W9-ART-926

MRI of the Musculoskeletal System

MRI of the Musculoskeletal System

Wing P. Chan, M.D.
Visiting Assistant Professor
Department of Radiology
University of California, San Francisco
San Francisco, California

Philipp Lang, M.D.
Resident in Diagnostic Radiology
Department of Radiology
University of California, San Francisco
San Francisco, California

Harry K. Genant, M.D.
Professor of Radiology, Medicine, and Orthopaedic Surgery
Chief, Skeletal Section, and Director, Osteoporosis Research Group
University of California, San Francisco
San Francisco, California

W.B. SAUNDERS COMPANY
A Division of Harcourt Brace & Company
Philadelphia ■ London ■ Toronto ■ Montreal ■ Sydney ■ Tokyo

W.B. Saunders Company
A Division of Harcourt Brace & Company

The Curtis Center
Independence Square West
Philadelphia, Pennsylvania 19106

Library of Congress Cataloging-in-Publication Data

MRI of the musculoskeletal system / [edited by] Wing P. Chan, Philipp Lang, Harry K. Genant.
p. cm.
Includes index.
ISBN 0-7216-4295-0
1. Musculoskeletal system—Magnetic resonance imaging. I. Chan, Wing P. II. Lang, Philipp. III. Genant, Harry K.
[DNLM: 1. Magnetic Resonance Imaging. 2. Musculoskeletal Diseases—diagnosis. WE 141 M93893 1994]
RC925.7.M7523 1994
616.7′07548—dc20
DNLM/DLC 93-26277

MRI of the Musculoskeletal System ISBN 0-7216-4295-0

Printed in the United States of America.

Last digit is the print number: 9 8 7 6 5 4 3 2 1

To my wife, Fiu-hua, my daughter Eve, and my mother, Fung-lan,
for their love and support,
and to the memory of my father, Kui-lam, and my daughter Joyce.

Wing P. Chan

To my dear parents, Walter and Lilo,
for a lifetime of love, support, and encouragement.

Philipp Lang

To my parents, Helen and Harry,
my wife, Gail,
and our children, Laura, Justin, and Jonathan.

Harry K. Genant

CONTRIBUTORS

W. Dillworth Cannon, M.D.
Clinical Professor of Orthopaedic Surgery, Department of Orthopaedic Surgery, University of California, San Francisco, San Francisco, California.
The Knee: Internal Derangements; The Knee: Other Pathologic Conditions

Wing P. Chan, M.D.
Attending Radiologist, Department of Radiology, Sun Yat-Sen Cancer Center, Taipei, Taiwan, Republic of China; formerly Visiting Assistant Professor, Department of Radiology, University of California, San Francisco, San Francisco, California.
Magnetic Resonance Imaging Anatomy and Cryomicrotome Atlas; Three-Dimensional Display in Magnetic Resonance Imaging; The Knee: Internal Derangements; The Knee: Other Pathologic Conditions; The Ankle and Foot; The Temporomandibular Joint; Disorders of Skeletal Muscle

Scott J. Erickson, M.D.
Assistant Professor and Chief, Body MRI, Department of Radiology, Medical College of Wisconsin, Milwaukee, Wisconsin.
The Ankle and Foot

James L. Fleckenstein, M.D.
Assistant Professor, Department of Radiology, and Director, The Algur H. Meadows Diagnostic Imaging Center, University of Texas Southwestern Medical Center at Dallas, Dallas, Texas.
Disorders of Skeletal Muscle

Bruce E. Fredrickson, M.D.
Associate Professor of Orthopedic Surgery and Neurosurgery, Department of Orthopedic Surgery, State University of New York, Health Science Center, Syracuse, New York.
The Cervical and Thoracic Spine; The Lumbar Spine

Russell C. Fritz, M.D.
Clinical Assistant Professor of Radiology, University of California, San Francisco, San Francisco, California; Staff Radiologist, Marin General Hospital; Medical Director, Marin Magnetic Resonance Imaging, Greenbrae, California.
The Elbow; The Wrist and Hand; The Knee: Internal Derangements; The Knee: Other Pathologic Conditions

Harry K. Genant, M.D.
Professor of Radiology, Medicine, and Orthopaedic Surgery, and Chief, Skeletal Section, and Director, Osteoporosis Research Group, Department of Radiology, University of California, San Francisco, San Francisco, California.
Magnetic Resonance Imaging Anatomy and Cryomicrotome

Atlas; Three-Dimensional Display in Magnetic Resonance Imaging; The Shoulder; The Hip; The Knee: Internal Derangements; The Knee: Other Pathologic Conditions; Musculoskeletal Neoplasm; Bone Marrow Disorders; Disorders of Skeletal Muscle

Clyde A. Helms, M.D.
Professor of Radiology in Residence and Clinical Professor of Restorative Dentistry, Department of Radiology, University of California, San Francisco, San Francisco, California.
The Ankle and Foot; The Temporomandibular Joint

Gordon Honda, M.D.
Department of Pathology, University of California, San Francisco, San Francisco, California.
Musculoskeletal Neoplasm

Harry E. Jergesen, M.D.
Associate Clinical Professor of Orthopaedic Surgery, Department of Orthopaedic Surgery, University of California, San Francisco, San Francisco, California.
The Hip

James O. Johnston, M.D.
Professor of Clinical Orthopaedic Surgery, Department of Orthopaedic Surgery, University of California, San Francisco, San Francisco, California.
Musculoskeletal Neoplasm

Philipp Lang, M.D.
Resident in Diagnostic Radiology, Department of Radiology, University of California, San Francisco, San Francisco, California.
Three-Dimensional Display in Magnetic Resonance Imaging; The Shoulder; The Hip; Musculoskeletal Neoplasm; Bone Marrow Disorders

Gin-Chung Liu, M.D.
Associate Professor and Chief, Department of Radiology, Kaohsiung Medical College, Kaohsiung, Taiwan, Republic of China.
Disorders of Skeletal Muscle

Sharmila Majumdar, Ph.D.
Assistant Professor in Residence, Department of Radiology, University of California, San Francisco, San Francisco, California.
Magnetic Resonance Imaging: Basic Principles and Imaging Technique; Bone Marrow Disorders

Steven Melnikoff, Ph.D.
Sausalito, California. Formerly Assistant Professor of Radiology, University of California, San Francisco, San Francisco, California.
Three-Dimensional Display in Magnetic Resonance Imaging

Sarah J. Nelson, Dr.rer.Nat.
Assistant Professor in Residence, Department of Radiology, University of California, San Francisco, San Francisco, California.
Magnetic Resonance Spectroscopy

Charles Peterfy, M.D., Ph.D.
Assistant Professor of Radiology, Department of Radiology, University of California, San Francisco, San Francisco, California.
The Ankle and Foot

Larry B. Poe, M.D.
Assistant Professor of Radiology, Department of Radiology, State University of New York, Health Science Center, Syracuse, New York.
The Cervical and Thoracic Spine; The Lumbar Spine

Wolfgang Rauschning, M.D., Ph.D.
Professor and Chief, Department of Orthopedic Surgery, Academic University Hospital, Uppsala, Sweden.
Magnetic Resonance Imaging Anatomy and Cryomicrotome Atlas

Arthur E. Rosenbaum, M.D.
Professor of Radiology and Director, Neuroradiologic Research, Department of Radiology, State University of New York, Health Science Center, Syracuse, New York.

The Cervical and Thoracic Spine; The Lumbar Spine

Lynne S. Steinbach, M.D.
Associate Professor of Clinical Radiology, University of California, San Francisco; Consultant, San Francisco Magnetic Resonance Center, and San Francisco Veterans' Administration Medical Center, San Francisco, California.
The Elbow; The Wrist and Hand; The Knee: Internal Derangements; The Knee: Other Pathologic Conditions

Martin Vahlensieck, M.D.
Assistant Professor of Radiology, Department of Radiology, Universitäts Klinik Bonn, Bonn, Germany.
Three-Dimensional Display in Magnetic Resonance Imaging; The Shoulder

Daniel B. Vigneron, Ph.D.
Assistant Professor in Residence, Department of Radiology, University of California, San Francisco, San Francisco, California.
Magnetic Resonance Spectroscopy

Chun Y. Wu, M.D.
Research Fellow in Radiology, Department of Radiology, University of California, San Francisco, San Francisco, California.
Magnetic Resonance Imaging Anatomy and Cryomicrotome Atlas; The Knee: Internal Derangements; The Knee: Other Pathologic Conditions

Shiwei Yu, M.D.
Research Fellow in Radiology, State University of New York, Health Sciences Center, New York, New York.
The Cervical and Thoracic Spine; The Lumbar Spine

DR. MARGULIS'S FOREWORD

Magnetic resonance (MR) imaging has established itself as the most versatile and precise approach for imaging much of the human body. Although in the beginning MR imaging was identified with studies of the central nervous system, its applications to the musculoskeletal system have by now equaled and surpassed in importance those to the central nervous system. Hardly any conditions of the spine, joints, and muscles in the body are diagnosed today without the help of MR imaging. The development of new techniques that improve spatial and soft tissue contrast resolution, the addition of gadopentetate dimeglumine contrast medium, and the continuous acquisition of experience have made this approach indispensable in modern rheumatology and orthopedic treatment. Magnetic resonance imaging has also proved to be very valuable in the study and evaluation of bone and soft tissue neoplasms and diseases that involve the bone marrow. Magnetic resonance spectroscopy, initially used only for studying muscular diseases that affect energy transfer, now is being applied also to the study of the biology of tumors, aging, and osteoporosis.

This book, containing contributions from many highly respected researchers, covers the global extent of the field, starting with MR anatomy and cryomicrotome image correlation and a simplified explanation of the basic principles and techniques of MR imaging. It brings the readers up to date on MR spectroscopy of the musculoskeletal system, and, after a three-dimensional display of the normal musculoskeletal system, it discusses and analyzes pathologic conditions of the spine, joints, bone marrow, skeletal muscles, and musculoskeletal neoplasms.

In a period when MR imaging has proved to be so important in the evaluation of patients with musculoskeletal disorders, this book should be of great help to the practicing clinician.

Alexander R. Margulis, M.D.
Emeritus Professor
Department of Radiology
University of California, San Francisco

DR. BRADFORD'S FOREWORD

Magnetic resonance (MR) imaging has been without question one of the most important and exciting advances in medical technology since the discovery and clinical use of diagnostic radiographs. Certainly, it is difficult, if not impossible, to imagine how we might approach the diagnosis and management of many of our patients without the use of this powerful tool. As we look back just in a short span over the past 10 years alone, we can see how rapidly this technology has made all but obsolete many of the previous procedures on which we in orthopedic surgery relied so heavily, for example, myelography and arthrography, to mention only two. Indeed, it would be hard to imagine at the present time how we could manage the diversity of patients with musculoskeletal disorders without the aid of this technology. In our subspecialties of orthopedic oncology, sports medicine, pediatric orthopedics, traumatology, adult reconstruction, and spinal disorders, the MR image has become indispensable.

Drs. Chan, Lang, and Genant have combined their efforts and presented this superb text, which should be the definitive word on musculoskeletal MR imaging for the radiologist, rheumatologist, and orthopedic surgeon. The text not only provides a careful review and an in-depth analysis for a thorough understanding of imaging techniques and the pathophysiology and correlation between the disease process and the image seen, but also is a useful and important resource for physicians and specialists interested in this subject. All physicians and surgeons involved in managing patients with musculoskeletal disease should have this well-written and comprehensive text available in their own personal libraries.

David S. Bradford, M.D.
Professor and Chairman
Department of Orthopaedic Surgery
University of California, San Francisco

DR. SACK'S FOREWORD

Ten years ago, if I were evaluating a patient with persistent pain and swelling in the knee and could not make a diagnosis after doing a careful examination and obtaining appropriate laboratory tests (including a plain radiograph of the knee), I would have consulted an orthopedic surgeon. The patient, in turn, likely would have undergone arthroscopic examination of the affected joint. That invasive procedure would have required an anesthetic, but it would have given me information regarding the hyaline and fibrocartilage, the cruciate ligaments, and the synovial tissues. Nowadays, under the same circumstances, I would order a magnetic resonance (MR) imaging examination of the patient's knee. This noninvasive procedure, which takes about 45 minutes, can accurately identify abnormalities in the cartilaginous and ligamentous structures as well as in the soft tissues and bones. All this at a cost of $800 to $1000 compared with $2000 to $3000 for arthroscopy.

In predicting the outcome of a disease such as rheumatoid arthritis, rheumatologists rely mainly on the presenting features (e.g., rapidity of onset, number of involved joints, systemic symptoms), the gross appearance of the joints (e.g., swelling, tenderness, limitation of motion, deformity), and the results of laboratory tests (e.g., hematocrit, erythocyte sedimentation rate, serum rheumatoid factor). In this regard, radiographs do not have much additional predictive value. By contrast, an MR imaging examination of involved joints has enormous potential for identifying hidden areas of synovial proliferation and early attrition of tendons and cartilage.

The ability of newer MR imaging techniques to identify and *quantify* cartilage loss enables us (at long last) to determine more accurately the effects of treatment on the outcome of rheumatic diseases. Indeed, the evolution of MR imaging in evaluating the musculoskeletal system seems limited only by the constraints of our ingenuity and resources.

Approximately one third of the adults in the United States have some form of arthritis, and rheumatologists caring for them face a number of vexing problems: What is the most cost-effective way of making the correct diagnosis? What are the predictors of disease severity and disability? How can we assess accurately the effects of treatment? This textbook, authored by superb musculoskeletal radiologists, shows how MR imaging helps to answer these questions.

KENNETH E. SACK, M.D.
Director, Clinical Program on Rheumatology
Division of Arthritis and Rheumatology, Department of Medicine
University of California, San Francisco

PREFACE

Disorders of the musculoskeletal system constitute the most common nonneurologic application of magnetic resonance (MR) imaging. The indications for MR imaging of the musculoskeletal system have expanded rapidly in recent years and will continue to grow. Development of new fast imaging sequences and special dedicated coils for imaging the axial skeleton and the joints will help to decrease scan times, reduce costs, increase patient throughput, and improve the image quality even further.

Superb tissue contrast, high spatial resolution, and multiplanar imaging capability are the hallmarks of MR imaging of the musculoskeletal system. Unlike computed tomography (CT), MR imaging provides high image contrast among tissue fat, fluid, effusion, tendons and ligaments, vascular structures, nerves, muscle, and bone marrow.

Magnetic resonance imaging is used for primary diagnosis as well as for monitoring disease progression or response to therapy. In evaluating the spine, MR imaging supplements and frequently replaces CT, including CT with intravenous or intrathecal contrast. Magnetic resonance imaging has revolutionized the assessment of the knee joint and has largely replaced arthrography. It represents a noninvasive alternative to arthroscopy for many joint disorders. In musculoskeletal oncology, MR imaging is more accurate than CT in the staging of bone and soft tissue neoplasms. Magnetic resonance imaging is unique in localizing and characterizing diffuse and focal marrow abnormalities. In addition, MR imaging affords a noninvasive assessment of the soft tissues of the locomotor system, which is particularly advantageous in imaging disorders of skeletal muscle or ligamentous injuries.

MRI of the Musculoskeletal System is written for the radiologist and the clinician who desire a complete comprehensive text on MR imaging of the musculoskeletal system. This book will help the newcomer understand the principles of MR imaging in orthopedics and rheumatology, and it will serve the experienced radiologist or clinician as an exhaustive reference text. The first section of the book is organized in atlas form and correlates MR images with color photographs of frozen cadaver sections. The second section describes the basic principles and MR imaging techniques, including MR spectroscopy, as they apply to musculoskeletal disorders. Perspectives on three-dimensional MR image displays that may gain more widespread use in the near future are outlined in a separate chapter. The third and largest section of the book focuses on the current clinical application of MR imaging to diseases of the spine, major joints, bone marrow, neoplasm, and skeletal muscle. Disease mechanisms and pathophysiology are initially addressed in each chapter.

This is followed by a detailed description of clinical indications, diagnostic criteria, advantages and limitations of MR imaging, and a comparison with other imaging modalities. In each chapter, a large number of correlative pathologic and histologic specimens obtained at surgery, biopsy, or arthroscopy are shown. This book will help radiologists, orthopedic surgeons, and rheumatologists who are involved with MR imaging in daily clinical practice or research.

WING P. CHAN
PHILIPP LANG
HARRY K. GENANT

ACKNOWLEDGMENTS

The editors thank the following people who gave full support to this project: the administrative staff of the Musculoskeletal Section, Denice Nakano and Claire Colangelo; the technologists of the Magnetic Resonance Science Center, Pauline Mattei and Evelyn Proctor; the photographers of the Illustration Resource Services at the University of California, San Francisco, Richard Wada and Tom Chan; and the permissions coordinator at the Radiological Society of North America, Mary Ann Morrissey.

A very special acknowledgment is given to Catherine F. Fix for her expert revision of manuscripts for presentation to the publisher.

Finally, the editors extend their appreciation to Lisette Bralow and the production staff, Peter Faber and Lee Ann Draud, at W.B. Saunders.

CONTENTS

Part I ANATOMY

Part II PRINCIPLES

Part III PATHOLOGY

Part I
ANATOMY

1 MAGNETIC RESONANCE IMAGING ANATOMY AND CRYOMICROTOME ATLAS

Wing P. Chan, Wolfgang Rauschning, Chun Y. Wu, and Harry K. Genant

Accurate interpretation of magnetic resonance (MR) images requires an appreciation of normal anatomy. In this chapter, the anatomic details as seen in axial, coronal, and sagittal planes about the joints and spine are shown by cryomicrotome sections from cadavers and MR images from asymptomatic adults. Table 1–1 summarizes the MR signal intensity characteristics of normal tissues about the joints.[2]

LUMBAR SPINE

Axial Images

Axial images provide the best views to show the spinal canal, nerve roots, and dorsal root ganglia. At the level of the neural foramina (Fig. 1–1), the dorsal ganglion is of low signal intensity and is surrounded by fat that is of high signal intensity. Also, the nerve root is well defined as a linear structure of low signal intensity. Careful attention should be given to the ventral internal longitudinal vein, which can easily be mistaken for a nerve ganglion. The spinal canal is round in the upper lumbar region but triangular in the lower lumbar region. Within the thecal sac, of low signal intensity, the nerve root demonstrates slightly higher intensity than the surrounding cerebrospinal fluid. When the subject is in a supine position, the nerve roots are clustered inferiorly, and cerebrospinal fluid is identified mainly in the superior layer. Posterior epidural fat is of high signal intensity, in contrast to the subarachnoid space of low signal intensity. On T2-weighted images, the nerve roots have low signal intensity and contrast with the cerebrospinal fluid, which has high signal intensity.

The bone marrow of the vertebral body, the pedicle, the lamina, and the transverse and spinous processes of the vertebra are of high signal intensity. The cortical bone margins decrease in signal intensity owing to the lack of resonating protons. The ligamenta flava, of intermediate signal intensity, parallel the inner surface of the lamina.

Differentiation between nucleus pulposus and anulus fibrosus cannot be made precisely on axial MR images regardless of pulse sequences. The intervertebral disk demonstrates overall intermediate signal intensity on T1-weighted images, high signal intensity on T2-weighted images, and very high signal intensity on T2* gradient-echo images.

The articulation of the facet joints is formed by the concave surface of the superior articular process and the convex surface of the inferior articular processes. The superior facet is located anterolaterally and faces posteromedially. The inferior facet is lo-

Table 1–1. RELATIVE SIGNAL INTENSITY OF ADULT NORMAL TISSUES

Tissue	T1-weighted spin-echo	Proton density-weighted spin-echo	T2-weighted spin-echo	T2* gradient-echo
Cortical bone	Low	Low	Low	Low
Ligament or tendon	Low	Low	Low	Low
Nerve	Low	Low	Low	Low
Fibrocartilage	Low	Low	Low	Low
Anulus fibrosus	Low	Low	Low	Low
Nucleus pulposus	Intermediate	Intermediate	High	High
Cerebrospinal fluid or vessel*	Low	Intermediate	High	High
Muscle	Intermediate	Intermediate	Low	High
Articular cartilage	Intermediate	Intermediate	Low	High
Fat	High	High	Intermediate	Low
Bone marrow	High	High	Intermediate	Low

* Generally speaking, the faster the flow velocity (measured in cm/sec, not in ml/sec) and the more perpendicular to the plane of section, the darker the signal.

cated posteromedially and faces anterolaterally. On T2* gradient-echo images, facet cartilage images have high signal intensity, in contrast to the low signal intensity of the cortical facet margins.

Sagittal Images

The lumbar spine is composed of five vertebral bodies, separated from one another by intervertebral disks. The anatomic level of these vertebrae is well appreciated on sagittal planes.

On midsagittal images (Fig. 1–2), the anterior and posterior longitudinal ligaments appear as an uninterrupted band of very low signal intensity on all pulse sequences. The posterior longitudinal ligament adheres to the fibers of the anulus fibrosus. It cannot be distinguished separately from the dura and anulus on T1-weighted images. On T2* gradient-echo images, a structure of high signal intensity, the venous plexus, separates the posterior longitudinal ligament (of low signal intensity) and the cortical bone of the vertebral body. Careful attention should be given to the basivertebral vein, which is seen as a structure of high signal intensity in the posterior aspect of the vertebral body on T2* gradient-echo images and should not be mistaken for a fracture.

On T1-weighted images, vertebral bodies demonstrate high signal intensity because of marrow fat, in contrast to the intermediate signal intensity of intervertebral disks. In younger age groups, fatty marrow of high signal intensity can be depicted as a linear area confined to the region along the basivertebral vein. With advancing age, this fatty marrow of high signal intensity can appear bandlike, triangular, or multifocal and can take up relatively large areas in patients older than 40 years.[16]

The disks of L1 and L2 have a straight posterior margin, but those of the L3, L4, and L5 vertebrae have a distinctly convex and bulging configuration posteriorly.[15] This slight posterior convexity of the lower lumbar disks should not be confused with pathologic bulging of the disks that results from their degeneration or instability. Differentiation of the nucleus and anulus may not be made precisely on T1-weighted images. On T2-weighted and T2* gradient-echo imaging, however, the central portion of the disk is of higher signal intensity, in contrast to the lower signal intensity from the surrounding, less hydrated anulus fibrosus.[3] Small concentric or transverse tears are identified frequently in normal adult disks. However, radial annular tears are considered an important hallmark of a degenerative disk.[20]

Within the thecal sac, on T1-weighted images, the lower signal intensity of the cerebrospinal fluid contrasts with the higher (intermediate) signal intensity of the nerve root bundles. The conus medullaris terminates at the L1–2 vertebral level. On T2-weighted images, cerebrospinal fluid demonstrates high signal intensity, in contrast to the low signal intensity of nerve root bundles. Posterior epidural fat is consistently present in the posterior part of the spinal canal, whereas the anterior epidural fat is most prominent in the L5–S1 region.

On paramedian sagittal images through the intervertebral foramina (Fig. 1–3), the dorsal ganglia and nerve roots of intermediate signal intensity occupy the upper portion of the intervertebral foramen (subpedicular notch). The radicular veins of low signal intensity are anterior to the neural elements. Both the neural elements and the veins are surrounded by and contrast with the fat of high signal intensity on T1-weighted images.

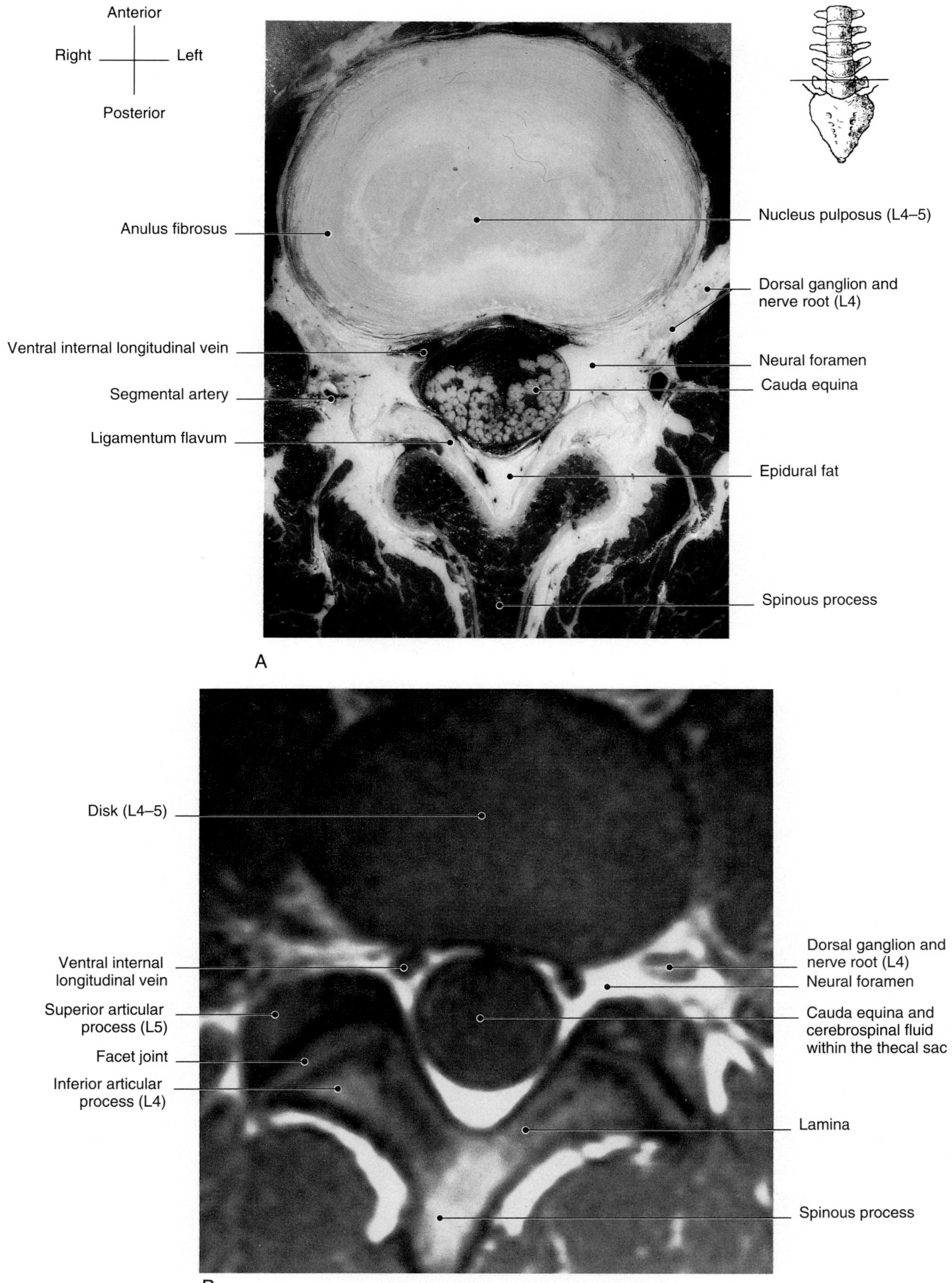

Figure 1–1. Axial cryoplaned section *(A)* and MR image (TR, 800 ms; TE, 20 ms) *(B)* of the lumbar spine at the L4–5 level through the neural foramen.

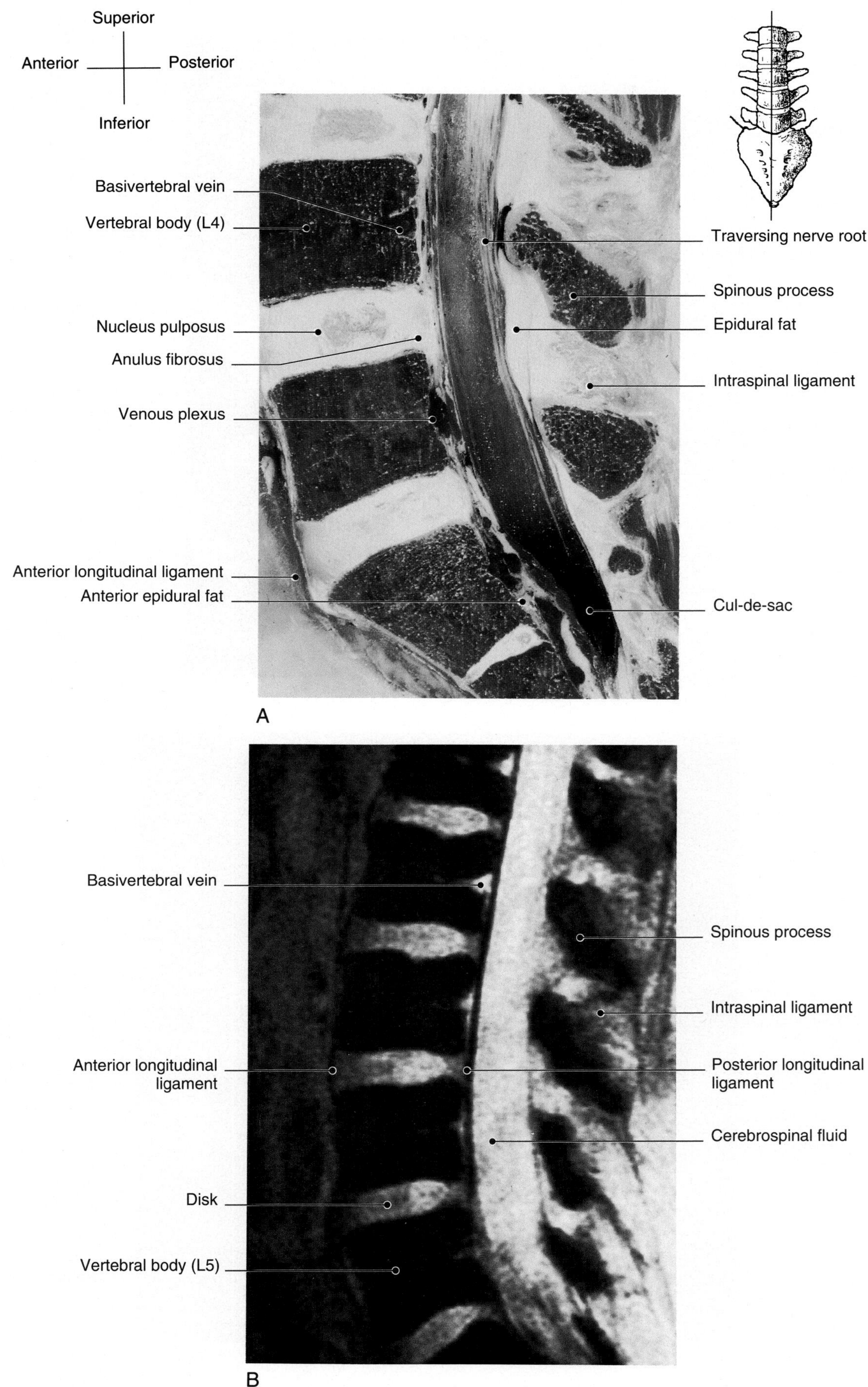

Figure 1–2. Midsagittal cryoplaned section *(A)* and MR image (TR, 400 ms; TE, 15 ms; flip angle, 15 degrees) *(B)* of the lumbar spine.

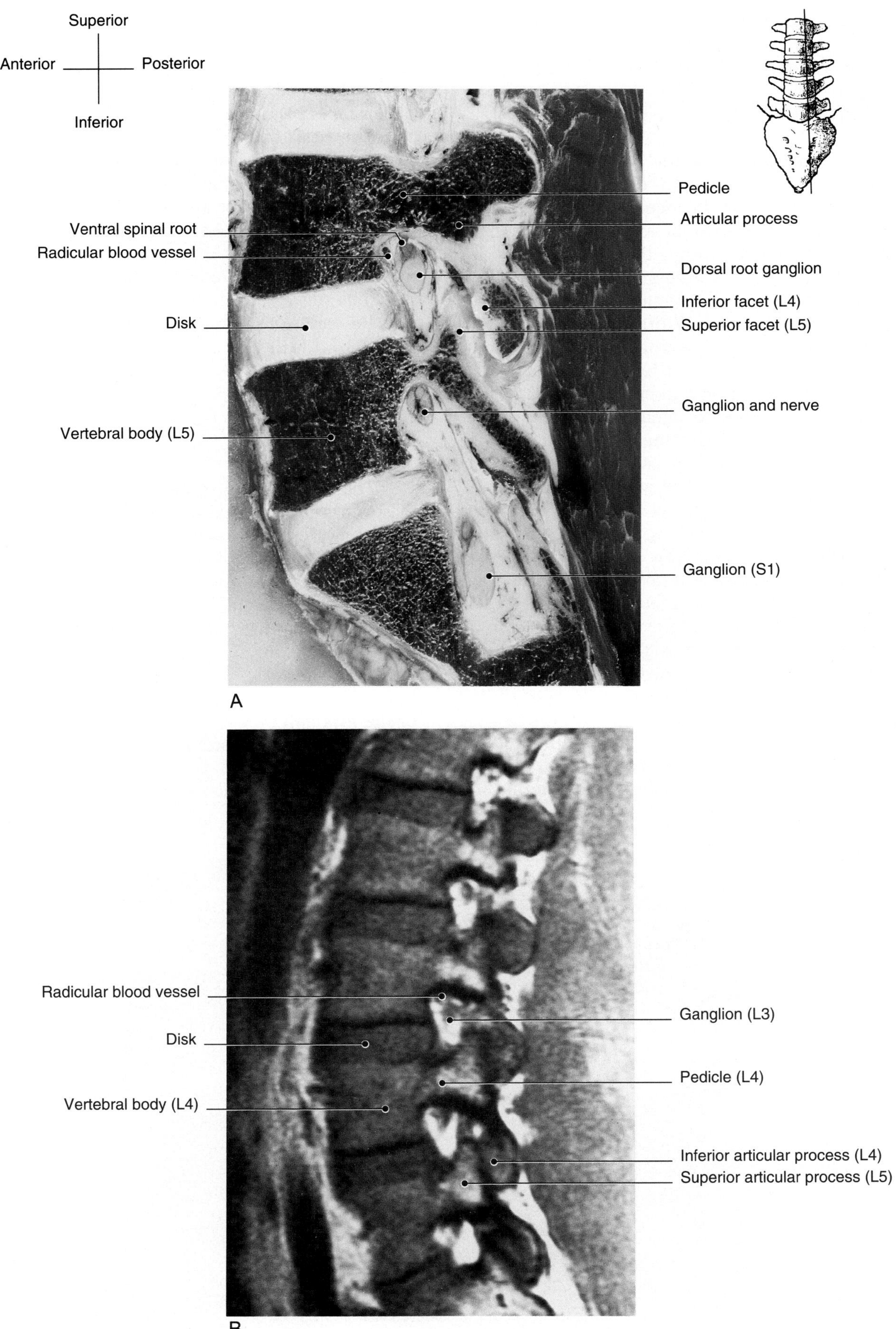

Figure 1–3. Paramedian sagittal cryoplaned section *(A)* and MR image (TR, 500 ms; TE, 20 ms) *(B)* of the lumbar spine.

CERVICAL AND THORACIC SPINE

Axial Images

Axial images provide the best views of the spinal canal, nerve roots, and dorsal root ganglion. Through the level of the neural foramina, on T1-weighted images, the high signal intensity of the neural foraminal fat outlines the nerve roots of low signal intensity. The vertebral arteries are seen as circular structures of low signal intensity owing to a flow-void phenomenon. Within the thecal sac, the cerebrospinal fluid of low signal intensity surrounds the spinal cord and roots of intermediate signal intensity. The central intervertebral disk has a nucleus pulposus of intermediate signal intensity.

On T2* gradient-echo images (Figs. 1–4, 1–5), the vascular structures and cerebrospinal fluid are high in signal intensity. The gray and white matter can be distinguished. The central gray matter appears as a butterfly-shaped region of high signal intensity. Distinction of nerve roots is difficult on these sequences, because both nerve roots and surrounding fat are intermediate in signal intensity. The nucleus pulposus shows high signal intensity on T2-weighted images and very high signal intensity on T2* gradient-echo images. The facet joints are shown in oblique orientation. The facet cartilage is of high signal intensity and contrasts well with the low signal intensity of cortical bone. The ventral internal venous plexus can be seen between the vertebrae and the spinal cord. The vertebral body, lamina, and transverse and spinous processes of the vertebrae are well delineated. The spinous processes of the cervical spine are short and carry bifid tips.

The first (atlas, C1) and second (axis, C2) segments of the cervical spine are distinct from the others. The atlas is a ring-shaped bony structure composed of anterior and posterior arches and lateral articular pillars. An odontoid process of the axis (dens) is seen posterior to the anterior arch of the atlas. In a normal adult, atlantoaxial instability might be suggested when the space between the dens and the anterior arch of the atlas exceeds 3 mm. In children, this distance could be as much as 4 mm.

Sagittal Images

The cervical and thoracic spines are composed of seven and 12 vertebrae, respectively. The atlantoaxial motion segment is formed by the fusion of the bodies of C1 and C2. The tip of the odontoid process abuts on the lower pons and the medulla oblongata. The transverse ligament (or more strictly, the transverse portion of the cruciate ligament), which is covered by the tectorial membrane, holds the odontoid process posteriorly.[14] The thin apical ligament of the dens directly anchors the tip of the dens to the clivus of the foramen magnum.

Sagittal images provide the best views to show the anatomic level of the vertebral body. The relationships between the osseous structures, posterior

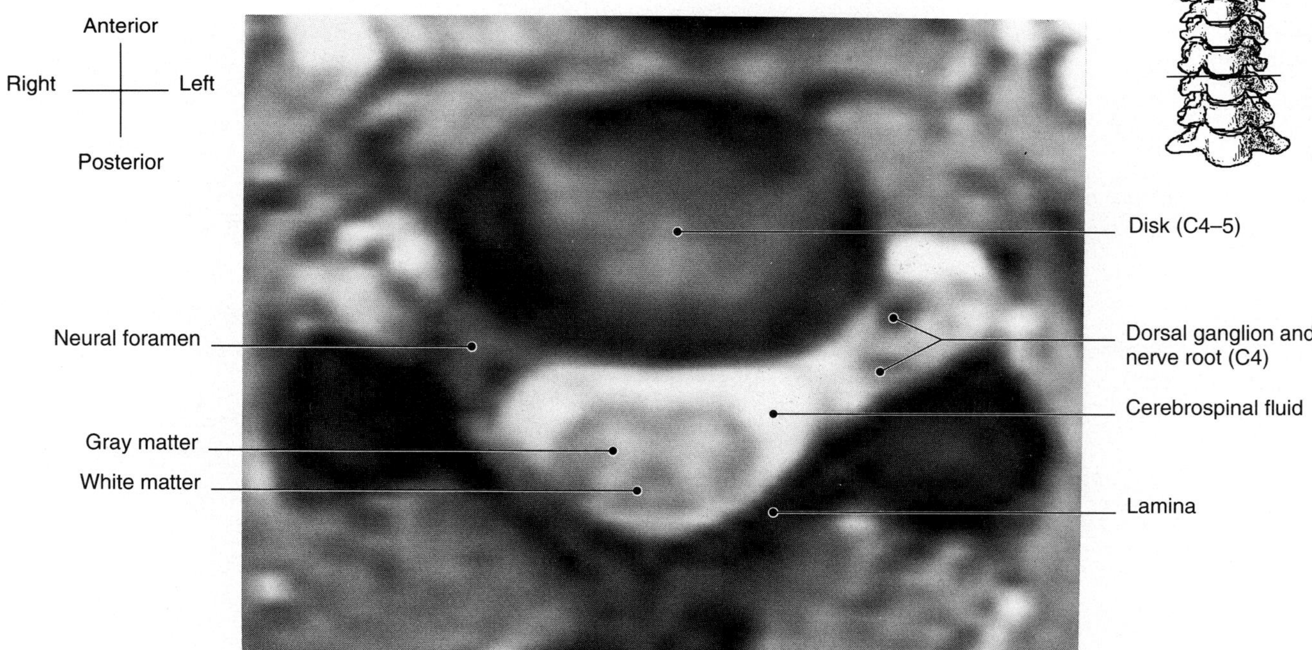

Figure 1–4. Axial MR image (TR, 600 ms; TE, 20 ms; flip angle, 20 degrees) of the cervical spine at the C4–5 level.

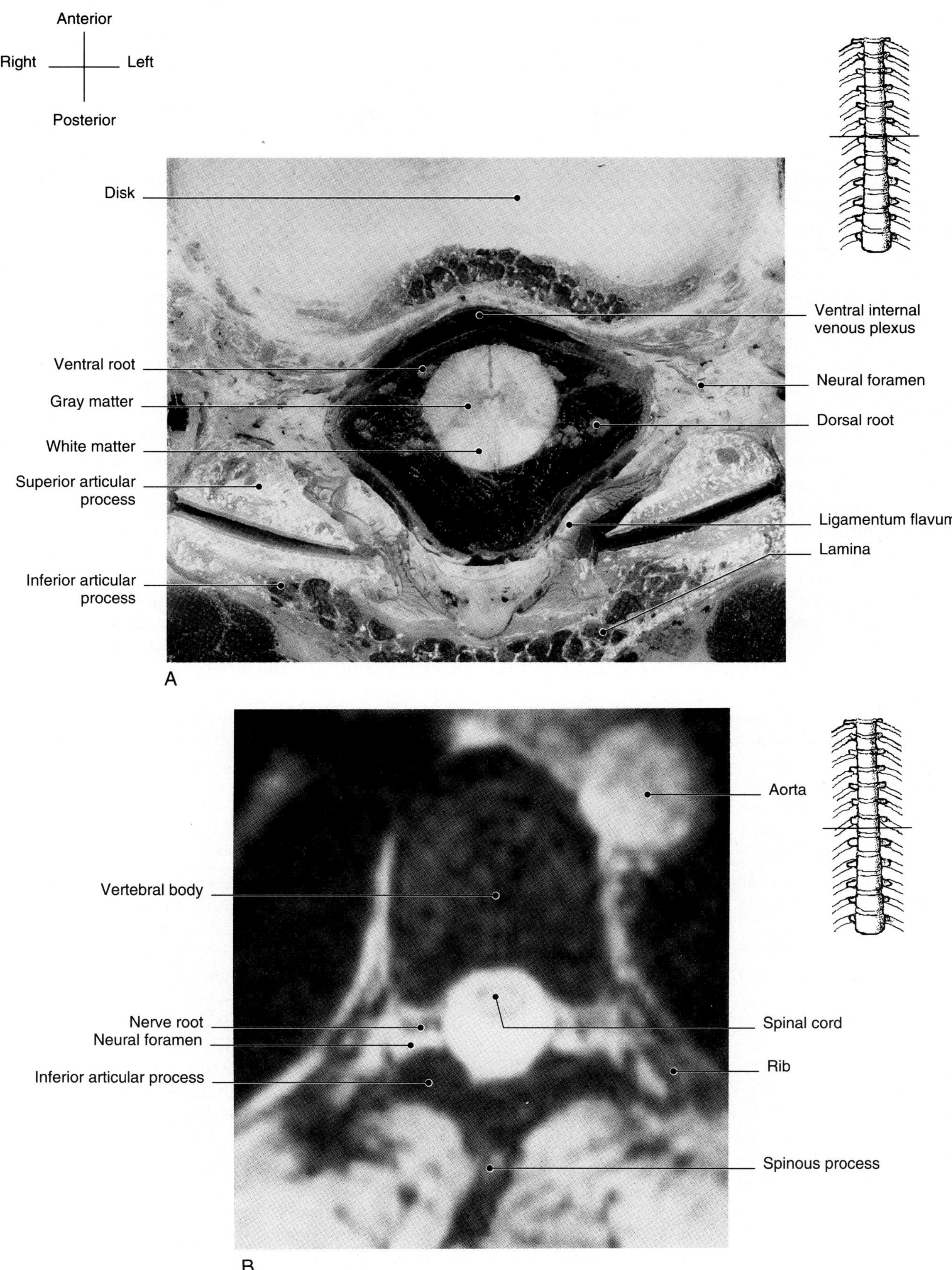

Figure 1–5. Axial cryoplaned section *(A)* and MR image (TR, 600 ms; TE, 20 ms; flip angle, 15 degrees) *(B)* of the midthoracic spine. Note that *A* and *B* are not of the same anatomic level.

longitudinal ligament, spinal canal, and nerve roots can be well depicted on sagittal views.

On midsagittal images (Figs. 1–6, 1–7), the anterior and posterior longitudinal ligaments are seen as a continued band of very low signal intensity on all sequences. On T1-weighted images, vertebral bodies show high signal intensity because of bone marrow fat, in contrast to the intermediate signal intensity of the intervertebral disks. The bone marrow fat may become inhomogeneous in signal intensity with advancing age, however. The intervertebral disk is of intermediate intensity. The gray and white matter may not be distinguishable owing to the cerebrospinal fluid flow and truncation artifacts. The truncation artifact (or Gibbs's phenomenon) is seen as bands parallel to the spinal cord. This occurs at the cerebrospinal fluid–spinal cord interface because of high-contrast boundaries and is related to field of view, acquisition size, and pixel size.

On T2* gradient-echo sequences, the intervertebral disks and cerebrospinal fluid show high signal intensity and contrast well with the posterior longitudinal ligament of low signal intensity. The alignment of this ligament and its relationships to the disk are well shown on these sequences. The spinal cord becomes of low signal intensity and is surrounded by cerebrospinal fluid of high signal intensity.

The cross-sectional view of the foramina is obtained by oblique paramedian sagittal scans. An axial scan is used to obtain an optimal angulation of the oblique sagittal planes through the foramina. The nerve roots of intermediate signal intensity are well depicted within the perineural fat of high signal intensity (Figs. 1–8, 1–9). On T2* gradient-echo images, nerve elements and perineural fat are difficult to distinguish. The articular cartilage facet is seen as a region of high signal intensity, separating the superior and inferior articular processes.

SHOULDER

Axial Images

Axial images provide the best views to show the glenohumeral joint (Fig. 1–10). On superior axial images the supraspinatus muscle courses obliquely. The supraspinatus tendon inserts onto the greater tuberosity of the humerus, which is posterior to the bicipital groove. The spine of the scapula, visualized as marrow fat of high signal intensity, runs posterior and parallel to the supraspinatus muscle. The spine of the scapula separates the supraspinatus and infraspinatus muscles. At the level of the coracoid process, the fusiform infraspinatus muscle is parallel and inferior to the supraspinatus muscle. The infraspinatus tendon inserts inferoposteriorly to the supraspinatus on the lateral aspect of the greater tuberosity.

Caudal to the coracoid process level is the glenohumeral articulation. Its anterior capsular mechanism consists of the synovial membrane, the glenohumeral ligaments, the glenoid labrum, the subscapularis bursa and related recesses, and the subscapularis muscle and tendon.[22] The anterior portion of the capsule, the glenohumeral ligaments, and the subscapularis tendon are difficult to distinguish from each other in normal adults. The anterior labrum is commonly a triangular structure of low signal intensity that is continuous with the capsule and the glenohumeral ligaments.[7] The anterior labrum is larger than the posterior labrum. The subscapularis is located anterior to the body of the scapula, and its tendon merges with the anterior aspect of the capsule before inserting onto the lesser tuberosity. The subscapularis recess usually is not identified unless glenohumeral joint fluid is present. The long head of the biceps tendon appears as a round area of low signal intensity within the bicipital groove between the greater and lesser tuberosities.

The glenohumeral articular cartilage can be distinguished as a region of intermediate signal intensity following the concave shape of the glenoid cavity. Posteriorly, the labrum usually is rounded in appearance. The collapsed subacromial-subdeltoid bursa appears of high signal intensity because of its associated fat; it is identified between the infraspinatus and deltoid muscles.

Coronal Images

Coronal oblique images provide the best views for visualizing the tendons of the rotator cuff muscles (Fig. 1–11). These sections are obtained parallel to the supraspinatus muscle and tendon. On anterior and midcoronal oblique images, these structures can be seen in continuity. The supraspinatus tendon of low signal intensity inserts on the superolateral aspect of the greater tuberosity of the humerus. The musculotendinous junction commonly is located superior to the head of the humerus. The subacromial and subdeltoid bursae are potential spaces separating the rotator cuff tendons from the acromioclavicular joint. These bursae are demarcated by peribursal fat and so are seen as bands of high signal intensity. The coracoclavicular ligament is composed of two portions. The anterolaterally located trapezoid ligament is best demonstrated on the anterior coronal images. The posteromedially located conoid ligament is visualized on the posterior coronal images.

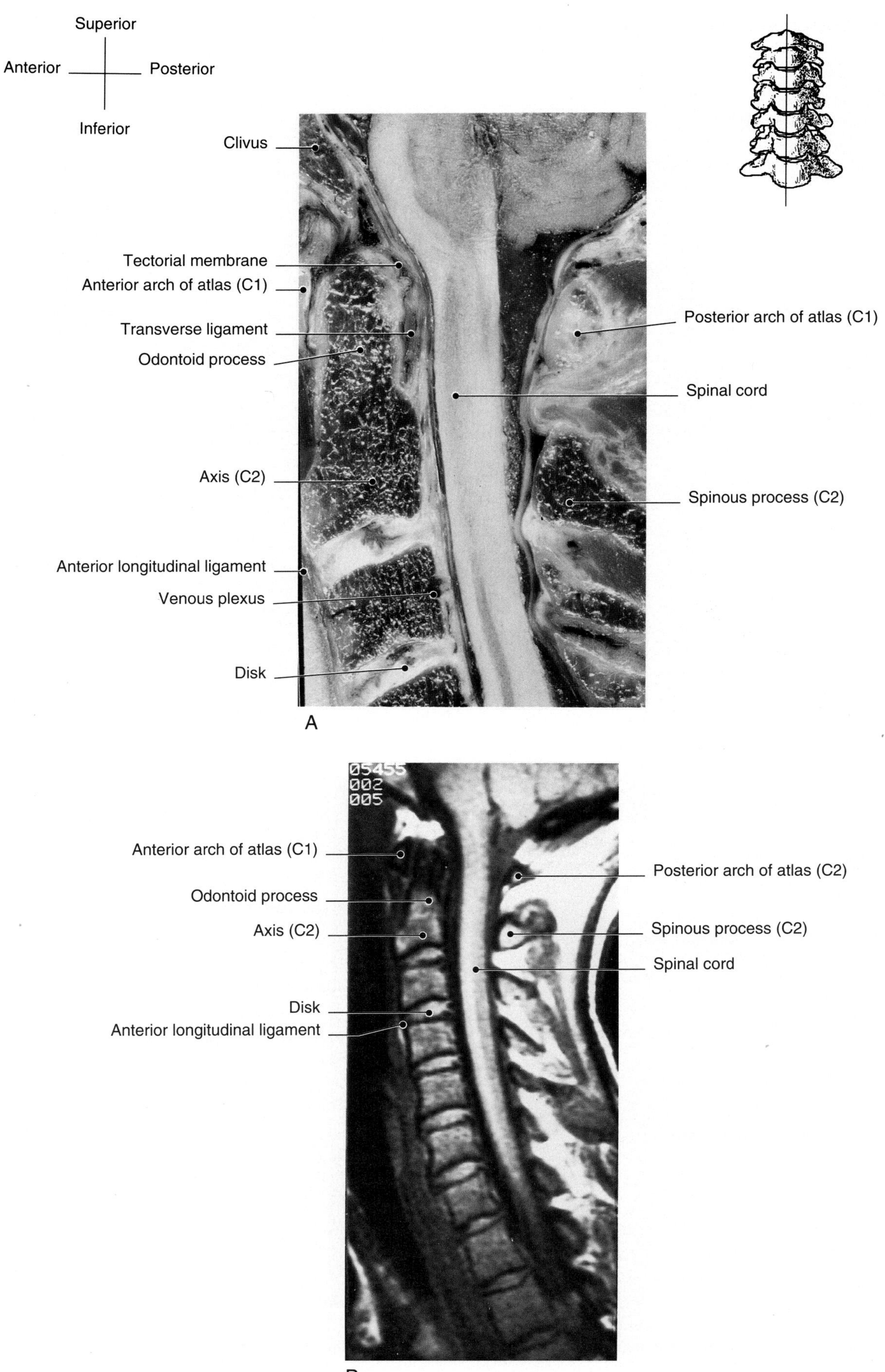

Figure 1-6. Midsagittal cryoplaned section *(A)* and MR image (TR, 500 ms; TE, 20 ms; flip angle, 15 degrees) *(B)* of the cervical spine.

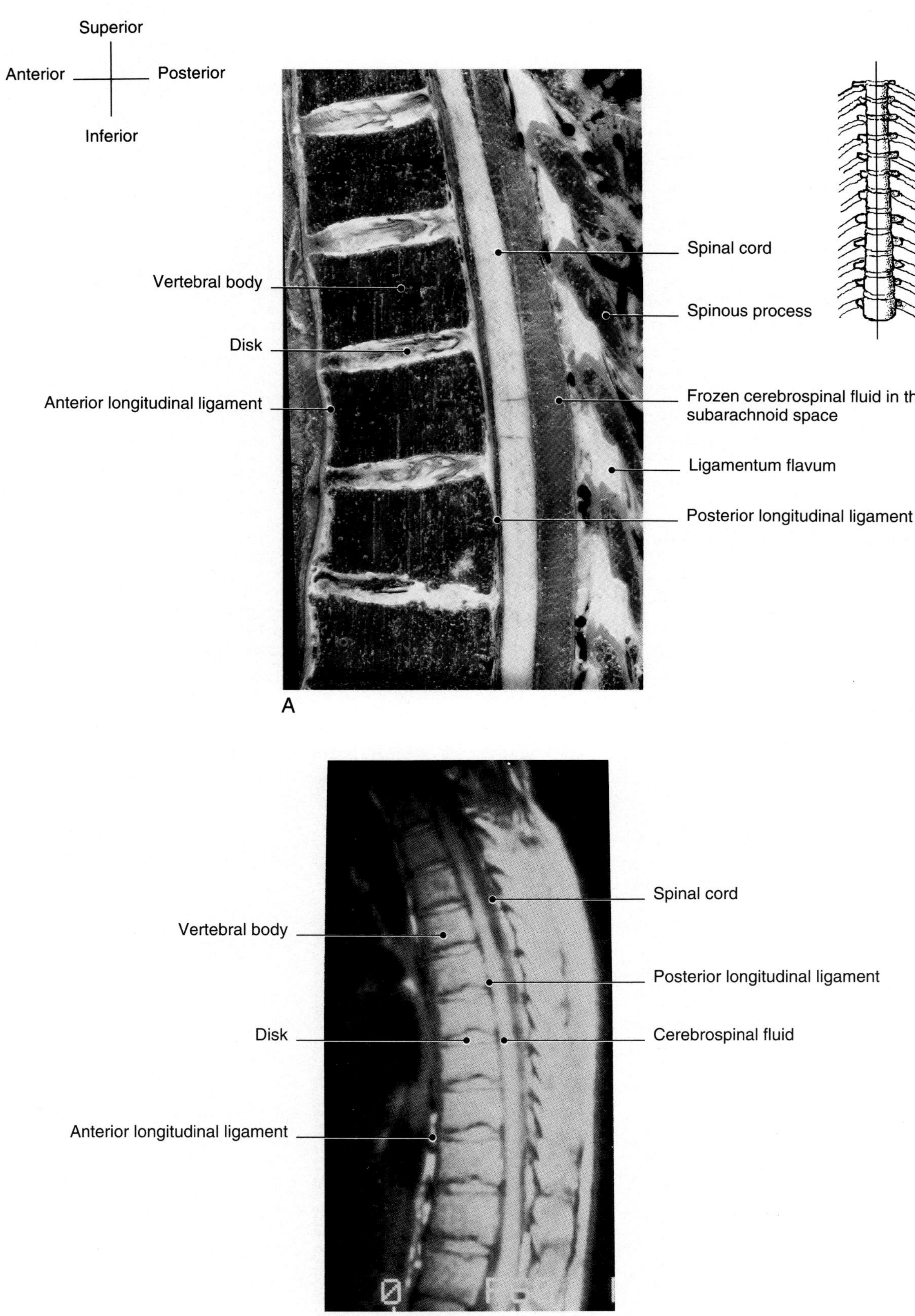

Figure 1–7. Midsagittal cryoplaned section *(A)* and MR image (TR, 600 ms; TE, 20 ms) *(B)* of the thoracic spine.

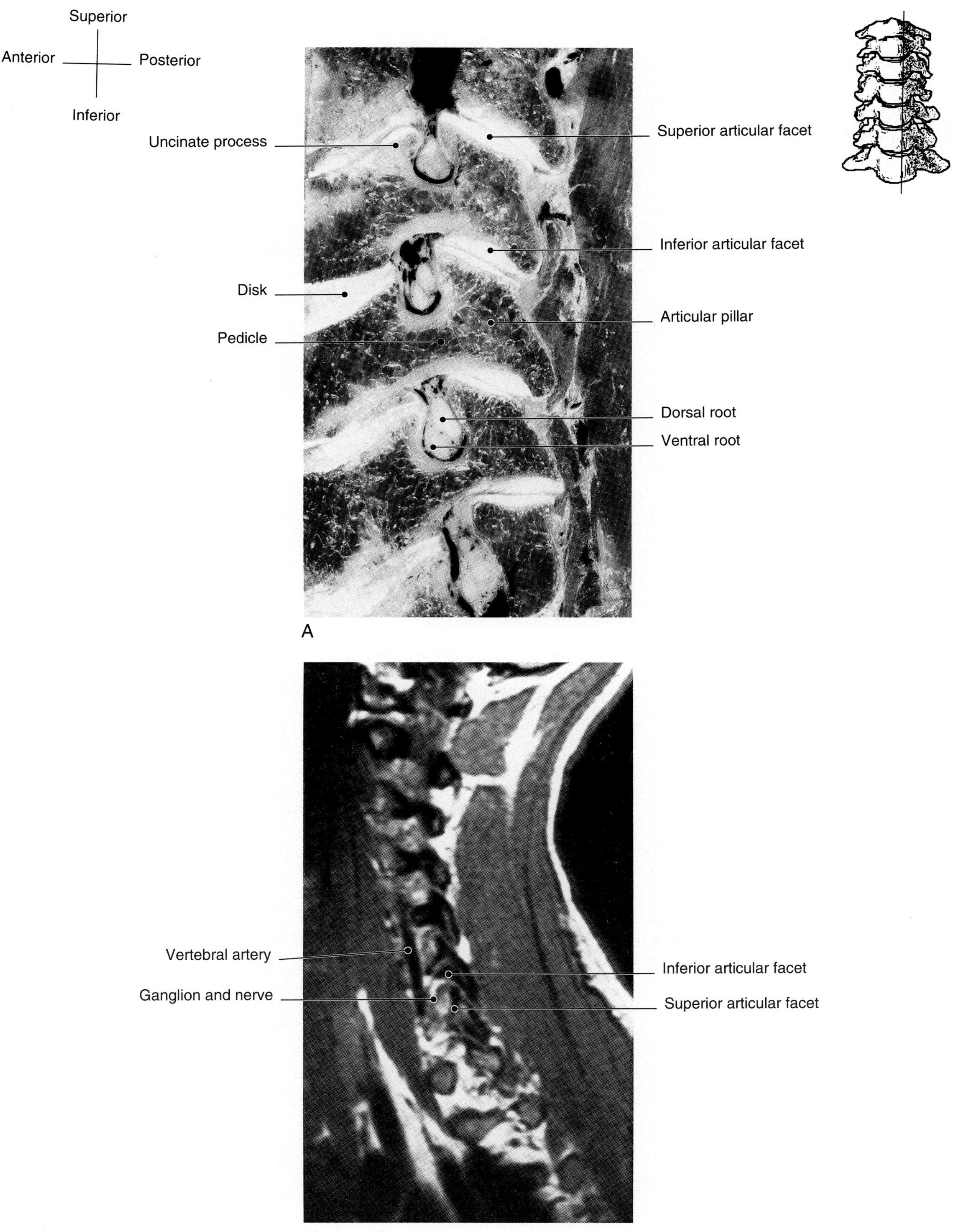

Figure 1–8. Paramedian oblique sagittal cryoplaned section *(A)* and MR image (TR, 500 ms; TE, 20 ms; flip angle, 15 degrees) *(B)* of the cervical spine.

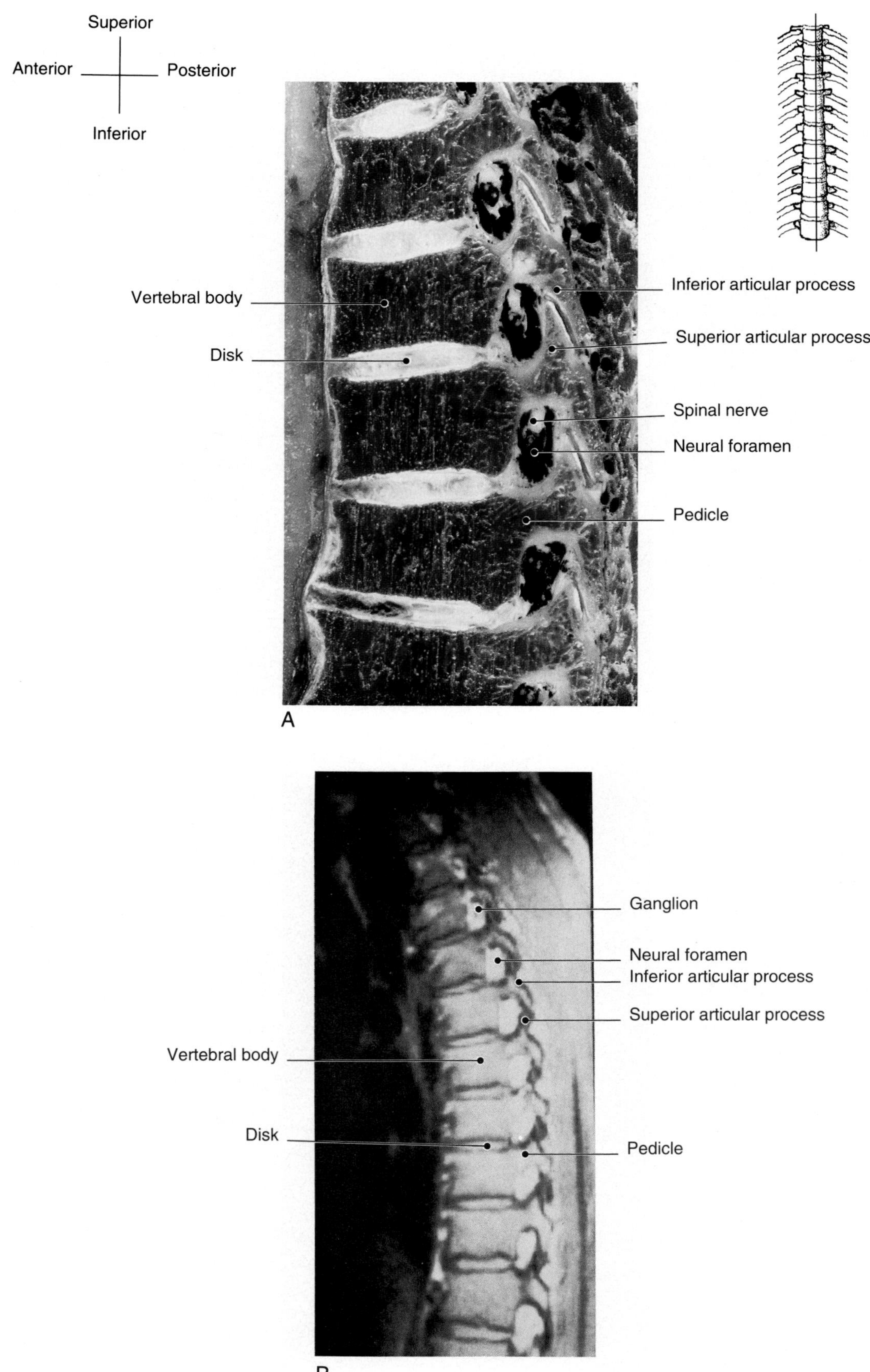

Figure 1-9. Paramedian sagittal cryoplaned section *(A)* and MR image (TR, 600 ms; TE, 20 ms) *(B)* of the thoracic spine.

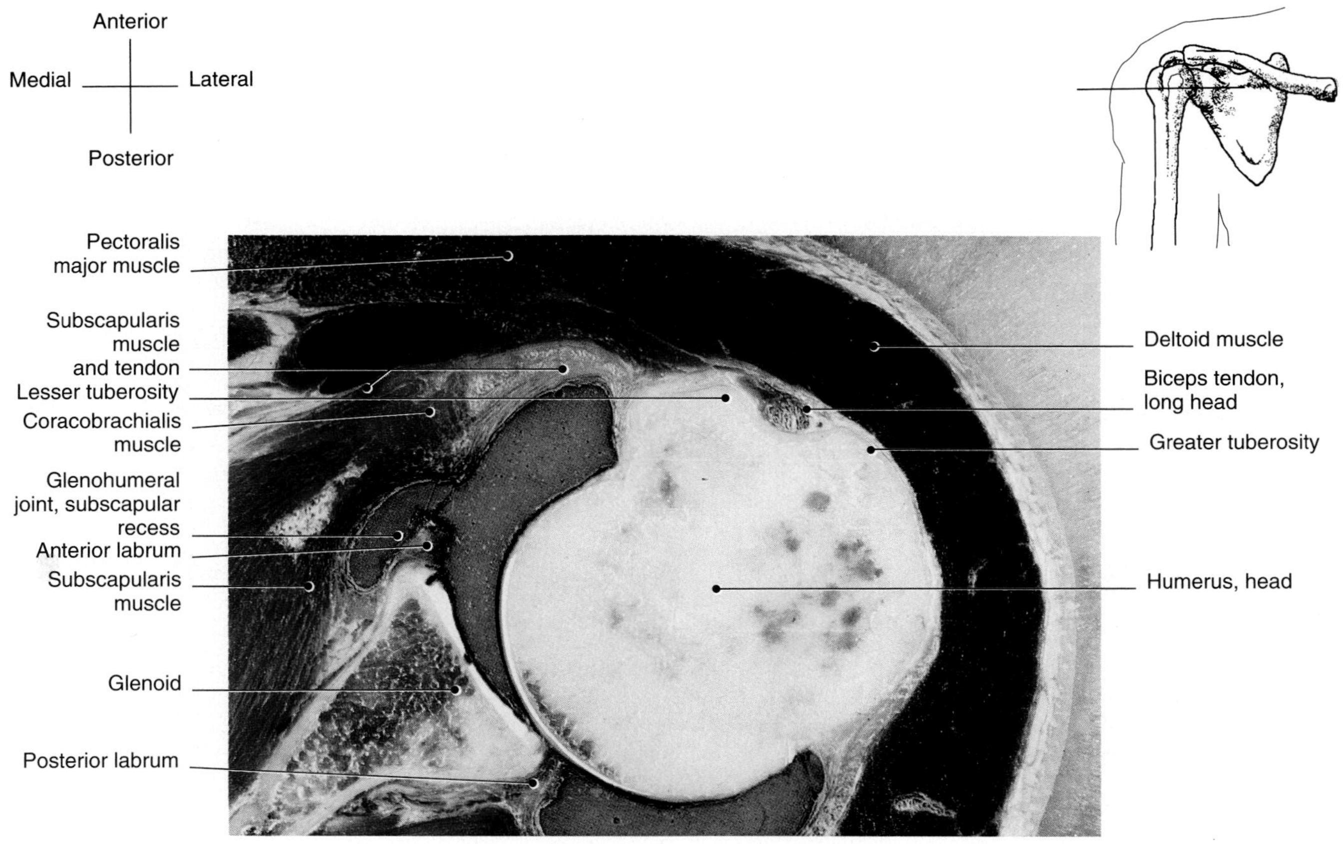

Anterior
Medial
Lateral
Posterior
Pectoralis major muscle
Subscapularis muscle and tendon
Lesser tuberosity
Coracobrachialis muscle
Glenohumeral joint, subscapular recess
Anterior labrum
Subscapularis muscle
Glenoid
Posterior labrum
Deltoid muscle
Biceps tendon, long head
Greater tuberosity
Humerus, head

A

Pectoralis major muscle
Subscapularis tendon
Coracobrachialis muscle
Anterior labrum
Subscapularis muscle and tendon
Glenoid
Posterior labrum
Infraspinatus muscle
Biceps tendon, long head
Greater tuberosity
Lesser tuberosity
Humerus, head
Deltoid muscle

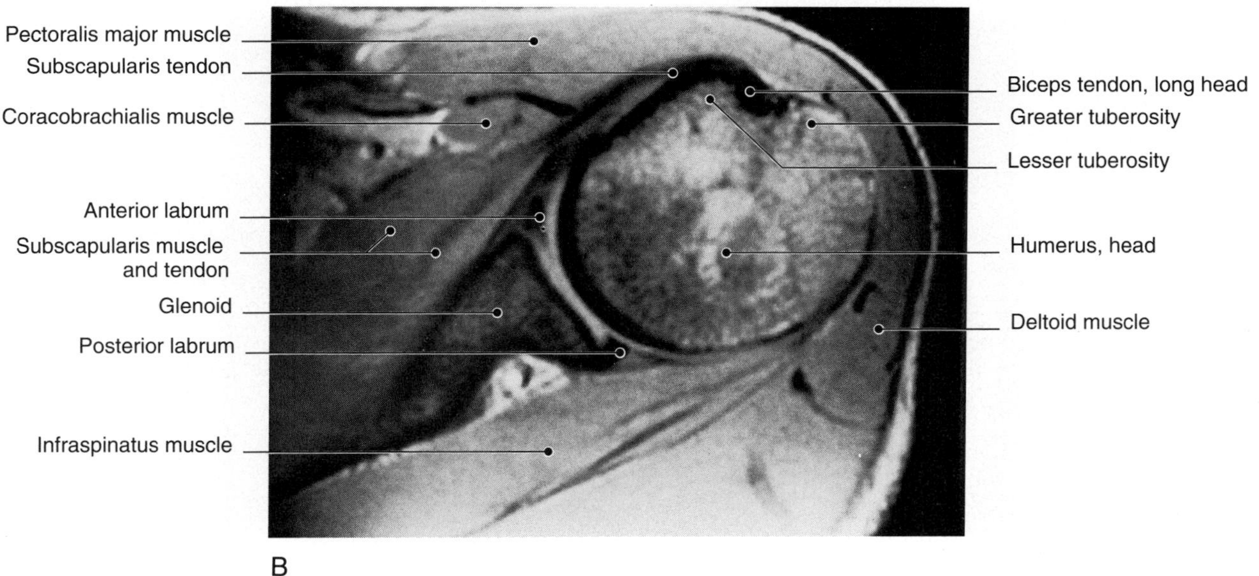

B

Figure 1–10. Axial cryoplaned section *(A)* and MR image *(B)* of the shoulder at the level of the glenohumeral joint.

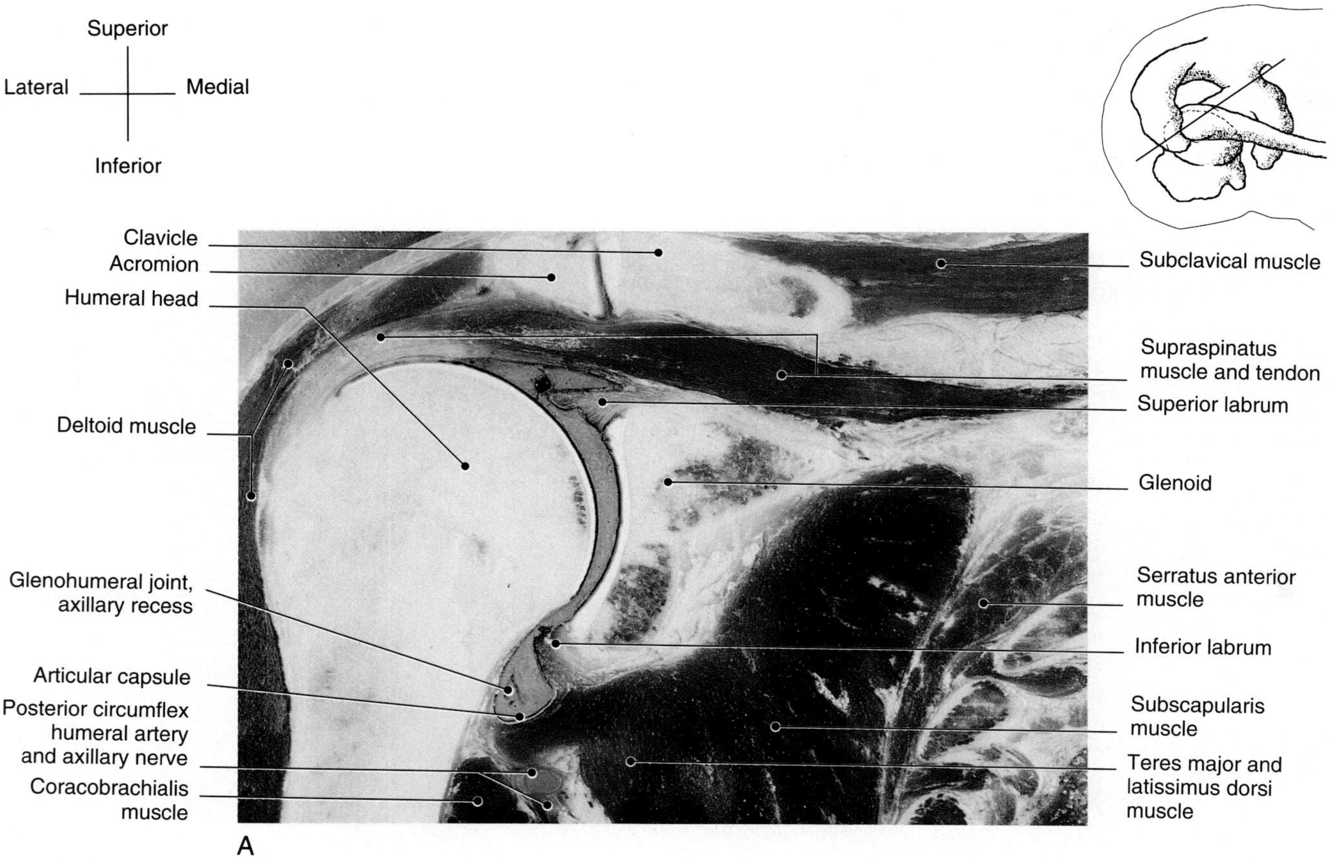

Superior
Lateral
Medial
Inferior
Clavicle
Acromion
Humeral head
Deltoid muscle
Glenohumeral joint, axillary recess
Articular capsule
Posterior circumflex humeral artery and axillary nerve
Coracobrachialis muscle
Subclavical muscle
Supraspinatus muscle and tendon
Superior labrum
Glenoid
Serratus anterior muscle
Inferior labrum
Subscapularis muscle
Teres major and latissimus dorsi muscle
A

Clavicle
Acromion
Supraspinatus tendon
Humeral head
Deltoid muscle
Subclavical muscle
Supraspinatus muscle
Glenoid
B

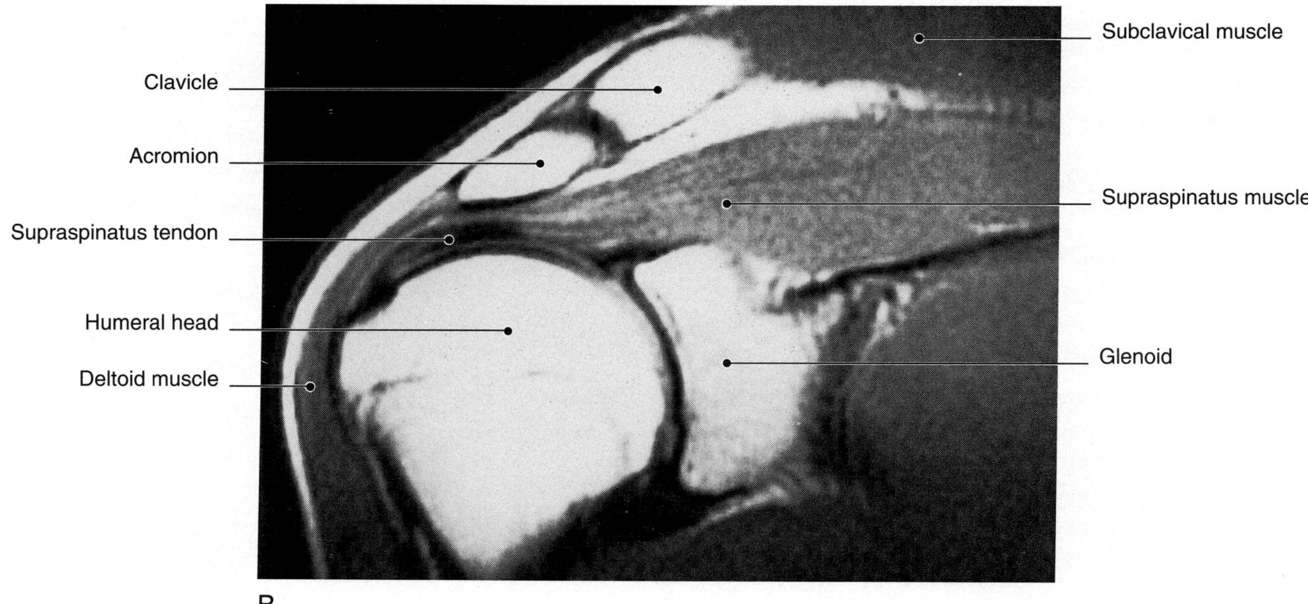

Figure 1–11. Oblique coronal cryoplaned section *(A)* and MR image (TR, 1000 ms; TE, 20 ms) *(B)* of the shoulder parallel to the supraspinatus muscle and tendon.

On posterior images, a transitional zone is seen representing the common insertion of the supraspinatus and infraspinatus tendons. The infraspinatus tendon is easily misinterpreted as the supraspinatus tendon, which is now out of the plane of section. Humeral head articular cartilage appears as a region of intermediate signal intensity between the cortex (low signal intensity) inferiorly and supraspinatus tendon superiorly.

Sagittal Images

Sagittal oblique images provide the best views for visualizing the supraspinatus outlet through which the supraspinatus muscle and tendon pass (Fig. 1–12). The most anterior of the rotator muscles is the subscapularis; this muscle is separated from the rest of the rotator cuff by the tendon of the long head of the biceps and the coracohumeral ligament. The supraspinatus is the most superior muscle of the rotator cuff. Posterior and inferior to the supraspinatus is the infraspinatus. The teres minor is located inferior to the infraspinatus. It is the smallest muscle of the rotator cuff and sometimes appears as part of the infraspinatus.

On midsagittal images, the coracoacromial ligament is seen as a band of low signal intensity that crosses superiorly and anteriorly to the rotator cuff. Midsagittal images also display the clavicle and acromioclavicular joint in profile.

ELBOW

Axial Images

Axial images provide the best views to show the relationship of the muscle groups and the neurovascular bundles. Muscles of the elbow are classified into four groups[1]: (1) The medial muscle group includes the pronator teres, the flexors of the fingers and wrist, and the palmaris longus; (2) the lateral muscle group includes the brachioradialis, the extensors of the fingers and wrist, and the supinator muscle; (3) the anterior muscle group consists of the brachialis and biceps; and (4) the posterior muscle group consists of the triceps and anconeus muscles.

On proximal axial images (Fig. 1–13), the radial nerve is of low signal intensity and lies between the brachioradialis (lateral) and brachialis (medial) muscles. The ulnar nerve is identified as a dot of low signal intensity just posterior to the medial humeral epicondyle, surrounded by perineural fat of high signal intensity. The median nerve runs medially along the superficial surface of the brachialis muscle. The brachial artery, seen as a signal void, runs between the biceps muscle and tendon and the medial nerve. Posteriorly, the capitellum and the trochlea are concave areas in the distal humerus that accommodate the radial and ulnar heads when the elbow is extended. The hyaline articular cartilage of these three osseous structures is well delineated.

At the proximal radioulnar joint level, the hy-

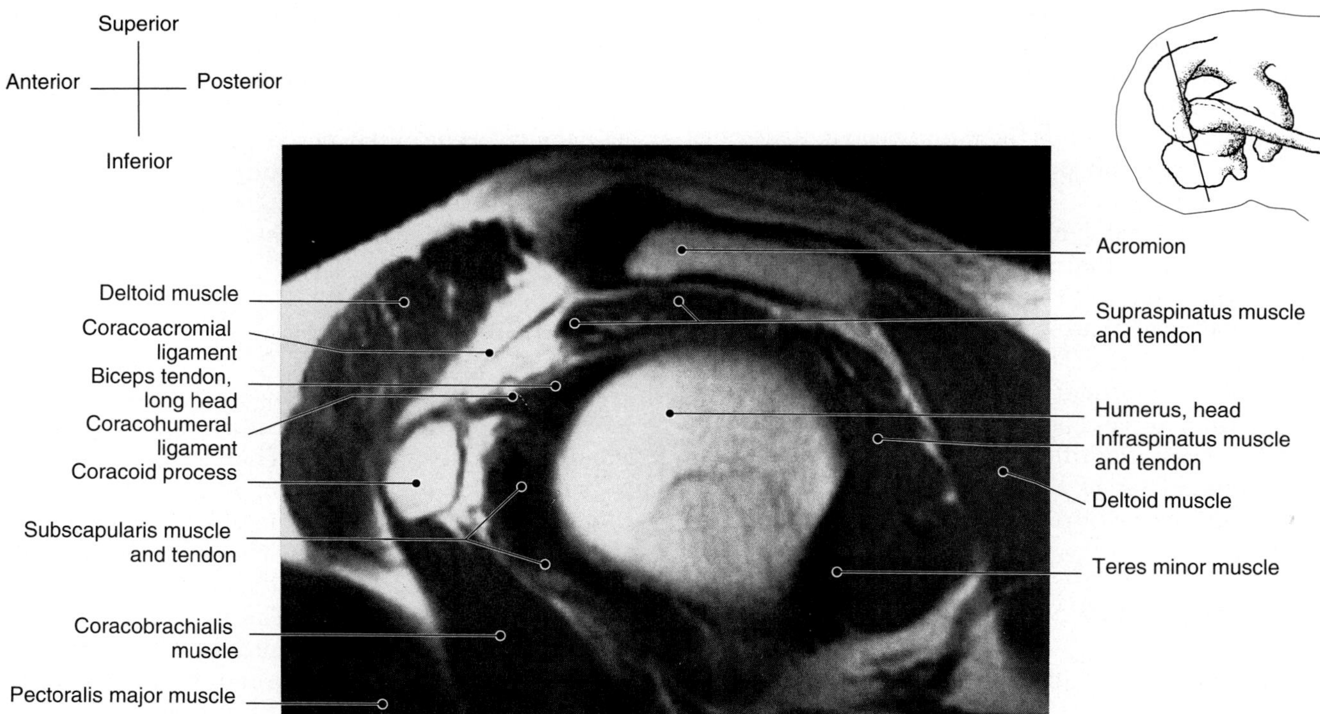

Figure 1–12. Oblique sagittal MR image (TR, 800 ms; TE, 20 ms) of the shoulder at the level of the coracoid process.

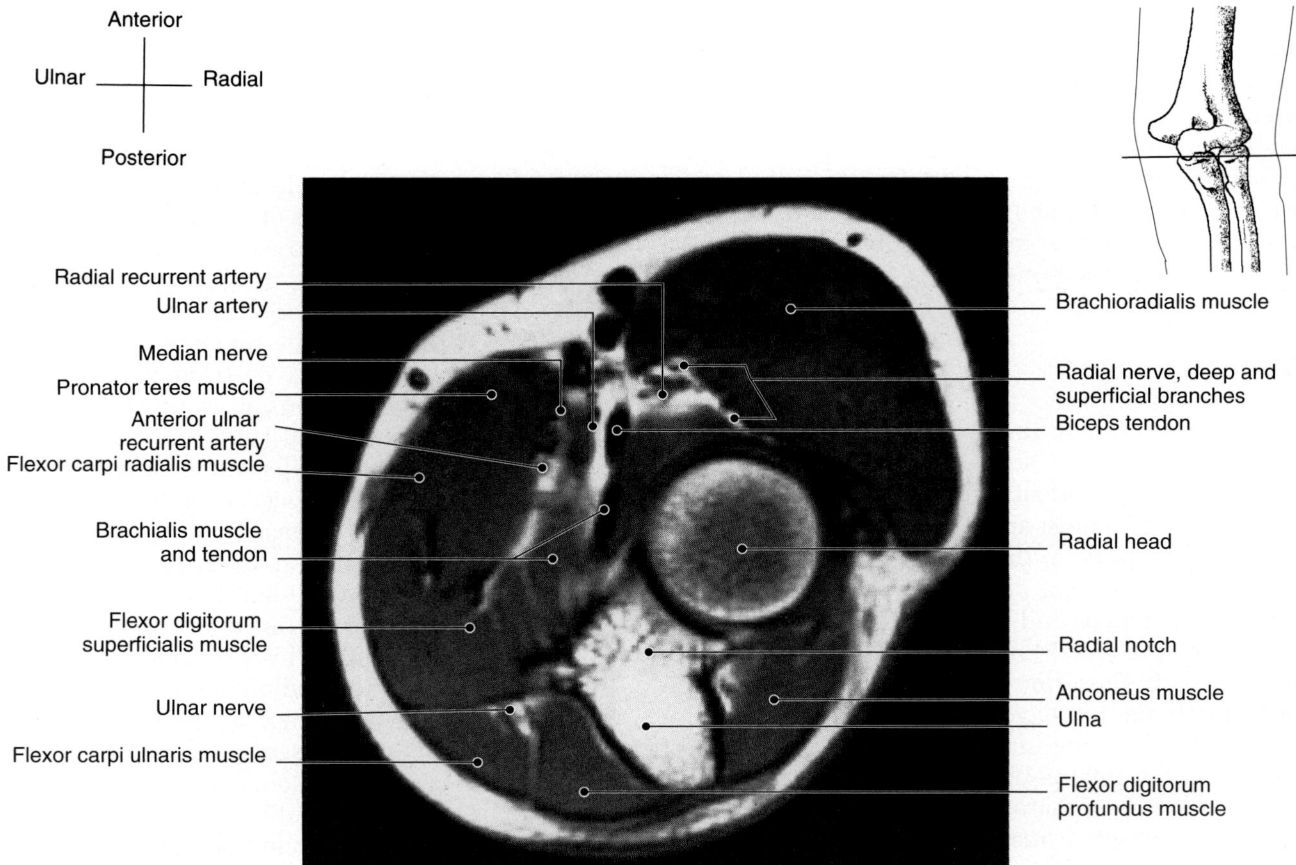

Figure 1–13. Axial MR image (TR, 600 ms; TE, 20 ms) of the elbow at the level of the proximal radioulnar joint.

aline articular cartilage of this joint is identified as a structure of intermediate signal intensity on T1-weighted images. The rounded portion of the radial head should be positioned within the concave radial notch of the ulna. The annular ligament, of low signal intensity, surrounds the radial head and attaches to the anterior and posterior portions of the radial notch of the ulna.

On distal axial images, the brachial artery divides into the radial and ulnar arteries. The ulnar artery runs deep to the pronator teres muscle, whereas the radial artery runs between the brachialis and pronator teres. The radial nerve divides into a deep and a superficial branch at this level. The ulnar and median nerves become difficult to identify more distally. Their identification depends on the amount of perineural fat of high signal intensity that surrounds the nerve of relatively low signal intensity.

Coronal Images

Coronal images provide the best views to show the low signal intensity, longitudinally oriented medial (ulnar) and lateral (radial) collateral ligaments (Fig. 1–14). The medial collateral ligament originates from the medial epicondyle of the humerus and inserts on the medial aspect of the trochlear notch of the ulna. The lateral ligament arises from the lateral epicondyle of the humerus and attaches to the annular ligament. The ulnar nerve courses immediately posterior to the medial epicondyle of the humerus. It is of relatively low signal intensity and is surrounded by perineural fat of high signal intensity.

Sagittal Images

Paramedian sagittal images provide the best views to show the trochlear-ulnar and capitellar-radial head joints (Fig. 1–15). Their articular surfaces are surrounded by cartilage of intermediate signal intensity on T1-weighted images. The anterior and posterior fat pads of the elbow joint are seen as bands of high signal intensity along the anterior and posterior surfaces of the humerus. The anterior fat pad lies proximal to the trochlea, whereas the posterior fat pad is located within the coronoid and olecranon fossae. The biceps muscle runs anterior to the brachialis, and its tendon inserts onto the radial tuber-

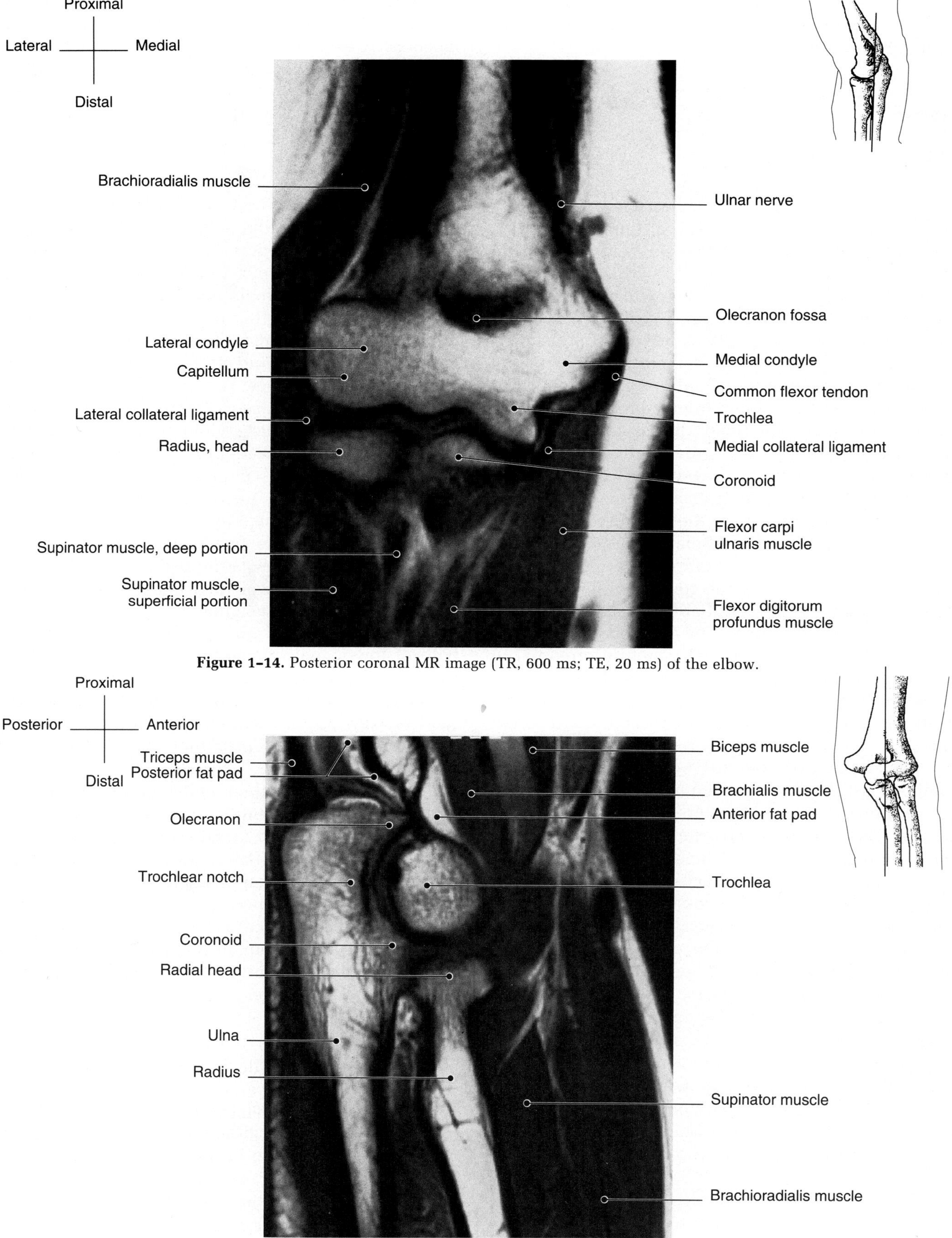

Figure 1-14. Posterior coronal MR image (TR, 600 ms; TE, 20 ms) of the elbow.

Figure 1-15. Paramedian sagittal MR image (TR, 600 ms; TE, 20 ms) of the elbow.

osity. The triceps muscle courses posteriorly over the elbow, and its joint tendon inserts onto the posterior surface of the olecranon.

WRIST

Axial Images

Axial images provide the best views of the carpal tunnel (Fig. 1–16). The carpal tunnel is an oval compartment, which is bordered ventrally by the flexor retinaculum and dorsally by the palmar surfaces of the carpal bones. On the palmar border of the tunnel, the flexor retinaculum is visualized as a thin band of low signal intensity extending from the pisiform and the hook of the hamate medially to the scaphoid and trapezium laterally.[21] On the dorsal or deep aspect of the tunnel, the flexor digitorum profundus (deep) and the flexor digitorum superficialis (superficial) muscles are individually identified as black, tubular structures outlined by investing sheaths of higher signal intensity.

In the neutral position, the median nerve is seen either anterior to the superficial flexor tendon of the index finger or interposed more posterolaterally between this tendon and the flexor pollicis longus.[9] The median nerve is identifiable as a structure of intermediate signal intensity on T1- and T2-weighted images. It appears as an oval structure proximal to and within the carpal tunnel, but it flattens at the level of the pisiform. The signal intensity of the normal median nerve is not increased on T2-weighted images.

The flexor carpi radialis and flexor carpi ulnaris tendons are clearly visualized in the radial and ulnar aspects, respectively, of the boundaries of the carpal tunnel. On the most radial aspect, the abductor pollicis longus and extensor pollicis brevis are identified. However, clear definition between them usually is not possible. Centrally, the osseous structures separate the flexor tendons ventrally and extensor tendons dorsally. Dorsally, the extensor tendons are depicted clearly deep to the extensor retinaculum. Like the flexor tendons, the extensor tendons can be identified individually because of the synovial sheaths of higher signal intensity that invest them. Vessels are usually black, but venous structures may have high signal intensity owing to even-echo rephasing or paradoxical enhancement secondary to slow flow. The flexor retinaculum attaches to the hook of the hamate. The three thenar and hypothenar muscles all arise from the flexor retinaculum. The adductor pollicis is located deep to the thenar muscles. Arterial and venous structures both show increases in signal intensity on gradient-recalled techniques.

Coronal Images

Coronal images best demonstrate the carpal bones, the triangular fibrocartilage, and the course of nerves, vessels, and tendons (Fig. 1–17). The triangular fibrocartilage is a curvilinear "bow tie" band of low, homogeneous signal intensity in the ulnocarpal space. It arises from the ulnar aspect of the distal portion of the radius and extends horizontally to the base of the ulnar styloid process.[19] In general, the triangular fibrocartilage is not well differentiated from the adjacent meniscus. The meniscus originates at the base of the ulnar styloid process and inserts onto the proximal aspect of the triquetrum. The meniscus helps prevent communication between the radiocarpal and the pisiform-triquetral compartments. The scapholunate ligament can be identified in some normal adults as a structure of low signal intensity. The lunatotriquetral ligament often cannot be recognized on T1-weighted images. On gradient-echo or spin-echo T2-weighted images, the lunatotriquetral ligament and the interosseous ligaments, both of low signal intensity, may be distinguished from the surrounding synovial fluid of high signal intensity. High-resolution images with small (8 cm) field of view permit demonstration of the cartilage surfaces between the proximal and distal carpal rows.

The radial and ulnar collateral ligaments are identifiable in dorsal coronal images. The radial ligament extends from the radial styloid process to the scaphoid. The ulnar collateral ligament extends from the ulnar styloid process to the triquetrum.

In the palmar aspect of the wrist, the flexor tendons are seen en face. They transverse the carpal tunnel between the hook of the hamate and the trapezium. The median nerve is not consistently identifiable on coronal images.

Sagittal Images

Sagittal images show the flexor and extensor tendons in profile (Fig. 1–18), although evaluation of these tendons is not common in the sagittal plane.

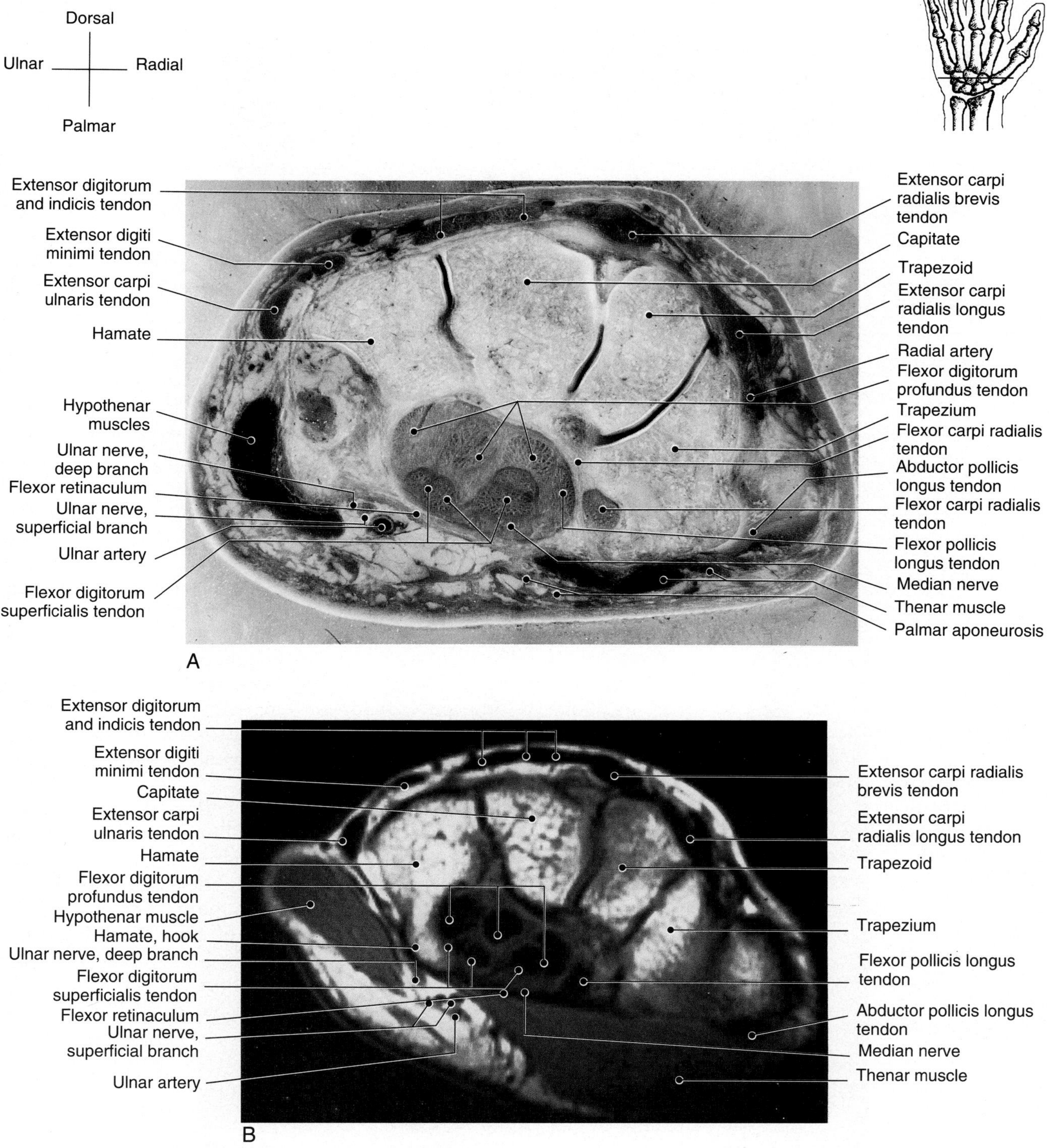

Figure 1–16. Axial cryoplaned section *(A)* and MR image (TR, 700 ms; TE, 20 ms) *(B)* of the wrist at the level of the hamate and trapezium.

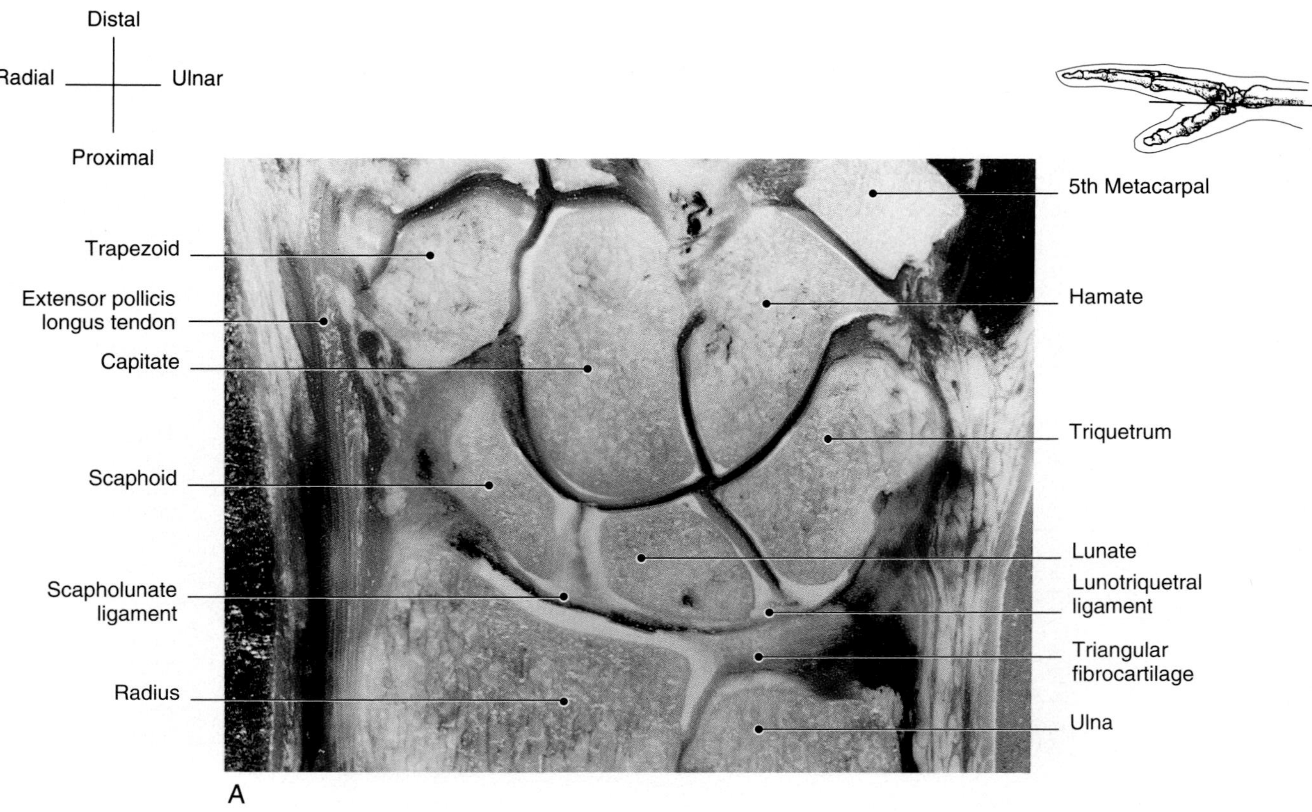

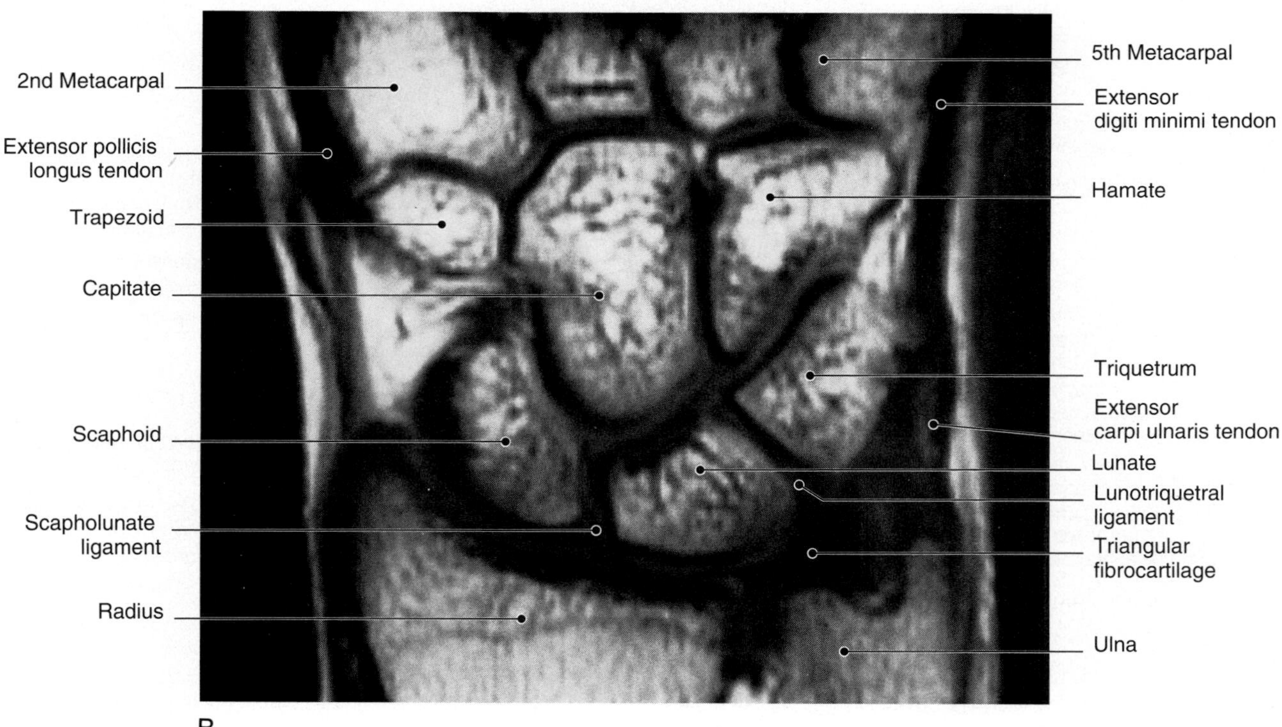

Figure 1–17. Midcoronal cryoplaned section *(A)* and MR image (TR, 700 ms; TE, 20 ms) *(B)* of the wrist.

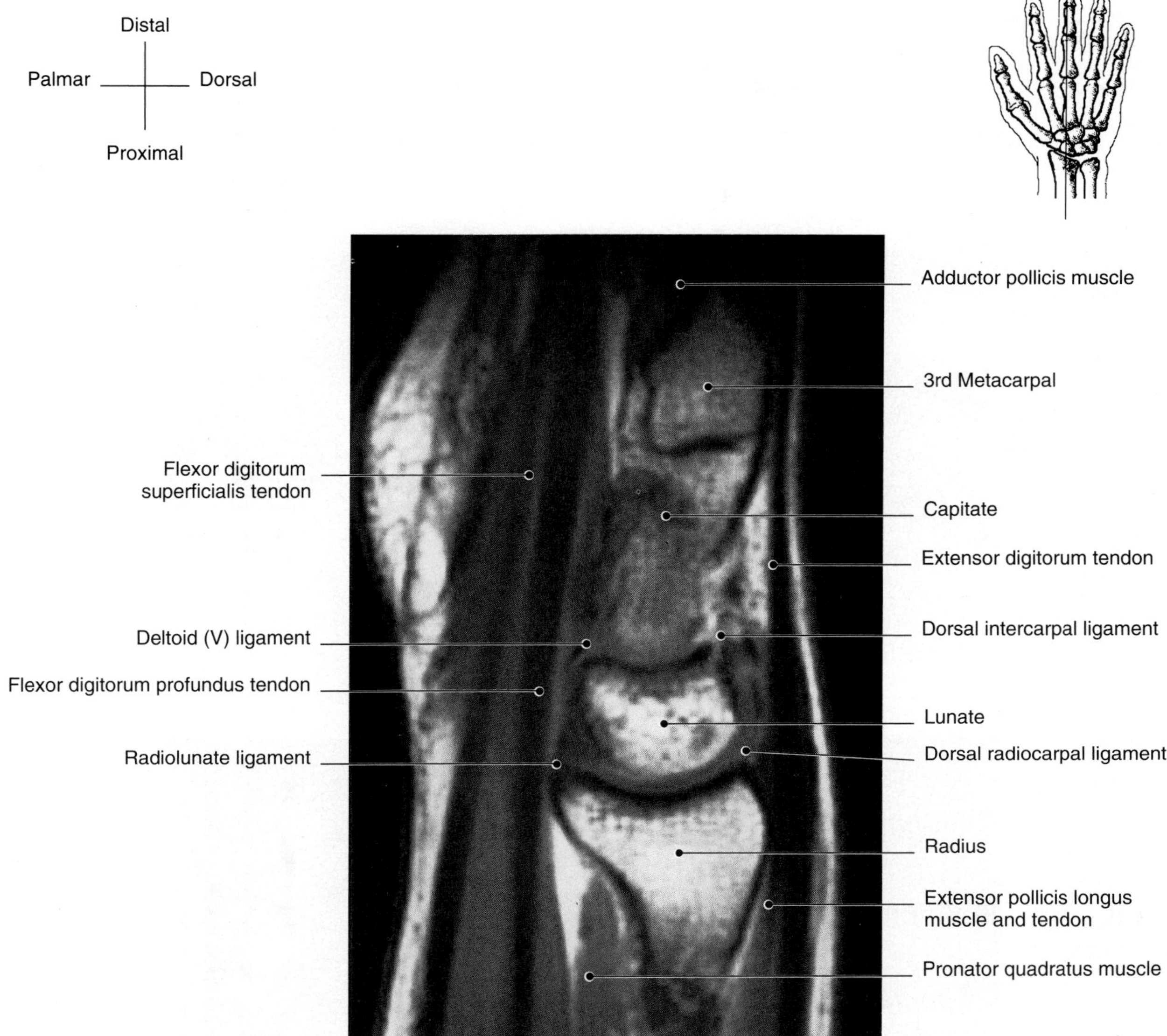

Figure 1–18. Sagittal MR image of the wrist (TR, 600 ms; TE, 20 ms) through the lunate.

FINGERS

Axial Images

The metacarpophalangeal joint is supported by a number of ligaments. The palmar plate (or volar plate) is a fibrocartilaginous plate extending from the base of the proximal phalanx to the metacarpal head (Fig. 1–19). Flexor digitorum profundus and superficialis tendons can be recognized as they run toward their respective digits. The flexor superficialis tendon is a single tendon superficial to the flexor profundus tendon at the level of the metacarpus. It subsequently diverges into two slips: one lies on each side of the profundus tendon at the level of the proximal phalanx and then runs deep to the profundus tendon at the level of the proximal interphalangeal joint, where the two slips reunite.[6] The extensor expansion can be clearly depicted on the dorsum of fingers. The collateral ligaments are seen along the sides of the metacarpophalangeal joints. Ligaments and tendons generally are of low signal intensity. The volar plate may have a higher (intermediate) signal intensity than the other ligaments.

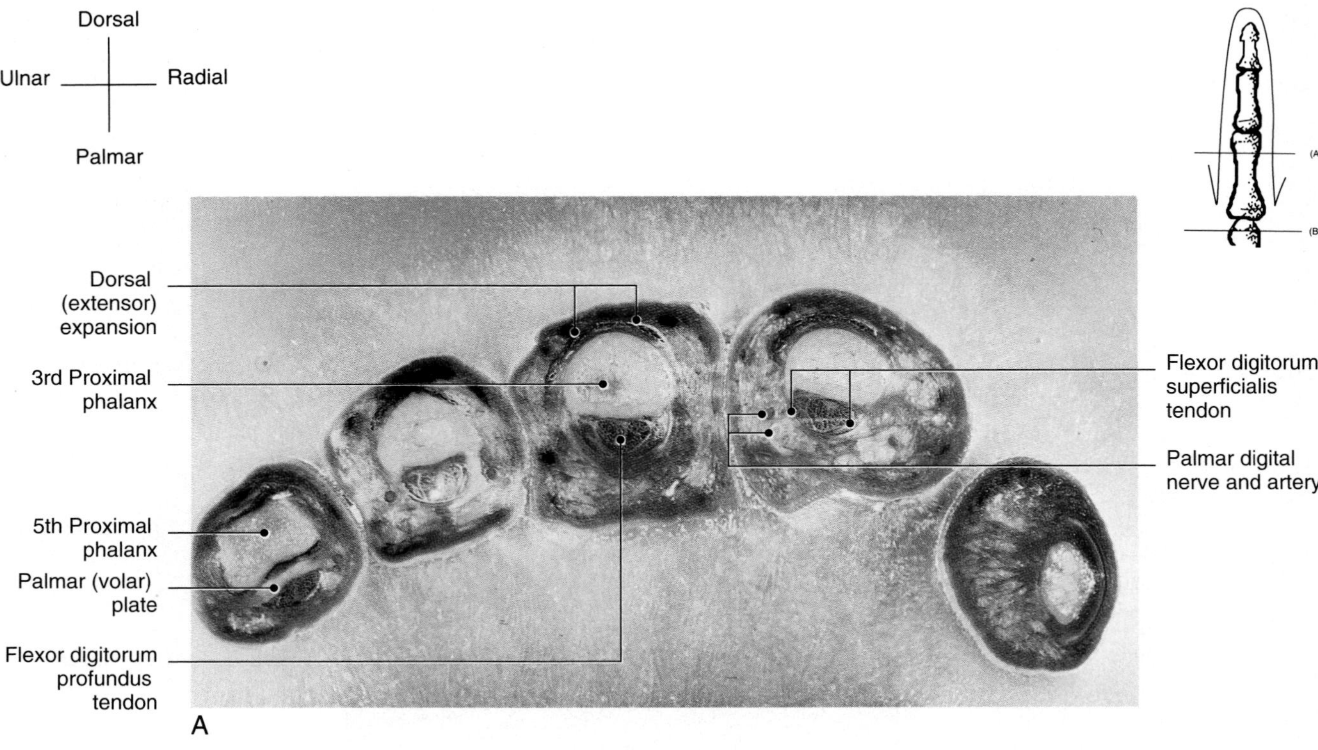

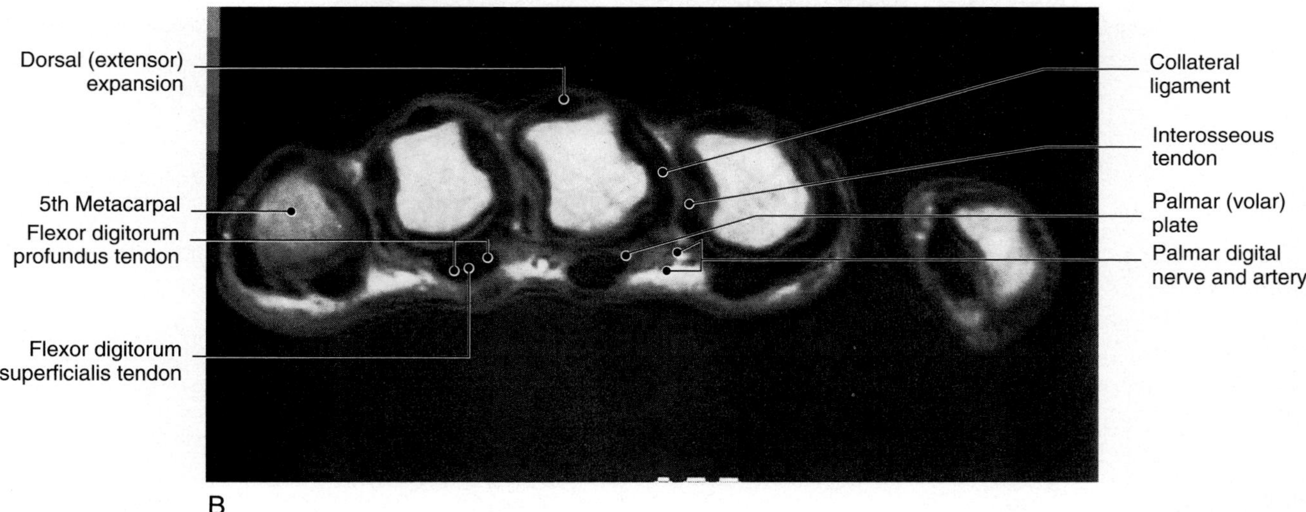

Figure 1-19. *A*, Axial cryoplaned section of the hand at the level of the third proximal phalanx. *B*, Axial MR image (TR, 700 ms; TE, 20 ms) of the hand at the level of the metacarpus. Note that *A* and *B* are not of the same level.

HIP

Axial Images

Axial images offer the best views of the anterior and posterior anatomic relationships of the hip joints (Fig. 1–20). Signal intensity in the acetabulum and femoral head is high owing to the presence of marrow fat. The stellate collection of low signal intensity represents weight-bearing trabeculae within the femoral head.[8] The sciatic nerve, a structure of low signal intensity, is located directly posterior to the posterior aspect of the ischium. The external iliac vessels, also of low signal intensity, course anterior to the pubis and medial to the iliopsoas muscle. The muscle groups exhibit intermediate signal intensity and are separated by fatty fascia of high signal intensity. The superficial muscles are the sartorius, tensor fasciae latae, gluteus medius, and gluteus maximus. The iliopsoas muscle is directly anterior to the femoral head. The gluteus minimus muscle is deep to the gluteus medius muscle. The obturator internus muscle lines the inner aspect of the acetabular column.

At the level of the femoral neck, the iliofemoral ligament crosses obliquely anterior to the femoral neck. The sciatic nerve runs posterolateral to the ischium, dorsal to the gemellus muscle, and deep to the gluteus maximus muscle.

Coronal Images

On anterior coronal images, the fovea capitis of the femoral head appears as a defect of low signal intensity (Fig. 1–21). This is an insertion site for the ligamentum teres. The iliofemoral ligament bends lateral to the femoral head and neck. On midcoronal images, from superior to inferior, the internal obturator, the external obturator, and the iliopsoas muscles fan out medial to the femoral head. The external obturator muscle crosses the femoral neck on posterior coronal images (Fig. 1–22).

The signal characteristics of the marrow and the articular cartilage of the femoral head vary with age. In children, the epiphysis of the femoral head is of high signal intensity owing to predominant marrow fat. The metaphyseal marrow contains hematopoietic red marrow and is of homogeneous, intermediate signal intensity on both T1- and T2-weighted images. The physeal plate, of low signal intensity, separates the epiphysis from the metaphysis. The thick, nonossified cartilage is intermediate in signal intensity.

With maturity and physeal closure, the red marrow in the metaphysis is gradually replaced by fatty marrow. Areas of uniformly high signal intensity appear in the epiphysis, greater and lesser trochanters, and inferomedial aspect of the femoral head. In middle-aged adults, many small areas of low signal intensity representing hematopoietic marrow are identified in the intertrochanteric region in addition to the fatty region found in younger adults. With increasing age, 89 per cent of normal adults older than 50 years have uniform high signal intensity throughout the proximal femur.[16]

In children and adults, linear striations of low signal intensity represent weight-bearing trabeculae, which course from superomedially in the femoral head to inferolaterally in the femoral neck. The physeal scar, a curvilinear band of low signal intensity, crosses the high signal intensity marrow of the femoral head transversely.[8] On T2-weighted images, the physeal scar cannot be seen owing to less contrast between scar and marrow, which also is of low signal intensity.

Sagittal Images

Sagittal images offer the best views of the articular cartilage of the acetabulum and femoral head. The transverse ligament, of low signal intensity, runs caudal to the acetabulum and forms the cotyloid ligament or acetabulum labrum. The iliofemoral ligament extends anterior to the joint. The iliopsoas muscle crosses anterior to the hip joint, and its tendon inserts into the lesser trochanter. The sartorius muscle lies most anterior.

KNEE

Axial Images

Axial images offer the best views of the patellofemoral compartment (Fig. 1–23). The articular cartilage of the patellar facets and femoral trochlea are depicted well on gradient-echo images. The vastus medialis and vastus lateralis are medial and lateral, respectively, to the patella and femur.

The entire extent of both menisci cannot be appreciated reliably on axial MR images. Morphologically, the medial meniscus has a more open C-like shape and a wider posterior horn than the lateral meniscus.[17] At the same level, the transverse ligament may be depicted as a band of low signal intensity connecting the anterior horns of both menisci. Although the axial view is not useful for assessing cruciate ligaments, the anterior cruciate ligament can be identified along the *lateral* femoral condyle in the intercondylar notch. The posterior

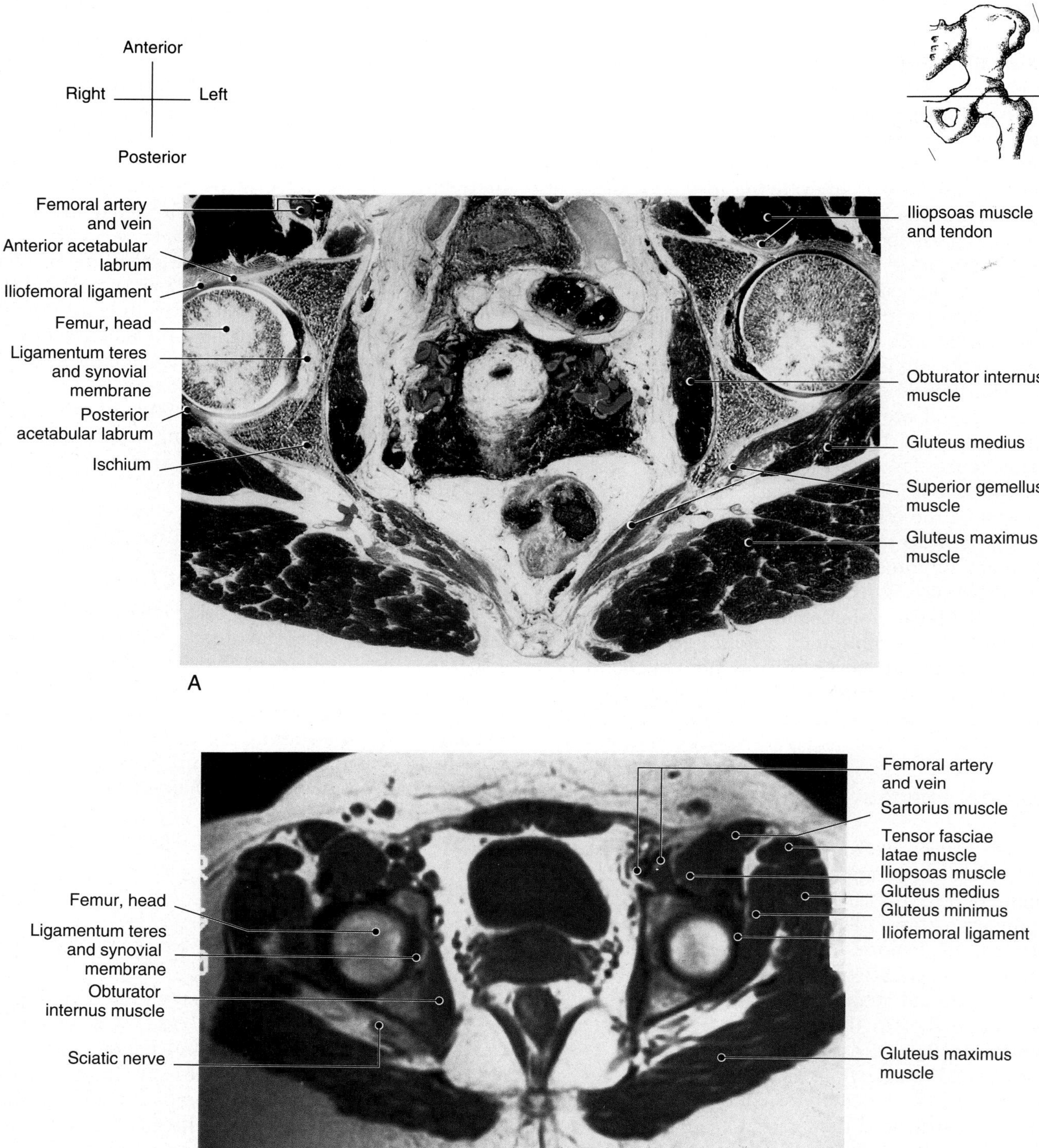

Figure 1–20. Axial cryoplaned section *(A)* and MR image (TR, 600 ms; TE, 20 ms) *(B)* of the hip at the level of the femoral head.

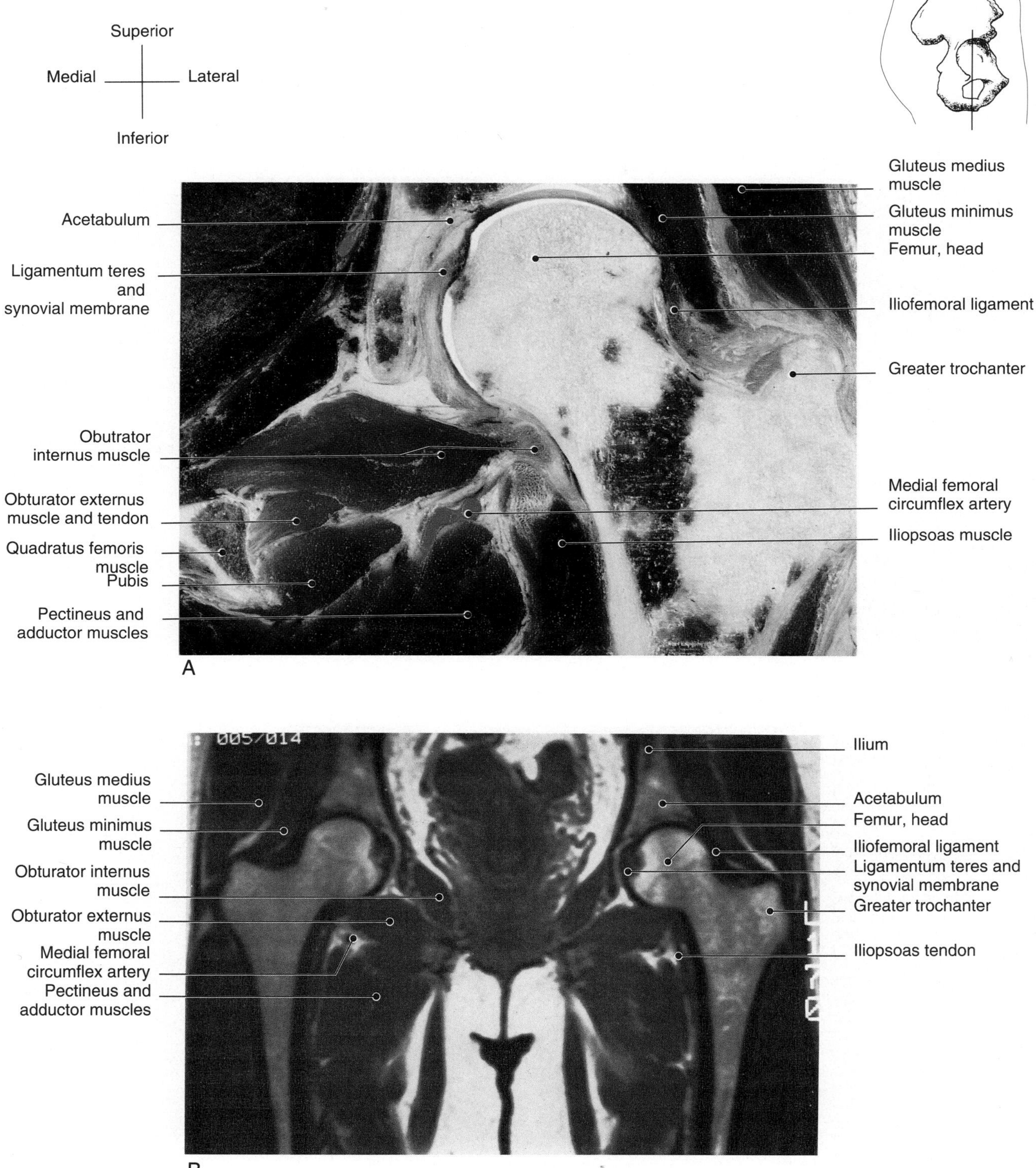

Figure 1-21. Anterior coronal cryoplaned section *(A)* and MR image (TR, 600 ms; TE, 20 ms) *(B)* of the hip.

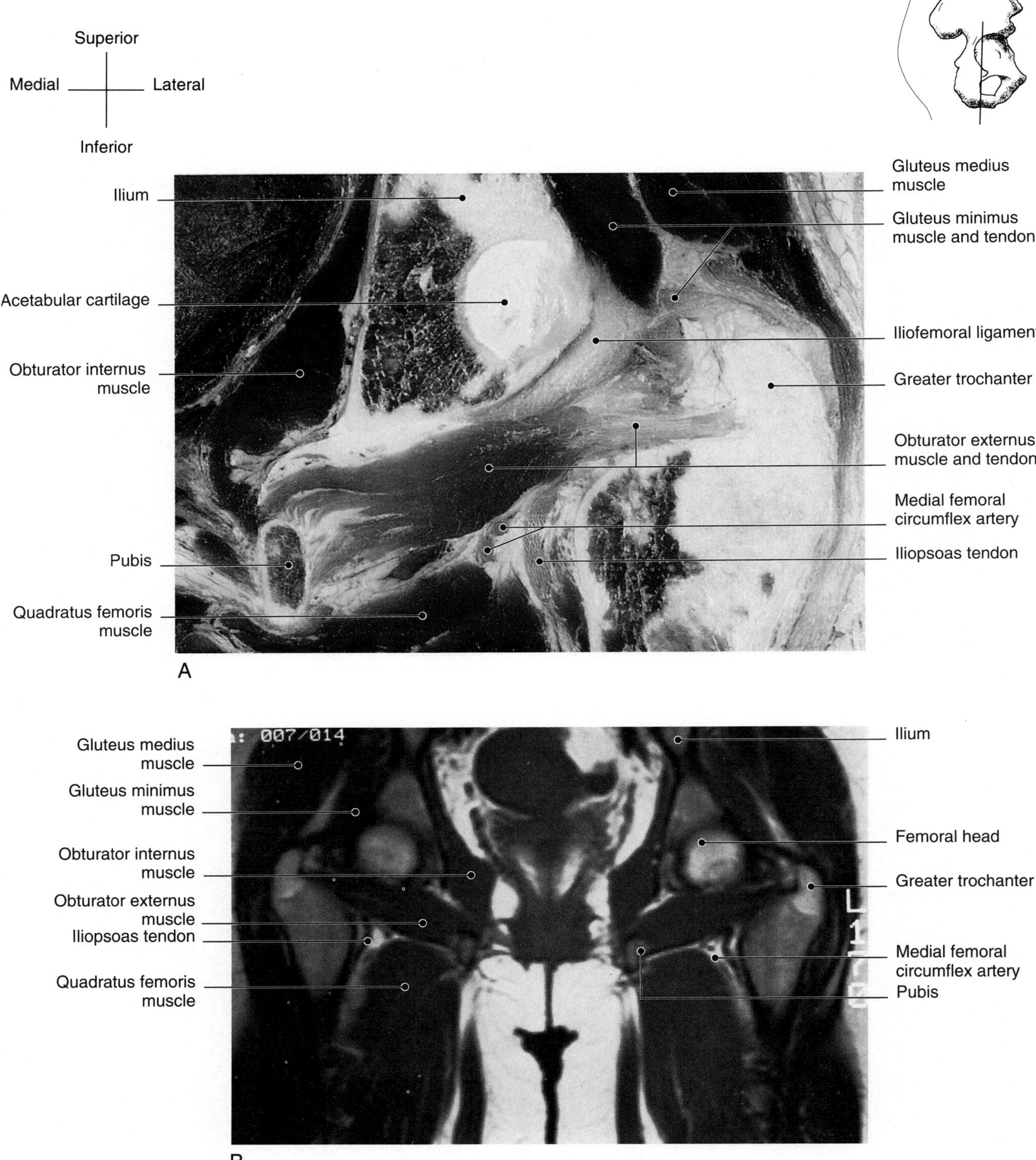

Figure 1-22. Posterior coronal cryoplaned section *(A)* and MR image (TR, 600 ms; TE, 20 ms) *(B)* of the hip.

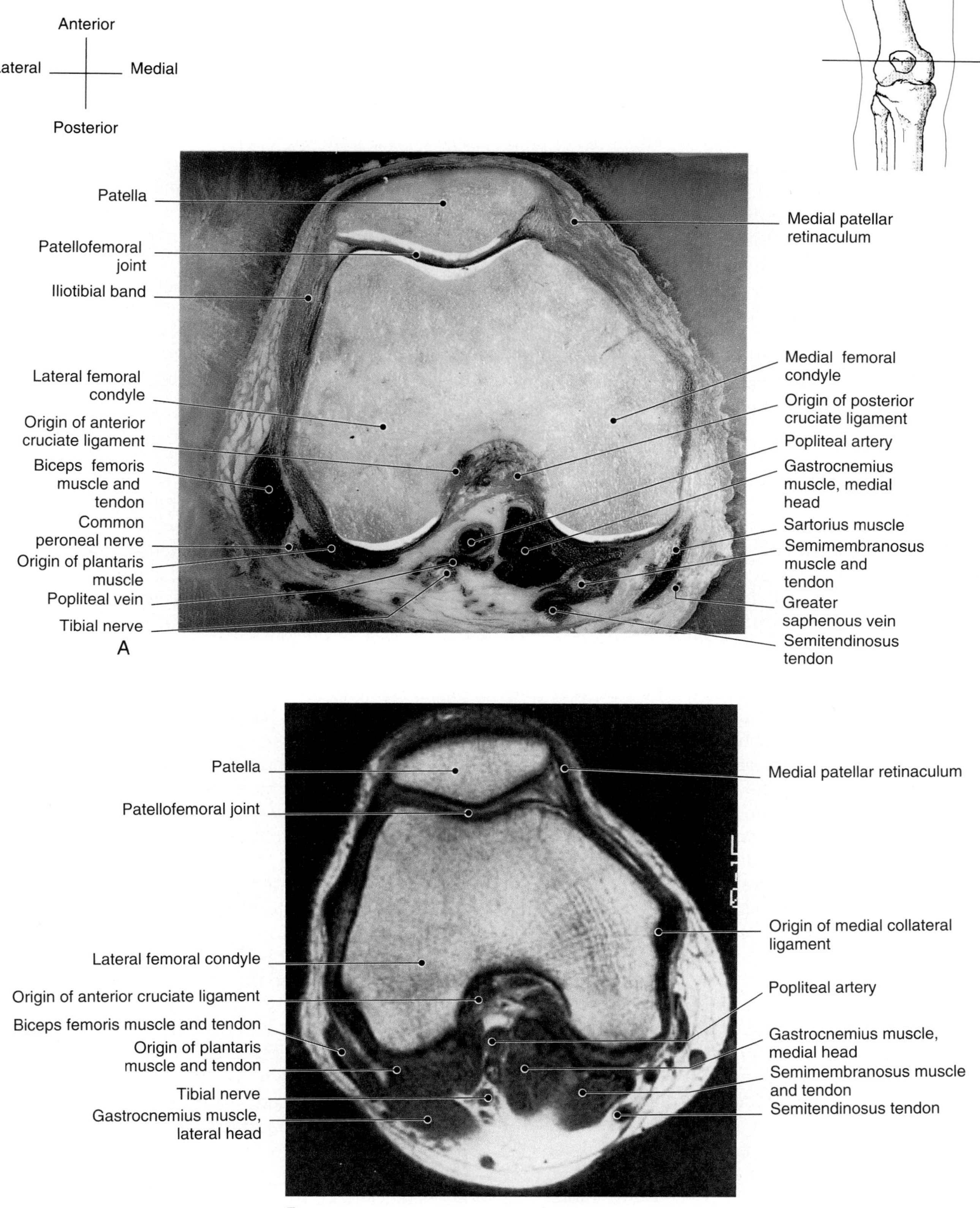

Figure 1–23. Axial cryoplaned section *(A)* and MR image (TR, 500 ms; TE, 20 ms) *(B)* of the knee at the level of the patellofemoral joint. (Case *A* courtesy of Murray A. Reicher, M.D., San Diego, Calif.)

cruciate ligament is circular in cross section and attaches to the *medial* femoral condyle in the intercondylar notch.

Posteriorly, the semimembranosus tendon and the medial and lateral heads of the gastrocnemius muscle can be identified separately. The popliteal artery and vein (deep and medial), tibial nerve, and peroneal nerve (lateral) also can be distinguished. For example, a popliteal cyst can be distinguished anatomically from a popliteal artery aneurysm. Medially are seen the medial collateral ligament and hamstring muscles. Laterally, from anterior to posterior, are the iliotibial band, lateral collateral ligament, and the biceps femoris tendon. The iliotibial tract continues anteriorly to the lateral patellar retinaculum. The medial and lateral patellar retinacula are Y-shaped structures that extend from the quadriceps tendon and fascia lata.

Coronal Images

Coronal gradient-echo images show uniquely the tibial (medial) and fibular (lateral) collateral ligaments (Fig. 1–24). Laterally, in the posterior coronal planes, the fibular collateral ligament and the biceps femoris tendon form a conjoined tendon before insertion on the fibula. The fibular collateral ligament stretches between the fibular head and the lateral femoral condyle. Because this ligament runs obliquely, a single image often cannot depict its entire course. Medial to the fibular collateral ligament, from superficial to deep, are the fibrous capsule and the popliteal tendon. The posterior horn of the lateral meniscus is not attached firmly from the peripheral soft tissues, through which the popliteal tendon passes. In the popliteal fossa, the meniscofemoral ligament is seen occasionally arising from the posterior horn of the lateral meniscus, coursing obliquely, and attaching to the lateral surface of the medial femoral condyle.[18] The course of the popliteal vein (lateral) and artery (medial) are vertically in the middle of the fossa. The coronal plane is parallel to the posterior curve of the C-shaped menisci; therefore, the posterior horn of each meniscus may be seen as a continuous horizontal band of low signal intensity.

Medially, in the midcoronal plane, the continuation of the tibial collateral ligament can be identified. It extends between the medial femoral condyle and the medial tibial metaphysis, approximately 9 cm below the joint line.[11] Unlike the lateral meniscus, the medial meniscus is firmly attached at its periphery to peripheral connective tissue. The inner border of the medial meniscus is free of attachment. The average width of the middle zone of the medial meniscus is 10 mm.[11] The cruciate ligaments cannot be seen in their entirety on one image because of their oblique course. The anterior cruciate ligament originates along the intercondylar surface of the lateral femoral condyle and extends obliquely to the anterior portion of the medial tubercle of the intercondyle eminence of the tibia. The posterior cruciate ligament is delineated well on the anterior and midcoronal images. It attaches to the intercondylar surface of the medial femoral condyle, passes obliquely, and attaches to the tibial plateau, just posterior and lateral to the intercondylar eminence.

Anterior coronal images show the iliotibial band blending with the lateral patellar retinaculum. The relatively high signal intensity of the fat between the anterior horn of the medial meniscus and the medial patellar retinaculum on T1-weighted images should not be mistaken for a meniscocapsular tear.[11]

Sagittal Images

Sagittal images are useful for showing menisci and cruciate ligaments. The medial and lateral menisci normally are of homogeneous low signal intensity regardless of the pulse sequence. After the age of 20 years, areas of increased signal intensity, not extending to the articular surface, frequently appear within the menisci, representing degeneration.[17] The bodies of both menisci have a homogeneous, black, and uninterrupted or bow-tie appearance between the convex femoral and tibial articular cartilages, which appear as thin lines of intermediate signal intensity interposed between the meniscus and cortical bone, which are of low signal intensity. However, the articular cartilage of the femur and tibia is well appreciated on gradient-echo images.

On approaching the midline of images, the anterior and posterior horns of both menisci appear as two opposing triangles (Fig. 1–25). The triangular tip should be sharp in appearance. The posterior horn is larger in both height and width than the anterior horn. The height of the medial meniscus peripherally is 3 to 5 mm.[11,17] This relation is important for evaluation of bucket-handle tears, which can appear as a normal-shaped but diminutive posterior horn.

Viewing the intercondylar notch from medially to laterally, the posterior cruciate ligament is first seen (Fig. 1–26). The entire posterior cruciate ligament is demonstrated in 95 per cent of patients when the knee is in slight external rotation.[11] Usually it appears as an arclike black band extending from the anterolateral side of the medial femoral condyle to the posterolateral intercondylar region of the tibia. With the leg in extension, it is posteriorly convex or buckled; it becomes taut when the knee is flexed.

Proceeding laterally, the anterior cruciate liga-

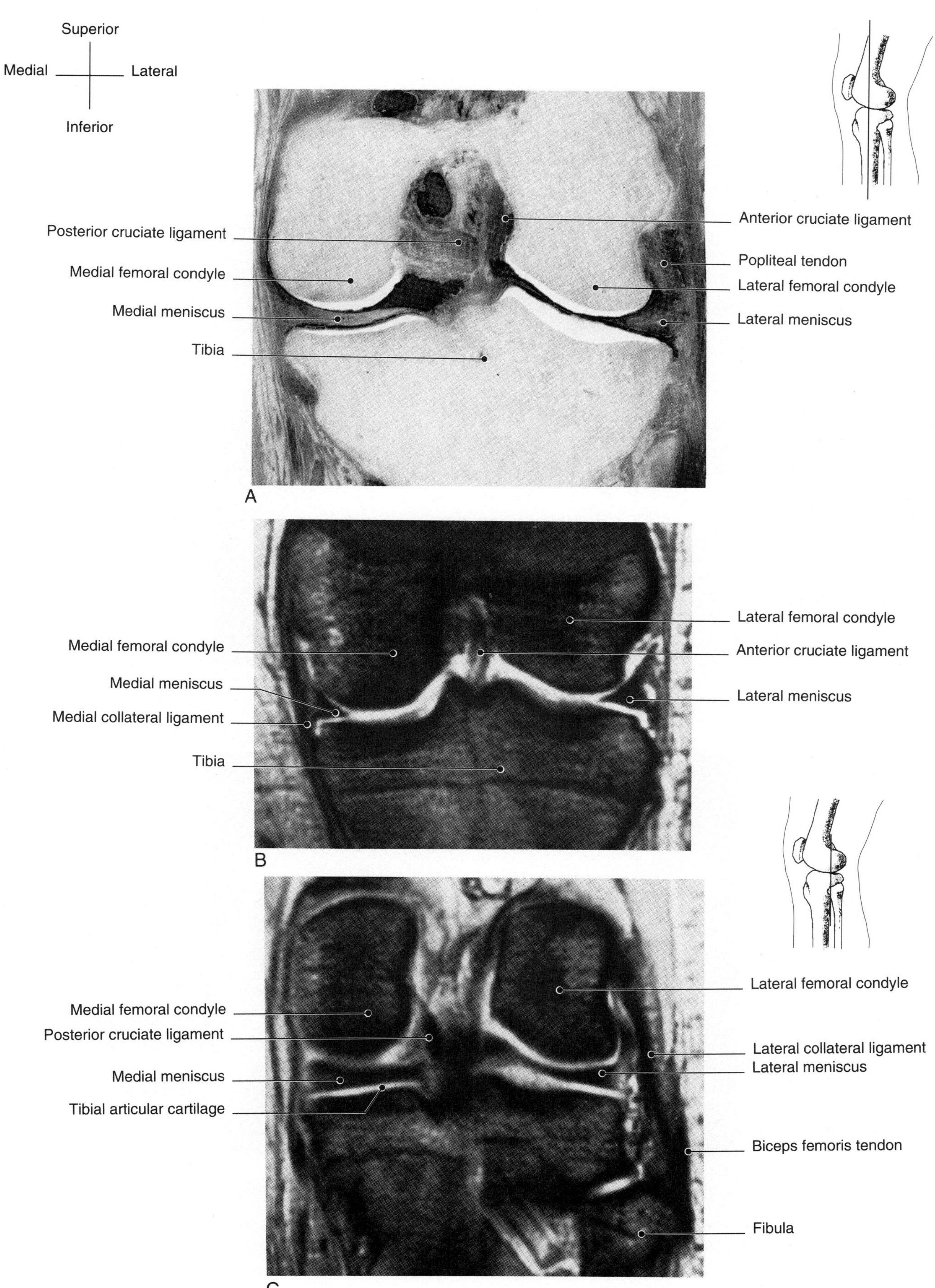

Figure 1–24. Midcoronal cryoplaned section of the knee *(A)* and midcoronal *(B)* and posterior coronal *(C)* MR images (TR, 400 ms; TE, 15 ms; flip angle, 30 degrees) of the knee.

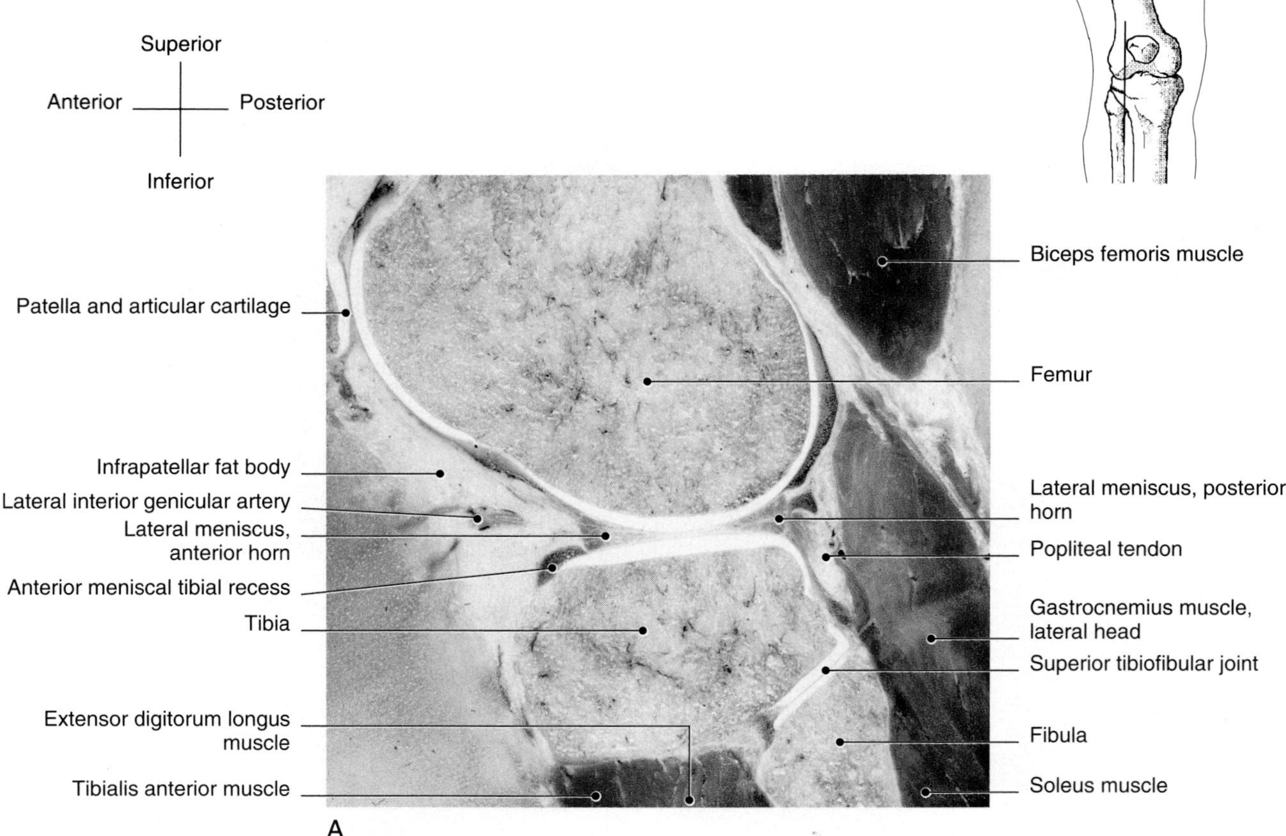

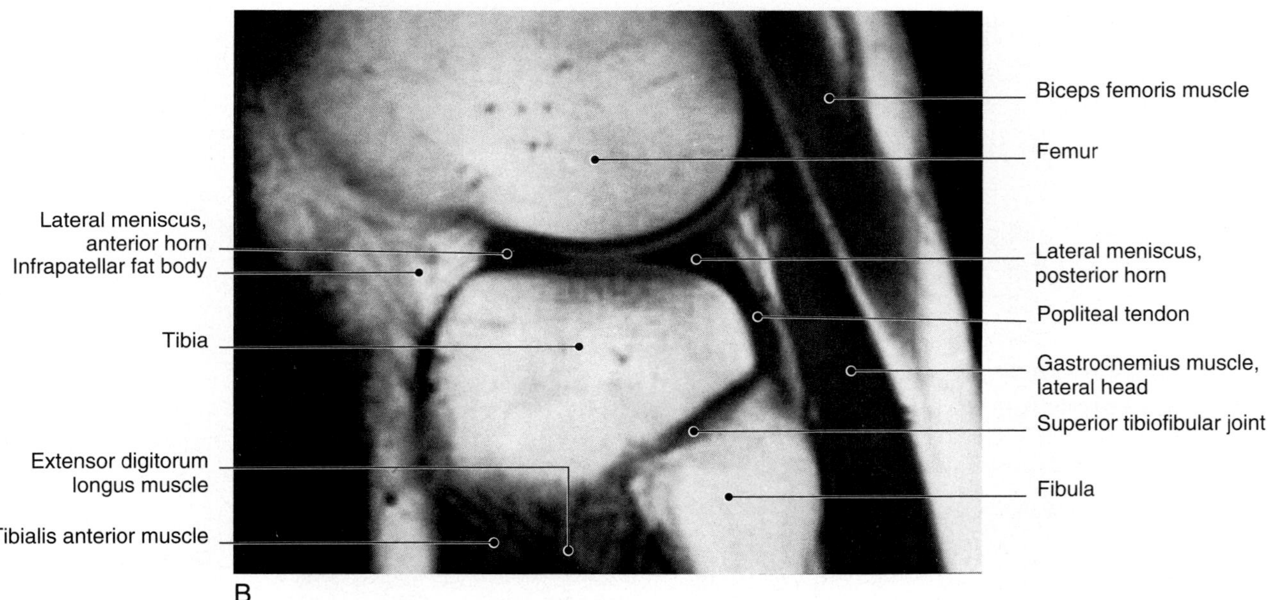

Figure 1–25. Paramedian sagittal cryoplaned section *(A)* and MR image (TR, 800 ms; TE, 20 ms) *(B)* of the knee at the level of the lateral meniscus.

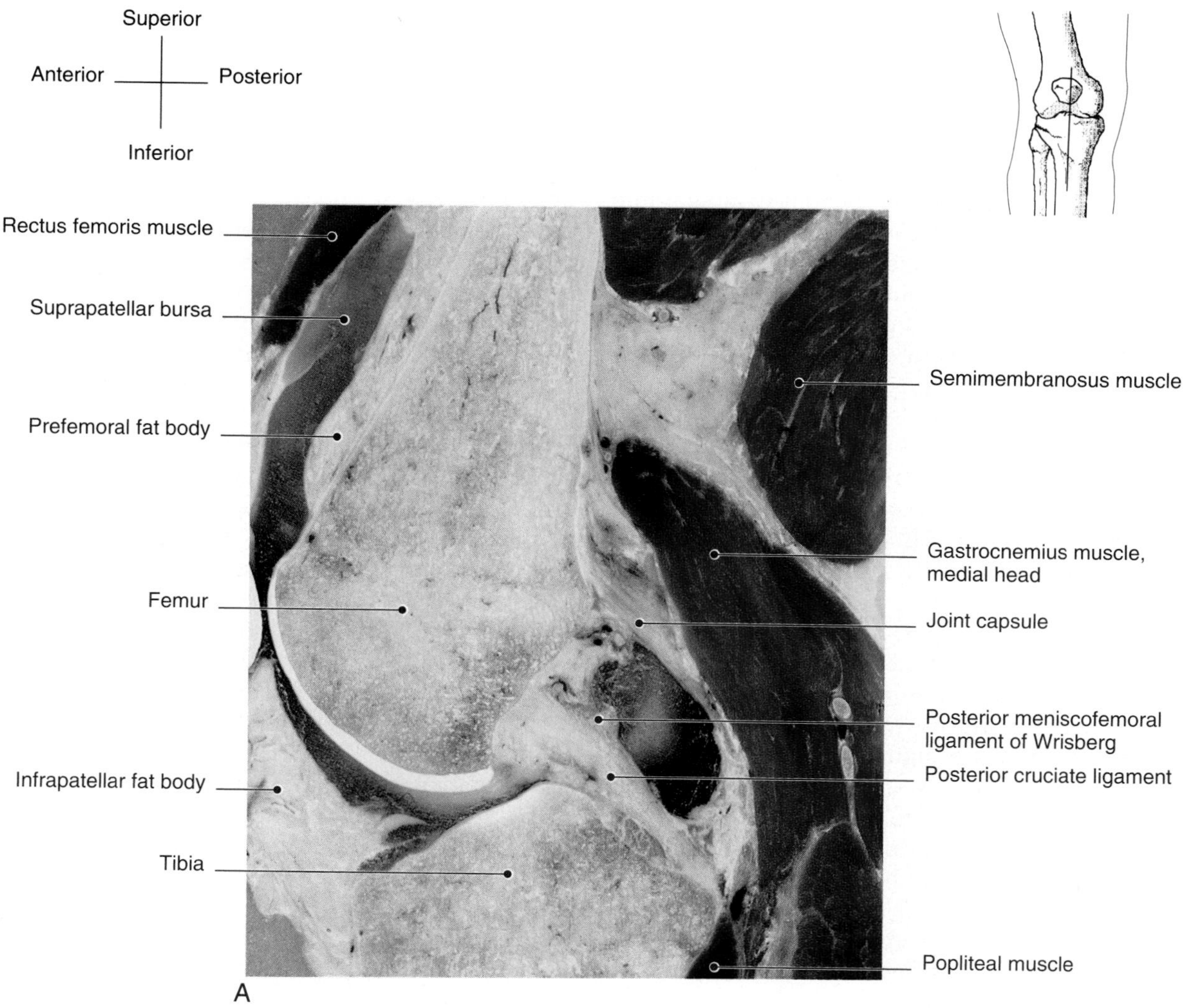

Superior
Anterior
Posterior
Inferior
Rectus femoris muscle
Suprapatellar bursa
Prefemoral fat body
Femur
Infrapatellar fat body
Tibia
Semimembranosus muscle
Gastrocnemius muscle, medial head
Joint capsule
Posterior meniscofemoral ligament of Wrisberg
Posterior cruciate ligament
Popliteal muscle
A

Suprapatellar fat body
Quadriceps tendon
Prefemoral fat body
Patella
Femur
Patellar ligament
Infrapatellar fat body
Tibia
Semimembranosus muscle
Gastrocnemius muscle, medial head
Posterior meniscofemoral ligament of Wrisberg
Posterior cruciate ligament
B

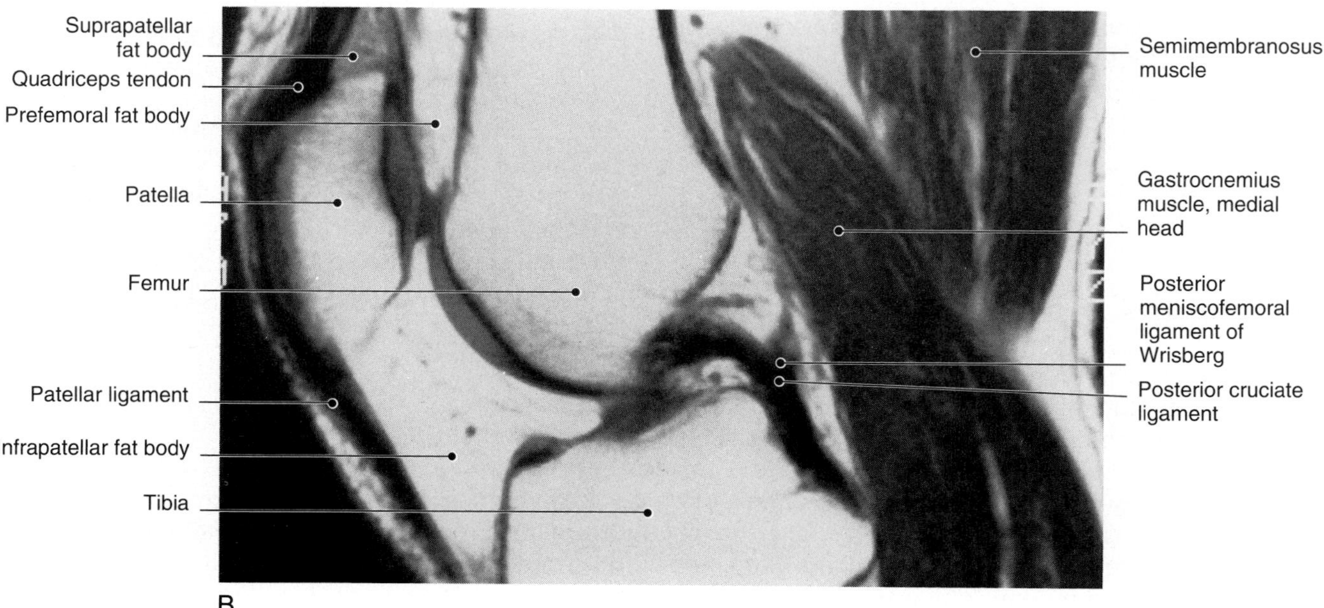

Figure 1–26. Sagittal cryoplaned section *(A)* and MR image (TR, 800 ms; TE, 20 ms) *(B)* of the knee through the posterior cruciate ligament.

ment is seen either as a single black band or as two or three bundles of fibers extending from the posteromedial aspect of the lateral femoral condyle to the anteromedial edge of the proximal tibia (Fig. 1–27). The normal thickness of the anterior cruciate ligament is 10 mm. The radiologist is advised to avoid interslice gaps in imaging the ligment when a 5-mm slice thickness is used.[11] With the knee properly positioned in 10 to 15 degrees of external rotation, the anterior cruciate ligament can be seen on a single sagittal image in at least 90 per cent of patients.[10,11] The anterior cruciate ligament is of slightly higher signal intensity than the posterior cruciate ligament on both T1- and T2-weighted images because of the partial volume averaging of the lateral femoral condyle. The fiber bundle striations of the anterior cruciate ligament are prominent at the femoral and tibial attachment sites. This ligament is relatively taut in the extended position and lax in the flexed position.

On more lateral images, the central ligamentous attachments of the lateral meniscus can be mistaken for tears of the anterior horn.[11] In contrast to the triangular meniscal horns, the anterior ligamentous attachment produces a signal and is rhomboid in shape, pointing obliquely upward. The transverse ligament is closely related to the anterior horn of the lateral meniscus. It may mimic a displaced anterior meniscal fragment. Farther laterally, the bow-tie appearance of the body is more symmetric in the lateral meniscus than in the medial meniscus. The obliquely orientated popliteal bursa is easily misinterpreted as a meniscal tear.

Anteriorly, the quadriceps tendon inserts into the anterior aspect of the superior pole of the patella. The patellar ligament attaches to the inferior pole of the patella. Hoffa's infrapetellar fat pad lies directly posterior to the patellar ligament. The posterior concave free border of the infrapatellar fat pad should be smooth.

ANKLE

Axial Images

On superior ankle images, the distal tibia and fibula divide the longitudinally oriented tendons of the ankle into anterior and posterior compartments (Fig. 1–28). From medial to lateral, the anterior compartment consists of the anterior tibial tendon, the extensor hallucis longus muscle and tendon, the extensor digitorum longus tendon, and the peroneus tertius muscle and tendon. The deep peroneal nerve and anterior tibial artery are positioned directly behind the extensor hallucis longus.

The posterior compartment is occupied by the posterior tibial tendon, the flexor digitorum longus tendon, and the flexor hallucis longus muscle and tendon. The peroneus brevis and peroneus longus tendons are seen directly posterior to the distal fibula. The sural nerve is located posteromedial to the peroneus brevis muscle. The tibial nerve and posterior tibial artery are found between the flexor digitorum longus and flexor hallucis longus. The Achilles tendon is seen in the most posterior aspect. It appears as a elliptic structure of low signal intensity, with a flattened or mildly concave anterior border and convex posterior margin.[13]

On midankle images, the deltoid ligament of low signal intensity (tibiocalcaneal and tibiotalar fibers) can be identified spanning the medial malleolus and talus. On inferior ankle images, the interosseous ligament, also of low signal intensity, runs between the superior aspect of the calcaneus and inferior aspect of the talus. Medial to the calcaneus lie the anteroposteriorly oriented muscles—quadratus plantae and abductor hallucis. The flexor tendons run anteromedially to the calcaneus, whereas the peroneus tendons run anterolaterally to the same bone. The calcaneofibular ligament is located between the peroneus tendons and calcaneus. The plantar calcaneonavicular (spring) ligament is seen on the most plantar images.

Coronal Images

On anterior coronal images, the tibiocalcaneal fibers of the deltoid ligament span the medial malleolus and the calcaneus.[12] The posterior tibial tendon is located medial to the deltoid ligament. The flexor tendons are defined medial to the calcaneus, whereas the peroneus tendons are identified lateral to this bone.

On midcoronal images, the tibiotalar fibers of the deltoid ligament appear as a fan-shaped structure (Fig. 1–29).[12] Posterior coronal images show the lateral collateral ligament, including the anterior talofibular ligament, the posterior talofibular ligament, and the calcaneofibular ligament. The posterior inferior talofibular ligament is identified in the most posterior plane.

Sagittal Images

On midsagittal images, the posterior tibial, flexor digitorum longus, and flexor hallucis longus tendons are seen extending posteriorly to the medial malleolus of the distal tibia (Fig. 1–30). The flexor hallucis longus is the most posterior of the three. It courses inferiorly to the sustentaculum tali.

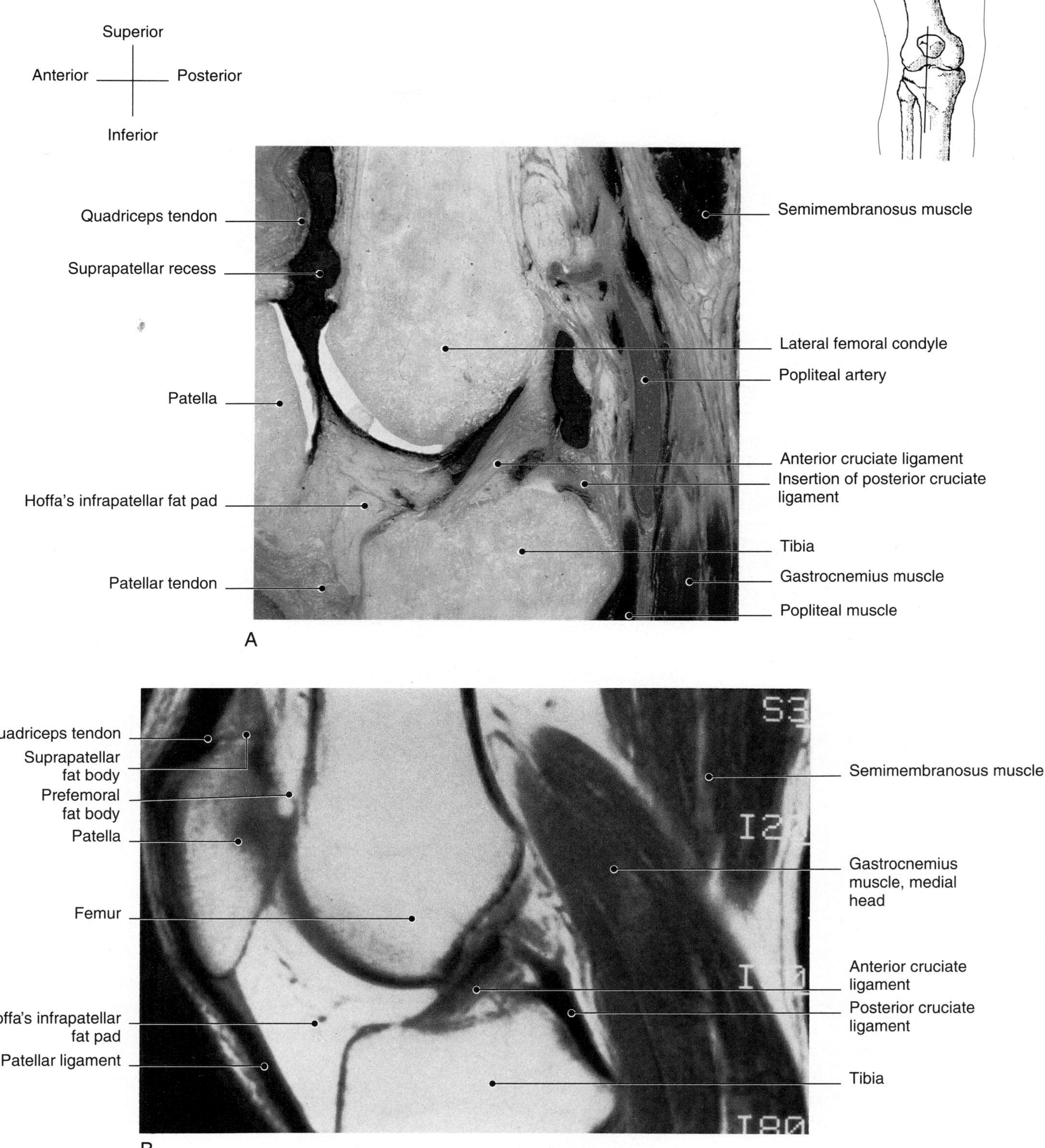

Figure 1–27. Sagittal cryoplaned section *(A)* and MR image (TR, 800 ms; TE, 20 ms) *(B)* of the knee through the anterior cruciate ligament. (Case *A* courtesy of Murray A. Reicher, M.D., San Diego, Calif.)

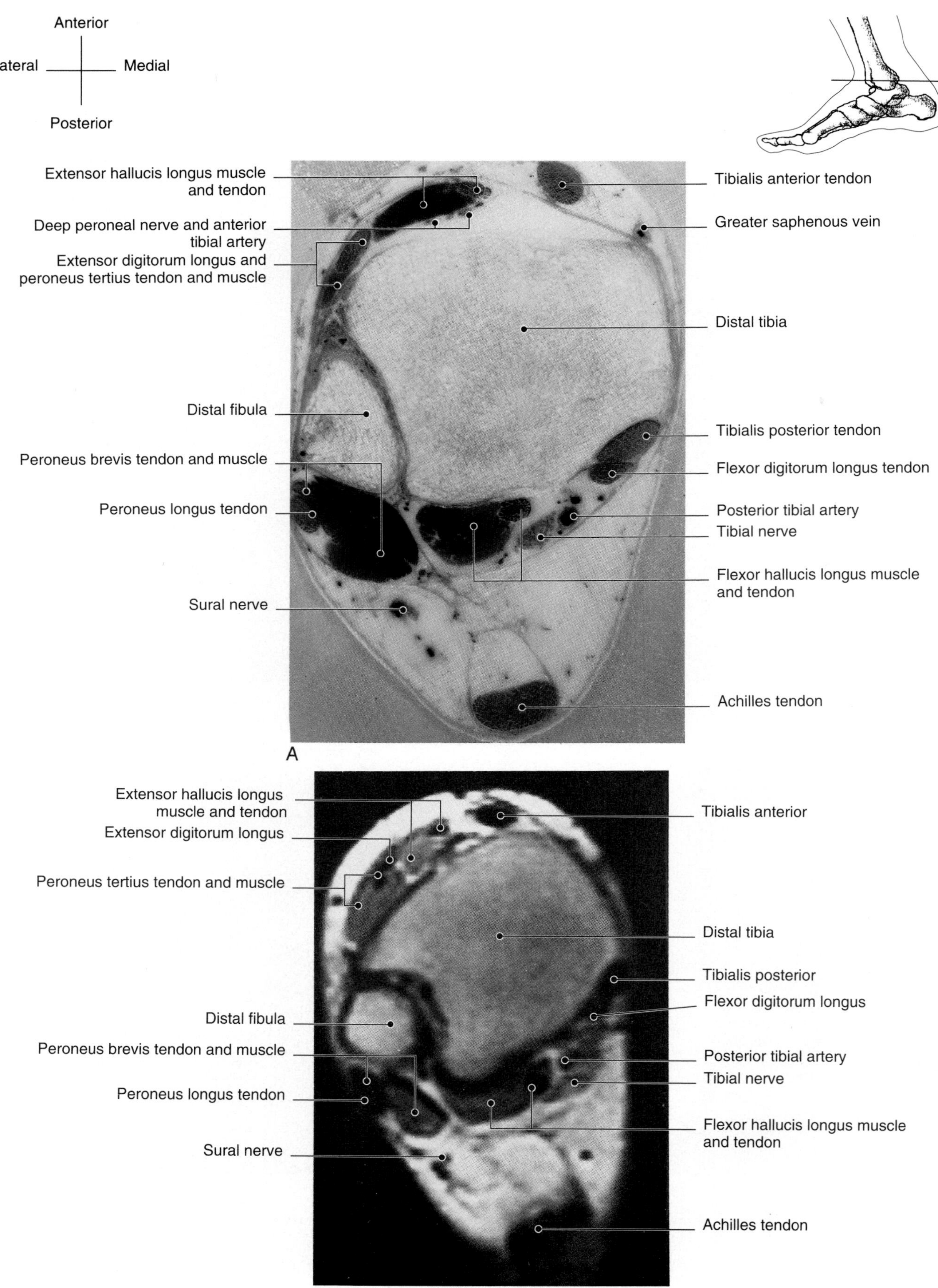

Figure 1–28. Axial cryoplaned section *(A)* and MR image (TR, 2000 ms; TE, 20 ms) *(B)* of the ankle at level of the distal tibia and fibula. (Case *A* courtesy of Scott J. Erickson, M.D., and G. F. Carrera, M.D., Milwaukee, Wis.)

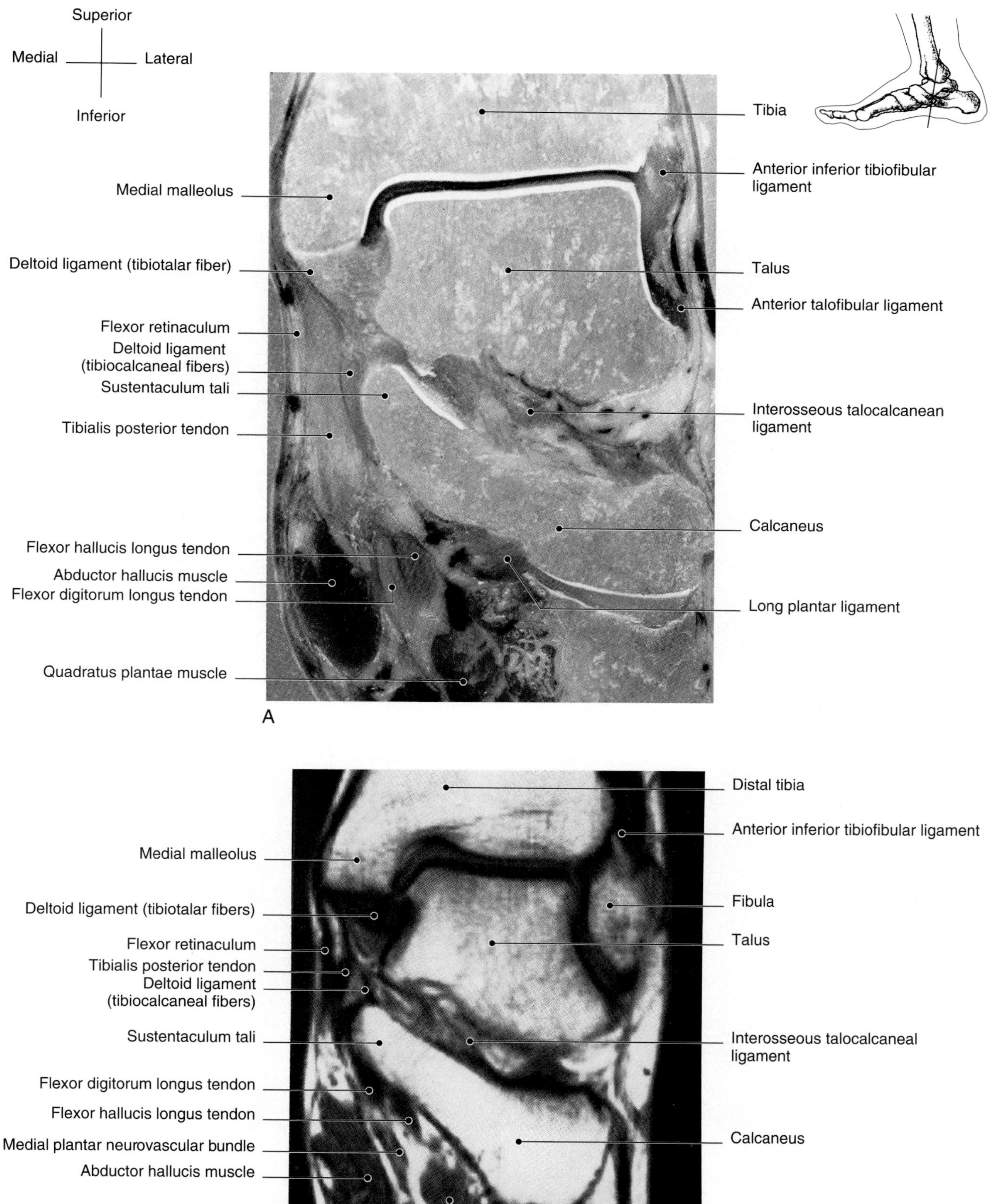

Figure 1–29. Midcoronal cryoplaned section *(A)* and MR image (TR, 500 ms; TE, 20 ms) *(B)* of the ankle. (Case *A* courtesy of Scott J. Erickson, M.D., and G. F. Carrera, M.D., Milwaukee, Wis.)

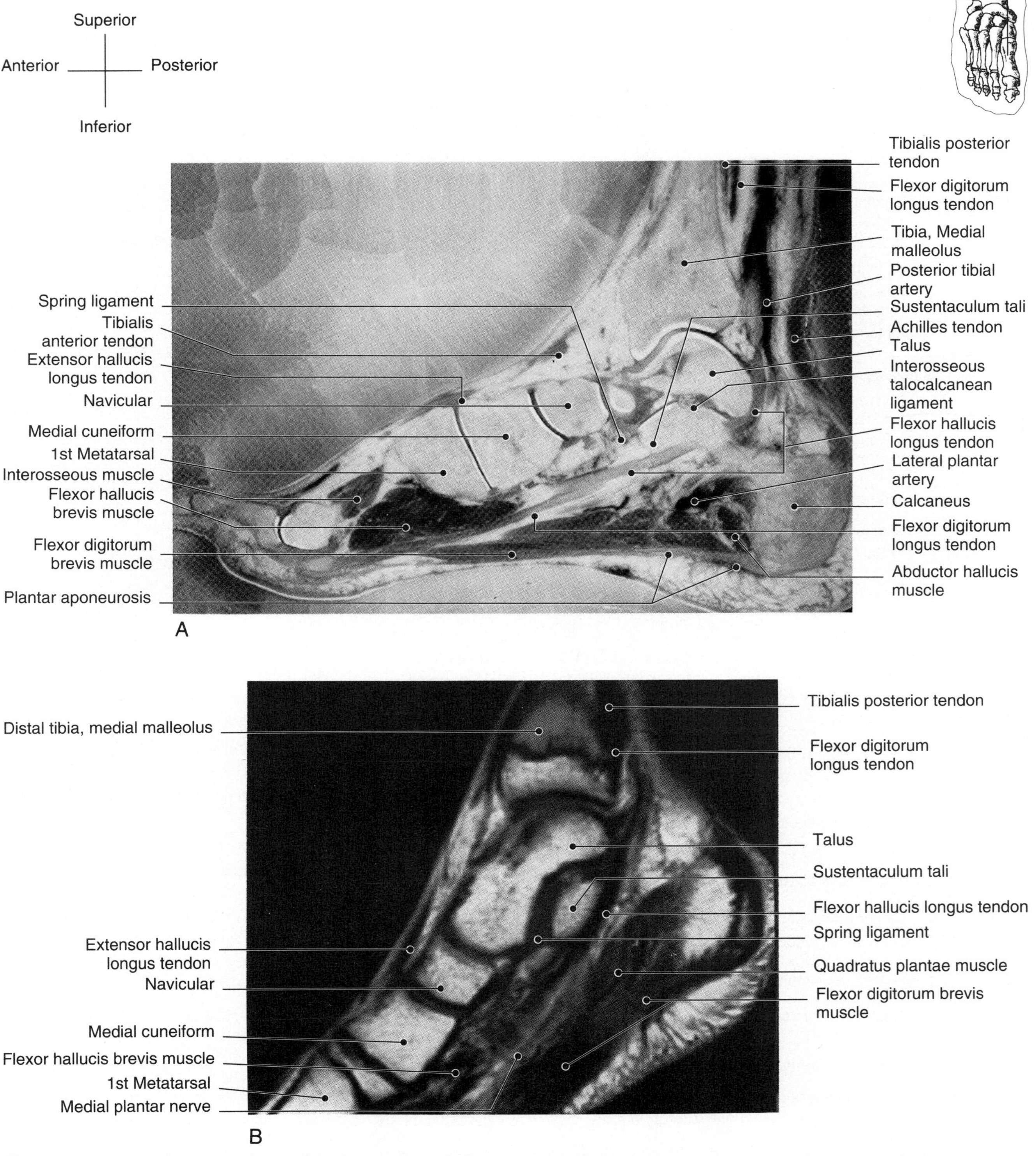

Figure 1–30. Paramedian sagittal cryoplaned section *(A)* and MR image (TR, 500 ms; TE, 20 ms) *(B)* of the ankle and foot through the medial compartment tendons. (Case *A* courtesy of Scott J. Erickson, M.D., and G. F. Carrera, M.D., Milwaukee, Wis.)

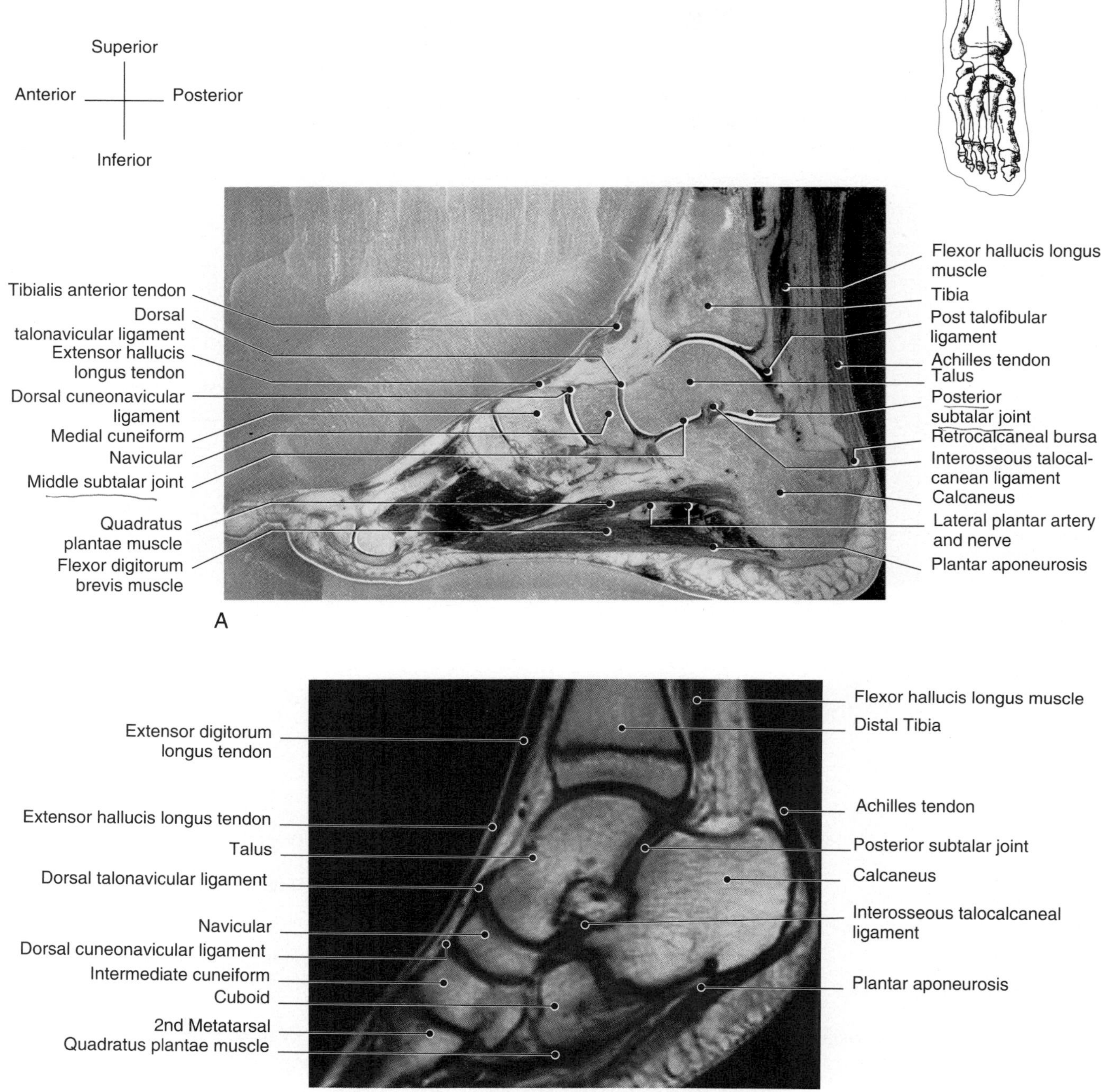

Figure 1–31. Midsagittal cryoplaned section *(A)* and MR image (TR, 500 ms; TE, 20 ms) *(B)* of the ankle. (Case *A* courtesy of Scott J. Erickson, M.D., and G. F. Carrera, M.D., Milwaukee, Wis.)

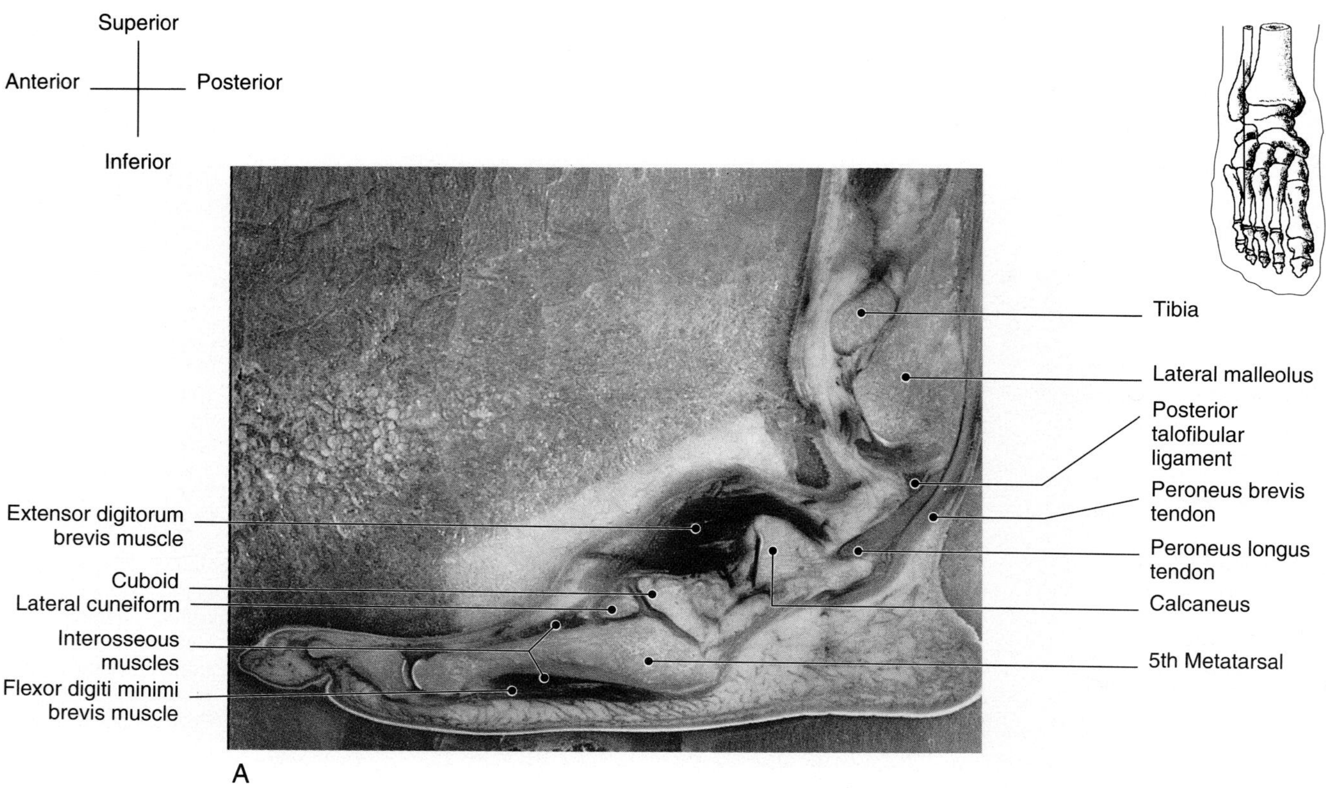

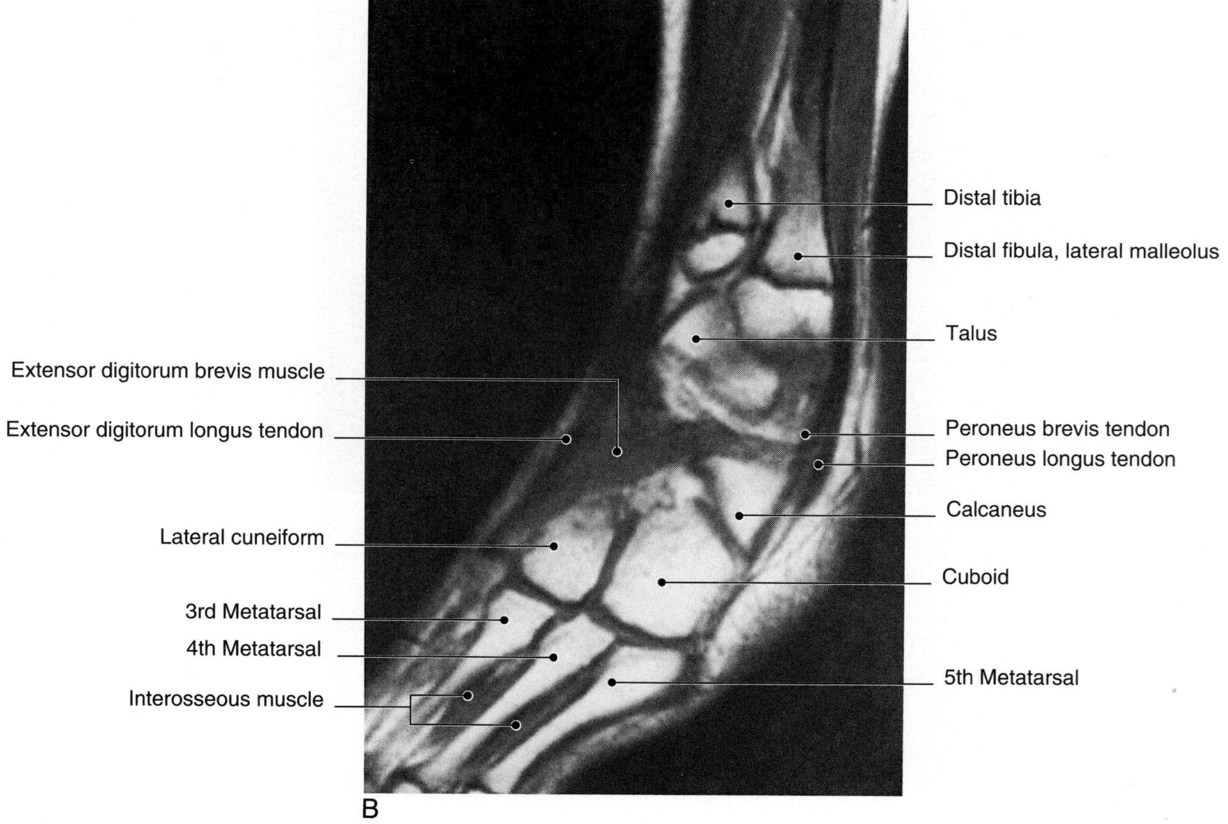

Figure 1–32. Paramedian sagittal cryoplaned section *(A)* and MR image (TR, 500 ms; TE, 20 ms) *(B)* of the ankle and foot through the lateral compartment tendons. (Case *A* courtesy of Scott J. Erickson, M.D., and G. F. Carrera, M.D., Milwaukee, Wis.)

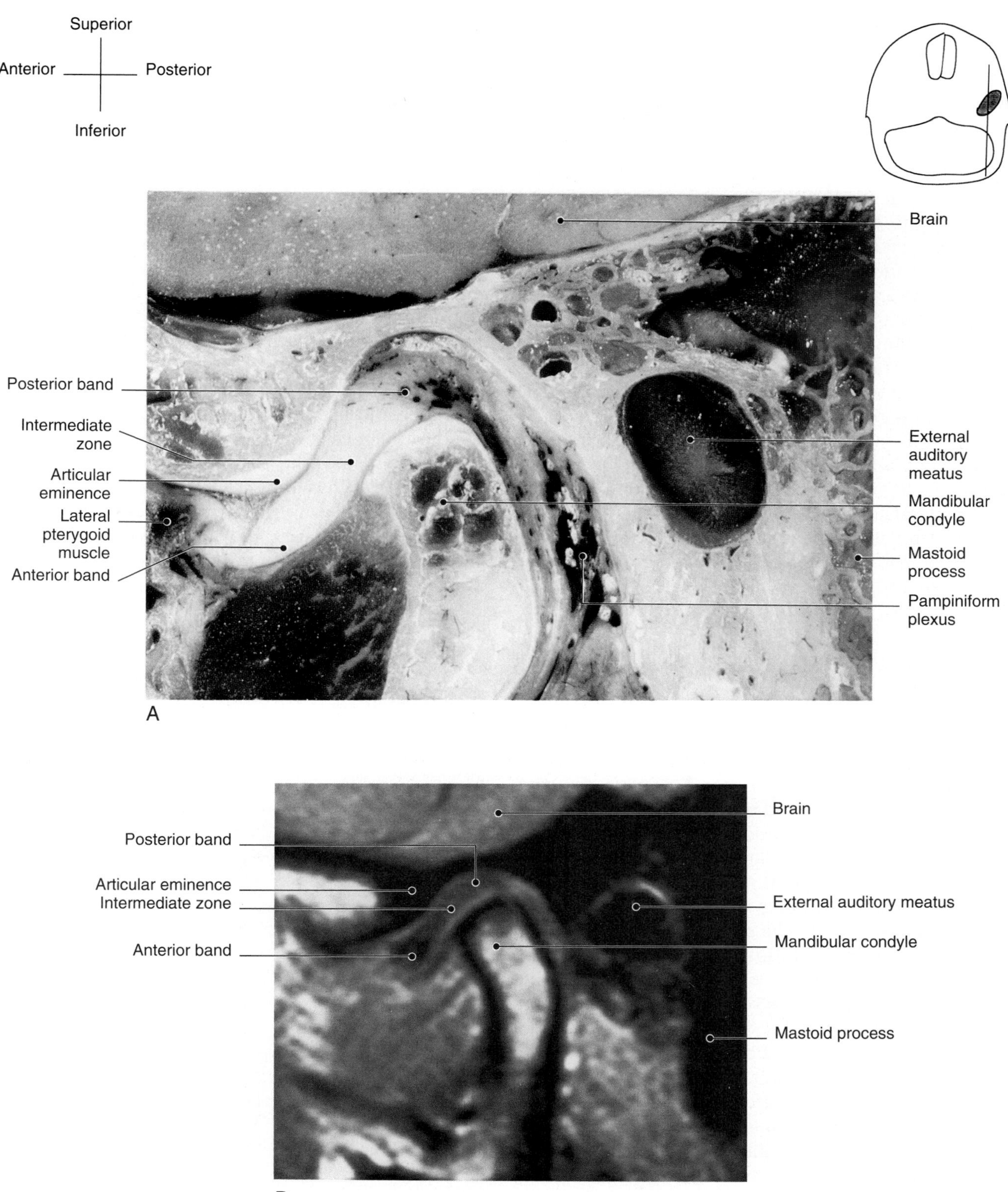

Figure 1–33. Sagittal cryoplaned section *(A)* and MR image (TR, 500 ms; TE, 20 ms) *(B)* of the temporomandibular joint.

On midsagittal images, the long axis of the Achilles tendon is imaged as a thin structure of low signal intensity originating from the gastrocnemius-soleus muscle complex and inserting onto the posterior aspect of the calcaneus (Fig. 1–31). The pre-Achilles fat pad is seen as an area of high signal intensity. On the plantar surface, the plantar aponeurosis attaches to the calcaneus and extends forward to the toes. Deep to the plantar fascia are muscles such as, from superficial to deep, the flexor digitorum brevis, quadratus plantae, and peroneus longus. The articular cartilage surfaces of the tibiotalar, talonavicular, and subtalar joints and the posterior talar dome are well delineated.

The lateral sagittal images show the interosseous ligament between its attachments. The peroneus brevis and peroneus longus tendons pass posterior to the lateral malleolus (Fig. 1–32). The peroneus brevis runs anterior to the peroneus longus muscle below the lateral malleolus and inserts onto the base of the lateral aspect of the fifth metatarsal. The posterior talofibular ligament is visualized as a dot of low signal intensity and is located immediately posterior to the talus.

TEMPOROMANDIBULAR JOINT

Sagittal Images

Two sequences of sagittal images, closed-mouth and partial open-mouth positions, provide the best views of the rotational and translational components of the temporomandibular joint. The bony structures of the temporomandibular joint include glenoid (condylar) fossa, temporal eminence, and mandibular condyle. The articular disk separates the bony structures of the joint. The disk is divided into three parts: an anterior band, thinner intermediate zones, and a thicker posterior band (Fig. 1–33). In the closed-mouth position, 95 per cent of the normal structure of the disk lies with the posterior band within 10 degrees of the 12 o'clock (vertical) position in relation to the mandibular condyle.[4] As the mouth opens, the condyle translates anteriorly to a variable degree, and the disk rotates posteriorly. The intermediate zone represents the weight-bearing zone of the disk. It maintains a consistent relationship between the closest cortical surfaces of the condyle and the eminence during translation.

The anterior band is anchored to the superior belly of the lateral pterygoid muscle. This muscle is obliquely oriented and tends to direct most disk displacements in an anteromedial direction. The posterior band is attached to the vascularized bilaminar zone or retrodiskal pad. A parallel elastic band of low signal intensity may be identified within the bilaminar zone. This band in the bilaminar zone opposes the anteromedial pull of the lateral pterygoid muscle. These opposing forces maintain the disk in its normal position on the condyle. A disrupted retrodiskal elastic band may lead to an unopposed lateral pterygoid muscle, resulting in anteromedial displacement of the disk. A synovial joint space separating the disk from the condylar head is called a lower joint. The upper joint thus separates the disk from the condylar fossa and the temporal eminence. However, the upper and lower joints do not communicate. Hence, a contrast agent injected into either joint should not fill both when the disk is intact.

The normal biconcave disk has a bow-tie appearance and is composed of proteoglycans, similar to the nucleus pulposus of the spine. On T1-weighted images, the central portion of the normal posterior band has a higher (intermediate) signal intensity than the low signal intensity of the anterior band or intermediate zone.[5] On T2-weighted images, the posterior band is of uniformly low signal intensity. The cortical bone of the bony component of the temporomandibular joint has a very low signal intensity in all pulse sequences.

References

1. Bunnell DH, Fisher DA, Bassett LW, et al: Elbow joint: Normal anatomy on MR images. Radiology 165:527, 1987.
2. Chan WP, Resendes M, Genant HK: The skeletal system. *In* Higgins CB, Hricak H, Helms CA (eds): Magnetic Resonance Imaging of the Body. 2nd ed. New York, Raven Press, 1992.
3. Coventry MB: Anatomy of the intervertebral disk. Clin Orthop 67:9, 1969.
4. Drace JE, Enzmann DR: Defining the normal temporomandibular joint: Closed-, partially open-, and open-mouth MR imaging of asymptomatic subjects. Radiology 177:67, 1990.
5. Drace JE, Young SW, Enzmann Dr: TMJ meniscus and bilaminar zone: MR imaging of the structure—diagnostic landmarks and pitfalls of interpretation. Radiology 177:73, 1990.
6. Erickson SJ, Kneeland JB, Middleton WD, et al: MR imaging of the finger: Correlation with normal anatomic sections. AJR 152:1013, 1989.
7. Holt RG, Helms CA, Steibach L, et al: Magnetic resonance imaging of the shoulder: Rationale and current applications. Skeletal Radiol 19:5, 1990.
8. Littrup PJ, Asien AM, Braunstein EM, et al: Magnetic resonance imaging of the femoral head development in roentgenographically normal patients. Skeletal Radiol 14:159, 1985.
9. Mesgarzadeh M, Schneck C, Bonakdarpour A: Carpal tunnel: MR imaging. Part I. Normal anatomy. Radiology 171:743, 1989.
10. Mesgarzadeh M, Schneck CD, Bonakdarpour A: Magnetic resonance imaging of the knee and correlation with normal anatomy. Radiographics 8:707, 1988.
11. Mink JH, Reicher MA, Crues JV III: Magnetic Resonance Imaging of the Knee. New York, Raven Press, 1987.
12. Noto AM, Cheung Y, Rosenberg ZS, et al: MR imaging of the ankle: Normal variants. Radiology 170:121, 1989.
13. Quinn SF, Murray WT, Clark RA, et al: Achilles tendons: MR imaging at 1.5-T. Radiology 164:767, 1987.
14. Rauschning W: Anatomy and pathology of the cervical spine. *In* Frymoyer JW (ed): The Adult Spine: Principles and Practice. New York, Raven Press, 1991.
15. Rauschning W: Anatomy and pathology of the lumbar spine. *In* Frymoyer JW (ed): The Adult Spine: Principles and Practice. New York, Raven Press, 1991.

16. Ricci C, Cova M, Kang YS, et al: Normal age-related patterns of cellular and fatty bone marrow distribution in the axial skeleton: MR imaging study. Radiology 177:83, 1990.
17. Stoller DW, Martin C, Crues JV III, et al: Meniscal tears: Pathologic correlation with MR imaging. Radiology 163:731, 1987.
18. Watanabe AT, Carter BC, Teitelbaum GP, et al: Normal variations in MR imaging of the knee: Appearance and frequency. AJR 153:341, 1989.
19. Weiss KL, Beltran J, Shamam OM, et al: High-field MR surface-coil imaging of the hand and wrist. Part I. Normal anatomy. Radiology 160:143, 1986.
20. Yu S, Haughton VM, Sether LA, et al: Criteria for classifying normal and degenerated lumbar intervertebral disks. Radiology 170:523, 1989.
21. Zeiss K, Skiem, Ebraheim N, et al: Anatomic relations between the median nerve and flexor tendons in the carpal tunnel: MR evaluation in normal volunteers. AJR 153:533, 1989.
22. Zlatkin MB, Bjorkengren AG, Gylys-Morin V, et al: Cross-sectional imaging of the capsular mechanism of the glenohumeral joint. Radiology 150:151, 1988.

Part II
PRINCIPLES

2 MAGNETIC RESONANCE IMAGING

Basic Principles and Imaging Technique

Sharmila Majumdar

BASIC PRINCIPLES

In the last decade magnetic resonance (MR) imaging has become increasingly popular as a diagnostic tool for assessing the musculoskeletal system. Its popularity is due largely to the fact that MR can produce images with superior soft tissue contrast and high spatial resolution and that it does not require the use of ionizing radiation. Moreover, images may be obtained in any plane, and by varying the timing and pulse sequences, different tissue contrast may be achieved. With recent developments in the field, MR imaging also has the potential to be used to obtain quantitative information pertaining to morphology, tissue function, and physiology. In this chapter the basic principles underlying MR imaging are discussed, with some emphasis on topics that are clinically relevant. The amount of material to be discussed is vast, and it is not possible to cover these topics in great depth in one chapter; however, sources in which in-depth technical discussions may be found are provided as often as possible.

In clinical MR imaging, the signal detected is most commonly that originating from the hydrogen atom. In addition to its presence in water, which constitutes 70 per cent of the human body, hydrogen also occurs in human body fat, cholesterol, and other constituents. It is the nucleus of the hydrogen atom that is responsible for the MR signal. Nuclei with an odd number of protons and neutrons have a net magnetic moment. The hydrogen atom, which has a single proton, therefore has a magnetic moment and hence behaves like a magnet when placed in a strong magnetic field. In addition to a net magnetic moment, all such nuclei spin about their axes (Fig. 2–1) and when placed in a strong magnetic field, they tend to align themselves in the direction of the field as well as to rotate about the axis of the main field at a precise frequency called the *Larmor frequency*. This frequency depends on the nucleus under consideration and also on the strength of the applied magnetic field. The mathematical formula for the Larmor frequency (ω) is given by the following equation:

$$\omega = \gamma H$$

where γ is the gyromagnetic ratio, which is different for different nuclei (e.g., 4.26×10^3 for the hydrogen nucleus and 1.72×10^3 for phosphorus). The magnetic field strength (H) is usually expressed in Tesla (T) using SI units. At a field strength of 1.5 T, the Larmor frequency is 63.5 MHz for hydrogen and 25.86 MHz for phosphorus.

The nuclei within a sample give rise to a net

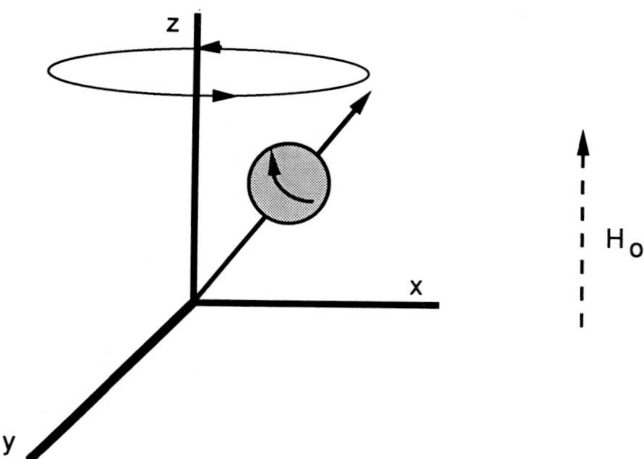

Figure 2–1. The nucleus of the hydrogen atom, which consists of one proton, has an associated magnetic moment and behaves like a magnet. Such hydrogen nuclei tend to align themselves in the direction of the field as well as rotate about the axis of the main field at a precise frequency called the Larmor frequency. X, Y, and Z represent the three orthogonal coordinates, and H_0 is the static magnetic field.

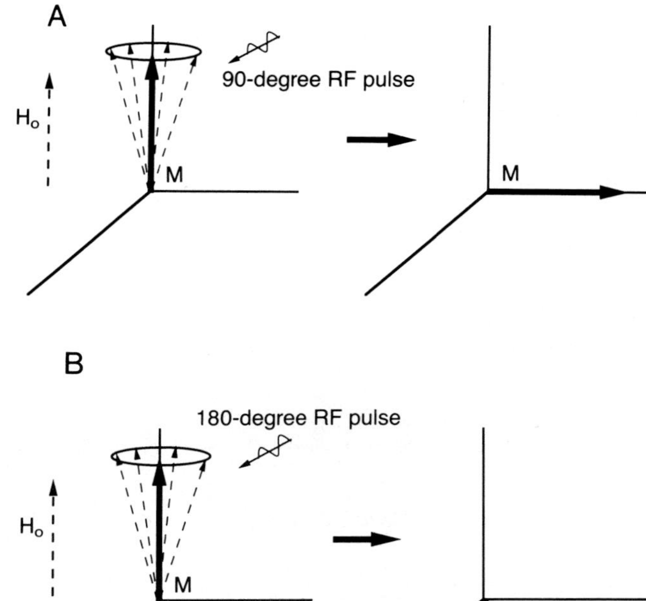

Figure 2–3. A radiofrequency (RF) pulse changes the position of the magnetization. *A*, A 90-degree pulse. A transmission of RF energy with the duration and power of the RF pulse rotating the magnetization, M, such that the new orientation of the net magnetization is at a 90-degree angle to the main field. *B*, A 180-degree pulse. A transmission of RF energy with the duration and power of the RF pulse rotating the magnetization, M, such that the new orientation of the net magnetization is antiparallel to the main field. H_0 is the static magnetic field.

magnetization (Fig. 2–2), which is the physical entity in MR imaging that is manipulated using radiofrequency (RF) pulses and gradients in the magnetic field. If an RF wave is transmitted at precisely the Larmor frequency, the nuclei absorb the RF energy. This absorption of the RF energy causes the nuclei and thus the net magnetization to change orientation relative to the main magnetic field. If the RF energy is on for a time such that the new orientation of the net magnetization is at a 90-degree angle to the main field, the RF pulse or energy transmission is termed a 90-degree pulse (Fig. 2–3*A*); similarly, if the orientation of the net magnetization after the cessation of the RF pulse is antiparallel to the main field, the RF pulse is termed a 180-degree pulse (Fig. 2–3*B*). After the RF energy transmission ceases, the nuclei tend to relax and reorient themselves. During this process, the nuclei emit radio waves and revert to their equilibrium position at characteristic rates termed *relaxation rates*. The absorption of the RF energy is an instantaneous process provided the RF energy is applied while tuned to the Larmor frequency. The relaxation process depends on the local magnetic field fluctuations within the molecule that are experienced by the nuclei and on whether these fluctuations occur at the Larmor frequency.[1] The local magnetic fields surrounding the nuclei depend on the chemical structure and the motion of the molecules to which these nuclei are bound. These relaxation processes occur over a longer time scale than the absorption of the RF energy and are dependent on several factors, such as the chemical environment of the nuclei. These relaxation times largely govern tissue contrast in MR imaging and are discussed in some detail in the following sections.

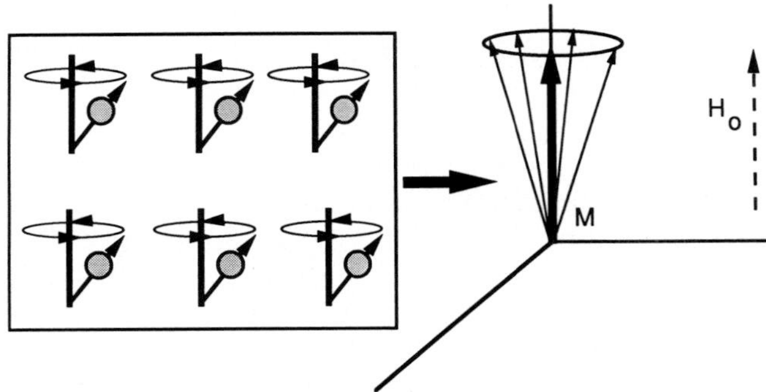

Figure 2–2. The combined effect of all hydrogen nuclei within a sample gives rise to a net magnetization, M, which is manipulated using radiofrequency pulses and gradients in the magnetic field in magnetic resonance imaging; H_0 is the static magnetic field.

RELAXATION TIMES

The signal intensity in an MR image results from a complex interaction of several intrinsic factors. In this section a few of the primary ones are addressed, namely, density of the nuclei (or spin density), spin-lattice relaxation time (T1), and spin-spin relaxation time (T2). The spin density merely reflects the total

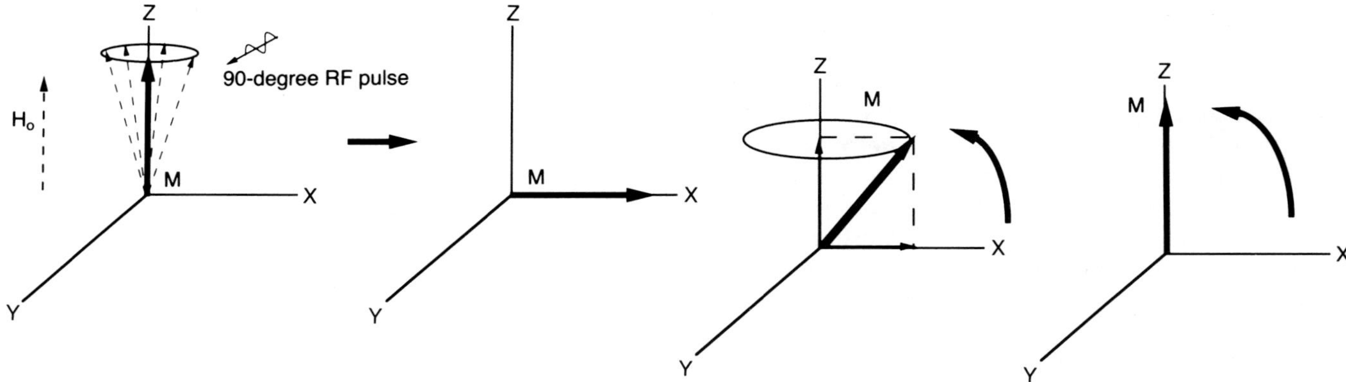

Figure 2–4. T1 relaxation. The equilibrium position of the magnetization is along the Z axis, with no components along the transverse plane. Application of a 90-degree radiofrequency pulse, for example, rotates the magnetization into the transverse plane. As the magnetization is recovering toward its equilibrium position (toward +Z), the net magnetization may be considered to be composed of two components, one along the Z axis, and one in the XY or transverse plane. Ultimately the magnetization acquires its equilibrium position along +Z. The total time taken for the magnetization to recover fully depends on the T1 relaxation time of the nuclei that constitute the magnetization.

number of nuclei that contribute to the net magnetization within a tissue sample. After the application of a 90-degree RF pulse, the nuclei, and hence the net magnetization in a sample, are rotated away from the direction of the static magnetic field, which is denoted as the Z axis (Fig. 2–3*A*), and into the transverse plane or the XY plane. The equilibrium position of the magnetization is along the direction of the applied magnetic field (i.e., along the Z axis). After being perturbed from its equilibrium position, in order to regain its equilibrium state, the magnetization tends to realign itself along the Z axis. As the equilibrium position is being recovered, a finite time interval occurs during which the magnetization is aligned neither in the transverse plane nor about the Z axis (Fig. 2–4). The net magnetization in such cases may be considered to be composed of two components, one along the Z axis, denoted as MZ, and one in the XY plane, denoted as MXY. The total time taken for the magnetization to recover fully and be realigned in the Z direction is dependent on the T1 relaxation time of the nuclei that constitute the magnetization. During this relaxation process the nuclei exchange energy with the surrounding lattice and, hence, T1 relaxation is also termed *spin-lattice relaxation.*

A second relaxation process, or the *spin-spin relaxation process,* also occurs after the nuclei are perturbed by an RF pulse. After the application of a 90-degree pulse, for example, the individual nuclei are all in phase (Fig. 2–5). However, owing to differences in local magnetic fields within the molecule, the nuclei may all precess at slightly different Larmor frequencies, exchanging energy with neighboring nuclei; thus, the net magnetization may be a summation of dephased nuclei and appear to decrease with time in the XY plane. Such a loss of signal intensity arising as a result of spin-spin interactions is governed by the T2 relaxation time or the spin-spin relaxation time. T1 relaxation occurs over a time period of several hundred to 1000 milliseconds (ms), whereas T2 decay occurs over a time range of 10 to 100 ms. By an appropriate manipulation of RF pulses and changing the times at which these pulses are applied in an imaging sequence, relaxation time–dependent tissue contrast can be widely varied in MR imaging.

PRINCIPLES OF MAGNETIC RESONANCE IMAGING

The preceding sections presented a discussion of the origin of the MR signal from tissue and the basic properties, such as tissue relaxation, that may

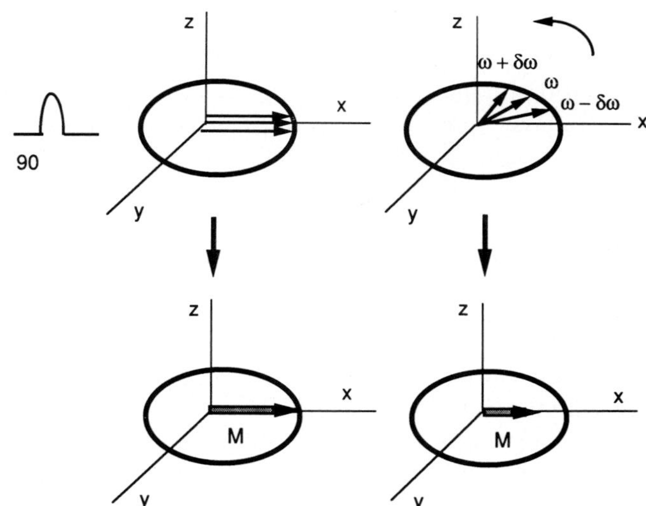

Figure 2–5. T2 or spin-spin relaxation process. After the application of a 90-degree pulse, the individual nuclei are all in phase, and the magnetization is at its maximum value. Differences in local magnetic fields within the molecule cause the nuclei to precess at slightly different Larmor frequencies ($\omega - \delta\omega$, ω, $\omega + \delta\omega$), and the net magnetization, which is a summation of dephased nuclei, decreases with time in the xy plane.

affect such signals. This section offers a simplified version of the imaging process. To obtain an MR signal and form an image, an array of magnetic field gradient coils are essential in addition to the strong magnetic field, the RF transmitter, and the RF receiver. These gradient coils are used for spatial localization of the MR signal. Although the MR images may be reconstructed using several different techniques, only the principles of the most commonly used Fourier transform techniques[8] are discussed here.

The Larmor frequency, or the frequency with which the nuclei precess, depends on the strength of the applied magnetic field (see previous equation). Thus, if a spatially varying magnetic field were superimposed on the main magnetic field, the Larmor frequency of the spins would vary spatially. This is the fundamental basis of spatial localization in MR imaging. As seen in Figure 2–6*A*, applying a strong magnetic field gradient across a sample results in a spatial variation of Larmor frequencies. If the gradient is strong, such that a large variation in the magnetic field strength occurs over a short distance, a distribution of frequencies ranging, for example, from $\omega - \delta\omega$ to $\omega + \delta\omega$ occurs over a small spatial region, as seen in Figure 2–6*A*. If the applied gradient is weaker, an equivalent range of Larmor frequencies may be seen extending over a larger spatial distance (Fig. 2–6*B* and *C*). Thus, by varying the strength of the applied gradient, it is possible to produce different thicknesses of tissue that consist of nuclei precessing within a specified frequency range. If now an RF pulse tuned to a narrow band of frequencies ($\omega \pm \delta\omega$) is applied, the RF pulse manipulates only the nuclei that are precessing within this narrow bandwidth of Larmor frequencies; if the RF pulse is on sufficiently long that all the magnetization within this slice is rotated by 90 degrees, all the nuclei that constitute this magnetization are now in the transverse or XY plane, in phase, and precessing at the same frequency. Thus, slice selection is achieved by applying a magnetic field gradient around the Z axis, such that the Larmor frequency varies along this axis. A pulse sequence diagram for slice selection in a standard imaging sequence is shown in Figure 2–7. After selecting the slice of interest, a signal can be detected; however, the signal originates from the entire slice, and a two-dimensional map cannot be obtained.

To spatially encode the other two dimensions, namely, X and Y, additional gradients are then applied; Figures 2–8 and 2–9 may be used in conjunction to understand the formation of the image after the process of slice selection. After the application of the 90-degree pulse, all the spins are in phase in the transverse plane, as shown in Figure 2–8*A*. If a gradi-

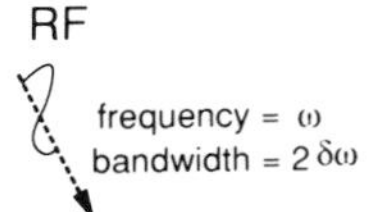

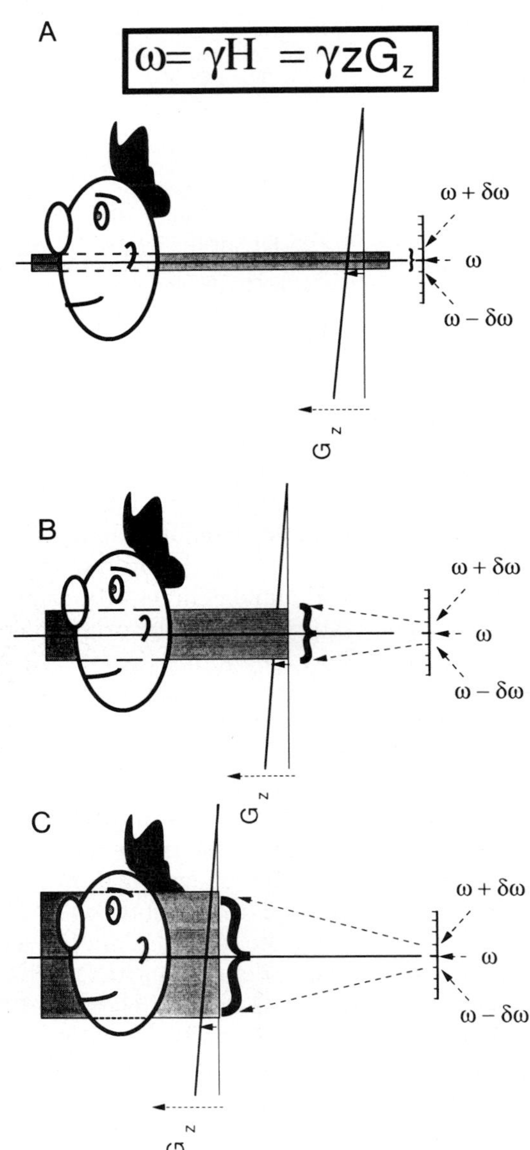

Figure 2–6. Varying the strength of the applied gradient, G_z, it is possible to produce different thicknesses of tissue that consist of nuclei precessing within a specified frequency range. An RF pulse applied within this selected frequency range thus excites a single slice. *A*, A strong magnetic field gradient is applied across a sample. This results in a spatial variation of Larmor frequencies ranging from $\omega - \delta\omega$ to $\omega + \delta\omega$. *B*, A magnetic field gradient weaker than that in (*A*) is applied across a sample. This results in a spatial variation of Larmor frequencies over a thicker slab of the sample, although the range of the frequencies still remains $\omega - \delta\omega$ to $\omega + \delta\omega$. *C*, A still weaker gradient results in the same range of Larmor frequencies occurring over a thicker slab. Thus, for the same RF pulse with constant bandwidth, $\omega - \delta\omega$ to $\omega + \delta\omega$, different thicknesses of tissue can be selected.

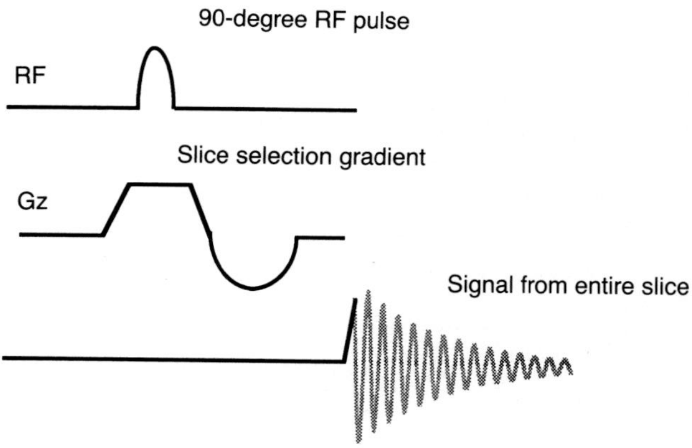

Figure 2–7. A pulse sequence diagram showing the RF pulse and slice selection gradient.

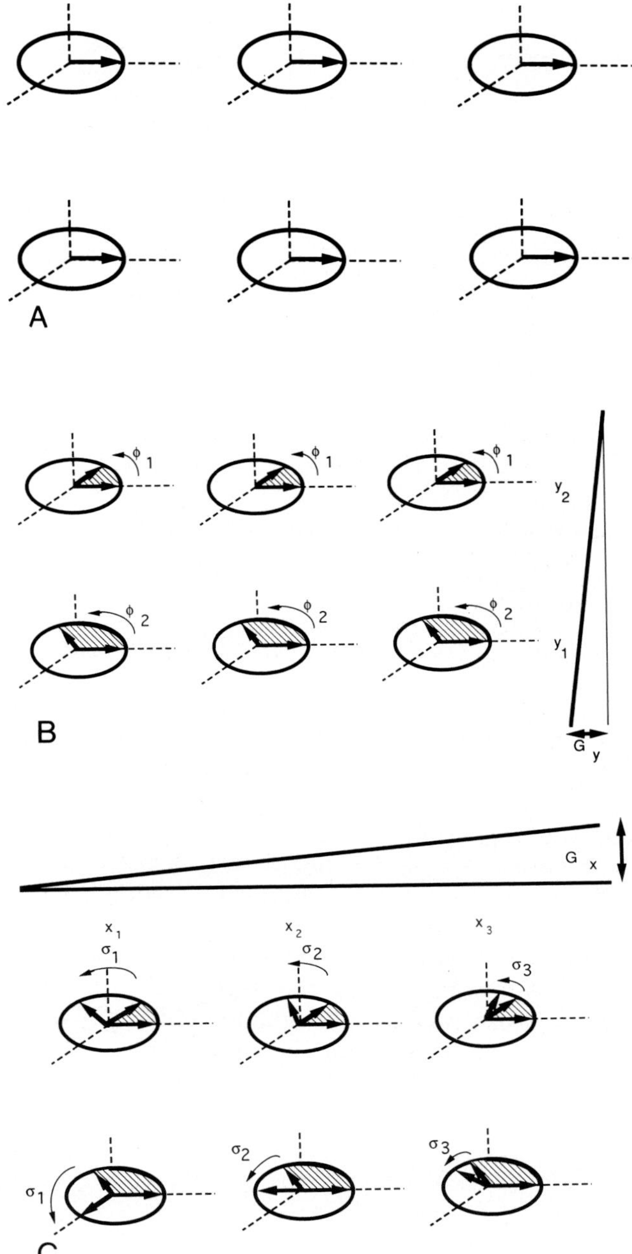

Figure 2–8. *A*, Immediately after the 90-degree pulse is applied, the nuclei constituting the magnetization are in phase. *B*, Application of a phase-encoding gradient along y results in the magnetization at each location along y having a unique phase ϕ associated with it. *C*, Application of a readout gradient along x results in the magnetization at each location along x having a unique frequency σ associated with it.

ent is then applied in the Y direction, the spins along the Y axis precess at slightly different frequencies; thus, spins in location Y_1 move faster and are ahead of those in Y_2, and so forth (see Fig. 2–8*B*). After the Y gradient ceases to act, the spins continue to precess at their original frequency; however, those in slice Y_1 are somewhat ahead of those in Y_2. When the spins are at different locations in their precessional cycle, they are said to be at different phases. Thus, the phase associated with the magnetization at location Y_1 is ϕ_1, and that at location Y_2 is ϕ_2. This phase information is used to discriminate different regions along the Y axis by mathematically processing the data using Fourier transforms. However, this procedure is repeated numerous times (usually 128, 256, or 512), depending on the spatial resolution desired, and these gradients are often termed *phase-encoding gradients* (see Fig. 2–9). The number of times the procedure is repeated is under operator control and is referred to as the *image matrix in Y*. To encode the signal in the third direction, gradients—usually called the readout gradients—are applied along the X direction (see Fig. 2–8*C*), and the signal is acquired during this time. The signal is acquired at discrete time intervals, and the total number of points that are acquired is often referred to as *image matrix in X*. The spins at different X locations precess at slightly different frequencies, σ_1 and σ_2, for example. By taking the Fourier transformation of the stack of data obtained for different phase-encoding gradients, the magnetization at each XY location may be determined uniquely for that location. For example, only the magnetization at location X_1Y_1 has a frequency σ_1 and phase ϕ_1 associated with it, the magnetization at location X_2Y_1 has a frequency σ_2 and phase ϕ_1 associated with it, and so forth. Thus, by identifying the unique phase and frequency, the net magnetization originating from a particular location may be identified, and an image can be generated. The pulse sequence for such a sequence is shown in Figure 2–9. However, there are technical limitations to acquiring the signal as shown in Figures 2–8 and 2–9. Hence one modification is to apply a 180-degree RF pulse at a time TE/2 after the 90-degree excitation and then acquire the data at time TE, the *echo time*. Such a sequence is called a *spin-echo sequence* (Fig. 2–10). To understand the role of the 180-degree pulse, consider the

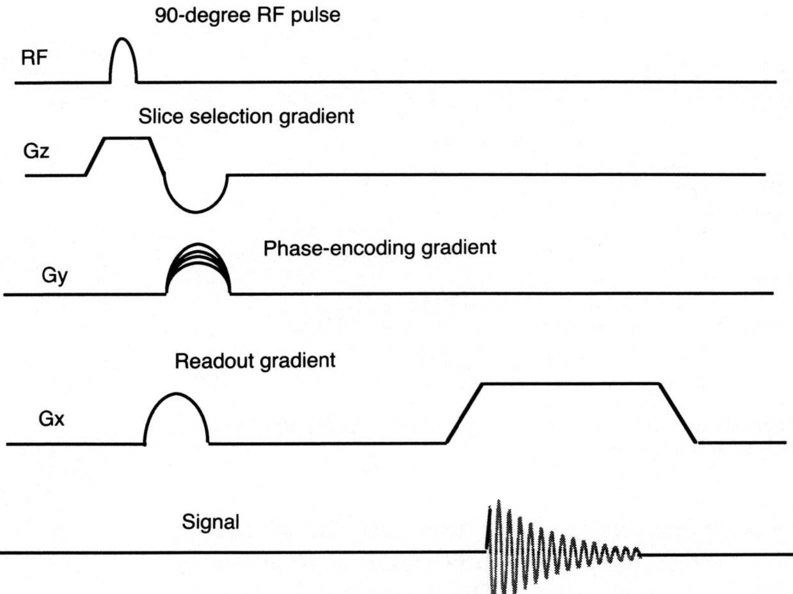

Figure 2-9. A pulse sequence diagram showing the RF pulse, and the slice selection, phase-encoding, and readout gradients for a standard imaging sequence.

behavior of the magnetization immediately after the 90-degree pulse (Fig. 2-11). At each point in space, the nuclei responsible for the magnetization precess in the transverse plane and dephase with respect to one another (see Fig. 2-11) for a period of time TE/2. This dephasing of the nuclei may occur owing to differences in the Larmor frequency arising from the application of additional magnetic field gradients or magnetic field inhomogeneities and may be reversed by the application of a 180-degree RF pulse. The fluctuations in the local magnetic field that give rise to T2 relaxation phenomena, however, are not reversible by the application of such RF pulses. The 180-degree RF pulse changes the position of the nuclei, as shown in Figure 2-11, and the spins continue to precess at the same frequency; at time TE the spins are back in phase, resulting in the formation of an echo (see Fig. 2-11). This echo is acquired, for different values of the phase-encoding gradient as shown in Figure 2-10, and is transformed using Fourier transform techniques to produce an image. The time between subsequent 90-degree pulses that are applied before each value of the phase-encoding gradient is known as the *repetition time* or TR. The times

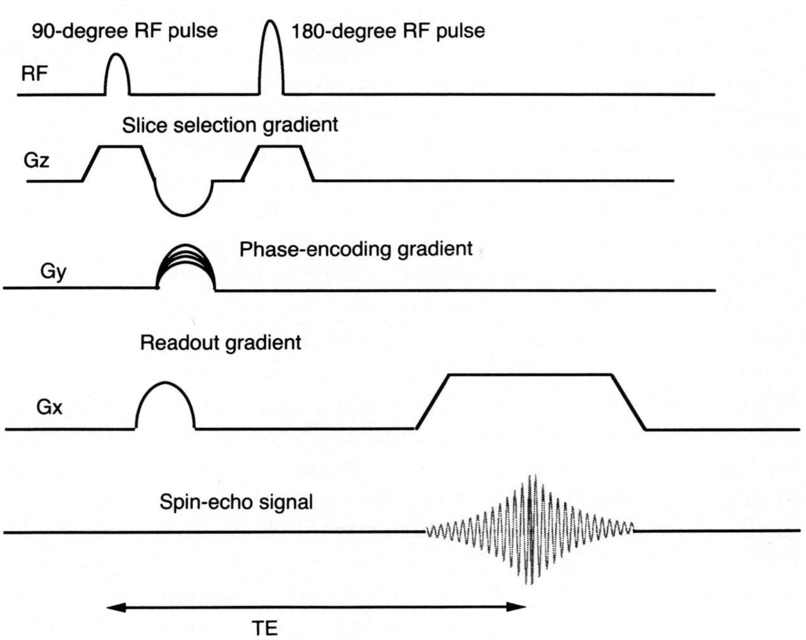

Figure 2-10. A pulse sequence diagram showing the RF pulse and the slice selection, phase-encoding, and readout gradients for a spin-echo imaging sequence. TE, echo time.

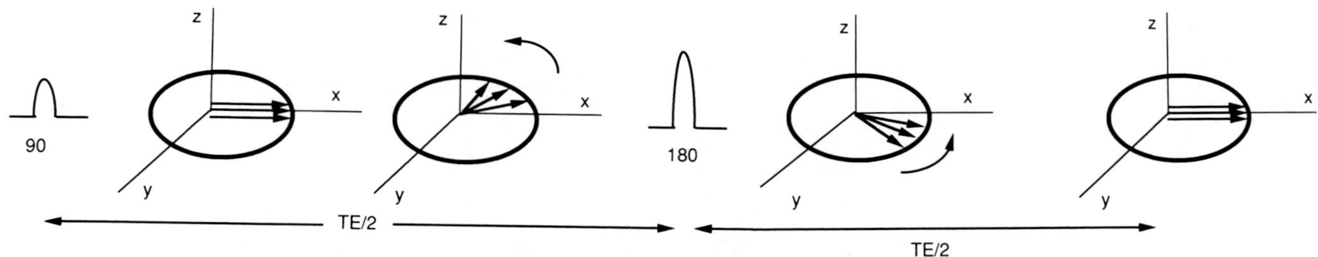

Figure 2–11. Formation of a spin-echo signal. Immediately after the 90-degree pulse is applied, the nuclei constituting the magnetization are in phase. The nuclei precess in the transverse plane and dephase with respect to one another as time goes on, for a total period TE/2, in which TE represents the echo time. The 180-degree RF pulse changes the position of the nuclei as shown. The spins continue to precess at the same frequency, and at time TE the spins are back in phase, resulting in the formation of an echo.

TR and TE are crucial for determining image contrast, as T1 relaxation occurs during TR and T2 relaxation during TE.

In MR imaging, fast imaging techniques have been and currently are being developed in an attempt to reduce the total imaging time. One way of reducing imaging time is to reduce the TR, which has been the basis of gradient-echo imaging sequences. In these sequences, the total image acquisition time is minimized by dispensing with the 180-degree pulse and forming an echo by using a sequence of gradients (Fig. 2–12). The excitation pulse in this case is typically less than 90 degrees and is called the flip angle (α). The choice of the TR, TE, and flip angle in these sequences produces images with widely varying image contrast. Several different acronyms have been used to describe such sequences, for example, GRASS, FLASH, and FISP. In these sequences, in addition to the phase-encoding, slice-selection, and readout gradients, additional gradients or RF pulses are used to destroy the coherent transverse magnetization that often builds up. Not all of these different techniques are discussed here; however, a very good discussion of the technical aspects may be found in a review by Wehrli and in references therein.[14]

Gradient-echo imaging sequences may be used to reduce the total imaging time to a few minutes. Developments such as the echo planar imaging techniques[9] and hybrid imaging sequences[4] have resulted in imaging times ranging from approximately 100 ms to 1 sec. The technical requirements and ramifications of these sequences are not discussed here; however, the important point is that the reduction of the total scan time does affect the image resolution and the signal-to-noise ratio, and it changes image contrast so that the radiologist must be careful when comparing these sequences to those produced using standard imaging techniques.[13] The fast spin-echo sequence[6] is another that has been

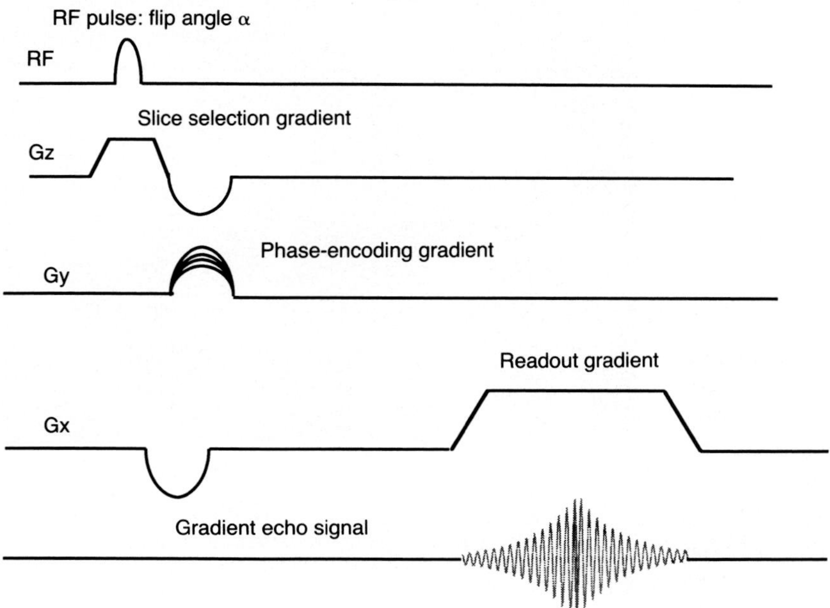

Figure 2–12. Gradient-echo imaging sequence.

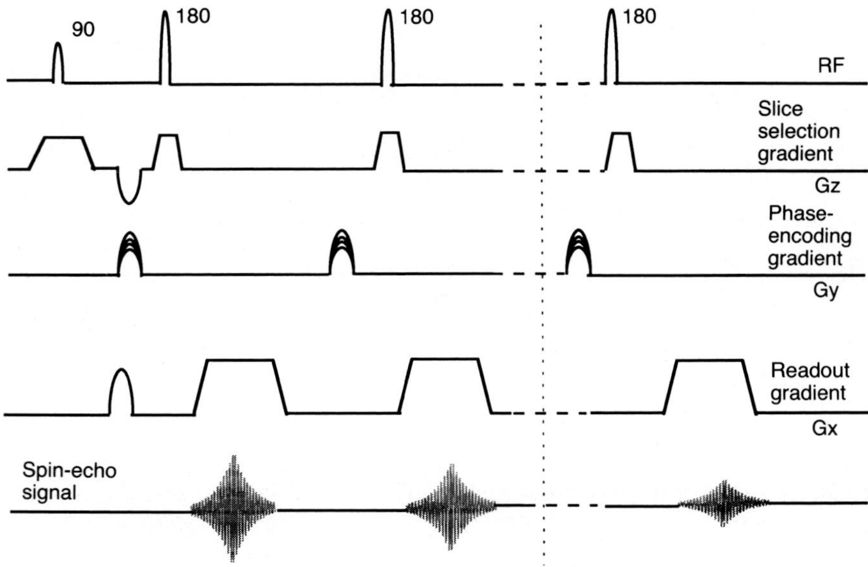

Figure 2–13. Fast spin-echo sequence.

used in several clinical applications. This sequence is a modification of the standard spin-echo sequence; however, rather than applying a single 180-degree pulse, a sequence of 180-degree pulses are applied in the same TR period, thus generating many echoes (e.g., 8 or 16). Phase-encoding gradients are turned on before each echo, thus reducing the total phase-encoding time by a factor of 8 or 16, depending on the total number of echoes that are generated (Fig. 2–13). Such sequences make it possible to obtain images with long TR times, in which the contrast mechanism is based primarily on differences in T2 relaxation times, at considerably shorter imaging times. Although the details are not discussed in this chapter, it suffices to say that this technique has gained a great deal of clinical popularity; a representative image is shown in Figure 2–14.

FACTORS THAT AFFECT IMAGE QUALITY

Importance of Coil Selection

Positioning the patient and choosing the optimal coils for the specific application is one of the most important considerations in MR imaging. The size of the patient, the part of the body to be imaged, and the total examination time are some of the factors that affect these decisions. It is best that the image of the organ to be studied be made with the coil that is most closely coupled to the body part to achieve the maximum signal-to-noise ratio and spatial resolution. In Figure 2–15, coronal images through the knee joint are obtained using the body coil and a transmit-receive knee coil. As is clearly seen, the close coupling of the knee coil (see Fig. 2–15*A*) produces images with a much better signal-to-noise ratio. In musculoskeletal imaging, the body coil typically is used to make images of the hip and thigh regions; the head coil may be used for imaging pediatric patients and for making images of both legs of children simultaneously (Fig. 2–16). In most cases, however, specialized coils for the knee joint and wrist are being used routinely to optimize the quality of images obtained.[7]

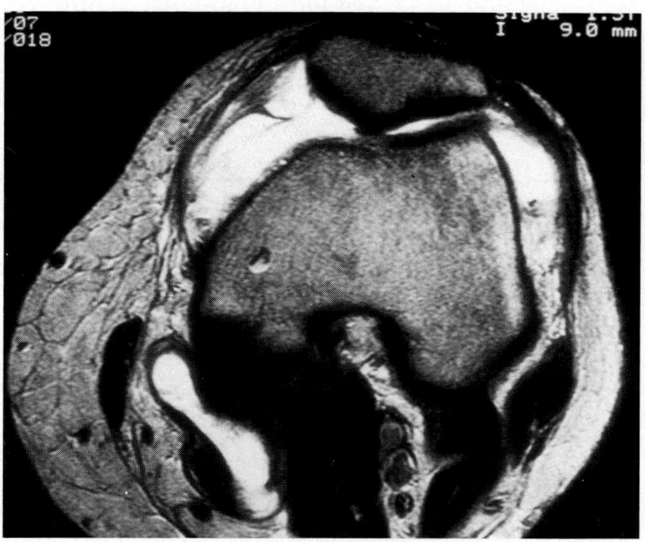

Figure 2–14. High-resolution image of the knee joint using the fast spin-echo sequence, phase-encoding eight echoes. TR, 3000 ms; TE, 95 ms; field of view, 16 cm; slice thickness, 5 mm; image matrix, 512 × 256; number of excitations, 2. Total time taken for this sequence was 3.2 minutes, compared with the 25.6 minutes it would have taken for a standard spin-echo sequence using the same repetition time.

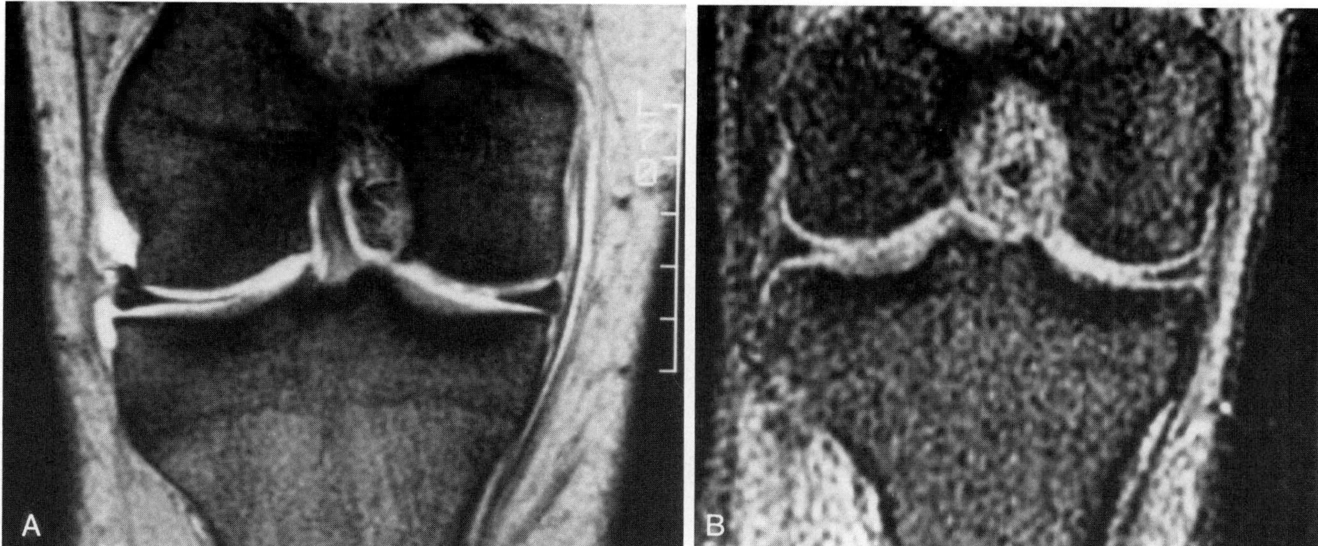

Figure 2–15. Coronal images through the knee joint using a transmit-receive knee coil *(A)* or a body coil *(B)*, which depict coil-dependent differences in the signal-to-noise ratio.

Many musculoskeletal applications require the use of surface coils—for example, in the shoulder, lumbar spine, and ankle. Surface coils are mainly receive-only coils (i.e., they only detect the RF signal from the tissue and do not transmit the excitation pulses). The signal intensity in images obtained from these coils may vary depending on the distance of the part being scanned from the coil (Figs. 2–17, 2–18). Although current software compensates for and minimizes these effects, the efficacy of these correction schemes depends significantly on the size of the subject and region of interest compared to the coil size. Newer coil developments include dual switchable coils for examining the temporomandibular joint bilaterally[5] and multiple coils[10] to increase signal-to-noise ratio and reduce examination time.

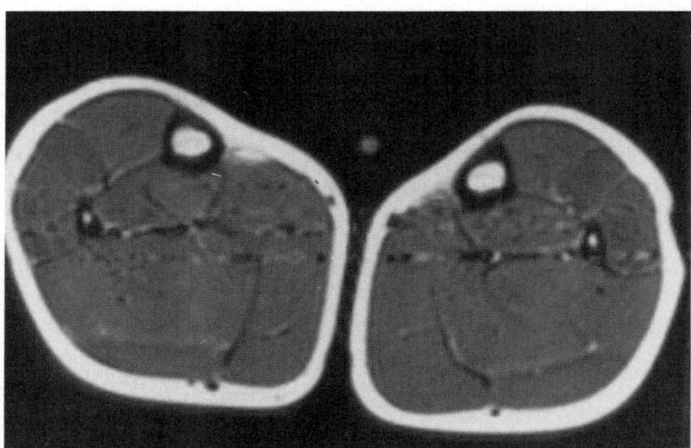

Figure 2–16. Axial image through the calves of a 10-year-old child obtained using a spin-echo sequence and a quadrature head coil.

ACQUISITION PARAMETERS THAT AFFECT IMAGE CHARACTERISTICS

The diagnostic quality of an MR image depends on several factors, such as spatial resolution, signal-to-noise ratio, contrast-to-noise ratio, and scanning time required to obtain the images. All of these fac-

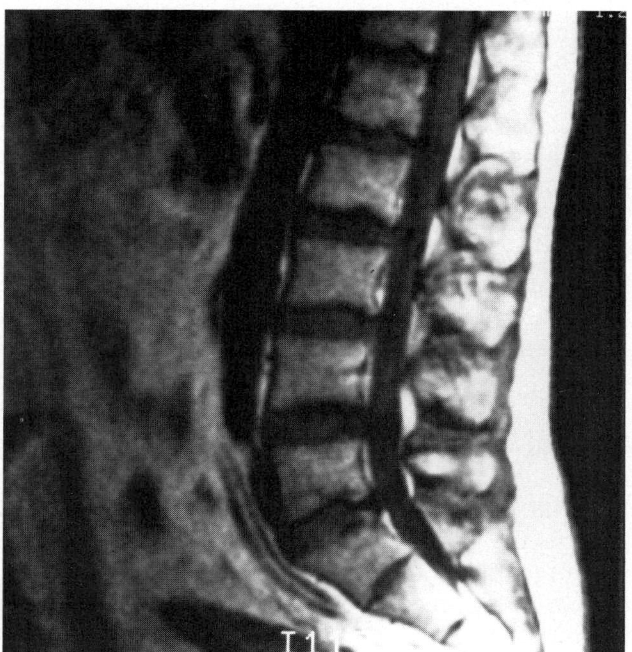

Figure 2–17. Sagittal image through the lumbar spine using a License Plate surface coil (Medical Advances, Milwaukee, Wis.). Note the variation in signal intensity of subcutaneous fat on the back and abdomen, demonstrating the drop off of signal intensity away from the coil surface.

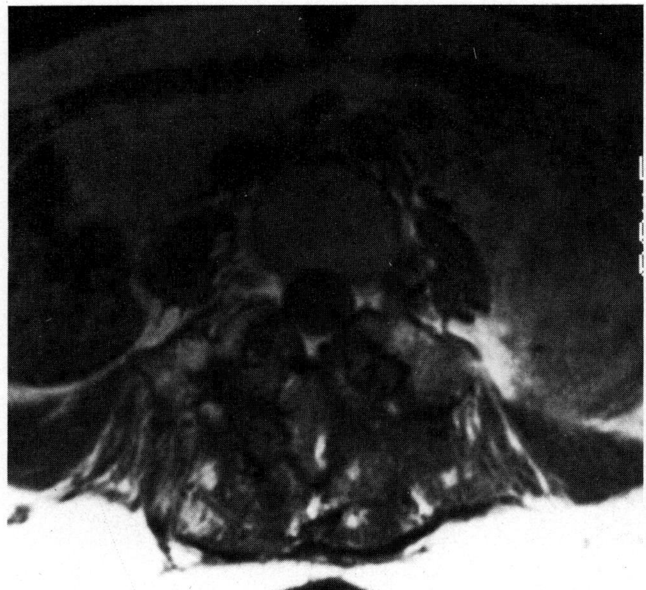

Figure 2–18. Axial image through the lumbar spine using a License Plate surface (Medical Advances, Milwaukee, Wis.) coil. As in Figure 2–17, note the variation in signal intensity of subcutaneous fat on the back and abdomen, demonstrating the drop off of signal intensity away from the coil surface.

tors are interrelated, and, depending on the study objective, either one or more may play a predominant role. Spatial resolution of an image is under operator control and may be varied by varying the thickness of the selected slice, number of phase encoding and readout steps, and imaging field of view. The signal-to-noise ratio is a measure of the signal intensity relative to the image noise, which is dependent on technical factors, such as preamplifier noise, coil characteristics, and physiologic noise, such as respiration or flow. However, a high signal-to-noise ratio does not ensure a high degree of discrimination between lesion and normal tissue. Detectability of pathologic conditions or lesions depends on the difference in signal intensity between the lesion and the surrounding tissue. Thus, in addition to optimal signal-to-noise ratio in an image, it is imperative to optimize contrast-to-noise ratio.

Spatial Resolution

The achievable voxel dimension in MR imaging governs the spatial resolution in an MR image. The voxel resolution depends on the slice thickness and the in-plane resolution and can be written as follows:

$$\text{Spatial Resolution} = \frac{\text{Slice thickness} \times \text{Field of view in X} \times \text{Field of view in Y}}{\text{Image matrix size in X} \times \text{Image matrix size in Y}}$$

Ideally the radiologist would like to obtain the highest spatial resolution possible in any image, which implies reducing the slice thickness and fields of view and increasing the image matrix. However, increasing the spatial resolution often leads to a reduction in the signal-to-noise or contrast-to-noise ratio, or both, and hence these factors must also be kept in mind when designing protocols for diagnostic purposes.

Slice Thickness. In an ideal setting, a true cross-sectional image would be one in which the slice thickness is infinitesimally small, so that the image obtained is not an average over a finite dimension, and overlapping structures do not lead to a blurring or obliterating of structural information. The slice thickness is governed by the strength of the slice-selection gradient (see Fig. 2–6) and apart from instrumental limitations in reducing slice thickness, the signal-to-noise ratio decreases as the slice thickness decreases. In going from a slice thickness of 10 mm to a slice thickness of 3 mm, there is a change in the signal-to-noise ratio as seen in the sagittal image through a normal knee joint in Figure 2–19. For example, the noise level in the muscle and bone marrow in the image with the thinner slice is apparent; from Table 2–1, the increase in signal-to-noise ratio in going from a slice thickness of 3 mm to one of 10 mm is seen to be a factor of 3.33. As is clear from the figure, at a slice thickness of 10 mm an averaging effect occurs, as a result of which the anterior cruciate ligament is not resolved, although it is clearly visible in the image with the slice thickness of 3 mm.

In-plane Resolution. The in-plane resolution in MR images is governed by the choice of field of view and image matrix size. The field of view is the area selected by the user that constitutes the image, and a combination of the image matrix and field of view governs the magnitude of the gradients that are applied in the phase-encoding and readout directions (see Fig. 2–10). Although it is optimal to choose the smallest field of view possible and the largest image matrix to optimize spatial resolution, which varies as

$$\text{Spatial Resolution} = \frac{\text{Field of view in X} \times \text{Field of view in Y}}{\text{Image matrix size in X} \times \text{Image matrix size in Y}}$$

it must be noted that signal-to-noise ratio varies as:

$$\text{Signal-to-Noise Ratio} = \frac{\text{Field of view in X} \times \text{Field of view in Y}}{\text{Image matrix size in X} \times \sqrt{\text{Image matrix size in Y}}}$$

Hence, in applications in which high spatial resolution is absolutely essential, the loss in signal-to-

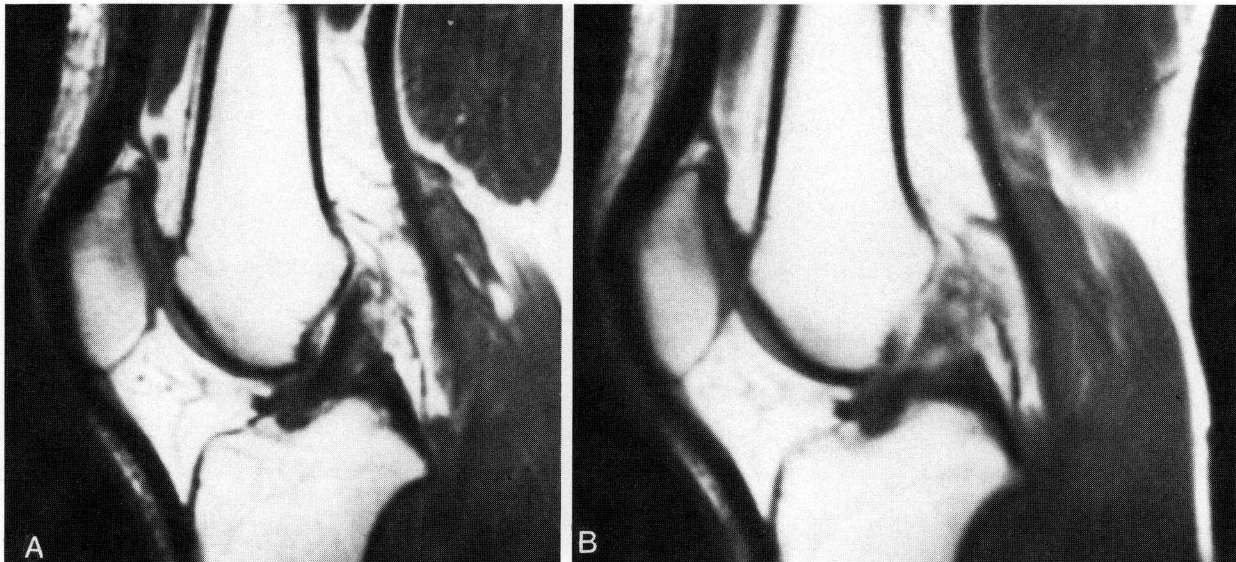

Figure 2–19. Sagittal section through the knee joint obtained using a spin-echo sequence at a TR of 400 ms and a TE of 20 ms. *A*, 3-mm slice thickness. *B*, 10-mm slice thickness.

noise ratio must be compensated for by an increase in the number of excitations used to obtain the MR image. In three-dimensional imaging, the signal-to-noise ratio is given by the following equation:

$$\text{Signal-to-Noise Ratio} = \frac{\text{Field of view in X} \times \text{Field of view in Y} \times \text{Field of view in Z}}{\text{Image matrix size in X} \times \sqrt{\text{Image matrix in Y}} \times \sqrt{\text{Image matrix in Z}}}$$

For a given slice thickness and spatial resolution, a three-dimensional imaging sequence may provide images with higher signal-to-noise ratios; however, there is likely to be a trade-off in terms of total scan time and contrast-to-noise ratios.

Number of Excitations. The number of excitations is another important factor that affects the signal-to-noise ratio in an MR image. It is the number of times the sequence is repeated and the signal averaged.

$$\text{Signal-to-Noise Ratio} \propto \sqrt{\text{number of excitations}}$$

Total Scan Time. An important consideration in clinical MR imaging is the total time required to acquire a dataset that may be used to reconstruct an image or a set of images. In two dimensional images

$$\text{Total time} = \text{Number of excitations} \times \text{TR} \times \text{Image matrix in Y (phase encoding)}$$

whereas in a three-dimensional image

$$\text{Total time} = \text{Number of excitations} \times \text{TR} \times \text{Image matrix in Y (phase encoding)} \times \text{Image matrix in Z (slice-selection direction)}.$$

Thus, it is quite possible to keep imaging time constant and vary the other parameters, keeping in mind, however, that image contrast and signal-to-noise ratios may differ depending on the factors that are varied.

Figure 2–20 shows a set of sagittal images obtained through the knee joint of a normal subject. In Figure 2–20*A* and *B*, images have been obtained with identical parameters for TE, TR, and slice thickness. However, the field of view in Figure 2–20*A* is 28 cm, and the image matrix size is 256 × 256, whereas in Figure 2–20*B* the field of view is 28 cm and the image matrix is 512 × 512. The increased signal-to-noise ratio in Figure 2–20*A* seen in the muscle and bone marrow, however, is at the cost of reduced spatial resolution (Table 2–2). In Table 2–2, the trade-off among the number of excitations and the ability to achieve a given signal-to-noise ratio, imaging time, and spatial resolution are shown. As seen in the table, for a doubling of the imaging time, only a 1.4-fold increase in signal-to-noise ratio is achieved when the number of excitations is doubled; however, if reduced spatial resolution is acceptable,

Table 2–1. EFFECT OF SLICE THICKNESS ON SIGNAL-TO-NOISE RATIO

Slice Thickness	Signal-to-Noise Ratio
3 mm	1
10 mm	Increases 3.33 times

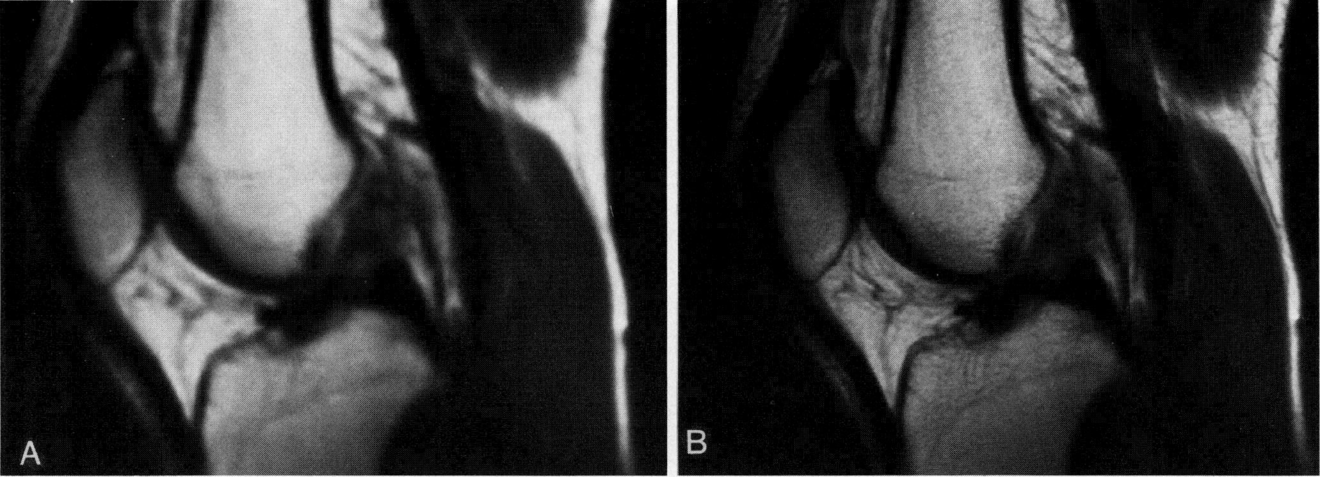

Figure 2–20. Sagittal section through the knee joint obtained using a spin-echo sequence at a TR of 400 ms and a TE of 20 ms. *A*, Field of view, 28 cm; image matrix, 256 × 256. *B*, Field of view, 28 cm; image matrix, 512 × 512.

for the same imaging time and number of excitations, a fourfold increase in signal-to-noise ratio is achievable when a larger field of view is selected.

Image Contrast

Image contrast depends on several intrinsic factors, such as spin density, spin-lattice (T1) relaxation time, and spin-spin (T2) relaxation times. In addition, the chemical environment of the nuclei being scanned (chemical shift) and variations in the magnetic susceptibility of tissues being scanned also can affect image contrast. Appropriate choice of pulse sequences, such as spin-echo or gradient-echo, or choice of imaging parameters, such as TR, TE, and flip angle, can govern the extent and mechanism of image contrast.

Repetition Time. The rate at which the magnetization regains its equilibrium position depends on the T1 relaxation time of the tissue. The magnitude of the magnetization that may have recovered between successive excitations depends on the ratio between the repetition time (TR) and the relaxation time (T1), or TR/T1. For a 90-degree excitation, the recovery of the signal intensity may be simplified as follows:

$$\text{Signal Intensity} = M_0\left(1 - e^{\frac{-TR}{T1}}\right)$$

where M_0 is the initial magnetization with TE-dependent contributions from the spin-spin relaxation time T2.

Thus, for all tissues when TR is much less than T1, the tissue contrast varies depending on TR/T1 if TR is sufficiently long and all the magnetization has recovered, after which the signal intensity does not change with TR time.

Echo Time. In most MR imaging experiments, in addition to the repetition time, the time between the

Table 2–2. IMPACT OF IMAGING PARAMETERS OF SIGNAL-TO-NOISE RATIO AND TOTAL TIME SCAN

Field of View (cm)	Matrix Size	Number of Excitations	Signal-to-Noise Ratio	Pixel Resolution (mm)	Scan Time (relative to first entry in table)
14	256 × 256	1	1	0.55 × 0.55	1
14	256 × 256	2	Increases 1.4 times	0.55 × 0.55	2
14	256 × 256	4	Increases 2 times	0.55 × 0.55	4
28	256 × 256	1	Increases 4 times	1.1 × 1.1	1
28	512 × 512	1	Increases 1.4 times	0.55 × 0.55	2

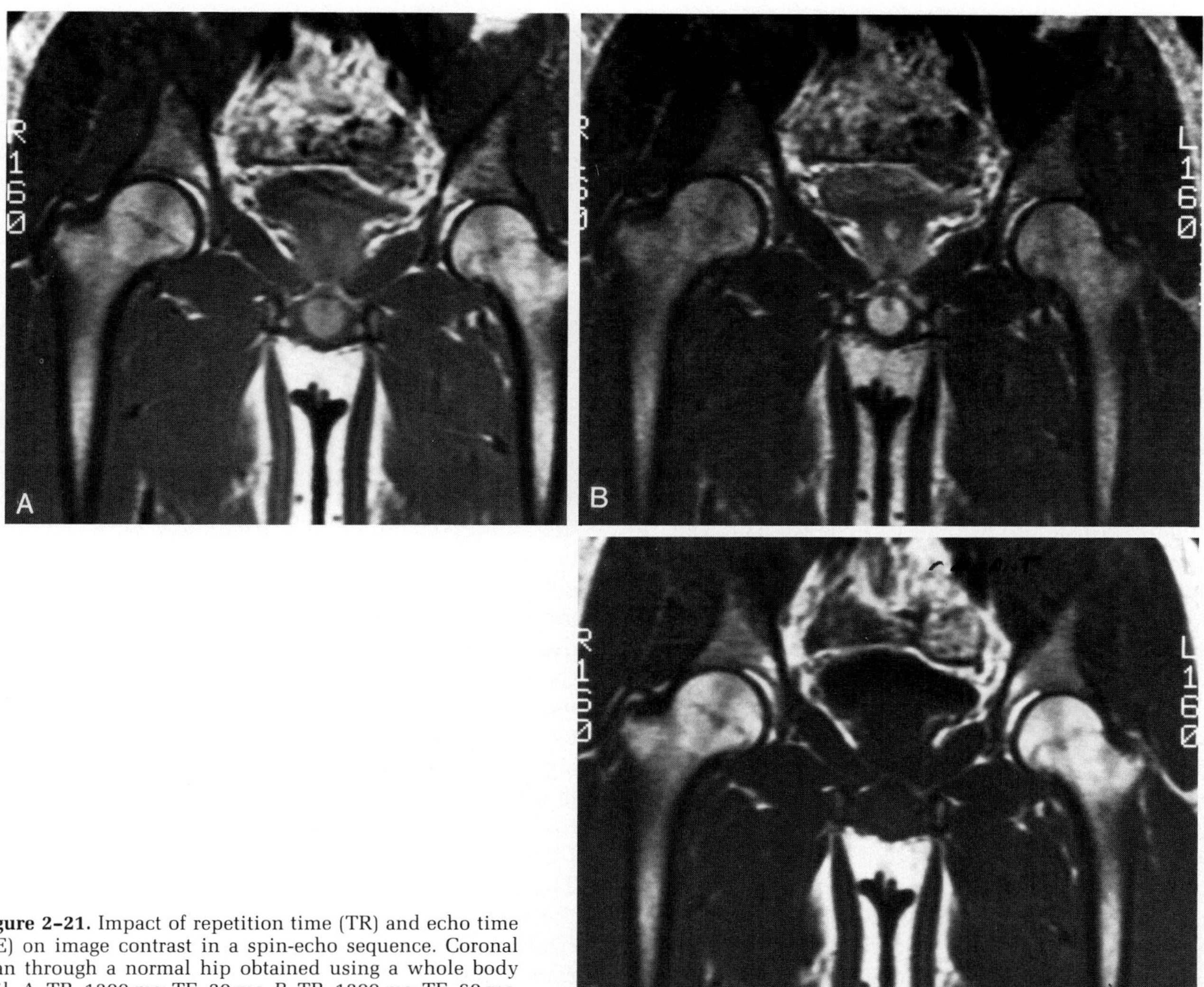

Figure 2–21. Impact of repetition time (TR) and echo time (TE) on image contrast in a spin-echo sequence. Coronal scan through a normal hip obtained using a whole body coil. *A*, TR, 1200 ms; TE, 20 ms. *B*, TR, 1200 ms; TE, 60 ms. *C*, TR, 300 ms; TE, 20 ms.

90- and 180-degree pulses governs the extent of spin-spin relaxation that occurs. The signal intensity decays as a function of echo time:

$$\text{Signal Intensity} = M_0\, e^{\frac{-TE}{T2}}$$

Figure 2–21 shows images obtained through the hip joint of a normal subject at a TR of 1200 ms and a TE of 20 and 60 ms, and at a TR of 300 ms and a TE of 20 ms. The contrast between muscle, subcutaneous fat, and fluid in the bladder shows considerable variation, both as a function of TR and TE. Figure 2–22 shows the quantitative variation in signal intensity that occurs between muscle and subcutaneous fat when regions of interest are selected. As seen in the figure, muscle and fat show increases in signal intensity with the TR time, although at different rates, which are governed by the T1 relaxation time of the tissues, whereas they decrease in signal intensity as a function of echo times, also at different rates, governed by the intrinsic T2 relaxation time.

Flip Angle. In gradient-echo images, the signal intensity and thus the image contrast in tissue vary not only as a function of TR and TE, but also as a function of the flip angle selected, as shown in Figure 2–23. Figure 2–23 shows a set of images through the hip joint of a normal subject that were obtained using a gradient-echo sequence at TR values ranging from 30 to 100 ms, TE values ranging from 15 to 45 ms, and flip angles varying from 10 to 90 degrees. These are by no means the only combinations of imaging parameters that may be used; however, images obtained using these values represent the wide variations in signal intensity and image contrast that are achievable. In addition to qualitative trends seen in Figure 2–23, regions of interest were selected in the muscle, subcutaneous fat, bone marrow, and fluid in the bladder; the resulting variations in signal

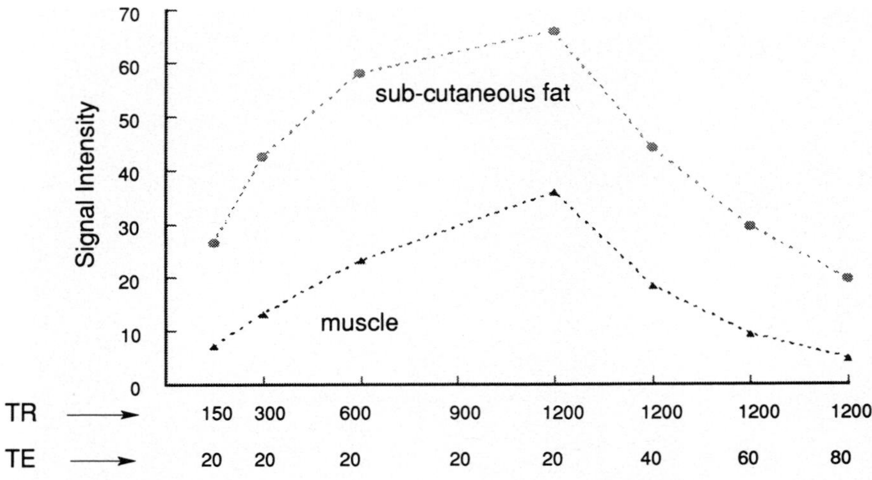

Figure 2–22. Variation of signal intensity in different tissues as a function of repetition time (TR) and echo time (TE).

intensity are shown in Figure 2–24. For example, the signal intensity for all the different tissue types sampled increases as a function of TR (TR ranging from 30 to 100 ms) for a TE of 15 ms and a flip angle of 10 degrees (see Fig. 2–24*A*) and decreases as a function of TE (TE ranging from 15 to 45 ms) (see Fig. 2–24*B*) at a TR of 60 ms and a flip angle of 60 degrees. However, for a TR of 100 ms and TE of 15 ms, the signal intensity of muscle, subcutaneous fat, fluid, and bone marrow shows considerable variation as a function of the flip angle (see Fig. 2–24*C*). In fact, on plotting the difference in signal intensity among fat-fluid, fat-muscle, and muscle-fluid as a function of flip angle (Fig. 2–25), it may be seen that for fat-fluid and fat-muscle there is a contrast reversal, in which the difference in image intensities ranges from negative to positive as a function of the flip angle. For muscle and fluid, on the other hand, although there is a variation in image contrast that is dependent on the flip angle, contrast reversal is absent. These combinations of TR, TE, and flip angle are not the only possible ones; it must be remembered that in addition to depending on the relaxation times of the tissue being scanned, image contrast differs between tissues depending on the values of these parameters.

Chemical Shift

In musculoskeletal MR imaging, the chemical shift effect between fat and water produces interesting variations in image contrast. The hydrogen nuclei bound to the fat molecules experience a different chemical environment from those bound to water molecules. Such differences in the chemical environment result in a shielding effect, and the hydrogen bound to water precesses at a slightly different Larmor frequency from that bound to fat. At 1.5 T, the difference in Larmor frequency between these two components is approximately 220 Hz. In gradient-echo images in which no 180-degree RF pulses are used or in spin-echo images in which the 180-degree pulse is not occurring at a time equal to TE/2, the fat and the water components of magnetization may be in phase or completely out of phase. For example, analyzing the behavior of fat and water signals immediately after a 90-degree excitation in a gradient-echo sequence (Fig. 2–26), the magnetization arising from the fat and water components are in phase, giving rise to a net maximum magnetization. The difference in precessional frequencies between these components, however, causes the net magnetization from fat and water to dephase with time, thus reducing the magnitude of the net magnetization (see Fig. 2–26). Depending on the time elapsed after the excitation pulse (e.g., at 1.5 T, at intervals of approximately 2.2 ms), the magnetization from these two components may be in opposite directions, and at this time the net resultant magnetization or signal is a difference between these two components (see Fig. 2–26). After a further period of approximately 2.2 ms at 1.5 T, the magnetization from the fat and water components rephases again, and the signal intensity is once again a summation of the individual fat and water signals. Thus, as a result of this phenomenon, the signal intensity of tissues containing both fat and water protons, at different echo times in gradient-echo images, may show an oscillatory behavior with echo time.[15] Similar signal intensity oscillations and maximum and minimum values in signal intensity may be seen between spin-echo images obtained at the same echo time but with the 180-degree RF pulse applied at a time TE/2, and at TE/2 ± 1.1 ms (at 1.5 T) (Fig. 2–27). It must be remembered, however, that this effect is independent of the effects of the dephasing that may occur

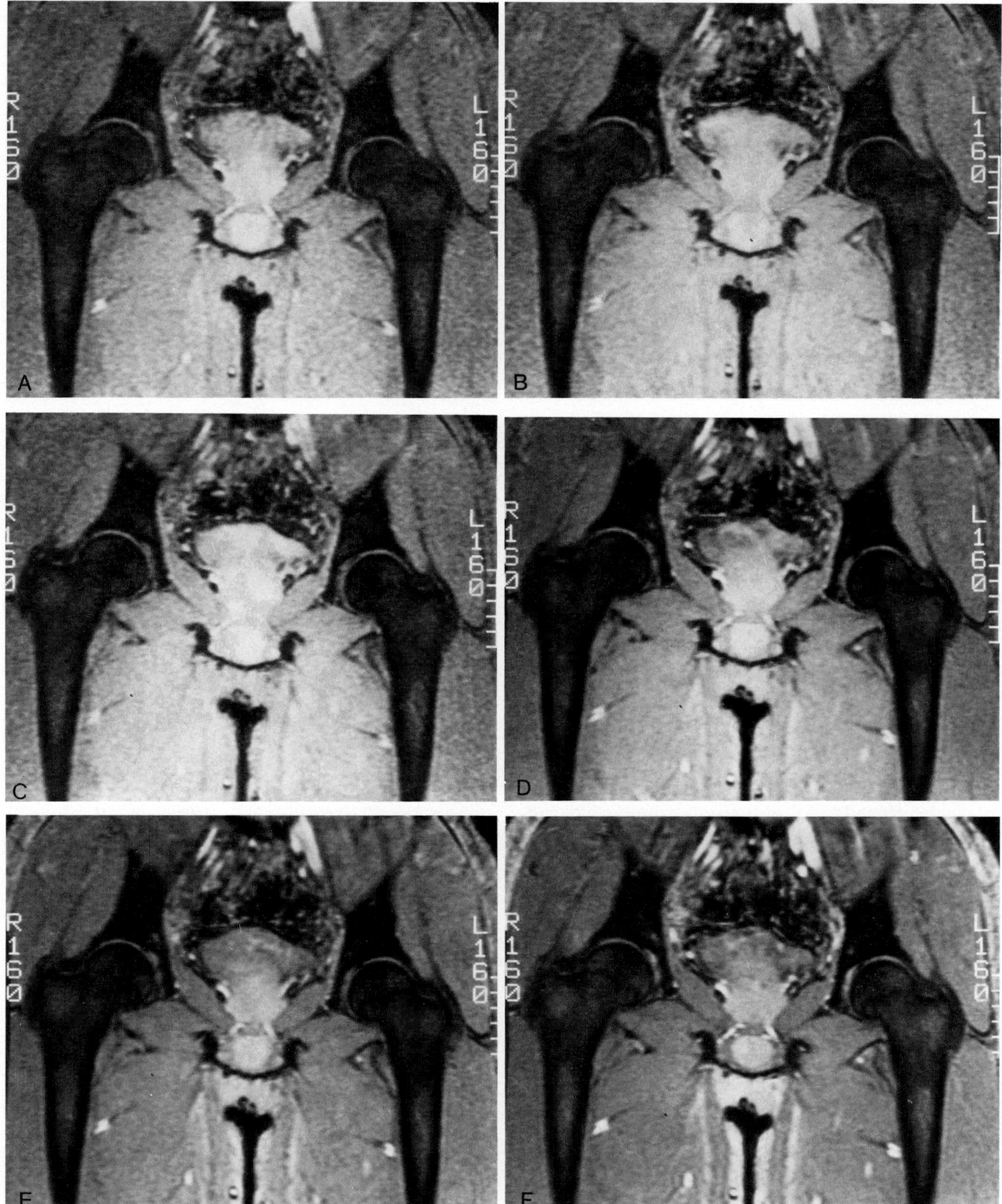

Figure 2–23. Coronal view of the hip joint of a normal subject obtained using a spoiled gradient-echo sequence. *A*, TR, 30; TE, 15; α, 10. *B*, TR, 60; TE, 15; α, 10. *C*, TR, 100; TE, 15; α, 10. *D*, TR, 100; TE, 15; α, 20. *E*, TR, 100; TE, 15; α, 30. *F*, TR, 100; TE, 15; α, 45.

Illustration continued on following page

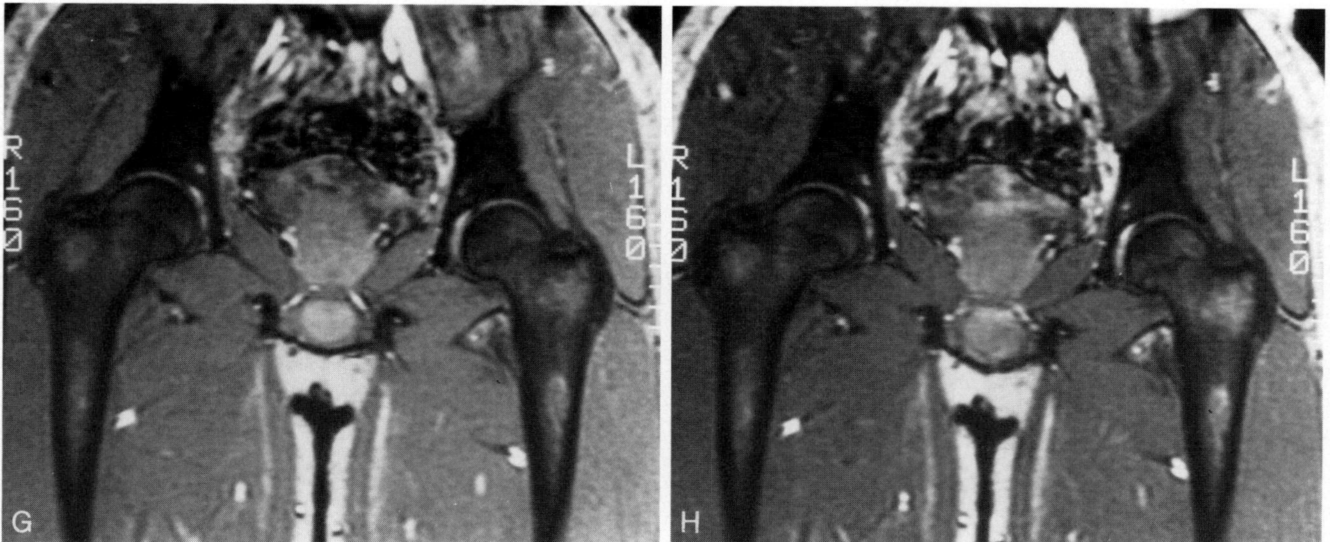

Figure 2–23 *Continued* G, TR, 100; TE, 15; α, 60. *H*, TR, 100; TE, 15, α, 90.

owing to the presence of magnetic field inhomogeneities and gradients.

The difference in the fat and water Larmor frequency has been used to develop imaging techniques that image only the fat or the water component of tissue. The chemical saturation technique, which employs an RF pulse to suppress one component,[11] the Dixon technique,[2] the hybrid technique,[12] and the multipoint Dixon technique[3] are some of the currently used methods that make use of the difference in the Larmor frequencies; the short tau inversion-recovery (STIR) technique is another method used to selectively image the water component only, making use of the differences in fat and tissue water relaxation time T1. All of these techniques have advantages and disadvantages and are not discussed in detail here; however, a short description of the commonly used fat saturation and STIR techniques is provided in the following paragraphs.

In the fat saturation technique, a narrow band RF pulse is used to first rotate only the fat magnetization by 90 degrees (Fig. 2–28) into the transverse plane. A subsequent application of a 90-degree pulse, which selects the desired slice, affects both the fat and water magnetization. The effect of the RF pulses may be understood from Figure 2–29. As seen in the figure, the chemically selective RF pulse rotates the fat component of the magnetization into the transverse plane; a subsequent 90-degree pulse that affects both the fat and the water components results in a rotation of the previously unaffected water magnetization onto the transverse or XY plane, while the fat magnetization is rotated by 90 degrees along the Z axis. The imaging sequence that follows thus affects only the magnetization present in the transverse plane, in the selected slice, and only an image of the water component is acquired. The amplitude of the chemical suppression pulse, its duration, and the frequency at which it is applied are crucial factors that dictate the effectiveness of such fat suppression techniques. Inhomogeneities in the main magnetic field may result in a shift of the resonant frequencies of fat and water; hence the application of a fat saturation pulse in many instances may lead to incomplete suppression or even complete suppression of the water component. For example, in Figure 2–30, an image through the ankle, the fat saturation technique resulted in suppression of the fat component in parts of the image and of the water component in other parts.

In the STIR technique, a 180-degree RF pulse is applied to rotate the transverse magnetization in a direction that opposes the main magnetic field (Fig. 2–31). After the application of such a pulse, T1 relaxation governs the recovery of the transverse magnetization toward equilibrium. The difference between the relaxation time T1 of fat and fluid, for example, governs the choice of the TI delay, the time after which a sequence of 90- to 180-degree pulses and the sequence of gradients are turned on. If the time TI is selected such that the fat component of the magnetization is along the transverse plane (see Fig. 2–31), the 90-degree excitation pulse returns the fat component along the direction of the main field; however, all other components that were not aligned entirely along the transverse plane but contained some component of magnetization along the longitudinal plane are rotated into the transverse plane and are subject to the imaging RF pulses and gradients. STIR images are not prone to magnetic field inhomogeneities (Fig. 2–32), unlike images obtained with the fat saturation techniques (see Fig. 2–30),

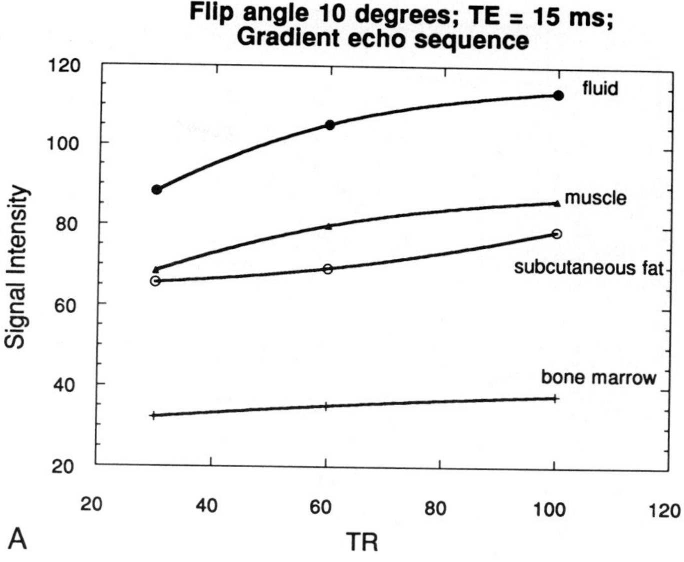

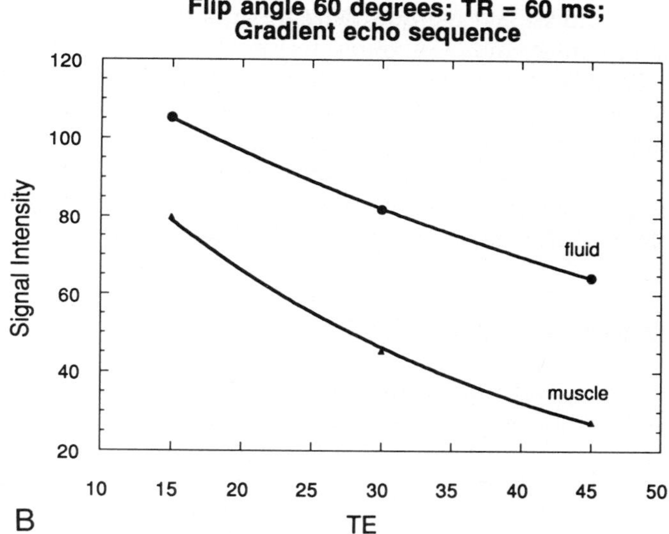

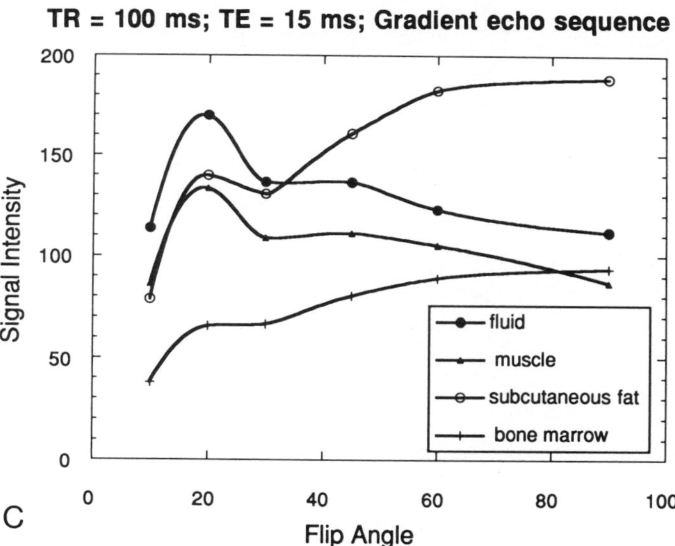

Figure 2–24. Signal intensity variations in tissues in images obtained at different repetition time (TR), echo time (TE), and flip angle (α) values in a spoiled gradient-echo sequence. *A*, Signal intensity as a function of TR. *B*, Signal intensity as a function of TE. *C*, Signal intensity as a function of α.

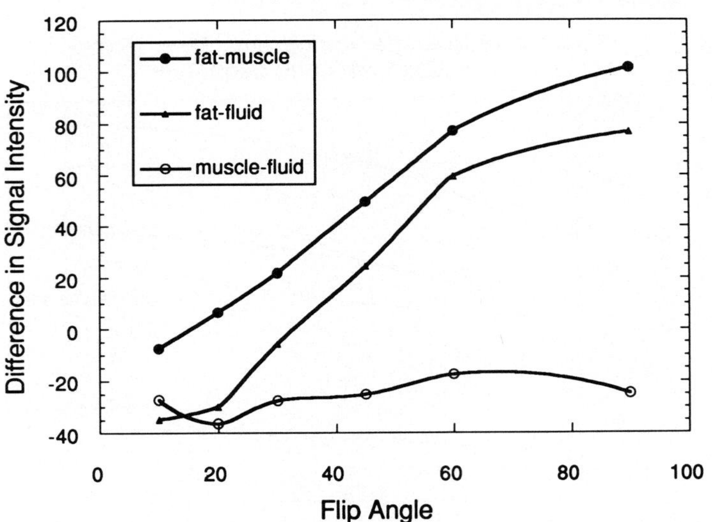

Figure 2–25. Difference in signal intensity or tissue contrast in a spoiled gradient-echo sequence as a function of flip angle, α.

since the STIR technique exploits the difference between the T1 relaxation time of fat and other tissues and in fact cannot distinguish between the fat component and any other tissue with similar T1 relaxation times. Moreover, the image contrast in a STIR sequence is different from that seen in a standard spin-echo sequence with identical TR and TE values. An additional 180-degree pulse in a STIR sequence also increases the RF deposition and hence the specific absorption rate in STIR sequences.

Magnetic Susceptibility

Magnetic susceptibility is the ability of a material to become magnetized. Differences in tissue magnetic susceptibility may be seen at air-tissue interfaces, such as in the lungs; at bone–bone marrow interfaces in the skeletal system; and in the presence of iron deposits as in hemorrhages. Such differences in tissue susceptibility give rise to inhomogeneities in the magnetic field, which cause differences in the Larmor frequency of the nuclei. As a result of such differences in the Larmor frequency, the transverse magnetization is dephased, leading to a loss of signal intensity, particularly in gradient-echo images. These effects are amplified at long echo times. For example, in Figure 2–33 a spin-echo image and gradient-echo image obtained at a TE value of 35 ms are shown. Susceptibility-induced dephasing due to calcification is apparent in the gradient-echo image and is amplified in the image obtained at a TE of

Text continued on page 68

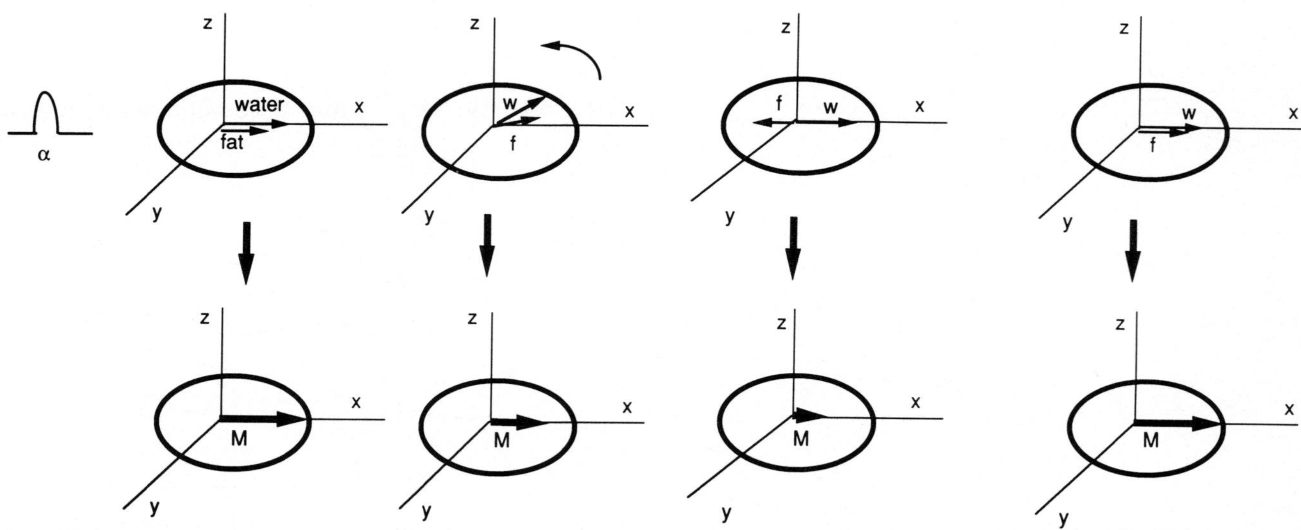

Figure 2–26. Behavior of fat and water magnetization in a gradient-echo sequence (see text for explanation). The fat component of magnetization is represented as *f*, the water component is represented as *w*, and the net magnetization is M.

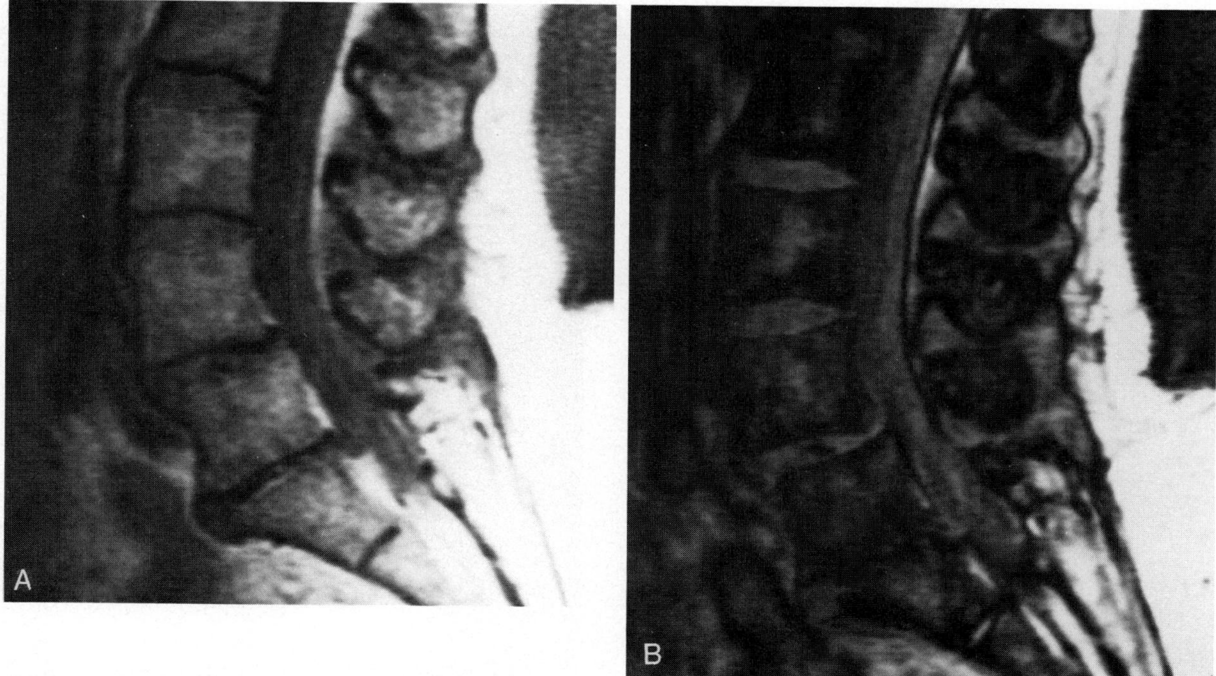

Figure 2–27. Spin-echo image at a repetition time (TR) of 400 ms and an echo time (TE) of 20 ms, obtained through the sagittal plane of the lumbar spine. *A*, 180-degree RF pulse is applied at a time TE/2 or 10 ms. *B*, 180-degree RF pulse is applied at a time TE/2 + 1.1 ms or 11.1 ms. Note the difference in signal intensity of vertebral bone marrow. In *A* the fat and water components of the magnetization are in phase and the signal intensity is at a maximum, whereas in *B* the fat and water signal are out of phase, causing a drop off in signal intensity in vertebral bone marrow.

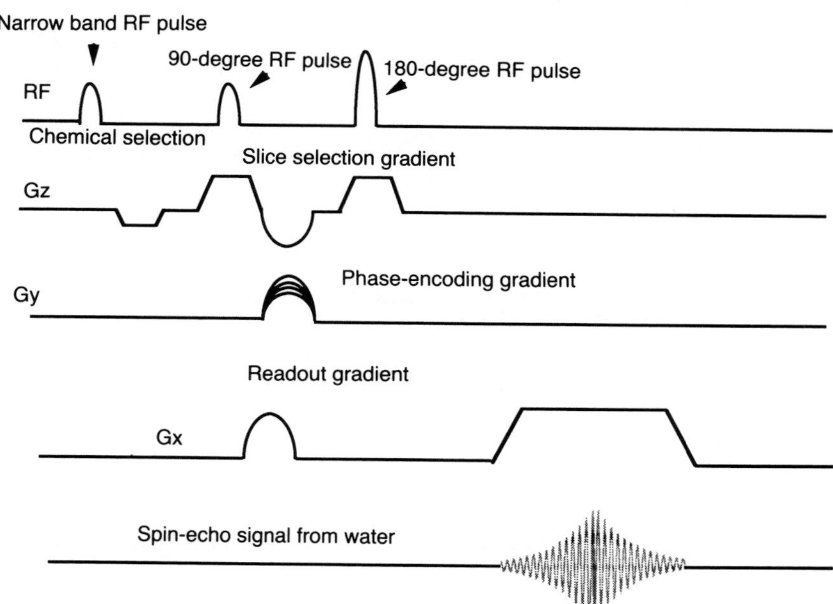

Figure 2–28. Chemically selective spin-echo sequence.

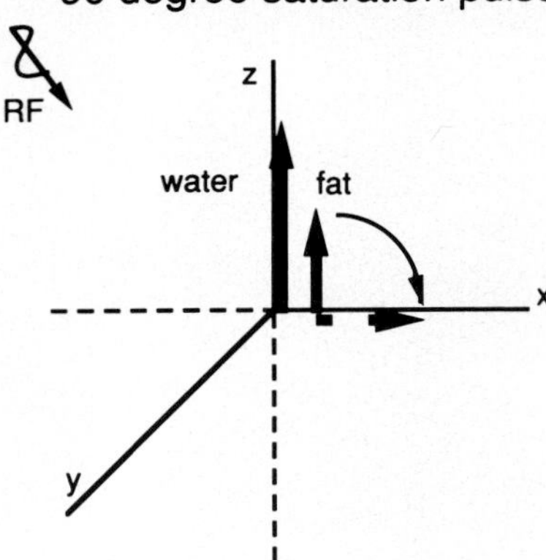

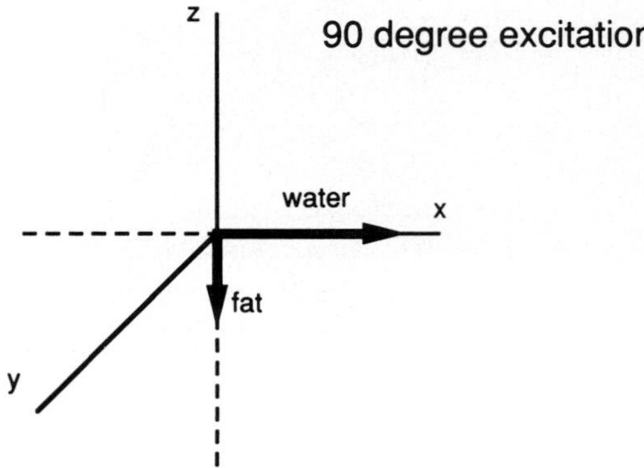

Figure 2–29. Impact of the chemically selective RF pulse on the fat and water magnetization.

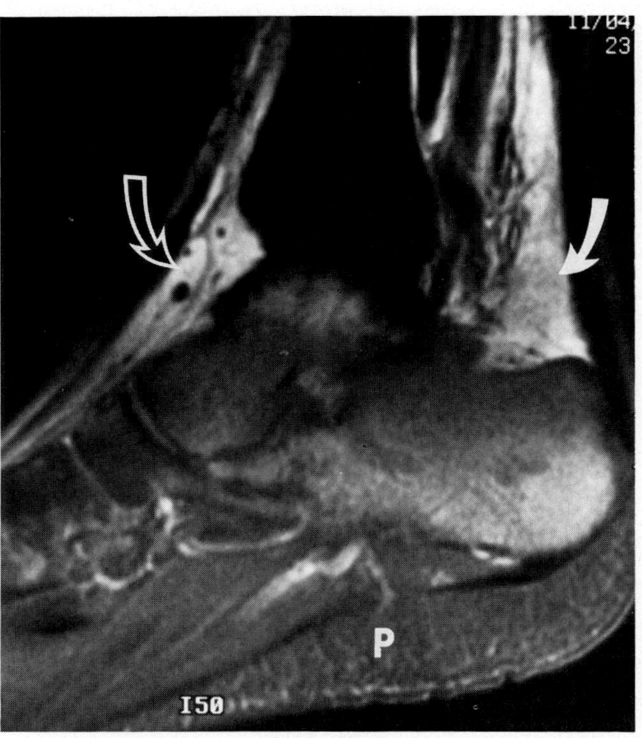

Figure 2–30. Sagittal view through the ankle of a patient with marrow reconversion, obtained using a fat saturation technique. Note that selective fat and water saturation are both present in this image owing to the magnetic field inhomogeneities generated by the shape of the foot. At the top of the image the hematopoietic red marrow is suppressed whereas the fat in the anterior *(open arrow)* and posterior compartments *(solid arrow)* is not suppressed; at the bottom of the image, the fat in the plantar aponeurosis (P) has been suppressed.

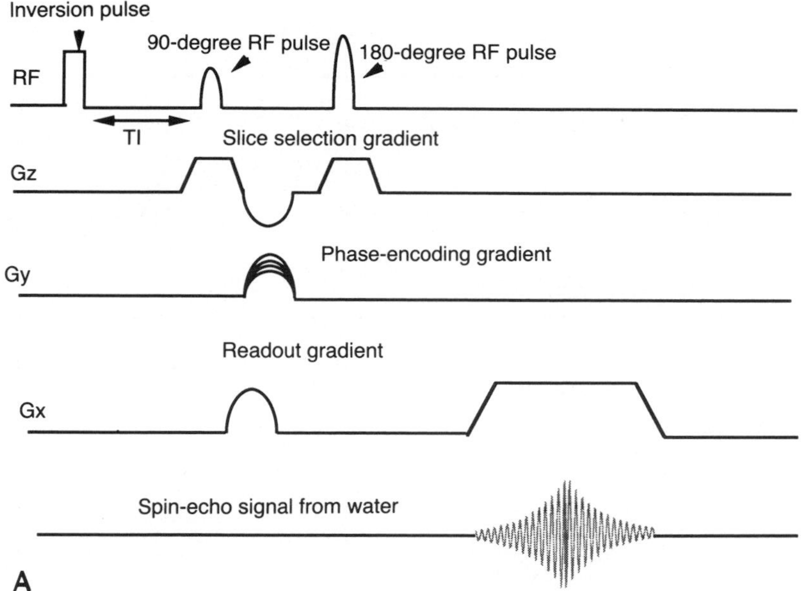

Figure 2–31. *A*, Pulse sequence for a short tau inversion-recovery (STIR) sequence. *B*, Impact of the inversion pulse on the fat and water magnetization. The dashed lines represent the transverse and longitudinal components of the water magnetization.

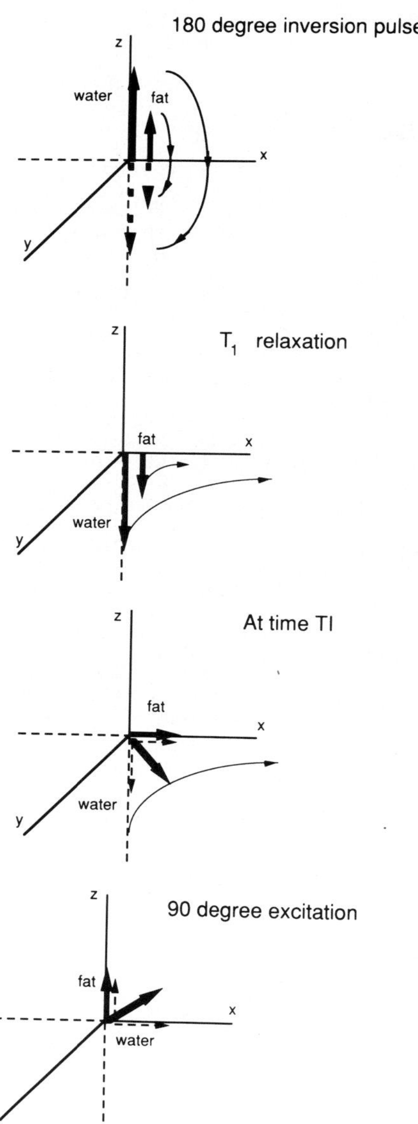

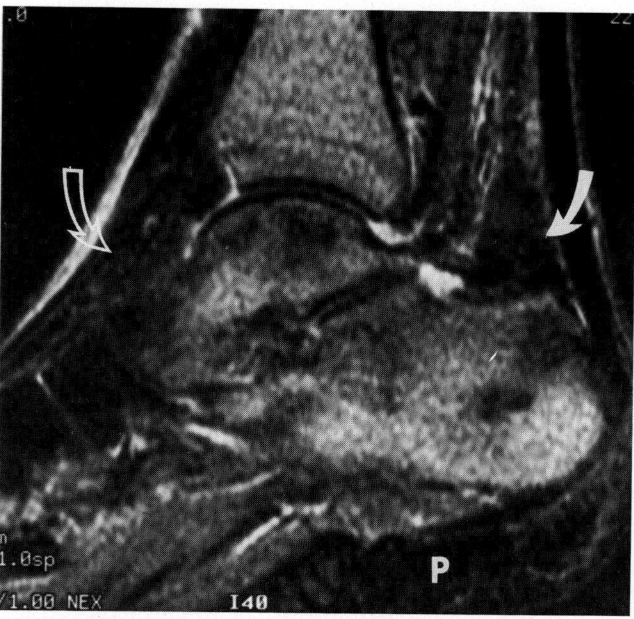

Figure 2–32. Short tau inversion-recovery (STIR) image obtained through the same section as that in Figure 2–30. Note the effectiveness of fat saturation. At the top of the image, the hematopoietic red marrow is not suppressed, whereas the fat in the anterior *(open arrow)* and posterior compartments *(solid arrow)* as well as in the plantar aponeurosis (P) has been suppressed.

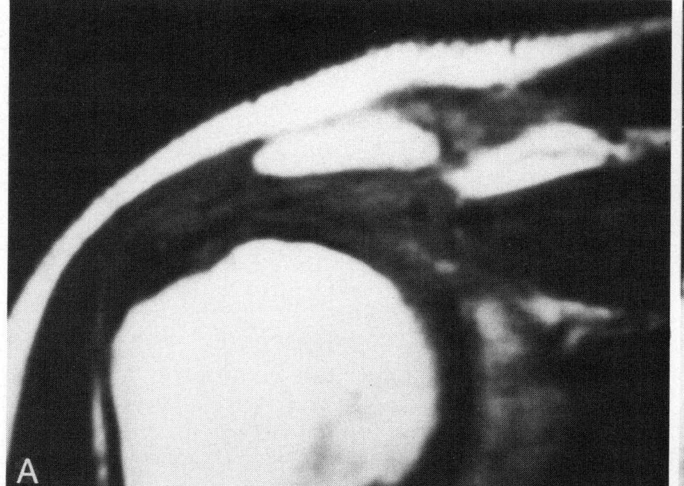

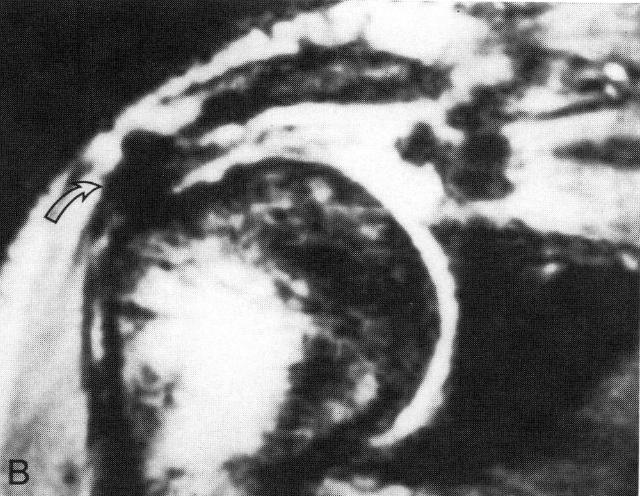

Figure 2–33. *A*, Spin-echo sequence obtained through the shoulder joint, showing no evidence of calcification. *B*, Gradient-echo image obtained through the same section at an echo time of 35 ms, showing severe image distortion near the calcification *(arrow)*.

35 ms; however, the spin-echo image through the same section shows no evidence of the apparent calcifications.

SUMMARY

The achievable image contrast, resolution, and characteristics depend on the complex interaction of several system-dependent, technique-dependent, and tissue-dependent factors. In this chapter only a few of these issues are addressed. In addition, factors such as patient respiration, blood flow, flow of cerebrospinal fluid, tissue perfusion, and diffusion of tissue water also affect image characteristics; however, such details are beyond the scope of this chapter. It is the complexity of the MR imaging technique that makes it a potentially valuable tool for diagnosing tissue pathologies and distinguishing different disease conditions.

References

1. Abragam A: The Principles of Magnetic Resonance. London, Oxford University Press, 1961.
2. Dixon WT: Simple proton spectroscopic imaging. Radiology 153:189, 1984.
3. Glover G: Multipoint Dixon technique for water and fat proton and susceptibility imaging. J Magn Reson Imag 1:521, 1991.
4. Haacke EM, Bearden FH, Clayton JR: Reduction of MR imaging scan time by the hybrid fast scan technique. Radiology 158:521, 1986.
5. Hardy CJ, Katzberg RW, Frey RL, et al: Switched surface coil system for bilateral imaging. Radiology 167:835, 1988.
6. Hennig J, Nauerth A, Friedburg H: RARE imaging: A fast method for clinical MR. Magn Reson Med 3:823, 1986.
7. Kneeland JB, Hyde JS: High resolution MR imaging with local coils. Radiology 17:1, 1989.
8. Kumar A, Welti D, Ernst RR: NMR Fourier Zeugmatography. J Magn Reson 18:69, 1975.
9. Mansfield P, Pykett IL: Biological and medical imaging by nuclear magnetic resonance. J Magn Reson 29:355, 1978.
10. Roemer PB, Edelstein WA, Hayes CE, et al: The NMR phased array. Magn Reson Med 16:192, 1990.
11. Rosen BR, Wedeen VJ, Brady TJ: Selective saturation NMR imaging. JCAT 8:813, 1984.
12. Szumowski J, Eisen JK, Vinitski S, et al: Hybrid methods for chemical shift imaging. Magn Reson Med 9:379, 1989.
13. Twieg DB: Acquisition and accuracy in rapid imaging methods. Magn Reson Med 2:37, 1985.
14. Wehrli FW: Fast scan magnetic resonance principles and applications. MR 6:165, 1990.
15. Wehrli FW, Perkins TG, Shimakawa A: Chemical shift-induced amplitude modulations in images obtained with gradient refocusing. Magn Reson Imaging 5:157, 1987.

3 MAGNETIC RESONANCE SPECTROSCOPY

Sarah J. Nelson and Daniel B. Vigneron

Magnetic resonance (MR) spectroscopy has been employed as a laboratory tool for assaying changes in chemical composition and structure for many years. Atomic nuclei in a magnetic field behave in a manner that depends on both the magnitude of the main magnetic field and the chemical environment of the nucleus. Thus, it is possible to distinguish signals from different atoms in the same molecule or from different compounds and, when applied to living tissue, to estimate the concentration of a number of important cellular metabolites. Magnetic resonance spectroscopy has been applied to clinical studies to assess tissue function to make a diagnosis, to monitor disease progression, or to follow response to therapy.

As has been detailed in several reviews,[2,10,41,45,52,63-65] MR spectroscopy has had numerous applications in both animal and human systems. Many groups have examined the bioenergetics of skeletal muscle, and MR spectroscopy is now established as a viable, noninvasive tool for measuring the function of superficial muscles in the forearm and calf. A large number of these studies have involved placing the limb in the bore of an experimental magnet and measuring the changes in metabolites such as phosphocreatine (PCr) and adenosine triphosphate (ATP). These studies have provided considerable data concerning normal physiology and abnormalities associated with different muscle diseases. With advances in the technology of whole body MR imaging equipment and improvements in techniques for obtaining MR spectra, it has now become feasible to perform MR imaging and obtain spectra that are localized to different regions of the structures in the same clinical examination. In this way, morphologic and functional measurements can be made on the same tissue; differences between superficial and deeper lying muscles can be observed; and results of sequential examinations can be interpreted more accurately. In the future, it is likely that MR imaging and MR spectroscopy will be combined in routine clinical examinations to aid in diagnosis and monitoring response to therapy. To understand the potential advantages and practical limitations of such studies, the biochemical basis of muscle contraction and the technical issues involved in measuring and quantifying MR spectra must be considered.

MUSCLE STRUCTURE AND FUNCTION

Muscle is a highly organized tissue.[34] The molecular architecture of muscle cells comprises a network of proteins, which are polymerized to form filamentous structures. Two main types of filaments have a structure that can be observed using light and electron microscopy. Actin is the main component of thin filaments, which also have tropomyosin and troponin subunits spaced at regular intervals along the strand. The actin filaments are joined at one end through the protein alpha-actinin to create the Z line. The number of connections at the Z line differs in different fiber types, fast muscles having a thin Z line with few connections and slow muscles having a much thicker Z line. Thick filaments comprise about 300 myosin molecules that are bound together. The cell architecture is arranged so that the thin filaments can slide between the thick filaments and form cross bridges. Other proteins that play a role in maintaining the structure of the arrays of thin and thick filaments are desmin, spectrin, and dystrophin.

At a higher level, bundles of 100 to 400 thick filaments are grouped to form myofibrils, which are in turn bound together to form individual fibers. Slow (type I) and fast (type II) fibers are distinguished on the basis of their contractile properties, with additional subclasses being determined by fiber size and fatigability. Muscle fibers vary considerably in number and length among different muscles in the body and among different persons. Some muscles have predominantly type I fibers; others have mainly type II fibers; and some have a mixture of fiber types. The composition of any particular muscle can be established by staining the tissue obtained by muscle biopsy and examining frozen sections under a microscope. Although it is reliable, biopsy is an extremely invasive procedure, and it would be a considerable advantage to have a noninvasive mechanism for assessing changes in the composition and function of different muscles.

The mechanisms of muscle contraction and extension are based on the formation and detachment of cross bridges between the thin and thick filaments. This requires ATP, which can be maintained at a constant level for short periods of exercise by the metabolism of PCr and adenosine diphosphate (ADP), catalyzed by the enzyme creatine kinase (CK). As ATP is used, inorganic phosphate (P_i) is formed and intracellular pH decreases. The basis for the great utility of phosphate-31 (^{31}P) MR spectroscopy in studying exercise response in skeletal muscle has been that it is able to measure all of these biochemical endpoints.[20,47] In the case of longer-term exercise and muscle fatigue, the level of ATP may be reduced, and other mechanisms for producing ATP, such as glycolysis and oxidative metabolism, may become operative. The buildup of lactate that occurs during exercise also can be measured using water suppressed hydrogen-1 (^{1}H) MR spectroscopy.[62] Owing to technical difficulties in excluding spectral contamination from fat, only a relatively small number of such studies have been performed.

Several different types of exercise may be used to evaluate muscle function. Isometric exercise involves holding a muscle contraction for a fixed period of time. At the maximum level of contraction (maximum voluntary contraction), this requires a great deal of effort, and, in practice, the force exerted usually declines by 30 to 60 sec after the initial contraction.[34] Monitoring the response to this form of exercise therefore requires that measurements be made in less than 1 min. Isokinetic exercise involves contractions at regular intervals, and these can be maintained for longer periods of time. Typically isokinetic exercise is performed by squeezing a bulb to a predetermined pressure every few seconds. Again the amount of effort exerted depends on the percentage of maximum voluntary contraction that is employed. To establish reproducibility for both isometric and isokinetic exercise, it is necessary to monitor the force exerted as a function of time during the experiment and, ideally, to provide a feedback to the subject so that he or she maintains the chosen level of effort. Thus, an important part of the design of an MR spectroscopy exercise study is the construction of a specialized exercise device that can operate in the magnet.

SPECTROSCOPY TECHNIQUES

When a sample is placed in a magnetic field, its atomic nuclei precess at characteristic (Larmor) frequencies about the main magnetic field. In the most simple form of MR spectroscopy experiment, a short radiofrequency (RF) pulse is applied to the sample to tip the nuclei at an angle to the main magnetic field. The shape of the pulse and the magnitude of the tip angle used depend on the details of the particular experiment. In some cases magnetic field gradients also are applied to selectively excite the nuclei in particular regions of space.[5] The response of the sample to the RF pulse typically is recorded as a complex valued time series or free induction decay. The contributions of different chemical species are then separated by taking the Fourier transform of the free induction decay to produce a spectrum. Consider first the situation of a uniform sample with a single magnetically resonant visible nucleus in a perfectly homogeneous magnetic field. This would provide a free induction decay with the form of a decaying sine wave, and the corresponding spectrum would comprise a single peak with a shape described by a lorentzian function. The width of the peak would correspond to the rate of decay of the free induction decay, its position to the frequency of the nuclei, and the area under the peak to the concentration of the compound in the sample (Fig. 3–1). When more than one chemical species is present in the sample, multiple peaks are obtained with different frequencies (chemical shifts). In practice, the data acquisition cannot always coincide precisely with the application of the RF pulse, and the nuclei may have a short period to relax before the first data point is acquired. This causes the phases of the peaks to be different, but it can be corrected by modifying the phase so that the real part of the spectrum has positive, symmetric peaks (Fig. 3–2).

In addition to the existence of multiple compo-

Figure 3-1. Simulation of a free induction decay *(A)* and spectrum *(B)* of a single chemical species in a homogeneous magnetic field. The width of the peak reflects the T2 relaxation time of the sample; the peak area is proportional to its concentration; and the peak position (or chemical shift) relative to a known standard is characteristic for the chemical composition of the sample.

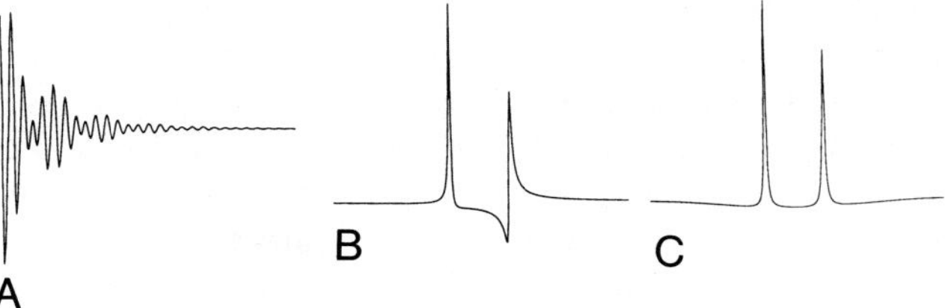

Figure 3-2. Simulation of a free induction decay *(A)*, and an unphased *(B)* and a phased spectrum *(C)* of a sample with two chemical species. The phase of the spectrum is adjusted using a function that is the sum of constant and linear frequency terms until both peaks are in positive orientation.

nents, the quality of the spectrum is affected by inhomogeneities in the main magnetic field. This broadens peaks and, in extreme circumstances, may totally distort their shape. Shimming (adjusting the homogeneity of the main magnetic field) is therefore extremely important for MR spectroscopy and determines whether components with similar Larmor frequencies can be resolved. In practice, the spectral resolution that can be achieved is limited by the strength of the main magnetic field, the magnet shim coil design, and the skill of the operator. Several reports have described different approaches to automatic shimming,[75,78] which have provided reproducible results within a defined region of interest. These techniques will be extremely valuable for clinical applications of MR spectroscopy, but they are not yet generally available. Whether the shim is manual or automatic, most in vivo MR spectroscopy studies are performed at field strengths of 1.5 T or higher to obtain sufficient spectral resolution to distinguish the components of interest.

The quantification of in vivo MR spectra is complicated by the presence of low signal-to-noise ratios, overlapping peaks with nonlorentzian peak shapes, and broad baseline components. For superficial skeletal muscle, the baseline component usually is not too severe, but it does become evident when larger regions of tissue that include bone are considered (Fig. 3-3). A range of different methods have been proposed for quantification of ^{31}P MR spectra. Many different approaches have been suggested, ranging from cutting out and weighing peaks, peak triangulation, maximum entropy,[29,39] linear prediction, singular value decomposition,[8,76] simulated annealing,[77] and the PIQABLE (Peak Identification Quantification and Automatic BaseLine Estimation) algorithm.[55,56] Currently the most common approach is to use maximum likelihood estimation in the time or frequency domain, assuming lorentzian or gaussian peak shapes.[21,71] These fitting routines generally are superior to manual integration,[48] but their accuracy depends on the validity of the peak shape assumptions and the method used to remove the baseline. An advantage of automated procedures, such as the PIQABLE algorithm, is that they provide reliable and reproducible peak parameter estimates, using nonparametric statistical methods to estimate both the baseline and the number of statistically significant peaks in the spectrum.[55,56]

To determine absolute concentrations of metabolites, more analysis is needed. The first step is to include a standard of known concentration within the sensitive region of the RF coil. For ^{31}P, this is commonly methylene diphosphoric acid, which gives a peak that is well separated from the metabolites of interest. The next, more difficult, problem is to estimate the T1 relaxation time of each metabolite to correct the peak intensities for partial saturation effects. This is necessary because the time between data acquisitions (repetition time, or TR) used for

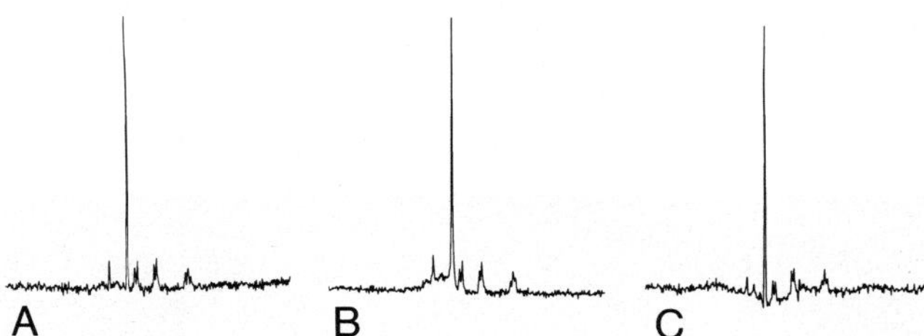

Figure 3-3. Examples of experimental ^{31}P spectra obtained from the calves of normal subjects. The peaks are (from left to right) inorganic phosphate (P_i), phosphocreatine (PCr), gamma-adenosine triphosphate (ATP), alpha-ATP, and beta-ATP. Note that each of the ATP peaks actually has a multiple structure owing to J coupling between the different atoms in the ATP molecule. These spectra were acquired on a General Electric Signa clinical imager operating at 1.5 T using a small surface coil on the anterior tibialis muscle *(A)*, with a volume coil that gave a signal from the calf using either no localization *(B)* or localization by selecting out a 4-cm thick slice through the calf *(C)*. The main peaks look very similar in all three cases, but the baseline components are quite different. For accurate quantification of peak areas it is thus extremely important to obtain a reproducible and reliable estimate of the baseline component.

most kinetic experiments is too short to allow complete relaxation between acquisitions. For calf muscle the T1 times at 1.5 T are as follows: P_i, 4.7 sec; PCr, 6.5 sec; alpha-ATP, 4.2 sec; beta-ATP, 4.1 sec; and gamma-ATP, 3.9 sec.[14] In practice, the restricted time available for most studies means that usually it is not feasible to measure T1 times during the course of a single experiment. Thus, in general, either relative concentrations are reported using peak area ratios or the T1 values are obtained from separate experiments.

An additional major problem in interpreting the results of MR spectroscopy experiments is identifying accurately the region of tissue that gives rise to the signals in the spectrum. Most of the early studies localized to the tissue of interest used a small surface coil placed directly on the muscle of interest.[1] This works for superficial tissue because it provides high sensitivity close to the coil (Fig. 3–4), which means that signals from deeper lying tissues are detected poorly. Another complication arises if a standard rectangular RF pulse is used, because tissues at different spatial positions receive different tip angles, and even for a region of homogeneous tissue, the overall spectrum is a sum of multiple signals. This problem has been partially overcome by the design of adiabatic pulses, which give approximately uniform excitation over a range of different RF fields,[12,28] or by using a relatively large coil for transmitting and a smaller surface coil for receiving the MR signals. Even in this case, the observed spectrum may still reflect contributions from heterogeneous tissue types.

Numerous localization techniques have been proposed to overcome such difficulties,[5] but only a small number have been implemented for human studies. These include methods that give spectra from a single slice (depth-resolved surface coil spectroscopy [DRESS]) or an approximately rectangular region of tissue (e.g., [ISIS]).[60] The size of the selected region is usually relatively large owing to the intrinsically low signal-to-noise ratio of ^{31}P MR spectroscopy. The shape of the region may be modified using, for example, conformal ISIS,[70] and it is possible to obtain spectra from two or more voxels at a time.[61] More recent applications of ^{31}P spectroscopy have used a multivoxel localization called *chemical shift* or *spectroscopic imaging*.[11,46] Here the spatial localization is provided by the same phase-encoding gradients as in MR imaging, allowing a direct comparison with the anatomy. The spatial resolution that can be achieved in muscle using chemical shift imaging is of the order of 1 to 10 cc. This is determined by the low sensitivity of the phosphorus nucleus and the limited time available for data acquisition. An additional complication arises because of the need to process and interpret the multidimensional arrays of data that are obtained with this technique (Fig. 3–5).

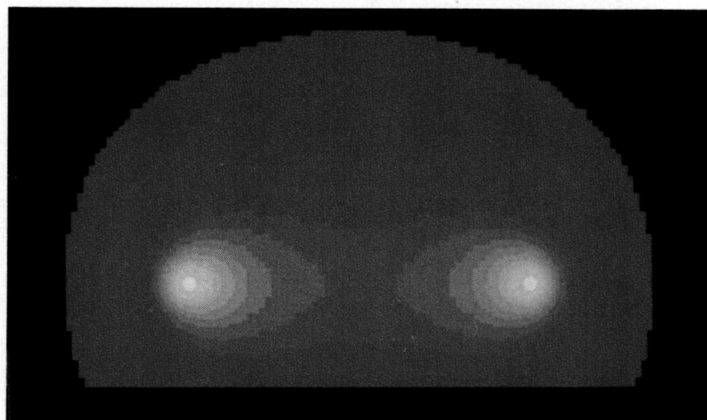

Figure 3–4. Map of the sensitivity profile of a circular surface coil calculated in a plane perpendicular to the coil. The bright spots represent the spatial location of the wires.

Water-suppressed localized ^{1}H spectroscopy techniques also have been developed for obtaining spectra from humans, mainly from within the brain,[23-25,42,51] but with a small number of applications to muscle. Although ^{1}H MR spectroscopy is much more sensitive than ^{31}P spectroscopy, significant technical hurdles must be overcome for in vivo applications. The first problem is the suppression of the 100-molar water resonance, which must be reduced to be able to visualize the much lower concentrations of cellular metabolites. To suppress the water resonance with minimal effect on the resonances nearby requires very good shimming within the region of interest and the use of saturation pulses that are restricted to a narrow range of frequencies. This is difficult in vivo because of variations in magnetic susceptibility that may lead to broad or multiple component water resonances.

The second problem, which is particularly relevant in muscle studies, is the presence of large lipid resonances from fat. For brain studies these can be substantially reduced by selecting a region that has no subcutaneous fat. The most commonly used techniques provide localization to a single voxel by selective 90°–90°–90° stimulated echo acquisition mode (STEAM) or 90°–180°–180° excitation point resolved spectroscopy (PRESS). STEAM can be used with shorter echo times and therefore allows the detection of ^{1}H metabolites with short T2. Its disadvantages are an inherent twofold decrease in signal-to-noise ratio and an increased sensitivity to signal loss owing to diffusion.[51] In addition to these volume selection techniques, several studies have applied phase encoding simultaneously to produce one-dimensional, two-dimensional, or three-dimen-

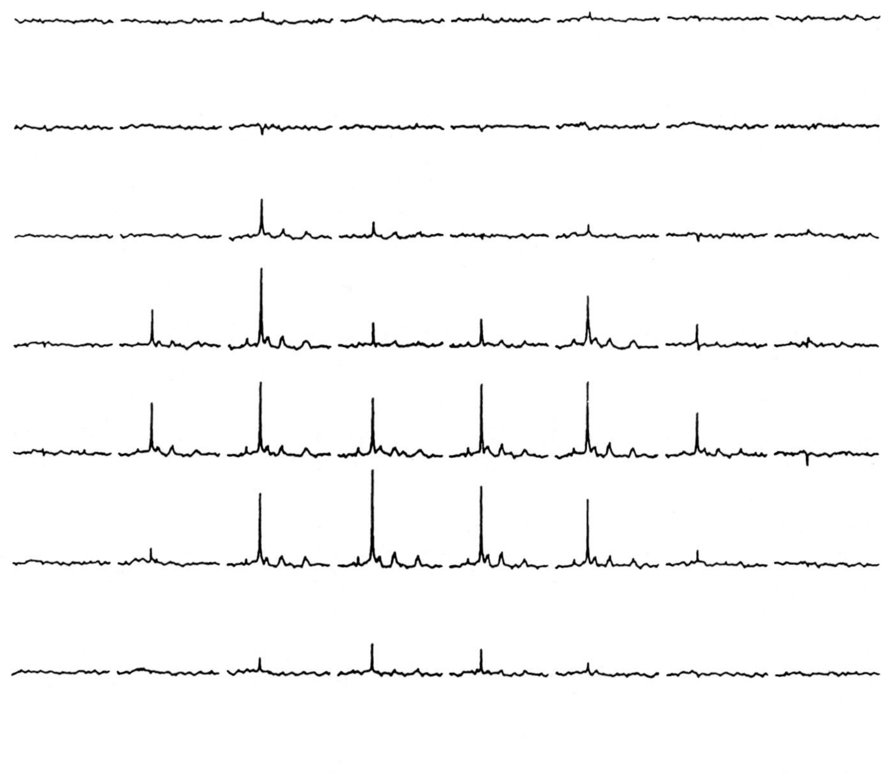

Figure 3-5. A two-dimensional array of ^{31}P spectra obtained from the calf of a normal subject using a General Electric Signa clinical imager operating at 1.5 T. These spectra represent the central slice from a three-dimensional chemical shift imaging dataset that was acquired using a 7 × 7 × 7 phase-encoding matrix and a 16 × 16 × 24-cm field of view in approximately 22 min of data acquisition. Each spectrum corresponds to a region of approximately 18 cm^3. Note that the relatively good signal-to-noise ratio of these data allows an estimation of the spatial distribution of inorganic phosphate (P_i) and adenosine triphosphate (ATP) resonances as well as that of the higher concentration phosphocreatine (PCr).

sional arrays of localized spectra[69] In muscle, the contribution of subcutaneous fat may be eliminated by selecting out an appropriate region, but considerable fat that has infiltrated into muscle tissue may still be present, particularly in older persons or in patients with degenerative diseases such as muscular dystrophy.

STUDIES OF ISOLATED MUSCLES AND ANIMAL SYSTEMS

Studies of isolated muscles and animal systems have established the validity of MR measurements in analyzing muscle function and have produced a sound basis for future clinical applications. One of the first demonstrations of MR spectroscopy from intact tissue was by Holt and co-workers,[36] who used a 7.4-T magnet to study muscle from the hind leg of a freshly killed rat. This showed peaks corresponding to the key ^{31}P metabolites, P_i, PCr, and the resonances of the three ATP species. The peaks were relatively broad compared with peaks from the spectra of corresponding compounds free in solution. In addition, if the muscle was stimulated before measurements were made, the PCr to P_i ratio was observed to be lower. Subsequent experiments extended these techniques to small animals or isolated muscle preparations in which the response to exercise could be measured in more detail.

Ackerman and colleagues[1] laid the foundation for many of these studies by designing surface coils to obtain spectra from localized regions of superficial muscle. The ^{31}P spectra were acquired from the leg of a rat, and they showed a dramatic difference between the spectra from ischemic muscle and that from normal muscle: an almost absent PCr, a slight reduction in ATP, a large increase in P_i and a change in the chemical shift of P_i by 0.61 ppm. This difference in chemical shift corresponded to a change in pH from 7.1 to 6.7. Since then, many applications of ^{31}P MR spectroscopy have been made to study ischemia in the skeletal muscle, in the myocardium, and in the brain. Regional differences in the extent of ischemia were demonstrated in the rat leg using localization by one-dimensional chemical shift imaging.[35]

In addition to estimating pH and concentrations of high-energy phosphates, ^{31}P MR spectroscopy can be used to determine the concentration of free intracellular magnesium (Mg). Cohen and Burt used ^{31}P spectroscopy to measure the relaxation parameters of PCr from intact frog gastrocnemius muscle.[18] Both T1 and T2 relaxation times were estimated. From a study of model solutions, the T2 of PCr was shown to depend on Mg concentrations, whereas both the T1 and chemical shift of PCr were constant. On the basis of the T2 value of PCr from muscle, the free intracellular Mg was estimated as 4.4 mM, the T1 as 4.8 sec, and the T2 as 912 ms. An alternative method for estimating Mg concentration is to use the chemical shift of beta-ATP. This was applied by Dillon and colleagues[22] and Kushmerick and associates[38] in studies of isolated rabbit muscle. Values obtained ranged from 0.4 mM to 1.5 mM depending on the type of muscle used.

Valuable information concerning muscle bioenergetics also was obtained with isolated muscle preparations. Kushmerick and associates used perfused cat biceps and soleus muscles to examine the response of muscles with different fiber composition: in the cat, the biceps muscle consists of 75 per cent fast glycolytic and 25 per cent fast oxidative glycolytic fibers whereas 95 per cent of the soleus is made up of slow oxidative fibers.[37] At rest, the PCr to P_i ratio differed between the two types of muscles, but the pH was similar at 7.0 to 7.2. The concentration of PCr was measured both by MR spectroscopy and chemical assay as 11 mmol/g in the soleus and 22 mmol/g in the biceps. The levels of P_i in the biceps measured by MR spectroscopy (1 μmol/g) were lower than those obtained by chemical assay (6 μmol/g), but the two methods agreed fairly well for the soleus muscle.

In addition to examining the resting state, Kushmerick and associates[37] also followed the metabolic changes that occurred as a result of stimulated isometric twitches in both muscles. They found that the oxygen consumption was different for the two muscles and that the biceps has a larger reduction in pH, by about 0.3 unit. This suggests a greater glycolytic lactate production in the fast twitch muscle fibers. The P_i peak in the biceps also tended to broaden during recovery, which was interpreted as being due to functional differences between the different fiber types in the muscle. If the muscles were made ischemic by stopping perfusion, the PCr levels fell, P_i increased, and acidification resulted but more slowly than by contractile activity. Recovery from ischemia was also slower than from exercise.

Although the precise mechanisms of force generation and muscle fatigue are still uncertain, clearly the changes in ^{31}P metabolites observed by MR spectroscopy are closely correlated with muscle function. For example, when frog muscles were stimulated repetitively in the anaerobic state, development of force declined rapidly and relaxation slowed. The decline in force appeared to be independent of the pattern of stimulation but was strongly correlated with the buildup of ADP, P_i, and hydrogen ions (H^+).[19] On the basis of these and similar experiments, the investigators suggested that the P_i that exists in the ionic species $H_2PO_4^-$ was an important factor in the formation of mechanical fatigue. They also reported that the time constants for mechanical relaxation correlated closely with the concentrations of ADP, P_i and H^+.

As discussed earlier, most of the MR spectroscopy studies of muscle have been performed under conditions of nonsaturation (e.g., TR of 1 to 2 sec) to follow rapid kinetic changes in ^{31}P metabolites. If the T1 of PCr or P_i changed during exercise or recovery, the estimates of relative PCr and P_i concentrations might well be in error. So far, the measurements that have been obtained in animals and isolated muscles suggest that there are no significant T1 changes. In one study, Meyer and Adams found no change in T1 times of ATP, PCr, or P_i in the gastrocnemius muscles of rats.[49] Although the peak areas of P_i and PCr decreased during stimulation, so did the sum of ATP peak areas, and the researchers found no evidence of a selective decrease in P_i during exercise, as had been suggested by some previous studies. Full recovery from exercise occurred within approximately 5 min of the end of stimulation.

STUDIES OF SUPERFICIAL HUMAN MUSCLES

A large number of MR spectroscopy studies of the superficial muscles of the arm and calf have been performed.[3,4,7,15-17,26,27,44,50,59,64,66-68,72,73] Taylor and co-workers[72] examined the changes in P_i, PCr, and pH both during and after forearm exercise. The exercise protocols required squeezing a rubber bulb to a pressure of between 100 and 300 mm Hg at a rate of once every 2 sec for 1 to 20 min. A surface coil 2.5 cm in diameter was used and placed by first palpitating the relevant muscle and marking the appropriate position. Relative concentrations of metabolites were determined by estimating peak areas of $P_i/(P_i + PCr)$, $PCr/(P_i + PCr)$, and $ATP/(P_i + PCr)$ and correcting for saturation effects by comparing the ratio of peak areas obtained from resting spectra with a TR of 2 sec with those obtained with a Tr of 10 or 20 sec. No difference was observed between the corrections estimated from the 10 or 20 sec spectra. The pH was calculated from the chemical shift difference of P_i and PCr. At rest the pH was 7.04, the PCr to P_i ratio was 10 to 1, and PCr/ATP was 4.4.

The response to exercise was qualitatively the same among subjects but varied in magnitude. Differences in the extent of change in pH depended on whether the coil was placed over the finger flexor (pH 6.4) or wrist (pH 6.8) muscles. It is not clear whether this variation was attributable to differences in fiber types or to differential work loads experienced by the muscle groups. Some researchers have suggested that the splitting of the P_i peaks is caused by contamination of the measured spectrum by signals from multiple muscle groups, whereas others have suggested that heterogeneity exists within a single muscle because of differences in fiber types. In the studies of Taylor and colleagues, single P_i peaks were observed, and the changes in pH occurred more slowly than did changes in PCr.[72] The time constant for 50 per cent recovery of PCr was 0.9 min, but for pH was between 4 and 5 min. If a 1-min period of light ischemic exercise was followed by 6 min of ischemia without exercise, the metabolic recovery did not occur until after the ischemic period ended. Recovery then occurred at the same rate as it did without ischemia.

In a later study with a similar experimental protocol, Taylor and colleagues used exhaustive dynamic exercise (4 to 20 min) to study the biochemical response to fatigue.[73] In six of the 12 subjects examined, ATP became depleted by approximately 55 per cent of its original value, PCr was reduced by 17 per cent, and the pH fell to 6.12. In the other six volunteers, no reduction in ATP was observed, the PCr reduction was less than 26 per cent, and the pH fell only to 6.37. When ATP was lost, the recovery rates for PCr and P_i were slower (5.3 and 1.5 min, as compared with 1.1 and 0.5 min) and the initial rate of oxidative phosphorylation was slower. The recovery of ATP was much slower than that of other metabolites, taking more than 40 min.

The enzymatic flux through the CK pathway in forearm muscles was examined in more detail by Rees and co-workers using ^{31}P saturation transfer techniques.[67] In this technique one metabolite is selectively irradiated to observe the effect on a chemical species that is exchanged with it. Thus, irradiation of the gamma-ATP causes a reduction in the PCr peak. In these experiments, the T1 of PCr was estimated to be 4.76 sec, the concentration of PCr was 35.7 mM, the P_i was 3.9 mM, and the flux through the CK reaction in exercising muscle was 64 per cent of its resting value.

Miller and co-workers examined the connection between muscle fatigue and metabolism in the adductor pollicis muscle.[50] By monitoring changes in MR spectra, muscle force, and electromyogram, they found evidence for three phases of recovery from fatiguing exercise: muscle membrane function (4 to 6 min), force and metabolic components (maximum voluntary contraction, pH, PCr, and P_i 10 to 12 min) and neuromuscular efficiency. Newham and Cady also studied fatigue and metabolism in humans.[59] They used the first dorsal interosseous muscle, which could be exercised both voluntarily and by stimulation. Although an overall correlation was noted between force generated and extent of metabolic changes, they did not find a special relationship between any single metabolite and the level of fatigue. A considerable difference existed between the level of force generated by a maximum voluntary contraction and an electrical stimulation, indicating that even motivated volunteers did not fully activate their muscles.

Magnetic resonance spectroscopy has been applied in studies that examined the variations in levels of ^{1}H metabolites in skeletal muscle. Bruhn and colleagues showed that considerable differences existed among subjects and also among muscles of different fiber types.[13] In a study of response to exercise, Pan and co-workers followed the production and removal of lactate.[62] Using a surface coil 2.5 cm in diameter placed next to the flexor digitorum superficialis muscle, they estimated that the decay of lactate was exponential (half-time of 10.6 min) but much slower than the rate of change of pH (half-time of 4.7 min). The amount of lactate accumulated after finger flexion against a 2.5-kg load until exhaustion was 25.3 mM with a drop in pH of 1.0. These results are consistent with previous reports of the rate of lactate recovery from animal systems.

Taken together, these studies of normal human physiology define a series of parameters that can be used as a basis for examining abnormalities that occur in various diseases. They can be summarized as follows. First, the ratio of the normal resting values of PCr to P_i is approximately 10 to 1, the value of PCr/ATP is approximately 4.4, and the pH is 7.0 to 7.1. With exercise, the PCr decreases; the P_i increases rapidly; the pH also decreases but relatively more slowly; and lactate is formed. The exercise response may vary owing to differences in fiber types, but it is not clear whether split P_i peaks result from heterogeneity within or between muscle groups. The magnitude of these changes depends on the extent of exercise as measured by the force exerted. Normally recovery from exercise has a half-time of approximately 1.0 min for PCr, 0.5 min for P_i, 4 min for pH, and 10 min for lactate. Under extremely strenuous, fatiguing exercise, the level of ATP may decrease, and if this occurs, recovery slows.

Studies of abnormal muscle function have been carried out in patients with mitochondrial myopathies, Duchenne's muscular dystrophy, Becker's muscular dystrophy, neuropathies, tumors, dermatomyositis, peripheral vascular disease, and heart failure.[3,4,7,16,17,26,27,44,64,66,67] The pattern of depletion of ^{31}P metabolites has been shown to be altered in patients who have peripheral vascular disease[27] and congestive heart failure.[44,66] Whether these observations are useful indicators of the severity of

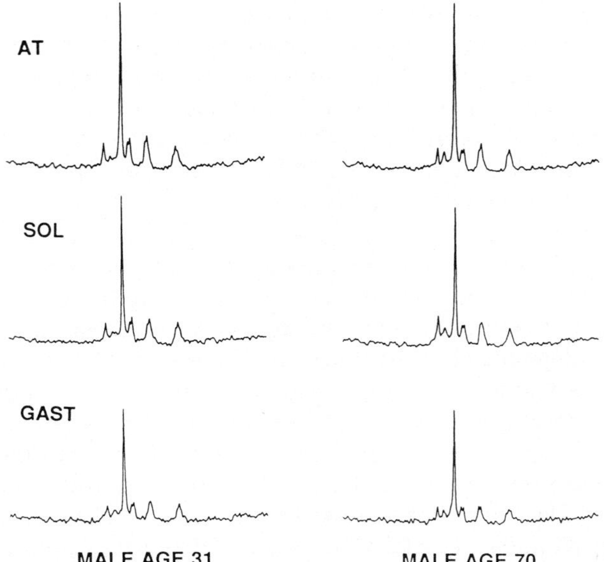

Figure 3–6. A comparison of ^{31}P spectra obtained from the anterior tibialis (AT), soleus (SOL), and gastrocnemius (GAST) muscles of male subjects, aged 31 and 70 years. The spectra were obtained by selecting groups of voxels from a three-dimensional chemical shift imaging dataset that corresponded spatially to each muscle group. The spectra shown are each the sum of spectra from three voxels and therefore represent data from approximately 54 cm^3 of tissue. The relative proportions of inorganic phosphate (P_i) and phosphocreatine (PCr) differ among the muscle groups, which may reflect variations in fiber type. In addition, the younger volunteer shows a very small phosphodiester (PDE) peak (between P_i and PCr), whereas the older volunteer demonstrates a much higher PDE peak.

disease or the response to therapy has not yet been established. In other disorders, such as Duchenne's muscular dystrophy, a difference is seen in resting levels of metabolites, specifically a relatively low PCr and an abnormally high peak in the phosphodiester region of the spectrum. This may well correlate with loss of strength and increase in fatty infiltration in subjects with these conditions. Our own studies of younger versus older persons have shown similar but less severe effects with aging (Fig. 3–6).

Magnetic resonance spectroscopy also can be applied readily in studying muscle metabolism in patients who have certain enzyme deficiencies, which, although they are not common, do show clear abnormalities. An early example was a patient with McArdle's syndrome,[68] who failed to develop any exercise-related acidosis. This was thought to be attributable to the patient's inability to mobilize glycogen to form lactic acid. Chance and associates described a patient with phosphofructokinase deficiency.[16] In this disorder the P_i increases slightly and then stays relatively constant during exercise and recovery but is overshadowed by a large increase in the phosphomonoester region of the spectrum. Patients who have soft tissue or bone tumors also demonstrate relative changes in resting concentrations of different ^{31}P metabolites.[54] These differences are attributable to increased phosphomonoester and phosphodiester resonances and allow the distinction of benign from malignant lesions with a sensitivity of 1.0 and a specificity of 0.93.

STUDIES COMBINING MAGNETIC RESONANCE IMAGING AND LOCALIZED MAGNETIC RESONANCE SPECTROSCOPY

Advances in hardware design and spectroscopy techniques have made it possible to implement chemical shift imaging on clinical imagers operating at 1.5 T. Thus, is now possible to obtain high-quality MR images and localized spectra within the same examination. This overcomes the problem of identifying the region of anatomy that gives rise to a particular spectrum, and, through the application of new data processing procedures, allows the estimation of peak parameters, which can be displayed as images of the distribution of ^{31}P metabolites. These techniques can be applied either by phase encoding alone (in one, two, or three dimensions) or by both selection and phase encoding (e.g., slice selection with two-dimensional phase encoding). The phase encoding uses the same gradient techniques as conventional imaging, and a direct correspondence exists among the different kinds of data.

Phase encoding in one dimension has been applied by several groups of researchers using a large surface coil to obtain spectra localized to slices parallel to the plane of the coil. Buchthal and co-workers employed this technique to estimate localized T1 times from both the calf and the liver, demonstrating the large differences between the values from muscle and liver tissue.[14] Bailes and colleagues also applied it to normal subjects and patients with tumors.[6] Tropp and co-workers used slice selection and two-dimensional phase encoding to obtain spectra from the calf of a normal subject at a 27-cm^3 spatial resolution and obtained a low-resolution image of the distribution of PCr.[74] Similar results were obtained by other groups of investigators.[9,40]

In a more extensive series of studies, Nelson and colleagues examined both the distribution of resting ^{31}P metabolites and the influence of exercise on forearm muscles.[57] Subjects inserted an arm through the center of a 12-cm surface coil and were placed in the right lateral position in a Siemens 1.5-T clinical imager. Imaging and two-dimensional ^{31}P chemical shift imaging were performed within a examination time of approximately 1 hour. The chemical shift imaging data were Fourier transformed, phased, and analyzed using specialized algorithms for baseline removal and peak area estimation. Arrays of peak areas were then used to produce images of the distribution of P_i, PCr, and ATP with an underlying spatial planar spatial resolution of 1 cm.[57] In this way the distribution of metabolites in both the superficial and the deep-lying muscles could be observed. Considerable variation in concentrations of metabolites could be seen, corresponding to the locations of bones and different muscle groups.[57] No statistically significant difference was found between the distributions of the different metabolites at the signal-to-noise ratio of the data.[58] The relatively higher concentration of PCr meant that good signal-to-noise PCr images could be obtained in as little as 1 min of data acquisition time or, depending on the time available, at relatively finer spatial resolutions.

The exercise response was tested with the same experimental protocol but with the volunteer squeezing a foam bar at either 1- or 4-sec intervals. In this protocol, dramatic changes in the distribution of P_i and PCr, corresponding to regions of muscle that were used in the particular exercise regimen, were observed. The results showed that the magnitude of response varied both among muscle groups and within the same muscle group when different fingers were exercised. As a result of this study, it became clear that the signals measured from the superficial regions of the flexor digitorum superficialis and flexor digitorum profundus correspond to the action of the ring and little fingers and not the other fingers.[32] This caused Jeneson and colleagues to change their surface coil protocols to measure the work done solely by these two fingers.[30,31] The resultant agreement between work done and change in the PCr to P_i ratio was much closer than in previous studies that measured the force exerted by all four fingers. In addition to the measurements made using ^{31}P chemical shift imaging, similar localization of response to different regions of muscle was observed by determining the postexercise changes in T2-weighted images from the forearm.[33]

Similar techniques have now also been applied to the calf using a volume $^{31}P/^{1}H$ coil that gives relatively uniform excitation over a region approximately 15 cm long. Both two-dimensional slice selection and three-dimensional chemical shift imaging experiments have been performed (Fig. 3–7). The distributions of ^{31}P metabolites in younger and older persons have been compared.[43] These have shown the existence of a significant phosphodiester component in all of the older persons examined, whereas the younger subjects had either no or scarcely detectable phosphodiester (see Fig. 3–6). Differences also were observed in the PCr to P_i ratios of the anterior tibialis, gastrocnemius, and soleus muscles. These are thought to reflect the differences in fiber types of these muscles. Exercise experiments have been performed, and either two-dimensional slice selection or three-dimensional images of PCr and P_i during exercise were obtained for plantar flexion of these muscles. In both cases the results showed complete removal of PCr in the anterior tibialis muscle and a significant shift in pH (Fig. 3–8). These experiments were performed with a time resolution of 6 min. To follow the kinetics of response and recovery, it is necessary to acquire spectra with a time resolution on the order of 10 sec. Our chemical shift imaging experiments have allowed us to design a protocol using a small surface coil, whereby we can be sure that we are indeed measuring a response specifically in the anterior tibialis muscle. Similar

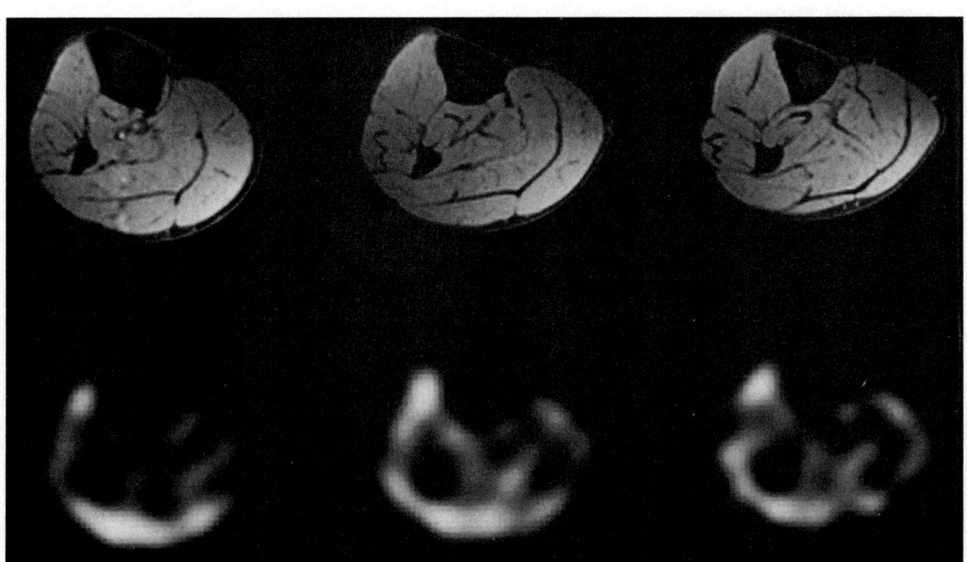

Figure 3–7. Proton and phosphocreatine (PCr) images from the calf of a normal subject. These images are slices from volume image datasets at positions 4.5 cm apart. The PCr image was obtained from a three-dimensional ^{31}P chemical shift imaging dataset with a 14 × 14 × 7 phase-encoding matrix and a 16 × 16 × 24-cm field of view. The acquisition time for the chemical shift imaging dataset was 45 min, although the signal-to-noise ratio would have been sufficient to obtain a good quality PCr image in 22 min. The proton images are slices from an SPGR volume dataset (60 slices, 4-mm thick, 16-cm field of view) obtained by saturating the fat resonance.

Figure 3-8. Axial and sagittal images demonstrating the effect of exercise on the distribution of ^{31}P metabolites. The data were obtained on a General Electric Signa clinical imager using a dual ^{1}H-^{31}P coil. The proton images are SPGR volume images obtained as described in Figure 3-7. The phosphocreatine (PCr) and inorganic phosphate (P_i) metabolite images were obtained from three-dimensional ^{31}P chemical shift imaging spectra using techniques described previously.[72,78] After baseline resting spectra and images were obtained, the subject performed plantar flexion once per second for a period of 8 minutes. A 7 × 7 × 7 three-dimensional ^{31}P chemical shift imaging dataset was acquired approximately 1 min after the beginning of exercise (6 min acquisition time). Before exercise the P_i is at a very low level and the PCr is relatively high in the anterior tibialis muscle (middle images). During exercise the PCr is depleted, and P_i builds up selectively in the anterior tibialis muscle (right images). These dramatic regional changes in ^{31}P metabolites underline the importance of accurate localization.

localization has been observed using T2-weighted imaging.

FUTURE PROSPECTS

In the future, it can be expected that the characterization of muscle function will involve both MR imaging and MR spectroscopy studies. MR imaging can measure size, T1 and T2 relaxation times, and extent of fatty infiltration, whereas spectroscopy can test the spatial distribution of ^{31}P metabolites at rest and either the utilization of different muscles (using chemical shift imaging) or the kinetics of response to exercise using surface coil spectroscopy. This combined approach will yield for the first time a true functional analysis of muscle performance using a noninvasive technique.

References

1. Ackerman JJH, Grove TH, Wong GG, et al: Mapping of metabolites in whole animals by ^{31}P NMR using surface coils. Nature 283:167–169, 1980.
2. Aisen AM, Chenevert TL: MR spectroscopy: Clinical perspective. Radiology 173:593–599, 1989.
3. Argov Z, Bank WJ, Ro YI, Chance B: Phosphorus magnetic resonance spectroscopy of partially blocked muscle glycolysis. An *in vivo* study of phosphoglycerate mutase deficiency. Arch Neurol 44:614–617, 1987.
4. Argov Z, Bank WJ, Maris J, Chance B: Muscle energy metabolism in McArdle's syndrome by in vivo phosphorus magnetic resonance spectroscopy. Neurology 37:1720–1724, 1987.
5. Aue WP: Localization methods for in vivo nuclear magnetic resonance spectroscopy. In Reviews of Magnetic Resonance in Medicine 1:21, 1986.
6. Bailes DR, Bryant DJ, Bydder GM, et al: Localized phosphorous-31 NMR spectroscopy of normal and pathological human organs *in vivo* using phase-encoding techniques. J Magn Reson 74:158–170, 1987.
7. Barany M, Siegel IM, Venkatasubramanian PN, et al: Human leg neuromuscular diseases, P-31 MR spectroscopy. Radiology 172:503–508, 1989.
8. Barkhuijsen H, de Beer R, Bovee WMMJ, van Ormondt D: Retrieval of frequencies, amplitudes, damping factors, and phases from time-domain signals using a linear least-squares procedure. J Magn Reson 61:465, 1985.
9. Bottomley PA, Charles HC, Roemer PB, et al: Human *in vivo* phosphate metabolite imaging with ^{31}P NMR. Magn Reson Med 7:319–336, 1988.
10. Bottomley PA: Human in vivo NMR spectroscopy in diagnostic medicine: Clinical tool or research probe? Radiology 170:1–15, 1989.
11. Brown TR, Kincaid BM, Ugurbil K: NMR chemical shift imaging in three dimensions. Proc Natl Acad Sci USA 79:3523–3526, 1982.
12. Brown TR, Buchthal S, Murphy-Boesch J, et al: A multi-slice sequence for ^{31}P *in vivo* spectroscopy: 1-D chemical shift imaging with an adiabatic-half passage pulse. J Magn Reson 82:629–633, 1988.
13. Bruhn H, Frahm J, Gyngell ML, et al: Localized proton NMR spectroscopy using stimulated echoes: Applications to human skeletal muscle in vivo. Magn Reson Med 17:82–94, 1991.
14. Buchthal SB, Taylor JS, Nelson SJ, Brown TR. *In vivo* T_1 values of phosphorous metabolites in human liver and muscle determined at 1.5 T by chemical shift imaging. NMR Biomed 2:298–304, 1989.
15. Chance B, Eleff S, Lee JS Jr: Noninvasive, nondestructive approaches to cell bioenergetics. Proc Natl Acad Sci USA 77:7430–7434, 1980.
16. Chance B, Eleff R, Bank W, et al: ^{31}P NMR studies of controls of mitochondrial function in phosphofructokinase-deficient human skeletal muscle. Proc Natl Acad Sci USA 79:7714–7718, 1982.
17. Chance B, Younkin DP, Kelley R., et al: Magnetic resonance spectroscopy of normal and diseased muscles. Am J Med Genet 25:659–679, 1986.
18. Cohen SM, Burt CT: ^{31}P nuclear magnetic resonance studies of phosphocreatine in intact muscles: Determination of intracellular free magnesium. Proc Natl Acad Sci USA 74:4271–4275, 1977.

19. Dawson MJ, Gadian DG, Wilke DR: Muscular fatigue investigated by phosphorus nuclear magnetic resonance. Nature 274:861–866, 1978.
20. Dawson JM: The relation between muscle function and metabolism studied by ^{31}P NMR spectroscopy. In Shu Chien, Chien Ho (eds): NMR in Medicine and Biology. New York, Raven Press, 1986, pp 185–200.
21. Derby K, Hawryszko C, Tropp J: Baseline deconvolution, phase correction and signal quantification in Fourier localized spectroscopic imaging. Magn Reson Med 12:235–240, 1989.
22. Dillon PF, Meyer RA, Kushmerick MJ, Brown TR: ^{31}P spectroscopy of smooth muscle. Biophys J 41:252a, 1983.
23. Frahm J, Merboldt K-D, Hanicke W: Localized proton spectroscopy using stimulated echoes. J Magn Reson 72: 502–508, 1987.
24. Frahm J, Bruhn H, Gyngell ML, et al: Localized high-resolution H-1 NMR spectroscopy using stimulated echoes; initial applications to human brain in vivo. Magn Reson Med 9:79–93, 1989.
25. Frahm J, Michaelis T, Merboldt KD, et al: Improvements in localized proton NMR spectroscopy of human brain, water suppression, short echo times, and 1 ml resolution. J Magn Reson 90:464–473, 1990.
26. Gadian DG, Radda GK, Ross BD, et al: Examination of a myopathy by phosphorus nuclear magnetic resonance. Lancet 2:774–775, 1981.
27. Hands LJ, Bore PJ, Galloway G, et al: Muscle metabolism in patients with peripheral vascular disease investigated by ^{31}P nuclear magnetic resonance spectroscopy. Clin Sci 71:283–290, 1986.
28. Heindrich K, Garwood M, Merkle H, Ugurbil K: Spectroscopic imaging using variable angle excitation from adiabatic plane-rotation pulses. Magn Reson Med 19:496–501, 1991.
29. Hore PJ: J Magn Reson 62:561, 1985.
30. Jeneson JAL, Nederveen D, Bakker CJG: Dynamic exercise of the human forearm muscles: A combined ^{1}H MRI and ^{31}P MR spectroscopy study. Proceedings of the Eighth Annual Meeting of the Society of Magnetic Resonance in Medicine, 1989, p 188.
31. Jeneson JAL, Wesseling MW, de Boer RW, Amelink HG: Peak splitting in organic phosphate during exercise: Anatomy or physiology? A MRI guided ^{31}P study of human forearm muscle. Proceedings of the Eighth Annual Meeting of the Society of Magnetic Resonance in Medicine, 1989, p 1030.
32. Jeneson, JAL, Nelson SJ, Vigneron DB, et al: ^{31}P two-dimensional chemical shift imaging of intramuscular heterogeneity in exercising human forearm muscle. Am J Physiol 263 (2 Part 1):C357–364, 1992.
33. Jeneson JAL, Taylor JS, Vigneron DB, et al: ^{1}H MR imaging of anatomical compartments within the finger flexor muscles of the human forearm. Magn Reson Med 15:491–496, 1990.
34. Jones DA, Round JM: Skeletal Muscle in Health and Disease. Manchester, Manchester University Press, 1990.
35. Hazelgrove J, Subramanian VH, Leigh JS, et al: In vivo one-dimensional imaging of phosphorus metabolites by phosphorus-31 nuclear magnetic resonance spectroscopy. Science 220:1170–1173, 1983.
36. Holt DI, Busby SJW, Gadian DG, et al: Observation of tissue metabolites using ^{31}P nuclear magnetic resonance. Nature 252:285–287, 1974.
37. Kushmerick MJ, Meyer RA, Brown TR: Phosphorus NMR spectroscopy of cat biceps and soleus muscles. In Bicher HI, Bruley DF (eds): Oxygen Transport to Tissue—IV. New York, Plenum, 1983.
38. Kushmerick MJ, Dillon PF, Meyer RA, et al: ^{31}P NMR spectroscopy, chemical analysis and free Mg2+ of rabbit bladder and uterine smooth muscle. J Biol Chem 261:14420–14429, 1986.
39. Laue ED, Skilling J, Staunton J, et al: Maximum entropy method in nuclear magnetic resonance spectroscopy. J Magn Reson 62:437, 1985.
40. Lenkinski RE, Holland GA, Allman T, et al: Integrated MR imaging and spectroscopy with chemical shift imaging of P-31 at 1.5 T: Initial clinical experience. Radiology 169:201–206, 1988.
41. Localized NMR spectroscopy in vivo: Problems, strategies and applications. Proceedings of Workshop, 29–30 June, 1989. NMR in Biomed 2:179–345, 1989.
42. Luyten PR, Marien AJH, den Hollander J: Acquisition and quantitation in proton spectroscopy. NMR Biomed 4:64–69 1991.
43. Majumdar S, Nelson SJ, Vigneron DB, Margulis AR: Quantitative assessment of muscle aging using magnetic resonance imaging and spectroscopy. Proceedings of the Tenth Annual Meeting of the Society of Magnetic Resonance in Medicine, 1991, p 310.
44. Massie BM, Conway M, Yonge R, et al: ^{31}P nuclear magnetic resonance evidence of abnormal skeletal muscle metabolism in patients with congestive heart failure. Am J Cardiol 60:309–315, 1987.
45. Matson GB, Weiner MW: MR spectroscopy *in vivo:* Principles, animal studies and clinical applications. *In* Stark DD, Bradley WG (eds): Magnetic Resonance Imaging. St. Louis, CV Mosby, 1988, p 201.
46. Maudsley AA, Hilal SK, Perman WH, Simon HE: Spatially resolved high resolution spectroscopy by four-dimensional NMR. J Magn Reson 51:147–152, 1983.
47. Meyer RA, Kushmerick MJ, Brown TR: Applications of ^{31}P NMR spectroscopy to the study of striated muscle metabolism. Am J Physiol 242:C1–C11, 1982.
48. Meyer RA, Fisher MJ, Nelson SJ, Brown TR: Evaluation of manual methods of integration of in vivo phosphorus NMR spectra. NMR in Biomed 1:131–135, 1988.
49. Meyer RA, Adams GR: Stoichiometry of phosphocreatine and inorganic phosphate changes in rat skeletal muscle. NMR Biomed 3:206–210, 1990.
50. Miller RG, Giannini D, Milner-Brown HS, et al: Effects of fatiguing exercise on high energy phosphates, force and EMG: Evidence for three phases of recovery. Muscle Nerve 10:810–821, 1987.
51. Moonen CTW, von Kienlin M, van Zijl PCM, et al: Comparison of single-shot localization methods (STEAM and PRESS) for in vivo proton NMR spectroscopy. NMR Biomed 2:201–208, 1989.
52. Moonen CTW, van Zijl PCM, Frank JA, et al: Functional magnetic resonance imaging in medicine and physiology. Science 250:53–61, 1990.
53. Narayana PA, Jackson EF, Hazle JD, et al: In vivo localized proton spectroscopic studies of human gastrocnemius muscle. Magn Reson Med 8:151–159, 1988.
54. Negendank WG, Crowley MG, Ryan JR, et al: Bone and soft-tissue lesions: Diagnosis with combined H-1 MR imaging and P-31 spectroscopy. Radiology 173:181–188, 1989.
55. Nelson SJ, Brown TR. A new method for automatic quantification of 1-D spectra with low signal-to-noise ratio. J Magn Reson 75:229–243, 1987.
56. Nelson SJ, Brown TR: A study of the accuracy of quantification which can be obtained from 1-D NMR spectra using the PIQABLE algorithm. J Magn Reson 84, 95–109, 1989.
57. Nelson SJ, Taylor JS, Vigneron D, et al: Metabolite images of the human arm: Changes in spatial and temporal distribution of high energy phosphates during exercise. NMR Biomed 4:268–273, 1991.
58. Nelson SJ, Vigneron DB, Brown TR: Detection of significant features and analysis of 2-D and 3-D images of ^{31}P metabolites. Magn Reson Med 25:85–93, 1992.
59. Newham DJ, Cady EB: A ^{31}P study of fatigue and metabolism in human skeletal muscle with voluntary contractions in different forces. NMR Biomed 3:211–219, 1990.
60. Ordidge RJ, Connnelly A, Lohman JAB: Image-selected in vivo spectroscopy (ISIS). A new technique for spatially resolved NMR spectroscopy. J Magn Reson 66:283–294, 1986.
61. Ordidge RJ, Bowley RM, McHale G: A general approach to selection of multiple cubic volume elements using the ISIS technique. Magn Reson Med 8:323–331, 1988.
62. Pan JW, Hamm JR, Hetherington HP, et al: Correlation of lactate and pH in human skeletal muscle after exercise by ^{1}H NMR. Magn Reson Med 20:57–65, 1991.
63. Physiological NMR Spectroscopy: From Isolated Cells to Man. New York, New York Academy of Sciences, 1987.
64. Radda GK, Bore PJ, Gadian DG, et al: ^{31}P NMR examination of two patients with NADH-coQ reductase deficiency. Nature 295:608–609, 1982.
65. Radda GK: The use of NMR spectroscopy for the understanding of disease. Science 233:640, 1986.
66. Rajagopalan B, Conway MA, Massie B, Radda GK: Alterations in skeletal muscle metabolism in humans studied by phosphorus 31 magnetic resonance spectroscopy in congestive heart failure. Am J Cardiol 62:53E–57E, 1988.
67. Rees D, Smith MB, Harley J, Radda GK: In vivo functioning of creatine phosphokinase in human forearm muscle studied by ^{31}P NMR saturation transfer. Magn Reson Med 9:39–52, 1989.
68. Ross BD, Radda GK, Gadian DG, et al: Examination of a case of suspected McArdle's syndrome by ^{31}P nuclear magnetic resonance. N Engl J Med 304:1338–1342, 1981.
69. Segebarth CM, Baleriaux DF, Luyten PR, den Hollander JA: Detection of metabolic heterogeneity of human intracranial tumors in vivo by H-1 NMR spectroscopic imaging. Magn Reson Med 13:62–76 1990.
70. Sharp JC, Leach MD: Conformal MR spectroscopy: Accurate localization to noncuboidal volumes with optimum SNR. Magn Reson Med 11:376–388, 1989.
71. Spielman D, Webb P, Macovski A: A statistical framework for in vivo spectroscopic imaging. J Magn Reson 79:66–77, 1988.
72. Taylor DJ, Bore PJ, Styles P, et al: Bioenergetics of intact human muscle: A ^{31}P nuclear magnetic resonance study. Mol Biol Med 1:7–94, 1983.
73. Taylor DJ, Styles P, Matthews PM, et al: Energetics of human muscle: Exercise-induced ATP depletion. Magn Reson Med 3:44–54, 1986.
74. Tropp JS, Sugina S, Derby KA, et al: Characterization of MR spectroscopic imaging of the human head and limb at 2.0 T. Radiology 169:207–212, 1988.
75. Tropp J, Derby KA, Hawryszko C, et al: Automated shimming of B0 for spectroscopic imaging. J Magn Reson 85:244–254, 1989.
76. Van der Veen JWC, de Beer R, Luyten PR, van Ormondt D: Accurate quantification of in vivo ^{31}P NMR signals using the variable projection method and prior knowledge. Magn Reson Med 6:92–98, 1988.
77. Webb S, Leach MO, Collins D, Bentley RE: A numerical fitting technique for quantitative MR spectroscopy. Book of Abstracts, Ninth Annual Meeting of Society of Magnetic Resonance in Medicine, 1990, p 208.
78. Webb P, Macovski A: Rapid, fully automatic, arbitrary volume in vivo shimming. Mag Reson Med 20:113–122, 1991.

4 THREE-DIMENSIONAL DISPLAY IN MAGNETIC RESONANCE IMAGING

Wing P. Chan, Martin Vahlensieck, Philipp Lang,
Steven Melnikoff, and Harry K. Genant

Complex pathoanatomic structures are difficult to interpret using conventional radiographic techniques. Although transaxial computed tomography (CT) provides images that facilitate accurate interpretation, the observer needs to integrate these into a three-dimensional (3-D) representation. The use of multiplanar two-dimensional CT data reformations adds to the understanding of complex pathoanatomy, but even these are limited, because they merely represent a two-dimensional display of a 3-D object.

Magnetic resonance (MR) imaging is unique because it provides multiplanar imaging capability. But it is hampered again by a two-dimensional display of what frequently are complex 3-D structures. Three-dimensional reconstruction has been widely used in CT for various applications.[5–10,14,20–24,29,31,32,35] Advances in computer science, image processing, and related engineering have made possible 3-D reconstruction of MR images.[33]

Three-dimensional MR image display in the musculoskeletal system is still a developing area. Reports of this technique have been few.[2,3,11,13,17,18,25,27,34] This chapter summarizes our experience with techniques and applications in this area, emphasizing the role each plays in specific circumstances.

TECHNICAL SURVEY

Scanning Technique

Scan protocols generally do not need to be altered for computing 3-D–rendered images. In this chapter, all MR images were generated on a 1.5-T system (SIGNA, General Electric, Milwaukee, Wisc.). Imaging studies for 3-D display included T1-weighted, T2-weighted spin-echo, and T2*-weighted gradient-echo pulse sequences. Other pulse sequences, such as short tau inversion-recovery (STIR) or fast spin-echo, also might be employed. Thin contiguous slices were obtained with intervals of 5 mm or less. The field of view was kept as small as technically feasible to obtain a high spatial resolution. The image matrix should be 256 × 192 to 256 × 256 pixels. Images were produced using circumferential and surface coils. Three-dimensional reconstruction could be performed on the scanner. Alternatively, many off-line work stations for 3-D reconstruction currently are offered from independent manufacturers.

Reconstruction Procedure

At the University of California, San Francisco, the 3-D image processing is performed off-line, using the

Maxiview work station developed by Dimensional Medicine (Minnetonka, Minn.). System configuration includes a 20-MHz pace frequency and an 8-megabyte random-access memory, a 766 megabyte hard disk, and a 4 megabyte graphics memory for display on a 19-inch color monitor. Computation speed ranges from 3 to 5 million instructions per second. Application programs designed for 3-D surface or volume displays from sequences of high-resolution images were supplied by the vendor.

Before any 3-D rendering, the MR imaging raw data are preprocessed in a series of several steps. Selection of the data (T1, T2, or T2*) type was determined by visual inspection of the images for the best contrast range. Pre-processing is necessary because MR imaging surface coils result in a falloff of the signal near the boundaries.[1] Similarly, images acquired with circumferential coils show a reduction in the signal that is proportional to the distance of the object from the center of the coil.[19] To minimize this problem, an unsharp mask of the original images is computed, and this is divided on a pixel-by-pixel basis. The overall signal intensity variations within an object are then normalized (Fig. 4–1).[18] If the different tissue types have only a slight difference in signal intensity, their contrast may be enhanced by adding or subtracting the normalized T2-weighted images from the T1-weighted images.[17] Linear low-pass filtering can be applied to the data to reduce the noise in the image. This step, however, may result in some loss of spatial resolution.

Final rendering is done using the same techniques as for CT images. The data flow for 3-D display is shown in Figure 4–2.

Linear Interpolation. An MR image display is composed of picture elements—pixels—which have intensities (gray values) proportional to the MR signal being acquired. Slices can be stacked on top of one another, so that planes of pixels form a 3-D mesh.[9] Each pixel in the mesh can then be considered to occupy a volume and is reinterpreted as a volume element or voxel. Because this procedure results in voxels that are not necessarily cubic, the data need to be reformatted into a volume of uniform dimensions using linear interpolation. The resultant interpolated voxels are smaller than the original volume elements, with the edge of the cube ranging from 0.2 to 0.9 mm.

Segmentation. The most difficult step in the rendering of MR imaging data is its segmentation. Segmentation is the process of extracting the voxels belonging to the object to be displayed from the total number of voxels in the 3-D volume. In CT scans, bone is characterized by a significantly higher density than surrounding soft tissue, thereby facilitating segmentation and 3-D reconstruction of osseous structures.[6,10,32] MR images demonstrate a markedly greater soft tissue contrast than CT scans. However, different anatomic structures may have similar, overlapping signal intensity. Thus, a segmentation process that is based on MR signal intensity only results in anatomically incorrect images with "fused" ob-

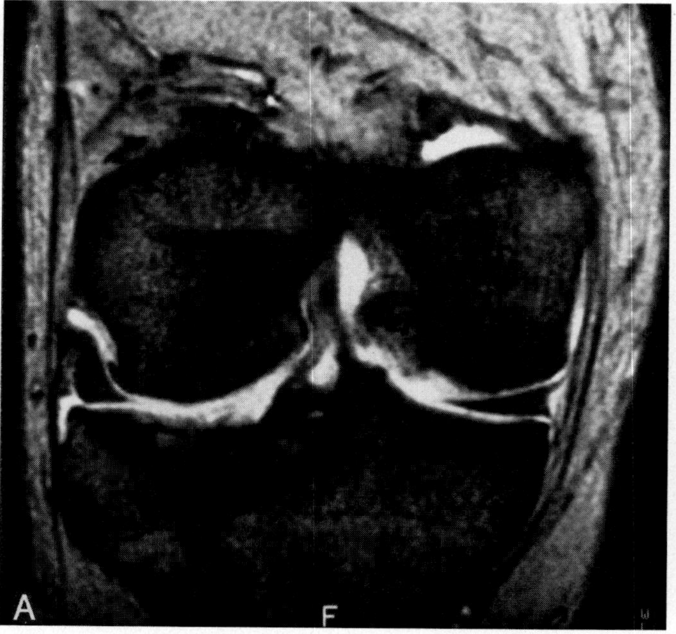

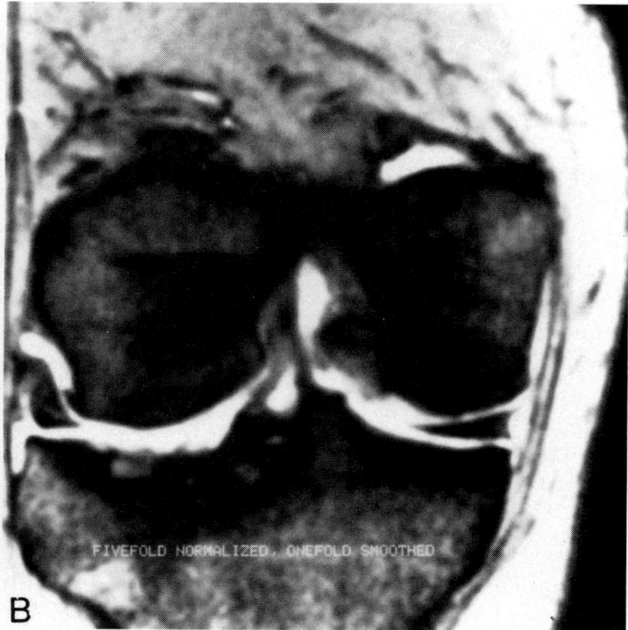

Figure 4–1. Preprocessing procedure. *A*, Original coronal T2*-weighted gradient-echo image shows an old depressed fracture of the tibial plateau. Associated degenerative joint changes and subchondral cyst formation are evident. *B*, Preprocessing image shows signal compensation that facilitates segmentation.

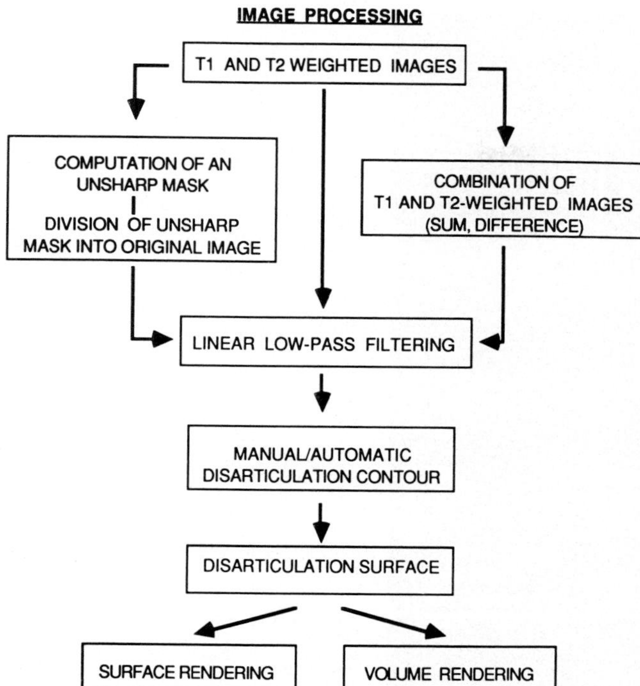

Figure 4–2. Data flow for three-dimensional MR image display in the work station.

jects (i.e., a display of different anatomic structures as part of a single 3-D object).

When the anatomic structures demonstrate similar signal intensities, the identification, or "disarticulation," of object boundaries requires interaction and guidance by an operator (Fig. 4–3). Two methods of disarticulation are available: (1) Tracking draws boundaries around certain regions either completely manually or partially manually by setting seedpoints or tracing lines along the area of interest; (2) region growing (autodisarticulation) identifies the boundary of a given structure with signal similarities and delineates it automatically.

The boundary in each slice is extracted, or isolated, using the disarticulated contour. Grouped together, these generate the desired volume of interest. Any procedure requiring manual intervention is very time consuming and represents a major limitation for 3-D display.

Using signal-intensity segmentation, a voxel satisfies a threshold criterion if its MR intensities fall within an operator-defined range. All voxels within the given range are included as part of the object (binary segmentation). Because the range of signal intensities is defined by the operator, incorrect thresholding can render the displayed image inaccurately.

Segmentation is a leading source of error in 3-D data processing and display. Partial volume effect of tissue interfaces can artificially increase or decrease the mean signal intensity of a voxel. If the signal intensity is artificially decreased, the voxel may fall outside the range of operator-defined signal intensities for 3-D display. This may lead to artificial surface gaps. Similarly, an artificially elevated voxel intensity may cause incorrect inclusion into the volume of interest.[14]

Binary Volume. After thresholding is applied, the voxel data can be used to create binary volumes. This representation simply assigns a binary value of one (1) to each element included after segmentation or thresholding and a binary value of zero (0) to elements that do not fulfill the criterion.[8] For the objects of interest, binary volumes are very compact and represent compressed 3-D representations.

Surface and Volume Rendering. For photorealistic object display, some sort of surface rendering must be done. In surface rendering, only visible boundaries or edges are shown.[33] Surfaces can then be represented using many small triangles or rectangles in a polygon tiling or directed-contour representation method or as a uniform matrix of noncubic (anisotropic) or cubic (isotropic) voxels in a binary voxel or cuberille representation method.[28] The advantage of the directed-contour representation method is very short computation time with reduced storage requirements. This enables all segments to be rotated in real time on the monitor along three axes.

Volume-rendering techniques differ from surface-rendering techniques by retaining all the voxel data, not just those from surface boundaries.[25,26] Frequently a ray tracing method is employed for volume data display.[4,30] The display is constructed either by summing all voxels along the project path or by computing the maximum value voxel along the rays. Images rendered with the summed projection in the ray-tracing method are reminiscent of radiographs. As a point-sampling process, ray tracing is sensitive to aliasing artifacts.[12] Volume-rendering techniques enable internal contours and structures to be viewed, but they fail to provide details of the surface. The surface irregularities, which may be of clinical significance, can be obscured owing to the high degree of transparency in volume-rendered images.

Surface rendering creates an artificial surface irregularity, and volume rendering yields an unsatisfactory display of surface details (Figs. 4–4 and 4–5). We suggest using a combination of surface and volume rendering (or hybrid rendering), which is thought to be superior to either technique alone (Fig. 4–6).

Shading. For a realistic representation of the anatomy, shading is applied to visible surfaces of the image immediately before projection. Shading also provides additional depth cues to the viewer for a

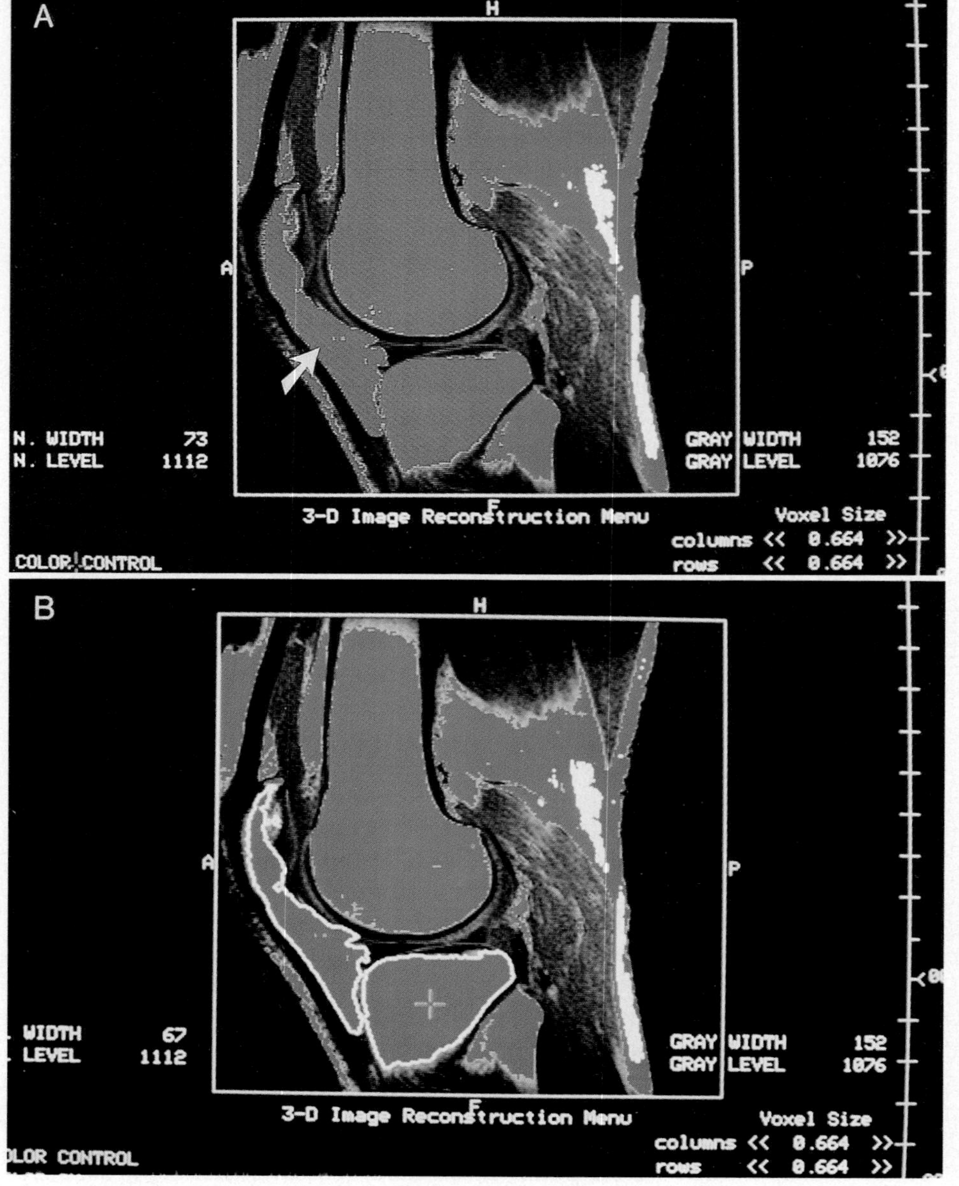

Figure 4–3. Segmentation steps. *A*, Thresholding. A range of signal intensities is selected by the operator. All voxels that fall within the range of predefined signal intensities (displayed in blue color) are considered part of the object. Because Hoffa's fat pad *(arrow)* and the fat located posterior to the femur have a signal intensity similar to that of the femur and tibia, they would be part of the three-dimensional display, which results in an anatomically incorrect view. The image therefore needs further "editing" (i.e., disarticulation), before a three-dimensional rendering is made. *B*, Disarticulation. The tibia and Hoffa's fat pad are surrounded by disarticulation lines (white), which allows separate three-dimensional displays as two different anatomic structures. Similar disarticulation can be performed for the menisci and cruciate ligaments if an overlap in signal intensity with other structures occurs.

greater sense dimension. This is accomplished by simulating the effect of an external light source illuminating the object. With the surface-shading model, the simulation of the diffusely reflected light depends on the location and material composition of a surface element (surfel) and its orientation (surface normal direction) with respect to the incident light and viewing direction.[12,30] Each element of the surface either is hidden or reflects light toward the viewer's eyes from the hypothetical source. Surface elements whose position generates speculate reflection are bright; those whose position creates glancing reflection are darker. Here the displayed gray scale of pixel intensities no longer has any relation to raw signal intensities. In comparison, the depth-shading model is nearly obsolete (see Fig. 4–4).

A transparent shading model is one that allows the display of several objects simultaneously. Transparent surfaces are attained by simulating different reflectivity on the surface. For a glassy appearance, the reflection coefficient must be high for light incidence at tangent and fall toward zero for normal incidence.

As the virtual light source can be emitted either in parallel (directional) or diffusely (isotropic), the associated computations for transparent surfaces are

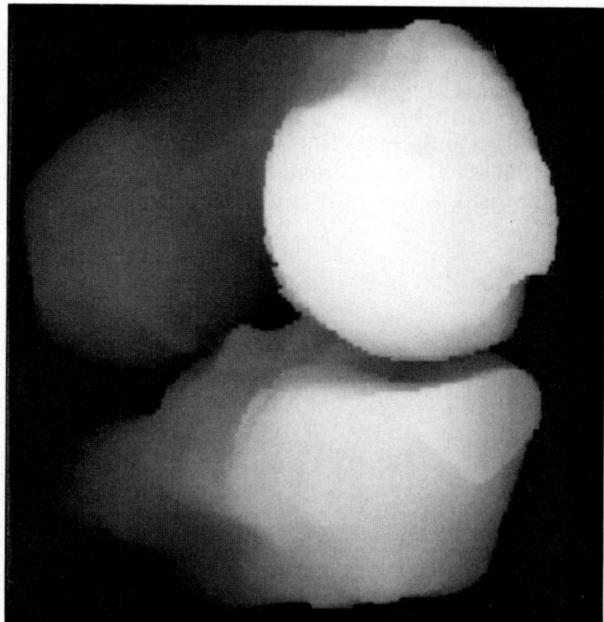

Figure 4-4. A single object display of the femur and tibia using a surface rendering with pure depth shading model. Note the lack of surface information regarding the bone.

Figure 4-6. Multiobject display using a hybrid rendering. A surface rendering of the femur (F) and tibia (T) with a coronal "cut" plane through the femur is shown. The patella (P) and the subchondral cyst *(curved arrow)* are rendered in terms of volume. A depressed tibial plateau fracture also is illustrated *(arrowhead)*.

relatively complex and time consuming. Satisfactory results can be obtained by using isotropic light and specular reflection. Overlapping objects are more apparent when they are displayed by color transparent shading, with the use of one color to represent one object so that several objects can be displayed without occlusion in one image, with different colors and transparencies.

Console operator time ranges from 20 min for single-object segmentation to 1.5 hours for multiobject segmentation.

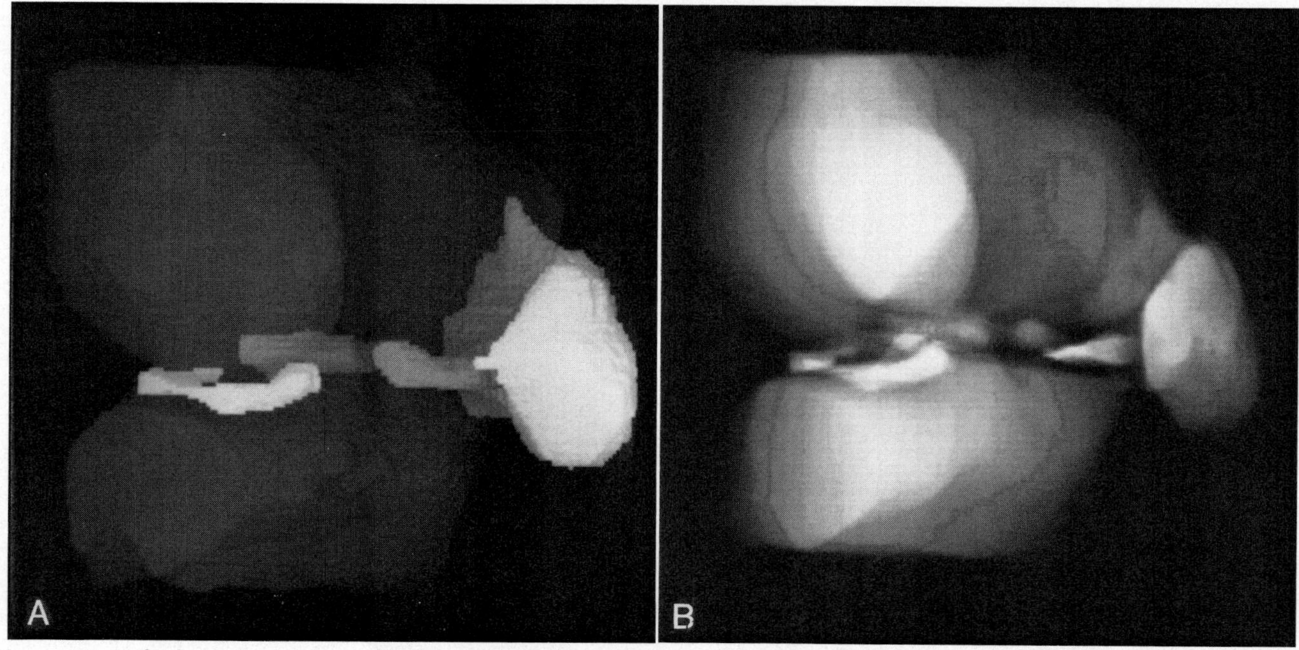

Figure 4-5. Multiobject display in a patient with a lateral meniscal cyst. *A*, Surface rendering with gray level gradient surface normal shading and different transparencies. *B*, Volume rendering. Note the artificial surface irregularity in *A* and the lack of surface details in *B*.

CLINICAL TRIALS

At the University of California, San Francisco, several clinical trials of 3-D MR imaging display have been performed in the musculoskeletal system, including the hip, the knee, and quantification of joint effusion.

Hip

Surgical decisions relating to congenital dysplasia of the hip depend on an understanding of the 3-D relationship between the femoral head and the acetabulum. Magnetic resonance imaging permits imaging without gonadal irradiation and, unlike CT

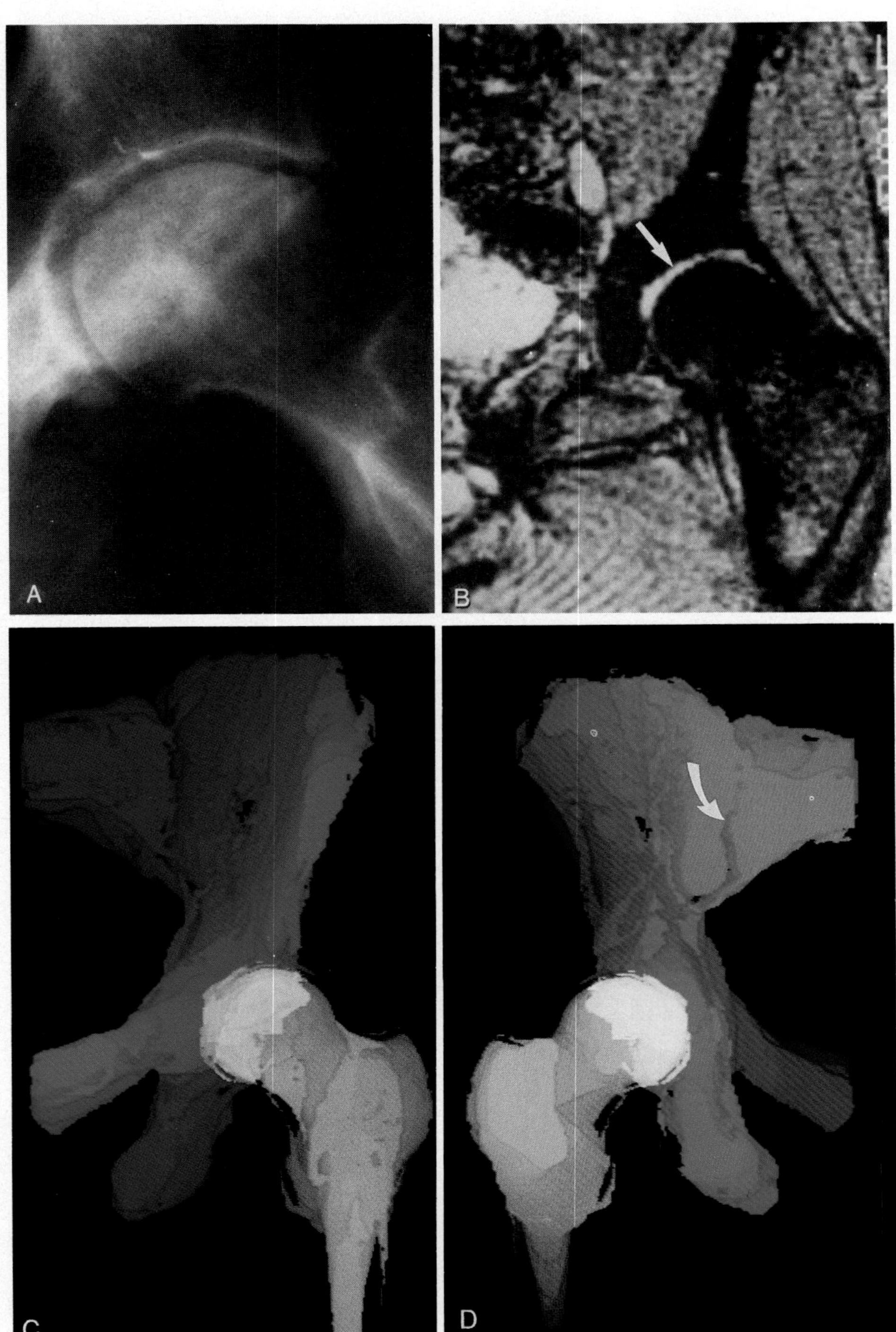

Figure 4–7. Avascular necrosis of the femoral head. *A*, Conventional radiograph shows late stage osteonecrosis with cystic and sclerotic changes of the femoral head, segmentation flattening, and subchondral fracture. *B*, T2*-weighted gradient-echo MR image. The femoral head and the acetabulum can be differentiated and segmented easily on this image owing to the interposed joint fluid of high signal intensity *(arrow)*, thereby facilitating three-dimensional reconstruction. The necrotic zone is not visualized on this sequence. Three-dimensional reconstruction of the necrotic area needed to be performed based on conventional spin-echo sequences (not shown). *C*, three-dimensional MR reconstruction in anteroposterior projection. The pelvis is displayed in blue, the femur in orange, and the necrotic zone in yellow. The position of the necrotic zone relative to the weight-bearing surface is directly demonstrated in all three dimensions. *D*, Three-dimensional MR reconstruction in posteroanterior projection in the same patient. The necrotic zone also occupies the entire posterior portion of the weight-bearing area. Note the sacroiliac joint *(arrow)*.

scanning, demonstrates the cartilaginous femoral head of patients under 6 months of age.[16] Three-dimensional MR imaging display allows the spatial relationships among the femoral head, the acetabulum, and (when present) the pseudoacetabulum to be shown.

In avascular necrosis of the femoral head, MR imaging is able to demonstrate tissue changes before these become obvious on radiograph. Three-dimensional MR imaging display can reveal the extent of involvement of the necrotic zone and areas of granulation tissue, which are computed on the basis of T2-weighted images.[15] Using different colors and transparencies, the extent of the necrotic zone, the adjacent granulation tissue, and their relationship to the weight-bearing surface can be identified (Fig. 4–7). This capability may be useful in preoperative and intraoperative localization in complex joint preserving procedures such as osteotomies.

Knee

The present classification of grade III meniscal tears does not describe accurately anatomic relationships with the various horizontal and vertical tear patterns as identified during arthroscopic surgery of the knee. Three-dimensional MR imaging displays of the meniscus are thought to facilitate subclassification of the morphology of tears. In a preliminary study by Stoller and Lindquist,[27] a series of 40 patients with grade III meniscal tears who underwent 3-D MR imaging displays had their tears subcategorized into vertical and horizontal tear planes as well as multiple tear patterns (Fig. 4–8). They found 100 per cent correlation between tear morphology on 3-D meniscus rendering and arthroscopic findings. This investigation is useful in patients who might benefit from primary meniscal repair versus partial meniscectomy.

Lang and colleagues[13] have shown several examples of 3-D MR imaging displays of the knee using different colors and transparencies to distinguish different objects (e.g., bone, ligaments, meniscus) (Fig. 4–9) and have compared surface- with volume-rendered images (Fig. 4–10). The decision regarding which structure is to be rendered by surface or volume technique differs from case to case. These investigators proposed that intra-articular lesions or anatomic structures should be rendered by surface technique, and the bone structure can be created by volume rendering.

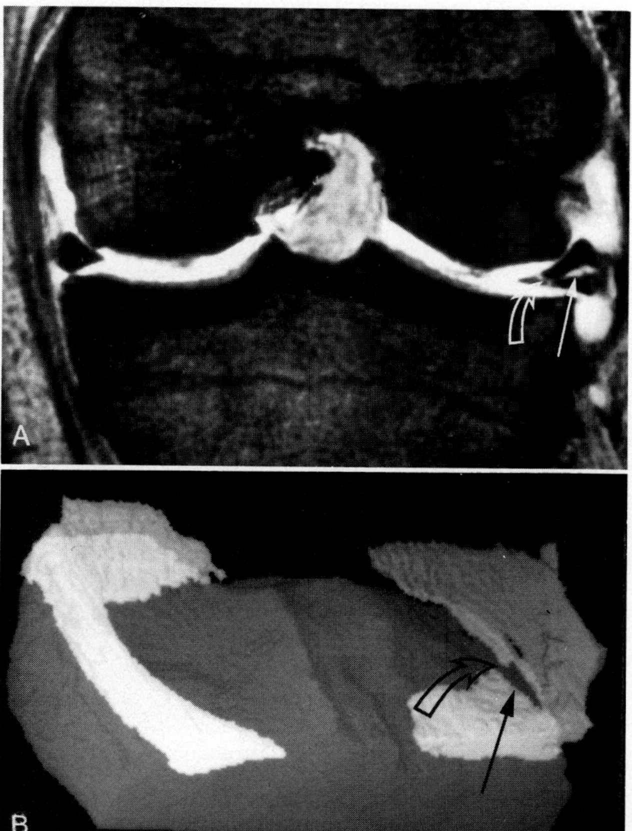

Figure 4–8. *A*, T2*-weighted coronal image shows a complex tear of the lateral meniscus *(arrows)*. An anterior cruciate ligament tear also is noted. *B*, Three-dimensional MR reconstruction image looking down from above. Note that the extent and morphology of the corresponding complex meniscal tear in *A* are depicted graphically in *B*.

Volume Quantification

In a preliminary study using cadavers, Heuck and colleagues[11] reported quantification of knee joint fluid using 3-D MR imaging data processing with satisfactorly accuracy and precision (Fig. 4–11). They found a mean accuracy error of −2.4 per cent (standard deviation of ± 5.1 per cent) in estimating the total knee joint fluid volume from T2-weighted MR images, compared with −12.5 per cent (standard deviation ± 16.9 per cent) when the joint fluid volume was reconstructed from CT scans. Volumetric quantification of joint fluid may be used as a noninvasive gauge for monitoring the effects of anti-inflammatory therapy in arthritides.[3]

To summarize, 3-D MR imaging display is feasible both from a technical and from a clinical perspective. The major limitation is the time factor in the segmentation of the two-dimensional images. Three-dimensional images provide information in a fashion that may be very useful for clinicians

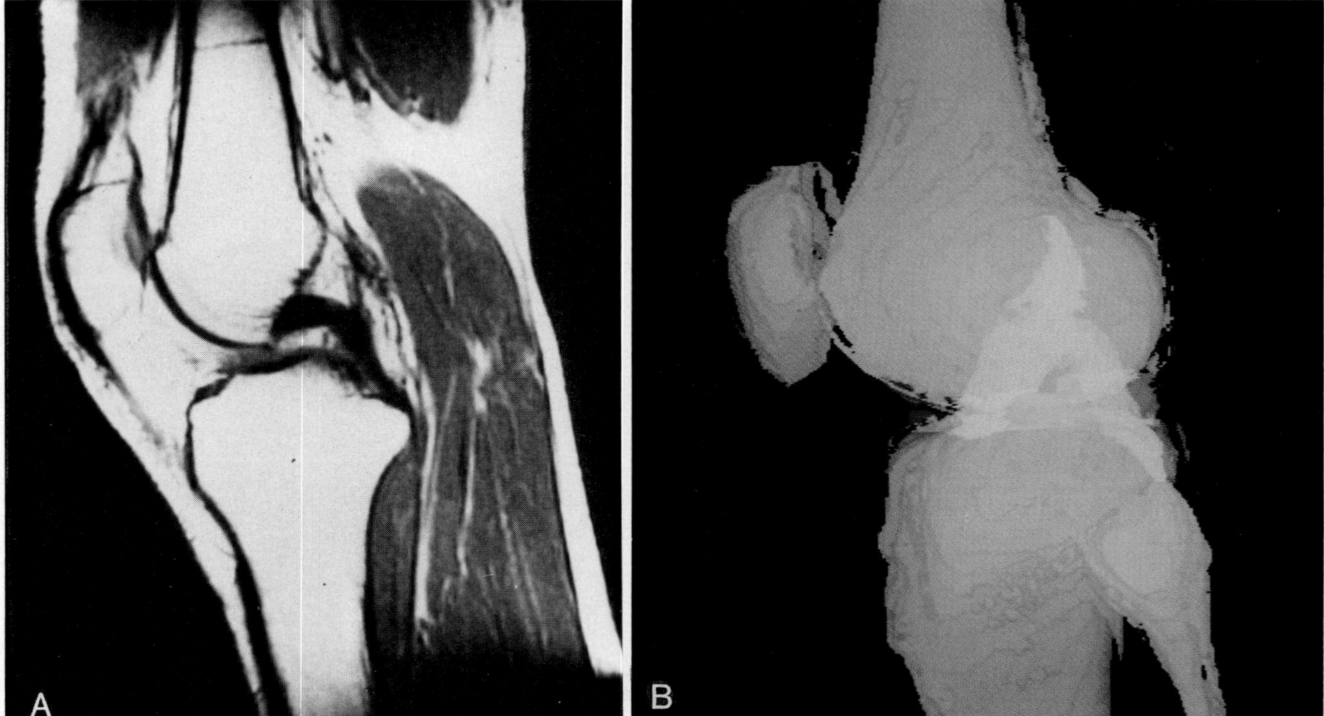

Figure 4–9. Three-dimensional reconstruction of a normal knee. *A*, T1-weighted spin-echo MR image shows the normal antaomy of the femur, tibia, and posterior cruciate ligament at this level. *B*, Three-dimensional reconstruction shows the femur, the tibia, and the patella (green); the menisci (yellow and red); and the anterior and posterior cruciate ligaments (light blue). The spatial orientation of both ligaments is shown simultaneously in one three-dimensional image. Each structure (e.g., menisci or cruciate ligaments) can be displayed separately if desired.

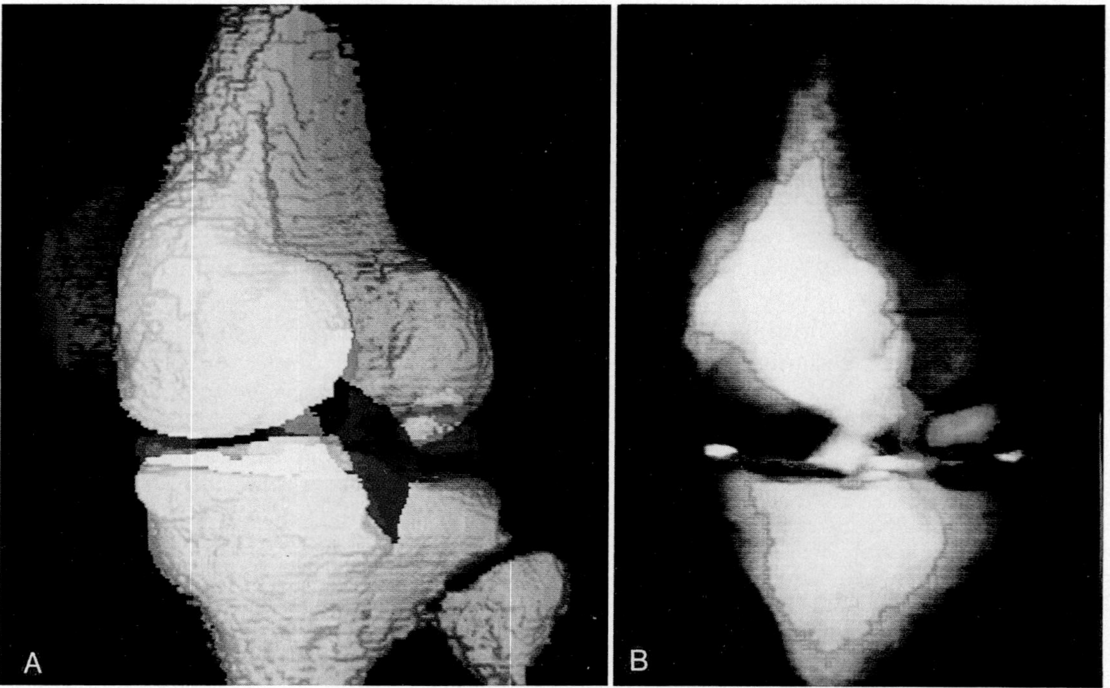

Figure 4–10. *A*, Three-dimensional surface reconstruction of the knee in a patient with spontaneous osteonecrosis of the right lateral femoral condyle. The normal bone is shown in white, the posterior cruciate ligament (PCL) is blue, the anterior cruciate ligament (ACL) is green, and the menisci are shown in magenta. The necrotic bone is displayed in red color. The position of the necrotic bone is demonstrated relative to the weight-bearing zone of the knee. Such displays may aid in preoperative planning of joint preserving surgery. *B*, Three-dimensional volume reconstruction in the same patient. The object surfaces are smoother in the volume reconstruction than in the surface reconstruction *(A)* and appear esthetically more pleasant. However, the anteroposterior orientation of the anterior cruciate ligament and the posterior cruciate ligament is not as clearly defined in the volume reconstruction when compared with that in the surface reconstruction.

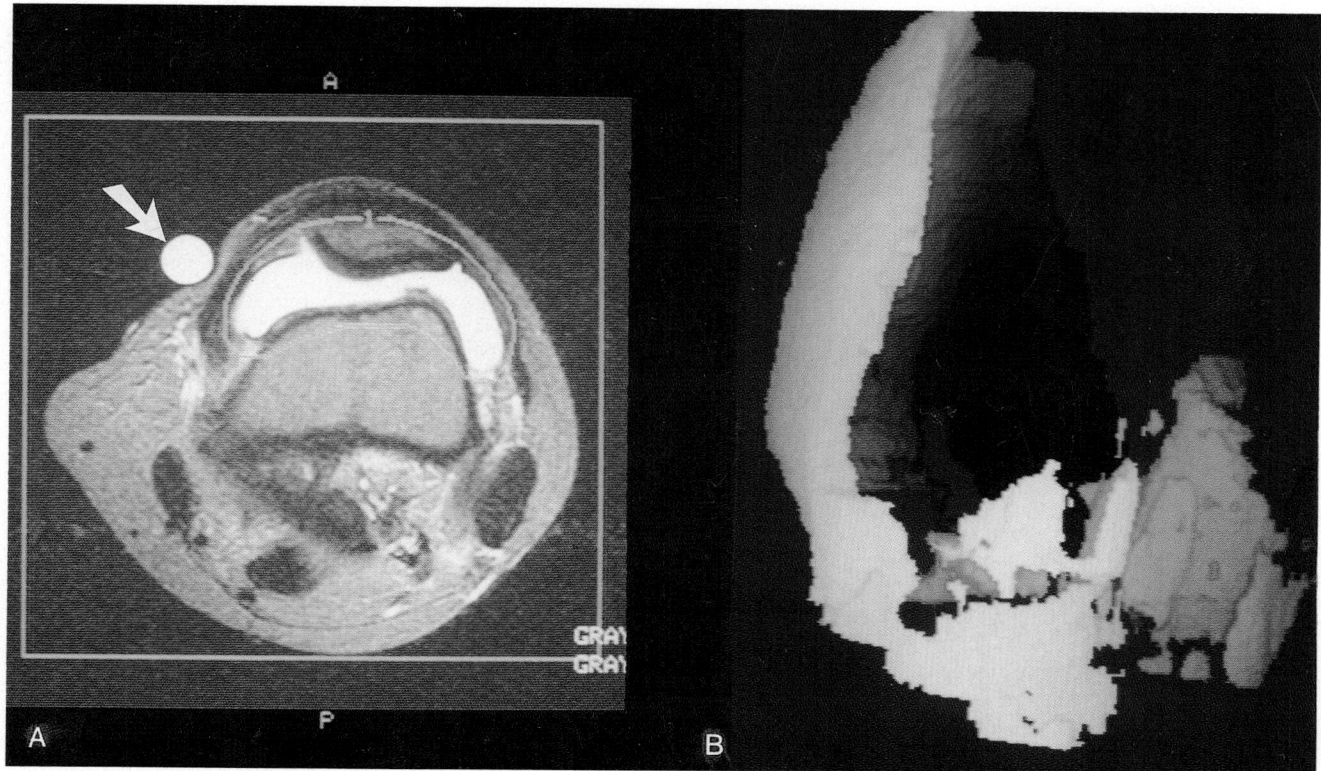

Figure 4–11. *A*, Axial T2-weighted image (TR, 2500 ms; TE, 120 ms) shows joint fluid of high signal intensity within the suprapatellar bursa. Joint fluid is surrounded by a disarticulated line (green). Using disarticulation procedure, other structures of high signal intensity that do not correspond to joint fluid are excluded from the three-dimensional rendering. A saline-filled test tube of known volume is placed lateral to the knee *(arrow)*. *B*, Three-dimensional MR image display of the joint fluid. The volume of joint fluid is computed on the basis of this display.

viewing the spatial relationship of the lesions and the surrounding anatomic structures. However, it should be noted that 3-D display is not used for primary diagnosis because only two-dimensional data reflect tissue characteristics in pathologic situations.

References

1. Axel L, Herman G, Udupa J, et al: Technical note. Three-dimensional reconstruction of nuclear magnetic resonance cardiovascular images. J Comput Assist Tomogr 7:172, 1983.
2. Chan WP, Lang P, Chieng P-U, et al: Three-dimensional imaging of the musculoskeletal system: An overview. J Formosan Med Assoc 90:713, 1991.
3. Chan WP, Wu CY, Majumdar S, et al: Quantification of knee synovial fluid with three-dimensional MR imaging data processing. Radiology 180(P):307, 1991.
4. Drebin RA, Carpenter L, Hanrahan P: Volume rendering. Comput Graphics 22:65, 1988.
5. Fishman EK, Drebin B, Magid D, et al: Volumetric rendering techniques: Applications for three-dimensional imaging of the hip. Radiology 163:737, 1987.
6. Fishman EK, Magid D, Ney DR, et al: Three-dimensional imaging and display of musculoskeletal anatomy. J Comput Assist Tomogr 12:465, 1988.
7. Gerber JD, Ney DR, Magid D, et al: Simulated femoral repositioning with three-dimensional CT. J Comput Assist Tomogr 15:121, 1991.
8. Hemmy DC, Lindquist TR: Optimizing 3-D imaging techniques to meet clinical requirements. National Computer Graphics Association, Conference Proceedings, Technical Sessions III:69, 1987.
9. Herman GT: Three-dimensional imaging on a CT or MR scanner. J Comput Assist Tomogr 12:450, 1988.
10. Herman GT, Udupa JK: Display of 3-D digital images: Computational foundations and medical applications. IEEE Comput Graphics Applications: 39, 1983.
11. Heuck A, Steiger P, Stoller DW, et al: Quantification of joint fluid volume by MR imaging and CT using 3D data processing: A preliminary study. J Comput Assist Tomogr 13:287, 1989.
12. Höhne KH, Bernstein R: Shading 3D-images from CT using grey-level gradients. Trans Med Imaging 5:45, 1986.
13. Lang P, Steiger P, Chan W, et al: Three-dimensional reconstruction of MR images of the knee: Combination of surface and volume displays. Radiology 177(P):223, 1990.
14. Lang P, Genant HK, Chafetz N, et al: Three-dimensional computed tomography and multiplanar reformations in the assessment of pseudoarthrosis in posterior lumbar fusion patients. Spine 13:59, 1988.
15. Lang P, Genant HK, Lindquist T, et al: Avascular necrosis of the hip: 3D reconstruction from spin-echo and gradient-echo MR images. Radiology 169(P):192, 1988.
16. Lang P, Genant HK, Steiger P, et al: Three-dimensional digital displays in congenital dislocations of the hip: Preliminary experience. J Pediatr Orthop 9:532, 1989.
17. Lang P, Genant HK, Steiger P, et al: 3-D reformatting asserts clinical potential in MRI. Diagn Imaging 12:100, 1989.
18. Lang P, Steiger P, Genant HK, et al: Three-dimensional CT and MR imaging in congenital dislocation of the hip: Clinical and technical considerations. J Comput Assist Tomogr 12:459, 1988.
19. Lufkin R, Sharpless T, Flannigan B, et al: Dynamic-range compression in surface-coil MRI. AJR 147:379, 1986.
20. Magid D, Michelson JD, Ney DR, et al: Adult ankle fractures: Comparison of plain films and interactive two- and three-dimensional CT scans. AJR 154:1017, 1990.
21. Ney D, Kuhlman JE, Hruban RH, et al: Three-dimensional CT-volumetric reconstruction and display of the bronchial tree. Invest Radiol 25:736, 1989.
22. Ney DR, Fishman EK: New algorithms extend utility of 3-D imaging. Diagn Imaging 11:220, 1990.
23. Ney DR, Fishman K, Magid D, et al: Volumetric rendering of computed tomography data: Principles and techniques. IEEE Comput Graphics & Application 16:24, 1990.
24. Pate D, Resnick D, Andre M, et al: Three-dimensional imaging of the musculoskeletal system. AJR 147:545, 1986.

25. Rusinek H, Mourino MR, Firooznia H, et al: Volumetric rendering of MR images. Radiology 171:269, 1989.
26. Sabella P: A rendering algorithm for visualizing 3D scalar fields. Comput Graphics 4:51, 1988.
27. Stoller DW, Lindquist TR: Three-dimensional rendering and classification of meniscal tears disarticulated from three-dimensional Fourier transform images. Radiology 177(P):263, 1990.
28. Tiede U, Hoehne KH, Bomans M, et al: Surface rendering: Investigation of medical 3D-rendering algorithms. IEEE Comput Graphics Applications 16:41, 1990.
29. Totty WG, Vannier MW: Complex musculoskeletal anatomy: Analysis using three-dimensional surface reconstruction. Radiology 150:173, 1984.
30. Upson C, Keeler M: VBUFFER: Visible volume rendering. Comput Graphics 22:59, 1988.
31. Vannier MW, Marsh JL, Gado MH, et al: Clinical applications of three-dimensional surface reconstruction from CT scans: Experience with 250 patient studies. Electromedica 51:123, 1983.
32. Vannier MW, Marsh JL, Warren GO: Three dimensional CT reconstruction images for craniofacial surgical planning and evaluation. Radiology 150:179, 1984.
33. Wallis JW, Miller TR: Three-dimensional display in nuclear medicine and radiology. J Nucl Med 32: 534, 1991.
34. Wrazidlo W, Brambs HJ, Lederer W, et al: An alternative method of three-dimensional reconstruction from two-dimensional CT and MR data sets. Eur J Radiol 12:11, 1991.
35. Zinreich SJ, Long DM, Davis R, et al: Three-dimensional CT imaging in postsurgical "failed back" syndrome. J Comput Assist Tomogr 14: 574, 1990.

Part III
PATHOLOGY

5 THE CERVICAL AND THORACIC SPINE

Shiwei Yu, Arthur E. Rosenbaum, Larry B. Poe, and Bruce E. Fredrickson

In contrast to the lumbar spine, imaging of the cervical and thoracic spine is more difficult because morphologic structures such as the intervertebral disks, intervertebral canals, and vertebrae in these regions are much smaller in volume. Computed tomography (CT) and myelography have been the mainstays for diagnosing cervical and thoracic degenerative disease; however, inaccuracies in imaging radiculopathy by these two methods are known.[8] Intrathecal administration of contrast material to render images that constitute the gold standard of CT myelography is now heavily challenged by advances in magnetic resonance (MR) imaging. The clear advantage of MR imaging over CT for detecting pathologic spinal conditions is owing to the excellent discrimination of soft tissues in nearly any desired plane when the patient is cooperative; cooperation is essential when imaging smaller structures. The relationships among degenerated intervertebral disks, the spinal cord, the intervertebral canals, and nerves can be well demonstrated by MR imaging. For higher resolution imaging, specialized techniques are needed for the cervical and thoracic spine.

MAGNETIC RESONANCE IMAGING TECHNIQUES

For imaging the cervical spine, specially designed surface coils are available, usually from the scanner's manufacturer, to achieve the maximum signal-to-noise ratio. Selecting appropriate scanning sequences and parameters significantly influences image contrast, spatial resolution, and signal-to-noise ratio. Spin-echo imaging is performed routinely in most medical centers despite the increasing use of fast scanning techniques.

The protocols of MR imaging of the cervical and thoracic spine are illustrated in Tables 5–1 and 5–2. T1-weighted spin-echo images depict well the bony structure of the cervical spine because the bone marrow of increased signal intensity, epidural fat, and inter-vertebral disk are contrasted against the quite dark cortical bone. The T1-weighted parameters for the cervical spine vary from a repetition time (TR) of 500 to 800 ms, an echo time (TE) of 10 to 30 ms, a slice thickness of 3 to 4 mm, 2 excitations, and a matrix of 256 × 256 or larger. Proton density images usually are obtained simultaneously with TRs of 2000 to 2500 ms (or even 3000 ms) and TEs of 20 and 80 to 90 ms. T2-weighted imaging detects the reduction in signal intensity from the decreased proton presence in disk degeneration. Proton density images add additional information by affording a high detail planar plain film for diagnosing spinal disease. However, the relatively poor distinction between the vertebral cortex and the epidural structures on spin-echo sequences is due to little anterior epidural fat in the cervical and thoracic spine compared with that in the lumbar spine. Spin-echo imaging also may poorly differentiate osteophytes from cerebrospinal fluid and disk material.

Gradient-recalled echo (GRE) imaging plays an important role in this situation.[6] Gradient-recalled echo imaging can provide higher signal-to-noise and contrast-to-noise ratios for a given acquisition time than are found in spin-echo sequences.[17] The desired contrast among bone, disk, cerebrospinal fluid,

Table 5-1. PROTOCOLS OF MAGNETIC RESONANCE IMAGING OF THE CERVICAL SPINE AT THE UNIVERSITY OF NEW YORK, HEALTH SCIENCE CENTER, SYRACUSE

Plane	TR (ms)	TE (ms)	FOV (cm)	Slice (mm)	Skip (mm)	NEX	Matrix (pixel)	Flip (degrees)	Time (min)
Sagittal (SE)	350–450	15	26	3	0.2	2	256 × 256		3~5
Sagittal (SE)	2500	20, 80	26	3	0.2	1	256 × 256		8~10
Sagittal (GRE)	350–500	12	30	4	0.2	3	256 × 256	15	3~4
Axial (SE)	350–600	15	25	4	0.2	2	256 × 256		4~5
Axial (GRE)	400–600	12	30	4	0.2	3	256 × 512	15	3~5

TR, Repetition time; TE, echo time; FOV, field of view; Skip, interslice gap; NEX, number of excitations; GRE, gradient-recalled echo; SE, spin echo.

and spinal cord can be obtained with selected flip angles. In general, images can be obtained with flip angles of 5 to 15 degrees for T2*-weighted images, of 20 to 50 degrees for proton density images, and of greater than 60 degrees for T1*-weighted images.[6] Flow compensation (gradient moment nulling) and occasionally gating techniques are used in MR imaging of the cervical and thoracic spine to minimize artifacts from the flow of cerebrospinal fluid. Sagittal and axial planes are routinely indicated for cervical and thoracic spine evaluation. Oblique foraminal plane images obtained perpendicular to the dorsal root ganglia and nerve roots are useful for visualizing the intervertebral canals and their contents, the uncovertebral joints, and the facet joints in the same image.[45] Coronal plane imaging may prove advantageous and often is helpful in demonstrating scoliosis (Fig. 5–1).

NORMAL ANATOMY—CRYOMICROTOME AND MAGNETIC RESONANCE IMAGING

This section emphasizes those aspects of the cervical and thoracic spine that differ from those of the lumbar spine and relate to MR imaging.

The cervical spine comprises the first seven vertebrae of the spinal column (Fig. 5–2). The first two vertebrae, the atlas (C1) and the axis (C2), are quite different from other vertebrae.[27] The atlas has no centrum. It consists of anterior and posterior arches and two bulky lateral masses. The atlas contains prominent tubercles along the lateral masses. Each of these osseous structures can be appreciated readily on high-resolution MR images. The fatty bone marrow in the cervical vertebrae of adults is slightly increased in signal intensity on T1-weighted spin-echo images and is decreased in signal intensity on T2-weighted sequences, whereas cortical bone is very hypointense on all MR sequences. The transverse ligament, which extends between the two osseous tubercles on the medial aspect of the lateral masses of C1, confines the odontoid process of C2 and delimits the anterior compartment. This relationship allows free rotation of C1 on C2 and provides for stability for the upper cervical spine in flexion, extension, and lateral bending. Other prominent ligaments in the upper cervical spine include the cruciform, alar, accessory atlantoaxial, apical, and anterior atlantoaxial ligaments.[31] These ligaments can be demonstrated on MR images by their location and relatively low signal intensity (Fig. 5–3).[46]

The other cervical vertebrae are similar in size

Table 5-2. PROTOCOLS OF MAGNETIC RESONANCE IMAGING OF THE THORACIC SPINE AT THE UNIVERSITY OF NEW YORK, HEALTH SCIENCE CENTER, SYRACUSE

Plane	TR (ms)	TE (ms)	FOV (cm)	Slice (mm)	Skip (mm)	NEX	Matrix (pixel)	Flip (degrees)	Time (min)
Sagittal (SE)	350–450	15	30	3	0.2	3	256 × 512		3~5
Sagittal (SE)	2200	20, 90	30	3	0.2	1	256 × 512		8~10
Sagittal (GRE)	350–500	12	30	4	0.2	3	256 × 512	15~18	3~5
Axial (SE)	400–700	15	25	4	0.2	2	256 × 256		4~6
Axial (GRE)	400–600	12	30	4	0.2	3	256 × 512	15~18	3~5

TR, Repetition time; TE, echo time; FOV, field of view; Skip, interslice gap; NEX, number of excitations; GRE, gradient-recalled echo; SE, spin echo.

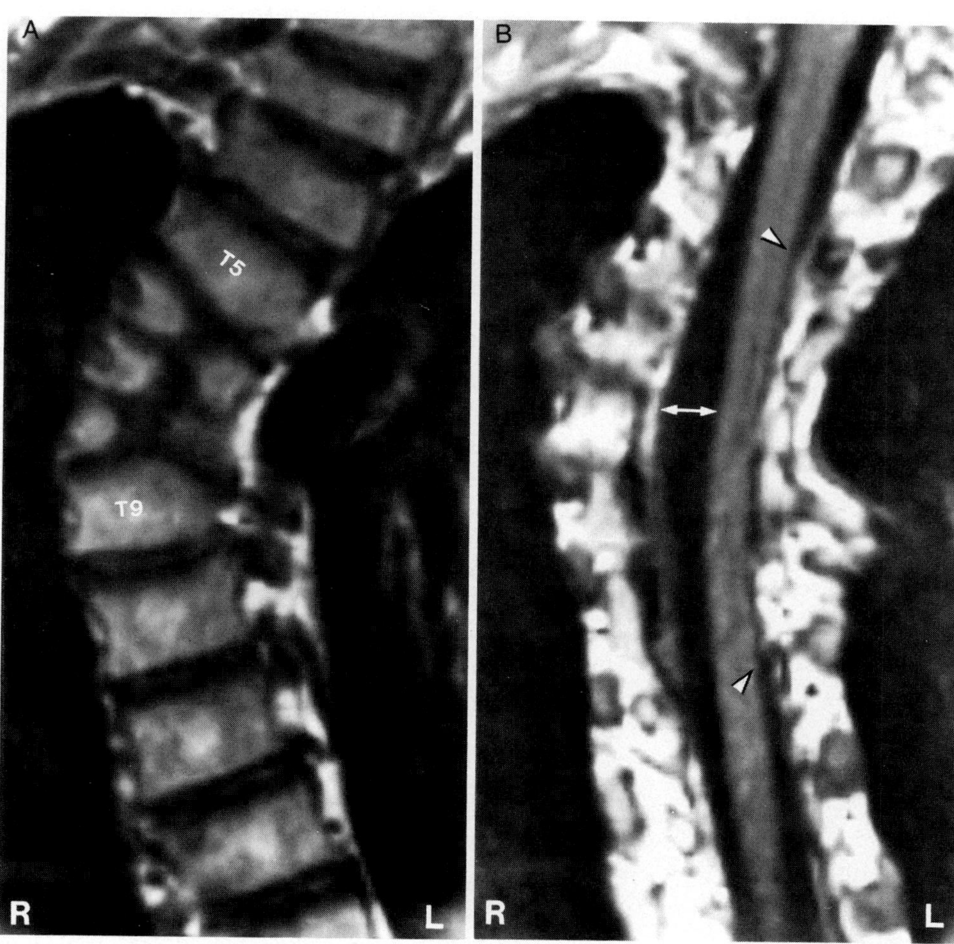

Figure 5–1. Scoliosis. Value of coronal sections in demonstrating one of the causes of scoliosis, a hemivertebra. The coronal plane in scoliosis is often more graphically informative because the vertebrae are better visualized than in the sagittal plane. *A*, T1-weighted coronal section anteriorly through vertebral body plane. Note that there are three portions of the vertebral bodies of T6 to T8 on the right but only two bony components on the left, indicating the agenesis of a hemivertebra. *B*, T1-weighted image. At the plane of the spinal cord, the cord has the smallest diameter *(arrowheads)*. Because there is a herniation and greater curvature focally, the subarachnoid space is widened *(opposing arrows)* contralaterally.

and configuration. Each vertebra has a small vertebral body. Except for C7, the other six vertebrae have foramina transversaria bilaterally for the passage of the vertebral arteries. The cervical intervertebral canal is bounded by the vertebral bodies, intervertebral disk, articular pillars, pedicles, anteriormost spinal processes, and the uncovertebral and facet joints.[46] It is oriented at about 45 degrees with respect to the sagittal plane and 10 to 15 degrees downward from the axial plane. Each intervertebral canal can be divided arbitrarily into superior and inferior portions. The ventral and dorsal roots are located in the inferior portion and thus lie below the disk level. The superior portion of the canal contains veins and epidural fat.[28] The intervertebral canal and the nerve root, of low signal intensity, within the canal are well visualized on T1-weighted spin-echo images owing to the high signal intensity of fat, which largely occupies the canal. The intervertebral canal and nerve root can be seen on axial T1-weighted images but are readily identified on GRE images also.

The facet joints are oriented in an oblique plane and angled 45 degrees with respect to the coronal planes as they parallel the vertical axis of the cervical spine.[28] The inferiorly located nerve root within the intervertebral canal is closely juxtaposed to the superior articular process. The articular processes and the facet joints can be well differentiated on T2-weighted images, especially on gradient-echo images. The anterior and posterior longitudinal ligaments in the lower cervical region at present are not readily differentiable from the dura mater, cortical bone, and, especially, the outermost portion of the anulus on MR images.

As it does in the lumbar spine, MR imaging of the cervical and thoracic spine allows distinctions to be made among the cortical bone, bone marrow, intervertebral disks, spinal canal, intervertebral canals, and ligamentum flavum. Sagittal and paramedian sagittal MR imaging planes are used routinely in thoracic spine evaluation. Axial images may be useful in some cases to visualize the relationship between the expanded vertebral bodies or protruded intervertebral disk and the subarachnoid space and spinal cord.

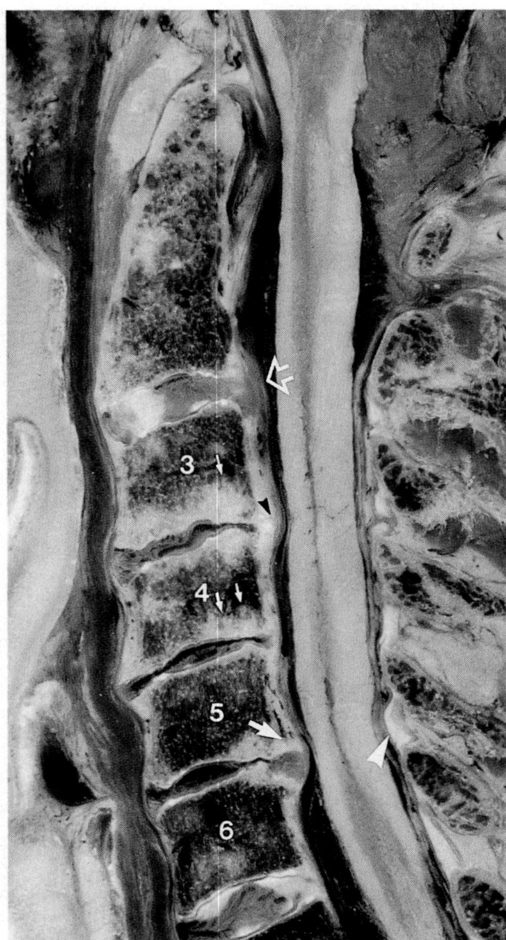

Figure 5-2. A cryomicrotome section in degenerative disk disease. The sagittal section shows the progression of intervertebral degenerative disease from thinning through end plate irregularity, thickening, and vertebral body marrow replacement *(small arrows)* by fibrous, fatty, and osseous tissue. Disks C3 to C5 are totally replaced by fluid and fibrous tissue. Anular bulging and spondylophytes *(large arrow)* are well seen also, together with the thickened ligamentum flavum *(white arrowhead)* that is compressing the spinal cord. One posterior osteophyte bridge is large enough to flatten the spinal cord anteriorly and slightly indent it *(black arrowhead)*. A disk herniation *(open arrow)* is seen at the C2-3 disk. The herniated disk extends to the epidural space behind the C3 body.

The intervertebral disks of the cervical and thoracic spine, like the lumbar intervertebral disks, consist of nucleus pulposus, anulus fibrosus, and hyaline cartilage endplates. Unlike the lumbar disks, however, these disks are much thinner, and the outermost portion of the anulus described in Chapter 6 is not as thick. Therefore, the internal structure of the cervical and thoracic intervertebral disks is not shown in as much detail as that of the lumbar spine. Nevertheless, the high signal intensity within the intervertebral disk that represents the inner anulus and nucleus pulposus and the relationship of the disk to its surroundings are well visualized.

INTERVERTEBRAL DISK DEGENERATION AND HERNIATION

Disk degeneration and osteophyte formation may result in spinal stenosis with myelopathy or radiculopathy or both. Degeneration of the cervical spine from degenerative disk disease is the most common indication for MR imaging. As with the lumbar disks, in cervical and thoracic disk degeneration, loss of water-binding capacity (water content), tears in the anulus fibrosus, and fibrous tissue replacement of the normal fibrocartilaginous tissue in the disks are seen. Disk degeneration is commonly followed by spondylotic changes (spondylosis deformity), including a decrease in disk height, osteophyte formation of the corners of vertebral bodies, malalignment of the cervical vertebral column, uncinate process hypertrophy, and reactive changes in the vertebral body marrow adjacent to the endplates of degenerative disks (see Fig. 5-2). A complete disruption of the outer anulus of the cervical disk leads to the escape of material from the nucleus pulposus through the anulus into the epidural space (Figs. 5-2, 5-4). Vertebrae C4-5, C5-6, and C6-7 are the more frequent sites for cervical disk herniation (Figs. 5-5 to 5-9). High-level disk herniations also can be found (Figs. 5-10, 5-11); two cases of C2-3 disk herniation are being reported. Very high level disk herniation can be associated with rapid axial loading. Lateral disk herniation that affects the intervertebral canal may cause intervertebral canal remodeling (Fig. 5-12).[4]

On MR images, loss of signal intensity of the disk on T2-weighted images is a common finding in disk degeneration (see Figs. 5-5, 5-9). The internal structure of the intervertebral disk is not easily shown, as it is in the lumbar region. Changes in the adjacent vertebral body marrow are well shown on spin-echo images, as in the lumbar region. The signal characteristics of bone marrow changes associated with disk degeneration on MR images are similar to those seen on lumbar MR images. Osteophytes can be readily appreciated on GRE images, although when they are subtle, they may go undetected on T1-weighted images because the signal intensities of the osteophyte and cerebrospinal fluid are similar and these structures merge (see Figs. 5-6, 5-8). A complete defect in the outermost portion of the anulus of the intervertebral disk indicates that the paradiskal or extradural mass is an extruded or sequestrated disk material. This is well shown on cryomicrotome sections (see Fig. 5-2). Proton density MR imaging may reveal the defect in the outermost portion of the anulus, but owing to the smallness of cervical disks and the marginal resolution at a matrix size of 256 ×

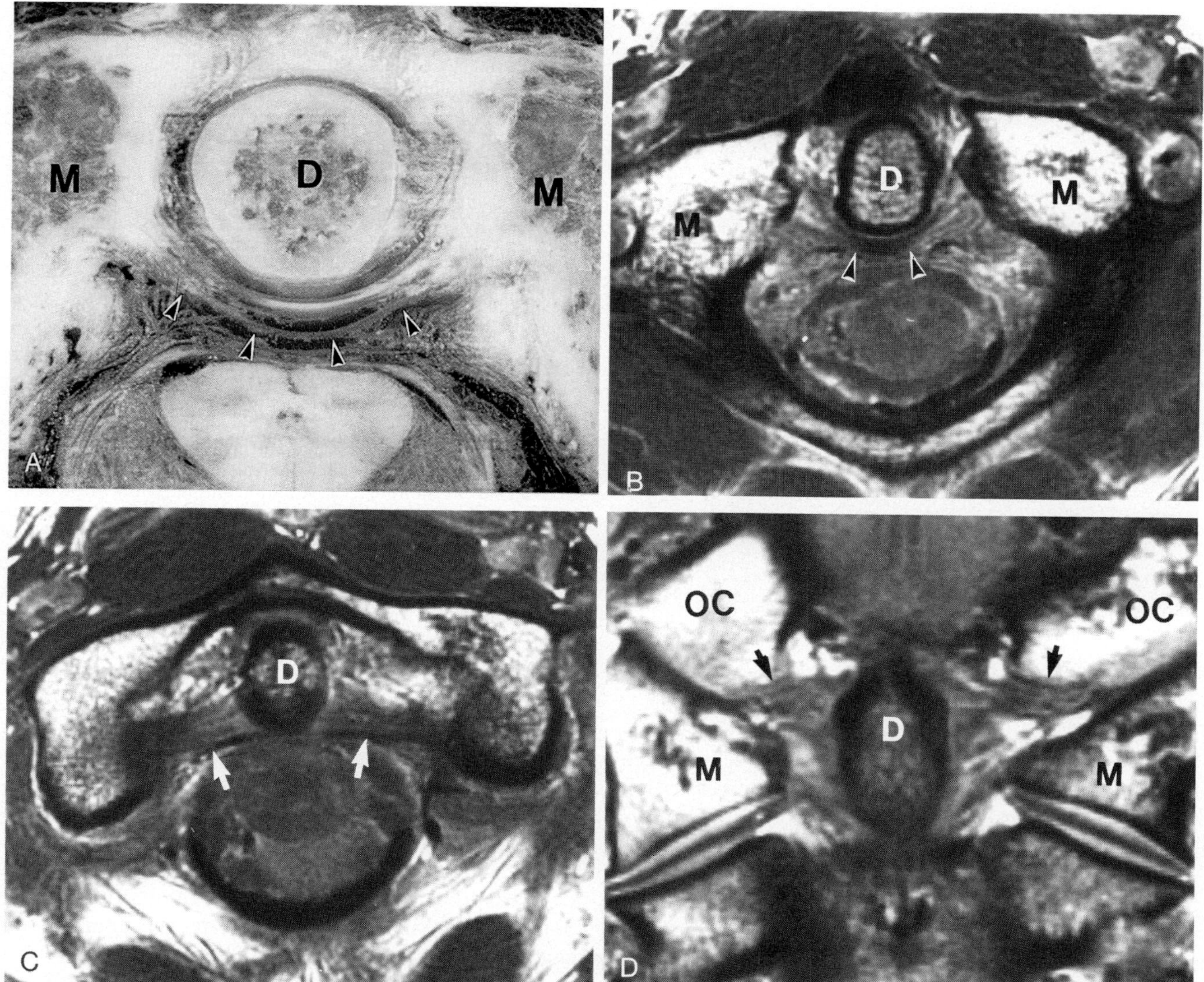

Figure 5–3. Normal anatomy of ligaments in the upper cervical spine. *A*, Axial plane cryomicrotome section at the level of the dens; *B*, T1-weighted image. The transverse ligament *(arrowheads)* bowed behind the dens (D) of C2 between the lateral masses (M) of C1 is well shown. *C* and *D*, Axial and coronal T1-weighted images. The paired winglike structures *(arrows)* are the alar ligaments, which connect the lateral aspects of the dens with the occipital condyles (OC).

256, such defects are difficult to identify at present in most of the cases.

Disk herniation is much less common in the thoracic spine than in the cervical and lumbar spine. It accounts for less than 5 per cent of all disk herniations[4,39]; however, calcification is much more common in thoracic disks than in those from other spinal regions. Most disk herniations in the thoracic spine occur in the lower portion (Fig. 5–13). The clinical manifestations of the herniation are usually pain and myelopathic manifestations rather than radiculopathy. Acute disk herniation may occur with thoracic spinal trauma (Figs. 5–14, 5–15). Conventional sagittal spin-echo T1-weighted, proton density, and T2-weighted images are usually used for evaluation of the thoracic disc herniation. Axial images are helpful for further confirmation of this diagnosis. Gradient-echo imaging becomes critical to differentiating disk material from bony mimickers. Disk degeneration, such as loss of disk height and gas accumulation within disks, can be seen (Fig. 5–16) but often requires increased imaging efforts to gain greater anatomic details because of the relative smallness of thoracic and cervical disks compared with the plump normal lumbar disks.

SPINAL STENOSIS

Congenital cervical spinal stenosis is a developmental disorder; severe forms are most commonly

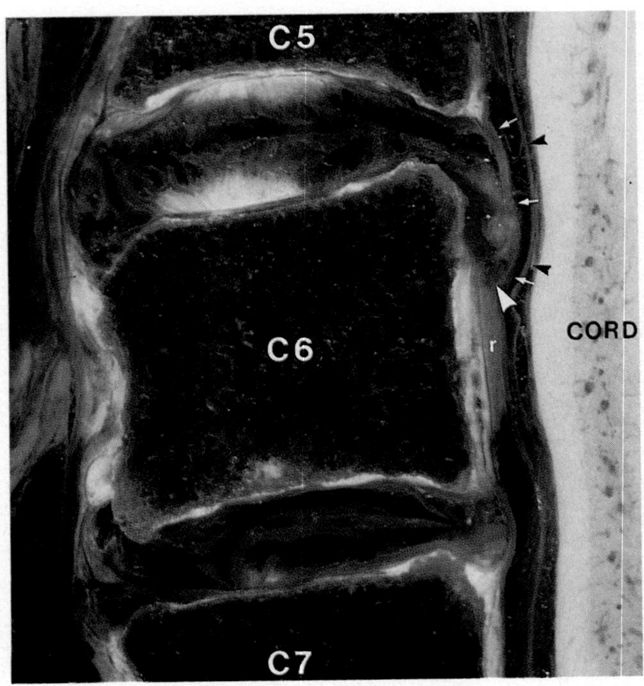

Figure 5–4. Disk degeneration and herniation. Sagittal cryomicrotome section of C5–6 and C6–7 demonstrates that C5–6 disk material extrudes through the ruptured anulus *(white arrowhead)* into the extradural space. The posterior longitudinal ligament *(small arrows)*, dura mater *(black arrowheads)*, and reactive tissue (r) to the herniated material are labeled. The C6–7 disk is also severely degenerated, as shown by the fissuring and fibers replacing the disk. An anterior disk herniation is also seen.

seen in achondroplasia (Fig. 5–17). Acquired cervical spinal stenosis usually is secondary to disk herniation, spondylosis deformans, and degenerative changes of the apophyseal joints (osteoarthritis). Among the causes of spinal stenosis, the most important factors are posterior disk expansion in combination with hypertrophy of the ligamentum flavum.[39] The narrowing of the intervertebral canal may result in nerve root compression. Osteophytic overgrowth at the posterior or posterolateral margins of the vertebral bodies and posterior protrusion or extrusion of disk material compromise the size of the subarachnoid space while displacing, indenting posteriorly, or deforming the spinal cord (Figs. 5–8, 5–9, 5–18). The position of the cervical spine is significant with respect to the severity of the spinal stenosis. The maximal canal size is achieved in slight flexion. The degree of the stenosis may appear more severe when the cervical spine is hyperextended.[39]

Spinal stenosis and its cause can be readily demonstrated on MR images. On the T1-weighted sagittal images, the cervical spinal stenosis often appears as a thinned or absent subarachnoid space anteriorly, posteriorly, or around the spinal cord (see Figs. 5–8, 5–9, 5–18). Spinal cord deformity or displacement is seen in more severe degrees of the spinal stenosis. The deformity of the spinal cord caused by a disk herniation or an osteophyte may be manifested as a focal indentation on the anterior surface of the cord (see Fig. 5–6). Spinal stenosis may go undetected on T1-weighted sagittal images in certain circumstances. For instance, on T1-weighted images, osteophytes and cerebrospinal fluid often are of similar signal intensity and are not readily distinguished. Sagittal T2-weighted and GRE images with a bright cerebrospinal fluid signal, in such circumstances, are very useful because the osteophyte of low signal intensity can be differentiated from the cerebrospinal fluid of high signal intensity (see Fig. 5–9). Axial images are important in corroborating the sagittal plane observations in spinal stenosis. On axial T1-weighted images, deformity and displacement of the normal elliptic shape of the spinal cord are common in stenosis. Reduction or lack of cerebrospinal fluid in this region also can be seen. Routine axial GRE images are recommended. Herniated or protruded material may have a higher signal intensity, whereas the osteophyte is of low signal intensity (see Figs. 5–6 to 5–8).

In the absence of severe spinal trauma, spinal stenosis in the thoracic region is much less common than in the cervical and lumbar spine regions because the spinal canal is relatively large with respect to the spinal cord in this region. Neoplastic disease, scoliosis, and infection with spinal deformity can all cause stenosis (Fig. 5–19).

SPONDYLOSIS

Schmorl and Junghanns[36] described spondylosis deformans as the most important pathologic

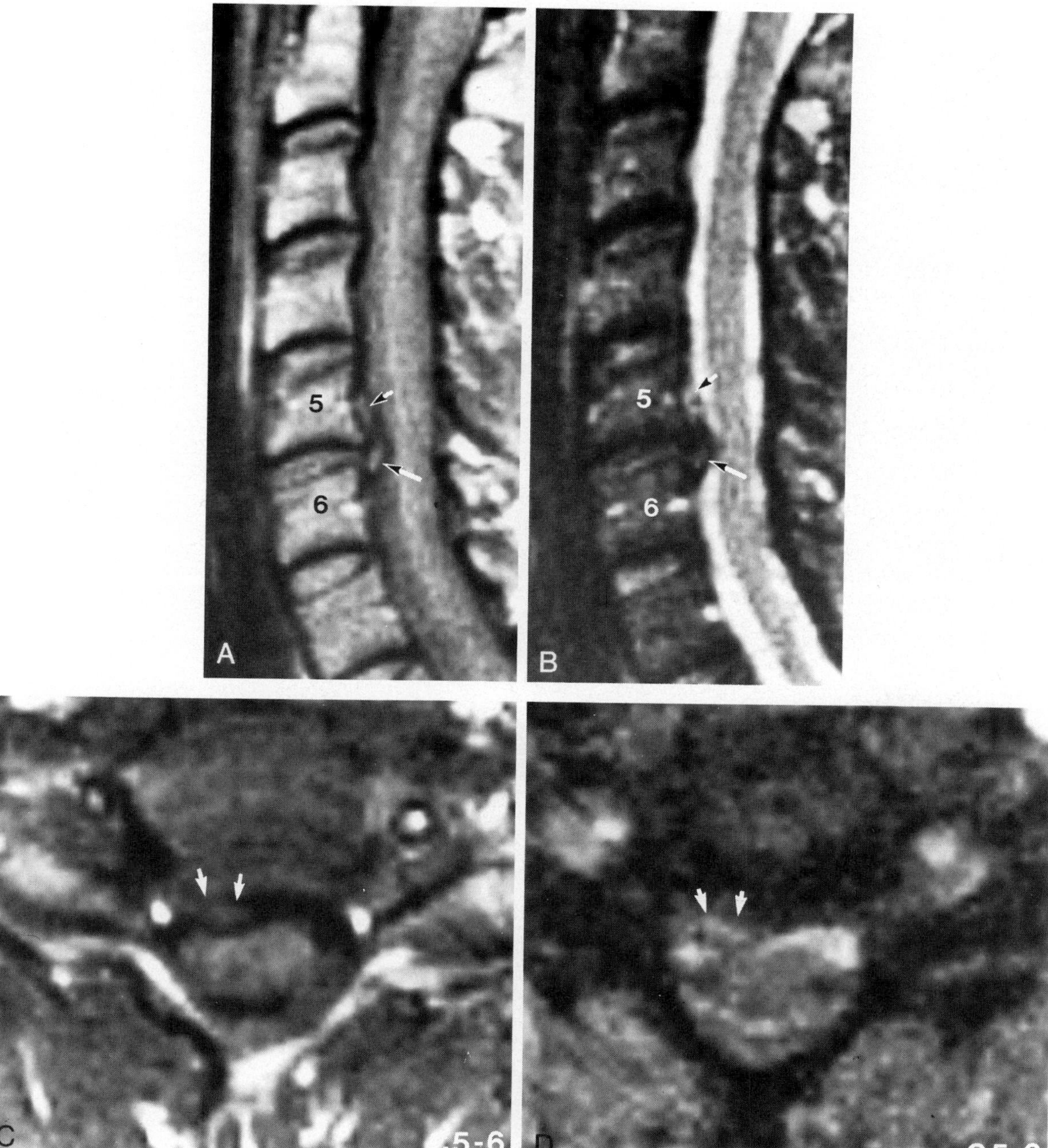

Figure 5-5. Sequestered ("free fragment") herniated disk. *A*, Sagittal motion density image. Two isolated extradural fragments of high signal intensity are seen *(arrows)*; one is precisely at the parent interspace C5-6, and the dissociated free fragment lies behind the C5 vertebral body inferiorly. *B*, Sagittal T2-weighted image. The C5-6 intervertebral disk is decreased in signal and height compared with that of the other disks. However, the connected herniated portion *(long arrow)* is similar in intensity to the parent disk, and the dissociated fragment *(short arrow)* is more intense. *C*, Axial T1-weighted image. The eccentric disk herniation *(arrows)* affects both the right anterior aspect of the spinal cord and the adjacent right intervertebral canal. *D*, Axial gradient-recalled echo, fast imaging steady state precession 12 degrees. The herniated fragment *(arrows)* is intermediate in signal intensity between the spinal cord and its surrounding cerebrospinal fluid. A thin rim of low signal intensity also surrounds the herniated fragment and usually represents the reactive immunologic or allergic response now resolvable morphologically.

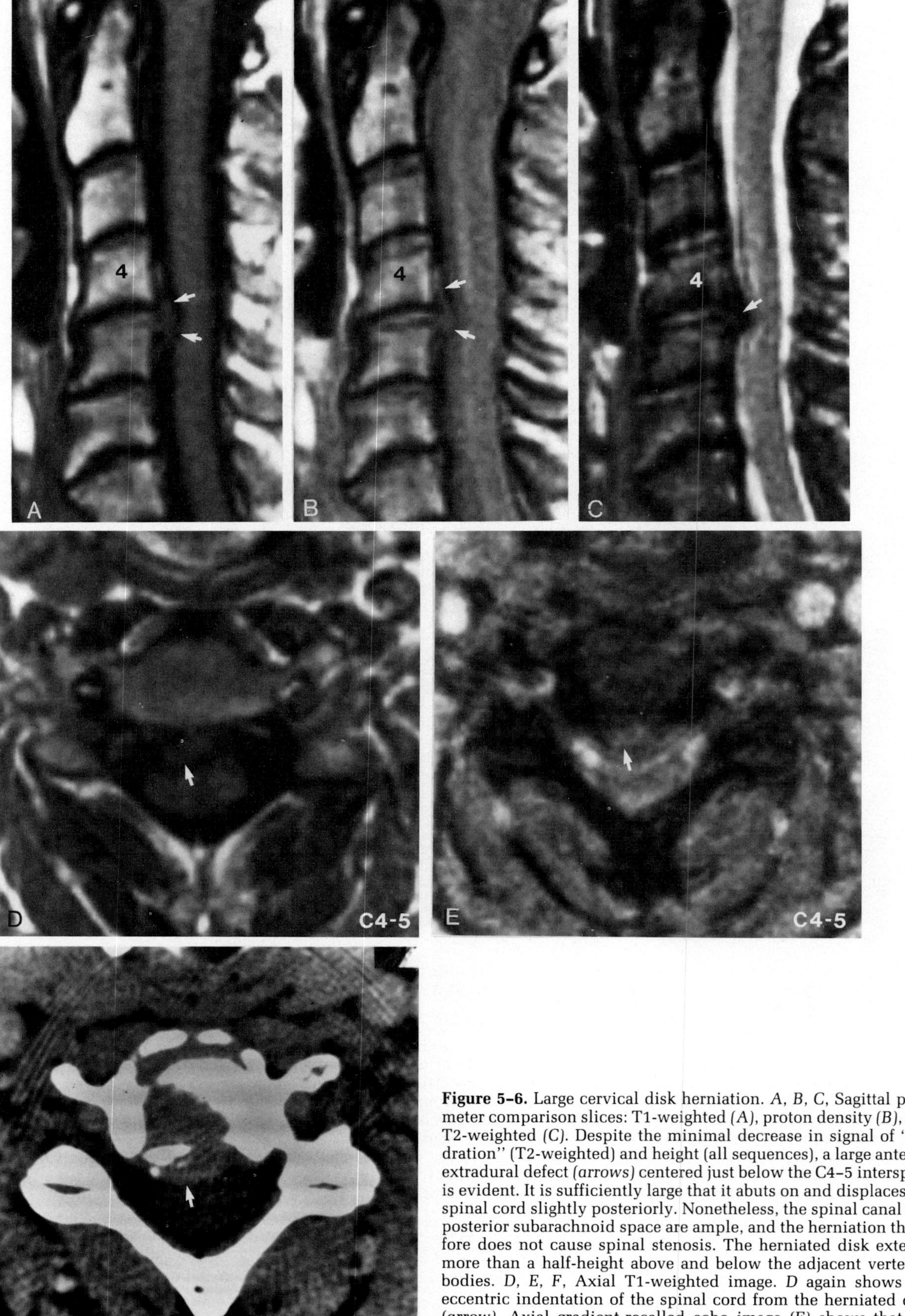

Figure 5–6. Large cervical disk herniation. *A, B, C,* Sagittal parameter comparison slices: T1-weighted *(A)*, proton density *(B)*, and T2-weighted *(C)*. Despite the minimal decrease in signal of "hydration" (T2-weighted) and height (all sequences), a large anterior extradural defect *(arrows)* centered just below the C4–5 interspace is evident. It is sufficiently large that it abuts on and displaces the spinal cord slightly posteriorly. Nonetheless, the spinal canal and posterior subarachnoid space are ample, and the herniation therefore does not cause spinal stenosis. The herniated disk extends more than a half-height above and below the adjacent vertebral bodies. *D, E, F,* Axial T1-weighted image. *D* again shows the eccentric indentation of the spinal cord from the herniated disk *(arrow)*. Axial gradient-recalled echo image *(E)* shows that the herniated disk *(arrow)* has a high signal intensity. Axial computed tomographic *(F)* image confirms the soft tissue nature *(arrow)*.

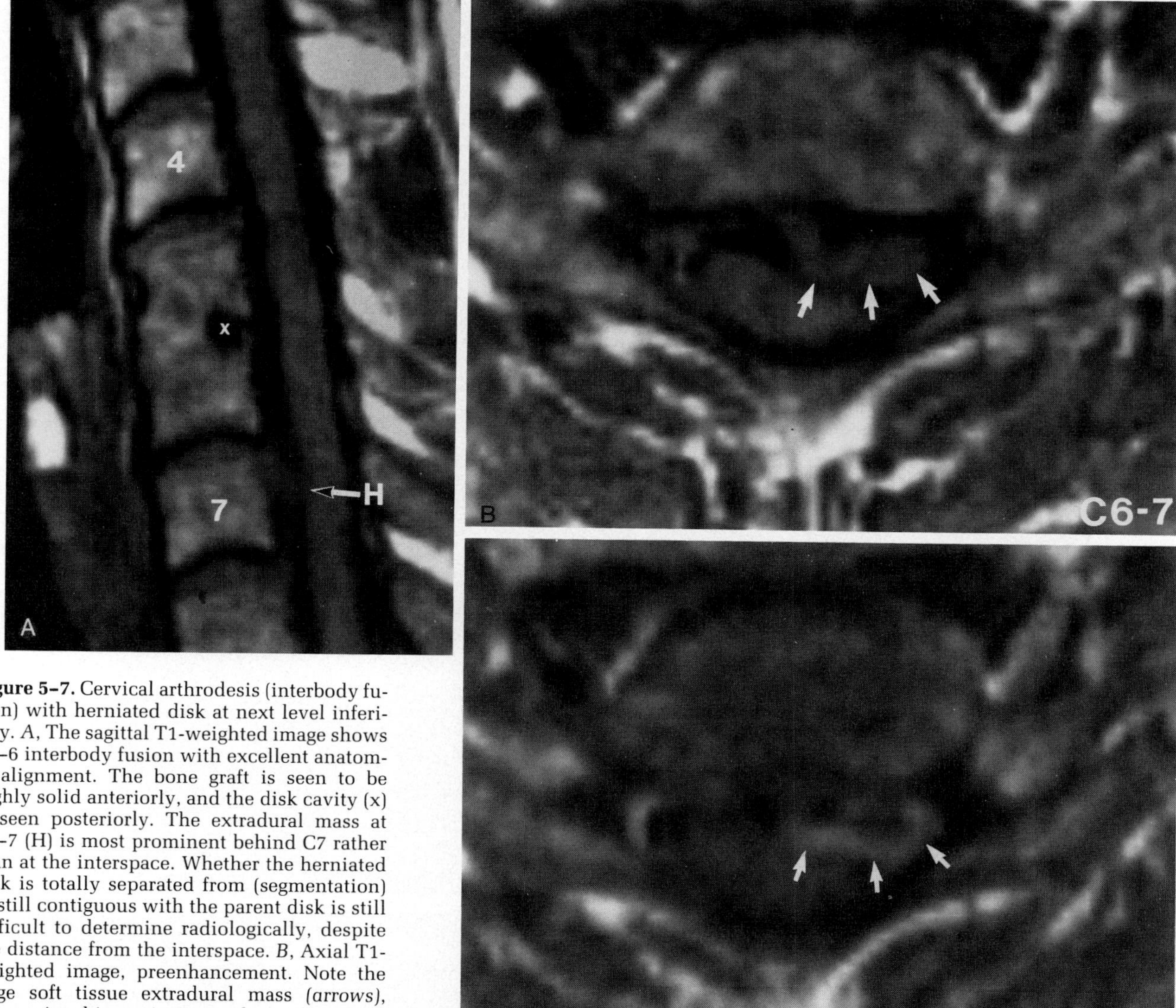

Figure 5–7. Cervical arthrodesis (interbody fusion) with herniated disk at next level inferiorly. *A*, The sagittal T1-weighted image shows C5–6 interbody fusion with excellent anatomic alignment. The bone graft is seen to be highly solid anteriorly, and the disk cavity (x) is seen posteriorly. The extradural mass at C6–7 (H) is most prominent behind C7 rather than at the interspace. Whether the herniated disk is totally separated from (segmentation) or still contiguous with the parent disk is still difficult to determine radiologically, despite the distance from the interspace. *B*, Axial T1-weighted image, preenhancement. Note the large soft tissue extradural mass *(arrows)*, whose signal intensity is similar to that of the spinal cord, which displaces and flattens the spinal cord posteriorly; this extends from the midline and is larger toward the left. *C*, Axial T1-weighted image after enhancement with gadopentetate. A rim of enhancement is seen *(arrows)* along the medial, posterior, and lateral aspects of the herniated disk. This contributed to a diagnosis of inflammation rather than to one of postsurgical change because no enhancement was present extradurally at the level of the surgery. Enhancement usually distinguishes the disk herniation from the regional response to it.

and radiographic changes within the spine. Intervertebral disk disease plays an inciting role in the development of spondylosis deformans. Spondylosis often takes the form of marginal osteophytes on the lips of the vertebral body, uncovertebral process hypertrophy, or facet joint disease. The osteophytes on the posteroinferior or posterosuperior margins of the vertebral bodies are the most important clinically. They may be located in the midline or extend laterally to the intervertebral canal.[30] Osteophytes may be silhouetted on T1-weighted scans because of the lack of difference in intensity between the osteophytes and the cerebrospinal fluid (see Fig. 5–18). Osteophytic spurs and osteophytic narrowing of the intervertebral canal are best visualized on axial images, especially on axial GRE images (see Figs. 5–8, 5–18).[12] Here the bony margins of the vertebral body and the intervertebral canal are well depicted. Sagittal oblique images also are helpful in visualizing the intervertebral canal.[5,24] Computed tomography still has high value for detecting foraminal narrowing, but both GRE and T1-weighted spin-echo imaging offer the advantage of showing both bony and soft tissue components very well (see Fig. 5–8).

Text continued on page 106

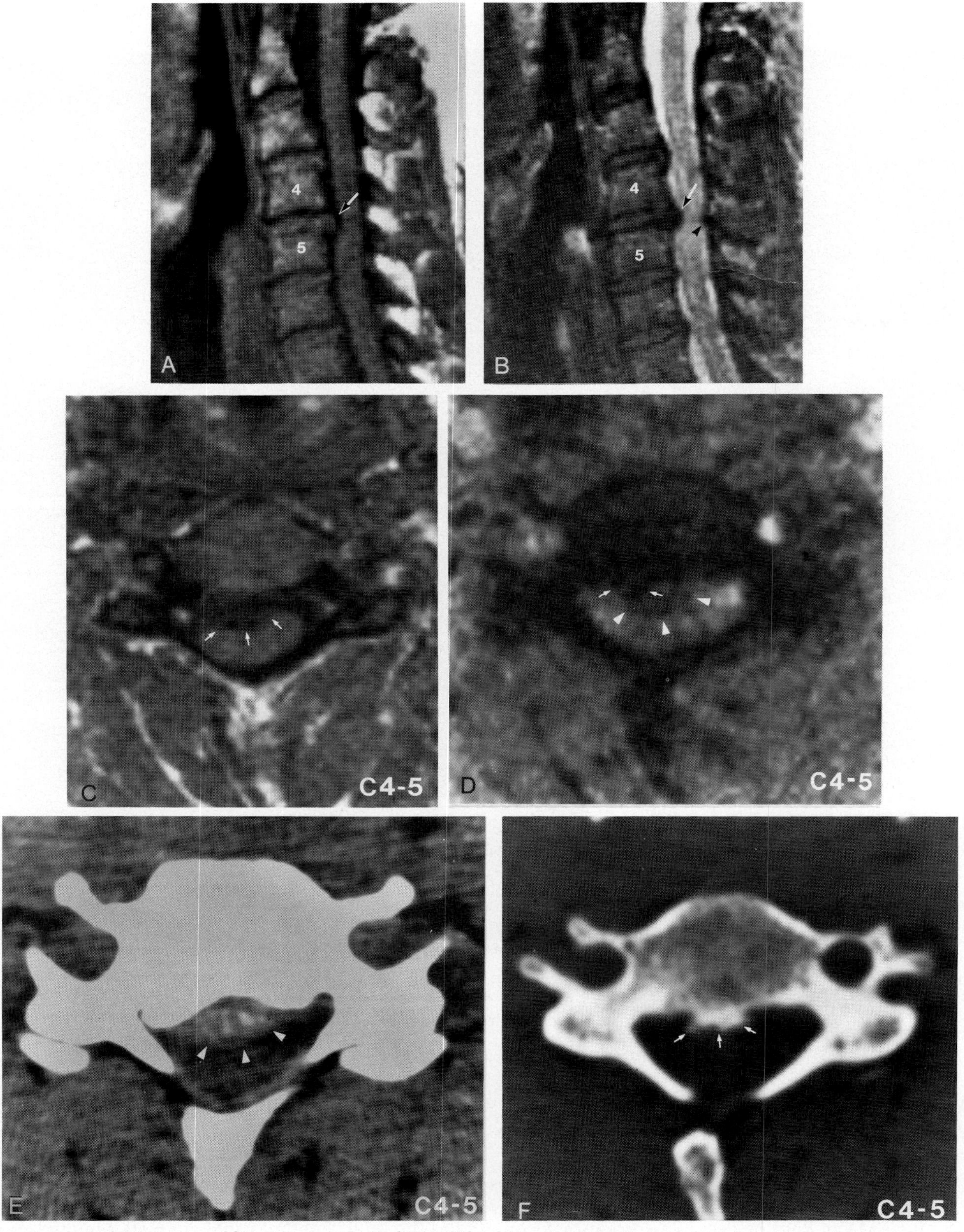

Figure 5–8. Spondylitic spur and herniation. *A*, Sagittal T1-weighted image showing an extradural defect *(arrow)* continuous with the C4–5 intervertebral disk. (There is a lesser example at C3–4). *B*, Sagittal T2-weighted image. The L4–5 defect *(arrow)* of low signal intensity obliterates the anterior subarachnoid space, reduces the space posteriorly on displacing the spinal cord, and is associated with mild degenerative changes of ligamentous thickening *(arrowhead)*. Note also the flattening and slight reversal of the cervical lordosis. *C*, Axial T1-weighted image showing a poorly defined mass *(arrows)* of low signal intensity compared with the intervertebral body and indented spinal cord. *D*, Axial gradient-recalled echo image. Like the vertebral body, the margin between the posterior aspect of the mass *(arrowheads)* and the anterior portion *(small arrows)*, which is less intense, is not well defined. *E*, Computed tomography (CT) scan with intravenous enhancement of a soft tissue window. There indeed is a well-defined mass indenting the spinal cord *(arrowheads)*. *F*, CT scan at the same level as *E*, with a slightly changed scanning angle, of a bone window. Note the exuberant calcified anterior component *(arrows)* of the mass, indicating a combination of soft tissue (herniated disk material) and hard material (spondylophyte).

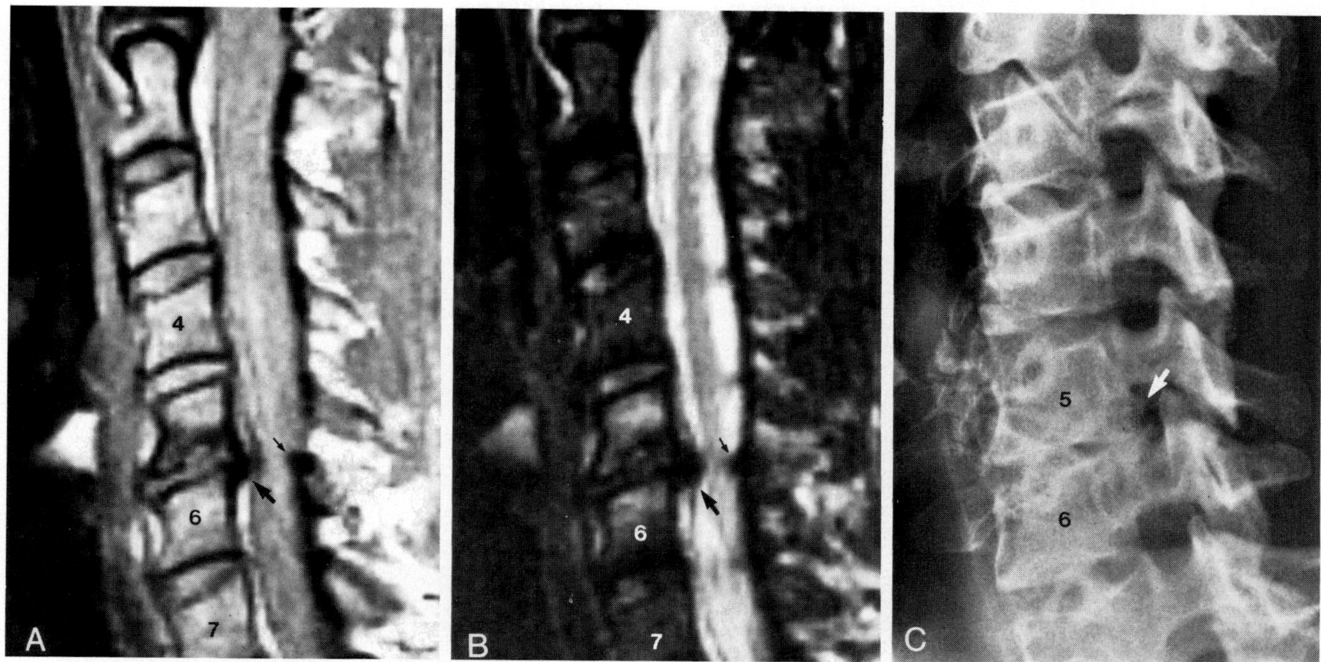

Figure 5–9. Osteophyte in a degenerated disk (C5–6). Sagittal proton density *(A)*, T2-weighted *(B)*, and plain radiographic *(C)* images show that the degenerated C5–6 disk is characterized by a loss of signal intensity, a narrowing of disk space, a response in the adjacent vertebral marrow, and an extradural mass of very low signal intensity *(arrow* in *A* and *B)*. The extradural mass and the thickened ligamentum flavum *(small arrow)* together narrow the spinal canal at this level. The plain film in *C* confirms that the extradural mass seen on the MR image is an osteophyte that is also laterally narrowing the intervertebral canal *(arrow)*.

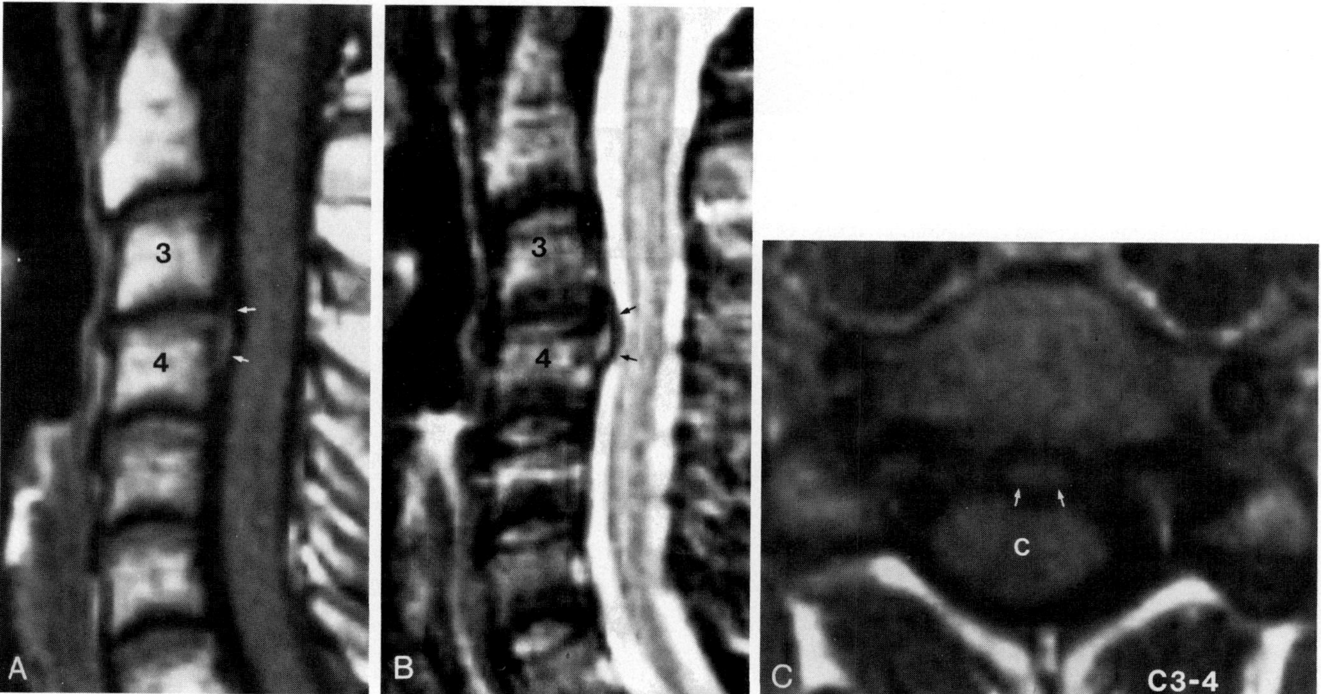

Figure 5–10. Intervertebral disk herniation at C3–4, an unusually high level. *A*, Sagittal T1-weighted image. On this sequence, spinal stenosis from the herniation with cord displacement appears to be present. Note the continuity of the herniated portion *(arrows)* with its parent disk. The disk has herniated more inferiorly than superiorly. *B*, Sagittal T2-weighted image. The herniated portion shows a higher signal intensity than its somewhat less intense parent disk. The black rim surrounding the herniated disk posteriorly is principally the posterior longitudinal ligament *(arrows)*, characteristically thick in the cervical region. The dura mater and the outermost portion of the anulus may participate in this displaced curvilinear region of low signal intensity. Chemical shift artifact is less contributory because phase encoding was in the horizontal direction and a relatively wide band width was used (0.67 for T2-weighted, 1.30 for T1-weighted images). *C*, Axial T1-weighted image shows that the herniated disk *(arrows)* in the midline compresses the spinal cord *(C)* anteriorly.

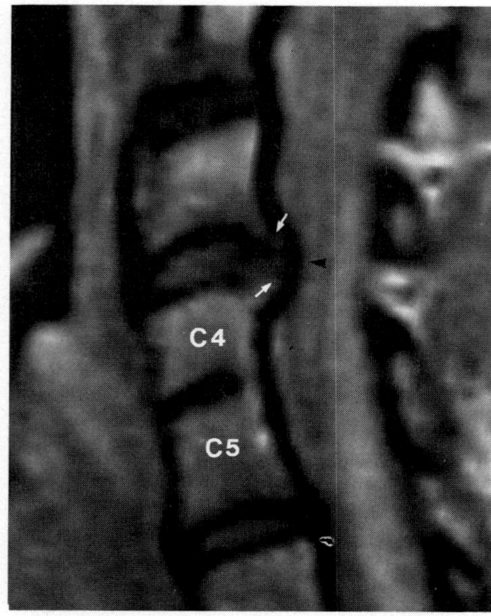

Figure 5–11. Congenital block vertebra with protruded disk above it. Sagittal proton density image reveals the fusion of C4 and C5 vertebral bodies posteriorly and the incompletely developed C4–5 disk anteriorly. The anteroposterior narrowing of C4 and C5 bodies is shown. Severe disk protrusion *(arrowhead)* with large osteophytes *(small arrows)* is present at the C3–4 disk level posteriorly. The spinal canal is narrowed significantly at that level.

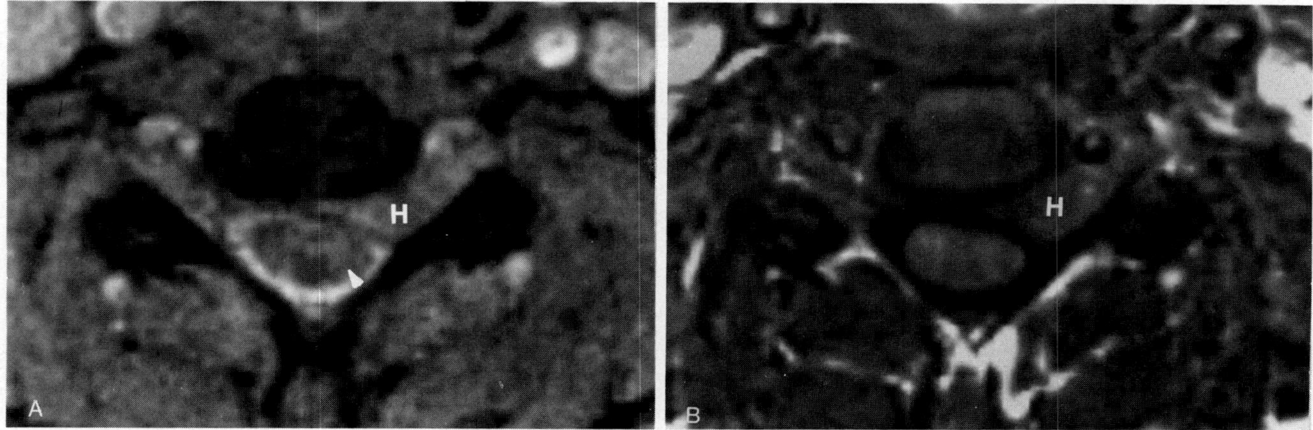

Figure 5–12. Huge intervertebral canal herniated disk at C4–5 of long duration. *A*, Axial gradient-recalled echo (GRE) image. Bone detail is well shown at GRE parameters. Thus, the remodeled, widened intervertebral canal is well illustrated with the mass (H) that is causing it. The canal is moderately bright and quite uniform in intensity, which is characteristic of disk material. The medial portion of the herniated disk extends sufficiently to the midline to displace the spinal cord posteriorly. The displacement and rotation of the cord are also inferred from the distorted configuration of the "H" *(arrowhead)* formed by the gray matter and the obliterated cerebrospinal fluid (brightest intensity) from the 12 to the 2 o'clock positions around the spinal cord. *B*, Axial T1-weighted image. This disk has a higher intensity signal than the parent intervertebral disk, which is characteristically less intense on T1-weighted images.

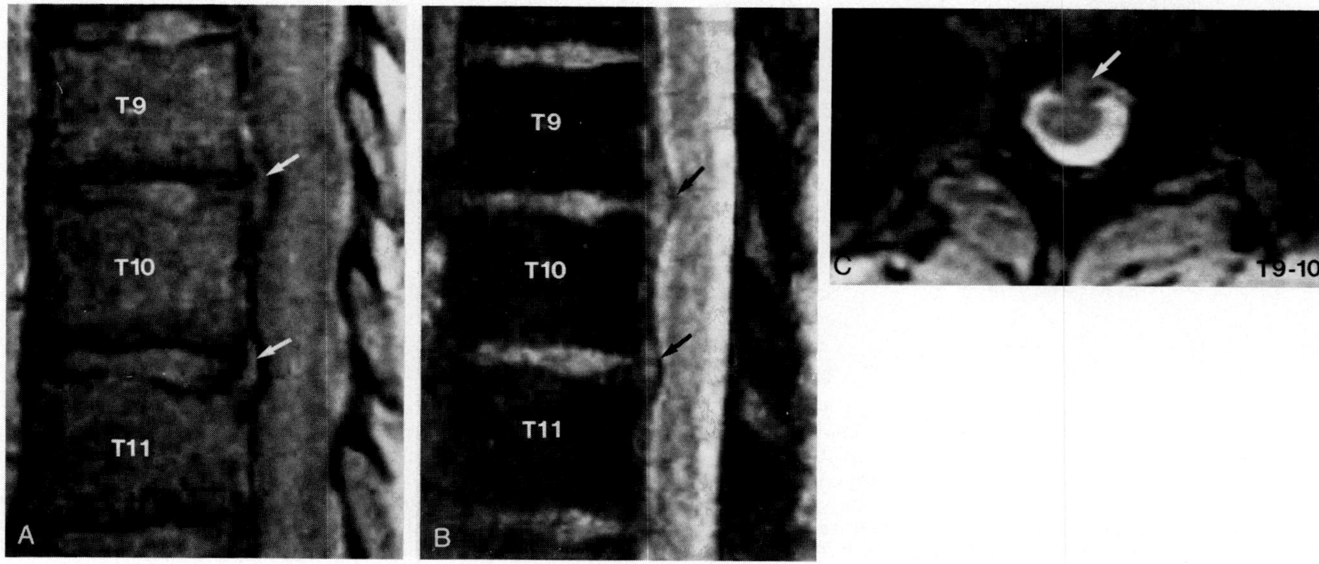

Figure 5–13 *See legend on opposite page*

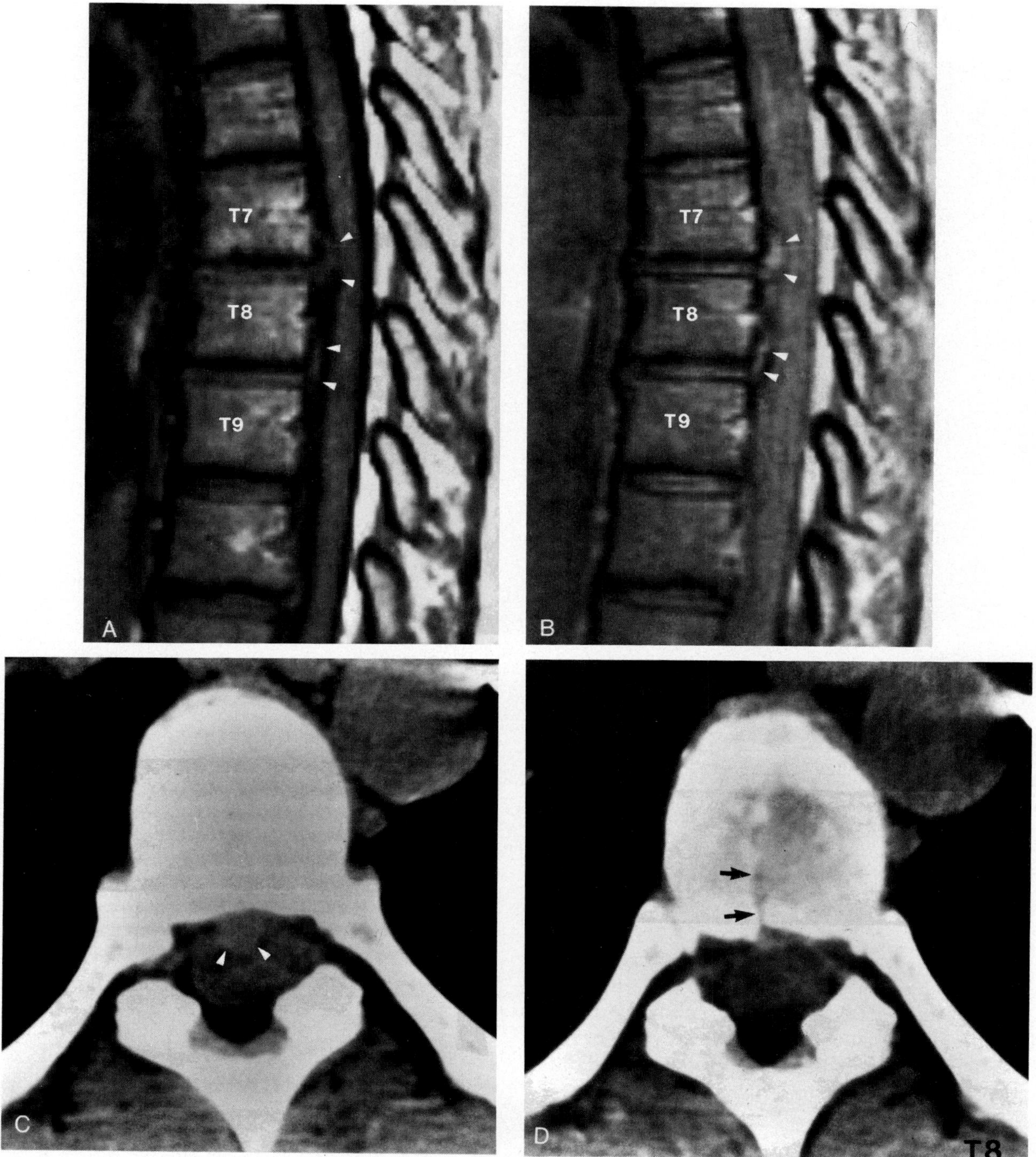

Figure 5-14. Large acute central thoracic disk herniation. *A*, Sagittal T1-weighted image. The anterior extradural defects are herniated disks *(arrowheads)* located at and above the interspaces of T7–8 and T8–9. *B*, Sagittal proton density image. Note the brightness and confirmation of the herniation with this sequence. (A T2-weighted image was noisy and provided no additional information.) *C*, Axial computed tomography (CT) scan. The midline location at C7–8 is well illustrated *(arrowheads)*. *D*, Axial CT scan of the T8 vertebral body. A fracture is noted in the posterior aspect of the T8 body *(arrows)*.

Figure 5-13. Two contiguous thoracic herniated disks, T9–10 and T10–11. *A*, Sagittal proton density image. The herniations *(arrows)* are largely confined to the level of the parent disks. Both have lower signal intensity than the two normal intervertebral disks above them. *B*, Sagittal gradient-recalled echo (GRE) image. Each of the intervertebral disks is bright on GRE images, so that altered signal intensity in degeneration cannot be detected. The contiguity of the herniation to the parent disks is well shown, especially in the upper disk. *C*, Axial GRE image. The herniated portion of the disk *(arrow)* is well distinguished from the low signal intensity of the bone (adjacent centrum) and the thecal sac. Note the "H" of gray matter within the spinal cord on this sequence.

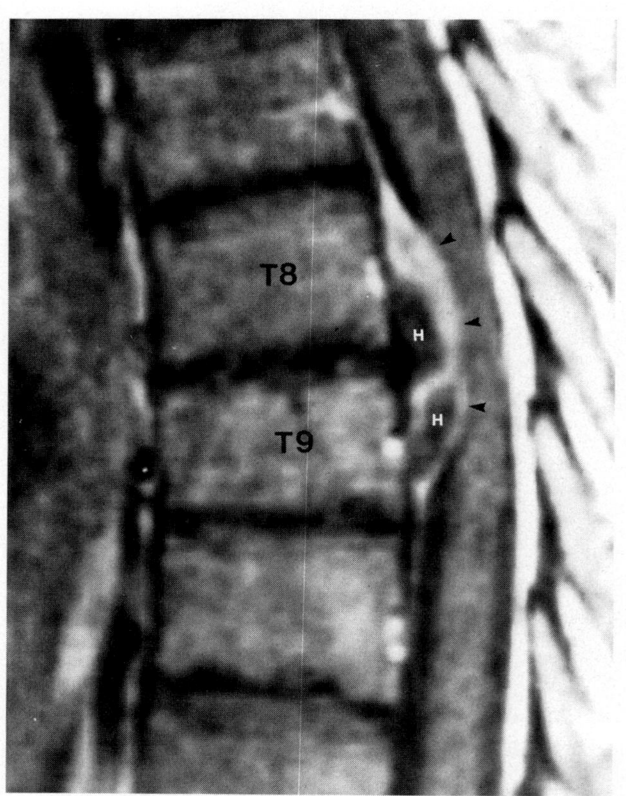

Figure 5-15. Sagittal T1-weighted image showing acute thoracic trauma with extradural hemorrhage and disk herniation. The anteriorly wedged, comminuted vertebral body fracture is associated with regional large anterior epidural hemorrhage *(arrowheads)* that displaces the spinal cord posteriorly and compresses it. The defects (H) of low signal intensity found within the hemorrhage proved to be herniated disk fragments at surgery.

Although thoracic herniation is a well-recognized entity, thoracic radiculopathy secondary to spondylosis is uncommon.[33]

SPINAL TRAUMA

In recent years, MR imaging has been advocated for the initial evaluation of spinal cord injury in spinal trauma.[19,23] Because most patients with potential spinal cord injuries are critically ill or clinically unstable, physiologic monitoring and life support systems are required during the MR imaging examination. The use of MR imaging in spinal trauma is limited by the incompleteness of physiologic monitoring.[23] The development of MR-compatible monitoring equipment has largely ameliorated this problem,[3,18,37] but CT is still much better with respect to physical access to patients.

Undoubtedly because of its superior contrast resolution and limited invasiveness, MR imaging is the best method currently available for imaging acute or chronic injury to the spinal cord and identifying a lesion that has the potential to be corrected surgically.[3,19,23] Magnetic resonance imaging, both spin-echo and GRE, affords direct and accurate imaging of ligamentous injuries. Recognition of ligamentous strain or disruption can be difficult, however, because this is highly dependent on the subject's lack of movement during scanning; physiologic movement, as in the thoracic spine, cannot be eliminated with current "high resolution" conventional spin-echo MR scanning times. Factors that contribute to this difficulty also include the relatively thick sections (e.g., 4 mm) needed to satisfy signal-to-noise requirements. The similarity of the signal intensities of interstitial water in fascial planes to pathologic edema, of fascial fat to hemorrhage, of soft tissues to bone marrow, and of ligaments to cortical bone in the various sequences all impede imaging. Ligaments are best separated from other tissues on proton density and small flip-angle GRE sequences. Ligamentous injury leads to increased signal intensity on long TR, long TE spin-echo (T2-weighted), or small flip angle GRE sequences or to abnormal signal intensity on short TR, short TE spin-echo (T1-weighted) sequences when the normal anatomic pattern is disrupted.[22,40] This may be less obvious when using newer "fast" spin-echo techniques, in which fat is not completely suppressed on T2-weighted sequences. Tears in the anterior or posterior longitudinal ligaments may be recognized by loss of the "line" of low signal intensity represented by these structures normally in the lumbar region[14]; however,

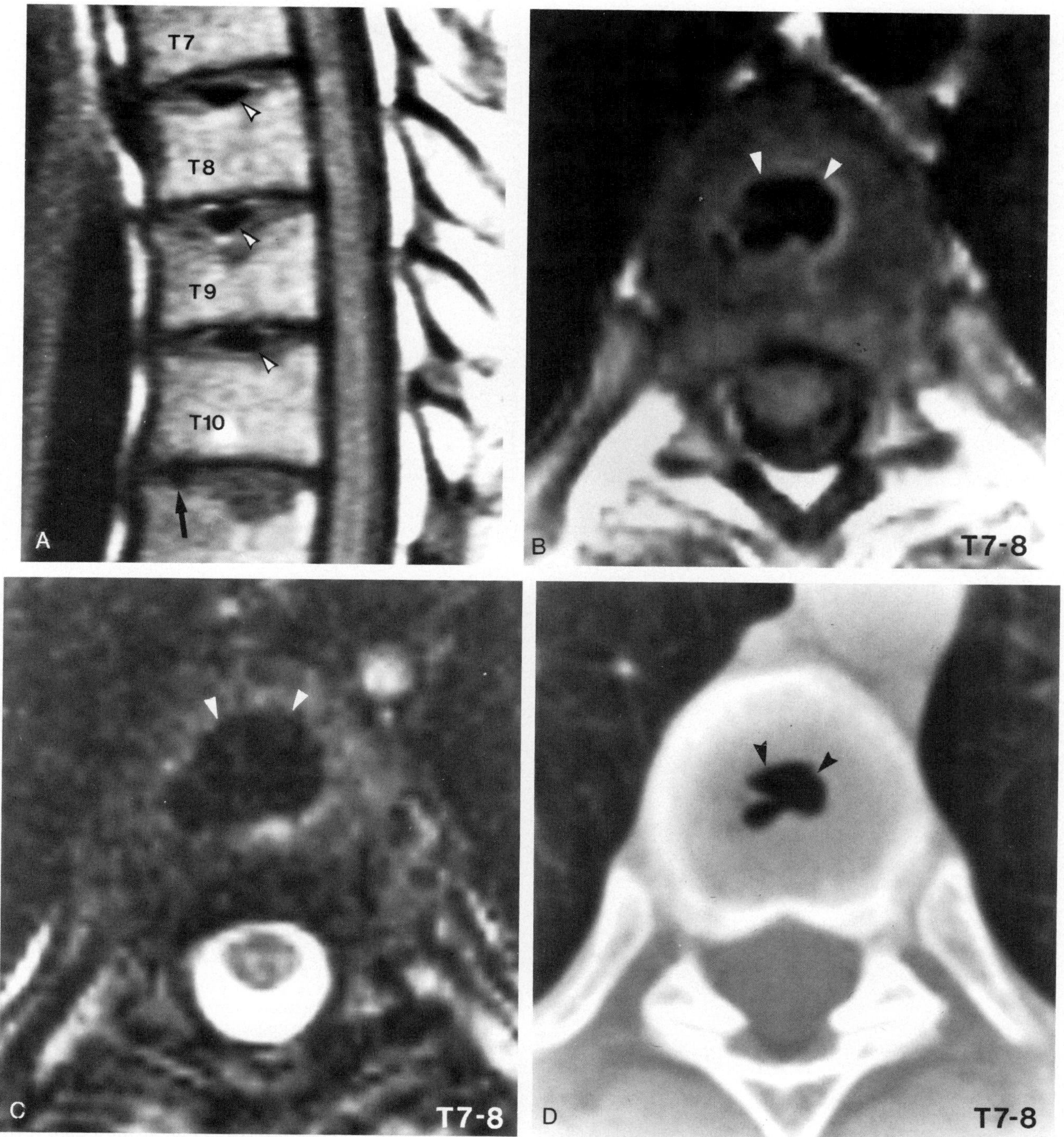

Figure 5–16. Consecutive degenerated thoracic disks that contain gas. *A*, Sagittal T1-weighted image. Note the central location of the gas collection *(arrowheads)* and that the middle of the three disk bulges inferiorly because of a Schmorl's node. A small gas collection *(arrow)* is seen in the anterior part of T10–11. *B*, Axial T1-weighted image. Note that the gas collection *(arrowheads)* is circumscribed by a rim of high signal intensity. *C*, Axial gradient-recalled echo image. This sequence defines less clearly both the margins of the gas *(arrowheads)* and the rim. *D*, Computed tomography scan affirming that the low signal intensity on MR image indeed represents gas and not calcification *(arrowheads)*.

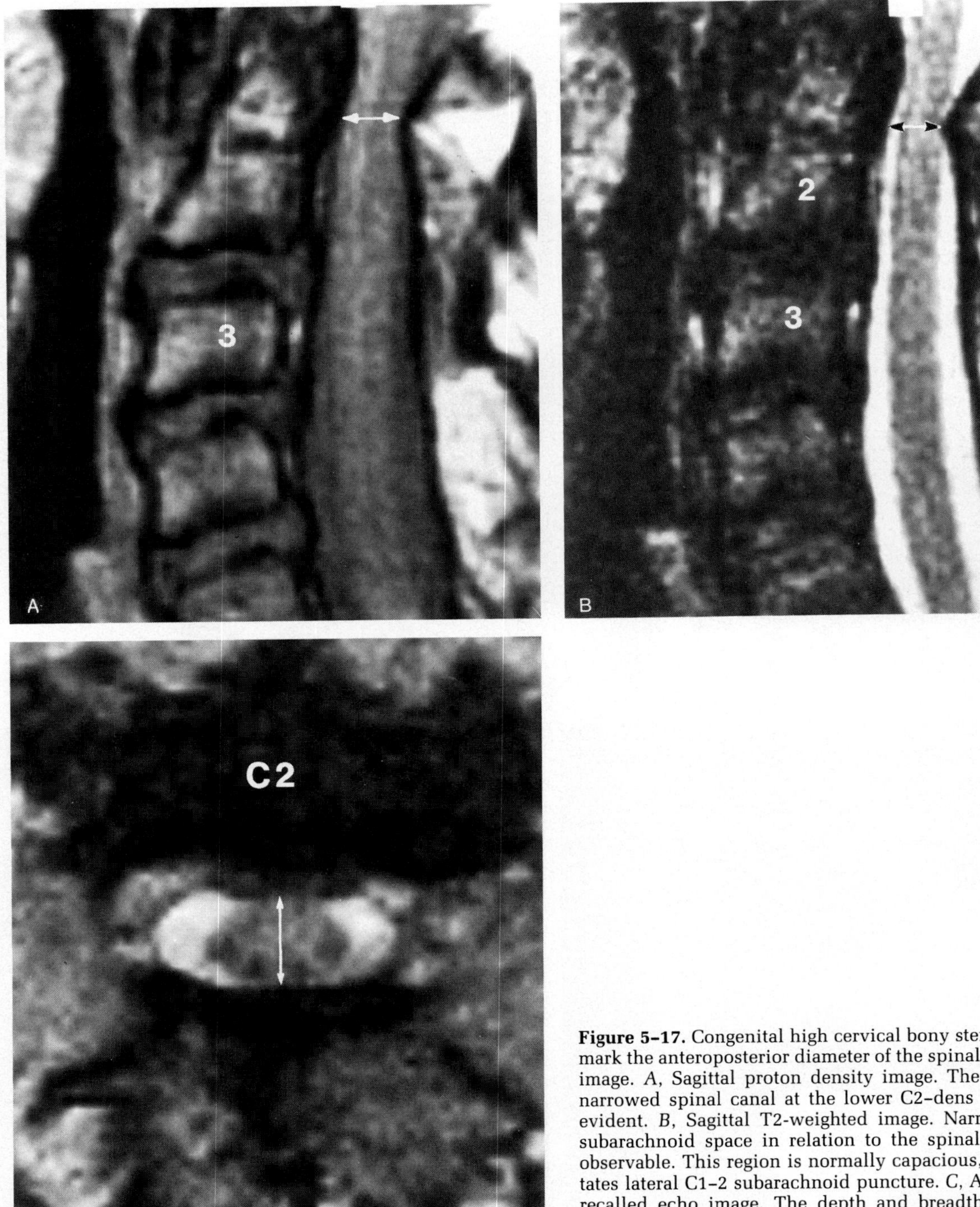

Figure 5-17. Congenital high cervical bony stenosis. Arrows mark the anteroposterior diameter of the spinal canal in each image. *A*, Sagittal proton density image. The detail of the narrowed spinal canal at the lower C2–dens level is quite evident. *B*, Sagittal T2-weighted image. Narrowing of the subarachnoid space in relation to the spinal cord is now observable. This region is normally capacious, which facilitates lateral C1–2 subarachnoid puncture. *C*, Axial gradient-recalled echo image. The depth and breadth of the bony narrowing that compromises the canal are seen both anteriorly and posteriorly.

in our experience, the posterior longitudinal ligament in the cervical region may be difficult to visualize accurately, so that its specific identification is best accomplished by observing the hemorrhage about it.

A distinct low signal intensity framing of the bone marrow on most sequences represents the bony cortex, although in some instances it may also represent an artifact of truncation.[10] Xu and colleagues[44] have demonstrated that the outer anulus, posterior longitudinal ligament, and dura mater all contribute to the contours of low signal intensity that are seen posterior to the vertebral bodies and intervertebral disks. We have made this distinction earlier in the

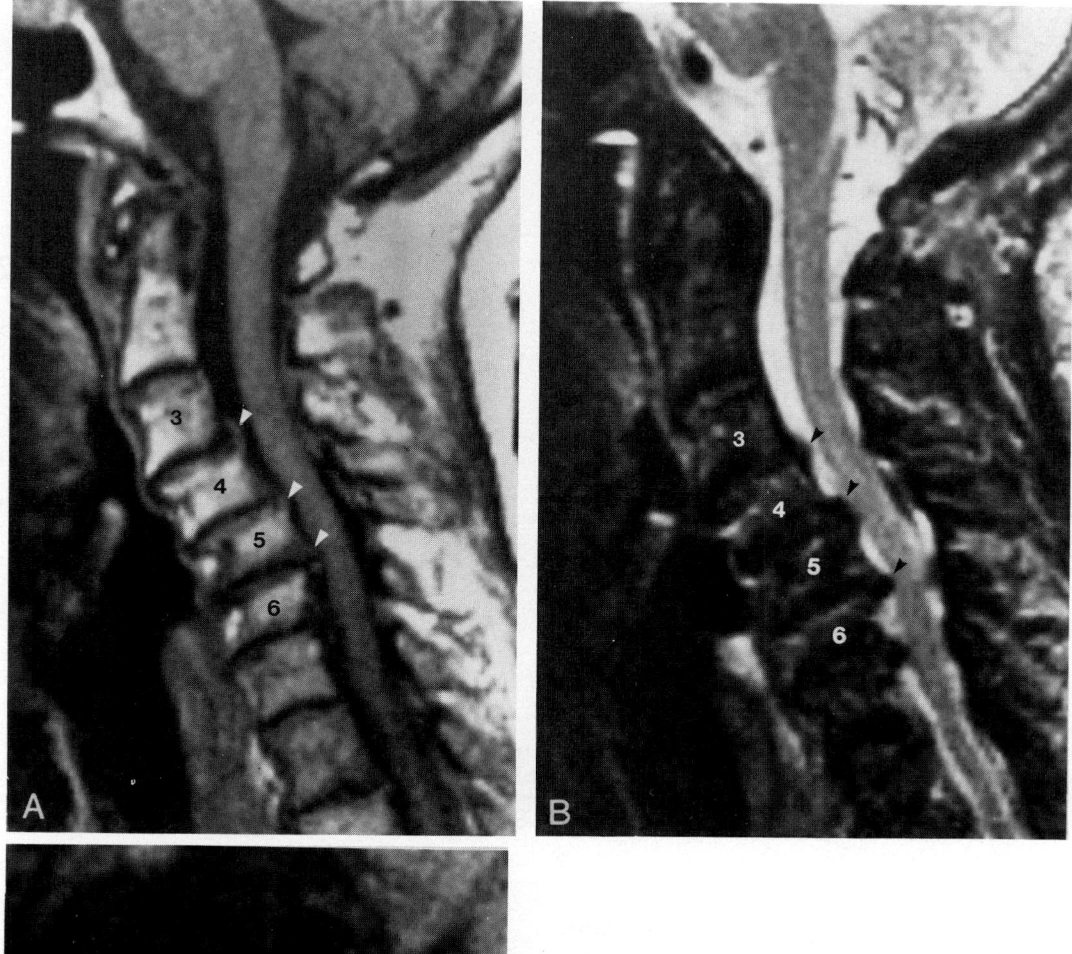

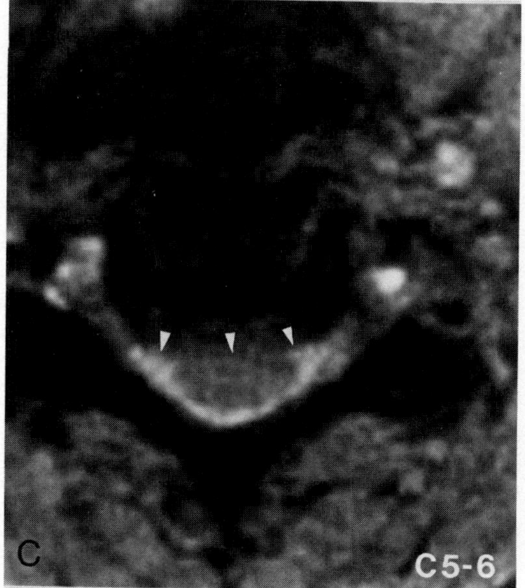

Figure 5-18. Cervical malalignment in advanced degenerative disk disease. *A* and *B*, T1- and T2-weighted sagittal images. The cervical lordosis is reversed at C4–6, and subluxation of the vertebral bodies is observed along this reversed lordosis. This indicates disk ligamentous laxity or stretching to allow for a disordered alignment. Bony bridging *(arrowheads)* and spondylophytes evolve to make this alignment permanent, as is better recognizable on the T2-weighted image. *C*, Axial gradient-recalled echo (GRE) image. The flattening and posterior displacement of the spinal cord *(arrowheads)* due to spondylosis is shown on the T2-weighted image. That this is caused by the spondylophyte and not by disk material is affirmed by the characteristic low signal intensity of bone on GRE images. In a spondylophyte, the intensity may be mixed (centrally), as is shown along the smooth arched posterior border of the spondylophyte at C5–6.

section on lumbar herniation (see Chapter 6) and now believe that higher resolution techniques and greater anatomic definition allow for their distinction. At the level of the intervertebral disk posteriorly, the low signal intensity is essentially that of the outermost portion of the anulus.[32] Acute hemorrhage, whether it is ligamentous or extra-axial or paraspinal in location, may be recognized as isointense in signal intensity to cord on T1-weighted images and low in signal intensity on T2-weighted and small flip-angle GRE sequences owing to the paramagnetic effect of deoxyhemoglobin. Because cartilage and facet capsular fluid and synovium are of similar intensity on T1-weighted images, the facet

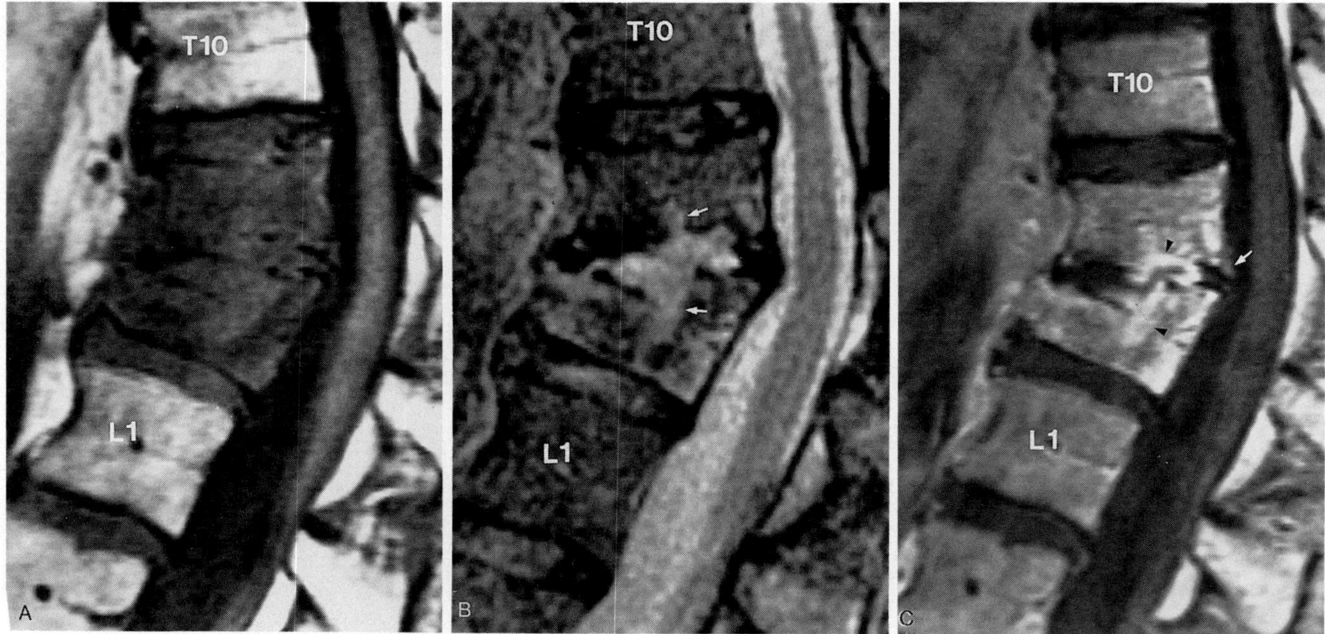

Figure 5–19. Osteomyelitis and diskitis in a 76-year-old woman. *A*, Sagittal T1-weighted image. The T11 and T12 vertebral bodies are markedly affected by edema and an anterior compression fracture of T12. The cortical margins are obscured, and the intevertebral disk and the marrow are of equal signal intensity. *B*, Sagittal T2-weighted image. The higher signal intensity corresponds to hyperemia and inflammatory edema. Very bright protrusions of abnormal material (disk, infection, or other) extend through the end plates *(arrows)*. The bright signal of edema on T2-weighted images is less obvious than the edema on T1-weighted images, perhaps because of the magnetic susceptibility of bone. *C*, T1-weighted gadopenetate dimeglumine–enhanced image. Prominent enhancement of intervertebral disks, end plates, and ameboid protrusions into the body *(arrowheads)* is seen. The small enhancing epidural mass *(arrow)* is consistent with granulation tissue. Cultures of tissue from multiple aspirations and biopsies failed to reveal organisms or abnormal cytologic findings.

joints are separable using small flip-angle GRE imaging (e.g., 10 to 15 fast imaging steady state precession).

There are several classifications of vertebral fractures and dislocations.[7,15] Most severe spinal injuries are a combination of hyperflexion or hyperextension with rotational, axial loading, or shearing forces. The three-column model reported independently by Denis[9] and McAfee and colleagues[21] is useful in assessing the probability of instability. The anterior column includes the anterior two thirds of the vertebral body and the anterior longitudinal ligament; the middle column, the posterior one third of the vertebral body and the posterior longitudinal ligament; and the posterior column, the osseous neural arch, facet joints and capsules, and the more posterior ligaments. Instability is deemed present if two or more columns are disrupted.

In a predominantly flexion injury, the anterior portion of the vertebral axis is compressed and the posterior portion is disrupted; in a predominantly extension injury, the opposite occurs.

Injuries with a flexion component are the most common. Flexion injuries typically disrupt the posterior elements. These injuries include vertebral body wedge compressions, burst fractures (with an associated axial loading component), facet distractions, and anterior subluxation (Figs. 5–20 to 5–24). Hallmarks of flexion injury may also include vertebral body fragmentation, teardrop avulsions, "fanning" of spinous processes with ligamentous edema or hemorrhage in the supra- and interspinous processes, subluxation or dislocation of facet joints with abnormal capsular signals, loss of height of the anterior portion and widening of the posterior intervertebral disk with possibly altered signal characteristics, subluxation of the anterior vertebral body, and disruption of the posterior longitudinal ligament. When a simple anterior wedge fracture is noted, these features can be of value in searching more specifically for ligamentous disruption, which would indicate a more serious and possibly unstable injury. In anterior subluxation, an abnormal signal from the supra- and interspinous and posterior longitudinal ligaments, outermost portion of the anulus, ligamentum flavum, or facet joint capsules may be obtained. The flexion teardrop fracture of the lower cervical spine with an acute anterior cord syndrome often is characterized by a triangular-shaped fracture of the anteroinferior vertebral body with complete

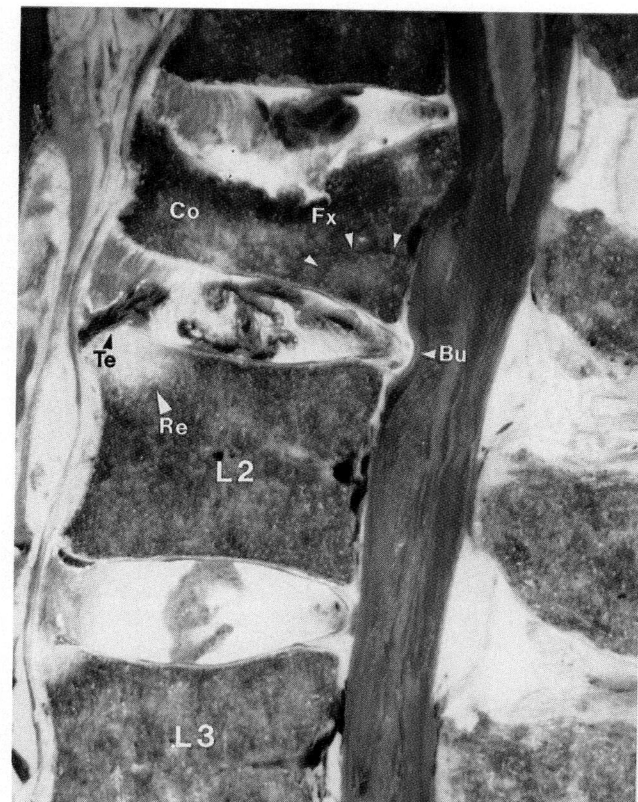

Figure 5–20. Vertebral body compression fracture with abnormalities of adjacent disks as seen by cryomicrotome. The traumatized L1 vertebral body reveals a compression (Co) fracture (Fx) with depression fraying and intravertebral herniation of the superior end plate of L1. There is consequent expansion of the T12–L1 disk. Note the fracture of the posteroinferior portion of L1 *(arrowheads)*. The L1–2 intervertebral disk bulges (Bu) as it is thinned by chronic degeneration (pigmentation). A transverse tear (Te) also affects the anteroinferior part of the middle and outer portions of the anulus. The fatty substance that replaced the adjacent L2 body is seen (Re).

disruption of all ligamentous complexes, facet joints, and the intervertebral disk (Fig. 5–25). The burst fracture is created by implosion of the nucleus pulposus into the vertebral body while the anulus fibrosus remains largely intact. There is comminution of the body, frequently with retropulsed fragments. The combination of acute edema, hemorrhage, and bony fragments contributes to the complex, heterogeneous signals from the central portion of the vertebral body.

Tearing or edema in the anterior or posterior longitudinal ligaments indicates associated hyperextension or hyperflexion forces, respectively. In facet joint dislocations, rotational forces also accompany flexion forces, particularly when they are unilateral. Disruption of the posterior longitudinal ligament and posterior ligament complex is an integral part of this injury. The anterior longitudinal ligament and disk also may be disrupted, depending on the severity of involvement and whether the disease is bilateral or unilateral. Foramen transversarium fracture usually involves the vertebral artery (Fig. 5–26).

Extension injuries are much more common in the cervical spine than in the thoracic or lumbar spine,[7] with the classic hangman's fracture (traumatic spondylolisthesis) being the most common cervical extension injury. Hyperextension fracture-dislocations are created by circular impacting forces, which cause disruption of the anterior longitudinal ligament, facet joint fracture, disk disruption, and vertebral body retrolisthesis. A special but poorly understood extension fracture is the hyperextension dislocation. A hyperextension dislocation injury produces central cord syndrome with possibly only diffuse prevertebral soft tissue swelling evident on plain films or CT scans.[11] Magnetic resonance imaging will very likely play a critical role in evaluating the injuries of these paralyzed patients by allowing visualization of the anterior longitudinal ligament disruption and the disk herniations that characterize this lesion. Rotary injury resulting in atlantoaxial subluxation requires disruption of the transverse and alar ligaments. Wiesel and Rothman[43] have stated that the atlanto-occipital and atlantoaxial joints depend on ligaments to maintain stability. High-resolution MR imaging affords visualization of the upper cervical ligaments[47] and potential ligamentous injuries associated with rotary subluxation (fixation) and odontoid fractures (Fig. 5–27).

Disk herniation after severe trauma was reported with a greater than 50 per cent incidence by Mirvis and co-workers[23] and Flanders and colleagues,[13] 40

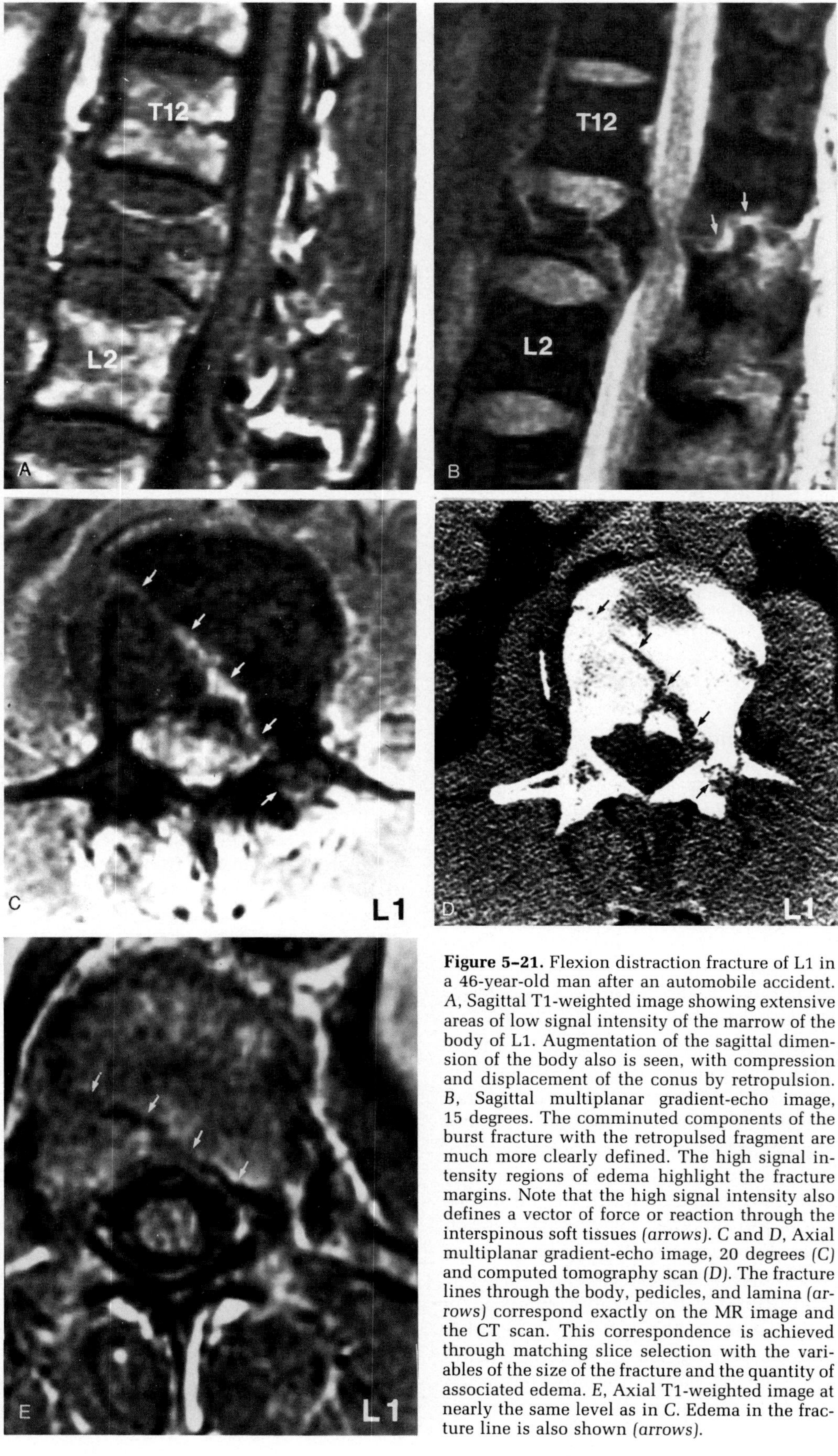

Figure 5–21. Flexion distraction fracture of L1 in a 46-year-old man after an automobile accident. *A*, Sagittal T1-weighted image showing extensive areas of low signal intensity of the marrow of the body of L1. Augmentation of the sagittal dimension of the body also is seen, with compression and displacement of the conus by retropulsion. *B*, Sagittal multiplanar gradient-echo image, 15 degrees. The comminuted components of the burst fracture with the retropulsed fragment are much more clearly defined. The high signal intensity regions of edema highlight the fracture margins. Note that the high signal intensity also defines a vector of force or reaction through the interspinous soft tissues *(arrows)*. *C* and *D*, Axial multiplanar gradient-echo image, 20 degrees *(C)* and computed tomography scan *(D)*. The fracture lines through the body, pedicles, and lamina *(arrows)* correspond exactly on the MR image and the CT scan. This correspondence is achieved through matching slice selection with the variables of the size of the fracture and the quantity of associated edema. *E*, Axial T1-weighted image at nearly the same level as in *C*. Edema in the fracture line is also shown *(arrows)*.

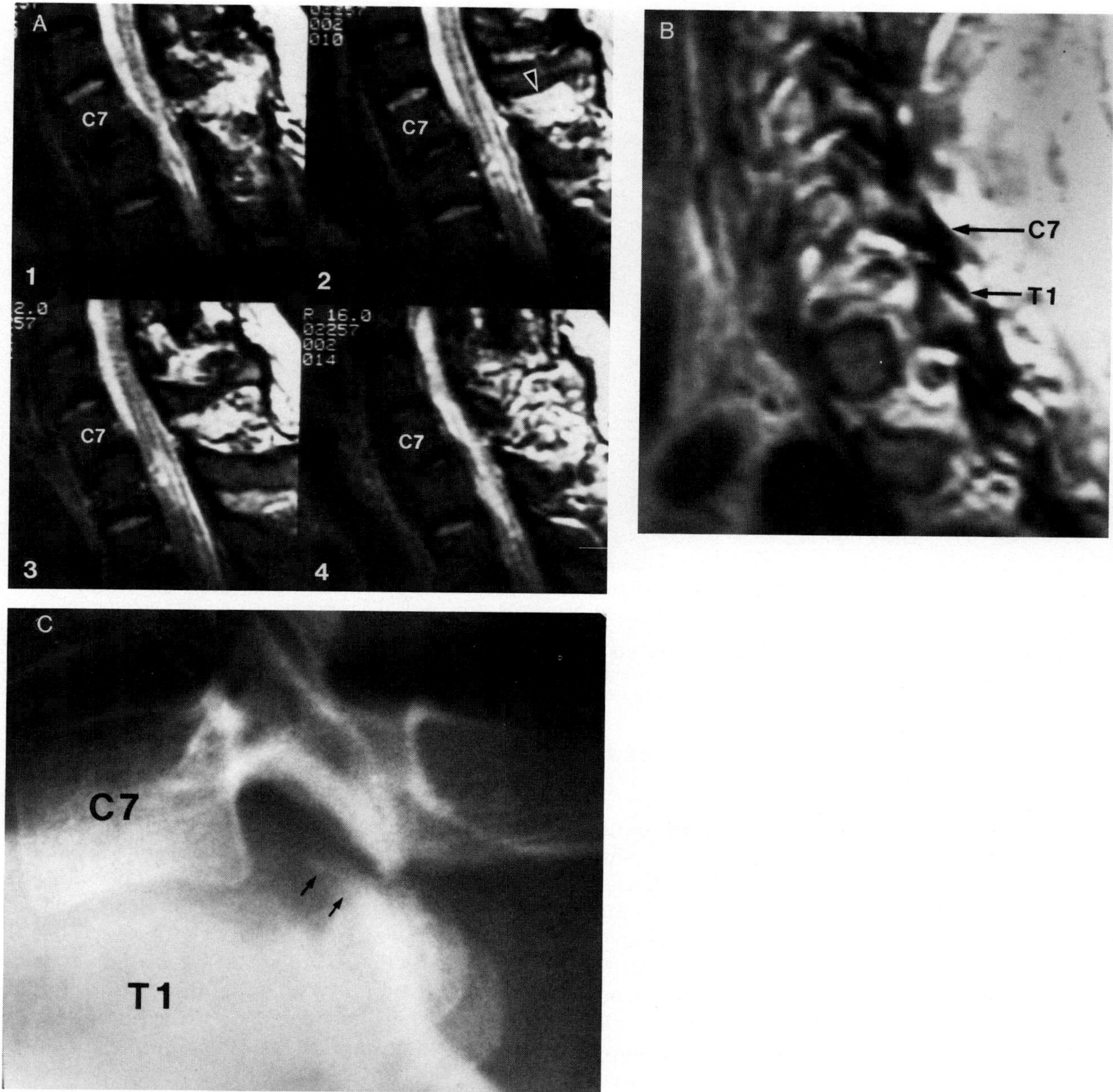

Figure 5–22. Hyperflexion injury of C7 with bilateral perched facets and spinal cord injury in a 25-year-old quadriparetic man who had been in a barroom fight. *A*, Montage of sagittal T2-weighted images. All findings indicate the hyperflexion mechanism of injury: *1*, Slight anterior subluxation of C7 on T1; *2*, splaying of the C7 and T1 spinous processes; edema and acute hemorrhage *(arrowhead)* within the interspinous and supraspinous ligaments; slight wedging of the superoanterior corner of the T1 vertebral body; hyperintensity (edema, injury [*2, 3*] of the cord in *3* and *4*). *B*, Sagittal T1-weighted image after traction. Persistent unilateral posterior subluxation of the superior facet of T1 is seen, as contrasted with C7. *C*, Plain radiograph substantiates subluxation near the perched facet *(arrows)*.

per cent by Cacayorin and associates,[3] and only 14 per cent (2 of 14) patients by McArdle and co-workers.[22] Neurologic dysfunction is frequent, and full function may not return despite expedited diskectomy. Regenbogen and colleagues[29] reported that cervical spondylosis increases the risk of traumatic cord injury.

Clinically significant spinal epidural and subdural hematomas are rare occurrences after trauma despite the frequency of regional hemorrhage in these compartments. Occasionally these hematomas may develop after minor trauma, particularly in the context of anticoagulation, coagulopathies, and vascular malformations.[34,35] Clinically unsuspected small extra-axial hemorrhages after vertebral trauma are likely to be uncovered by MR imaging in acute

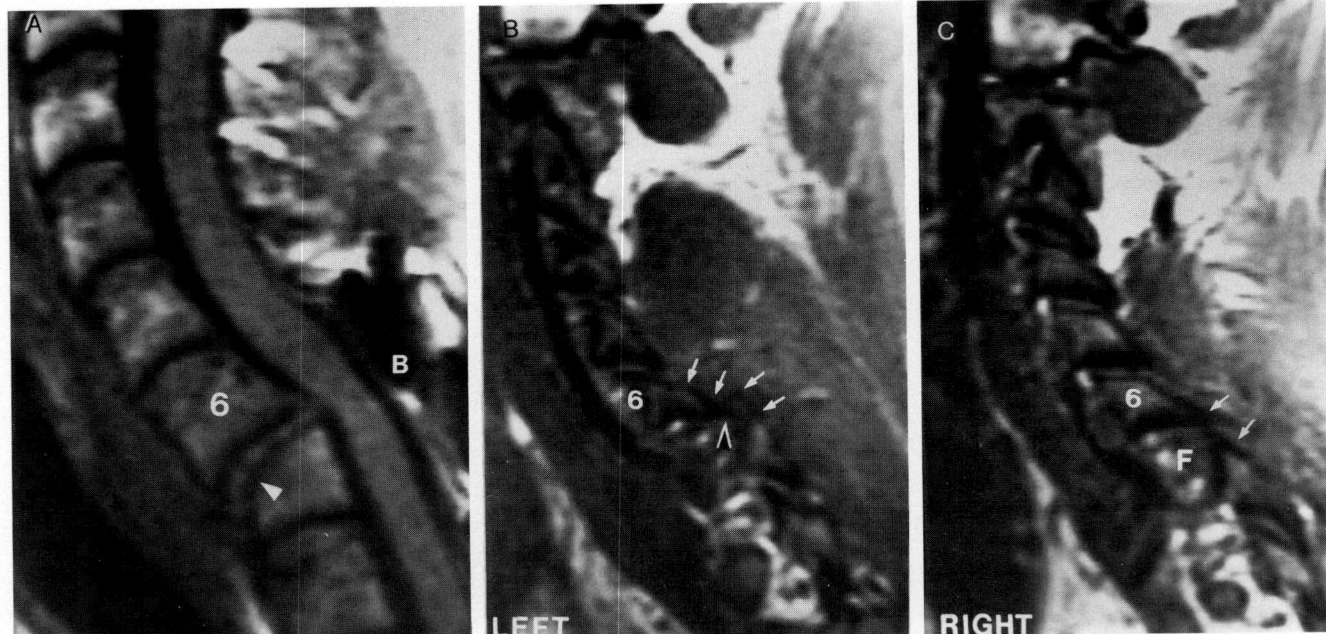

Figure 5–23. The spine of a 34-year-old man who suffered a gunshot wound and recent blunt neck trauma with locked cervical facet. *A*, Sagittal T1-weighted image. Focal kyphotic deformity of the spinal canal is associated with cord narrowing. Slight anterior subluxation of C6 and C7 is seen, with an anterosuperior wedge fracture of C7 *(arrowhead)*. The C6–7 disk is thinned and bulges posteriorly. The low signal intensity of the bullet fragment with artifact (B) can be seen; the fragment has disrupted the posterior elements, predisposing the spine to instability. *B* and *C*, Far lateral sagittal T1-weighted image. In *B*, the left superior facet of C7 is dislocated posteriorly *(arrows)* and "locked" *(arrowhead)*; in *C*, the right superior facet of C7 is posteriorly dislocated and "perched" *(arrows)*, with widening of the adjacent neural canal (F).

trauma as the use of MR imaging increases and GRE imaging sequences in trauma become more routine. Conversely, for displaying extra-axial blood, small collections, evolving blood from the hyperdense to isodense stage, and the inability to differentiate blood in a subacute stage from edema are disadvantages of CT (and myelography) compared with MR imaging. Small flip-angle gradient-echo sequences prove useful, particularly in this situation, because the magnetic susceptibility of deoxyhemoglobin is greater in these sequences than it is with conventional spin-echo techniques.[13,19]

NEOPLASTIC CONDITIONS AND METASTATIC DISEASE

Hemangiomas occur most commonly in the thoracic spine.[20,25] Although usually they are incidental findings during radiologic examination and are of little clinical significance, occasionally they expand and break through the bone cortex to cause myelopathic or radiculopathic symptoms. Hemangioma can be well demonstrated by MR imaging (Fig. 5–28). On both T1-weighted and T2-weighted spin-echo images, the hemangioma is of high signal intensity owing to the fatty component and the hemangiomatous tissue. A distinction between focal fat replacement in the bone marrow and hemangiomas should be possible on conventional T2-weighted spin-echo sequences, because the fat becomes lower in signal intensity whereas the hemangioma maintains its high signal intensity.[38]

Aneurysmal bone cysts are benign, expansive lesions of unknown cause with eggshell-thin bony margins. They most commonly involve both the posterior elements and the vertebral bodies. Twenty per cent of the tumors occur in the spine, but less than 5 per cent of the tumors occur in the sacrum.[41] Internally these lesions are typified by large cavernous spaces, which may be filled with unclotted blood.[41] They may be cystic or solid. Trabeculae may contain giant cells, creating confusion with giant cell tumor histologically. Aneurysmal bone cysts rarely occur after 25 years of age (unlike the giant cell tumor); that age can be very helpful in differentiating these lesions with similar imaging appearances. On MR imaging, distinguishing features include a thin, low-intensity region of the bony periphery, which may be incomplete, may be of an expansive cystic nature, and may have multiple fluid-fluid levels (Fig. 5–29). These are characteristically of varying signal intensities because they contain hemorrhage in different stages of resolution. Unfortunately, this charac-

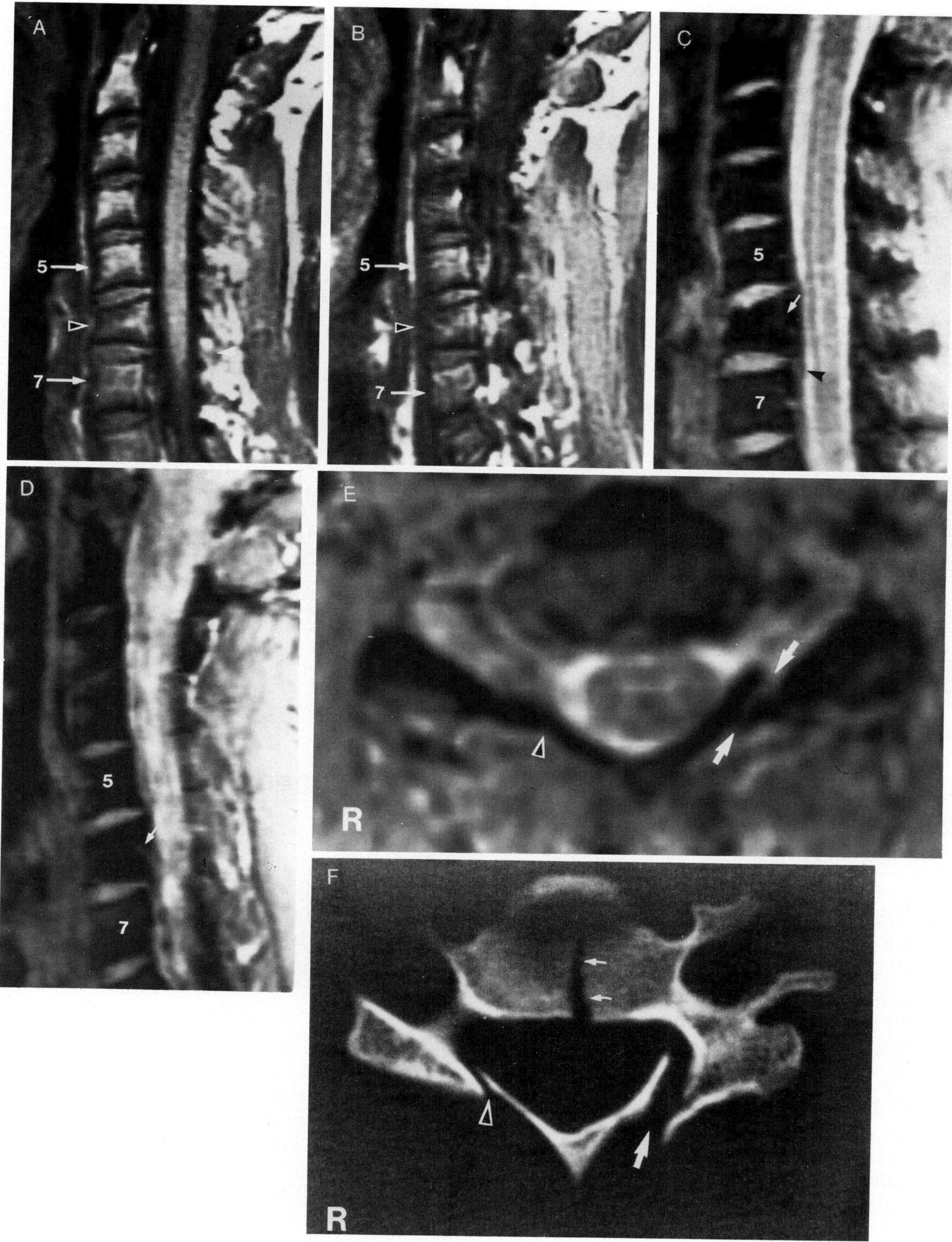

Figure 5–24. C6 fracture and subluxation in a 37-year-old man involved in an automobile accident. *A* and *B*, Sagittal and paramedian sagittal T1-weighted images. Mild anterior compression of C6 is present *(arrowhead)*. Slight posterior subluxation of the C6 vertebral body on C7 also is seen, indicating anterior and middle column instability from a flexion injury and implying a tear in the posterior longitudinal ligament (not seen on these MR images). *C* and *D*, Sagittal and paramedian sagittal multiplanar gradient-echo images, 20 degrees. A subtle high-intensity signal *(arrows)* is seen along the posterior aspect of the C6 vertebral body on this multiplanar gradient-echo sequence *(D)*. This abnormal signal was not detectable on T1-weighted images. Loss of the normal low signal intensity line representing the posterior longitudinal ligament at C6–7 also is seen (compare with other levels retrovertebrally), implying a tear in the posterior ligament. In *C* the outermost portion of the anulus is shown to be stretched *(arrowhead)*. Definite cord edema is present. *E*, Axial multiplanar gradient-echo image, 20 degrees. Bilateral lamina fractures *(arrows and arrowhead)* are present. The left is characterized by a disrupted cortex and a high-intensity signal *(arrows)* between the fragments. The right fracture is characterized primarily by "buckling" of the laminar margin *(arrowhead)*. *F*, Axial computed tomography scan at the bone window. Both fractures *(large arrow and arrowhead)* are well shown, as is the fracture in the midline of the vertebral body *(small arrows)*.

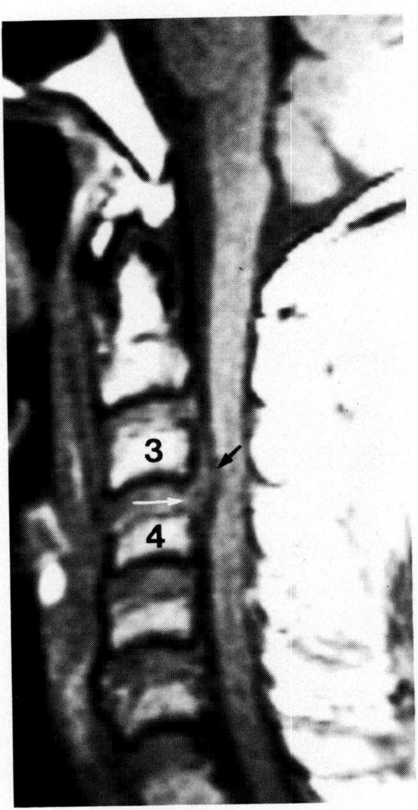

Figure 5–25. Compression fracture and disk herniation occurring after a fall in a 60-year-old woman. Sagittal T1-weighted image shows the compression fracture of C4 anterosuperiorly with prevertebral soft tissue swelling that causes cord compression. A large posttraumatic disk herniation *(small arrow)* is present also. Traumatic herniation is described here because (1) herniation at this high level is rare, and (2) more objectively, the outermost portion of the anulus is torn *(tip of long arrow)*, allowing for so-called subligamentous herniation to occur. Here the herniation is high and posterior to the vertebral body of C3 and anterior to the posterior longitudinal ligament at C3.

teristic is not unique to aneurysmal bone cysts, as hemorrhagic fluid levels may also be found in telangiectatic osteosarcoma, giant cell tumor, and chondroblastoma.[2,26] Aneurysmal bone cysts may also exhibit the unusual feature of crossing the intervertebral disk to involve two adjacent vertebral bodies and even create a vertebral plana.[16]

The thoracic spine is a common site for metastatic disease, and the metastases are readily detectable by MR imaging (Figs. 5–30 to 5–34). On T1-weighted images, metastasis results in decreased signal intensity in the bone marrow of the vertebrae owing largely to the neoplastic tissue or the abnormal increase in extracellular fluid.[42] Because of the variety of tissue components in the metastatic tumor, the appearance on MR T2-weighted images varies. Low, intermediate, and high signal intensity may be present. Pathologic fractures are common findings in metastatic disease.

Other primary tumors occurring in the cervical and thoracic spine, such as osteoblastoma (Fig. 5–35), are relatively uncommon.

OTHER DISORDERS

Rheumatoid arthritis that affects the cervical spine causes pathologic changes in the ligaments with distention and rupture, articular cartilage destruction, and bone erosion, cyst formation, or osteoporosis.[1] The deformity or instability that occurs in the cervical spine depends on the structures involved. The upper cervical region is a common site for the occurrence of rheumatoid arthritis. The fine details of the soft tissue reaction and bony destruction can be appreciated by MR imaging (Fig. 5–36).

Paget's disease, a metabolic bone disease,[47] can be evaluated by MR imaging and scintigraphy. A heterogeneous low signal intensity of the bone marrow is shown on T1-weighted images (Fig. 5–37). An increased osteoclastic activity demonstrated by scintigraphy confirms this diagnosis.

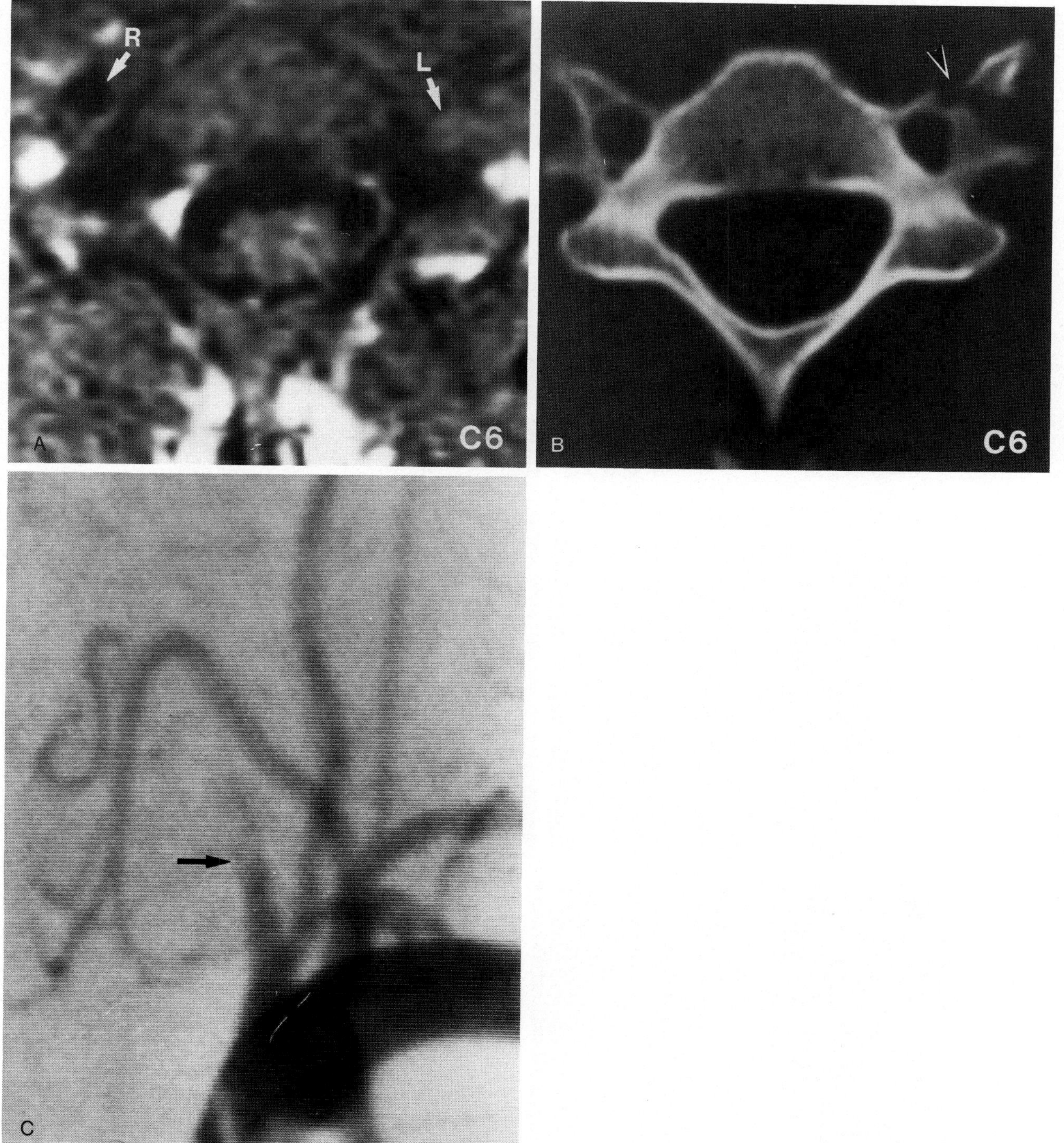

Figure 5–26. Fracture about and through the foramen transversarium of C6 with vertebral artery involvement in a 33-year-old man. *A*, Axial T1-weighted image of C6. This initial MR image reveals disparity in the expected signals of the vertebral arteries; the left (L) is of abnormally low intensity, indicating slow or absent flow. *B*, Axial computed tomography scan of C6 shows a distracted fracture of the anterior tubercle of the left transverse process *(arrowhead)* with the fracture line extending through the foramen transversarium. *C*, The high density of the left vertebral artery is affirmed on other windows (not included), and dissection of the vertebral artery was substantiated by angiography *(arrow)*.

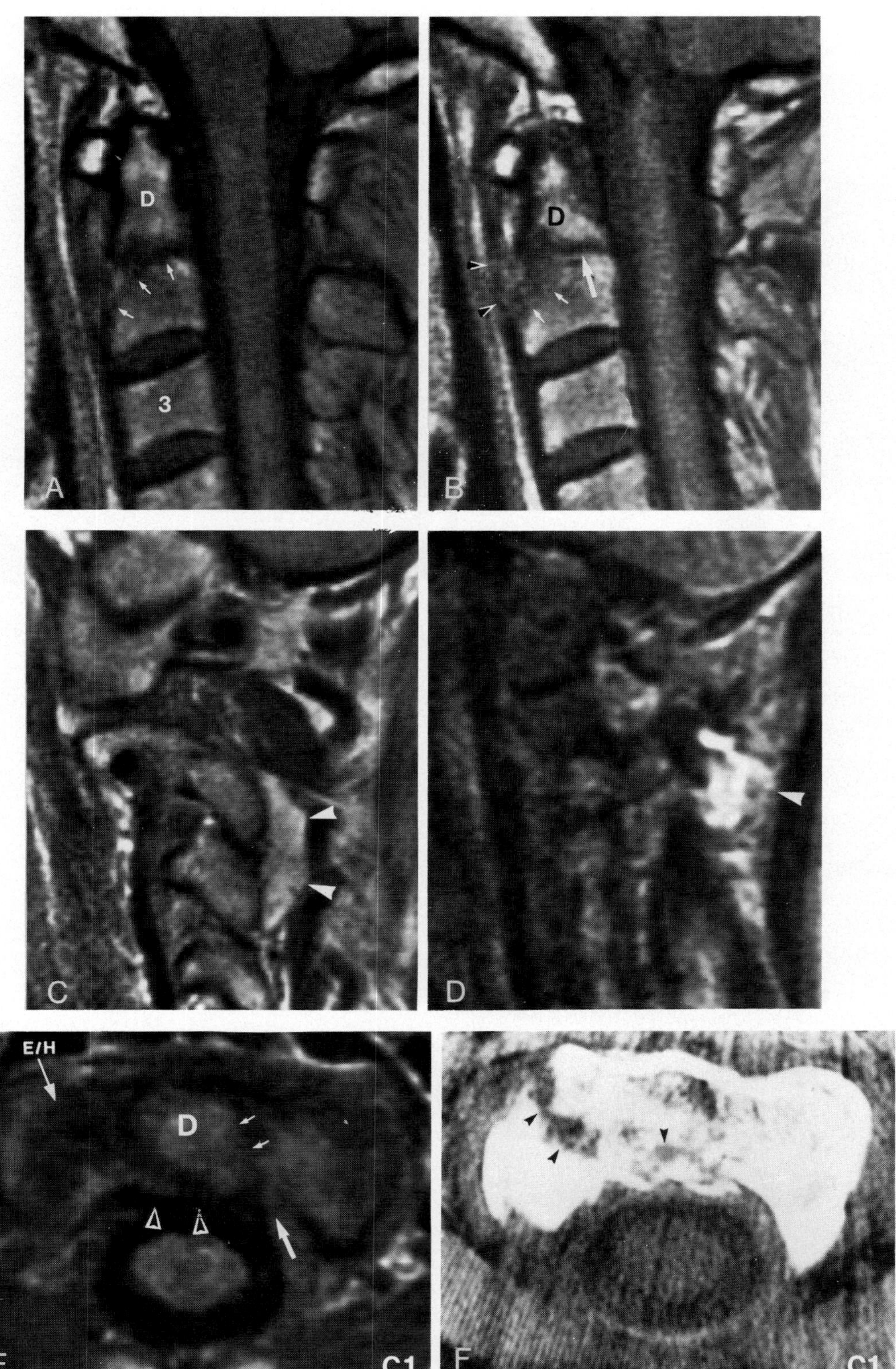

Figure 5–27. Type III C2 fracture in a 45-year-old woman after a motor vehicle accident. *A*, Sagittal T1-weighted image. The low signal intensity of edema or hemorrhage *(arrows)* is noted in the base of the odontoid (dens, D) fracture. No displacement of the fragment is seen other than the fracture. *B*, Paramedian sagittal T1-weighted image. The ondontoid (dens, D) fracture is seen only posteriorly *(large arrow)*, but the involvement of the body of C2 is now defined *(small arrows)*. Loss of the very low signal intensity of the anterior atlantoaxial ligament is seen, indicating that it is torn *(arrowheads)*. *C* and *D*, Paramedian sagittal T1- and T2-weighted images. Material of high signal intensity *(arrowheads)* in the paraspinal soft tissues on both sequences is consistent with extracellular methemoglobin in acute hemorrhage. *E*, Axial T1-weighted image. Loss of the normally very low-intensity signal *(arrowheads)* of the transverse ligament on the left side *(arrow)* is seen, representing its disruption also. The left margin of the dens (D) *(small arrows)* is less clearly defined than its normal right side. Moreover, the right lateral mass shows greater low signal intensity, indicating involvement. E/H, edema, hemorrhage, or both (see Figure 5–3*A* for comparison). *F*, Axial computed tomography scan. The oblique course of the fracture *(arrowheads)* through the right lateral mass and odontoid posteriorly parallels the indistinctness of the dens on the MR image in *E*.

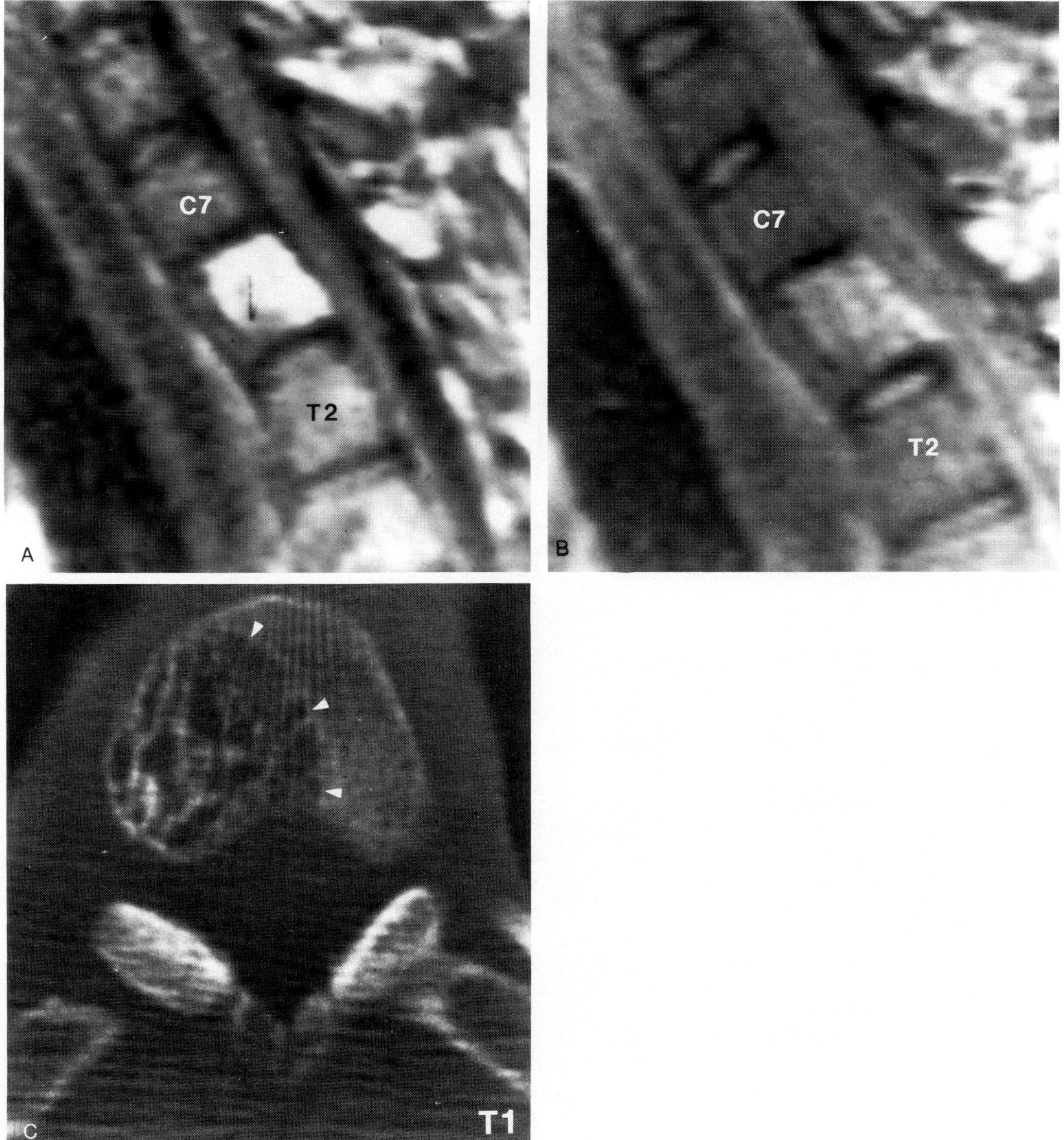

Figure 5–28. Hemangioma of a T1 vertebral body in a 40-year-old man. *A*, Sagittal T1-weighted image showing a T1 vertebral body of homogeneously high signal intensity as a result of significant T1 shortening by the presumably fatty components of this tumor. *B*, Sagittal proton density image. The high signal intensity of the tumor within the vertebral body is maintained on the proton density image, which is typical for hemangioma. Partitions within the lesions are suggested. *C*, Axial computed tomography scan. The multiloculated appearance of a hemangioma *(arrowheads)* is seen on this scan.

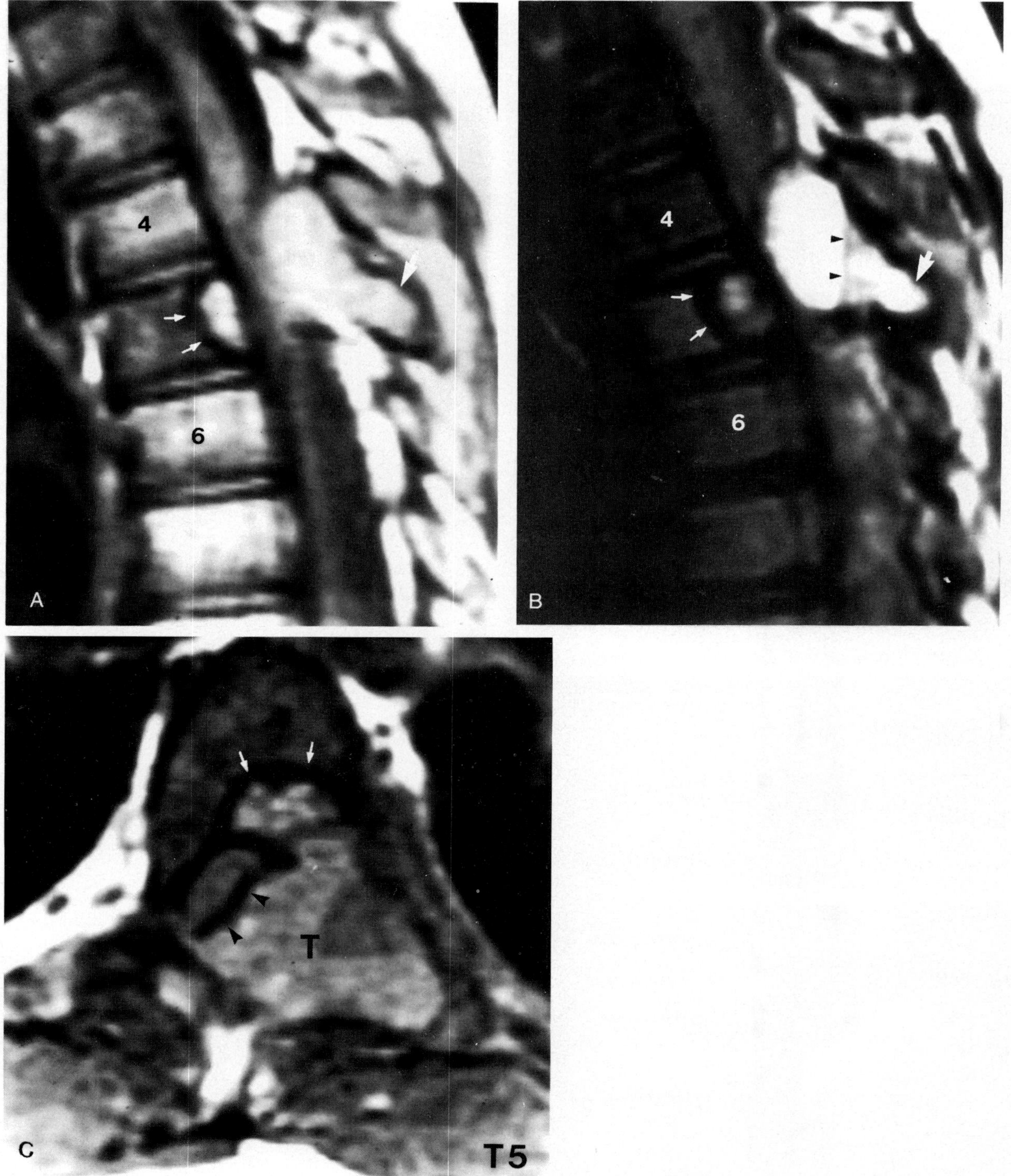

Figure 5–29. Aneurysmal bone cyst in a 7-year-old boy. *A*, Sagittal T1-weighted image. A mass expands the posterior elements on the right side *(arrow)*. The tumor also has a vertebral body component *(small arrows)*. The spinal cord is displaced. The high signal intensity of the tumor implies a high protein content and the presence of methemoglobin, melanin, or fat. *B*, Sagittal proton density image. A fluid-fluid level *(arrowheads)* is evident in the posterior aspect of this mass *(arrow)*, highly suggestive in this age group of an aneurysmal bone cyst. The signal intensity of the most posterior fluid material is slightly higher on the T1-weighted image and is lower on this proton density image, consistent with evolving blood products. *C*, Axial T1-weighted image of T5. The expansile nature of the mass is better defined. The dura mater of very low signal intensity *(arrowheads)* appears to define the malpositioned thecal sac. The high signal intensity of the tumor (T) prevents confusion of the tumor with the thecal sac. However, gadopentetate dimeglumine enhancement can readily separate an enhancing tumor from the thecal sac. Small arrows indicate the vertebral body component of the tumor.

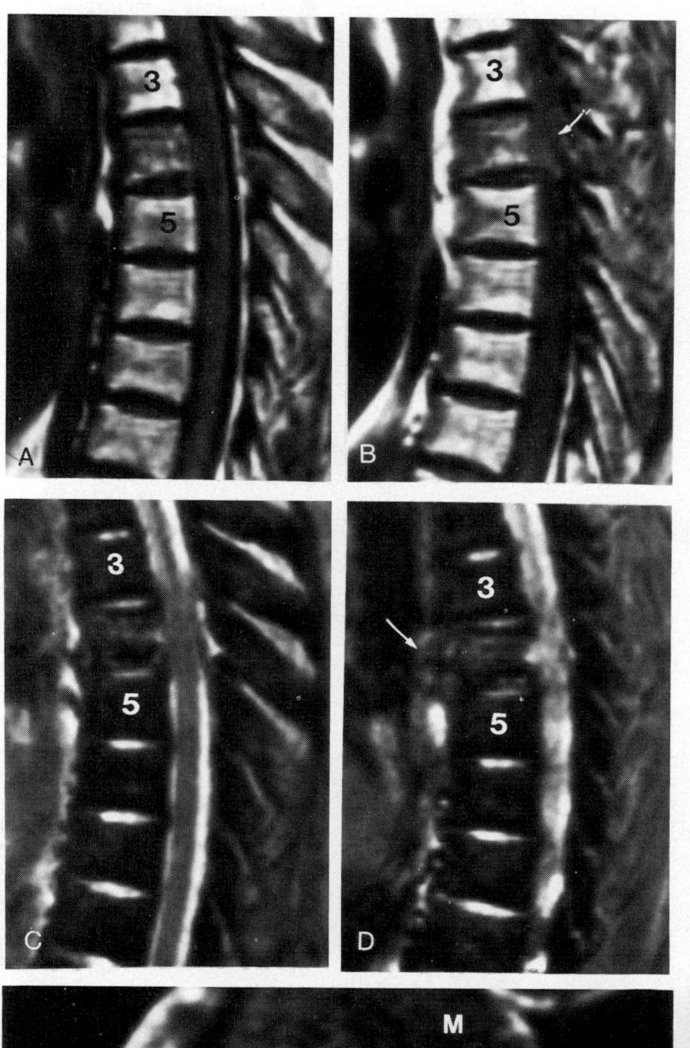

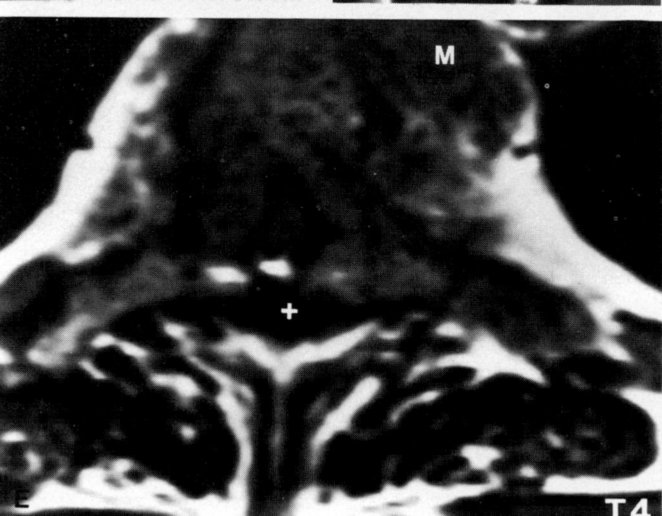

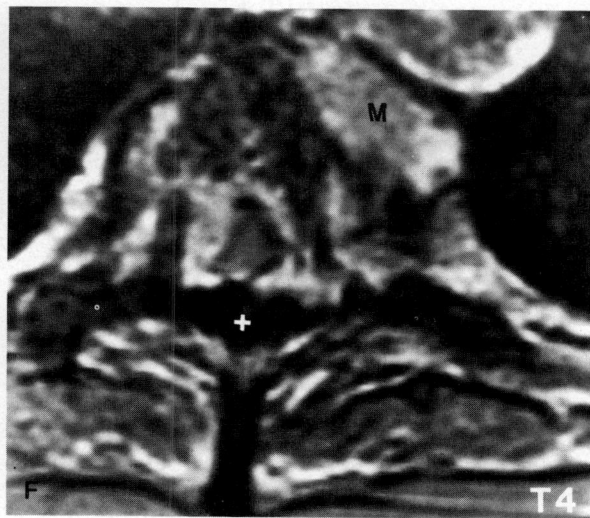

Figure 5–30. Solitary metastatic breast carcinoma to T4 in a 62-year-old woman. *A* and *B*, Midsagittal and paramedian sagittal T1-weighted images. The tumor has created a more heterogeneous low-intensity marrow signal of T4 than of the other vertebral bodies. The adjacent disks and superior and inferior T4 cortical margins are less involved, whereas the vertebral body does not appear intact anteriorly. In *B*, an anterior epidural component of this metastasis parasagittally is suggested *(arrow)*. *C* and *D*, Midsagittal and paramedian sagittal multiplanar gradient-echo images, 20 degrees. Heterogeneity is clearly depicted, with multiple foci of high signal intensity among the bone, which normally is of low signal intensity. Although the spinal cord is not displaced posteriorly, the subarachnoid space is narrowed both anterior and posterior to the spinal cord. Also prevertebral extension is seen in the paramedian sagittal image in *D (arrow)*. *E*, Axial T1-weighted image of T4. The marrow of the vertebral body and the cortex, the spinal canal (+), and the pedicles are distorted morphologically, but all of these structures are of relatively similar signal intensity on this sequence. A prominent left paraspinal mass (M) is observed contiguous to the vertebral body. *F*, Axial multiplanar gradient-echo images of T4, 20 degrees. As seen in many cases, the gradient-echo sequence is superior to T1-weighted images in separating pathologic (M, mass) from normal structures. The epidural tumor of high signal intensity reduces the spinal canal (+) by more than 50 per cent circumferentially.

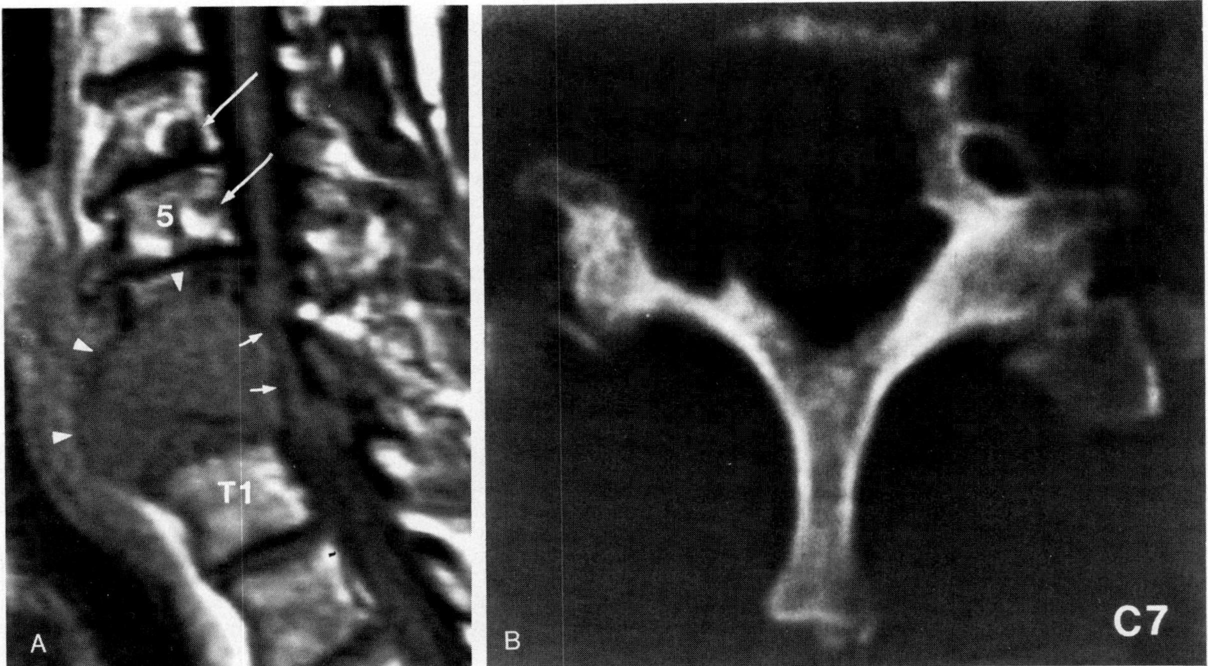

Figure 5–31. Metastatic bronchogenic carcinoma of C6 and C7 in a 70-year-old man. *A*, Sagittal T1-weighted image. A large bulky mass *(arrowheads)* has infiltrated and expanded the C7 vertebral body, crossed the disk space, invaded the C6 body, and largely affected the C7–T1 disk also. Extradural cord compression at C6 and C7 *(small arrows)* and focal abnormal signals within the C4 and C5 bodies also are seen *(long arrows)*. *B*, Axial computed tomography scan through the C7 vertebral body. Gross destruction of the vertebral body, right pedicle, and adjacent lamina has occurred, with dissolution of the right foramen transversarium inferiorly.

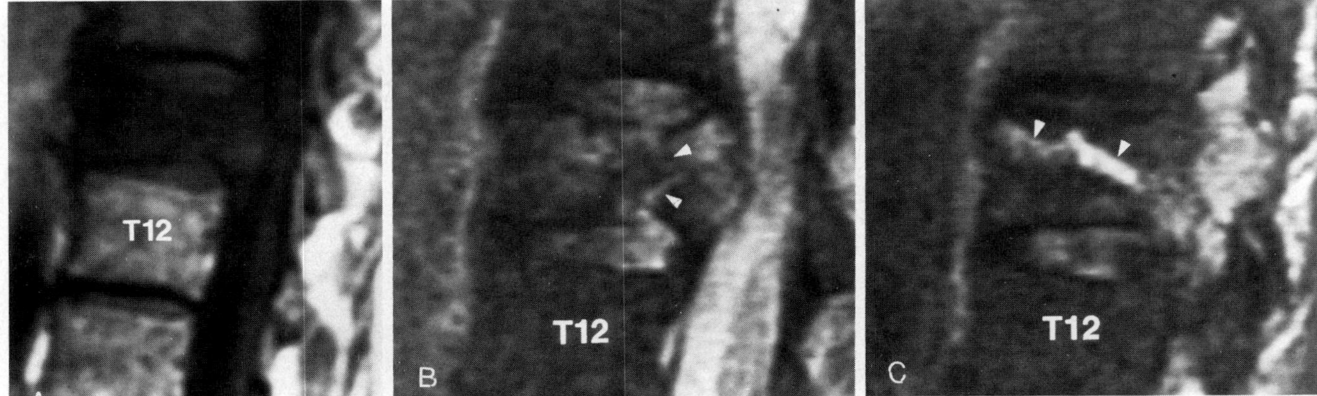

Figure 5–32. Adenocarcinoma metastatic to T11 with a pathologic fracture in a 67-year-old man. *A*, Sagittal T1-weighted image. The entire T11 vertebral body is replaced by a tumor of low signal intensity. The vertebral body has expanded in the sagittal plane, and partial loss of height of the vertebral body is seen as the disks bulge into it. Note the spinal cord compression from the bulging of the adjacent disks into the expanded body. The T10 body also is of low signal intensity but less so than the T11. The vertebral body similarly is made more evident by the slight low signal intensity of the disks. *B* and *C*, Sagittal and paramedian sagittal T2-weighted images. These sequences afford visualization of the fracture line of high signal intensity *(arrowheads)*. Note the heterogeneous but primarily high signal intensity marrow. In this case, the T1-weighted image is more suggestive of tumor replacement than is the T2-weighted sequence. Most tumors replace the marrow signal on T1-weighted images completely before a fracture is visualized. However, an acute nonpathologic or osteoporotic fracture may be of high signal intensity on T2-weighted images.

Figure 5–34. Large cell lymphoma of T6–7 in a 65-year-old man. *A*, Sagittal T1-weighted image. A large soft tissue mass (T) of low signal intensity affects the posterior elements and soft tissues. Its anterior component compresses the spinal cord *(arrow)* at the T6 level. *B*, Sagittal proton density image. The extracellular water content of this tumor (T) has a long T2 relaxation time, causing elevated signal intensity within this sequence. The posterior extradural component of the tumor is more evident, but its relationship to the spinal cord is less obvious than on T1-weighted sequences. *C*, Axial T1-weighted image of T6. The mass of low signal intensity is confined to the posterior bony and soft tissue structures despite the compression of the thecal sac and widening of the spinal cord *(arrows)*). Although the inferior facets of T6 are not visible, the superior facets of T7 are still identifiable *(arrowheads)*.

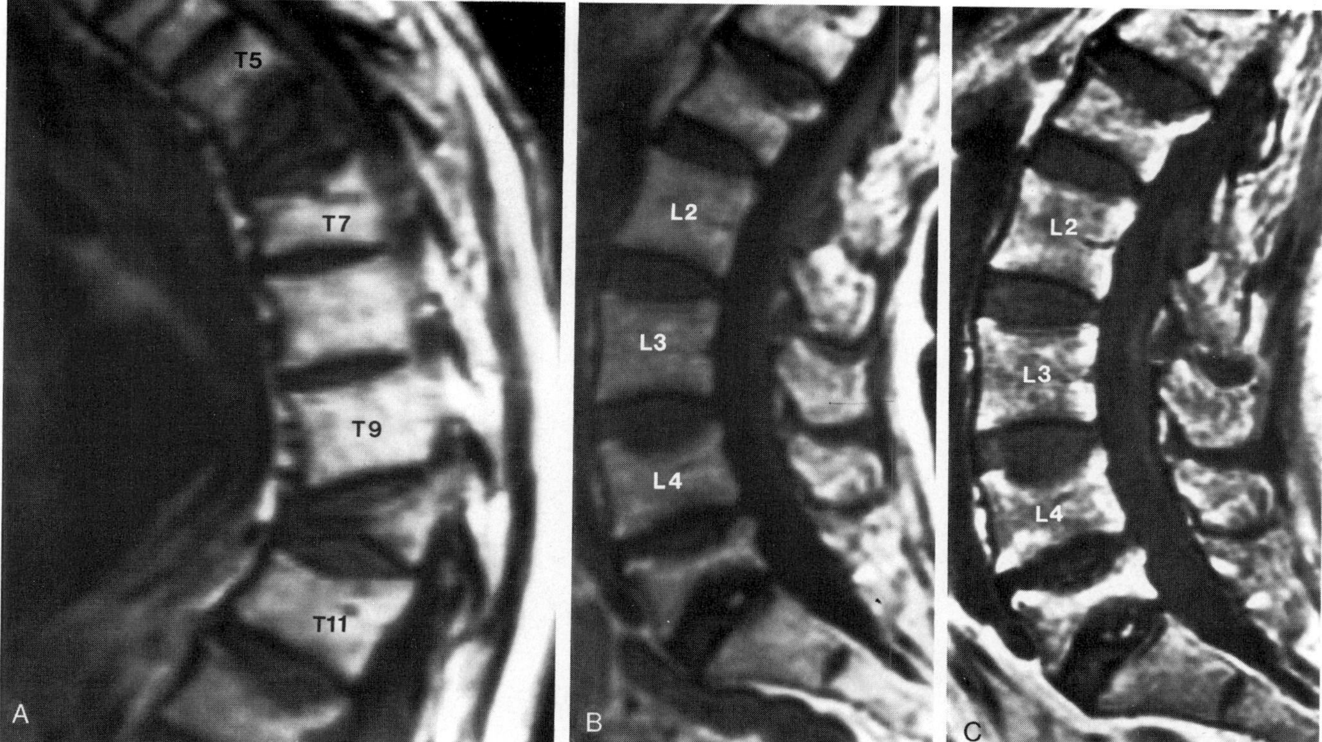

Figure 5–33. Metastatic squamous cell carcinoma and chronic osteoporotic fractures in a 75-year-old woman. *A*, Paramedian sagittal T1-weighted image of thoracic spine showing an almost complete collapse of the T6 and anterior T10 vertebral bodies with severe kyphosis. The marrow signal in T6 is of very low signal intensity. Skeletal scintigraphy showed abnormal findings in both of these locations, but the marrow signal in T10 is maintained compared with that of adjacent vertebral bodies, which is more consistent with osteoporotic fracture than with metastasis. *B*, Sagittal T1-weighted image of the lumbar spine obtained at the same time as *A*. Several compressed vertebral bodies are noted, with their signal intensities consistent with osteoporotic fractures rather than metastasis. Chronic osteoporotic fractures should maintain their "normal" signal intensity on all sequences. *C*, Paramedian sagittal T1-weighted image obtained 8 months after that in *B*. No significant change in the signal intensity or the compression of the abnormal vertebral bodies has occurred, further substantiating the view that their collapse is due to osteoporosis rather than to metastasis. (The images appear to be of different intensities owing to use of a high pass filter; radiation therapy was not given to this region.)

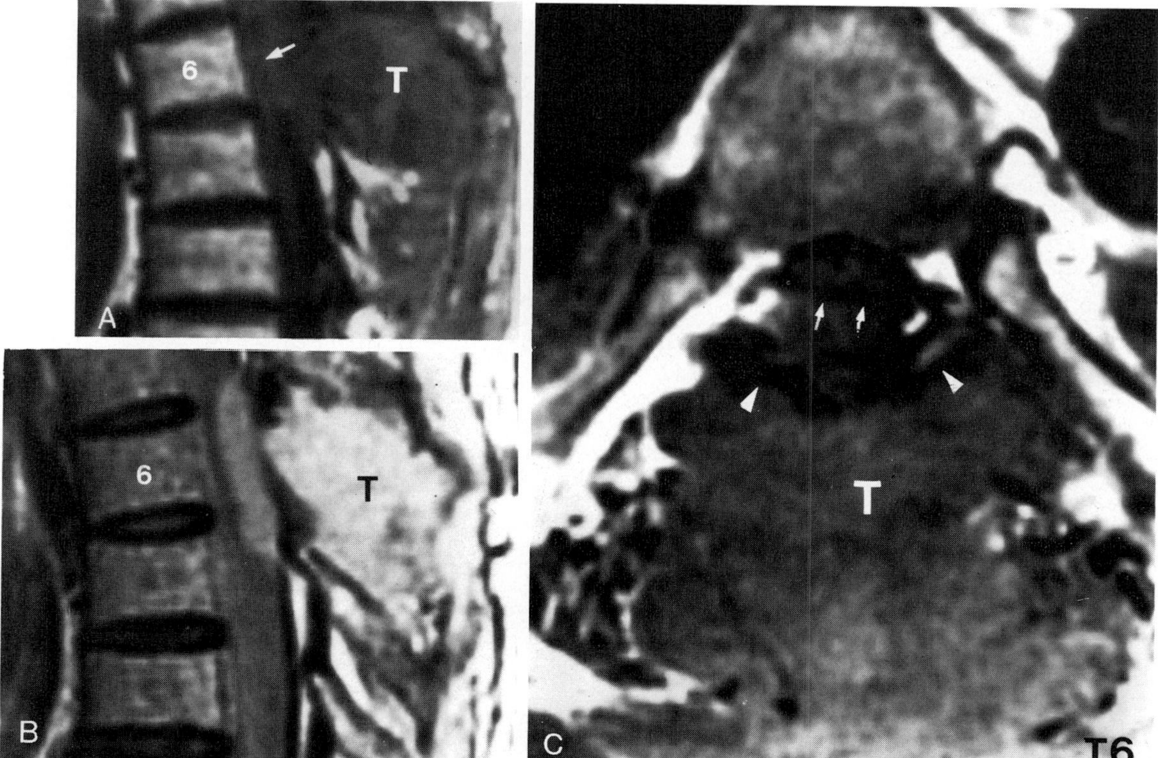

Figure 5–34 *See legend on opposite page*

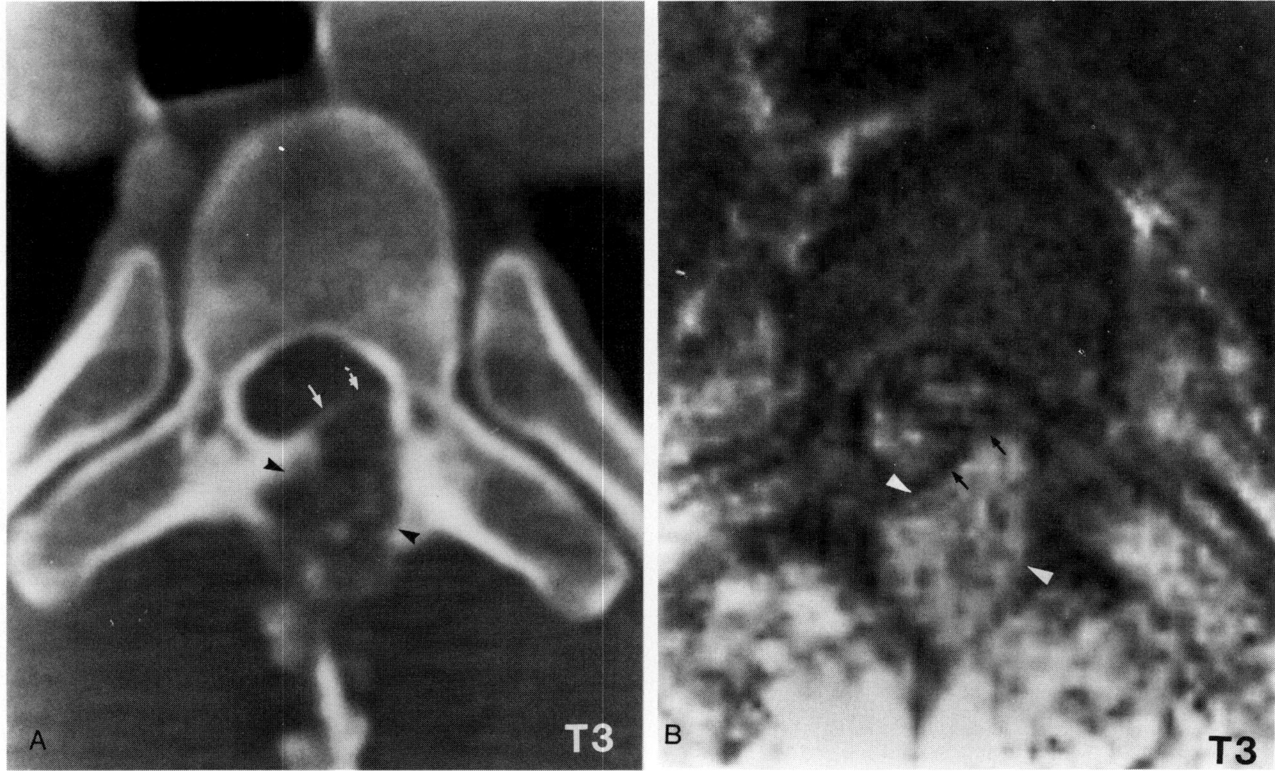

Figure 5-35. Osteoblastoma at T3 in a 20-year-old man. *A*, Axial computed tomography (CT) scan. An expansile lytic lesion *(arrowheads)* of the lamina and adjacent spinous process has a largely benign appearance (cortical disruption is questioned posteriorly). Hyperostotic changes can be seen along the distal aspect of the spinous process. The most anterior part of the expanding mass bulges into the spinal canal *(small arrows)*. *B*, Axial T1-weighted image. The low-density portion of the mass on CT is of high signal intensity on the T1-weighted image *(arrowheads)*, probably reflecting blood products within the lesion. The thecal sac is mildly compressed on the left side *(small arrows)*.

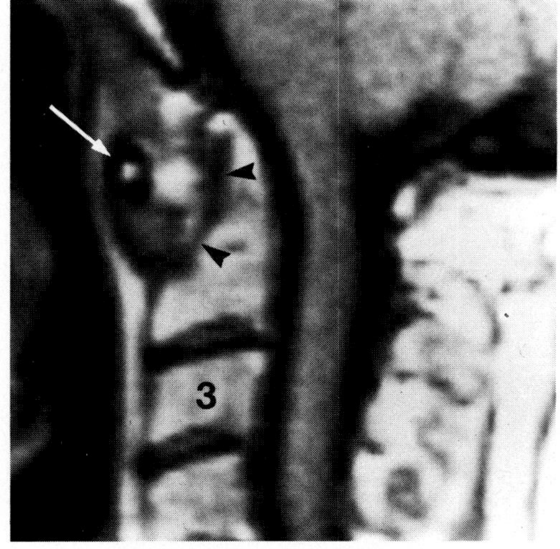

Figure 5-36. Rheumatoid arthritis involving C2. Sagittal T1-weighted image showing erosion of the anterosuperior portion of the odontoid by pannus *(arrowheads)*. This inflammatory focus increases the bursal space, lending itself to ligamentous instability manifested through flexion and extension of the neck. An arrow indicates the anterior arch of C1.

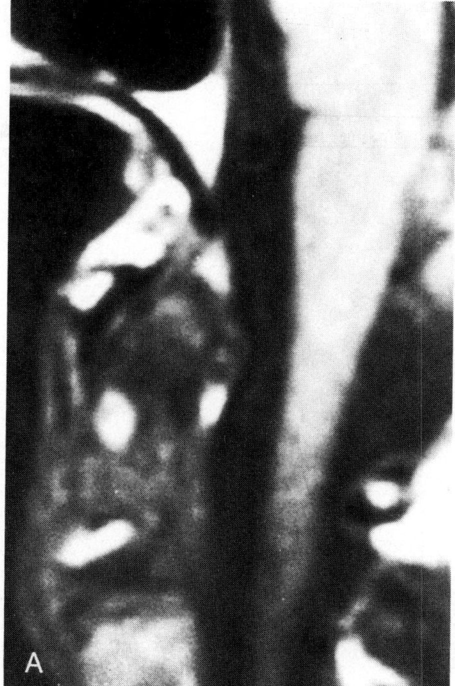

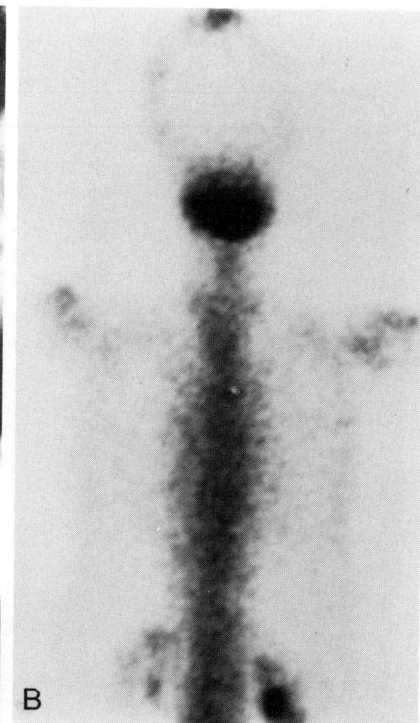

Figure 5-37. Paget's disease of C2 in a 70-year-old patient with headache, who was scheduled for a brain MR image. *A*, This T1-weighted sagittal image of the upper cervical region was included on MR imaging of the brain. A heterogeneous low signal intensity area of the marrow of the dens and body of C2 is seen. C2 also appears enlarged, which was confirmed on plain radiographs. *B*, Frontal tomographic image from ^{99m}Tc methylene disphosphonate bone scintigraphy. The only area of abnormality was in the C2 area. One year later, no clinical or further imaging changes were found.

References

1. Ashby Grantham S, Lipson SJ: Rheumatoid arthritis and other noninfectious inflammatory diseases. *In* Sherk HH, et al: The Cervical Spine. Philadelphia, JB Lippincott, 1989, p 564.
2. Beltran J, Simon DC, Levy M, et al: Aneurysmal bone cyst: MR imaging at 1.5T. Radiology 158:689, 1986.
3. Cacayorin ED, Poe LB, Hochlauser L, et al: MR imaging in spine trauma: Preliminary experience. AJNR 11:412, 1986.
4. Castillo M: Neural foramen remodeling caused by a sequestered disk fragment. AJNR 12:566, 1991.
5. Czervionke LF, Daniels DL, Ho PSP, et al: Cervical neural foramina: Correlative anatomic and MR imaging study. Radiology 169:753, 1988.
6. Czervionke LF, Haughton VM, Daniels DL, et al: Cervical spine. *In* Cranial and Spinal Magnetic Resonance Imaging: An Atlas and Guide. New York, Raven Press, 1987, p 235.
7. Daffner RH: Imaging of vertebral trauma. Rockville, Md, Aspen Publications, 1988.
8. Daniels DL, Grogan JP, Johansen JG, et al: Cervical radiculopathy: Computed tomography and myelography compared. Radiology 151:109, 1984.
9. Denis F: Spinal instability as defined by the 3 column spine concept in acute spinal trauma. Clin Orthop 189:65, 1984.
10. Dick BW, Mitchell DG, Burk DL, et al: The effect of chemical shift misregistration on cortical bone thickness on MR imaging. AJR 151:537, 1988.
11. Edeiken-Monroe B, Wagner LK, Harris JH Jr: Hyperextension-dislocation of the cervical spine. AJNR 7:135, 1986.
12. Enzmann DR, Rubin JB: Cervical spine: MR imaging with a partial flip angle, gradient-refocused pulse sequence. Part I. General considerations and disk disease. Radiology 166:467, 1988.
13. Flanders AE, Schaefer DM, Doan HT, et al: Acute cervical spine trauma: Correlation of MR imaging with degree of neurological deficit. Radiology 177:25, 1990.
14. Grenier N, Greselle JF, Vital JM, et al: Normal and disrupted lumbar longitudinal ligaments: Correlative MR and anatomic study. Radiology 171:197, 1989.
15. Harris JH, Edeiken-Monroe B: The radiology of acute cervical spine trauma. Baltimore, William & Wilkins, 1987.
16. Hay MC, Paterson D, Tayloe TK: Aneurysmal bone cysts of the spine. J Bone Joint Surg [Am] 60B:406, 1978.
17. Hendrick RE, Kneeland JB, Stark DD: Maximizing signal-to-noise and contrast-to-noise ratios in FLASH imaging. Magn Reson Imaging 5:117, 1987.
18. Holshouser BA, Hinslow DB Jr, Shellock FG: Sedation, anesthesia and physiologic monitoring during MRI. *In* Hasso AN, Stark DD (eds): Spine and Body Magnetic Resonance Imaging Categorical Course Syllabus. Boston, American Roentgen Ray Society, 1991, p 9.
19. Kulkarni MV, McArdle CB, Kopanicky D, et al: Acute spinal cord injury: MR imaging at 1.5T. Radiology 164:837, 1987.
20. Laredo JD, Reizine D, Bard M, et al: Vertebral hemangiomas: Radiologic evaluation. Radiology 161:183, 1986.
21. McAfee PC, Yuan HA, Fredrickson BE, et al: Value of computed tomography in thoraco-lumbar fractures. J Bone Joint Surg [Am] 65:461, 1983.
22. McArdle CB, Crofford MJ, Mirfakhraee M, et al: Surface coil MR of spinal trauma: Preliminary experience. AJNR 7:885, 1986.
23. Mirvis SE, Geisler FH, Jelineck JJ, et al: Acute cervical spine trauma: Evolution with 1.5T MR imaging. Radiology 166:807, 1988.
24. Modic MT, Masaryk TJ, Ross JS, et al: Cervical radiculopathy: Value of oblique MR imaging. Radiology 163:227, 1987.
25. Mohan V, Gupta SK, Tuli SM: Symptomatic vertebral hemangiomas. Clin Radiol 31:575, 1980.
26. Munk PL, Helms CA, Holt RG, et al: MR imaging of aneurysmal bone cysts. AJR 153:99, 1989.
27. Parke WW, Sherk HH: Normal adult anatomy. *In* Sherk HH, et al: The Cervical Spine. Philadelphia, JB Lippincott, 1989, p 11.
28. Pech P, Daniels DL, Williams AL, et al: The cervical neural foramina: Correlation of microtomy and CT anatomy. Radiology 155:143, 1985.
29. Regenbogen VS, Rogers LF, Atlas SW, et al: Cervical spinal cord injuries in patients with cervical spondylosis. AJR 146:277, 1986.
30. Resnick D, Niwayama G: Degenerative disease of the spine. *In* Resnick D, Niwayama G: Diagnosis of Bone and Joint Disorders. 2nd ed. Philadelphia, WB Saunders, 1988, p 1480.
31. Resnick D, Niwayama G: Calcification and ossification of the posterior spinal ligaments and tissues. *In* Resnick D, Niwayama G: Diagnosis of Bone and Joint Disorders. 2nd ed. Philadelphia, WB Saunders, 1988, p 1603.
32. Rosenbaum AE, Yu SF: The outer anulus: The final barrier to containing intervertebral disc herniation. Part II. MR imaging. Presented at the 29th annual meeting of American Society of Neuroradiology, Washington, DC, June 1991.
33. Ross JS, Perez-Reyes N, Masaryk TJ, et al: Thoracic disk herniation: MR imaging. Radiology 165:511, 1987.
34. Rothus WE, Chedid M, Deeb ZL, et al: MR imaging in the diagnosis of spontaneous spinal epidural hematomas. J Comput Assist Tomogr 11:851, 1987.
35. Russell NA, Benoit BG: Spinal subdural hematoma. Surg Neurol 20:133, 1983.

36. Schmorl G, Junghanns H: The Human Spine in Health and Disease. New York, Grune & Stratton, 1959, p 187.
37. Shellock FG: Biological effects and safety aspects of magnetic resonance imaging. Magn Reson Q 5:243, 1989.
38. Shoukimas GM, Hesselink JR: MR imaging of the cervical and thoracic spine. *In* Edelman RR, Hesselink JR (eds): Clinical Magnetic Resonance Imaging. Philadelphia, WB Saunders, 1990, p 667.
39. Stollman A, Pinto R, Benjamin V, et al: Radiologic imaging of symptomatic ligamentum flavum thickening with and without ossification. AJNR 8:991, 1987.
40. Tarr TW, Drolshagen LF, Kerner TC, et al: MR imaging of recent spinal trauma. J Comput Assist Tomogr 11:412, 1987.
41. Tillman BP, Dahlin DC, Lipscomb PR, et al: Aneurysmal bone cyst: An analysis of 95 cases. Mayo Clin Proc 43:478, 1968.
42. Vogler JB, Murphy WA: Bone marrow imaging. Radiology 168:679, 1988.
43. Wiesel SW, Rothman RH: Occipital-atlantal hypermobility. Spine 4:187, 1979.
44. Xu GL, Haughton VM, Carrera GE: Lumbar facet joint capsule: Appearance at MR imaging and CT. Radiology 177:415, 1990.
45. Yenerich DO, Haughton VM: Oblique plane MR imaging of the cervical spine. J Comput Assist Tomogr 10:823, 1986.
46. Yu S, Haughton VM, Rosenbaum AE: Magnetic resonance imaging and anatomy of the spine. Radiol Clin North Am 29:691, 1991.
47. Yu S, Haughton, VM, Poe LB, et al: Upper cervical spinal ligaments: Correlative MR imaging and anatomic study. Radiology 177(P): 119, 1990.

6 THE LUMBAR SPINE

Shiwei Yu, Arthur E. Rosenbaum, Larry B. Poe, and Bruce E. Fredrickson

Magnetic resonance (MR) imaging is the best imaging modality for evaluation of spinal diseases. It has several advantages over computed tomography (CT) in the diagnosis of diseases of the spine. However, the excellent anatomic details achieved by high-resolution MR imaging have not been elucidated, and its potential for further detailed study of spinal imaging has not been fully developed. Correlation of fine anatomic details with cross-sectional imaging of the spine has been a very useful approach that has greatly improved the quality of imaging interpretation. This chapter discusses and illustrates the spectrum of disk degeneration and musculoskeletal pathologic conditions that affect the spine.

TECHNICAL CONSIDERATIONS IN MAGNETIC RESONANCE IMAGING

Gross anatomically equivalent detail of the lumbar spine can be achieved with high-resolution MR imaging. Optical MR technical performance, however, varies considerably depending on the imaging sequences chosen, the anatomic region and pathologic process studied, and—very important—the cooperation of the patient to minimize motion.

T1-weighted, proton density, and T2-weighted spin-echo pulse sequences are commonly used in clinical MR imaging of the lumbar spine. The protocol of MR imaging of the lumbar spine at the State University of New York, Syracuse, is illustrated in Table 6–1. In physical diagnosis, a conclusion is reached by combining the results of observation, palpation, percussion, and auscultation. In MR imaging, differential sampling through T1-weighted, proton density, and T2-weighted spin-echo and often gradient-recalled echo (GRE) imaging is used to reach conclusions. The high signal intensity of fatty tissue is a major contributor to the success of T1-weighted images in evaluating the lumbar spine owing to the abundance of fat in this region, especially in the epidural space. On T1-weighted spin-echo images, the spinal cord and cerebrospinal fluid are distinguished readily by their contrast differences. The intervertebral disks have a relatively low, uniform signal intensity on T1-weighted images, and, the in-

Table 6–1. PROTOCOLS OF MAGNETIC RESONANCE IMAGING OF THE LUMBAR SPINE AT THE UNIVERSITY OF NEW YORK, HEALTH SCIENCE CENTER, SYRACUSE

Plane	TR (ms)	TE (ms)	FOV (cm)	Slice (mm)	Skip (mm)	NEX	Matrix (pixel)	Flip (degrees)	Time (min)
Sagittal (SE)	350–600	15	25~30	4	0.2	2	256 × 256		3~5
Sagittal (SE)	2500	20, 80	25~30	4	0.2	1	256 × 256		8~10
Sagittal (GRE)	350–450	12	30	4	0.2	3	256 × 512	25	4~5
Axial (SE)	500~650	15	25	4	0.2	2	256 × 256		3~5
Axial (GRE)	350~650	12	30	4	0.2	3	256 × 512	25	3~5

TR, Repetition time; TE, echo time; FOV, field of view; Skip, interslice gap; NEX, number of excitations; GRE, gradient-recalled echo; SE, spin echo.

ternal architecture of the intervertebral disks is not well appreciated on this pulse sequence. The use of intravenous gadopentetate dimeglumine has proved very valuable in differentiating recurrent or residual disk herniation from postoperative scarring in which the granulation tissue is enhanced.[7,70] In general, T2-weighted imaging reveals greater contrast differences among structures than T1-weighted imaging. The normal nucleus pulposus and the inner portion of the anulus fibrosus have a high signal intensity on T2-weighted images.[101] The usual decrease in signal intensity of the intervertebral disks in disk degeneration is particularly well shown on T2-weighted images. The spectrum of intervertebral disk disease is largely demonstrable on T2-weighted images. However, the acquisition time of the T2-weighted imaging technique is two to three times longer than that of the T1-weighted technique, which makes it susceptible to patient motion and also to greater noise, which may reduce its value in the evaluation of spinal disease. Proton density images, derived from an earlier (first) echo in generating T2-weighted images, provide valuable information concerning normal and pathologic spinal morphology (Fig. 6–1). Sagittal proton density images are always obtained with T2-weighted images in our MR examinations to evaluate spinal disease.

An important complement to long repetition time spin-echo imaging and even primary technique is GRE imaging with its relatively faster scanning pulse sequences. Gradient-recalled echo imaging often has minimal motion artifacts and increased contrast, whereas the signal-to-noise ratio is increased in comparison with spin-echo imaging.[29,94] When appropriate flip angles and shorter repetition times and echo times are selected, GRE provides good differentiation of the cerebrospinal fluid (bright) from the spinal cord.[94,95] Moreover, bone and adjacent soft tissues also can be distinguished readily, which they often are not on T2-weighted images. However, fat is of low signal intensity on most GRE sequences, unlike its appearance on T1-weighted spin-echo imaging, so that morphologic detail defined by fat is not so well demonstrated on GRE images as on spin-echo images.

Among the other factors that affect spinal imaging significantly are the matrix size, field of view, scanning planes selected, gaps or overlaps on angled slices, surface coils, gradient moment nulling motion compensation, pulse triggering and gatting,

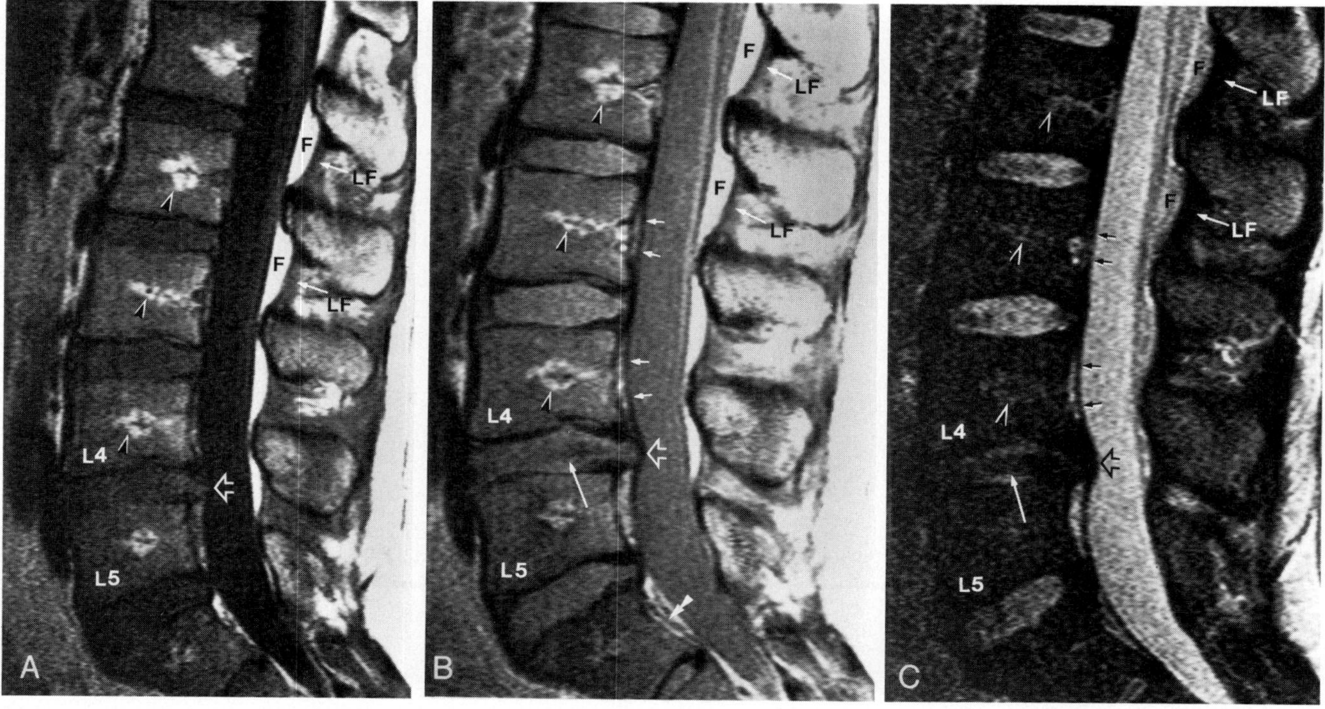

Figure 6–1. L4–5 disk herniation. Sagittal T1-weighted *(A)*, proton density *(B)*, and T2-weighted *(C)* images show the fat *(arrowheads)* in the vertebral venous channels of the vertebrae. They are of high signal intensity on T1-weighted and proton density images and of intermediate signal intensity on T2-weighted sequences. The combination of posterior longitudinal ligament and dura mater *(small arrows)* is shown in *B* and *C*. The posterior longitudinal ligament behind the ST vertebral body is well visualized in *B (stacked arrowheads)*. Disk herniation is seen at the L4–5 disk *(open arrow)*. Reduced signal intensity is noted with a dark band in the equator of the L4–5 disk *(long arrow)*. In *B*, the equator of the L4–5 disk is characterized by a stacked V appearance of low signal intensity *(long arrow)*, which is the inverted anterior inner anulus and fibrous replacement within the disk (see Fig. 6–9). F, Fat; LF, ligamentum flavum.

band width, phase-encoding axis, use or nonuse of laser camera photography, and postprocessing blurring when magnification is used.

NORMAL ANATOMY OF THE SPINE: CRYOMICROTOME AND MAGNETIC RESONANCE IMAGES

The segmental osseous structures of the lumbar spine defined by MR imaging include the vertebral bodies and their appendages, which include the pedicles, the articular pillars, the laminae, and the transverse and spinous processes. The contiguous osseous segments interrelate through the intervertebral disks, facet joints, and ligaments. The predominant ligaments in the lumbar spine are the anterior longitudinal ligament, the posterior longitudinal ligament, and the ligamentum flavum. The osseous structures, intervertebral disks, and ligaments form two canals: the spinal or vertebral canal, containing the thecal sac enclosed by the dura mater; and the intervertebral canals (or neural foramina), which contain the nerve root, dorsal root ganglion, fat, and vessels (Figs. 6–2 to 6–5).

The vertebral body is approximately rectangular when viewed in the sagittal and coronal sections (cryomicrotome or MR images) and ovoid with a flat or slightly concave posterior border when viewed in the axial section (see Figs. 6–1, 6–2, 6–5). The periphery of the vertebral body is composed of dense cortical bone, which has a low signal intensity on MR imaging regardless of the pulse sequence chosen. The central portion of the vertebral body is composed of trabecular bone and bone marrow. The signal intensity from the interior of the vertebral body depends largely on the bone marrow and its relative fat content and on the pulse sequence chosen. The fat content of the bone marrow varies with age; degeneration of adjacent intervertebral disks; therapy (e.g., radiotherapy produces atrophy); and increased hematopoiesis, with possible infarcts in sickle cell disease, agnogenic myeloid metaplasia, and similar conditions.[47] A canal containing the basivertebral vein is well seen on sagittal (see Fig. 6–2) and axial cryomicrotome sections. In the adult, along the peripheries of the vertebral bodies superiorly and inferiorly are the residua of the bony ring apophyses; cartilaginous end plates cover the more central portion of the superior and inferior surfaces of the vertebral body (see Fig. 6–3).

In general, the signal intensity of the marrow varies from moderately high on T1-weighted spin-echo images, to intermediate to low on T2-weighted spin-echo images, to low on GRE images. The channel of the basivertebral vein is usually of high signal intensity on T1-weighted spin-echo images and of intermediate signal intensity on T2-weighted spin-echo images owing to the fat surrounding the vein (see Fig. 6–1). The hyaline cartilage end plate usually is difficult to recognize on high field spin-echo MR images largely because of the overlap of chemical shift artifact.

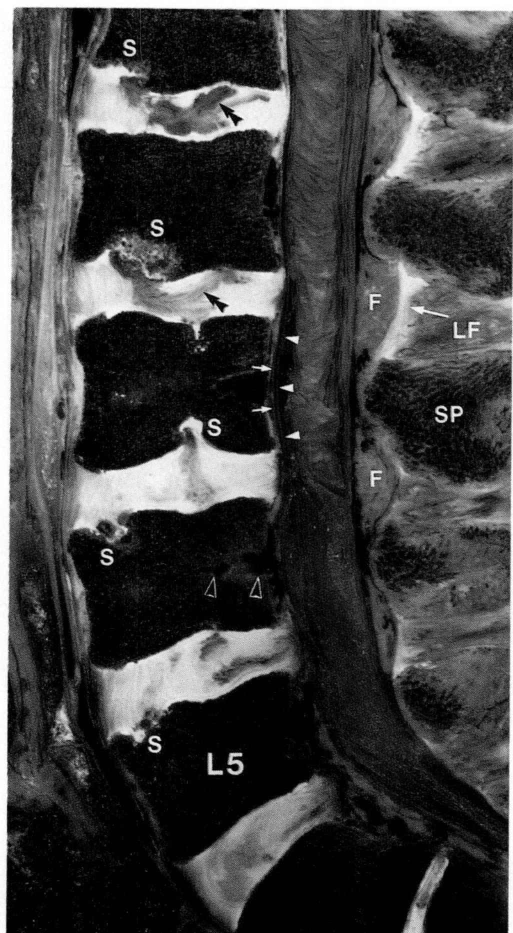

Figure 6–2. Sagittal cryomicrotome section from a 20-year-old man. Schmorl's nodes (S) are present at L1–5 vertebral bodies. Regions of pigmentation *(stacked arrowheads)* are noted within the nucleus pulposus. The posterior longitudinal ligament *(small arrows)*, dura mater *(arrowheads)*, epidural fat (F), ligamentum flavum (LF), spinous process (SP), and basivertebral vein channel *(open arrowheads)* are labeled.

The osseous spinal canal of the lumbar region is formed by the vertebral bodies, pedicles, articular pillars, and laminae. The spinal canal contains ligaments, fat, and a large venous plexus in the epidural space; it also contains the dural envelope enclosing cerebrospinal fluid in the subarachnoid space, the lower spinal cord, and the conus medullaris and cauda equina.

The neural or intervertebral canal is bounded anteriorly by the adjoining vertebral bodies and their interspaced intervertebral disk, superiorly and infe-

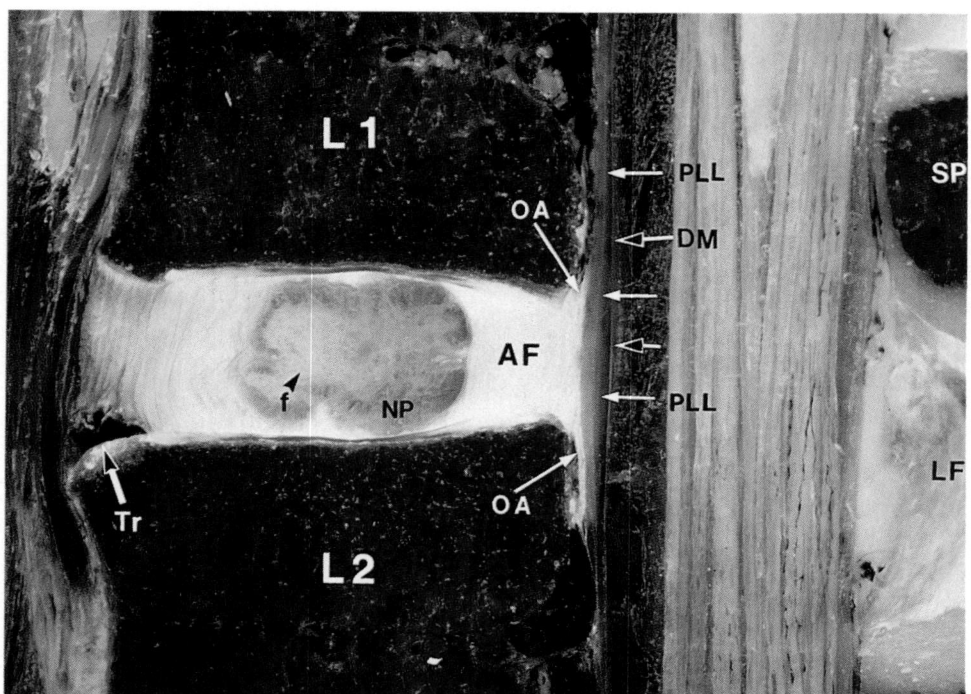

Figure 6–3. Sagittal cryomicrotome section of a largely normal intervertebral disk in a 52-year-old woman. The nucleus pulposus (NP) is oval, large, and compact in appearance, with some internal fibrous replacement (f). The woman has suffered trauma, as manifested by a transverse tear (Tr). The attachments of the outermost portion of the anulus (OA) to the posterior aspect of the adjacent L1 and L2 vertebral bodies can be seen. The relationship of the thin posterior longitudinal ligament (PLL) and dura mater (DM) to the OA is evident. AF, Middle portion of the anulus fibrosus; SP, spinous process; LF, ligamentum flavum.

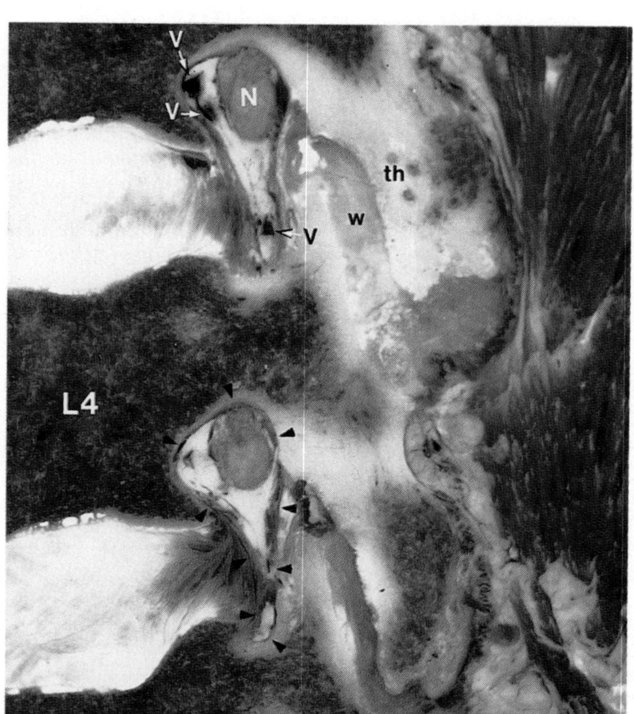

Figure 6–4. Paramedian sagittal cryomicrotome section of normal neural foramen (L3–4 and L4–5). Arrowheads depict the pearlike shape of the neural foramen. The nerve root (N) is located in the upper portion within the neural foramen. Slightly degenerative changes of the facet joint (thickening of the cortical bone [th] and widening of the facet joint space [w]) are also visible. V, Vessels.

riorly by the pedicles of the two adjoining vertebrae, and posteriorly by the facet joints (see Fig. 6–4). The intervertebral canals have an orientation that is nearly directly lateral. They contain the ipsilateral nerve root, dorsal root ganglion, vascular structures, fat, and ligamentum flavum, which covers the facet joint.

Magnetic resonance imaging can demonstrate the osseous structures that form the spinal and intervertebral canals and the contents within these canals. As with the signal intensity of the vertebral body, that of the osseous appendages of the vertebral body varies depending on the fat content, the thickness of the compact cortical bone, the cancellous trabecular structure and marrow, and the pulse sequence chosen. Conversely, the ligamentum flavum has an intermediate to lower signal intensity, in contrast to the higher signal intensity of fat anterior to it, and this is the case for all spin-echo pulse sequences (see Fig. 6–1). The ligamentum flavum is better shown on axial images. A combination of the dura mater and the posterior longitudinal ligament is well seen behind the vertebral bodies as a band of low signal intensity (see Fig. 6–1) in which these structures contrast with the higher signal intensity of the fat situated anterior to them in the concavities of the posterior vertebral bodies. Occasionally, only in

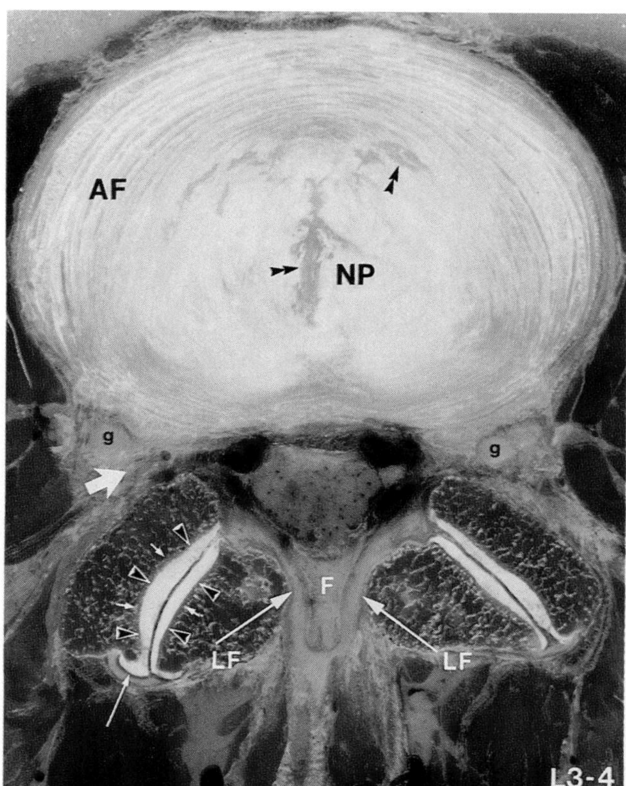

Figure 6–5. Axial cryomicrotome section through the L3–4 disk showing a normal intervertebral disk and facet joints. The nucleus pulposus (NP) is located slightly posteriorly. Small regions of pigmentation *(stacked arrowheads)* are noted within the NP. The anulus fibrosus (AF) is thicker anteriorly and laterally than posteriorly. The intervertebral canal *(large arrow)* is capacious. Facet cartilage *(arrowheads)* is relatively uniform in thickness and extends beyond the opposite surfaces of the articular processes, covering the posterior median surfaces of articular processes *(long thin arrows)*. A uniform thin bony cortex *(small arrows)* of articular processes can be seen underneath the facet cartilage. The ligamentum flavum (LF) is relatively thin and uniform. F, Epidural fat; g, dorsal nerve root ganglion.

the regions of the lower lumbar spine and sacrum can the posterior longitudinal ligament be differentiated from the dura mater because sufficient fat is distributed on both the anterior and posterior aspects of this ligament (see Fig. 6–1).[103] The intervertebral canal portion of the ligamentum flavum is well seen on MR images as a band of lower signal intensity covering the facet joints.

INTERVERTEBRAL DISK DISEASES

Disease of the intervertebral disk is the most common cause of low back pain,[44] and in most medical centers it is the most common clinical indication for MR imaging. The chain of events of degeneration of the intervertebral disk is manifested on gross morphologic specimens by fissures and tears, thinning, herniation, malalignment, and spondylosis; physicochemically it is characterized by a reduction in the amount of proteoglycan, a transformation of the collagen, and a substitution of keratan sulfate for chondroitin sulfate. Evidence of degenerative disk disease was found at autopsy in 85 to 95 per cent of adults by the age of 50 years.[63] Disk herniation is more frequent in middle-aged adults and is infrequent in children and adolescents; in older patients the disk becomes even less hydrated than it is in middle age. A strong understanding of the ways the process of aging leads to degenerative changes within the intervertebral disk is extremely useful in interpreting radiologic images for referring physicians who treat patients with low back pain. This important responsibility may be complicated by the inability to distinguish clearly between usual and unusual aging and degeneration of the respective intervertebral disks.

Normal Changes in the Intervertebral Disk as a Result of Aging

Intervertebral disks are composed of highly specialized connective tissue that forms two major compartments: the central gelatinous nucleus pulposus and the peripheral anulus fibrosus.[11,12,31,60,72] In the neonate, the nucleus pulposus is a highly gelatinous translucent, relatively large, ovoid, and amorphous structure that is readily distinguishable from the anulus fibrosus; in contrast, the anulus fibrosus consists of dense fibers organized as concentric lamellae not unlike tree rings in appearance. In the second decade of life, the disk is characterized by a substantial replacement in the outer portion of the nucleus pulposus by solid tissue.[31,72,101] The anulus fibrous also becomes denser at this age than it is in the neonate. The nucleus pulposus begins to lose its translucency as fibrous replacement continues; moreover, a distinctive fibrous band can be identified in the ventral or dorsal aspect of the nucleus. This type of disk is called the *transitional disk*,[38,60,101] but its appearance, characteristic of adolescents, can still be found in adults (Fig. 6–6).

In the adult, the nucleus pulposus of the disk is composed of amorphous fibrocartilage represented by the collagen fibril enmeshed in a mucoprotein gel (see Fig. 6–3). The anulus is now composed of denser lamellated fibrous and fibrocartilaginous tissue. Collagen as the major structural protein of the body accounts for one half the dry weight of the

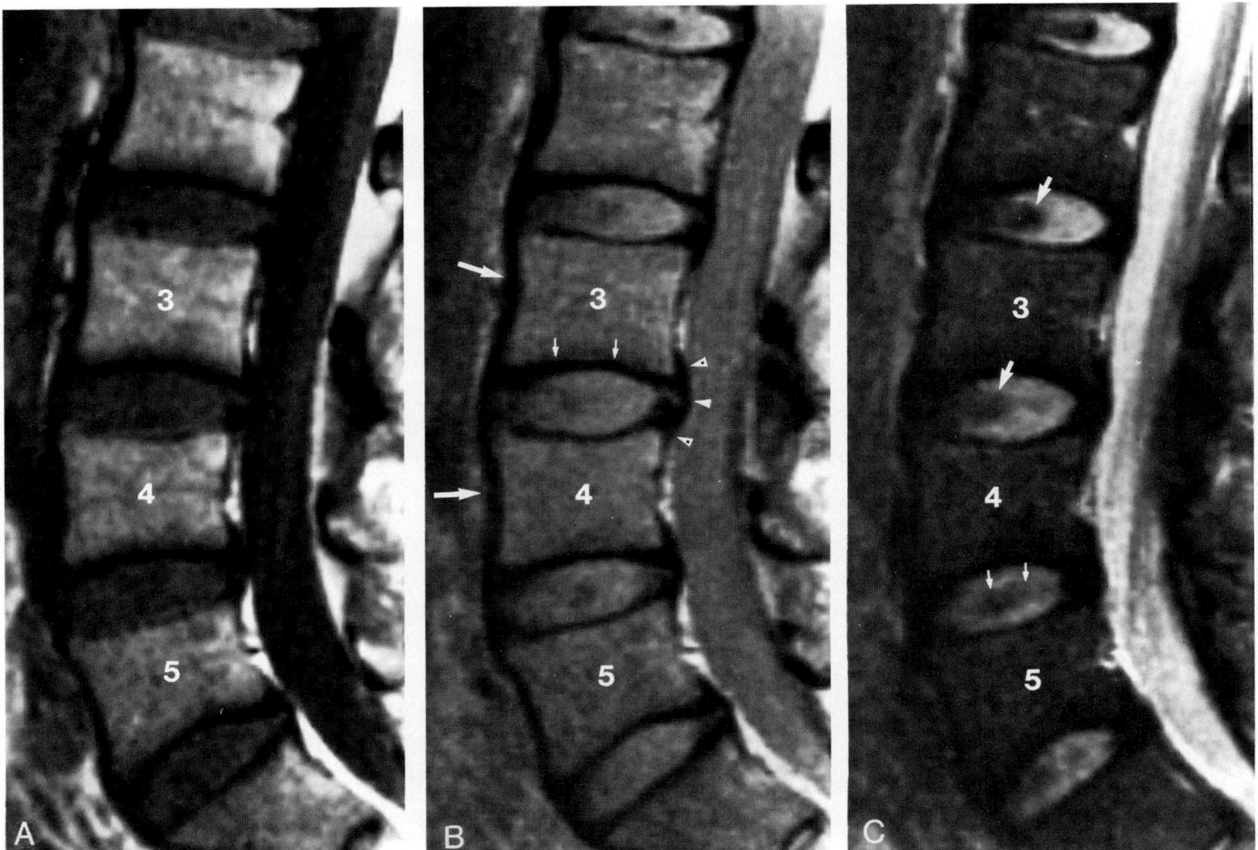

Figure 6–6. Normal intervertebral disk of a youth or young adult (patient's age 29 years). *A*, Sagittal T1-weighted image. The intervertebral disks are well visualized with respect to height and overall intensity, but fine detail of the disks is not visible. *B*, Sagittal proton density image. The bony cortex adjacent to the cartilaginous end plates *(small arrows)*, the outermost portion of the anulus *(arrowhead)* with its posterior upper and lower vertebral body attachment *(dotted open arrowheads)*, and the anterior longitudinal ligament (and its continued intensity with cortical bone) *(larger arrows)* are very well seen. Between the outermost portions of the anulus, fat fills the concavity behind the vertebral bodies anterior to the thecal sac. The internal disk architecture becomes more visible. *C*, Sagittal T2-weighted image. The fibrous components within the hydrated nucleus pulposus become evident. Like the transitional disk [101] usually seen in the early second decade of life, the fibrous replacement *(arrows)* of the nucleus pulposus is very prominent in the nucleus pulposus anteriorly. At L4–5, the fibrous replacement *(small arrows)* of the disk is broader anteroposteriorly, and to a lesser extent, it can be seen at L5–S1. This is the normal adult appearance, and this configuration can be resolved well before the onset of degenerative changes. The appearance of the fibrous replacement at L4–5 and at L5–S1 precedes the degenerative changes, which usually begin at L4–5 and L5–S1. Note also that the "hydration" of the nucleus pulposus is quite broad, larger than the normal nucleus compartment of discography. Therefore, the hydration involves at least the inner and probably the middle anulus here also.

anulus and 20 to 30 per cent of the dry weight of the nucleus.[91] Both type I and type II collagens are found in the disk. The fine fibers within the nucleus are virtually all type II collagen, whereas the outer anulus is predominantly type I collagen. The proportion of type I collagen in the anulus decreases toward the nucleus as the proportion of type II increases. Both types of collagen have a mechanical role. Although their mechanical properties are not yet understood, type II collagen apparently has fibrils that are much finer than those of type I. In keeping with this observation, the fibrils of the nucleus are finer than those of the anulus.[18] With increasing age, hydration of the disk decreases.[11,12,15,31,60]

On axial cryomicrotome sections, the posterior margins of the L1–4 disks are slightly concave or flat (see Fig. 6–5); the posterior margin of the L5 disk is flat or slightly convex.[103] On sagittal cryomicrotome sections the disks have a biconvex appearance, especially at L5–S1 (see Fig. 6–2). The nucleus pulposus may be stained by a lipofuscin pigment, giving it a dark brown color (see Figs. 6–2, 6–3). The pigmentation denoting degeneration, however, is not necessarily confined by the boundaries of the nucleus pulposus; furthermore, the demarcation between the nucleus and anulus becomes less distinct with age.

The anulus can be divided into four portions—the inner, middle, outer, and outermost anulus—instead of the usual three. Proceeding peripherally, the lamellae of the anulus become more densely compacted. The middle and outer portions of the anulus have highly organized lamellae, and the outermost portion contains very dense fibers, which appear darker in color than the more internal layers. These outermost anular fibers insert into the ring apophyses of the adjacent vertebrae and along the inferior and superior cortices of the vertebral bodies (see Fig. 6–3). Transverse tears in the outer anulus near the ring apophyses and concentric tears between the lamellae are common findings in normal adult disks (see Fig. 6–3).[104]

On spin-echo MR images, the internal architecture of intervertebral disks is better shown on proton density and T2-weighted sequences (see Figs. 6–1, 6–6). The nucleus pulposus and the inner anulus have a relatively high signal intensity and are surrounded by a low signal intensity of the middle and outer anulus on proton density images; their signal intensity is very high on T2-weighted images.[101] The outermost anulus has the lowest signal intensity (see Fig. 6–6). In the nucleus in adolescence, the developing fibrous tissue can be identified as a region of low signal intensity in the anterior or posterior aspects of the nucleus pulposus.[101] As described previously, a similar MR imaging appearance of the disk can also be found in adults (see Fig. 6–6). In adults, a transversely oriented band of low signal intensity in the midportion of the disk represents the fibrous plate that is demonstrable on cryomicrotome sections (see Figs. 6–1, 6–6).[102] The outer and middle portions of the anulus are thicker anteriorly than posteriorly. The outermost portion of the anulus appears as a curvilinear band of very low signal intensity in the periphery of the disk, bordering the adjacent inferior and superior vertebral bodies (see Fig. 6–6). A small area of high signal intensity can be seen in the outer anulus near the ring apophyses of the vertebrae,[104] which indicates the transverse tear shown on cryomicrotome sections (see Fig. 6–3).

Disk Degeneration

Disk degeneration refers to the biochemical and structural changes that occur within the intervertebral disks, predominantly dehydration of the nucleus and the anulus as the collagen fibrous content of the disk increases.[11] Disk degeneration is attributable to the repeated trauma to the disk and end plate disruption that results in nutritional disturbance of the disk.[12] As the intervertebral disk degenerates, the nucleus pulposus dehydrates and fissures occur within it. Fibrous tissue gradually replaces the normal fibrocartilaginous tissue. The inner anulus changes from convex to concave toward the center of the disk (Figs. 6–7, 6–9). Tears in the anulus fibrosus take three forms: concentric, transverse, and radial.[104] Concentric and transverse tears are not thought to be significant. A radial tear is a fissure that extends from the central portion to the periphery of the disk, forming a channel from the nucleus through the anulus (Figs. 6–7 to 6–9). It is an important sign of early disk degeneration.[104] The outermost anulus remains intact (in the absence of herniation), but it thins and bulges (see Fig. 6–8). Loss of disk height and anular bulging are common findings in this phase of degeneration. Magnetic resonance imaging shows that in the early stage of disk degeneration, disks have lower signal intensity owing to the loss of water content and the replacement by collagen fi-

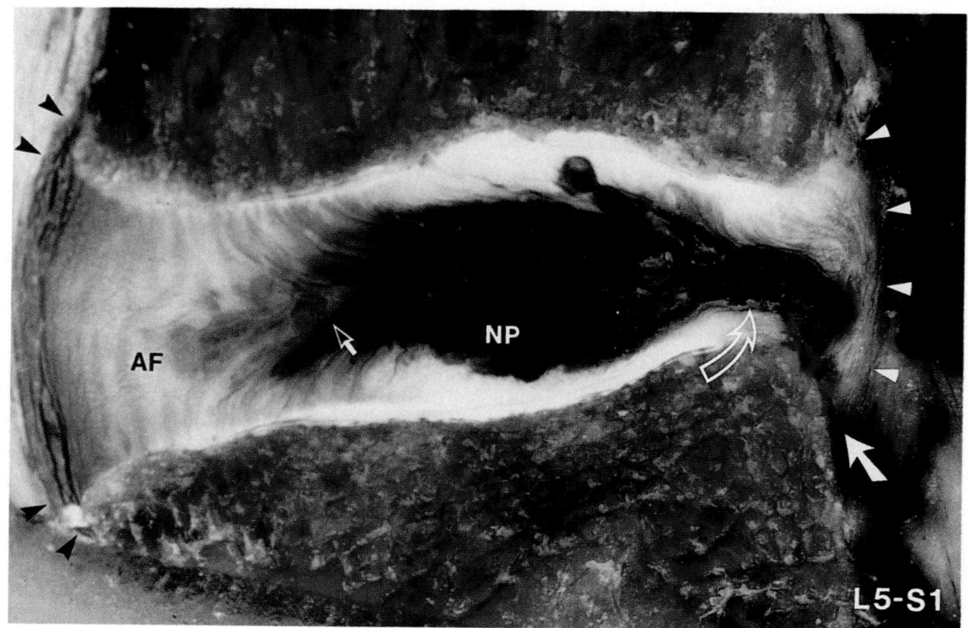

Figure 6–7. Sagittal cryomicrotome section of the L5–S1 disk illustrating a defect of the outermost portion of the anulus fibrosus (AF) in disk herniation. A severe degree of pigmentation in the nucleus pulposus (NP) is seen. The anterior inner anulus *(small open arrow)* has invaginated toward the center of the disk. The posterior inner, middle, and outer anulus are totally disrupted *(curved open arrow)*. A defect *(large arrow)* inferiorly at the site of attachment of the outermost portion of the anulus *(white arrowheads)* to the posterior bone cortex of S1 is shown. The attachments of the anterior outermost portion of the anulus to the anteroinferior and anterosuperior surfaces of the vertebral bodies *(black arrowheads)* also are demonstrated.

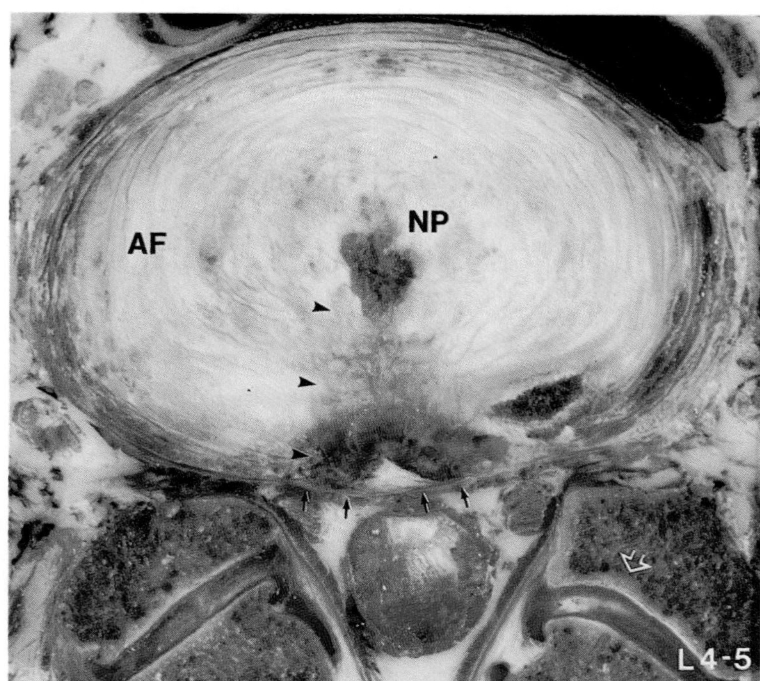

Figure 6–8. Axial cryomicrotome section through the L4–5 disk illustrating central disk protrusion. Disruption *(arrowheads)* of the inner, middle, and outer anulus is seen. The outermost portion of the anulus *(small arrows)* remains intact but bulges. Thickening of the cortical bone of the facet *(open arrow)* is noted. The nucleus pulposus (NP) and anulus fibrosus (AF) are shown.

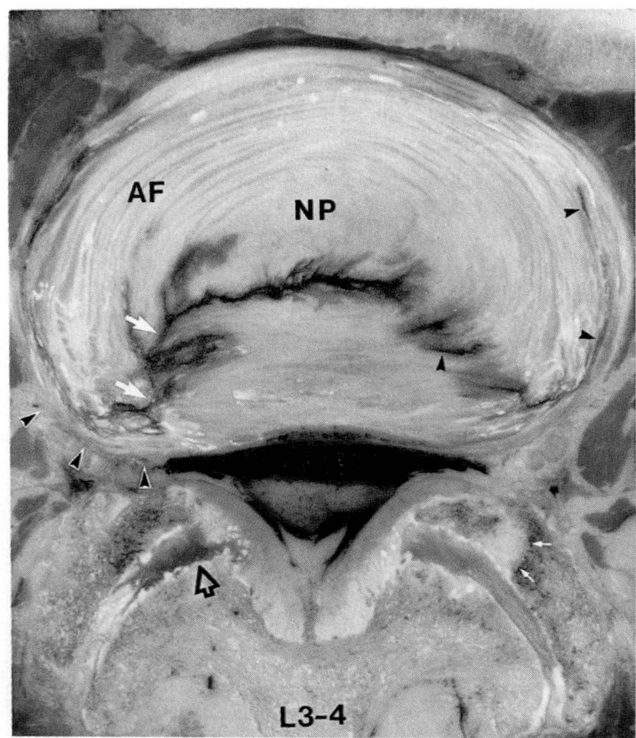

Figure 6–9. Axial cryomicrotome section of L3–4 disk injected with fast green FCF. There is a radial tear *(arrows)* in the right posterolateral aspect of the anulus fibrosis (AF). The outermost portion of the anulus at this location is still intact but has expanded *(arrowheads)*. The circumferential tears in the anulus *(black arrowheads)* also are noted. Fat replacement *(small arrows)* in the articular facet and loss of facet cartilage *(open arrow)* are shown. NP, Nucleus pulposus.

bers.[101] Radial tears also can be visualized on MR images, especially on proton or T2-weighted images (Fig. 6–10)[101] and after gadopentetate dimeglumine enhancement.[70]

Late disk degeneration is characterized by a severe loss of disk height as the disk space is almost replaced by fibrous tissue, gas, and fluid (see Fig. 6–27). Magnetic resonance imaging shows that the severely degenerated disks usually have low signal intensity owing to the replacement of fibrous tissue and occasionally high signal intensity on T2-weighted images due to the presence of fluid within the disk. Changes in the juxtaposed vertebrae include fibrosis or fat infiltration, irregularity of the end plates, fibrous replacement, and osteophytosis (Fig. 6–11). Changes in the marrow of the adjoining vertebral bodies between the degenerated disk take three forms[51]: Type I changes are characterized by the disruption and fissuring of the end plates and the presence of vascularized fibrous tissue within the adjacent marrow. These changes demonstrate a decrease in signal intensity on T1-weighted images and an increase in signal intensity on T2-weighted images (Fig. 6–12). Type II changes consist predominantly of fat replacement of the bone marrow, which results in increased signal intensity on T1-weighted sequences and uniform or slightly higher signal intensity on T2-weighted images (Fig. 6–13). Type III changes represent extensive bone sclerosis within

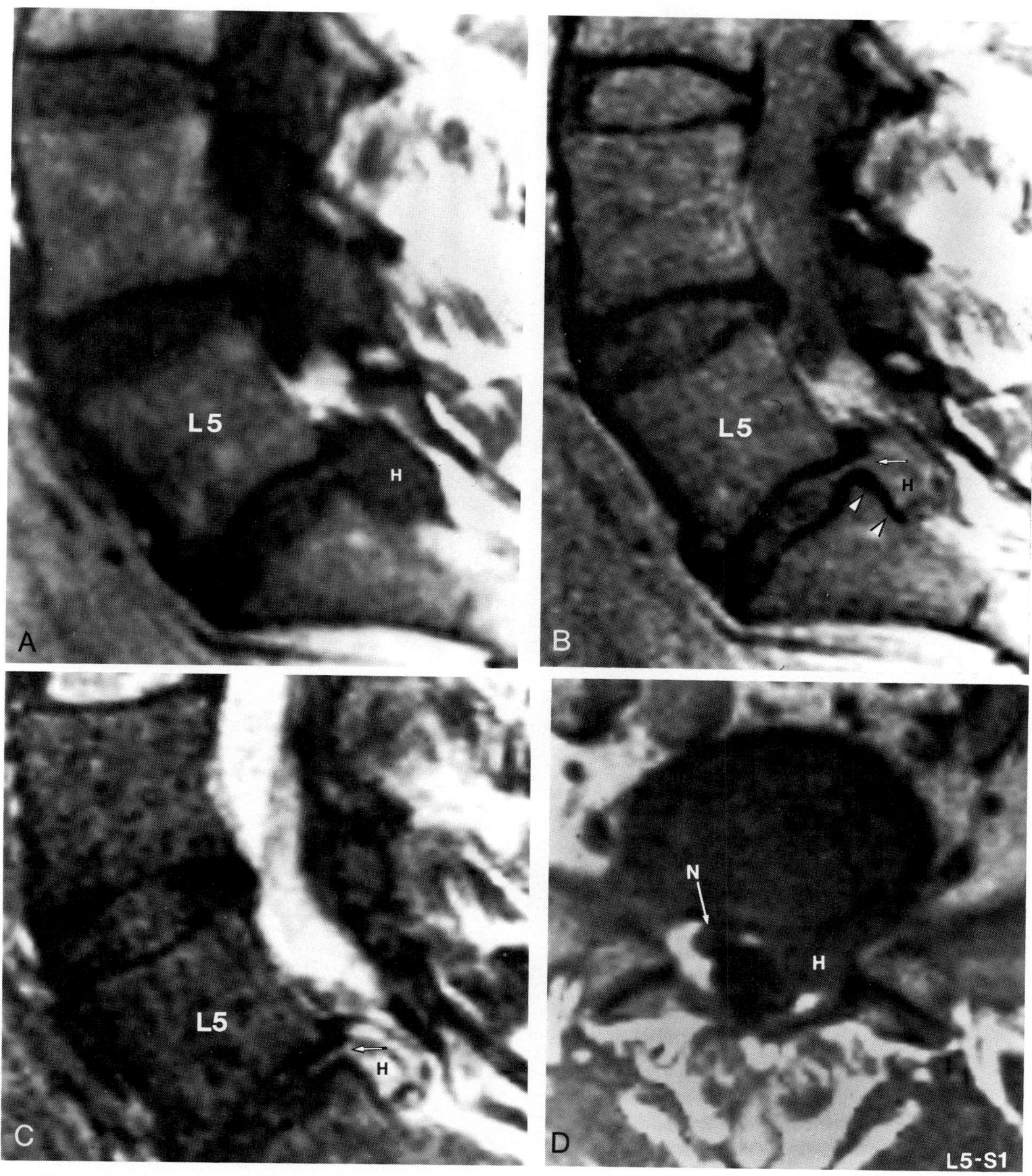

Figure 6–10. Disrupted outermost portion of the anulus folded back by disk extrusion or sequestration. *A*, Sagittal T1-weighted image. The L5–S1 herniation (H) is well shown. In this acute herniation, the extruding mass is of higher signal intensity relative to the parent disk. *B*, Sagittal proton density image. Disruption *(arrow)* of the outermost portion of the anulus of L5–S1 can be seen easily. The vertebral attachments of this outermost portion are well shown, especially inferiorly *(arrowheads)*. The herniated disk (H) is quite large. The disrupted outermost portion of the anulus with herniated disk together form an "open fish mouth" sign. *C*, Sagittal T2-weighted image. The degenerated, thinned L5–S1 disk of low signal intensity, whereas the extruded or sequestered portion (H) is quite bright. The arrow indicates the disruption of the outermost portion of the anulus. L4–5 shows early degeneration characterized by loss of signal intensity. *D*, Axial T1-weighted image. The L5–S1 herniated mass (H) abuts on the thecal sac from the 1 o'clock to the 3 o'clock positions, affects the left S1 nerve root, and extends into the lateral intervertebral canal. N, Contralateral S1 nerve root.

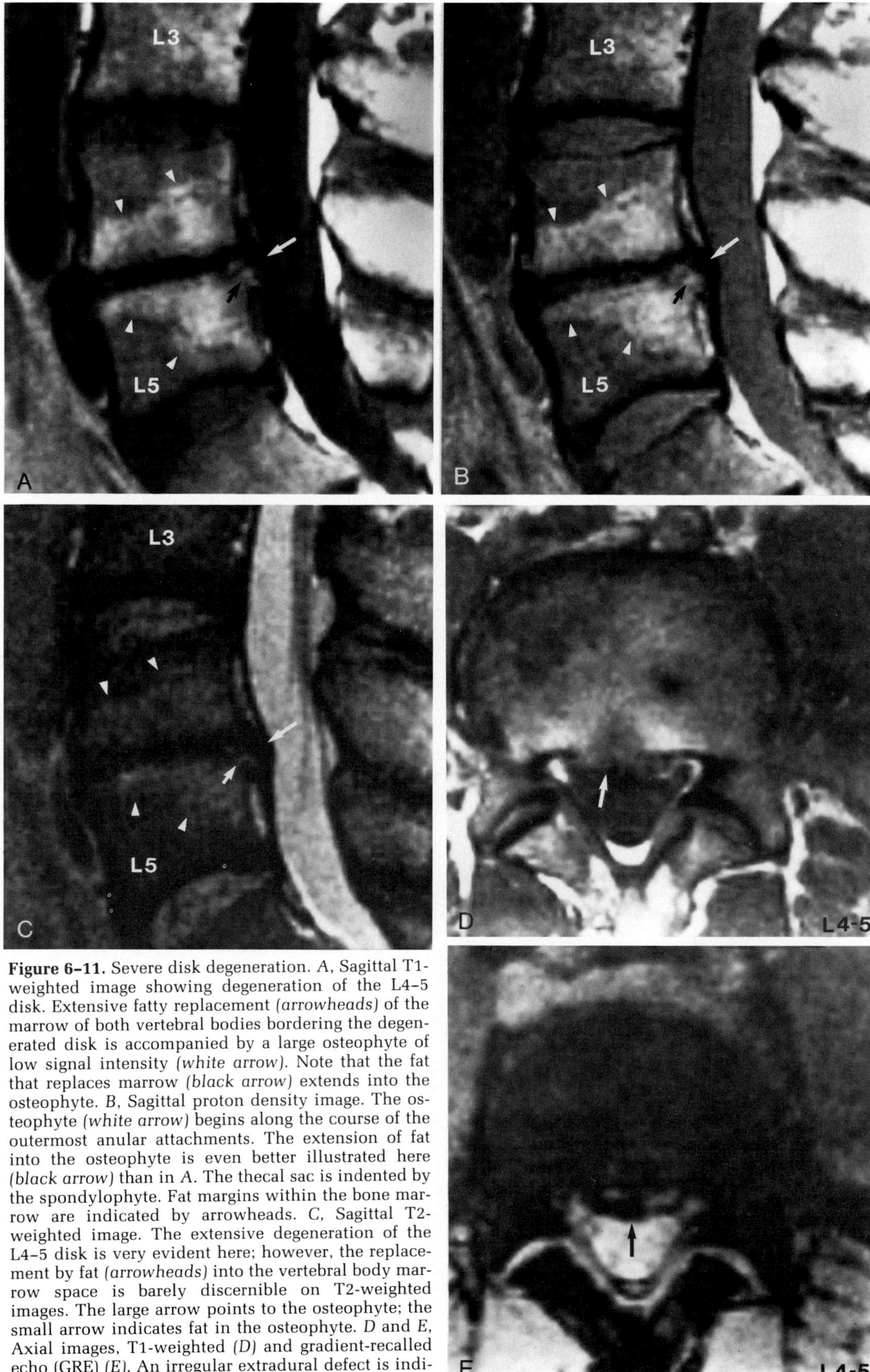

Figure 6–11. Severe disk degeneration. *A*, Sagittal T1-weighted image showing degeneration of the L4–5 disk. Extensive fatty replacement *(arrowheads)* of the marrow of both vertebral bodies bordering the degenerated disk is accompanied by a large osteophyte of low signal intensity *(white arrow)*. Note that the fat that replaces marrow *(black arrow)* extends into the osteophyte. *B*, Sagittal proton density image. The osteophyte *(white arrow)* begins along the course of the outermost anular attachments. The extension of fat into the osteophyte is even better illustrated here *(black arrow)* than in *A*. The thecal sac is indented by the spondylophyte. Fat margins within the bone marrow are indicated by arrowheads. *C*, Sagittal T2-weighted image. The extensive degeneration of the L4–5 disk is very evident here; however, the replacement by fat *(arrowheads)* into the vertebral body marrow space is barely discernible on T2-weighted images. The large arrow points to the osteophyte; the small arrow indicates fat in the osteophyte. *D* and *E*, Axial images, T1-weighted *(D)* and gradient-recalled echo (GRE) *(E)*. An irregular extradural defect is indicated by an arrow in *D*. Its dense bony nature (calcified) is corroborated by the GRE image (arrow in *E*).

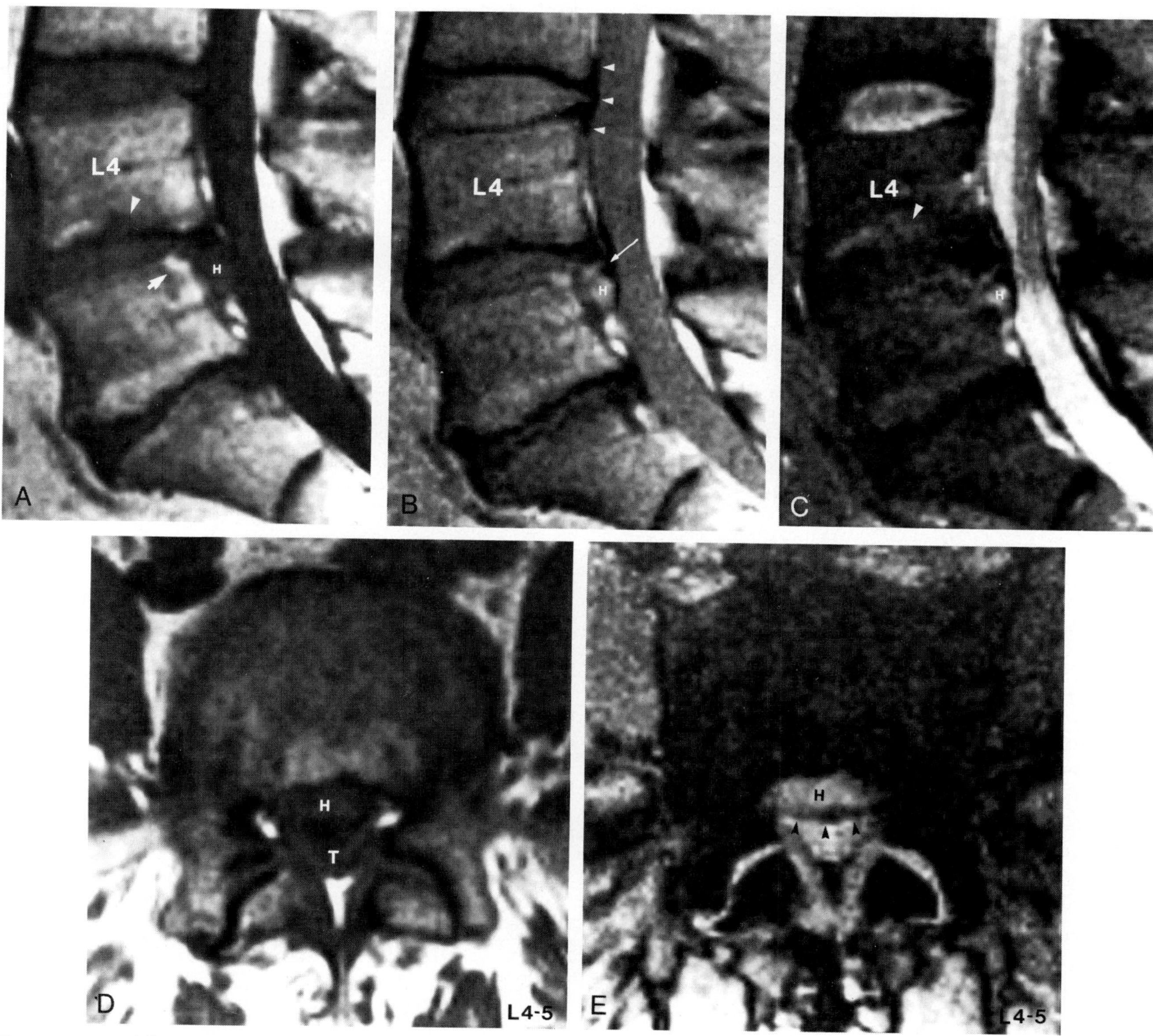

Figure 6–12. Disk herniation causing dissection and avulsion of the outermost portion of the anulus extending inferiorly. *A*, Sagittal T1-weighted image. The herniated disk (H) extends as a mass that is thicker than the parent intervertebral disk and is situated behind the upper portion of the L5 vertebral body. It fills the epidural space sufficiently so that it indents the thecal sac anteriorly. The bright signal of fat replacement *(arrow)* is seen in the adjacent bone marrow of L5, where the anular avulsion has occurred. Vascularized fibrous tissue within L4 *(arrowhead)*, which has low signal intensity in this sequence, is seen (type I). *B*, Sagittal proton density image. The brightness of the herniated fragment has increased in this sequence. Note that the outermost anular attachment is lifted off the vertebral body *(arrow)* when the herniation (H) is juxtaposed to the vertebral body. The normal appearance of the outermost portion of the anulus is seen at L3–4 *(arrowheads)*. *C*, Sagittal T2-weighted image. Degeneration of the L4–5 and L5–S1 disks is quite evident (loss of signal intensity and thinning). The L4–L5 herniation (H) is quite bright. The vascularized fibrous tissue in the marrow of L4 *(arrowhead)* is of high signal intensity. *D*, Axial T1-weighted image. The herniated fragment (H) is barely separable from the thecal sac (T). *E*, Axial gradient-recalled echo image. A curvilinear band *(arrowheads)* of low signal intensity separates the herniated disk (H) from the thecal sac at this level. The anteroposterior dimension of the herniated disk varies with the axial slice level and slice angle so that as the herniation thins inferiorly, the curvilinear band of low signal intensity at a more inferior location becomes less separable from the vertebral body. Here the outermost portion of the anulus appears intact axially but is not intact sagittally (dissected off the vertebral body).

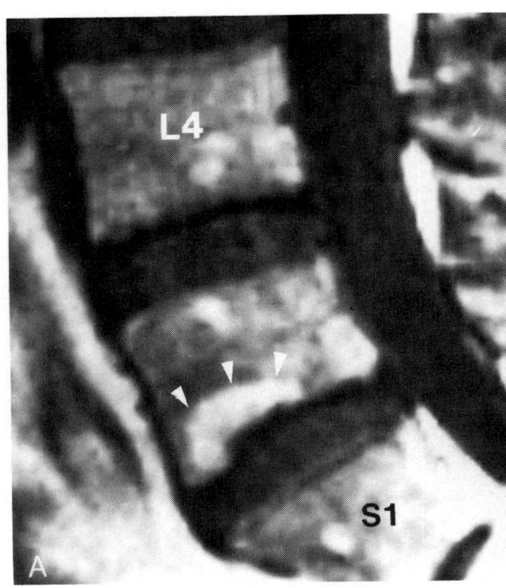

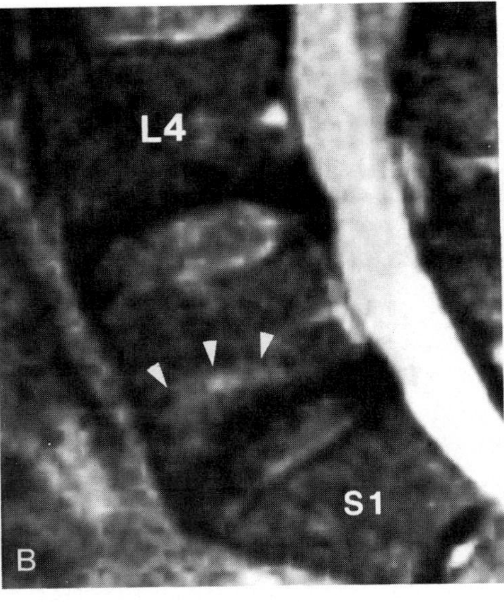

Figure 6–13. Replacement of bone marrow of the vertebral body by fat secondary to disk degeneration. *A* and *B*, Sagittal T1-weighted (*A*) and T2-weighted (*B*) images. High signal intensity (*arrowheads*) is noted in the anteroinferior portion of L5 in *A*. Signal intensity appears slightly low in *B*, indicative of type II marrow change.[51] Diminished disk height and loss of signal intensity at the L5–S1 disk, which indicate disk degeneration, are seen.

the adjacent bone marrow; thus, consistently low signal intensity is seen on both T1-weighted and T2-weighted images.

Disk Herniation

Disk material may herniate in several directions: into the vertebral body (intravertebral) or posterior, lateral, or anterior to the parent disk. Intravertebral herniation occurs when nuclear material disrupts or remodels the end plates and may extend into the bone marrow of the vertebral body (see Fig. 6–2). Such herniations are typically referred to as Schmorl's nodes.[72] This type of herniation is presumed to have little clinical significance, and it has been observed as early as the second decade of life. The type of herniation that is highly important clinically is intervertebral herniation, that is, when material from the nucleus pulposus (and sometimes from the anulus) gains access to the subligamentous (preligamentous) and epidural spaces through tears in the anular lamellae and has the potential to displace and compress nerve roots. Depending on its precise site of expansion, herniated disk material may compress not only the nerve root but the dorsal root ganglion, conus medullaris, or spinal cord and thus cause neurologic manifestations. The process of herniation originates with the radial tear in the disk, which allows the egress of the more central contents peripherally.[19,79]

The terms used by surgeons, radiologists, and other health care professionals to describe herniated disks often are confusing.[86] Bulged, protruded, prolapsed, extruded, sequestered, and slipped disks, as well as free fragment, herniated nucleus pulposus, and "disk" are among the commonly used terms or synonyms of disk ectopia.[44] The following definitions represent a consensus of current thought and usage.

Bulging of the anulus implies a *nonfocal*, diffuse, circumferential disk expansion sufficiently beyond the contours of the vertebral bodies secondary to loss in disk height.[15,30,44,49] A *protruded* disk represents a herniation in which nuclear substance has dissected through the inner, middle, and outer anular fibers[86] and has caused thinning of the outermost anular fibers, resulting in focal expansion (usually involving the posterior aspect of the disk) without a breach in the outermost portion of the anulus. An *extruded* disk indicates that the nucleus pulposus (and perhaps fragments of the disrupted anulus) extends through all layers of the anulus,[30,49,86] including the outermost anulus. Extrusion usually appears as an irregularly marginated soft tissue mass that focally deforms the contour of the parent disk and that may impinge on or distort adjacent regional structures (nerves, fat, and thecal sac). *Sequestration* or *free fragment* refers to the stage beyond an extrusion, in which the fragment has become disconnected from the parent disk and may migrate into the subligamentous (between the outermost anulus and posterior longitudinal ligament) or retroligamentous (posterior to the posterior longitudinal ligament extradurally) compartments. Rarely, the fragment may be intradural. Macnab and McCulloch refer to protrusion as a *contained* herniation, as opposed to a *noncontained* herniation: extrusion and sequestration.[44] Schellinger and coworkers[71] define containment to be in relationship to the posterior longitudinal ligament, whereas we believe containment to be by the outermost anulus (see Fig. 6–7).

Correlating multisectional anatomy and high-resolution MR images of the intervertebral disks makes it possible to understand these distinctions more readily and precisely and affords more accurate communication among radiologists, referring physicians, and other health care professionals. On cryomicrotome sections the outermost portion of the anulus appears as a curvilinear dense fibrous band attaching to and beyond the superior and inferior margins of the vertebral bodies (see Figs. 6–3, 6–9). This fibrous band, called the *outermost anulus*, forms the final barrier in the containment of disk material. It prevents herniation of the nucleus through channels within the inner, middle, and outer portions of the anulus fibrosus. Thus, disk protrusion, the mildest form of herniation, represents contained disk herniations (i.e., contained by the thinned but focally expanded outermost anulus) (see Figs. 6–8, 6–9). Complete breakdown of the outermost anulus is indicative of the occurrence of a noncontained disk herniation.

The vast majority of lumbar disk herniations occur in the midline or posterolaterally and at the L4–5 and L5–S1 disks.[49] Although intraforaminal or far lateral disk herniation occurs, initially it was only by diskography and later by cross-sectional or multisectional imaging—first CT and now MR—that lateral and far lateral herniations could be detected preoperatively (Fig. 6–14). Upper lumbar disk herniations and anterior disk herniations, although less common than lower lumbar posterior disk herniations, also occur (Figs. 6–15 to 6–17).

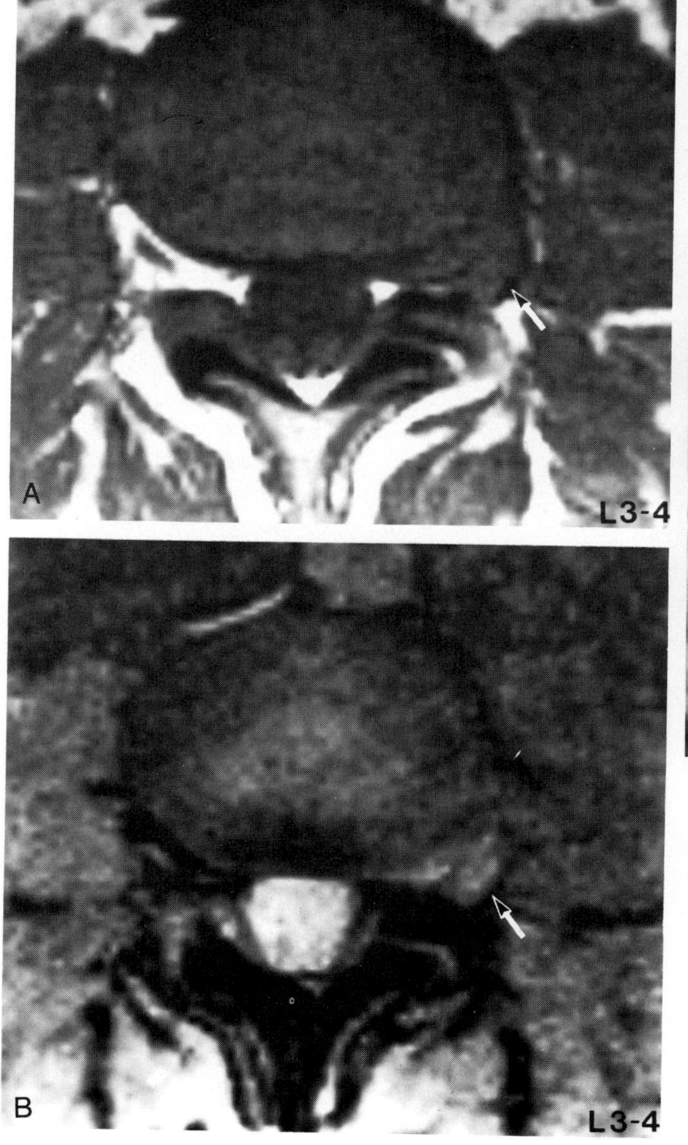

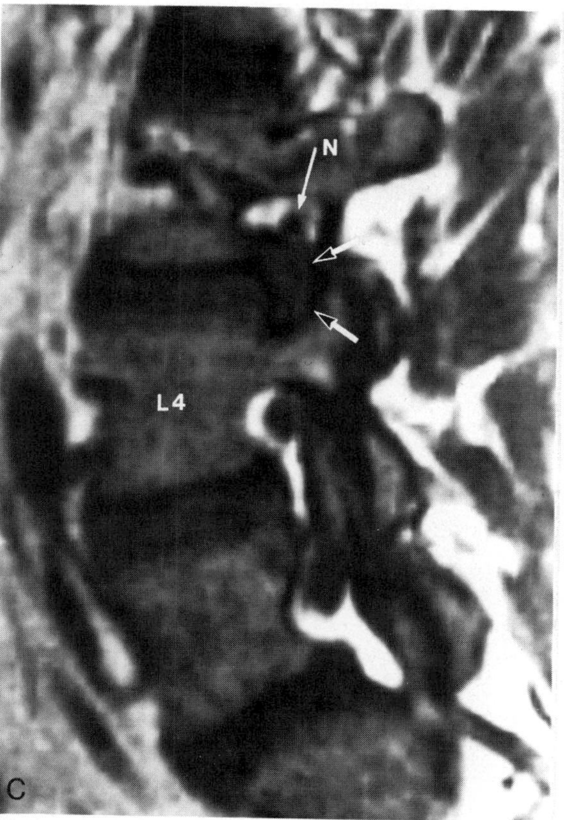

Figure 6–14. Far lateral disk herniation. *A*, Axial T1-weighted image. The lateralmost portion of the left intervertebral canal is reduced by a mass *(open arrow)* adjacent to the intervertebral disk and of the same signal intensity. This reduction is along the characteristic course of the nerve as it leaves and enters the intervertebral canal. *B*, Axial gradient-recalled echo image. The herniated disk *(open arrow)* is brighter than the parent disk, a difference that is usually ascribed to the high water content of the herniated part. *C*, Paramedian sagittal T1-weighted image, foraminal plane. The inferior two thirds of the L3–4 intervertebral canal is occupied by the herniated disk *(open arrows)* (fat not visualized). The nerve (N) within the canal is abutted but not displaced upward.

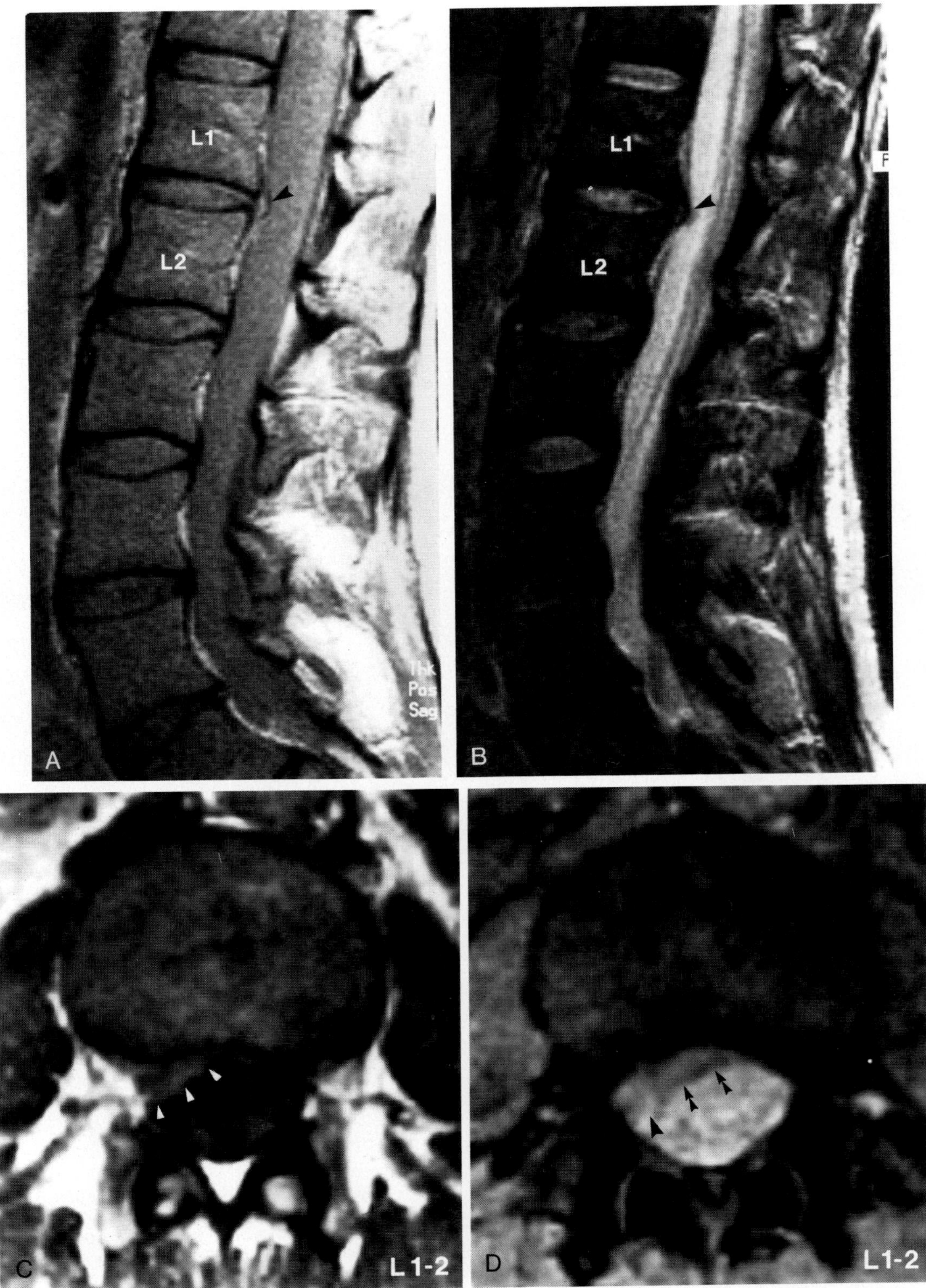

Figure 6–15. Upper lumbar disk herniation (L1–2). *A*, Sagittal proton density image. Note the carat or sideward V-shaped shadow of low signal intensity of the stretched outermost portion of the anulus (and potentially the posterior longitudinal ligament) at L1–2 *(arrowhead)*. *B*, Sagittal T2-weighted image. The lifting of the outermost portion of the anulus *(arrowhead)* by the herniation is confirmed. *C*, Axial T1-weighted image. The herniated disk *(arrowheads)* is located posterolaterally. *D*, Axial gradient-recalled echo image. The herniated portion is now better defined, not only by its higher signal intensity of fibrocartilage and nucleus but also by the low signal intensity rim of outwardly bowing outermost portion of the anulus *(stacked arrowheads)* and the anular disruption manifested by the lack of low signal intensity *(single arrowhead)*.

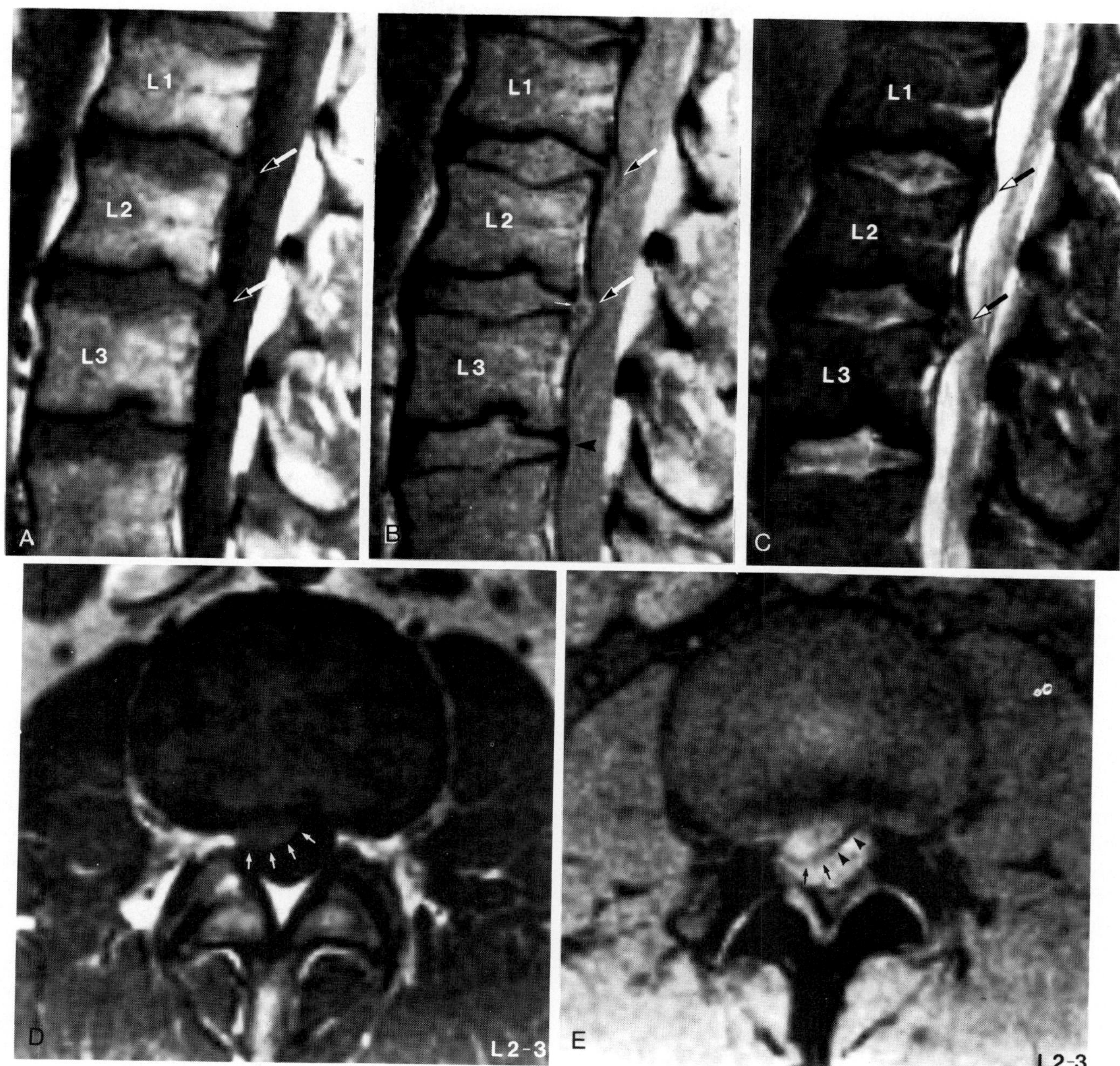

Figure 6–16. Two upper lumbar disk herniations. *A*, Sagittal T1-weighted image of L1–2 and L2–3 extradural defects *(arrows)* that are of equal signal intensity with and centered at the respective intervertebral disk levels. *B*, Sagittal proton density image. Imprinting of the thecal sac is better detected with this sequence *(arrows)*. Proton density imaging, however, also affords visualization of the disrupted outermost portion of the anulus at L2–3 *(small arrow)*; the normal intact outermost portion of the anulus of L3–4 is defined well *(arrowhead)*. At L1–2 the outermost portion of the anulus is thinned. *C*, Sagittal T2-weighted image showing nearly normal hydration of the parent intervertebral disk (L1–2 and L2–3). This sequence allows better visualization of the heterogeneous character of the herniation (especially at L2–3) and the displacement of the posterior longitudinal ligament and dura mater by the herniated disks *(arrows)*. *D*, Axial T1-weighted image. The extruded eccentric disk material of L2–3 is readily seen (arrows define its posterior limits). *E*, Axial gradient-recalled echo image. Not only is the L2–3 extruded disk well shown, but in addition the bulged outermost portion of the anulus shows less signal intensity on the right side *(arrows)*, indicating that it is thinner than that on the left *(arrowheads)*.

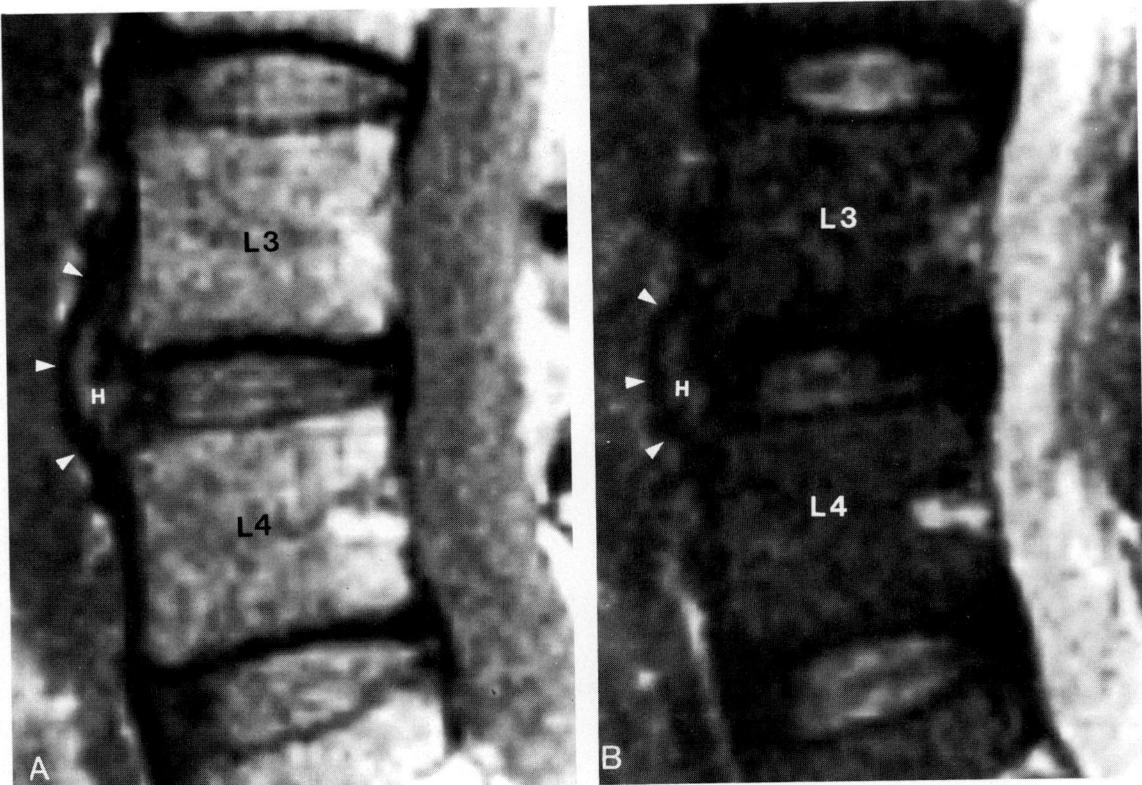

Figure 6–17. Anterior disk herniation of L3–4. *A*, Sagittal proton density image. The herniated nucleus pulposus (H) has a signal intensity similar to that of its parent disk. It extends along the anteroinferior surface of L3 and anterosuperior surface of L4 and elevates the anterior longitudinal ligament *(arrowheads)*. *B*, Sagittal T2-weighted image. The herniated disk (H) has a high signal intensity. The L3–4 disk space narrows, and the signal intensity decreases in comparison with that of L2–3 and L4–5. The anterior longitudinal ligament is indicated by arrowheads.

Magnetic resonance imaging is the most comprehensive and noninvasive imaging method at present of detecting disk herniation. High-resolution MR imaging now is able not only to differentiate anular bulging from different forms of herniation but also to show the relationship of the intervertebral disk to the adjacent thecal sac, neural elements, and osseous structures. These relationships characteristically require a compilation of spin-echo (T1-weighted, proton density, and T2-weighted) and GRE images. Magnetic resonance imaging also can demonstrate internal structural detail within normal and abnormal intervertebral disks (see Figs. 6–1, 6–6). In general, a decrease in signal intensity of an intervertebral disk on a T2-weighted image is indicative of disk degeneration.[52] Determining whether disk herniation is present usually requires a more detailed scrutiny of multiple images in at least two planes. In our experience, sagittal proton density imaging combined with axial GRE imaging (FISP [fast imaging with steady precession], FLASH [fast low-angle shot], GRASS [gradient-recalled acquisition of steady state], MPGR [multiplanar gradient-recalled acquisition of steady state]) are very important, and some believe mandatory, for demonstrating disk herniation most accurately. T1-weighted imaging usually does not differentiate the nucleus pulposus from the anulus fibrosus, whereas conventional T2-weighted imaging, which highlights the nucleus pulposus and the inner anulus, is noisy and usually does not reveal detailed disk morphology. Proton density imaging, in our experience, can often differentiate readily the nucleus pulposus–inner anulus from the middle and outer anuli. Moreover, the outermost anulus is readily identifiable on sagittal proton density images (Figs. 6–6, 6–18). The outermost anulus appears as a curvilinear thick line of low signal intensity that represents the most peripheral boundary of the intervertebral disk and extends beyond the adjacent inferior and superior margins of the vertebral bodies (see Figs. 6–6, 6–15, 6–18).

On MR imaging, differentiation between a contained and a noncontained disk herniation is possible because the defect in the outermost portion of the anulus is demonstrable, especially on proton density images. Because the outermost anulus can be well shown as a low signal band on proton density images, loss of continuity of this band indicates that a defect is present within it (Figs. 6–10, 6–16, 6–18 to 6–21). When an extradural mass is associated with a

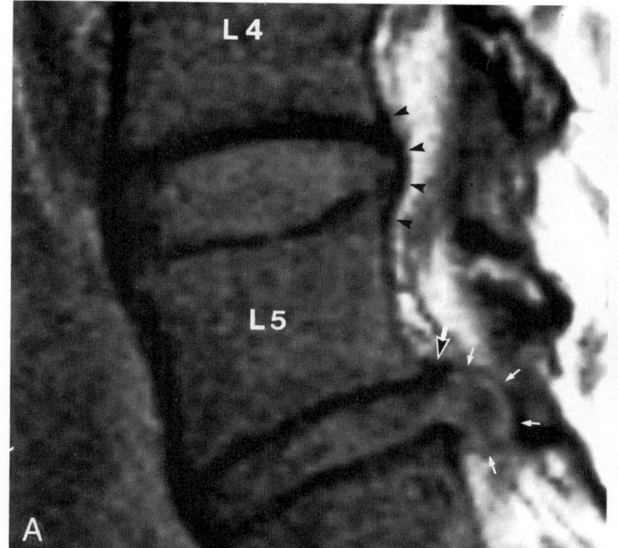

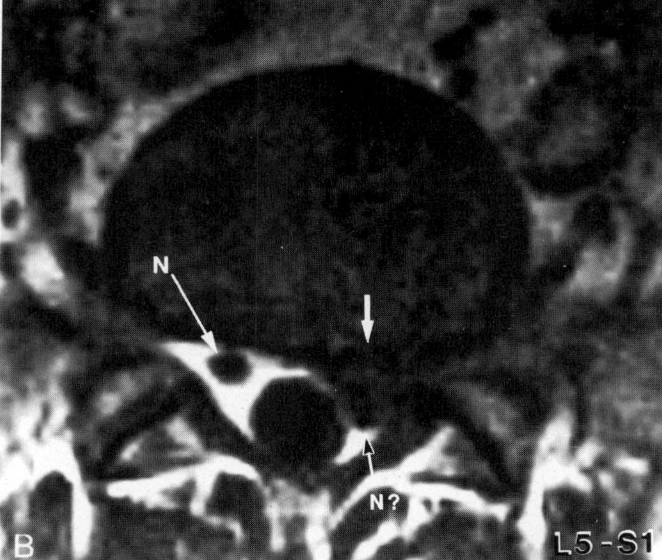

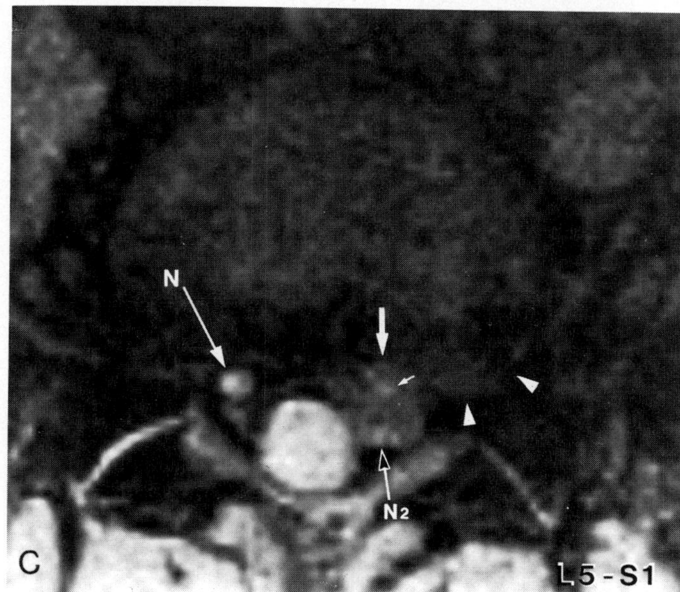

Figure 6–18. Large herniated disk, which is the unequivocal cause of the extradural defect. *A*, Paramedian sagittal proton density image. The intraspinal mass at L5–S1 *(small arrows)* is of the same intensity as that of the parent disk. In addition, one can see the disruption and splaying *(open arrow)* of the outermost portion of the anulus through which the herniation occurred. Arrowheads indicate the normal outermost portion of the anulus of L4–5. *B*, Axial T1-weighted image. The herniated disk *(arrow)* occupies the left lateral recess and has a signal intensity that merges with that of the S1 nerve root so that the latter is not definitely visualized (N?). The herniation also reduces the fat within the adjacent intervertebral canal, but this component of the herniation is not resolved directly. N, Right S1 nerve root. *C*, Axial gradient-recalled echo image. The prethecal and parathecal herniation *(arrow)* as seen on the T1-weighted image is confirmed; moreover, a large component of less high signal intensity does occupy the left intervertebral canal confirming greater herniation *(arrowheads)*. The normal right S1 nerve (N) is well shown, as it is on T1-weighted images, but the position of the left S1 nerve is better identified (N2). Although there is a bright focus *(small arrow)* in the normal position of the left S1 root, such a large disk herniation would displace the nerve root posteriorly, corresponding to the N2 brightness and the hypointensity (N?) on the T1-weighted image.

defect in the outermost portion of the anulus, it tends to splay the outermost anulus, offering a picture called the *open fish mouth sign*. The combination of the mass and outermost anulus disruption indicates unequivocally that the mass is a herniated disk (see Figs. 6–10, 6–18). Moreover, a second sign, but one that is not invariably able to be imaged, is direct visualization on axial sections of the defect in the outermost portion of the anulus that allowed herniation to occur (see Figs. 6–15, 6–16). Discontinuity of the band of low signal intensity on axial images is evidence that extrusion or sequestration has occurred. Disruption of the outermost portion of the anulus on axial MR images is an especially important sign for differentiating noncontained from contained disk herniation; again in the case of contained disk herniation, the band of low signal intensity representing the outermost portion of the anulus remains intact. These features have important implications for surgical treatment and disk exploration as well as for safe and successful chemonucleolysis and automated diskectomy. A distinction between extruded and sequestered (free fragment) disks (see Fig. 6–21) also is important when surgery is planned because migration of the sequestered fragment may produce misleading localizing signs and symptoms and change the surgical procedure.[49] The disk herniation (extradural mass) often, but not invariably, manifests a signal intensity similar to that of the parent disk. Herniation is usually well shown on sagittal MR images, but axial MR corroboration is very helpful, and reaching a match fit is important

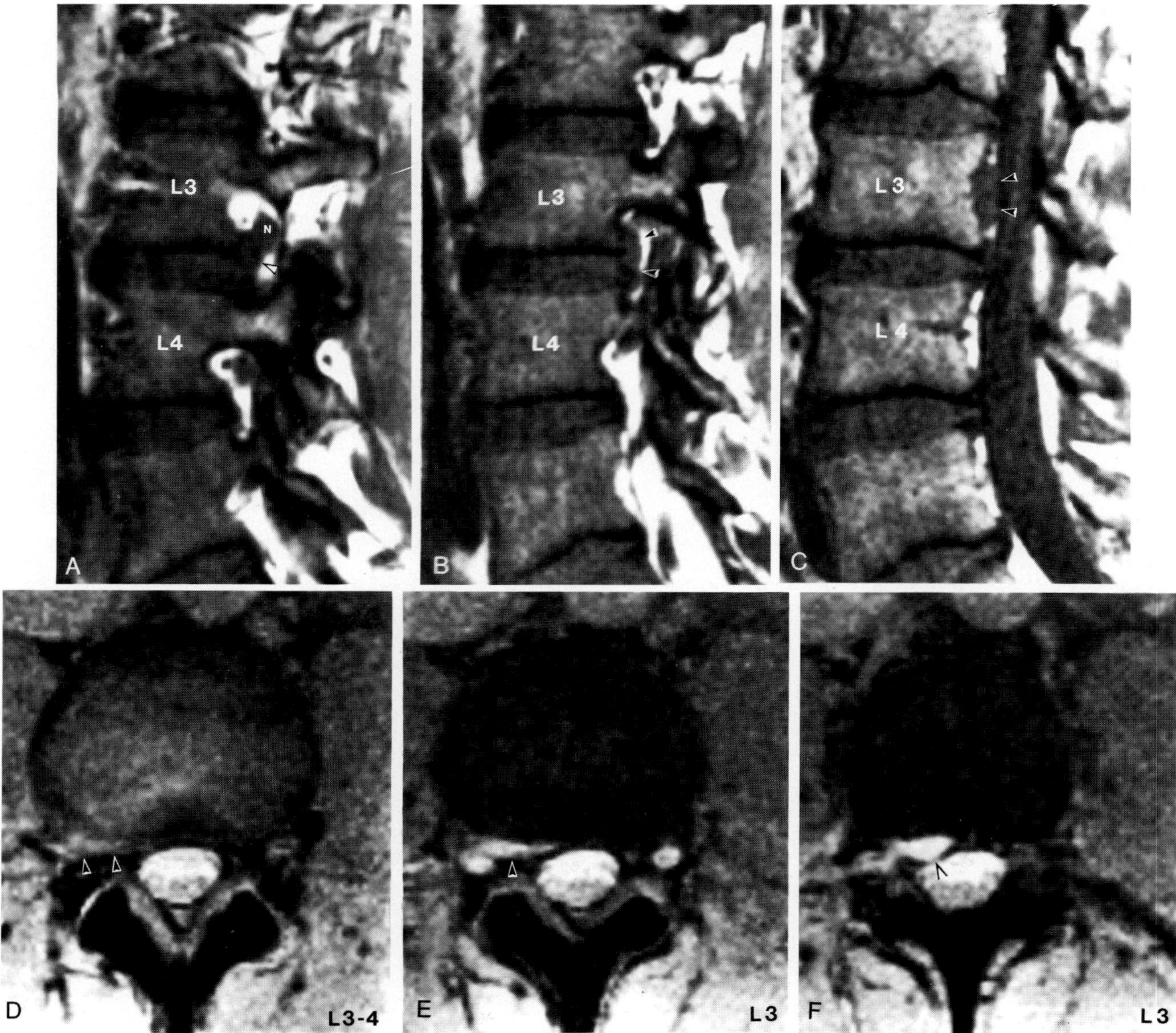

Figure 6–19. Noninterspace intervertebral disk herniation that mimics sequestration. *A–F*, Serial sagittal T1-weighted and axial gradient-recalled echo (GRE) images. The largest segment of disk herniation *(arrowheads)* lies posterior to the vertebral body and continues with its parent disk. This herniation extends from laterally on the right at the level of the intervertebral disk to superiorly and medially at the site behind L3. This herniation was not proved surgically, but a subsequent follow-up MR examination was performed. N, Nerve root.

diagnostically, particularly in the planning of surgery. The true herniation abnormality usually is visualized in more than one plane (see Fig. 6–19).

SPINAL STENOSIS

Spinal stenosis can be divided into two major categories: congenital (developmental) and acquired. The congenital form is typically bony stenosis due to underdevelopment of the spinal canal and most typically the short pedicles, for example, achondroplasia.[30] In congenital stenosis even minimal degenerative changes of the intervertebral disk, facet, or ligamentum flavum can result in neurologic manifestations or pain as the stenosis increases. This section focuses on acquired lumbar stenosis, which is different from congenital spinal stenosis. Although the adult form may be bony in its confinement of structures, it may be caused by solely abnormal soft tissue or a combination of bony and soft tissue abnormalities. With MR and CT imaging, the scans are presumed to be from the nonaxially loaded spine, so that the findings may be very different from the true findings in the erect person.

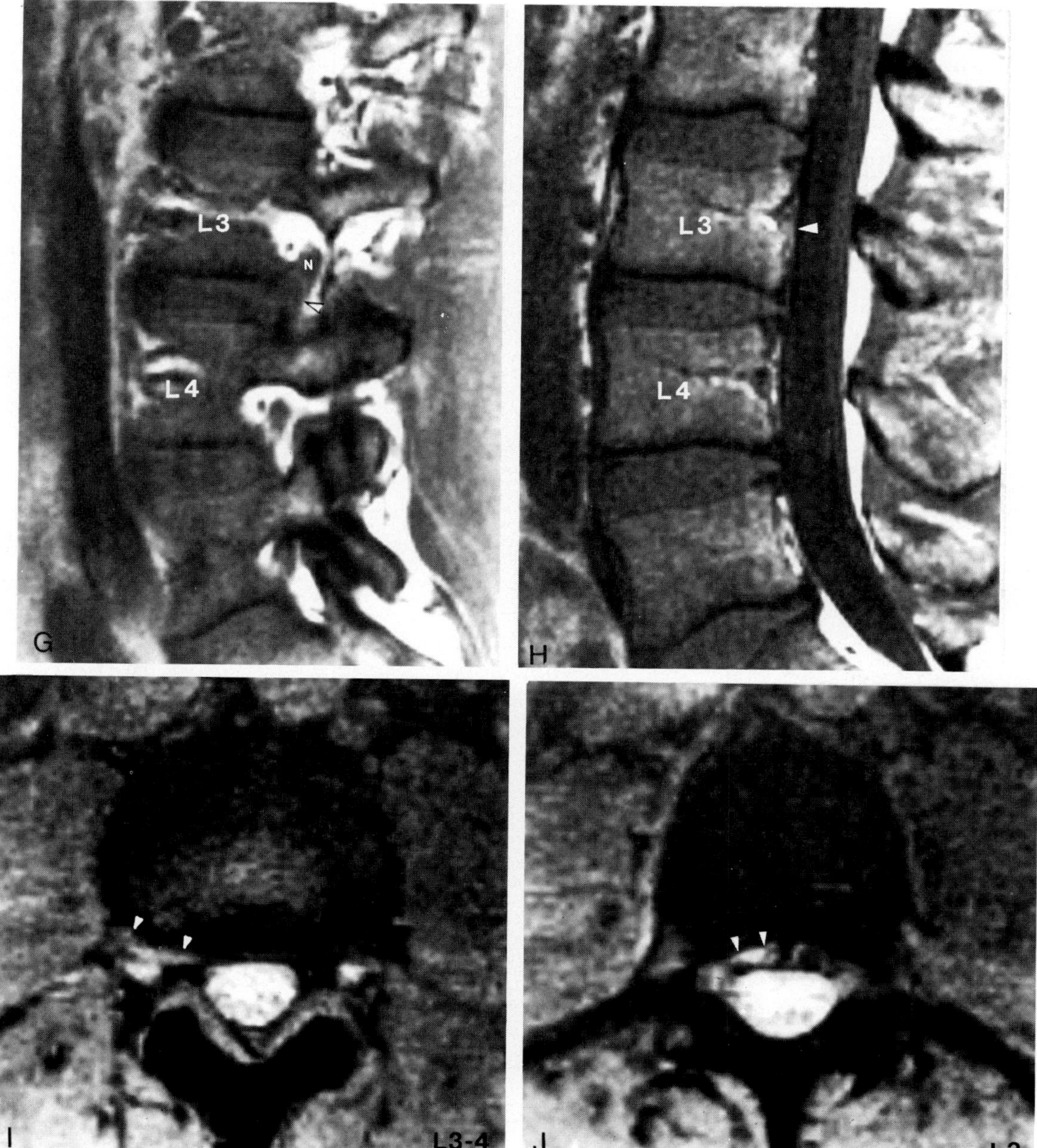

Figure 6–19 *Continued G–J,* Serial sagittal T1-weighted and axial GRE images. Four months later, the patient's symptoms subsided, and the disk herniation *(arrowheads)* became much reduced. N, Nerve root.

Stenosis in the acquired form may involve the spinal canal, the lateral recess, or the intervertebral canal. In these locations the stenosis may develop simultaneously or independently. Depending on its location—whether compression is of the spinal cord, nerve roots, or dorsal root ganglion—stenosis may cause radiculopathy or myelopathy. Common causes of acquired stenosis include degenerative diseases (e.g., bulging of the disk, contained or noncontained disk herniation, spondylosis, apophyseal osteoarthropathy, thickened ligamentum flavum) (Fig. 6–22); malalignment of the spinal column; postoperative scarring; trauma; and neoplastic and similar conditions that cause disk expansion, such as Paget's disease, metastatic disease, and pathologic fracture.

Herniation of a disk, evidence of osteophyte, hypertrophy of the ligamentum flavum, and apophyseal osteoarthritis are the common causes of stenosis of the spinal canal. Spinal canal stenosis occurs most commonly at the L2–3, L3–4, and L4–5 disk levels.[57] It is characterized by a diminution of the cross-sectional area, especially by a decrease in the anteroposterior diameter of the spinal canal. The average

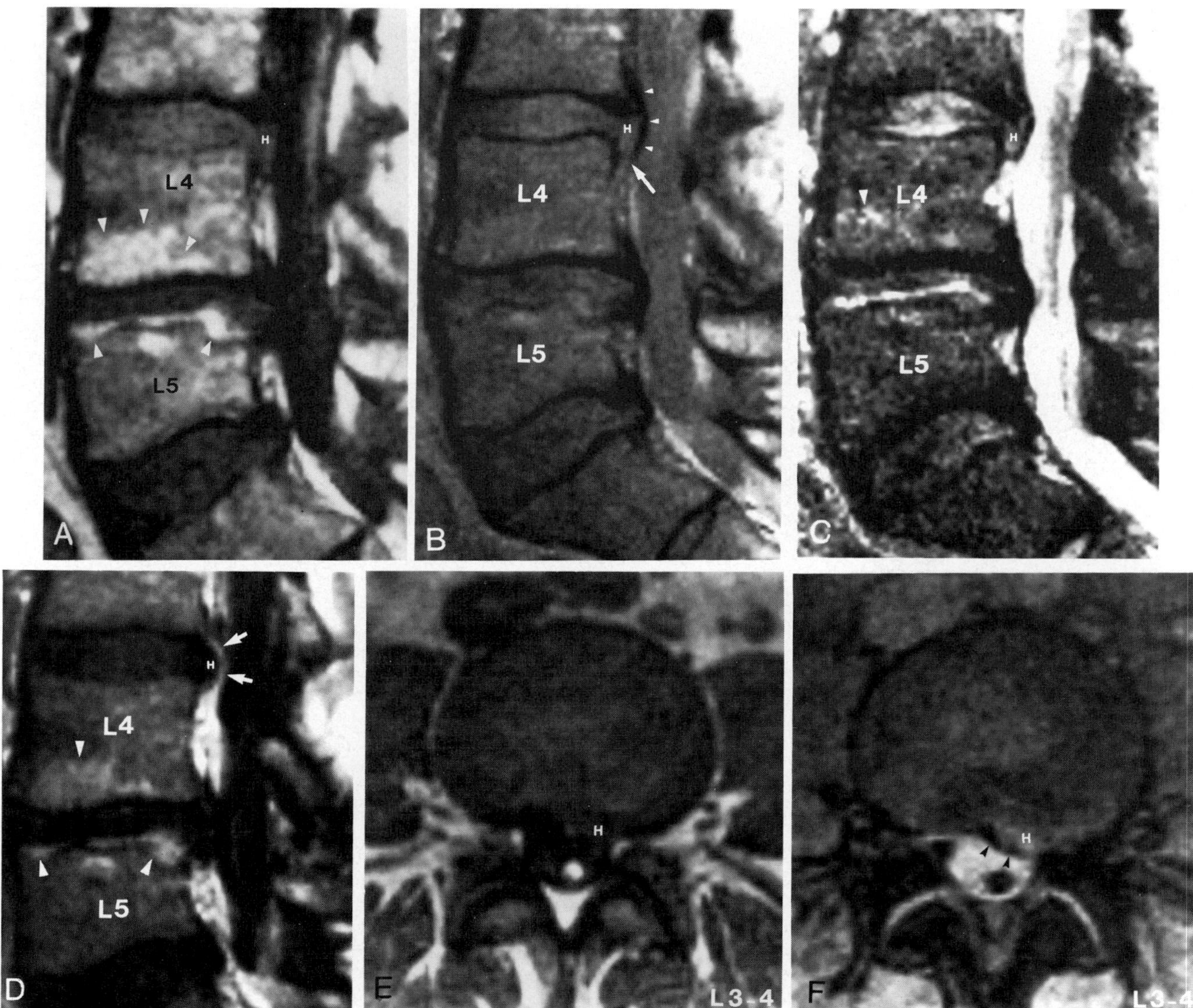

Figure 6-20. Multiple features of disk herniation (L3-4): Torn outermost portion of the anulus, hydrated herniated material, and enhancement along the outermost portion of the anulus. *A*, Sagittal T1-weighted image. The L3-4 posterior herniation (H) is shown but is not well defined in this sequence. Arrowheads point to type II bone marrow changes.[51] *B*, Sagittal proton density image. This sequence defines the outermost portion of the anulus as a structure of low signal intensity *(arrowheads)*, bridging the vertebral bodies circumferentially; here its posterior aspect is seen and is shown to be disrupted inferiorly *(arrow)*, allowing for herniation (H). *C*, Sagittal T2-weighted image. The herniated segment (H) demonstrates less signal intensity than that of the parent disk. Most herniations occur in degenerated disks so that the parent disk usually is of low signal intensity on T2-weighted images also; here the parent L3-4 disk is of normal signal intensity and normal hydration (note the reduced signal of degenerated L5-S1 and the less intense, thinned L4-5 intervertebral disk). *D*, Sagittal T1-weighted, gadopentetate dimeglumine-enhanced image. Capping the outermost portion of the anulus is a curvilinear band of enhancement that defines the posterior limits of the herniation even better *(arrows)* (contrast with *A*, a pre-enhanced image). Arrowheads indicate type II bone marrow change. *E*, Axial T1-weighted image. The eccentrically positioned (left-sided) herniated segment (H) has an intensity similar to that of the parent disk. The high signal density within the thecal sac is fat (lipome) within the filum terminale. *F*, Axial gradient-recalled echo (GRE) image. The focally herniated disk (H) of high signal intensity on the GRE image further distinguishes the thecal sac and the rim of low signal intensity (the outermost portion of the anulus and the reaction to herniation are emphasized by enhancement *[D]*).

anteroposterior dimension of the midsagittal spinal canal in adults is 11.5 mm.[89] Because this dimension varies considerably among persons and at different disk levels, morphologic rather than direct measurement of the spinal canal in stenosis is of much higher radiologic importance in our opinion. Stenosis of the spinal canal can be well appreciated in most patients on T1-weighted sagittal MR images, because the expanding abnormality can be well depicted by the surrounding cerebrospinal fluid. Demonstrating and confirming the presence of an osteophyte, an expanded anulus of the intervertebral disk, and a herniation may require a variety of sequences. Although proton density and T2-weighted images often are helpful at this time, GRE imaging, in our opinion, seems required for this differentiation, because bony

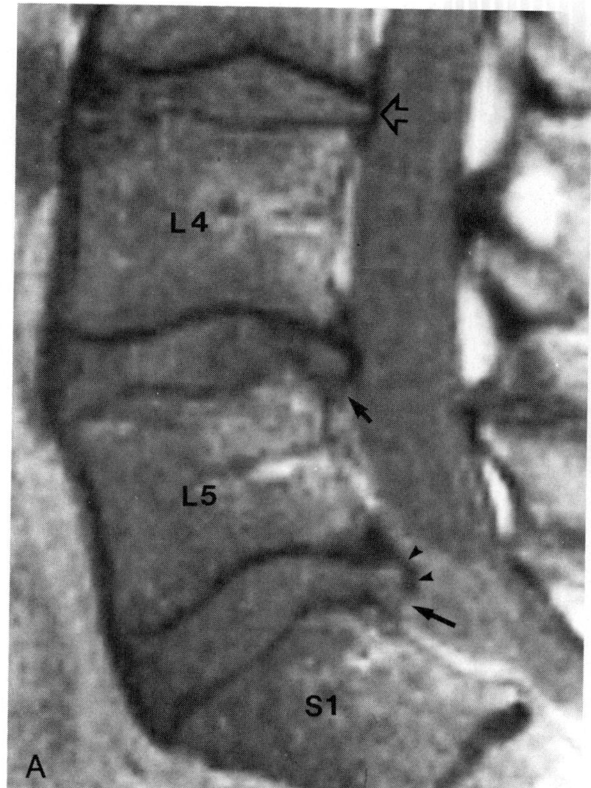

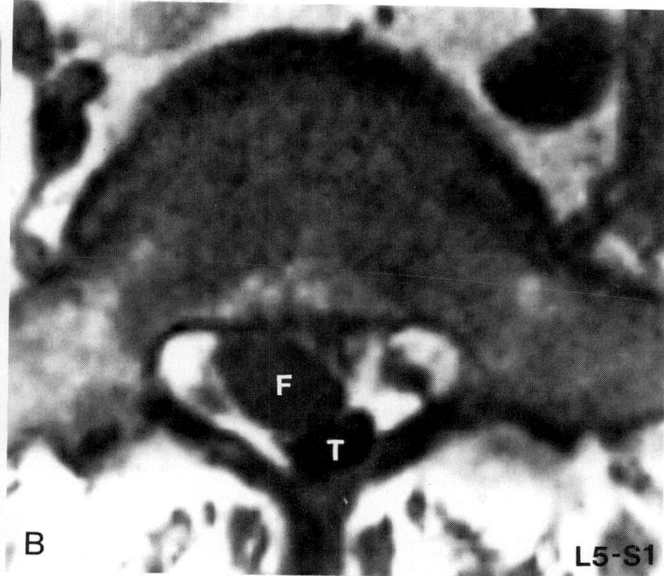

Figure 6–21. L5–S1 disk sequestration (free fragment). *A,* Sagittal proton density image. A defect *(long arrow)* is seen in the outermost portion of the anulus *(arrowheads)* of L5–S1. Another more subtle tear in the posterior outermost portion of the anulus is present at L4–5 *(short arrow).* The open arrow points to the normal appearance of the posterior outermost portion of the anulus at L3–4. *B,* Axial T1-weighted image. The sequestered L5–S1 disk material (free fragment, F) displaces the thecal sac (T) posteriorly.

spurs are of low signal intensity, whereas herniated disks and cerebrospinal fluid are of higher signal intensity (Fig. 6–23). Hypertrophy of the ligamentum flavum also may cause spinal stenosis and is readily demonstrated on MR imaging (Fig. 6–24).

Herniation of the posterolateral disk, postoperative scar formation, or hypertrophy of the superior articular process may result in stenosis of the lateral recess (see Fig. 6–23). The hypertrophic, enlarged superior articular process of the vertebral body can encroach on the lateral recess (the subarticular zone)[44] and often compresses the nerve root before it

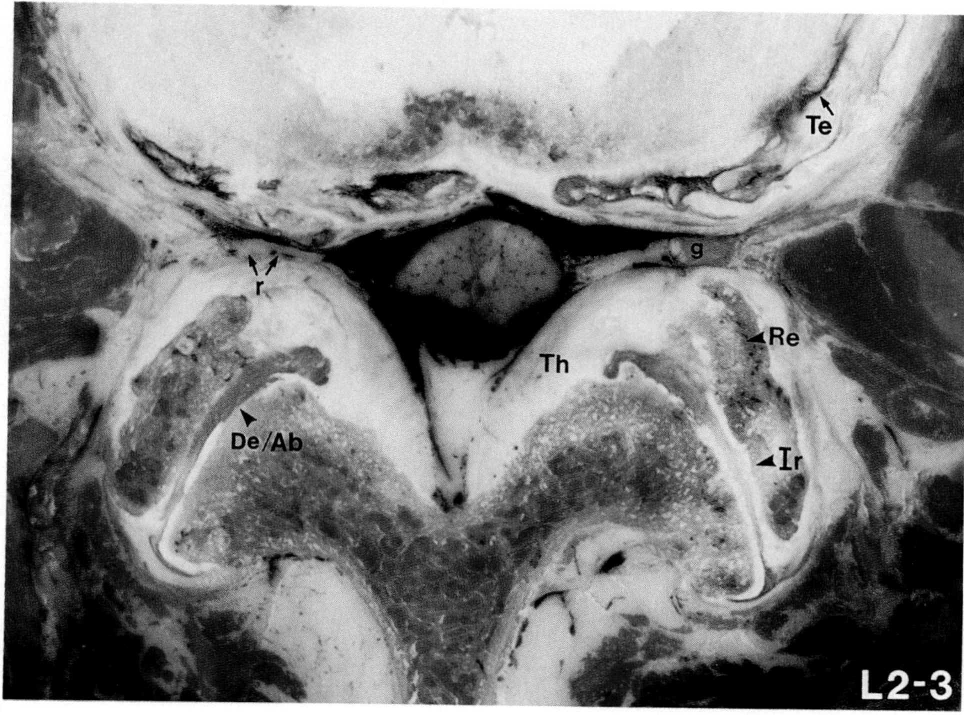

Figure 6–22. Degenerative changes of facets and intervertebral disk leading to neural compression, as seen in the cryomicrotome section (axial section, L2–3). The abnormal findings are as follows: facet joint thinning with irregularity of the cartilage (Ir), fatty replacement of bone within the superior articular facet (Re) and enlargement of the facet, loss of cartilage along the contralateral facet joint (De/Ab), thickening of the ligamentum flavum (Th), and a circumferential tear in the outer anulus (Te). The dorsal root ganglion (g) is flattened by compression between the bulging outermost portion of the anulus and the superior articular facet. Contralaterally, the ramus (r) is compressed between the thickened ligamentum flavum and the bulging disk as manifested by the leakage of injected dye into the outermost portion of the anulus.

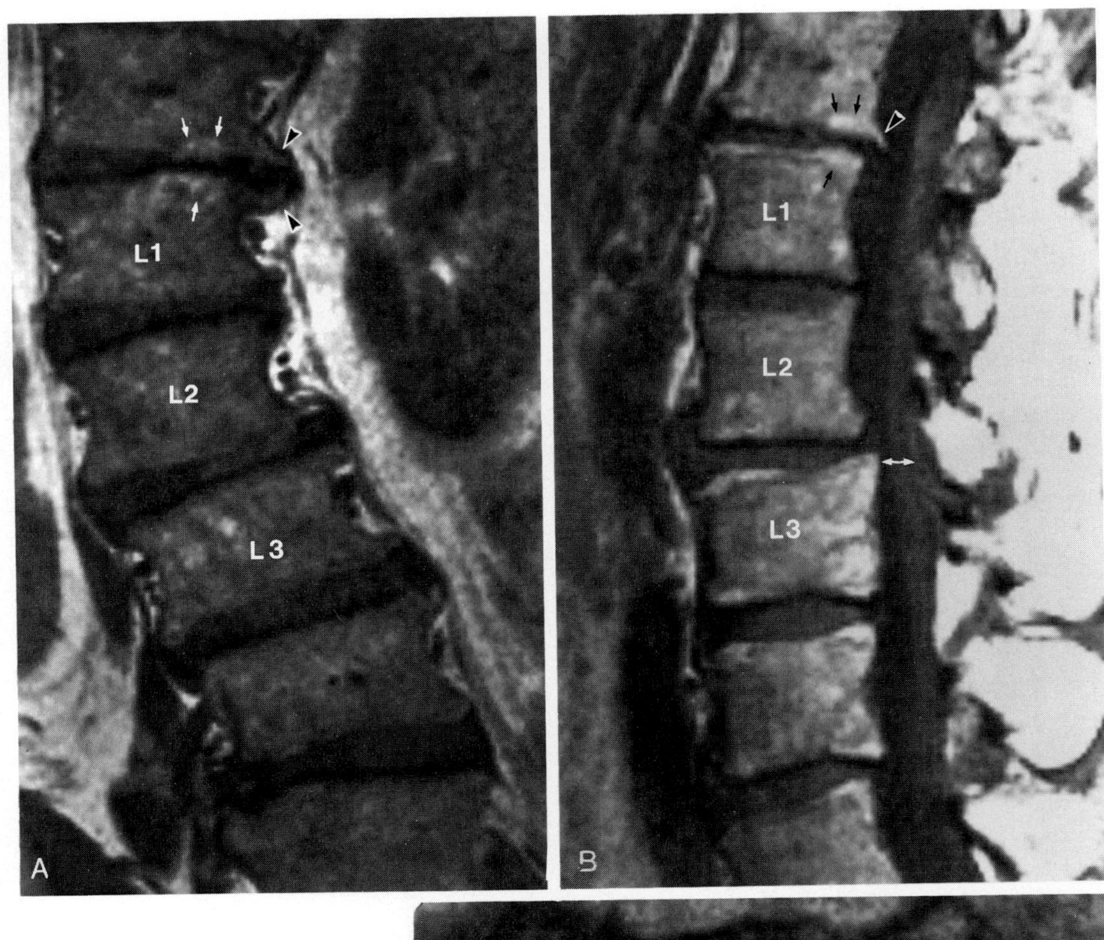

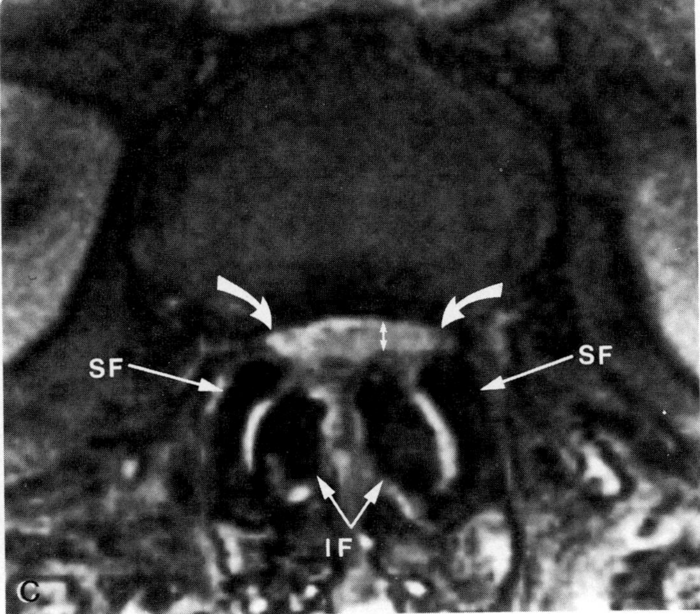

Figure 6–23. Scoliosis with spinal stenosis. *A*, Coronal T1-weighted image. In this example of scoliosis, the osteophyte *(arrowheads)* is more evident at T12–L1, and the degenerative vertebral body responses *(small arrows)* are shown. *B*, Sagittal T1-weighted image. This plane also shows the subluxation associated with scoliosis, regional spinal canal narrowing *(white arrows)*, the spondylophyte *(arrowhead)*, and marrow changes *(black arrows)*. *C*, Axial gradient-recalled echo image. The stenosis of the spinal canal *(small arrows)*, including lateral recesses *(curved arrows)*, is evident. Canal stenosis is caused by hypertrophy of reoriented L2–3 facets and L2–3 subluxation, as shown in the sagittal section *(B)*. SF, Superior facet; IF, inferior facet.

exits through the intervertebral canal. Lateral recess stenosis can be readily identified on T1-weighted axial spin-echo and GRE images. Normally, the lateral recess is filled with fat, which is of high signal intensity and contrasts with its surroundings on T1-weighted images. Its displacement or diminution caused by an epidural mass is indicative of stenosis. Gadopentetate dimeglumine administered intravenously is needed for differentiating postoperative scarring from residue or recurrent disk herniation.

Stenosis of the intervertebral canal usually is caused by the presence of bony growth or abnormal soft tissue within the intervertebral canal. Lateral disk herniation is a very important and common

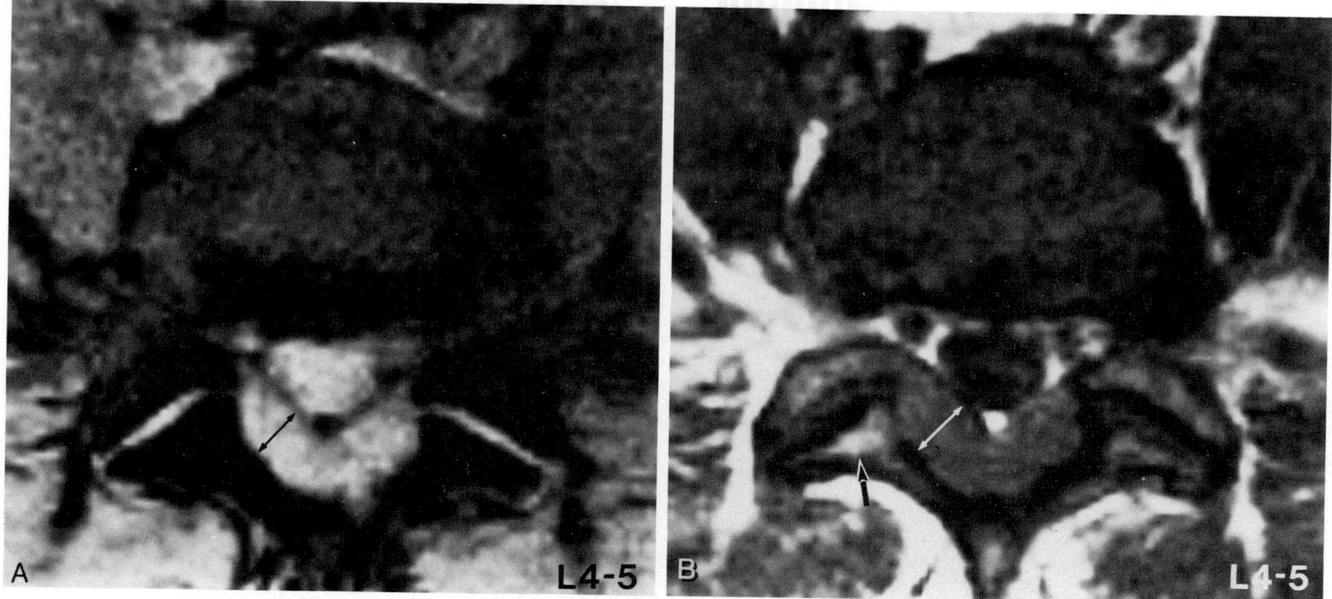

Figure 6–24. Very large ligamentum flavum at the L4–5 disk level. Axial gradient-recalled echo *(A)* and spin-echo *(B)* images. The thickness of the ligamentum flavum, marked by arrows, measures 8 mm. This thickening causes spinal stenosis. The fatty replacement *(large open arrow)* within the right facet of L4 is visible in *B*.

cause of diminished caliber of the intervertebral canal anteroposteriorly in young and middle-aged patients. However, a slight disk bulging or a small disk herniation may not cause compression of the nerve root because the root within the intervertebral canal medially is located in the upper portion (see Fig. 6–4).[61,103] In later middle age and in older patients, osteoarthritis of the facets and degenerated, sagging disks that bulge into the neural canal are the common causes of lateral spinal stenosis. Enlargement of the superior articular process and its anterior subluxation can significantly narrow the intervertebral canal anteroposteriorly, producing compression of the nerve root (Fig. 6–25).[44] The in-

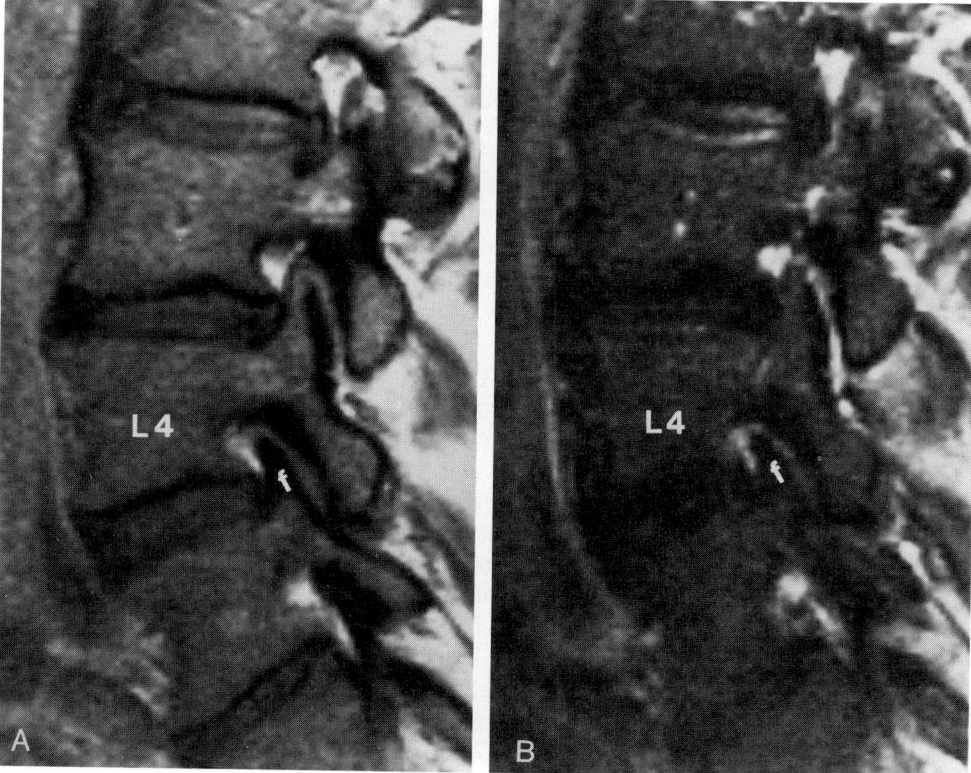

Figure 6–25. Facet subluxation that narrows the intervertebral foramina bilaterally (L4–5), causing nerve root compression. Radiculopathy was present clinically. Sagittal proton density *(A)* and T2-weighted *(B)* images show the superior facet of L5 (f) spearing the upper portion of the L4–5 foramen, compressing the nerve root anterosuperiorly.

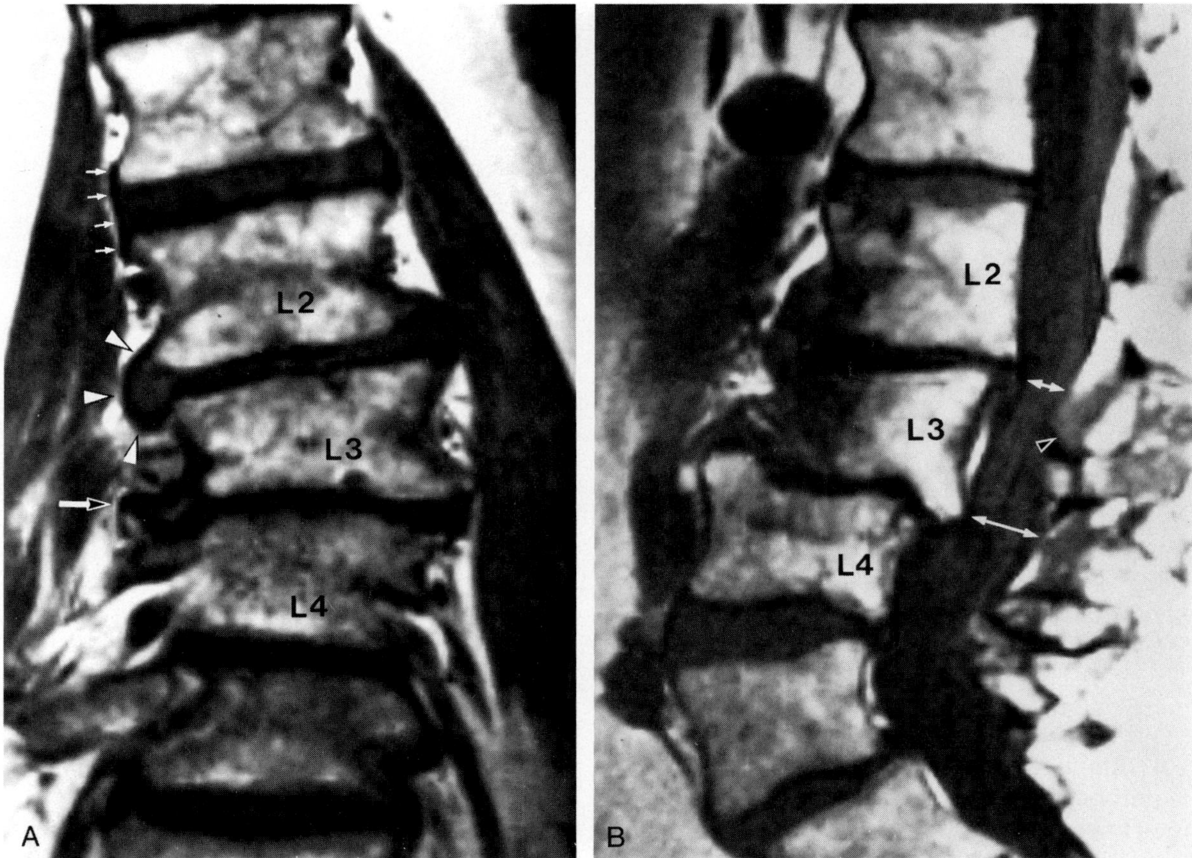

Figure 6–26. Scoliosis accompanied by retrolisthesis. *A*, Coronal T1-weighted image. The most severe degenerative changes of the disks are shown at L2–3, L3–4, and L4–5, where the lumbar spine is most concave. A big bony spur *(arrow)* is noted at L3. The lateral outermost portion of the anulus *(arrowheads)* of L2–3 of low signal intensity is stretched by the protruded disk, but it is still intact. Small arrows point to the normal outermost portion of the anulus of low signal intensity at L1–2. *B*, Sagittal T1-weighted image. Retrolisthesis is noted at the L2–3 and L3–4 disks. The L3 vertebral body is subluxated posteriorly with respect to L4, as is L2 on L3. The spinal canal is narrowed significantly in this region *(double-headed arrows)*. Ligamentum flavum thickening at L2–3 level is seen *(arrowhead)*.

tervertebral canal is best shown on paramedian sagittal T1-weighted spin-echo images, so much so that this plane can be called the plane of the intervertebral canal. On this image, the nerve roots, with intermediate signal intensity, contrast well with the high signal intensity of fat within the intervertebral canal. Axial GRE imaging also is advantageous for evaluating the intervertebral canal because of its good depiction of bony margins.

Malalignment of the spinal column in spondylolisthesis and retrolisthesis leads to narrowing of the central spinal canal (Fig. 6–26) and the intervertebral canal anteroposteriorly. On paramedian sagittal anatomic and MR sections through the intervertebral canal, the vertically or slightly obliquely oriented configuration of the intervertebral canals are converted, in spondylolisthesis, to a horizontal orientation with a configuration simulating an infinity sign or a horizontal figure-of-eight shape (Figs. 6–27, 6–28). Compression of the nerve root in this circumstance may be due to the upward bulging and buckling of the intervertebral disk (see Fig. 6–27). On sagittal images, the anteroposterior dimension of the spinal canal is usually reduced at the level of subluxation owing largely to the listhesis. Spondylophyte or ligament hypertrophy, if present, significantly increases the degree of spinal stenosis.

MALALIGNMENT OF THE LUMBAR SPINE

Of the various types of abnormalities of spinal alignment, spondylolisthesis is the most common.[49] Spondylolisthesis is characterized by the forward displacement of one vertebra over another. This malalignment, which occurs most commonly in the lower lumbar region,[30,49] most often with the forward slippage of L4 on L5 or L5 on S1,[72] may be secondary to developmental or acquired abnormalities. Spondylolisthesis secondary to a defect in the pars interarticularis is thought to be a developmental ab-

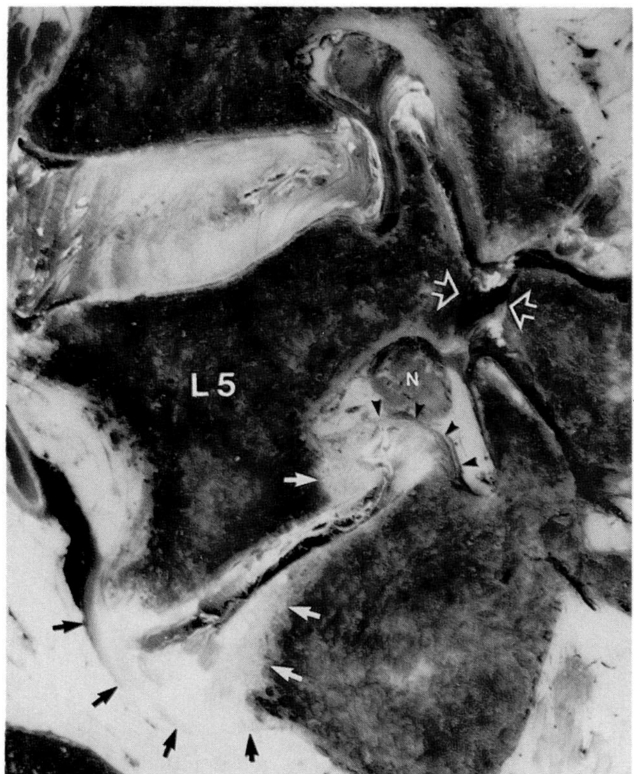

Figure 6–27. Spondylolysis with spondylolisthesis compression of the neural canal is visualized in cryomicrotome section. At the plane of the intervertebral canal medially (the nerve [N] is in the uppermost portion of canal), the forward displacement of L5 on S1 resulting from spondylolysis *(open arrows)* produces buckling of the nondegenerated posterosuperior portion *(arrowheads)* of the degenerated L5–S1 intervertebral disk. Interestingly, the long duration of the spondylolisthesis is manifested by the fatty replacement of the marrow around the buckled upwardly herniating disk *(single white arrow)*, as contrasted with the degenerated anterior part of the intervertebral disk *(double white arrows)* with its spondylophyte *(black arrows)*.

normality and usually occurs at L5–S1. However, a defect in the pars interarticularis is also thought to be traumatic because it is never present at birth.[96] The disk between the slipped vertebrae may undergo accelerated degeneration secondary to loss of protection by the posterior elements. All the stresses, and especially torsional stresses, are exerted on the disk, which results in the breakdown of the anulus. Bulging of the disk is usually present, but true herniation of the disk is uncommon (see Figs. 6–27, 6–28).[96]

Degenerative spondylolisthesis is secondary to a combination of factors, including reluxation of the disk complex, orientation of the facets, and action of the forces of gravity. Herniation of a disk may be associated with degenerative spondylolisthesis (Fig. 6–29). This type of spondylolisthesis occurs most often between the L4 and L5 vertebrae.[72] Retrolisthesis is characterized by posterior subluxation of a given vertebral body with respect to the vertebral body below and is usually related to similar factors as in degenerative spondylolisthesis. This occurs more often at the upper lumbar spine.[64]

Plain radiography and CT are traditionally used for the evaluation of abnormalities in alignment of the spine. MR imaging, however, not only shows the occurrence of the alignment abnormality but in addition can better demonstrate the causes and morphologic consequences, such as degenerative changes of the disk or facets, defects in the pars interarticularis, compromise of the spinal and intervertebral canal, and neural effects. What cross-sectional imaging (CT and MR) currently does not show are the effects of gravitational loading, because patients can only be scanned while recumbent. On MR images, the deformity of the intervertebral canal is well depicted as a horizontal figure-of-eight (see Fig. 6–28). The upwardly bulging disk may abut against the nerve root, which normally is located in the upper portion of the intervertebral canal. The intact or disrupted outermost anulus of low signal intensity is well shown, in contrast to the fat in the canal and the higher signal intensity of the protruded nucleus material. The degenerated disk appears to be of decreased signal intensity on T2-weighted images. Gas of low signal intensity or fluid of high signal intensity may be found within the disk. Usually the disk bulges downwardly in the anterior aspect associated with the spondylolisthesis.

POSTOPERATIVE SPINE

The occurrence of "failed back surgery" syndrome is reported in from 10 to 40 per cent of patients after spinal surgery.[64] This syndrome commonly occurs in younger patients and is characterized by pain, radiculopathy, and functional incapacitation. Recurrent disk herniation, lateral (subarticular) recess or intervertebral canal stenosis, epidural scarring, and arachnoiditis are the more common causes.[8] Less common etiologic factors in failed back surgery syndrome include mechanical facet subluxation and meningocele formation.[49]

The evaluation of patients who have symptoms after spinal surgery can be a difficult diagnostic problem. Clinical assessment is difficult because the clinical manifestations may be nonspecific and nonfocal. Findings visualized on conventional radiography, film-screen myelography, and plain CT of the spine also are nonspecific in differentiating between epidural scarring and recurrent disk herniation or lateral spinal stenosis. Computed tomographic myelography and intravenously enhanced CT represent major advances in diagnosing the consequences of and residual diseases occurring after spinal surgery.[6,87,98] The ability of MR imaging to demonstrate

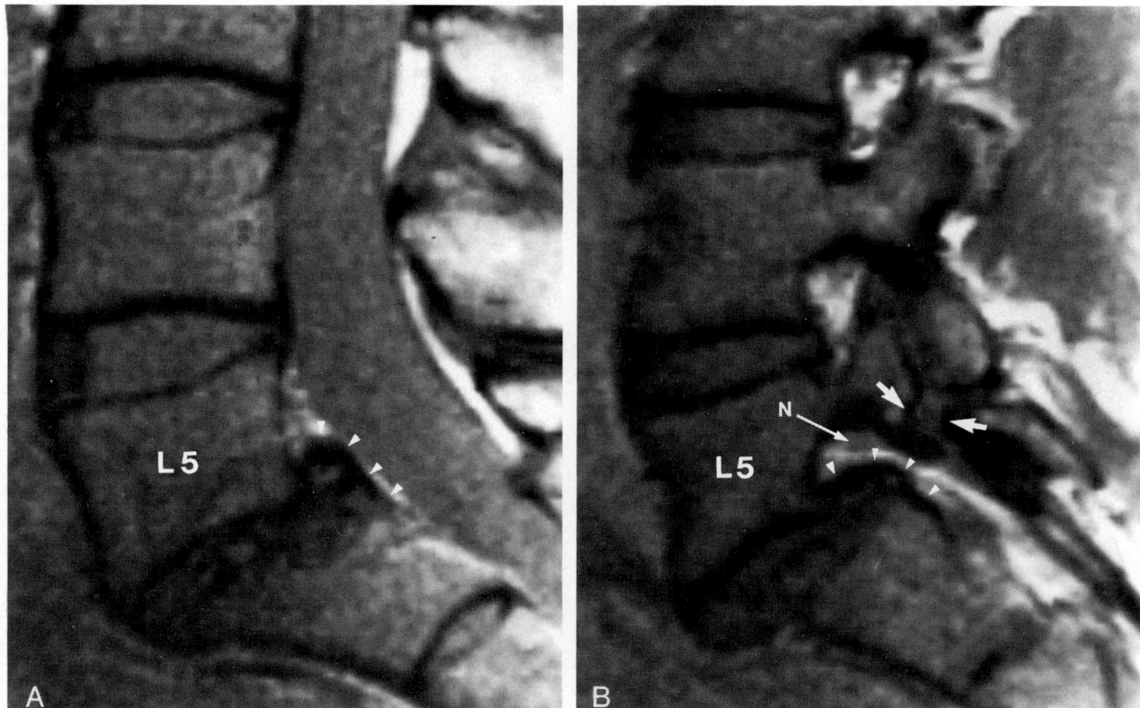

Figure 6–28. Spondylolisthesis of L5–S1. *A*, Sagittal proton density image shows the grade I–II spondylolisthesis of L5 on S1 and the degeneration of the L5–S1 intervertebral disk (lower signal intensity, thinning, and altered intensities of the end plate and adjacent centra). *B*, The paramedian sagittal proton density section at the foraminal plane shows the transverse figure-of-eight configuration of the elongated intervertebral canal with the L5 nerve (N) in its uppermost portion and the wide defect in the pars interarticularis axis *(opposing arrows)* proportional to the slippage. The thickened outermost portion of the anulus is stretched but remains intact *(arrowheads)*.

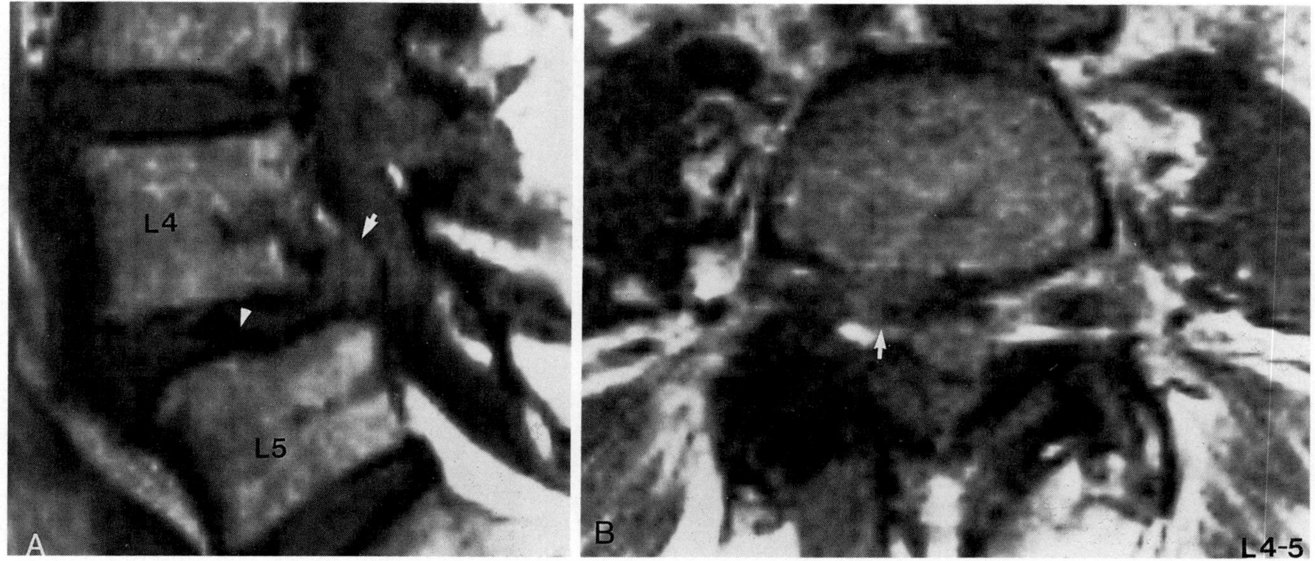

Figure 6–29. Spondylolisthesis with disk herniation. Sagittal *(A)* and axial *(B)* T1-weighted images. The finding of a herniated disk in spondylolisthesis is very uncommon. Nonetheless, it is demonstrated here *(arrows)*. Degeneration of the disk on T1-weighted images in grade I–II spondylolisthesis (L4–5) is indicated by the large, irregular gas collection *(arrowhead)* within the thinned disk.

detailed soft tissue morphology of the spine has further helped clarify the specific cause of an individual patient's symptoms in failed back surgery syndrome. Gadopentetate dimeglumine–enhanced MR imaging, not dissimilar to intravenously enhanced CT, has shown unparalleled sensitivity and specificity in evaluating the spine postoperatively, particularly in differentiating postoperative scarring from a recurrent or residual herniated disk.[32,69] Gadopentetate dimeglumine should always be used in the postoperative spine patient because of its ability to differentiate among normal tissue, postoperative scarring, and recurrent or residual disk herniation.

Laminectomy

Laminectomy or laminotomy is the most common surgical technique used to approach the herniated disk.[56] Laminectomy (in which the spinal processes are removed also) affords greater decompression, as is required in spinal stenosis. The laminectomy site is usually readily identifiable on MR images, not necessarily because of the bony changes but because of those of the soft tissues. The absence of the lamina and ligamentum flavum, the fat graft placed along the operative tract of the lamina, the scar within the laminectomy site, the distortion of the normal anatomic structures, and the fat replacement of the posterior musculature often are well appreciated on T1-weighted and T2-weighted images. The congenital absence of the facet may mimic the configuration of postlaminotomy (Fig. 6–30). Scar formation along the tract and perithecal scarred tissues are much better defined on comparing the pre-enhanced T1-weighted axial (and sagittal) slices with T1-weighted contrast-enhanced images in the same plane or planes.

Spinal Fusion

Spinal fusion is a surgical procedure used for stabilizing painful movement and for treating intervertebral joint disease.[36] A bone graft is placed across a spinal segment (e.g., between the transverse processes, facets, or lamina of bodies) to maintain an anatomic realignment and stabilization of the spine. The graft may be an autograft or an allograft. The appearance of the graft on MR images varies with its origin (Fig. 6–31), which may be from the fibula, the iliac crest, or the rib. The autograft usually demonstrates a rim of decreased signal intensity on both T1-weighted and T2-weighted images, representing cortical bone. The signal intensity of the bone marrow in the graft also varies, depending on the relative quantity of marrow within the graft. Postoperative stresses placed on the fusion mass, graft revascularization, and manipulation required during the surgical procedure are other factors that can alter the signal intensity of the bone graft. Understanding the complexity of grafts from cross-sectional imaging is facilitated by three-dimensional postprocessing, especially three-dimensional CT.

Diskectomy

Diskectomy with or without spinal fusion is an increasingly successful and popular procedure used to treat painful and unstable degenerative disk disease.[56] This procedure is carried out in conjunction with laminotomy and laminectomy, in which the degenerated disk is partially or totally excised. Sagittal and axial T1-weighted images obtained immediately after diskectomy show soft tissue of intermediate signal intensity at the original site of disk herniation and diskectomy. To reach an accurate interpretation for the immediate postdiskectomy MR images may be very difficult owing to the variable appearance manifested by the reactive granulation tissue, blood, and fluid at the site of surgery and the surgical bed within the disk. The variable appearance may mimic that of a disk herniation. Two to three months after surgery, the scar tissue becomes more mature so that the distinction between scar tissue and recurrent disk herniation becomes possible with the administration of gadopentetate dimeglumine. Images made after the intravenous gadopentetate dimeglumine in this circumstance may show a significant enhancement in the site of diskectomy that represents scar tissue (Fig. 6–32). The routine use of gadopentetate dimeglumine for an accurate interpretation of the postdiskectomy MR images is necessary. Another characteristic finding in the case of diskectomy is the absence of the low signal band of the outermost anulus on sagittal proton density MR images.

Residual or Recurrent Disk Herniation and Postoperative Scarring

Postoperative patients often were referred to the radiologist for differentiating recurrent (or residual) disk herniation from epidural scarring. This distinction is of considerable importance because the removal of recurrent or residual disk herniation generally entails acceptable results, whereas removal of epidural scarring, even with the employment of the operating microscope and expert technique, usually leads to a re-formation and a potential increase

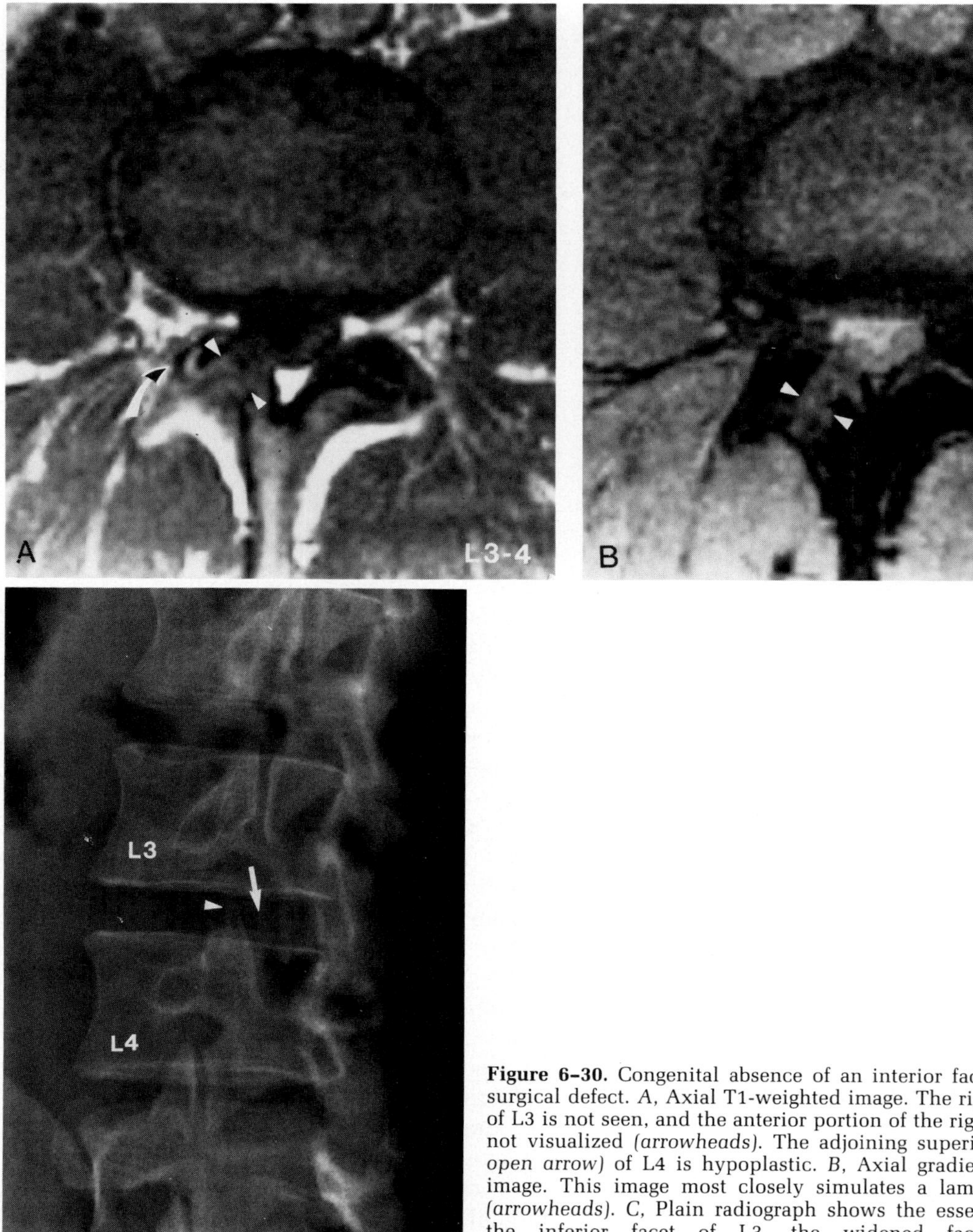

Figure 6–30. Congenital absence of an interior facet, simulating a surgical defect. *A*, Axial T1-weighted image. The right interior facet of L3 is not seen, and the anterior portion of the right lamina also is not visualized *(arrowheads)*. The adjoining superior facet *(curved open arrow)* of L4 is hypoplastic. *B*, Axial gradient-recalled echo image. This image most closely simulates a laminectomy defect *(arrowheads)*. *C*, Plain radiograph shows the essential absence of the inferior facet of L3, the widened facet "space" in its absence *(arrow)*, and the hypoplastic superior facet of L4 *(arrowhead)*.

in scar formation.[49] On nonenhanced MR images, herniated disks are relatively well defined and usually exhibit decreased signal intensity on both T1-weighted and T2-weighted images. A sequestered fragment may have slightly increased signal intensity compared with that of its parent disk and anterior epidural scarring on T1-weighted images; it may have high signal intensity, similar to that of scarring, on T2-weighted images.[69] Epidural scarring, however, often has poorly defined margins, is of low or intermediate signal intensity on T1-weighted images, and is either of low or relatively high signal intensity on T2-weighted sequences. Scar tissue usually is distributed along the lateral margin of the spinal canal along sites of the surgical procedure. Mass effect on the thecal sac is common to recurrent disk herniation and less common to epidural scarring.

Bundschuh and colleagues reported that scarring and herniated disk may be differentiated on nonenhanced MR imaging with an accuracy comparable to that of the intravenous contrast-enhanced

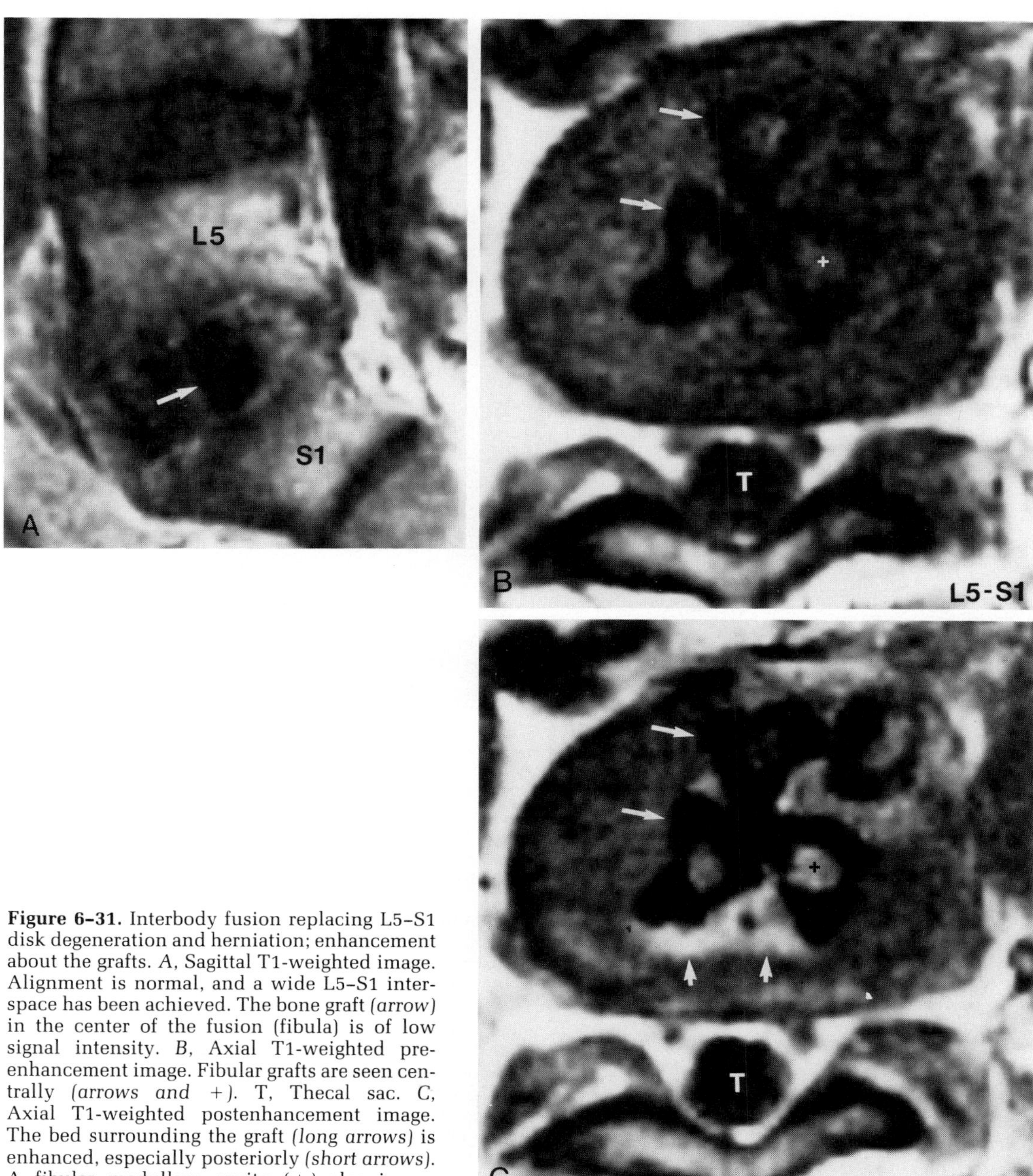

Figure 6–31. Interbody fusion replacing L5–S1 disk degeneration and herniation; enhancement about the grafts. *A*, Sagittal T1-weighted image. Alignment is normal, and a wide L5–S1 interspace has been achieved. The bone graft *(arrow)* in the center of the fusion (fibula) is of low signal intensity. *B*, Axial T1-weighted preenhancement image. Fibular grafts are seen centrally *(arrows and +)*. T, Thecal sac. *C*, Axial T1-weighted postenhancement image. The bed surrounding the graft *(long arrows)* is enhanced, especially posteriorly *(short arrows)*. A fibular medullary cavity *(+)* also is enhanced.

CT scan.[7] However, the intravenous administration of gadopentetate dimeglumine has improved markedly the ability to distinguish accurately recurrent disk herniation from postoperative scarring (Figs. 6–33, 6–34).[68] When images are obtained immediately after the administration of gadopentetate dimeglumine, the scar tissue is enhanced significantly, whereas the herniated disk usually remains the same.[16] The lack of early enhancement of the herniated disk is attributed to its relative avascularity. Occasionally, herniated disks also may enhance early (Fig. 6–35).

SPINAL TRAUMA

Spinal trauma is discussed in Chapter 5.

MARROW DISEASE

The axial skeleton acts as a major site of red marrow hematopoiesis throughout life. T1-weighted images convey well the conversion of red marrow, which is of intermediate signal intensity, to fatty marrow, which is of high signal intensity, in the

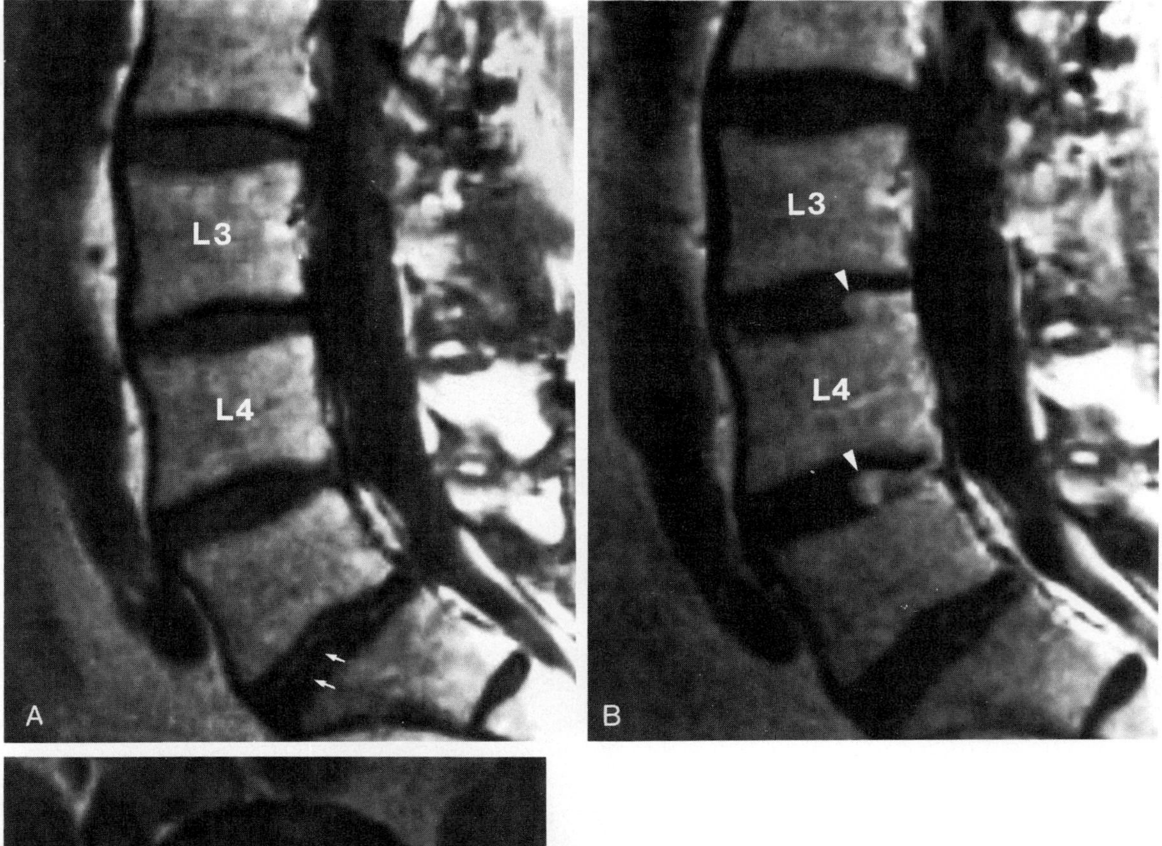

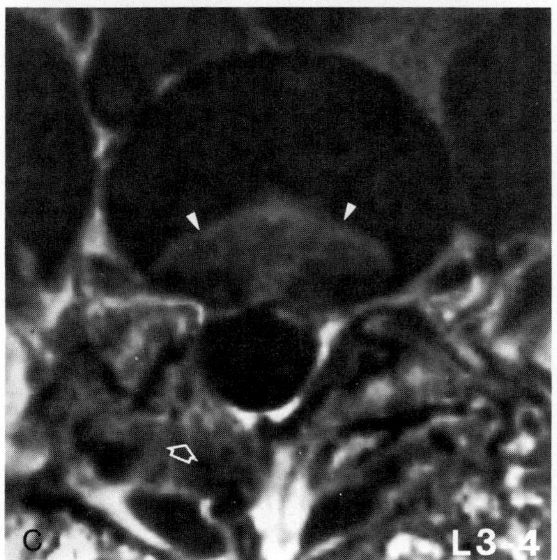

Figure 6–32. Enhancement of diskectomy bed (fibrous tissue) after surgery. *A*, Sagittal T1-weighted pre-enhancement image. Thinning of the L3–4, L4–5, and L5–S1 lumbar disks with gas *(arrows)* from the more degenerated L5–S1 disk is seen. T1-weighted sagittal *(B)* and axial *(C)* enhanced images. The L3–4 and L4–5 disks enhance posteriorly after gadopentetate dimeglumine administration. After cutting of the outermost portion of the anulus, the posterior portions of the disks were curetted to the same depth. Scar tissue replacement of the anulus is the likely cause for this enhancement *(arrowheads)*. (It is not diskitis because the signal intensity of the enhanced structure is identical to that of the region that underwent curettage; moreover, in diskitis, enhancement of the entire disk occurs.) Axial sections showed a symmetric wedge with sharp, well-defined borders of the disk curettage *(arrowheads)*. Open arrow points to laminectomy site.

appendicular skeleton that begins in childhood and is completed by approximately 25 years of age. Despite acting as a major hematopoietic center, even the red marrow in the vertebral bodies undergoes conversion to fatty marrow (Fig. 6–36). The changes are more subtle in the axial skeleton. Adult patterns, as described by Ricci and co-workers,[66] include bandlike and trianglelike fatty infiltrates along the end plates (Fig. 6–37); small and large areas of well-marginated marrow fat create considerable heterogeneity (Fig. 6–38). Variability is great in the marrow pattern among adults and is exemplified by differences among the marrow of the cervical, thoracic, and lumbar regions in the same person.

Even though MR imaging has proved to be very sensitive in cases of pathologic marrow, the findings often are nonspecific and usually require close clinical correlation to derive an accurate diagnosis. Focal disease of the vertebral bodies can be discerned more easily than diffuse disease. Focal disease usually is manifested primarily as high signal intensity (within the predominantly fatty character of the centrum in the adult) on T1-weighted images and as low signal intensity on T2-weighted (Fig. 6–39), small flip-an-

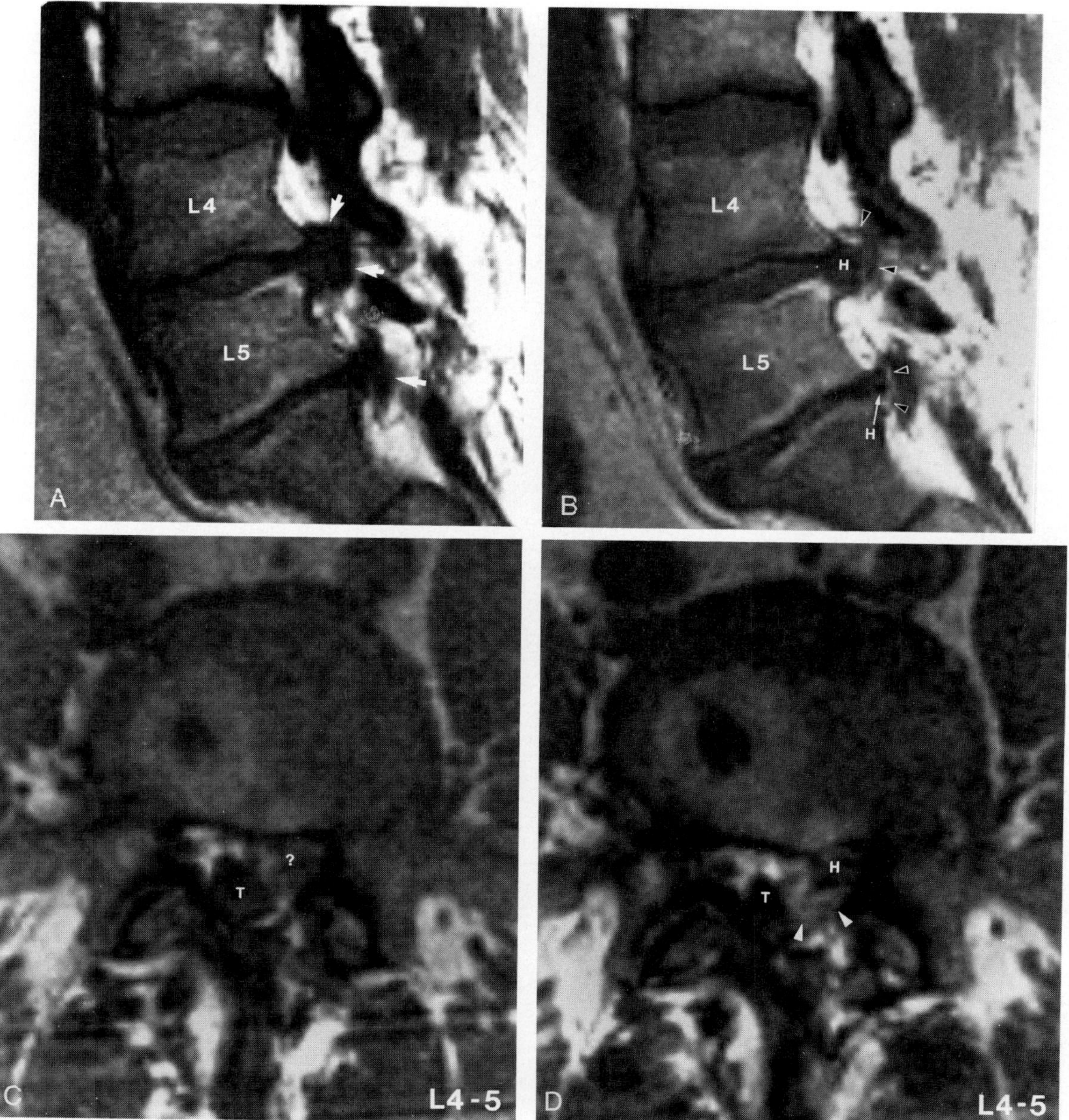

Figure 6–33. Recurrent disk herniation with extradural fibrosis. *A* and *B*, Sagittal T1-weighted pre-enhancement and postenhancement images. Degenerative disease of L4–5 and L5–S1 is associated with the large extradural defects *(arrows at both levels)*. Despite subtle motion, enhancement *(arrowheads)* can be detected about the posterior portion of the L4–5 herniation (H) and within portions of the L5–S1 herniation (H). *C* and *D*, Axial T1-weighted pre-enhancement and postenhancement images. A characteristic extradural defect *(?)* is seen on the left side anteriorly; however, the patient had had a lumbar diskectomy previously. The enhancement pattern of scar tissue *(arrowheads)* and the unenhanced pattern consistent with residual or recurrent disk herniation (H) are observed. The thecal sac (T) is displaced to the right side posteriorly.

gle GRE, or short tau inversion-recovery (STIR) sequences. The STIR sequences (which suppress signal from fat, fibrosis, and bone) and the fat suppression sequences can be especially sensitive to focal pathology. On gradient-echo sequences, bone marrow has low signal intensity owing to intravoxel dephasing secondary to discrepant diamagnetism of trabecular bone and marrow cellular elements. This holds true for either small or large flip-angle GRE sequences.[73] Although diffuse disease of the marrow may be difficult to recognize, whenever an intervertebral disk appears brighter than bone marrow on T1-weighted images, the adjacent vertebral body should be considered abnormal in an adult (Fig. 6–40).[9] The sensitivity of MR imaging in detecting neoplastic involvement of bone has been confirmed in the results of a large series. It may be particularly useful in detecting myeloma, in which abnormalities apparent on MR images may appear normal on a bone scan.[14]

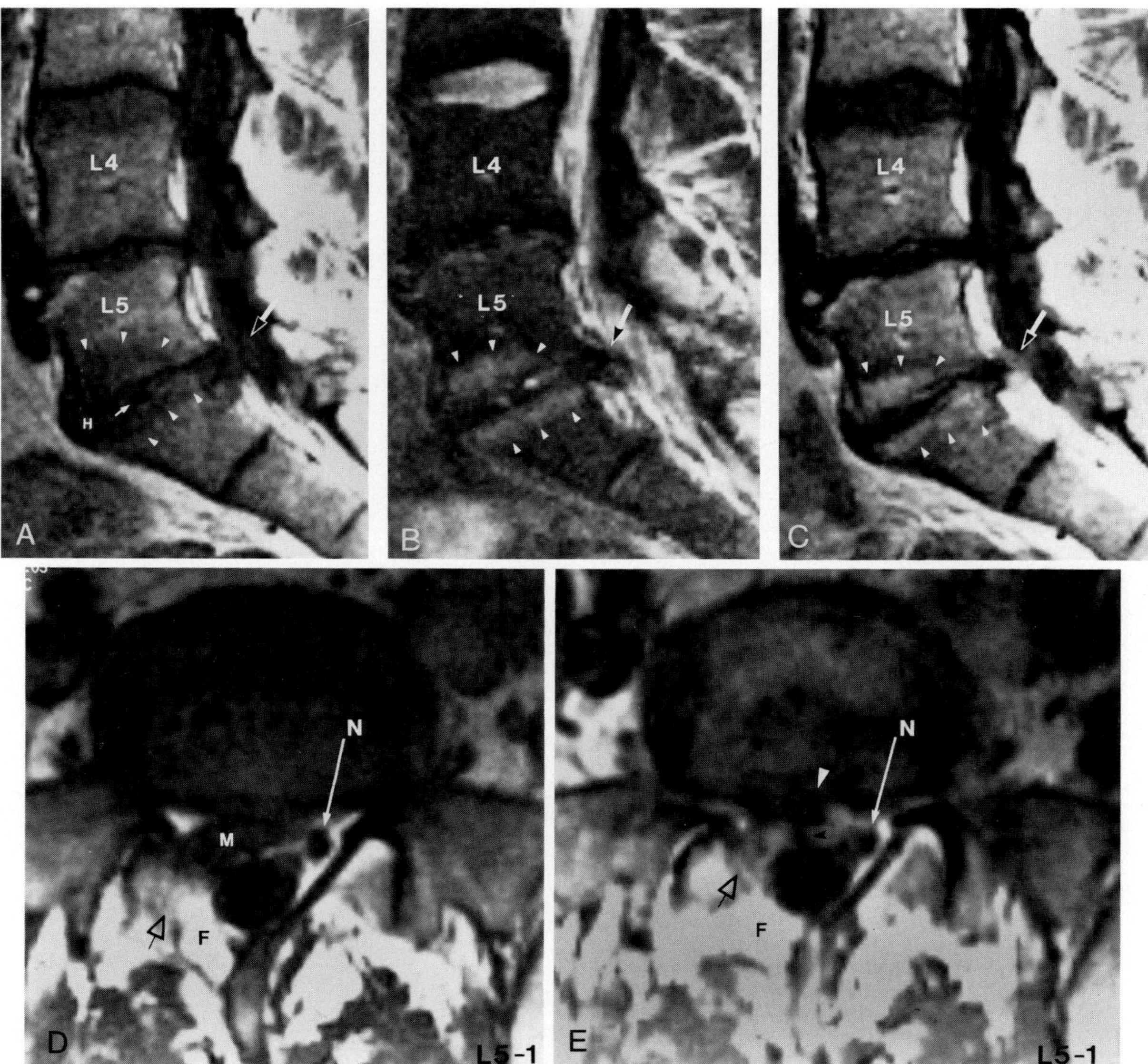

Figure 6–34. Extradural mass from a scar, simulating herniation. *A*, Sagittal T1-weighted, pre-enhanced image. Degeneration of the disks is present: thinning, gas, or calcification within the L5–S1 *(small arrow)*; type I bone marrow changes of the adjacent vertebral bodies *(arrowheads)*[51]; possible herniation as the cause of the larger posterior mass *(large arrow)*; and anterior disk herniation (H). *B*, Sagittal T2-weighted image. The anterior extradural mass *(large arrow)* is well outlined by cerebrospinal fluid in this sequence. Note the brightness *(arrowheads)* of the intervertebral bodies adjacent to the degenerated disk, indicating type I changes.[51] *C*, Sagittal T1-weighted image intravenously enhanced with gadopentetate dimeglumine. Note that the cap of the large extradural mass has enhanced almost completely. The vascularized fibrous tissue (type I bone marrow change, *arrowheads*) is also enhanced. *D*, Axial T1-weighted, pre-enhanced image. The right S1 nerve root is probably obscured by the large extradural mass (M). The left S1 nerve root (N) is well shown. Fat (F) was used at the site of laminectomy *(open arrow)*. *E*, Axial T1-weighted, postenhancement image. Note how the mass (see *C)* has enhanced nearly completely, but the right S1 nerve root is not definitely identified. The midline septum of Schallinger and colleagues[71] is slightly canted. The region of low signal intensity *(white arrowhead)* between the septum and vertebral body in the midline is an osteophyte and explains the low signal intensity beneath the enhancing cap in *C*. F, Fat; *open arrow*, laminectomy site.

Bone marrow disorders may be divided into four main categories: reconversion (e.g., hyperplastic anemias), infiltration (e.g., tumors, infection, myeloproliferative disorders, storage disorders), ischemia or edema (e.g., avascular necrosis, reflex sympathetic dystrophy, migratory osteoporosis), and depletion (e.g., aplastic anemia, radiation therapy, chemotherapy).[55,90,93]

It may be especially difficult to distinguish regeneration of normal bone marrow after therapy from recurrent tumor or other pathologic conditions. For this reason, research has centered on quantitative T1 and T2 relaxation times and quantitative chemical shift imaging to differentiate these conditions; unfortunately, the results are mixed.[54,67,78,84,97] It is possible that each specific marrow disease has its own appearance in response to therapy and regeneration. Aplastic anemia is asso-

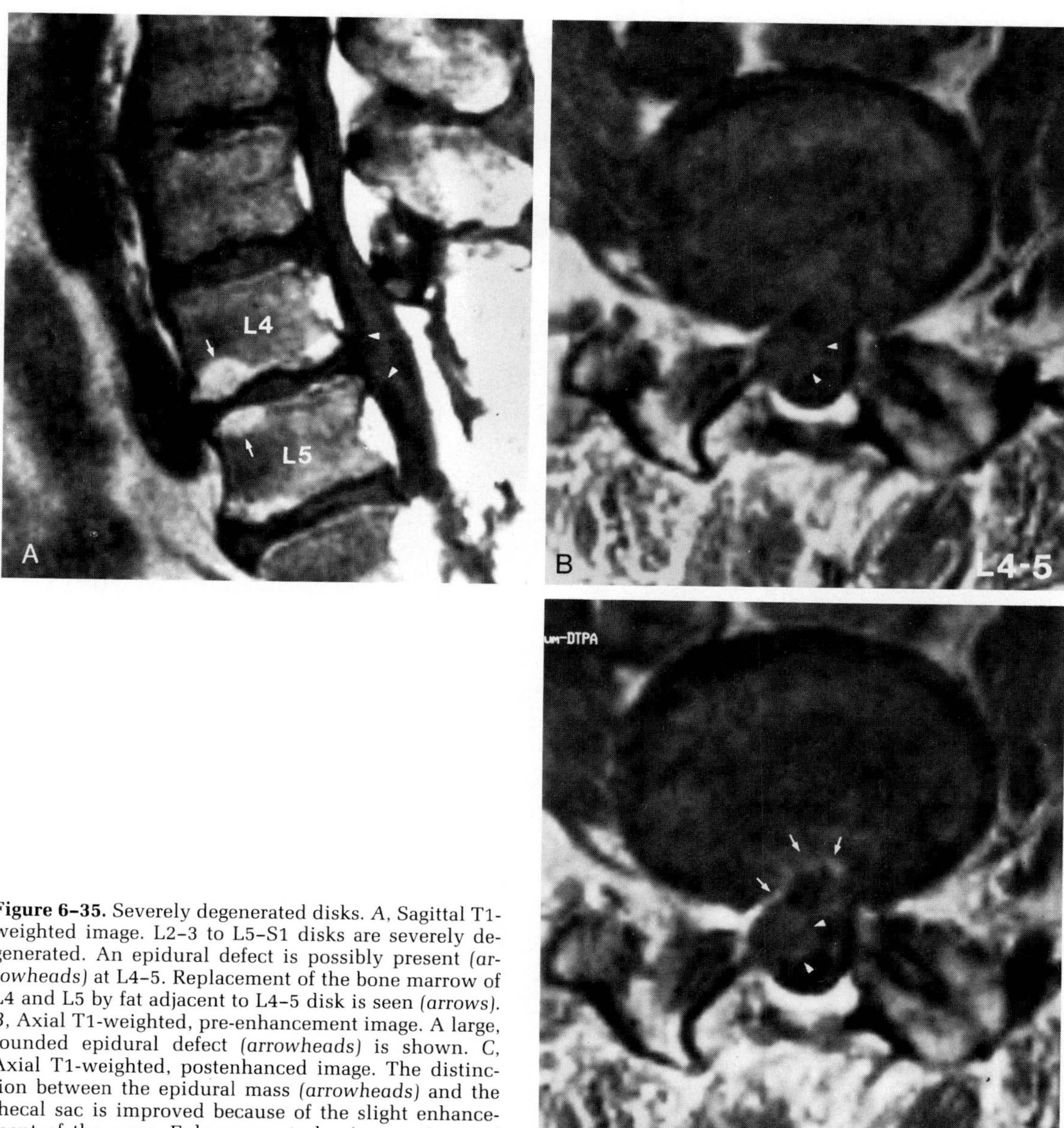

Figure 6–35. Severely degenerated disks. *A*, Sagittal T1-weighted image. L2–3 to L5–S1 disks are severely degenerated. An epidural defect is possibly present *(arrowheads)* at L4–5. Replacement of the bone marrow of L4 and L5 by fat adjacent to L4–5 disk is seen *(arrows)*. *B*, Axial T1-weighted, pre-enhancement image. A large, rounded epidural defect *(arrowheads)* is shown. *C*, Axial T1-weighted, postenhanced image. The distinction between the epidural mass *(arrowheads)* and the thecal sac is improved because of the slight enhancement of the mass. Enhancement also is seen *(arrows)* within the ruptured anulus.

ciated with markedly hypocellular marrow with increased fat content (Fig. 6–41). Although initially this may be difficult to identify in adults, after successful therapy foci of low to average signal intensity representing reconverted red marrow would be expected to be found and can be followed qualitatively.[47] Quantitative measurements of either chemical shift or T1 relaxation times have been necessary to evaluate correctly the therapeutic responses of various leukemias.[47,54] Hanna and colleagues[28] performed a prospective study in children treated with radiation or chemotherapy for bone metastases from neuroblastoma, lymphoma, sarcoma, or leukemia; biopsies of abnormal sites suggestive of neoplasm were done. These investigators were unable to differentiate consistently recurrent or residual neoplasm from posttherapy changes using T1-weighted, T2-weighted, STIR, or contrast-enhanced (gadopentetate dimeglumine) T1-weighted images. These nontumor marrow abnormalities may be correlated pathologically with necrosis, serous atrophy, or granulation tissue. Marrow signal intensities after chemotherapy may be increased, decreased, or normal, depending on when the study is performed.[55]

Repopulation after bone marrow transplantation characteristically appears as material of intermediate signal intensity along the periphery of the body that gradually replaces the entire internal signal of the fatty marrow.[80]

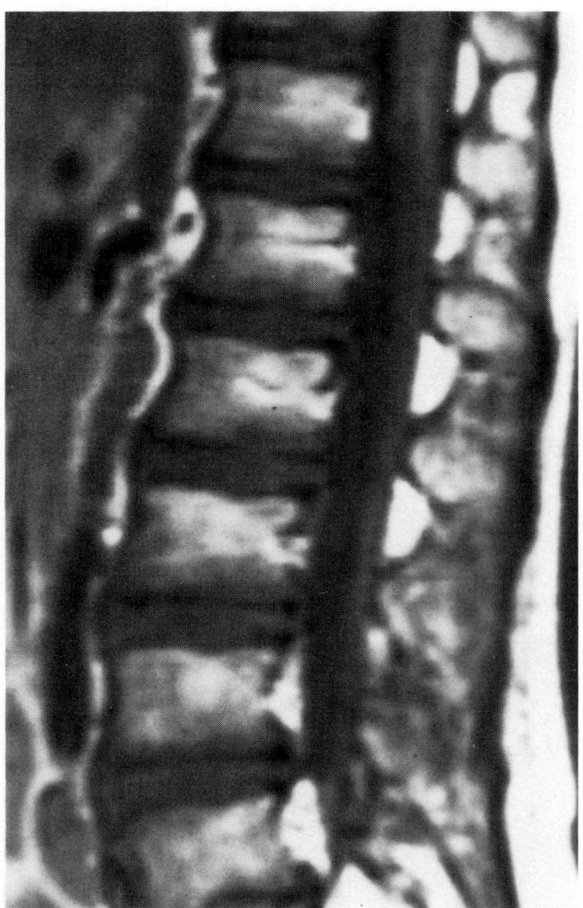

Figure 6–36. Fourteen-year-old boy with nonspecific back pain. Sagittal T1-weighted image shows the typical juvenile pattern of fat distribution in the lumbar spinal vertebrae as described by Ricci and co-workers.[66] Fat clusters around the base of the vertebral vein at the middle of the centrum.

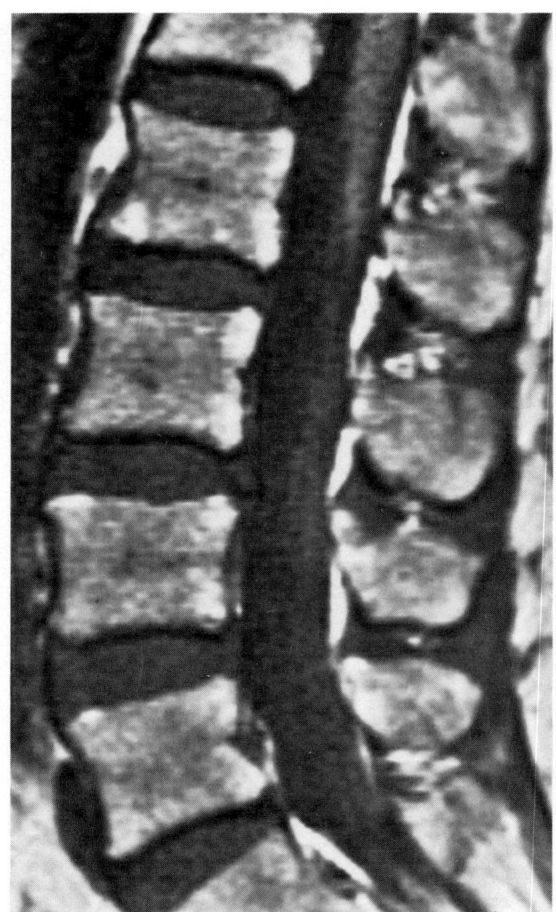

Figure 6–37. Nonspecific low back pain in a 61-year-old woman. Sagittal T1-weighted image. Linear and triangular areas of fatty infiltration are seen, which Ricci and colleagues[66] described as being typical in an older patient.

Radiation effects on vertebral marrow are now well known. Usually significant change does not occur in the first 2 weeks after radiotherapy, although some edema may be appreciated on T2-weighted or STIR sequences. Between 3 and 6 weeks after therapy, more apparent heterogeneity and elevated signal intensity (on T1-weighted images) are seen in the central portion of the body. Between 6 and 14 weeks and continuing for as long as 2 years, the fatty conversion of the marrow predominates. Peripheral intermediate signal intensity, presumably reflecting regenerating tissue, may be present adjacent to the endplates.[81,99]

NEOPLASTIC CONDITIONS AND METASTATIC DISEASE

Hemangiomas are the most common benign tumors of the spine. Schmorl and Junghanns[72] found them in approximately 11 per cent of persons in an autopsy series. However, they are rarely symptomatic.[42,53] Hemangiomas are angiomatoid fibroadipose tissue interspersed among tortuous thin-walled sinuses. They are most common in the thoracic spine. On MR images, these lesions tend to be well circumscribed tumors within the vertebral body and appear to be of high signal intensity on both T1-weighted and T2-weighted sequences (Fig. 6–42). The T1 shortening is produced by the fatty component and the T2 prolongation by the angiomatous component. The very low signal intensity of bone trabeculae is overshadowed by the signals of the matrix. Focal fatty infiltration, a common marrow variant, may be confused with hemangiomas on T1-weighted

Figure 6–39. Unusual location for a lipoma of a vertebral body. Paramedian sagittal T1-weighted (*A*) and T2-weighted (*B*) images show that the region developed from the ring apophysis contains this lesion (*arrowheads*), which is of high signal intensity on T1-weighted image and is less intense on the T2-weighted sequences, a finding that is consistent with the presence of a fatty lesion (i.e., a lipoma). It is mildly expansile, as its upper half bulges into the intervertebral disk, and no cortical rim is resolved.

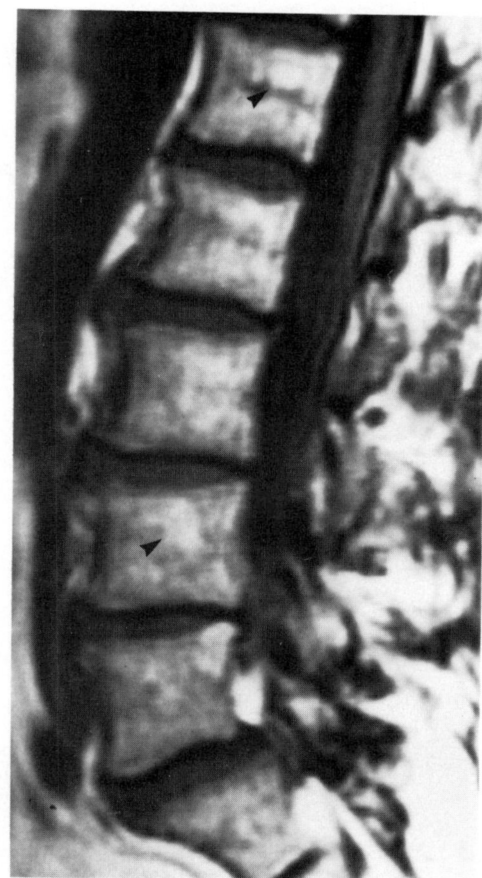

Figure 6–38. Midsagittal T1-weighted image in a 65-year-old woman. Typical large areas of well-circumscribed fatty infiltration *(arrowheads)* of the marrow are present, which are described by Ricci and colleagues[66] as an older adult pattern. The wavy lines are artifact created from intrathecal iohexol (Omnipaque) for myelography that was performed within the previous 6 hours.

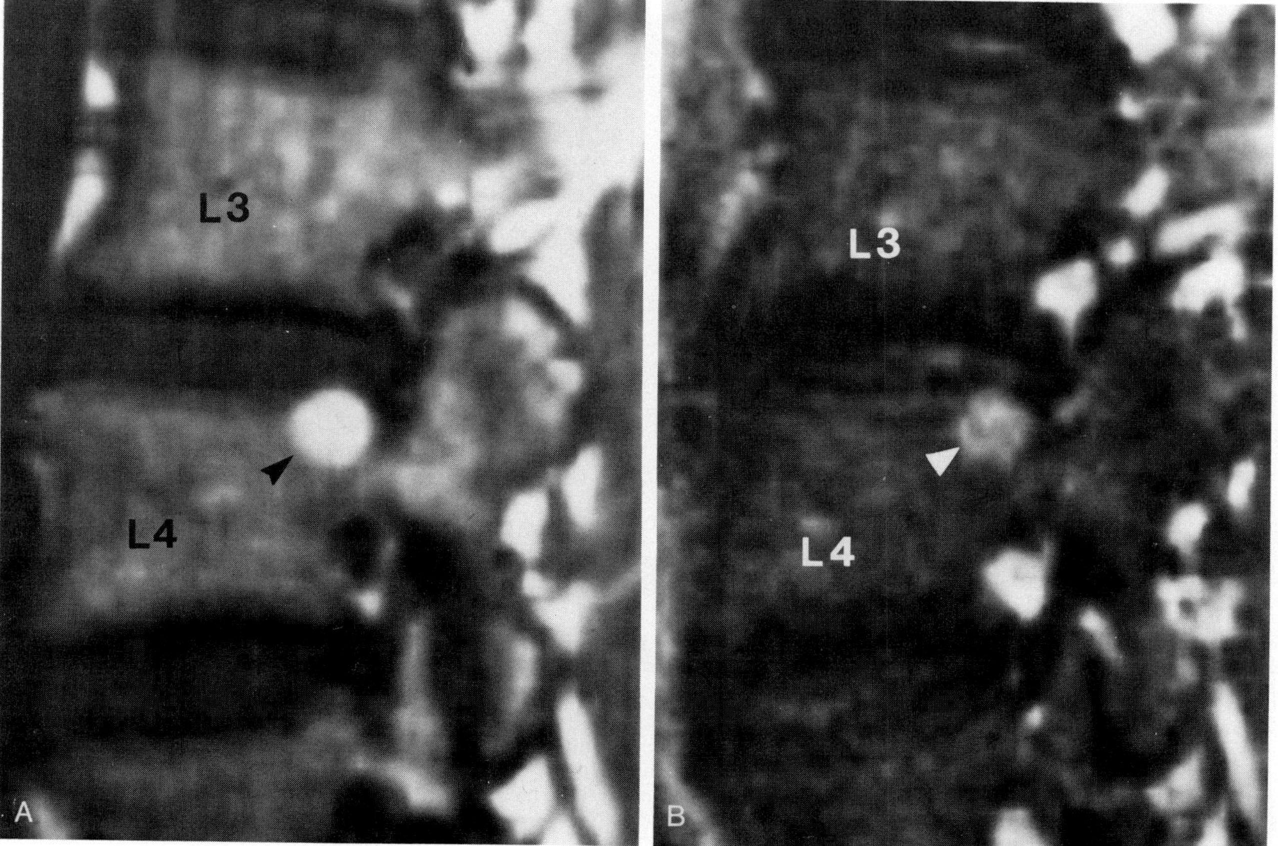

Figure 6–39 See legend on opposite page

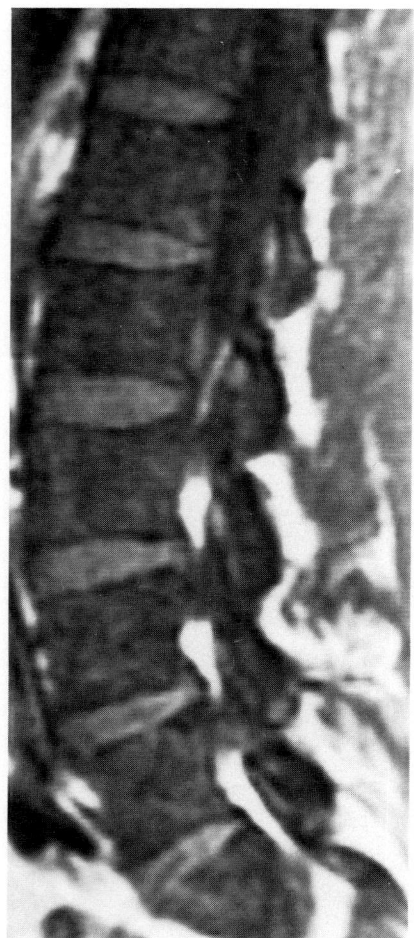

Figure 6–40. Sagittal T1-weighted image from a 58-year-old man with prostate carcinoma shows diffusely decreased signal intensity within all of the vertebral bodies, which is atypical of metastatic disease. As Castillo and associates noted,[9] whenever the disks are brighter than the vertebral bodies on T1-weighted images, the marrow is always abnormal. This pattern would be indistinguishable from myelofibrosis or diffuse leukemic infiltration.

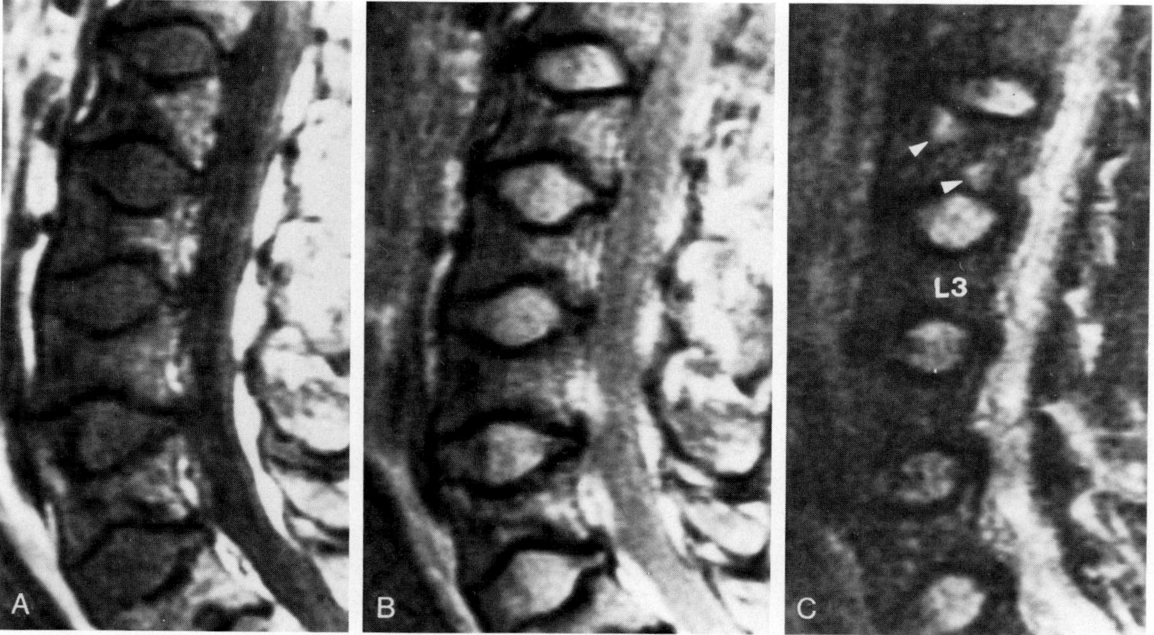

Figure 6–41. Sickle cell crisis with back pain in a 28-year-old man. *A*, The sagittal T1-weighted image shows markedly heterogeneous marrow for a patient of this age, which in this instance is indicative of anemia with consequent red marrow hyperplasia. *B*, The sagittal proton density image best depicts the "fish mouth" deformity of the spine commonly described on plain radiographs that represents vertebral body infarctions with compression from the intervertebral disks. *C*, The T2-weighted image demonstrates focal areas of increased signal intensity in the L2 body *(arrowheads)* consistent with acute infarctions. Also note the triangular or wedge-shaped areas of high signal intensity in the posterior aspect of the lumbar vertebral bodies in *A* and *B* that represent fatty replacement of marrow in subacute hemorrhages.

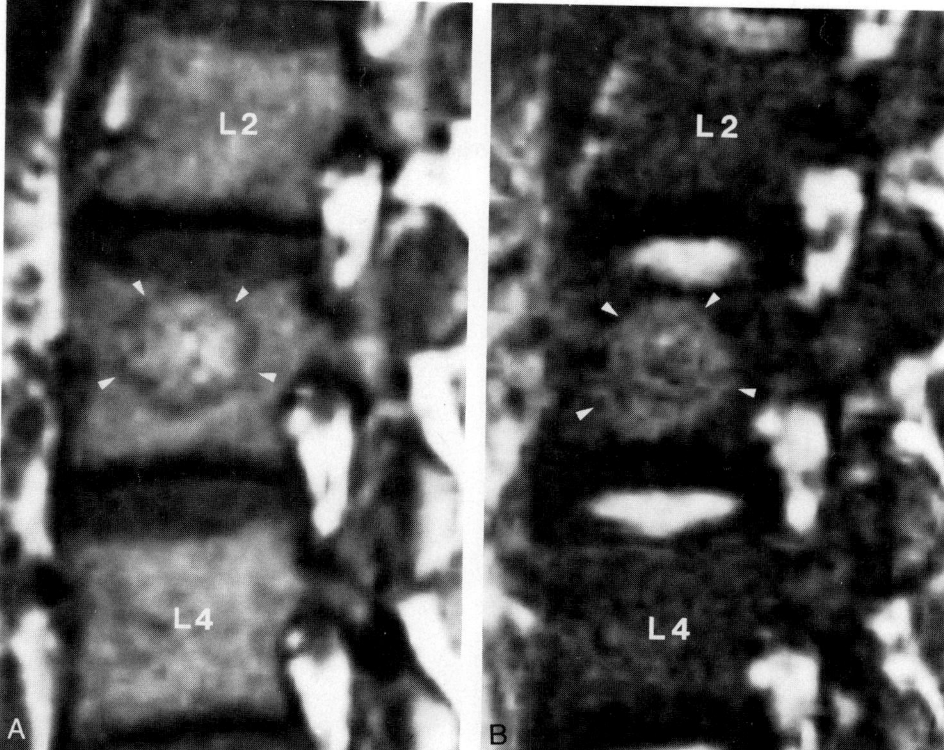

Figure 6–42. Evidence of hemangioma of L3. *A*, Paramedian sagittal T1-weighted image. A well-circumscribed lesion is seen *(arrowheads)*, which is primarily of high signal intensity with a low-intensity convoluted rim surrounding it. *B*, Paramedian sagittal T2-weighted image. The high signal intensity of the lesion *(arrowheads)* is maintained, which is typical of hemangiomas.

images; however, the expected decrease in signal intensity of fat on T2-weighted imaging serves to distinguish it from the high signal intensity of hemangiomas on T2-weighted images (see Fig. 6–39).[27]

Only 6 per cent of osteoid osteomas occur in the spine. This tumor is found predominantly in the lumbar area in males between 5 and 20 years of age.[40] The nidus is classically described as less than 1 cm in diameter, and the tumor characteristically arises from the posterior elements.[33] The vertebral body alone is involved in only 7 per cent of cases.[23] The calcification in the nidus and the surrounding sclerosis on plain radiographs are often accompanied by a signal of low intensity on T1-weighted and T2-weighted spin-echo images (Fig. 6–43). Computed tomographic scanning is described as more sensitive in distinguishing the central nidus from the surrounding sclerosis, which typify this lesion on CT scans. The nidus may be obscure on MR images. An inflammatory response of edema may be obvious on T2-weighted images. Benign osteoblastomas (giant osteoid osteomas) are osteoblastic lesions that are similar microscopically to osteoid osteomas and are those lesions arbitrarily defined as having a nidus greater than 1 to 1.5 cm.[3] This expansive tumor usually has a thin cortical rim and may have associated soft tissue masses and an epidural extension.[45] These tumors typically are heterogeneous because of both hemorrhagic and calcific components. In 50 per cent of lesions, the tumor matrix may not be heavily mineralized, such that this tumor may have an appearance similar to that of an aneurysmal bone cyst on MR imaging. On CT scans, they may be differentiated by demonstrating a surrounding sclerosis in the osteoblastoma that is not present in aneurysmal bone cysts.[3] Typically osteoblastomas have a reactive component, which would lower the signal intensity of adjacent marrow on T1-weighted images. Crim and associates described a case in which the inflammatory response involved two adjacent vertebral bodies.[13]

Eosinophilic granuloma is part of the spectrum of Langerhans's cell histiocytosis that occurs traditionally in bone.[58] This is a nonneoplastic condition (previously classified as neoplastic) that is associated with infiltrating granulomas. The disease is uncommon in persons older than 30 years of age. Symptoms are variable, but systemic signs such as fever and weight loss are not uncommon. In the spine, the lesions may be single or multiple.[46,59] In the early stages, the lesions appear as a cystic mass within the vertebral body, of low signal intensity on T1-weighted images and of high signal intensity on T2-weighted sequences. Frequently the lesions have progressed to vertebral plana by the time of presentation. Magnetic resonance imaging has been used to demonstrate the normalization of the morphology of the vertebral body with healing.[26] Meningeal infil-

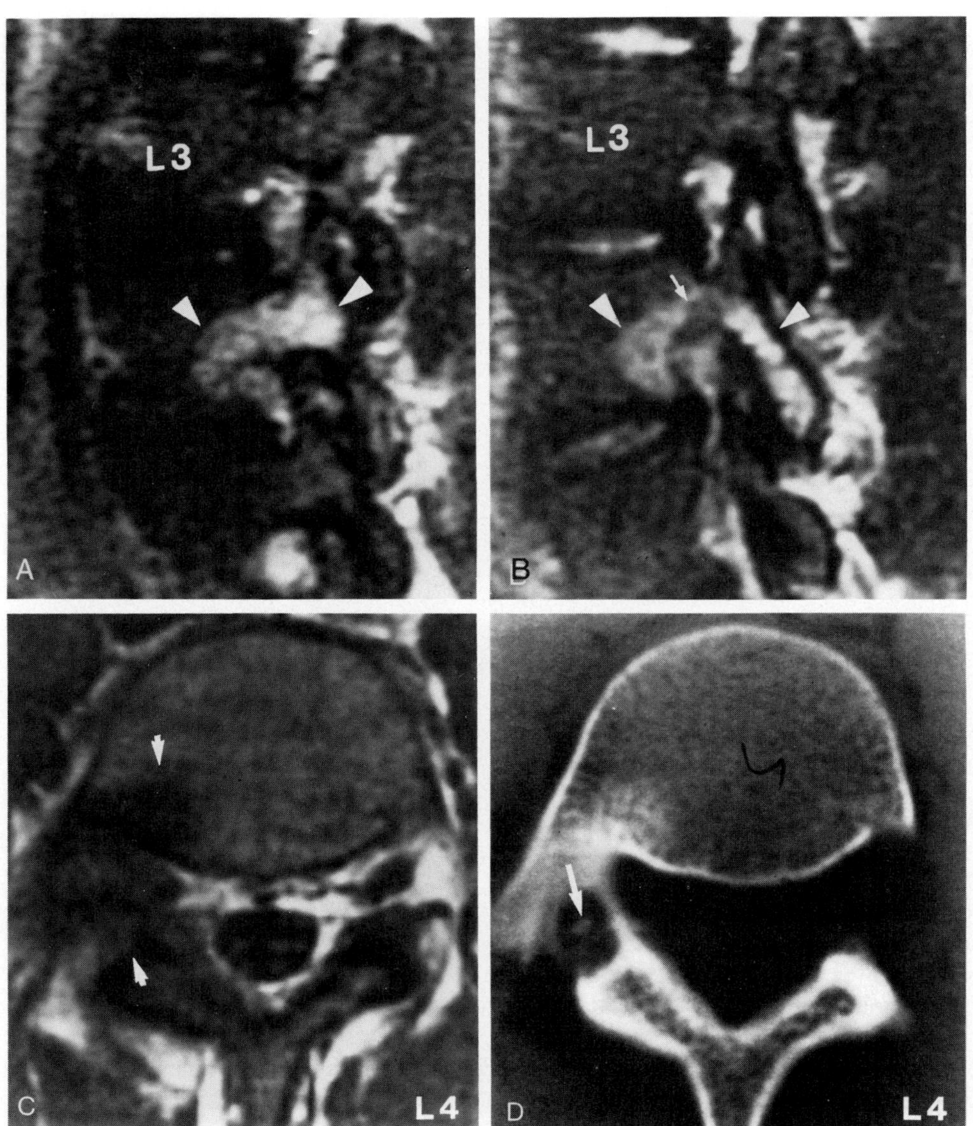

Figure 6–43. Signs of osteoid osteoma of L4 in a 21-year-old man. *A* and *B*, Paramedian sagittal T2-weighted images. The dramatic, well-circumscribed high signal intensity *(arrowheads)* of the left inferior facet, pedicle, and posterior portion of the body of L4 is evident. The lesion does not expand these structures. A small focus of low signal intensity, the calcified nidus, is seen *(arrow)*. *C*, Axial T1-weighted image of L4. The low signal intensity edema of the inflammatory response from this tumor extends into the right lamina and the right posterolateral portion of the vertebral body *(arrows)*. Note that the patient's scoliosis has created an axial slice with the right pedicle and the left neural foramen in the same plane. *D*, Axial computed tomography scan of L4. The calcified nidus *(arrow)* is evident within the right pedicle. Note an asymmetry in the axial cut similar to that evident in *C*.

tration of the central nervous system is not uncommon when other central nervous system lesions of Langerhans's cell histiocytosis are present.[37] With the inherent sensitivity of MR imaging with gadopentetate dimeglumine in vivo, demonstration of meningeal involvement is now possible.[17]

It has been stated that all metastatic foci are of decreased intensity on T1-weighted images when compared with that of the surrounding bone marrow in adults.[14,93] This feature exemplifies a basic advantage that MR imaging has over bone scintigraphy in detecting metastases. Because most hematogenously spreading metastases infiltrate the bone marrow and then destroy trabecular and cortical bone later, MR imaging would be expected to be able to discover infiltration earlier and define its extent more accurately than scintigraphy. This assumption may be particularly true in cases of myeloma, renal cell carcinoma, and melanoma, in which extensive disease may exist without creating the factors necessary for phosphate chemisorption.[20] However, CT is better able to detect cortical erosions, sclerotic changes, and calcification than MR imaging. Magnetic resonance imaging is superior to plain CT and equal to CT with intrathecal contrast for determining important anatomic relationships before surgery.[4] Because it is now accepted that MR imaging and CT findings complement each other in any given neoplasm,[105] the goal is to determine whether either alone can provide sufficient diagnostic answers.

There are no specific measures to distinguish various metastatic processes from lymphoma (Fig. 6–44) or myeloma (Fig. 6–45) on MR images. Typically each creates a signal of low intensity within the marrow on T1-weighting. The appearance of metastasis varies considerably on T2-weighted im-

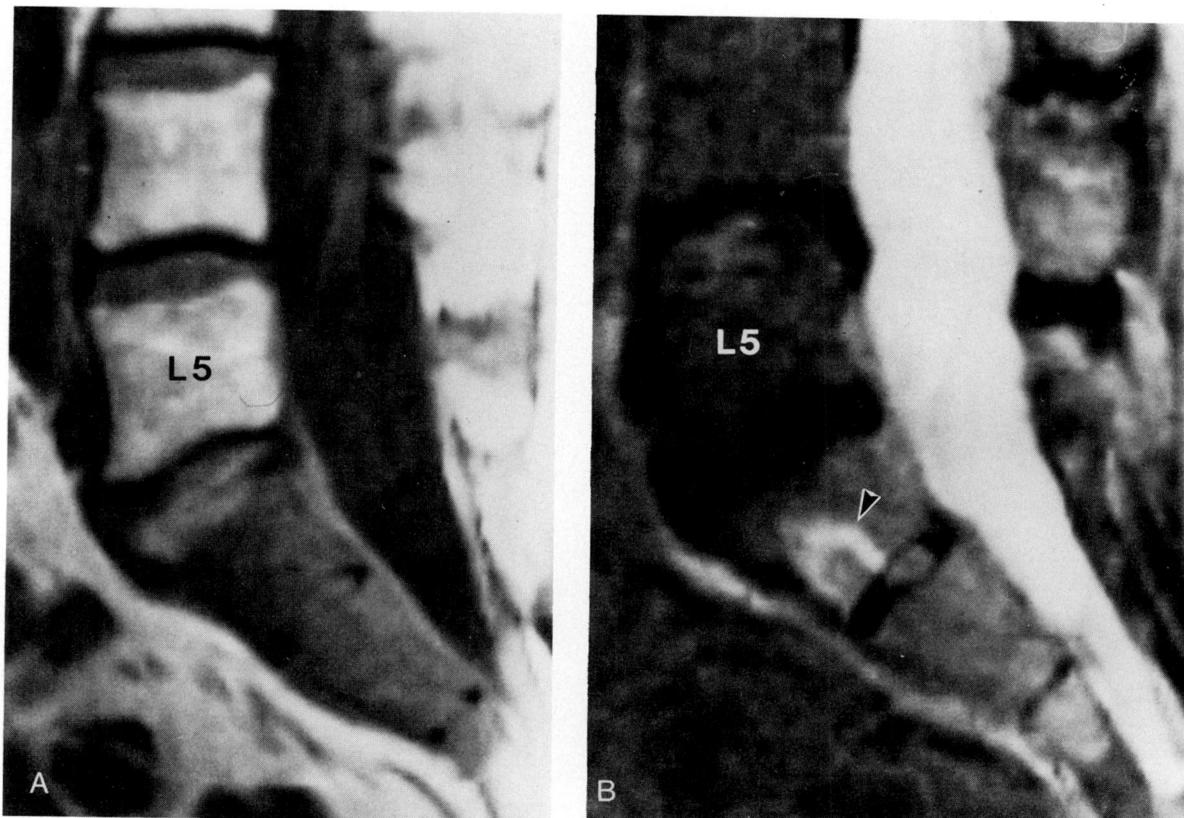

Figure 6–44. Evidence of lymphoma of S1, S2, and S3 in a 73-year-old woman. *A*, Sagittal T1-weighted image. A tumor of low signal intensity infiltrates S1 and to a lesser extent S2 and S3, sparing the disk spaces as myeloproliferative disorders would be expected to do. The bright signal of the adjacent vertebral bodies is created as a result of the very fatty marrow seen in the elderly. This patient has not received radiation therapy. *B*, Sagittal T2-weighted images. Lymphoma involving S1 and S3 is manifested as the more focal, very high signal intensity area, which may represent necrosis. The foci of involvement in S1 *(arrowhead)* and S3 are seen on the T2-weighted image, as is the generalized S1–3 increase in intensity.

ages, however, reflecting differences in tissue type, cellularity, water content, and quantity of fibrotic, hemorrhagic, calcific, necrotic, or inflammatory components that are present.[93,105] Signals of low, intermediate, or high intensity may be visible on T2-weighted spin-echo, STIR, fat suppression techniques, or small flip-angle GRE sequences (Fig. 6–46).[22,82,85] Therefore, T2-weighted images alone are not particularly useful for identifying metastases. The administration of gadopentetate dimeglumine may make some lesions more conspicuous in certain circumstances (e.g., leukemia)[54]; it may also obscure other lesions if fat suppression is not applied. Blastic metastases are of very low signal intensity on T1-weighted and T2-weighted images. Moreover, lesions often lose signal intensity on T2-weighted images after therapy.

A common clinical problem is differentiating neoplastic pathologic compression fractures in the elderly from nonneoplastic fractures. Unfortunately, there is no easy way to make this distinction in most malignant fractures. On spin-echo images, the bone marrow, which is completely replaced by tumor, can be identified. Unfortunately, complete tumor replacement is not present in a significant number of cases.[100] The issue is further complicated in acute fractures because edema, blood products, and compressed bone collectively contribute to the observed signal intensities. Regardless of cause, in acute trauma the marrow initially may be abnormal and appear to have been replaced as manifested by low signal intensity on T1-weighted images and high signal intensity on T2-weighted sequences.[2,100] In acute osteoporotic compression fractures, the pedicles typically are spared[100]; we have noted, as Schmorl and Junghanns[72] also found, the bowed end plates, "enlarged" disks, and intact bone cortical surfaces in osteoporosis are uncomplicated by metastatic disease. Typically, chronic benign compression fractures have marrow that is of similar signal intensity to that of adjacent vertebral bodies on all sequences.[2,34] In a prospective study, Baker and colleagues[2] concluded that benign nonacute compression fractures could be differentiated from pathologic fractures by using a variety of sequences: Neoplastic involvement led to increased signal intensity

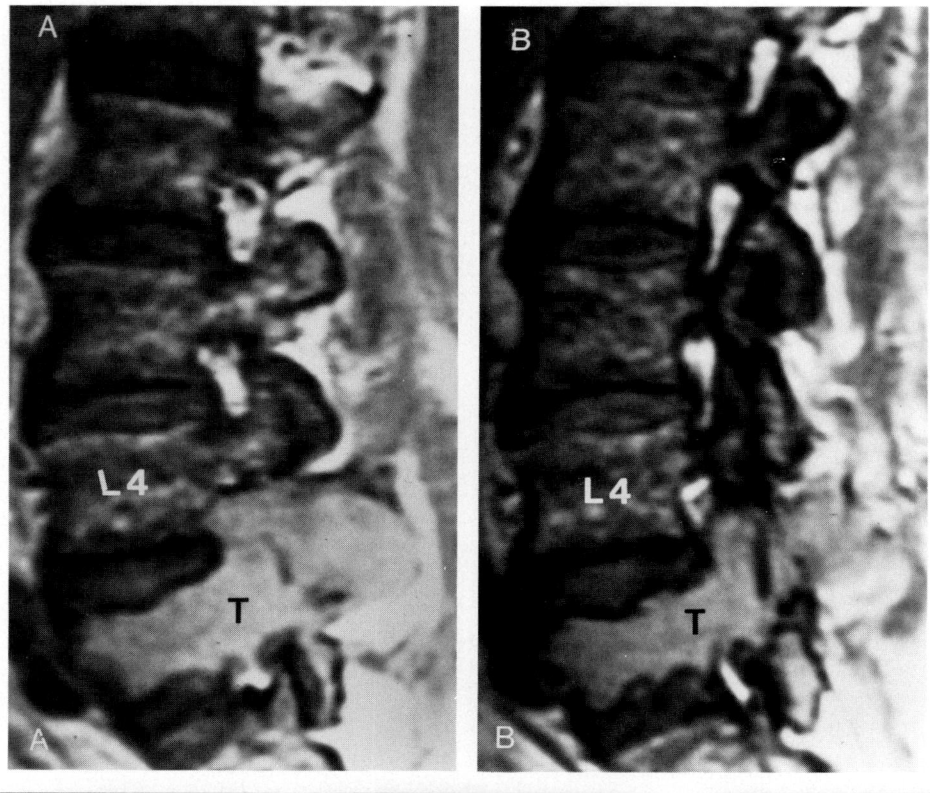

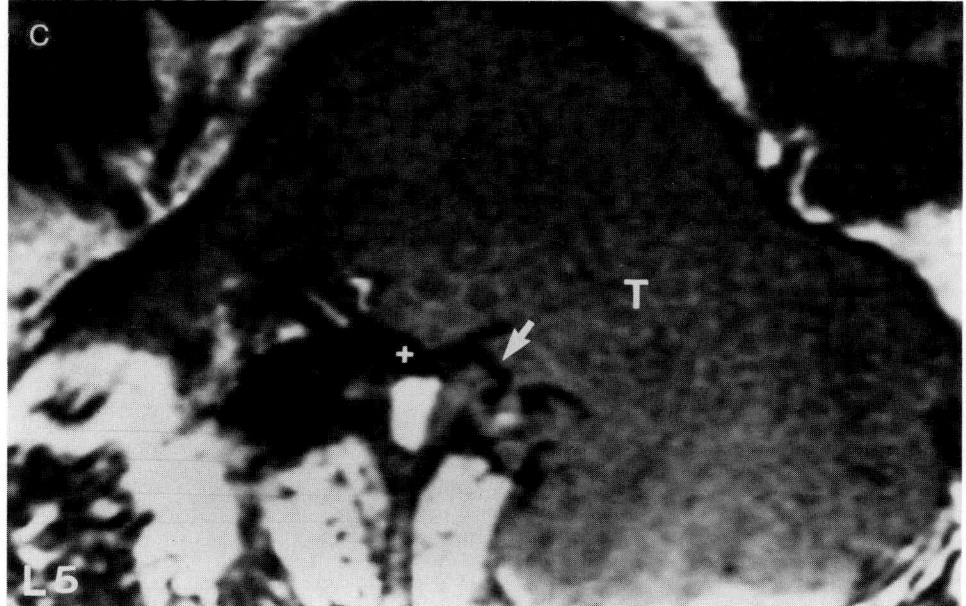

Figure 6–45. Indications of multiple myeloma in a 70-year-old man. *A* and *B*, Paramedian sagittal proton density images. A tumor (T) of high signal intensity has replaced the marrow of L5. The softening of bone that results has led to the relative expansion of the L4 and L5 disks, which are spared. The pedicle, superior facet of L5, and inferior facet of L4 cannot be seen owing to tumor tissue replacement. *C*, Axial T1-weighted image of L5. The axial view better highlights the large paraspinal component of this tumor (T). The tumor has replaced the left lamina, superior facet of L5, and most of the inferior facet of L4. A remnant *(arrow)* of the inferior facet of L4 can be seen. The spinal canal is almost completely occupied by tumor, and the thecal sac *(+)* is displaced posteriorly.

within the collapsed vertebra on T1-weighted images with fat suppression techniques, on T2-weighted images, or on STIR sequences.

Metastatic disease to the spine that is spread hematogenously can involve both the primary vertebral body and the epidural space. Tumors that frequently metastasize to the spine in adults include lung carcinoma, breast carcinoma, prostate carcinoma, myeloma, and lymphoma.[24] Metastases from leukemia also are seen in adults (Fig. 6–47). In children, metastases commonly arise from leukemia, sarcoma, neuroblastoma, lymphoma, and germ cell tumor.[39] In patients with lymphoma, MR imaging findings may lead to a reconsideration of marrow involvement when marrow biopsy findings are not in agreement with MR findings.[43]

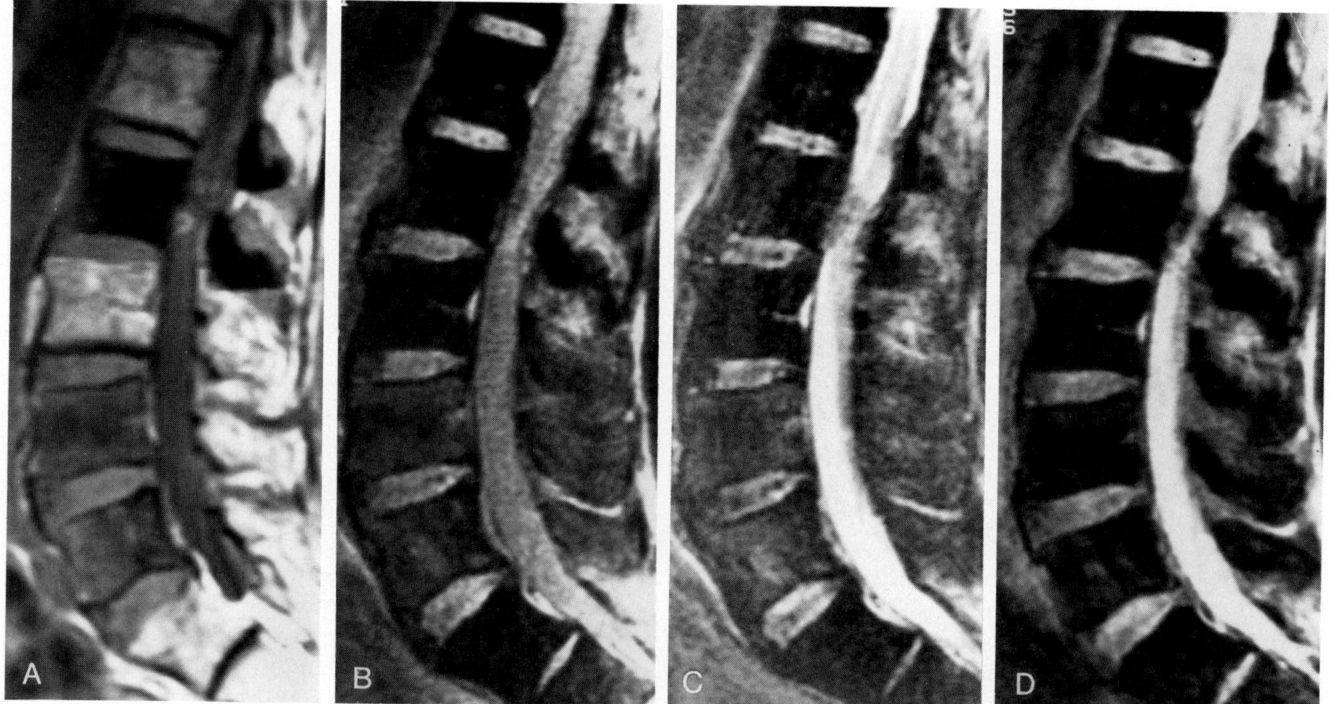

Figure 6–46. Evidence of blastic prostate carcinoma in a 66-year-old man. *A*, Sagittal T1-weighted image. Metastases here run the entire gamut of MR signal intensities even though they all appear similarly highly dense on plain radiographs. Which vertebral bodies are normal? None. They are all abnormal: of very low signal intensity at L2, of moderately low signal intensity at L4 and L5, and of intermediate intensity at L1 and L3. Note the extradural compression of the thecal sac. *B*, Sagittal proton density image. The L4 and L5 vertebral bodies have increased in signal intensity, presumably as a result of increased water content. S1, which might appear normal on a T1-weighted image *(A)*, has a signal intensity as low as that of L2 from metastatic ossification, more noticeable on longer TR images. *C*, Sagittal T2-weighted image. A further shift in signal intensities is now seen. Note that the L2 vertebral body is intermediate in intensity anteriorly. *D*, Sagittal multiplanar gradient-recalled acquisition of steady state (MPGR), 20 degrees. The MPGR sequences are more sensitive to the magnetic susceptibility effects of bone, resulting in a reduction in intensity.

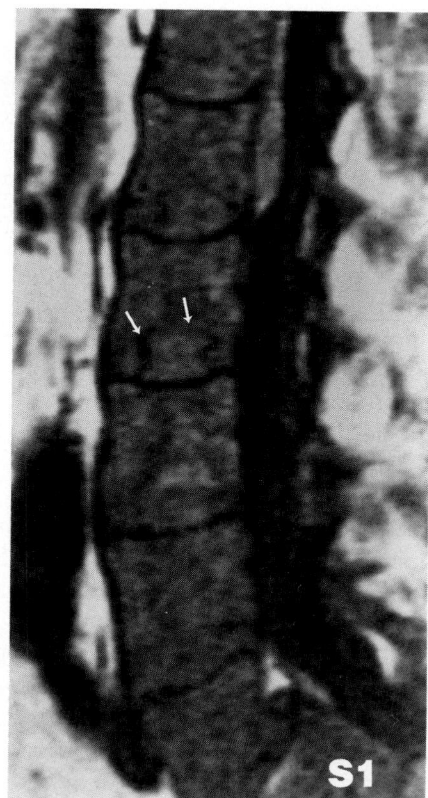

Figure 6–47. The spine of a 60-year-old patient with chronic lymphocytic leukemia, thrombocytopenia, and acute back pain. Sagittal T1-weighted image shows diffusely decreased signal intensity, owing to the replacement of marrow by fat at this age. The intervertebral disks are of equal to lower signal intensity in comparison with the marrow, indicating cellular leukemic infiltrates. (A prevertebral hematoma at T12–L1; a large Schmorl's node at L2–3*(arrows)*; and a grade I–II spondylolisthesis at L5–S1 are also seen.)

INFECTION

Infectious spondylitis primarily affects the vertebral body in adults, with secondary spread to other spinal components, most commonly the intervertebral disk. In pediatric patients, the infection primarily affects the disk space.[65] Magnetic resonance imaging is ideal for pediatric assessment because it does not require the use of ionizing radiation. Modic reported that MR imaging evaluation alone was more accurate (94 per cent) than bone scintigraphy alone (86 per cent) in detecting vertebral osteomyelitis.[50] For detecting experimentally induced osteomyelitis, Chandnani and co-workers found scintigraphy to be just as sensitive on MR imaging, with both of these being more sensitive than CT.[10]

By far the most common organism to infect the spine is *Staphylococcus aureus*. Other organisms include *Streptococcus viridans*, *Escherichia coli*, and *Klebsiella*. An increased frequency of *Salmonella* infection, tuberculosis, and fungal infection are seen in immunocompromised persons; tuberculosis, brucellosis, and echinococcosis can be considered in endemic areas or when the patient has traveled to such areas. The earliest sign of osteomyelitis is marrow edema manifested by low signal intensity on T1-weighted images and high signal intensity on T2-weighted images (Fig. 6–48). Marginal sclerosis observed on CT scans may be seen on MR images as a thin hypointense line. A classic feature of spondylitis in adults, attributable to the pattern of collateral blood flow, is involvement of the adjacent vertebral bodies and the intervening disk with sparing of the posterior elements. Diffusely decreased signal intensity of the vertebral body marrow is seen with poor distinction between the disk and the adjacent vertebral body, especially on T1-weighted images with absent or poorly defined cortical margins and increased signal intensity of the body or end plates on T2-weighted or small, flip-angle GRE images. The disk may protrude into the softened end plate or become flattened or both; the disk has very high signal intensity on T2-weighted images in adults but not always in children.[92] Prominent circumferential, inflamed prevertebral soft tissues also may be noted.[49,75,77,88]

The characteristic description just provided may be countered by many exceptions.[48,92] Only a solitary vertebral body may be involved, or adjacent vertebral bodies may be abnormal without changes in the intervertebral disk, or the posterior elements may be included. These exceptions are particularly true of tuberculous spondylitis or brucellosis in which in the early stage cortical margins may be maintained. A destructive crystal-induced spondyloarthropathy has been recognized in patients undergoing long-term hemodialysis; this mimics infection on plain radiographs and CT scans[35,41] and presumably on MR images. Tumors such as chordoma, lymphoma, and myeloma may mimic spondylitis by invading the disk. Systemic manifestations (e.g., fever and elevated erythrocyte sedimentation rate) in lymphoma and myeloma may further complicate the clinical picture such that aspiration biopsy or isolation of organisms from the blood becomes necessary for diagnosis and management.[10,24,39,41,50,65,77] Differentiation of infection from neoplasm may be a problem in some instances using MR imaging alone. We have seen a case in which a benign mild compression fracture, intensely enhanced on T1-weighted fat suppression, led to a biopsy. Determining the presence of a postoperative infection may be difficult. In our experience, gadopentetate dimeglumine enhancement in the post-diskectomy patient may occasionally mimic the pattern of early or mild disk infection; namely, there is enhancement of the disk and adjacent end plates, sometimes with extra-axial granulation tissue.[50] Enhancement with gadopentetate dimeglumine is expected in inflamed tissue.[50] Often contrast is decreased within the fatty marrow-laden vertebral body, making less conspicuous the inflammation, which is readily recognized as of low signal intensity on pre-enhancement T1-weighted images. Enhancement within a paraspinal soft tissue mass also frequently is less than expected, the exception being suggestive of tuberculosis, in which a thick peripheral rim of enhancement is characteristic.[75,88] Gadopentetate dimeglumine is very useful for demonstrating the subligamentous and epidural extension of infection. It may also be useful in depicting the most favorable site for aspiration.[62,75,88] Fat suppression techniques may improve the visualization of pathologic tissue.[76,88] Some investigators have advocated the use of STIR sequences to improve contrast among tissues,[5,88] partly because of its tendency to suppress the signal from the fat. Imaging in the coronal plane may simultaneously show the relationship of the vertebral end plates, the disk, and the paraspinal soft tissues to good advantage. Bone erosion is best depicted on proton density sequences or T1-weighted images with fat suppression sequences. For all these reasons, multiple sequences are necessary for an appropriate evaluation. Because gadopentetate dimeglumine has a tendency to cause elevation in signal intensity even in normal marrow, preenhanced T1-weighted sequences are needed. Vertebral body healing after therapy typically results in increased signal intensity on T1-weighted sequences because fatty marrow accumulates and in decreased signal intensity on T2-weighted or small flip-angle GRE images as edema, hyperemia, and

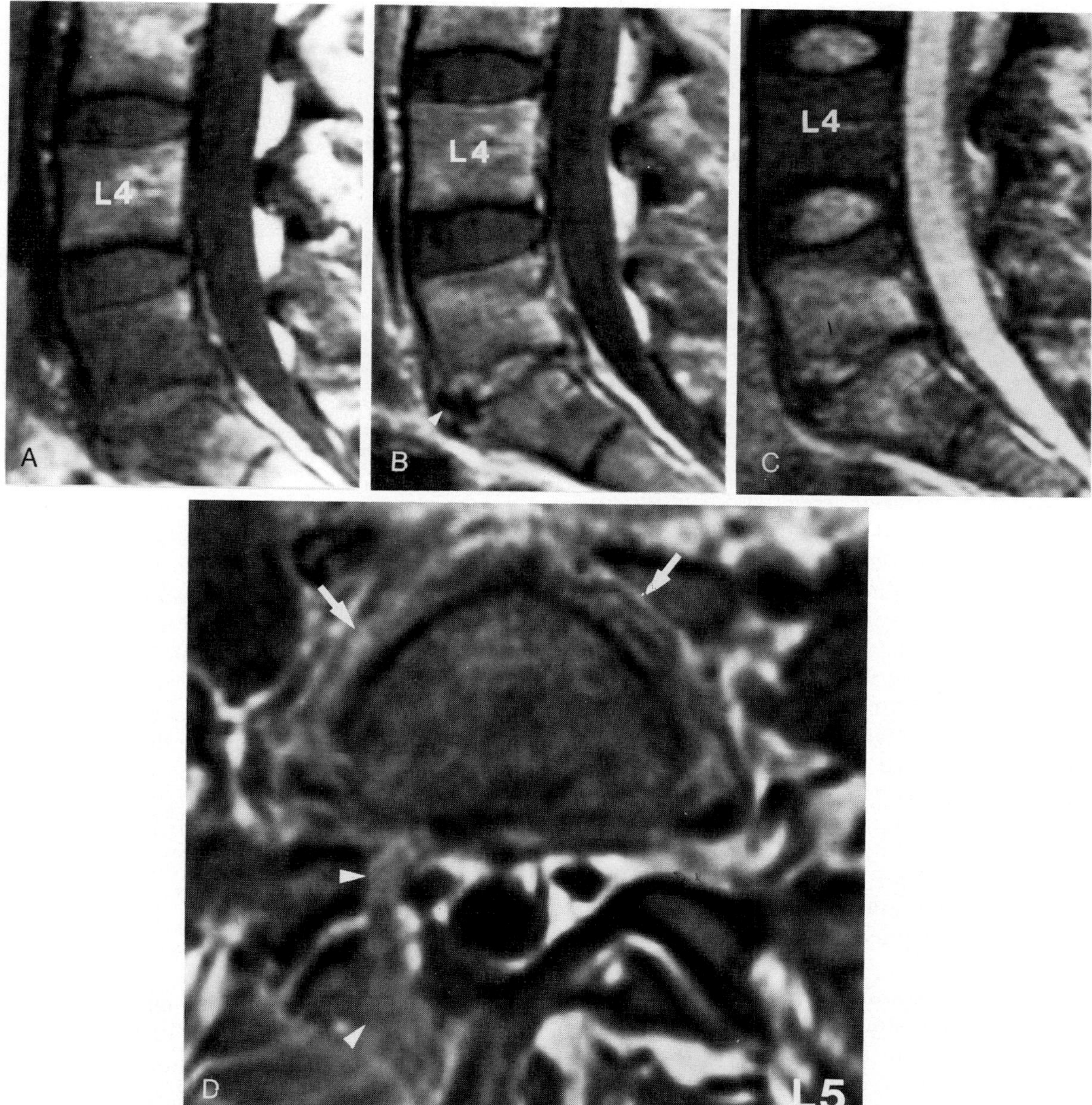

Figure 6–48. Evidence of *Enterobacter* diskitis and osteomyelitis in a 36-year-old woman. *A*, Sagittal T1-weighted image. Decreased signal intensity is seen in the inferior aspect of L5 and superior aspect of S1 with a smudged disk-cortical interface. *B*, Sagittal T1-weighted postenhancement image. Enhancement of two thirds of the disk occurs. Enhancement of the L5 and S1 marrow exceeds that of other vertebral bodies. A peculiar low signal intensity of the disk also is seen anteriorly *(arrowhead)*, indicating the residuum of the anulus. *C*, T2-weighted sagittal image. The "edema" within the L5 and S1 bodies again is apparent, as is the decrease in the L5 disk height. In this case, the T1-weighted contrast-enhanced and the T2-weighted studies complement each other because the edema (high water content) is not masked. The bright signal of the disk anteriorly confirms that it is due to the preserved segment of the disk. *D*, Axial T1-weighted postenhancement image at L5. A relatively small circumferential region of enhancing inflammatory tissue is observed in the prevertebral soft tissues *(arrows)*, typical of pyogenic osteomyelitis. Enhancing "granulation" or "inflammatory" tissue also is noted *(arrowheads)* extending medially to the right facets and laminectomy site in this postsurgical patient.

inflammation subside. Although this technique is not yet well established, it seems prudent to assume that gadopentetate dimeglumine enhancement would continue well beyond clinical evidence of healing (as it does in the postoperative spine) with improvement in the appearance of the vertebral body as osteogenesis proceeds.

Pyogenic spondylitis appears to be increasing in frequency.[65] It typically creates diffusely abnormal signals within the vertebral body. The paravertebral soft tissue inflammatory component tends to be small, anterior, and circumferentially located. Cortical margins tend to be eroded and posterior elements spared.[50,77,88] This phenomenon aids in differentiating pyogenic infection from tuberculosis and neoplasm.

Tuberculosis has a predilection for the thoracic spine. It creates typical destructive changes of the vertebral bodies, especially anteriorly, leading to a gibbous deformity. Classically, skip lesions are seen as the infection spreads beneath the anterior longitudinal ligament, sparing some disks. Large paraspinal masses may exhibit a thick rim of enhancement. Posterior elements show a tendency to erode and to induce epidural granulomas, which may produce neurologic manifestations. Cortical margins are frequently lost even when the disk is uninvolved. Detailed bony changes are better visualized on CT scans than on MR images.[74,77]

Brucella spondylitis can be manifested in two distinct forms. Infection classically occurs as focal involvement of the anterior aspect of one of the lower lumbar end plates and is usually self-limited. Alternately, the spondylitis may be a diffuse process involving several vertebral bodies and intervening disk spaces but maintaining the cortical margins and the height of the centra.[74]

In the United States fungal infections that affect the spine usually occur in immunocompromised hosts and may be associated with bizarre pathologic patterns, including large osteophytes, indicating indolence. Some organisms such as *Aspergillus* may produce a picture that mimics tuberculosis, whereas others, such as *Actinomyces*, may create only patchy changes in the vertebral body that may be lytic or sclerotic.[65] It is common for aspergillosis to recur despite what seems to be adequate surgical and antibiotic therapy.

Epidural abscesses are rare, but they are more common in granulomatous than in pyogenic infections. Rarely they may arise directly from hematogenous spread without osteomyelitis. An epidural abscess often is difficult to distinguish from granulation tissue. It would appear to be best seen on T1-weighted images, on which it is of intermediate intensity. Proton density sequences should allow differentiation from adjacent cerebrospinal fluid. Gadopentetate dimeglumine enhancement should increase its conspicuousness. Extra-axial masses may have mixed and complex signals and can be difficult to distinguish from dural inflammation.[1,62] In this situation, CT myelography or intraoperative ultrasonography may be superior to MR imaging[62] in detecting the fluid component of a frank abscess.

Most cases of pyogenic spondylitis respond to antibiotic therapy. Signs of spinal instability and neurologic compromise are usually followed by the regimen of traction, needle aspiration for culture, antibiotic therapy, débridement, and spinal fusion. Débridement is especially useful in countering granulomatous infection.[21,25,83]

References

1. Angtuaco EJC, McConnell JR, Chadduck WM, et al: MR imaging of spinal epidural sepsis. AJR 149:1249, 1987.
2. Baker LL, Goodman SB, Perkash I, et al: Benign versus pathologic compression fractures of vertebral bodies and assessment with conventional spin echo, chemical shift, and STIR MR imaging. Radiology 174:495, 1990.
3. Beabout JW, McLead RA, Dahlin DC: Benign tumors. Semin Roentgenol 14:33, 1979.
4. Beltran J, Noto AM, Chakeres DW, et al: Tumors of the osseous spine: Staging with MR imaging versus CT. Radiology 162:565, 1987.
5. Bertino RE, Poiter BA, Simac GK, et al: Imaging spinal osteomyelitis and epidural abscess with short inversion recovery (STIR). AJNR 9:563, 1988.
6. Braun IF, Hoffman JC, Davis PC, et al: Contrast enhancement in CT differentiation between recurrent disk herniation and postoperative scar: Prospective study. AJNR 6:607, 1985.
7. Bundschuh CV, Modic MT, Ross JS, et al: Epidural fibrosis and recurrent disk herniation in the lumbar spine: Assessment with MR. AJNR 9:169, 1988.
8. Burton CV, Kirkaldy-Willis WH, Young-Hing K, et al: Causes of failure of surgery on the lumbar spine. Clin Orthop 157:191, 1981.
9. Castillo M, Melko JA, Hoffman JC: The bright intervertebral disk: An indirect sign of abnormal spine bone marrow on T1-weighted MR images. AJNR 11:23, 1990.
10. Chandnani VP, Beltron J, Morris CS, et al: Acute experimental osteomyelitis and abscesses: Detection with MR imaging vs. CT. Radiology 174:233, 1990.
11. Coventry MB, Ghormley RK, Kernohan JW: The intervertebral disc: Its microscopic anatomy and pathology. Part I. Anatomy, development, and physiology. J Bone Joint Surg 27:105, 1945.
12. Coventry MB, Ghormley RK, Kernohan JW: The intervertebral disc: Its microscopic anatomy and pathology. Part II. Changes in the intervertebral disc concomitant with age. J Bone Joint Surg 27:233, 1945.
13. Crim JR, Mirra JM, Eckardt JJ, et al: Widespread inflammatory response to osteoblastoma: The flare phenomenon. Radiology 177:835, 1990.
14. Daffner RH, Lupetin AR, Dash N, et al: MRI in the detection of malignant infiltration of bone marrow. AJR 146:353, 1986.
15. DeLaPaz RL: Physical basis and anatomic correlates of MR signal in the spine and cord. *In* Enzmann DR, DeLaPaz RL, Rubin JB: Magnetic Resonance of the Spine. St Louis, CV Mosby, 1990, p 115.
16. Djukic S, Lang P, Morris J, et al: The postoperative spine: Magnetic resonance imaging. Orthop Clin North Am 21:603, 1990.
17. Drolshegen LF, Kessler R, Partain CL: Cervical meningeal histiocytosis demonstrated by magnetic resonance imaging. Pediatr Radiol 17:63, 1987.
18. Eyre DR: The intervertebral disc. *In* Frymoyer JW, Gordon SL (eds): New Perspectives on Low Back Pain. Park Ridge, Ill, American Academy of Orthopedic Surgery, 1989, p 147.
19. Fernstrom V: A discographic study of ruptured lumbar intervertebral discs. Acta Chir Scand 258(suppl):11, 1960.
20. Frank JA, Ling A, Patronas NJ, et al: Detection of malignant bone tumors: MR imaging vs. scintigraphy. AJR 155:1043, 1990.
21. Fredrickson B, Yuan H, Olans R: Management and outcome of pyogenic vertebral osteomyelitis. Clin Orthop 131:160, 1978.
22. Fruehwald FX, Tscholakoff D, Schwaighofer B, et al: Magnetic resonance imaging of the lower vertebral column in patients with multiple myeloma. Invest Radiol 23:183, 1988.
23. Gamba JL, Martinez S, Apple J, et al: CT of axial skeletal osteoid osteomas. AJR 142:769, 1984.
24. Gilbert RW, Kim JH, Posner JB: Epidural spinal cord compression from metastatic tumor: Diagnosis and treatment. Ann Neurol 3:40, 1987.
25. Govender S, Rejoo R, Goga IE: *Aspergillus* osteomyelitis of the spine. Spine 16:746, 1991.
26. Haggstrom JA, Brown JC, Marsh PW. Eosinophilic granuloma of the spine: MR demonstration. J Comput Assist Tomogr 12:344, 1988.
27. Hajeck PC, Baker LL, Goober JE, et al: Focal fat deposition in axial bone marrow: MR characteristics. Radiology 162:245, 1987.
28. Hanna SL, Fletcher BO, Fairclough DL, et al: Magnetic resonance imaging of disseminated bone marrow disease in patients treated for malignancy. Skeletal Radiol 20:79, 1991.
29. Hendrick RE, Kneeland JB, Stark DD: Maximizing signal-to-noise and contrast-to-noise ratios in Flash imaging. Magn Reson Imaging 5:117, 1987.
30. Hesselink JR, Shoukimas GM: MR imaging of the lumbar spine. *In* Edelman RR, Hesselink JR: Clinical Magnetic Resonance Imaging. Philadelphia, WB Saunders, 1990, p 705.
31. Hirsch C, Schajowicz F: Studies on structural changes in the lumbar annulus fibrosus. Acta Orthop Scand 22:185, 1952.
32. Hueftle M, Modic MT, Ross JS, et al: Lumbar spine: Postoperative MR imaging with Gd-DPTA. Radiology 167:817, 1988.
33. Jackson RP, Reckling FW, Mantz FA: Osteoid osteoma and osteoblastoma. Clin Orthop 128:303, 1977.

34. Kaplan PA, Orton DF, Asleson RJ: Osteoporosis with vertebral compression fractures, retropulsed fragments and neurologic compromise. Radiology 165:533, 1987.
35. Kaplan P, Resnick D, Murphy M: Destructive noninfectious spondyloarthropathy in hemodialysis patients. A report of four cases. Radiology 162:241, 1987.
36. Kaufman HH, Jones E: The principles of bony spinal fusion. Neurosurgery 24:264, 1989.
37. Kepes JJ, Kepes M: Predominantly cerebral forms of histiocytosis X. Acta Neuropathol 14:77, 1969.
38. Kieffer SA, Stadlan EM, Mohandas A, et al: Discographic-anatomical correlation of developmental changes with age in the intervertebral disc. Acta Radiol 9:733, 1969.
39. Klein SL, Sanford RA, Muhlbauer MS: Pediatric epidural metastases. J Neurosurg 74:70, 1991.
40. Kransdorf MJ, Stull MA, Gilkey FW, et al: Osteoid osteoma. Radiographics 11:671, 1991.
41. Kuntz D, Nareau B, Bardin T, et al: Destructive spondyloarthropathy in hemodialyzed patients. Arthritis Rheum 27:369, 1984.
42. Laredo JD, Reizine D, Bard M, et al: Vertebral hemangiomas: Radiologic evaluation. Radiology 161:183, 1986.
43. Linden A, Zankovich R, Theissen P, et al: Malignant lymphoma: Bone marrow imaging versus biopsy. Radiology 173:335, 1989.
44. Macnab I, McCulloch J: Backache. Baltimore, Williams & Wilkins, 1990, pp 131, 335.
45. McCleod RA, Dahlin DC, Beabout JW: The spectrum of osteoblastoma. AJR 126:321, 1976.
46. McGavran MH, Spady HA: Eosinophilic granuloma of bone, a study of 28 cases. AJR 97:719, 1966.
47. McKinstry CS, Steiner RE, Young AJ, et al: Bone marrow in leukemia and aplastic anemia: MR imaging before, during and after treatment. Radiology 162:701, 1987.
48. Michael AS, Mikhael MA: Spinal osteomyeliis: Unusual findings on magnetic resonance imaging. Comput Med Imaging Graph 12:329, 1988.
49. Modic MT: Degenerative disorders of the spine. *In* Modic MT, Masaryk TJ, Ross JS: Magnetic Resonance Imaging of the Spine. Chicago, Year Book Medical Publishers, 1989, p 75.
50. Modic MT, Feiglin DH, Piraino DW, et al: Vertebral osteomyelitis: Assessment using MR. Radiology 157:157, 1985.
51. Modic MT, Masaryk TJ, Ross JS, et al: Imaging of degenerative disk disease. Radiology 168:177, 1988.
52. Modic MT, Pavlicek W, Weinstein MA, et al: Magnetic resonance imaging of intervertebral disk disease: Clinical and pulse sequence considerations. Radiology 152:103, 1984.
53. Mohan V, Gupta SK, Tuli SM: Symptomatic vertebral hemangiomas. Clin Radiol 31:575, 1980.
54. Moore SG, Gooding CA, Brasch RC, et al: Bone marrow in children with acute lymphocytic leukemia: MR relaxation times. Radiology 160:237, 1986.
55. Moore SG: Pediatric marrow and musculoskeletal MRI. *In* Hasso AN, Stark DD: Spine and Body Magnetic Resonance Imaging. Categorical Course Syllabus, Boston: American Roentgen Ray Society, 1991, p 145.
56. Morris JM: Overview of surgical management of lumbar disc disease. *In* Genant HK (ed): Spine Update 1987. San Francisco, Radiology Research and Education Foundation, 1987, p 87.
57. Newton TH, Potts DG: Computed Tomography of the Spine and Spinal Cord. San Anselmo, Calif, Clavadel Press, 1983.
58. Osband ME, Pochedly C: Histiocytosis-X: An overview. Hematol Oncol Clin North Am 1:1, 1987.
59. Oschner SF: Eosinophilic granuloma of bone: Experience with 20 cases. AJR 97:719, 1966.
60. Peacock A: Observations on the postnatal structure of the intervertebral disc in man. J Anat 86:162, 1952.
61. Pech P, Haughton VM: Lumbar intervertebral disc: Correlative MR and anatomic study. Radiology 156:699, 1985.
62. Post MJD, Quencer RM, Montalvo BM, et al: Spinal infection: Evaluation with MR imaging and intraoperative US. Radiology 169:765, 1988.
63. Quinet RJ, Hadler NM: Diagnosis and treatment of backache. Semin Arthritis Rheum 8:261, 1979.
64. Resnick D, Niwayama G: Degenerative diseases of the spine. *In* Resnick D (ed): Bone and Joint Imaging. Philadelphia, WB Saunders, 1989, p 413.
65. Resnick D, Niwayama G: Osteomyelitis, septic arthritis and soft tissue infection in the axial skeleton. *In* Resnick D, Niwayama G (eds): Diagnosis of Bone and Joint Disorders. Philadelphia, WB Saunders, 1988, p 2619.
66. Ricci C, Cova M, Kang YS, et al: Normal age-related patterns of cellular and fatty bone marrow distribution in the axial skeleton: MR imaging study. Radiology 77:83, 1990.
67. Rosen BR, Fleming DM, Kushner DC, et al: Hematologic bone marrow disorders: Quantitative chemical shift MR imaging. Radiology 169:799, 1988.
68. Ross JS, Hueftle MG: Postoperative spine. *In* Modic MT, Masaryk TJ, Ross JS: Magnetic Resonance Imaging of the Spine. Chicago, Year Book Medical Publishers, 1989, p 135.
69. Ross JS, Hueftle MG: Postoperative spine. *In* Modic MT, Masaryk TJ, Ross JS: Magnetic Resonance Imaging of the Spine. Chicago, Year Book Medical Publishers, 1989, p 120.
70. Ross JS, Modic MT, Masaryk TJ: Tears of the anulus fibrosus: Assessment with Gd-DTPA-enhanced MR imaging. AJR 154:156, 1990.
71. Schellinger D, Manz HJ, Vidic B, et al: Disk fragment migration. Radiology 175:831, 1990.
72. Schmorl G, Junghanns H: The Human Spine in Health and Disease. New York, Grune & Stratton, 1959, p 12.
73. Sebag GH, Moore SG: Effect of trabecular bone on the appearance of marrow in gradient-echo imaging of the appendicular skeleton. Radiology 174:855, 1990.
74. Sharif HS, Aideyon OA, Clark DC, et al: Brucellar and tuberculous spondylitis: Comparative imaging features. Radiology 171:419, 1989.
75. Sharif HS, Clark DC, Aabed MY, et al: Granulomatous spinal infections: MR imaging. Radiology 177:101, 1990.
76. Simon JH, Szumowski J: Chemical shift imaging with paramagnetic contrast material enhancement for improved lesion depiction. Radiology 171:539, 1989.
77. Smith AS, Weinstein MA, Mizushima A, et al: MR imaging characteristics of tuberculous spondylitis vs. vertebral osteomyelitis. AJNR 10:619, 1989.
78. Smith SR, Williams CE, Davies JM, et al: Bone marrow disorders: Characterizations with quantitative MR imaging. Radiology 172:805, 1989.
79. Steindler A: Lectures on the Interpretation of Pain in Orthopedic Practice. Springfield, Ill, Charles C Thomas, 1959.
80. Stevens SK, Moore SG, Amylon MD: Repopulation of marrow after transplantation: MR imaging with pathologic correlation. Radiology 175:213, 1990.
81. Stevens SK, Moore SG, Kaplan ID: Early and late bone marrow changes after irradiation: MR evaluation. AJR 154:745, 1990.
82. Stimac GK, Porter BA, Olson DO, et al: Gadolinium-DTPA-enhanced MR imaging of spinal neoplasms: Preliminary investigation and comparison with unenhanced spin-echo and STIR sequences. AJNR 9:839, 1988.
83. Stone JL, Cybulski GR, Rodriguez J, et al: Anterior cervical debridement and strutgrafting for osteomyelitis of the cervical spine. J Neurosurg 70:879, 1989.
84. Suginmura K, Yamasaki K, Kitagoki H, et al: Bone marrow diseases of the spine: Differentiation of T1 and T2 relaxation times in MR imaging. Radiology 165:541, 1987.
85. Sze G, Twohig M: Neoplastic disease of the spine and spinal cord. *In* Atlas SW: Magnetic Resonance Imaging of the Brain and Spine. New York, Raven Press, 1991, p 921.
86. Taveras JM: Herniated intervertebral disk: A plea for more uniform terminology. AJNR 10:1283, 1989.
87. Teplick JG, Haskin ME: Intravenous contrast-enhanced CT of the postoperative lumbar spine: Improved identification of recurrent disk herniation, scar, arachnoiditis, and diskitis. AJNR 5:373, 1984.
88. Thrush A, Enzmann D: MR imaging of infectious spondylitis. AJNR 11:1171, 1990.
89. Ullrich CG, Binet EF, Sanecki MG, et al: Quantitative assessment of the lumbar spinal canal by computed tomography. Radiology 134:137, 1980.
90. Unger EC, Summers TB: Bone marrow topics. Magn Reson Imaging 1:31, 1989.
91. Urban J, Holm SH, Lipson SJ: Disc biochemistry in relation to function. *In* Weinstein JN, Wiesel SW (eds): The Lumbar Spine. Philadelphia, WB Saunders, 1990, p 232.
92. VanLon KJ, Kellerhouse LE, Pathria MN, et al: Infection versus tumor in the spine: Criteria for distinction with CT. Radiology 166:851, 1988.
93. Vogler JB, Murphy WA: Bone marrow imaging. Radiology 168:679, 1988.
94. Wasenko JJ, Rosenbaum AE, Yu SF, et al: The importance of gradient echo for lumbar spine imaging. Presented at the 29th annual meeting of American Society of Neuroradiology, Washington, DC, June 1991.
95. Wehrli FW, Atlas SW: Fast imaging: Principles, techniques, and clinical applications. *In* Atlas SW: Magnetic Resonance Imaging of the Brain and Spine. New York, Raven Press, 1991, p 1016.
96. Wiltse LL: The effect of the common anomalies of the lumbar spine upon disc degeneration and low back pain. Orthop Clin North Am 2:573, 1971.
97. Wisner GL, Rosen BR, Buxton R, et al: Chemical shift imaging of bone marrow: Preliminary experience. AJR 145:1031, 1985.
98. Yang PJ, Seeger JF, Dzioba RB, et al: High-dose IV contrast in CT scanning of the postoperative lumbar spine. AJNR 7:703, 1986.
99. Yankelevitz DF, Henschke CI, Knapp PH: Effect of radiation therapy in thoracic and lumbar bone marrow: Evaluation with MR imaging. AJR 157:87, 1991.
100. Yuh WTC, Zachar CK, Barloon TJ, et al: Vertebral compression fractures: Distinction between benign and malignant causes with MR imaging. Radiology 172:215, 1989.

101. Yu S, Haughton VM, Ho PSP, et al: Progressive and regressive changes in the nucleus pulposus. Part II. The adult. Radiology 169:93, 1988.
102. Yu S, Haughton VM, Lynch KL, et al: Fibrous structure in the intervertebral disk: correlation of MR appearance with anatomic sections. AJNR 10:1105, 1989.
103. Yu S, Haughton VM, Rosenbaum AE: Magnetic resonance imaging and anatomy of the spine. Radiol Clin North Am 29:691, 1991.
104. Yu S, Sether LA, Ho PSP, et al: Tears of the anulus fibrosus: Correlation between MR and pathologic findings in cadavers. AJNR 9:367, 1988.
105. Zimmer WD, Berquist TH, McLeod RA, et al: Bone tumors: Magnetic resonance imaging versus computed tomography. Radiology 155:709, 1985.

7 THE SHOULDER

Martin Vahlensieck, Philipp Lang, and Harry K. Genant

Imaging of the shoulder has been subject to significant changes in the last few years. Conventional radiographs display osseous disorders and soft tissue calcification well. Arthrography is a powerful tool in diagnosing rotator cuff tears and adhesive capsulitis. Computed tomography (CT) has facilitated the evaluation of bony pathoanatomy. Computed tomography arthrography is a most sensitive method of evaluating the glenoid labrum. Ultrasonography detects fluid and, to a lesser extent, tendinous tears. Owing to the complex anatomy of the shoulder, however, only a limited region can be assessed sonographically. Magnetic resonance (MR) imaging is unique in that it is capable of multiplanar imaging and is incomparable in the differentiation of soft tissue. It may therefore become the most important diagnostic tool in evaluating the painful shoulder.[21,24,30,47,48,51]

Nevertheless, conventional radiographs should still be obtained before the MR study is done, because comparison of MR images with the radiographs is essential for a complete evaluation.[6]

IMAGING PROTOCOLS FOR THE SHOULDER

To achieve good MR image quality, off-axis examinations using a small field of view with local coils are the procedures of choice. Various coil configurations are available for MR imaging of the shoulder.[27,28] In many institutions, a pair of loop gap resonators in Helmholtz configuration is used. In our experience, a curved 6.5-inch loop surface coil (Medical Advances, Milwaukee, Wis.) helps to reduce sensitivity to motion, raises the signal-to-noise ratio, and facilitates handling. However, the signal falloff with increasing distance from the coil remains a disadvantage, as signal inhomogeneities occur within the image. Other coils have been suggested for shoulder imaging, such as flexible wraparound local coils[21] or a combination of several small surface coils (coil array).[28] Coil arrays may become the method of choice in the future, because signal-to-noise ratios and spatial resolution can be improved. A slice thickness of 4 mm with an interslice gap of 0.4 mm to minimize cross-talk effects represents a compromise between sufficient spatial resolution and the signal-to-noise ratio. We use a field of view of 16 cm^2 with a matrix size of 256 $\times$ 192 elements, resulting in a pixel size of 0.52 mm^2 and four exitations. Oversampling in the direction of phase and frequency encoding helps to avoid aliasing (wraparound) artifacts. Respiratory compensation and spatial presaturation are used to decrease blood flow motion and ghost artifacts.

Three well-defined planes (axial, oblique-coronal, and oblique-sagittal) have been established as standard planes for MR imaging of the shoulder.[22] Patients are in the supine position with their arms along their sides in the neutral position. The series begins with an axial plane image, which is most useful for evaluating the anterior and posterior portions of the glenoid labrum, the capsular structures, and the humeral head contour. This image serves as a localizer for the following two planes (Fig. 7–1). The oblique-coronal plane, which lies parallel to the course of the supraspinatus muscle, best depicts the supraspinatus tendon, the subacromial-subdeltoid bursa, the undersurface of the acromion, the acromioclavicular joint, and the superior and inferior portions of the glenoid labrum (Fig. 7–2). The third plane is an oblique-sagittal plane, located perpendicular to the supraspinatus muscle, which best as-

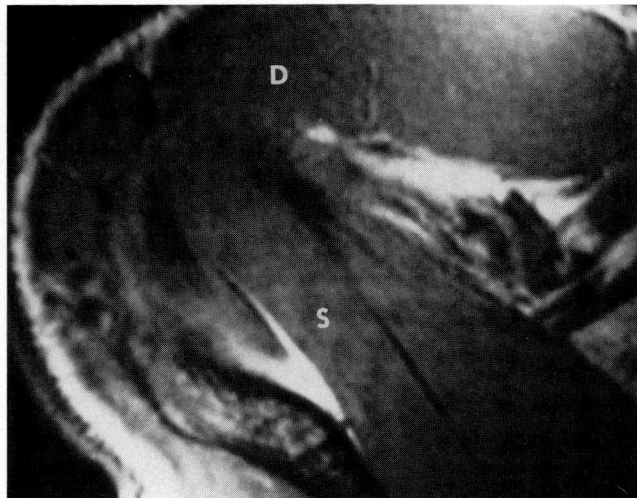

Figure 7–1. Image in the axial plane through the supraspinatus muscle and tendon serving as localizer for the oblique sequences. S, Supraspinatus muscle; D, deltoid muscle. Spoiled volumetric gradient-echo sequence (TR, 34 ms; TE, 5 ms; flip angle, 30 degrees).

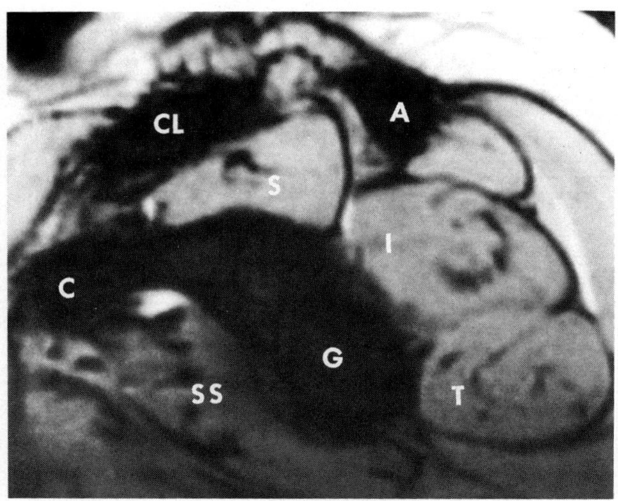

Figure 7–3. Multiplanar proton density–weighted gradient-echo sequence (TR, 500 ms; TE, 15 ms; flip angle, 30 degrees) in the oblique-sagittal plane through the glenoid. S, Supraspinatus muscle; I, infraspinatus muscle; T, teres minor muscle; SS, subscapularis muscle; C, coracoid process; A, acromion; CL, clavicle; G, glenoid.

sesses the shape and slope of the lateral acromion and its relationship to the supraspinatus tendon and muscle and the presence of fluid in the subacromial-subdeltoid bursa (Fig. 7–3). The total image acquisition time for such a protocol is approximately 30 minutes. The total examination time (i.e., image acquisition, prescanning, and reconstruction time) is approximately 50 minutes.

To achieve high contrast between normal and pathologic structures, T1- and T2-weighted sequences should be obtained in each plane. A T1-weighted spin-echo sequence in the oblique-coronal plane provides high spatial resolution for assessing anatomic details of the rotator cuff. A T2-weighted spin-echo sequence provides the desired contrast because most pathologic lesions cause prolonged T2 relaxation times. Fast gradient-recalled echo sequences (GRE) with proton density or T2*-weighting can be used to replace time-consuming, low signal-to-noise, and motion-sensitive long echo time (TE), long repetition time (TR) (T2-weighted) spin-echo sequences.[54] A representative sequence protocol is demonstrated in Table 7–1.

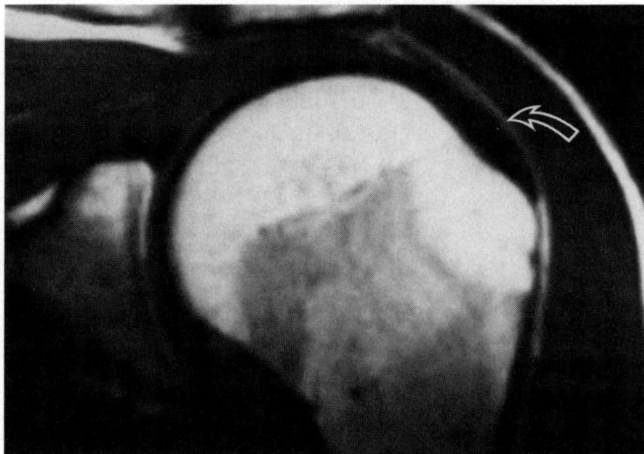

Figure 7–2. T1-weighted spin-echo sequence (TR, 600 ms; TE, 20 ms) in the oblique-coronal plane through the supraspinatus muscle and tendon. The peribursal fat plane (*arrow*) and normal distribution of fatty bone marrow within the humeral epiphysis and the superior glenoid are seen in this 35-year-old man.

Susceptibility artifacts caused by local inhomogeneities in the magnetic field are emphasized on GRE sequences as a gradient inversion rather than a 180-degree radiofrequency (RF) pulse rephases the spins. Magnetic susceptibility effects can occasionally help differentiate between calcified and noncalcified tissue. The effect is TE dependent; the shorter the TE, the smaller are the effects resulting from inhomogeneous magnetic susceptibility. We therefore recommend the acquisition of two echoes in one sequence. The first echo, with a short TE, emphasizes anatomic details with only slight susceptibility effects. The second echo is generated at a long TE to yield a high T2* contrast, thus aiding in the separation of fluid.

TE-dependent changes in signal behavior must be considered carefully when assessing a GRE image. Because dephasing effects in gradient-echo images demonstrate an oscillation, fat and fat-water interfaces demonstrate a TE-dependent signal intensity.

Table 7–1. PROTOCOLS OF MAGNETIC RESONANCE IMAGING OF THE SHOULDER AT THE UNIVERSITY OF CALIFORNIA, SAN FRANCISCO

Plane	TR (ms)	TE (ms)	FOV (cm)	Slice (mm)	Skip (mm)	NEX	Matrix (pixel)	Flip (degrees)	Time (min)
Axial (GRE)	600	11, 36	16	5	0	4	192	20	8
Oblique-coronal (SE)	600	20	16	4	0	4	192		8
Oblique-coronal (GRE)	600	11, 36	16	4	0	4	192	20	8
Oblique-sagittal (GRE)	600	11, 36	16	5	0	2	192	20	8
Oblique-coronal* (FSE)	3500	30, 90	16	4	0.5	4	256		8
Oblique-sagittal* (FSE)	3500	30, 90	16	4	0.5	4	256		8

TR, Repetition time; TE, echo time; FOV, field of view; Skip, interslice gap; NEX, number of excitations; GRE, gradient-recalled echo; SE, spin echo; FSE, fast spin echo; * optional sequence.

These in-phase and out-of-phase effects might be used to increase the contrast of fat-containing tissues to surrounding structures.[53] Fat stripes appear as lines of low signal intensity with certain TE values (Fig. 7–4).

Another alternative to conventional T2-weighted spin-echo sequences are fast spin-echo sequences, or rapid acquisition with relaxation enhancement (RARE).[20] This technique uses information from several spin echoes (echo train) of variable length for spatial encoding and thus reduces acquisition time. The factor of reduction in the imaging time equals the number of echoes in the echo train. T2-weighted images may thus be generated within 4 minutes or less.

In certain cases, subtle marrow abnormalities may be better detected by techniques based on the signal of water or fat protons (chemical shift imaging), such as fat presaturation (CHESS), out-of-phase imaging (Chopper-Dixon), and short tau inversion-recovery (STIR) sequences. Fat-saturated images were not found to be necessary to improve the soft tissue contrast in the evaluation of the soft tissues of the shoulder.[54]

A radial imaging technique using a fast GRE sequence[39] may provide additional information about the glenoid labrum, as the labrum can be visualized in cross section on each image. However, the large field of view necessary for this technique results in a lower spatial resolution.

Intra-articular injection of paramagnetic contrast medium is thought to improve the assessment of the glenoid labrum and rotator cuff.[13,17] This technique is still experimental and not yet in routine use. The value of three-dimensional acquisition, three-dimensional rendering, and intravenous contrast medium application also is not yet well established.

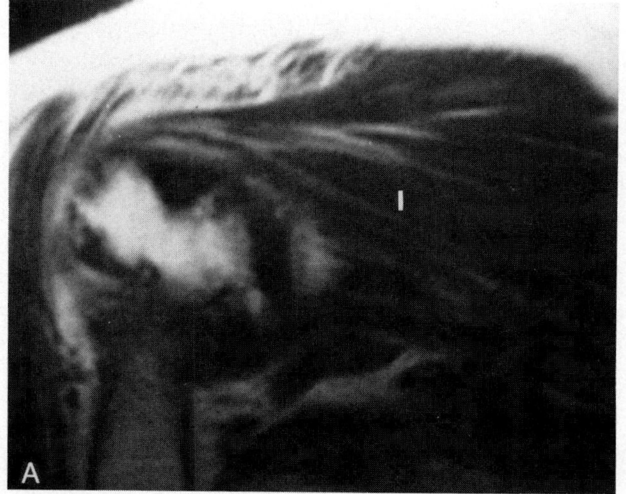

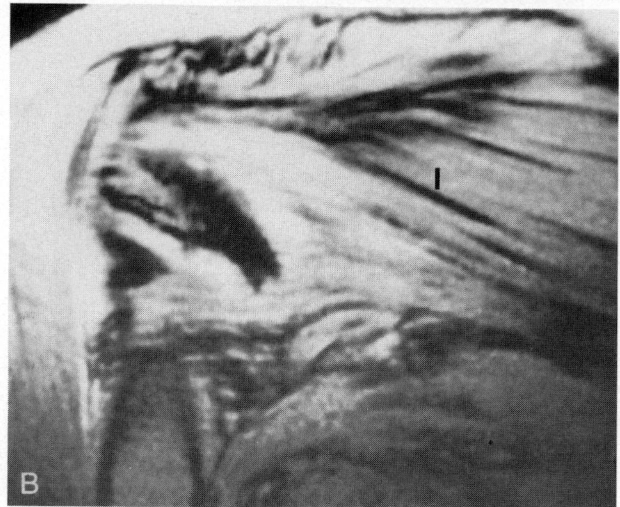

Figure 7–4. Oblique-coronal image of an overweight patient with fat enrichment between the muscle fibers of the infraspinatus muscle (I), appearing as bright fat stripes on the T1-weighted spin-echo sequence *(A)* and as black lines on the gradient-echo sequence *(B)*.

MAGNETIC RESONANCE IMAGING ANATOMY

Anatomy of the Rotator Cuff and Other Muscles

An overview of the anatomy of the shoulder is given in Chapter 1. However, because several anatomic peculiarities of the shoulder exist on MR imaging and critical knowledge of this anatomy is essential in assessing MR images, we describe it here in more detail and with correlations to clinical considerations.

The shoulder is stabilized primarily by four muscles and their tendons, which form the rotator cuff. These are the supraspinatus, infraspinatus, subscapularis, and teres minor muscles. The components of the rotator cuff are well demonstrated on the oblique-sagittal plane. The supraspinatus muscle is best demonstrated on oblique-coronal and axial images as a thick structure of intermediate signal intensity tapering into a tendon of low signal intensity that inserts onto the superolateral aspect of the greater tuberosity of the humerus. The musculotendinous junction generally is located superior to the head of the humerus. The infraspinatus and teres minor muscles are best demonstrated on axial images as fusiform structures of intermediate signal intensity parallel and inferior to the supraspinatus. The infraspinatus muscle inserts inferoposteriorly to the supraspinatus on the greater tuberosity, and the teres minor muscle inserts farther inferiorly. The subscapularis muscle is located anterior to the body of the scapula and appears on T1-weighted axial images as a structure of intermediate signal intensity. Multiple low-intensity tendinous portions join to form one broad main tendon anteriorly. This tendon merges with the anterior aspect of the capsule before inserting onto the lesser tuberosity. The rotator cuff in the region of the supraspinatus tendon is about 10 mm thick.[50]

The long head of the biceps muscle originates from the superior glenoid labrum and the glenoid tubercle. Its tendon passes within the joint superiorly and obliquely under the rotator cuff (Fig. 7–5) between the supraspinatus tendon and the subscapularis tendon. It then passes through the intertubercular groove of the humerus into the muscle belly. The tendon is covered by a synovial sheath that communicates with the joint capsule. It is best seen on axial images as a circular structure devoid of signal in the intertubercular groove. The tendon of the short head of the biceps muscle lies anterior to the humeral head. Together with the coracobrachialis muscle tendon it originates from the coracoid process and is well demonstrated on axial planes.

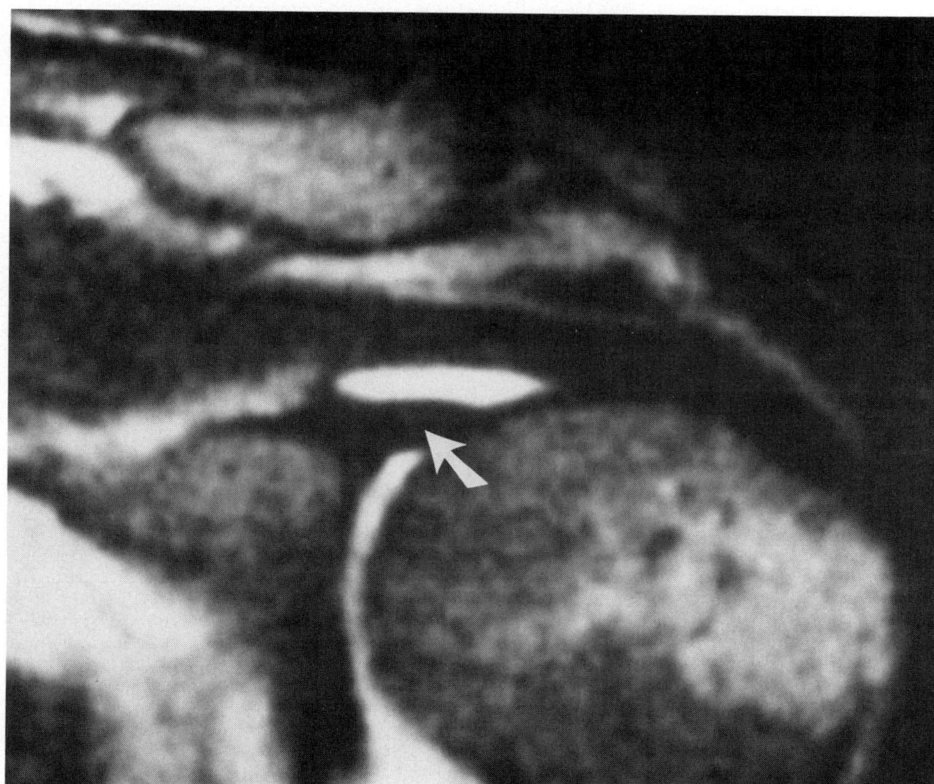

Figure 7–5. Oblique-coronal T2-weighted spin-echo image from a patient with a small joint effusion that emphasizes the insertion of the long head of the biceps onto the superior labrum *(arrow)*.

The other muscles can be summarized briefly. The deltoid muscle originates from the lateral clavicle, acromion, and scapular spine and inserts onto the deltoid tuberosity of the humerus. The teres major muscle originates from the inferior lateral scapula and inserts onto the medial intertubercular humeral groove. The trapezius muscle originates from the thoracic spinous processes and inserts onto the distal clavicle, acromion, and scapular spine. The latissimus dorsi muscle originates from the spinous processes T6 through T12 and inserts onto the medial intertubercular humeral groove. Finally, the pectoralis major muscle originates from the inferomedial clavicle, sternum, and costochondral junctions and inserts onto the lateral intertubercular humeral groove.

Anatomy of the Bursae

The largest bursa of the human body[50] lies between the acromion and the subdeltoid muscle on one side and the rotator cuff and the humeral head on the other. It consists of three portions: the subacromial, subdeltoid, and subcoracoid portions (Fig. 7–6). The subcoracoid portion lies inferior to the coracoid process and anterior to the subscapularis muscle, between the combined tendons of the short head of the biceps and the coracobrachialis muscle anteriorly and the subscapularis tendon posteriorly. It is variable in its extent and communicates with the subacromial portion in about 10 per cent of cases.[23,50] As an individual bursa it may also be called the coracobrachialis or subcoracoid bursa.[9,40]

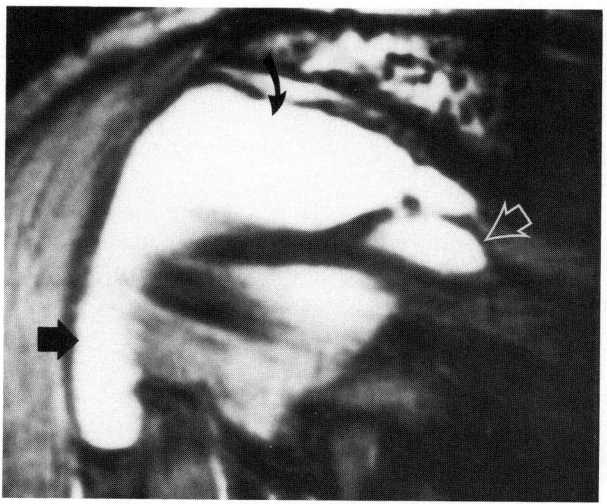

Figure 7–6. Oblique-coronal gradient-echo image of a large fluid collection within the subacromial-subdeltoid bursa. The subdeltoid portion *(arrow)*, subacromial portion *(curved arrow)*, and connection to the subcoracoid portion *(white arrow)* are visible.

The subacromial portion lies between the acromion and the supraspinatus tendon and muscle. It is attached to the undersurface of the acromion and the coracoacromial ligament as well as to the supraspinatus tendon.[50] The large subdeltoid portion lies below the deltoid muscle, superior to the rotator cuff and lateral to the greater tuberosity. It is firmly attached to the upper outer part of the greater tuberosity (approximately 0.75 inch ≅ 1.9 cm) and the rotator tendon.[32] Part of the bursa covers the bicipital groove. The subacromial and subdeltoid portions usually communicate with each other and are constant findings. Thus, we prefer the term *subacromial-subdeltoid bursa*, as was also used by others.[44] Occasionally (8 to 20 percent of cases), the subacromial and subdeltoid portions are separated by a septum.[18,50] The bursa does not communicate with the glenohumeral joint. Its inner layer consists of synovial tissue and its outer layers of fat. This extrasynovial fat may be seen on radiographs as a radiolucent stripe, 1 to 2 mm wide, whereas the potential space of the bursa is usually not visualized.[57] Mitchell and colleagues[35] could identify radiographically the fat stripe of the subacromial-subdeltoid bursa in 60 per cent of normal subjects. They claimed better visibility on internal rotation of the arm.

On T1-weighted MR images, the peribursal fat appears as a bright stripe (see Fig. 7–2). A direct relationship has been demonstrated between its thickness, the amount of subcutaneous fat,[22,35] and increasing age.[8] The peribursal fat plane continues as the intermuscular fat between the trapezius and supraspinatus muscles medially and a fat plane adjacent to the lateral aspect of the proximal humeral metaphysis.[35]

A second bursa of major interest for MR imaging is the subscapular bursa, located between the subscapularis tendon and the middle glenohumeral ligament, which extends over the neck of the scapula. The subscapular bursa normally communicates with the glenohumeral joint between the three glenohumeral ligaments. Therefore, it may also be called the subscapular recess of the shoulder capsule. It may surround the subscapularis tendon and extend along the anterior surface of the subscapularis tendon for a variable distance (Fig. 7–7).[59] The subscapular bursa may be indistinguishable from a large subcoracoid portion of the subacromial-subdeltoid bursa or the coracobrachialis bursa.[37] In the superior aspect of the bursa, folds of capsular tissue can be seen and should not be mistaken for pathologic structures. A good way to assess this bursa is in the axial plane.

Several other bursae are present in the shoulder, but these are smaller and inconstant and thus have not been important on MR imaging. They do not

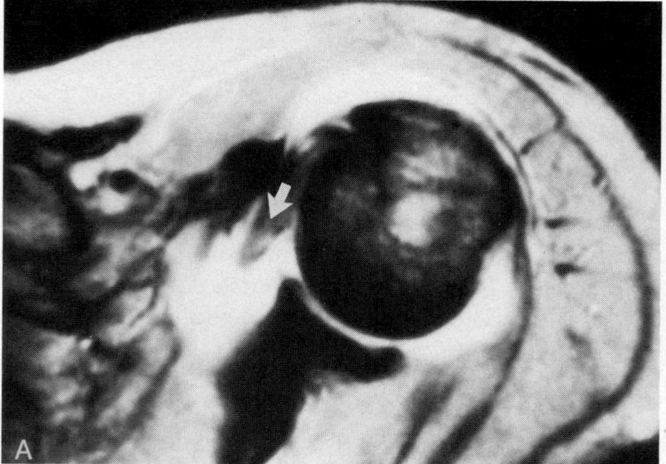

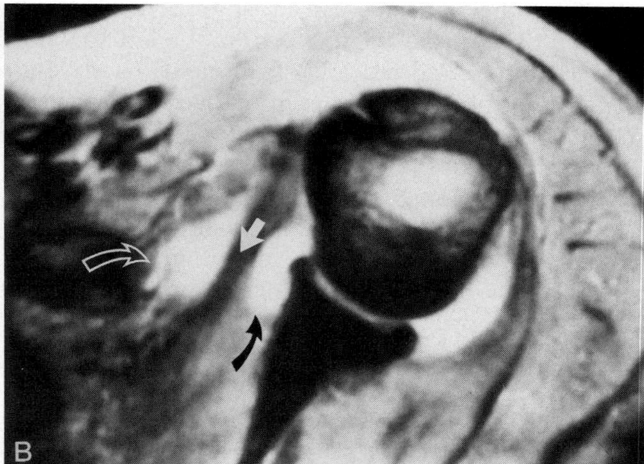

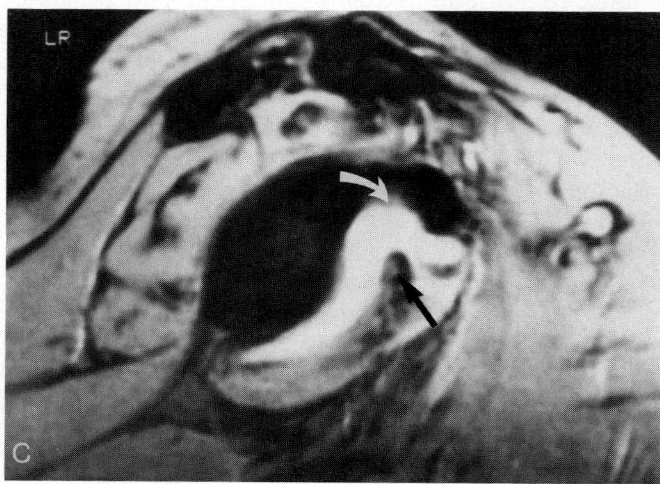

Figure 7–7. Axial gradient-echo images from a patient with a large joint effusion. *A*, The superior level cuts through the extension of the subscapular recess of the joint capsule anterior to the subscapularis tendon, which is only partially shown *(arrow)*. *B*, The inferior level cuts through the subscapularis tendon totally *(arrow)*, revealing two spaces of fluid *(curved arrows)*. *C*, The oblique-sagittal plane of the same patient also shows the subcoracoid extension of the subscapular recess *(curved arrow)* anterior to the subscapularis muscle and tendon *(arrow)*.

commonly communicate with the joint. Among them are the infraspinatus bursa, between the capsule and infraspinatus tendon; the teres major bursa, between the teres major tendon and its humeral insertion; and the pectoralis major bursa, between the pectoralis major tendon and its humeral insertion.

Anatomy of the Glenoid Labrum and Joint Capsule

The glenoid labrum is a thick, fibrous extension of the joint capsule, which forms a rim that serves to deepen the glenoid fossa. On axial and oblique-coronal images, it is a triangular structure of low signal intensity, continuous anteriorly with the capsule and the glenohumeral ligaments. The anterior portion of the labrum appears larger than the more rounded posterior portion. The base is attached to the hyaline cartilage of the glenoid fossa, causing a linear area of high signal intensity on gradient-echo sequences and T1-weighted spin-echo sequences. This line is not apparent on T2-weighted spin-echo sequences. It must not be mistaken for a labral tear. Occasionally, a second linear area of high signal intensity that is parallel to the base of the labrum and slightly lateral to the sublabral hyaline articular cartilage may be noted (Fig. 7–8). The exact nature of this structure has not been established, but it is not believed to be pathologic. Possible explanations include a zone of vascularity, cartilaginous reinforcement, or insertion of individual bands of the inferior glenohumeral ligament.[22]

The capsular insertion is best seen on axial planes at a midglenoid level. Because of the complex anatomy of the anterior capsular complex, use of standard levels in the axial plane can facilitate comparison of different studies. Conversely, the posterior capsular insertion is consistently seen at the base of the fibrous labrum. The anterior capsular insertion seems to demonstrate a wide variation in its attachment.[10]

Three types of anterior capsular insertions have been suggested (Fig. 7–9).[29] In type I, the capsule inserts near or on top of the glenoid labrum. In types II and III, the insertion along the scapular neck is

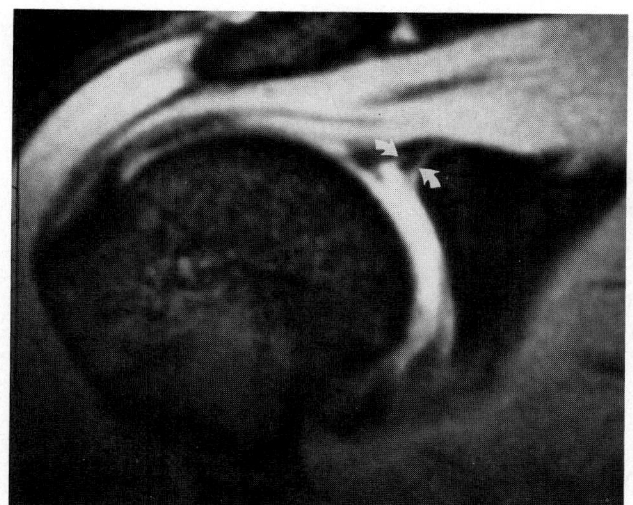

Figure 7–8. Oblique-coronal gradient-echo image of the normal superior glenoid labrum with two linear areas of high signal intensity *(arrows)*.

located farther medially. A type III capsule is thought to be susceptible to recurrent glenohumeral dislocations.[59] The variations in the attachment of the anterior glenohumeral joint capsule may lead to misinterpretation: Types II and III capsules should not be referred to as a stripped capsule or anterior instability. Anterior stability is provided by the anterior capsular complex. The anterior capsular complex consists of the synovial membrane; the superior, middle, and inferior glenohumeral ligaments; the coracohumeral ligament; the subscapularis bursa; and the subscapularis muscle and tendon.[52] The glenohumeral ligaments are variable in their anatomy and usually are not demonstrated on MR images unless fluid is within the joint. Posterior stability is provided by the capsular complex and the infraspinatus and teres minor muscles.

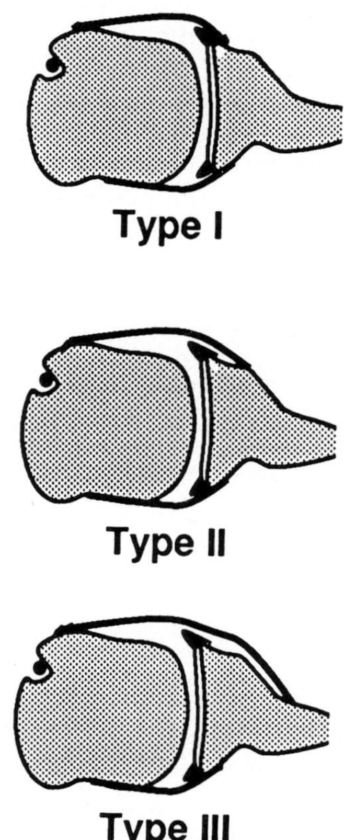

Figure 7–9. Schematic drawings of the axial plane of the variable anterior insertion of the shoulder capsule (types I–III).

Anatomy of the Bones

Cortical bone appears as a region devoid of signal because of its lack of resonating protons. Trabecular bone is of high or intermediate signal intensity on T1-weighted images, depending on its rates of fatty to hematopoietic marrow (see Fig. 7–2). The epiphysis often shows fatty marrow, whereas the apposing metaphysis shows variable hematopoietic marrow. This distribution causes a typical pattern, and, especially on axial planes, it should not be mistaken as pathologic infiltration.

The shape and slope of the acromion are important in determining the height of the supraspinatus outlet, which is the space below the coracoacromial arch (anterior acromion, coracoacromial ligament, and acromioclavicular joint). The shape and slope are best seen on oblique-sagittal planes (Fig. 7–10).

The shape of the undersurface of the acromion may appear linear (flat), curved, or overhanging (elongated or hooked)[42] (Fig. 7–11). In one study of 71 cadavers these variations were referred to as types I to III.[4] The linear or flat acromion (type I) was found in 17.1 per cent, the curved acromion (type II) in 41.9 per cent, and the overhanging (or hooked) acromion (type III) in 39.3 per cent.

The slope of the acromion relative to the horizontal axis can be well assessed on the oblique-sagittal plane. It seems to vary within a range of about 5 to 40 degrees.[1,4] An acromion with a small slope angle has been described as a flat[42] or a downsloping acromion. To avoid confusion of the slope with the shape of the acromion, the term *flat* should not be used to describe a linear shape of the acromion.

On the oblique-coronal plane the relative location of the acromion with respect to the distal clavicle can be well assessed. An inferior location of the inferior margin of the anterior acromion relative to the undersurface of the distal clavicle has been described as a low-lying acromion.[49] This appearance might be due to a small slope angle of the acromion

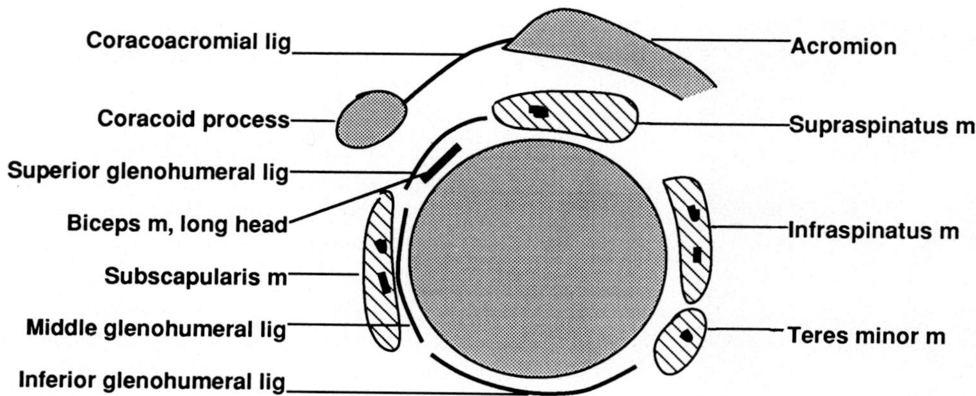

Figure 7-10. Schematic drawing of the oblique-sagittal plane of the shoulder. Lig, ligament; m, muscle.

(i.e., the oblique-coronal appearance of a downward-sloping acromion).

A deep notch in the humerus of varying size, posterior to the greater tubercle, which is best visualized on axial planes, should not be mistaken for a Hill-Sachs infraction (see Fig. 7-7A).

DISEASES OF THE ROTATOR CUFF

Diseases of the rotator cuff include inflammation, degeneration with or without calcification, and partial and complete ruptures.

Rotator Cuff Tears

Except for a small percentage of cases attributable to acute trauma, most tears of the rotator cuff are the result of attritional change and tendon degeneration. As the pathophysiology of these changes is not completely understood, different theories have been developed.

One model proposes an intrinsic degeneration process within the tendon,[7] starting in the "critical zone," a zone of relative avascularity in the supraspinatus tendon approximately 1 cm from its insertion.[33,46] Other contributing factors to cuff degeneration are aging and overuse of the shoulder (e.g., excessive athletic activity).

A clinically based theory has been introduced by Neer.[41] Initially, impingement of the rotator cuff tendons, especially the supraspinatus tendon, causes acute inflammation as a consequence of microtrauma with edema and hemorrhage (Neer stage I). This stage typically occurs in patients younger than 25 years. Continuous impingement leads to fibrosis and chronic tendinitis (Neer stage II), causing symptoms mainly in patients 25 to 40 years old. This leads to an increased vulnerability and makes the tendons liable to tear (Neer stage III). Stage III usually is seen in patients older than 40 years and is manifested by bony spurs. Impingement most frequently is caused by narrowing of the supraspinatus outlet. Narrowing occurs with anterior acromial spurs, an overhanging or curved shape of the acromion, a flat slope of the acromion, or a prominent acromioclavicular joint. Joint prominence can be caused by capsular distention or hypertrophy and spurs.[42] In one study, 69.9 per cent of the rotator cuff

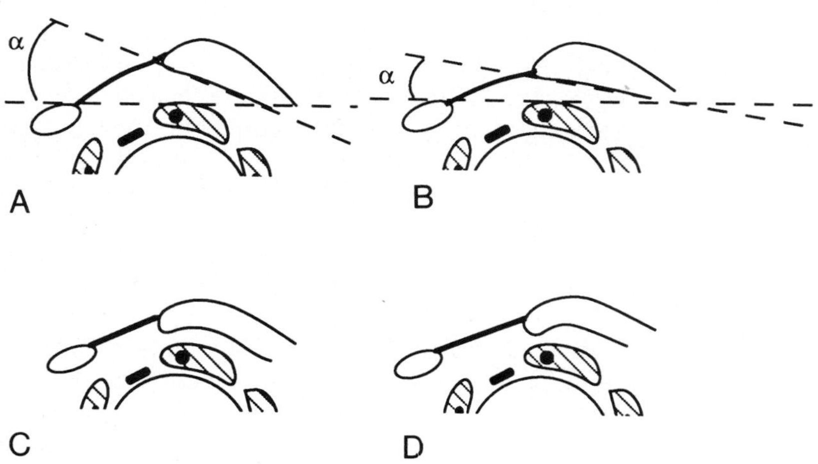

Figure 7-11. Schematic drawing in the oblique-sagittal plane of large *(A)* and small *(B)* acromial slope (α) and flat *(A, B)*, curved *(C)*, and hooked *(D)* acromial shape.

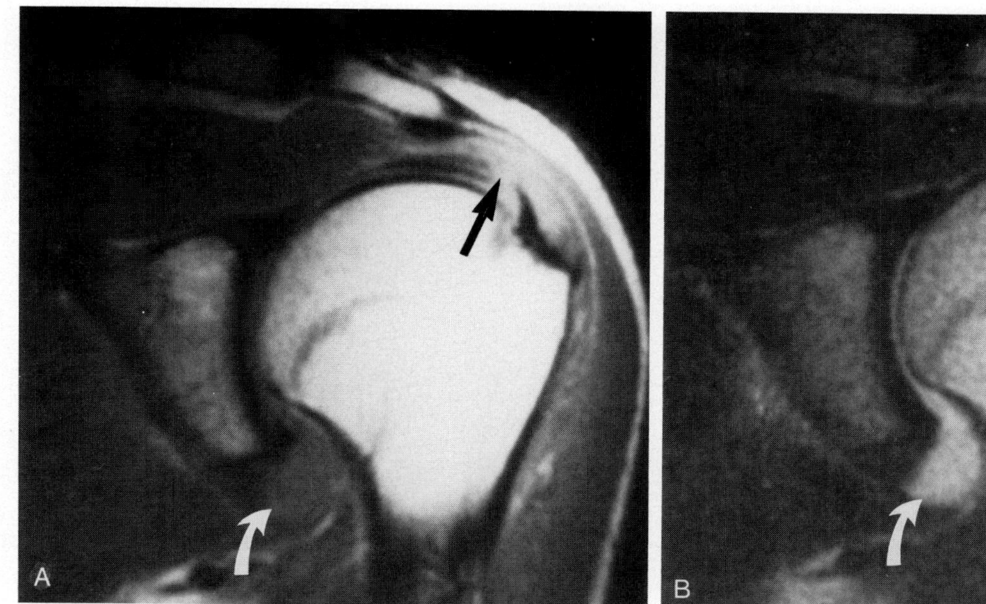

Figure 7-12. Oblique-coronal T1-weighted *(A)* and T2-weighted *(B)* spin-echo images. Note the rotator cuff tear with a gap in the supraspinatus tendon *(arrows)*, the abnormal signal intensity that becomes brighter on the T2-weighted image, and the fluid in the subacromial-subdeltoid bursa and the glenohumeral joint space *(curved arrows)*.

tears occurred in shoulders with a hooked acromion.[36] It is presumed that 95 per cent of rotator cuff tears are caused by impingement mechanisms.[61]

Less frequent impingement mechanisms (nonoutlet impingement) include a prominent greater tuberosity, loss of head depressors, loss of glenohumeral fulcrum, loss of suspensory mechanism, defects of the acromion, thickened bursa, or abnormal use of the extremities (e.g., paraplegia).

A study of 200 cadavers[43] revealed bony changes at the undersurface of the acromion only in complete and superior partial tears. In contrast, cadavers with inferior partial tears showed a smooth undersurface of the acromion. These findings indicate that bony changes of the acromion resulting from rotator cuff disorders may also be secondary and not the main cause of the tear.

The intensity of the MR imaging signal of normal tendons is low on all sequences. Signs of a complete rotator cuff tear are (1) a gap within the tendon on T1-weighted images that becomes brighter on T2-weighted images (Figs. 7-12, 7-13), (2) free fluid in the subacromial-subdeltoid bursa, (3) obliteration of the peribursal fat stripe, and (4) muscle retraction and fatty atrophy (depending on the extent and chronicity of the tear). Fatty atrophy of retracted muscles is characterized by bands of bright signal on T1-weighted images within the muscle and by diminished bulk of the muscle.

The most sensitive sign of a complete tear is the presence of fluid in the subacromial-subdeltoid bursa.[12] However, this sign lacks specificity. Criteria highly specific for a complete tear are the presence of a tendinous gap and musculotendinous retraction (Fig. 7-14). Obliteration of the peribursal fat stripe is not a very useful sign for diagnosing tears.[12]

However, the literature reports cases of proved complete tears without high signal on T2-weighted images.[44] Accurate diagnosis in these cases was based primarily on marked contour abnormalities and associated secondary findings such as intrabursal fluid and muscle retraction. The bright signal on

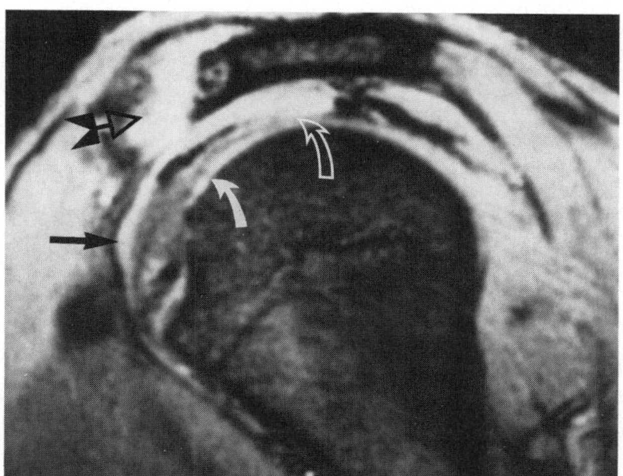

Figure 7-13. Oblique-sagittal gradient-echo image. Note the rotator cuff tear with fluid in the subacromial-subdeltoid bursa *(arrow)* and joint space *(curved arrow)*, the abnormal elevated signal intensity within the supraspinatus muscle and tendon *(open arrow)*, and the fluid in the acromioclavicular joint *(flagged arrow)*. The acromial shape is curved (type II).

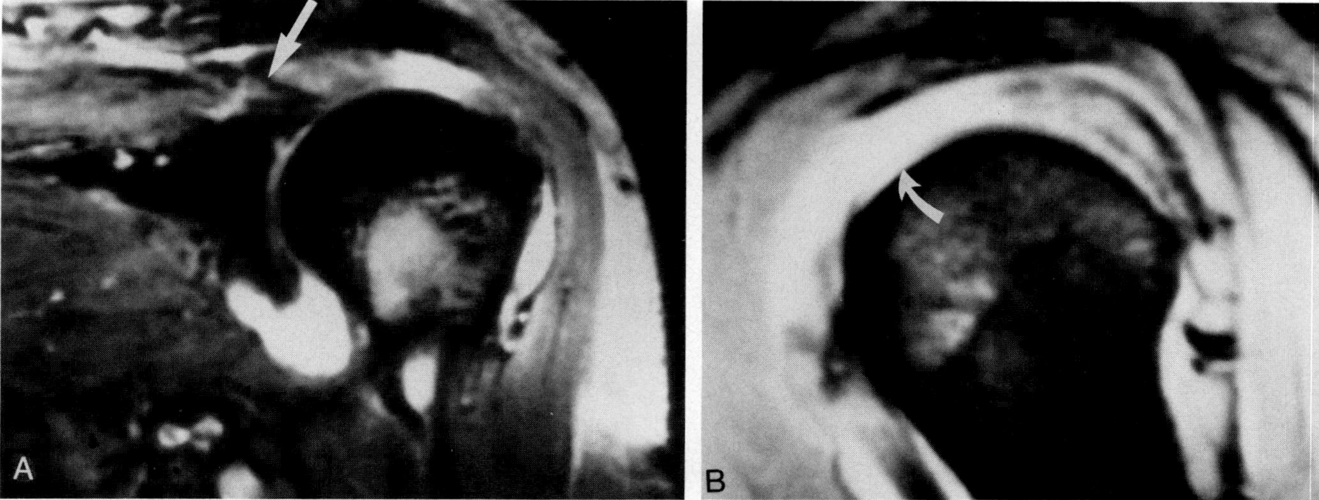

Figure 7–14. Oblique-coronal (*A*) and oblique-sagittal (*B*) gradient-echo images. Note the extensive rotator cuff tear with the large gap in the supraspinatus tendon and the muscle retraction (*arrow*). Supraspinatus muscle tissue is missing on the oblique-sagittal image (*curved arrow*).

T2-weighted images in these cases was probably missing owing to fibrotic scar that represented an old tear.[44]

Although the supraspinatus tendon is attached to the inferior layer of the subacromial-subdeltoid bursa[50] and the glenohumeral joint capsule, not every torn tendon leads to defects in these synovial tissue layers. This has been demonstrated by intraoperative injection of saline solution into the joint space in one case of complete supraspinatus tendon tear.[44] This finding could explain the absence of fluid in the subacromial-subdeltoid bursa in some cases of cuff tear. The sensitivity of MR imaging in detecting complete tears was found to be very high (about 75 to 90 per cent).[5,60]

For suitable treatment planning of rotator cuff tears, the following information may be helpful for the surgeon: (1) the condition of the ends of the tendon, (2) the size of the tear, (3) the specific tendons involved, (4) the degree of atrophy of the muscle, and (5) the degree of tendon retraction. Depending on this information, rotator cuff tears can be treated by suturing the torn tendons together, suturing the tendon to bone, and moving local tissue (biceps, subscapularis, or other tendons) or prosthetic material into the deficient area. Impingement can be treated by decompression (division of the coracoacromial ligament and acromioplasty with or without bursectomy).

Postoperatively, approximately 25 per cent of the patients have recurrent symptoms.[3] Magnetic resonance imaging helps to detect recurrent tears as well as inadequate acromial resection. Frequently, the subacromial-subdeltoid bursal fat plane is obliterated postoperatively or was removed, and so its assessment is not helpful in evaluation of the cuff. Signal abnormalities within skin and subcutanous fat frequently are observed postoperatively and reveal a typical pattern (Fig. 7–15).

Acute or Chronic Tendinitis, Degeneration, and Partial Tears

The differential diagnosis of acute or chronic tendinitis, degeneration, and partial tears is difficult. No definite model provides a reasonable correlation with the surgical changes. The contrast behavior on

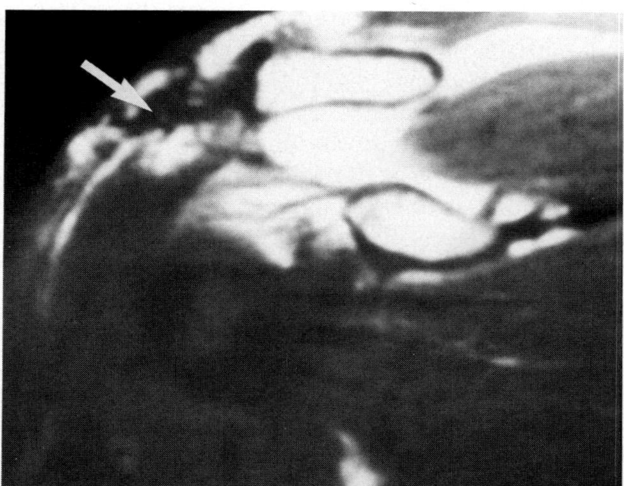

Figure 7–15. Oblique-coronal T1-weighted image after acromioplasty in a patient with impingement syndrome. A signal intensity decrease is observed in the skin and in the subcutaneous and muscular tissue layers (*arrow*) owing to scar, hemosiderin, and bone chips.

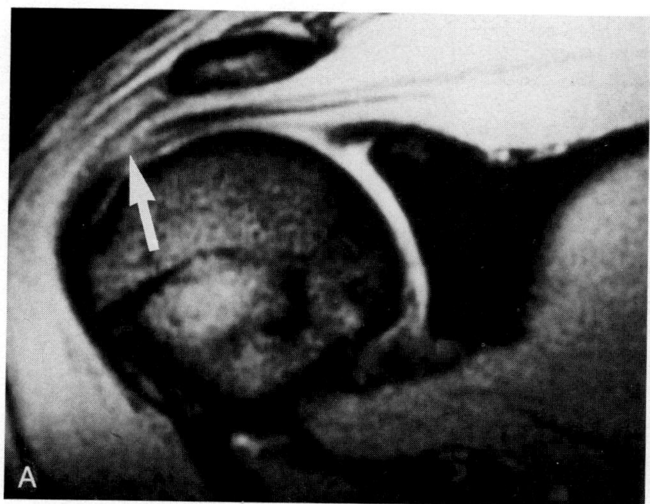

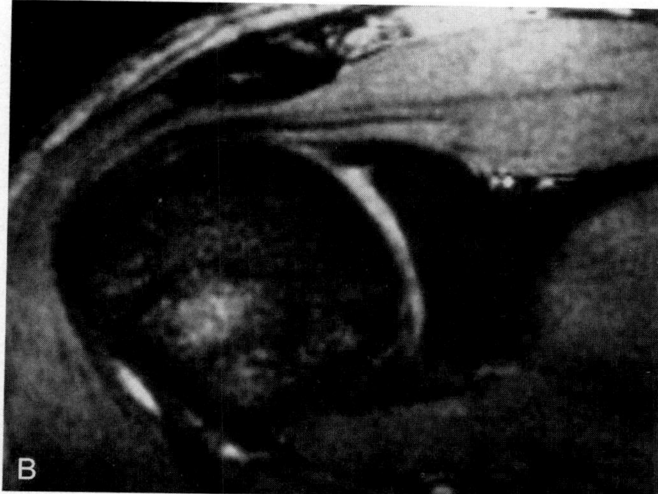

Figure 7–16. Oblique-coronal gradient-echo images of a healthy, 27-year-old subject. *A*, proton density image (TR, 300 ms; TE, 11 ms; flip angle, 30 degrees). *B*, T2*-weighted image (TR, 300 ms; TE, 35 ms; flip angle, 30 degrees). Focally increased signal intensity within the supraspinatus tendon is seen on the proton density image *(arrow)*. Note the homogeneous low signal intensity of the tendon on the T2*-weighted image.

T1- versus T2-weighted images and the tendon morphology, such as thickening and thinning, are important in making the diagnosis. Acute tendinitis, for example, might cause increased signal intensity on T1- and T2-weighted images without contour abnormalities or slight enlargement of the tendons.[44] This increased signal intensity is due to edema with or without acute hemorrhage or myxoid degeneration.[25] Chronic tendinitis is thought to cause an increase in signal intensity on T1-weighted images without an increase in signal intensity on T2-weighted images or contour irregularities. A partial tear might be diagnosed by signal abnormality (increased signal intensity) within the tendon on T1-weighted images, increasing signal intensity on T2-weighted images, and contour irregularities.

It should be noted that abnormal signal within the supraspinatus tendon near its insertion also has been described in asymptomatic young adults (Fig. 7–16). This is seen particularly on proton density–weighted and to a lesser extent on T1-weighted images. The reason for this increased signal intensity remains unclear. Different theories have been proposed supporting signal averaging with surrounding fat or muscle fibers, signal averaging with the intra-articular portion of the biceps tendon, or differences in the histoanatomic composition of that region (Fig. 7–17).

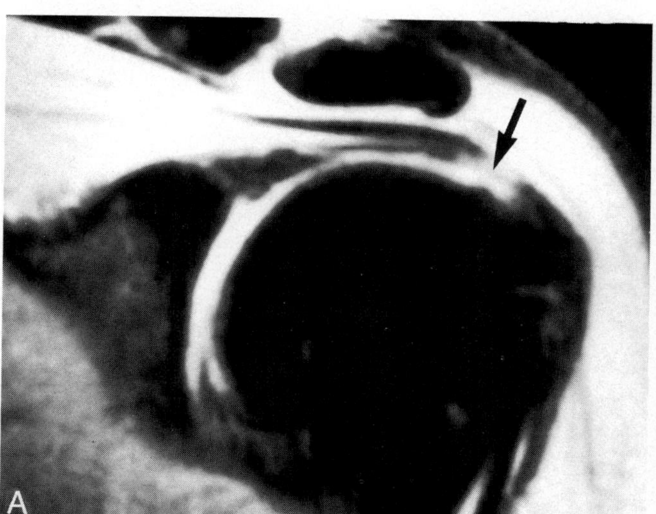

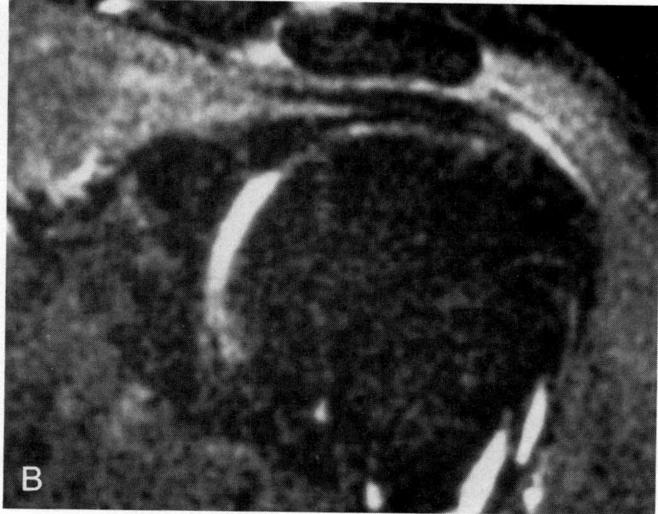

Figure 7–17. Oblique-coronal proton density *(A)* and T2-weighted *(B)* spin-echo, fat-saturated images from a patient with a small rotator cuff tear. The signal abnormality on the proton density–weighted image *(arrow)* appears larger than that on the T2-weighted image. Note the black appearance of the superior glenoid owing to fatty bone marrow.

Table 7–2. DIAGNOSTIC MAGNETIC RESONANCE IMAGING SIGNS OF THE INSERTION OF THE SUPRASPINATUS TENDON ONTO THE HUMERAL HEAD AND THE PERIBURSAL FAT STRIPE IN THE OBLIQUE-CORONAL PLANE

T1 →	T2	Signs	Suggested Diagnosis	Neer Stage	Possible Secondary Signs
		Form: Normal Contour: Regular Signal*: Void	Normal tendon		
		Form: Normal or thick Contour: Variable Signal: Brightening	Acute tendinitis or partial tear	I	Tendon thickening
		Form: Normal or thin Contour: Variable Signal: Indifferent	Degeneration Chronic tendinitis, Artificial	II	Chronic bursitis with thickening of bursal walls, tendon thinning
		Form: Different Contour: Gap Signal: Brightening	Complete tear	III	Tendon and muscle retraction, bursal and joint effusion

The first two columns illustrate the signal intensity of the tendinous insertion on T1- and T2-weighted MR images. The stripe above the tendon represents the subacromial-subdeltoid peribursal fat.

* Signal changes from T1- to T2-weighting.

A summary of the different MR imaging patterns of the rotator cuff is given in Table 7–2.

Additional Findings in Rotator Cuff Disease

Acromioclavicular joint degenerative disease may represent a cause or a result of impingement of the supraspinatus muscle and tendon (Fig. 7–18). Magnetic resonance imaging can help to depict degenerative changes in earlier stages than radiographs. A joint distention with intermediate signal intensity on T1- and T2-weighted images may result from capsule hypertrophy. This joint distention can cause obliteration of the subacromial-subdeltoid fat stripe and impingement of the rotator cuff.[49] Occasionally a cystic or ganglionlike distention of the acromioclavicular joint with high signal intensity on T2-weighted images may be present. Joint effusion typically brightens on T2-weighted images. Later stages of degeneration reveal the different types of acromial and clavicular spurs.

Cystic changes such as erosions or resorptions and sclerosis of the humeral head and acromion are evident on MR imaging and are common findings in rotator cuff disease (Fig. 7–19). Cystic resorptions are located frequently along the lateral aspect of the

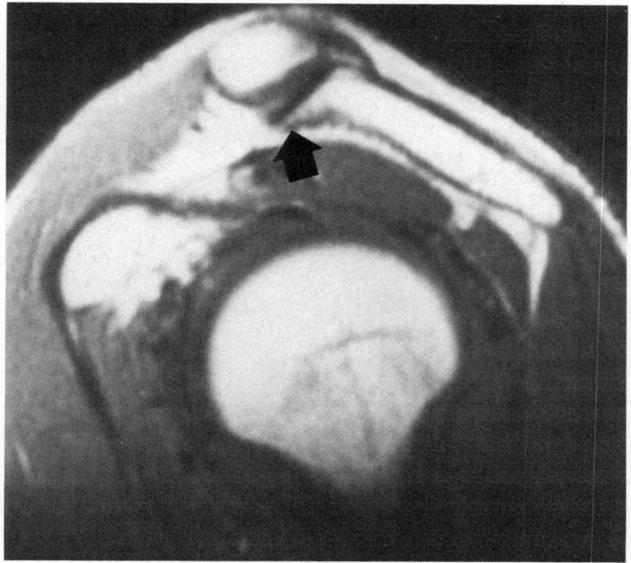

Figure 7–18. Oblique-sagittal proton density–weighted image of the anterior bone marrow–containing spur of the acromioclavicular joint *(arrow)*.

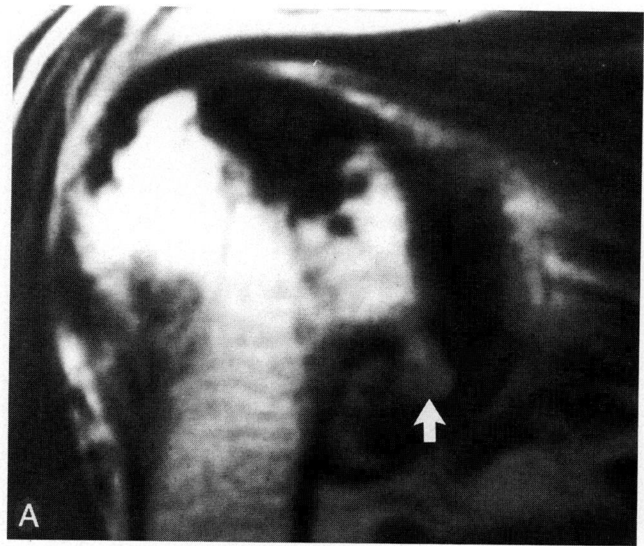

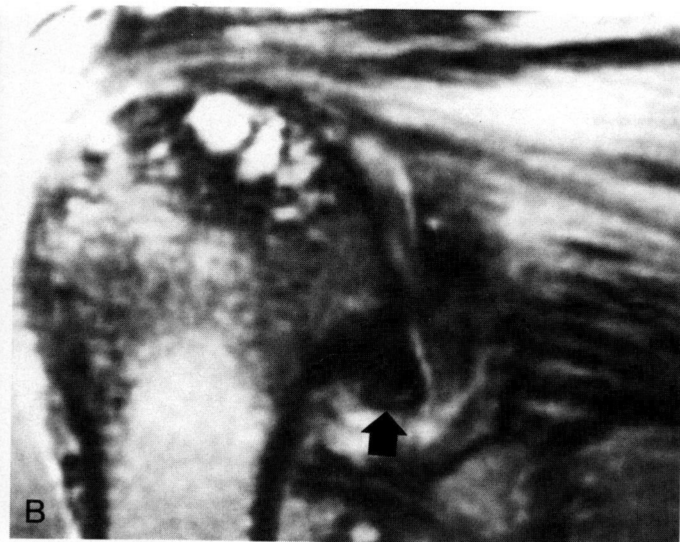

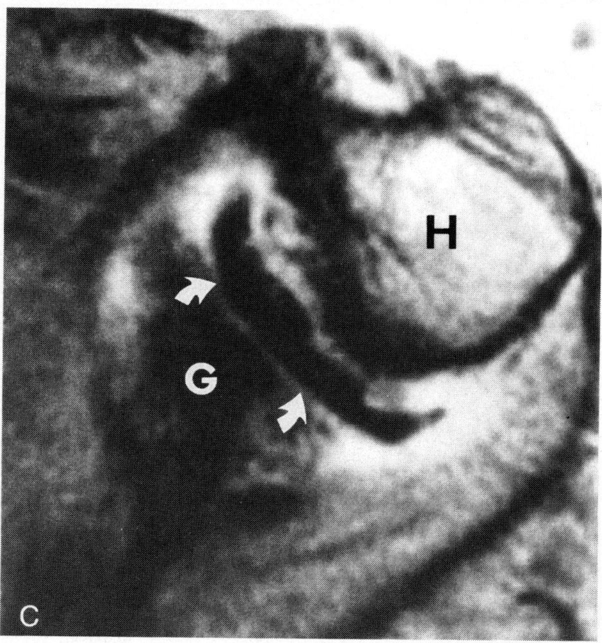

Figure 7–19. Oblique-coronal T1-weighted spin-echo *(A)* and proton density–weighted gradient-echo *(B)* images. There is severe osteoarthritis with subchondral cysts and a large axillary spur *(arrow)*, which shows a bandlike signal void between the humeral head (H) and the glenoid (G) on the axial gradient-echo image *(C)*.

greater tuberosity. Sclerotic areas may demonstrate signal void on all sequences. Cysts show low signal intensity on T1-weighted images and high signal intensity on T2-weighted images. However, owing to its multiplanar imaging capability, MR imaging is able to depict sclerotic and cystic changes at uncommon localizations.

The biceps tendon or its synovial sheath may become inflamed as a result of impingement syndrome. Occasionally, rheumatoid arthritis, calcific deposits, trauma, or primary tenosynovitis causes biceps tendinitis. A nonspecific finding with inflammation consists of increased fluid within the tendon sheath. This appears as decreased signal intensity on T1-weighted images and increased signal intensity on T2-weighted images. Common complications of biceps tendinitis that accompany impingement syndrom are tears of the long head of the biceps, which show an absence of the circular signal void in the intertubercular groove. Another complication is a dislocation of the long head of the biceps after disruption of the retinaculum of the bicipital groove. In these cases, the biceps tendon may be demonstrated medially, displaced toward the minor tubercle.

Tendinous and peritendinous calcifications are seen in calcific tendinitis (dystrophic calcification) caused by impingement syndrome and calcifying crystal deposition diseases (calcium hydroxyapatite deposition or calcium pyrophosphate dihydrate deposition disease). They occur most commonly in the supraspinatus tendon (Fig. 7–20). Occasionally involvement of the infraspinatus, teres minor, sub-

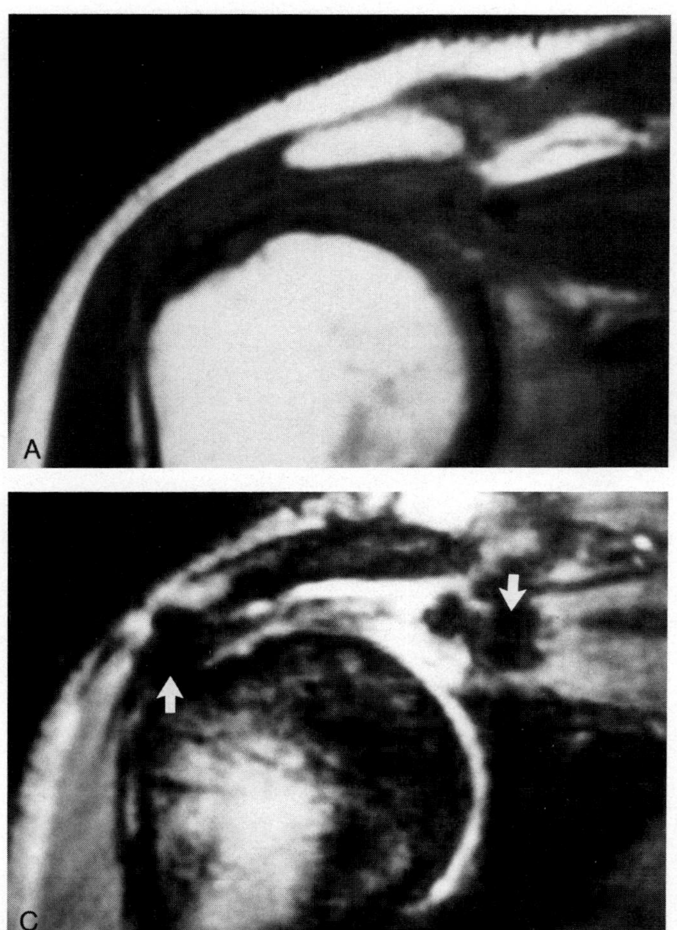

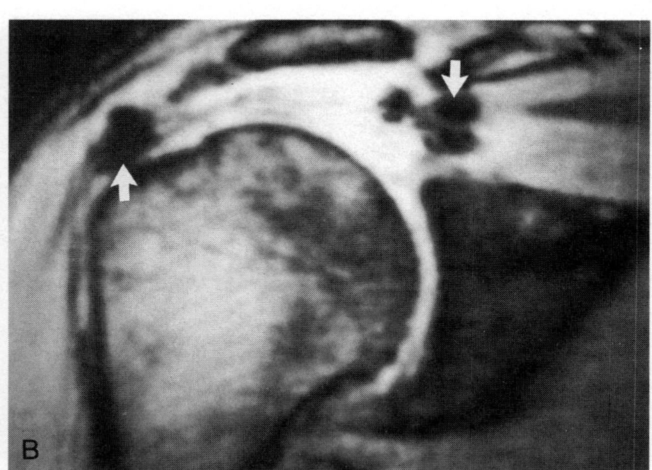

Figure 7–20. Oblique-coronal T1-weighted spin-echo *(A)*, proton density *(B)*, and T2*-weighted gradient-echo *(C)* images of calcified tendinitis and a rotator cuff tear. Tendinous and calcified tissue are indistinguishable on the T1-weighted image. However, the calcified depots demonstrate a large area of signal void on the gradient-echo images *(arrows)*, which is emphasized even more with longer echo time.

scapularis, and biceps tendons also is seen.[19] Calcifications are devoid of signal on MR images. They may be missed on spin-echo sequences within the low signal intensity tendon; however, owing to susceptibility artifacts they may be even better appreciated on GRE sequences. Comparison with radiographs is essential in making the diagnosis. Different stages of this disorder as described on radiographs (i.e., silent, mechanical, and rupture phases)[38] have not been demonstrated on MR images.

Involvement of the subacromial-subdeltoid bursa is a common finding in impingement syndrome. The typical finding with a rotator cuff tear is fluid collection within the bursa. Whereas the radiographic diagnosis of bursal pathology is based on indirect signs, such as displacement and obliteration,[56,58] MR imaging gives additional information about its content. An important sign in complete rotator cuff tears is an intrabursal fluid collection within the subacromial-subdeltoid bursa. Synovial proliferation in these cases may be seen as nodular filling defects of low signal intensity within the effusion of high signal intensity on T2-weighted images[44] that is comparable to that seen in bursography.[56] Adhesions in bursitis can complicate the anatomic situation with atypical fluid distribution.[32]

SHOULDER INSTABILITY

Tears of the glenoid labrum can be either acute or a result of recurrent dislocations or subluxations due to capsular laxity. Bankart first described a tear of the inferior aspect of the anterior labrum due to inferior humeral dislocation, which is referred to as a Bankart lesion.[2] Direct signs of labral tears on cross-sectional images are a detached and displaced labrum (Fig. 7–21) or an absent labrum (Fig. 7–22), a blunted labral edge (Fig. 7–23), and labral fragmentation. On MR images increased signal intensity within the fibrous cartilage continuing to its surface also is considered to represent a tear, even though to date the variant signal patterns of the normal and abnormal labrum are not completely known. Magnetic resonance imaging can detect anterior labral tears better than superior labral tears and far better than posterior and inferior labral tears.[16,31] Sensi-

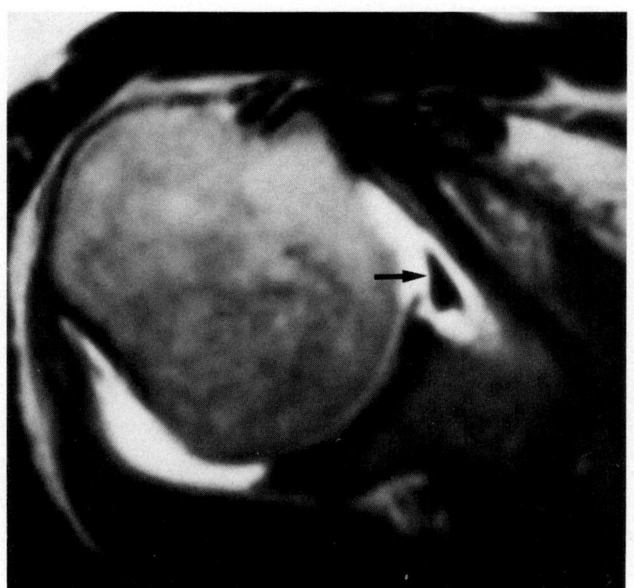

Figure 7-21. Axial T1-weighted image with intra-articular injection of gadopentetate dimeglumine. An anterior glenoid labral tear with labral displacement *(arrow)* is seen.

tivity in detecting labral abnormalities is improved in the presence of intra-articular fluid, either naturally or iatrogenically induced.[59] Because MR imaging is noninvasive and most labral tears occur anteriorly, this method might gain more importance in the future in assessing the glenoid labrum. However, one study revealed MR imaging to have only a somewhat low overall sensitivity (44 to 78 per cent) in assessing labral abnormalities.[14] As the role of MR imaging in labral abnormalities is still controversial, larger studies with improved techniques (e.g., volume acquisition, radial imaging) have to define its value. Cystic changes within the glenoid labrum are readily appreciated as bright areas on T2-weighted images (Fig. 7-24).

Infractions of the humeral head resulting from glenohumeral dislocation are depicted with high sensitivity on axial images at the level of the coracoid process. A Hill-Sachs defect is located posterosuperiorly after anteroinferior dislocation (Fig. 7-25). A posterior dislocation may cause an anteromedial defect (reverse Hill-Sachs lesion). Hill-Sachs lesions appear on images as mild, flattened to wedge-shaped defects of the humeral contour. An additional type of fracture of the glenoid fossa is called a bony Bankart lesion.

Abnormalities of the subscapularis tendon include increased signal intensity on T2-weighted images along the course of the tendon, interruption of the tendon, and retraction of the musculotendinous junction.

SYNOVIAL DISEASES

Inflammatory processes of the synovial tissue of the joint capsule include pigmented villonodular synovitis. In this condition, deposition of nodular hemosiderin can be demonstrated on MR imaging as areas of low signal intensity on T1- and T2-weighted images owing to its superparamagnetic effects. In synovial osteochondromatosis, calcified and ossified intra-articular loose bodies appear as signal voids.

The glenohumeral joint is commonly involved in rheumatoid arthritis. In the early stages of the disease, synovitis may lead to soft tissue edema and effusion. With progressive and persistent inflamma-

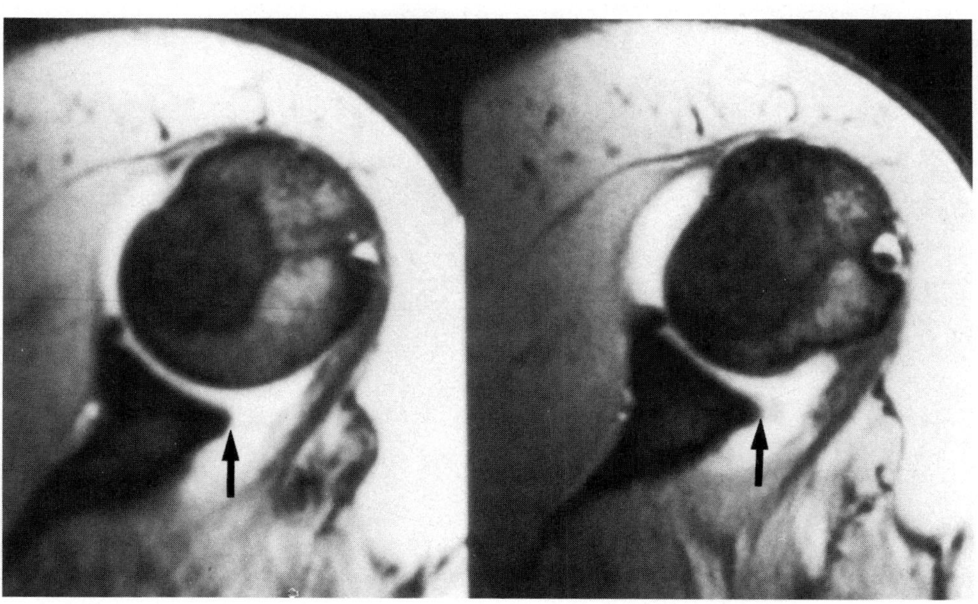

Figure 7-22. Axial gradient-echo images, showing anterior shoulder instability with blunted anterior glenoid labrum *(arrows)* and a large joint effusion of high signal intensity.

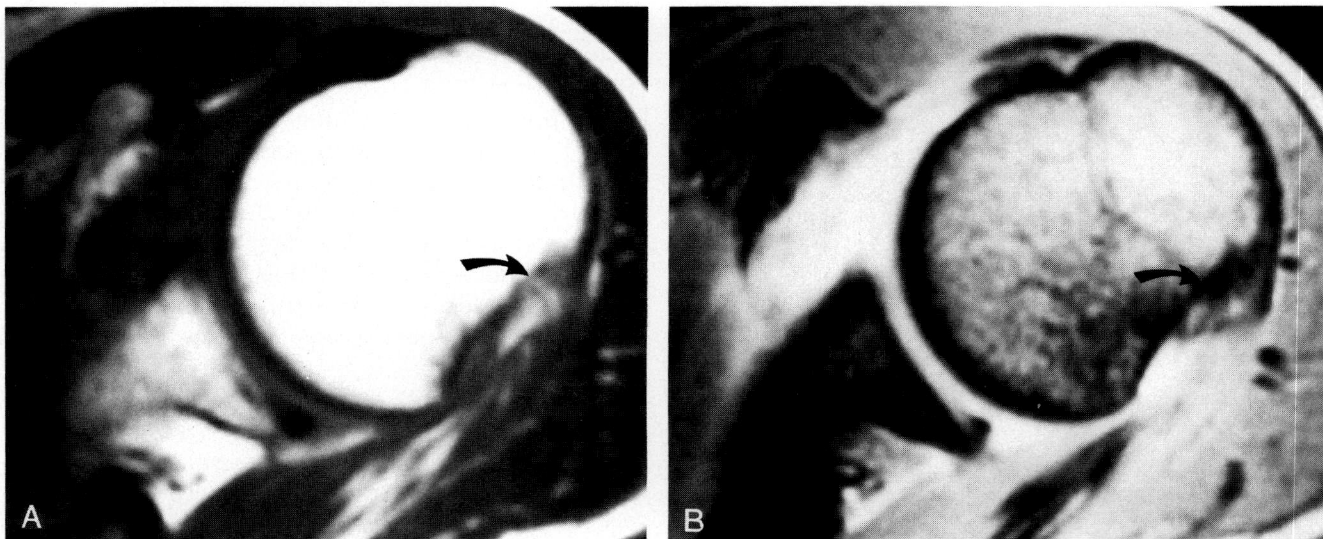

Figure 7–23. Axial T1-weighted spin-echo *(A)* and proton density gradient-echo *(B)* images of an anterior glenoid labral tear with absent anterior glenoid labrum, capsular stripping, impression fracture of the posterior humeral head (Hill-Sachs lesion) *(curved arrows)*, and joint effusion.

tion, a proliferative synovitis develops, which destroys the articular cartilage; erodes the underlying bone; and may disrupt the ligaments, tendons, and joint capsule. With disruption of the capsule, synovial cysts may develop, which can increase in size, dissect through soft tissue planes, or rupture. Magnetic resonance imaging may demonstrate all of these findings in a noninvasive manner. Contrast-enhanced studies are able to demonstrate the size and location of the inflammatory pannus.[55] Dynamic studies with gadopentetate dimeglumine further allow the differentiation between effusion and inflammatory pannus. Although loss of articular cartilage and osseous erosions may be seen with conventional radiographs, they may be detected at an earlier stage and their extent may be better assessed with MR imaging.[26] Erosions appear as defects of low signal intensity within the bone surface, occasionally filled with pannus, and show high signal intensity on T2-weighted images. Another advantage of MR imaging is its ability to reveal the muscular atrophy that accompanies arthritis. Although no definite role has been defined yet for shoulder MR imaging in rheumatoid arthritis, it may help to identify and differentiate complications related to this disease.

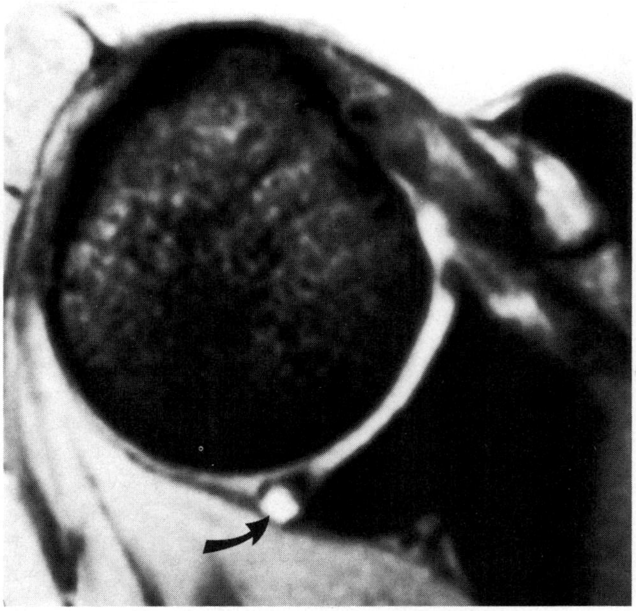

Figure 7–24. Axial T2*-weighted gradient-echo image of a cystic lesion within the posterior glenoid labrum *(curved arrow)*.

Another entity, most likely caused by synovial proliferation, is posttraumatic osteolysis of the clavicle. The MR imaging appearance of this process has been described as consisting of low signal intensity on T1- weighted images and high signal intensity on T2-weighted images.[11]

FRACTURES

Magnetic resonance imaging, to date, has not played a significant role in evaluating osseous trauma of the shoulder. However, because of its multiplanar tomographic capabilities this technique can assess fractures in locations that are otherwise difficult to visualize. The sensitivity of MR imaging to changes in the signal intensity of bone marrow also makes it a valuable technique for assessing bony

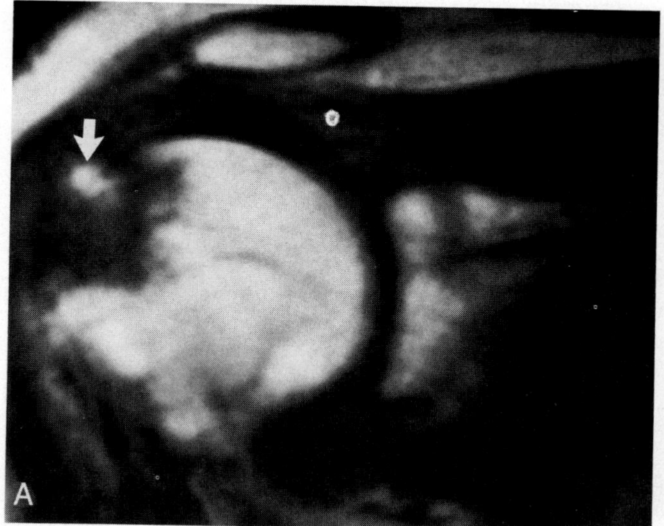

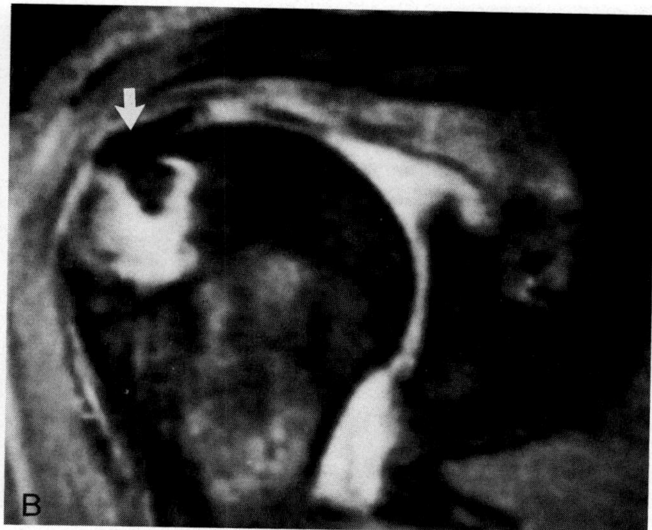

Figure 7–25. Oblique-coronal T1-weighted spin-echo *(A)* and T2*-weighted gradient-echo *(B)* images showing severe fracture of the humeral head with displaced fragment *(arrow)* and bone marrow edema.

injuries, such as bone contusions, stress fractures, and osteochondral fractures. The appearance of fractures on MR imaging usually consists of linear regions of low signal intensity that represent the fracture line, which may be surrounded by poorly defined regions of intermediate signal intensity on T1-weighted images that represent marrow edema. These regions of intermediate signal intensity usually brighten on T2-weighted images, owing to the high water content of the edema (see Fig. 7–25).

OSTEONECROSIS

The humeral head is the second most common site of avascular osteonecrosis after the femoral head. Causes of osteonecrosis include fractures interrupting the blood supply, Caisson's disease, vasculitis, steroid medication, and hemoglobinopathies such as sickle cell anemia or Gaucher's disease. If no cause can be identified, the osteonecrosis is termed idiopathic. Magnetic resonance imaging is a very sensitive method for the depiction of bone necrosis. In one study of avascular necrosis of the hip the sensitivity was 97 per cent and the specifity was 98 per cent.[15] The findings of avascular necrosis in the hip are similar to those found in the shoulder.[45] On T1-weighted images, areas of low signal intensity can be found below the articular surface. Another pattern on T1-weighted images shows a band or bands of low signal intensity surrounding a central area of higher signal intensity (Figs. 7–26, 7–27). On T2-weighted images, areas of low signal intensity can become bright, and regions of high signal intensity remain high.

The double-line sign can be seen in up to 50 per cent of cases.[45] This sign consists of a linear area of bright signal juxtaposed to an area or band of low signal intensity on T2-weighted images. The double-line sign is a well-known feature of avascular necrosis of the femoral head and is considered to be pathognomonic.[34] Bilateral lesions are common. Ancillary findings include joint effusion and articular collapse. An important indication for MR imaging is depiction of the early stages of this disease.

PITFALLS

Because most pitfalls in shoulder MR imaging are discussed in the previous sections, the main sources of misinterpretation are summarized only briefly here.

Increased signal intensity within the supraspinatus tendon on T1-weighted and proton density-weighted images within a normal tendon (pseudogap) probably represents the most important finding that could cause misdiagnosis. The physiologic basis for this signal variation remains unclear. However, the diagnosis of tendinitis or partial tears should be considered carefully.

A small band of high signal intensity at the base of the glenoid labrum can be due to the adjoining hyaline cartilage. A second bandlike region of increased signal intensity within the center of the labrum might be the result of normal vascularization. The reason for this second band remains uncertain; however, it can be considered as a normal variation. Neither signal variation should be mistaken for a labral tear.

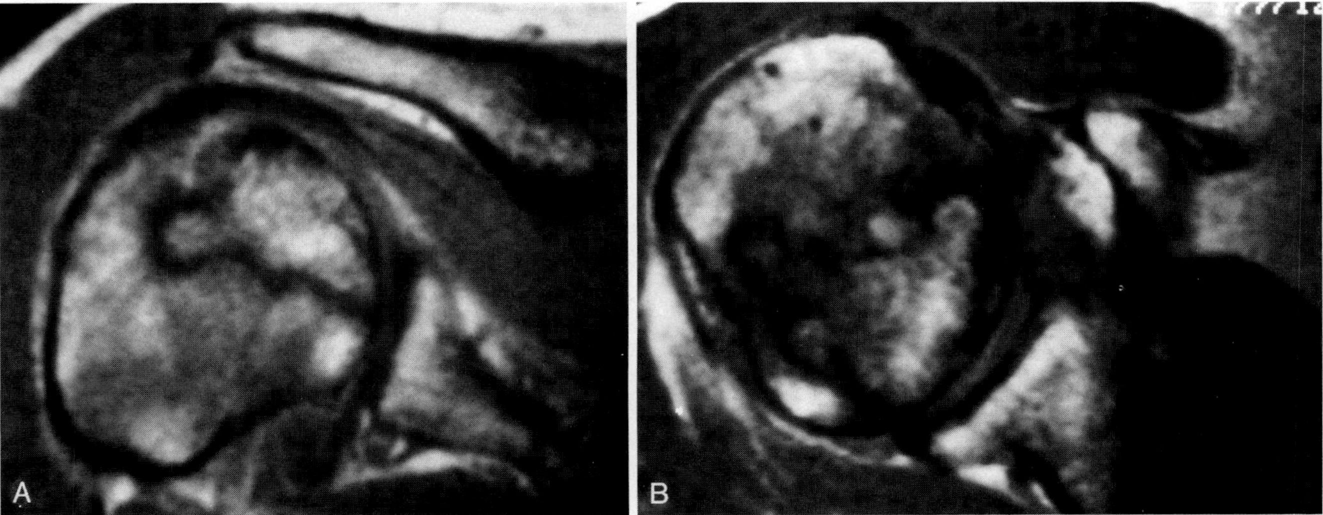

Figure 7-26. Oblique-coronal *(A)* and axial *(B)* T1-weighted spin-echo images of avascular osteonecrosis of the humeral head with a band of low signal intensity surrounding a subchondral area of intermediate to high signal intensity.

The surface of the humerus may vary in appearance. An irregularity and indentation along the posterolateral aspect of the humeral metaphysis are frequently observed. These abnormalities should not be mistaken for a Hill-Sachs lesion, which is located near the posterosuperior surface.

Several pitfalls are possible when gradient-echo sequences are used. It is thus critical to be familiar with the peculiarities of this technique. Because gradient-echo sequences are sensitive to spin dephasing, signal variations of fat and fat-muscle interfaces are observed. Consequently, pitfalls include overestimation or underestimation of certain signal intensities or overestimation of the size of particles that cause magnetic field variations, such as metal or bone fragments. The visualized signal void of these substances increases with increasing TE. Knowledge of these technical aspects can provide additional information.

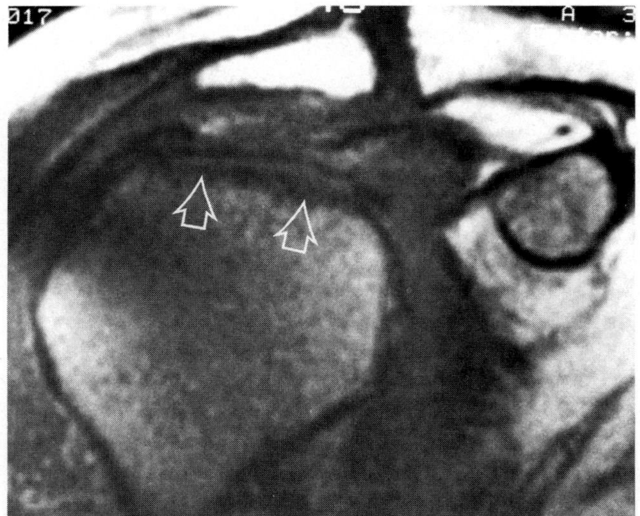

Figure 7-27. Oblique-coronal T1-weighted spin-echo image showing avascular osteonecrosis with a subchondral bandlike dark area *(arrows)*.

References

1. Aoki M, Ishii S, Usui M: The slope of the acromion and rotator cuff impingement. Orthop Trans. 10:228, 1986.
2. Bankart A: The pathology and treatment of recurrent dislocation of the shoulder joint. Br J Surg 26:23-29, 1938.
3. Berquist TH: Imaging of Orthopedic Trauma and Surgery. Philadelphia, WB Saunders, 1986.
4. Bigliani LU, Morisson DS: The morphology of the acromion and its relationship to rotator cuff tears. Orthop Trans 10:228, 1986.
5. Burk DL, Karasick D, Kurtz AB, et al: Rotator cuff tears: Prospective Comparison of MR imaging with arthrography, sonography and surgery. AJR 153:87-92, 1989.
6. Burk DL, Karasick D, Mitchell DG, et al: MR imaging of the shoulder: Correlation with plain radiography. AJR 154:549-553, 1990.
7. Codman EA: The Shoulder, Rupture of the Supraspinatus Tendon and Other Lesions in or about the Subacromial Bursa. Boston, Thomas Todd, 1934.
8. Deichgräber E, Olsson B: Soft tissue radiography in painful shoulder. Acta Radiol 16:393-400, 1975.
9. DeSment AA: Arthrographic demonstration of the subcoracoid bursa. Skeletal Radiol 7:275-276, 1982.
10. Deutsch AL, Resnick D, Mink JH, et al: Computed and conventional arthrotomography of the glenohumeral joint: Normal anatomy and clinical experience. Radiology 153:603-612, 1984.
11. Erickson SJ, Kneeland JB, Komorowski RA, et al: Post-traumatic osteolysis of the clavicle: MR features. J Comput Assist Tomogr 14:835-837, 1990.
12. Farley TE, Neumann CH, Steinbach LS, et al: Full-thickness tears of the rotator cuff of the shoulder: Diagnosis with MR imaging. AJR 158: 347-351, 1992.
13. Flannigan B, Kursunoglu-Brahme S, Snyder S, et al: MR arthrography of the shoulder. AJR 155:829-832, 1990.
14. Garneau RA, Renfrew DL, Moore TE, et al: Glenoid labrum: Evaluation with MR imaging. Radiology. 179:519-522, 1991.
15. Glickstein MF, Burk DL, Schiebler ML, et al: Avascular necrosis versus other diseases of the hip: Sensitivity of MR imaging. Radiology 169:213-215, 1988.
16. Gross ML, Seeger LL, Smith JB, et al: Magnetic resonance imaging of the glenoid labrum. Am J Sports Med 18:229-234, 1990.
17. Hajek PC, Sartoris DJ, Neumann CH: Potential contrast agents for MR arthrography: In vitro evaluation and practical observations. AJR 149:97-104, 1987.
18. Hara B: Studies about the periarticular tissue of the shoulder joints with advancing age. J Japanese Orthop Assn 16:833-876, 1941.

19. Hayes CW, Conway WF: Calcium hydroxiapatite deposition disease. Radiographics 10:1031–1048, 1990.
20. Hennig J, Nauerth A , Friedburg H: RARE imaging: A fast imaging method for clinical MR. Magn Reson Med 3:823–833, 1986.
21. Heuck A, Appel M, Kaiser E, et al: Magnetresonanztomographie (MRT) der Schulter: Möglichkeiten der Überinterpretation von Normalbefunden. Fortschr Röntgenstr 152:587–594, 1990.
22. Holt RG, Helms CA, Steinbach L, et al: Magnetic resonance imaging of the shoulder: Rationale and current applications. Skeletal Radiol 19:5–14, 1990.
23. Horwitz MT, Tocantins LM: An anatomical study of the role of the long thoracic nerve and the related scapular bursa in the pathogenesis of local paralysis of the serratus anterior muscle. Anat Rec 71:375–385, 1938.
24. Iannotti JP, Zlatkin MB, Esterhai JL, et al: Magnetic resonance imaging of the shoulder. J Bone Joint Surg 73-A:17–29, 1991.
25. Kieft GJ, Bloem JL, Rozing PM, et al: Rotator cuff impingement syndrome: MR imaging. Radiology 166:211–214, 1988.
26. Kieft GJ, Dijkmans BAC, Bloem JL, et al: Magnetic resonance imaging of the shoulder in patients with rheumatoid arthritis. Ann Rheum Dis 49:7–11, 1990.
27. Kneeland JB, Carrera GF, Middleton WD, et al: Rotator cuff tears: Preliminary application of high-resolution MR imaging with counter rotating current loop-gap resonators. Radiology 160:695–699, 1986.
28. Kneeland JB, Hyde JS: High-resolution MR imaging with local coils. Radiology 171:1–7, 1989.
29. Kummel BM: Spectrum of lesions of the anterior capsular mechanism of the shoulder. Am J Sports Med 7:111–120, 1979.
30. Kursunoglu-Brahme S , Resnick D: Magnetic resonance imaging of the shoulder. Radiol Clin North Am 28:941–954, 1990.
31. Legan JM, Burkhard TK, Goff WB, et al: Tears of the glenoid labrum: MR imaging of 88 arthroscopically confirmed cases. Radiology. 179:241–246, 1991.
32. Lie S, Mast WA: Subacromial bursography: Technique and clinical application. Radiology 144:626–630, 1982.
33. Ling SC, Chen CF , Wan RX: A study on the vascular supply of the supraspinatus tendon. Surg Radiol Anat 12:161–165, 1990.
34. Mitchell DG, Rao VM, Dalinka MK, et al: Femoral head avascular necrosis: Correlation of MR imaging, radiographic staging, radionuclide imaging, and clinical findings. Radiology 162:709–715, 1987.
35. Mitchell MJ, Causey G, Berthoty DP, et al: Peribursal fat plane of the shoulder: Anatomic study and clinical experience. Radiology 168:699–704, 1988.
36. Morrison DS, Bigliani LU: The Clinical significance of variations in acromial morphology. Orthop Trans 11:234–244, 1987.
37. Moseley HF, Overgaard B: The anterior capsular mechanism in recurrent anterior dislocation of shoulder. Morphological and clinical studies with special reference to the glenoid labrum and the gleno-humeral ligaments. Br J Bone Joint Surg 44-B:913–927, 1962.
38. Moseley HF: Shoulder Lesions. Baltimore, Williams & Wilkins, 1969.
39. Munk PL, Holt RG, Helms CA, et al: Glenoid labrum: Preliminary work with use of radial-sequence MR imaging. Radiology 173:751–753, 1989.
40. Naimark A, Baum A: Pitfall to avoid injection of the subcoracoid bursa: A cause of technical failure in shoulder arthrography. J Can Assoc Radiol 40:170–171, 1989.
41. Neer C: Impingement lesions. Clin Orthop 173:70–77, 1982.
42. Neer CS: Shoulder Reconstruction. Philadelphia, WB Saunders, 1990.
43. Ozaki J, Fujimoto S, Nakagawa Y, et al: Tears of the rotator cuff of the shoulder associated with pathological changes in the acromion. J Bone Joint Surg 70:1224–1230, 1988.
44. Rafii M, Firooznia H, Sherman O, et al: Rotator cuff lesions: Signal patterns at MR imaging. Radiology 177:817–823, 1990.
45. Randall M: MR Image of humeral head AVN similar to that of femoral head. Radiology Today 8:18, 1991. Abstract.
46. Rathbun JB, MacNab I: The microvascular pattern of the rotator cuff. J Bone Joint Surg 52B:541–553, 1970.
47. Reiser M, Erlemann R, Bongartz G, et al: Möglichkeiten der magnetischen resonanz Tomographie (MRT) in der Diagnostik des Schultergelenkes. Radiologe 28:79–83, 1988.
48. Reiser MF: Imaging of the joint and soft-tissue disorders. Curr Opin Radiol 2:684–690, 1990.
49. Seeger LL, Gold RH, Basset LW, et al: Shoulder impingement syndrome: MR findings in 53 shoulders. AJR 150:343–347, 1988.
50. Strizak AM, Danzig TL, Jackson DW, et al: Subacromial bursography: An anatomical and clinical study. J Bone Joint Surg 64-A:196–201, 1982.
51. Tsai JC , Zlatkin MB: Magnetic resonance imaging of the shoulder. Radiol Clin North Am 28:279–291, 1990.
52. Turkel SJ, Panio IMW, Marshall JL, et al: Stabilizing mechanisms preventing anterior dislocation of the glenohumeral joint. J Bone Joint Surg 63A:1208–1217, 1981.
53. Vahlensieck M, Majumdar S, Lang P, et al: Fast gradient-echo imaging of the shoulder: TE dependent effects on image contrast. SMRM 10th Annual Scientific Meeting San Francisco, Book of Abstracts. 1:316, 1991.
54. Vahlensieck M, Majumdar S, Lang P, et al: Shoulder MRI: Routine examinations using gradient recalled and fat-saturated sequences. Eur Radiol 2:142–147, 1992.
55. Vestring T, Bongartz G, Konermann W, et al: Stellenwert der Magnetresonanztomographie in der Diagnostik von Schultererkrankungen. Fortschr Röntgenstr 154:143–149, 1991.
56. Weston WJ: The enlarged subdeltoid bursa in rheumatoid arthritis. Br J Radiol 42:481–486, 1969.
57. Weston WJ: Soft Tissues of the Extremities: A Radiologic Study of Rheumathic Disease. Berlin, Springer, 1978.
58. Weston WJ: The subdeltoid bursa. Aust Radiol 17:214–215, 1973.
59. Zlatkin MB, Bjorkengren AG, Gylys-Morin V, et al: Cross-sectional imaging of the capsular mechanism of the glenohumeral joint. AJR 150:151–158, 1988.
60. Zlatkin MB, Iannotti JP, Roberts MC, et al: Rotator cuff tears: Diagnostic performance of MR imaging. Radiology 172:223–229, 1989.
61. Zlatkin MB, Reicher MA, Kellerhouse LE, et al: The painful shoulder: MR imaging of the glenohumeral joint. J Comput Assist Tomogr 12:995–1001, 1988.

8 THE ELBOW

Russell C. Fritz and Lynne S. Steinbach

Magnetic resonance (MR) imaging provides clinically useful information for assessing the elbow joint. Clinicians who are familiar with the advantages of MR imaging in evaluating other joints have been requesting examinations of the elbow with increasing frequency. Experience has demonstrated the utility of this technique in detecting and characterizing disorders of the elbow in a noninvasive fashion.[2,16] The superior depiction of muscles, ligaments, and tendons as well as the ability to visualize bone marrow and hyaline cartilage directly are advantages that MR imaging has over conventional imaging techniques. These features of MR imaging may help to establish the cause of elbow pain and to determine accurately the extent of bone and soft tissue disease.

IMAGING TECHNIQUE

The patient is typically scanned in a supine or a posterior oblique position with the arm at the side. Depending on the size of the patient relative to the bore of the magnet, it may be necessary to scan in a prone position with the arm extended overhead. In general, the prone position is less well tolerated and results in a greater number of motion-degraded studies. Taping a vitamin E pill to the skin at the site of tenderness or at the site of a palpable mass is useful to ensure that the area of interest is imaged, especially when no pathologic lesion is identified on the images. The protocol of MR imaging of the elbow at the University of California, San Francisco, is listed in Table 8–1.

T2-weighted images are usually obtained in the axial and sagittal planes using spin-echo technique. T1-weighted and short tau inversion-recovery (STIR) sequences usually are obtained in the coronal plane. Although the STIR sequence has a relatively poor signal-to-noise ratio because of the suppression of the signal from fat, pathologic structures often are more conspicuous owing to the effects of additive T1 and T2 contrast.

Additional sequences may be added or substituted depending on the clinical problem to be solved. T2*-weighted gradient-echo sagittal se-

Table 8–1. PROTOCOLS OF MAGNETIC RESONANCE IMAGING OF THE ELBOW AT THE UNIVERSITY OF CALIFORNIA, SAN FRANCISCO

Plane	TR (ms)	TE (ms)	FOV (cm)	Slice (mm)	Skip (mm)	NEX	Matrix (pixel)	Time (min)
Coronal (SE)	600	15	10	4	1	2	192	4
Coronal (STIR)	1800	30, T1 = 160	10	4	1	1	128	4
Axial (FSE)	3500	30, 90	12	4	1	2	256	8
Sagittal (FSE)	3500	30, 90	12	4	1	2	256	8
Sagittal* (SE)	600	15	10	4	1	2	192	4
Axial* (SE)	600	15	10	4	1	2	192	4

TR, Repetition time; TE, echo time; FOV, field of view; Skip, interslice gap; NEX, number of excitations; STIR, short tau inversion-recovery; T1, inversion time; SE, spin echo; FSE, fast spin echo; * optional sequence.

quences provide useful supplemental information in identifying loose bodies. Gradient-echo volume sequences allow acquisition of a sequence of very thin axial images, which may be reformatted subsequently in any plane using a computer work station. These volume sequences have not been routinely useful in our experience because of their relatively poor soft tissue contrast compared with that of the spin-echo technique. In general, gradient-echo sequences are to be avoided after surgery because magnetic susceptibility artifacts associated with micrometallic debris may obscure the images and may also be mistaken for loose bodies. Furthermore, the degree of artifact surrounding orthopedic hardware is most prominent on gradient-echo sequences owing to the lack of a 180-degree refocusing pulse and is least prominent on fast spin-echo sequences owing to the presence of multiple 180-degree pulses. Fast spin-echo sequences may be substituted for the conventional T2-weighted spin-echo sequences if they are available; these newer sequences allow greater flexibility in imaging the elbow while continuing to provide information that is comparable to that of the spin-echo sequences. The speed of fast spin-echo sequences may be used to obtain higher resolution T2-weighted images in the same amount of time as the conventional spin-echo sequences or may simply be used to increase the speed of the examination. The ability to shorten the examination with fast spin-echo has been useful when scanning claustrophobic patients or when scanning patients who become uncomfortable in the prone position with the arm overhead.

Fat suppression may be added to various pulse sequences to improve visualization of the hyaline articular cartilage. Avoidance of chemical shift artifact at the interface of cortical bone and fat-containing marrow permits a more accurate depiction of the overlying hyaline cartilage. T1-weighted images with fat suppression are useful whenever gadopentetate dimeglumine is administered. Intravenous gadopentetate dimeglumine may provide additional information in the assessment of neoplastic or inflammatory processes about the elbow. Injection of the joint with dilute gadopentetate dimeglumine may be useful in patients without a joint effusion to detect loose bodies, to determine if the capsule is disrupted, or to determine if an osteochondral fracture fragment is stable.

ANATOMY

A thorough understanding of the anatomy and function of the elbow is essential for interpretation of the MR images. The anatomic structures of the elbow are reliably depicted with MR imaging.[6,13] Knowledge of the relative functional significance of these structures allows the radiologist to identify and assess the clinically important anatomy. Focusing on the relevant anatomic structures leads to more meaningful interpretation of the images and facilitates clinical problem solving.

The elbow is composed of three articulations contained within a common joint cavity. The radial head rotates within the radial notch of the ulna, allowing supination and pronation distally. The radial head is surrounded by the annular ligament, which is best seen on the axial images. Disruption of the annular ligament results in proximal radioulnar joint instability. The radius articulates with the capitellum, and the ulna articulates with the trochlea in a hinge fashion. The anterior and posterior portions of the joint capsule are relatively thin, whereas the medial and lateral portions are thickened to form the collateral ligaments. The ulnar collateral ligament (UCL) complex consists of anterior and posterior bundles as well as an oblique band also known as the transverse ligament. The functionally important anterior bundle of the UCL extends from the medial epicondyle to the medial aspect of the coronoid process and is well seen on coronal images. The anterior bundle provides the primary restraint to valgus stress and is commonly damaged in throwing athletes.[8,17,18] The radial collateral ligament (RCL) complex is more variable and less well understood than the UCL.[17,32] The RCL proper arises from the lateral epicondyle anteriorly and inserts onto the annular ligament. A more posterior bundle known as the lateral ulnar collateral ligament is present in 90 per cent of anatomic specimens and is thought to provide the primary restraint to varus stress.[32] Disruption of the lateral ulnar collateral ligament results in the pivot shift phenomenon and posterolateral rotatory instability of the elbow.[23] Both the RCL proper and the lateral ulnar collateral ligament are well seen on the coronal images and should be evaluated separately because of the difference in functional significance of these structures.

The muscles of the elbow are divided into anterior, posterior, medial, and lateral compartments. The anterior compartment contains the biceps and brachialis muscles, which are best evaluated on sagittal and axial images. The brachialis extends along the anterior joint capsule and inserts on the ulnar tuberosity. The biceps lies superficial to the brachialis and inserts on the radial tuberosity. The posterior compartment contains the triceps and anconeus muscles, which are best evaluated on sagittal and axial images. The triceps inserts on the proximal aspect of the olecranon. The anconeus arises from the posterior aspect of the lateral epicondyle and inserts more distally on the olecranon. The anconeus provides dynamic support to the RCL complex in resisting varus stress. The medial and lateral

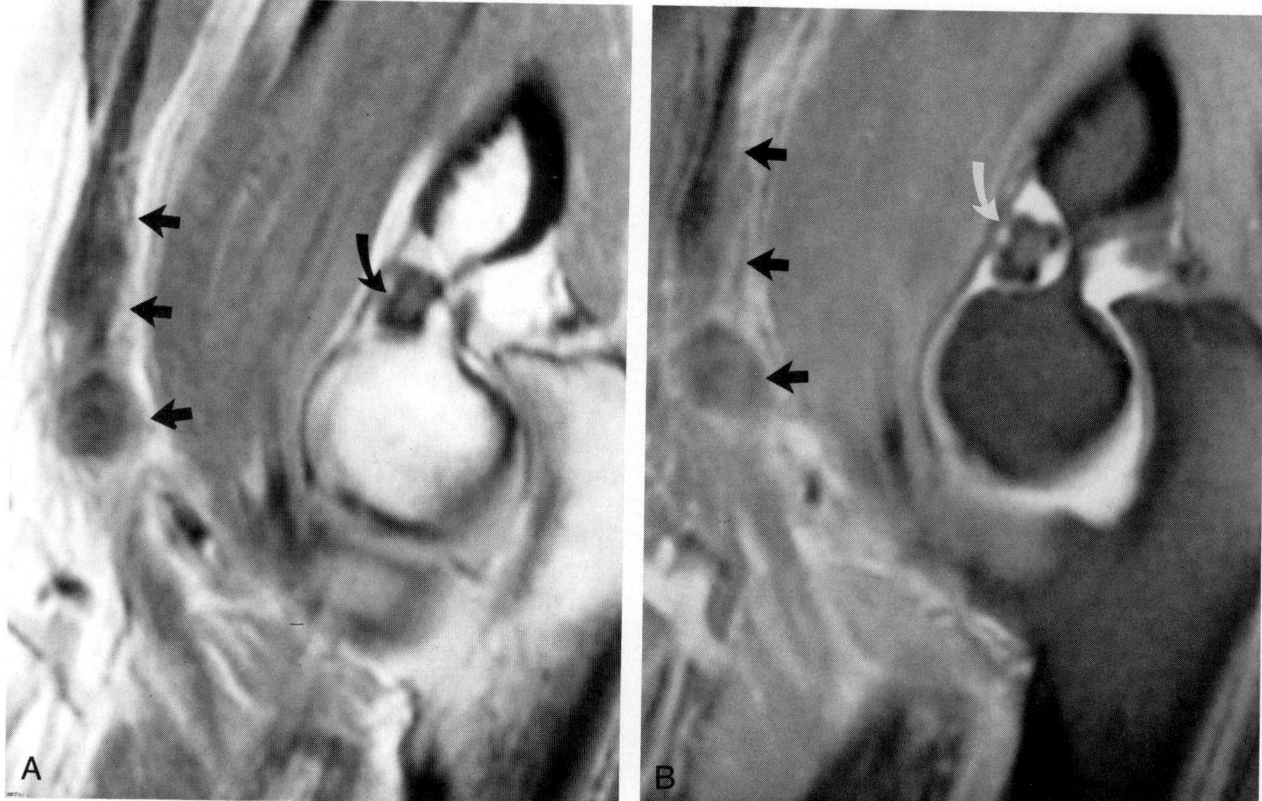

Figure 8–1. *A*, Fast spin-echo proton density sagittal image reveals a high-grade partial tear of the biceps tendon *(arrows)*. A loose body is also noted in the coronoid fossa anteriorly *(curved arrow)*. *B*, Fast spin-echo proton density sagittal image with fat suppression again shows thickening and retraction of a large portion of the biceps tendon *(arrows)*. The loose body is more conspicuous *(curved arrow)* on the fat-suppressed image owing to readjustment of the gray scale, which makes the joint fluid relatively bright.

compartment muscles are best seen on coronal and axial images. The medial compartment structures include the pronator teres and the flexors of the wrist and hand, which arise from the medial epicondyle as the common flexor tendon. The common flexor tendon provides dynamic support to the UCL complex in resisting valgus stress. The lateral compartment structures include the supinator, the brachioradialis, and the extensors of the wrist and hand, which arise from the lateral epicondyle as the common extensor tendon.

The ulnar, median, and radial nerves are subject to entrapment in the elbow region. These nerves are normally surrounded by fat and are best seen on the axial images.

PATHOLOGY

Biceps Tendon

Rupture of the distal biceps tendon is a relatively uncommon injury, which accounts for 3 to 10 per cent of all biceps tendon tears. Clinical diagnosis may be difficult as the bicipital aponeurosis usually remains intact and proximal retraction of the muscle is minimal. Flexion power at the elbow may be preserved; however, supination of the forearm is usually weakened.

Magnetic resonance imaging is useful in evaluating these injuries because partial and complete ruptures may be distinguished (Fig. 8–1). The T2-weighted axial images are most useful for determining the degree of tearing. The axial images must extend from the musculotendinous junction to the insertion of the tendon on the radial tuberosity. Magnetic resonance imaging provides useful information regarding the size of the gap and the location of the tear for preoperative planning. Complete tears of the distal biceps are thought to be more common than partial tears. The vast majority of distal biceps ruptures occur in men, with the injury involving the dominant arm in 80 per cent.[3] These tears usually occur as a result of forceful elbow flexion against resistance.[3,19] Other injuries that may occur via the same mechanism include avulsion and strain of the brachialis as well as disruption of the annular ligament with anterior dislocation of the radial head. These less common injuries may also be identified with MR imaging.

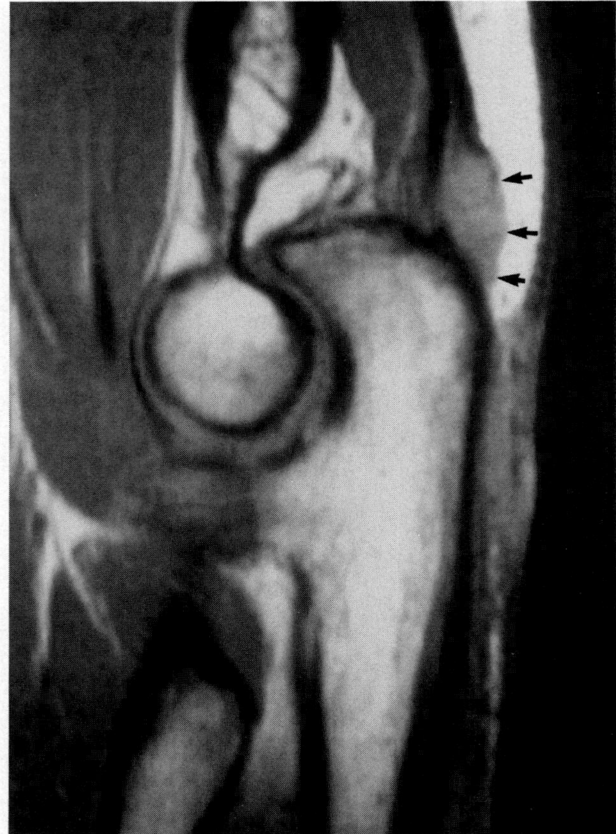

Figure 8–2. T1-weighted sagittal image of a partial tear of the triceps tendon. The superficial fibers are separated from their insertion on the olecranon by a fluid-filled gap *(arrows)*. The deep fibers remain attached.

Triceps Tendon

Injuries of the triceps tendon are well seen with MR imaging (Fig. 8–2). Complete avulsion of the distal triceps has been considered one of the least common tendon injuries in the body.[32] Partial tears are also considered uncommon.[26] The usual mechanisms of injury include a direct blow to the tendon or a decelerating counterforce during active extension.[19] The normal triceps tendon often appears wavy with foci of increased signal intensity within the tendon on the sagittal images. These foci normally decrease in signal intensity on T2-weighted images and should not be mistaken for pathologic lesions. The apparent laxity of the tendon is normal when the elbow is imaged in full extension and resolves when the elbow is imaged in mild degrees of flexion.

Lateral Epicondylitis

Lateral epicondylitis, also referred to as tennis elbow, is caused by degeneration and tearing of the common extensor tendon.[19,20] This condition often occurs as a result of repetitive sports-related trauma to the tendon, although it is seen far more commonly in nonathletes.[4,7,32] In the typical case, the extensor carpi radialis brevis tendon is partially avulsed from the lateral epicondyle.[20] Scar tissue forms in response to this partial avulsion, which is then susceptible to further tearing with repeated trauma. Local steroid injections are commonly used to treat lateral epicondylitis and may increase the risk of tendon rupture.[29] Histologic studies have clearly demonstrated angiofibroblastic hyperplasia with a lack of inflammation in the surgical specimens of patients with lateral epicondylitis; this finding suggests that the abnormal signal intensity seen on MR images is secondary to tendon degeneration and repair rather than an indication of tendinitis.[32]

Overall, 4 to 10 per cent of cases of lateral epicondylitis are resistant to conservative therapy[19]; MR imaging is useful in assessing the degree of tendon damage in such cases (Fig. 8–3). Tendon degeneration (also referred to as tendinopathy or tendinosis) is manifested by normal to increased tendon thickness with increased signal intensity on T1-weighted images, which does not further increase in signal intensity on T2-weighted images. Partial tears are characterized by thinning of the tendon, which is outlined by adjacent fluid on the T2-weighted images. Complete tears may be diagnosed on MR images by identifying a fluid-filled gap that separates the tendon from its adjacent bony attachment site (Fig. 8–4).

At surgery, 97 per cent of the tendons appear scarred and edematous, and 35 per cent have macroscopic tears.[20] Magnetic resonance imaging is useful in identifying high-grade partial tears and complete tears, which are unlikely to improve with rest and repeated steroid injections. In addition to determining the degree of tendon damage, MR imaging also provides a more global assessment of the elbow and therefore is able to detect additional pathologic conditions that may explain the lack of a therapeutic response. For example, unsuspected ruptures of the RCL complex may occur in association with tears of the common extensor tendon (Fig. 8–5). Moreover, the lack of a significant abnormality involving the common extensor tendon on MR imaging may prompt consideration of an alternative diagnosis, such as radial nerve entrapment, which may mimic lateral epicondylitis.[19,31,33]

Medial Tension–Lateral Compression Syndromes

Repetitive valgus stress injury in the elbow is well described and is commonly seen in baseball and javelin throwing. This spectrum of injuries may

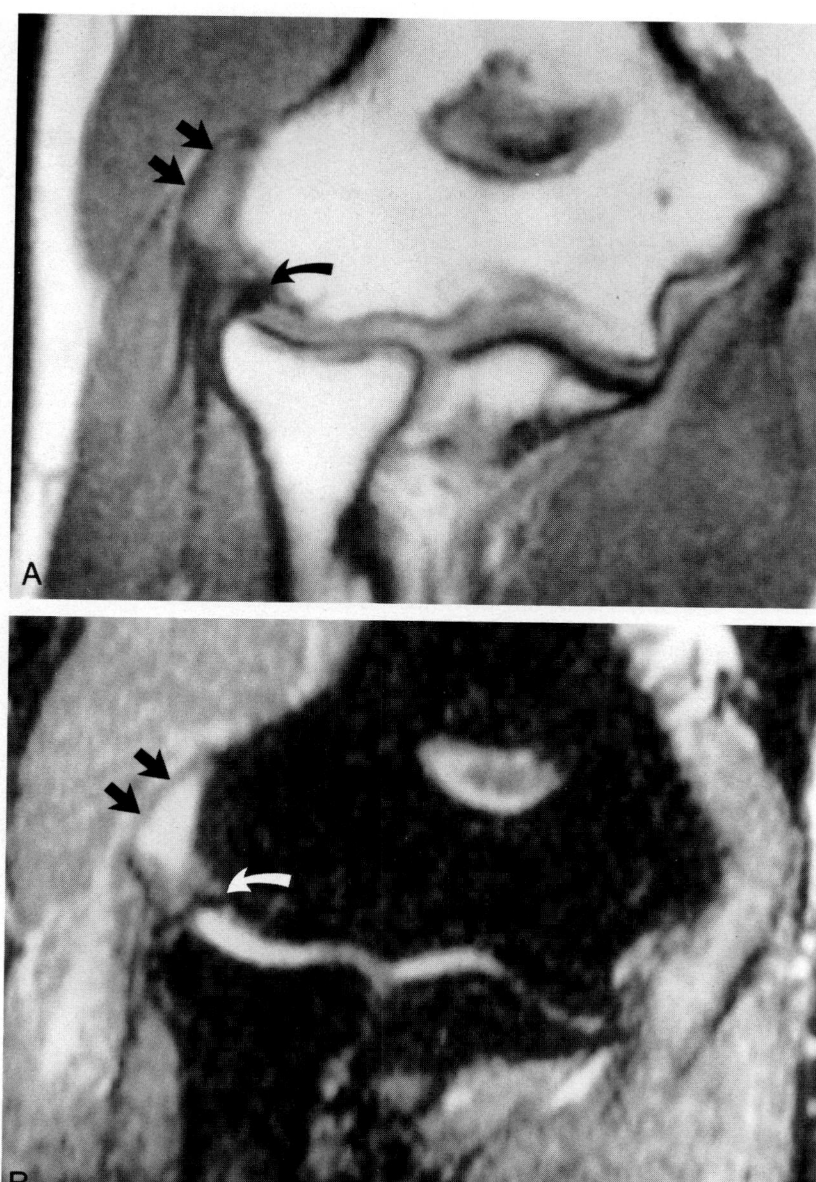

Figure 8–3. T1-weighted *(A)* and short tau inversion-recovery (STIR) *(B)* coronal images of a ruptured common extensor tendon in a 50-year-old tennis player with pain that did not respond to rest and local steroid injections. A fluid-filled gap separates the tendon from the lateral epicondyle *(arrows)*. The underlying radial collateral ligament is intact *(curved arrows)*.

also occur in quarterbacks, wrestlers, gymnasts, golfers, weightlifters, tennis players, and violinists. A number of different conditions may occur in addition to the repeated valgus stress of the throwing motion. Medial tension overload typically produces extra-articular injury, such as flexor-pronator strain, UCL sprain, ulnar traction spurring, and ulnar neuropathy. Lateral compression overload typically produces intra-articular injury, such as osteochondritis dissecans of the capitellum or radial head, degenerative arthritis, and loose body formation. Magnetic resonance imaging can assess for each of these pathologic processes that is associated with repeated valgus stress.[16] The information provided by MR imaging can be quite helpful in formulating a logical treatment plan, especially when surgery is being considered.

Medial Epicondylitis

Medial epicondylitis, also known as golfer's elbow, pitcher's elbow, or "medial tennis elbow," is less common than lateral epicondyitis.[28,32] This condition is caused by overload of the flexor-pronator muscle group, which arises from the medial epicondyle.[30,32] The spectrum of injuries seen in medial epicondylitis that may be characterized with MR imaging includes tendon degeneration (Fig. 8–6), muscle strain (Fig. 8–7), and macroscopic tendon disruption (Fig. 8–8). The coronal and axial sequences are useful for assessing the degree of tendon injury. Magnetic resonance imaging facilitates surgical planning by delineating and grading tears of the common flexor tendon as well as evaluating the underlying UCL complex. The increased preoperative

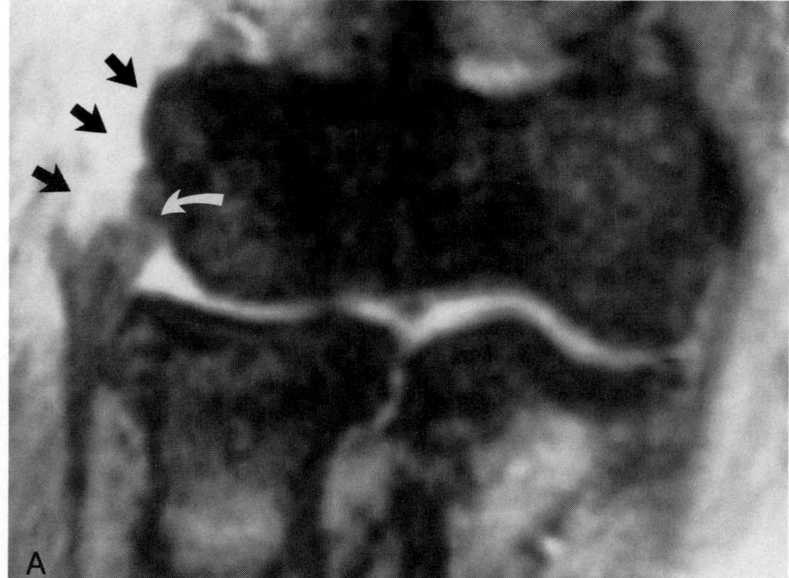

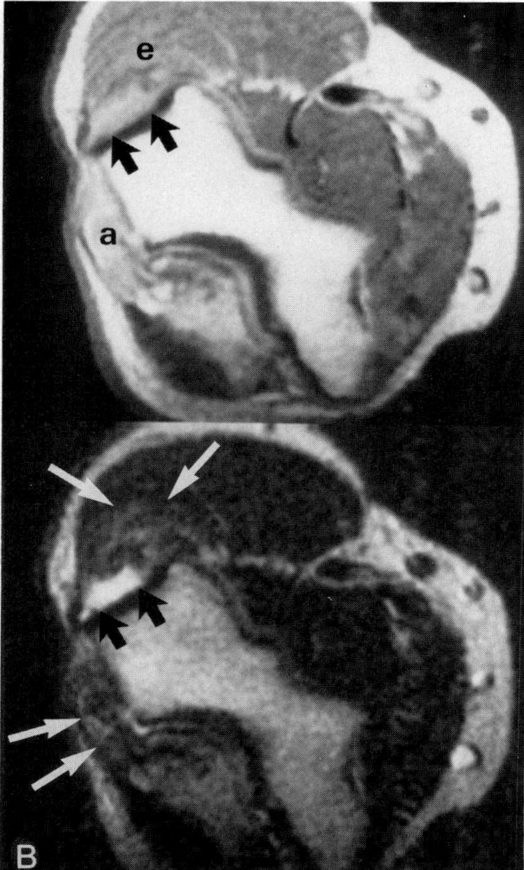

Figure 8–4. *A*, T2*-weighted gradient-echo coronal image of a complete tear of the common extensor tendon *(arrows)*. The underlying radial collateral ligament is intact *(curved arrow)*. *B*, Proton density and T2-weighted axial images just inferior to the lateral epicondyle allow the extent of the tendon tear to be quantified *(black arrows)*. Mild increased signal intensity is also noted in the adjacent anconeus and extensor musculature *(white arrows)*. a, Anconeus; e, extensor musculature.

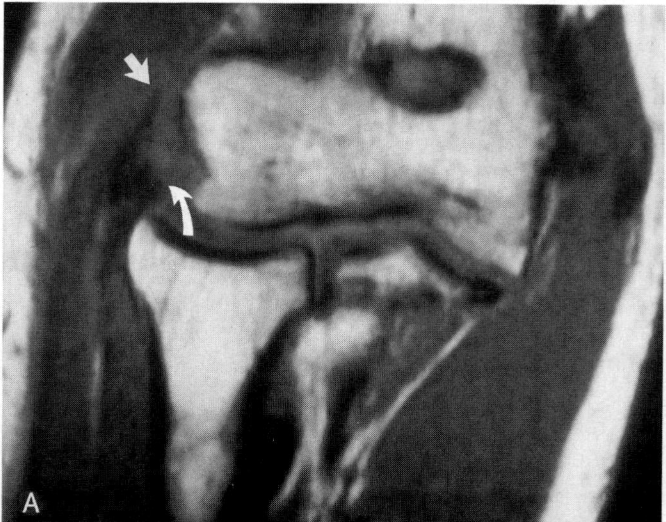

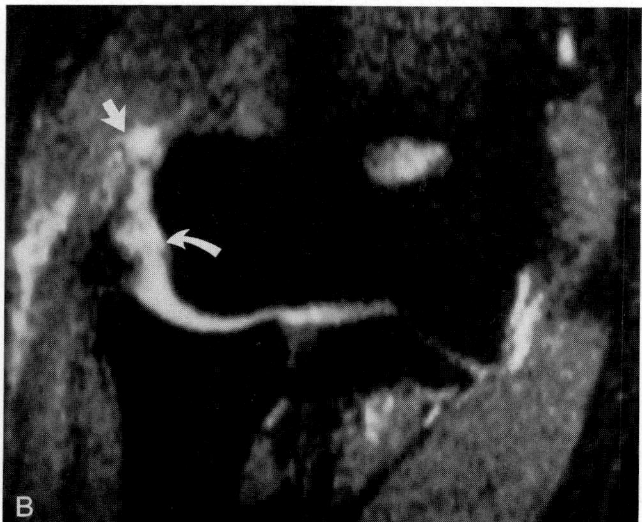

Figure 8–5. T1-weighted *(A)* and short tau inversion-recovery (STIR) *(B)* coronal images of a ruptured common extensor tendon *(arrows)* and a ruptured radial collateral ligament *(curved arrows)*. These findings were confirmed at surgery.

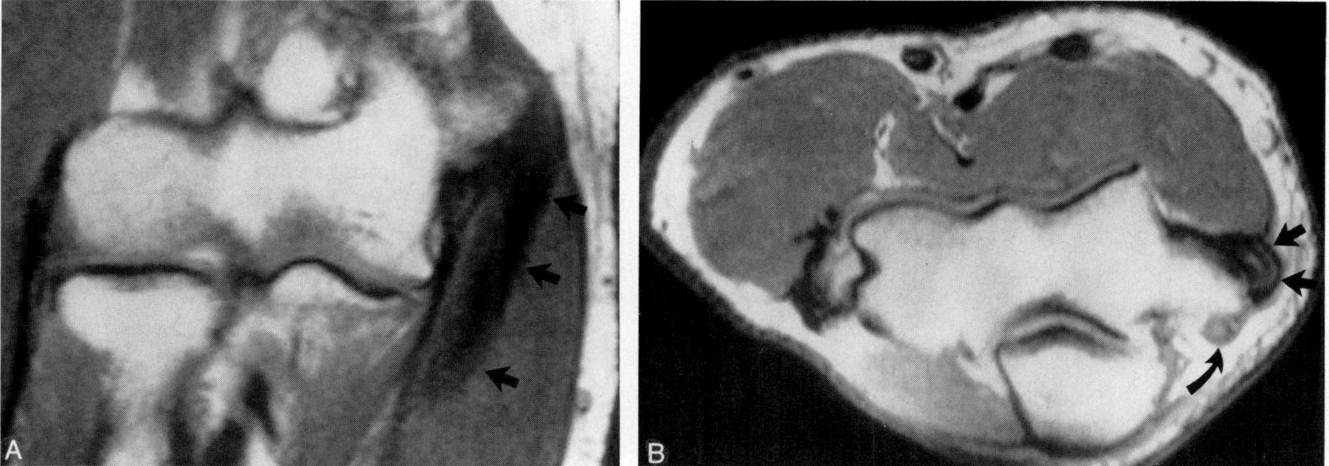

Figure 8–6. T1-weighted coronal *(A)* and proton density axial *(B)* images reveal thickening of the common flexor tendon *(arrows)* compatible with degeneration. There is no evidence of thinning of the tendon or of a fluid-filled gap to suggest a partial or complete tear. The ulnar nerve is mildly enlarged *(curved arrow)*. The overlying cubital tunnel retinaculum is absent, allowing anterior dislocation of the ulnar nerve during elbow flexion and friction neuritis.

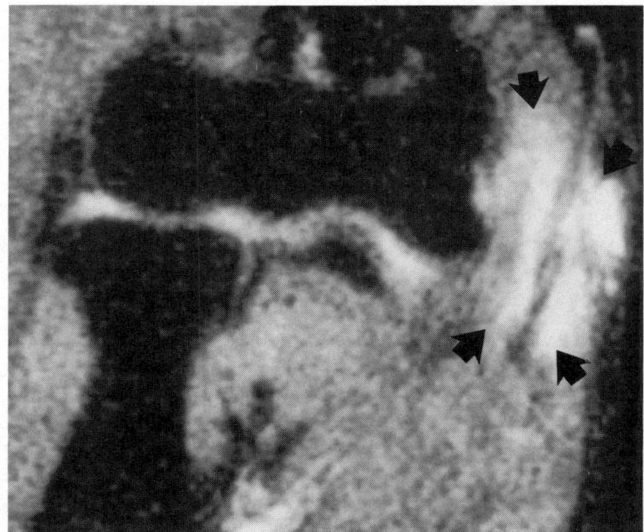

Figure 8–7. The short tau inversion-recovery (STIR) coronal image reveals increased signal intensity throughout the pronator teres and adjacent flexor musculature that is compatible with a strain secondary to overuse *(arrows)*. These findings were extremely subtle on the spin-echo T2-weighted sequence.

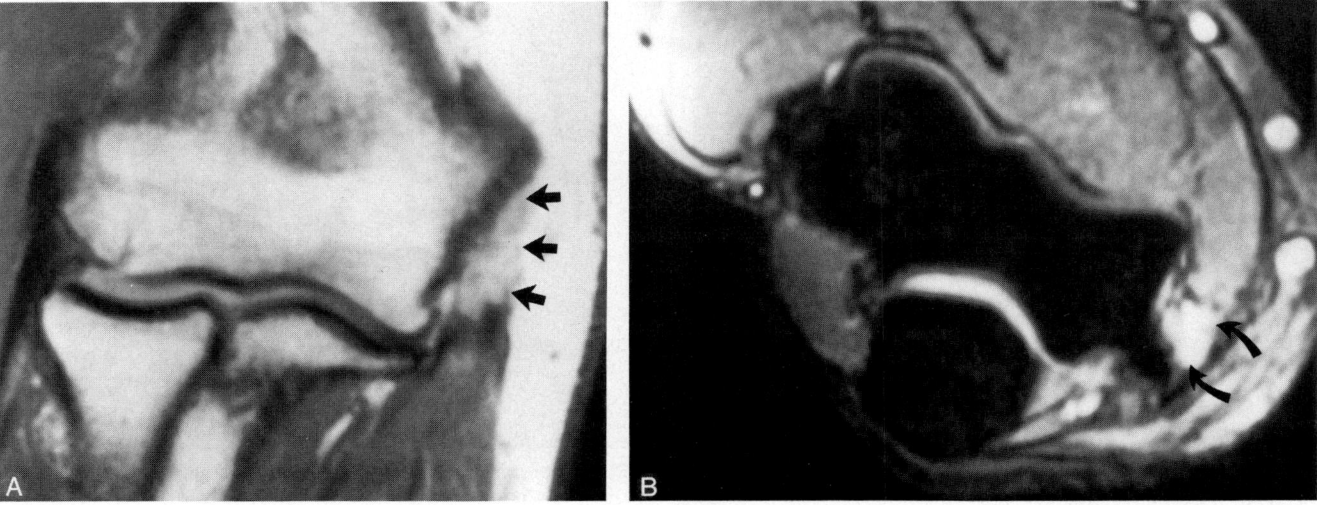

Figure 8–8. Proton density coronal *(A)* and T2*-weighted gradient-echo axial *(B)* images of a rupture common flexor tendon *(arrows)*. A fluid-filled gap separating the tendon from the medial epicondyle was found at surgery.

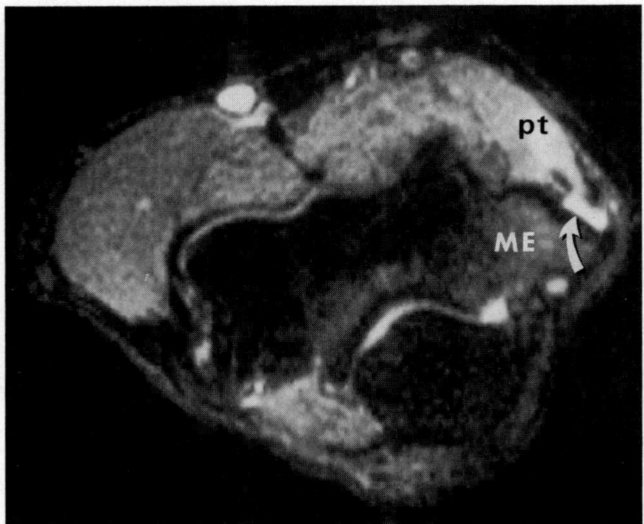

Figure 8–9. A short tau inversion-recovery (STIR) axial image of a 38-year-old professional tennis player with chronic medial epicondylitis that did not respond to steroid injections and rest. A complete tear of the common flexor tendon is seen, outlined by a fluid-filled gap along the anterior margin of the medial epicondyle *(curved arrow)*. Increased signal intensity is also noted in the adjacent pronator teres muscle and within the cancellous bone of the medial epicondyle. ME, Medial epicondyle; pt, pronator teres.

diagnostic information may lessen the need for extensive surgical exploration in cases in which the UCL is clearly intact.

The STIR sequence is the most sensitive for detecting pathologic conditions in the muscle and marrow (Fig. 8–9). Increased signal intensity on STIR and T2-weighted sequences may be seen after an intramuscular injection and may persist for as long as 1 month.[24] Abnormal signal intensity within a muscle may simply be the effect of a therapeutic injection for epicondylitis rather than an indication of muscle strain. Inquiring if and when a steroid injection was given may be useful in recognizing this phenomenon; this information may be obtained easily from the patient questionnaire at the time of MR imaging.

Stress fractures or avulsion of the medial epicondylar apophysis may occur in skeletally immature baseball players (little leaguer's elbow).[5] Magnetic resonance imaging may detect these injuries before complete avulsion and displacement by revealing soft tissue or marrow edema about the medial epicondylar apophysis on the STIR images.

Ulnar Collateral Ligament

In older throwing athletes, degeneration and tearing of the UCL complex with or without injury of the common flexor tendon may be seen. Acute injury of the UCL can be detected and graded. Lateral compartment bone contusions are usually seen in association with acute UCL tears. The status of the functionally important anterior bundle of the UCL complex may be determined by assessing the axial and coronal images (Fig. 8–10). The majority of the tears occur in the middle and proximal fibers of the anterior bundle, whereas 10 per cent of tears are distal avulsions. Chronic degeneration of the UCL is characterized by thickening of the ligament, which may contain foci of calcification or heterotopic bone (Fig. 8–11); these findings are similar to those seen

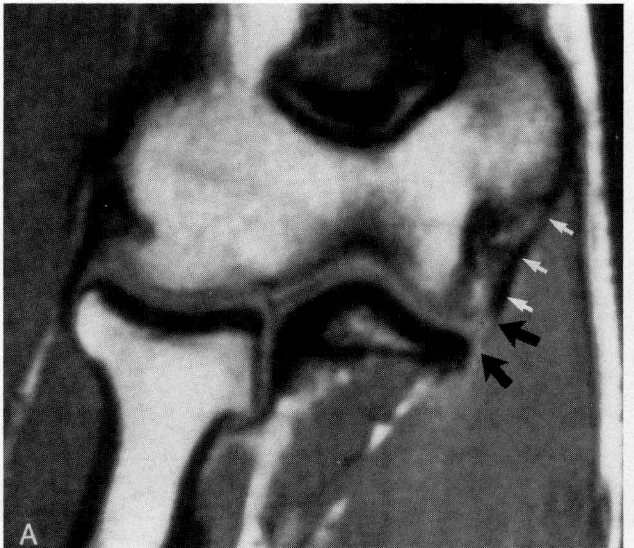

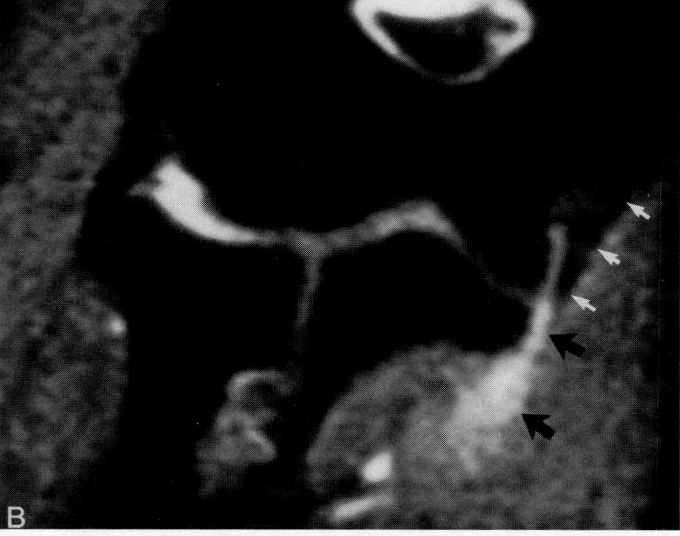

Figure 8–10. T1-weighted *(A)* and short tau inversion-recovery (STIR) *(B)* coronal images of a complete tear of the ulnar collateral ligament in a 23-year-old professional baseball player. The functionally important anterior bundle of the ligament *(white arrows)* is detached from its distal insertion on the medial aspect of the coronoid process. The detachment is outlined by fluid in the adjacent soft tissues *(black arrows)*.

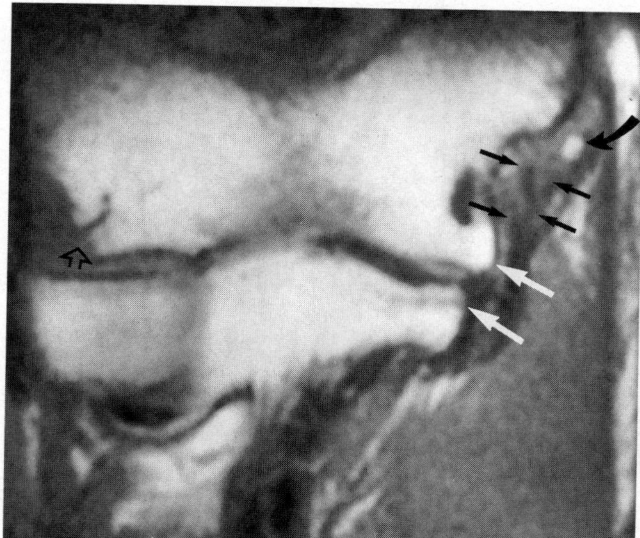

Figure 8–11. The T1-weighted coronal image reveals thickening and increased signal intensity throughout the proximal aspect of the ulnar collateral ligament *(small black arrows)* in a 36-year-old professional baseball player. Heterotopic ossification is noted within the ligament just distal to the medial epicondyle *(curved arrow)*; this finding is similar to that seen in the Pellegrini-Stieda phenomenon after tears of the medial collateral ligament in the knee. Irregularity of the capitellum laterally *(open arrow)* as well as bony hypertrophy and traction spurring medially *(white arrows)* is also noted.

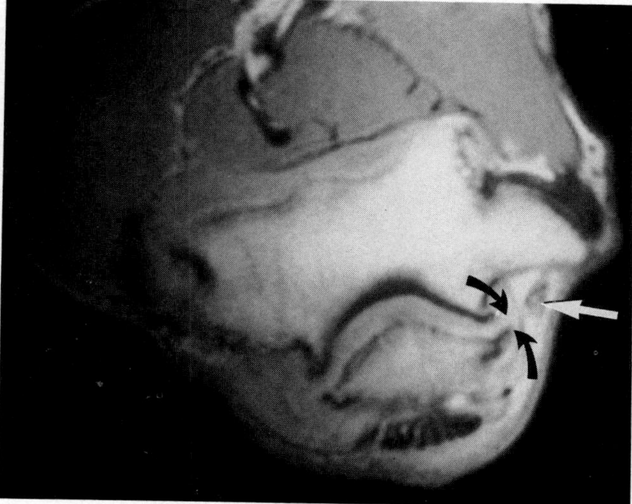

Figure 8–12. The T1-weighted axial image reveals changes of osteoarthritis with medial compartment osseous ridging *(curved arrows)* undermining the floor of the cubital tunnel. Although the volume of the cubital tunnel is diminished, the ulnar nerve *(white arrow)* appears normal and remains surrounded by fat.

after tears of the medial collateral ligament in the knee (Pellegrini-Stieda phenomenon). Ulnar traction spurs are commonly seen at the insertion of the UCL on the coronoid process owing to repetitive valgus stress, occurring in 75 per cent of professional baseball pitchers.[10]

Ulnar Nerve Entrapment

The ulnar nerve is well seen on axial MR images as it passes through the cubital tunnel. The floor of the tunnel is formed by the capsule of the elbow and the posterior and transverse portions of the UCL. Thickening of the UCL and medial bony spurring may undermine the floor of the cubital tunnel, resulting in ulnar neuropathy (Fig. 8–12).[12,22] The roof of the tunnel is formed by the flexor carpi ulnaris aponeurosis distally and the cubital tunnel retinaculum proximally. Anatomic variations of the cubital tunnel retinaculum may contribute to ulnar neuropathy. The cubital tunnel retinaculum may be absent in 10 per cent of cases, allowing anterior dislocation of the nerve over the medial epicondyle during flexion, with subsequent friction neuritis (see Fig. 8–6).[32] The retinaculum may be thickened in 22 per cent of cases, resulting in dynamic compression of the ulnar nerve during elbow flexion. The retinaculum may be replaced by an anomalous muscle, the anconeus epitrochlearis, in 11 per cent of cases, resulting in static compression of the ulnar nerve.[22] These variations in the cubital tunnel retinaculum as well as the ulnar nerve itself may be evaluated on the axial MR images.

Osteochondritis Dissecans

Chronic lateral impaction may lead to osteochondritis dissecans of the capitellum or radial head in young pitchers.[14,15,19,25] The repeated valgus stress and the relatively tenuous blood supply within the capitellum have been proposed to explain the frequent occurrence of osteochondritis dissecans in this location. Stable osteochondral lesions are usually treated with rest and splinting, whereas unstable lesions are either pinned or excised.[25,27] Magnetic resonance imaging can reliably detect and stage these lesions (Fig. 8–13). Unstable lesions are characterized by fluid encircling the osteochondral fragment on T2-weighted images. Loose in situ lesions may also be diagnosed by identifying a cystlike lesion beneath the osteochondral fragment.[9] These apparent cysts typically are found to contain loose granulation tissue at surgery, explaining why they may enhance after intravenous administration of gadopentetate dimeglumine (Fig. 8–14).

Osteochondritis dissecans should be distinguished from osteochondrosis of the capitellum, which is known as Panner's disease. Osteochondritis dissecans is typically seen in the 13- to 16-year age range, whereas Panner's disease is typically seen in the 5 to 10-year range. Loose body formation and significant residual deformity of the capitellum are concerns in osteochondritis dissecans but are usu-

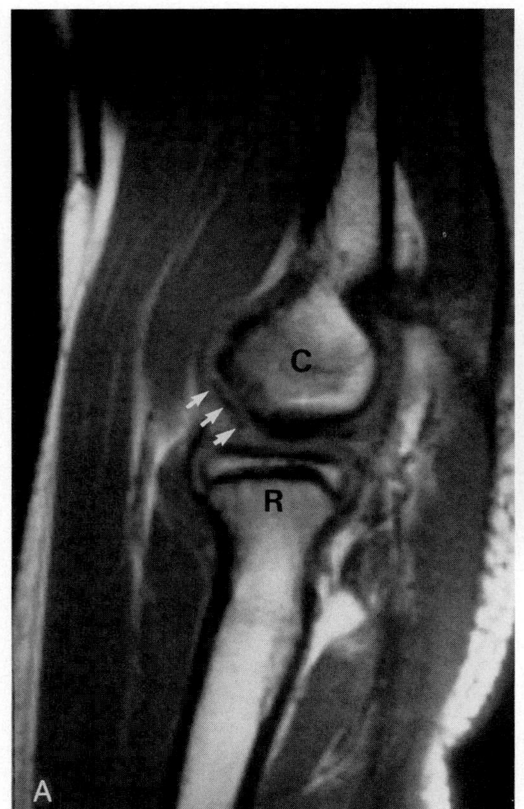

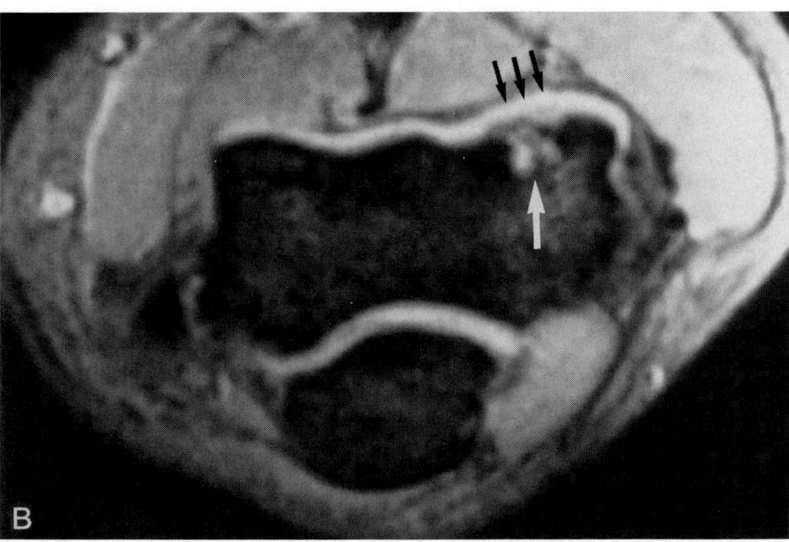

Figure 8–13. *A*, A T1-weighted sagittal image of an in situ osteochondral fracture fragment *(arrows)* in a 15-year-old baseball pitcher. C, Capitellum; R, radius. *B*, The T2*-weighted gradient-echo axial image reveals a small cystic-appearing area in the anterior aspect of the capitellum *(large white arrow)*, which suggests instability of the overlying osteochondral fragment *(black arrows)*.

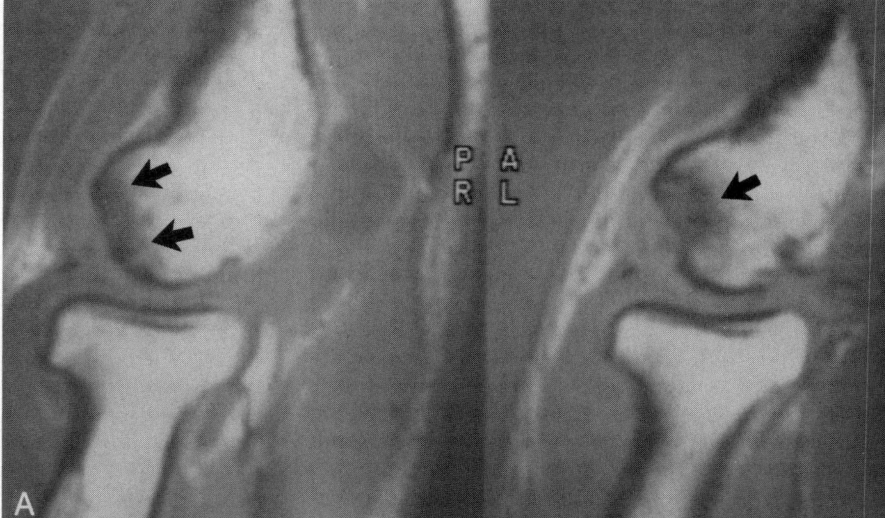

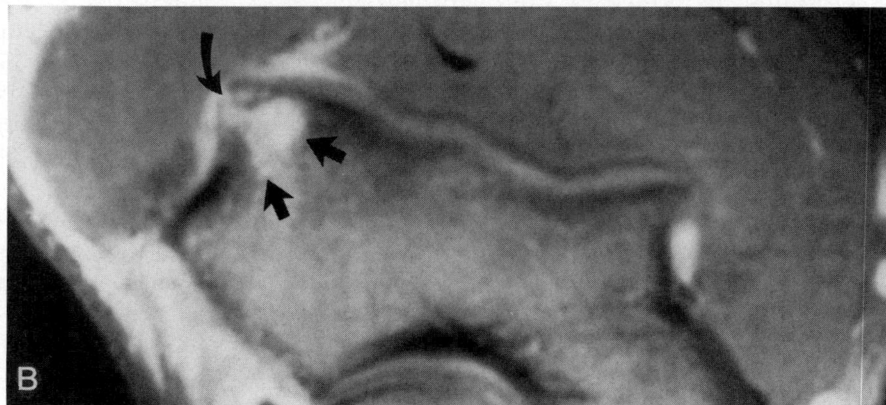

Figure 8–14. T1-weighted sagittal images *(A)* and a fat-suppressed axial image *(B)* after intravenous administration of gadopentetate dimeglumine reveal an enhancing cystic-appearing area *(arrows)* compatible with granulation tissue in the anterior aspect of the capitellum. A partially detached osteochondral flap *(curved arrow)* is noted anteriorly.

ally not seen in Panner's disease. Panner's disease is characterized by fragmentation and abnormally decreased signal intensity within the ossifying capitellar epiphysis on the T1-weighted images, similar in appearance to Legg-Calvé-Perthes disease in the hip; subsequent scans reveal normalization of these changes with little or no residual deformity of the capitellar articular surface.

Loose Bodies

Unstable osteochondral lesions may fragment and migrate throughout the joint as loose bodies. These loose bodies may attach to the synovial lining of the joint and undergo gradual laminar growth with variably thick hyaline cartilage (Fig. 8–15). Loose bodies may become quite large and result in mechanical symptoms, such as locking and limitation of motion (Fig. 8–16). Loose bodies usually are removed arthroscopically when detected, as they may lead to premature degenerative arthritis in addition to their effects on joint function.

Laxity of the UCL in throwing athletes may lead to incongruity between the medial olecranon and the medial aspect of the olecranon fossa, resulting in loose body formation.[1,19] These posterior compartment loose bodies frequently occur in baseball pitchers and may be well seen when joint fluid is present. Small loose bodies may be difficult to exclude with MR imaging in the absence of an effusion.

Bone Injury

Radiographically occult or equivocal fractures may be assessed with MR imaging. Magnetic resonance imaging can be performed with the patient in a cast without significant degradation of the image quality. A large cast may require using a larger surface coil, such as the head coil. In general, the findings of bone injury are conspicuous on T1-weighted or STIR sequences but may be subtle on proton density, T2-weighted, and T2*-weighted gradient echo sequences.

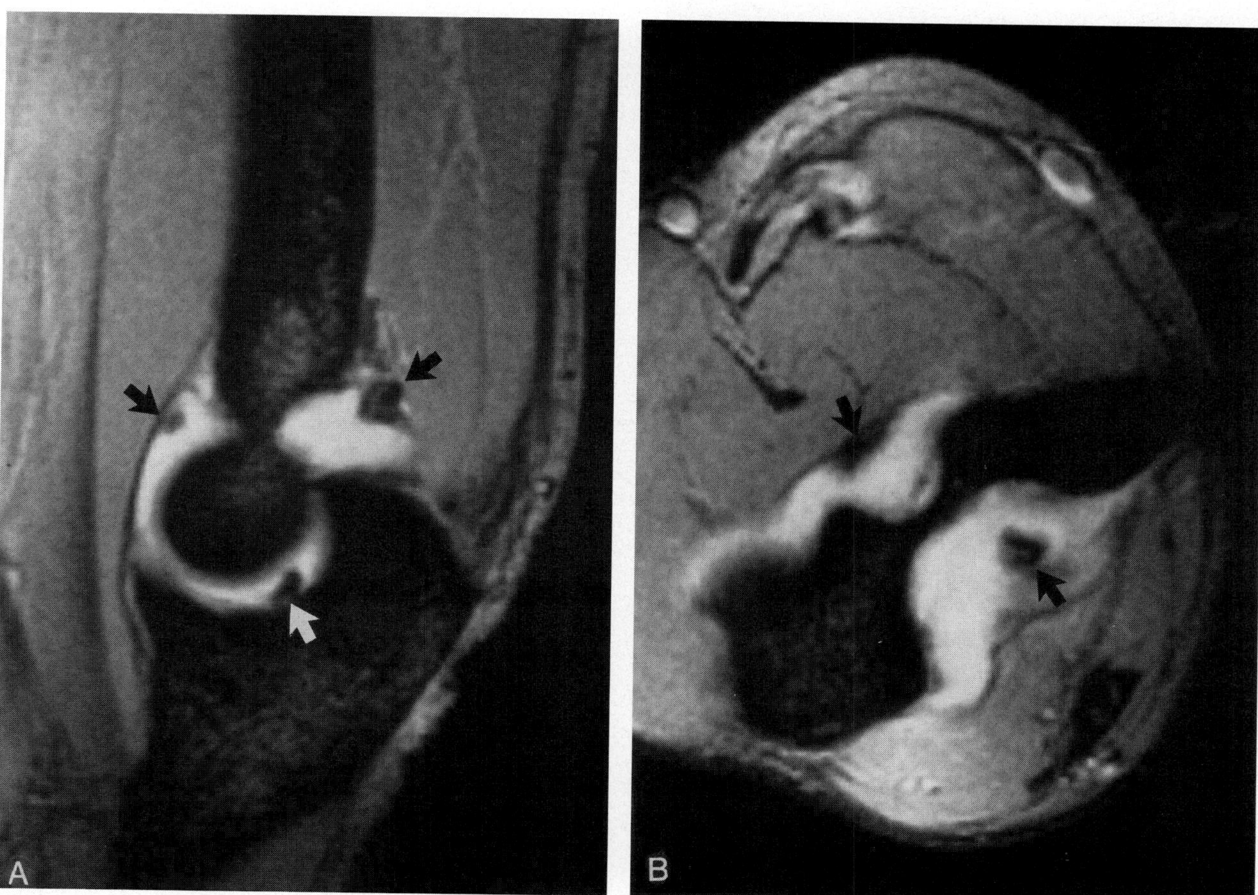

Figure 8–15. T2*-weighted gradient-echo sagittal *(A)* and axial *(B)* images reveal multiple loose bodies *(arrows)* outlined by a joint effusion. These osteochondral loose bodies also have a rim of variably thick hyaline cartilage, which often is of equal signal intensity to the surrounding joint fluid. These loose bodies may seemingly "defy gravity" by embedding within the synovial lining, as demonstrated by the anterior loose body in this case.

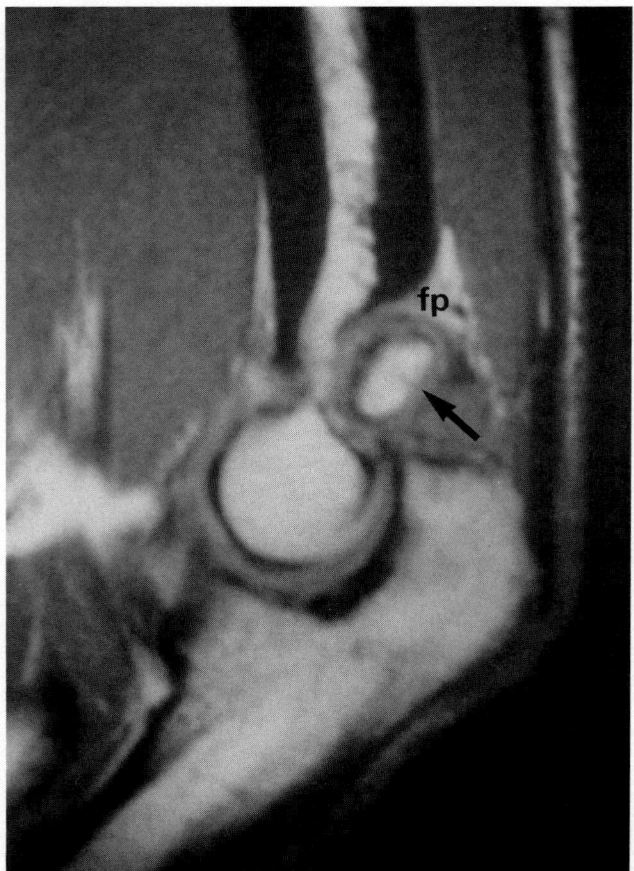

Figure 8–16. The T1-weighted sagittal image reveals a large loose body in the olecranon fossa *(arrow)*, displacing the posterior intracapsular fat pad (fp) superiorly. The patient is a 20-year-old pitcher who was unable to extend his elbow fully.

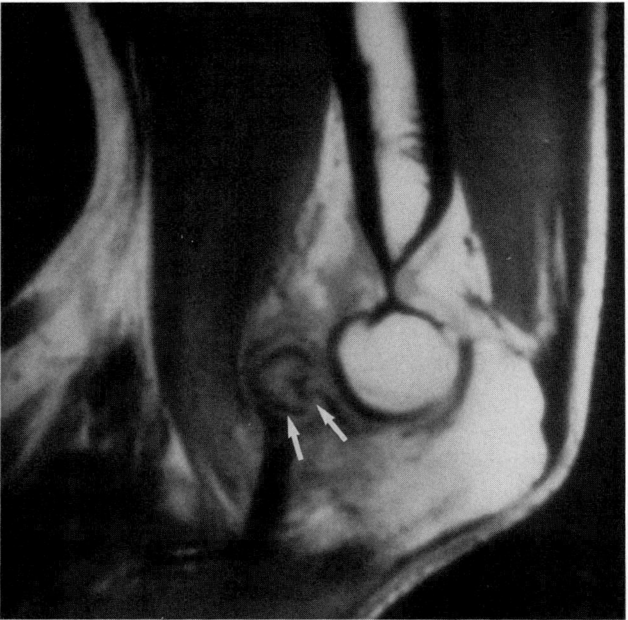

Figure 8–17. The T1-weighted sagittal image reveals a minimally displaced fracture of the anterior coronoid process *(arrows)* after a posterior dislocation of the elbow.

An anterior coronoid process fracture or "contusion" may be seen after posterior elbow dislocation or subluxation, which may occur as a consequence of a hyperextension injury (Fig. 8–17); anterior capsule injury and strain of the adjacent brachialis muscle are associated MR imaging findings in this setting.

Magnetic resonance imaging may identify or exclude radial head fractures in adults and supracondylar fractures in children when radiographic evidence of a joint effusion is present and a fracture is not visualized (Fig. 8–18). Injury to the physis as well as the unossified epiphyseal cartilage may be assessed with MR imaging in difficult pediatric cases when the plain film findings are equivocal (Fig. 8–19).[11]

Stress fractures of the middle third of the olecranon may occur in throwing athletes as a consequence of overload by the triceps mechanism.[21] These fractures have the potential to become displaced and are usually treated surgically. Similar chronic extension overload in adolescent baseball pitchers leads to nonunion of the olecranon physeal plate, which may also require surgery. Magnetic res-

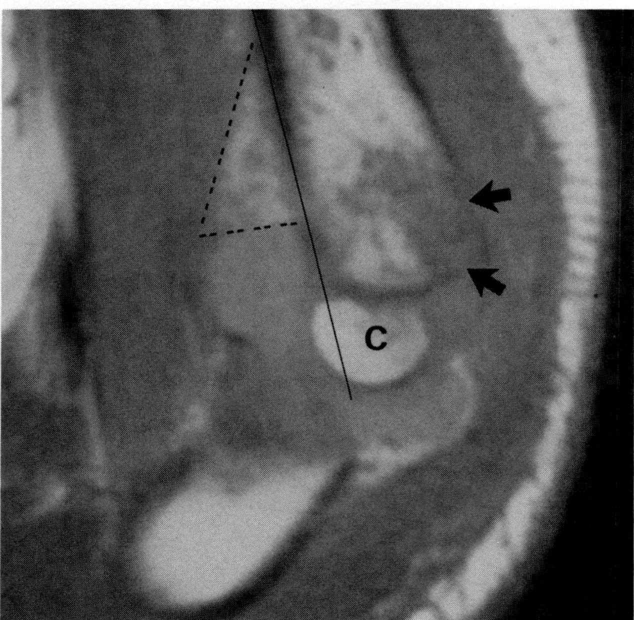

Figure 8–18. The T1-weighted sagittal image reveals elevation of the anterior fat pad *(dashed lines)* secondary to a hemorrhagic effusion occurring after trauma in this 4-year-old child. The capitellum is posteriorly displaced relative to the anterior humeral line *(straight line)*. The anterior humeral line normally passes through the central one third of the capitellum. A radiographically occult supracondylar fracture of the humerus *(arrows)* accounts for the posterior displacement of the capitellum in this case. C, Capitellum.

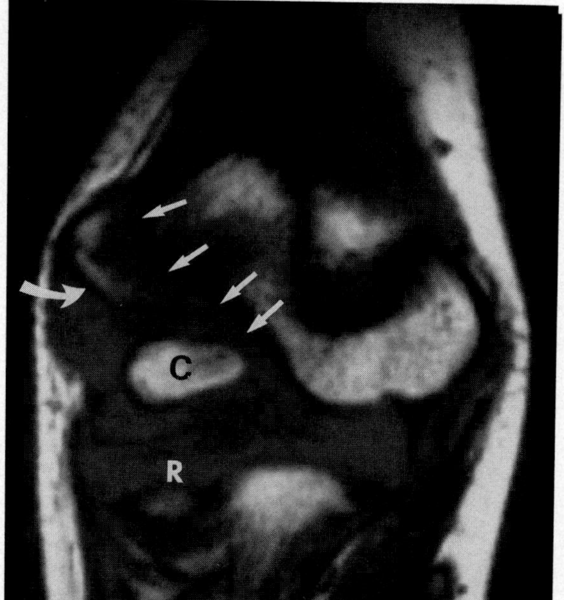

Figure 8–19. The T1-weighted coronal image reveals a Salter-Harris type II physeal fracture *(arrows)* with mild lateral displacement of the lateral humeral condyle *(curved arrow)* in a 5-year-old child; the plain film findings were equivocal. C, Capitellum; R, unossified radial head cartilage.

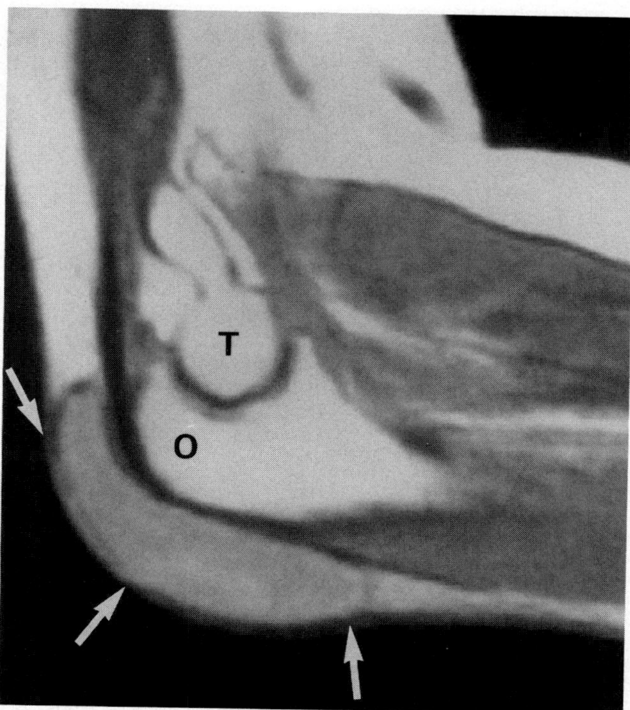

Figure 8–20. The T1-weighted sagittal image shows enlargement of the olecranon bursa *(arrows)*. No evidence is seen of an abnormal signal in the underlying olecranon to indicate osteomyelitis. O, Olecranon; T, trochlea.

onance imaging may detect these lesions, which may be difficult to diagnose clinically and radiographically.

Miscellaneous Conditions

Applications for MR imaging are continually evolving. It is often requested when a clinical problem has not been adequately defined with conventional imaging techniques. Infections, neoplasms, and cystic masses about the the elbow may be evaluated with MR imaging. Examples of problem-solving applications include assessing whether osteomyelitis is present (Fig. 8–20) and determining the cause and site of peripheral nerve entrapment (Fig. 8–21). Magnetic resonance imaging may also be useful in evaluating unusual arthridites and synovial disorders, such as pigmented villonodular synovitis or synovial chondromatosis (Fig. 8–22).

CONCLUSION

Magnetic resonance imaging provides clinically useful information in patients with a variety of traumatic and degenerative disorders that result in elbow pain. It is perhaps most useful when patients have not responded to conservative therapy, and therefore surgery as well as additional diagnoses are being considered.

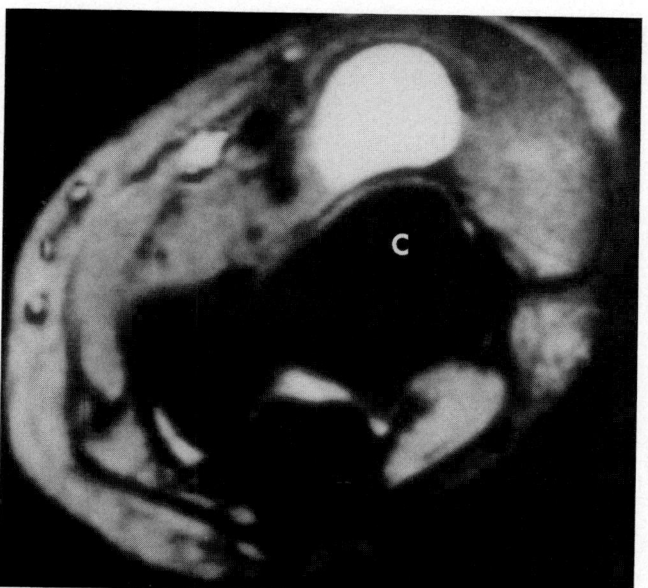

Figure 8–21. The T2*-weighted gradient-echo axial image reveals a large ganglion cyst anterior to the capitellum, causing entrapment of the radial nerve. C, Capitellum.

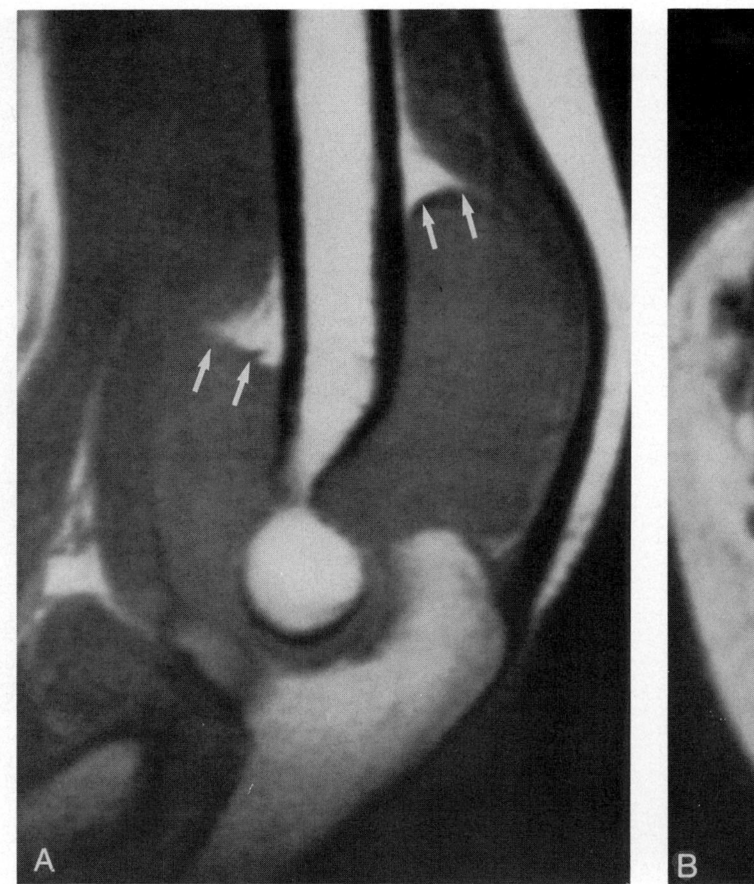

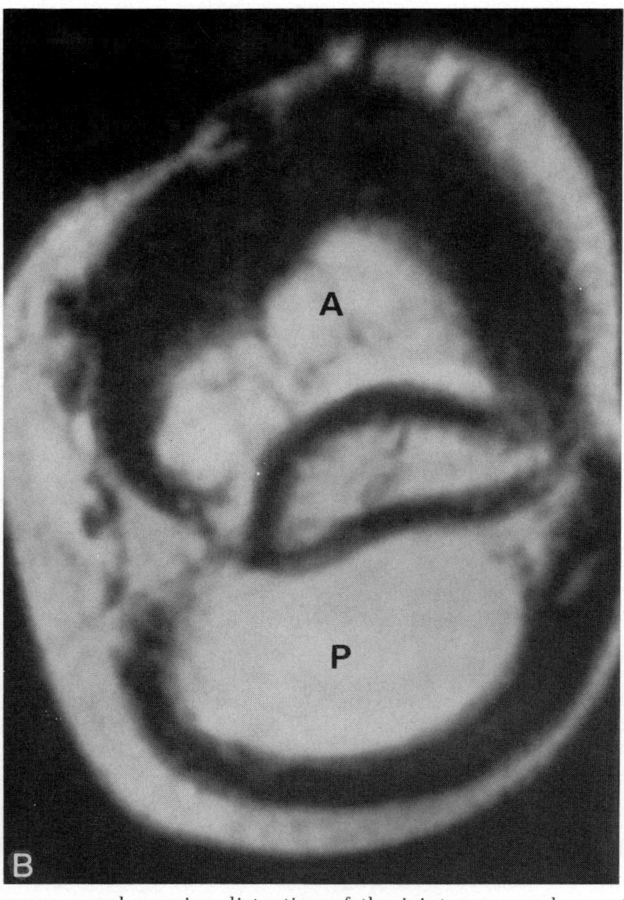

Figure 8-22. T1-weighted sagittal *(A)* and T2-weighted axial *(B)* images reveal massive distention of the joint space and superior elevation of the fat pads *(arrows)* secondary to synovial chondromatosis. Linear foci of low signal intensity delineate the margins of numerous cartilaginous loose bodies in the anterior joint space on the T2-weighted image. A, Anterior; P, posterior.

References

1. Andrews JR, Craven WM: Lesions of the posterior compartment of the elbow. Clin Sports Med 10:637–652, 1991.
2. Berquist TH: The elbow and wrist. Top Magn Reson Imaging 1:15–27, 1989.
3. Bourne MH, Morrey B: Partial rupture of the distal biceps tendon. Clin Orthop 271:143–148, 1991.
4. Boyd HB, McLeod AC: Tennis elbow. J Bone Joint Surg [Am] 55:1183–1187, 1973.
5. Brogdon BG, Crow NE: Little leaguer's elbow. AJR 83:671–675, 1960.
6. Bunnell DH, Fisher DA, Bassett LW, et al: Elbow joint: Normal antaomy on MR images. Radiology 162:527–531, 1987.
7. Coonrad RW, Hooper WR: Tennis elbow: Its course, natural history, conservative and surgical management. J Bone Joint Surg [Am] 55:1177–1182, 1973.
8. Conway JE, Jobe FW, Glousman RE, Pink M: Medial instability of the elbow in throwing athletes. Treatment by repair or reconstruction of the ulnar collateral ligament. J Bone Joint Surg [Am] 74:67–83, 1992.
9. DeSmet AA, Fisher DR, Burnstein MI, et al: Value of MR imaging in staging osteochondral lesions of the talus (osteochondritis dissecans). AJR 154:555–558, 1990.
10. Gore RM, Rogers LF, Bowerman, et al: Osseous manifestations of elbow stress associated with sports activities. AJR 134:971–977, 1980.
11. Jaramillo D, Hoffer FA, Shapiro F, Rand F: MR imaging of fractures of the growth plate. AJR 155:1261–1265, 1990.
12. McPherson SA, Meals RA: Cubital tunnel syndrome. Orthop Clin North Am 23:111–123, 1992.
13. Middleton WD, Macrander S, Kneeland JB, et al: MR imaging of the normal elbow: Anatomic correlation. AJR 149:543–547, 1987.
14. Milgram J, Rogers L, Miller H: Osteochondral fractures: Mechanism of injury and fate of fragments. AJR 130:651–658, 1978.
15. Mitsunaga MM, Adishian DA, Bianco AJ Jr: Osteochondritis dissecans of the capitellum. J Trauma 22:53–55, 1982.
16. Murphy BJ: MR imaging of the elbow. Radiology 184:525–529, 1992.
17. Morrey BF, An KN: Functional anatomy of the ligaments of the elbow. Clin Orthop 201:84–90, 1985.
18. Morrey BF, An KN: Articular and ligamentous contributions to the stability of the elbow joint. Am J Sports Med 11:315–319, 1983.
19. Morrey BF: The Elbow and Its Disorders. Philadelphia, WB Saunders, 1985.
20. Nirschl RP, Petrone FA: Tennis elbow: The surgical treatment of lateral epicondylitis. J Bone Joint Surg [Am] 61:832–839, 1979.
21. Nuber GW, Diment MT: Olecranon stress fractures in throwers. Clin Orthop 278:58–61, 1992.
22. O'Driscoll SW, Horii E, Carmichael SW, Morrey BF: The cubital tunnel and ulnar neuropathy. J Bone Joint Surg [Br] 73:613–617, 1991.
23. O'Driscoll SW, Bell DF, Morrey BF: Posterolateral rotatory instability of the elbow. J Bone Joint Surg [Am] 73:440–446, 1991.
24. Resendes M, Helms CA, Fritz RC, Genant HK: MR appearance of intramuscular injections. AJR158:1293–1294, 1992.
25. Ruch DS, Poehling GG: Arthroscopic treatment of Panner's disease. Clin Sports Med 10:629–636, 1991.
26. Tarsney FF: Rupture and avulsion of the triceps. Clin Orthop 83:177–183, 1972.
27. Tivnon MC, Anzel SH, Waugh TR: Surgical management of osteochondritis dissecans of the capitellum. Am J Sports Med 4:121–128, 1976.
28. Thorson EP, Szabo RM: Tendinitis of the wrist and elbow. Occup Med 4:419–431, 1989.
29. Unverferth LJ, Olix ML: The effect of local steroid injections on tendon. J Sports Med 1:31–37, 1973.
30. Vangsness CT, Jobe FW: Surgical treatment of medial epicondylitis. J Bone Joint Surg [Br] 73:409–411, 1991.
31. Verhaar J, Spaans F: Radial tunnel syndrome. J Bone Joint Surg [Am] 73:539–544, 1991.
32. Wilkins KE, Morrey BF, Jobe FW, et al: The elbow. Instr Course Lect 40:1–87, 1991.
33. Wittenberg RH, Schaal S, Muhr GG: Surgical treatment of persistent elbow epicondylitis. Clin Orthop 278:73–80, 1992.

9 THE WRIST AND HAND

Lynne S. Steinbach and Russell C. Fritz

In the past, significant anatomic information about the painful wrist was obtained from conventional radiographs supplemented by arthrography, tomography, bone scintigraphy, or computed tomography (CT). At present, high-resolution magnetic resonance (MR) imaging can be used to enhance detection and evaluation of several wrist and hand disorders, allowing for discrimination of soft tissue structures including marrow, ligaments, tendons, cartilage, muscles, nerves, and blood vessels. Magnetic resonance imaging can be of aid in the evaluation of carpal instability, disorders of the triangular fibrocartilage, ulnar impaction syndrome, distal radioulnar joint instability, fracture, avascular necrosis, tendinopathy, nerve entrapment syndromes, synovial abnormalities, and soft tissue masses.

IMAGING PROTOCOLS

We are currently using a General Electric 1.5 T superconducting magnet (GE, Milwaukee, Wis.). Most patients are able to place the arm at the side with the wrist or hand surrounded by a surface coil. Larger patients may need to have the arm placed above the head, which is less comfortable and, as a result, tends to be associated with increased motion artifact.

A variety of surface coils can be used for wrist and hand imaging. We have used the following coils for wrist imaging: (1) paired circular 3-inch coils, (2) a shoulder coil, and (3) a small parts transmit-receive volume coil (Medical Advances, Inc., Milwaukee, Wis.). Ideally, the wrist should be in neutral alignment; however, we have found it difficult for the patient to maintain the wrist in this position during scanning without the aid of a dorsal splint. Instead, we image the wrist prone, parallel to the table top with the fingers extended.

When imaging the wrist we use axial images as a guide with cursors aligned parallel and then perpendicular to the radioulnar joint, providing coronal and sagittal images, respectively. An axial scout sequence is followed by coronal T1-weighted and gradient-echo three-dimensional Fourier transform (3DFT) sequences, an axial fast spin-echo sequence, and a sagittal T1-weighted sequence (Table 9–1). In some cases, we will also obtain coronal fast spin-echo images to further evaluate the triangular fibrocartilage and ligaments. The field of view should be as small as possible, ranging between 8 and 10 cm. Although they are not routinely obtained at our institution, cine motion studies can be performed with fast imaging techniques such as gradient-echo sequences for evaluation of ligamentous abnormality and carpal instability.

For hand imaging, we begin with an axial T1-weighted sequence, followed by coronal T1-weighted, axial and sagittal dual-echo fast spin-echo, and optional axial T1-weighted images after administration of gadopentetate dimeglumine for evaluation of ligamentous abnormalities and soft tissue tumors (Table 9–2). The field of view ranges between 10 and 12 cm.

Additional sequences with thinner sections (2 to 3 mm) can be used to image the metacarpophalangeal joints and phalanges. Abnormalities of the metacarpophalangeal joints are well seen on coronal and axial images. Phalangeal pathology is best seen in axial and sagittal planes.

Table 9–1. PROTOCOL FOR MAGNETIC RESONANCE IMAGING OF THE WRIST AT UNIVERSITY OF CALIFORNIA, SAN FRANCISCO

Plane	TR	TE	Flip Angle (degree)	FOV (cm)	Slice (mm)	Skip (mm)	NEX	Matrix
Axial	500	15		16	4	1	1	128
Coronal	500	15		10	3	0	2	256
Coronal MPGR 3DFT	35	15	20	10	2	0	2	256
Axial FSE	3500	30/102		10	4	1	2	256
Sagittal	500	15		10	3	0.5	2	192
Coronal FSE (opt)	3500	80		10	3	0.5	4	256

TR, Repetition time; TE, echo time; FOV, field of view; Skip, interslice gap; NEX, number of excitations; FSE, fast spin-echo; MPGR 3DFT, multiplanar gradient-echo 3-dimensional Fourier transform; opt, optional.

Table 9–2. PROTOCOL FOR MAGNETIC RESONANCE IMAGING OF THE HAND AT UNIVERSITY OF CALIFORNIA, SAN FRANCISCO

Plane	TR	TE	Flip Angle (degree)	FOV (cm)	Slice (mm)	Skip (mm)	NEX	Matrix
Axial	600	15		10	5	1.0	2	192
Coronal	500	15		10	4	0.5	2	192
Axial FSE	3500	30/102		12	4	1.0	2	192
Sagittal FSE	3500	30/102		12	4	1.0	2	192
Axial Post-gad (opt)	600	15		10	5	1.0	2	192

TR, repetition time; TE, echo time; FOV, field of view; Skip, interslice gap; NEX, number of excitations; FSE, fast spin-echo; gad, gadolinium; opt, optional.

LIGAMENTOUS DISRUPTION AND CARPAL INSTABILITY

Two types of ligaments are present in the wrist—intrinsic and extrinsic. Both serve to maintain proper carpal alignment and when torn can result in carpal instability. With normal carpal alignment, the volar tubercle of the scaphoid lies approximately 47 degrees volar to the longitudinal axis of the wrist. This is best assessed on sagittal MR images (Fig. 9–1). On sagittal images, the radius, lunate, capitate, and third metacarpal are colinear with the wrist in neutral position (Fig. 9–2). When a ligament on the radial aspect of the wrist is stretched or disrupted, the lunate may tilt dorsally, producing a dorsiflexed intercalated segment instability (DISI) deformity (Figs. 9–2, 9–3). Presumably, abnormalities of ligaments on the ulnar side of the wrist, including the lunotriquetral and ulnocarpal ligaments, produce a volar-flexed intercalated segment instability (VISI) pattern in which the lunate is flexed volarly. The DISI and VISI patterns are well seen on sagittal MR images (Figs. 9–3, 9–4). The radiographer evaluates the angle formed between the long axis of the capitate and a perpendicular from a line drawn between the distal dorsal and volar aspect of the lunate. This angle should not measure more than 15 degrees in either direction. The physician must be sure that the wrist is not in ulnar deviation when evaluating for DISI, as this position itself can produce some dorsal tilt of the lunate (see Fig. 9–2). With ulnar deviation, however, the lunate also shifts

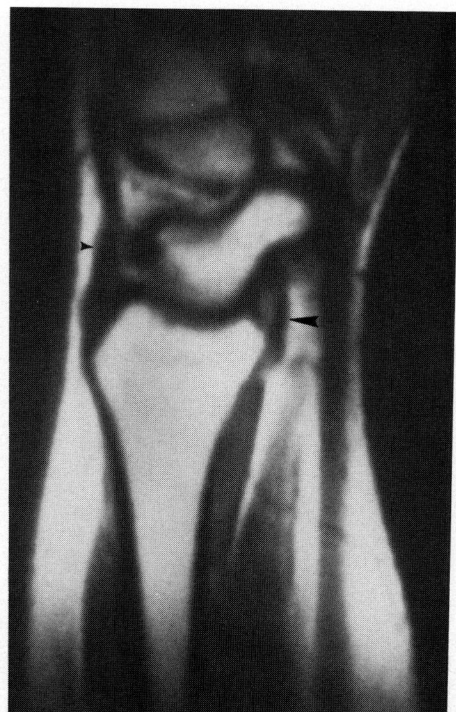

Figure 9–1. Normal T1-weighted sagittal MR image of the wrist through the scaphoid. The long axis of the scaphoid lies approximately 47 degrees volarly to the longitudinal axis of the wrist. The palmar *(large arrowhead)* and dorsal *(small arrowhead)* radiocarpal ligaments are well seen as structures of low signal intensity.

volarly with relation to the radius, which does not occur with a true DISI. Sequelae of malalignment from ligamentous tears include wrist shortening, which produces crowding in the carpal tunnel, and degenerative arthritis.

Intrinsic Wrist Ligaments (Intercarpal Ligaments)

The radiocarpal joint is separated from the midcarpal joint by interosseous ligaments that extend proximally between the scaphoid, lunate, and triquetrum and are best seen on thin-section coronal MR images. It is important to use T2- or T2*-weighting to fully evaluate these ligaments. Three-dimensional volume Fourier transform gradient-echo techniques may be used to study the wrist for ligamentous disruption in suspected cases of instability. The wrist also can be studied in various positions with MR imaging (ulnar and radial deviation, flexion and extension) when ligamentous disruption or carpal instability is suspected.[48]

Normally, the ligaments are of low signal intensity on MR imaging (Fig. 9–5). The scapholunate and lunotriquetral ligaments are associated with age-related degeneration, similar to that seen in the triangular fibrocartilage. With degeneration, signal intensity increases on T1-weighted images, but decreases on T2-weighted images.[24]

Tears are common in both the scapholunate and lunotriquetral ligaments. Tears of the scapholunate ligament can be manifested as nonvisualization of the ligament; focal areas of high signal intensity within the ligament on T1- and T2-weighted images; or abnormal morphology, including irregular margins (Fig. 9–6). In general, the lunotriquetral ligament is not visualized reliably with MR imaging, and arthrography should be employed to evaluate patients with suspected lunotriquetral ligament tears. Lunotriquetral tears can be diagnosed by MR imaging only when the ligament is seen and demonstrates focal high signal intensity on all imaging sequences or abnormal morphology. Fluid may be seen in the midcarpal joint in patients with either scapholunate

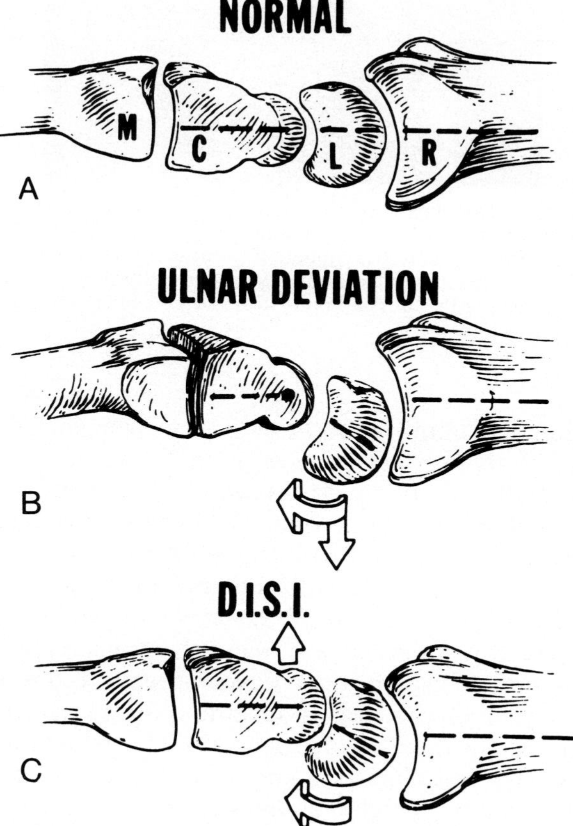

Figure 9–2. Sagittal views showing normal wrist *(A)*, ulnar deviation *(B)* and dorsiflexed intercalated segment instability (DISI) *(C)*. *A*, A normal colinear relationship of the central column of the wrist between the radius (R), lunate (L), capitate (C), and third metacarpal (M) is seen. *B*, With ulnar deviation of the normal wrist, the capitate and radius remain colinear while the lunate tilts dorsally and shifts volarly. *C*, The lunate tilts dorsally but does not shift volarly when DISI is present. The capitate may shift dorsally in wrists with DISI. (Illustration by Elizabeth Roselius, © 1985. Reprinted with permission from Taleisnik J. The Wrist. New York, Churchill Livingstone, 1985.)

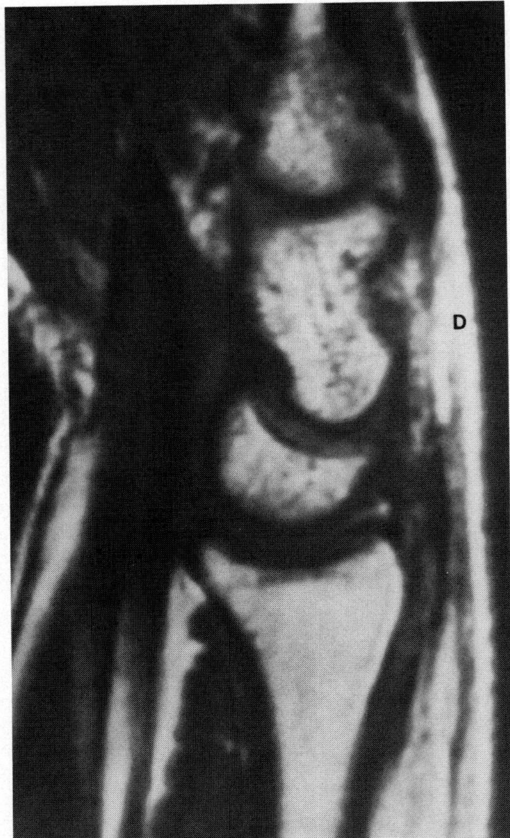

Figure 9–3. Dorsiflexed intercalated segment instability (DISI). A sagittal T1-weighted image demonstrates the dorsal tilt of the lunate with respect to the capitate (D, dorsal aspect of wrist). No evidence is noted of palmar displacement of the lunate, which may be seen with a dorsal lunate tilt produced solely with ulnar deviation of the wrist.

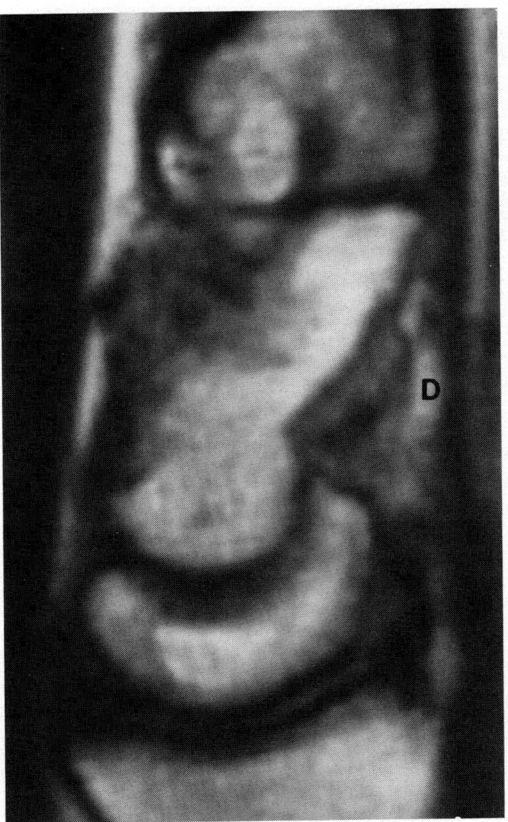

Figure 9–4. Volar-flexed intercalated segment instability (VISI). A sagittal T1-weighted image demonstrates the volar tilt of the lunate with respect to the capitate (D, dorsal aspect of wrist). This patient had a lunotriquetral ligament tear—an abnormality frequently associated with VISI.

or lunotriquetral ligament tears; however, this finding can also be seen in asymptomatic wrists.[47]

Zlatkin and co-workers reported accuracy of 90 per cent and 80 per cent for MR imaging evaluation of scapholunate and lunotriquetral ligament tears, respectively.[63] Magnetic resonance imaging demonstrated a sensitivity of 50 per cent, a specificity of 86 per cent, and an accuracy of 76 per cent for detection of scapholunate ligament tears, and a sensitivity of 50 per cent, a specificity of 46 per cent, and an accuracy of 49 per cent for detection of lunotriquetral tears in a study of 15 patients using arthrography as a gold standard.[47]

The distal carpal interosseous ligaments do not usually extend completely from the dorsal to the volar aspects of the carpal bones, allowing communication between the midcarpal and carpometacarpal joints. The capitohamate ligament, which is of low signal intensity, extends obliquely from the ulnar aspect of the mid capitate to the upper radial aspect of the hamate, and is routinely visualized on coronal MR images (see Fig. 9–6).

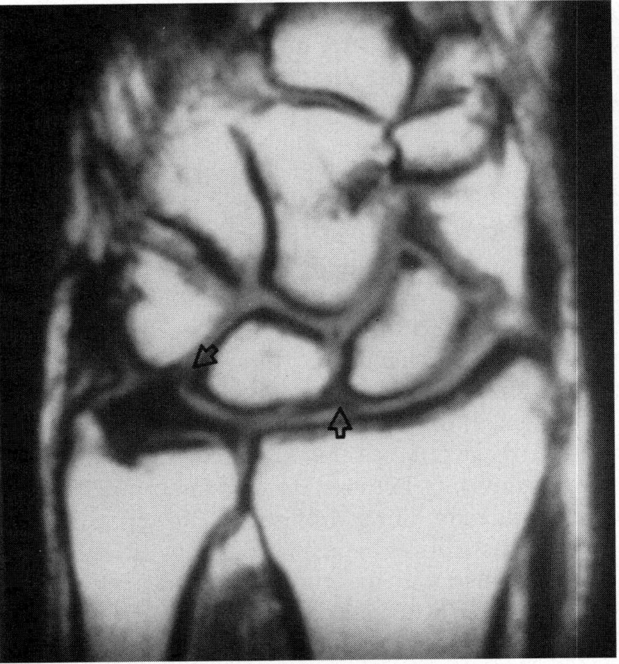

Figure 9–5. The scapholunate and lunotriquetral ligaments of low signal intensity are well delineated between the proximal aspects of the scaphoid, lunate, and triquetrum on this normal T1-weighted coronal MR image *(arrows)*. The lunotriquetral ligament is not always seen on a normal coronal MR image.

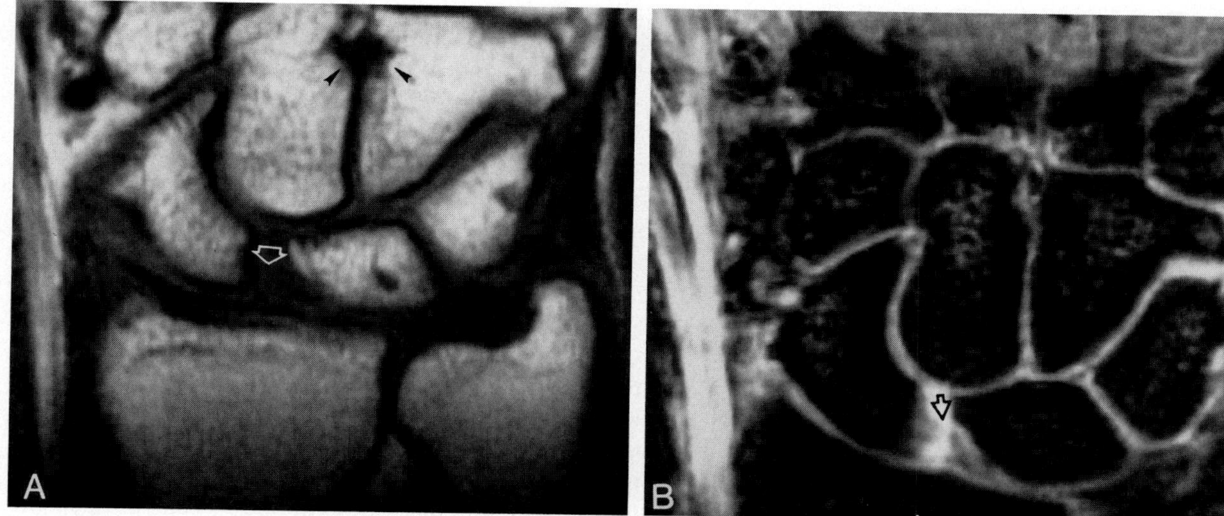

Figure 9–6. Scapholunate ligament tear. *A*, Coronal T1-weighted MR image reveals abnormal widening between the proximal scaphoid and the lunate *(arrow)*. A normal capitohamate ligament is demonstrated *(arrowheads)*. Incidentally noted are bone islands of low signal intensity in the lunate and triquetrum. *B*, The scapholunate ligament, normally of low signal intensity, is disrupted by fluid of high signal intensity on this gradient-echo image *(arrow)*.

Extrinsic Wrist Ligaments

The articular capsule is strengthened by dorsal and palmar ligaments, which provide wrist stability. Alignment of the carpal bones is maintained principally by the extrinsic ligaments, which may be difficult to identify and follow in their entirety on MR imaging. The strong palmar ligaments include the radial collateral, radioscaphocapitate, radiolunate, radioscapholunate (also termed the ligament of Kuenz and Testut), ulnolunate, and ulnar collateral ligaments (Fig. 9–7). The less clinically important dorsal radiocarpal ligament has three components—the radioscaphoid, radiolunate, and radiotriquetral fascicles (Fig. 9–8).

Arthrography provides little information about these ligaments. They are of low signal intensity and are best seen on sagittal and coronal MR images. A wavy contour of the unstressed ligament may be seen in normal persons. Currently MR imaging for evaluation of disruption of the extrinsic carpal ligaments is limited. In our experience, because of the obliquity of these ligaments, they are not always seen in their entirety and should not be called abnormal unless there is obvious discontinuity of the ligament, with high signal intensity within the gap on T2-weighted images (Fig. 9–9). Laxity of these ligaments may occur with stretching due to trauma.

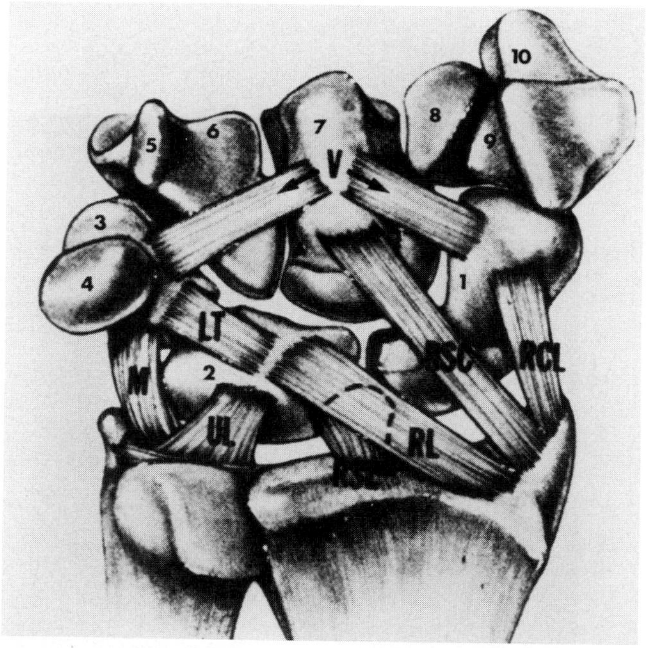

Figure 9–7. Volar view of the carpal bones and volar carpal ligaments. 1, Scaphoid; 2, lunate; 3, triquetrum; 4, pisiform; 5, hook of hamate; 6, body of hamate; 7, capitate; 8, trapezoid; 9, groove within trapezium for flexor carpi radialis tendon; 10, distal articular surface of trapezium; V, deltoid ligament; RCL, radial collateral ligament; RSC, radioscaphocapitate ligament; RL, radiolunate ligament; LT, lunotriquetral ligament; RSL, radioscapholunate ligament; UL, ulnolunate ligament; M, medial collateral ligament. (Illustration by Elizabeth Roselius, © 1985. Reprinted with permission from Taleisnik J: The Wrist. New York, Churchill Livingstone, 1985.)

Ulnar Collateral Ligament of the Thumb

Injury to the ulnar collateral ligament of the thumb (gamekeeper's thumb) is quite common, espe-

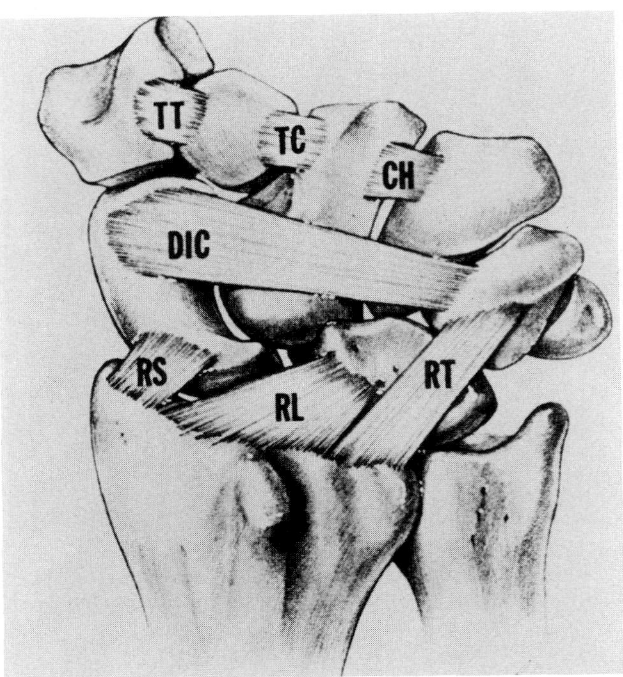

Figure 9–8. Dorsal intrinsic and extrinsic ligaments of the wrist. TT, trapeziotrapezoid; TC, trapeziocapitate; CH, capitohamate; DIC, dorsal intercarpal; RS, radioscaphoid; RL, radiolunate; RT, radiotriquetral. (Illustration by Elizabeth Roselius, © 1985. Reprinted with permission from Taleisnik J: The Wrist. New York, Churchill Livingstone, 1985.)

cially in skiers. It was first described as an occupational hazard in English game wardens, who killed rabbits by twisting their necks between the thumb and second digit. The injury is significant when the ligament is ruptured distally, particularly if the ligament is displaced significantly, with interposition of the adductor aponeurosis and extensor hood, which prevents spontaneous healing.[52] Adductor aponeurosis interposition cannot occur in partial ruptures; therefore, it is important to distinguish acute complete ruptures from partial ruptures. Complete rupture with retraction requires surgical intervention. Magnetic resonance imaging is useful for patients with ambiguous physical findings of complete ulnar collateral ligament rupture (Fig. 9–10). The ligament is of low signal intensity and is well seen in the coronal plane. High signal intensity is seen on T2-weighted images in the region of the tear. Noting the presence of proximal retraction of the ligament can aid in the evaluation of these injuries.

DISORDERS OF THE TRIANGULAR FIBROCARTILAGE

The triangular fibrocartilage is an important structure that cushions and stabilizes the ulnocarpal and distal radioulnar joints. Tears of the triangular

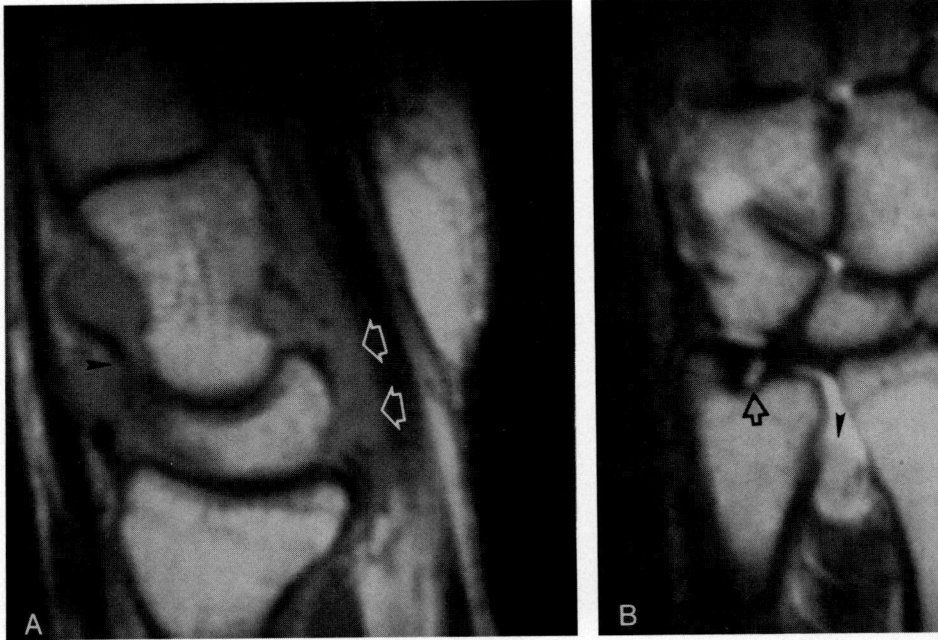

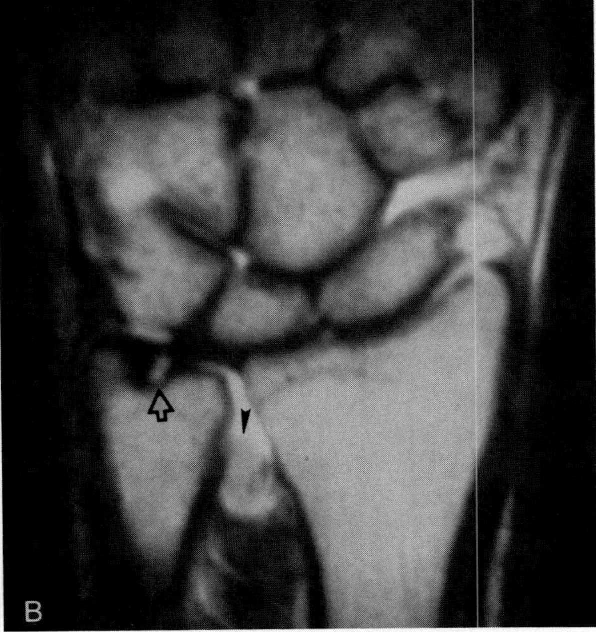

Figure 9–9. MR images of a patient who came for evaluation of wrist pain after a car accident. *A,* There is discontinuity of the palmar radiocarpal ligament consistent with disruption on this T1-weighted sagittal MR image *(arrows).* The dorsal radiocarpal ligament is wavy but intact *(arrowhead).* The distention produced by hemorrhage in the midcarpal and radiocarpal compartments makes it easier to evaluate these ligaments. *B,* The hemorrhagic fluid is seen in all three wrist compartments (midcarpal, radiocarpal, and distal radioulnar) on this coronal T2-weighted MR image. Hemosiderin, which is of low signal intensity and is produced by the hemorrhage, lies in the distal radioulnar joint fluid *(arrowhead).* The triangular fibrocartilage tear, of high signal intensity, is well demonstrated *(arrow).*

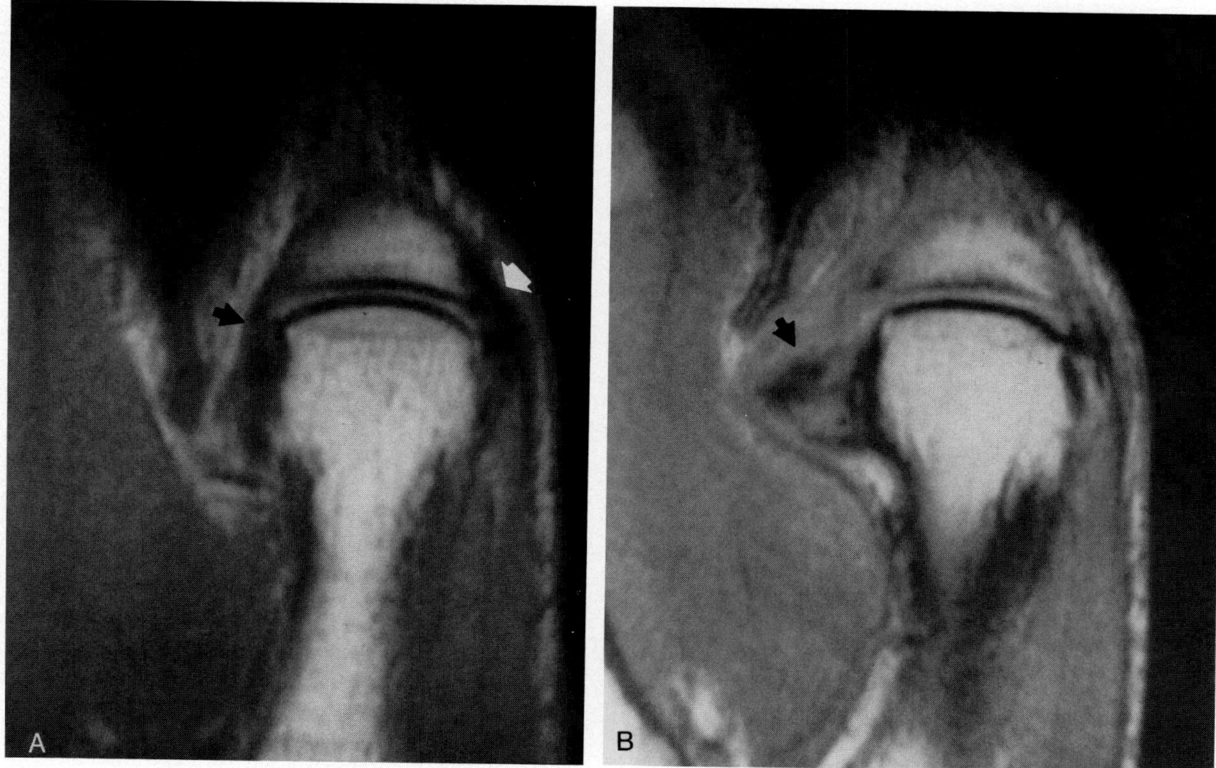

Figure 9–10. *A*, Normal ulnar *(black arrow)* and radial *(white arrow)* collateral ligaments of low signal intensity lie on either side of the first metacarpophalangeal joint on this coronal T1-weighted MR image. *B*, Proton density coronal MR image reveals a distal disruption of the ulnar collateral ligament with proximal retraction *(arrow)* in a patient with gamekeeper's thumb. The ligament was repaired surgically rather than by immobilization in a cast on the basis of this MR imaging finding.

fibrocartilage result in nonspecific pain, crepitus, and weakness that can be difficult to distinguish from injuries to the lunotriquetral ligament, extensor carpi ulnaris tendon, or the pisotriquetral or distal radioulnar joints.[10,55] Diagnostic imaging usually is performed in these patients because clinical examination may not give a precise diagnosis for ulnar wrist pain.

Palmer and Werner have defined the triangular fibrocartilage complex (TFCC) as a composite of five structures: (1) the triangular fibrocartilage (also referred to as the articular disc) and its two dorsal and volar capsular reinforcements (the volar and dorsal distal radioulnar ligaments); (2) the ulnocarpal meniscus, not always identified in the human wrist; (3) the ulnocarpal ligaments, which add stability to the ulnar aspect of the midcarpal joint; (4) the ulnar collateral ligament, which extends from the base of the ulnar styloid to the carpus; and (5) the sheath of the extensor carpi ulnaris tendon, which inserts along the dorsal margin of the base of the fifth metacarpal (Fig. 9–11).[39]

The triangular fibrocartilage frequently undergoes degeneration, which often is asymptomatic. Histologic studies have demonstrated degeneration with increasing prevalence in older persons. For the first two decades of life, no degenerative changes were detected in Mikic's anatomic study of 180 wrist joints from 100 cadavers. Such changes were seen in 38 per cent and 55 per cent of persons in the third and fourth decades of life, respectively, and in up to 100 per cent of persons during the sixth decade.[35] More than 40 per cent of these wrists had perforations of the triangular fibrocartilage.

Degeneration tends to be more severe on the proximal surface because more intensive biomechanical forces are operating on this surface.[35,39] Progressive degeneration of the proximal surface leads to erosion, thinning, and perforation of the triangular fibrocartilage. Degenerative perforations are more common in the thinner central portion of the triangular fibrocartilage, whereas traumatic tears tend to occur in the radial portion.[38]

Arthrography has been useful for excluding complete tears of the triangular fibrocartilage, demonstrating a lack of communication between the radiocarpal joint and distal radioulnar joint. Communication may occur, however, between these wrist compartments in 7 to 35 per cent of asymptomatic persons, presumably owing to degenerative perforation, which produces false-positive arthrograms.[21,26] The prevalence of this finding increases

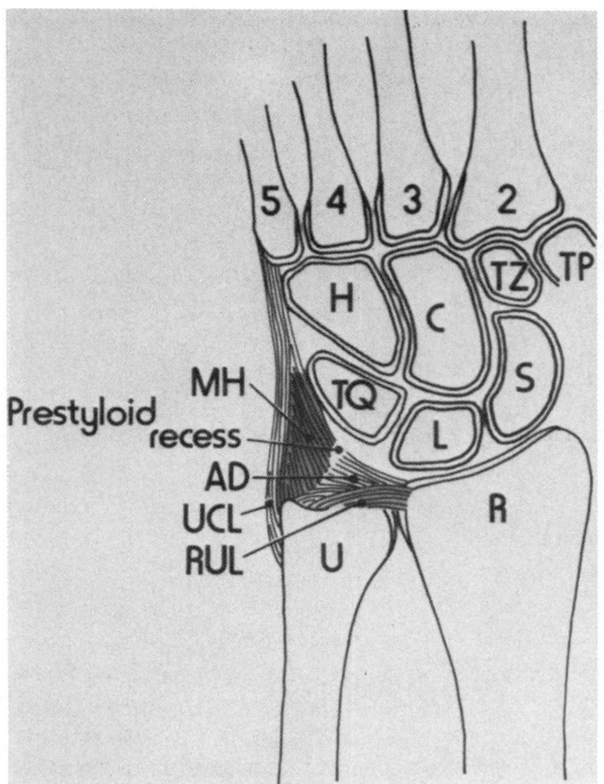

Figure 9–11. Diagrammatic representation of some components of the triangular fibrocartilage complex: the ulnocarpal meniscus, also referred to as the meniscus homologue (MH); prestyloid recess; triangular fibrocartilage or "articular disk" (AD); ulnar collateral ligament (UCL); and radioulnar ligaments (RUL). The sheath of the extensor carpi ulnaris tendon is not shown. H, hamate; C, capitate; TZ, trapezoid; TP, trapezium; S, scaphoid; L, lunate; TQ, triquetrum; R, radius; U, ulna. (From Palmer AK, Werner FW: The triangular fibrocartilage complex of the wrist: Anatomy and function. J Hand Surg [Am] 6:153–162, 1981.)

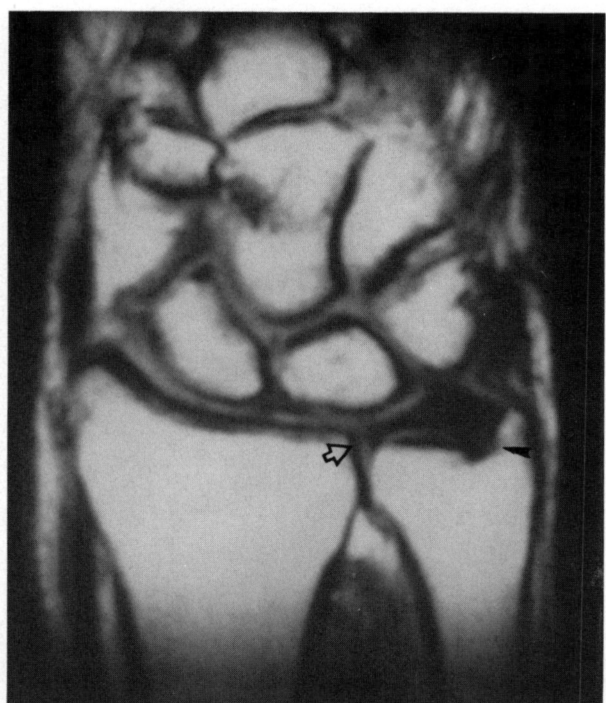

Figure 9–12. Normal triangular fibrocartilage (T1-weighted coronal MR image). The triangular fibrocartilage, of low signal intensity, extends from the hyaline cartilage (intermediate signal intensity) of the distal radius *(arrow)* to the ulnar styloid *(arrowhead)*. The margins of the triangular fibrocartilage are smooth.

in older persons. Therefore, a test is needed that can eliminate these false positives. Magnetic resonance imaging, with its potential to separate degenerative from traumatic abnormalities of the triangular fibrocartilage, can aid in this distinction.

Magnetic resonance imaging has been established as a noninvasive technique that evaluates the entire triangular fibrocartilage directly.[4,14,18,20,22,45,49,63] The triangular fibrocartilage is best seen on coronal images obtained with a field of view between 8 and 12 cm. It is a bow tie–like structure of low signal intensity that extends radially from the dorsal ulnar aspect of the lunate fossa, where it attaches to the hyaline articular cartilage of the radius, which is of intermediate to high signal intensity (Fig. 9–12). Its ulnar attachments are to the fovea at the base of the radial aspect of the ulnar styloid and to the ulnar styloid process. The ulnar attachment often is obscured by surrounding loose vascular connective tissue, which is of intermediate signal intensity. The prestyloid recess is an extension of the radiocarpal joint that also lies near the ulnar attachment of the triangular fibrocartilage. Fluid in this recess is seen as a region of increased signal intensity on T2-weighted images. The dorsal and volar distal radioulnar ligaments, of low signal intensity, are most easily seen on axial images as they extend from the radius to the ulna.

When degeneration of the triangular fibrocartilage is present, it is manifested on MR images as a region of intermediate signal intensity on short echo time (TE) images that does not increase in intensity on T2- or T2*-weighted images (Fig. 9–13).[24] This is believed to be caused by synovial fluid diffusing into areas of degeneration or an alteration of chemical binding components for "free water" or both.[6,24,42]

Traumatic tears are seen as a region of intermediate signal intensity within the triangular fibrocartilage on T1-weighted and proton density spin-echo images. The signal intensity increases on T2-weighted images (Figs. 9–14 to 9–16). Therefore, to fully evaluate the triangular fibrocartilage, MR images must be obtained with some type of T2 weighting. If the tear is complete, it extends to the edges of the fibrocartilage. Some tears are partial and may only extend to the superior or inferior surface. With a traumatic triangular fibrocartilage tear, fluid is usually present in the distal radioulnar joint.[47] This finding, however, is not specific for triangular

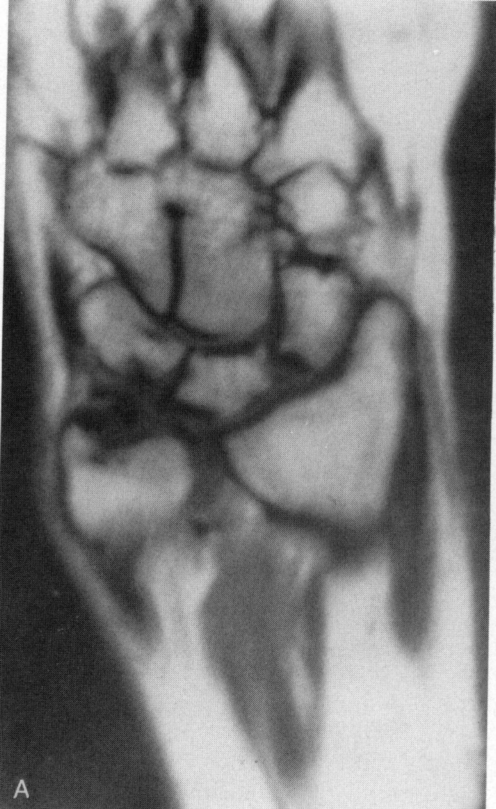

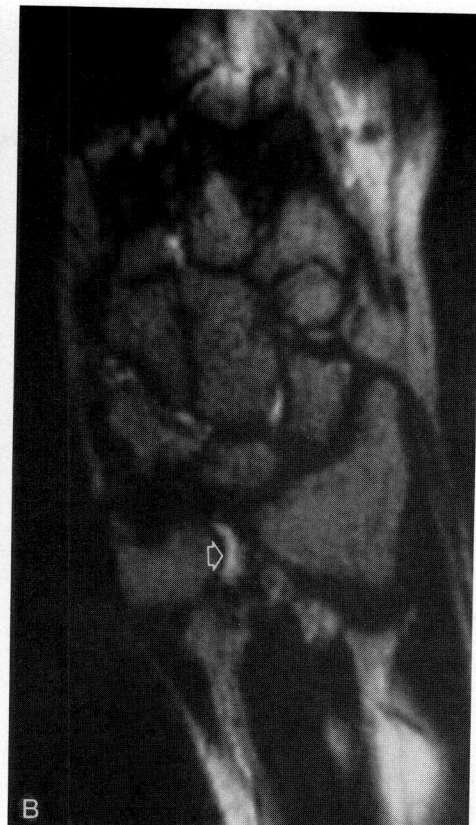

Figure 9–13. Degeneration of the triangular fibrocartilage as seen on images obtained from a 0.35 T Diasonics imager. *A,* Foci of intermediate signal intensity are present diffusely throughout the triangular fibrocartilage on this coronal proton density MR image. *B,* The entire triangular fibrocartilage is of low signal intensity on the T2-weighted MR image. This pattern is characteristic for degeneration of the triangular fibrocartilage, a finding that was confirmed surgically. Fluid of high signal intensity is seen in the distal radioulnar joint, even in the absence of a true tear *(arrow),* possibly as a consequence of overuse.

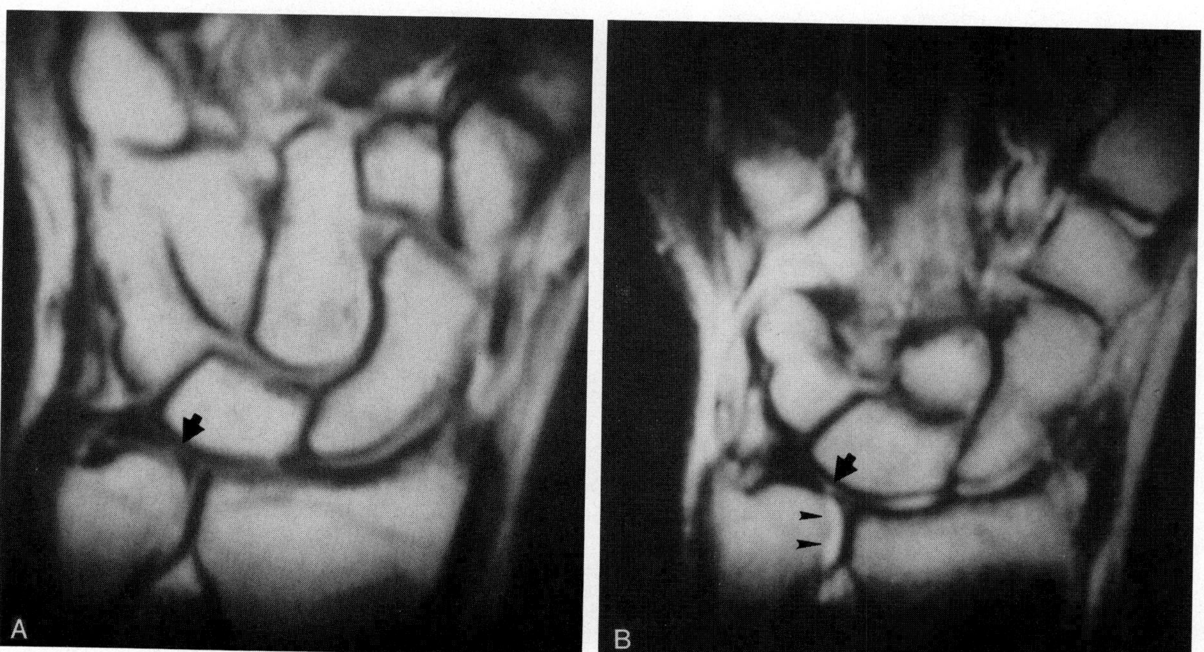

Figure 9–14. A triangular fibrocartilage tear is well seen several millimeters from the radial insertion *(arrow)* on coronal proton density *(A)* and T2-weighted *(B)* MR images. In *B,* fluid of high signal intensity in seen in the distal radioulnar joint *(arrowheads)*—a common abnormality associated with triangular fibrocartilage tears.

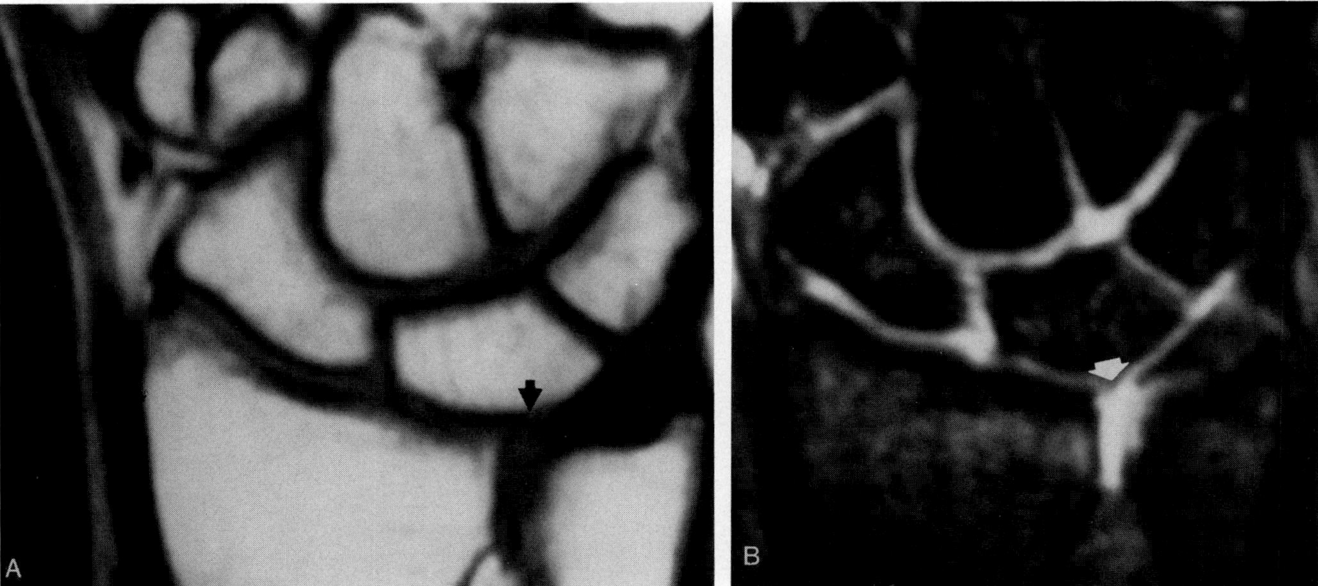

Figure 9–15. Tear of the triangular fibrocartilage at the radial insertion, seen only with the aid of a T2*-weighted gradient-echo MR image. *A*, The tear, of intermediate signal intensity, is not evident on the T1-weighted coronal MR image because it is of equal intensity with the adjacent hyaline cartilage, of intermediate signal intensity *(arrow)*. *B*, Fluid of high signal intensity traverses the tear *(arrow)* on this T2*-weighted gradient-echo MR image. The fluid extends into the distal radioulnar joint.

fibrocartilage tear and can be seen in patients with synovitis or mechanical irritation of the distal radioulnar joint.

Detection of triangular fibrocartilage tears using MR imaging has been investigated by Zlatkin and coworkers, who reported a sensitivity of 100% and a specificity of 93% in 41 patients when compared with results using arthrography, and a sensitivity of 89%, a specificity of 92%, and an accuracy of 90% when compared with results using arthroscopy and

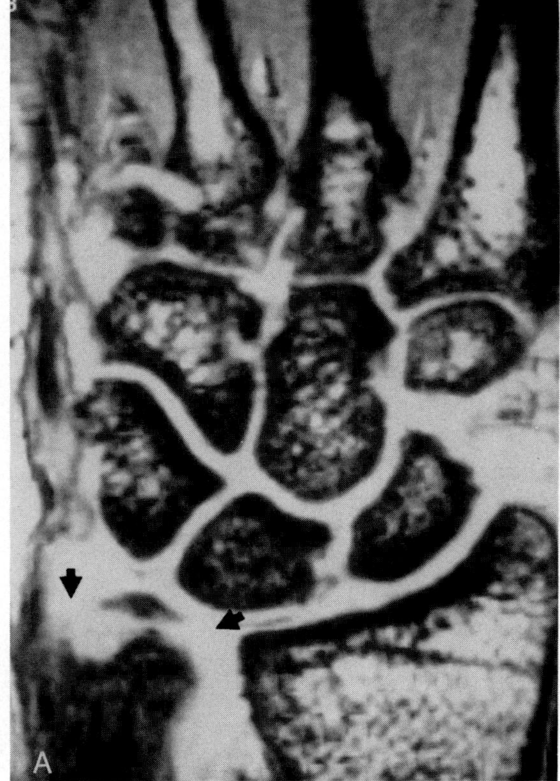

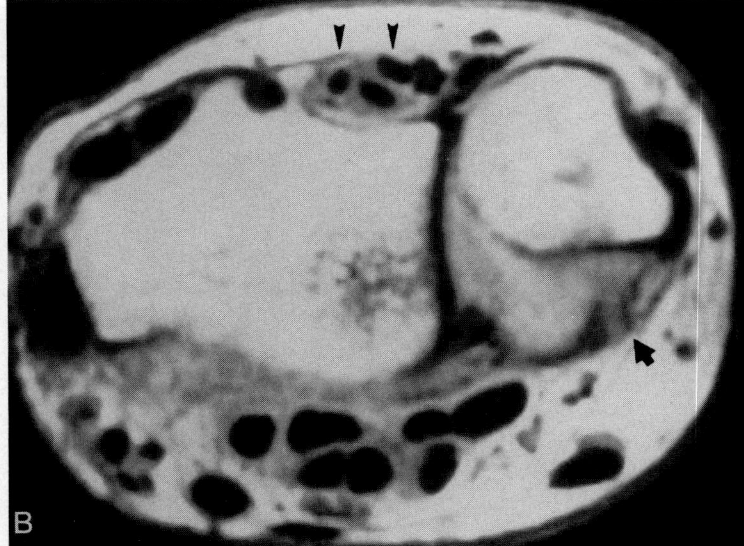

Figure 9–16. Triangular fibrocartilage tear in a patient with rheumatoid arthritis. *A*, A coronal T2*-weighted three-dimensional Fourier transform (3DFT) gradient-echo MR image shows high signal intensity at the radial and ulnar attachments of the triangular fibrocartilage *(arrows)*. Fluid also is present in the distal radioulnar joint. *B*, A tear of the volar radiocarpal ligament *(arrow)*, a component of the triangular fibrocartilage complex, with dorsal subluxation of the ulna is seen on this T1-weighted axial MR image. The patient also had tenosynovitis in the fourth dorsal tendon compartment surrounding the extensor digitorum tendons *(arrowheads)*. The tendons are separated by fluid of intermediate signal intensity.

arthrotomy.[63] Another study by Golimbu and colleagues found a sensitivity of 93% and an accuracy of 95% in 20 patients with surgical correlation.[18] Schweitzer and associates compared MR imaging with arthrography as a gold standard in 15 patients with chronic wrist pain and found MR imaging to have a 72% sensitivity, a 94% specificity, and an 89% accuracy.[47] Because the triangular fibrocartilage also stabilizes the distal radioulnar joint, triangular fibrocartilage tear is occasionally associated with subluxation of the ulnar head, which is well seen on axial MR images (see Fig. 9–16).

ULNAR IMPACTION SYNDROME

Forces in the wrist can cause the ulna and lunate to abut each other.[31] This occurs more frequently with positive ulnar variance. With positive ulnar variance, the carpal surface of the ulna extends farther distal than that of the radius, which can produce rotational contact between the ulna and lunate with subsequent perforation of the triangular fibrocartilage, a tear in the lunotriquetral ligament, and erosion of the articular cartilage of the ulna and lunate. In some cases, MR imaging is able to detect abnormalities in the articular cartilage of these patients. In Kang and associates' series, MR imaging was able to demonstrate cartilaginous abnormalities only in two of the four cadaveric wrists with histologically proved erosion of the articular cartilage caused by ulnolunate impingement.[24] Sclerosis or cysts, or both, are often found on the proximal ulnar aspect of the lunate or the proximal radial aspect of the triquetrum in ulnar impaction syndrome, and, in our experience, may be detected by MR imaging before their detection by conventional radiography. The marrow is of low signal intensity in areas of sclerosis and of fluidlike signal intensity in areas of cyst formation on MR imaging (Fig. 9–17).

DISTAL RADIOULNAR JOINT INSTABILITY

The distal radioulnar joint is located between the semicircular convex ulnar head and the ulnar concavity in the distal radius—the sigmoid notch.

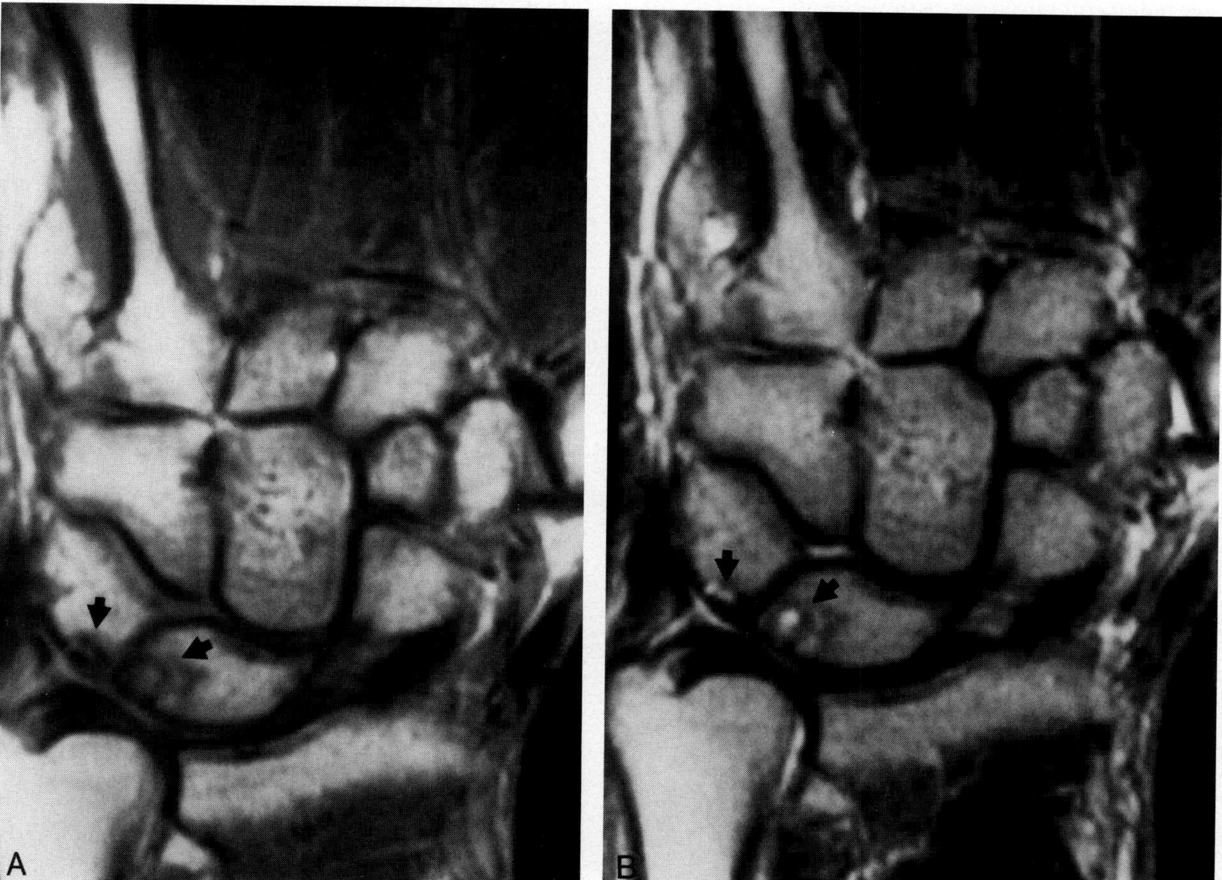

Figure 9–17. Ulnar impaction syndrome. Cystic changes with surrounding sclerosis are seen in characteristic locations along the proximal ulnar aspect of the lunate and proximal radial aspect of the triquetrum *(arrows)*. The sclerosis is of low signal intensity on both the coronal proton density MR image *(A)* and the T2-weighted MR image *(B)*. The cysts are of high signal intensity in *B*. Positive ulnar variance is present, and the triangular fibrocartilage is intact.

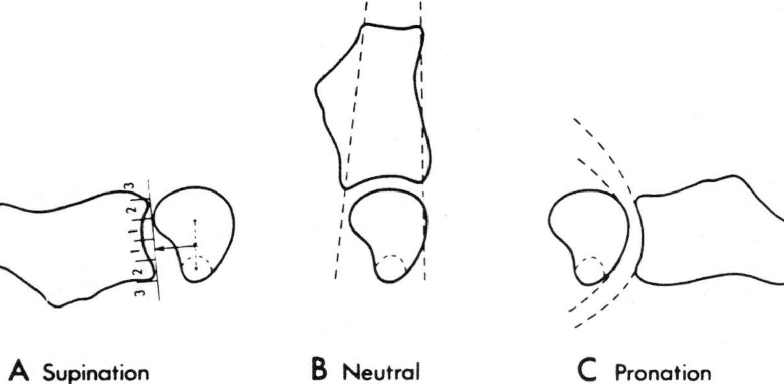

Figure 9–18. Methods for assessing radioulnar subluxation. *A*, Supination; epicenter method. A perpendicular line is drawn from the center of rotation of the distal radioulnar joint (a point halfway between the ulnar styloid process and center of the ulnar head) to the chord of the sigmoid notch. The joint is considered normal if this line is in the middle of the sigmoid notch. The dashed lines represent the location of the styloid process. *B*, Neutral; radioulnar line method. Articulation of the ulnar head with the radius is normal if the head falls between the two pictured lines. *C*, Pronation; congruity method. Note congruity of the arc of the ulnar head with that of the sigmoid notch. (From Wechsler RJ, Wehbe MA, Rifkin MD, et al: Computed tomography diagnosis of distal radioulnar subluxation. Skel Radiol 16:1, 1987.)

Stability for this joint is provided by the interosseous membrane and to a greater extent by the triangular fibrocartilage complex.

The diagnosis of distal radioulnar joint subluxation can be difficult. Symptoms and physical examination findings are often nonspecific, and conventional radiographs generally are unreliable. Axial images from both CT scanning and MR imaging delineate the cross-sectional anatomy of this joint and can be used for evaluating instability.

Wechsler and co-workers have described several methods for the diagnosis of distal radioulnar joint subluxation by CT, which can also be applied to MR imaging (Fig 9–18).[57] When evaluating the distal radioulnar joint on axial MR images it is important to be familiar with the small degree of subluxation that can occur with changes in wrist positioning in the normal wrist (Fig 9–19). On images of the wrist obtained in neutral position, the distal portion of the ulna articulates with the radius in the sigmoid notch. When the wrist is pronated, the ulna can move dorsally, and when the wrist is supinated, the ulnar head can move in a volar direction. When evaluating distal radioulnar joint instability, it is best to include axial images of both wrists in pronation, supination, and neutral positioning. These images are obtained with the arms above the patient's head and the wrists placed in a head or neck coil. T1-weighted images provide a fast and accurate method for determining distal radioulnar joint instability. Magnetic resonance imaging has an advantage for demonstrating soft tissue abnormalities, including triangular fibrocartilage tears, which may be associated with distal radioulnar joint instability (Figs. 9–16, 9–20).

FRACTURE

Although MR imaging is not used for initial diagnosis of carpal fracture, it can be used for detecting subtle or occult fractures (Figs. 9–21, 9–22).[59] The fracture line typically is of low signal intensity on all imaging sequences with surrounding edema and hemorrhage in the marrow in acute situations,

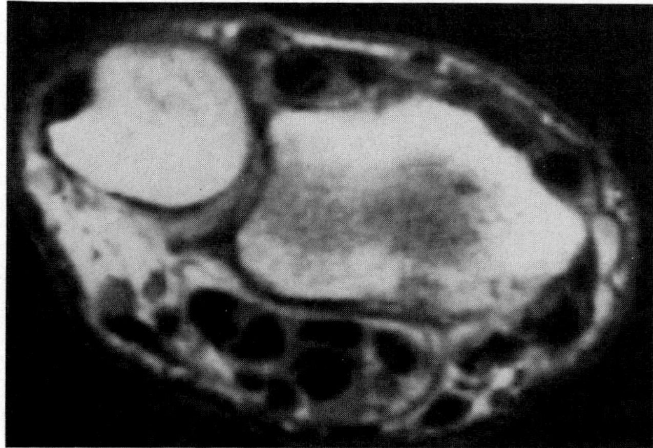

Figure 9–19. Pronation of the normal wrist may produce a mild dorsal subluxation of the ulnar head, as demonstrated on this axial T1-weighted MR image.

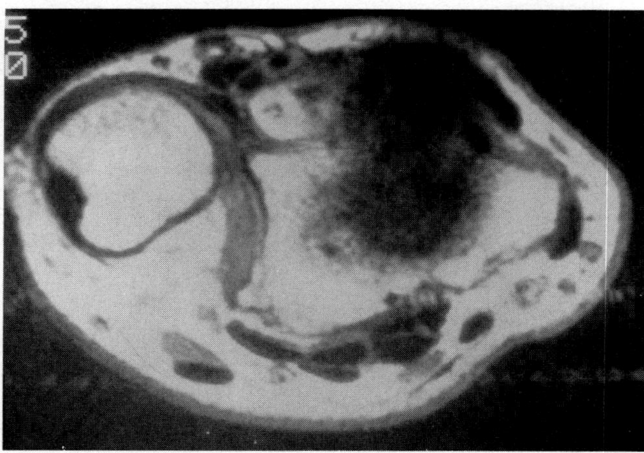

Figure 9–20. True dorsal subluxation of the ulnar head associated with a triangular fibrocartilage tear. The volar radiocarpal ligament is absent on this axial T1-weighted MR image.

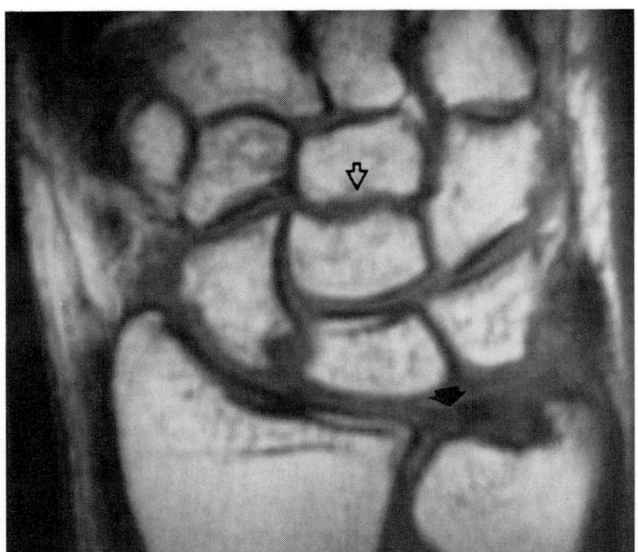

Figure 9-21. Transverse fracture of low signal intensity through the midportion of the capitate on the T1-weighted coronal MR image *(open arrow)*, which was not visible on conventional radiographs. The triangular fibrocartilage also is torn *(solid black arrow)*.

which leads to low to intermediate signal intensity on T1-weighted images and high signal intensity on T2-weighted images. Stress fractures also can be identified using MR imaging.[29]

Multipartite carpal bones are not uncommon and can be mistaken for a fracture. The most common site for this is the scaphoid.[37] Multipartite carpals may be difficult to distinguish from true fractures with MR imaging, except if surrounding marrow contusion is present. Therefore, in the absence of surrounding marrow abnormality, a suspected carpal fracture on MR imaging should be correlated with clinical history and symptoms.

It is important to identify scaphoid nonunion, which commonly occurs when a diagnosis of fracture is delayed or when activity is resumed too soon after fracture. Sequelae of nonunion include avascular necrosis and advanced degenerative arthritis. T2-weighted images are useful for detecting fracture nonunion. This entity is diagnosed when increased signal intensity is observed in the region of the fracture line, caused by fluid collecting in the area of nonunion. Granulation tissue may also have this ap-

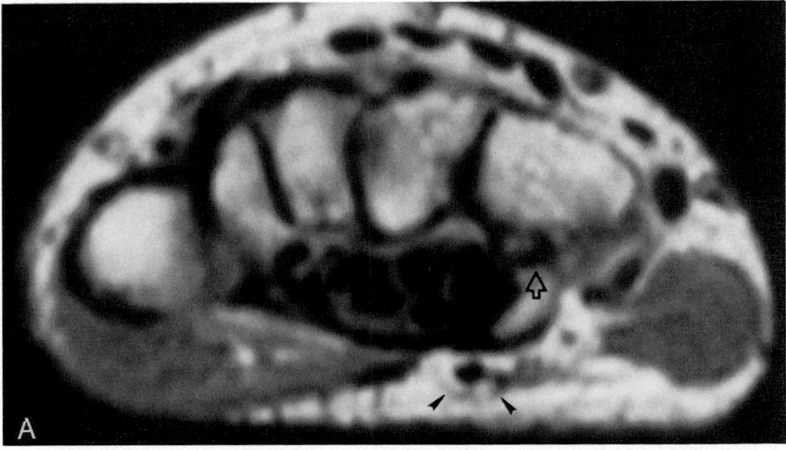

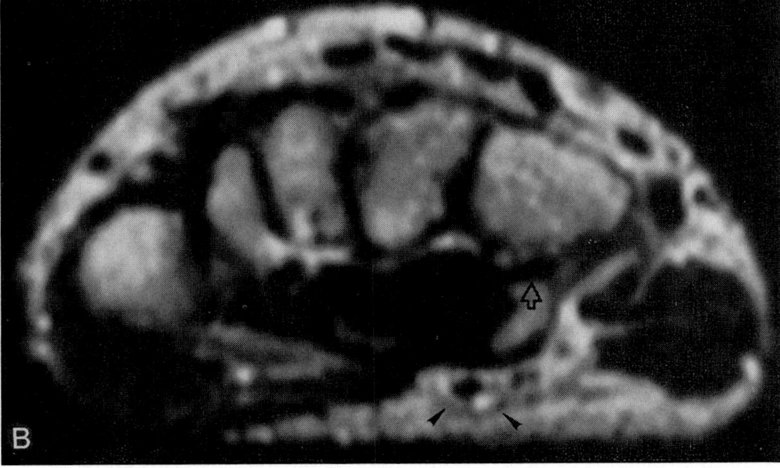

Figure 9-22. T1-weighted *(A)* and T2-weighted *(B)* axial MR images demonstrate a fracture of low signal intensity at the base of the hook of the hamate *(arrow)*. This type of fracture is evaluated more easily on axial CT images. Note Guyon's canal, which contains the ulnar nerve and vessels, lying just below the hook of the hamate *(arrowheads)*.

pearance. If increased signal intensity in the suspected site of nonunion is not demonstrated, the physicians cannot presuppose that nonunion does not exist, as the sensitivity and specificity of this finding have not been determined. Although MR imaging also can be used for evaluation of fracture and bone graft healing, CT generally is more useful.

AVASCULAR NECROSIS

Carpal avascular necrosis (AVN) is most frequent in the proximal scaphoid, with the lunate being the second most common site. Avascular necrosis has proved to be difficult to both diagnose and treat.[15] It is best to identify AVN early, before the onset of collapse and fragmentation of the bone that results in a progressive arthritic process owing to loss of congruity of the joints and disorganization of the normal relationship between other intact carpal bones. MR imaging is useful for the early diagnosis of carpal AVN.[43]

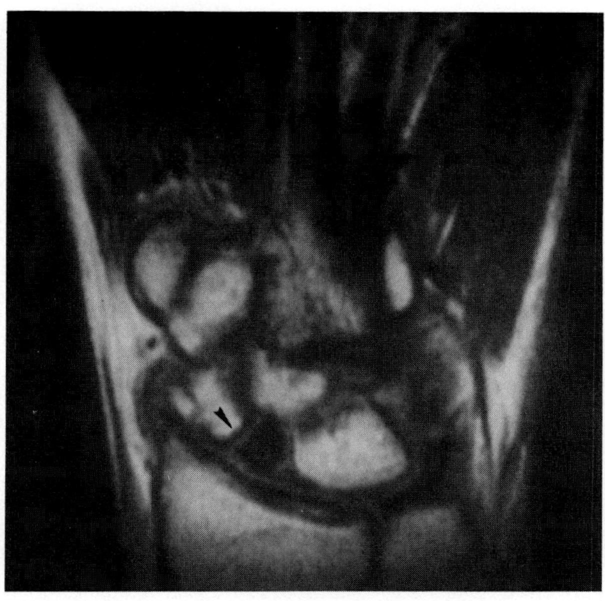

Figure 9-23. There is a fracture of the scaphoid *(arrowhead)* associated with avascular necrosis (low signal intensity) in the proximal fragment on this proton density coronal MR image.

Scaphoid Necrosis

Because the scaphoid receives its blood supply from branches of the radial artery, which enters the bone distally, fractures of the scaphoid often result in AVN of the proximal pole. Avascular necrosis also can be seen in the proximal scaphoid in patients without evidence of fracture. It has been reported in 16 per cent of patients with scaphoid nonunion.[16] The presence of AVN in a proximal fragment suggests a poorer prospect for fracture healing.[56] Knowledge of AVN in the proximal fragment of a nonunited fracture influences treatment and may result in selection of a surgical procedure, such as internal fixation or bone grafting.

Early AVN can be difficult to detect in the scaphoid on conventional radiographs. Bone scintigraphy often is nonspecific. Avascular necrosis of the scaphoid is easily detected on coronal MR images (Fig. 9-23).[56] Magnetic resonance imaging can accurately predict the vascularity of the nonunited scaphoid.[41] The signal intensity can vary according the degree of fibrosis, sclerosis, and revascularization. Osteonecrosis commonly appears as a region of low signal intensity on all MR imaging sequences. Other patterns of signal intensity in osteonecrosis have been described in the femoral head, which are also applicable to AVN in the wrist.[36] These include fatlike, fluidlike, and subacute hemorrhage-like signal intensities.

Kienböck's Disease

Kienböck's disease (also known as lunatomalacia) refers to AVN of the lunate. It is most common in men 20 to 40 years old. Presenting features usually are dorsal wrist pain, decreased range of motion of the wrist, and loss of grip strength. The cause is uncertain, although some authorities have suggested that chronic low-grade trauma is the most important factor. Kienböck's disease is often seen in wrists with negative ulnar variance, in which the ulna lies proximal to the distal radius (Fig. 9-24). This configuration is believed to increase shear stress on the lunate, resulting in microfracture and necrosis.[23,31]

Lichtman defined four stages of Kienböck's disease.[30] Stage 1 represents an early form when radiographs show no evidence of AVN. In this stage, the wrist can be immobilized, with reversal of the AVN.[1] In stage 2, the lunate is sclerotic, without evidence of collapse. Fragmentation and collapse of the lunate are seen in stage 3, with proximal migration of the capitate, resulting in wrist shortening. This can predispose a patient to scapholunate dissociation with disruption of the scapholunate ligament as well as carpal tunnel syndrome. In stage 4 the patient develops degenerative changes in the radiocarpal joint as well as subarticular cysts and carpal instability.

In our experience, MR imaging has been able to demonstrate all stages of Kienböck's disease, including stage 1, when conventional radiographs appear

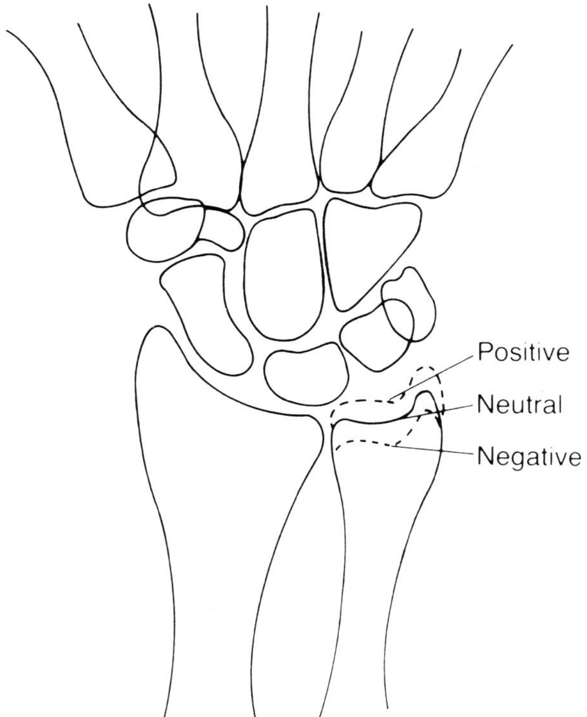

Figure 9–24. Schematic representation of the radiographic appearance of ulnar variance. Neutral variance occurs when the carpal surfaces of the radius and ulna are equal. If the ulna is shorter than the radius, negative ulnar variance exists; if the ulnar is longer than the radius, the relationship is termed as positive ulnar variance. (From Lichtman DM [ed]: The Wrist and Its Disorders. Philadelphia, WB Saunders, 1988.)

normal. On T1-weighted coronal and sagittal MR images, an area of low signal intensity is seen within the lunate (Fig. 9–25).[2,50] Low signal intensity is usually identified within the lunate on T2-weighted images, although early necrosis may be manifested as intraosseous edema, hemorrhage, and hyperemia, producing areas of higher signal intensity.

Abnormal signal intensity produced by necrosis in the lunate can be diffuse or focal. When focal, the radiologist should also consider other abnormalities, including a bone island, an intraosseous ganglion or cyst, or an ulnocarpal impingement. Evaluation of conventional radiographs is essential to aid in characterizing the lesion.

Follow-up MR images are useful to monitor treatment of patients with stage I necrosis. After successful immobilization, the marrow signal intensity may return to normal.[40] Many different treatments are available for advanced Kienböck's disease, all having variable success rates. If the lunate is replaced with a spacer, such as a rolled tendon or prosthesis, MR imaging can be used to evaluate spacer position periodically. It can also aid in the identification of prosthesis fracture, displacement, or synovitis when a Silastic implant is used (Fig. 9–26).

TENDINOPATHY

The extensor tendons of the wrist are divided into six compartments that overlie the carpal bones and interosseous ligaments along the dorsal aspect of the wrist (Fig. 9–27). The first compartment lies lat-

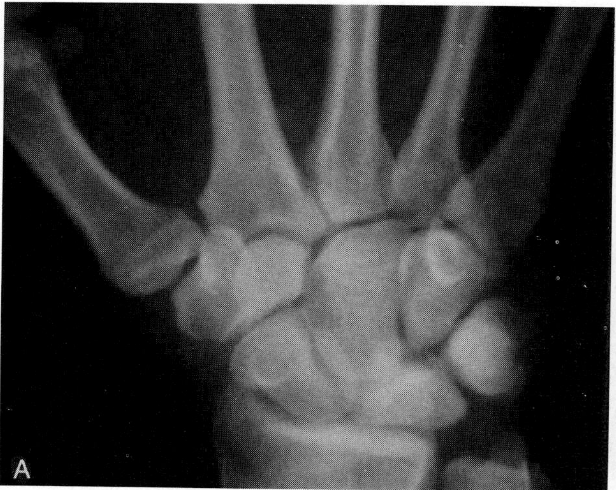

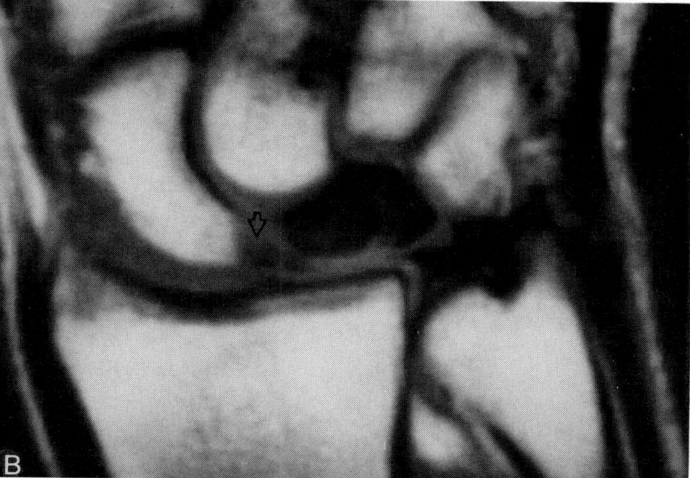

Figure 9–25. Kienböck's disease. *A*, Sclerosis of the lunate is observed on the conventional radiograph. *B*, The lunate is of low signal intensity throughout on this coronal T1-weighted MR image. A scapholunate ligament tear *(arrow)* and negative ulnar variance also are present—two abnormalities that are commonly seen with Kienböck's disease.

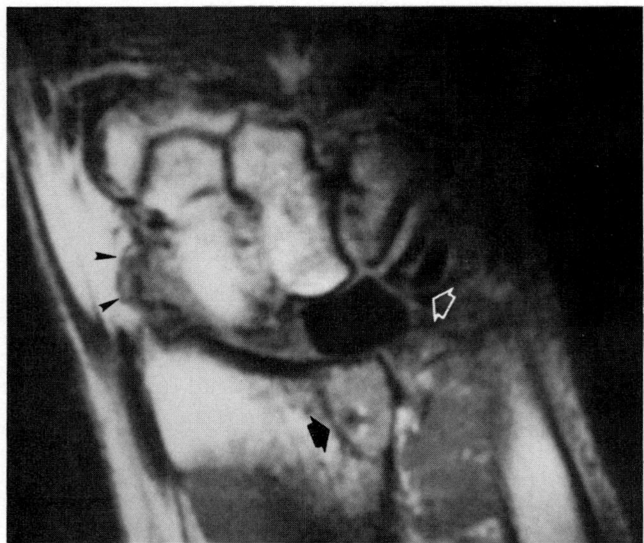

Figure 9–26. A fracture is present at the base of the triquetral peg of this Silastic prosthesis *(white arrow)*, which replaces the lunate in a patient with Kienböck's disease. This type of prosthesis may or may not be seen on conventional radiographs. The irregularity of the synovium *(arrowheads)* is suggestive of synovitis, which is commonly seen in patients with a fractured Silastic prosthesis. The patient has also had a fusion of the scaphoid, trapezium, trapezoid, and capitate, as well as packing of a subcortical cyst in the distal radius *(black arrow)*.

eral to the radius and contains the abductor pollicis longus and extensor pollicis brevis tendons. The second compartment, lateral to Lister's tubercle, contains the extensor carpi radialis brevis and extensor carpi radialis longus tendons. The third compartment, located medial to Lister's tubercle, contains the extensor pollicis longus tendon. The fourth compartment contains the extensor indicis proprius and extensor digitorum communis tendons. The extensor digiti minimi lies in the fifth compartment, and the extensor carpi ulnaris is located in the sixth compartment, medial to the ulnar styloid. Volar to the palmar ligaments lie the deep and superficial flexor tendons of the digits and the flexor pollicis longus tendon, which traverse the wrist through the carpal tunnel.

Tendinopathy can be seen easily on MR imaging. It may be manifested as fluid in the tendon sheath (tenosynovitis), thickening of the tendon, abnormal elevation of signal intensity within the tendon (which may be produced by tendinitis or tear) on T1- or T2-weighted images, or both, or as complete disruption of the tendon.

When evaluating tendons of the wrist and hand, it is important to keep in mind the "magic angle" phenomenon, which can be seen in normal tendons if they lie at an angle of approximately 55 degrees from the direction of the static magnetic field.[17] At this angle, normal anisotropic structures such as tendons may have intermediate signal intensity on short TE images. Signal intensity observed on short TE images decreases with increasing TE. Thus, increased signal intensity due to the magic angle effect may be misdiagnosed as tendinous degeneration or tendinitis.

Tenosynovitis is seen in both flexor and extensor compartments. Tenosynovitis in the sheaths of the abductor pollicis longus and extensor pollicis brevis tendons at the level of the radial styloid, or de Quervain's syndrome, can be identified with MR imaging. A frequent source of pain along the ulnar aspect of the wrist is tenosynovitis of the extensor carpi ulnaris tendon sheath (Fig. 9–28). Subluxation of the extensor carpi ulnaris also is common and can best be demonstrated with the forearm in supination

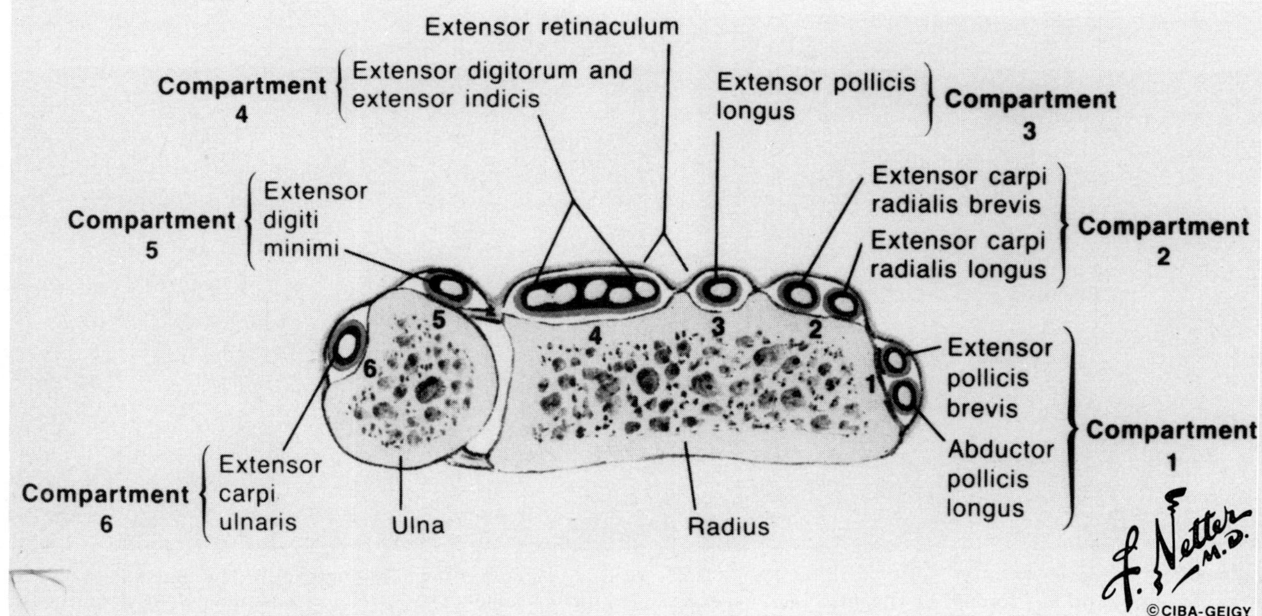

Figure 9–27. Extensor tendons of the wrist. Six dorsal tendon compartments are present in the wrist, each with an overlying extensor retinaculum. (From Netter F: The Ciba Collection of Medical Illustrations. Summit, NJ: Ciba-Geigy Corporation, 1987.)

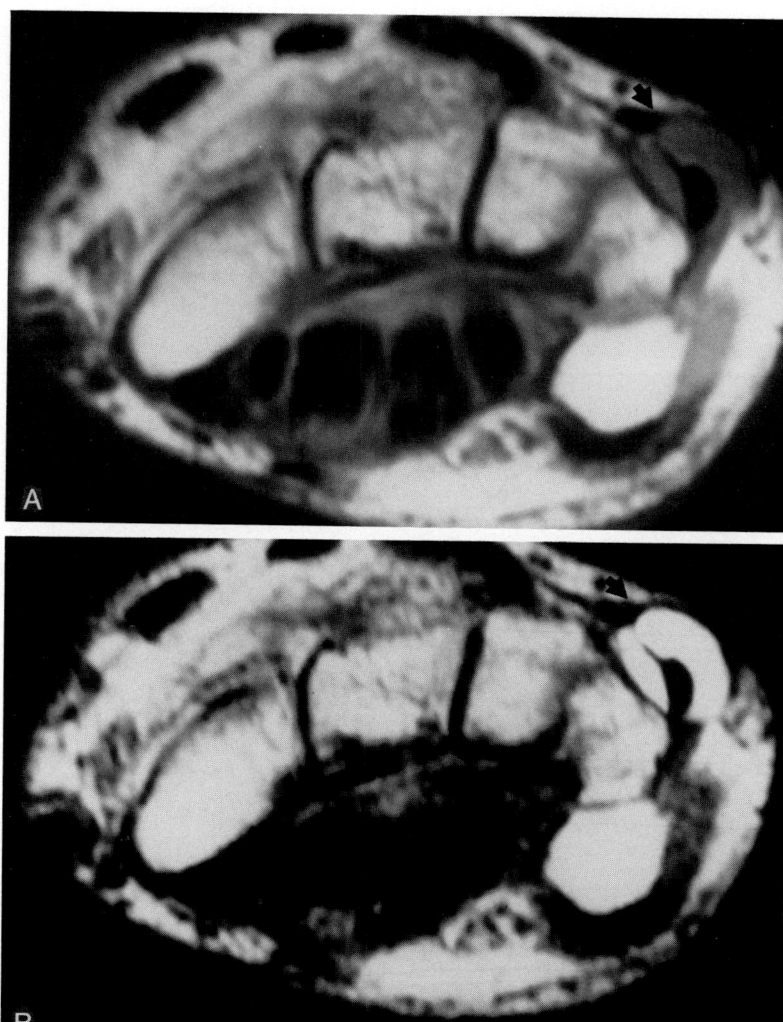

Figure 9–28. Tenosynovitis of the extensor carpi ulnaris tendon sheath *(arrows)* is shown on axial proton density *(A)* and T2-weighted *(B)* MR images at the level of the pisiform.

and the wrist deviated ulnarly. Inflammation of flexor tendons within the carpal tunnel is a frequent cause of carpal tunnel syndrome, as discussed later. The presence and extent of acute suppurative tenosynovitis of the hand can be assessed with MR imaging (Fig. 9–29). This has been called the most disastrous of all hand infections and requires prompt surgical drainage.[13]

Magnetic resonance imaging is useful for demonstrating the presence and extent of tendon rupture, which may be difficult to diagnose clinically (Fig. 9–30).[5] This information also can be utilized for planning tendon repair, relocation, or transfer. The sagittal plane is best for a demonstration of flexor and extensor tendons of the fingers. Complete tendon ruptures are manifested as a loss of continuity in the tendon, which is of low signal intensity. Incomplete ruptures or chronic tears are depicted by MR imaging as irregular thickening or thinning of the tendon, which may contain areas of high signal intensity on T2-weighted images. Gadopentetate dimeglumine is taken up in the area of tendon rupture, but this contrast medium probably is not necessary in most cases (Fig. 9–31). Postoperative adhesions and scar tissue around a tendon can lead to functional impairment. Magnetic resonance imaging can demonstrate the scar tissue as an area of low signal intensity around the tendon.

NERVE ENTRAPMENT SYNDROMES

Carpal Tunnel Syndrome

Carpal tunnel syndrome, produced when the median nerve is compressed in the wrist, is increasing in frequency with the increasing number of jobs that require repetitive motion of the wrist. Patients with carpal tunnel syndrome have pain and tingling of the fingers along the distribution of the median nerve. Most patients with this disorder are 30 to 50 years old. Carpal tunnel syndrome is more common

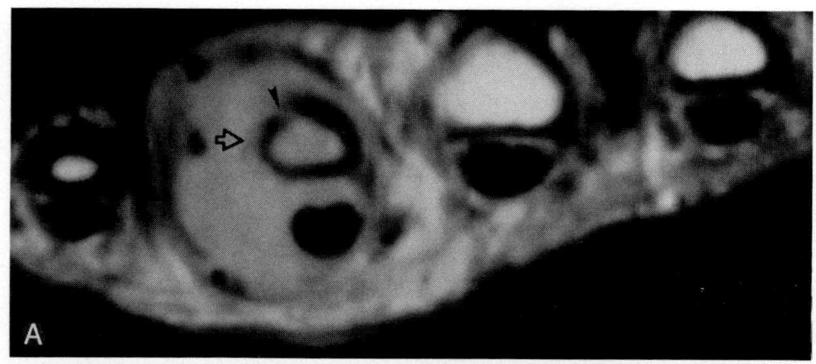

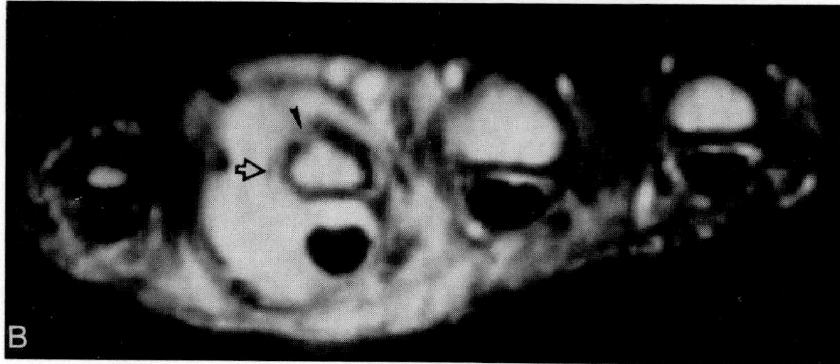

Figure 9–29. Suppurative tenosynovitis of the flexor digitorum tendon *(arrows)* is seen in association with osteomyelitis in the proximal phalanx of the ring finger on axial proton density *(A)* and T2-weighted *(B)* MR images. The marrow demonstrates abnormal intermediate signal intensity on the proton density image, and the signal increases in intensity on the T2-weighted image, consistent with osteomyelitis. Cortical disruption *(arrowheads)* and periostitis *(open arrow)* also are present.

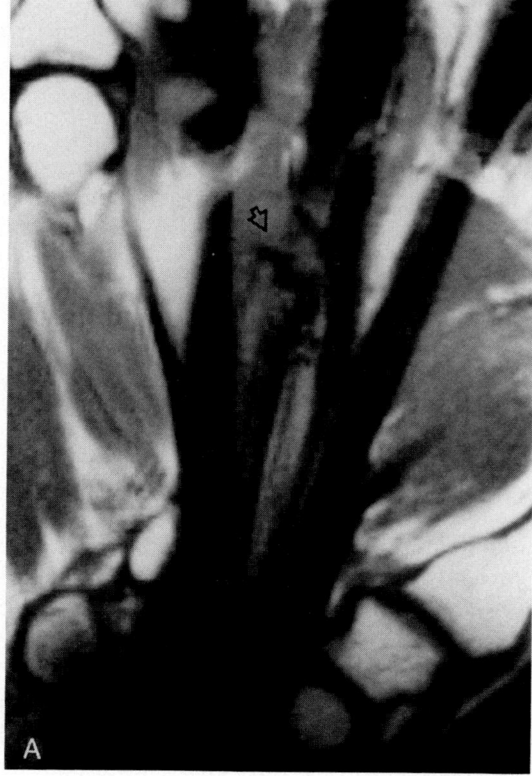

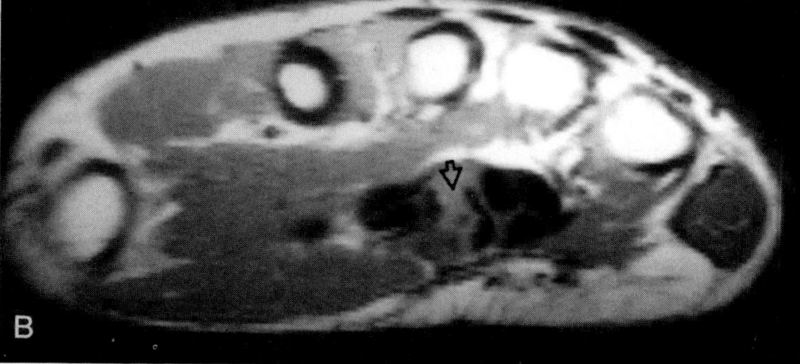

Figure 9–30. Flexor tendon rupture in the hand is well seen on coronal *(A)* and axial *(B)* T1-weighted MR images *(arrows)*. The low signal intensity tendon shows loss of continuity on the coronal image, and the tendon is absent on the axial image at the level of the metacarpals.

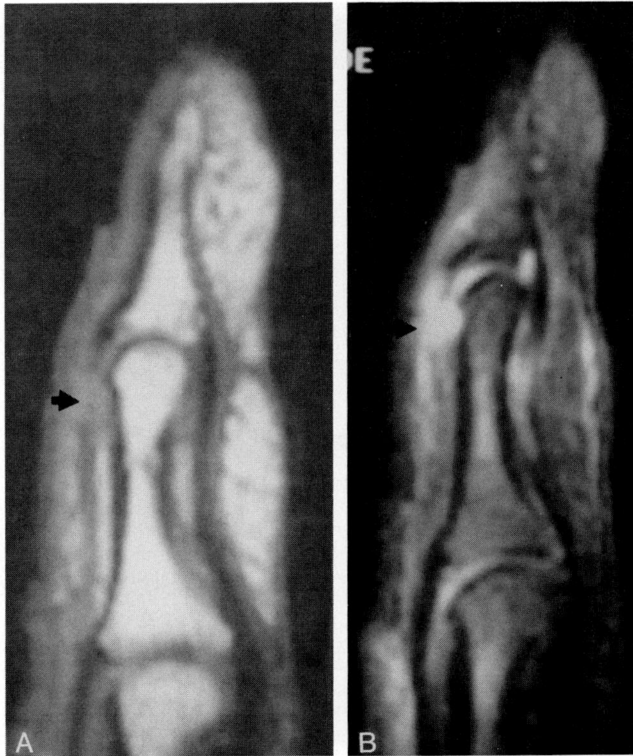

Figure 9–31. Extensor tendon rupture in a finger. *A*, The extensor tendon is interrupted by a region of intermediate signal intensity near the tear on a sagittal T1-weighted MR image *(arrow)*. *B*, The site of the tendon rupture enhances with the administration of gadolinium contrast agent on a sagittal T1-weighted MR image obtained with fat saturation *(arrow)*.

in women and is bilateral in up to 50 per cent of cases.[33]

The anatomic causes of median nerve compression can be assessed by MR imaging. Causes of carpal tunnel syndrome can be divided into two main categories: (1) abnormalities that compress the carpal tunnel from outside, such as a mass or malalignment of osseous structures as a consequence of fracture, Kienböck's disease, or carpal instability producing abnormal volar narrowing, and (2) increased volume within the carpal tunnel caused by inflammation, arthritis, masses, excess fat along the dorsal aspect of the tunnel, persistent median artery, a large adductor pollicis muscle, and edematous and infiltrative disorders. Flexor tenosynovitis and tendinitis resulting from repetitive wrist flexion are the most common causes of carpal tunnel syndrome.[54]

Magnetic resonance imaging is accepted as a useful method for assessment of carpal tunnel syndrome.[32-34,62] The carpal tunnel is bordered inferiorly by the flexor retinaculum, a broad ligament of low signal intensity that extends from the hook of the hamate medially to the tuberosities of the scaphoid and trapezium laterally, holding the flexor tendons in place during wrist flexion. The rigid roof of the carpal tunnel is formed by the carpal bones. The flexor digitorum profundus and flexor digitorum superficialis tendons, of low signal intensity, lie within the tunnel, as does the flexor pollicis longus tendon, which lies radial to the other tendons. The tendon sheaths are of intermediate signal intensity. Only several millimeters of distance should be present between each of the tendons (Fig. 9–32).

The median nerve is easily seen on axial MR images as a rounded or occasionally flattened structure of intermediate signal intensity. The caliber of the median nerve is relatively constant at the level of the distal radioulnar joint, pisiform, and hook of the hamate. This nerve usually is located along the superficial radial aspect of the carpal tunnel just deep to the flexor retinaculum and anterior to the superficial flexor tendon of the index finger (see Fig. 9–32).[32,58] Occasionally, the nerve can lie deeper in the carpal tunnel, perpendicular to the flexor retinaculum (Fig. 9–33). Patients with this normal variation are more prone to develop carpal tunnel syndrome. The position of the nerve also can vary between flexion and extension, and this can be observed with MR imaging.[62] Flexion of the wrist produces anatomic crowding in the carpal tunnel, seen as flattening or interposition of the nerve between the flexor tendons, whereas wrist extension increases the distance between the median nerve and flexor tendons.

Carpal tunnel abnormalities can be visualized with MR imaging; however in our experience, the MR image is not always abnormal in patients with carpal tunnel syndrome. Because of the lack of sensitivity and the expense of MR imaging, we believe that MR imaging should be reserved for certain situations. Current clinical applications for MR imaging of carpal tunnel syndrome include (1) equivocal cases when the electromyogram does not correlate with clinical symptoms of carpal tunnel syndrome; (2) identification of the cause of carpal tunnel syndrome, such as an intrinsic or extrinsic mass, which may obviate flexor retinaculum release; (3) preoperative assessment of the position of the median nerve in patients who are scheduled for endoscopic retinacular release; and (4) postoperative assessment of patients whose symptoms recur after surgery.

In cases of tenosynovitis, MR images demonstrate distention of the tendon sheaths (Fig. 9–34). In acute tenosynovitis, fluid is seen within the tendon sheaths on MR images. Tendinitis may be manifested as enlargement and abnormal increase in signal intensity within the tendon. It is important to exclude the presence of a magic angle phenomenon in tendons that are of high signal intensity on short TE images.[17] Tenosynovitis and tendinitis can be localized or diffuse and may be caused by overuse

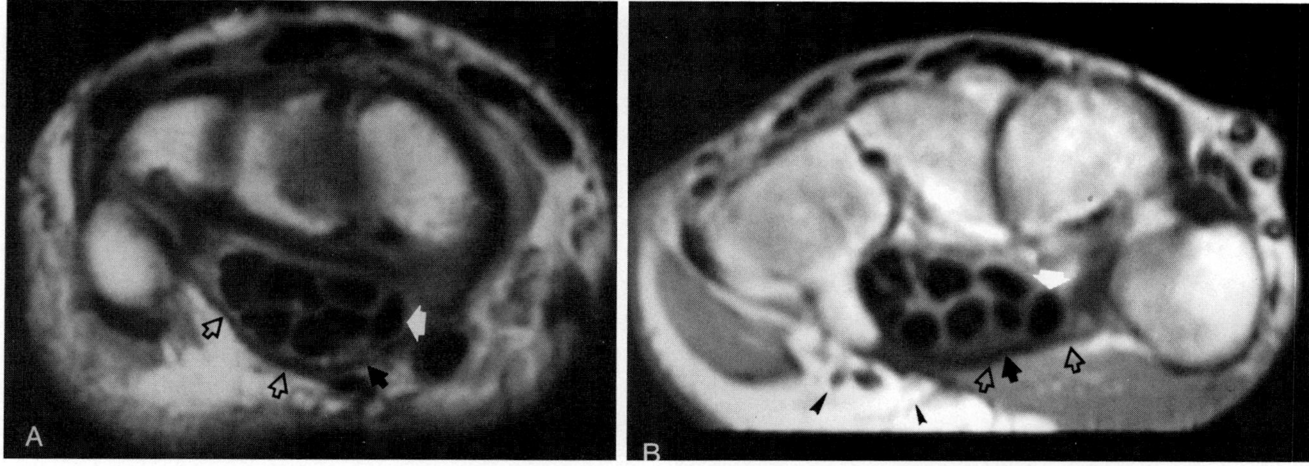

Figure 9–32. Normal carpal tunnel on axial T1-weighted MR images obtained at the level of the pisiform *(A)* and hamate *(B)*. The carpal tunnel is bordered superiorly by the carpal bones and palmar extrinsic and intrinsic ligaments of the wrist. The flexor retinaculum (low signal intensity) *(open arrows)* extends across the palmar aspect of the carpal tunnel. The flexor digitorum profundus and superficialis tendons, also of low signal intensity, as well as the flexor pollicis longus tendon *(white arrow)*, lie within the carpal tunnel. The median nerve is a structure of intermediate signal intensity that usually lies in a radial volar location *(black arrow)*. The nerve flattens out at the level of the hook of the hamate. Also note the location of Guyon's canal *(arrowheads)*.

syndromes or an inflammatory arthropathy, such as rheumatoid arthritis. Other abnormalities that can be identified by MR imaging in patients with carpal tunnel syndrome include palmar bowing of the flexor retinaculum (best identified at the level of the hamate) (Fig. 9–35); swelling, flattening, or attenuation of the median nerve; and elevated signal intensity within the nerve on T2-weighted images, presumably from compression-induced edema.[33] Swelling of the median nerve often is seen proximal to the carpal tunnel at the level of the pisiform (Fig. 9–36). With long-standing carpal tunnel syndrome, the nerve may be of low signal intensity on all imaging sequences, presumably owing to fibrosis. Thenar muscle atrophy also may be seen. Occasionally, muscles can hypertrophy within the carpal tunnel, particularly the lumbricals, which lie distally and originate from the flexor digitorum profundus tendons. It is important that the fingers be kept extended when evaluating the carpal tunnel to avoid overdiagnosing lumbrical muscle hypertrophy, as the lumbrical muscles may slip into the location of the carpal tunnel with finger flexion.[34]

Carpal tunnel syndrome usually is treated conservatively with splinting and medication. If symptoms continue despite conservative treatment, the

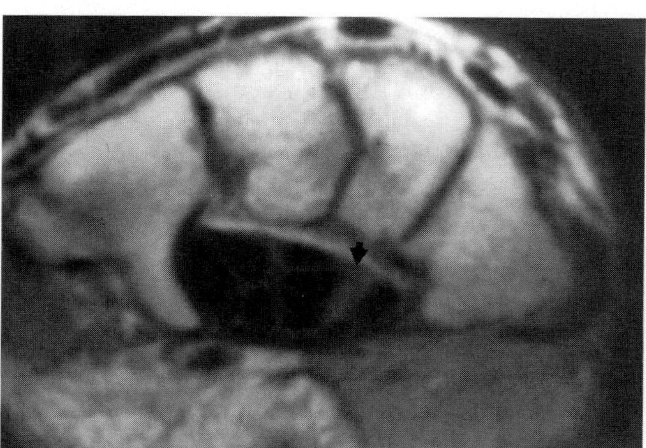

Figure 9–33. Normal variation of the median nerve, which lies deeper in the carpal tunnel perpendicular to the flexor retinaculum *(arrow)*. When the nerve is deeper in the carpal tunnel, it is more prone to compression.

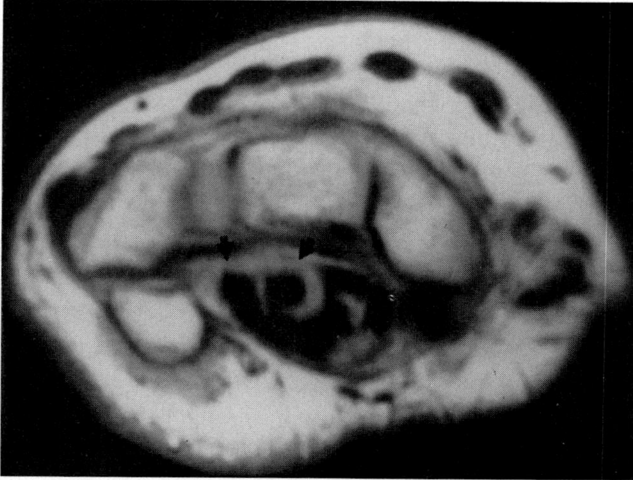

Figure 9–34. Focal tenosynovitis is seen along the ulnar aspect of the carpal tunnel on this axial proton density MR image obtained at the level of the pisiform. Two of the flexor digitorum profundus tendons are surrounded by fluid of intermediate signal intensity *(arrows)*.

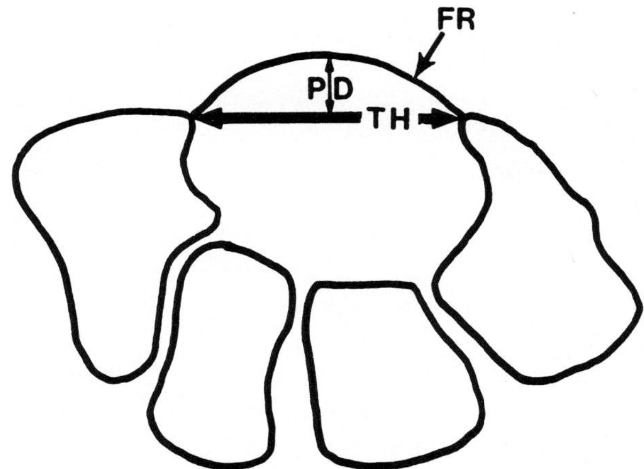

Figure 9–35. The bowing ratio of the flexor retinaculum (FR) is derived by dividing the amount of palmar displacement (PD) by the unbowed length measured as the shortest distance between the trapezium and the hook of the hamate (TH). (From Mesgarzadeh M, Schneck CD, Bonakdarpour A: Carpal tunnel: MR imaging. Part I. Normal anatomy. Radiology 171:743–748, 1989.)

flexor retinaculum is released surgically. This can be done as an open procedure or by blind endoscopic release. Because the endoscopic technique is done with a limited field of the view, the nerve is more prone to injury. After successful carpal tunnel release, the flexor retinaculum and contents of the carpal tunnel are displaced volarly (Fig. 9–37).[8] Magnetic resonance imaging can be used after surgical decompression for those patients who are symptomatic to assess for incomplete retinacular excision or reattachment (Fig. 9–38), scarring in the region of the carpal tunnel, neuritis of the median nerve, or development of a neuroma, which can occur when the median nerve is cut inadvertently.[33]

Guyon's Canal Syndrome

Magnetic resonance imaging can demonstrate the area of Guyon's canal reliably.[61] It is in this fascial tunnel that the ulnar nerve enters the palm from the forearm and extends distally from the level of the pisiform, where it divides into two sensory branches and a deep motor branch. Guyon's canal extends approximately 4 cm from the palmar carpal ligament at the proximal edge of the pisiform to the origin of the hypothenar muscles at the level of the hamate. Besides the ulnar nerve, the ulnar artery and occasional veins pass through this region, and abundant adipose tissue is present surrounding these structures, which allows them to be well visualized by MR imaging (see Fig. 9–32).

The deep motor branch of the ulnar nerve is subject to compression by repetitive motion; adjacent masses such as ganglia, lipomas, and anomalous muscles compressing the canal; ulnar artery false aneurysm; hypertrophy of the flexor carpi ulnaris tendon; hypertrophy of the palmar carpal ligament; osteoarthritis of the pisiform-triquetral joint; or fractures at the bases of the metacarpals, hook of the hamate, and pisiform.[12,19,53,60,61] Magnetic resonance imaging can reliably demonstrate surrounding masses as well as anomalous muscles (Fig. 9–39).

SYNOVIAL ABNORMALITIES

Effusions, inflammation, and edema are well seen by MR imaging, displaying signal intensities characteristic of fluid (long T1 and T2). Progression or improvement of synovitis can be monitored by

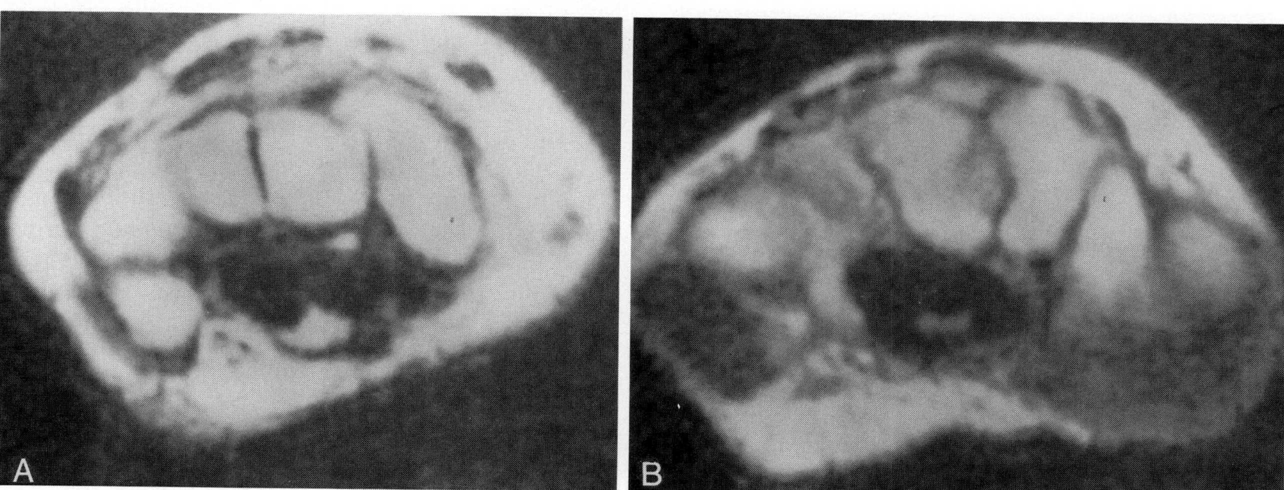

Figure 9–36. Carpal tunnel syndrome. The T2-weighted axial MR images obtained at the level of the pisiform *(A)* and hamate *(B)* demonstrate that the median nerve is swollen at the level of the pisiform but normal in size at the level of the hamate, a finding often seen in patients with carpal tunnel syndrome. The nerve also is of high signal intensity, consistent with edema and inflammation. Some degradation of the image has occurred from wrist movement related to pain produced by the carpal tunnel syndrome.

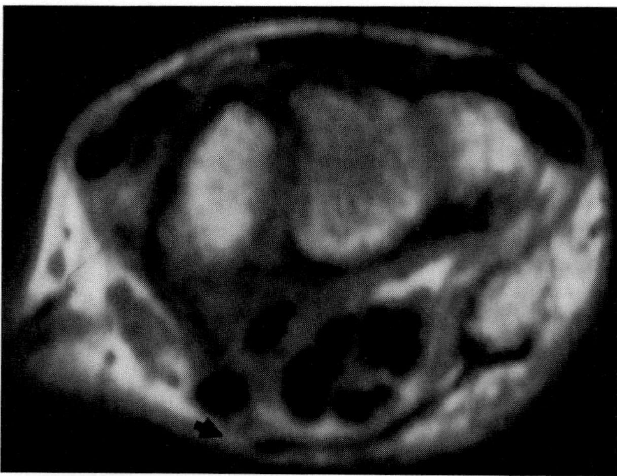

Figure 9–37. Normal appearance of the carpal tunnel after release of the flexor retinaculum. The axial T1-weighted image at the level of the pisiform reveals volar displacement of the contents of the carpal tunnel. The flexor retinaculum also is displaced volarly and is incomplete, with evidence of release along the radial aspect *(arrow)*.

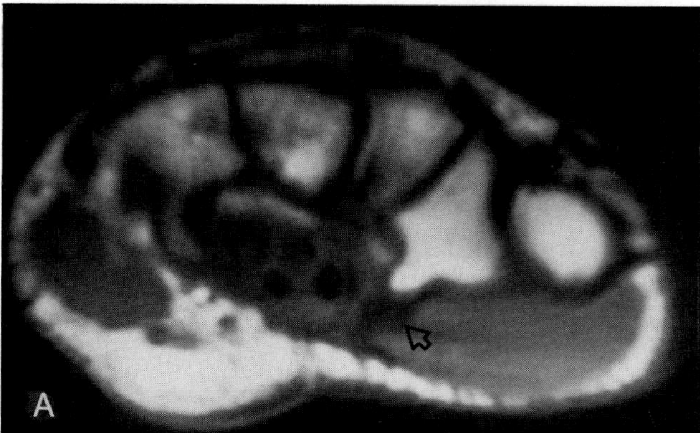

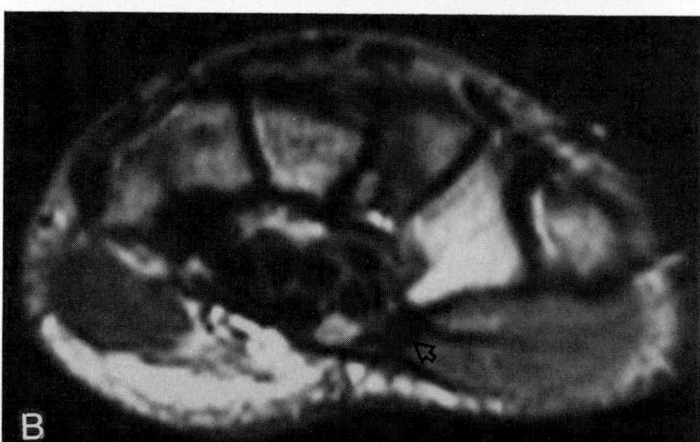

Figure 9–38. Scarring and reattachment of the flexor retinaculum in a symptomatic patient after carpal tunnel release. Axial proton density *(A)* and T2-weighted *(B)* MR images show an area of thickening of low signal intensity adjacent to the triquetrum, consistent with scarring *(arrows)*. On the T2-weighted image *(B)*, the flexor retinaculum is seen spanning the floor of the carpal tunnel without interruption.

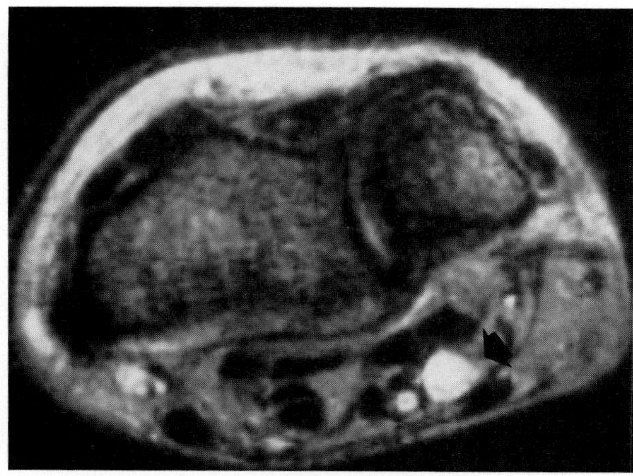

Figure 9–39. A neuroma of high signal intensity is involving the ulnar nerve just proximal to Guyon's canal on this axial T2*-weighted gradient-echo MR image *(arrow)*.

MR imaging, allowing for presurgical evaluation and assessment of therapy. Characteristics of synovitis on MR images include irregular synovium that can be thickened in long-standing cases of arthritis. Joint effusion can be difficult to distinguish from hypervascular pannus without the aid of dynamic gradient-echo imaging after administration of an intravascular contrast agent such as gadopentetate dimeglumine (Fig. 9–40).[27,28,44] Using dynamic sequential MR imaging immediately after the administration of gadopentetate dimeglumine, the radiologist can differentiate inflamed synovium, which takes up the gadolinium within 30 sec, from synovial fluid, which does not become enhanced during the first 10 min after injection. In our experience, chronically thickened, fibrotic pannus has a shorter T2 relaxation time than synovial fluid, demonstrating intermediate to low signal intensity on T2-weighted MR images. Pannus may extend into the prestyloid recess and distal radioulnar joint, causing erosion of these structures. Magnetic resonance imaging also is useful for the early identification of osseous and cartilaginous erosion in joints affected by inflammatory arthritis, such as rheumatoid arthritis. The synovium and articular cartilage around the metacarpophalangeal joints are sites of major involvement for many arthropathies and can be evaluated with high-resolution MR imaging for such abnormalities as effusion, erosion, and cartilaginous narrowing.

BENIGN SOFT TISSUE MASSES

Benign tumors occur much more frequently than malignant tumors in the wrist and hand.[11] Magnetic resonance imaging is useful for identifying and

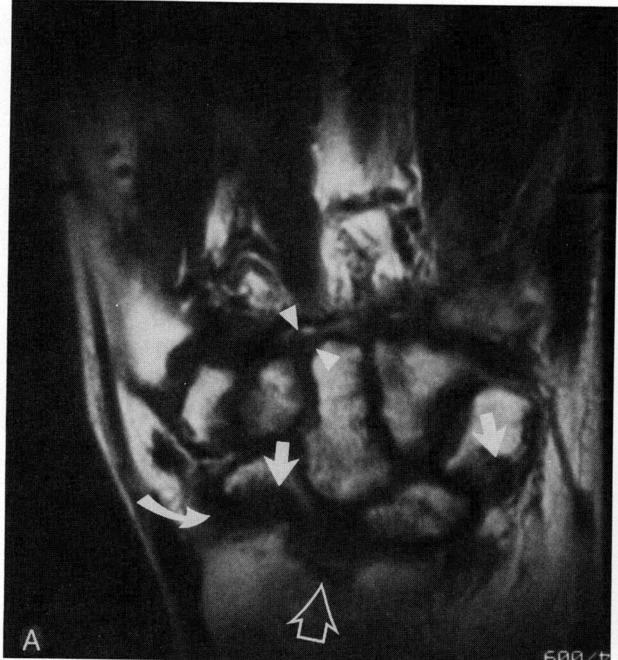

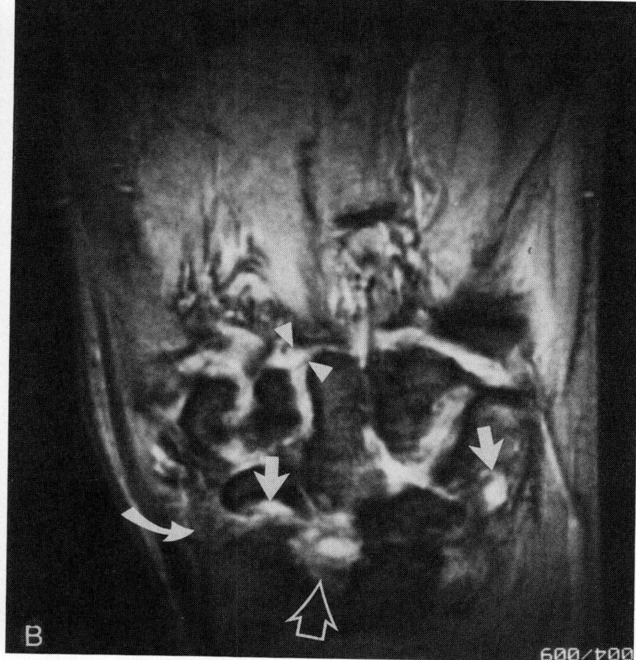

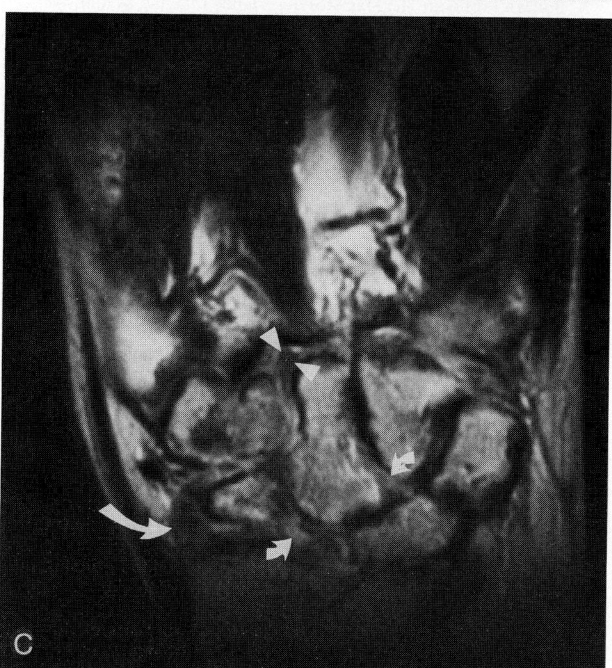

Figure 9–40. Rheumatoid arthritis. *A*, A T1-weighted MR image demonstrates erosion of low signal intensity in the distal radius *(open arrow)* and multiple erosions of low signal intensity in the carpal bones *(straight arrows)*. Note the soft tissue swelling (low signal intensity) between the distal radius and carpals *(curved arrow)* and between the carpals *(arrowheads)*. It is not certain whether this represents effusion, inflammatory granulation tissue associated with rheumatoid arthritis, or fibrotic pannus. *B*, T2*-weighted gradient-echo MR image demonstrates radial *(open arrow)* and carpal *(straight arrows)* erosions of intermediate to high signal intensity. Portions of the intercarpal soft tissue swelling are of high signal intensity *(arrowheads)*. These may reflect effusion or inflammatory granulation tissue with a high water content. Other portions lateral and distal to the radius remain of intermediate to low signal intensity. These probably correspond to fibrotic pannus *(curved arrow)*. *C*, T1-weighted postcontrast MR image demonstrates contrast enhancement in some areas of intercarpal soft tissue swelling *(small curved arrows)* when compared with the precontrast T1-weighted image *(A)*. These are likely to reflect perfused, hyperemic inflammatory granulation tissue. Other portions of the intercarpal soft tissue swelling do not enhance, indicating that they represent nonperfused joint effusion *(arrowheads)*. The soft tissue lateral to the radius *(large curved arrow)*, which was of intermediate to low signal intensity on T2*-weighting, shows slight enhancement when compared with the precontrast T1-weighted image. This likely corresponds to fibrotic pannus (as shown in *B*). (Courtesy of Philipp Lang, M.D., and Harry K. Genant, M.D.)

demonstrating the osseous and soft tissue extent of tumors in the wrist and hand.[8,51] Because malignant tumors and benign osseous tumors are discussed in another chapter, this section focuses only on the characteristic benign soft tissue masses. The common benign soft tissue tumors of the wrist and hand include ganglion, giant cell tumor of the tendon sheath, and lipoma.[8,25,51,58]

Ganglions are the most common wrist tumor. Usually they are located dorsally in the wrist, although they can occur anywhere. Magnetic resonance imaging can provide information preoperatively regarding their extent and relationship to adjacent structures, such as tendons, nerves, and the joint capsule. Ganglions may be multiple and multiloculated. They contain mucinous material that has long T1 and T2 relaxation times, similar to those of fluid (Fig. 9–41). Differentiation from other benign lesions is not always possible on the basis of MR imaging features alone. Because the recurrence rate

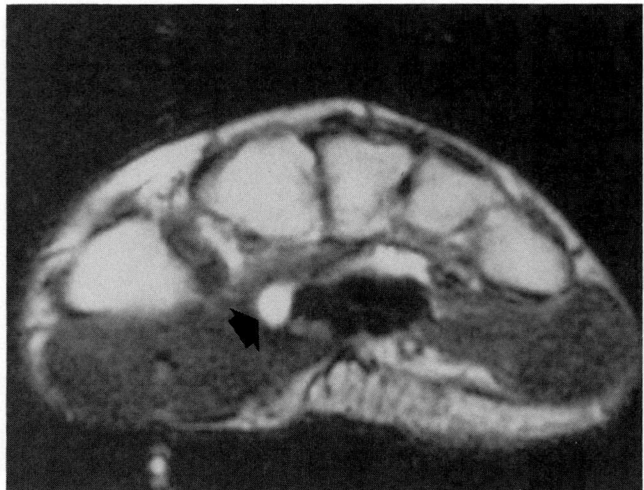

Figure 9–41. A ganglion of high signal intensity is present adjacent to the median nerve *(arrow)* on this T2-weighted axial MR image obtained at the metacarpal level. This ganglion was producing a carpal tunnel syndrome more proximally.

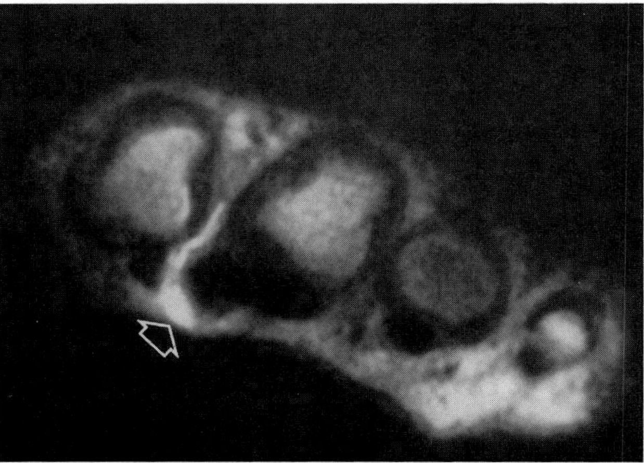

Figure 9–42. A metallic foreign body is producing an artifact adjacent to the flexor tendon of the third metacarpal on this T2-weighted axial MR image *(arrow)*.

after ganglion resection is high, MR imaging can be utilized to assess for the presence of ganglia postoperatively.

Giant cell tumor of the tendon sheath is the second most common soft tissue mass in the hand. It can produce erosion of neighboring bone and appear on MR images as homogeneous or inhomogeneous areas of low signal intensity on all imaging sequences as a consequence of the paramagnetic effect of hemosiderin.[7,25]

Lipomas commonly are located in the thenar eminence and have signal characteristics identical to that of subcutaneous fat. They may contain streaks of low signal intensity and usually are well circumscribed. Lipomas and ganglions, as well as other masses, can cause compression of the ulnar nerve in Guyon's canal along the ulnovolar aspect of the wrist.[3,9]

Magnetic resonance imaging also is useful for identification of a variety of foreign bodies in the wrist and hand.[46] This is particularly true of nonradiopaque foreign bodies, such as plastic and wood, which are not well seen with xeroradiography. Ultrasonography also can be used for localization. Metallic foreign bodies produce some artifact, particularly with high-field MR imagers (Fig. 9–42).

References

1. Alexander AH, Lichtman DM: Kienbock's disease. Orthop Clin North Am 17:461, 1986.
2. Amadio PC, Hanssen AD, Berquist TH: The genesis of Kienböck's disease: Early diagnosis by magnetic resonance imaging. J Hand Surg [Am] 12:1044, 1987.
3. Angelides AC: Ganglions of the hand and wrist. In Green DP (ed): Operative Hand Surgery, 2nd ed. New York, Churchill Livingstone, 1988.
4. Baker LL, Hajek PC, Bjorkengren A, et al: High-resolution magnetic resonance imaging of the wrist: Normal anatomy. Skel Radiol 16:128–132, 1987.
5. Beltran J, Mosure JC: Magnetic resonance imaging of tendons. Crit Rev Diagn Imaging 30:111, 1990.
6. Beltran J, Noto AM, Mosure JC, et al: Meniscal tears: MR demonstration of experimentally produced injuries. Radiology 158:691, 1986.
7. Binkovitz LA, Berquist TH, McLeod RA: Masses of the hand and wrist: Detection and characterization with MR imaging. AJR 154:323, 1990.
8. Binkovitz LA, Cahill DR, Ehman RL, et al: Magnetic resonance imaging of the wrist: Normal cross-sectional imaging and selected abnormal cases. Radiographics 8:1171, 1988.
9. Bogumill GP: Tumors of the wrist. *In* Lichtman DM (ed): The Wrist and Its Disorders. Philadelphia, WB Saunders, 1988.
10. Bowers WH: The distal radioulnar joint. *In* Green DP (ed): Operative Hand Surgery. 2nd ed. New York, Churchill Livingstone, 1988.
11. Butler ED, Hamell JP, Seipel RS, deLorimier AA: Tumors of the hand. A 10 year survey and report of 437 cases. Am J Surg 100:293, 1960.
12. Carlson CS, Clark GL: False aneurysm of ulnar artery in Guyon's canal. J Hand Surg [Am] 8:223, 1983.
13. Carter SJ, Mersheemer WL: Infections of the hand. Orthop Clin North Am 1:455, 1970.
14. Cerofolini E, Luchetti R, Pederzini L, et al: MR evaluation of the triangular fibrocartilage complex tears in the wrist: Comparison with arthrography and arthroscopy. J Comput Assist Tomogr 14:963, 1990.
15. Cooney WP, Dobyns JH, Linscheid RL: Complications of Colles' fractures. J Bone Joint Surg [Am] 62:613, 1980.
16. Cooney WP, Linscheid RL, Dobyns JH: Scaphoid fractures: Problems associated with nonunion and avascular necrosis. Orthop Clin North Am 15:381, 1981.
17. Erickson SJ, Cox IH, Hyde JS, et al: Effect of tendon orientation on MR imaging signal intensity: A manifestation of the "magic angle" phenomenon. Radiology 181:389, 1991.
18. Golimbu CN, Firooznia H, Melone CP, et al: Tears of the triangular fibrocartilage of the wrist: MR imaging. Radiology 173:731, 1989.
19. Guliani G, Poppi M, Pozzati E, Forti A: Ulnar neuropathy due to a carpal ganglion: The diagnostic contribution of CT. Neurology 40:1001, 1990.
20. Gundry CR, Kursunoglu-Brahme S, Schwaighofer B, et al: Is MR better than arthrography for evaluating the ligaments of the wrist? In vitro study. AJR 154:337, 1990.
21. Harrison MO, Freiberger RH, Ranawat CS: Arthrography of the rheumatoid wrist joint. AJR 112:480, 1971.
22. Heuck A, Steinbach L, Neumann C, et al: Möglichkeiten der MR-Tomographie bei Erkrankungen von Hand und Handgelenk. Radiologe 29:53, 1989.
23. Hulten O: Über anatomische Variationen der Handgelenkknochen. Acta Radiol Scand 9:155, 1928.
24. Kang HS, Kindynis P, Brahme SK, et al: Triangular fibrocartilage and intercarpal ligaments of the wrist: MR imaging-cadaveric study with gross pathologic and histologic correlation. Radiology 181:401, 1991.
25. Karasick D, Karasick S: Giant cell tumor of tendon sheath: Spectrum of radiologic findings. Skel Radiol 21:219, 1992.
26. Kessler I, Silberman Z: Experimental study of the radiocarpal joint by arthrography. Surg Gynecol Obstet 112:33, 1961.
27. Konig H, Sieper J, Wolf KJ: Rheumatoid arthritis: Evaluation of hyper-

vascular and fibrous pannus with dynamic MR imaging enhanced with Gd-DTPA. Radiology 176:473, 1990.
28. Kursunoglu-Brahme S, Riccio T, Weisman MH, et al: Rheumatoid knee: Role of gadopentetate-enhanced MR imaging. Radiology 176:831, 1990.
29. Lee JK, Yao L: Stress fractures: MR imaging. Radiology 169:217, 1988.
30. Lichtman DM: Kienböck's disease. *In* Lichtman DM (ed): The Wrist and Its Disorders. Philadelphia, WB Saunders, 1988.
31. Linscheid R: Ulnar lengthening and shortening. *In* Taleisnik J (ed): Management of Wrist Problems. Hand Clinics III:69, 1987.
32. Mesgarzadeh M, Schneck CD, Bonakdarpour A: Carpal tunnel: MR imaging. Part I. Normal anatomy. Radiology 171:743, 1989.
33. Mesgarzadeh M, Schneck CD, Bonakdarpour A, et al. Carpal tunnel: MR imaging. Part II: Carpal tunnel syndrome. Radiology 171:749, 1989.
34. Middleton WD, Kneeland JB, Kellman GM, et al: MR imaging of the carpal tunnel: Normal anatomy and preliminary findings in the carpal tunnel syndrome. AJR 148:307, 1987.
35. Mikic ZD: Age changes in the triangular fibrocartilage of the wrist joint. J Anat 126:367, 1978.
36. Mitchell DG, Rao VM, Dalinka MK, et al: Femoral head avascular necrosis: Correlation of MR imaging, radiographic staging, radionuclide imaging, and clinical findings. Radiology 162:709, 1987.
37. O'Rahilly R: A survey of carpal and tarsal anomalies. J Bone Joint Surg [Am] 35:626, 1953.
38. Palmer AK: Triangular fibrocartilage complex lesions: A classification. J Hand Surg [Am] 14:594, 1989.
39. Palmer AK, Werner FW: The triangular fibrocartilage complex of the wrist: Anatomy and function. J Hand Surg [Am] 6:153, 1981.
40. Pennes DR, Louis DS, Fechner K: Bone marrow imaging (letter to the editor). Radiology 170:894, 1989.
41. Perlik PC, Guilford WB: Magnetic resonance imaging to assess vascularity of scaphoid nonunions. J Hand Surg [Am] 16:479, 1991.
42. Reicher MA, Hartzman S, Duckwiler GR, et al: Meniscal injuries: Detection using MR imaging. Radiology 159:753, 1986.
43. Reinus WR, Conway WF, Totty WG, et al: Carpal avascular necrosis: MR imaging. Radiology 160:689, 1986.
44. Reiser MF, Bongartz GP, Erlemann R, et al: Gadolinium-DTPA in rheumatoid arthritis and related diseases: First results with dynamic magnetic resonance imaging. Skel Radiol 18:591, 1989.
45. Reuther G, Erlemann R, Grunert J, Peters PE: Untersuchungstechnik und ligamentare Binnenmorphologie in der MRT des Handgelenks. Radiologe 30:373, 1990.
46. Russell RC, Williamson DA, Sullivan JW, et al: Detection of foreign bodies in the hand. J Hand Surg [Am] 16:2, 1991.
47. Schweitzer ME, Brahme SK, Hodler J, et al: Chronic wrist pain: Spin-echo and short tau inversion recovery MR imaging and conventional and MR arthrography. Radiology 182:205, 1992.
48. Shellock FG, Mandelbaum BP: Kinematic MR Imaging of the joints. *In* Mink JH, Deutsch AL (eds): MR Imaging of the Musculoskeletal System—A Teaching File. New York; Raven Press, 1990.
49. Skahen JR, Palmer AK, Levinsohn EM, et al: Magnetic resonance imaging of the triangular fibrocartilage complex. J Hand Surg [Am] 15:552, 1990.
50. Sowa DT, Holder LE, Patt PG: Application of magnetic resonance imaging to ischemic necrosis of the lunate. J Hand Surg [Am] 14:1008, 1989.
51. Steinbach LS: Tumors and synovial processes in the wrist and hand. *In* Reicher MA, Kellerhouse LE (eds): MRI of the Wrist and Hand. New York: Raven Press, 1990.
52. Stener B: Displacement of the ruptured ulnar collateral ligament of the metacarpo-phalangeal joint of the thumb: A clinical and anatomical study. J Bone Joint Surg [Br] 44:869, 1962.
53. Subin GD, Mallon WJ, Urbaniak JR: Diagnosis of ganglion in Guyon's canal by magnetic resonance imaging. J Hand Surg [Am] 14:640, 1989.
54. Sunderland S: Carpal tunnel syndrome. *In* Nerves and Nerve Injuries. 2nd ed. New York, Churchill Livingstone, 1981.
55. Taleisnick J: Pain on the ulnar side of the wrist. *In* Taleisnik J (ed): Management of Wrist Problems. Vol 1. Philadelphia, WB Saunders, 1987.
56. Trumble TE: Avascular necrosis after scaphoid fracture: A correlation of magnetic resonance imaging and histology. J Hand Surg [Am] 15:557, 1990.
57. Wechsler RJ, Wehbe MA, Rifkin MD, et al: Computed tomography diagnosis of distal radioulnar subluxation. Skel Radiol 16:1, 1987.
58. Weiss KL, Beltran J, Lubbers LM: High field MR surface coil imaging of the hand and wrist. Part II. Pathologic correlations and clinical relevance. Radiology 160:147, 1986.
59. Yao L, Lee JK: Occult intraosseous fracture: Detection with MR imaging. Radiology 167:749, 1988.
60. Zahrawi F: Acute compression ulnar neuropathy at Guyon's canal resulting from lipoma. J Hand Surg [Am] 9:238, 1984.
61. Zeiss J, Jakab E, Khimji T, Imbriglia J: The ulnar tunnel at the wrist (Guyon's canal): Normal MR anatomy and variants. AJR 158:1081, 1992.
62. Zeiss J, Skie M, Ebraheim N, et al: Anatomic relations between the median nerve and flexor tendons in the carpal tunnel: MR evaluation in normal volunteers. AJR 153:533, 1989.
63. Zlatkin MB, Chao PC, Osterman AL, et al: Chronic wrist pain: Evaluation with high-resolution MR imaging. Radiology 173:723, 1989.

10 THE HIP

Philipp Lang, Harry E. Jergesen, and Harry K. Genant

Magnetic resonance (MR) imaging has greatly advanced the assessment of hip disorders. This is largely because of its great sensitivity and specificity in detecting osteonecrosis of the hip. Other applications for MR imaging of the hip joint have evolved in recent years, and, at the present time, this imaging technique supplements and frequently replaces computed tomography (CT) in evaluating hip disorders.

Magnetic resonance images are characterized by excellent soft tissue contrast and direct multiplanar imaging in the axial, sagittal, and coronal planes. The most important factors influencing signal intensity in MR imaging of the hip joint are proton or spin density, T1 relaxation time, and T2 relaxation time. Magnetic susceptibility effects have an effect on the imaging appearance of the hip in gradient-echo sequences. Other factors that influence MR signal intensity, such as flow phenomena, play a less important role in imaging the hip joint.

In this chapter, we summarize our experience with MR in imaging of the hip joint. Applications and limitations of MR in assessing osteonecrosis, transient osteoporosis, congenital dysplasia, Legg-Calvé-Perthes disease, hip trauma, effusion, osteoarthritis, rheumatoid arthritis, juvenile chronic arthritis, septic arthritis, pigmented villonodular synovitis, synovial osteochondromatosis, bursae, and herniation pits of the hip are described. The influence of metallic hardware on MR images of the hip joint also is discussed.

IMAGING PROTOCOLS AND HARDWARE

Magnetic resonance imaging protocols for evaluating patients with hip disorders vary depending on the clinical situation. A standard imaging protocol starts with a T1-weighted axial localizer scan (short repetition time [TR], short echo time [TE]; e.g., TR of 300 to 500 ms, TE of 20 to 30 ms, 1 excitation 256 × 128 matrix). The axial localizer sequence is followed by T1-weighted (TR, 400 to 800 ms; TE, 20 to 30 ms) and T2-weighted (TR, 1600 to 2500 ms; TE, 80 to 120 ms) scans in the coronal plane. These scans are acquired with a finer matrix consisting of 256 × 192 or 256 × 256 elements. Because the number of slices obtained with each acquisition is a function of the TR, a TR should be chosen on the T1-weighted spin-echo sequence that is long enough to provide enough images to cover the entire anatomic region in question.

Spin-echo sequences have proved advantageous for imaging the hip. T2-weighted spin-echo images help to improve diagnostic specificity.[59,63] In selected cases, T2-weighted spin-echo images may be replaced by a fast T2*-weighted gradient-echo sequence that helps to reduce the total imaging time. Short T1 inversion-recovery (STIR) sequences can be employed to suppress the signal that arises from normal, fatty marrow.[91,110] Structures with long T1 and T2 relaxation times (e.g., granulation tissue or joint effusion) produce high signal intensity with this imaging sequence, thereby improving contrast between normal marrow and pathologic tissues (Figs. 10–1, 10–2).[91,110] Similarly, fat-selective and water-selective chemical shift imaging techniques help to improve tissue specificity when detailed morphologic information is needed by accurately depicting areas that are characterized by a high fat or water content.[79]

Additional T1-weighted axial sequences are recommended when the exact anteroposterior position and extent of a lesion within the femoral head must be evaluated (Fig. 10–3). Sagittal imaging is useful in assessing acetabular coverage in congenital dyspla-

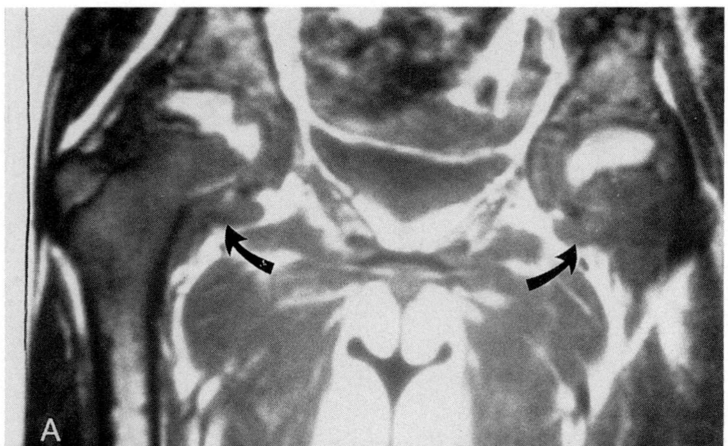

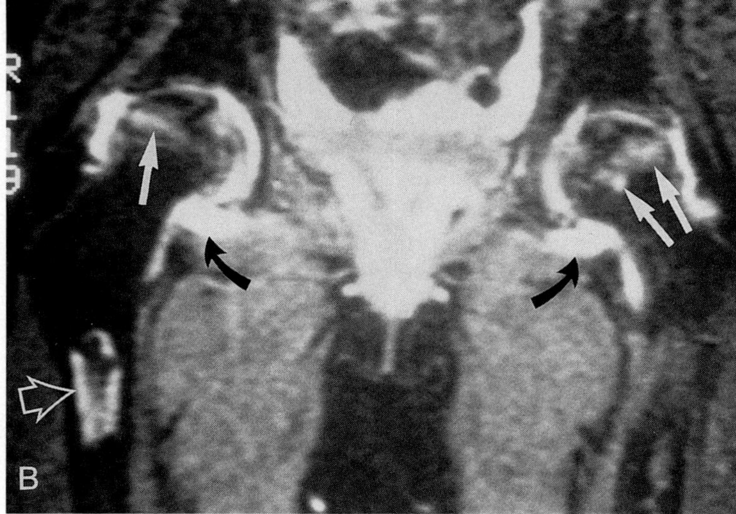

Figure 10–1. A comparison of spin-echo and short tau inversion-recovery (STIR) imaging of an adolescent patient with leukemia and chemotherapy and steroid-induced bone infarcts. *A*, T1-weighted spin-echo image demonstrates bilateral distention of the joint capsule *(curved arrows)*, reflecting joint effusion. The femoral epiphyses appear normal with this imaging sequence. *B*, STIR image. The normal medullary cavity is of low signal intensity with this imaging sequence. Both proximal epiphyses demonstrate irregularly shaped areas of abnormal, high signal intensity corresponding to chemotherapy- and steroid-induced bone infarcts *(straight arrows)*. A large bone infarct is also seen in the right femoral shaft *(open arrow)*. Note large bilateral joint effusions displaying high signal intensity with this sequence *(curved arrows)*. Contrast between normal marrow (low intensity) and pathologic tissues (i.e., necrotic areas and effusion) (high intensity) is very high.

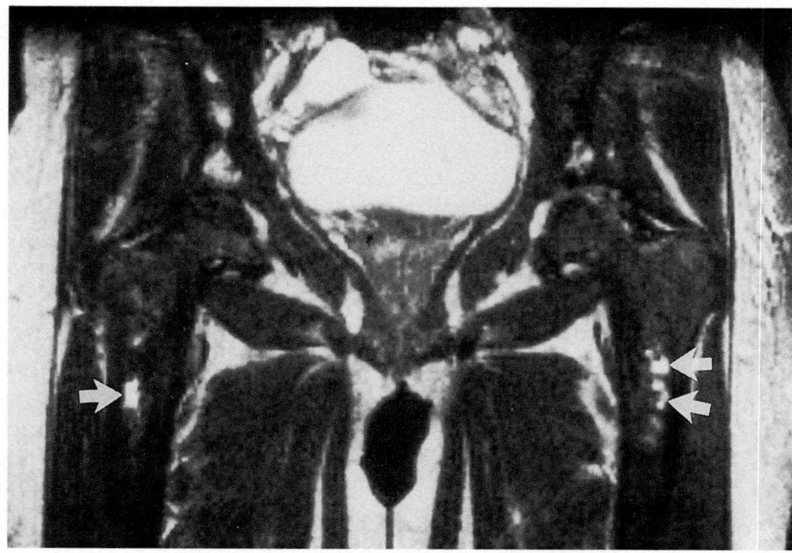

Figure 10–2. A short tau inversion-recovery (STIR) image of a patient with sickle cell anemia. The normal marrow demonstrates low signal intensity on STIR. Hemorrhagic bone infarcts which are of high signal intensity are seen in both femoral shafts *(arrows)*.

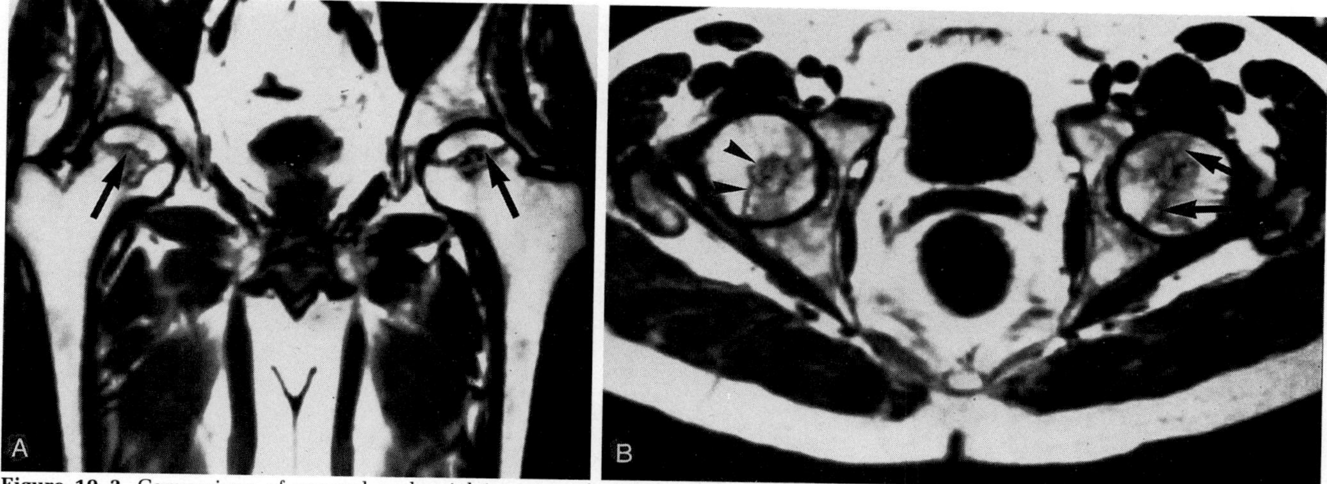

Figure 10–3. Comparison of coronal and axial imaging of osteonecrosis of the hip. *A*, Coronal T1-weighted spin-echo MR image demonstrates bilateral areas of signal abnormality in a patient with radiographically proved osteonecrosis. Bands of low signal intensity *(arrows)* are shown to surround subchondral areas (high signal intensity). *B*, Axial T1-weighted spin-echo MR image demonstrates the anteroposterior extent of necrotic lesion in both hips. On the right side, the necrosis involves primarily the posterior portions of the femoral head *(arrowheads)*. On the left side, both anterior and posterior portions are affected *(arrows)*.

sia of the hip. Sagittal images also are of use when an osteotomy is contemplated.

Most hip MR imaging studies are obtained with circumferential body coils. When high anatomic detail is required, single or dual surface coils can be employed with small fields of view (16 to 20 cm). When surface coil images are obtained, sagittal images can provide valuable information about the condition of the articular cartilage and the underlying subchondral bone.

IMAGING IN THE PRESENCE OF ORTHOPEDIC IMPLANTS

Computed tomography is severely limited and often nondiagnostic in patients who have been treated with metallic implants, owing to beam hardening and resultant streaklike artifacts. These streak artifacts may be limited to the area adjacent to the metallic implant, but, more frequently, the entire image is degraded. Technologic advances have made possible reduction of metallic artifacts in CT using special postprocessing algorithms.[53] However, these algorithms are time consuming, not widely available, and not always successful in eliminating metallic artifacts.[53] In evaluating the hip, metallic artifacts represent a major obstacle in CT because many hips disorders (e.g., osteonecrosis, osteoarthritis, and rheumatoid arthritis) occur bilaterally. Thus, when the patient has a metallic implant for treatment of one of these disorders in one hip and returns for follow-up with a symptomatic contralateral hip, CT frequently cannot be used for diagnosis because the entire CT image may be degraded from streak artifacts.

The presence of certain metallic implants (e.g., vascular clips and pacemakers) frequently represents a contraindication for body MR imaging. Several of the considerations that may prohibit MR imaging if the patient has a metallic implant are, however, not valid in deciding whether to image orthopedic hip implants. Prosthetic devices of the hip or plates for internal fixation are made of mostly nonferromagnetic alloys. In addition, they are attached rigidly to the bone so that motion of the implant secondary to the magnetic forces generally would not represent a problem. Local heating of the implant may occur in gradient-echo sequences or any other sequence with high radiofrequency (RF) power deposition. The temperature increase of the implant is, however, negligible overall in particular when RF is within the limits recommended by the U.S. Food and Drug Administration (FDA).[108,109] The only significant adverse effects of orthopedic implants on MR imaging of the hip are artifacts that degrade the image quality. Local eddy currents are induced in both nonferromagnetic and ferromagnetic orthopedic implants when they are exposed to the RF pulse. These eddy currents cause dephasing of spins in the adjacent tissue, with resultant artifactual signal loss. In most patients, however, these artifacts are limited to the area immediately adjacent to the implant. This is particularly true when nonferromagnetic materials have been used, as is the case with most prosthetic and other orthopedic devices.

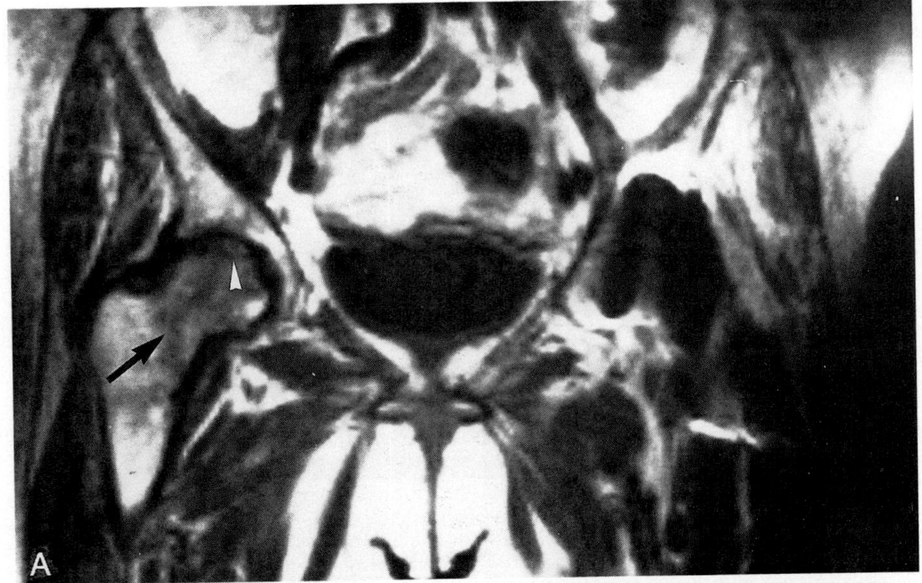

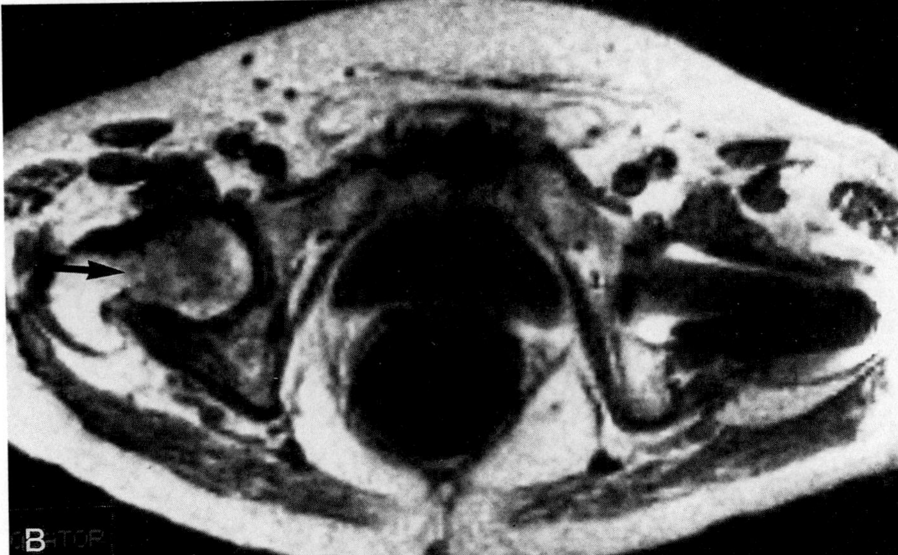

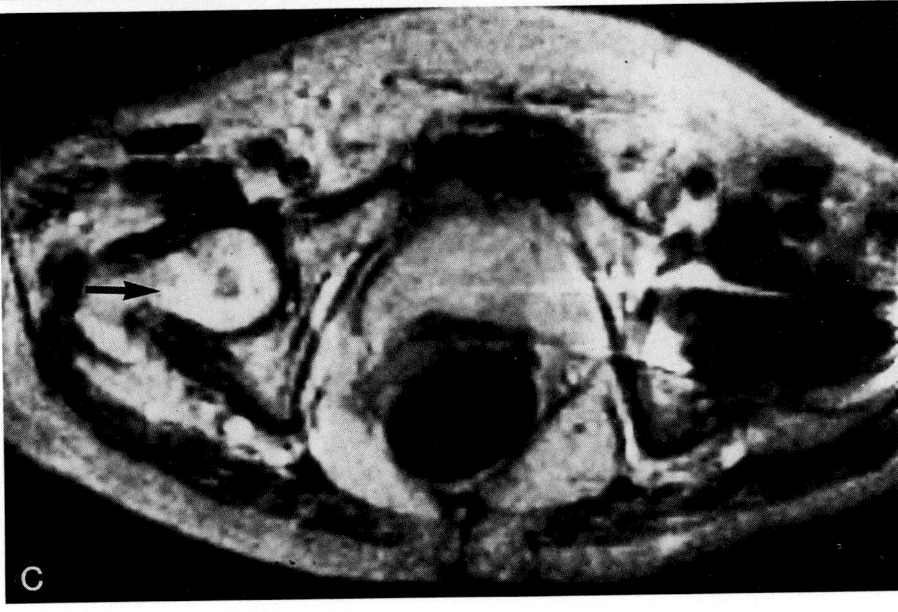

Figure 10–4. MR imaging in the presence of orthopedic implants. The patient was a 38-year-old man who had had a left total hip arthroplasty for treatment of osteonecrosis. Eighteen months after surgery, the patient developed symptoms in the other hip. *A*, Coronal T1-weighted spin-echo MR image demonstrates metal artifact in the left hip. The artifact is limited to the area immediately adjacent to the prosthesis. The right hip is not obscured by the artifact. The right hip shows a diffuse signal reduction *(arrow)*. A small focal, rimlike area of low signal intensity is demonstrated in the subchondral marrow consistent with early osteonecrosis *(arrowhead)*. *B*, Axial T1-weighted spin-echo image. The artifact is limited to the left hip. Diffuse loss of signal is seen in the right hip *(arrow)*. *C*, On T2-weighting, a marked increase in signal intensity is seen in the femoral head, extending into the femoral neck, consistent with diffuse bone marrow edema *(arrow)*. Bone marrow edema is an occasional finding in early cases of osteonecrosis.[123]

In patients with osteonecrosis of the femoral head who had unilateral total hip arthroplasty previously, follow-up MR imaging is therefore still useful for detecting osteonecrosis in the contralateral hip (Fig. 10–4). With few exceptions, the metal-induced artifact on MR images is significantly less pronounced than that on CT scans.

OSTEONECROSIS

Osteonecrosis of the hip results in partial or complete destruction of the femoral head. Osteonecrosis can be caused by local trauma to the hip, but it also can occur in absence of any prior traumatic condition. Traumatic osteonecrosis may be caused by femoral neck fracture or hip dislocation with resultant disruption or obliteration of the femoral head arteries, in particular the lateral epiphyseal arteries.[107] Nontraumatic osteonecrosis is a process with multifactorial bases that leads to local ischemia and subsequent necrosis of the cellular components of the bone and bone marrow.[54] Nontraumatic osteonecrosis is associated with a variety of disorders and predisposing factors, such as corticosteroid therapy, alcohol abuse, and hematologic and metabolic disorders.

Nontraumatic osteonecrosis is observed most frequently between the third and fifth decades of life. Bilateral involvement is found in up to 70 per cent of patients.[13,76] Osteonecrosis most often involves the anterosuperior portions of the femoral head.[73] The severity of the clinical course and its implications with respect to the patient's social life are further aggravated by the young patient age and frequent bilateral involvement.

Treatment modalities in osteonecrosis of the femoral head include joint-preserving surgery, hemiarthroplasty or total joint replacement, and, in rare cases, joint fusion. Joint-preserving surgery includes procedures such as core decompression, bone grafting, and osteotomy.[4,29,46,77,118] Most joint-preserving procedures must be performed at an early stage of osteonecrosis. This underlines the need for imaging modalities that allow early detection and delineate the morphology of an osteonecrotic lesion sufficiently.

Several different clinical and radiographic staging systems have been reported for osteonecrosis of the hip.[29,73,77] The most frequently used classification has been introduced by Ficat.[29] This system denotes five successive stages of osteonecrosis; stage 0 is preradiographic and preclinical. In stage 1, the patient has definite but usually preradiographic disease in one hip. In stage 2, the patient has pain, whereas radiographs demonstrate diffuse or localized areas of sclerosis or cysts. Stage 3 is characterized by the appearance of a bony sequestrum on the radiograph. A transition between stages 2 and 3 is evidenced by a crescentic line due to subchondral fracture and segmental flattening. In stage 4, progressive loss of articular cartilage and the development of acetabular osteophytes are observed.

In CT, early stages of osteonecrosis (i.e., Ficat stages 1 and 2) are characterized by a change of the asterisk sign.[24,25] The asterisk sign is a starlike figure that is formed by the weight-bearing trabeculae, seen on cross-sectional axial images through the femoral head. Typical alterations in the asterisk sign in early osteonecrosis are central or peripheral clumping of the asterisk and focal sclerosis.[24,25] In early cases of osteonecrosis, in which conventional radiography may be equivocal, CT has been used to verify or rule out osteonecrosis. Computed tomography affords improved demarcation of necrotic sequestra, detection of loose bodies in the hip joint as a result of sequestration, and delineation of cystic changes in the center of the femoral head (Fig. 10–5).[24,25,44] Computed tomography is, however, limited because

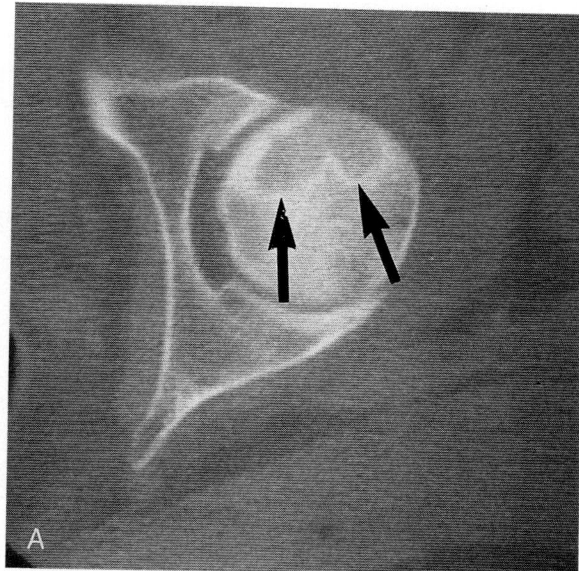

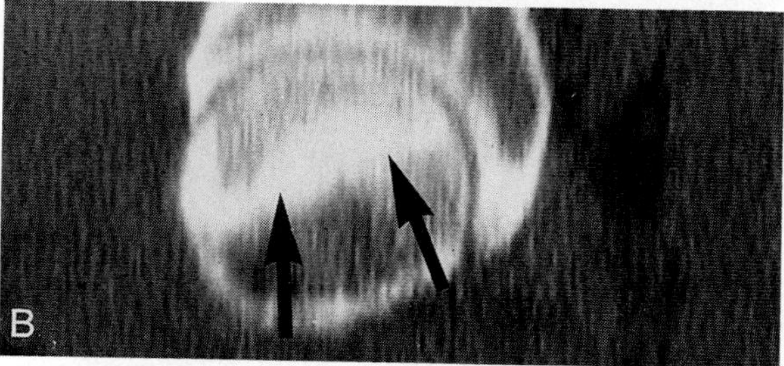

Figure 10–5. Computed tomography (CT) in osteonecrosis. *A*, Axial CT scan demonstrates a sclerotic rim *(arrows)* demarcating a radiolucent area in the anterior portion of the femoral head. *B*, Sagittal CT reformation. The sclerotic rim is clearly demonstrated *(arrows)*. Delineation of the position of the demarcated subchondral area *(arrows)* relative to the weight-bearing zone is improved.

it demonstrates only osseous (i.e., trabecular or subchondral) changes, which occur relatively late in the disease process.[34-36] Early medullary alterations, such as marrow necrosis and infiltration of granulation tissue, remain undetected on CT scans.

Nuclear scintigraphy is characterized by an excellent sensitivity but a relatively low specificity in diagnosing osteonecrosis of the femoral head.[9,74,81]

With MR, the normal nonnecrotic portions of the femoral head are characterized by a marrow space of high signal intensity on both T1- and T2-weighted spin-echo sequences. Linear striations of low signal intensity extending longitudinally from the top of the femoral head into the femoral neck correspond to the weight-bearing trabeculae. The medullary cavity of high signal intensity is surrounded by a sharply defined line of subchondral bone of low signal intensity.

In patients with manifest osteonecrosis, MR imaging demonstrates localized areas of signal abnormality that extend from the subchondral portions of the femoral epiphysis into the distal marrow cavity. The pattern and extent of MR imaging abnormalities display marked variation.[31,48,49,59,63,74,79,82,84,85,98,122] The abnormal areas may have the form of bands of low signal intensity or rings of low signal intensity with higher signal intensity centrally (Figs. 10–6 to 10–8). Segmental or wedge-shaped areas of low signal intensity have been described[74] (Figs. 10–9, 10–10). On both T1 and T2 weighting, these areas of low intensity may be homogeneous or heterogeneous, with patches of high signal intensity within a zone of low signal intensity[6,122] (see Fig. 10–9). In some patients, these areas may demonstrate an increase in signal intensity on T2-weighted MR images (see Fig. 10–9).[49,59,63,98]

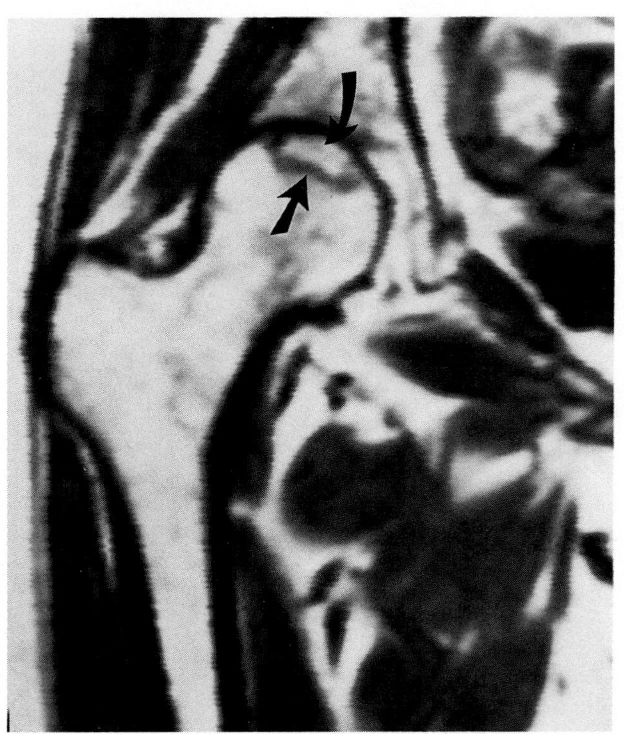

Figure 10–6. Band pattern of osteonecrosis. T1-weighted spin-echo MR image demonstrates a band of low signal intensity *(straight arrow)* surrounding a subchondral area of high signal intensity *(curved arrow)*.

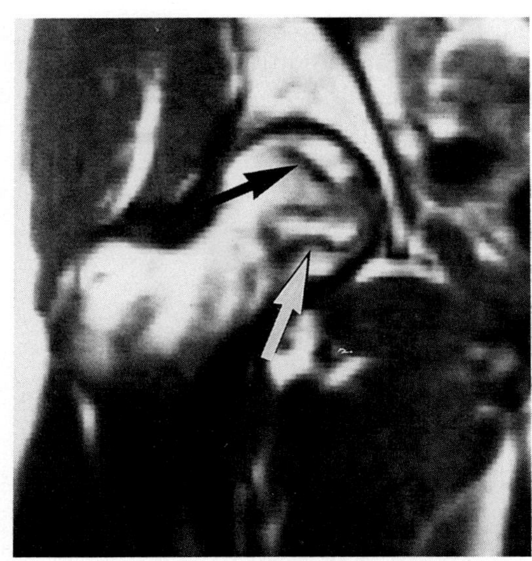

Figure 10–7. Ring pattern of osteonecrosis in a renal transplant patient who had a long-term history of corticosteroid use. T1-weighted spin-echo MR image demonstrates two rings of low signal intensity *(arrows)* surrounding central areas of high signal intensity.

Mitchell and co-workers described a "double-line sign" consisting of a band of low signal intensity with an inner border of high signal intensity on long TR-TE images (Fig. 10–11).[82] They believed that this sign may add specificity to the diagnosis of osteonecrosis by MR imaging.

Mitchell and colleagues employed an MR imaging classification scheme that identified four different types of signal abnormality in imaging osteonecrosis.[82] Class A regions demonstrate high signal intensity on T1-weighted images and intermediate signal intensity on T2-weighted images; class A signal is considered "fatlike." Class B lesions have high signal intensity on both T1- and T2-weighted images; they are classified as "bloodlike." Class C areas are characterized by low signal intensity on T1-weighted images and high signal intensity on T2-weighted images, corresponding to a "fluid-like" abnormality. Class D abnormalities have low signal intensity on T1 and T2 weighting, analogous to "fibrouslike" lesions.[82] Although this approach represents a first step toward a more tissue-specific evaluation of MR images, it does not entirely reflect the morphologic changes associated with osteonecrosis.

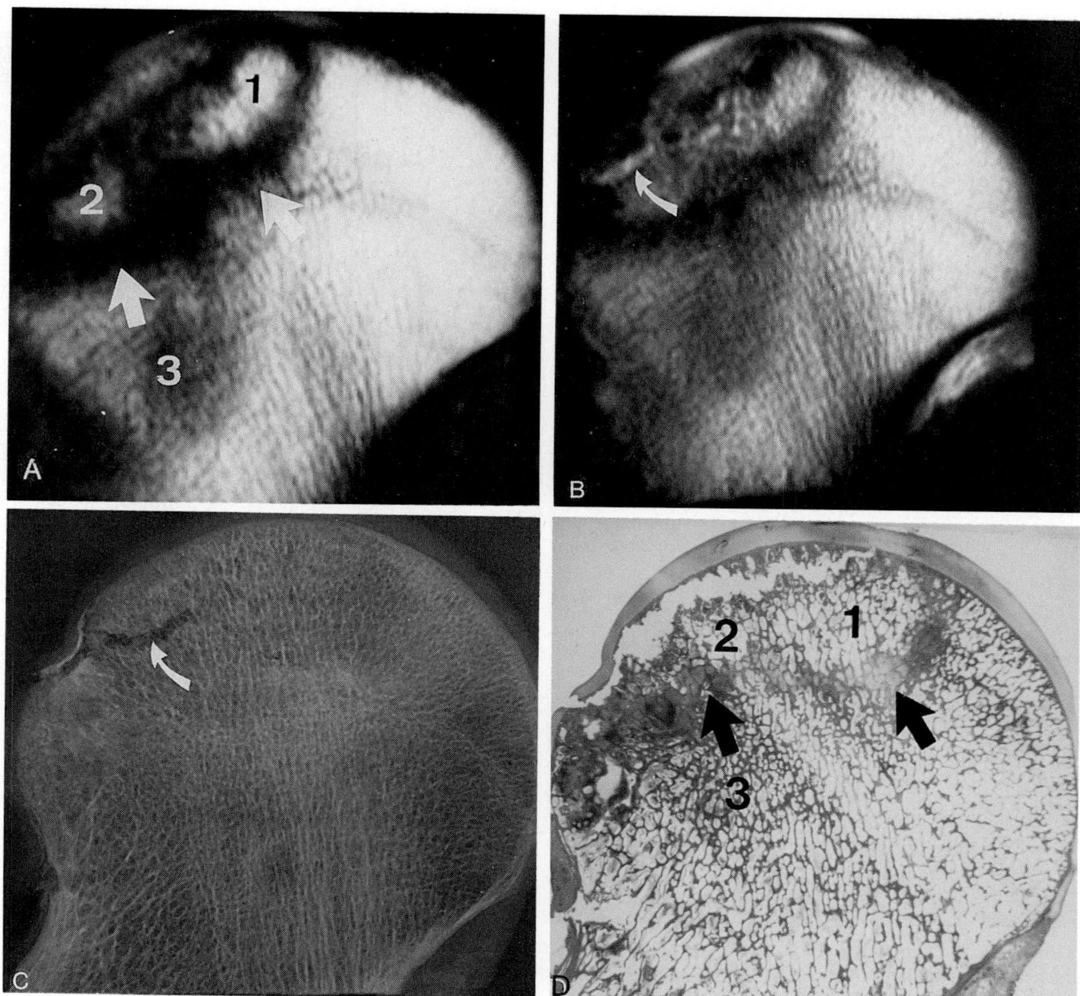

Figure 10–8. Alcohol-induced osteonecrosis of the femoral head in a 37-year-old man. *A*, T1-weighted spin-echo MR image demonstrates the ring pattern of osteonecrosis. A ring of low signal intensity *(arrows)* surrounds areas of intermediate to high signal intensity (1, 2). A large area of intermediate signal intensity is shown inferiorly (3). There are no signs of subchondral fracture. *B*, On the intermediate to T2-weighted spin-echo MR image, a line of high signal intensity *(curved arrow)* not visible on the T1-weighted image *(A)* is shown. *C*, Contact radiograph demonstrates Ficat stage 2 to 3 osteonecrosis with subchondral fracture *(curved arrow)* that corresponds to the line of high signal intensity on the intermediate-weighted image. Cystic and sclerotic changes are shown. *D*, Histologic section (hematoxylin-eosin). The ring of low signal intensity is composed of thickened trabecular bone, mesenchymal granulation and fibrous tissue, and amorphous cellular debris *(arrows)*. The zones of intermediate to high signal intensity inside the ring (1, 2) consist of necrotic bone and marrow. The inferior intermediate-intensity area (3) is composed of normal marrow and thickened trabecular bone. The subchondral fracture is artifactually enlarged from cutting.

When MR images were correlated with histologic macrosections, excellent agreement was found between imaging patterns and different histologic stages of osteonecrosis.[7,50,63] The initial phase of osteonecrosis is characterized by cellular ischemia with death of osteocytes and marrow cells (Fig. 10–12). With increasing morphologic degradation, necrotic debris appears in the intertrabecular spaces. Marked proliferation of mesenchymal cells and capillaries occurs in the marrow spaces of the living bone adjacent to the necrotic segment (see Fig. 10–12). Subsequently, mesenchymal cells differentiate into osteoblasts on the surface of the dead trabeculae. The osteoblasts commence to synthesize layers of new viable bone on top of the necrotic bone with resultant trabecular thickening. Osteoclastic activity leads to remodeling of repaired cancellous bone (see Fig. 10–12). During this period, structural failure of the weight-bearing trabeculae may occur, resulting in progressive collapse of the femoral head (see Fig. 10–12). As the repair process permeates the necrotic marrow spaces superiorly, different stages of repair are found in zones that lie between normal, viable bone and marrow and unrepaired necrotic tissue. In some cases, areas of necrotic bone and marrow are completely invaded by repair tissue. In other cases, the repair process ceases before permeating all necrotic zones.[34-36]

Magnetic resonance images reflect these histologic changes very well. A central area of high signal

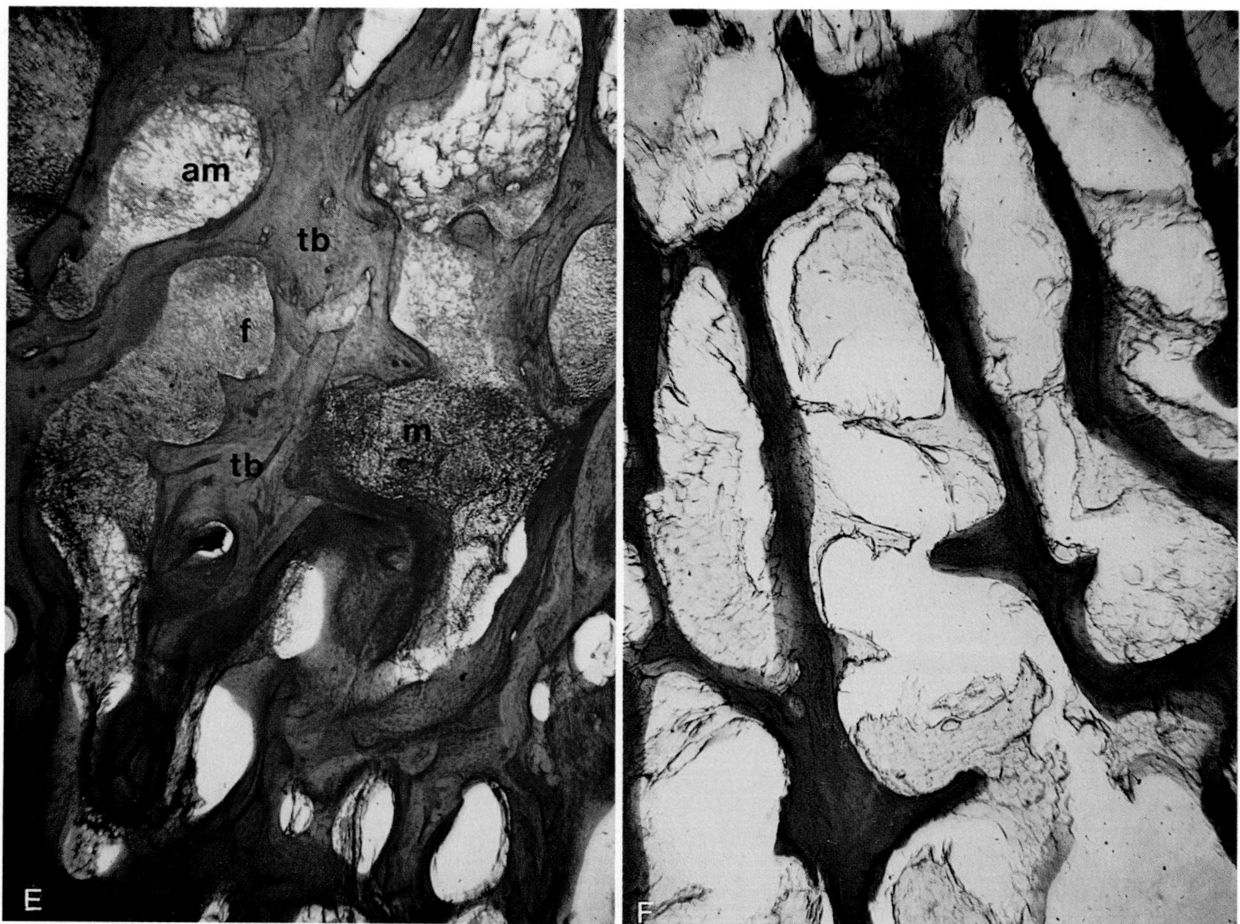

Figure 10–8 *Continued E,* Photomicrograph (hematoxylin-eosin, original magnification 150×) demonstrates the composite tissue of thickened trabecular bone (tb), amorphous cellular debris (am), mesenchymal (m) granulation, and fibrous (f) tissue corresponding to the ring of low signal intensity that separates viable from necrotic marrow. *F,* Photomicrograph (hematoxylin-eosin, original magnification 150×) shows the necrotic bone and marrow located within the ring of low signal intensity (areas 1 and 2 in *A* and *D*). The marrow structure is completely disrupted. (From Lang P, Jergesen HE, Moseley ME, et al: Avascular necrosis of the femoral head: High-field-strength MR imaging with histologic correlation. Radiology 169:517, 1988.)

intensity above or within a ring of low signal intensity corresponds to necrotic bone and marrow that have not been reached by invading capillaries and mesenchymal tissue (see Fig. 10–8). The band or ring of low signal intensity represents the repair tissue interface of mesenchymal and fibrous tissue, amorphous cellular debris, and thickened trabecular bone adjacent to the necrotic zone[50,63] (see Fig. 10–8).

When amorphous debris appears in the intertrabecular spaces and marked proliferation of mesenchymal cells and capillaries is present, segmental patterns of osteonecrosis with low signal intensity on T1-weighted images and increase in signal intensity in their distal portions on T2-weighted images are seen[50,63] (see Fig. 10–9). In this imaging pattern, areas of residual necrotic bone and marrow with amorphous intertrabecular debris and, sometimes, collapsed trabeculae are seen in the most superior portion of the femoral head, lending to low signal intensity on both T1 and T2 weighting (see Fig. 10–9*A*). Adjacent, more distal areas composed of mesenchymal granulation tissue and thickened trabecular bone demonstrate low signal intensity on T1-weighted images and high signal intensity on T2-weighted images (see Fig. 10–9*B*), owing to their high water content.[50,63]

When trabecular thickening from osteoblastic activity is the predominant histologic finding, areas of low signal intensity on both T1- and T2-weighted images are observed (see Fig. 10–10). Progressive collapse of the femoral head and necrotic debris in the intertrabecular spaces also contribute to this signal pattern.[50,63]

Subchondral fractures most often appear as gaps of low signal intensity on the MR images because of the lack of resonating protons (see Fig. 10–9*A*). In some patients, however, a fracture not apparent on T1-weighted images (see Fig. 10–8*A*) may be demonstrated by the presence of a line of high signal intensity on T2-weighted images (see Fig. 10–8*B*). The

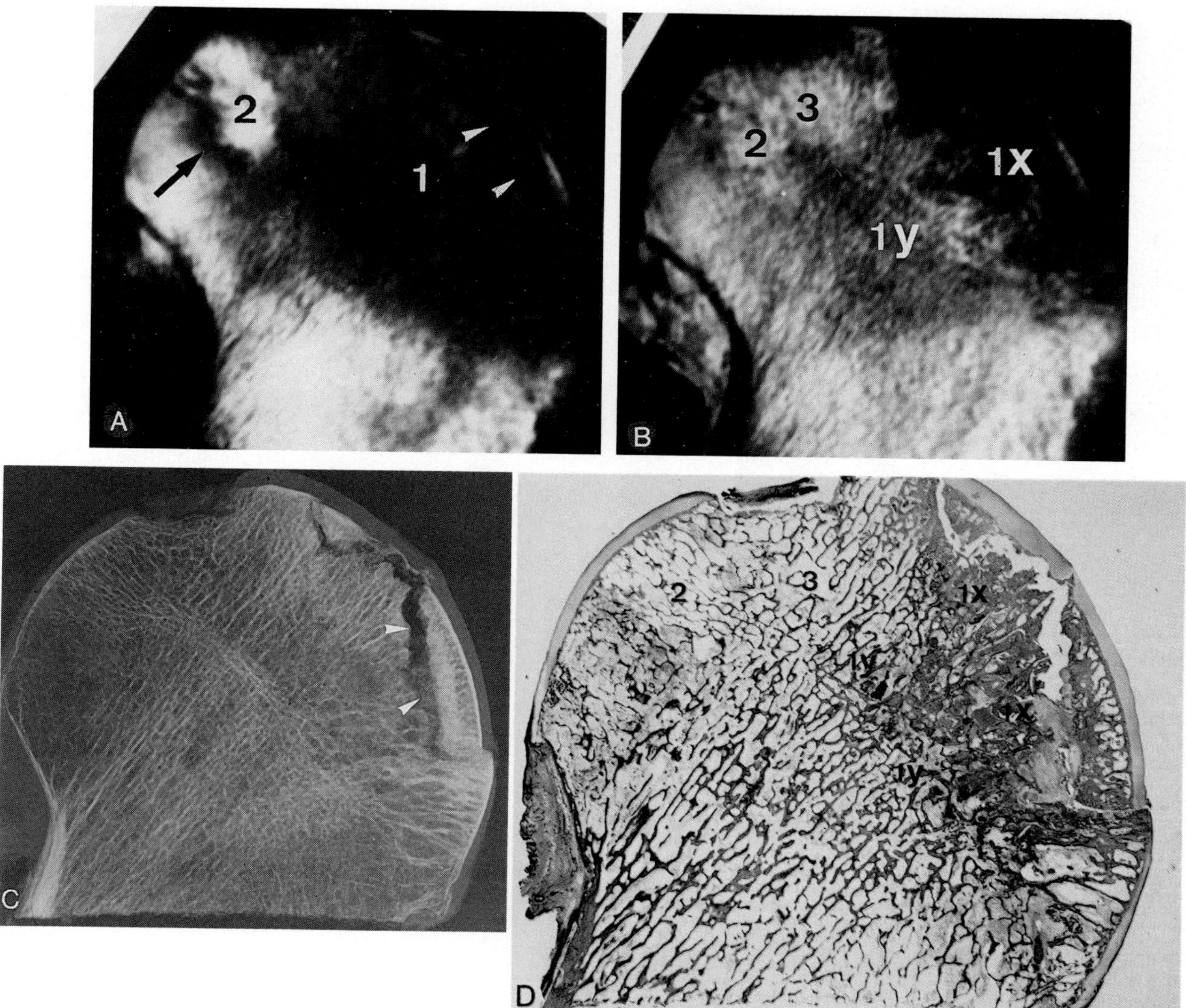

Figure 10–9. Osteonecrosis following steroid treatment in a 30-year-old man. *A*, T1-weighted spin-echo MR image demonstrates a large area of low signal intensity (1). Adjacent to it, but still demarcated from the marrow (normal-intensity) by a thin line of low signal intensity line *(arrow)*, is a small area of high signal intensity (2). A black gap in the superior portion of the specimen corresponds to a subchondral fracture *(arrowheads)*. *B*, The large area of low signal intensity on the T1-weighted image *(A)* changes on the intermediate- to T2-weighted image, leaving a superior area of low signal intensity (1x) and an inferior area of intermediate signal intensity (1y). An area of intermediate signal intensity (3) is shown below the fovea capitis. *C*, Contact radiograph demonstrates Ficat stage IV osteonecrosis with subchondral fracture *(arrowheads)* and collapse of the femoral head. *D*, Histologic section (hematoxylin-eosin). The area of low signal intensity on the intermediate-weighted image (1x) is composed of necrotic bone and marrow; amorphous cellular debris; and thickened, collapsed trabecular bone. The area of intermediate signal intensity on the intermediate-weighted image (1y) consists of thickened trabecular bone and mesenchymal granulation tissue. The area of high signal intensity on both sequences is composed of necrotic bone and marrow (2). The area below the fovea capitis is a mixture of normal bone marrow and granulation tissue (3).

high intensity of the signal in the fracture cleft on T2 weighting may correspond to intercellular or joint fluid filling the fracture cleft.[50,63]

Magnetic resonance imaging is unique in that it allows detailed evaluation of soft tissue changes in the marrow spaces in osteonecrosis. Different tissues can frequently be distinguished on the basis of their relative signal intensity characteristics and location (Table 10–1). The value of MR imaging in osteonecrosis is further enhanced by the fact that medullary changes occur histologically long before any alterations of the osseous components of the femoral head can be observed.[34,35] When information from signal intensities, configuration, and anatomic location is evaluated carefully, MR imaging may provide a detailed assessment of osteonecrosis that closely correlates with histologic alterations. On the basis of results of different studies that compared histologic and MR imaging findings, the following classification system of MR signal abnormalities has been developed that may help to improve the evaluation of osteonecrosis[59]:

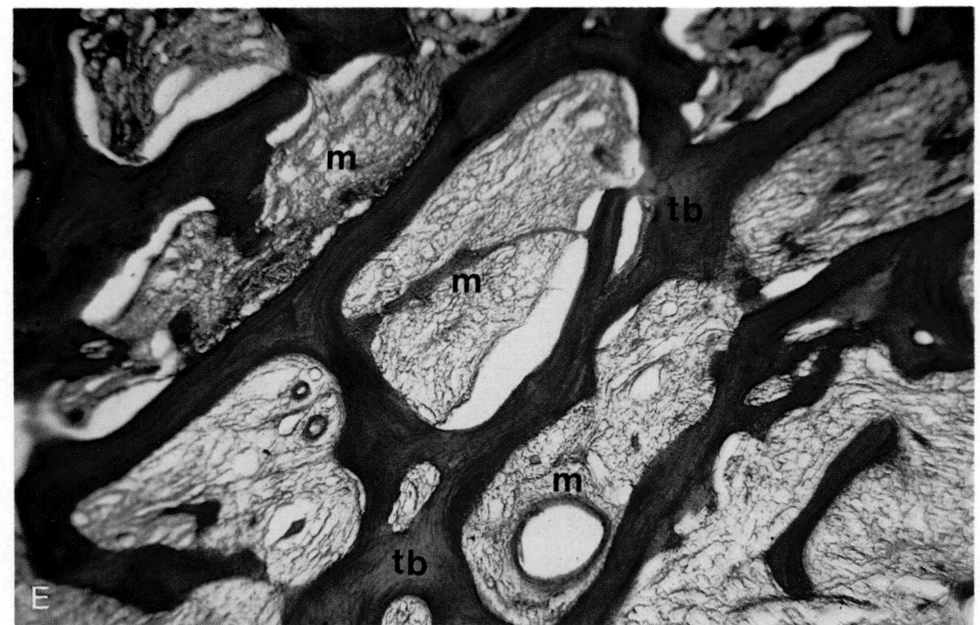

Figure 10–9 *Continued E,* Photomicrograph (hematoxylineosin, original magnification 150×) from the center of the femoral head demonstrates the mixture of mesenchymal (m) granulation tissue and thickened trabecular bone (tb) (area 1y in *B*). (From Lang P, Jergesen HE, Moseley ME, et al: Avascular necrosis of the femoral head: High-field-strength MR imaging with histologic correlation. Radiology 169:517, 1988.)

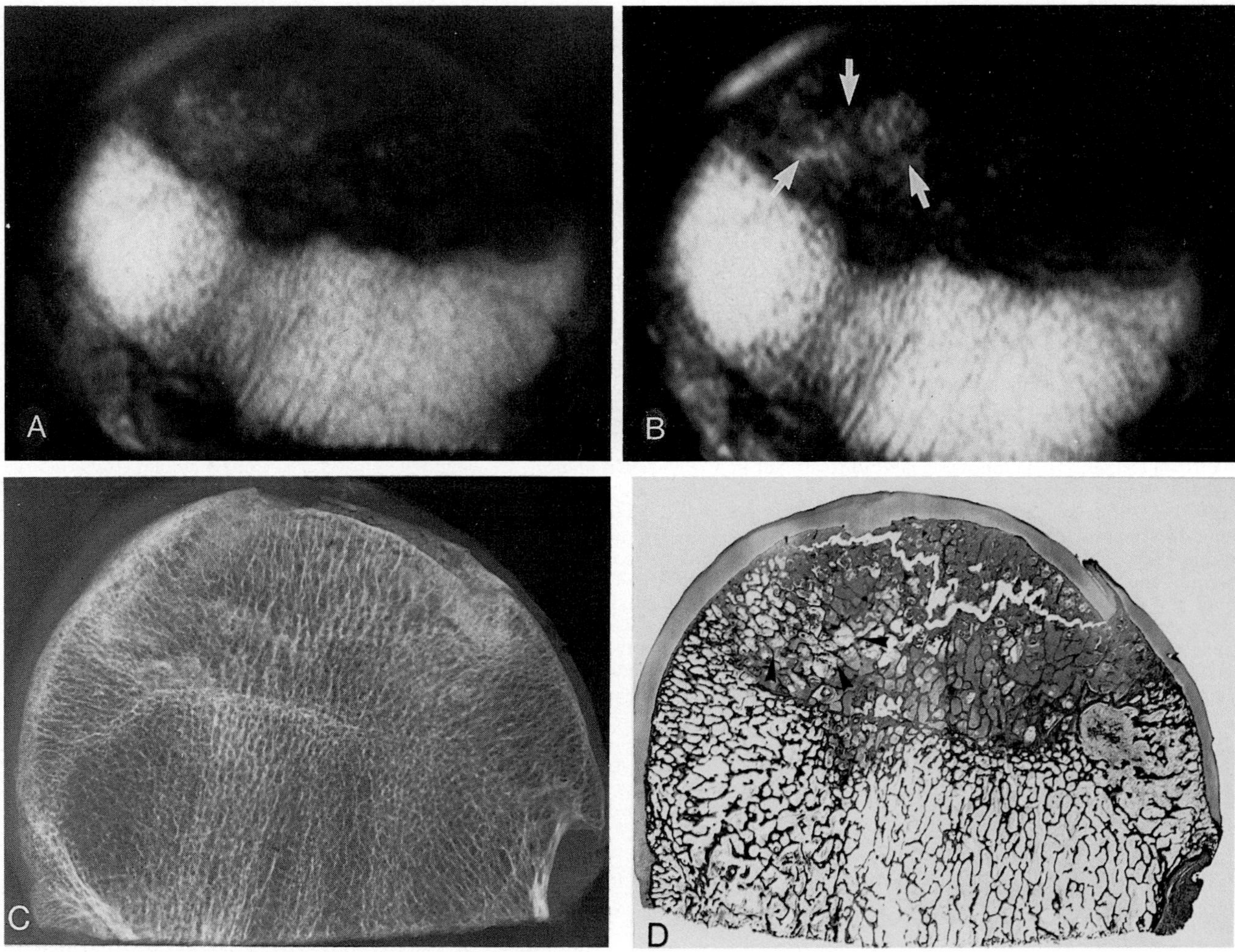

Figure 10–10. Steroid-induced osteonecrosis in a 31-year-old man. *A,* T1-weighted spin-echo MR image demonstrates a large segmental area of low signal intensity. *B,* On the intermediate- to T2-weighted image, small, spotty areas of intermediate signal intensity are demonstrated *(arrowheads).* The overall signal intensity, however, remains low. *C,* Contact radiograph demonstrates severe sclerotic changes with collapse of the femoral head. *D,* Histologic section (hematoxylin-eosin). The large area of low signal intensity on both T1- and intermediate-weighted images consists of thickened, collapsed trabecular bone with amorphous cellular debris in the intertrabecular spaces. The small areas of intermediate signal intensity on the intermediate-weighted image show some mesenchymal infiltration *(arrowheads).* (From Lang P, Jergesen HE, Moseley ME, et al: Avascular necrosis of the femoral head: High-field-strength MR imaging with histologic correlation. Radiology 169:517, 1988.)

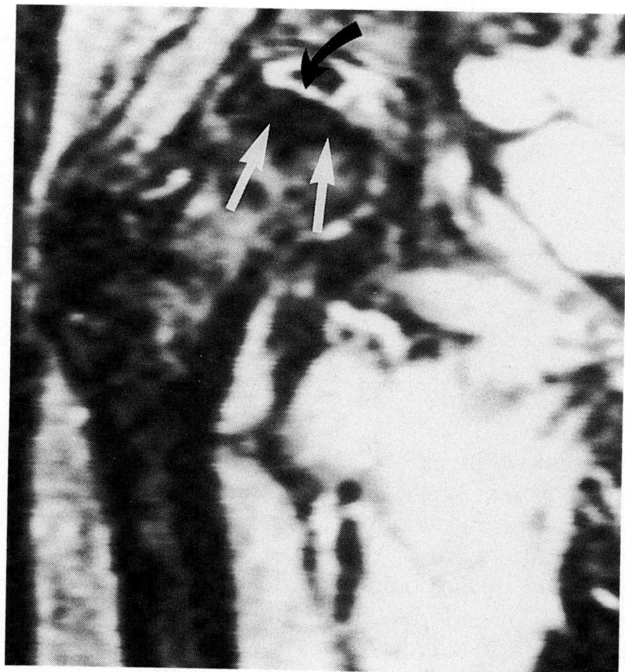

Figure 10–11. Osteonecrosis: double line sign. The short tau inversion-recovery (STIR) image demonstrates a band of low signal intensity *(straight arrows)* adjacent to an inner border of high signal intensity *(curved arrow)*.

1. **Type I:** Band or ringlike pattern of low signal intensity that surrounds a central zone of high signal intensity (see Fig. 10–8)
2. **Type II:** Segmental pattern with low signal intensity on T1-weighted images and increased signal intensity in the distal portions on T2-weighting (see Fig. 10–9)
3. **Type III:** Segmental pattern with low signal intensity on both T1- and T2-weighted images (see Fig. 10–10)

The three different types of MR signal abnormality identified in this classification reflect the different histologic stages observed in osteonecrosis of the femoral head. Because osteonecrosis is a dynamic disease that may show different pathohistologic stages in different portions of the femoral head, a transition between different types of MR signal abnormality can be expected in some patients. In general, the identification of any such type of MR signal abnormality may provide a more detailed and specific evaluation of the stage of osteonecrosis that

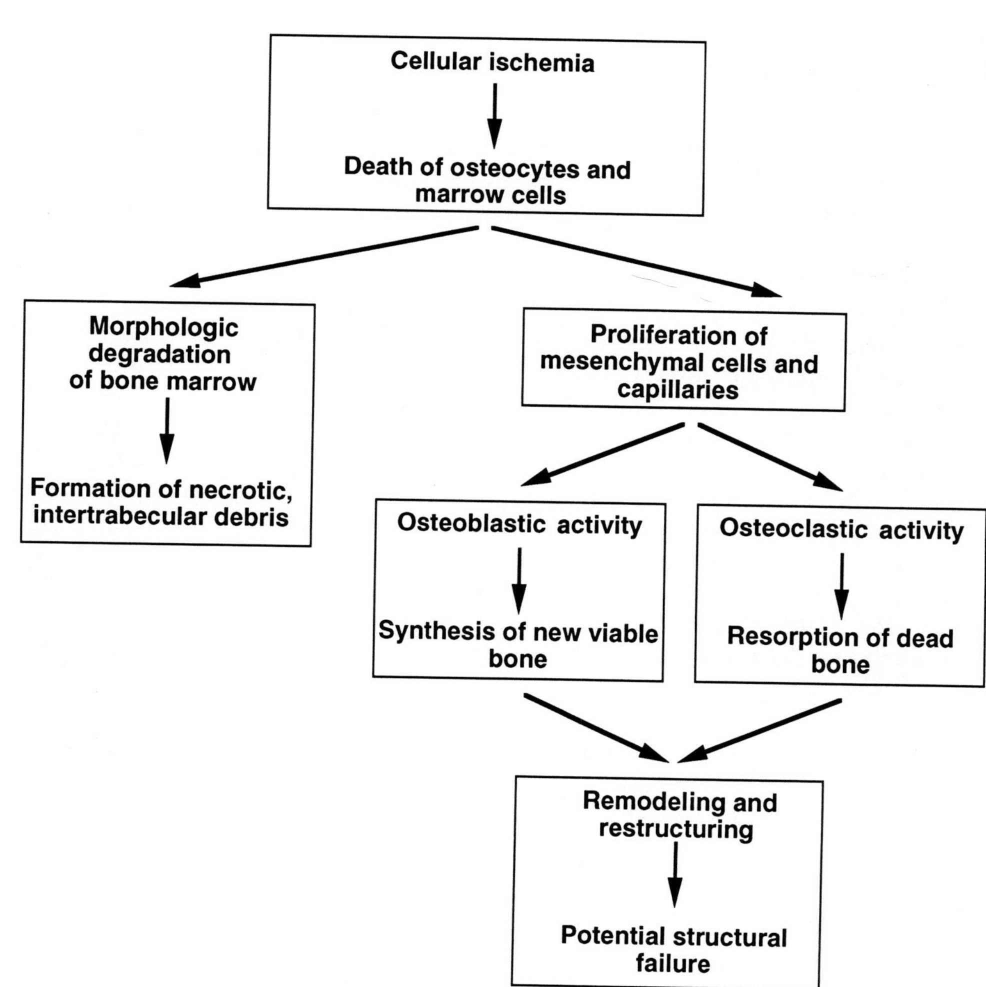

Figure 10–12. Histologic changes in osteonecrosis of the hip.

Table 10–1. MR SIGNAL INTENSITY CHARACTERISTICS OF DIFFERENT TISSUES IN OSTEONECROSIS

Tissue	Signal Intensity: T1 Weighting	Signal Intensity: T2 Weighting
Normal marrow	High	Intermediate to high
Necrotic marrow	High	Intermediate to high
Necrotic marrow and amorphous cellular debris	Low	Low
Mesenchymal granulation tissue (repair tissue)	Low	High
Sclerotic tissue	Low	Low

should ultimately benefit the patient by aiding in the selection of the most appropriate treatment method.

Turner and colleagues have described an MR imaging pattern in six patients with early osteonecrosis of the femoral head that differed from the three types that are typically seen in manifest osteonecrosis.[123] They found diffuse signal abnormality in the marrow of the femoral head and neck, which —unlike the case in manifest osteonecrosis— extended into the intertrochanteric region (see Fig. 10–4). The abnormal regions were of low signal intensity on T1-weighted and high signal intensity on T2-weighted images. This finding was attributed to bone marrow edema. No focal MR imaging abnormalities characteristic of osteonecrosis were observed. In our own experience, they may, however, be present, although discrete (see Fig. 10–4*A*). At core biopsy, Turner and co-workers found osteonecrosis in all six patients.[123] Conventional radiographs were normal or showed mild osteopenia. Although diffuse marrow abnormalities in the proximal femur are not specific and may be related to other disorders, such as transient osteoporosis (see later discussion), they may indicate the presence of osteonecrosis of the femoral head.[123] Diffuse marrow abnormalities of low signal intensity on T1 weighting and of high signal intensity on T2 weighting may thus represent an additional MR imaging pattern of osteonecrosis (i.e., type 0 in the classification scheme). However, further studies that include larger patient populations are needed to determine how strong the association is between this imaging pattern and osteonecrosis of the femoral head.

Several studies report the intravenous use of gadopentetate dimeglumine in the MR imaging of osteonecrosis.[65,125] Both nonenhancing and enhancing areas may be seen on postcontrast T1-weighted images. The zones with contrast enhancement demonstrate low signal intensity on T1-weighted precontrast scans and typically correspond to areas that show an increase in signal intensity on T2-weighted precontrast images.[65,125] Histologically, enhancing areas reflect viable hypervascular mesenchymal granulation tissue. Nonenhancing areas correspond to necrotic tissue. At present, it appears that T1-weighted images obtained after intravenous administration of gadopentetate dimeglumine do not add specificity to standard imaging protocols employing precontrast T1- and T2-weighted scans. However, in patients who are in pain and who cannot tolerate prolonged imaging, postcontrast MR images may be used to replace precontrast T2-weighted images, thereby decreasing the total imaging time.[65]

The great sensitivity of MR imaging in detecting osteonecrosis has been documented in many studies.[6,9,22,31,44,48,74,78,81,82,84,85,97,122] MR imaging is often positive for osteonecrosis when conventional radiography, and even scintigraphy and CT, are still negative.[22,31,48,74,84,85,97] In patient populations with symptomatic hip disorders, the sensitivity of MR imaging for detecting osteonecrosis of the femoral head has been found to vary between 88 and 100 per cent (Table 10–2). Markisz and colleagues[74] and Beltran and co-workers[9] reported that MR imaging is significantly more sensitive than scintigraphy in detecting osteonecrosis; the sensitivity of scintigraphy in diagnosing osteonecrosis was 10 to 20 per cent lower than that of MR imaging in these studies. Unlike scintigraphy, MR imaging also demonstrates the exact location and extent of the necrotic zone. Magnetic resonance imaging may, however, be limited in demonstrating subchondral fractures. Computed tomography can more accurately identify subchondral fractures and thus remains important for staging despite its lower diagnostic sensitivity.[80] Mitchell and colleagues compared the sensitivity of CT, scintigraphy, and MR imaging using receiver operating characteristic curves.[81] Diagnostic ability was measured as the area under the receiver operating characteristic curve. In their study, MR imaging was better than both CT and scintigraphy. In the patients with early osteonecrosis, the difference between MR imaging

Table 10–2. SENSITIVITY AND SPECIFICITY OF DIFFERENT DIAGNOSTIC MODALITIES FOR ASSESSING OSTEONECROSIS OF THE FEMORAL HEAD

	Sensitivity (Per Cent)					Specificity (Per Cent)				
	X-Ray	*CT*	*RNI*	*BMP*	*MR*	*X-Ray*	*CT*	*RNI*	*BMP*	*MR*
Beltran et al[9]	—	—	77.5	92	88.8	—	—	75	57	100
Genez et al[32]	—	—	11*/ 50*†	—	46	—	—	—	—	—
Glickstein et al[33]	—	—	—	—	—	—	—	—	—	98
Markisz et al[74]	—	—	81	—	100	—	—	100	—	100

X-ray, Conventional radiography; CT, computed tomography; RNI, radionuclide imaging; BMP, bone marrow pressure measurements; MR, magnetic resonance imaging.
* Bone scan with ^{99m}Tc methylene diphosphonate.
† Bone marrow scan with ^{99m}Tc sulfur colloid.

and CT exceeded 2 standard errors and that between MR imaging and scintigraphy exceeded 3 standard errors. Based on the results of this[81] and other studies,[6] MR imaging clearly is the most sensitive test for diagnosing early osteonecrosis.

Genez and associates studied seven patients who were at high risk for developing osteonecrosis.[32] Risk factors included steroid therapy and femoral neck fracture. The sensitivity of MR imaging for detecting osteonecrosis in these patients was only 46 per cent. These investigators concluded that the disease may be evident histologically before it is detected by MR imaging. The discrepancy with other studies[6,9,22,31,44,48,74,78,81,82,84,85,97,122] may be explained by a relatively long time interval between MR imaging and subsequent core biopsy. This interval reached up to 9 weeks, so that histologic changes may have well developed after MR imaging. More important, the histologic criteria for bone and marrow necrosis had not yet been defined clearly. Absence of osteocyte nuclei in the osteocytic lacunae alone, as observed in this study, does not represent proof of osteonecrosis, as it can be the result of the tissue fixation process.

MR imaging also is characterized by a very high specificity in differentiating osteonecrosis from other hip disorders. The specificity of MR approaches 100 per cent (see Table 10–2).

Tervonen and co-workers[120] used an abbreviated MR imaging protocol to determine the prevalence of clinically occult osteonecrosis of the hip in a series of 100 asymptomatic renal transplant patients treated with corticosteroids. All patients were older than 18 years, had been treated with corticosteroids for at least 6 months, and had no symptoms suggesting osteonecrosis before MR imaging. Of the 100 patients screened, six were found to have clinically occult osteonecrosis at MR imaging.[120] Tervonen and colleagues concluded that there is a significant prevalence of asymptomatic, clinically occult osteonecrosis among renal transplant patients treated with corticosteroids.[120]

Because MR imaging is expensive, we recommend that it be used in a cost-efficient manner in imaging osteonecrosis. Magnetic resonance imaging is best suited for diagnosing early cases of osteonecrosis, for which less expensive imaging modalities, such as conventional radiography and scintigraphy, may be equivocal. It appears also indicated in those patients in whom making a distinction between different hip disorders is difficult using other techniques. Magnetic resonance imaging may be employed before more costly interventional procedures, such as core biopsy and decompression, are performed.[6] An abbreviated, low-cost screening MR imaging examination consisting only of a T1-weighted spin-echo scan in the coronal plane can be used to detect clinically occult, asymptomatic osteonecrosis in high-risk patients. In the future, low-cost, screening MR imaging may improve patient care by detecting osteonecrosis early in its course, when it is still amenable to conservative treatment or joint-preserving surgical procedures, rather than waiting until total hip arthroplasty is required.[120] Prospective, longitudinal studies are needed to evaluate the clinical utility of low-cost screening MR imaging in osteonecrosis.

TRANSIENT OSTEOPOROSIS

Transient osteoporosis of the hip is a rare disorder of uncertain origin.[94] It may be related to the reflex sympathetic dystrophy syndrome and is commonly seen in young and middle-aged adults, particularly men. Hip pain related to transient

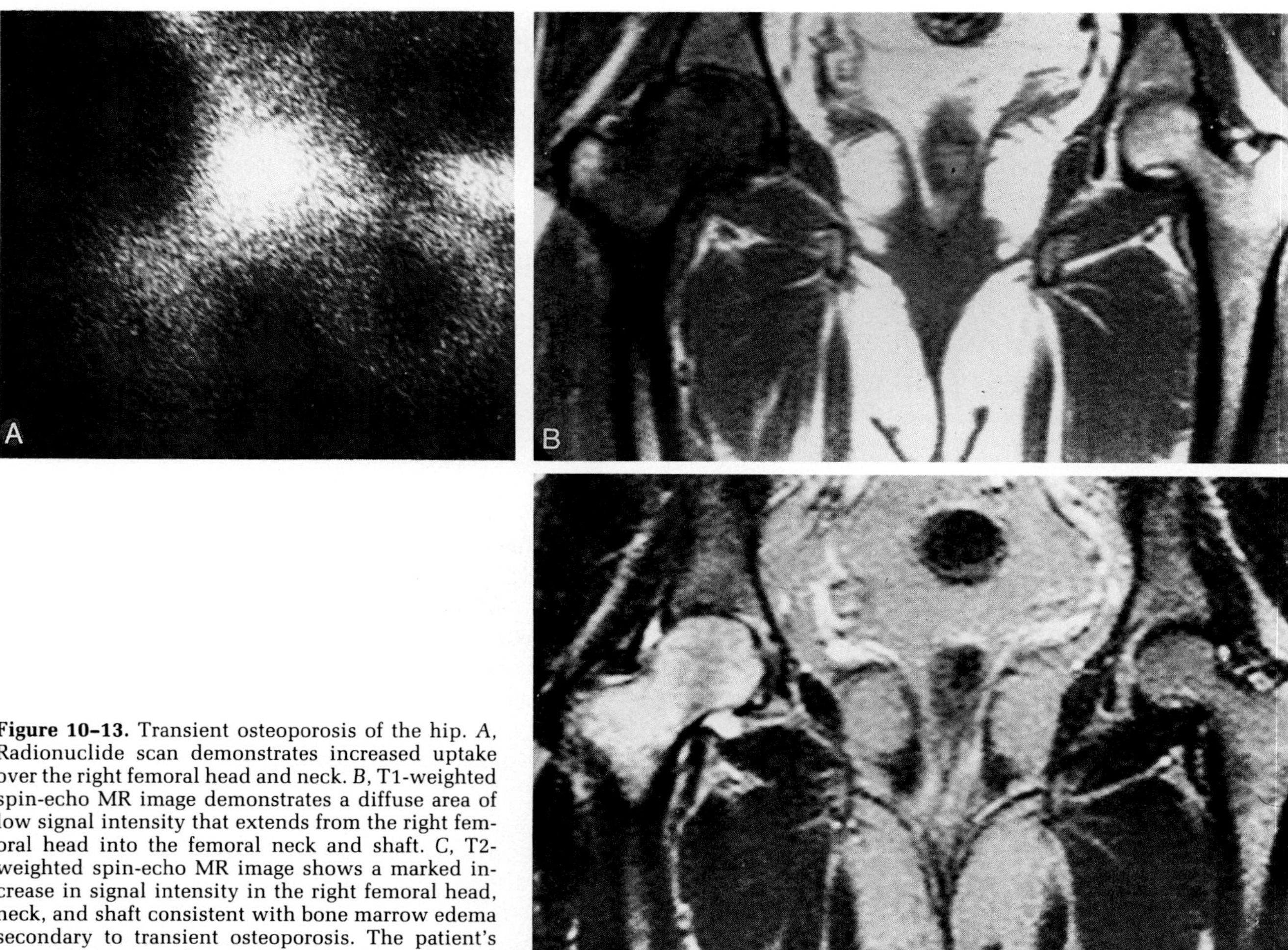

Figure 10–13. Transient osteoporosis of the hip. *A*, Radionuclide scan demonstrates increased uptake over the right femoral head and neck. *B*, T1-weighted spin-echo MR image demonstrates a diffuse area of low signal intensity that extends from the right femoral head into the femoral neck and shaft. *C*, T2-weighted spin-echo MR image shows a marked increase in signal intensity in the right femoral head, neck, and shaft consistent with bone marrow edema secondary to transient osteoporosis. The patient's symptoms regressed spontaneously 5 months after MR imaging was performed.

osteoporosis begins without a history of trauma or infection. Conventional radiographs usually demonstrate osteoporosis of the femoral head and neck.[45] Scintigraphy shows increased radionuclide uptake[30] (Fig. 10–13). Bone biopsy in patients with transient osteoporosis may demonstrate an increased bone turnover and inflammatory changes.[45] Synovial biopsy may be normal or show mild, chronic inflammatory changes.[5] Typically, the clinical and radiographic findings regress spontaneously within 6 to 12 months without any obvious morphologic defects.[94]

On MR imaging, diffuse signal abnormalities are observed in the femoral head and neck, which frequently extend into the femoral shaft (Figs. 10–13, 10–14).[11,39,40,124,128] The bone marrow in the abnormal areas demonstrates low signal intensity on T1-weighted images and high signal intensity on T2 weighting (see Figs. 10–13, 10–14). The degree of signal abnormality has been reported to vary markedly among patients.[11,128] The MR imaging findings are thought to reflect transient bone marrow edema associated with transient osteoporosis[128] (see Figs. 10–13, 10–14). Bloem reported that the MR imaging changes regressed completely after 6 to 10 months.[11]

The differential diagnosis of the MR imaging changes seen in transient osteoporosis includes osteomyelitis, neoplasm, septic arthritis, and avascular osteonecrosis. In most of the patients, clinical and laboratory findings, and eventually synovial cultures, help to differentiate transient osteoporosis from other disorders. Spontaneous regression of clinical, radiographic, and MR imaging findings will, however, help to verify the diagnosis. Follow-up MR imaging will help identify patients with transient osteoporosis because, unlike what occurs with neoplasm, signal abnormalities can be expected to regress.[11] At present, it is not certain whether transient osteoporosis may represent an abortive form of osteonecrosis that resolves spontaneously a few months after the initial marrow insult. This would explain the overlap of MR signal patterns between very early osteonecrosis without focal subchondral changes[123] and transient osteoporosis.[11,39,40,124,128]

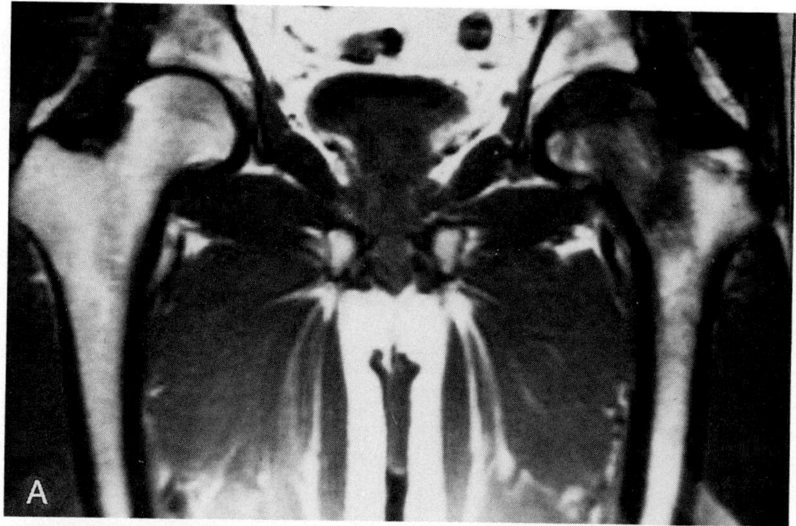

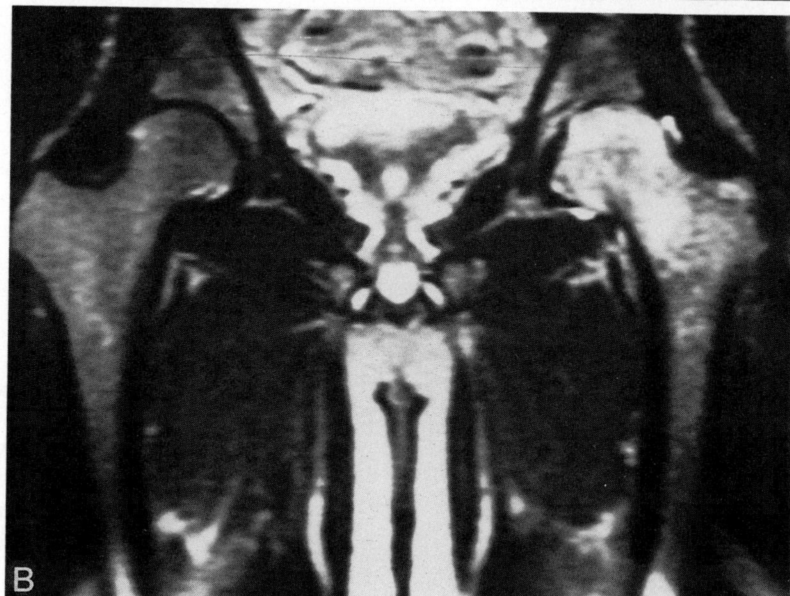

Figure 10–14. Transient osteoporosis of the hip. *A*, Coronal T1-weighted spin-echo MR image demonstrates a diffuse area of signal reduction in the left femoral head, neck, and shaft. The right proximal femur has normal, high signal intensity. *B*, T2-weighted spin-echo MR image demonstrates a marrow signal of high signal intensity in the left femoral head, neck, and shaft. This signal pattern—low intensity marrow signal on T1-weighting and high intensity marrow signal on T2-weighting—is thought to reflect bone marrow edema in transient osteoporosis. The patient's symptoms resolved 5 months after the MR examination.

CONGENITAL DYSPLASIA

Congenital dysplasia of the hip (CDH) or developmental dysplasia of the hip is a disturbance in the relationship of the femoral head and the acetabulum in three dimensions.[41] Congenital dysplasia of the hip is more frequent in females than in males. The rate of occurrence of CDH is increased in children with a family history of CDH. Three different categories of CDH have been defined[27,87]: In type I, the hip can be subluxed or is positionally unstable. In type II, the hip can be dislocated, and the labrum is everted. In type III, the femoral head is dislocated posterosuperiorly, the labrocapsular complex is infolded, and a pseudoacetabulum has formed.

Conventional radiographs frequently fail to display the anteroposterior relationships between the femoral head and the acetabulum before ossification of the capital epiphysis.[42,103] The adequacy of acetabular coverage is difficult to determine by routine conventional radiography, although it is of great importance for treatment decisions. Choices about splinting devices and pelvic or femoral osteotomies are better made when the extent of anterior and posterior coverage is known.

Although ultrasonography makes possible early detection of the anatomic relationships of the femoral head and acetabulum in infants with uncomplicated CDH without gonadal irradiation and is ideally suited as a screening tool,[38] acetabular architecture is frequently difficult to assess.

Axial CT has improved the understanding of anatomic relationships in CDH[19,42,43,89] (Fig. 10–15). However, although CT has been demonstrated to be more accurate than conventional radiography in detecting CDH, CT does not provide a clear distinction

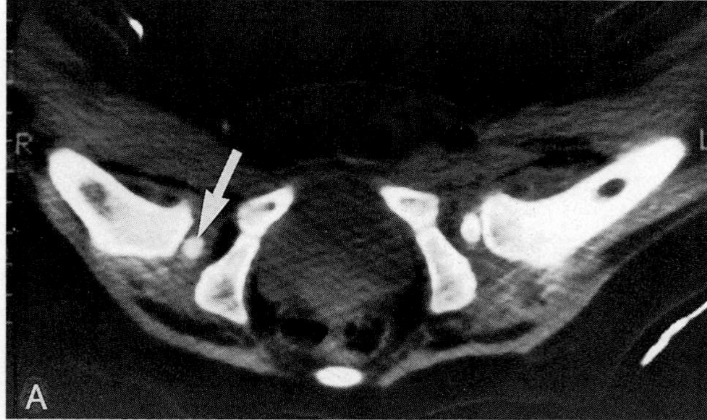

Figure 10–15. Congenital dysplasia of the hip. *A,* Axial computed tomography (CT) scan demonstrates mild subluxation of the right femoral head *(arrow).* Note that articular cartilage and surrounding soft tissues (e.g., joint capsule and ligaments) cannot be clearly differentiated on CT. *B,* Three-dimensional reconstruction in inferior projection confirms the presence of mild right-sided subluxation *(arrow).* The position of the femoral head relative to the acetabulum is shown. (From Lang P, Steiger P, Genant HK, et al: Three-dimensional CT and MR imaging in congenital dislocation of the hip: Clinical and technical considerations. J Comput Assist Tomogr 12:459, 1988.)

between the articular cartilage and the surrounding soft tissues (i.e., joint capsule, ligaments, muscles, and tendons)[19,42,43,89] (see Fig. 10–15). Thus, the cartilaginous, nonossified femoral heads in children younger than 6 months old or in children with delayed ossification cannot be directly demonstrated by CT, and their positions must be inferred by indirect signs. The ossified nuclei, however, must be present if imaging of anatomic relationships between the femur and the acetabulum is to be meaningful. Until adequate ossification has occurred, the diagnostic accuracy for detecting and evaluating CDH is suboptimal with CT[89] and even poorer with conventional radiography, in which displacement of the femoral head may not even be projected onto the plane of the radiograph.

Magnetic resonance imaging demonstrates both the nonossified cartilaginous femoral head and the ossified femoral head (Figs. 10–16 to 10–18). It is unique in that it demonstrates femoral head subluxation and dislocation as well as soft tissue changes that may prevent reduction (see Figs. 10–16, 10–17). The cartilaginous femoral head is shown as an area of intermediate signal intensity that is clearly separated from the surrounding tissue. Unlike any other imaging modality, MR imaging demonstrates the fibrocartilaginous labrum demarcated from the acetabular hyaline cartilage as a triangular structure with low signal intensity on both T1- and T2-weighted images[51] (see Fig. 10–16). Magnetic resonance imaging also displays the capsular insertion at the bone margin of the ilium just superior and lateral to the base of the labrum.[51] It can be used to assess labral hypertrophy or inversion, which may cause failure of closed reduction.[15,17,51] The ligamentum teres is demonstrated as a structure of low signal intensity that originates from the transverse acetabular ligament and inserts on the fovea capitis. The transverse ligament also is of low signal intensity on both T1- and T2-weighted images. In CDH, the transverse ligament may infold, thus impeding reduction.[15] Other structures that may be interposed between the femoral head and the acetabulum are the pulvinar, hypertrophied transverse ligament, and iliopsoas tendon, which also are delineated on MR imaging.

Osteonecrosis can be a crippling complication in the patient with CDH.[92] Magnetic resonance imaging is unique in detecting early osteonecrosis[22,31,48,74,84,85,97] and, therefore, makes early initiation of therapy possible. In the patient with CDH, MR imaging can be helpful in preventing the debilitating changes associated with advanced osteonecrosis.

Magnetic resonance imaging is also advantageous in assessing acetabular coverage (see Figs. 10–16 to 10–18). Conventional radiography demonstrates only osseous coverage, but MR imaging also displays coverage by acetabular cartilage and the labrum. It can be employed to evaluate the result of closed or open reduction. Magnetic resonance imaging appears particularly indicated when a series of follow-up studies are needed in female patients, as the ovaries, located at the level of the superior acetabulum, cannot be shielded.

Magnetic resonance imaging is limited in cases of CDH in that it can show only a two-dimensional display of what frequently are complex three-dimensional (3-D) relationships between the femoral head and the acetabulum. Conventional planar two-dimensional MR has been supplemented by 3-D reconstruction,[60,61,66] which demonstrates femoral subluxation and dislocation in three dimensions (see Fig. 10–18). The position of the femoral head after closed or open reduction is also evaluated easily. Three-dimensional displays are particularly useful in demonstrating the extent of acetabular cartilage,

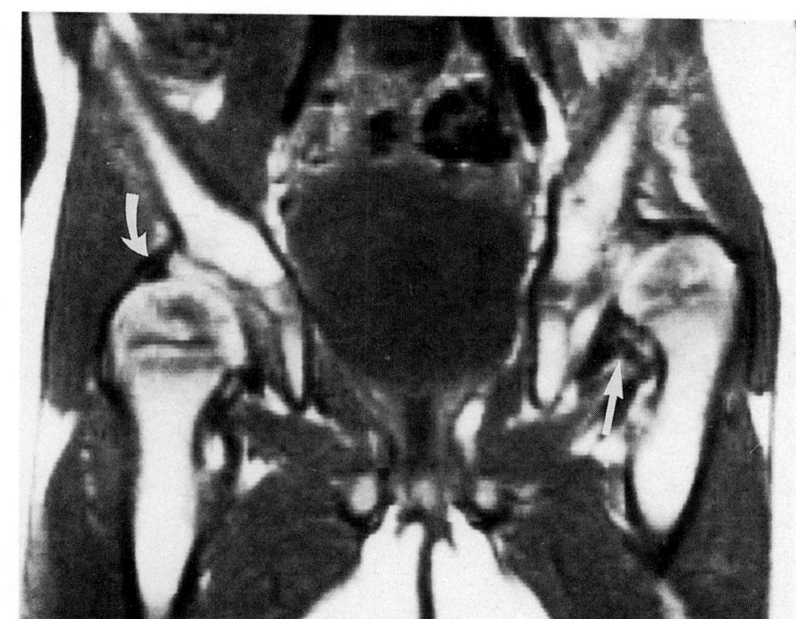

Figure 10–16. Congenital dysplasia of the hip. T1-weighted spin-echo MR image demonstrates lateral and superior displacement of the left femoral head. The joint capsule is infolded *(straight arrow)*. Note normal, triangular labrum of low signal intensity in the right hip *(curved arrow)*.

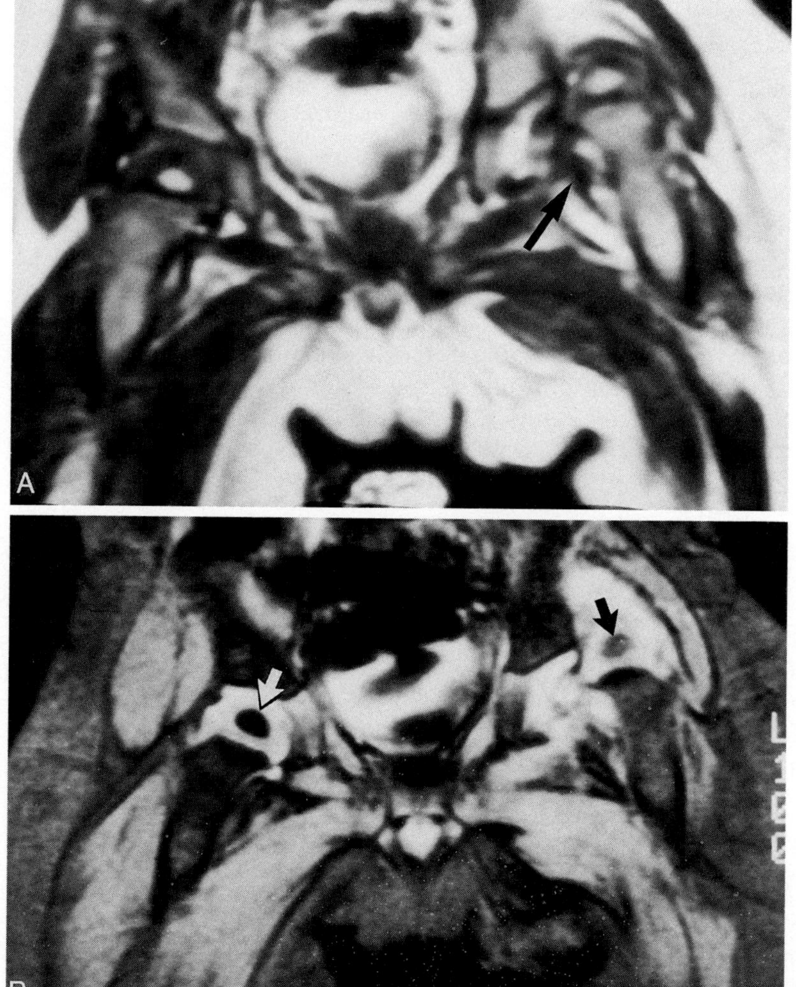

Figure 10–17. Congenital dysplasia of the hip. *A*, T1-weighted spin-echo MR image demonstrates lateral and superior displacement of the left femoral head. The joint capsule is hypertrophied and slightly infolded *(arrow)*. *B*, T2*-weighted gradient-echo image shows cartilaginous femoral heads of high signal intensity with ossified nuclei of low signal intensity *(arrows)*.

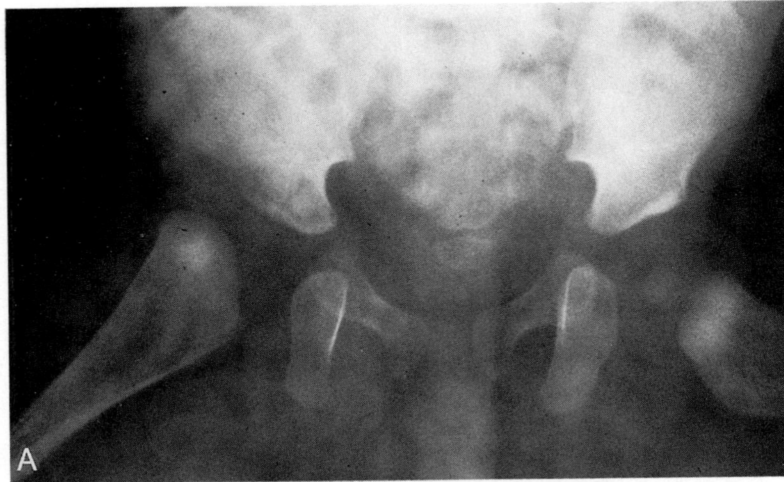

Figure 10–18. Congenital dysplasia of the hip in a 5-month-old patient. *A,* Anteroposterior radiograph. The acetabular roof is steeply inclined on the right side. Shenton's line is broken owing to subluxation. The right femoral head is not yet ossified. *B,* Three-dimensional reconstruction of MR images demonstrates both the cartilaginous femoral head and the osseous portions of the proximal femur. The right femoral head *(arrow)* is dislocated posterosuperiorly. The left femoral head is located in the acetabular fossa. (From Lang P, Steiger P, Genant HK, et al: Three-dimensional CT and MR imaging in congenital dislocation of the hip: Clinical and technical considerations. J Comput Assist Tomogr 12:459, 1988.)

on which the mode of therapy may be based. In children younger than 6 months old, 3-D MR imaging reconstructions allow visualization of the cartilaginous femoral head, which is not possible with 3-D CT[60,61,66] (see Fig. 10–18). In the future, 3-D MR imaging reconstructions may be useful in problem cases of CDH when complex corrective surgery is contemplated.

LEGG-CALVÉ-PERTHES DISEASE

Legg-Calvé-Perthes disease is an idiopathic form of osteonecrosis of the femoral head that is frequently encountered in children between the ages of 4 and 8 years.[95] Boys are affected more often than girls. Radiographically, fissuring, fracture, lateral displacement, flattening (Fig. 10–19), and sclerosis of the incompletely ossified epiphysis may be observed.[95]

Magnetic resonance imaging is useful for evaluating Legg-Calvé-Perthes disease.[12,16,28,96,99,104,121] It can detect this disorder when conventional radiographs and scintigraphy are still negative.[12,16] Similar to osteonecrosis in the adult patient, infarction of the capital femoral epiphysis in Legg-Calvé-Perthes disease is manifested by signal-deficient areas in the proximal part of the femur (Figs. 10–20, 10–21). Magnetic resonance imaging demonstrates the cartilaginous portions of the femoral capital epiphysis and acetabulum directly and noninvasively and can thus be used to evaluate the congruity of the acetabular and femoral articular surfaces.

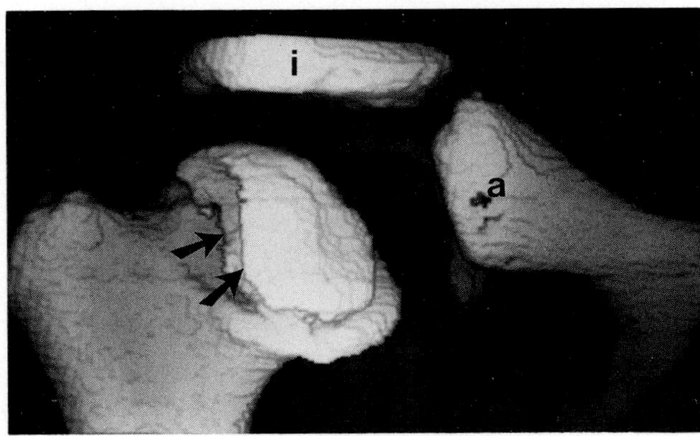

Figure 10–19. Legg-Calvé-Perthes disease. Three-dimensional computed tomography (CT) reconstruction demonstrates fissuring *(arrows)* and flattening of the femoral head. a, Anterior branch of acetabulum; i, ilium.

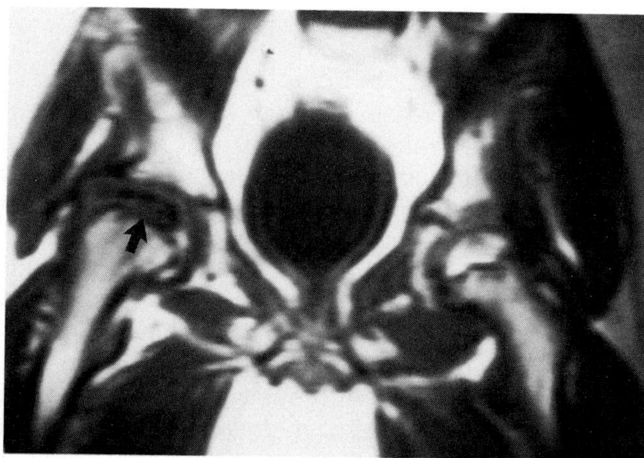

Figure 10–20. Legg-Calvé-Perthes disease. T1-weighted spin-echo MR image demonstrates abnormal low signal intensity in the right femoral epiphysis *(arrow)*. The right femoral head is flattened. The left femoral epiphysis has normal high signal intensity.

Rush and colleagues[99] demonstrated a statistically significant increase in both acetabular and femoral cartilage thickness in a study involving 20 patients with Legg-Calvé-Perthes disease. The increase in cartilage thickness was thought to reflect a response of the cartilage of the hip to the local event of bone infarction and subsequent subchondral fracture or a generalized disorder of cartilage in this disease. Loss of containment of the femoral head in the acetabulum was observed in most patients. A frondlike structure adjacent to the inferomedial joint space seen in seven patients may have represented villous hypertrophy of the synovium within the iliopsoas recess of the joint capsule, causing subluxation.

Magnetic resonance imaging is ideally suited to evaluate the acetabular and femoral cartilage in Leggs-Calvé-Perthes disease and provides unique information on femoral head containment.[28] Complications such as chondrolysis, synovitis, and deformity of the femoral head as well as soft tissue irregularities associated with this disorder are well demonstrated using MR imaging.

TRAUMA

Fractures of the acetabulum and proximal femur frequently are complex. Evaluation of joint stability and configuration and extent of the fracture is critical for treatment planning. Conventional radiography is often limited in evaluating the condition of the femoral head and the acetabulum in patients with hip trauma. Assessment of the anterior and posterior acetabular rim may be improved using special views described by Judet and co-workers.[52] Evaluation of the joint space and the medial acetabulum remains difficult, however. In addition, the presence of severe pain and the condition of the patient may prevent getting oblique and lateral views to further evaluate the acetabulum and femoral head.

Computed tomography has been reported to reveal fractures not seen on the radiographs.[111] It demonstrates the anterior and posterior acetabular rim, the medial acetabulum, and the femoral head and neck (Fig. 10–22). When the posterior lip of the acetabulum is fractured, CT is helpful in determining the size and location of the fragment and assessing the stability of the remaining posterior wall of the acetabulum (see Fig. 10–22). Computed tomography detects loose fragments in the hip joint that may be obscured on conventional radiography.[111] The major limitation of CT is, however, its inability to demonstrate cartilaginous fragments.[101]

Magnetic resonance imaging can detect fractures when the results from conventional radiography are negative[23,68] (Fig. 10–23). Patients who have clinical suspicion of fracture of the proximal femur but normal radiographs may be screened by MR imaging for hip fracture[23] (see Fig. 10–23). Because T1-weighted MR images are usually sufficient for fracture detection, MR imaging can be used in a rapid, cost-effective manner that provides anatomically precise diagnosis of hip fracture. The fracture is demonstrated on T1-weighted MR images as an intramedullary line of low signal intensity that is continuous at some point with the cortex (see Fig. 10–23). Adjacent

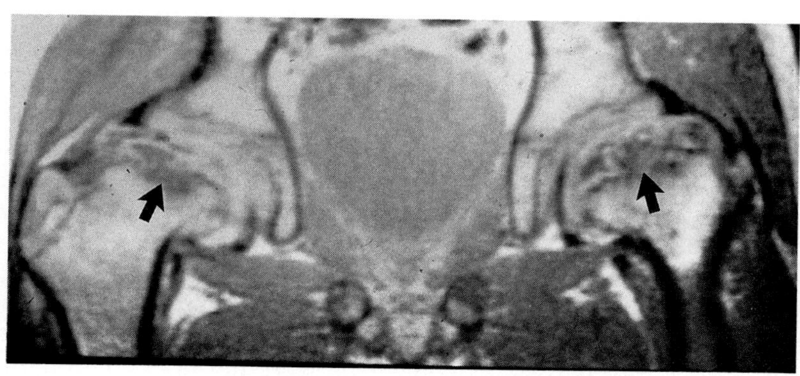

Figure 10–21. Legg-Calvé-Perthes disease. T1-weighted spin-echo MR image demonstrates bilateral Legg-Calvé-Perthes in this patient. The right femoral head is more deformed and flattened than the left femoral head. The signal-deficient areas correspond to infarction of the capital femoral epiphysis *(arrows)*.

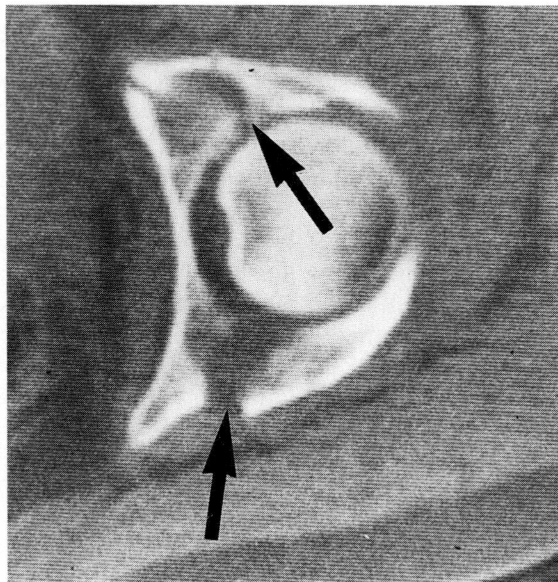

Figure 10–22. Acetabular fracture. Axial computed tomography (CT) scan demonstrates fracture of the anterior and posterior acetabulum in this patient *(arrows)*.

to the fracture, a surrounding area of mildly to moderately decreased signal intensity is present on T1-weighted images (see Fig. 10–23*A*). This area demonstrates high signal intensity on T2-weighting, reflecting edema or hemorrhage, or both (see Fig. 10–23*A*).

Also, MR imaging has the potential to be employed to assess the viability of the femoral head after intracapsular femoral neck fracture or hip dislocation. Both conditions may cause obliteration or rupture of the arteries supplying the femoral head. Patients in whom the femoral head is viable are most likely candidates for osteosynthesis. In patients with impaired perfusion, however, other, more invasive procedures—in the extreme, total hip replacement—may be preferable.

Scintigraphy is limited in detecting marrow ischemia because the ratio of radionuclide uptake between the injured and the contralateral side is an uncertain measure for the perfusion of the femoral head.[2]

Lang and co-workers performed sequential MR imaging of femoral heads in a porcine model for a total of 72 hours after the induction of ischemia.[62] They found a decrease in signal intensity during the first 24 hours of ischemia; signal intensities returned to baseline values after 72 hours. T1 relaxation times increased most significantly between baseline and 24 hours and then decreased to near baseline levels between 48 and 72 hours of ischemia. T2 relaxation times did not change significantly during the study interval. The MR imaging patterns occurring in clinical cases with manifest osteonecrosis were not observed. Such changes appear to require the presence of intact and vigorous repair response within adjacent viable bone. These investigators concluded that the transient decrease in signal intensity and prolongation of T1 relaxation times at 12, 24, and 48 hours may represent potential diagnostic indicators of early ischemia. However, clinically such studies may be limited because they require sequential follow-up MR imaging of the patient at different intervals after the injury to document any time-related signal intensity and relaxation time changes.

Speer and colleagues performed an in vivo study on 15 patients with subcapital intracapsular fracture using MR imaging to determine whether the technique can be used to evaluate the viability of the femoral head.[113] A single MR imaging study was performed 48 hours after the injury. In 11 patients, they found decreased signal intensity on T1-weighting at the base of the femoral head, immediately adjacent to the fracture. The decreased signal was thought to represent hemorrhage next to the site of the fracture. No signal intensity changes were observed in the other portions of the femoral head. They concluded that conventional MR imaging is inadequate to determine the viability of the femoral head in patients who sustain femoral neck fractures.[113]

Lang and associates studied 10 patients with acute femoral neck fractures before and after intravenous administration of gadopentetate dimeglumine (Magnevist).[64] Magnetic resonance imaging findings were compared with superselective digital subtraction angiography (Fig. 10–24). The femoral head demonstrated normal high signal intensity in all patients on the precontrast sequences. In the patients with intact femoral head arteries, the femoral head, neck, and shaft demonstrated an increase in signal intensity after administration of the contrast agent that was comparable to that of the unaffected side. In the patients with impaired blood supply to the femoral head, the signal intensity of the marrow cavity distal to the fracture increased, also markedly, on postcontrast images. The signal intensity of the femoral head, however, did not change on the postcontrast images and was lower than that of the marrow distal to the fracture or that of the femoral head on the unaffected side (see Fig. 10–24). These investigators concluded that gadopentetate dimeglumine–enhanced MR imaging may be a unique noninvasive imaging method for the assessment of femoral head perfusion after femoral neck fracture.[64]

At present, it appears that gadopentetate dimeglumine–enhanced MR imaging may eventually be used clinically for evaluating femoral head perfusion and viability after femoral neck fracture or hip dislocation. However, studies entailing larger patient populations are needed to validate these findings.

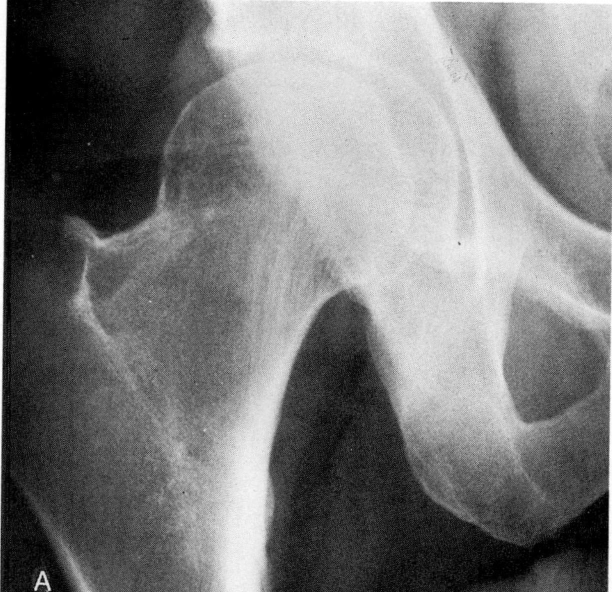

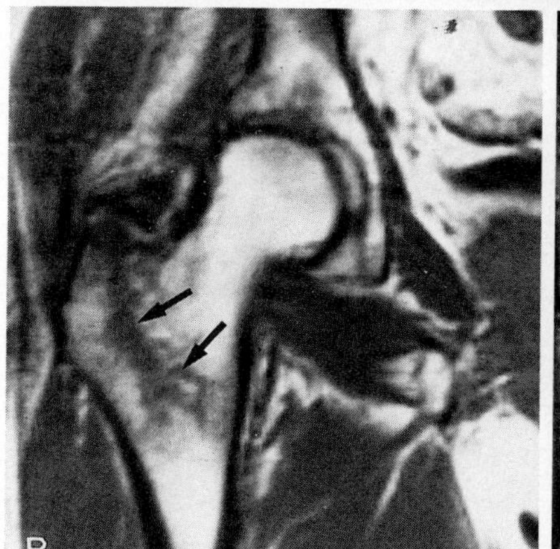

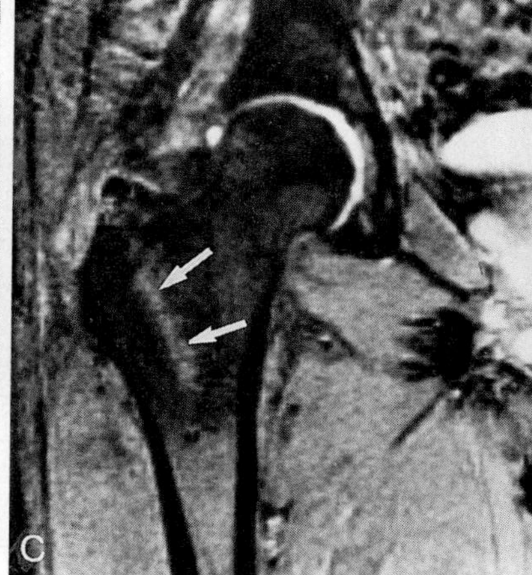

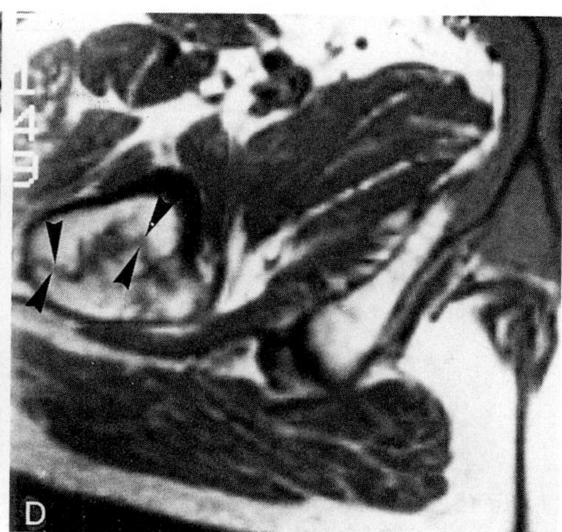

Figure 10–23. Occult femoral neck fracture detected by MR imaging. *A*, The patient had right hip pain after trauma. The conventional radiograph does not demonstrate a fracture. *B*, Coronal T1-weighted spin-echo MR image demonstrates an intramedullary band of low signal intensity *(arrows)*. *C*, T2*-weighted gradient-echo image shows marked increase in signal intensity of the intramedullary band *(arrows)*. The area of signal intensity increase is thought to reflect edema or acute hemorrhage surrounding the fracture. *D*, Axial T1-weighted spin-echo image demonstrates an intramedullary line of low signal intensity *(arrowheads)* that is continuous with the cortex and corresponds to the fracture.

EFFUSION

Conventional radiography is limited in detecting effusions of the hip joint. Diagnostic signs that may indicate effusion are displacement of fat layers; widening of the teardrop distance; and, in rare cases, subluxation and dislocation of the femoral head. Small amounts of joint fluid are often seen in normal hip joints on routine MR imaging. Hip joint effusion is found in trauma, synovial disorders, tumor, infection, arthritis, and aseptic osteonecrosis. Most effusions demonstrate low signal intensity on T1-weighted images and high signal intensity on T2-weighted images owing to their long T1 and T2 relaxation times (Figs. 10–1, 10–25). Fresh hemorrhage into the joint demonstrates low signal intensity on T1 weighting and high signal intensity on T2 weighting and cannot be differentiated from regular joint fluid on MR imaging.[10] However, when hemorrhage is subacute and methemoglobin has formed, it demonstrates high signal intensity on T1-weighted images and retains high signal intensity on T2-weighted sequences. Similarly, joint effusions containing large amounts of proteinaceous debris show high signal intensity on both T1- and T2-weighted images. Hemosiderin, in contrast, demonstrates low signal intensity with all sequences.[37,105] Hemosiderin deposits may be found in hemophilia, rheumatoid arthritis, pigmented villonodular synovitis, and synovial hemangioma.

OSTEOARTHRITIS

Osteoarthritis frequently affects the hip joint and can cause profound structural changes. These changes include cartilage loss, subchondral sclerosis, osteophytosis, and subchondral cyst formation.[3,127] In most patients, conventional radiography is sufficient to evaluate osteoarthritis of the hip. Ra-

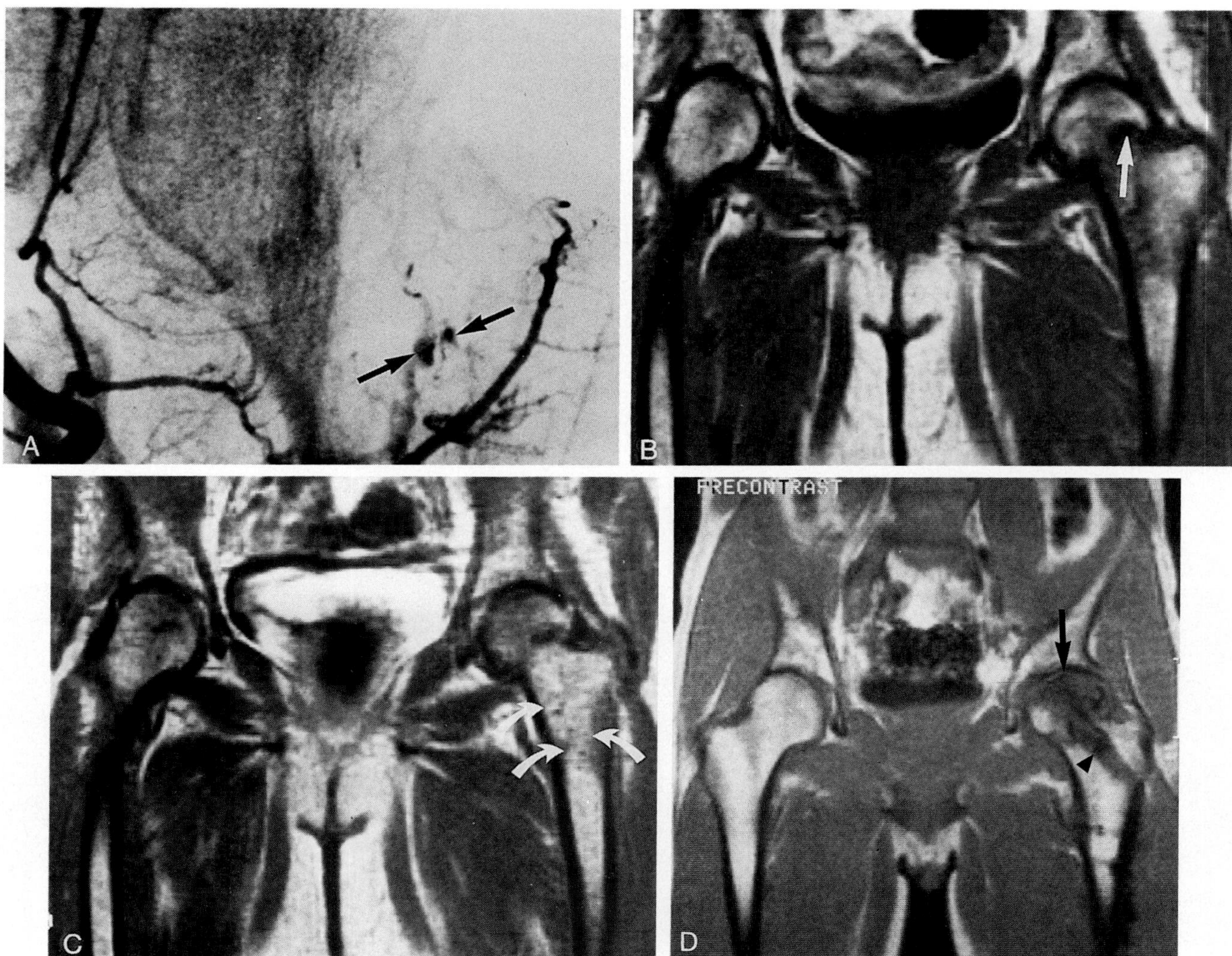

Figure 10–24. Femoral neck fracture: Assessment of femoral head perfusion using gadopentetate dimeglumine–enhanced MR imaging. The patient was a 60-year-old man with a femoral neck fracture, impaired blood supply on digital subtraction angiography, and capital osteonecrosis on long-term follow-up. *A,* Digital subtraction angiogram of the medial circumflex femoral artery demonstrates contrast extravasation *(arrows)* in the femoral neck resulting from vascular injury. The lateral epiphyseal arteries do not fill with contrast medium indicating an impaired blood supply. *B,* Precontrast gradient-echo MR image (TR, 314 ms; TE, 14 ms; flip angle, 90 degrees) made 24 hours after the fracture demonstrates a fracture of low signal intensity *(straight arrow).* No focal signal abnormalities indicating osteonecrosis are observed in the femoral head on the fractured side. *C,* Postcontrast gradient-echo MR image (TR, 314 ms; TE, 14 ms; flip angle, 90 degrees) shows uniform contrast enhancement in the femoral shaft and neck *(curved arrows).* The femoral head on the fractured side, however, does not demonstrate enhancement and is lower in signal intensity than the marrow distal to the fracture or the femoral head on the contralateral normal side, indicating impaired perfusion. *D,* Follow-up gradient-echo MR image (TR, 314 ms; TE, 14 ms; flip angle, 90 degrees) made 12 months after the initial injury, after the removal of surgical hardware, demonstrates manifest osteonecrosis *(arrow).* The band of low signal intensity in the femoral shaft and neck *(arrowhead)* corresponds to the area where the internal fixation device had been located. (From Lang P, Mauz M, Schörner W, et al: Acute fracture of the femoral neck: Assessment of femoral head perfusion using gadopentetate dimeglumine–enhanced MR imaging. AJR 160:335–341, 1993.)

diographic evaluation can detect structural changes of the bone; however, soft tissue involvement may not be recognized. Computed tomography is superior to conventional radiography and polytomography because it provides a tomographic assessment of both osseous and soft tissue changes. It is particularly useful for evaluating subchondral sclerosis and subchondral cyst formation. Osteophytes, which typically form at the junction of the femoral head and neck, can be detected by CT[55] even in the absence of radiographic findings. Loose bodies can be demonstrated by CT when they are calcified. Cartilaginous loose bodies, however, cannot be delineated on CT.

Magnetic resonance imaging is unique in that it affords direct visualization of cartilage.[69,83,117] High-resolution MR imaging may demonstrate three different zones in normal cartilage.[83] A zone of low signal intensity near the articular surface corresponds histologically to dense, tangentially oriented layers in the superficial zone of cartilage.[83] A region of higher signal intensity deep to the superficial zone

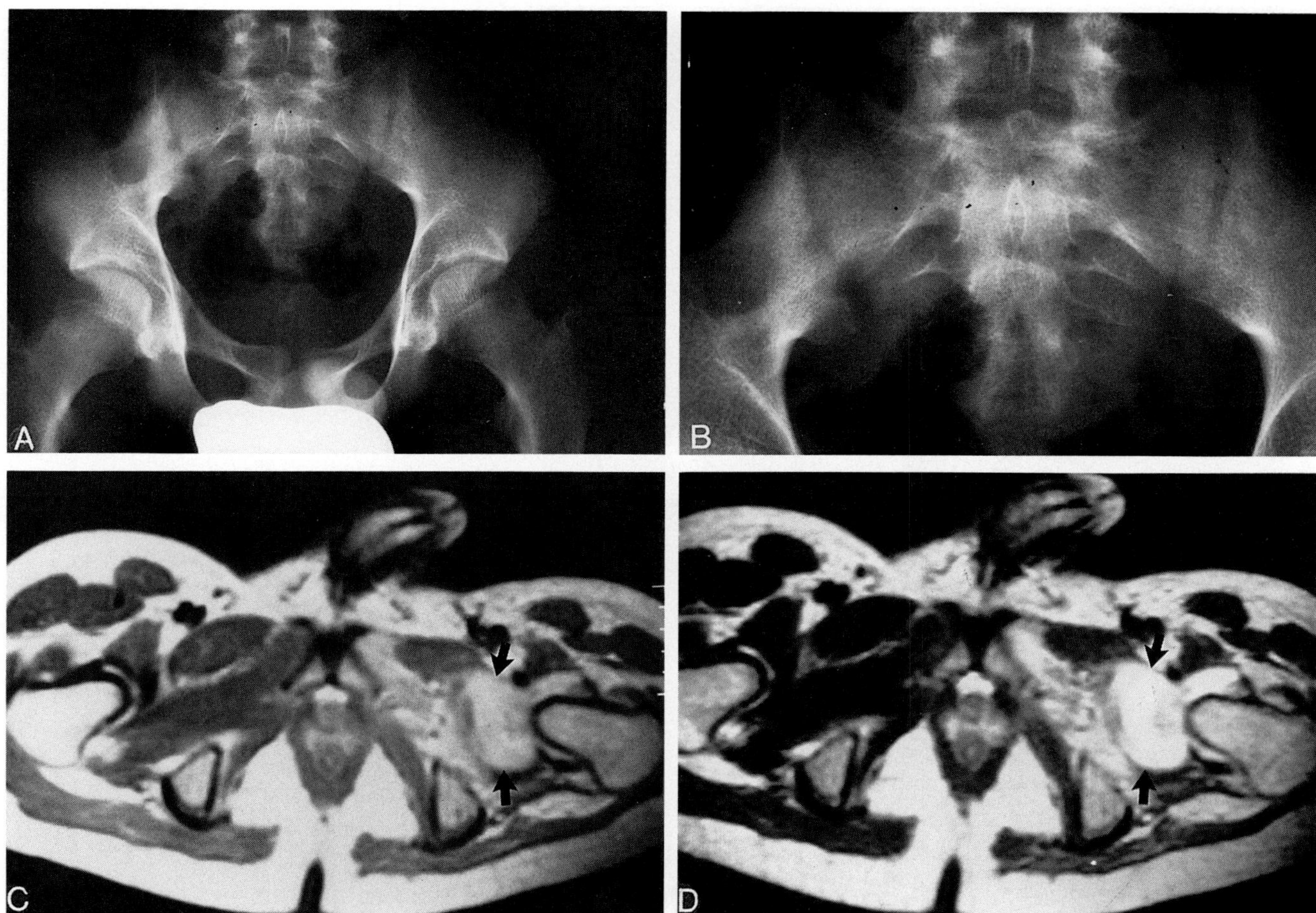

Figure 10–25. Joint effusion in a patient with ankylosing spondylitis. *A*, Conventional radiograph demonstrates the loss of the normal cortical outline of the right sacroiliac joint. Both hips appear normal. *B*, Close-up radiograph confirms the presence of pathologic changes in right sacroiliac joint. *C*, Proton density spin-echo MR image demonstrate joint effusion of intermediate intensity in the left hip *(arrows)* with distention of the joint capsule. *D*, The effusion has high signal intensity on T2-weighting *(arrows)*.

of low signal intensity reflects cartilage in the transitional histologic zone. A deep zone of low signal intensity correlates with a combination of deep radiate and calcified cartilage and subchondral bone.[83] Focal thinning of cartilage, irregularity of the cartilage surface, and inhomogeneity of signal intensity are MR imaging signs of cartilage loss in osteoarthritis (Fig. 10–26). Changes in cartilage signal on T2*-weighted gradient-echo sequences reflect degeneration with reduction of water content.[117] Magnetic resonance imaging may demonstrate cartilage loss before joint space narrowing is evident radiographically.[14,20,21,70] It also demonstrates osseous as well as cartilaginous loose bodies in the joint space. Subchondral sclerosis causes a region of low signal intensity within the marrow space of high signal intensity on all sequences owing to the lack of resonating protons. Subchondral cysts produce low signal intensity on T1-weighted images and high signal intensity on T2 weighting. The signal intensity most likely reflects joint fluid or other fluids filling the cyst. Magnetic resonance imaging has been reported to be more sensitive than conventional radiography and CT in demonstrating the joint changes observed in osteoarthritis[20,21] and may be used as a noninvasive tool for monitoring progression of early osteoarthritis. It may be particularly useful for longitudinal studies that assess the influence of drug therapy on the long-term course of this disease.[20,21]

RHEUMATOID ARTHRITIS

Rheumatoid arthritis (RA) affects approximately 1.0 to 1.5 per cent of the population of the United States and is observed in 4.5 per cent of people older than 55 years.[75] Synovial hypertrophy, cartilage defects, bone erosions, weakening, and rupture of ligaments and tendons are frequent findings.[72] Cartilage loss, bone erosion, osteopenia, and soft tissue swelling are the most frequent findings on conventional radiography in RA. Conventional radiography in RA

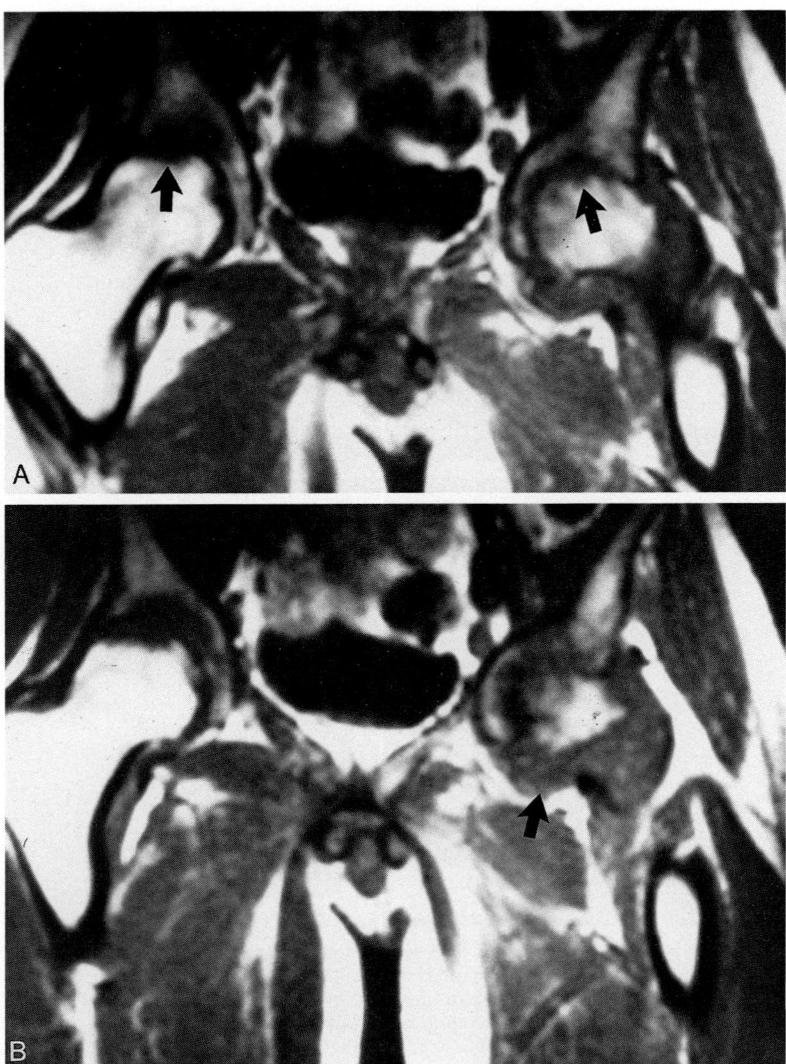

Figure 10–26. Osteoarthritis of the hip. *A*, T1-weighted spin-echo MR image demonstrates bilateral cartilage loss and inhomogeneous signal intensity subchondrally. Low signal intensity *(arrows)* is probably related to subchondral sclerosis. *B*, In a more anterior slice, the changes are even more evident in the left hip. Note the joint effusion of low signal intensity on the left *(arrow)*.

of the hip is limited because osseous changes occur relatively late in the disease process. Early changes such as inflammatory soft tissue proliferation are difficult if not impossible to determine by conventional radiography.

Magnetic resonance imaging provides excellent delineation of soft tissue changes, cartilaginous defects, and osseous erosions associated with RA (Fig. 10–27). Magnetic resonance imaging has been reported to be superior to conventional tomography and CT in demonstrating the joint changes related to RA, in general because of its great soft tissue contrast and multiplanar imaging capability.[112,115] The signal intensity of the inflammatory soft tissue that penetrates the synovial lining of the joint has been decribed to vary markedly on T1- and T2-weighted MR images. Beltran and co-workers[8] observed signal intensities in proliferative synovial tissue that were greater than that of joint fluid on T1- and T2-weighted images in five patients. In five other cases, pannus had lower signal intensity than joint fluid on T1 and T2 weighting; these investigators attributed the low signal intensity of synovial proliferation in these patients to hemosiderin deposits associated with RA. This signal inhomogeneity has also been reported by others and was thought to reflect different degrees of inflammation and fibrosis in the proliferative tissue.[129]

In evaluating inflammatory soft tissue changes in RA, MR imaging may be supplemented by the use of paramagnetic contrast agents such as gadopentetate dimeglumine. Preliminary studies indicate that intravenous administration of gadopentetate dimeglumine may provide even better distinction of inflammatory soft tissue alterations from joint effusion than that with unenhanced MR imaging.[1,56,58,93,119]

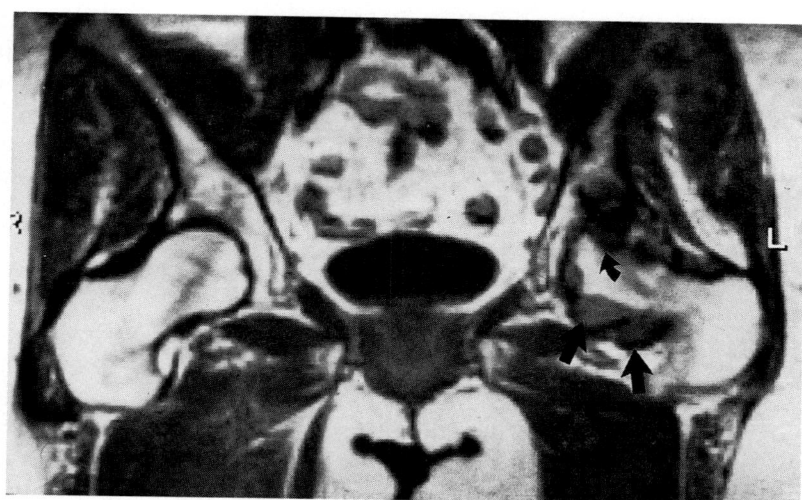

Figure 10–27. Rheumatoid arthritis of the hip. *A*, Coronal T1-weighted spin echo MR image demonstrates lesions of intermediate to low signal intensity *(curved arrow)* in the femoral head, reflecting erosive disease. Pathologic soft tissue extension is seen inferior to the femoral head *(straight arrows)*.

In our own observations in RA patients in a preliminary pilot project,[102] we were able to confirm that gadopentetate dimeglumine–enhanced MR imaging may allow a distinction between joint fluid and inflammatory soft tissue. In addition to improved morphologic assessment of the rheumatoid joint changes, fast sequential MR imaging after bolus intravenous administration of gadopentetate dimeglumine also provides quantitative information on the degree of synovial inflammation.[67] In patients with chronic synovial inflammation, contrast enhancement in the inflamed synovium tends to be slow, whereas rapid and marked enhancement is seen in acute synovial inflammation.[67] This faster and greater enhancement in acute inflammation may be related to hyperemia. This technique may be used in longitudinal drug trials to evaluate the efficacy of anti-inflammatory drugs in suppressing synovial inflammation.

JUVENILE RHEUMATOID ARTHRITIS

Juvenile rheumatoid arthritis (JRA) is a disease of childhood that may involve the hips. It is characterized by synovitis that causes joint effusion, synovial hypertrophy, and destruction of cartilage and bone. Conventional radiography does not provide information about cartilage loss until narrowing of the joint space occurs and the disease is irreversible.[106,129] Magnetic resonance imaging is particularly advantageous in detecting early cartilaginous lesions before they become manifest on radiography. It also shows synovial hypertrophy, overgrowth of the epiphyses, osseous erosions, and joint effusions in JRA. Reports on MR imaging in JRA have, however, been scarce,[106,129] and more experience is needed to define the role of MR imaging in JRA more clearly.

SEPTIC ARTHRITIS

Septic arthritis is not characterized by distinctive MR imaging patterns. The definitive diagnosis is based on clinical findings and joint aspiration. Frequent MR imaging findings in septic arthritis include hip joint effusion, monoarticular involvement, and abnormal signal intensity in the bone marrow of the acetabulum and the femoral head, indicating a possible ostemyelitis. If osteomyelitis is present, the bone marrow typically demonstrates low signal intensity on T1 weighting and high signal intensity on T2 weighting. However, this signal pattern may also result from reactive hyperemia adjacent to the inflammatory process in absence of osteomyelitis. The hyaline cartilage may be destroyed rapidly, resulting in progressive joint space narrowing. In more advanced, untreated stages, erosion of the osseous structures can be observed. Delineation of the extent of the inflammatory process is facilitated by multiplanar imaging.[100] In the hip, axial images are particularly helpful to demonstrate the extent of a periarticular abcess or inflammatory myositis.[100]

SYNOVIAL-BASED DISORDERS

Pigmented Villonodular Synovitis

Pigmented villonodular synovitis (PVNS) is a benign proliferative lesion of the synovial joint lining. It is most often found in the knee joint, but it may also affect the hip. Its cause remains uncertain.

Hypotheses about the origin of this lesion include a disorder of lipid metabolism, recurrent intra-articular hemorrhage, repeated trauma, postinflammatory changes, and neoplasia.[26,114] Pigmented villonodular synovitis is seen most frequently in adults 30 to 60 years old, and the distribution between the sexes is even. It causes pain, decreased range of motion, and locking sensation of the joint.

Macroscopically, PVNS has a typical yellow appearance with areas of rustlike, brownish discoloration. Histologically, the lesion consists of markedly hyperplastic synovium filled with lipid-laden foam macrophages and scattered hemosiderin deposits.

Detection of PVNS may be difficult because the findings of conventional radiographs frequently are unremarkable.[47,71] Three-phase bone scintigraphy typically demonstrates increased flow and blood pool in the region of the soft tissue mass. Computed tomography may demonstrate lytic defects of bone that are sharply defined and surrounded by sclerotic borders. Focal areas of hyperdensity may be observed in the soft tissue mass on CT.

The MR imaging appearance of PVNS demonstrates marked variability.[18,47,57] The lesions contain areas of intermediate and low signal intensity in comparison to skeletal muscle on T1- and T2-weighted images.[47,57] Areas with low signal intensity on both T1- and T2-weighted sequences are thought to represent the presence of hemosiderin.[57] Portions of the lesion may be cystic with low signal intensity on T1 weighting and increased signal intensity on T2-weighted sequences. Magnetic resonance imaging frequently detects joint effusion accompanying PVNS. Absence of osteopenia and subchondral cysts and preservation of joint space are other MR imaging findings in PVNS of the hip. The differential diagnosis of this imaging pattern includes early rheumatoid arthritis and lesions that result in intra-articular hemorrhage with hemosiderin formation.

Synovial Osteochondromatosis

Synovial osteochondromatosis is a monoarticular process resulting from cartilaginous metaplasia of the synovium. Synovial osteochondromatosis is found most frequently in the hip and in the knee joint in adults between the ages of 20 and 40 years. It may be difficult to differentiate from PVNS. Synovial osteochondromatosis is manifested by multiple intra-articular nodules that can calcify and ossify. Primary synovial osteochondromatosis is idiopathic. Primary synovial osteochondromatosis requires synovectomy for treatment. Recurrence is common. Secondary osteochondromatosis may result from trauma or from cartilage degeneration. Secondary synovial osteochondromatosis is managed by removal of loose bodies and treatment of the underlying condition.

Conventional radiography can be used to identify calcified loose bodies in synovial osteochondromatosis. The presence of calcified loose bodies on conventional radiographs helps to differentiate synovial osteochondromatosis from PVNS, as PVNS is usually not associated with calcification. However, both conventional radiography and CT fail to demonstrate cartilaginous loose bodies.

Magnetic resonance imaging is unique in demonstrating cartilaginous loose bodies. Osteochondromas may vary markedly in signal intensity, depending on their composition.[100] Cartilaginous osteochondromal bodies typically are of intermediate to high signal intensity on both T1- and T2-weighted images. Calcified loose bodies are of low signal intensity on all imaging sequences (Fig. 10–28). Osteochondromal bodies that contain fat may demonstrate high signal intensity on T1 weighting and intermediate to high signal intensity on T2 weighting.[100] The loose bodies are usually surrounded by a joint effusion of high signal intensity on T2-weighted sequences (see Fig. 10–28).[100]

AMYLOIDOSIS

Amyloidosis is the accumulation and infiltration of body tissues by a protein polysaccharide complex. Primary amyloidosis has to be differentiated from secondary amyloidosis; secondary amyloidosis occurs in chronic disease. Bone lesions are seen with amyloid deposits in and around the joints or within the marrow spaces. Osteolytic lesions are frequently present in the proximal humerus and the proximal femur (Fig. 10–29). The borders of the osteolytic lesions can be sclerotic. On MR imaging, amyloid typically is of low signal intensity on T1-weighted MR images (see Fig. 10–29) and of low to intermediate signal intensity on T2-weighted images. The amyloid may cause lysis of cortical bone and progressive joint destruction. Pathologic fracture may occur.

BURSAE

The hip joint is surrounded by multiple bursae (e.g., iliopsoas, iliopectinal, ischiotrochanteric, and trochanteric bursae).[88,116] The iliopsoas bursa is the largest normally occurring bursa in the human body. It is present in 98 per cent of adults and occurs bilaterally in most cases. The bursa is located

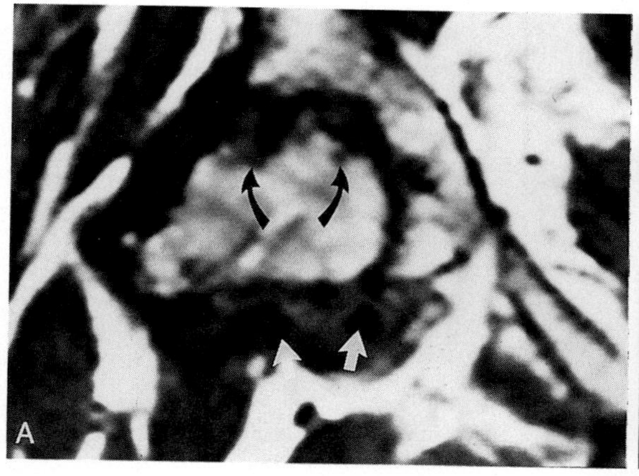

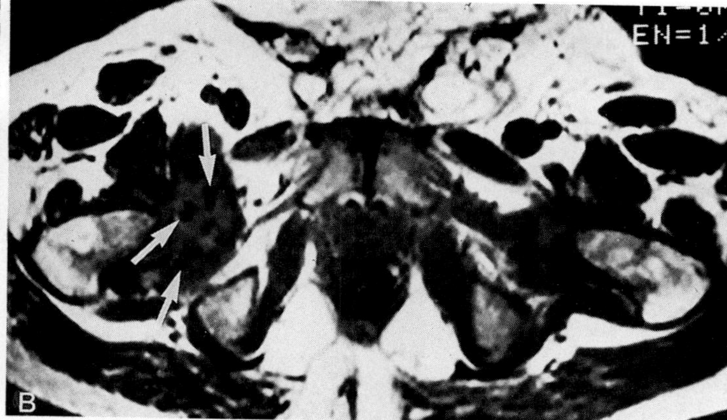

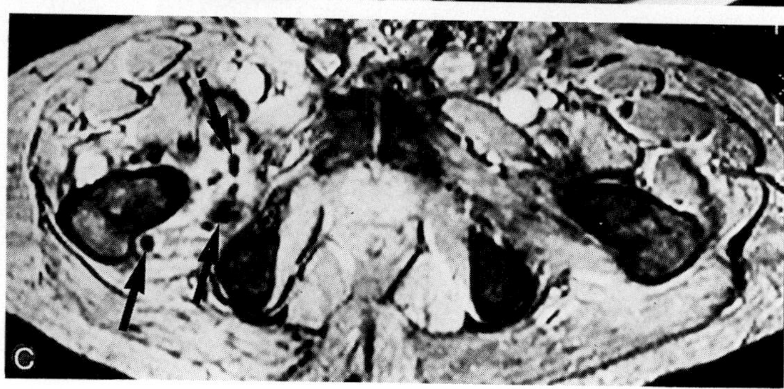

Figure 10–28. Synovial osteochondromatosis of the hip. *A,* Coronal T1-weighted spin-echo MR image demonstrates multiple loose bodies of low signal intensity surrounding the femoral head *(straight arrows).* Note the destruction of the articular surface and subchondral signal irregularity *(curved arrows). B,* T1-weighted axial spin-echo MR image demonstrates multiple osteochondromas of low signal intensity *(arrows)* in the right hip joint. The joint capsule is extended by a joint effusion of intermediate signal intensity. *C,* On T2-weighting, the joint effusion shows high signal intensity, thus highlighting the osteochondromas of low signal intensity *(arrows).*

between the psoas tendon and the hip joint capsule, deep to the musculotendinous part of the iliopsoas muscle and lateral to the femoral vessels. The iliopsoas bursa communicates with the hip joint in 15 per cent of adults. The communication may be secondary to trauma, degenerative joint disease, RA, gout, syphilis, PVNS, and synovial osteochondromatosis. Bursae can become distended with fluid in conditions that weaken the capsule and promote communication between the hip joint and bursa (e.g., friction of tendons on the base of the bursa, with subsequent thinning of the synovium).

Clinically, patients may reveal a soft tissue mass that is slowly enlarging. Often patients remain asymptomatic. Symptoms may arise from compression of adjacent structures, superimposed infection, or trauma. Fluid within the bursa demonstrates low signal intensity on T1-weighted sequences and high signal intensity on T2-weighted images. MR imaging is particularly useful in demonstrating the exact location and extent of an enlarged bursa, thus facilitating preoperative planning when resection is necessary. Magnetic resonance imaging also is helpful in differentiating bursae from other soft tissue masses surrounding the hip such as inguinal hernias, arteriovenous aneurysms, femoral artery aneurysm, inguinal lymphadenopathy, or undescended testes. However, MR imaging cannot distinguish uninfected from infected bursae. Aspiration of bursal fluid is indicated when infection is suspected clinically.

HERNIATION PITS

Herniation pits of the femoral neck represent an incidental finding on conventional radiographs. They represent subcortical pits or cavities that are formed by mechanical, abrasive effects of the anterior hip capsule on the anterior aspect of the femoral neck.[126] The radiologic diagnosis of a herniation pit is based on its location in the proximal anterior superior quadrant of the femoral neck.[90] The lesion is typically round to oval and usually measures less than 1 cm in diameter. A central radiolucency is sharply demarcated by an adjacent narrow zone of sclerosis. The sclerotic margin may show peripheral unsharpness. In some cases, peripheral extensions from the central radiolucency may be demonstrated, causing a lobular appearance. Herniation pits are seen in 5 per cent of the adult population.[90]

On MR imaging, herniation pits appear as round, well-marginated defects in the superior lateral quadrant of the femoral neck.[86] Herniation pits demonstrate variable signal intensity, depending on their composition. Low signal intensity is observed on both T1- and T2-weighted images when fibroconnective tissue fills the defect.[86] The low signal

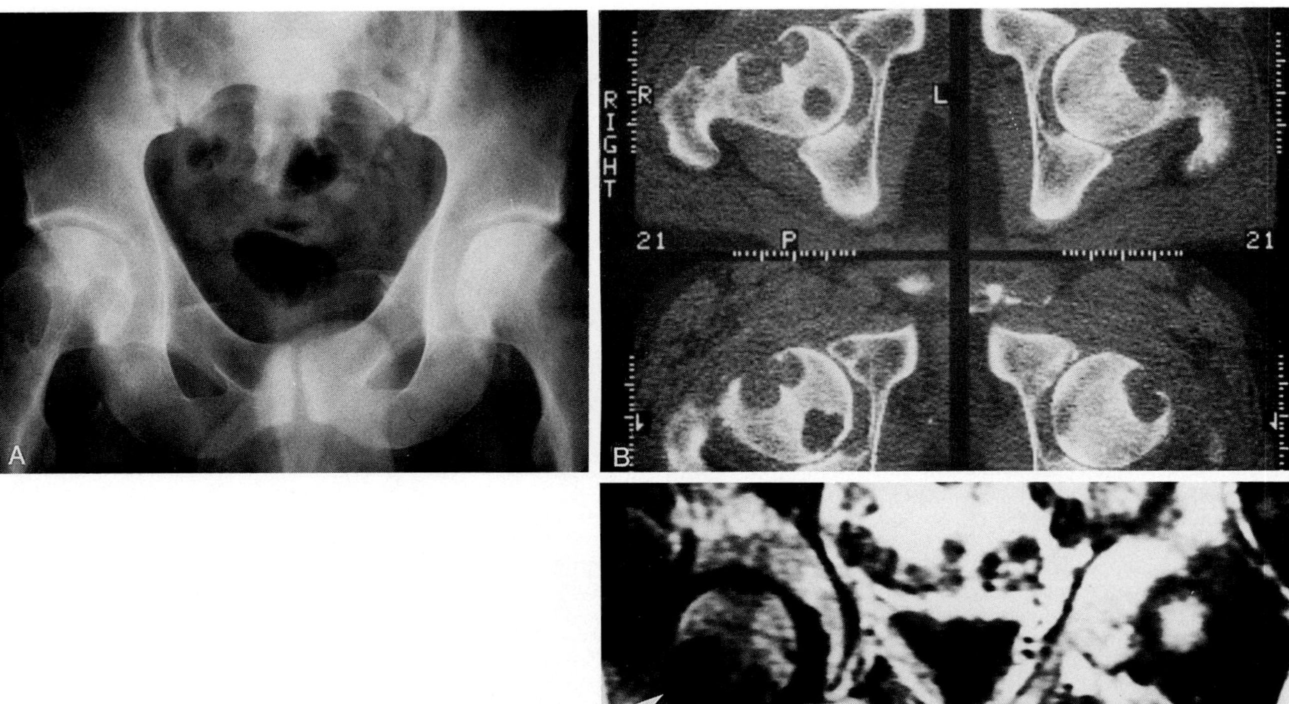

Figure 10–29. Amyloidosis. *A*, Conventional radiograph demonstrates a large lytic lesion in the right hip. Lytic areas also are seen in the lateral aspects of the left hip. *B*, Computed tomography (CT) scan shows multiple lytic defects of cortical and trabecular bone in both hips. *C*, The lytic defects are of low signal intensity on T1-weighted MR images *(arrows)*.

intensity is caused by the low proton density of fibroconnective tissue. If fluid is present, high signal intensity can be seen on T2-weighted images.[86] The lesion is frequently surrounded by a ring with low signal intensity on all sequences, which reflects the sclerotic rim on conventional radiographs. The cortical defect may be identified on axial images.[86] Differential diagnoses for these MR imaging patterns include osteoid osteoma, chronic abcesses, intraosseous ganglion, avascular necrosis, and atypical metastatic disease.

References

1. Adam G: Gadolinium-DTPA enhanced MR imaging of rheumatoid arthritis. Radiology 173 (P):389, 1989.
2. Alberts KA, Dahlborn M, Glas JE, et al: Radionuclide scintigraphy of femoral head specimens removed at arthroplasty for failed femoral neck fractures. Clin Orthop 205:222–229, 1986.
3. Alexander CJ: Osteoarthritis: A review of old myths and current concepts. Skel Radiol 19:327–333, 1990.
4. Arlet J, Ficat RP: Diagnostic de l'ostéo-nécrose fémoro-capitale primitive au stade I. Rev Chir Orthop 54:637–648, 1968.
5. Arnstein AR: Regional osteoporosis. Orthop Clin North Am 3:585–600, 1972.
6. Bassett LW, Gold RH, Reicher M, et al: Magnetic resonance imaging in the early diagnosis of ischemic necrosis of the femoral head. Clin Orthop 214:237–248, 1989.
7. Bassett LW, Mirra JM, Cracchiolo A, Gold RH: Ischemic necrosis of the femoral head: Correlation of magnetic resonance imaging and histologic sections. Clin Orthop 223:181–187, 1987.
8. Beltran J, Caudill JL, Herman LA: Rheumatoid arthritis: MR imaging manifestations. Radiology 165:153–157, 1987.
9. Beltran J, Herman LT, Burk JM, et al: Femoral head avascular necrosis: MR imaging with clinical-pathologic and radionuclide correlations. Radiology 166:215–220, 1988.
10. Beltran J, Noto AM, Herman LJ, et al: Joint effusions: MR imaging. Radiology 158:133–137, 1986.
11. Bloem JL: Transient osteoporosis of the hip: MR imaging. Radiology 167:753–755, 1988.
12. Bluemm RG, Falke THM, des Plantes BGZ, Steiner RM: Early Legg-Perthes disease (ischemic necrosis of the femoral head) demonstrated by magnetic resonance imaging. Skel Radiol 14:95–98, 1985.
13. Boettcher WG, Bonfiglio M, Hamilton HR, et al: Non-traumatic necrosis of the femoral head. Part I—Relation of altered hemostasis to etiology. J Bone Joint Surg [Am] 52:312–321, 1970.
14. Bongartz GE, Bock E, Horbach T, Requardt H: Degenerative cartilage lesions of the hip—magnetic resonance evaluation. Magn Reson Imaging 7:179–186, 1989.
15. Bos CF, Bloem JL: Treatment of dislocation of the hip, detected in early childhood, based on magnetic resonance imaging. J Bone Joint Surg [Am] 71:1523–1529, 1989.
16. Bos CF, Bloem JL, Bloem RM: Sequential magnetic resonance imaging in Perthes' disease. J Bone Joint Surg [Br] 73:219–224, 1991.
17. Bos CFA, Bloem JL, Obermann WR, et al: Magnetic resonance imaging in congenital dislocation of the hip. J Bone Joint Surg [Br] 70:174–178, 1988.
18. Boyd AD Jr, Sledge CB: Evaluation of the hip with pigmented villonodular synovitis. A case report. Clin Orthop 275:180, 1992.
19. Browning WH, Rosenkrantz H: Computed tomography in congenital hip dislocation. J Bone Joint Surg [Am] 64:27–31, 1982.
20. Chan W, Stevens M, Lang P, et al: Structural changes of osteoarthritis of the knee: Radiography, CT, and MR imaging correlation. Radiology 177(P): 183, 1990.

21. Chan WP, Lang P, Stevens MP, et al: Osteoarthritis of the knee: Comparison of radiography, CT, and MR imaging to assess extent and severity. AJR 157:799–806, 1991.
22. Coleman BG, Kressel HY, Dalinka MK, et al: Radiographically negative avascular necrosis: Detection with MR imaging. Radiology 168:525–528, 1988.
23. Deutsch AL, Mink JH: Occult fractures of the proximal femur: MR imaging. Radiology 170:113–116, 1989.
24. Dihlmann W: Koxale Computertomographie (KCT). ROFO 135:333–342, 1981.
25. Dihlmann W: CT analysis of the upper end of the femur: The asterisk sign and ischemic bone necrosis of the femoral head. Skel Radiol 8:251–258, 1982.
26. Dorwart RH, Genant HK, Johnston WH, et al: Pigmented villonodular synovitis of synovial joints: Clinical, pathologic, and radiologic features. AJR 143:877–885, 1984.
27. Dunn PM: Perinatal observations on the etiology of congenital dislocation of the hip. Clin Orthop 119:11–22, 1976.
28. Egund N, Wingstrand H: Legg-Calvé-Perthes disease: Imaging with MR. Radiology 179:89–92, 1991.
29. Ficat RP: Treatment of avascular necrosis of the femoral head. *In* The Hip. Proceedings of the 11th Open Scientific Meeting of the Hip Society. St Louis, CV Mosby, 1983, pp 279–295.
30. Gaucher A, Colomb JH, Naoun AR, et al: The diagnostic value of Tc-99m diphosphonate bone imaging in transient osteoporosis of the hip. J Rheumatol 6:574–583, 1979.
31. Genant HK, Moon KL, Heller M, et al: Nuclear magnetic resonance: Musculoskeletal applications. *In* Margulis AR, Higgins CB, Kaufman L, et al (eds): Clinical Magnetic Resonance Imaging. San Francisco, Radiology Research and Education Foundation, 1983, pp. 359–381.
32. Genez BM, Wilson MR, Houk RW, et al: Early osteonecrosis of the femoral head: Detection in high-risk patients with MR imaging. Radiology 168:521–524, 1988.
33. Glickstein MF, Burk DL, Schiebler ML, et al: Avascular necrosis versus other diseases of the hip: Sensitivity of MR imaging. Radiology 169:213–215, 1988.
34. Glimcher MJ, Kenzora JE: The biology of osteonecrosis of the human femoral head and its clinical implications. I. Tissue biology. Clin Orthop 138:284–309, 1979.
35. Glimcher MJ, Kenzora JE: The biology of osteonecrosis of the human femoral head and its clinical implications. II. The pathological changes in the femoral head as an organ and in the hip joint. Clin Orthop 139:283–312, 1979.
36. Glimcher MJ, Kenzora JE: The biology of osteonecrosis of the human femoral head and its clinical implications. III. Discussion of the etiology and genesis of the pathologic sequela: comments on treatment. Clin Orthop 140:273–312, 1979.
37. Gomori JM, Grossman RI, Goldberg HI, et al: Intracranial hematomas: Imaging by high-field MR. Radiology 157:87–93, 1985.
38. Graf R: New possibilities for the diagnosis of congenital hip joint dislocation by ultrasonography. J Pediatr Orthop 3:354–359, 1983.
39. Grimm J, Higer HP, Benning R, Meairs S: MRI of transient osteoporosis of the hip. Arch Orthop Trauma Surg 110:98–102, 1991.
40. Hauzeur JP, Hanquinet S, Gevenois PA, et al: Study of magnetic resonance imaging in transient osteoporosis of the hip. J Rheumatol 18:1211–1217, 1991.
41. Hensinger RN: Congenital dislocation of the hip. Clin Symp 31:1–31, 1979.
42. Hernandez RJ, Poznanski AK: CT evaluation of pediatric hip disorders. Orthop Clin North Am 16:513–542, 1985.
43. Hernandez RJ, Tachdijan MO, Dias LS: Hip CT in congenital dislocation: Appearance of tight iliopsoas tendon and pulvinal hypertrophy. AJR 139:335–337, 1982.
44. Heuck A, Lehner K: Bildgebende Diagnostik bei Hüftkopfnekrosen. Röntgenpraxis 40:245–251, 1987.
45. Hunder GG, Kelly PJ: Roentgenologic transient osteoporosis of the hip: A clinical syndrome? Ann Intern Med 68:539–552, 1968.
46. Hungerford DS, Lennox DW: The importance of increased intraosseous pressure in the development of osteonecrosis of the femoral head: Implications for treatment. Orthop Clin North Am 16:635–652, 1985.
47. Jelinek JS, Kransdorf MJ, Utz JA, et al: Imaging of pigmented villonodular synovitis with emphasis on MR imaging. AJR 152:337–342, 1989.
48. Jergesen HE, Heller M, Genant HK: Magnetic resonance imaging in osteonecrosis of the femoral head. Orthop Clin North Am 16:705–716, 1985.
49. Jergesen HE, Heller M, Genant HK: Signal variability in magnetic resonance imaging of femoral head osteonecrosis. Clin Orthop 253:137, 1990.
50. Jergesen HE, Lang P, Moseley M, Genant HK: Histologic correlation in magnetic resonance imaging of femoral head osteonecrosis. Clin Orthop 253:150–163, 1990.
51. Johnson ND, Wood BP, Jackman KV: Complex infantile and congenital hip dislocation: Assessment with MR imaging. Radiology 168:151–156, 1988.
52. Judet R, Judet J, Letournel E: Fractures of the acetabulum: Classification and surgical approaches for open reduction. J Bone Joint Surg [Am] 46:1615–1646, 1964.
53. Kalender WA, Hebel R, Ebersberger J: Reduction of CT artifacts caused by metallic implants. Radiology 164:576–577, 1987.
54. Kenzora JE, Glimcher MJ: Accumulative cell stress: The multifactorial etiology of idiopathic osteonecrosis. Orthop Clin North Am 16:669–679, 1985.
55. Kindynis P, Haller H, Kang HS, et al: Osteophytosis of the knee: Anatomic, radiologic, and pathologic investigation. Radiology 174:841–846, 1990.
56. Konig H, Sieper J, Wolf KJ: Rheumatoid arthritis: Evaluation of hypervascular and fibrous pannus with dynamic MR imaging enhanced with Gd-DTPA. Radiology 176:473–477, 1990.
57. Kottal RA, Vogler JB, Matamoros A, et al: Pigmented villonodular synovitis: A report of MR imaging in two cases. Radiology 163:551–553, 1987.
58. Kursunoglu-Brahme ST, Riccio T, Weisman MH, et al: Rheumatoid knee: Role of gadopentetate-enhanced MR imaging. Radiology 176: 831–835, 1990.
59. Lang P: Magnetic Resonance Imaging in Avascular Necrosis of the Femoral Head. Stuttgart, Enke, 1990.
60. Lang P, Genant HK, Steiger P, et al: Three-dimensional digital displays in congenital dislocation of the hip: Preliminary experience. J Pediatr Orthop 9:532–537, 1989.
61. Lang P, Genant HK, Steiger P, et al: 3-D reconstruction asserts clinical potential in magnetic resonance imaging. Diagn Imaging 5:80–84, 1989.
62. Lang P, Jergesen HE, Genant HK, et al: Magnetic resonance imaging of the ischemic femoral head in pigs. Dependency of signal intensities and relaxation times on elapsed time. Clin Orthop 244:272–280, 1989.
63. Lang P, Jergesen HE, Moseley ME, et al: Avascular necrosis of the femoral head: High-field-strength MR imaging with histologic correlation. Radiology 169:517–524, 1988.
64. Lang P, Mauz M, Schörner W, et al: Acute fracture of the femoral neck: Assessment of femoral head perfusion using gadopentetate dimeglumine-enhanced MR imaging. AJR 160:335–341, 1993.
65. Lang P, Schwetlick G, Mauz M, et al: Avascular necrosis of the hip: Unenhanced and Gd-DTPA enhanced MR imaging. 90th Annual Meeting of the American Roentgen Ray Society, New Orleans, 1990.
66. Lang P, Steiger P, Genant HK, et al: Three-dimensional CT and MR imaging in congenital dislocation of the hip: Clinical and technical considerations. J Comput Assist Tomogr 12:459–464, 1988.
67. Lang P, Stevens M, Vahlensieck M, et al: Rheumatoid arthritis of the hand and wrist: Evaluation of soft-tissue inflammation and quantification of inflammatory activity using unenhanced and dynamic Gd-DTPA enhanced MR imaging. Society of Magnetic Resonance in Medicine, San Francisco, Calif, 1991, p 66.
68. Lee JK, Yao L: Stress fractures: MR imaging. Radiology 169:217–220, 1988.
69. Lehner KB, Rechl HP, Gmeinwieser JK, et al: Structure, function, degeneration of bovine hyaline cartilage: Assessment with MR imaging in vitro. Radiology 170:495–499, 1989.
70. Li KC, Higgs J, Aisen AM, et al: MRI in osteoarthritis of the hip: Gradations of severity. Magn Reson Imaging 6:229–236, 1988.
71. Mandelbaum BR, Grant TT, Hartzmann S, et al: The use of MRI to assist in diagnosis of pigmented villonodular synovitis of the knee joint. Clin Orthop 231:135–139, 1988.
72. Mannerfelt L, Norman O: Attrition rupture of flexor tendons in rheumatoid arthritis. J Bone J Surg [Br] 51:270–277, 1969.
73. Marcus ND, Enneking WP, Massam RA: The silent hip in idiopathic necrosis. Treatment by bone grafting. J Bone Joint Surg [Am] 55:1351–1366, 1973.
74. Markisz JA, Knowles JR, Altchek DW, et al: Segmental patterns of avascular necrosis of the femoral heads: Early detection with MR imaging. Radiology 162:717–720, 1987.
75. McDuffie FC: Morbidity impact of rheumatoid arthritis in society. Am J Med 78:1–5, 1985.
76. Merle d'Aubigne RM, Postel M, Mazabraud A, et al: Idiopathic necrosis of the femoral head in adults. J Bone Joint Surg [Br] 47:612–633, 1965.
77. Meyers MH: The treatment of osteonecrosis of the hip with fresh osteochondral allografts and with the muscle pedicle graft technique. Clin Orthop 130:202–209, 1978.
78. Mikhael MA, Paige ML, Widen AL: Magnetic resonance imaging and the diagnosis of avascular necrosis of the femoral head. Comput Radiol 11:157–163, 1987.
79. Mitchell DG, Joseph PM, Fallon M, et al: Chemical-shift MR imaging of the femoral head: An in vitro study of normal hips and hips with avascular necrosis. AJR 148:1159–1164, 1987.
80. Mitchell DG, Kressel HY, Arger P, et al: Avascular necrosis of the femoral head: Morphologic assessment by MR imaging with CT correlation. Radiology 161:739–742, 1986.
81. Mitchell DG, Kundel HL, Steinberg ME, et al: Avascular necrosis of the hip: Comparison of MR, CT, and scintigraphy. AJR 147:67–71, 1986.
82. Mitchell DG, Rao VM, Dalinka MK, et al: Femoral head avascular necrosis: Correlation with MR imaging, radiographic staging, radionuclide imaging, and clinical findings. Radiology 162:709–715, 1987.

83. Modl JM, Sether LA, Haughton VM, et al: Articular cartilage: Correlation of histologic zones with signal intensity at MR imaging. Radiology 181:853–855, 1991.
84. Moon KL, Genant HK, Davis P, et al: Nuclear magnetic resonance imaging in orthopaedics. Principles and applications. J Orthop Res 1:101–108, 1983.
85. Moon KL, Genant HK, Helms CA, et al: Musculoskeletal applications of nuclear magnetic resonance. Radiology 147:161–171, 1983.
86. Nokes SR, Vogler JB, Spritzer CE, et al: Herniation pits of the femoral neck: Appearance at MR imaging. Radiology 172:231–234, 1989.
87. Ogden JA: Dynamic pathobiology of congenital hip dysplasia. *In* Tachdjian MO (ed): Congenital dislocation of the hip. New York, Churchill Livingstone, 1982, pp 93–144.
88. Peters JC, Coleman BG, Turner ML, et al: CT evaluation of enlarged iliopsoas bursa. AJR 135:392–394, 1980.
89. Peterson HA, Klassen RA, McLeod RA, et al: The use of computerized tomography in dislocation of the hip and femoral neck anteversion in children. J Bone Joint Surg [Br] 63:198–207, 1981.
90. Pitt MJ, Graham AR, Shipman JH, et al: Herniation pit of the femoral neck. AJR 138:1115–1121, 1981.
91. Porter BA, Olson DO, Stimack GK: STIR imaging of marrow malignancies. Society of Magnetic Resonance in Medicine, San Francisco, Calif, 1987, p 146.
92. Powell EN, Gerratana FJ, Gage JR: Open reduction for congenital hip dislocation: The risk of avascular necrosis with three different approaches. J Pediatr Orthop 6:127–132, 1986.
93. Reiser M, Bongartz GP, Erlemann R, et al: Gadolinium-DTPA in rheumatoid arthritis and related diseases: First results with dynamic magnetic resonance imaging. Skel Radiol 18:591–597, 1989.
94. Resnick D, Niwayama G: Transient osteoporosis of the hip. *In* Resnick D, Niwayama G (ed): Diagnosis of Bone and Joint Disorders. Philadelphia, WB Saunders, 1981, p 1653.
95. Resnick D, Niwayama G: Diagnosis of Bone and Joint Disorders. 2nd Ed. Philadelphia, WB Saunders, 1988.
96. Rix J, Maas R, Eggers SG, Bruns J: Legg-Calvé-Perthes disease. The value of MRT in its early diagnosis and the assessment of its course. ROFO 156:77–82, 1992.
97. Robinson HJ, Hartleben PD, Lund G, Schreiman J: Evaluation of magnetic resonance imaging in the diagnosis of osteonecrosis of the femoral head. J Bone Joint Surg [Am] 71:650–663, 1989.
98. Rupp N, Reiser M, Hipp E, et al: Diagnostik der Knochennekrose durch magnetische Resonanz- (MR-) Tomographie. ROFO 142:131–137, 1985.
99. Rush BH, Bramson RT, Ogden JA: Legg-Calvé-Perthes disease: Detection of cartilaginous and synovial changes with MR imaging. Radiology, 167:473–476, 1988.
100. Sanchez RB, Quinn SF: MRI of inflammatory synovial processes. Magn Reson Imaging 7:529–540, 1989.
101. Sauser DD, Billimoria PE, Rouse GA, Mudge K: CT evaluation of hip trauma. AJR 135:269–274, 1980.
102. Schörner W, Lang P, Bittner R, et al: Gadolinium-DTPA enhanced MR imaging: Applications in body MRI. Diagn Imaging Vol. 4, 1990.
103. Schuster W: Röntgenologische Beurteilung der dysplastischen Hüftpfanne. Orthopäde 2:219, 1973.
104. Scoles PV, Yoon YS, Makley JT, Kalamchi A: Nuclear magnetic resonance imaging in Legg-Calvé-Perthes disease. J Bone Joint Surg [Am] 66:1357–1363, 1984.
105. Scott WA, Mark AS, Grossman RI, Gomori JM: Intracranial hemorrhage: Gradient-echo MR imaging at 1.5T. Radiology 168:803–807, 1988.
106. Senac M, Deutsch D, Bernstein B, et al: MR imaging in juvenile rheumatoid arthritis. AJR 150:873–878, 1988.
107. Sevitt S, Thompson RF: The distribution and anastomoses of arteries supplying the head and neck of the femur. J Bone Joint Surg [Br] 47:560–573, 1965.
108. Shellock FG, Schaefer DJ, Crues JV: Alterations in body and skin temperatures caused by magnetic resonance imaging: Is the recommended exposure for radiofrequency radiation too conservative? Br J Radiol 62:904–909, 1989.
109. Shellock FG, Schaefer DJ, Crues JV: Exposure to a 1.5-T static magnetic field does not alter body and skin temperatures in man. Magn Reson Med 11:371–375, 1989.
110. Shields AF, Porter BA, Churchley S, et al: The detection of bone marrow involvement by lymphoma using magnetic resonance imaging. J Clin Oncol 5:225–230, 1987.
111. Shirkoda A, Brashear R, Staab EV: Computed tomography of acetabular fractures. Radiology 134:683–688, 1980.
112. Sims RE, Genant HK: Magnetic resonance imaging of joint disease. Radiol Clin North Am 24:179–188, 1986.
113. Speer KP, Spritzer CE, Harrelson JM, et al: Magnetic resonance imaging of the femoral head after acute intracapsular fracture of the femoral neck. J Bone Joint Surg [Am] 72:98–103, 1990.
114. Srinvasa RA, Vigorita VJ: Pigmented villonodular synovitis (giant cell tumor) of the tendon sheath and synovial membrane. J Bone Joint Surg [Am] 66:76–94, 1984.
115. Steinbach L, Hellman D, Petri M, et al: Magnetic resonance imaging: A review of rheumatologic applications. Semin Arthritis Rheum 16:77–91, 1986.
116. Steinbach LS, Schneider R, Goldman AB, et al: Bursae and cavities communicating with the hip. Radiology 156:303–307, 1985.
117. Stoller DW, Helms CA, Genant HK: Gradient-echo MR imaging of the knee. Radiology 165(P):176, 1987.
118. Sugioka Y: Transtrochanteric anterior rotational osteotomy of the femoral head in the treatment of osteonecrosis affecting the hip. Clin Orthop 130:191–201, 1978.
119. Terrier F, Revel D, Reinhold C, et al: Contrast-enhanced MRI of periarticular soft-tissue changes in experimental arthritis of the rat. Magn Reson Med 3:385–396, 1986.
120. Tervonen O, Mueller DM, Matteson EL, et al: Clinically occult avascular necrosis of the hip: Prevalence in an asymptomatic population at risk. Radiology 182:845–847, 1992.
121. Toby EB, Koman LA: Magnetic resonance imaging of pediatric hip disease. J Pediatr Orthop 5:665–671, 1985.
122. Totty WG, Murphy WA, Ganz WI, et al: Magnetic resonance imaging of the normal and ischemic femoral head. AJR 143:1273–1280, 1984.
123. Turner DA, Templeton AC, Selzer B, et al: Femoral capital osteonecrosis: MR findings of diffuse marrow abnormalities without focal lesions. Radiology 171:135–140, 1989.
124. Urbanski SR, deLange EE, Eschenroeder HJ: Magnetic resonance imaging of transient osteoporosis of the hip. A case report. J Bone Joint Surg [Am] 73:451–455, 1991.
125. Vande Berg B, Malghem J, Labaisse MA, et al: Avascular necrosis of the hip: Comparison of contrast-enhanced and non-enhanced MR imaging with histologic correlation. Radiology 182:445–450, 1992.
126. Walmsley T: Observations on certain structural details of the neck of the femur. J Anat 49:305–315, 1915.
127. Watt I, Dieppe P: Osteoarthritis revisited. Skel Radiol 19: 1990.
128. Wilson AJ, Murphy WA, Hardy DC, et al: Transient osteoporosis: Transient bone marrow edema? Radiology 167:757–760, 1988.
129. Yulish B, Lieberman JM, Newman AJ, et al: Juvenile rheumatoid arthritis: Assessment with MR imaging. Radiology 165:149–152, 1987.

11 THE KNEE

Internal Derangements

Wing P. Chan, Russell C. Fritz, Lynne S. Steinbach, Chun Y. Wu, Harry K. Genant, and W. Dillworth Cannon

In most institutions, knee examinations have become the most common nonneurologic application of magnetic resonance (MR) imaging.[1,10] Magnetic resonance imaging is gaining recognition as an alternative to knee arthroscopy and as a substitute to knee arthrography in evaluating internal derangements of the knee.

IMAGING PROTOCOLS

At the University of California, San Francisco, the routine knee protocol for evaluation of the internal derangement at a field strength of 1.5 Tesla includes T1-weighted spin-echo sequences acquired in the sagittal plane and gradient-echo sequences obtained in three orthogonal planes (Table 11–1). The patient is placed in the supine position with the knee in extension. The examined leg is rotated externally 15 to 20 degrees throughout the examination. This position places the obliquely oriented anterior cruciate ligament parallel to the sagittal imaging plane.[39] A transmit-and-receive circumferential extremity coil is employed to improve signal-to-noise ratio and make it uniform across the knee. The use of this type of coil avoids the limitations of signal dropoff encountered with flat surface coils.

An initial axial gradient-echo acquisition serves as a localizer for subsequent sagittal and coronal planes. It is the best sequence for showing structural changes in the patellofemoral compartment. Sagittal T1-weighted sequences provide information regarding the menisci, cruciate ligaments, and osseous structures. They are generated with a repetition time (TR) of 800 ms and an echo time (TE) of 20 ms. Gradient-echo techniques are acquired by using a TR of 600 ms, a TE of 20 ms, and a flip angle of 30 degrees. Sagittal gradient-echo images are valuable for evaluating abnormalities in the cruciate ligaments and are useful for detecting cartilaginous

Table 11–1. PROTOCOLS OF MAGNETIC RESONANCE IMAGING OF THE KNEE AT THE UNIVERSITY OF CALIFORNIA, SAN FRANCISCO

Plane	TR (ms)	TE (ms)	FOV (cm)	Slice (mm)	Skip (mm)	NEX	Matrix (pixel)	Flip (degrees)	Time (min)
Axial (localizer) (GRE)	600	20	14	5	0	2	192	30	4
Sagittal (SE)	800	20	14	4	0	2	192		5
Sagittal (GRE)	600	20	14	4	0	2	192	30	4
Coronal (GRE)	600	20	14	5	0	2	192	30	4

TR, Repetition time; TE, echo time; FOV, field of view; Skip, interslice gap; NEX, number of excitations; GRE, gradient-recalled echo; SE, spin echo.

changes, joint effusions, and popliteal cysts. Coronal gradient-echo images, however, serve to confirm findings when sagittal images are equivocal. They are also essential sequences for evaluating collateral ligaments or meniscocapsular separations.

Other parameters include a 256 × 192 acquisition matrix, 14-cm field of view, and two excitations. The slice thickness is 4 mm for imaging in the sagittal plane and 5 mm for the coronal and axial planes. There is no interslice gap. The total imaging time is approximately 20 minutes. A 14-cm field of view and a 256 × 192 acquisition matrix provide a spatial resolution of 0.73 mm in the phase-encoded direction and 0.55 mm in the frequency-encoded direction. A 14-cm field of view also provides adequate visualization of the quadriceps-patellar mechanism in the sagittal plane. A 12-cm field of view is suggested for pediatric patients. A slice thickness of 4 to 5 mm is optimal for evaluating menisci and ligaments. Thicker slices improve the signal-to-noise ratio but significantly decrease the spatial resolution owing to partial volume effects. Thin slices may be useful when the status of the anterior cruciate ligament is uncertain on the standard 4- or 5-mm thick images. However, a slice thickness of 3 mm is suggested for pediatric patients.

In selected cases of acute trauma and neoplasia, T2-weighted multiecho sequences are employed with a TR of 2000 ms and a TE of 20 and 70 ms combined with a 256 × 128 acquisition matrix. Conventional T2 weighting is helpful in highlighting occult bone contusions and ligamentous edema. Associated edema and hemorrhage seen with bone contusions and microfracture of bony trabeculae are more apparent on spin-echo sequences. Short tau inversion-recovery (STIR) sequences are particularly useful in detecting subtle changes of bone marrow. These STIR images show a suppression of marrow fat signal with markedly increased signal intensity in lesions such as osteomyelitis, bone contusion, and neoplasms. Magnetic susceptibility effects often obscure pathologic marrow lesions on gradient-echo images. Techniques for evaluating neoplastic lesions are discussed in Chapter 15.

Three-dimensional Fourier transform volumetric imaging techniques permit rapid ultrathin (0.7 mm) slice acquisitions that facilitate reformation into any orthogonal plane.[53] Because 64 thin slices are obtained with full coverage of the knee, a very short TR is required to prevent long acquisition times. These techniques may provide a powerful tool in the evaluation of the articular cartilage, because joint effusions are of very high signal intensity, whereas articular cartilage is intermediate in signal intensity with the use of a moderately low flip angle of 30 degrees. The ultrathin slice acquisition may improve specificity in evaluating anterior cruciate ligament tears. This sequence provides increased sensitivity but decreased specificity for meniscal tears.[37]

Radial imaging techniques commonly are acquired using long TR, multislice gradient-echo sequences. Radial scans allow the meniscus to be viewed along its entire length, which may improve detection of tears near the free edge of the meniscus (e.g., parrot beak tears).[9,32] However, evaluation of cruciate ligaments is inadequate with use of these techniques.

MENISCI

Suspected meniscal lesions are the reason for approximately two thirds of the requests for MR imaging of the knee. Clinically, a history of specific injury may or may not be obtained. Although it is not pathognomic, locking of the knee is highly suggestive of a bucket-handle tear of the meniscus.[47] When locking is absent, the diagnosis is relatively difficult. A sensation of giving way is nonspecific for meniscal injuries. The most frequent physical finding is tenderness over the joint line, which is believed to be related to synovitis in the adjacent capsular and synovial tissues. Although the meniscus is innervated only at its periphery, tears in the central zone of the meniscus that produce symptoms are not uncommon. This may be caused by tugging of innervated parameniscal tissues by a compressed meniscus when it is buckling at the tear site.[7]

Meniscal injuries often are not isolated findings. Concomitant injuries of the capsule, ligaments, and articular cartilage must always be sought. Magnetic resonance imaging permits examination of the entire meniscus and the parameniscal structures instead of being limited to surface abnormalities, as in arthroscopic and arthrographic studies.

Anatomy and Function

Magnetic Resonance Imaging Anatomy. The normal meniscus appears as a homogeneous C-shaped structure of low signal intensity on all pulse sequences because of the low proton density of fibrocartilage, which contains high levels of type I collagen.[3,25] The lateral meniscus is smaller in diameter and more mobile than the medial meniscus. In the medial meniscus, the average width of the anterior horn is 7.5 mm and that of the posterior horn is 10.5 mm.[30] In the lateral meniscus, the horns are symmetric in size and shape and measure 11 mm in width. The middle zone or body segment of both the medial and lateral menisci is on average 11.6 mm wide.[12] On sagittal images (Fig. 11–1), through the

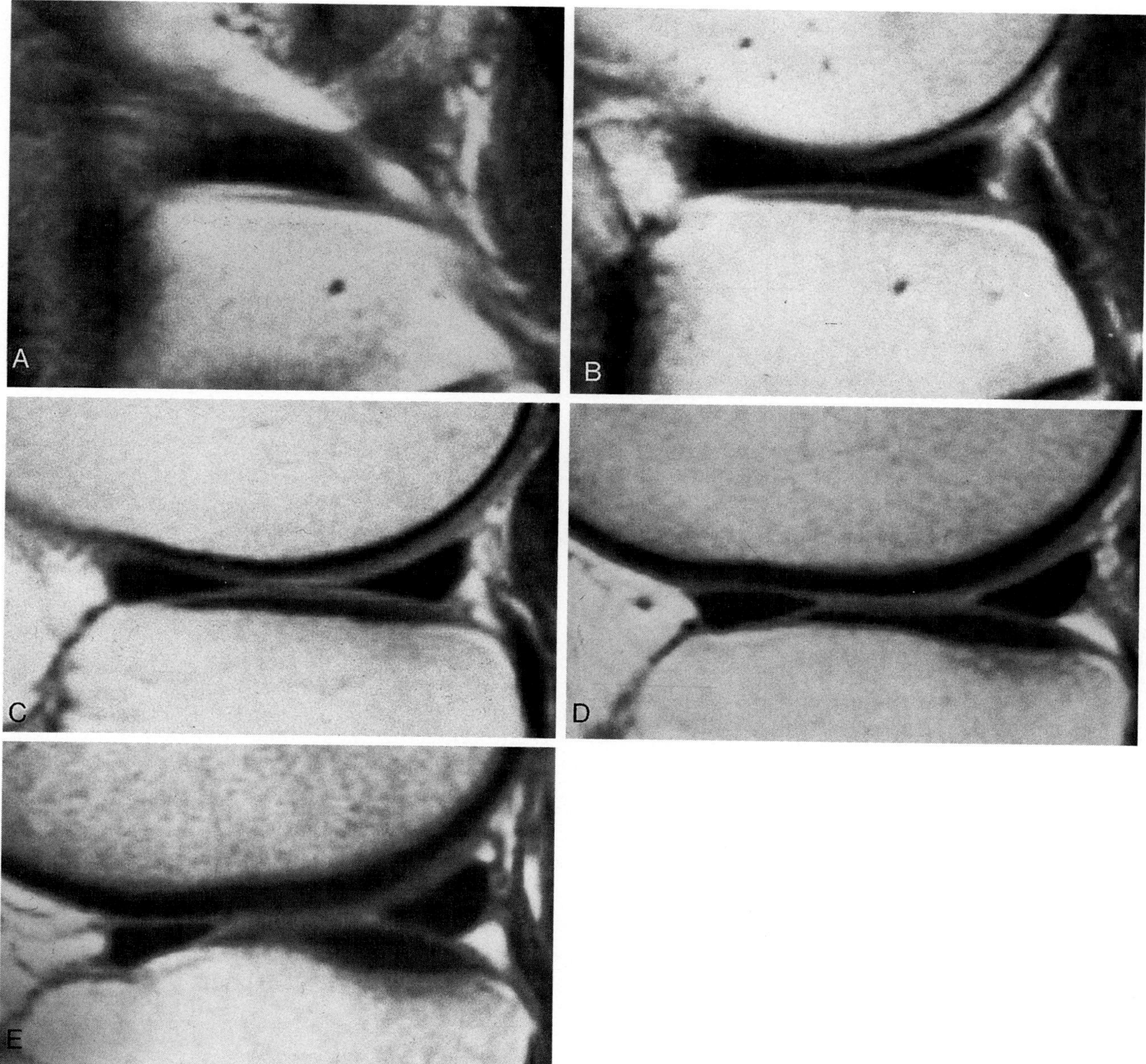

Figure 11–1. Normal appearance of the lateral meniscus on consecutive 5-mm thick sagittal T1-weighted images (TR, 800 ms; TE, 20 ms). Two "bow ties" of meniscal bodies and three to four "opposing triangles" of meniscal horns are captured on these planes.

body segment, the meniscus resembles a bow-tie in configuration. This bow-tie appearance is normally seen on two consecutive 5-mm thick images. The meniscus changes shape to two opposing triangles as the images approach the midline. The height of the normal meniscal body tapers from 4 to 5 mm at its periphery to the paper-thin free edge, with the posterior horn being slightly thicker (higher as seen on sagittal planes) than the anterior horn.[30] Sagittal views are best for depicting both the horns and the body of the meniscus. High-resolution coronal images allow more graphic depiction of the width of the meniscus. Because of partial volume effects, standard axial images cannot depict meniscal abnormalities reliably.

Attachments. The periphery of the medial meniscus is attached to the medial collateral ligament and the tibial plateau (Figs. 11–2 and 11–3). The inner edge is free. The anterior and posterior horns are fixed at their central tips anterior and posterior, respectively, to the tibial eminence. In the posterolateral portion of the lateral meniscus, the popliteal tendon separates the periphery of the meniscus from the joint capsule and the fibular collateral ligament. It also is loosely attached to the tibia via coronary ligaments. The inner edge is free. The

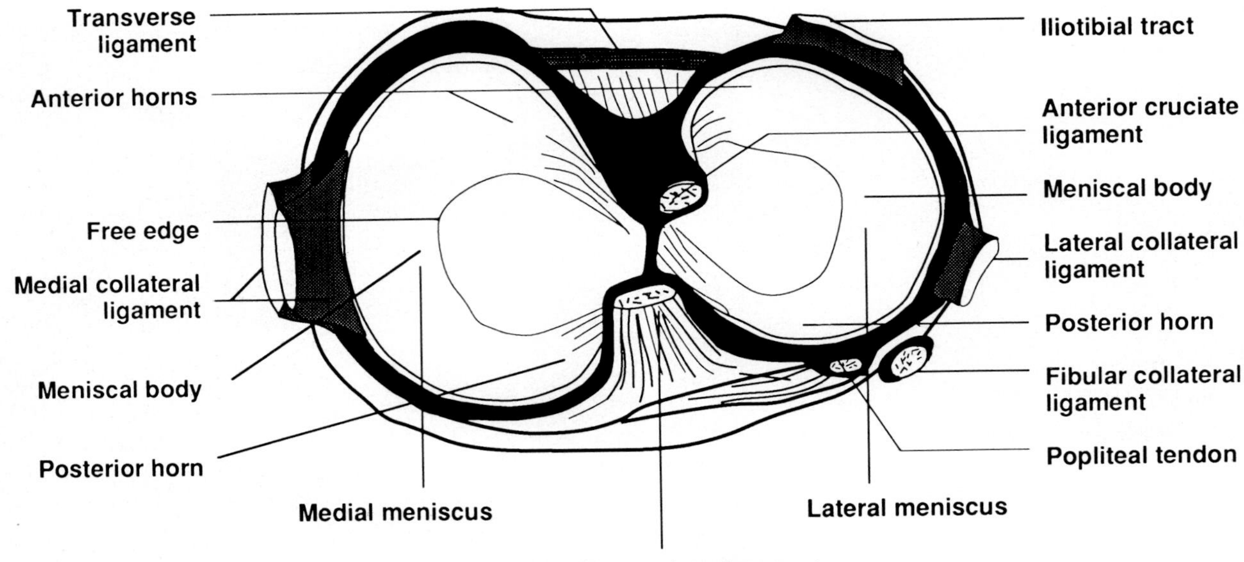

Figure 11–2. Superior view of tibial condyles shows the meniscoligamentous structure of the knee.

anterior and posterior horns are loosely anchored to the anterior and posterior aspects of the tibial eminence, respectively.

Function. In a standing adult, 40 to 60 per cent of the weight-bearing or load-transmitting forces are carried by the menisci.[47] Healthy menisci reduce the compressive loads or stress on the articular cartilage and protect the joint from osteoarthritic changes. The attachments and shape of the menisci tend to keep them from moving toward the center of the joint after the tibial condyles during extension and flexion. Injuries of the menisci commonly are produced by rotational forces that lead them to follow the femur and move on the tibia during rotation.

Mechanism of Tears

Traumatic Tears. During vigorous internal rotation of the femur with the knee in flexion, the femur tends to force the medial meniscus posteriorly and toward the center of the joint.[47] The medial meniscus has strong ligamentous attachments to the tibia at its central tips in the anterior and posterior horns. Therefore, its anterior and posterior attachments follow the tibia, but its intervening part tends to follow the femur. As a result of the tight peripheral attachments, which limit meniscal excursion, the meniscus may easily be trapped between the femur and the tibia. A *longitudinal tear* may be pro-

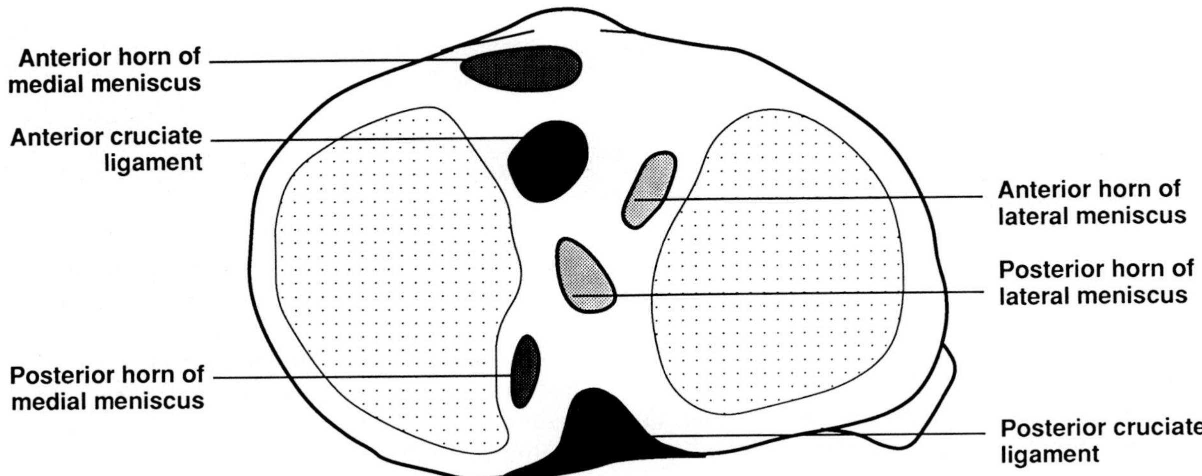

Figure 11–3. Sites of tibial attachments of menisci and ligaments.

duced when the joint is extended suddenly. Further extension of a longitudinal tear may lead to a *bucket-handle tear,* with intermittent locking of the joint occurring when the displaced inner segment is caught by the intercondylar notch.

The lateral meniscus is easily injured by vigorous external rotation of the femur on the tibia with the knee in partial flexion.[47] The lateral meniscus has loose attachments to the tibia at its anterior and posterior horns. Because of its small radius of curvature and loose peripheral attachments, a distractive force may separate the anterior and posterior horns, with the body segment being trapped by the femur and the tibia.[30] This mechanism leads to stretching of the inner concave border and results in *transverse* or *radial tears* of the lateral meniscus.

Degenerative Tears. A number of factors are thought to contribute to degenerative changes of the menisci. However, with regard to the biomechanics of an adult in the standing position, the weight-bearing force generally is transmitted across the posterior medial compartment of the knee. In an older meniscus, the collagen fibers may lose their elasticity and be unable to withstand as much trauma as a healthy meniscus.[47] Continued abnormal mechanics in a joint may lead to injury and weakening of these collagen fibers. With increasing age, the inferior surface of the weight-bearing meniscus degenerates, owing to wear and tear as it slides over the tibial plateau during joint motion. This may explain in part the frequent occurrence of the *horizontal cleavage tears* on the undersurface of the meniscus in elderly patients.

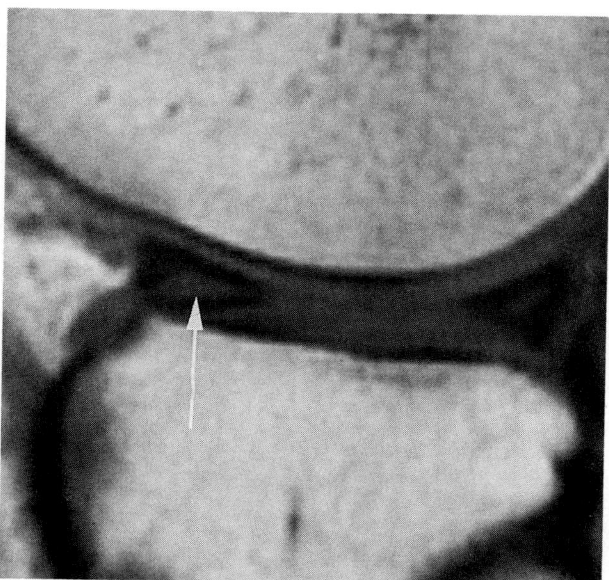

Figure 11–4. Meniscal degeneration. Sagittal T1-weighted image shows grade I signal intensity of the anterior horn of the lateral meniscus *(arrow)*.

Histology and Grading System for Magnetic Resonance Imaging

The normal meniscus on MR imaging appears as a homogeneous structure of low signal intensity on all pulse sequences because of the low proton density of fibrocartilage. In adults, increased intrameniscal signal intensity represents meniscal degeneration or intrasubstance tears, which may occur in either asymptomatic or symptomatic persons.[21,33] A grading system for MR imaging allows differentiation between meniscal degeneration and tear on the basis of intrameniscal signal pattern. Lotysch and colleagues[26] first divided intrameniscal signal into three grades according to the signal intensity with respect to the articular surface (Figs. 11–4 to 11–6). Stoller and co-workers[49] in an autopsy study of 12 menisci reported a one-to-one correspondence between histologic stages and MR imaging grades (Table 11–2). The grading system for MR imaging reduces false-positive diagnoses from intrameniscal degenerative signal (grades I and II) because only grade III signal is considered to indicate a meniscal tear. Subsequently, Frija and co-workers[14] in an autopsy study of 31 grossly normal menisci, reported a close correlation between intrameniscal signal intensity and severity and extent of histologic anomalies (Figs. 11–7 to 11–9).

Spin-echo T1-weighted and proton density-

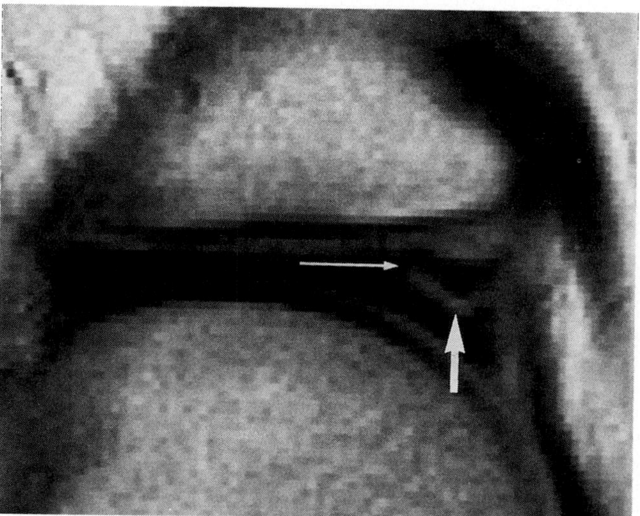

Figure 11–5. Meniscal degeneration. Sagittal T1-weighted image shows grade II signal intensity of the posterior horn of the medial meniscus *(thick arrow)*. Note the linear region of high signal intensity does not communicate with the tibial articular surface *(thin arrow)*.

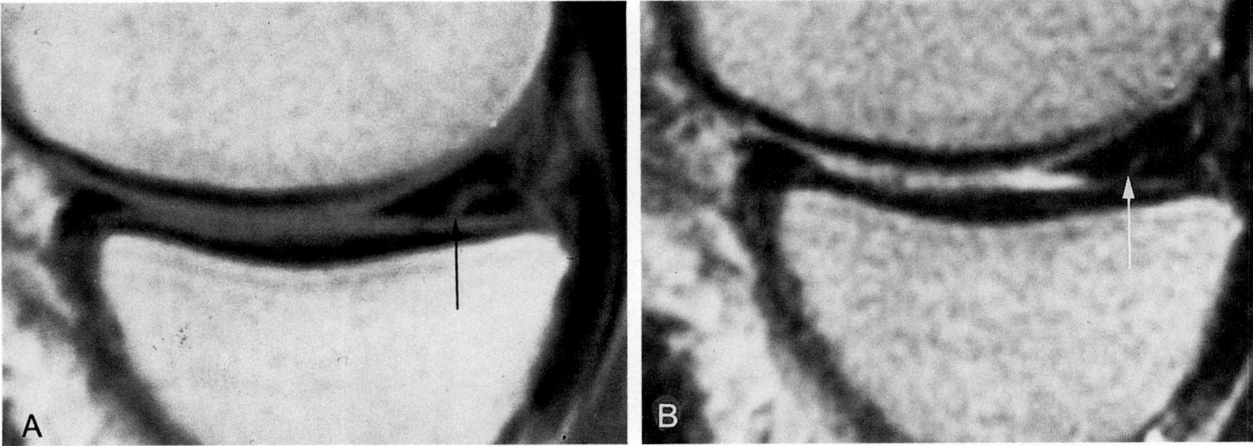

Figure 11-6. Meniscal tear with grade III signal intensity. *A*, Sagittal T1-weighted (TR, 800 ms; TE, 20 ms) image shows a linear intrameniscal area of high signal intensity communicating with the tibial articular surface *(arrow)*. *B*, The signal intensity in the intrameniscal region is less apparent on the T2-weighted (TR, 2000 ms; TE, 80 ms) image.

weighted images are the standard sequences to depict degeneration and tears. Spin-echo T2-weighted images do not offer any advantage in diagnosing tears (see Fig. 11-6). The linear increased signal intensity in meniscal tears has been recognized as imbibed synovial fluid within meniscal gaps.[49] The foci of increased signal intensity in early degenerative menisci may be the result of trapped water being proportionately absorbed onto boundary layers of macromolecules, as well as an absolute increase in local proton density that results in T1 shortening.[29] Although the presence of fat might shorten the TR value, Stoller and co-workers[49] emphasized that no lipids were found in the degenerative foci of the menisci. The decreased signal intensity on T2-weighted images is the result of the decrease in translational motion, leading to T2 shortening.[29]

Table 11-2. CORRELATION OF MAGNETIC RESONANCE IMAGING AND HISTOLOGIC FINDINGS IN MENISCAL TEARS

Grade	MR Imaging	Histologic Finding
I	One or several punctate intrameniscal signals not contiguous with an articular surface	Chondrocyte-deficient area and mucinous degeneration
II	A linear intrameniscal signal without articular extension	Extensive bands of mucinous degeneration; no distinct cleavage plane
III	A linear intrameniscal signal extending to at least one articular surface	Fibrocartilage separation with or without macroscopic extension to an articular surface

Results based on 12 cadaveric menisci in the study of Stoller DW, Martin C, Crues JV III, et al: Meniscal tears: Pathologic correlation with MR imaging. Radiology 163:731, 1987.

It is important to note that the grade II signals in menisci of children and adolescents do not carry the same significance as grade II abnormalities in adults. In children or adolescents, the central and inner menisci have a rich vascular supply, which accounts for the increased signal intensity seen on MR images.[34] In adult menisci, however, the capillary plexus is confined to the peripheral zone.[49] Areas of increased signal intensity should be interpreted as degeneration in adults.

In our experience, not all meniscal patterns fit definitely into one of the standard categories. Meniscal tears are considered equivocal when the signal is neither clearly confined to the substance of the meniscus nor definitely communicating with the articular surface (Fig. 11-10). The prevalence of equivocal meniscal tears is less than 14 per cent.[15,19] Of these cases up to 42 per cent have been reported to have tears at arthroscopy.[15]

Patients with meniscal degeneration or tear do not necessarily have symptoms. In the asymptomatic population, grade II signal intensity in the posterior medial meniscus accounts for most of the signal changes in all decades of life.[33] Intrameniscal signal changes begin in the second decade, progress with age, and occur in both sexes and in both inactive and exercised knees.[21]

Morphology of Tears

The present MR imaging grading system does not provide information on the morphology of tears, which is of particular value for surgical planning

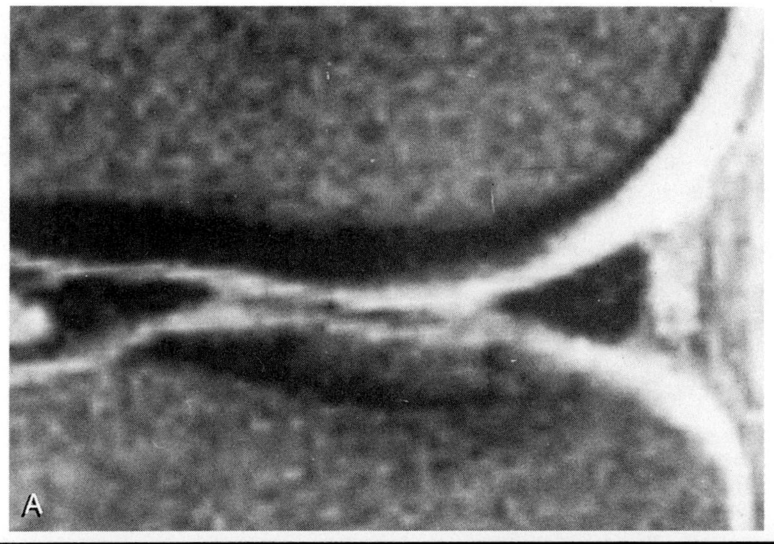

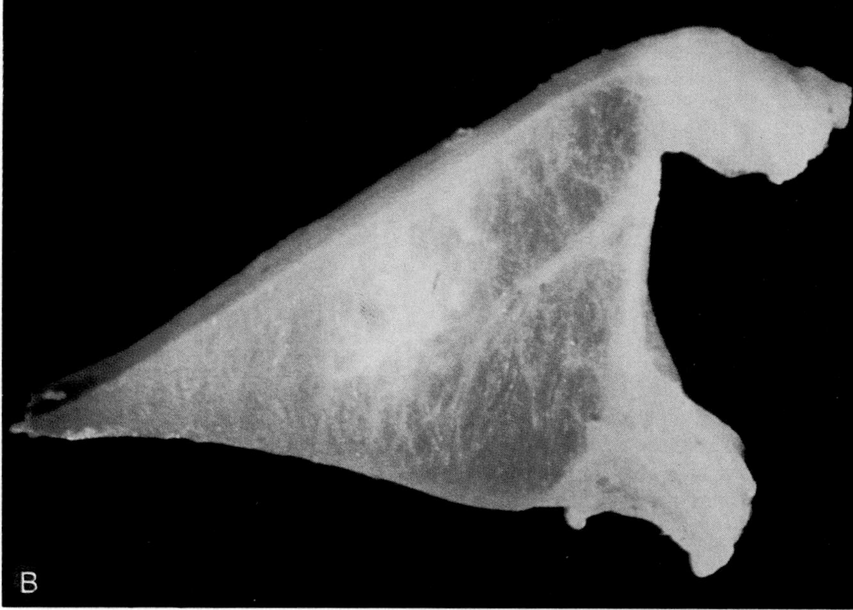

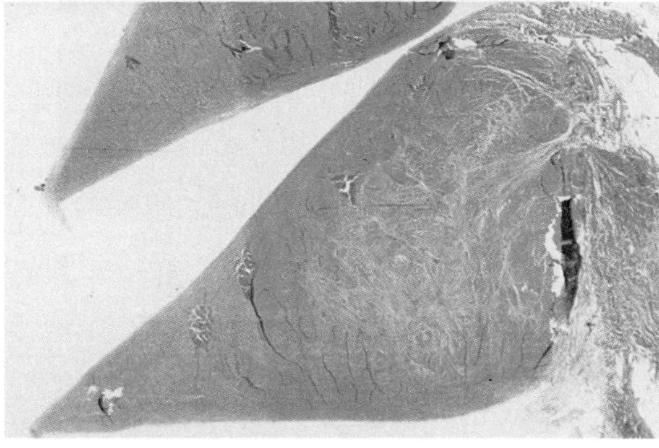

Figure 11-7. Punctate signal. *A*, A small punctate signal located intrameniscally that does not communicate with the articular surface can be seen. *B*, Macroscopic section showing a small whitish zone at the same level as in *A*. *C*, Histologic section with trichrome stain showing grade I fissures in the outer third combined with degenerative lesions. (Photographs courtesy of Professor G. Frija, Paris, France.)

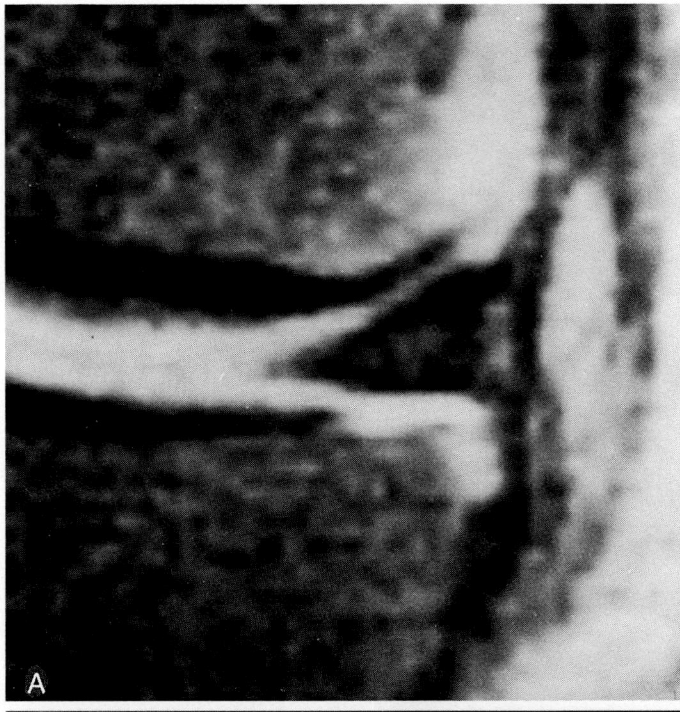

Figure 11–8. Diffuse signal. *A*, A homogeneous intrameniscal region of increased signal intensity is present that does not clearly extend to the articular surface. *B*, Macroscopic section showing a whitish central meniscal zone at the same level as in *A*. *C*, Histologic section with trichrome stain showing a central meniscal zone of myxoid degeneration corresponding to the MR signal distribution in *A*. (Photographs courtesy of Professor G. Frija, Paris, France.)

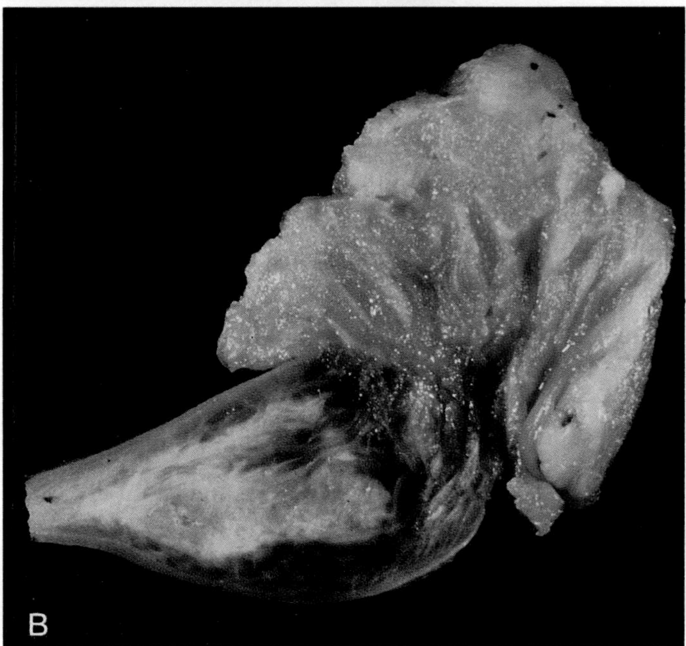

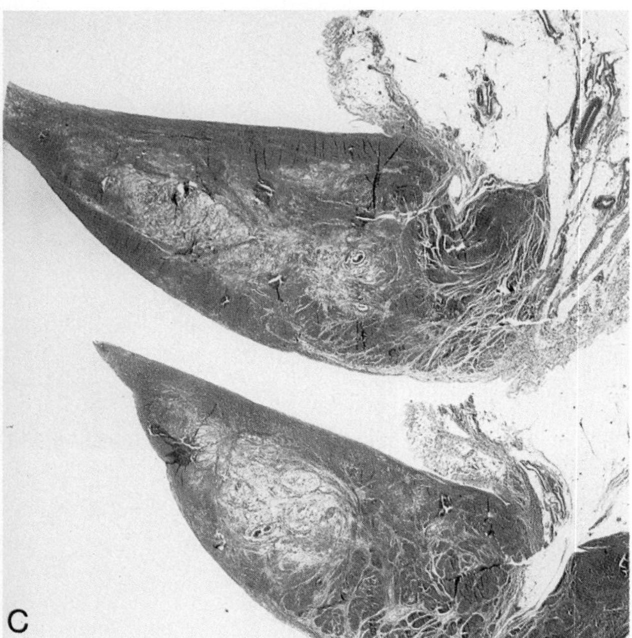

and prognosis. Meniscal tears are generally categorized into three types with reference to the triangular cross section of the meniscus (Fig. 11–11): (1) vertical tears, (2) horizontal tears, and (3) complex tears.[28,47] Vertical tears can be subclassified as (a) simple vertical tear, (b) peripheral tear, (c) meniscocapsular separation, and (d) bucket-handle tear. Horizontal tears can be subdivided into (a) cleavage or (b) radial, oblique, or parrot-beak tears. Complex tears are a combination of both vertical and horizontal tears. It is important to note that numerous classification systems have been devised for meniscal tears on the basis of surgical findings, mechanism, and other factors. The prevalence and significance of a different morphology of tears thus may vary widely in different reports. Various diagnostic modalities also yield different results. In our experience, horizontal cleavage tears of the posterior medial

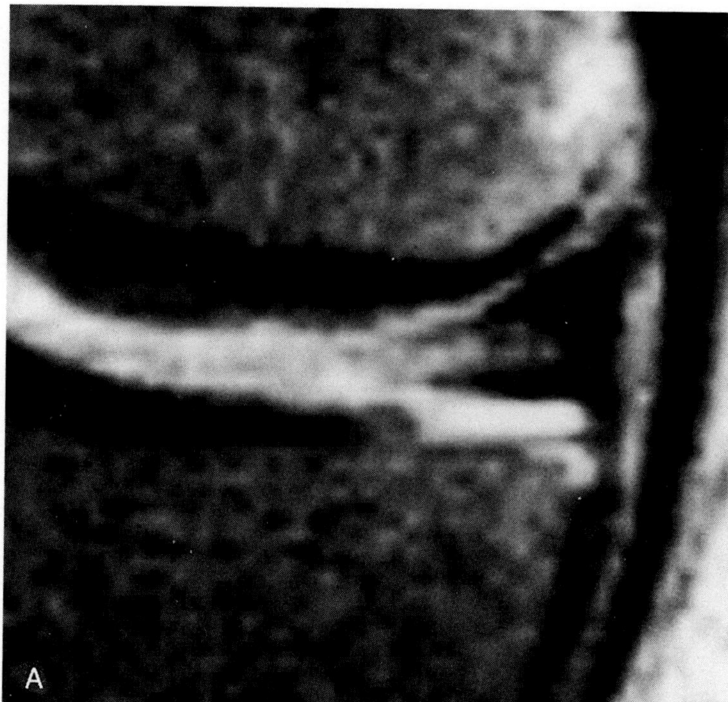

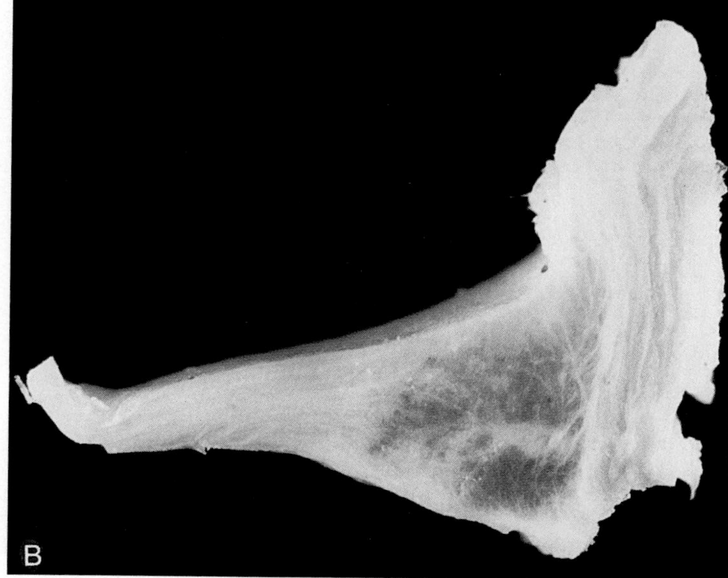

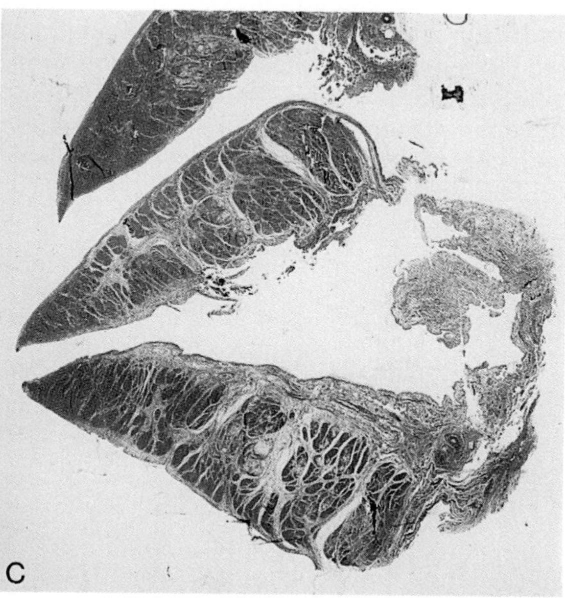

Figure 11-9. Linear signal extending to the articular surface. *A*, A horizontal intrameniscal linear region of increased signal intensity that communicates with the articular surface. *B*, Macroscopic section showing a thick whitish line in the outer third continuous with a diffuse whitish appearance of the inner third. *C*, Histologic section with trichrome stain showing complex lesions consisting of degeneration and fissuring. (Photographs courtesy of Professor G. Frija, Paris, France.)

meniscus are the most frequent type, commonly occurring in older patients as a consequence of degeneration. Acute traumatic episodes frequently result in vertical tears and occur slightly more often in the lateral meniscus. Complex tears can take place after degeneration and repeated traumatic episodes. The following description offers further explanation regarding tear morphology.

The vertical tear is a tear that extends parallel to the meniscal margins and frequently communicates with both superior and inferior articular surfaces. It commonly occurs in the middle portion of the meniscus; vertical tears located in the outer third of the meniscus are called *peripheral tears*.

The peripheral tear is a tear that involves the peripheral one third of the meniscus and frequently occurs in the posterior horn of the medial meniscus (see Fig. 11-27). This is the type of tear that can be repaired and may be expected to heal because of the blood supply in this region provided by the peri-

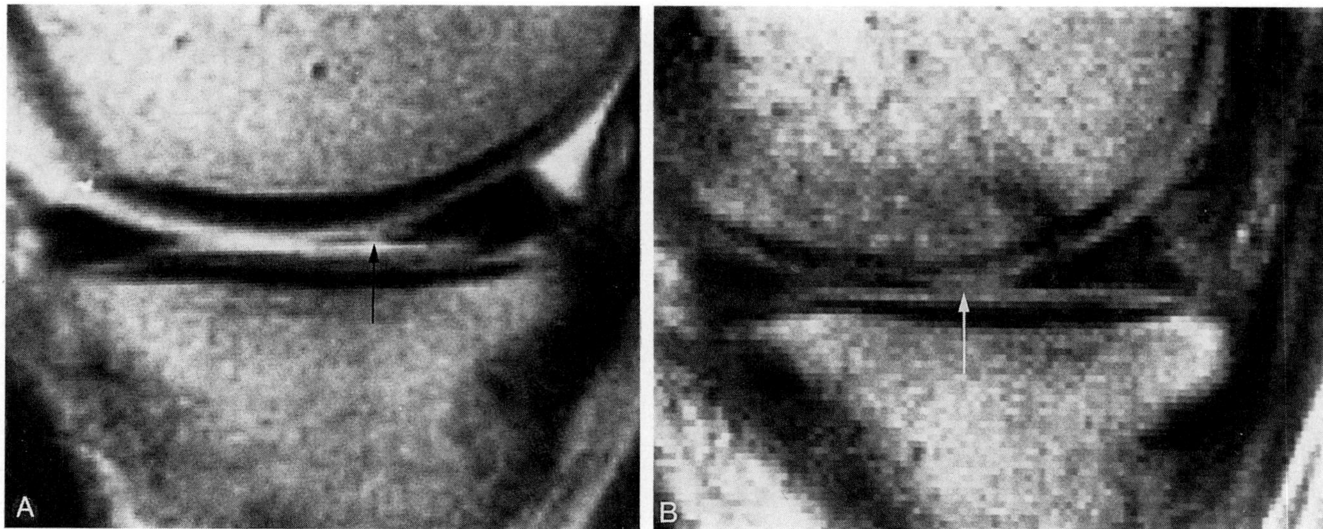

Figure 11–10. Equivocal meniscal tears in two different patients. Sagittal T2-weighted images show a suspicious region of grade III signal intensity in the posterior horn of the medial meniscus, connecting both superior and inferior articular surfaces *(arrow)*. Arthroscopy revealed a meniscal tear in the patient in *A*, but no tear was found in the patient in *B*.

meniscal capillary plexus, which arises from the geniculate arteries and supplies the outer 10 to 30 per cent of the meniscal tissue.[2] This vascular area has been termed the *red zone*, in contrast to the avascular central *white zone* of the meniscus.[8] Orthopedic surgeons are interested in red zone injuries because nondisplaced peripheral tears less than 1 cm long may heal with conservative management. Larger peripheral tears may be treated by arthroscopic repair.

Normally, there is no fluid between the medial meniscus and the joint capsule. The presence of fluid along the entire interface between the medial meniscus and its capsular attachment is diagnostic of meniscocapsular or meniscosynovial separation

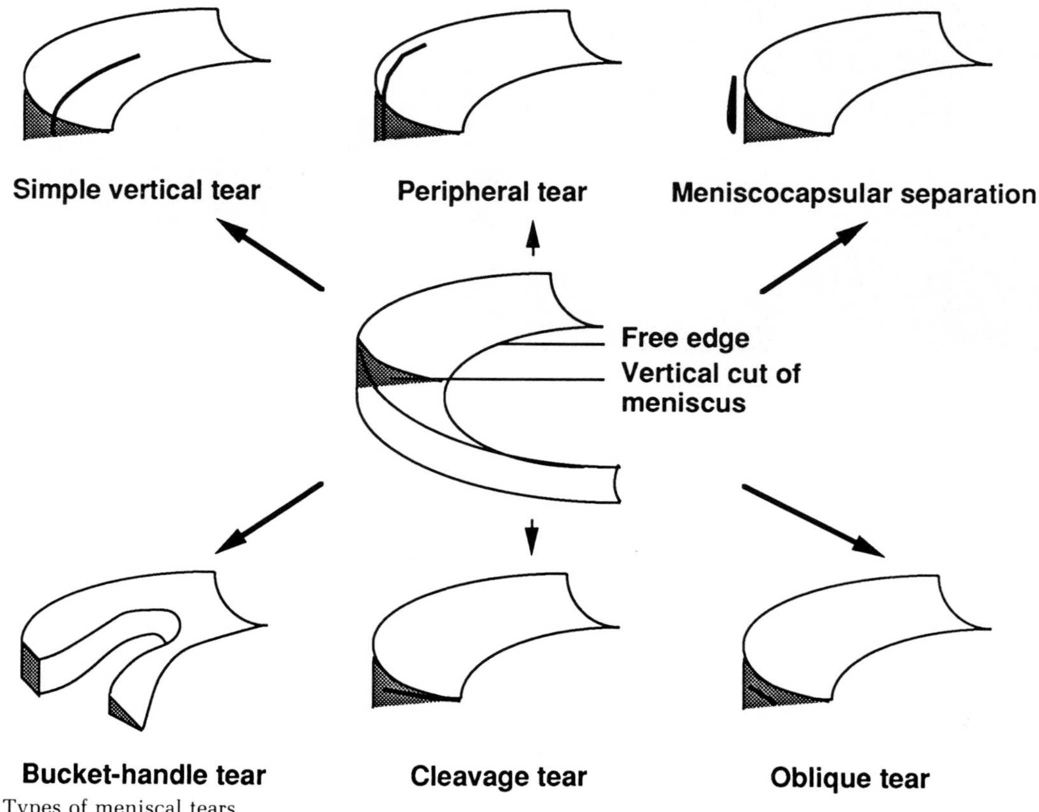

Figure 11–11. Types of meniscal tears.

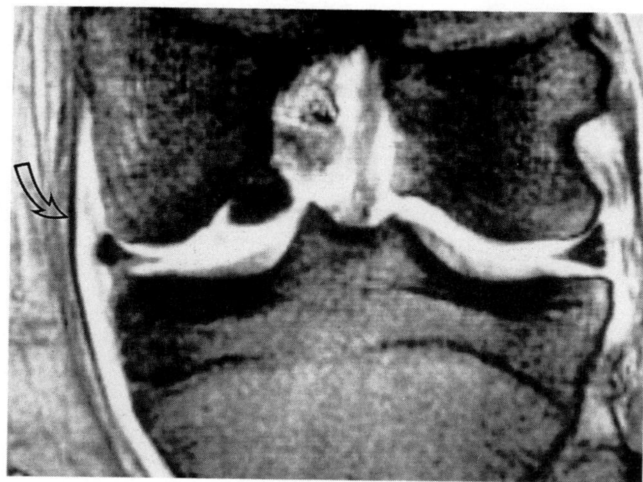

Figure 11–12. Meniscocapsular separation in a 50-year-old-man with osteochondritis dissecans. Coronal T2* gradient-echo image shows an extensive fluid tract separating the medial collateral ligament *(curved arrow)* and the medial meniscus. An osteochondral defect with a well-defined fluid interface from parent bone is seen in the intercondylar aspect of the medial femoral condyle. Attenuated anterior cruciate ligament is shown.

and most readily is depicted on coronal spin-echo T2-weighted or gradient-echo images (Fig. 11–12). It also can be depicted on sagittal images as a continuous band of fluid behind the posterior margin of the posterior horn of the medial meniscus.[30] However, fluid normally is present in the superior and inferior capsular recesses and should not be mistaken for a meniscocapsular separation. A band of low signal intensity between these two recesses represents the intact meniscal attachment. Occasionally, displacement of the meniscus may be a sign of meniscocapsular separation on MR imaging if a joint effusion is not present.[7] It should be noted that signs of the meniscocapsular separation on sagittal images may be unreliable because inaccuracy can be caused by inadequate window and level settings.

The bucket-handle tear often involves the medial meniscus and is a vertical or oblique tear with a full-thickness longitudinal extension that propagates anteriorly and posteriorly within the meniscus. The inner fragment frequently is displaced into the intercondylar notch in a way that resembles a handle; the peripheral nondisplaced fragment is the "bucket" (Fig. 11–13). Magnetic resonance imaging may show changes of meniscal morphology without visible abnormalities in intrameniscal signal intensity. These findings include (1) the nondisplaced fragment showing a truncated triangular appearance or decreased height (Fig. 11–14); (2) nonvisualization of the meniscal body (absence of a bow tie); (3) two fragments of the meniscus seen on coronal images (Fig. 11–15); and (4) a "double posterior cruciate ligament" sign indicating the displaced fragment of the meniscus (Figs. 11–16, 11–17). The double posterior cruciate ligament sign occurs when a band of low signal intensity of the displaced handle fragment is seen beneath the posterior cruciate ligament within the intercondylar notch on both sagittal and coronal MR images.[46,57]

The horizontal cleavage tear is a tear that runs at an acute angle with respect to the free edge of the meniscus and commonly is located in the inner third of the posterior horn of the medial meniscus (Figs. 11–18, 11–19; see also Figs. 11–6, 11–9).

The radial, oblique, or parrot-beak tear is a tear

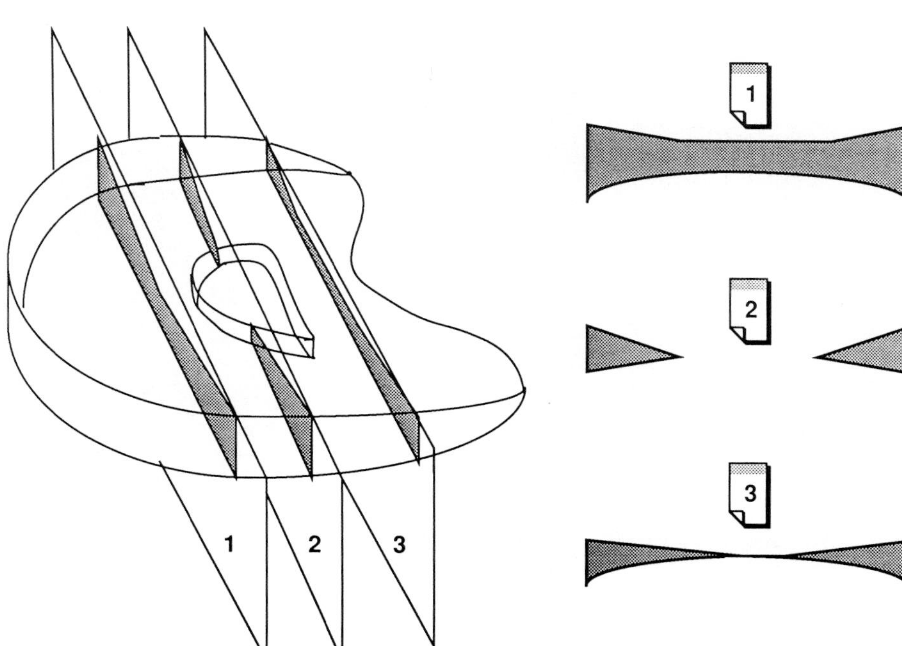

Figure 11–13. Bucket-handle tear with diagrams illustrating imaging findings.

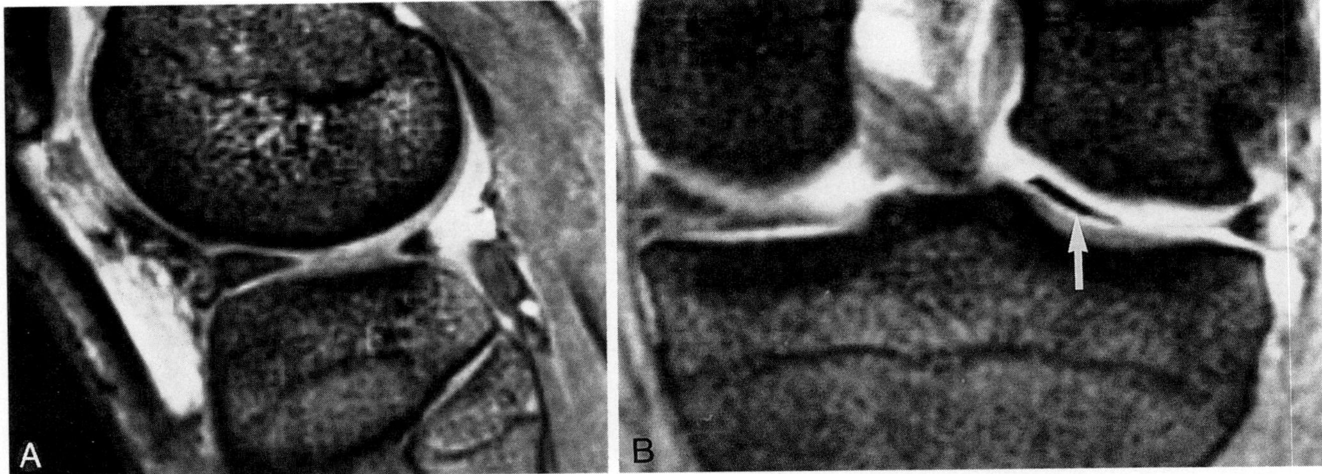

Figure 11–14. Bucket-handle tear of the posterior horn of the lateral meniscus. *A*, Sagittal, gradient-recalled acquisition of steady state (GRASS) image (TR, 35 ms; TE, 15 ms; flip angle, 30 degrees) shows the truncated triangular appearance of the posterior horn of the lateral meniscus. An abnormally large "anterior horn" is noted. *B*, Coronal T2* gradient-echo image (TR, 600 ms; TE, 20 ms; flip angle, 30 degrees) confirms a displaced meniscal fragment toward the intercondylar notch *(arrow)*.

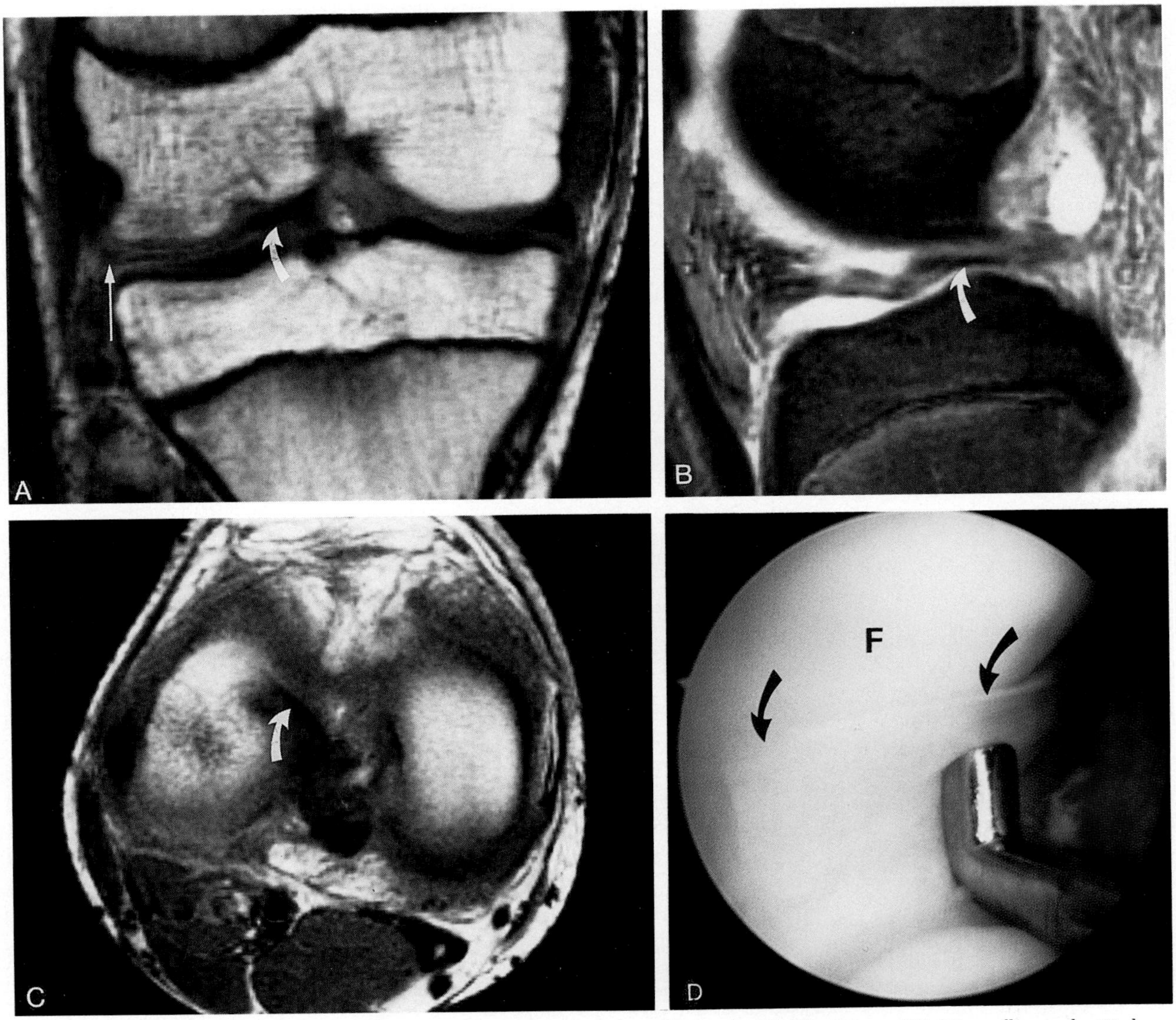

Figure 11–15. Coronal T1-weighted (*A*, TR, 600 ms; TE, 20 ms), sagittal gradient-echo (*B*, TR, 600 ms; TE, 20 ms; flip angle, 25 degrees), and axial T1-weighted (*C*, TR, 600 ms; TE, 20 ms) images show a bucket-handle tear of the lateral meniscus of the right knee. A displaced fragment is dislodged toward the intercondylar notch *(curved arrow)*. Note the truncated nondisplaced meniscal fragment on the coronal image in *A (arrow)*. *D*, A probe pushing the displaced fragment *(arrows)* in the intercondylar notch is demonstrated in an arthroscopic view. F, Lateral femoral condyle.

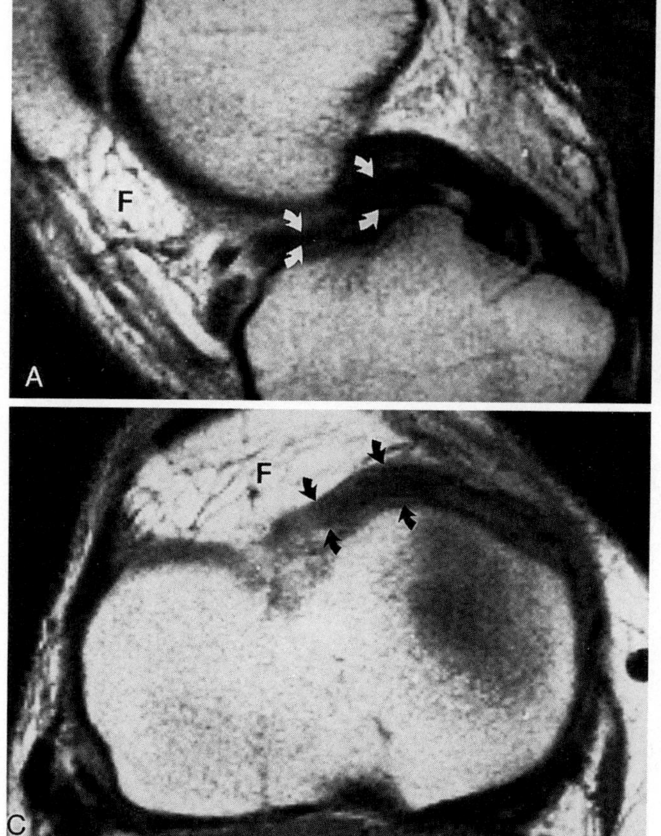

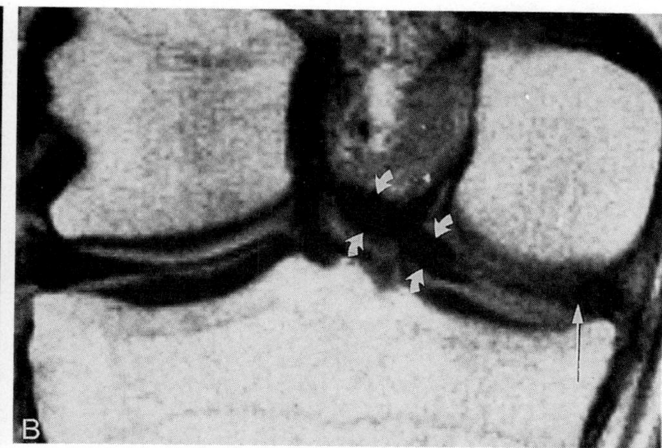

Figure 11–16. Bucket-handle tear of the posterior horn of the medial meniscus in a 31-year-old woman with locked knee (arthroscopically proved). Sagittal T1-weighted *(A)* and coronal T1-weighted *(B)* images show a band of low signal intensity of the displaced fragment *(curved arrows)* underneath and parallel to the posterior cruciate ligament. A nondisplaced fragment demonstrates a shortened meniscus *(straight arrow, B)*. *C*, Axial T1-weighted image shows that the displaced fragment is dislodged toward the infrapatellar fat pad (F) anteriorly and centrally.

that runs at an oblique angle with respect to the free edge of the meniscus and can penetrate from the inner to the outer third of the meniscus (Figs. 11–20, 11–21). A parrot-beak tear is a tear with both vertical and horizontal components in an oblique plane. Commonly, this type of tear is located at the junction of the body and posterior horn of the lateral meniscus. Magnetic resonance imaging may show a blunted free margin of the meniscus without changes in signal intensity.

Complex tears consist of both a vertical and a horizontal component or of two or more of these types of tears (Figs. 11–22 to 11–24).

Sagittal images are the best views to evaluate meniscal injuries of the anterior horn, body (Fig. 11–25), and posterior horn. Coronal images often are useful to confirm the location and morphology of meniscal injuries. Occasionally, axial images may show tear morphology. Determination of the tear morphology and location within the inner, middle, and outer third of the meniscus may be valuable for surgical planning.

Accuracy and Limitations

Using arthroscopic findings as a standard, the reported accuracy for MR imaging in the evaluation of meniscal tears varies from 72 to 94 per cent (Table 11–3). Overall, approximately 80 to 94 per cent of MR grade III menisci have been reported to have tears at arthroscopy.[7,38] False-positive MR imaging studies, seen in about 10 per cent of cases, often are related to lack of familiarity with interpretative pitfalls at MR imaging. However, arthroscopy is itself an imperfect gold standard in evaluating the knee joint. The reported accuracy for arthroscopy in the evaluation of meniscal tears ranges between 69.8 and 98.0 per cent,[24,27,43,50] indicating that arthroscopy does not necessarily provide diagnostic advantage over MR imaging. It is well known that without extensive probing, arthroscopists often underestimate tears in the periphery and inferior surface of the posterior medial meniscus.[24] Interestingly, nearly 70 per cent of so-called false-positive MR imaging results are found in the posterior medial menis-

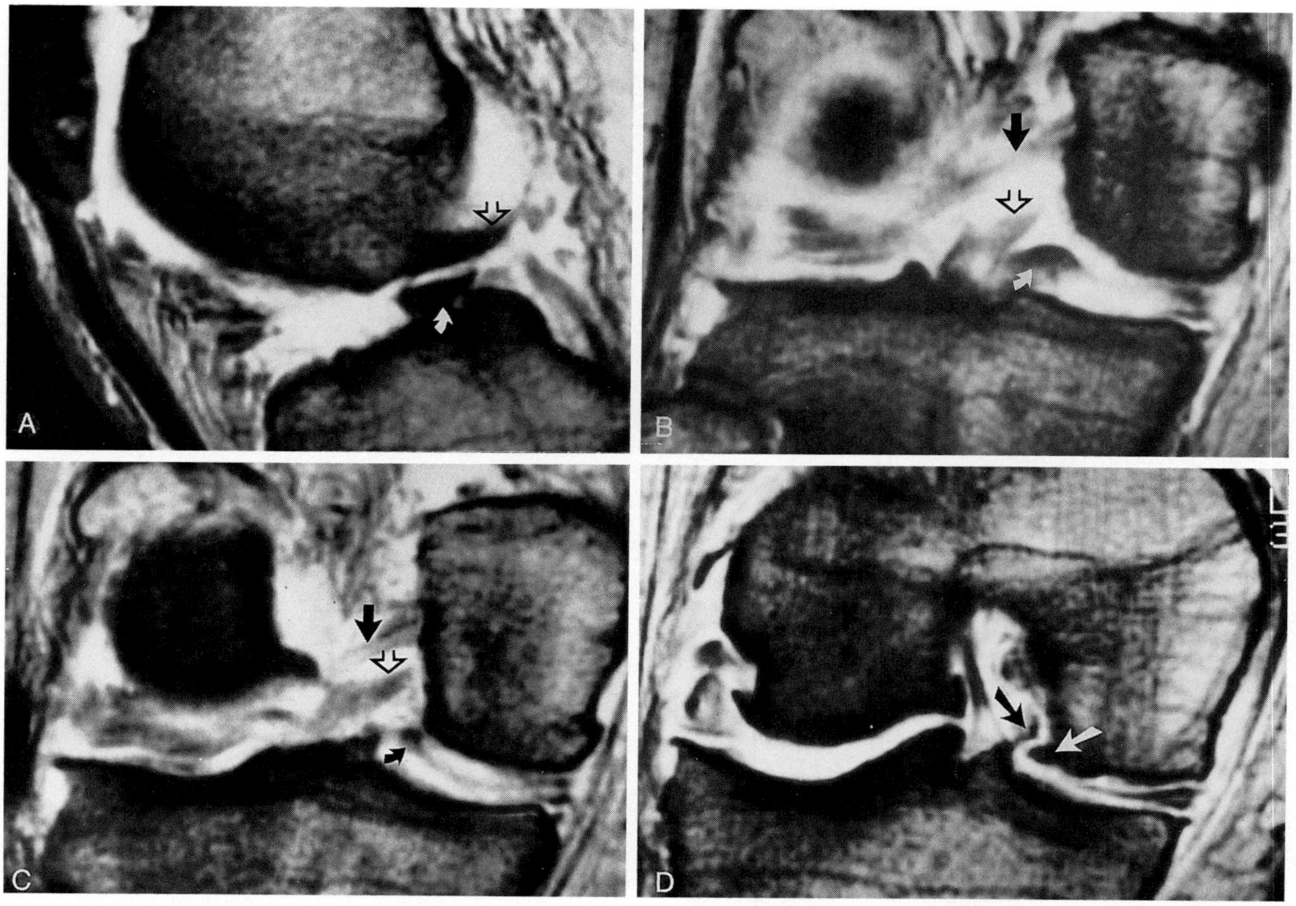

Figure 11–17. Bucket-handle tear of the posterior horn of the medial meniscus in a 59-year-old woman with degenerative joint disease. *A*, Sagittal gradient-echo image shows the double posterior cruciate ligament sign—a free meniscal fragment of low signal intensity *(curved arrow)* parallel to and beneath the posterior cruciate ligament *(open arrow)*. *B–D*, Posterior coronal gradient-echo images, from posterior to anterior, show the displaced meniscal fragment *(curved arrow)* beneath the posterior cruciate ligament *(open arrow)* and toward the intercondylar notch. The band of low signal intensity superior to the posterior cruciate ligament is a meniscofemoral ligament *(solid arrow)*. Note in *D* that osteophytes protruding from the tibial spine and femoral condyle should not be mistaken for free meniscal fragments *(arrows)*.

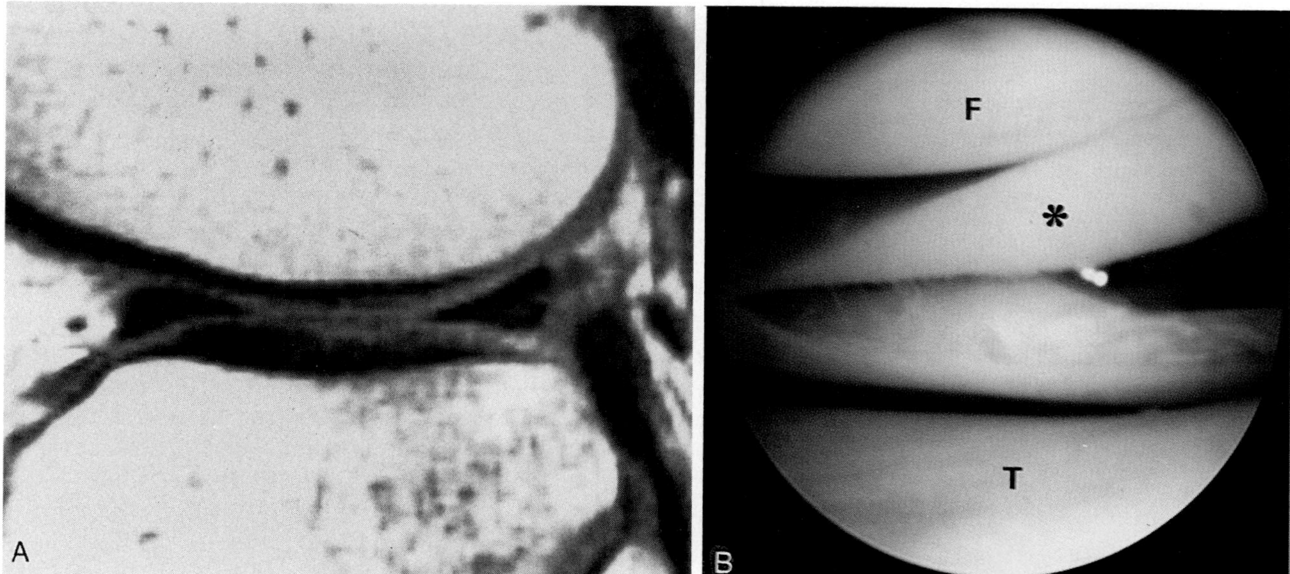

Figure 11–18. *A*, Sagittal T1-weighted image (TR, 1100 ms; TE, 20 ms) shows a horizontal cleavage tear in the posterior horn of the lateral meniscus of the right knee. *B*, Arthroscopic view from an anterolateral portal and probe from an anteromedial portal demonstrate a large flap *(asterisk)* extending from the free edge of the lateral meniscus. F, Lateral femoral condyle; T, tibial plateau.

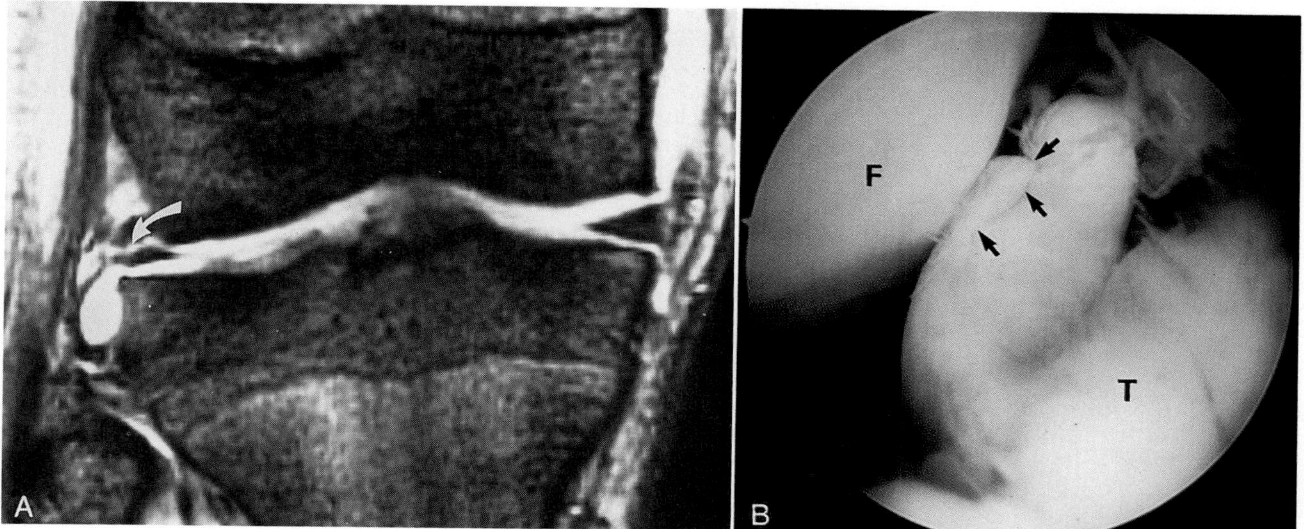

Figure 11–19. *A*, Coronal gradient-echo image (TR, 600 ms; TE, 20 ms; flip angle, 25 degrees) shows an oblique tear in the posterior horn of the lateral meniscus of the right knee, connecting the femoral articular surface *(arrow)*. *B*, Arthroscopic view through the intercondylar notch confirms the split *(arrows)* extending to the superior surface of the meniscus. F, Lateral femoral condyle; T, tibial plateau.

cus.[7,29] This may indicate that what appear to be false-positive MR imaging results may actually represent false-negative arthroscopic findings.

Perhaps more importantly, Fischer and co-workers reported that 17 per cent of MR imaging grade II menisci were found to have tears at arthroscopy.[13] Crues and associates also found that 17 tears in 154 menisci (11 per cent) prospectively were graded as I or II.[7] These data indicate a relatively low reliability for a negative MR imaging study. This can be explained either by an underestimation of grade III signal intensity or by a progression of definite small or intrasubstance tears to complete tears by the time of arthroscopic performance. Underestimation may occur when tears are oriented parallel to the plane of the image and they are manifested as abnormal morphology of the meniscus without changes in signal intensity.[3]

With regard to magnetic field strength, the 1.5 T unit may be more reliable than the 0.35 T unit.[7,13,29,38,44] However, a lower degree of accuracy for the 1.5 T system also has been reported.[36] At a

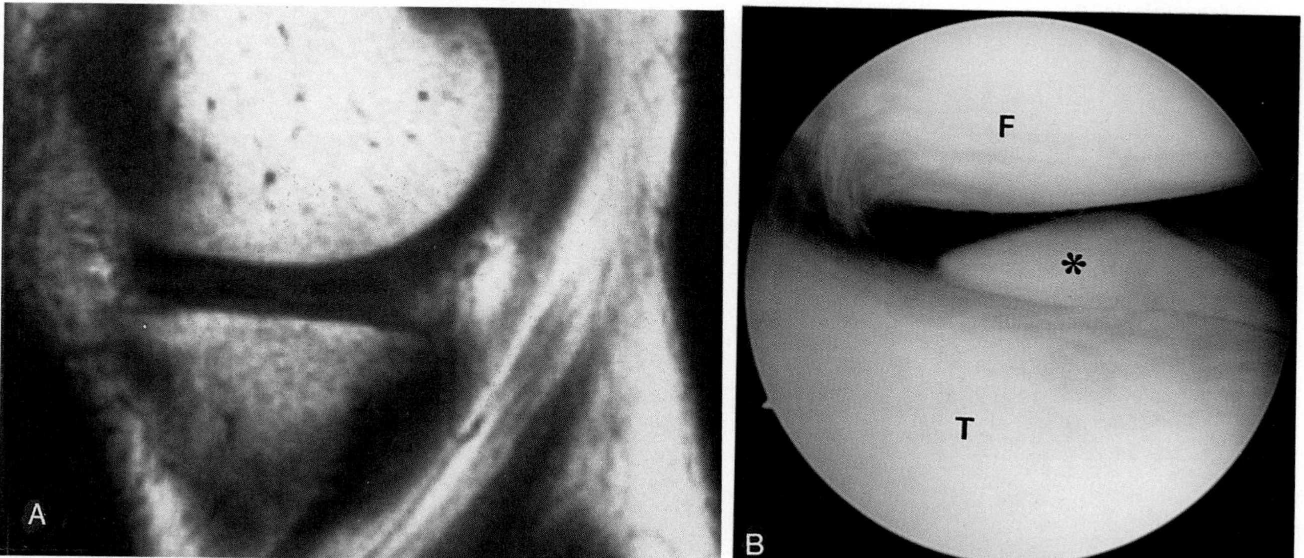

Figure 11–20. *A*, Sagittal T1-weighted image (TR, 600 ms, TE, 20 ms) shows an oblique tear in the body of the medial meniscus of the right knee. *B*, Arthroscopic view shows a mobile flap tear *(asterisk)* in the medial meniscus between the medial femoral condyle (F) and the tibial plateau (T).

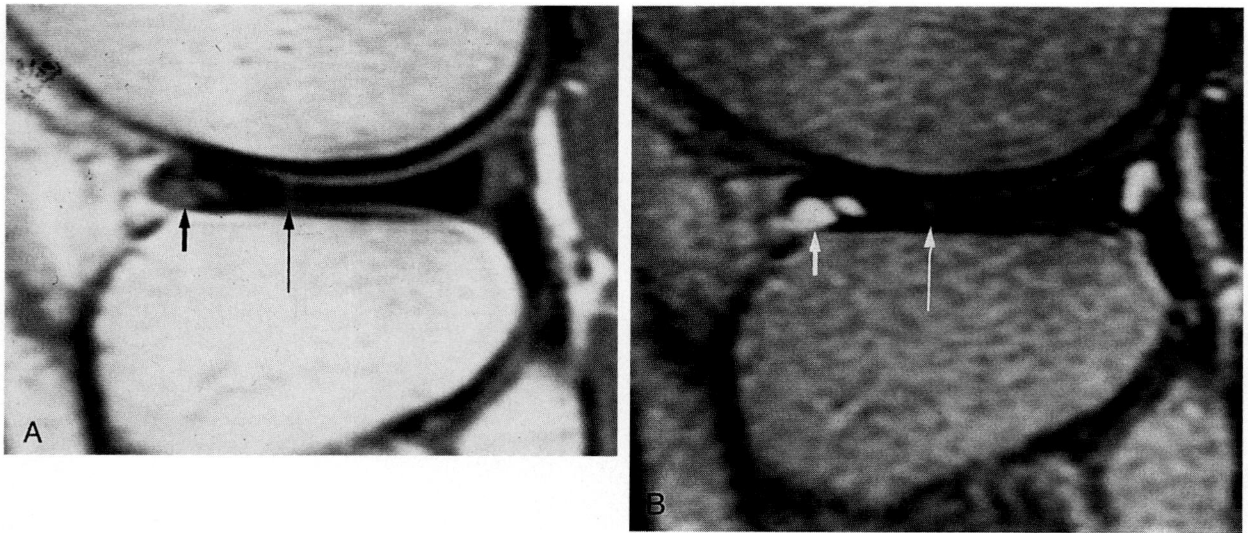

Figure 11-21. Radial (parrot beak) tear of the lateral meniscal body in a 26-year-old woman. Sagittal images (*A*, TR, 2000 ms; TE, 20 ms; *B*, TR, 2000 ms; TE, 80 ms) show a typical appearance and location of such a tear (*long arrow*). A small lateral meniscal cyst is also noted anteriorly (*small arrow*).

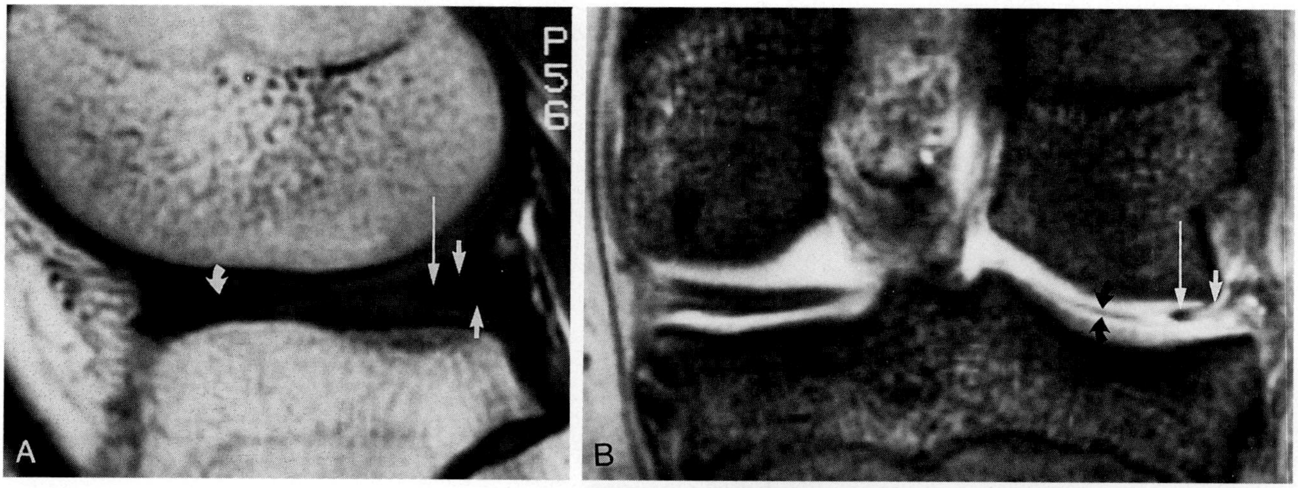

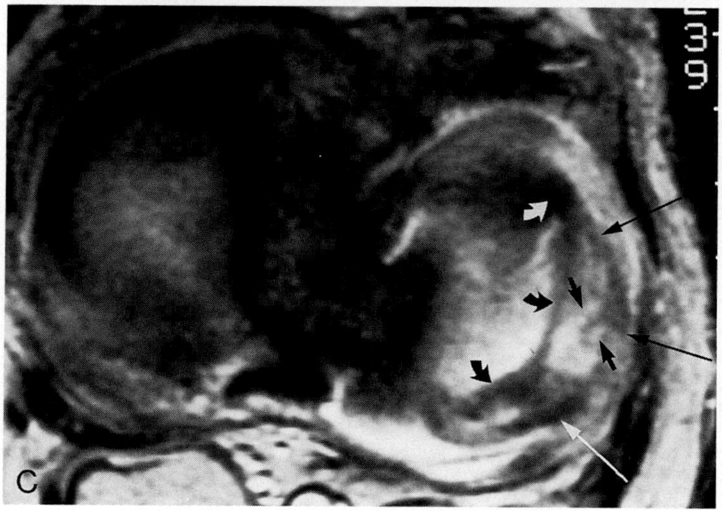

Figure 11-22. Traumatic complex tear and bucket-handle tear in a 38-year-old man. Sagittal T1-weighted (*A*), coronal gradient-echo (*B*), and axial T1-weighted (*C*) images show a complex tear (*small arrows*) at the point of break in the bucket-handle segment of the lateral meniscus. The complex tears are of high signal intensity on gradient-echo images because of imbibed fluid (*small arrows* in *B* and *C*). A fragment from the bucket-handle tear is displaced toward the intercondylar notch (*curved arrows*). Shortening of the nondisplaced fragment of the lateral meniscus is identified (*long arrow*). This patient's left knee had been twisted while playing soccer 3 months earlier. Arthroscopy confirmed all of the MR findings. A partially lateral meniscectomy was performed subsequently.

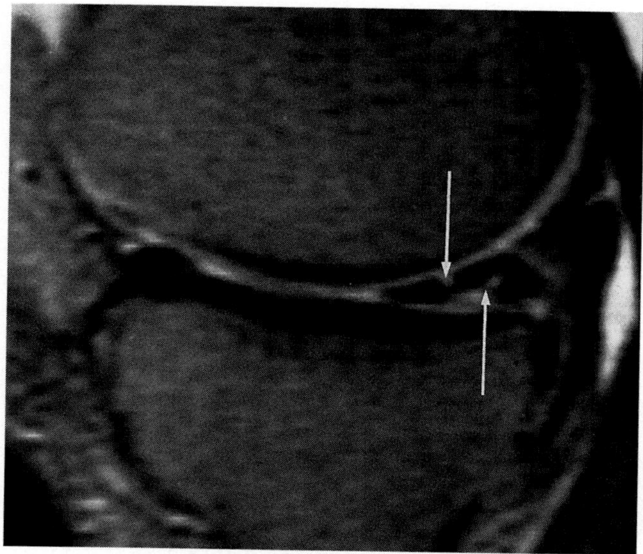

Figure 11–23. Degenerative, complex cleavage tears in the posterior horn of the lateral meniscus in a 60-year-old man (arthroscopically proved). Sagittal T2-weighted image (TR, 2000 ms; TE, 80 ms) shows meniscal tears of high signal intensity *(arrows)*.

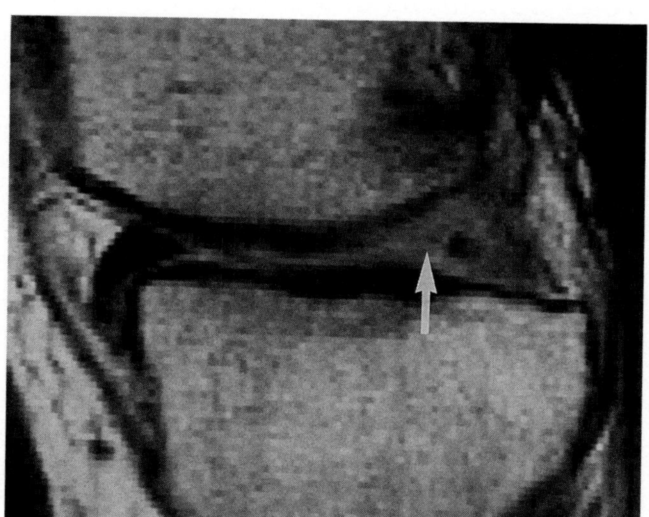

Figure 11–24. Sagittal T1-weighted image (TR, 800 ms; TE, 20 ms) shows maceration of the posterior horn of the medial meniscus *(arrow)* in a 65-year-old woman (arthroscopically proved). This patient was subsequently treated by arthroscopic partial medial meniscectomy.

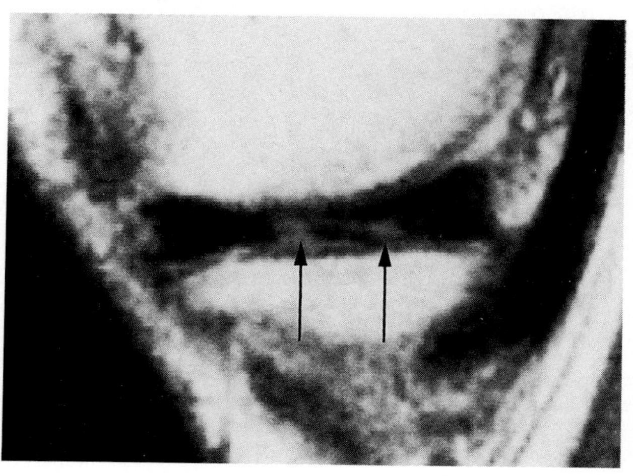

Figure 11–25. Sagittal T1-weighted image shows degenerative, complex cleavage tears of the medial meniscal body *(arrows)* in a 54-year-old man. This patient was subsequently treated by arthroscopic partial meniscectomy.

Table 11–3. COMPARISON OF MAGNETIC RESONANCE IMAGING RESULTS IN THE DIAGNOSIS OF MENISCAL TEARS

	Reicher 1986 (0.3 T)		Crues 1987 (1.5 T)		Mink 1988 (1.5 T)		Fischer 1991*		Raunest 1991 (1.5 T)	
	MM	LM	MM	LM	MM	LM	MM	LM	MM	LM
Number of cases	69	69	144	144	459		911	971	50	50
Accuracy (%)	77	78	89	94	94	92	89	88	72	72
Sensitivity (%)	84	58	87	88	97	92	93	69	94	78
Specificity (%)	80	100	91	98	89	91	84	94	37	69
PPV (%)	84	92	93	96			86	76	71	58
NPV (%)	69	86	84	92			92	92	78	85

*A multicenter and multi-unit analysis.

PPV, Positive predictive value; NPV, negative predictive value; MM, medial meniscus; LM, lateral meniscus.

very low field strength of 0.064 T, agreement of MR imaging and arthroscopy was found in 79 per cent of cases in the diagnosis of meniscal tears.[42]

Other Conditions of Menisci

Postoperative Mensicus. The outer (peripheral) third of the meniscus is relatively well vascularized.[2] Small tears within this zone can be treated conservatively, whereas large tears may be repaired arthroscopically with a high rate of success. However, a signal of grade III intensity from both conservatively treated and repaired menisci may persist long after the tear has become asymptomatic and presumably has healed, as documented in a report of 15 asymptomatic patients 3 to 27 months after injury.[8] It is important to note that such persistent signal intensity should not necessarily be interpreted as meniscal re-tear. Magnetic resonance imaging currently is unable to distinguish between meniscal healing and re-tear *at the same location* when a signal of grade III intensity is seen. To solve this problem, arthrography should be considered for an assessment of the symptomatic, previously repaired meniscus.[11] A tear at a location different from that which has been repaired arthroscopically, however, may be diagnosed accurately using conventional criteria (Figs. 11–26, 11–27).

Treatment of tears in the inner two thirds (avascular zone) of the meniscus are controversial. A current surgical principle is to preserve as much meniscal tissue as possible in an attempt to improve the outcome. Poor prognosis with a high prevalence of osteoarthritis is often found in patients who have undergone total meniscectomy. In patients who have had a partial meniscectomy, Smith and Totty[48] found that meniscal remnants in which more than two thirds of the normal meniscal length had been preserved have an appearance similar to that of menisci on which no surgery has been performed,

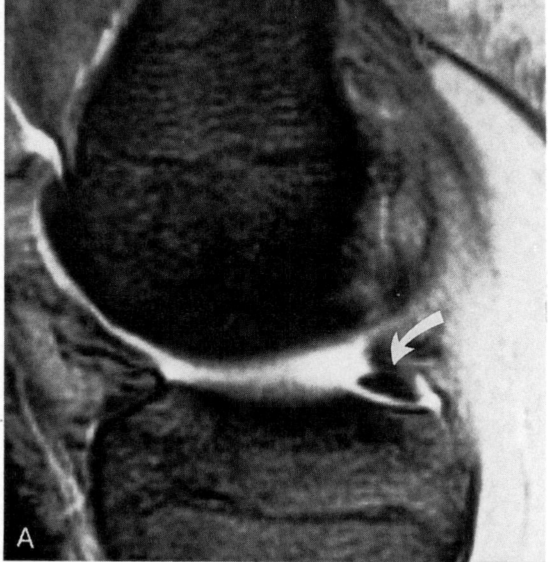

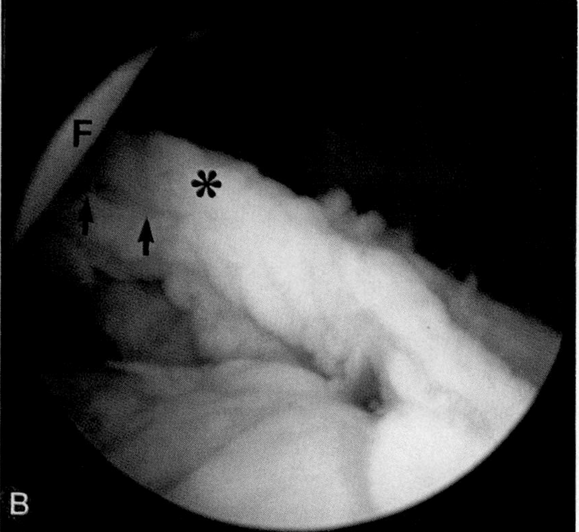

Figure 11–26. *A*, Sagittal T2* gradient-echo image (TR, 533 ms; TE, 15 ms; flip angle, degrees) shows a residual horizontal tear in the posterior horn of the medial meniscus of the right knee *(arrow)*. Previously this patient had had a partial meniscectomy. *B*, Arthroscopic view demonstrates a fibrillated superior fragment *(asterisk)*, which is torn from the posterior horn of the medial meniscus *(arrows)*. F, Medial femoral condyle.

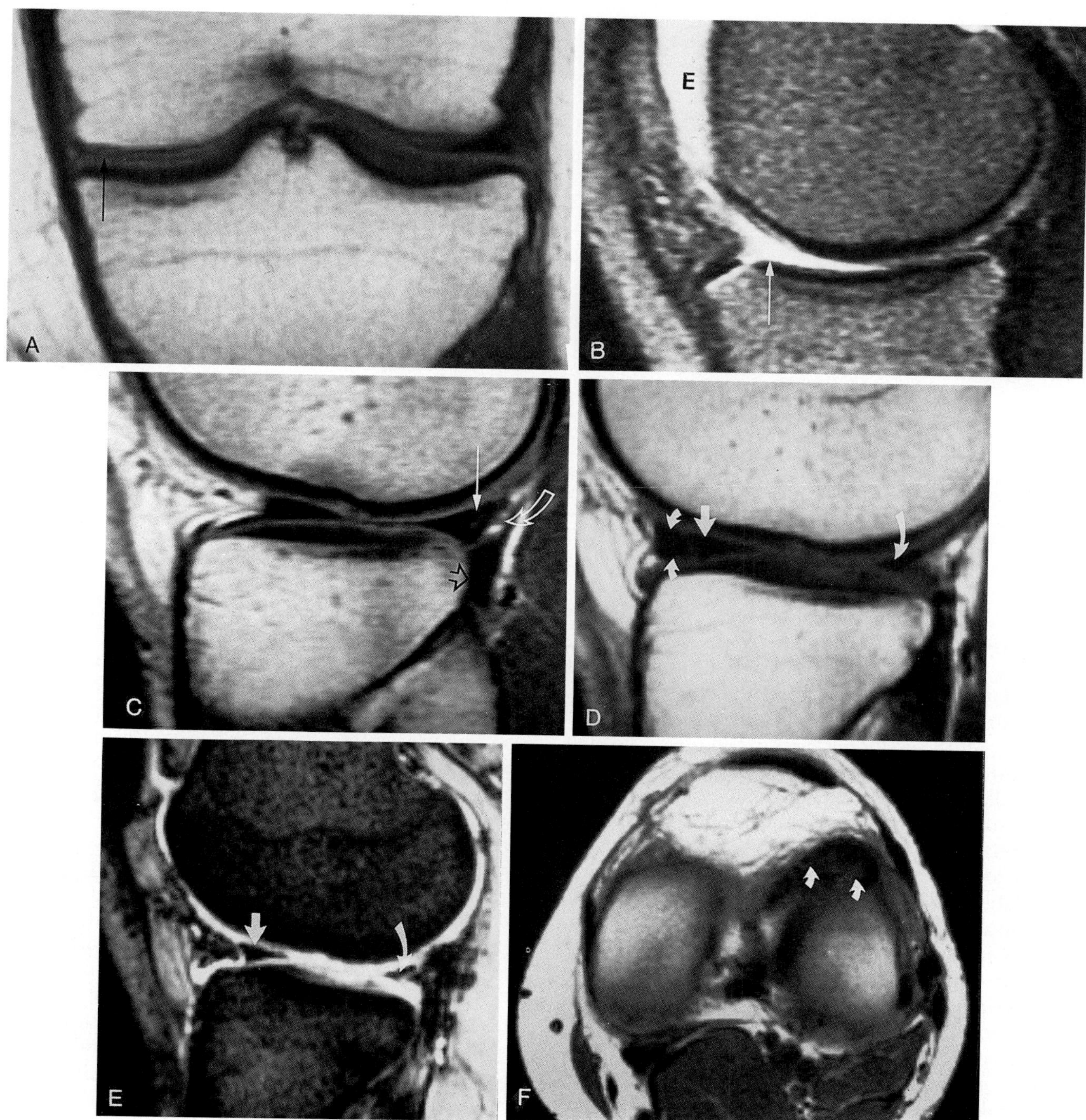

Figure 11–27. Postoperative meniscus in 26-year-old woman. Ten years earlier, this patient had had an anterior cruciate ligament tear and open medial meniscectomy of her left knee. MR imaging of her knee was requested because of her recurrent episodes of knee locking. Coronal T1-weighted *(A)* and sagittal T2-weighted *(B)* images show the absence of the medial meniscus *(arrows)* after medial meniscectomy. Associated effusion (E) replacing the resected anterior horn is shown *(arrow* in *B). C,* Sagittal T1-weighted image shows a peripheral tear of the *lateral* meniscus *(arrow)*. The popliteal tendon sheath *(curved open arrow)* should not be mistaken for a tear. *Open arrow,* popliteal tendon. Arthroscopic repair of her lateral meniscus was performed subsequently. *D–F,* After 7 months, the patient developed recurrent left knee pain and locking. Sagittal T1-weighted *(D)*, gradient-echo *(E)*, and axial T1-weighted *(F)* images show apparent re-tearing of the lateral meniscus (bucket-handle tear). A truncated triangular appearance of the posterior horn is identified, representing a nondisplaced fragment *(large curved arrow)*. The displaced fragment *(small curved arrows)* is seen anteriorly to the anterior horn *(straight arrow)*. The microscopic metallic particles present after the previous surgical meniscal repair create multiple small artifacts of low signal intensity. These artifacts are more prominent on the gradient-echo image in *E*.

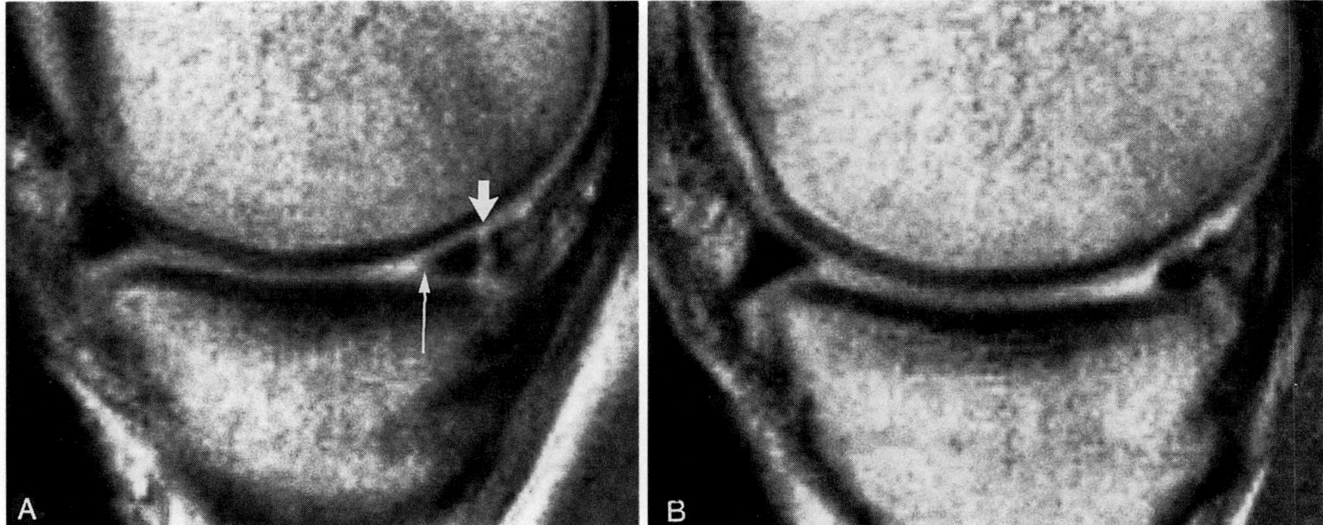

Figure 11–28. Postoperative view of the meniscus of a 29-year-old woman. This patient had had a bucket-handle tear of the medial meniscus and surgery on her left knee 4 years previously. MR imaging of her knee was requested because of her anterior cruciate ligament–deficient knee. *A*, Sagittal T1-weighted image (TR, 800 ms; TE, 20 ms) shows the postoperative shortened posterior horn of the medial meniscal remnant *(thin arrow)* with a grade III signal extending to both superior and inferior articular surfaces *(thick arrow)*. *B*, A marked irregularity of the superior articular surface is noted on the consecutive image, probably resulting from the previous operation. Arthroscopy revealed a new meniscal tear in the posterior horn of the medial meniscus. The shortened meniscal rim was the result of an excision of a bucket-handle tear.

although a blunted meniscal tip was seen in most patients (Fig. 11–28). With more extensive resection, marked contour irregularity simulating meniscal fragmentation appears frequently on MR images, seen in 30 per cent of segments with residual meniscal tissue. Of these, 86 per cent proved to be normal at second-look arthroscopy. Smith and Totty concluded that standard MR imaging criteria can be used to diagnose tears in the absence of marked contour irregularity; however, in the presence of marked contour irregularity, the standard criteria are not reliable.

Discoid Meniscus. A discoid meniscus is an uncommon, morphologically enlarged meniscus having an abnormal width or height. The prevalence is approximately 3 per cent for the lateral meniscus[47] and 0.1 to 0.3 per cent in the medial meniscus.[30] Most discoid menisci are asymptomatic unless they degenerate or become injured. However, a loud snap occasionally is heard during flexion and extension of the knee. An intact discoid meniscus seen as an incidental finding does not necessarily require treatment. A discoid meniscus is more susceptible to degeneration and tearing (Fig. 11–29); however, tears of discoid menisci that produce symptoms usually are treated by resection of the excess central meniscal tissue.

As mentioned previously, the diagnosis of discoid meniscus should be suspected when a bow-tie configuration of the meniscal body is seen on more than two contiguous 5-mm thick sagittal images (Fig. 11–30). Similarly, a discoid meniscus should be considered when the anterior and posterior horns are depicted on fewer than two slices.[30] This can be explained by the abnormal width of the discoid body, which is seen on three or more 5-mm thick contiguous sagittal images, relatively reducing the configuration of both triangular horns. Coronal images allow confirmation of the presence of a discoid meniscus, and they also allow comparison of meniscal height. A discoid meniscus often is increased in height by 2 mm compared with the opposite normal meniscus.[45] Silverman and colleagues[45] reported that MR imaging and arthroscopic findings were in agreement in 9 of 10 cases in the diagnosis of discoid menisci.

Counting the number of bow ties on sagittal images helps to determine the width of the body segment and thereby to avoid overlooking a discoid meniscus. Discoid menisci are more prone to tear, and MR imaging is reliable in evaluating this sequela.

Parameniscal Cysts. A parameniscal cyst is an accumulation of fluid within the meniscocapsular margin, with lateral cysts predominating 3 to 1.[47] Most observers believe that meniscal cysts develop after trauma, as a result of mucinous joint fluid being forced out through underlying tears. Other causes include degeneration, congenital predisposition, or postoperative effect, which occurs in 1 per cent of meniscectomies.[40] Pain is the predominant symptom, and it is accentuated by activity.

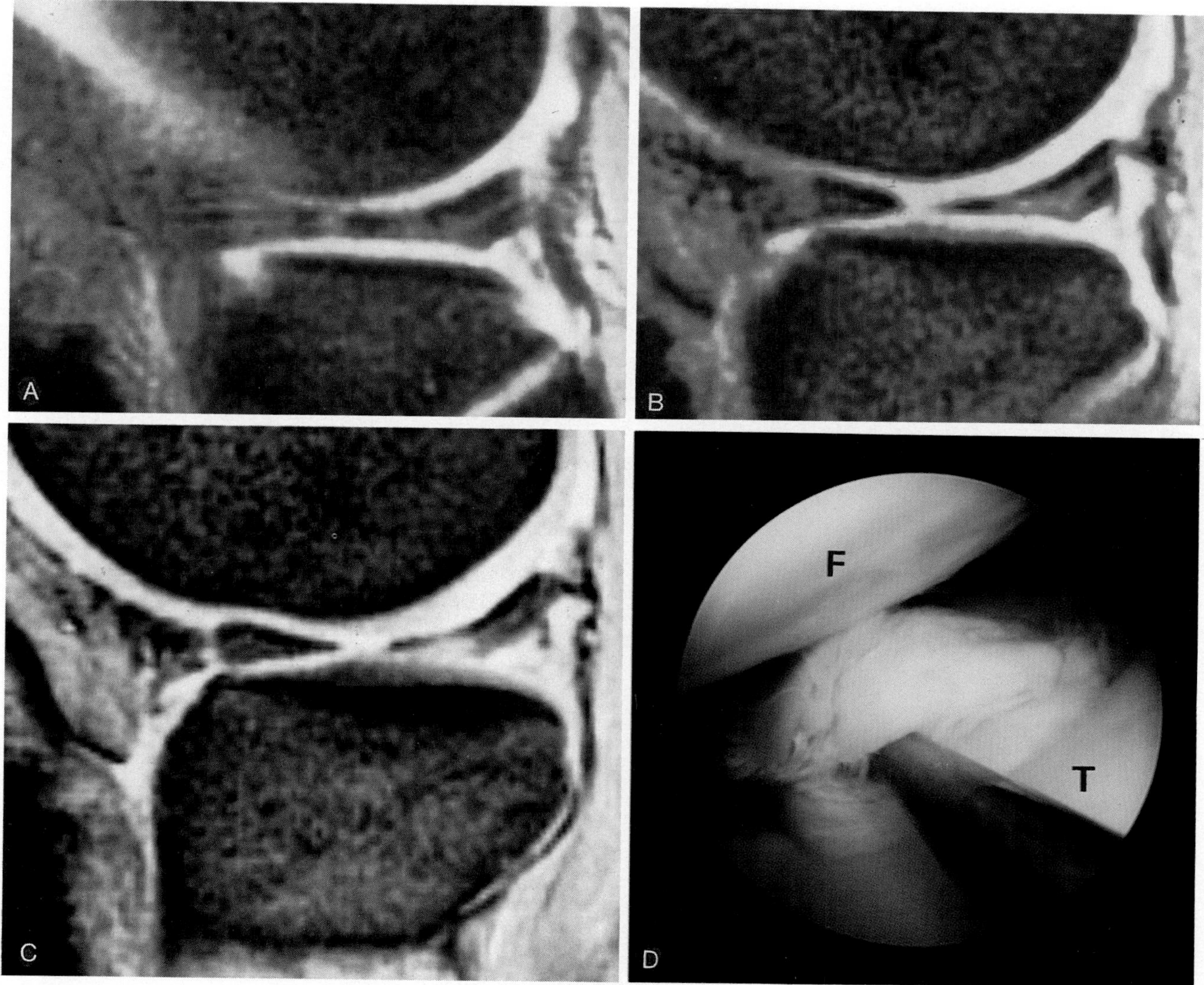

Figure 11–29. *A–C,* Sagittal gradient-echo images (TR, 600 ms; TE, 15 ms; flip angle, 20 degrees) show a complex tear of the central portion of a discoid lateral meniscus in the right knee (in a normal lateral meniscus the anterior and posterior horns are symmetric in size and shape). *D,* Arthroscopic view from an anterolateral portal demonstrates the lateral meniscus, which is thick and irregular in appearance. A probe has elevated the fragment and points to the tear. F, Medial femoral condyle; T, tibial plateau.

Burk and co-workers[5] reported 11 parameniscal cysts as being well circumscribed in appearance, of low or equal signal intensity to muscle on T1-weighted images, and of high signal intensity on T2-weighted or gradient-echo images (Fig. 11–31). These cysts may extend anteriorly to Hoffa's fat pad or laterally similar to a ganglion cyst. Parameniscal cysts are almost always associated with underlying horizontal cleavage or complex tears of the menisci, whereas ganglion cysts are not. Distinction between ganglion and meniscal cysts is essential before surgery because meniscal cysts may recur unless the underlying meniscal tear is treated.[5] Magnetic resonance imaging often depicts fluid tracking from the tear into the parameniscal cyst. Ganglion cysts frequently are multiloculated and may connect with the joint capsule. Unlike meniscal cysts, ganglion cysts usually do not communicate with the joint, and therefore they must be removed by an extra-articular surgical approach.

The Role of Magnetic Resonance Imaging for Meniscal Tears

Magnetic resonance imaging allows detection of intrasubstance (closed) tears and intrameniscal degeneration before visible open tears are seen arthroscopically.[29] Because of the limitations of arthroscopy, it is not surprising that a small number of false-positive MR imaging findings are reported. A positive MR imaging report, however, may alert the

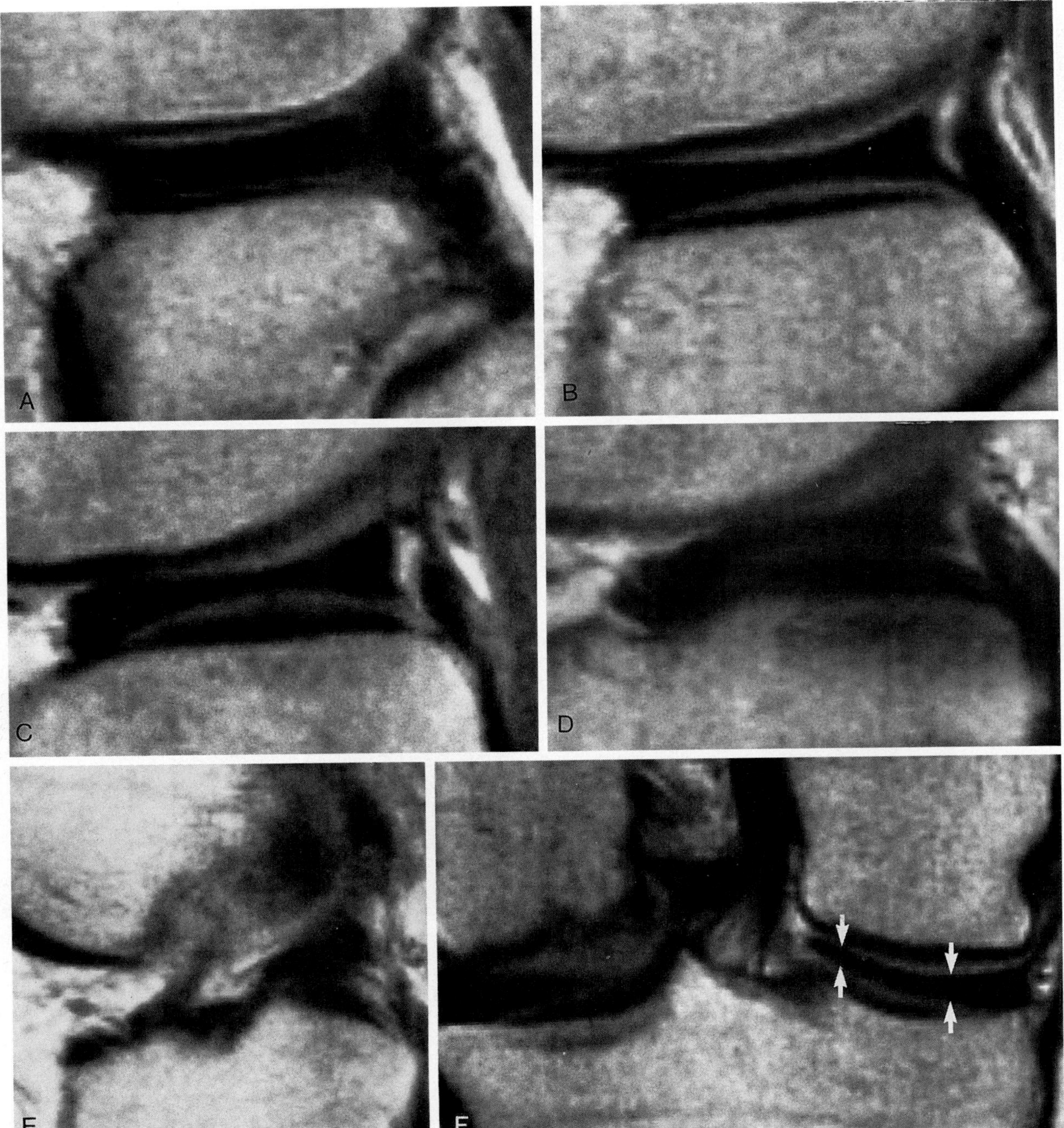

Figure 11–30. Discoid lateral meniscus in a 40-year-old woman. *A–D,* The bow-tie appearance of the meniscal body is seen on four contiguous 5-mm thick sagittal T1-weighted images, suggesting abnormal width of the lateral meniscal body. *E,* No meniscal horns can be identified in the fifth consecutive image. *F,* Coronal T1-weighted image shows the lack of normal tapering of the meniscal height as it extends toward the intercondylar notch *(arrows).*

arthroscopist to probe the meniscus extensively to avoid a false-negative arthroscopic result. The question of whether intrasubstance degeneration may progress to complete tear is controversial. Certainly there is no treatment for lesions that do not extend to the articular surface of the meniscus, and therefore signal patterns on MR images that do not clearly extend to the surface should not be referred to as tears, which require treatment.

CRUCIATE LIGAMENTS

Anterior Cruciate Ligament

The anterior cruciate ligament (ACL) arises from the posterior part of the lateral femoral condyle in the intercondylar notch and inserts onto a wide depressed area anterolateral to the anterior tibial spine, usually in close association with the anterior horn of

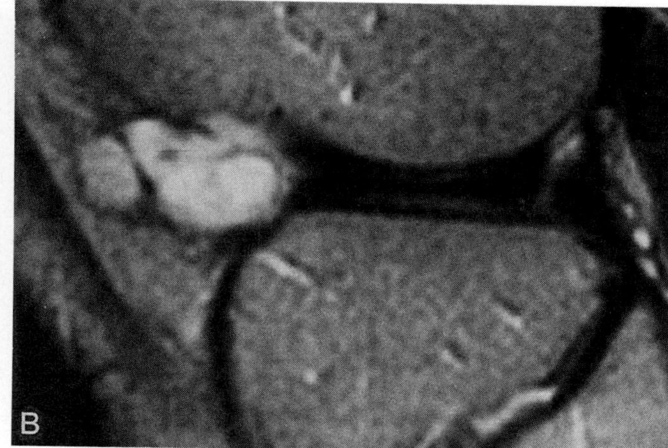

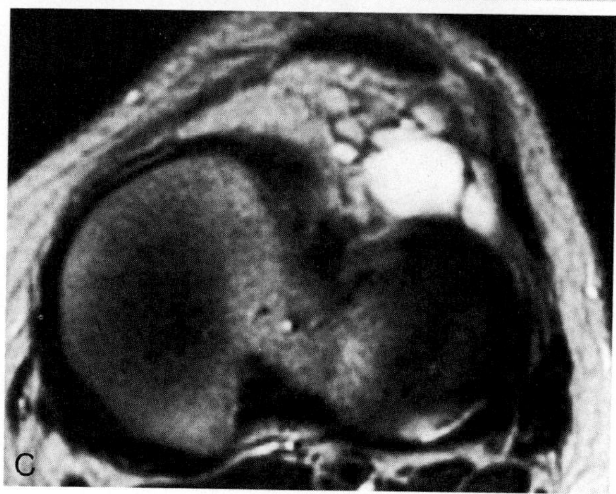

Figure 11–31. Sagittal T1-weighted *(A)* (TR, 1800 ms; TE, 20 ms), sagittal T2-weighted *(B)* (TR, 1800 ms; TE, 80 ms), and axial T2-weighted *(C)* (TR, 1800 ms; TE, 80 ms) images of a multiloculated lateral meniscal cyst in a 28-year-old woman. The cyst extends anteriorly to Hoffa's fat pad and posteriorly to the anterior horn of the lateral meniscus, which is truncated, suggesting tearing.

the lateral meniscus (Fig. 11–32). It is an intra-articular and extrasynovial structure and is composed of three fascicles: a small anteromedial bundle, an intermediate bundle, and a larger, bulky posterolateral bundle. The tibial attachment of the ACL is tighter than the femoral attachment. With the knee in flexion, the anteromedial fascicle becomes taut and the posterolateral bundle is relatively lax.[47] With extension, the entire ligament is equally loaded and tight. The average width of the ACL is about 11 mm.[29] It is a primary anterior stabilizer, preventing anterior translation of the tibia on the femur and serving as a rotational guide for the femoral condyles.

Injury to the ACL commonly is associated with injuries of the medial collateral ligament (MCL) and medial supporting structures. This type of injury is caused by deceleration and forceful valgus-external rotation of the knee (Fig. 11–33).[47] Isolated tears of the ACL are produced by deceleration, internal rotation, and excessive hyperextension. These injuries often occur in football players and downhill skiers. Tears of the ACL result in anterolateral instability of the knee. Physical examination often is difficult in the acute setting because of severe pain and muscle spasm.

With 15 to 20 degrees of external rotation and neutral extension of the leg, the oblique sagittal course of the ACL can be well seen on a single image or, more commonly, two adjacent-sagittal images in 95 per cent of patients.[39] Interslice gaps are not advised in imaging the ligament. The ACL is seen either as a single dark band or as two or three fiber bundles with prominent fiber striations at its tibial attachments (Fig. 11–34). A normal ligament can be seen as an intact structure extending from the femur to the tibia. Tears of the ACL may be suggested when one or more of the following signs are present[6,6a,18a,22,29,33a,55]: (1) a discontinuity of the ligament, with fluid filling the defect (Figs. 11–35, 11-36); (2) a cloudlike edematous mass replacing the ACL or a portion of its length (see Figs. 11–35, 11–36); (3) an anterior translation of the tibia in relation to the femur (Fig. 11–37); (4) an acute angulation or buckling of the proximal posterior cruciate ligament (PCL) (Figs. 11–37, 11–38); (5) a failure to visualize the ligament fibers on routine sagittal and coronal images (Fig. 11–39); (6) a fragment of the ACL with abnormal orientation (scarred down to the PCL) (Fig. 11–40); (7) a continuous band of the ACL with focal angulation; (8) a tear of the ACL at its

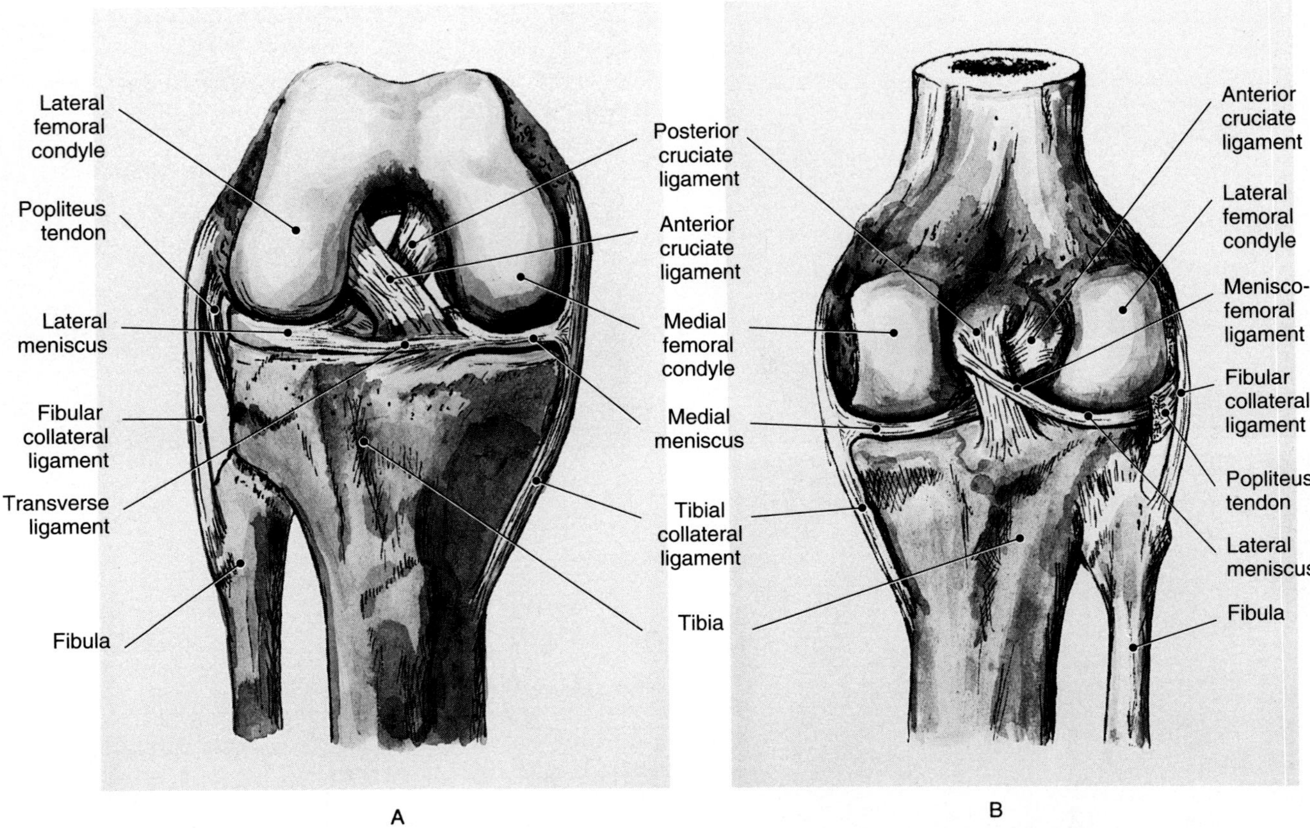

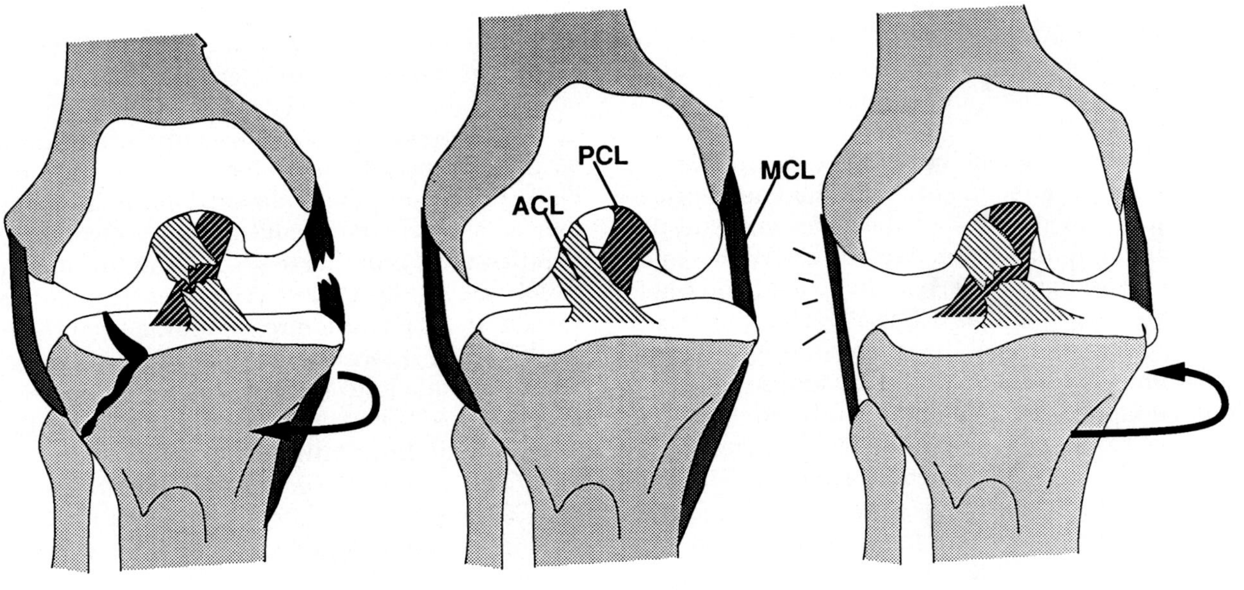

Figure 11–33. Mechanism of a tear in the anterior cruciate ligament.

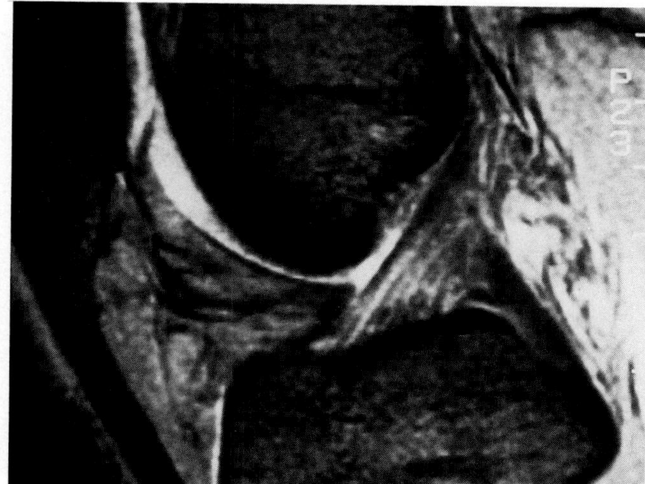

Figure 11–34. Normal anterior cruciate ligament. Prominent fiber bundles of low signal intensity are well shown on the gradient-echo image.

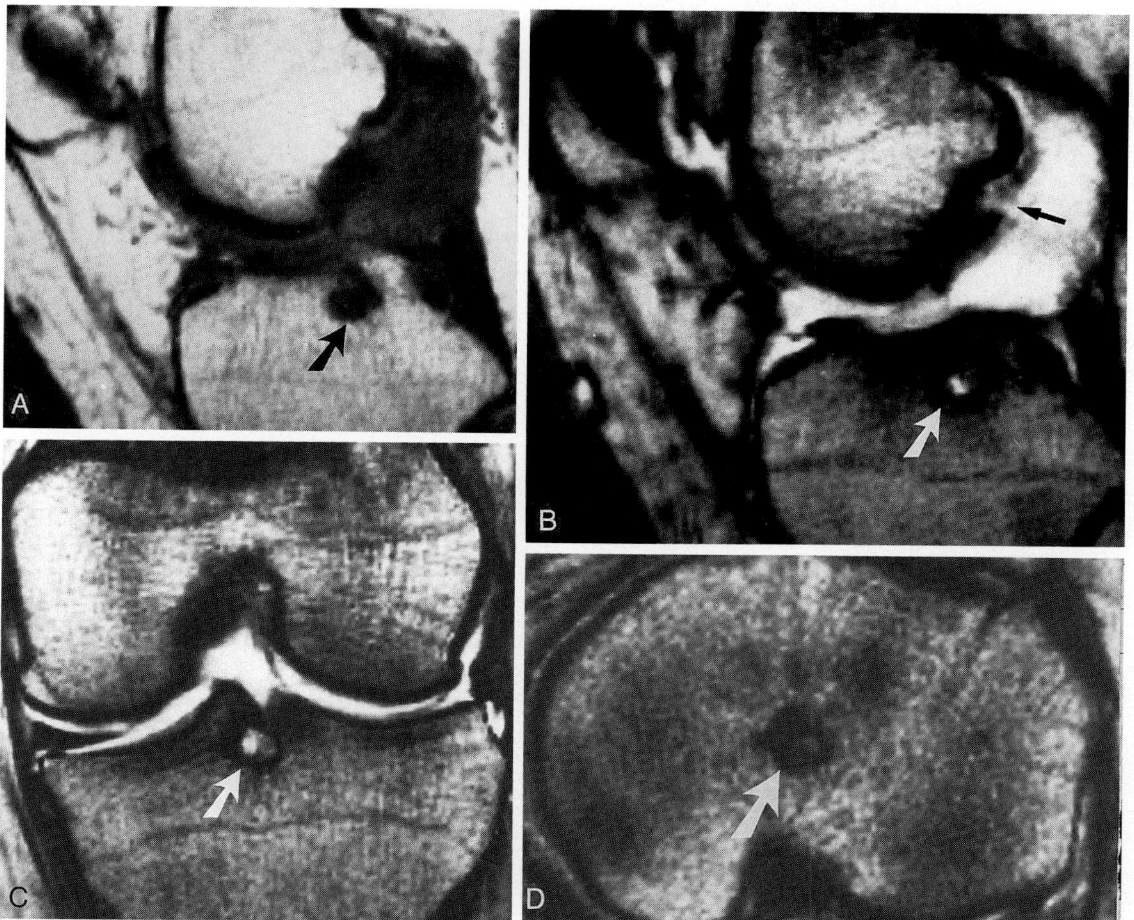

Figure 11–35. Complete tear in the anterior cruciate ligament (ACL) of a 56-year-old woman with a 3-year history of left knee pain. The ACL is torn at its midportion *(thin arrow)* and is accompanied by edema. Cystic resorption adjacent to the tibial attachment of the ACL is noted *(thick arrow)*. *A*, Sagittal T1-weighted image; *B*, T2-weighted image; *C*, coronal T2-weighted image; *D*, axial proton density-weighted image.

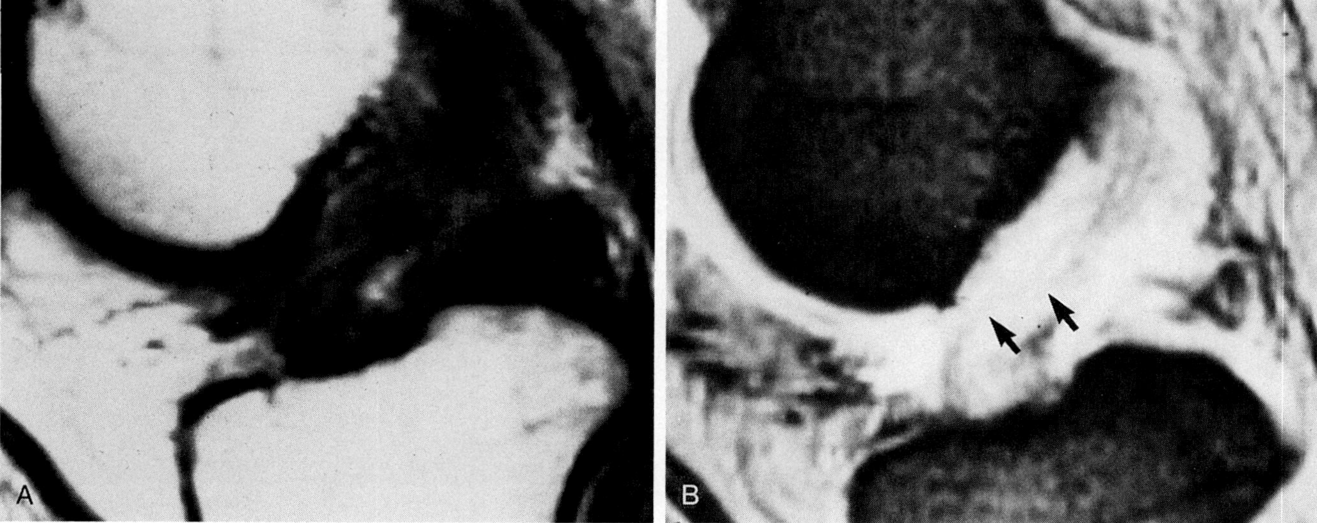

Figure 11–36. Acute, complete tear in the anterior cruciate ligament (ACL) of a 32-year-old woman (arthroscopically proved). *A*, Sagittal T1-weighted image (TR, 800 ms; TE, 20 ms) shows a cloudlike edematous mass replacing the ACL. *B*, The disruption of the ligamentous fibers is better seen on the T2* gradient-echo image (TR, 600 ms; TE, 20 ms; flip angle, 30 degrees) *(arrows)*.

tibial attachment (Fig. 11–41), although this is seen less frequently than a tear at its femoral attachment; (9) a bone contusion of the lateral compartment of the knee (see Figs. 11–38, 12–30); or (10) a deepened lateral femoral notch (see Fig. 12–32). The MR imaging findings of sprain or incomplete tears of the ACL have been less well defined and studied (Fig. 11–42). Mink and co-workers reported that T2-weighted images were significantly more sensitive in detecting tears of the ACL than T1-weighted sequences (100 versus 85 per cent).[29] Coronal images often are helpful in assessing the ACL when findings are equivocal on sagittal images. Attention should be paid to commonly associated injuries such as tears of the MCL and menisci in addition to a lateral compartment bone contusion.[41]

Acute tears of the ACL commonly have a well-defined edematous mass, not fluid, that is of homogeneous low to intermediate signal intensity on proton density images and becomes inhomogeneous and with slightly increased signal intensity on T2-weighted images.[55] The lateral femoral condyle may produce a partial volume averaging effect with the proximal half of the ACL, which should not be misinterpreted as a tear (see Fig. 11–70); the coronal images or repeat thin sections help to avoid this

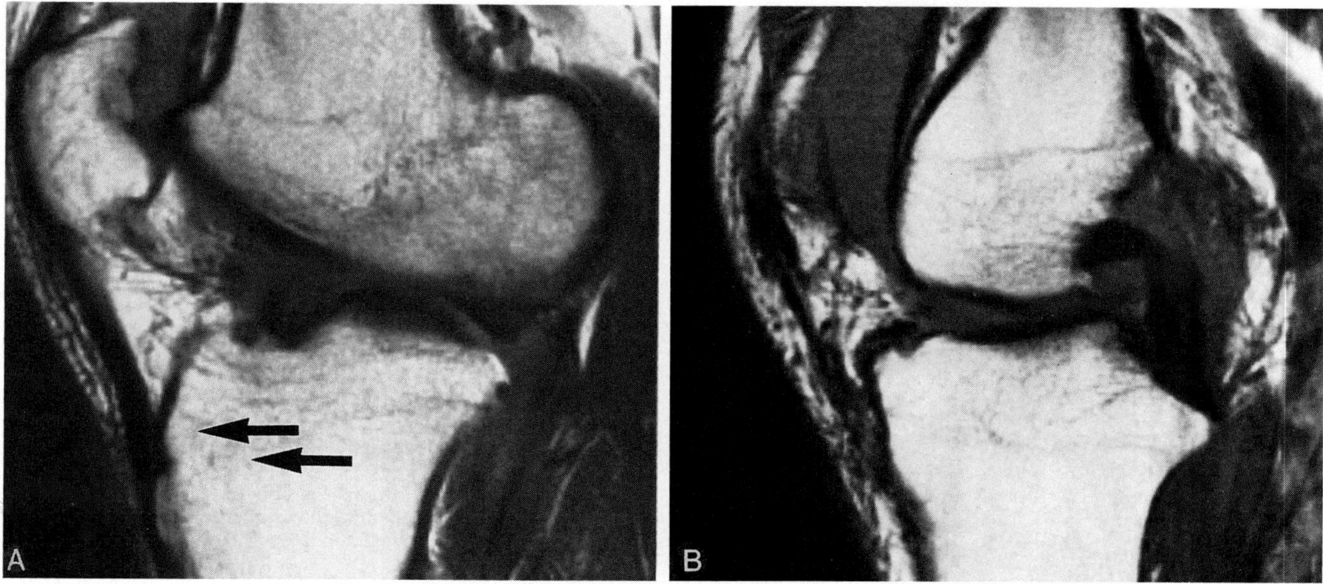

Figure 11–37. *A* and *B*, Sagittal T1-weighted images. The tibia is translated forward under the femur owing to insufficiency of the anterior cruciate ligament *(A)* *(arrows)*, resulting in proximal posterior cruciate ligament buckling *(B)*.

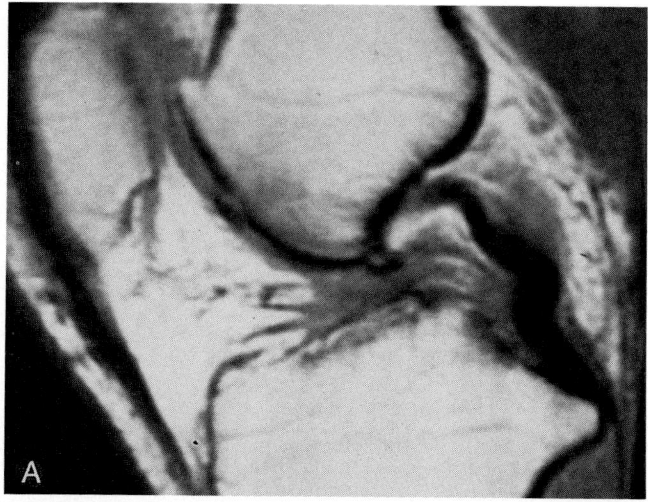

Figure 11–38. Buckled and lax posterior cruciate ligament (PCL) in a 38-year-old man with lateral compartment contusion. *A*, Lax PCL is a sign of increased laxity of the anterior cruciate ligament. *B* and *C*, Contusion of the lateral posterior tibial plateau appears as a nonlinear area of decreased signal intensity on T1-weighted images and increased signal intensity on T2-weighted images *(flagged arrow)*. *A* and *B*, Sagittal T1-weighted images (TR, 1800 ms; TE, 20 ms); *C*, sagittal T2-weighted image (TR, 1800 ms; TE, 80 ms).

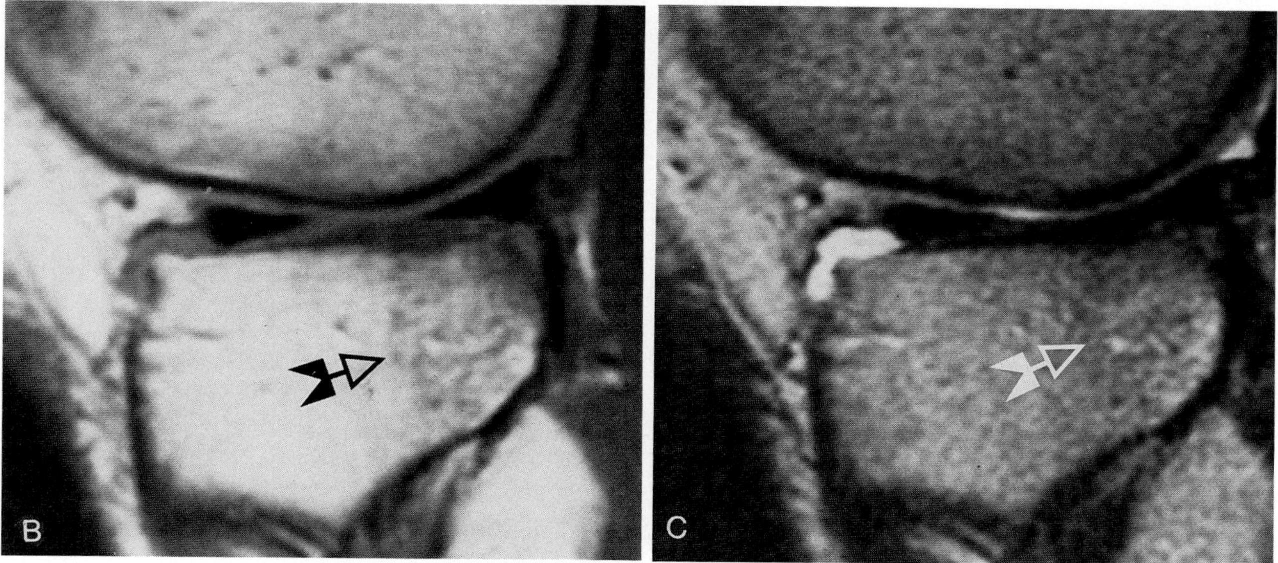

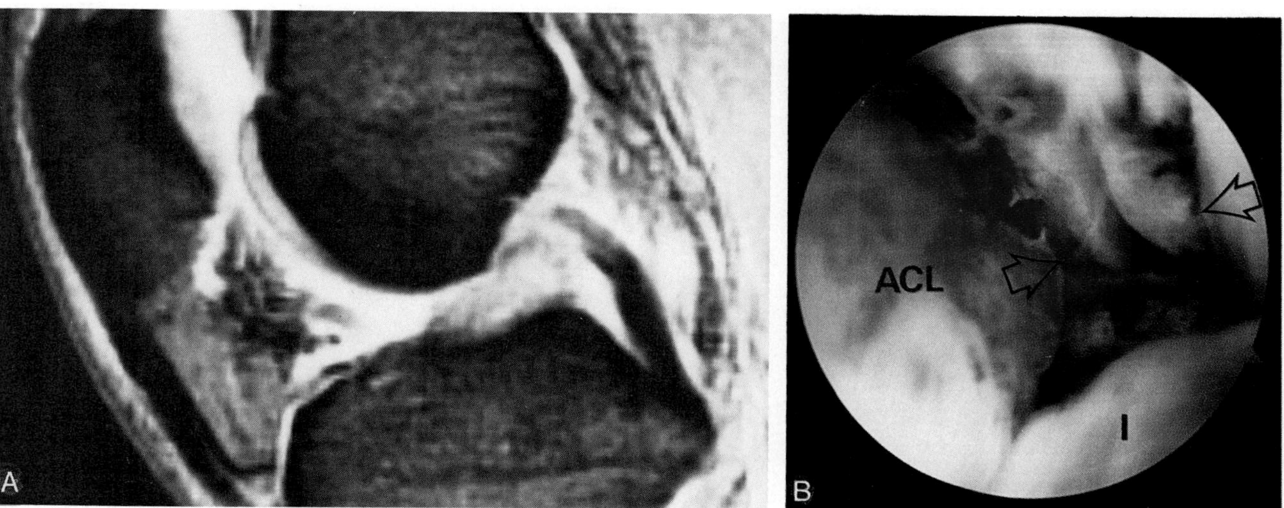

Figure 11–39. Acute, complete tear in the anterior cruciate ligament (ACL) of a 47-year-old woman. *A*, Sagittal gradient-echo image (TR, 500 ms; TE, 15 ms; flip angle, 20 degrees) shows lack of ACL fibers in continuity in the intercondylar notch. An associated effusion is noted. *B*, Arthroscopic view demonstrates a complex tear of the ACL and a discontinuity of its fibers heading toward the lateral femoral condyle in the left knee. ACL, Stump of anterior cruciate ligament; I, intercondylar eminence; *arrows*, the void area where the ACL should be.

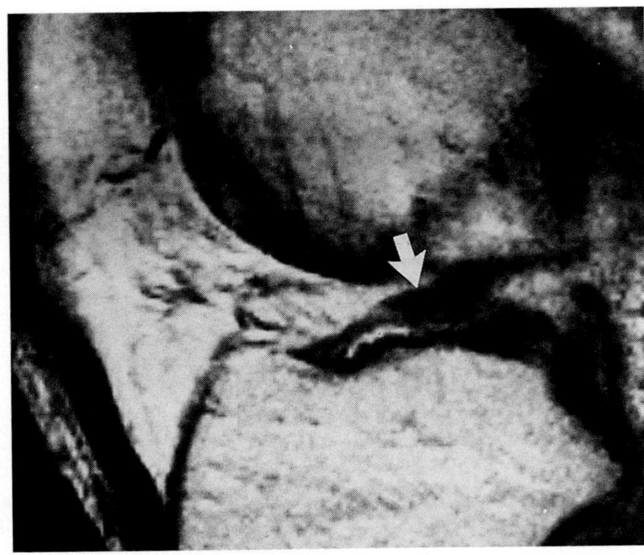

Figure 11–40. A longstanding tear in the anterior cruciate ligament (ACL) of a 27-year-old policeman. The sagittal T1-weighted image shows a reduced angulation of the ACL. There is no associated joint effusion, as may be seen with an acute injury. Arthroscopy revealed a scar tissue remnant of the ACL.

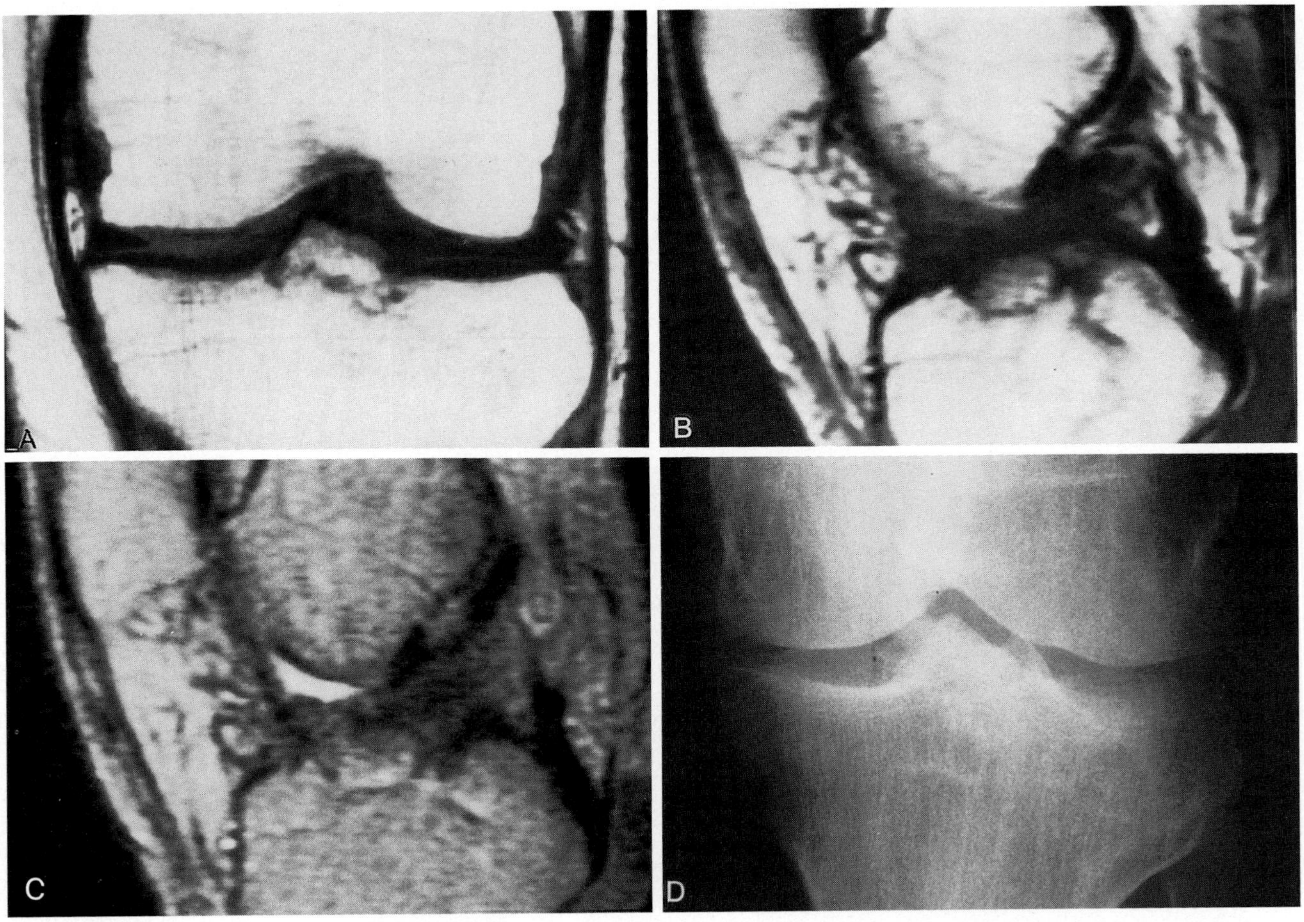

Figure 11–41. Avulsion of the anterior cruciate ligament (ACL) in a 28-year-old man. *A*, Coronal T1-weighted (TR, 1000 ms; TE, 25 ms) image shows a tibial spine fracture. *B*, Sagittal T1-weighted (TR, 200 ms; TE, 20 ms) and *C*, T2-weighted (TR, 2000 ms; TE, 60 ms) images show the avulsed ACL at its tibial insertion, with a bony fragment. The ACL is poorly recognized, indicating a tear. *D*, Initial radiograph does not show the avulsed tibial spine clearly.

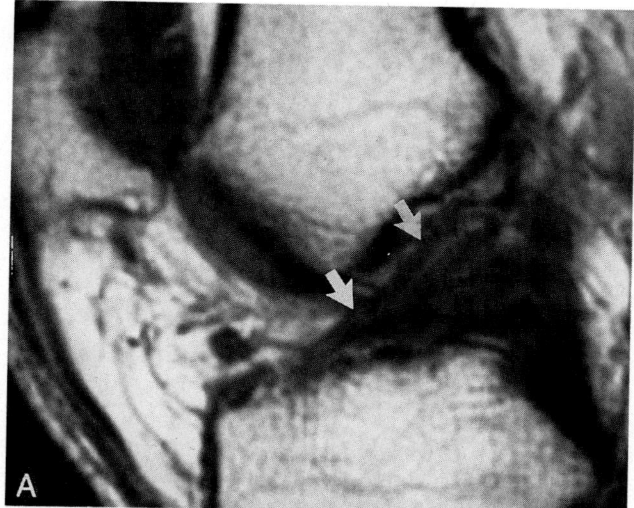

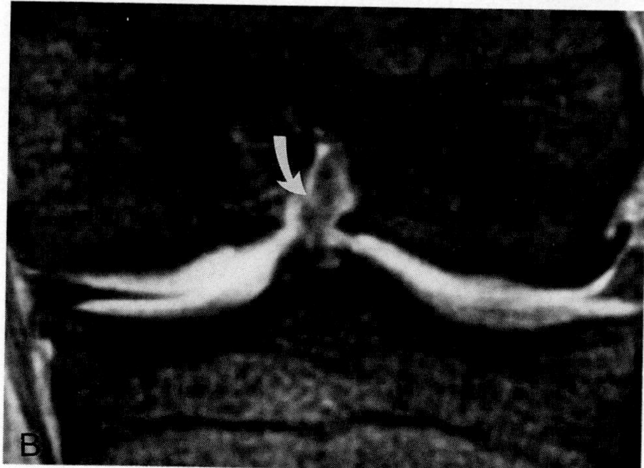

Figure 11–42. Sprain of the anterior cruciate ligament (ACL) in a 31-year-old basketball player. *A*, Sagittal T1-weighted image shows that the course of the ACL is relatively lax, but the continuity of the fiber bundles is maintained *(arrows)*. Arthroscopy revealed a frayed and loose anteromedial band of the ACL. *B*, Gradient-echo image shows a focal area of high signal intensity in the midportion of the anteromedial band.

potential pitfall. In chronic tears, a band with decreased signal intensity as a consequence of bridging fibrous scars may be misintepreted as an intact ligament. This pattern occurs in 30 per cent of patients with chronic ACL tears as reported by Vahey and colleagues.[55] The scar tissue, however, frequently is angulated, bowed, or focally thickened. No associated edema or effusion is identified.

Mink and co-workers[29] reported 10 false-positive and three false-negative cases in diagnosing ACL tears from 242 MR imaging studies of the knee. The overall sensitivity, specificity, and accuracy of MR imaging for the diagnosis of ACL tears were 92 per cent, 95 per cent, and 95 per cent, respectively. In a multicenter analysis of 1014 cases, Fischer and colleagues[13] reported MR imaging results to be true positive in 173 cases, true negative in 757, false positive in 54, and false negative in 13 cases. Accuracy, specificity, and sensitivity each were 93 per cent.

To restore the stability of the knee, a torn ACL may be reconstructed by using an autograft. Such an autograft may be derived from the the patellar tendon, the semitendinosus tendon, or the fascia latae.[2] Synthetic materials such as Gore-Tex also may be used. Grafts are inserted into holes 8 to 10 mm in diameter that have been drilled in the posterolateral femur and tibial tubercle, and are placed between these bone tunnels within the intercondylar notch (Fig. 11–43).[35] Attention should be given to the placement of the tibial tunnel.[18] A tibial tunnel that is posterior to the slope of the intercondylar roof appears to impinge the graft less frequently. Graft impingement is caused by the placement of the tibial tunnel anterior to the intercondylar roof, which may rub against the anterior surface of the graft during knee extension.

Reconstructed ACLs are best depicted on sagittal images. The graft is normally of low signal intensity and may be of variable thickness. This normal variation in thickness is thought to be due in part to vascular ingrowth and cellular proliferation within the reconstructed ligament.[35] Localized fluid collections within or around a graft suggest a re-tear (Fig. 11–44), which may also be confirmed by anterolateral laxity of the reconstructed ligament, resulting in buckling of the PCL. Impinged grafts are high in signal intensity in the distalmost third of the grafts. This could be explained by the replacement of the ligament by synovium and the increase in water content. The increased signal intensity in impinged grafts suggests that graft strength may be weaker.[18]

Posterior Cruciate Ligament

The PCL arises from the medial femoral condyle in the intercondylar notch and inserts onto a depression of the posterior intercondylar region in the tibia. With the leg in extension, it is posteriorly convex or buckled; the PCL becomes taut when the knee is in flexion.[47] The PCL is 30 per cent larger in size and two times stronger than the ACL in tensile force.[20] Therefore, isolated tears of the PCL are relatively rare, occurring in 3.4 to 20.0 per cent of ligamentous injuries. Direct posterior violence to a flexed knee causes injury to both ACL and PCL (Fig. 11–45).

The PCL accounts for 89 per cent of the resistance to posterior translation of the tibia on the femur. Isolated tears of the PCL are rare. They can be

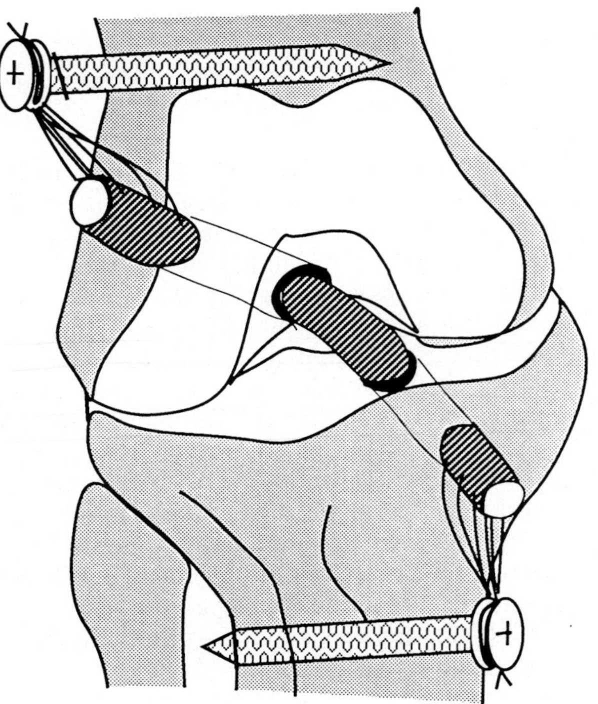

Figure 11–43. Reconstruction of the anterior cruciate ligament (ACL). Anterior view shows the course of the reconstructed ACL autograft through the intercondylar notch and bone tunnels and its attachment into the femur and tibia with two staples.

produced by forceful translation of the tibia in a flexed knee (e.g., dashboard injury). This type of injury also can be caused by a fall on the flexed knee. Tears of the PCL result in posterior instability of the knee. Patients may experience unsteadiness when the knee is in a semiflexed position, as in descending stairs. However, many untreated patients do not have severe symptoms until years later, when osteoarthritic changes develop in the knee. Eighty per cent of patients have reported osteoarthritic changes of the knee when repair or reconstruction of the PCL has been delayed beyond 4 years.[47] The PCL is an extrasynovial structure (see Fig. 11–2). An unsuspected PCL injury may be overlooked during routine arthroscopy.

Sagittal images are best to display the continuity of the PCL, which is seen as an arclike structure of low signal intensity on all pulse sequences. The entire PCL can be displayed on a single sagittal image in 95 per cent of patients in a routine MR imaging study. Sprain of the PCL is manifested as increased signal intensity from edema and hemorrhage within the ligament on T1-weighted images, without evidence of anatomic disruption.[16] Ruptures of the PCL appear as (1) fluid or blood filling the gaps within the ligaments (Fig. 11–46), (2) detachment of the PCL at bony insertion sites (Fig. 11–47), or (3) increased signal intensity over the entire ligament (Figs. 11–48, 11–49). Acute or subacute tears of the PCL may lead to increased signal intensity at or around the point of disruption on T2-weighted images as a result of hemorrhage and edema. The hemorrhage usually is less masslike in appearance than it is in ACL tears. In a chronic tear, the injured ligament may not have increased signal intensity because of fibrous scar formation. The site of disruption frequently appears in the center of the ligament.The tibial insertion is the second most common location for injuries (e.g., avulsion). Tears at or near the femoral origin of the ligament are much less frequent. The area should be searched carefully for commonly associated injuries, such as tears of the ACL, MCL, and menisci.

Grover and colleagues[16] reported 11 cases of PCL injuries in 202 MR imaging studies of the knee with surgical correlation, with no false-negative MR diagnoses. In a multicenter analysis of 1014 cases, Fischer and co-workers[13] reported true positive diagnoses by MR imaging in eight cases, true negative in 998, false positive in six, and false negative in two. Accuracy, specificity, and negative predictive value of MR imaging for the diagnosis of PCL tears were each 99 per cent.

COLLATERAL LIGAMENTS

Medial Collateral Ligament

The MCL is composed of a superficial and a deep band. The superficial band, also called the tibial collateral ligament (TCL), extends from the medial femoral epicondyle to the medial aspect of the tibia, approximately 5 cm below the joint line.[30] The deep band, also known as the medial capsular ligament or deep capsular ligament, is firmly attached to the midportion of the medial meniscus. The TCL and the deep capsular ligament are separated by a TCL bursa and fat, which allows for their movement.[23] Both the TCL and the deep capsular ligament provide medial stability that resists external rotation, valgus stress, and anterior forces to the knee. The medial joint capsule also is supported by the pes anserinus, which is formed by the sartorius, the gracilis, and the semitendinosus muscles that attach to the medial tibia.[23]

Injuries to the MCL are far more frequent than injuries to the lateral ligamentous structures. Often MCL tears are produced by a violent valgus force with associated external rotation (clipping injury). The deep capsular ligament usually is torn first, followed by the TCL. Once the deep capsular ligament tears, the medial meniscus may become free of peripheral attachments, resulting in a meniscocapsular

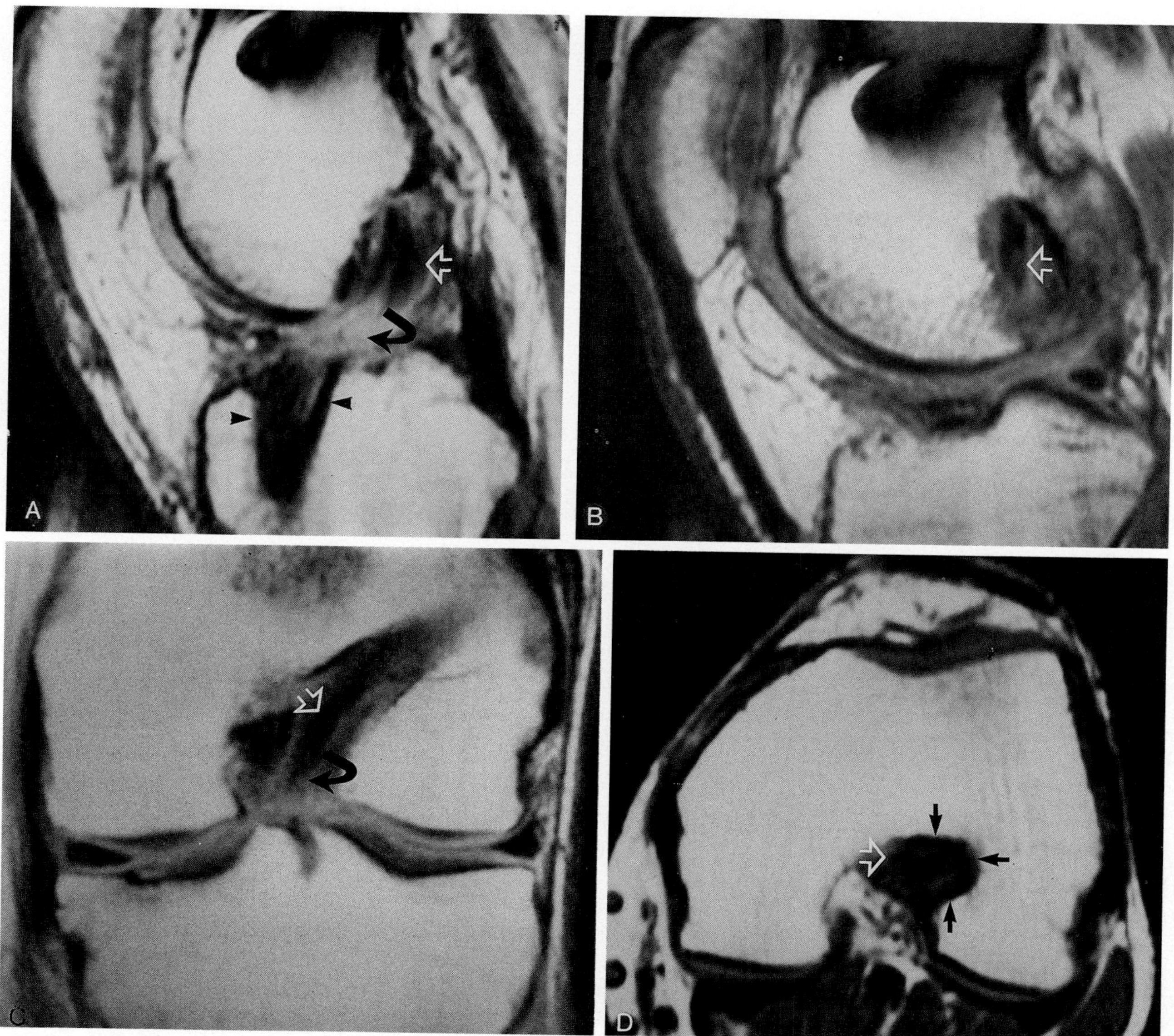

Figure 11–44. Re-tear of an anterior cruciate ligament (ACL) graft. *A–C,* Sagittal and coronal T1-weighted images show a high signal intensity gap *(curved arrow)* in the midportion of the ACL graft (which is of low signal intensity) *(open arrow)*, suggesting a complete tear of the ACL reconstruction. The tibial bone tunnel is placed in a right position *(arrowheads)* (i.e., posterior to the roof of the intercondylar notch). *D,* The axial T1-weighted image shows the tunnel-graft exits the posterosuperior aspect of the lateral condyle at the 1:30 clock position in this left knee *(arrows).*

separation. Tears of the MCL frequently coexist with tears of the ACL and lateral bone contusions.

Clinically, collateral ligament injuries are categorized into three grades. Grade I is a sprain or stretching of the ligament with localized tenderness but no clinical instability. Grade II is a partial disruption of the fibers with localized tenderness, mild to moderate laxity, and a firm endpoint to laxity on valgus stress. Grade III is a complete disruption of the ligament with significant instability and an indistinct endpoint.[58] Most complete MCL tears can be managed conservatively. Surgery is indicated when tears of the posterior capsular ligament are present as well.

Midcoronal images are the best views to display the continuity of the MCL, which is seen as a thin band of low signal intensity on all pulse sequences. T2-weighted spin-echo or gradient-echo sequences are the best images to evaluate collateral ligament injuries. Sprain of the collateral ligament (clinical grade I) shows poorly defined areas of increased signal intensity from edema and hemorrhage in the subcutaneous fat superficial to the intact ligament (Fig. 11–50). Although a few fibers of the deep cap-

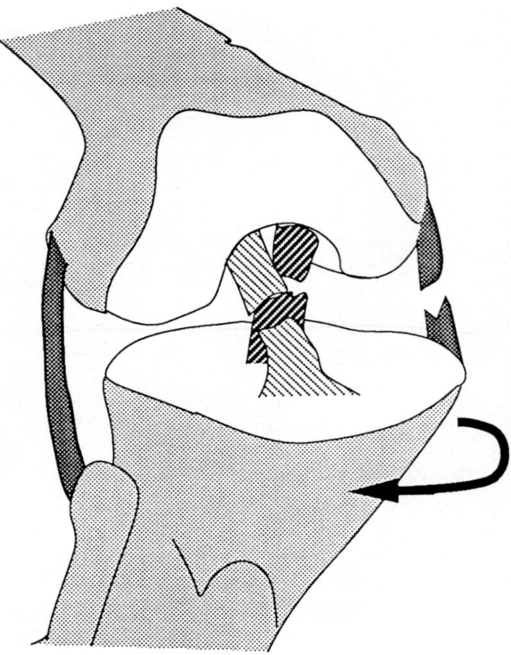

Vigorous External Rotation in a Flexed Tibia

Figure 11–45. Mechanism of a tear in the posterior cruciate ligament.

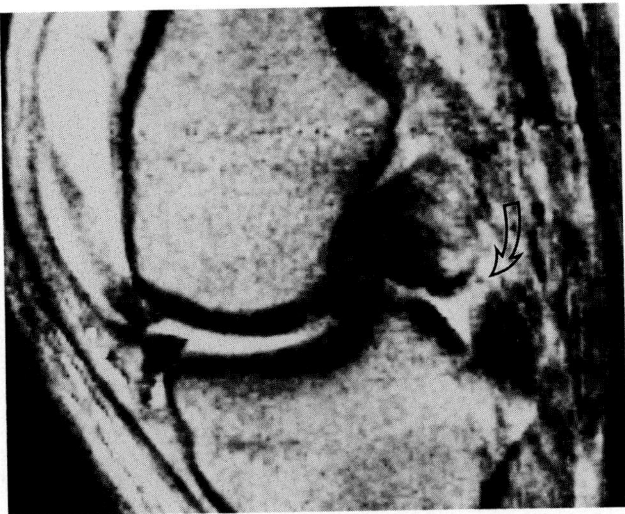

Figure 11–46. An acute tear in the posterior cruciate ligament (PCL) with a loss of continuity. Sagittal T2-weighted image (TR, 2000 ms; TE, 80 ms) shows a gap of high signal intensity transecting the PCL, indicating an acute ligamentous rupture with hemorrhage or edema.

sular ligament may actually tear, on MR images at this stage the ligament is neither disrupted nor displaced from bony attachments. Ruptures of the collateral ligament (clinical grade III) appear as fiber discontinuity (Figs. 11–51, 11–52), wavy contours, or detachment of the bony attachments of the injured ligament on MR images. Careful attention should be paid to associated injuries, such as contralateral bone contusions, tears of the ACL, and meniscocapsular separation of the medial meniscus. Magnetic resonance imaging diagnosis usually is classified as a sprain or torn collateral ligament. A clinical grade II MCL injury is difficult to define precisely on MR images. Not surprisingly, extra-articular edema or hemorrhage may be absent in chronic injuries of the MCL.

Bursitis of the TCL is noted when a focal area of high signal intensity occurs between the deep and superficial fibers on T2-weighted images.[23] A small amount of fat is normally seen between the TCL and the medial meniscus. When fluid is present between the fibers of the TCL and the medial meniscus, meniscocapsular separation may be diagnosed. Bursitis may be difficult to distinguish from meniscocapsular separation, although bursitis usually is more focal and produces isolated medial joint line pain (see Chapter 12).

Lateral Collateral Ligament

The lateral collateral ligament (LCL) is an imprecise term because the lateral ligamentous support is extremely complex. On posterior coronal images, the fibular collateral ligament (FCL) appears as a cord of low signal intensity extending between the fibular head and the lateral epicondyle of the femur. This ligament normally is not seen on a single image because of its oblique posteroinferior course. Before its insertion onto the fibular head, the FCL joins with the biceps femoris tendon, forming a conjoined tendon that is seen on posterior coronal images and extreme lateral sagittal images (Fig. 11–53). The popliteal tendon runs between the lateral meniscus and the FCL, attaches onto the femur superiorly, and joins with its muscle belly inferiorly.[51] On anterior coronal images, the iliotibial band runs distally as a dark vertical band and inserts onto the anterolateral tibia (Fig. 11–54). It blends with the lateral patellar retinaculum as seen on axial images. The FCL, popliteal tendon, and iliotibial band are all extra-articular structures. These structures, along with the arcuate ligament, provide the lateral support of the knee.

A violent varus stress may produce injury to the lateral ligamentous complex. However, injuries in this complex occur much less commonly than medial collateral injuries. Injuries of the lateral ligamentous complex have MR imaging patterns similar to those of other ligamentous injuries (Fig. 11–55), including (1) discontinuity of the ligament, (2) abnormal signal intensity within or around the injured

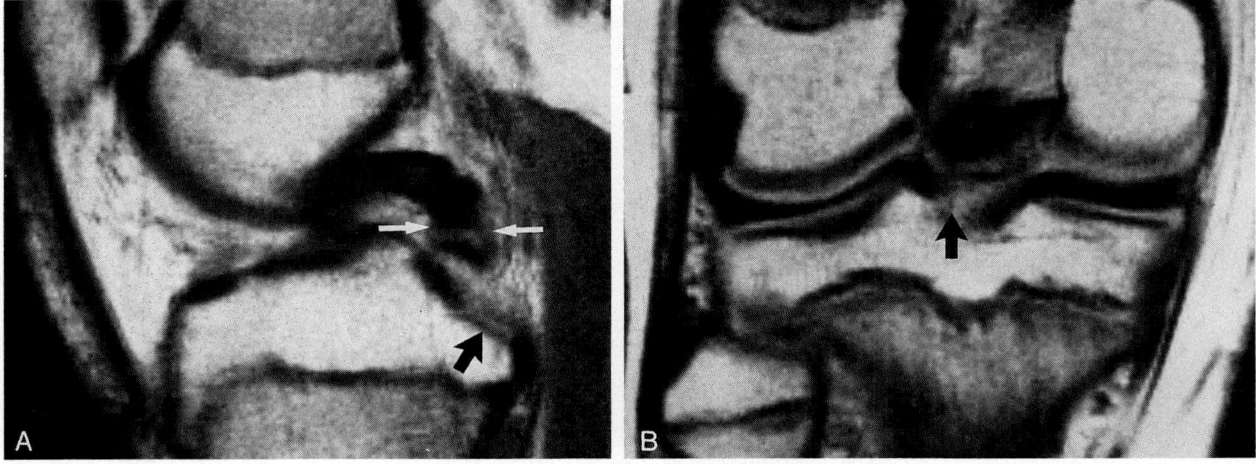

Figure 11–47. Sagittal *(A)* and coronal *(B)* T1-weighted images show a complete tear of the posterior cruciate ligament *(white arrows)* with detachment of its tibial insertion site *(black arrow)*. No associated effusion is noted.

ligament, (3) wavy appearance of the ligament, and (4) detachment from sites of bony insertion. Medial compression fractures and PCL tears also may be seen as associated injuries due to varus stress.

PITFALLS

Many pitfalls in MR imaging of the knee contribute in part to false-positive and false-negative diagnoses. These sources of error can be caused by normal anatomic structures or artifacts unique to MR imaging of the knee. Awareness of their existence is essential for accurate diagnosis.

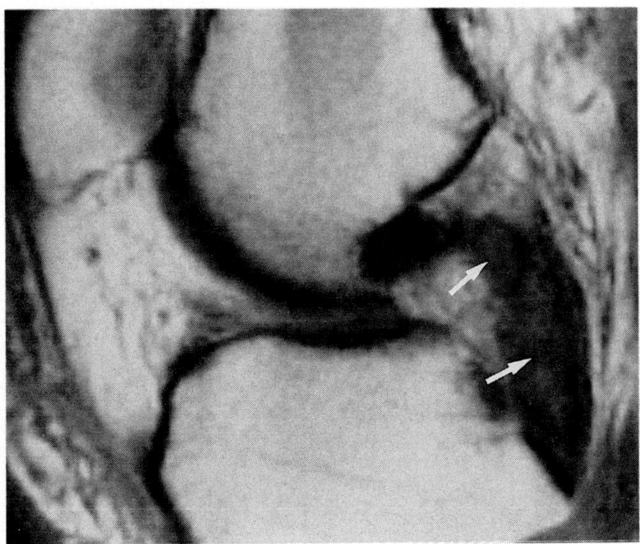

Figure 11–48. Tear in the posterior cruciate ligament. The sagittal T1-weighted image shows edema and hemorrhage of intermediate signal intensity almost within the entire length of the ligament, which should be devoid of signal.

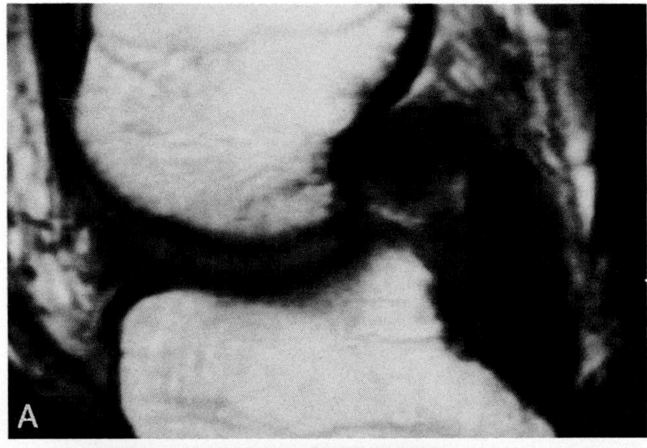

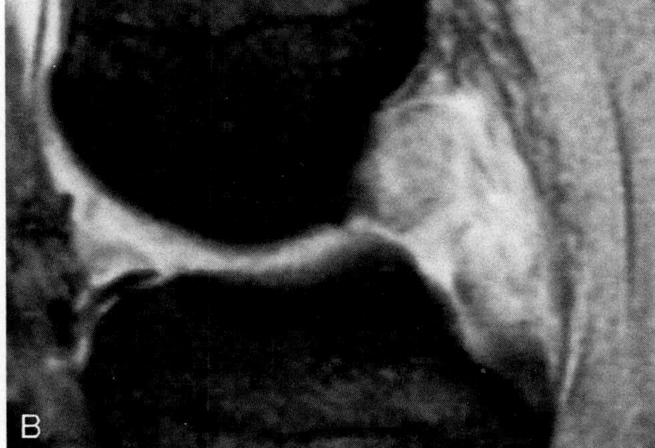

Figure 11–49. An acute tear in the posterior cruciate ligament of a 38-year-old man. Sagittal images show a pathognomonic edematous mass with low to intermediate signal intensity on a T1-weighted image *(A)* (TR, 600 ms; TE, 20 ms), which becomes of high signal intensity in gradient-echo (TR, 500 ms; TE, 20 ms; flip angle, 30 degrees) image *(B)*.

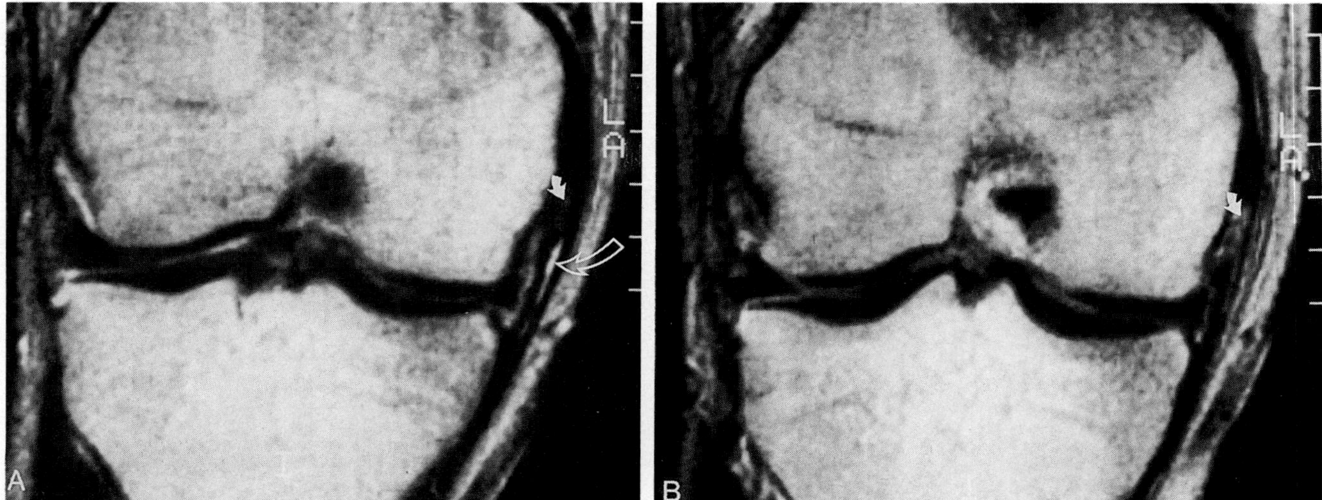

Figure 11–50. Presumed sprain of the medial collateral ligament (MCL) in a 22-year-old man. *A* and *B*, Coronal T2-weighted images (TR, 1500 ms; TE, 70 ms) show a small amount of fluid collected between the deep and superficial layers of the MCL *(open arrow)*. The adjacent deep layer (deep capsular ligament) is thought to be partially disrupted *(solid arrow)*.

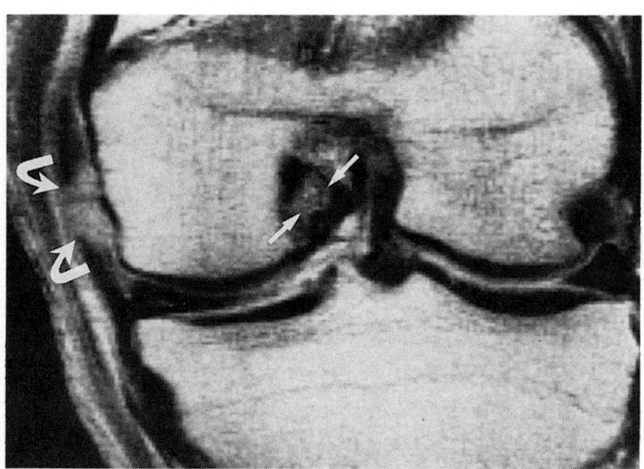

Figure 11–51. Coronal image (TR, 100 ms; TE, 40 ms) shows complete transection of the medial collateral ligament *(curved arrows)* and the posterior cruciate ligament *(arrows)* with hemorrhagic gaps of high signal intensity near their femoral insertions.

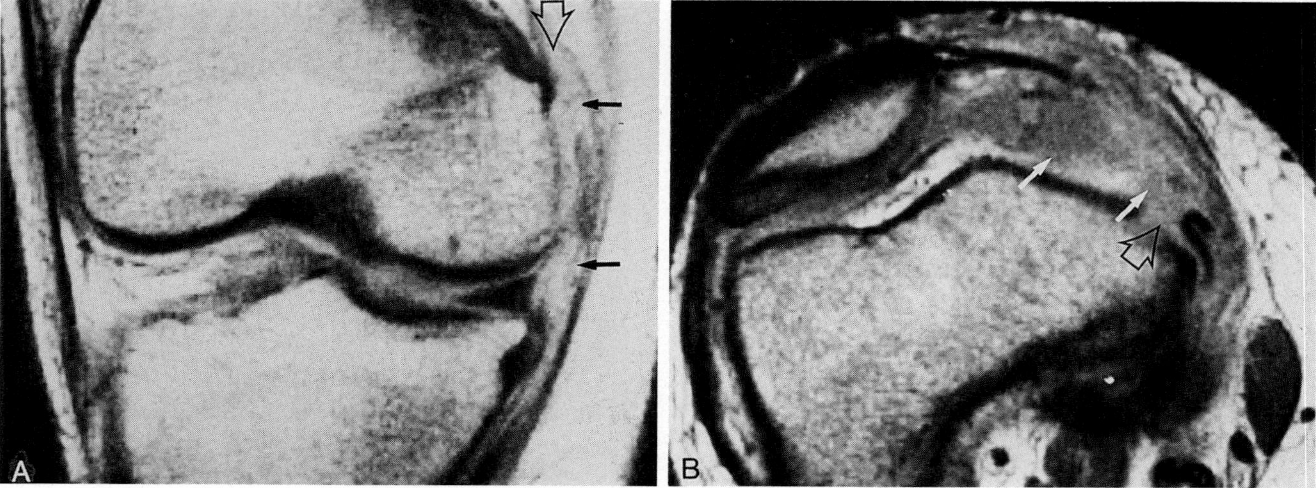

Figure 11–52. Avulsion of the medial collateral ligament (MCL). Acute avulsion of the MCL from its femoral epicondylar attachment *(open arrow)* associated with extensive hemorrhage and synovial fluid of inhomogeneous signal intensity *(arrows)*. *A*, Coronal image (TR, 1000 ms; TE, 40 ms); *B*, axial image (TR, 1000 ms; TE, 40 ms).

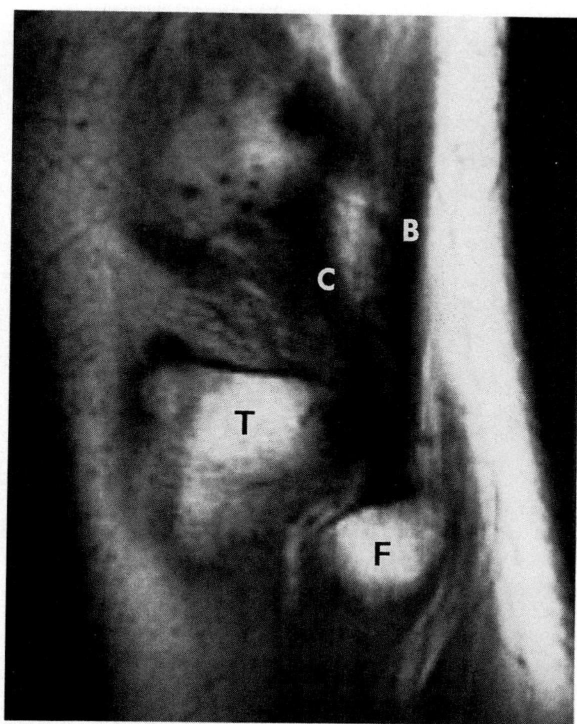

Figure 11-53. The conjoined insertion on the fibular head of the separate biceps femoris tendon (B) and fibular collateral ligament (C) is demonstrated on a peripheral sagittal T1-weighted image. T, Tibia; F, fibula.

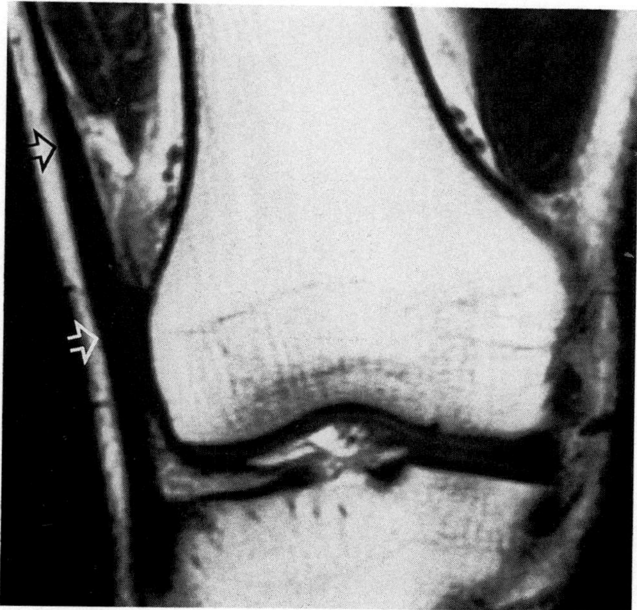

Figure 11-54. Anterior coronal T1-weighted image (TR, 600 ms; TE, 20 ms) shows the attachment of the iliotibial band onto the anterolateral tibia *(arrows)*.

Menisci

Posterior Horn, Lateral Meniscus. ***Popliteal Tendon.*** The popliteal tendon attaches to the lateral femoral condyle laterally and runs between the lateral meniscus and the lateral joint capsule as it passes medially and inferiorly to join its muscle belly on the posterior tibia. Its tendon sheath appears as an oblique line of intermediate to high signal intensity passing between the dark meniscus and tendon on sagittal T1- or T2-weighted images. The popliteal tendon sheath mimics grade III signal intensity and extends to both articular surfaces, with the tendon simulating the peripheral rim of the meniscus (Fig. 11-56). Watanabe and co-workers reported in an MR imaging study of 200 patients that 28 per cent had such an appearance.[56] Furthermore, in patients with a resected posterior horn of the lateral meniscus, the popliteal tendon can be mistaken for a retained posterior horn.[30]

Several clues help distinguish a pseudotear from a true peripheral tear: (1) The popliteal tendon and its sheath almost always can be identified on adjacent slices to allow excluding a tear with confidence. The orientation of the linear structure of increased signal intensity in the posterior horn is helpful but is not a reliable clue for differentiation of these structures. (2) The posterior horn of the lateral meniscus is the same size as the anterior horn. If it appears larger, the popliteal tendon may be contributing to this appearance, which can be mistaken for a meniscal fragment with its tendon sheath simulating a tear. If the meniscus appears smaller, a true peripheral tear should always be considered (see Fig. 11-27C). Coronal images may further help confirm or rule out a tear.

Meniscofemoral Ligament. The meniscofemoral ligament arises from the posterior horn of the lateral meniscus and passes cephalad and medially to insert onto the medial condyle of the femur (Fig. 11-57). The ligament is composed of two branches: Humphry's ligament and Wrisberg's ligament. Humphry's ligament runs anteriorly to the PCL and is visualized in 24 to 33 per cent of MR images.[4,56] Wrisberg's ligament passes posteriorly to the PCL and is identified in a similar percentage of MR images. Either one of these ligaments can be seen on 62 per cent of sagittal MR images, and both ligaments can be depicted concomitantly in 3 to 12 per cent of MR images (Fig. 11-58).[16,56] The pseudotear is caused by a cleft between the meniscal insertion of the meniscofemoral ligament and the posterior horn of the lateral meniscus. The cleft appears as a linear band of intermediate to high signal intensity on sagittal images, and it may a mimic a grade III signal that extends to the superior articular surface of the posterior horn (Fig. 11-59). In this instance the meniscofemoral ligament simulates the peripheral rim of the meniscus. Vahey and colleagues[54] reported that 39 per cent of 109 patients showed such an appearance.

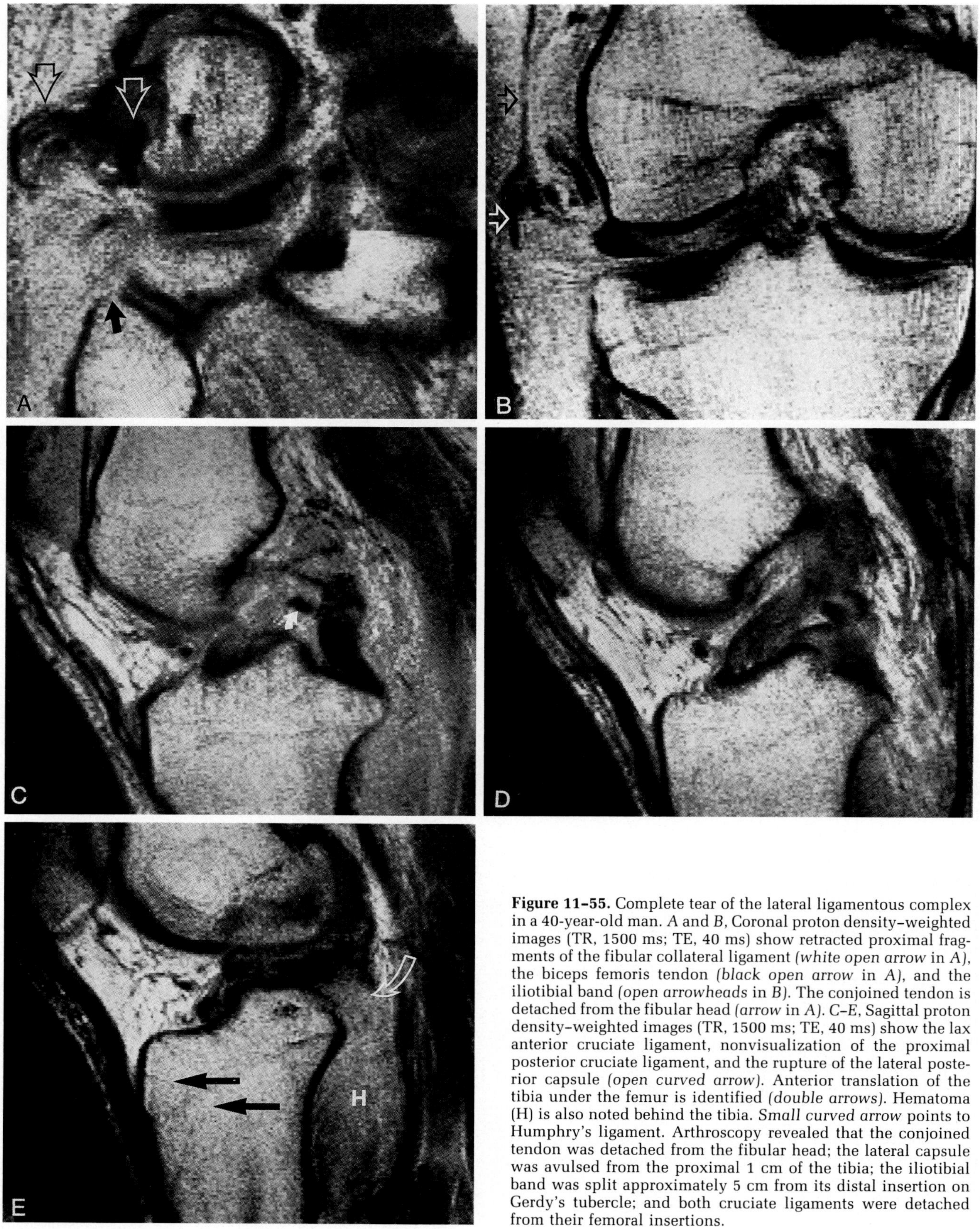

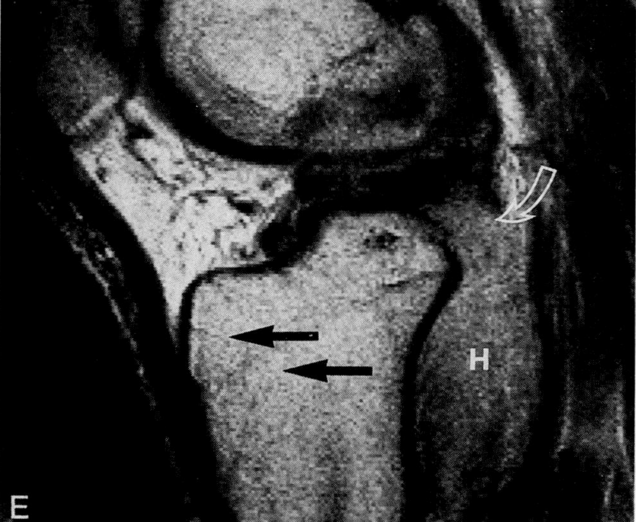

Figure 11–55. Complete tear of the lateral ligamentous complex in a 40-year-old man. *A* and *B*, Coronal proton density–weighted images (TR, 1500 ms; TE, 40 ms) show retracted proximal fragments of the fibular collateral ligament *(white open arrow* in *A)*, the biceps femoris tendon *(black open arrow* in *A)*, and the iliotibial band *(open arrowheads* in *B)*. The conjoined tendon is detached from the fibular head *(arrow* in *A)*. *C–E*, Sagittal proton density–weighted images (TR, 1500 ms; TE, 40 ms) show the lax anterior cruciate ligament, nonvisualization of the proximal posterior cruciate ligament, and the rupture of the lateral posterior capsule *(open curved arrow)*. Anterior translation of the tibia under the femur is identified *(double arrows)*. Hematoma (H) is also noted behind the tibia. *Small curved arrow* points to Humphry's ligament. Arthroscopy revealed that the conjoined tendon was detached from the fibular head; the lateral capsule was avulsed from the proximal 1 cm of the tibia; the iliotibial band was split approximately 5 cm from its distal insertion on Gerdy's tubercle; and both cruciate ligaments were detached from their femoral insertions.

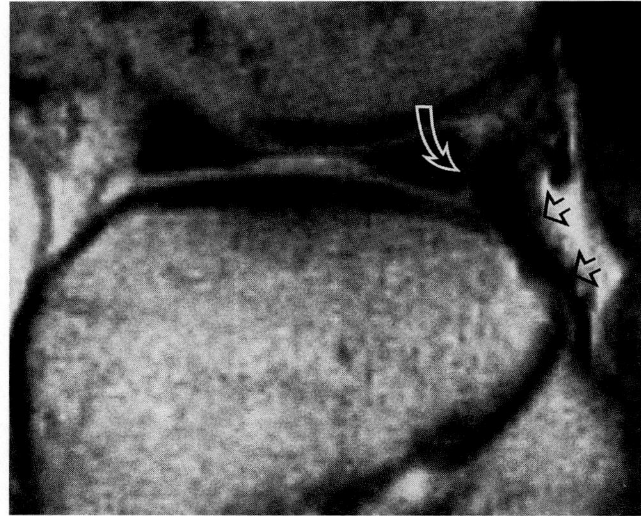

Figure 11–56. Sagittal image (TR, 800 ms; TE, 20) of the popliteal tendon sheath simulating a tear in the posterior horn of the lateral meniscus *(white arrow)*. Note that normal anterior and posterior horns of the lateral meniscus are symmetric in size and shape. *Black arrowheads* point to the popliteal tendon.

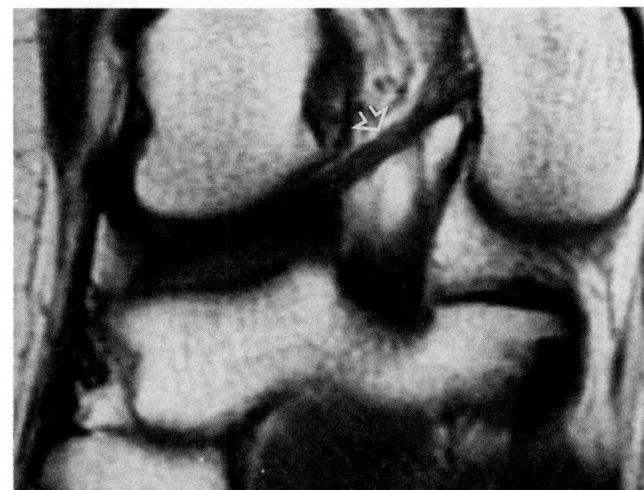

Figure 11–57. Normal meniscofemoral ligament illustrated on a posterior coronal T1-weighted image *(open arrow)*.

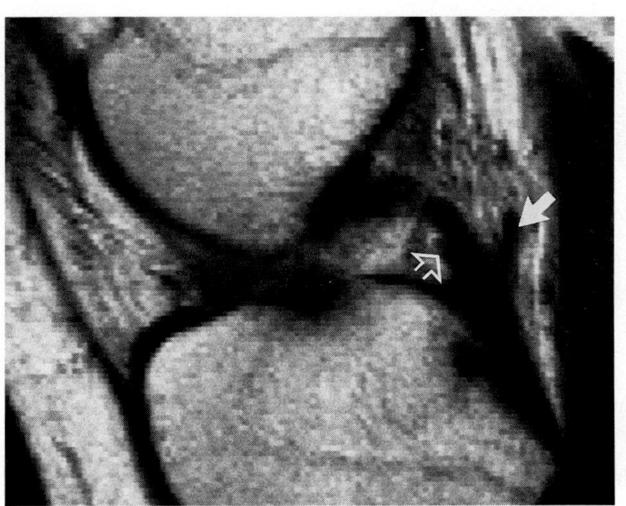

Figure 11–58. Wrisberg's ligament *(arrow)* and Humphry's ligament *(open arrow)*.

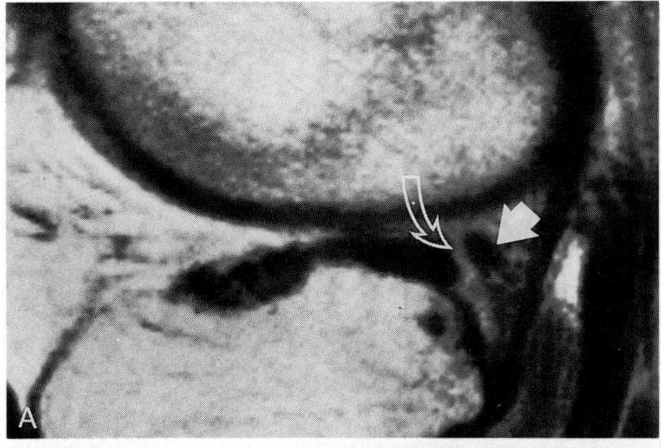

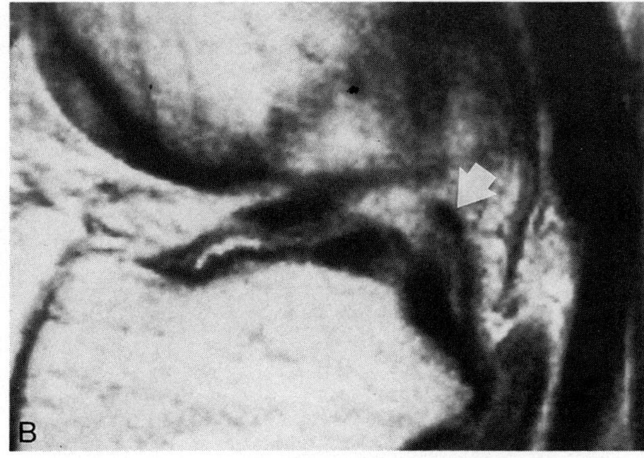

Figure 11–59. *A*, Sagittal T1-weighted image (TR, 800 ms; TE, 20 ms) shows a cleft *(open arrow)* between the meniscofemoral ligament *(solid arrow)* and the posterior horn of the lateral meniscus, simulating a meniscal tear. *B*, A consecutive sagittal image shows the continued course of the ligament *(solid arrow)*.

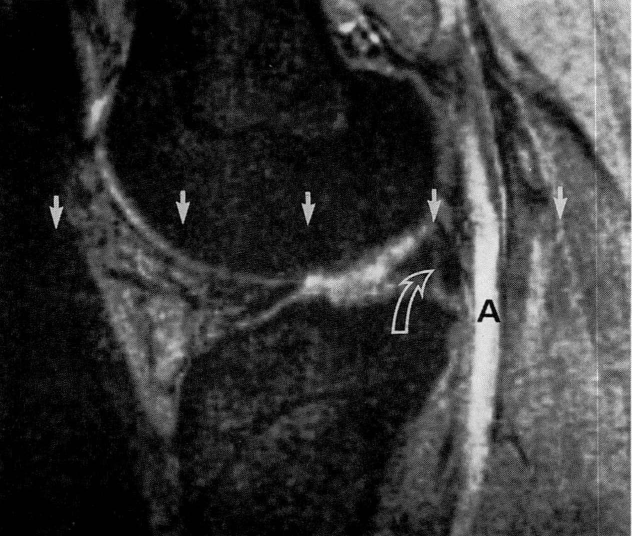

Figure 11–60. Pulsation artifact simulating a meniscal tear *(curved arrow)*. Pulsation and flow of the popliteal artery create artifacts in the phase-encoded direction *(arrows)* on this gradient-echo image. No meniscal tear was found at arthroscopy. A, popliteal artery.

Again, it is important to trace the ligament on consecutive sagittal images to avoid misinterpreting the normal meniscofemoral ligament origin as a tear in the posterior horn of the lateral meniscus.

Pulsation Artifact. A pulsation artifact is created by the regular motion of the popliteal arterial walls and blood flow when the phase-encoding gradient is directed from anterior to posterior (Fig. 11–60).[30] In particular, a posteriorly placed surface coil can magnify the strength of the pulsation artifacts. To decrease the magnitude of this artifact, the surface coil is better placed circumferentially around the knee. The posterior horn of the lateral meniscus may appear macerated owing to the signal from the popliteal artery, which is mismapped onto the meniscus, which normally is of low signal intensity. In contradistinction, a true tear may be missed when the pulsation artifact is present. Coronal images are helpful in establishing a correct diagnosis.

Anterior Horn, Lateral Meniscus. ***Transverse Geniculate Ligament.*** The transverse geniculate ligament runs horizontally posterior to Hoffa's fat pad and connects the anterior horns of the medial and lateral menisci (Fig. 11–61). The pseudotear is caused by a cleft between the meniscal insertion of the transverse geniculate ligament and the central tendinous attachment of the anterior horn of the lateral meniscus. This cleft appears as a band of intermediate signal intensity on sagittal T1- or T2-weighted images, which may mimic a grade III signal, with the ligament simulating the peripheral frag-

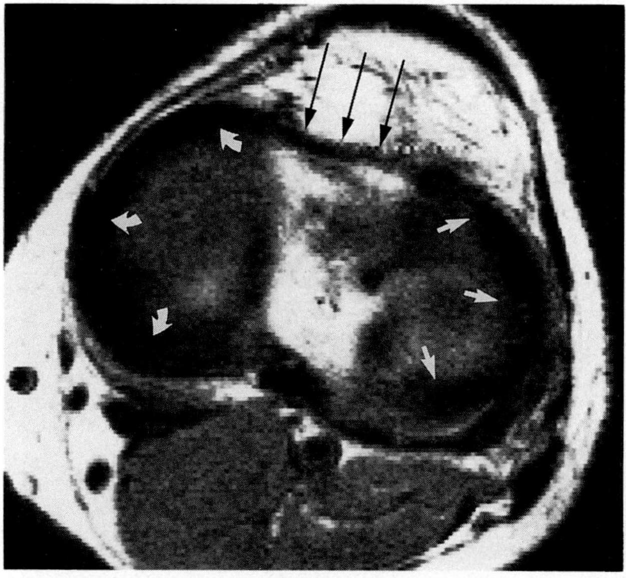

Figure 11–61. Axial image (TR, 1000 ms; TE, 40 ms) of the transverse geniculate ligament connecting both anterior horns of the medial and lateral menisci *(white arrows)*. The lateral meniscus *(black straight arrows)* has a smaller radius of curvature than the medial meniscus *(curved arrows)*.

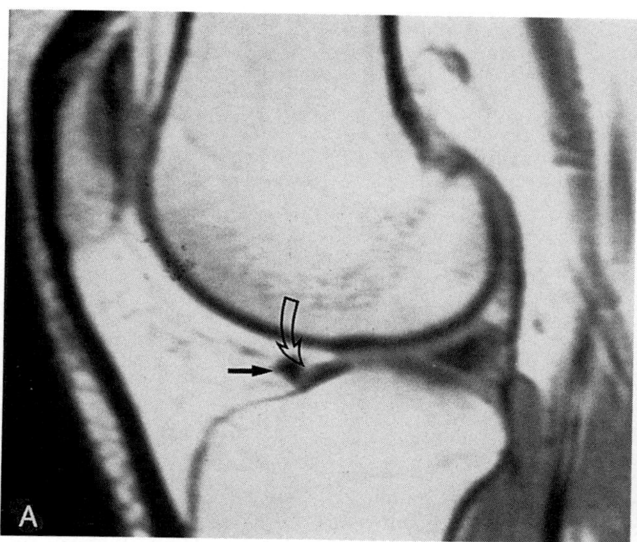

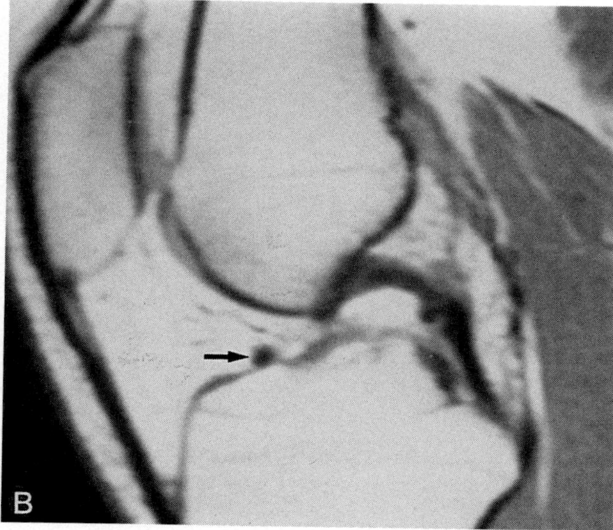

Figure 11–62. *A*, Sagittal image (TR, 2000 ms; TE, 20 ms) shows a cleft *(curved arrow)* between the transverse geniculate ligament *(straight arrow)* and anterior horn of the lateral meniscus, simulating a meniscal tear. *B*, The ligament can be seen along its course to the intercondylar notch *(arrow)*.

ment of the meniscus (Fig. 11–62). Twenty-two to 38 per cent of MR imaging examinations of the knee have been reported to have such an appearance.[17,30,56] Less commonly, the transverse geniculate ligament may mimic a tear of the anterior horn of the medial meniscus.

Diagnostic errors can be avoided by tracing the course of the ligament and the meniscus on consecutive sagittal images. As the observer proceeds centrally, the transverse geniculate ligament becomes a more rounded structure of low signal intensity, coursing anteriorly to the ACL, while the rhomboid central attachment of the anterior horn of the lateral meniscus gradually disappears after it inserts onto the tibia (Fig. 11–63).

Lateral Inferior Genicular Artery. This structure arises from the popliteal artery at the level of the tibiofemoral joint and courses laterally to the anterior aspect of the knee, where it and other arteries compose the genicular anastomosis.[17] The pseudotear is caused by a cleft between the artery and the anterior horn of the lateral meniscus. The cleft may mimic a grade III signal on sagittal images, with the artery simulating the peripheral fragment of the meniscus. Herman and Beltran reported 21 per cent of cases showing such an appearance on sagittal MR images, although it is debatable whether these pseudotears are actually caused by the artery or due to the transverse geniculate ligament.[17]

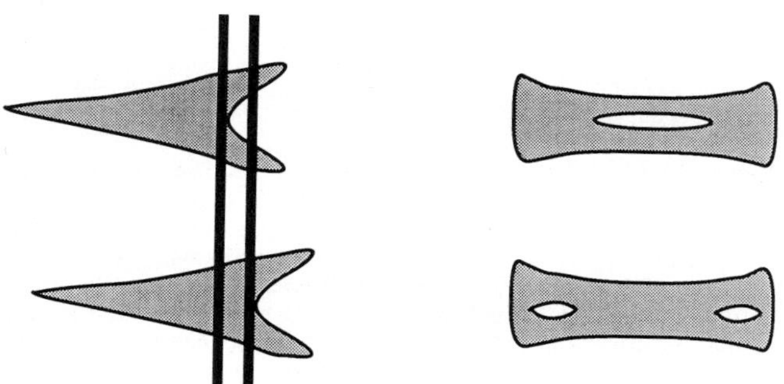

Figure 11–63. Concave meniscal edge, which may produce a linear artifact of high signal intensity on the most peripheral sagittal cut (upper drawing). A subsequent medial cut may create a short artifact in the anterior and posterior horns of the meniscus (lower drawing).

Both Horns, Both Menisci. ***Volume-Averaging Artifact.*** The concavity of the outer margin of the meniscus is filled with fat and neurovascular structures that create a linear artifact of high signal intensity within the dark meniscus, simulating a grade II meniscal signal on the most peripheral sagittal image (Figs. 11–63, 11–64). A coronal image can be used to exclude the presence of a true intrameniscal signal. Herman and Beltran[17] reported that 29 per cent of the medial menisci and 6 per cent of lateral menisci had such an appearance on sagittal images. This volume-averaging artifact is less apparent on thicker slices (e.g., 5 mm) than on thinner slices (e.g., 3 mm).

Truncation Artifact. A truncation artifact is seen as a subtle line that is uniform in thickness and runs parallel to the surfaces of the menisci. It simulates a grade III signal and is located approximately 2 pixels above the tibial articular surface (Fig. 11–65).[52] Turner and colleagues reported that the artifacts are more apparent when the acquisition matrix is 128 × 256 and the 128-pixel (phase-encoded) axis is in a superoinferior orientation.[52] The anteroposterior orientation of the phase-encoding gradient or the use of

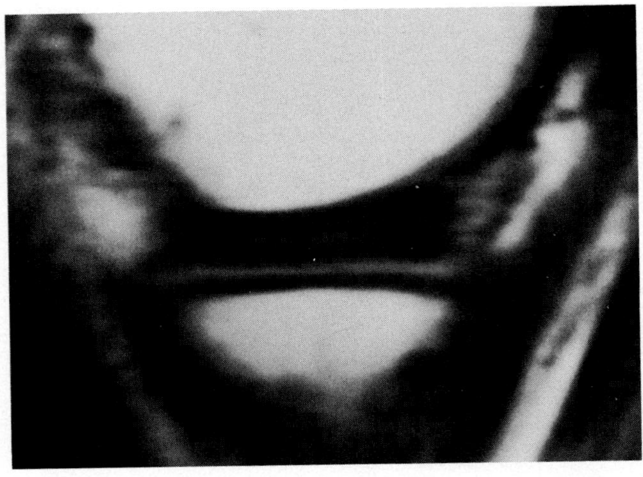

Figure 11-64. Sagittal T1-weighted image (the most peripheral medial cut) shows a linear horizontal artifact in the meniscus. This is caused by volume averaging of fat in the concave meniscal edge.

a 256 × 256 acquisition matrix can virtually eliminate these artifacts (Fig. 11-66). Diagnostic errors can be avoided by careful inspection, as the linear artifact may extend beyond the boundaries of the menisci.

Pseudodiscoid Meniscus. Coronal images through the far posterior horn can mimic a discoid meniscus (Fig. 11-67). This occurs because the posterior horn appears as a continuous horizontal band when the coronal plane is parallel to the posterior curve of the C-shaped menisci.[31] Therefore, a diagnosis of discoid meniscus should be avoided when the fibula is seen on coronal images, representing the far posterior cuts. A diagnosis of a true discoid meniscus is more reliable when made on the sagittal and midcoronal images (Fig. 11-68).

Ligaments

Medial Collateral Ligament. A small amount of fat can normally be identified between the TCL and the medial meniscus. The increased signal intensity of fat on T1-weighted images should not be mistaken for a sign of meniscocapsular separation (Fig. 11-69).

Anterior Cruciate Ligament. A cloudlike edematous appearance at the femoral attachments of the ACL is created by partial volume averaging of the lateral femoral condyle and may occasionally mimic ACL tears on the sagittal images (Fig. 11-70). This artifact can be reduced by a slice thickness of 3 mm. T2-weighted sagittal and coronal images are helpful to confirm whether a true tear exists. In addition, the linear increased signal intensity within the ACL, particularly near the tibial attachment, should not be mistaken for abnormality (Fig. 11-71). This represents the gaps between the fiber bundles of the distal ACL, which normally decrease in signal intensity with T2 weighting.

Meniscofemoral Ligament. The anterior or posterior branches of the meniscofemoral ligament appear as round structures of low signal intensity that can be mistaken for free fragments of a torn meniscus or loose bodies composed of cortical bone (Fig. 11-72; also see Fig. 11-58).

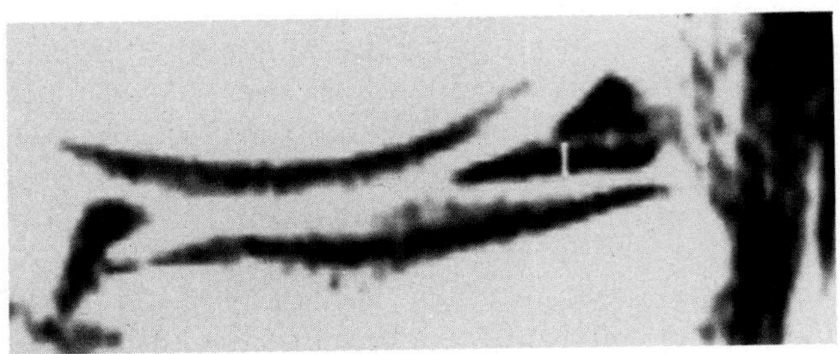

Figure 11-65. A linear intrameniscal signal communicating with the superior articular surface, simulating a meniscal tear. This image was acquired with a 128-pixel (phase-encoded) axis in the superoinferior orientation. The vertical white bar represents a distance of 2 pixels. No meniscal tear was found at arthroscopy. (From Turner DA, Rapoport MI, Erwin WD, et al: Truncation artifact: A potential pitfall in MR imaging of the menisci of the knee. Radiology 179:629, 1991. © Radiological Society of North America.)

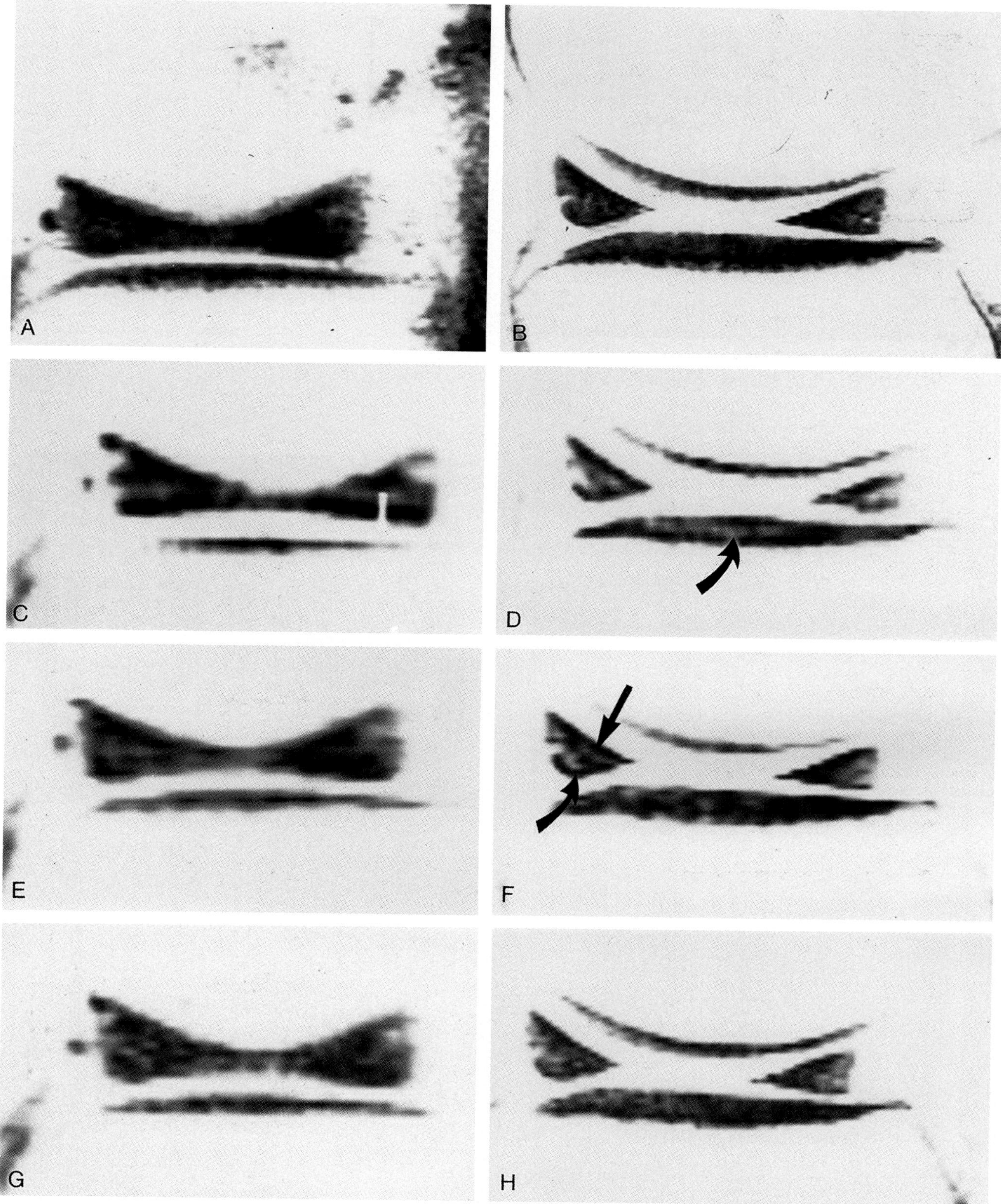

Figure 11–66. Truncation artifact simulating a meniscal tear in the knee of a cadaver. *A* and *B*, Witha 256 × 256 matrix, an intrameniscal signal is seen, but there is no grade III signal intensity. *C* and *D*, With a 128 × 256 matrix and a 128-pixel axis in the superoinferior orientation,a subtle line of high signal intensity appears, mimicking a meniscal tear. The vertical white bar in *C* represents a distance of 2 pixels. A subelt line also is seen in the subchondral bone plate of the tibia. *E* and *F*, With a 128 × 256 matrix and a 128-pixel axis in the anteroposterior orientation, the two subtle lines are less apparent than in *C* and *D*. However, lines parallel to the oblique femoral articular surface and posteroinferior margin of the meniscus seen in *D* remain apparent in *F* (*arrow*). *G* and *H*, With a 256 × 256 matrix and phase-encoded axis (superoinferor orientation), all oblique lines are much less prominent than in *C* to *F*. (From Turner DA, Rapoport MI, Erwin WD, et al: Truncation artifact: A potential pitfall in MR imaging of the menisci of the knee. Radiology 179:629, 1991. © Radiological Society of North America.)

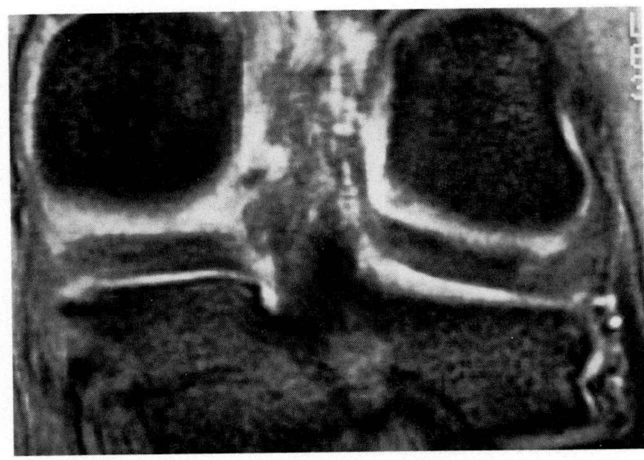

Figure 11-67. Gradient-echo coronal image shows the lack of normal tapering of the meniscal height of both menisci, simulating discoid menisci. Note that this image is the far posterior coronal cut, which may create a pseudodiscoid meniscus.

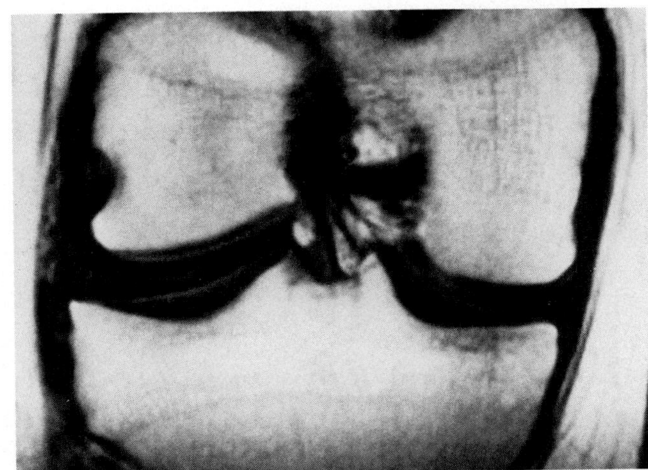

Figure 11-68. Discoid lateral meniscus. Coronal T1-weighted image shows the lack of normal tapering of the lateral meniscal height as it extends toward the intercondylar notch. A linear horizontal intrameniscal signal is noted, indicating meniscal degeneration. Note that the diagnosis is more reliable in the mid-coronal cut, in which the intercondylar notch is seen.

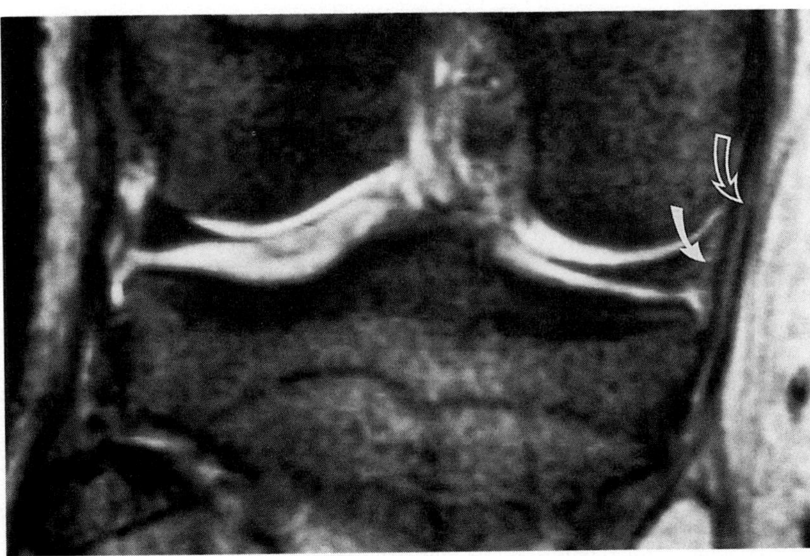

Figure 11-69. Coronal gradient-echo image shows a linear band of intermediate signal intensity between the deep and superficial layers of the medial collateral ligament (MCL) *(open arrow)*. This should not be mistaken for an abnormal finding. Also, a linear band of fat that separates the deep layer of the MCL and the periphery of the medial meniscus does not represent a meniscocapsular separation *(solid arrow)*.

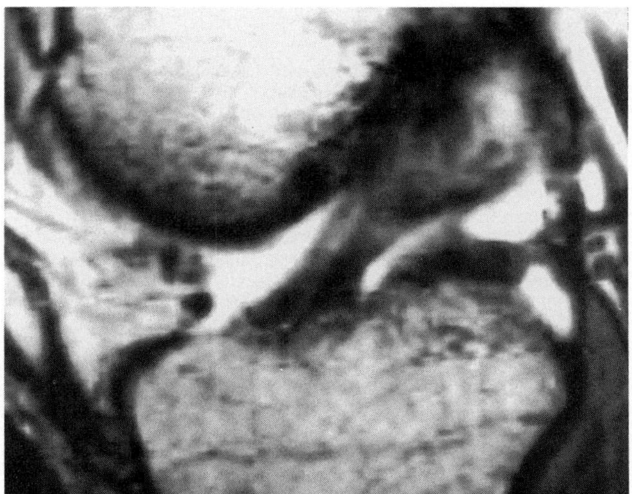

Figure 11–70. Partial volume averaging of the lateral femoral condyle simulating a tear of the anterior cruciate ligament at the femoral attachment. Sagittal T2-weighted image.

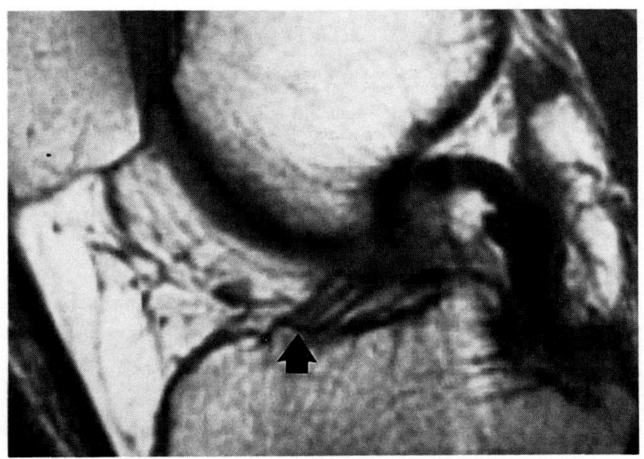

Figure 11–71. Sagittal T1-weighted image showing the normal tibial attachment of the anterior cruciate ligament *(arrow)*. The high signal intensity between the three separate fiber bundles should not be mistaken for a ligamentous sprain or partial tear.

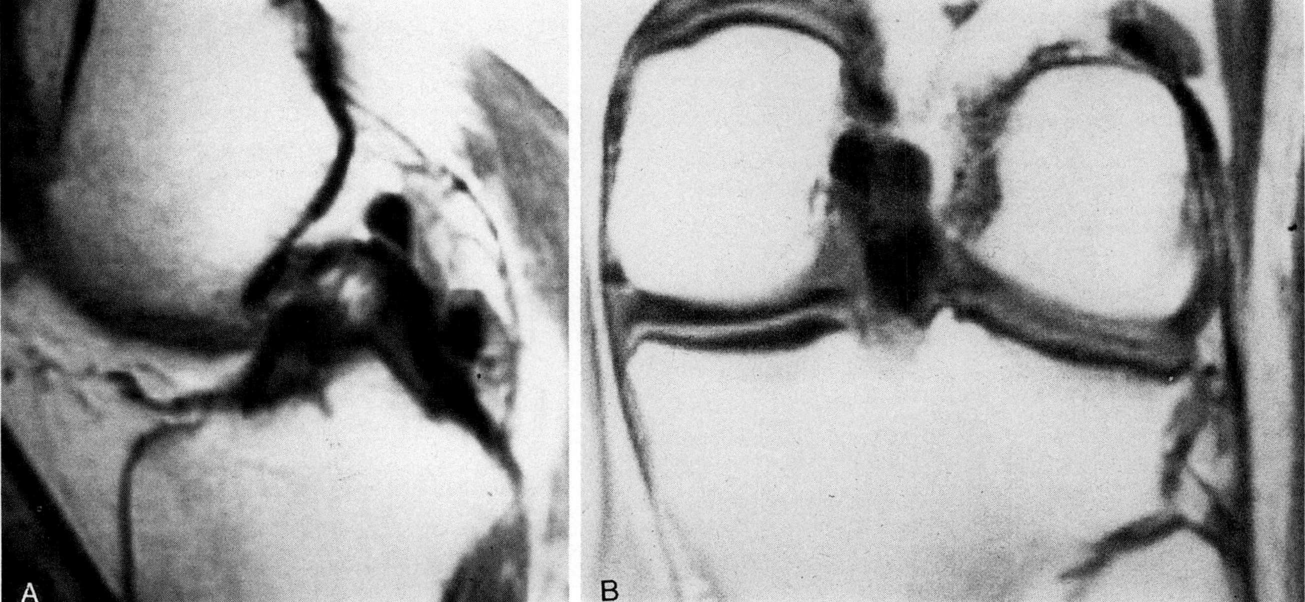

Figure 11–72. *A*, Sagittal T1-weighted image (TR, 800 ms; TE, 25 ms) shows two round structures of low signal intensity posterior to the posterior cruciate ligament. Does either one simulate Wrisberg's ligament? *B*, Coronal image (TR, 1000 ms; TE, 20 ms) shows an absence of the lateral meniscus. Two free fragments originally from the lateral meniscus were removed at surgery.

References

1. Council on Scientific Affairs: Musculoskeletal applications of magnetic resonance imaging. JAMA 262:2420, 1989.
2. Arnoczky SP, Warren R: The microvasculature of the meniscus and its response to injury: An experimental study in the dog. Am J Sports Med 11:131, 1983.
3. Beltran J, Noto AM, Mosure JC, et al: The knee: Surface-coil MR imaging at 1.5-T. Radiology 159:747, 1986.
4. Brantigan OC, Voshell AF: The relationships of the ligament of Humphry's to the ligament of Wrisberg. J Bone Joint Surg 28:66, 1945.
5. Burk DL, Dalinka MK, Kanal E, el at: Meniscal and ganglion cysts of the knee: MR evaluation. AJR 150:331, 1988.
6. Chan WP, Fritz RC, Stoller DW, et al: MR "anterior drawer" sign: A useful sign in the diagnosis of complete anterior cruciate ligament tears. Radiology 181(P):178, 1991.

6a. Cobby MJ, Schweitzer ME, Resnick D: The deepened lateral femoral notch: An indirect sign of a torn anterior cruciate ligament. Radiology 184:855, 1992.

7. Crues JV III, Mink J, Levy TL, et al: Meniscal tears of the knee: Accuracy of MR imaging. Radiology 164:445, 1987.
8. Deutsch AL, Mink JH, Fox JM, et al: Peripheral meniscal tears: MR findings after conservative treatment or arthroscopic repair. Radiology 176:485, 1990.
9. Drace J, Smathers R, Enzmann D: Prospective comparison of radial knee MR imaging with standard sagittal and coronal planes. Radiology 165(P):176, 1987.
10. Ehman RL, Berquist TH, McLeod RA: MR imaging of the musculoskeletal system: A 5-year appraisal. Radiology 166:313, 1988.
11. Farley TE, Howell SM, Love KF, et al: Meniscal tears: MR and arthrographic findings after arthroscopic repair. Radiology 180:517, 1991.
12. Ferrer-Rocca O, Vilalta C: Lesions of the menisci. Part I. Macroscopic and histologic findings. Clin Orthop 146:289, 1980.
13. Fischer S, Fox J, Pizzo W, et al: Accuracy of diagnoses from magnetic resonance imaging of the knee. J Bone Joint Surg 73:2, 1991.
14. Frija G, Schouman-Claeys E, d'Anthouard F, et al: Grossly normal knee menisci: Correlation with pathology and magnetic resonance imaging. Diagn Interv Radiol 1:29, 1989.
15. Fritz R, Helms C, Genant HK, et al: Frequency and significance of equivocal meniscal and anterior cruciate ligament tears in MR imaging of the knee. Radiology 177(P):225, 1990.
16. Grover JS, Bassett LW, Gross ML, et al: Posterior cruciate ligament: MR imaging. Radiology 174:527, 1990.
17. Herman LJ, Beltran J: Pitfalls in MR imaging of the knee. Radiology 167:775, 1988.
18. Howell SM, Berns GS, Farley TE: Unimpinged and impinged anterior cruciate ligament grafts: MR signal intensity measurements. Radiology 179:639, 1991.

18a. Kaplan PA, Walker CW, Kilcoyne RF, et al: Occult fracture patterns of the knee associated with anterior cruciate ligament tears: Assessment with MR imaging. Radiology 183:835, 1992.

19. Kaplan P, Nelson N, Garvin K, et al: MR of the knee: The significance of high signal in the meniscus that does not clearly extend to the surface. AJR 156:333, 1991.
20. Kennedy JC, Hawkins RJ, Willis EB, et al: Tension studies of human knee ligaments: Yield point, utimate failure, and disruption of the cruciate and tibial collateral ligaments. J Bone Joint Surg [Am] 58:350, 1976.
21. Kornick J, Trefelner E, McCarthy S, et al: Meniscal abnormalities in the asymptomatic population at MR imaging. Radiology 177:464, 1990.
22. Lee JK, Yao L, Phelps CT, et al: Anterior cruciate ligament tears: MR imaging compared with arthroscopy and clinical tests. Radiology 166:861, 1988.
23. Lee JL, Yao L: Tibial collateral ligament bursa: MR imaging. Radiology 178:855, 1991.
24. Levinsohn E, Baker B: Prearthrotomy diagnostic evaluation of the knee: Review of 100 cases diagnosed by arthroscopy and arthrotomy. AJR 131:107, 1980.
25. Li KC, Henkelman M, Poon PY, et al: MR imaging of the normal knee. J Comput Assist Tomogr 8:1147, 1984.
26. Lotysch M, Mink J, Crues JV III, et al: Magnetic resonance in the detection of meniscal injuries (Abstr). Magn Reson Imaging 4:185, 1986.
27. Mandelbaum B, Finerman G, Reicher M, et al: Magnetic resonance imaging as a tool for evaluation of traumatic knee injuries. Am J Sports Med 14:361, 1986.
28. Mink JH, Deutsch AL: MRI of the Musculoskeletal System: A Teaching File. New York, Raven Press, 1990.
29. Mink JH, Levy T, Crues JV III: Tears of the anterior cruciate ligament and menisci of the knee: MR imaging evaluation. Radiology 167:769, 1988.
30. Mink JH, Reicher MA, Crues JV III: Magnetic Resonance Imaging of the Knee. New York, Raven Press, 1987.
31. Munk PL, Helms CA: MRI of the Knee. Rockville, Md, Aspen, 1991.
32. Munk PL, Helms CA, Genant HK, et al: Magnetic resonance imaging of the knee: Current status, new directions. Skeletal Radiol 18:569, 1989.

32a. Murphy BJ, Smith RL, Uribe JW, et al: Bone signal abnormalities in the posterolateral tibia and lateral femoral condyle in complete tears of the anterior cruciate ligament: A specific sign? Radiology 182:221, 1992.

33. Negendank WG, Fernandez-Madrid FR, Heilbrun LK, et al: Magnetic resonance imaging of meniscal degeneration in asymptomatic knees. J. Orthop Res 8:311, 1990.
34. Quinn S, Muus C, Sara A, et al: Meniscal tears: Pathologic correlation with MR imaging (letter). Radiology 166:580, 1988.
35. Rak KM, Gillogly SD, Schaefer RA, et al: Anterior cruciate ligament reconstruction: Evaluation with MR imaging. Radiology 178:553, 1991.
36. Raunest J, Oberle K, Loehnert J, et al: The clinical value of magnetic resonance imaging in the evaluation of meniscal disorders. J Bone Joint Surg [Am] 73:11–16, 1991.
37. Reeder JD, Martz SO, Becker L, et al: MR imaging of the knee in the sagittal projection: Comparison of three-dimensional gradient-echo and spin-echo sequences. AJR 154:537, 1989.
38. Reicher MA, Hartzman S, Duckwiller GR, et al: Meniscal injuries: Detection using MR imaging. Radiology 159:753, 1986.
39. Reicher MA, Rauschning W, Gold RH, et al: High-resolution magnetic resonance imaging of the knee joint: Normal anatomy. AJR 145:859, 1985.
40. Resnick D, Madewell JE, Sweet DE: Tumors and tumor-like lesions of bone in or about joints. *In* Resnick D, Niwayama G (eds): Diagnosis of Bone and Joint Disorders. Philadelphia, WB Saunders, 1981.
41. Rosen MA, Jackson DW, Berger PE: Occult osseous lesions documented by magnetic resonance imaging associated with anterior cruciate ligament ruptures. Arthroscopy 7:45, 1991.
42. Rothschild PA, Domesek JM, Kaufman L, et al: MR imaging of the knee with a 0.064-T permanent magnet. Radiology 175:775, 1990.
43. Selesnick F, Noble H, Bachman D, et at: Internal derangement of the knee: Diagnosis by arthrography, arthroscopy, and arthrotomy. Clin Orthop 198:26, 1985.
44. Silva I, Silver DM: Tears of the mensicus as revealed by magnetic resonance imaging. J Bone Joint Surg [Am] 70:199, 1988.
45. Silverman JM, Mink JH, Deutsch AL: Discoid menisci of the knee: MR imaging appearance. Radiology 173:351, 1989.
46. Singson RD, Feldman F, Staron R, et al: MR imaging of displaced bucket-handle tear of the medial meniscus. AJR 156:121, 1991.
47. Sisk TD: Knee injuries. *In* Crenshaw A (ed): Campbell's Operative Orthopaedics. St. Louis, CV Mosby, 1987.
48. Smith DK, Totty WG: The knee after partial meniscectomy: MR imaging features. Radiology 176:141, 1990.
49. Stoller DW, Martin C, Crues JV III, et al: Meniscal tears: Pathologic correlation with MR imaging. Radiology 163:731, 1987.
50. Thijn C: Accuracy of double-contrast arthrography and arthroscopy of the knee joint. Skeletal Radiol 8:187, 1982.
51. Tria AJ, Johnson CD, Zawadsky JP: The popliteus tendon. J Bone Joint Surg [Am] 71:714, 1989.
52. Turner DA, Rapoport MI, Erwin WD, et al: Truncation artifact: A potential pitfall in MR imaging of the menisci of the knee. Radiology 179:629, 1991.
53. Tyrrell RL, Gluckert K, Pathria M, et al: Fast three-dimensional MR imaging of the knee: Comparison with arthroscopy. Radiology 166:865, 1988.
54. Vahey TN, Bennett HT, Arrington LE, et al: MR imaging of the knee: Pseudotear of the lateral meniscus caused by the meniscofemoral ligament. AJR 154:1237, 1990.
55. Vahey TN, Broome DR, Kayes K, et al: Acute and chronic tears of the anterior cruciate ligament: Differential features at MR imaging. Radiology 181:251, 1991.
56. Watanabe AT, Carter BC, Teitelbaum GP, et al: Normal variations in MR imaging of the knee: Appearance and frequency. AJR 153:341, 1989.
57. Weiss KL, Morehouse HT, Levy IM: Sagittal MR images of the knee: A low-signal band parallel to the posterior cruciate ligament caused by a displaced bucket-handle tear. AJR 156:117, 1991.
58. Zarins B, Nemeth VA: Acute knee injuries in athletes. Orthop Clin North Am 16:285, 1985.

12 THE KNEE
Other Pathologic Conditions

Wing P. Chan, Russell C. Fritz, Lynne S. Steinbach, Chun Y. Wu, Harry K. Genant, and W. Dillworth Cannon

This chapter deals with magnetic resonance (MR) imaging of the diverse pathologic conditions involving the knee. For some of the entities described in this chapter, MR imaging is still an investigational approach that is not employed routinely in the diagnosis and follow-up of patients with joint disorders. However, MR imaging is more accurate than other imaging modalities and may be of particular value when selecting patients for preoperative planning and in monitoring response to therapy.

PATELLOFEMORAL JOINT AND EXTENSOR MECHANISM

Patellar Instability

Patellar tracking abnormalities are a major cause of anterior knee pain and often result in chondromalacia patellae. Factors precipitating subluxation or dislocation of the patella include repetitive trauma, torn retinacular attachments, patella alta, femoral condyle hypoplasia, genu valgum or genu recurvatum, and abnormal insertion of the patellar tendon at the tibial tuberosity. Clinical evaluation is difficult, as signs and symptoms are nonspecific and may mimic those of other knee joint disorders. Many radiographic techniques have been devised to evaluate patellar malalignment. However, a major drawback of these radiographic techniques is the technical difficulty in obtaining tangential views of the patellofemoral joint.

The normal patellar position is displayed on axial MR images with the posterior patellar apex centered in the femoral trochlear groove. In a study of patellar tracking with kinematic MR imaging, Shellock and colleagues[48] proposed the following terminology: (1) *lateral subluxation*—the apex of the patella is displaced laterally relative to the femoral trochlear groove or the centermost part of the femoral trochlea, and the lateral facet of the patella overlaps the lateral aspect of the femoral trochlea (Fig. 12–1); (2) excessive *lateral pressure syndrome*—the lateral facet of the patella is tilted toward the lateral aspect of the femoral trochlear groove with little or no lateral subluxation of the patella; the relative degree of tilting typically increases with increased knee flexion; (3) *medial subluxation*—the apex of the patella is displaced medially relative to the femoral trochlear groove (Fig. 12–2); (4) *lateral-to-medial subluxation of the patella*—the patella starts in a laterally displaced position and moves to a medially displaced position with increasing knee flexion; (5) *dislocation*—the patella is completely displaced from its normal position relative to the femoral trochlear groove (Fig. 12–3).

The measurements of the congruence angle and patellar tilt angle have been employed for assessing patellar alignment radiographically. Such measurements have been described on axial computed tomography (CT) and MR images,[13] but their reliabilities have not been reported.

Chondromalacia Patellae

Chondromalacia patellae is a descriptive term for pathologic patellar cartilage softening, resulting

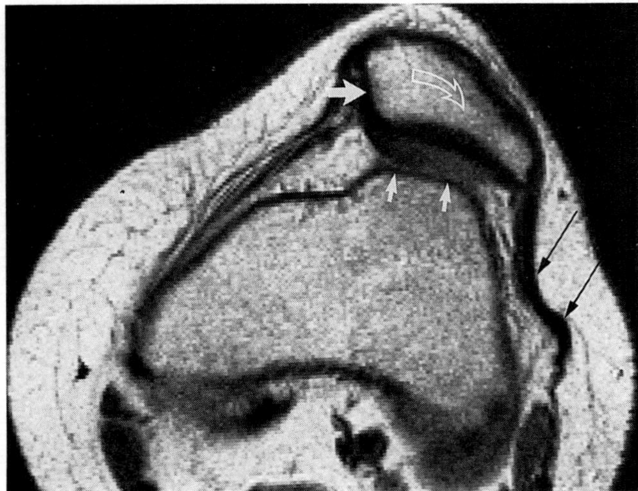

Figure 12–1. Lateral patellar subluxation in a 28-year-old woman with a congenital dysplastic patellofemoral joint. Axial image (TR, 1500 ms; TE, 20 ms) shows a dominant lateral patellar facet and an absence of the medial patellar facet *(thick arrow)*. The trochlear groove is replaced by a convex lateral femoral condyle, indicating the presence of a dysplastic femoral trochlea *(small arrows)*. The patella is displaced laterally *(curved arrow)*, and the lateral retinaculum is markedly redundant *(long arrows)*. This patient was treated by an arthroscopic lateral retinacular release.

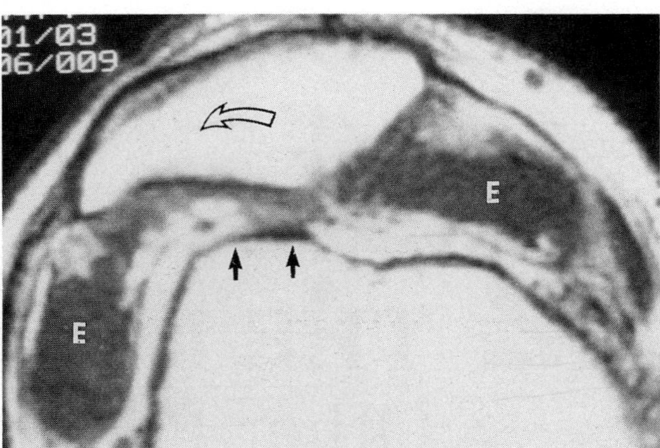

Figure 12–3. Lateral patellar dislocation in a 60-year-old man with degenerative joint disease. Axial T1-weighted image (TR, 600 ms; TE, 20 ms) shows a laterally dislocated patella *(curved open arrow)* with its apex riding on the dysplastic femoral condyle *(arrows)*. A giant marrow-containing osteophyte protruding from the lateral facet is seen, as is associated effusion (E).

in patellofemoral pain, frequently encountered in adolescents and young adults. These findings can be produced by acute injury to the knee or by repeated minor trauma. On the basis of arthroscopic findings, Shahriaree divided chondromalacia patellae into four grades.[47] Grade I shows cartilage softening due to disruption of the vertical collagen fibers. Grade II shows blister formation due to separation of the more superficial layers of cartilage from the deeper layers (Fig. 12–4). Grade III shows surface ulceration and fragmentation due to blister ruptures (Fig. 12–5). Grade IV shows subchondral bone exposure (Fig. 12–6). This mechanism of chondromalacia has been termed *basal degeneration*[20] and typically affects younger age groups. It is thought to occur secondary to trauma and perhaps as an inherited metabolic disturbance. Conversely, superficial degeneration affects older age groups, involves the superficial cartilage, and is caused by repetitive minor trauma (wear and tear mechanism) and may result in patellofemoral osteoarthritis. Stages I and II superficial degeneration consist of fine surface fibrillation, discoloration, and minor fissuring involving the superficial layer of the cartilage. Stages III and IV have imaging features similar to those of basal degeneration grades III and IV.

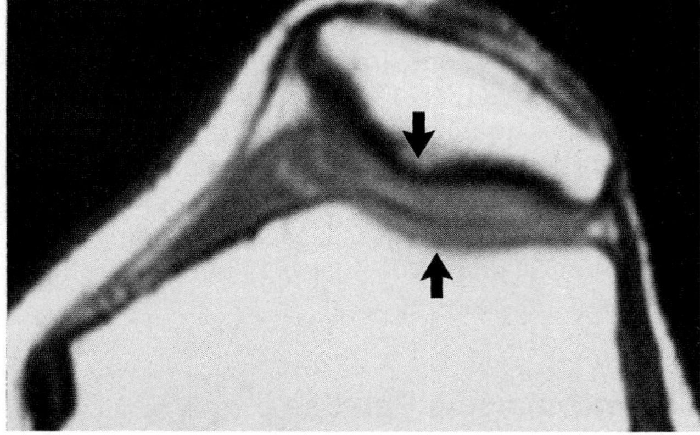

Figure 12–2. Medial patellar subluxation in a 46-year-old man. The patella is tilted and medial to the trochlear groove *(arrows)* on this axial T1-weighted image.

Magnetic resonance imaging allows the depiction of cartilage defects, internal chondral signal changes, and subchondral defects. Yulish and coworkers developed an MR imaging grading system for chondromalacia patellae.[62] MR grade I corresponds to arthroscopic grades I and II and shows focal areas of swelling with areas of decreased signal intensity on T1- and T2-weighted images. MR grade II corresponds to arthroscopic grade III and shows irregularity of the articular surface of the posterior patellar cartilage with focal thinning. MR grade III corresponds to arthroscopic grade IV and shows absence of cartilage with exposure of subchondral bone. Using the arthroscopic criteria described by Shahriaree,[47] Hayes and colleagues[21] reported that MR imaging offers an accurate means of detecting grade II to IV lesions in their study of 14 freshly disarticulated knee specimens (see Figs. 12–4 to 12–6, 12–7). However, both Yulish[62] and Hayes and their co-workers[21] found that MR imaging was not able to detect grade I lesions. To avoid false-negative diagnoses, patellar cartilage is best evaluated by both sagittal and axial images. T1-weighted and gradient-

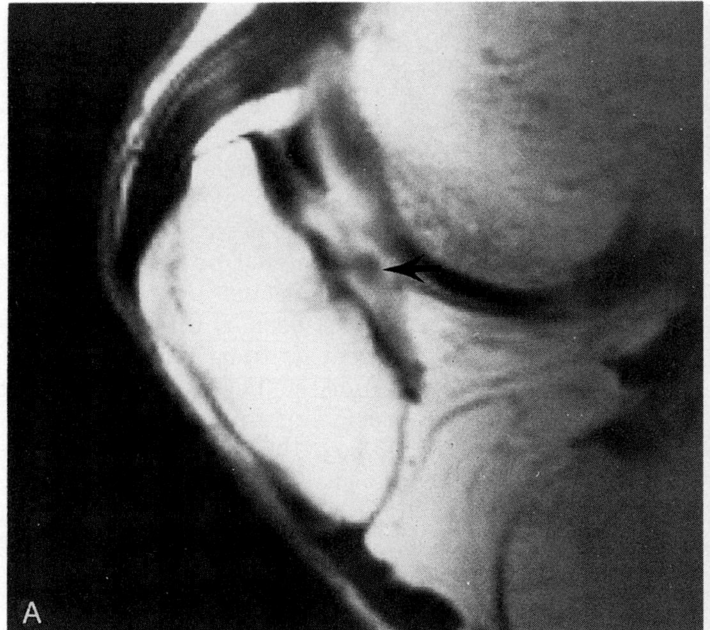

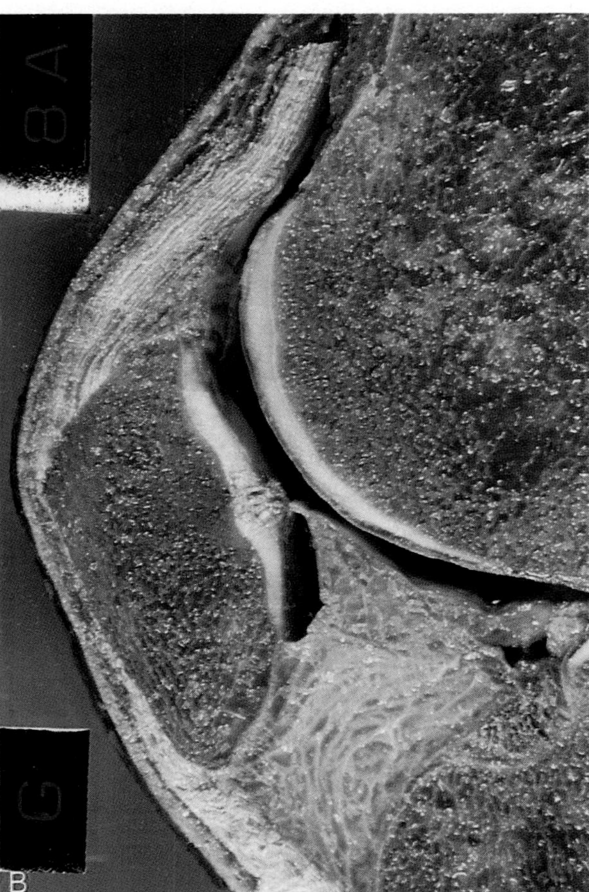

Figure 12–4. Grade II chondromalacia patellae in a freshly disarticulated knee specimen. *A*, Sagittal T1-weighted image (TR, 700 ms; TE, 20 ms) shows a focal area of decreased signal intensity extending to and deforming the articular cartilage surface *(arrow)*. *B*, Corresponding gross specimen shows a focal blisterlike lesion, which was found to be filled with vertical cartilage fibrils. (From Conway WF, Hayes CW, Loughran T, et al: Cross-section imaging of the patellofemoral joint and surrounding structures. Radiographics 11:195, 1991. © Radiological Society of North America.)

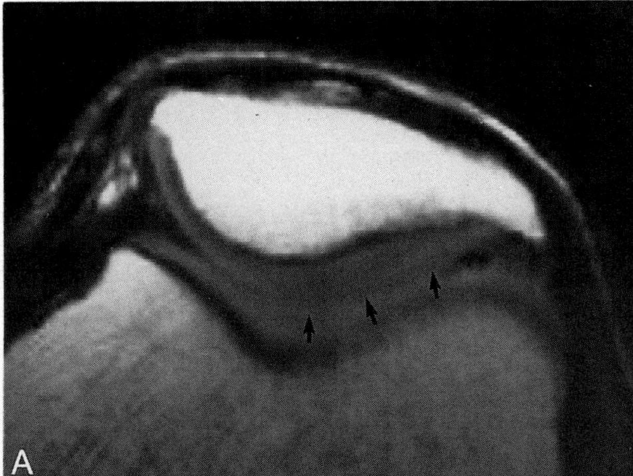

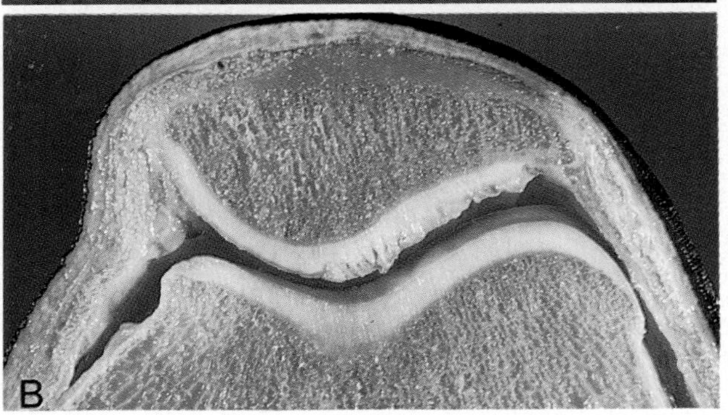

Figure 12–5. Grade III B chondromalacia patellae in a freshly disarticulated knee specimen. *A*, Axial T1-weighted image (TR, 700 ms; TE, 20 ms) shows a complete loss of the line between the patellar and trochlear cartilages *(arrows)*. B, Corresponding gross specimen shows a large area of "crabmeat" fibrillation involving the entire lateral patellar facet. (From Conway WF, Hayes CW, Loughran T, et al: Cross-section imaging of the patellofemoral joint and surrounding structures. Radiographics 11:195, 1991. © Radiological Society of North America.)

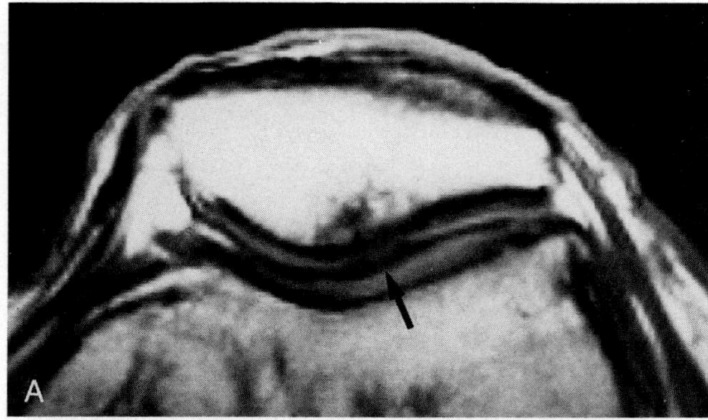

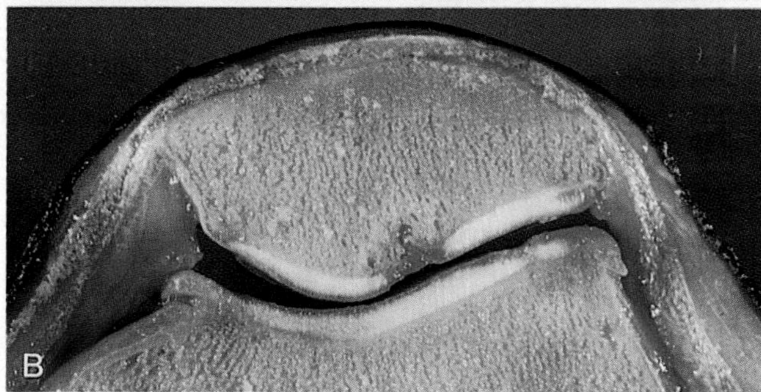

Figure 12–6. Grade IV chondromalacia patellae in a freshly disarticulated knee specimen. *A*, Axial T1-weighted image (TR, 700; TE, 20 ms) shows focal decreased signal intensity of both the articular cartilage *(arrow)* and the subchondral bone. *B*, Corresponding gross specimen shows a focal ulceration on the lateral patellar facet extending to subchondral bone. Early subchondral cystic changes are present. (From Conway WF, Hayes CW, Loughran T, et al: Cross-section imaging of the patellofemoral joint and surrounding structures. Radiographics 11:195, 1991. © Radiological Society of North America.)

echo sequences are useful in depicting patellar cartilage changes.

Retinacula

The retinacula are stabilizers of normal patellar tracking and are best seen on axial images. The normal lateral retinaculum appears as a dark single band extending from the vastus lateralis, and the medial retinaculum normally appears as somewhat thinner bands extending from the vastus medialis. The medial retinaculum is torn more frequently than the lateral one, reflecting the tendency for the patella to sublux and dislocate laterally (Fig. 12–8). Retinacular tears appear as lax structures of low signal intensity with or without surrounding edema or hemorrhage, depending on the severity of the injury.

Patellar Tendon

The "patellar tendon-to-patella ratio" has been used to gauge abnormalities of patellar tendon length. A normal patellar tendon is the same length as the patella.[53] A high position of the patella with a relatively long patellar tendon, termed *patella alta*, is associated with subluxation, chondromalacia, and cerebral palsy.[57] A high-riding patella most commonly results from patellar tendon tears and avulsions. *Patella baja* is defined as an inferiorly positioned patella and is seen in patients with quadriceps tendon rupture, poliomyelitis, achondroplasia, and juvenile rheumatoid arthritis. These abnormalities are best depicted on sagittal images (Fig. 12–9). Foci of high signal intensity on T1-weighted images may be seen within the patellar tendon, particularly near its tibial insertion, in patients without symptoms at that site. This may be attributed to the "magic angle phenomenon" as reported by Erickson and co-workers (see Chapter 13). Increased signal intensity on T2-weighted images, particularly near the patellar aspect of the patellar tendon, is characteristic of a patellar tendinitis commonly referred to as *jumper's knee*.

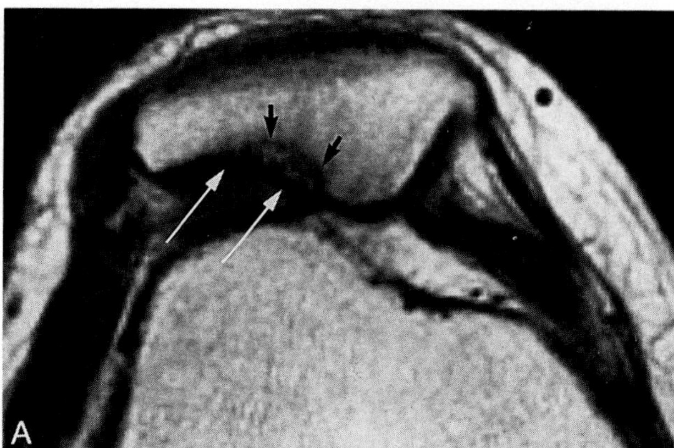

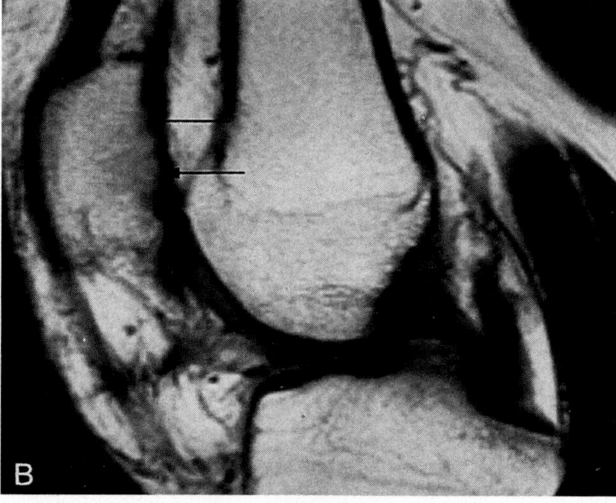

Figure 12–7. Lateral facet chondromalacia in a 32-year-old man. This patient has a history of lateral patellar dislocation and lateral retinaculum release. *A*, Axial and *B*, sagittal T1-weighted images (TR, 800 ms; TE, 20 ms) show lateral patellar subluxation, subchondral irregularity *(long arrows)*, and low signal intensity subchondrally, which is suggestive of sclerosis *(small arrows)*. Arthroscopy confirmed the presence of grade IV chondromalacia patellae.

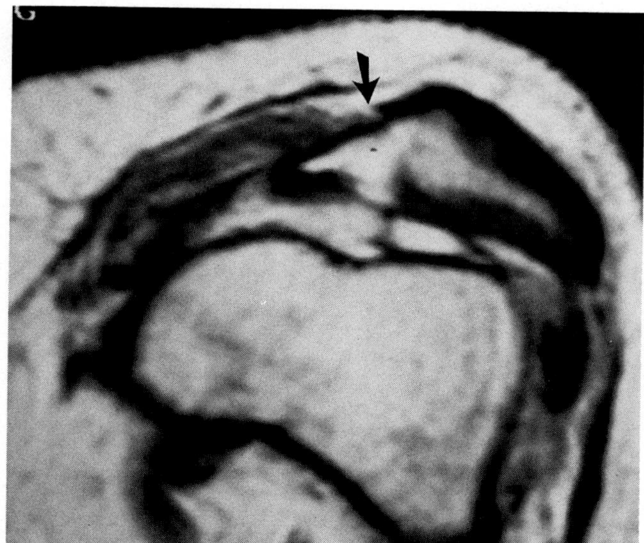

Figure 12–8. Axial T1-weighted image shows a tear of the medial retinaculum *(arrow)* and associated lateral subluxation of the patella. A patellar fracture at the medial facet is noted.

Synovial Plicae

Plicae are persistent remnants of embryonic synovial septa seen in 18 to 60 per cent of adults.[16] The medial patellar plica runs vertically along the medial capsule of the joint adjacent to the medial facet of the patella (Fig. 12–10). The suprapatellar plica can separate the suprapatellar pouch from the knee joint if it is complete. The infrapatellar plica extends from Hoffa's fat pad to the intercondylar notch just in front of the anterior cruciate ligament; if it is complete, it may rarely divide the knee into medial and lateral compartments. The medial patellar plica is the most common type to cause symptoms.

The medial patellar plica can be seen on both axial and sagittal images (Figs. 12–11, 12–12), whereas the suprapatellar plica is best identified on sagittal images (Fig. 12–13). The infrapatellar plica is not commonly seen on MR images. Normal plicae appear as a line of low signal intensity on T1- and T2-weighted images and are apparent only in the presence of knee effusions. We have seen cases in which thickening of the medial patellar plica was associated with cartilaginous erosions of the medial patellar facet and femoral trochlea (Figs. 12–14, 12–15). This finding has been observed by arthroscopy and arthrography as reported by Patel[40] and Lupi and colleagues,[29] respectively.

SYNOVIAL AND ARTHRITIC DISORDERS

Arthritides are joint disorders that cause swelling, pain, stiffness, and early disability. Clinical evaluation, laboratory studies, and conventional radiography commonly are employed in the routine diagnosis and follow-up of these diseases. Structural changes in the knee such as cartilage loss and osseous changes are predominant findings on radiography; however, these changes are manifested late in the disease process. Magnetic resonance imaging is emerging as a valuable noninvasive tool to evaluate early soft tissue changes before arthritic changes are evident radiographically.

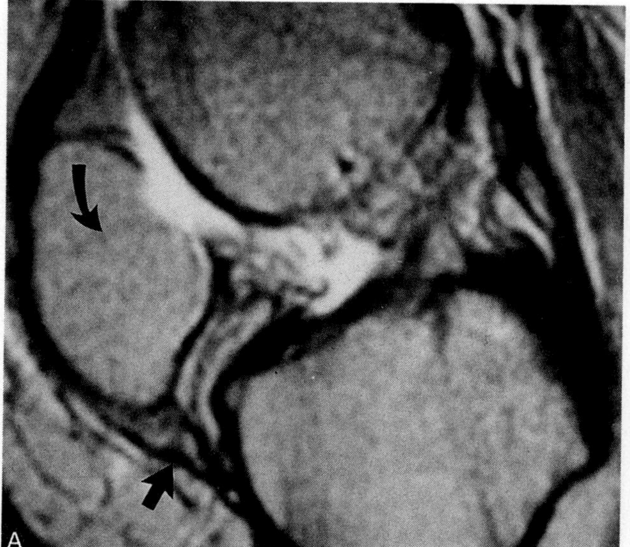

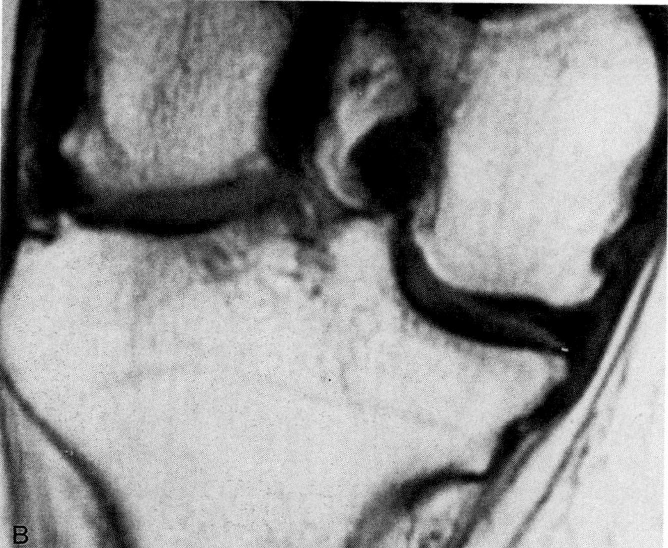

Figure 12–9. Patella baja in a 40-year-old woman with poliomyelitis. *A*, Sagittal T2-weighted image (TR, 1800 ms; TE, 80 ms) shows a low-riding patella *(curved arrow)* and a shortening of the patellar tendon *(straight arrow)*. *B*, Coronal T1-weighted image (TR, 800 ms; TE, 20 ms) shows asymmetric tibial epiphyses and femoral condyles. Hypoplastic menisci are noted.

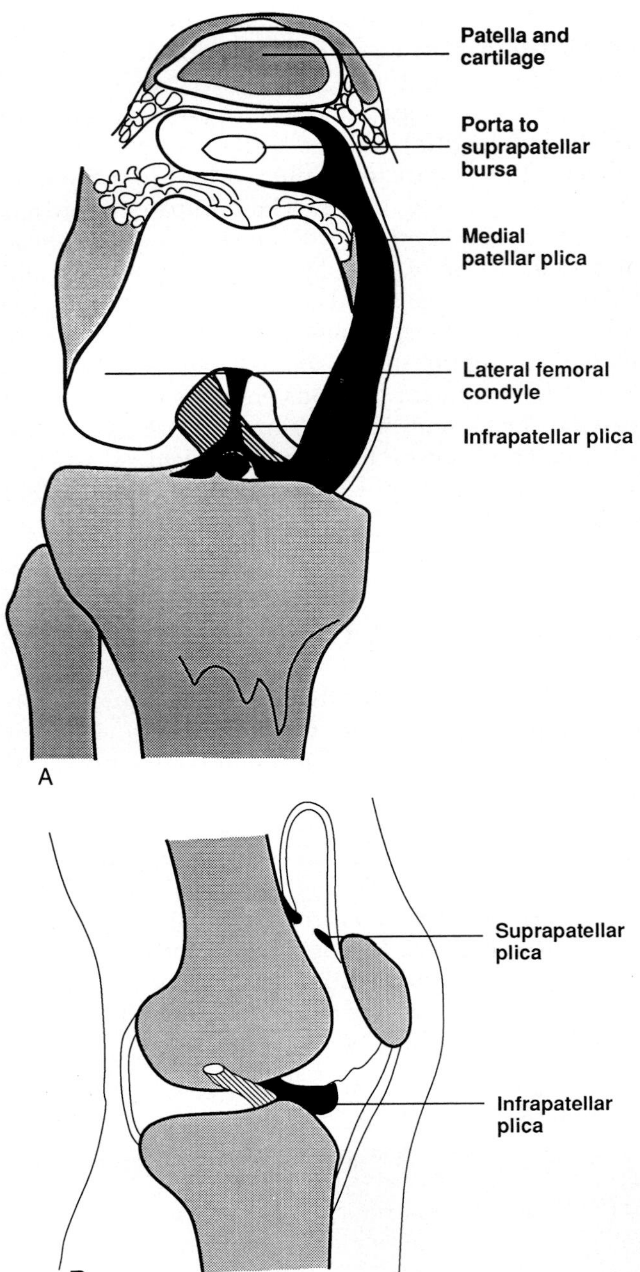

Figure 12–10. Anterior *(A)* and lateral *(B)* views of synovial plicae of the knee.

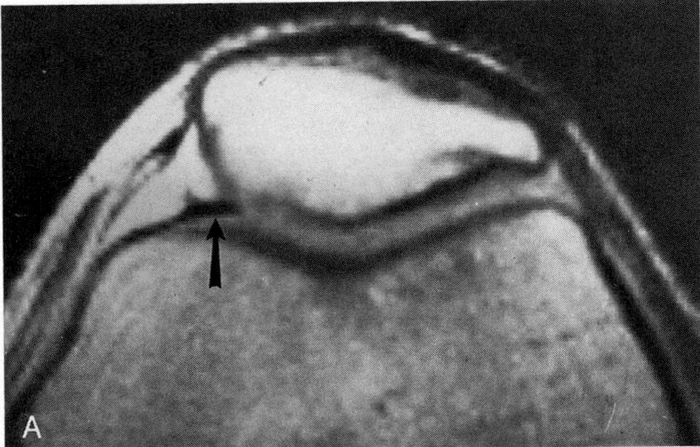

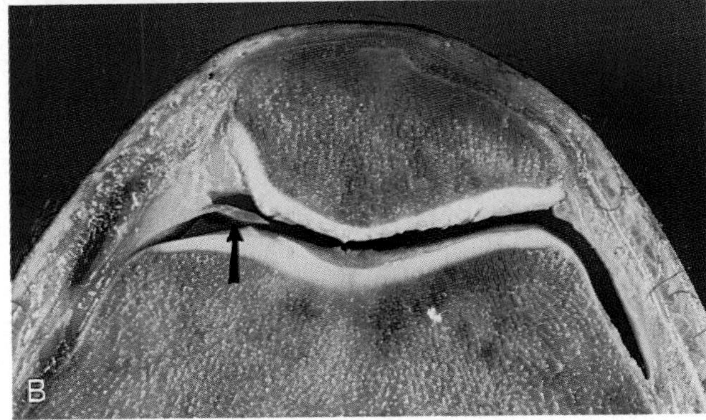

Figure 12–11. Medial plica of the knee in a freshly disarticulated knee specimen *(arrow)*. *A*, Axial T1-weighted image (TR, 700 ms; TE, 20 ms). *B*, Corresponding gross specimen. (From Conway WF, Hayes CW, Loughran T, et al: Cross-section imaging of the patellofemoral joint and surrounding structures. Radiographics 11:195, 1991. © Radiological Society of North America.)

Synovitis

The posterior free edge of Hoffa's fat pad normally is smoothly marginated. The *irregular infrapatellar fat pad sign* refers to the loss of the smooth posterior free border of the fat pad and indicates synovitis, which can be appreciated on sagittal T1-weighted images (see Figs. 12–21, 12–22).[53] This sign reflects synovial irritation, which may be caused by rheumatoid arthritis, hemophilia, septic arthritis, pigmented villonodular synovitis, or re-

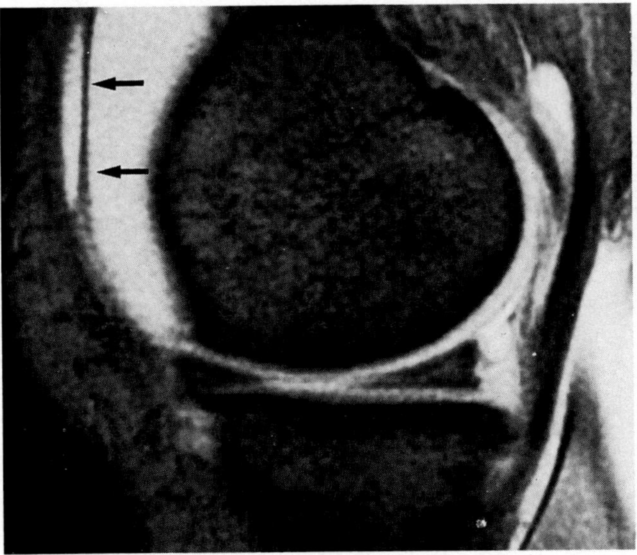

Figure 12–12. Sagittal image (TR, 850 ms; TE, 15 ms; flip angle, 30 degrees) shows a normal medial patellar plica *(arrows)*.

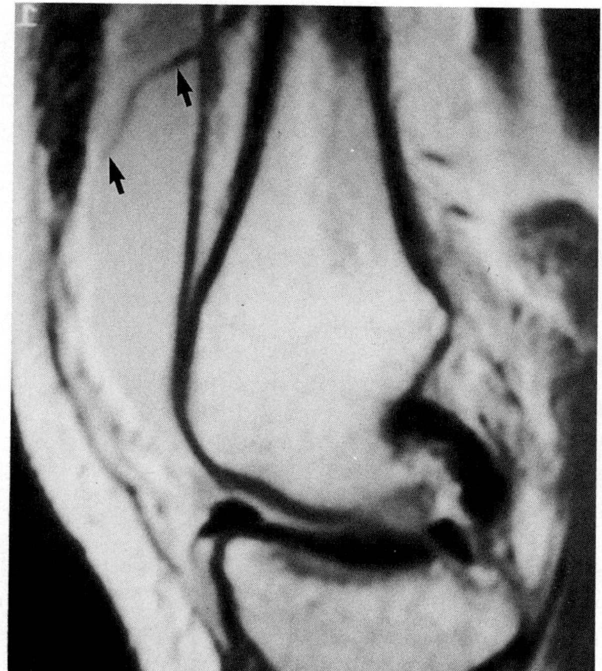

Figure 12–13. Sagittal image (TR, 1600 ms; TE, 40 ms) shows a normal suprapatellar plica *(arrows)*.

peated hemarthrosis from either osteoarthritis or trauma.

Articular Cartilage Changes

Articular cartilage changes, including focal erosions, contour irregularity, and thinning can be seen in patients with traumatic conditions such as osteochondral fractures, chondromalacia patellae, degenerative osteoarthritis, and synovial inflammatory

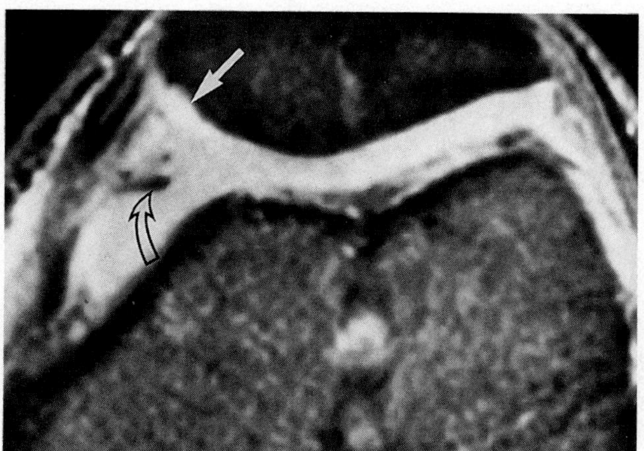

Figure 12–14. Thickened medial patellar plica *(curved arrow)* associated with joint effusion and medial facet cartilage erosion *(straight arrow)*. Axial image (TR, 600 ms; TE, 20 ms; flip angle, 30 degrees).

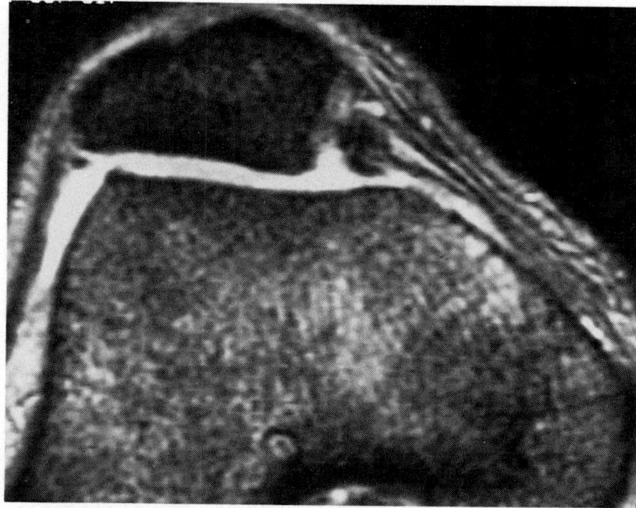

Figure 12–15. Plica syndrome in a 68-year-old man. Axial gradient-echo image (TR, 600; TE, 20 ms; flip angle, 30 degrees) shows a thickened medial patellar plica impinging on the medial patellar facet.

processes (e.g., rheumatoid arthritis). Magnetic resonance imaging demonstrates cartilaginous abnormality before it is evident radiographically. Erosion or thinning of the cartilage is more apparent in children than in adults on MR images because of the increased thickness of hyaline cartilage in children.

On T1-weighted images, the articular cartilage normally appears as a line of intermediate signal intensity that parallels the cortical bone (low-signal-intensity). Hyaline articular cartilage is normally increased in signal intensity relative to meniscal fibrocartilage on T1- and T2-weighted images owing to its higher levels of type II collagen and hydroxylysine, which result in a more hydrophilic state.[4] In a study of knees from cadavers, chondral defects as small as 3 mm could be appreciated on T2-weighted images in the presence of fluid.[11] Gradient-echo images show the best contrast between the articular cartilage of high signal intensity and the cortical bone of low signal intensity. The interface between joint fluid and cartilage could be differentiated satisfactorily with the use of moderate flip angle (20 to 30 degrees) gradient-echo imaging techniques (Fig. 12–16). One study[11] of knees from cadavers revealed that a T1-weighted hybrid fat suppression sequence provided superior visualization, with cartilage being of very high signal intensity in contrast to the low signal intensity of cortical bone and the intermediate signal intensity of joint fluid. Magnetic resonance imaging can detect articular cartilage changes, showing thinning, irregular articular surface (Fig. 12–17), or focal erosion (see Fig. 12–23). Chondromalacia appears as a region of decreased signal intensity because of the loss of hydration.

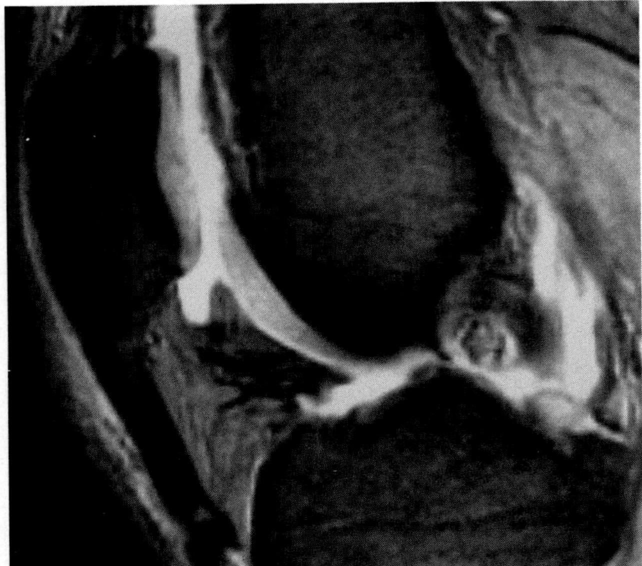

Figure 12-16. Sagittal gradient-echo image (TR, 500 ms; TE, 15 ms; flip angle, 20 degrees) shows a cartilage-fluid interface. Joint effusion has higher signal intensity than adjacent cartilage surface.

In patients with articular chondrocalcinosis, multiple punctate signal void calcium salts may be seen within the articular cartilage of high signal intensity on gradient-echo images (Fig. 12–18). This disease process can be present without symptoms or may be coincident with other types of arthritis.[19] Detection of this disease on MR images often is incidental to other arthritic disorders. However, MR imaging is unable to detect the deposition of calcium in the fibrocartilage (meniscus) of joints. Radiography is helpful in confirming the deposition of calcium salts in joints.

Rheumatoid Arthritis

Rheumatoid arthritis is a systemic and progressive inflammatory disease that affects synovial joints in which the synovium is the target tissue. It affects about 1 per cent of the adult population in the United States; approximately three fourths of those affected are women.[31] A majority of adults (80 per cent) with this disease have elevated levels of serum rheumatoid factor.[44] Rheumatoid factor is composed of immunoglobulin M (IgM) antibodies that are directed against immunoglobulin G (IgG). The inflamed synovium is characterized by edema, hypertrophied cells, and villous transformation within the synovial membrane. Progressive disease results in pannus formation, which is a pathogenetic factor in joint destruction. The pannus is hypervascular in the active phase and becomes fibrous in the nonactive phase.[25,41] Pannus formation cannot be differentiated from joint effusions radiographically. Nonspecific findings of soft tissue swelling may be observed clinically and radiographically when a large joint effusion or pannus formation is present.

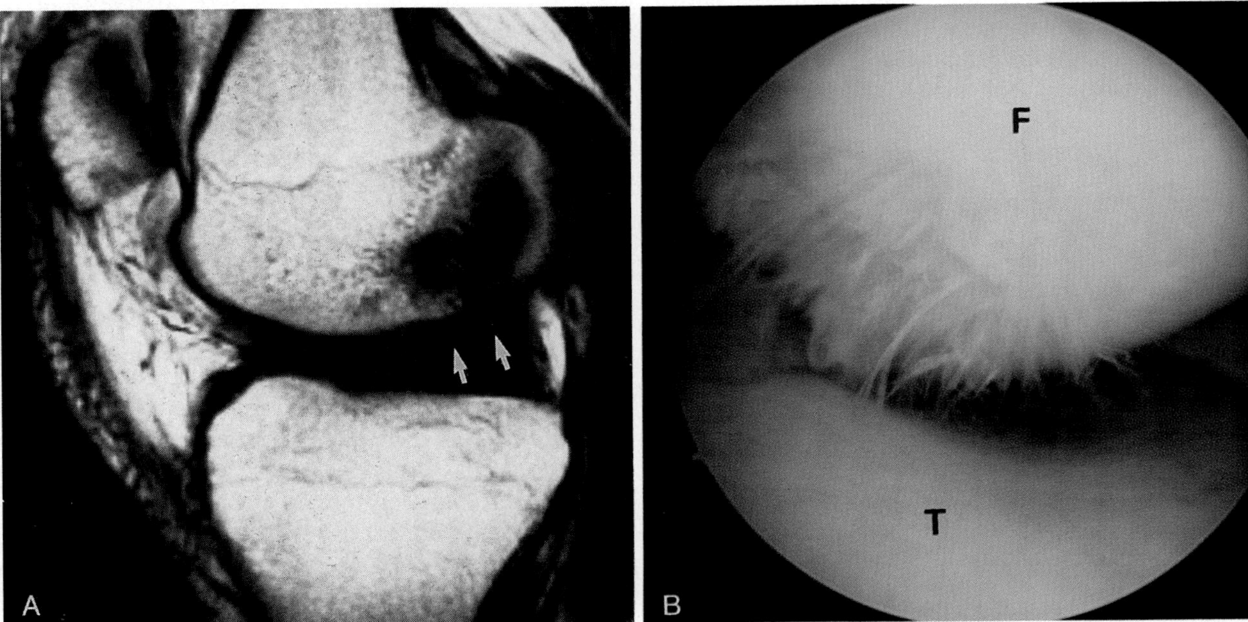

Figure 12-17. *A*, Sagittal T1-weighted image (TR, 600 ms; TE, 20 ms) shows irregular medial femoral articular surface of this right knee. *B*, Corresponding arthroscopic view demonstrates a wide area of fibrillation of the femoral articular surface. Note the smooth articular surface of the tibia for comparison. F, Femoral condyle; T, tibial plateau.

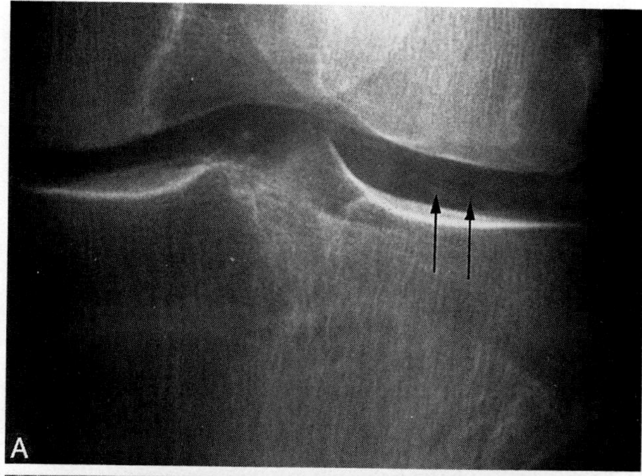

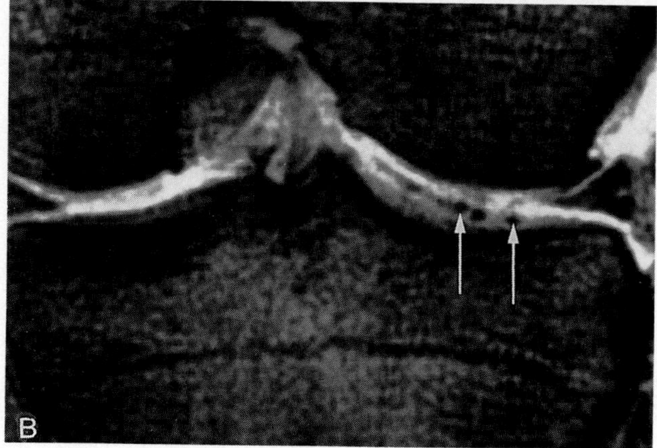

Figure 12–18. Calcium pyrophosphate dihydrate in a 62-year-old man with osteoarthritis of the knee. *A*, Initial radiograph shows narrowing of the joint space in the medial compartment. A distinct linear and punctate calcification is visualized in the lateral joint space *(arrows)*. *B*, Coronal gradient-echo image (TR, 750 ms; TE, 20 ms; flip angle, 30 degrees) shows many ovoid signal void lesions within the tibial articular cartilage of high signal intensity, corresponding to calcification as shown on the radiograph. Note the diffuse thinning and irregularity of the articular cartilage in both compartments, which could not be assessed on radiographic technique.

This finding precedes the marginal bone erosions that appear during the first 2 years after the onset of symptoms in about 90 per cent of patients.[8] Symmetric tricompartmental cartilage loss, which is caused by the invasion of destructive pannus and enzymatic degradation is characteristically seen on radiography.

Magnetic resonance imaging permits detection of active, hypervascular pannus, which is valuable for the early diagnosis of rheumatoid arthritis, planning of surgery, and follow-up after drug therapy. Fibrous pannus shows decreased signal intensity on T1- and T2-weighted images, whereas active pannus with increased vascularity has decreased signal intensity on T1 weighting and increased signal intensity on T2 weighting. However, hypervascular pannus cannot be differentiated from joint effusions on all pulse sequences. Reiser and colleagues first reported that the use of intravenous gadopentetate dimeglumine resulted in rapid and marked enhancement of the synovial pannus on T1-weighted spin-echo images, allowing differentiation from joint effusions.[41] Subsequently, König and associates[25] reported that the tissue that enhances rapidly on postcontrast images represents histologically proved hypervascular pannus (Fig. 12–19). Only minor and delayed enhancement is found in fibrous pannus (Fig. 12–20), effusions (Fig. 12–21), or normal structures. The average time required to reach maximum enhancement is 64 sec for the hypervascular pannus, in contrast to 120 sec and 108 sec for the joint effusion and fibrous pannus, respectively. The enhancement factor for the hypervascular pannus is 145 per cent, in contrast to 15 per cent and 25 per cent for joint effusions and fibrous pannus, respectively. Therefore, differentiation between hypervascular pannus and joint effusion is possible after bolus injection of gadopentetate dimeglumine on T1-weighted spin-echo images with rapidly acquired pulse sequences.

Beltran and co-workers found that MR imaging depicts bony erosions more reliably than radiographs.[5] This may be due to the tomographic nature of MR imaging as well as to the high contrast between the erosions filled with pannus of intermediate or low signal intensity and the marrow fat of high signal intensity on T1-weighted images. Tricompartmental cartilage loss can be seen directly on MR imaging. Magnetic resonance imaging may depict cartilage loss well before joint space narrowing is seen on radiographs. Popliteal cysts, commonly associated with rheumatoid arthritis, are well seen on MR images. Magnetic resonance imaging may also demonstrate the rupture of ligaments and tendons as well as bursitis in patients with rheumatoid arthritis.

Juvenile Rheumatoid Arthritis

In addition to the same features as in adult rheumatoid arthritis, atrophic or hypoplastic menisci are found in most patients with juvenile rheumatoid arthritis (Fig. 12–22).[46] This may occur as a result of factors that impair normal development of fibrocartilage, such as nutritional disturbances in joint fluid secondary to pannus or destruction by proteolytic enzymes. Common findings in juvenile rheumatoid arthritis also include enlarged epiphyses, hypoplastic inferior patellae, and popliteal cysts.[46,60] In a minority of cases, avascular necrosis, intra-articular fragments, and medullary infarcts also may be seen.

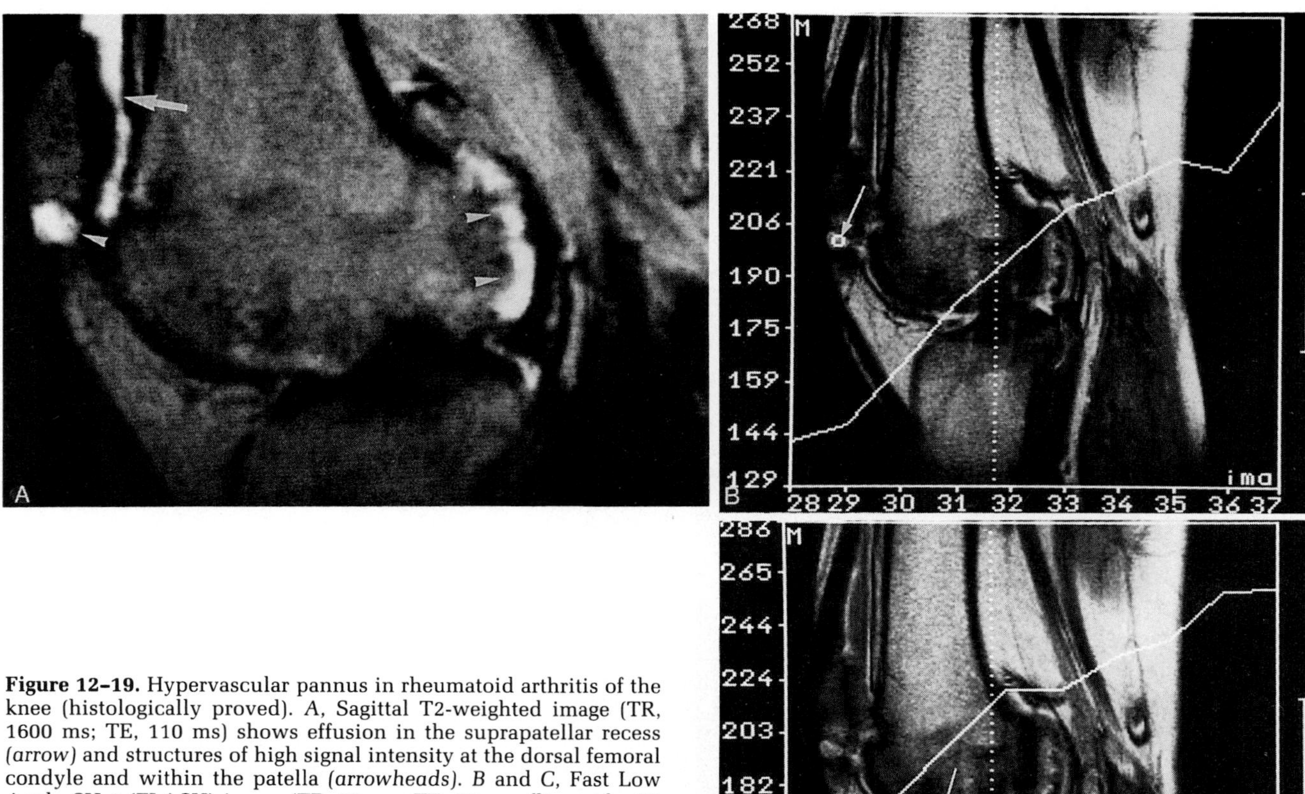

Figure 12–19. Hypervascular pannus in rheumatoid arthritis of the knee (histologically proved). *A,* Sagittal T2-weighted image (TR, 1600 ms; TE, 110 ms) shows effusion in the suprapatellar recess *(arrow)* and structures of high signal intensity at the dorsal femoral condyle and within the patella *(arrowheads)*. *B* and *C,* Fast Low Angle SHot (FLASH) image (TR, 30 ms; TE, 12 ms; flip angle, 70 degrees) obtained 72 sec after injection of gadopentetate dimeglumine. Note that enhancements are found in the hypervascular subchondral pannus in *B* and flat pannus layer covering the articular cartilage in *C (arrow)*. Effusion remains of low signal intensity. X axis, image numbers; Y axis, signal intensity values. Signal-intensity time curves of the corresponding areas of interest *(circles)* are shown. (From König H, Sieper J, Wolf K-J: Rheumatoid arthritis: Evaluation of hypervascular and fibrous pannus with dynamic MR imaging enhanced with Gd-DTPA. Radiology 176:473, 1990.)

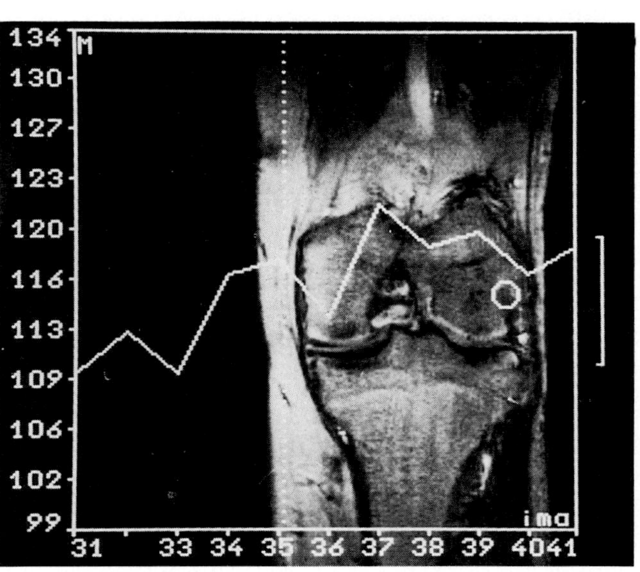

Figure 12–20. Fibrous pannus in rheumatoid arthritis of the knee (histologically proved). Fast Low Angle SHot (FLASH) image (TR, 30 ms; TE, 12 ms; flip angle, 70 degrees) obtained 80 sec after injection of gadopentetate dimeglumine. Signal-intensity time curve shows no significant enhancement. X axis, image numbers; Y axis, signal-intensity values. (From König H, Sieper J, Wolf K-J: Rheumatoid arthritis: Evaluation of hypervascular and fibrous pannus with dynamic MR imaging enhanced with Gd-DTPA. Radiology 176:473, 1990.)

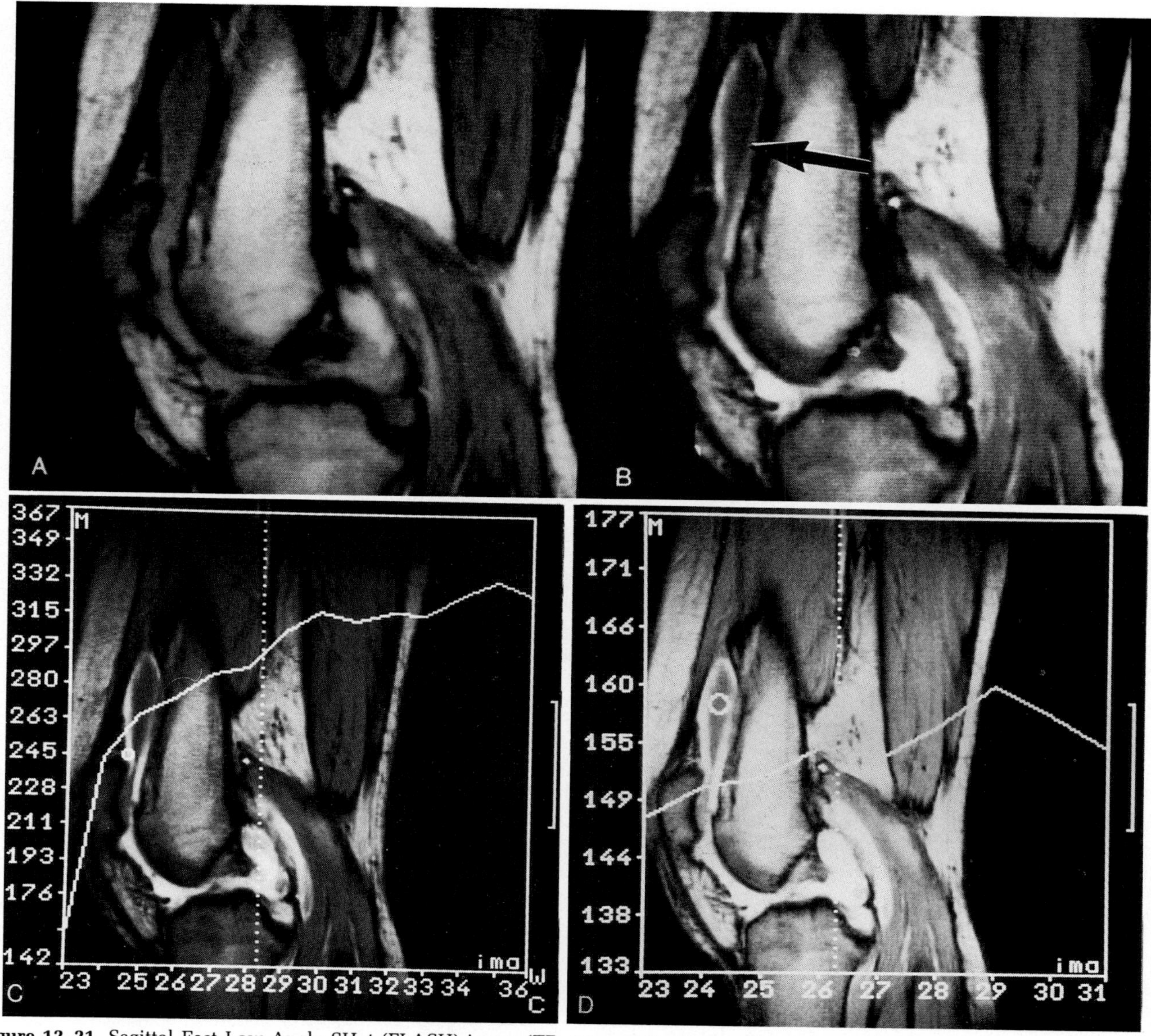

Figure 12–21. Sagittal Fast Low Angle SHot (FLASH) image (TR, 30 ms; TE, 12 ms; flip angle, 70 degrees) shows joint effusion and histologically proved surrounding pannus masses. *A*, Precontrast image shows indistinguishable border between joint effusion and pannus. *B*, Image obtained 96 sec after injection of gadopentetate dimeglumine shows rim enhancement of the hypervascular pannus *(arrow)*. Effusion was eventually enhanced 160 sec after injection, resulting in an appearance that is indistinguishable from pannus. *C* and *D*, Signal-intensity time curve of hypervascular pannus enhancement in *C* and joint effusion enhancement in *D*. X axis, image numbers; Y axis, signal intensity values. (From König H, Sieper J, Wolf K-J: Rheumatoid arthritis: Evaluation of hypervascular and fibrous pannus with dynamic MR imaging enhanced with Gd-DTPA. Radiology 176:473, 1990.)

Osteoarthritis

Osteoarthritis is a slowly progressive degenerative disease, in which the *articular cartilage* is the target tissue. It afflicts 10 to 20 per cent of the elderly in the United States and affects women twice as frequently as men.[39] Osteoarthritis is characterized pathologically by cartilage loss, osteophytosis, subchondral sclerosis, and cyst formation.[1] Cartilage loss often occurs in areas of increased load. Osteophytosis is believed to be a reparative response associated with early osteoarthritis. Subchondral sclerosis is the result of new bone formation from increased numbers of osteoblasts and increased osteoblastic activity. Subchondral cysts result from focal bone resorption in an area of high intra-articular pressure that is frequently associated with cartilage loss.

Many important findings in osteoarthritis of the knee remain undetected by conventional radiographic techniques, and even by CT. Magnetic resonance imaging frequently detects tricompartmental disease when radiography only shows bicompartmental involvement.[10] A radiographically normal

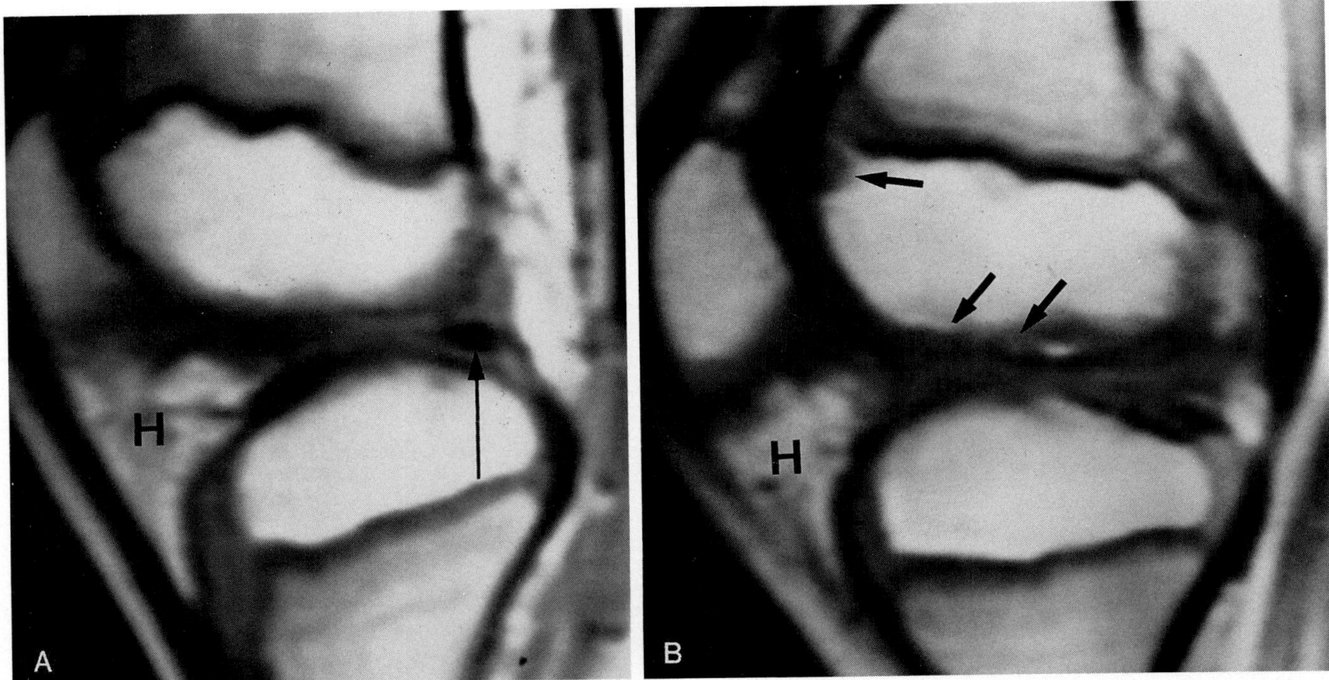

Figure 12–22. Juvenile rheumatoid arthritis in a 9-year-old boy. *A* and *B*, Sagittal T1-weighted images (TR, 800 ms; TE, 25 ms) show erosion of the articular cartilage and subchondral sclerosis, which contrasts well with the marrow fat of high signal intensity *(arrows)*. Irregularity of the superior margin of Hoffa's fat pad (H) is identified, indicating reactive synovitis. Hypoplastic meniscus is shown *(long arrow)*.

joint space does not exclude cartilage loss, which may be depicted directly on MR imaging (Figs. 12–23, 12–24). The ability to depict cartilage changes gives MR imaging an advantage over radiographs in preoperative planning for joint arthroplasty procedures. Osteophytes can be overlapped by adjacent anterior and posterior bony structures with radiography. However, the tomographic, thin-section, and multiplanar imaging capabilities of MR imaging significantly improve the sensitivity for detecting osteophytes (Fig. 12–25). Experimentally, osteophytes can be detected by MR imaging as early as 4 weeks after induction of osteoarthritis in dogs, which is 8 weeks earlier than osteophytes can be seen radiographically.[43]

Radiography and CT depict subchondral sclerosis as areas of increased density and attenuation (Fig. 12–26). On MR images, these areas of increased bone density are of low signal intensity on all pulse sequences owing to a lack of resonating protons. Subchondral cysts are frequently underestimated on radiography. On MR, subchondral cysts are of high signal intensity on T2-weighted and gradient-echo sequences and are easily differentiated from normal marrow and subchondral bone.

Chan and associates found a high association of meniscal and anterior cruciate ligament tears with osteoarthritis on MR imaging.[10] The posterior medial meniscus is the most common site of meniscal degeneration or tears in patients with osteoarthrits. Tears of the anterior cruciate ligament may result in increasing mechanical stress in the medial compartment of the unstable knee. The ability of MR imaging to assess tears of this ligament may alert the clinician to consider ligament reconstruction and possibly

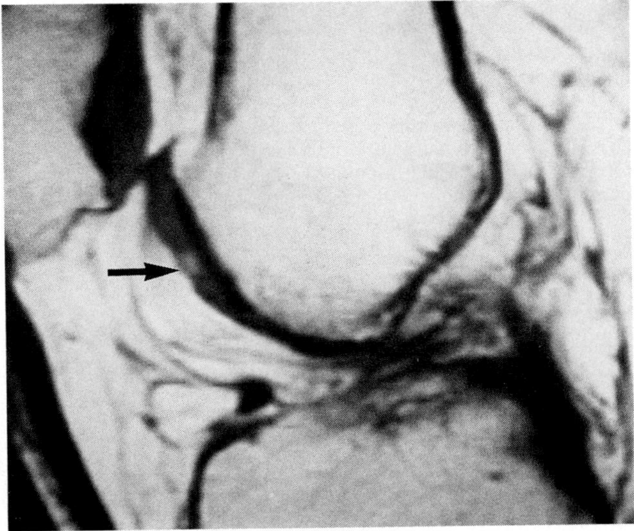

Figure 12–23. Sagittal T1-weighted image shows a focal ulceration of the trochlear hyaline cartilage *(arrow)* in a 73-year-old man who had a history of osteoarthritis of the knee. Arthroscopy confirmed the presence of focal fissuring and ulceration of the cartilage.

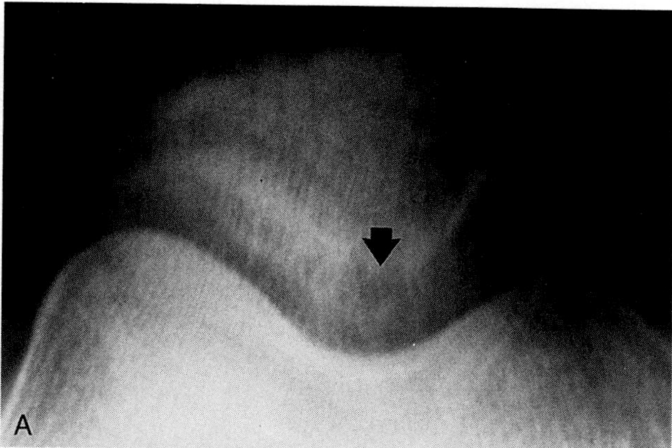

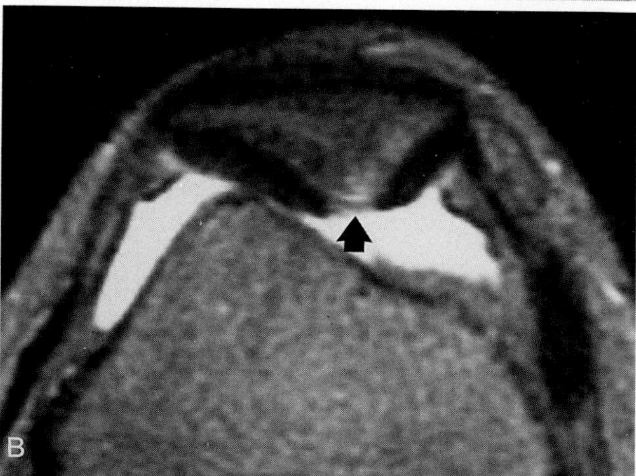

Figure 12–24. *A*, Radiograph of the knee (sunrise patellar projection) does not show narrowing of the joint space in the patellofemoral compartment. A radiolucent area at the apex of the patella is noted *(arrow)*. *B*, Axial T2-weighted image (TR, 2000 ms; TE, 60 ms) shows cartilage loss and cortical erosion at the apex of the patella *(arrow)* that cannot be detected on radiograph. Intrusion of the joint fluid into the patellar apex is noted. (From Chan WP, Lang P, Stevens M, et al: Osteoarthritis of the knee: Comparison of radiography, CT, and MR imaging to assess extent and severity. AJR 157:799, 1991.)

protect the knee from developing meniscal tears and osteoarthritis.

Pigmented Villonodular Synovitis

Pigmented villonodular synovitis (PVNS) is a slowly progressive disorder resulting in villous or nodular proliferation, or both, in the synovial lining of large joints, tendon sheaths, and bursae. It afflicts 1.8 persons per million in the United States, occurring most frequently in adults 20 to 50 years old. Sixty to 80 per cent of patients have monoarticular involvement of the knee.[38] The stimulus for the histologic proliferation in PVNS is unknown. This disorder may be either localized or diffuse, with the diffuse form accounting for 75 per cent of cases.[17,50] The affected synovium is highly vascular and has a tendency to bleed, resulting in hemosiderin deposition. Pigmented villonodular synovitis often invades the articular cartilage and underlying subchondral bone, resulting in progressive cartilage erosion and subchondral cyst formation.

Distention of the joint capsule, irregularity of the synovium, and diffuse hemosiderin deposits are the usual findings of PVNS on MR images (Fig. 12–27).[23,26,50,52] The "synovial mass" of PVNS often exhibits varying signal intensities because histologically it consists of a combination of lipid, hemosiderin, and fluid. An irregular infrapatellar fat pad sign is seen, reflecting synovitis. Areas of low signal intensity predominate on both T1- and T2-weighted images, which represent hemosiderin deposits. The paramagnetic effects of hemosiderin have been ex-

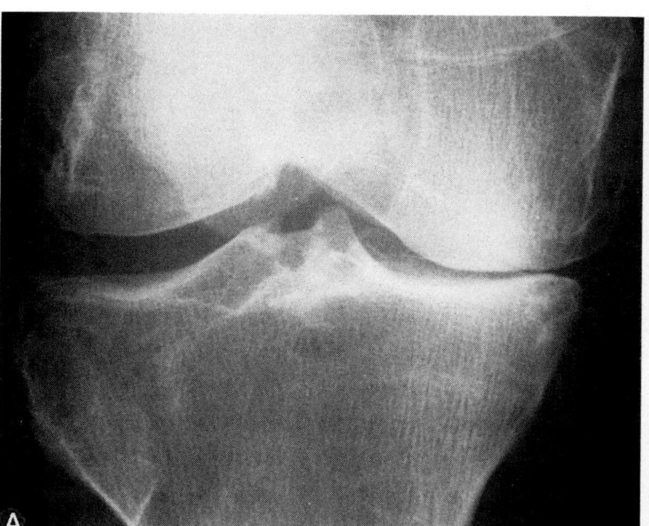

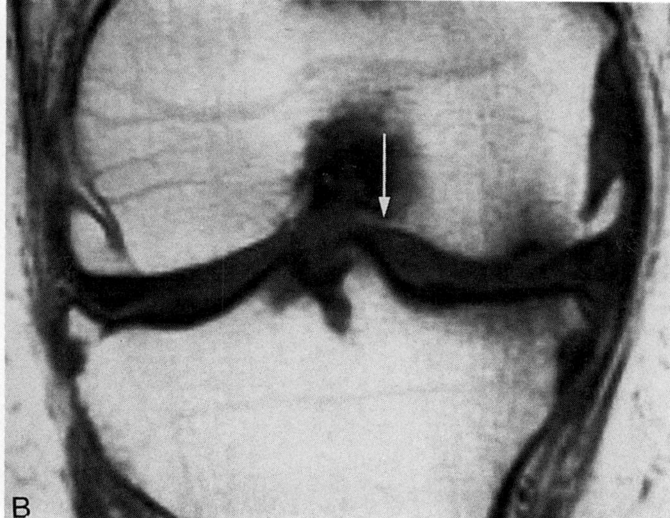

Figure 12–25. *A*, Radiograph of the knee (posteroanterior standing projection) shows formation of osteophytes in the medial and lateral compartments. *B*, Coronal T1-weighted image (TR, 600 ms; TE, 20 ms) confirms the presence of advanced osteophytosis in the medial and lateral femoral condyles and in the tibial plateau. Note the beaklike osteophyte at the medial femoral condyle in the intercondylar notch *(arrow)*, which cannot be seen on radiograph. (From Chan WP, Lang P, Stevens M, et al: Osteoarthritis of the knee: Comparison of radiography, CT, and MR imaging to assess extent and severity. AJR 157:799,1991.)

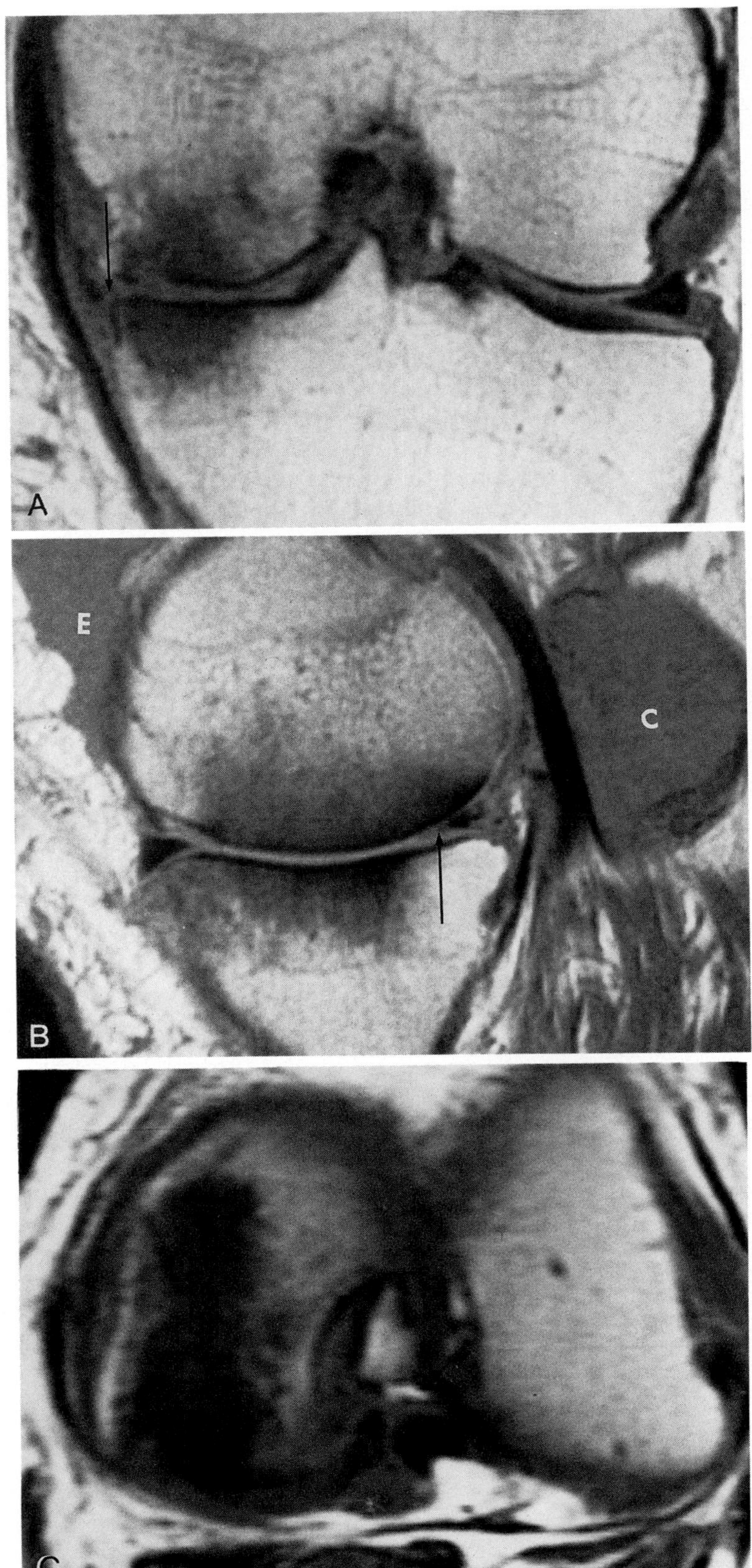

Figure 12-26. *A*, Coronal (TR, 1000 ms; TE, 20 ms); *B*, sagittal (TR, 2000 ms; TE, 20 ms); and C, axial (TR, 800 ms, TE, 20 ms) images showing osteoarthritis of the knee with extensive subchondral sclerosis. The weight-bearing medial compartment of the knee shows diffuse thinning of the articular cartilage, a meniscal tear in the posterior horn *(arrow)*, and extensive subchondral areas of low signal intensity consistent with sclerosis. Associated effusion (E) is shown. Note a large Baker's cyst (C) in the popliteal fossa.

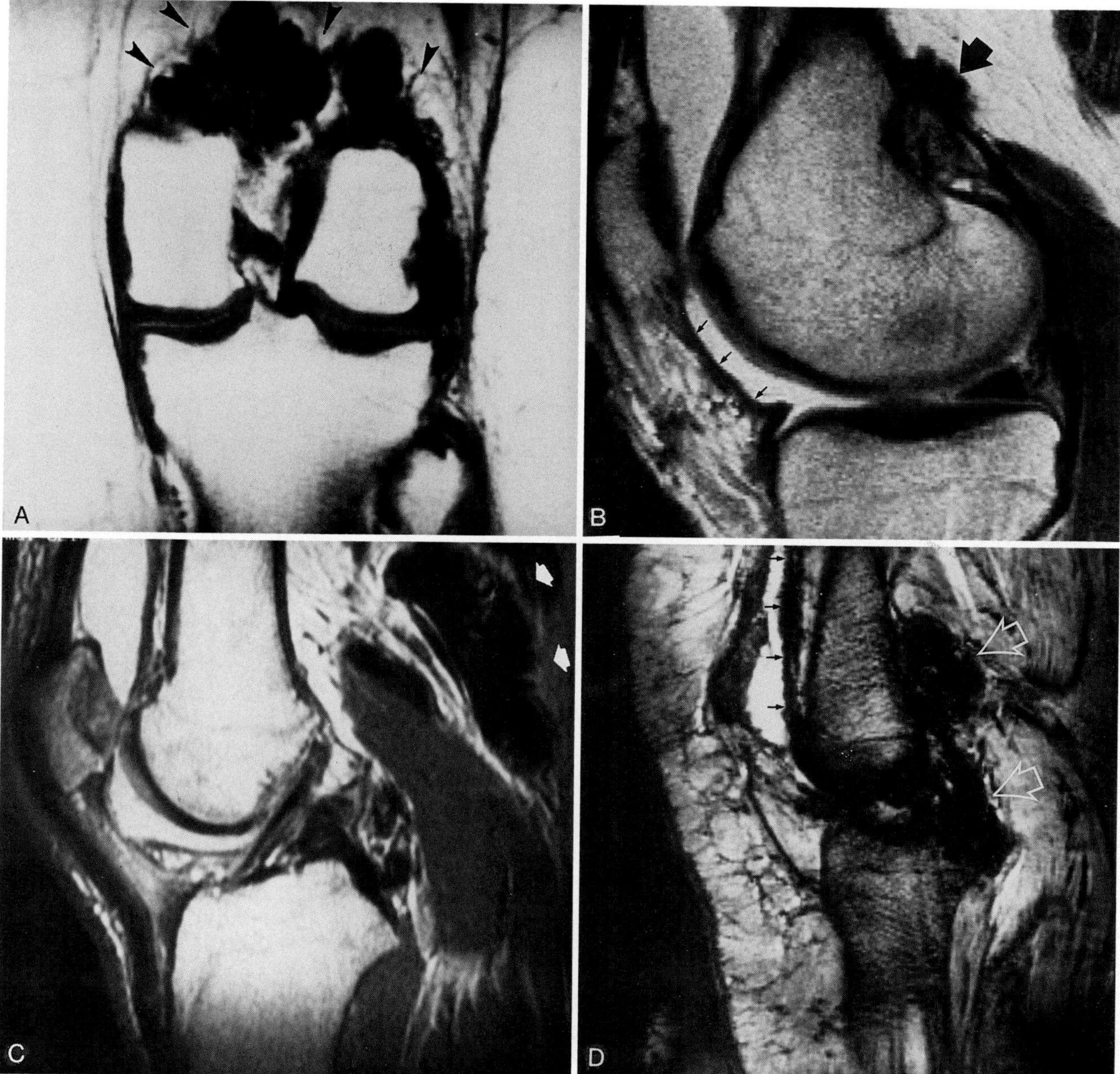

Figure 12–27. Pigmented villonodular synovitis in four different patients. *A*, Posterior coronal T1-weighted image shows a large lobulated mass of low signal intensity in the popliteal fossa *(arrowheads)*. *B*, Sagittal T2-weighted image shows thickened synovium of low signal intensity *(small arrows)*, in contrast to effusion of high signal intensity. A small popliteal mass of low signal intensity is noted *(thick arrow)*. *C*, Sagittal T2-weighted image shows a large hemosiderin-laden popliteal mass *(arrows)*. *D*, Sagittal gradient-echo image shows a hemosiderin mass of low signal intensity in the popliteal fossa *(open arrows)*. Thickened, hemosiderin-laden synovium is indicated *(small arrows)*.

plained as a preferential T2-weighted proton relaxation enhancement attributable to a portion of the iron being in the ferric (Fe^{+++}) state and interacting with adjacent water molecules to shorten the T2 relaxation time of water.[52] This effect is further enhanced with the use of the gradient-echo technique because of increased magnetic susceptibility. The magnetic susceptibility decreases signal intensity in proportion to the square of the magnetic field strength.

Other entities associated with hemorrhage and hemosiderin deposition such as recurrent trauma, hemophilia, synovial hemangioma, and rheumatoid arthritis, should be included in the differential diagnosis of PVNS. However, the clinical history often may exclude other causes of hemosiderotic synovitis. Generally, PVNS is characterized by the absence of osteophyte formation, calcification, and osteoporosis. When a soft tissue mass demonstrates a calcification, synovial chondromatosis (Fig. 12–28)

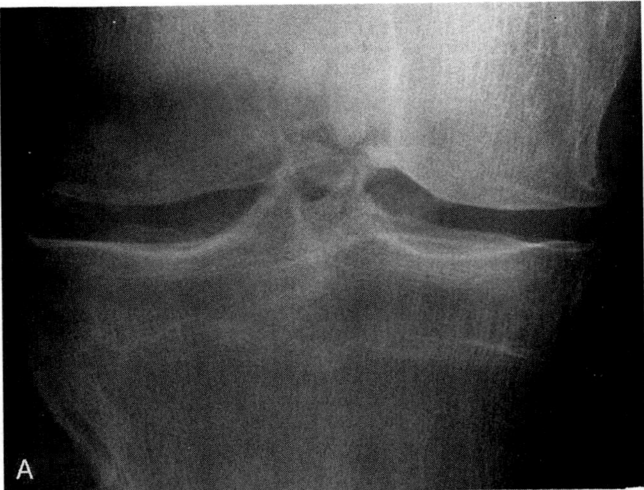
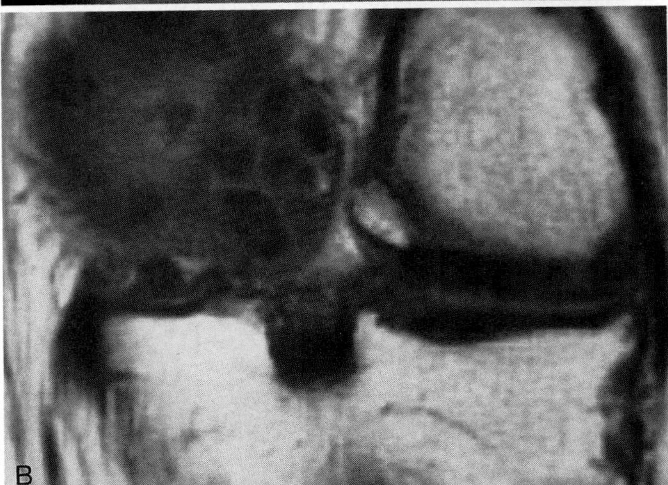
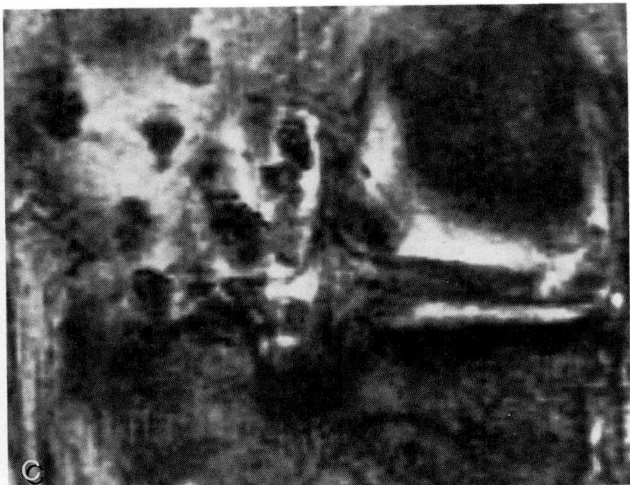

Figure 12-28. Synovial osteochondromatosis in a 47-year-old man. *A*, Initial radiograph shows multiple calcified loose bodies superimposed on the lateral femoral condyle. *B*, Coronal T1-weighted image (TR, 800 ms; TE, 20 ms) and *C*, gradient-echo image (TR, 500 ms; TE, 28 ms; flip angle, 4 degrees) show multiple loose bodies devoid of signal surrounded by synovial fluid. Fluid is of high signal intensity on the gradient-echo image.

and synovial sarcoma should be considered as additional differential diagnoses.

Magnetic resonance imaging signal characteristics can be explained by the tissue components of PVNS. Magnetic resonance imaging can demonstrate uniquely the extent and distribution of the disease, which is valuable for preoperative planning and postoperative follow-up for residual or recurrent disease.

Hemophilic Arthropathy

Hemophilia is an X-linked recessive disorder that is manifested by an abnormality of the coagulation mechanism in males resulting from deficiency of factor VIII (hemophilia A) or IX (hemophilia B, Christmas disease).[2] Repeated hemarthroses with secondary hemosiderin deposition result in hypertrophied, hypervascular, and inflamed synovium, which subsequently invades the articular cartilage. The proteolytic enzymes in the abnormal synovium destroy chondrocytes and cartilage matrix directly. Therefore, synovectomy performed before the stage of cartilage destruction may be effective in slowing the progress of hemophilic arthropathy.

The hypertrophic synovium is marked by areas of low to intermediate signal intensity on T1- and T2-weighted images that represent hemosiderin deposits (Fig. 12-29).[61] Cartilage destruction appears as areas of focal erosions, diffuse thinning, or complete cartilage loss that can be better appreciated on gradient-echo images. In contrast to osteoarthritis, subchondral cysts in the hemophilic joint may represent "subchondral hematoma" and may not be associated with overlying cartilage loss. A disorganized joint with enlarged epiphyses results from intraosseous hemorrhage, indicating a moderate to advanced stage of the disease.

Magnetic resonance imaging should not replace clinical, laboratory, and radiographic examinations for the routine approach and follow-up of patients with joint disorders. However, MR imaging is a unique noninvasive method of evaluating arthritis, which may prove useful in selecting patients for synovectomy, ligament reconstruction, or joint arthroplasty and for monitoring response to therapy.

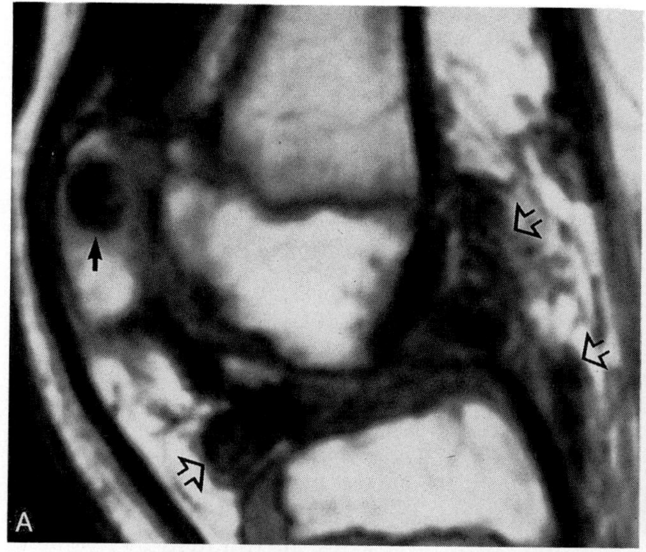

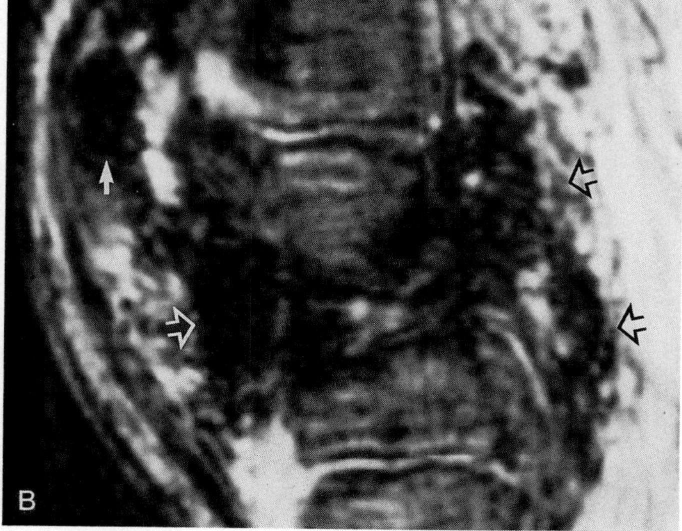

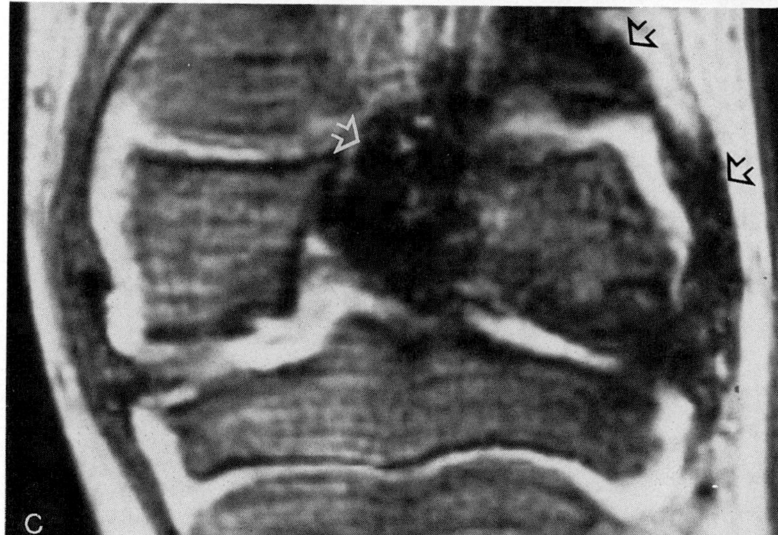

Figure 12–29. *A*, Sagittal (TR, 800 ms; TE, 15 ms); *B*, sagittal (TR, 600 ms; TE, 15 ms; flip angle, 30 degrees); and *C* coronal (TR, 600 ms; TE, 15 ms; flip angle, 30 degrees) images showing hemophilic arthropathy in a 9-year-old boy. Multiple areas of signal void are observed within the knee joint, consistent with hypertrophic hemosiderin-laden synovium *(open arrows)*. Note that the hemosiderin deposition is more pronounced on gradient-echo images in *B* and *C* owing to its magnetic susceptibility effect. Also, markedly destructive articular cartilage and menisci are more apparent on gradient-echo images. On the T1-weighted image *(A)*, the subchondral sclerosis and cysts imaged with low signal intensity contrast well with the marrow fat of high signal intensity. A large subchondral cyst in the upper pole of the patella remains of low signal intensity on gradient-echo images *(B* and *C)*, suggesting intraosseous hemosiderin deposition *(arrow)*. Mild enlargement of the epiphyses and lack of joint effusion are identified, indicating a late stage of hemophilia.

BONE INJURIES

Bone Contusion

A bone contusion refers to radiographically occult bone injuries that are commonly seen on MR imaging of the knee. These injuries are usually secondary to valgus stress or result from direct trauma to the knee. Because they are nonarticular in nature, bone contusions cannot be detected at arthroscopy.

A bone contusion is seen on T1-weighted images as a geographic, nonlinear area of signal loss involving the subchondral bone, which increasses in signal intensity on T2-weighted images (Fig. 12–30).[33] The majority of cases are confined to the epiphysis with varying degrees of extension to the metaphysis. Although the MR imaging features of bone contusions have been documented, histologic verification is lacking. The marrow signal changes are thought to represent trabecular microfracture, blood, edema, or hyperemia.[33,59] A bone contusion commonly accompanies an injury to the anterior cruciate ligament occurring in 16 of 27 patients reported by Mink and Deutsch,[33] nine of 22 patients reported by Yao and Lee,[59] and 41 of 87 patients reported by Lynch and colleagues.[30] The medullary signal changes of bone contusions on MR images usually resolve in 6 to 8 weeks without sequelae; the pain associated with these injuries resolves gradually with time.

T-1 weighted spin-echo and short tau inversion-recovery (STIR) sequences are the most sensitive methods for identifying these injuries. Gradient-echo images do not reliably show abnormal signal intensity in most bone contusions owing to the effects of magnetic susceptibility on the appearance of bone marrow. An amorphous stress fracture may occasionally mimic a bone contusion on MR imaging. A history of a single traumatic incident with involvement of a large segment of the epiphysis and

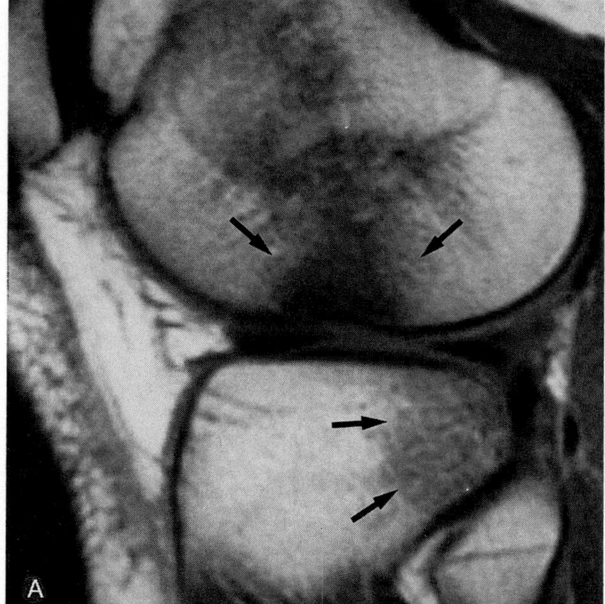

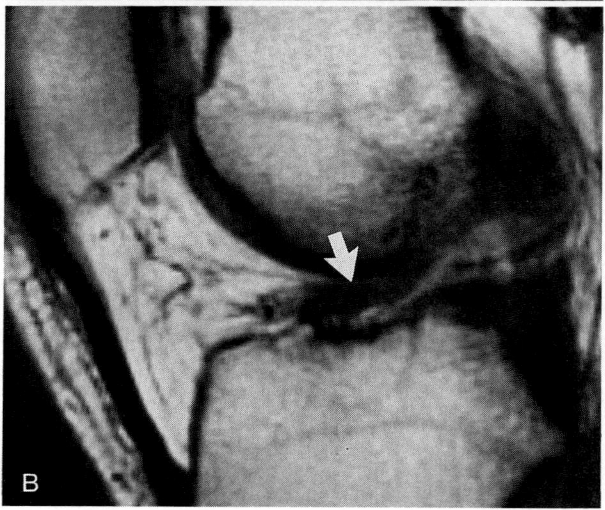

Figure 12–30. Anterior cruciate ligament tear and bone contusion in a 20-year-old woman who had a history of a valgus stress to her left knee 2 months earlier. Sagittal T1-weighted image *(A)* shows contusions in the lateral femoral condyle and posterior tibial plateau, resulting from injury with violent valgus forces or axial loading *(arrows)*. *B*, The anterior cruciate ligament shows reduced angulation *(white closed arrow)*, corresponding to a complete tear with fibrotic scarring found at arthroscopy. Kissing contusions is a unique pattern associated with a complete tear of the anterior cruciate ligament.

metaphysis is typical in bone contusions.[33] Conversely, patients with stress fractures usually have a history of unusual physical activity, and the lesion usually involves the metaphysis alone.

Osteochondral Fracture

Osteochondral fractures are injuries to hyaline articular cartilage and subchondral bone that usually result from compressive forces on the lateral joint compartment. In contrast to bone contusions, arthroscopy may detect these lesions.

Mink and Deutsch[33] categorized osteochondral fractures into two subgroups on the basis of their MR imaging appearance. The *displaced type* appears frequently in the inferior pole of the patella and the anterolateral femoral condyle (Fig. 12–31). The fracture is best depicted on T2-weighted images, especially in the presence of an effusion. The displaced fragments of the articular cartilage and bone could be identified in the intercondylar notch or in the medial and lateral reflections of the suprapatellar bursa. The *impacted type* of osteochondral fracture is a focal bone depression measuring 1 to 4 mm, which appears frequently in the anterior aspect of the lateral femoral condyle adjacent to the anterior horn of the lateral meniscus (Fig. 12–32).[55] The depressed cartilage and cortical bone are surrounded by a zone of edema. A persistent contour deformity has been reported on follow-up MR images.[55]

Tibial Plateau Fracture

The important feature in tibial plateau fractures is that the integrity of a major bearing joint is disrupted to a variable degree. In a surgical study by Hohl[22] of 915 tibial plateau fractures, 24 per cent were not displaced, 33 per cent had local compression, 16 per cent had split compression, 8 per cent had total compression, 4 per cent were split, and 14 per cent were comminuted (Fig. 12–33). Fractures of the compression type are found in almost half of the patients. The mechanism of injury is a combined valgus strain and axial loading force on the extended knee, causing the lateral femoral condyle to drive into the lateral tibial plateau. This usually is caused by a fall from a height and tends to occur in the elderly patient with osteoporosis. The surgeon should document the fracture completely and clearly before planning treatment. In general, when the depression exceeds 7 mm, surgery with elevation and restoration of the joint surface is indicated.[49] Also, in a split fracture with significant displacement, open reduction and internal fixation are advised.

Tibial condylar fractures appear as single or multiple linear areas of low signal intensity extending to the articular surfaces on both T1- and T2-weighted images (Figs. 12–34 to 12–36). In the acute or subacute phase, a zone of presumed edema, hemorrhage, or hypervascular response is seen surrounding the fracture line. This nonlinear component is of low signal intensity on T1-weighted sequences and of high signal intensity on T2-weighted images. Magnetic resonance imaging provides high contrast between tissues and offers high-resolution multi-

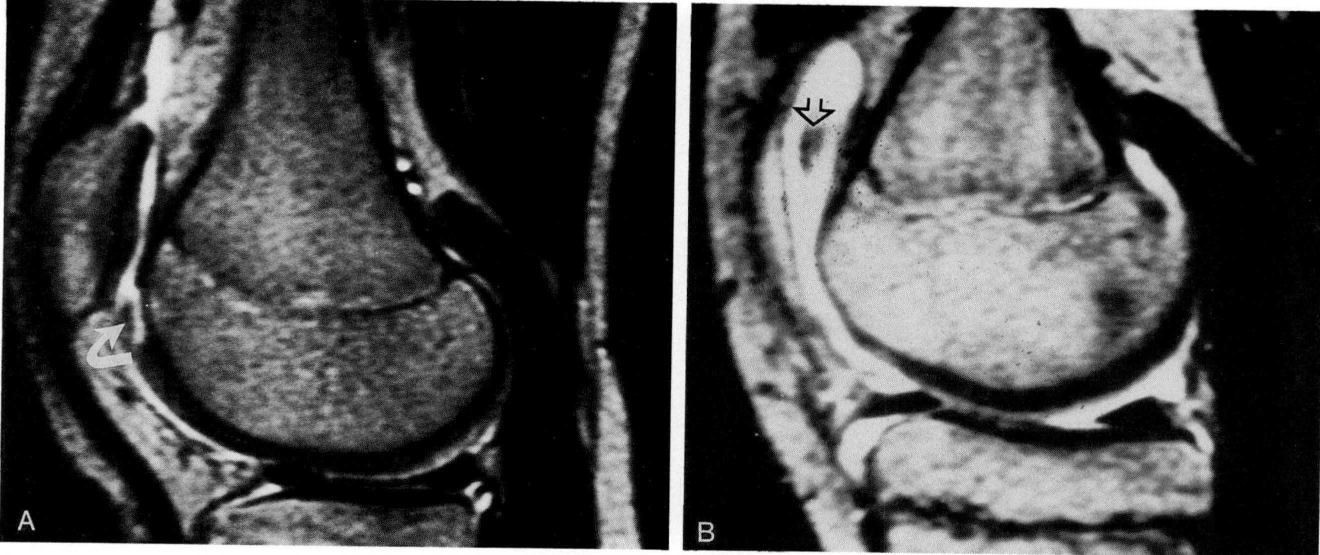

Figure 12–31. Osteochondral fracture, displaced type. *A*, Sagittal T2-weighted image shows a large osteochondral fracture of the femoral trochlea *(curved arrow)*. *B*, An associated effusion is present. The displaced osteochondral fragment of low signal intensity is identified in the suprapatellar bursa, surrounded by the effusion of high signal intensity *(open arrow)*.

planar imaging capability. Tissue contrast is particularly advantageous for detecting associated injuries of menisci, cruciate and collateral ligaments, and articular cartilage. Because of the multiplanar imaging capability, the degree of comminution, depression, displacement, and orientation of fracture fragments are clearly depicted on MR images. Magnetic resonance imaging can detect subtle tibial plateau fractures not evident on conventional radiographs. Sagittal or coronal MR images have better spatial resolution than reformatted CT images. All of these factors are essential in the determination of operative versus nonoperative therapy. However, small fragments are not as apparent on MR images as they are on CT scans.

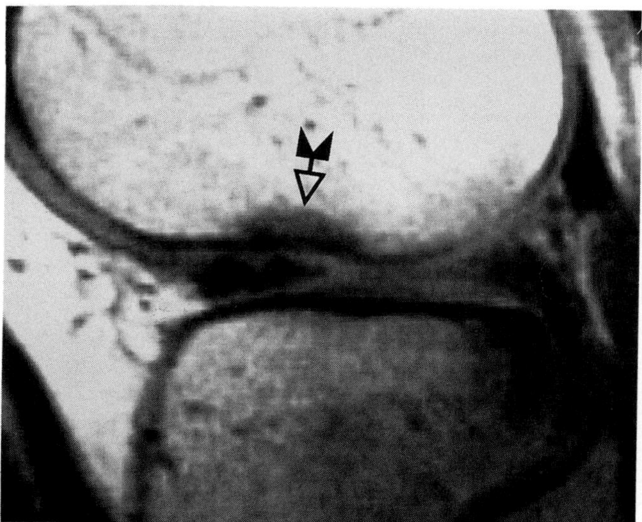

Figure 12–32. Impaction fracture in a 30-year-old man. Sagittal T1-weighted image (TR, 700 ms; TE, 30 ms) shows a focal impaction of the lateral femoral articular surface and an adjacent subcortical lesion of low signal intensity *(flagged arrow)*. Contusion of low signal intensity also is noted in the tibial plateau. Note that a focal impaction (or deepened) lateral femoral notch is suggestive of a complete anterior cruciate ligament tear.

The *Segond fracture* is a small vertical avulsion fracture involving the proximal lateral tibia just distal to the lateral plateau, resulting from excessive internal rotation and varus stress.[45] A focal bone contusion along the lateral tibial rim due to the injury to the lateral capsular junction is suggestive of this type of fracture on MR imaging (Figs. 12–37, 12–38). However, the fracture fragment was apparent on MR imaging in only four in 12 cases as reported by Weber and colleagues.[56] The Segond fracture is associated with anterior cruciate ligament tears in 75 to 100 per cent of cases and with meniscal tears in 66 to 70 per cent. Magnetic resonance imaging is particularly important for identifying these associated abnormalities when the Segond fracture is recognized on plain films of the knee.

Yao and Lee[58] reported two cases with small fractures at the posteromedial tibial plateau, resulting from avulsion of the semimembranosus tendon. This tendon gives rise to five distal fibrous expansions, of which the central one contributes to an avulsion injury. The semimembranosus serves as a flexor and internal rotator of the tibia. External rotation and abduction with knee flexed may violate the medial capsular structures, resulting in this avulsion injury. The avulsion fractures show marginal areas of high signal intensity on T2-weighted images, consistent with hemorrhage, edema, and joint fluid (Fig. 12–39). The semimembranosus tendon may be

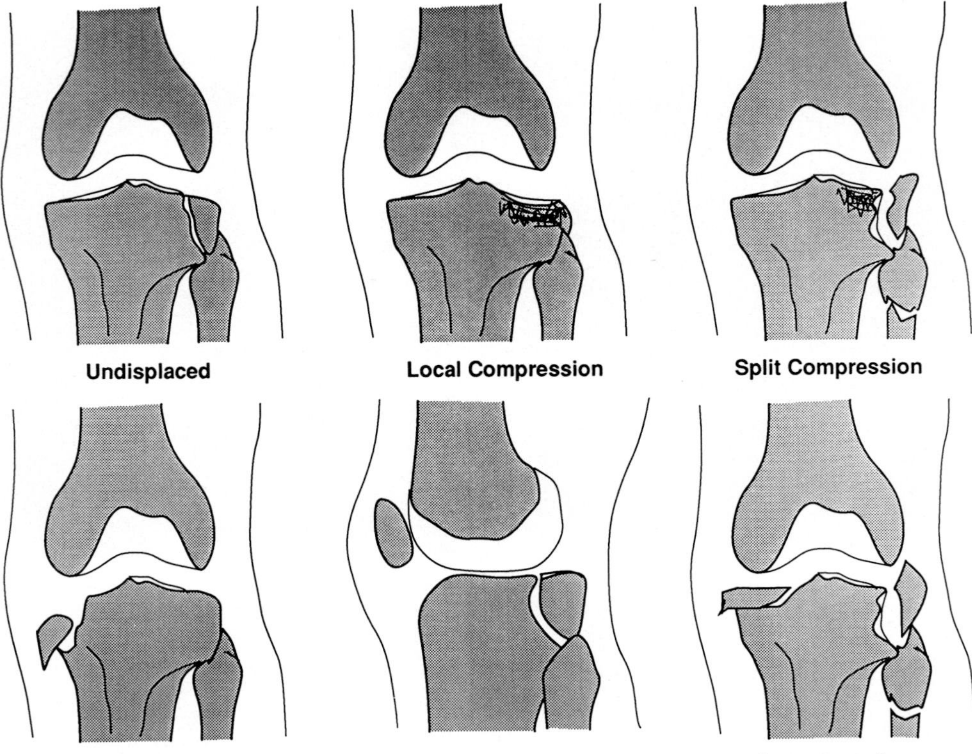

Figure 12–33. Hohl's classification of tibial condylar fractures.

intact. Tears of the anterior cruciate ligament and medial meniscus have been noted.

Stress Fracture

Stress fractures have been categorized into two types: fatigue and insufficiency fractures. Fatigue fractures result from *abnormal* stresses on a *normal* bone; *normal* stresses on an *abnormal* bone result in insufficiency fractures.[14] Patients with a stress fracture may initially have knee pain after exercise, then pain during exercise, and finally pain in the absence of exercise. Stress fractures often are not detected initially because of the lack of a single traumatic incident. A radiographically occult fracture is identi-

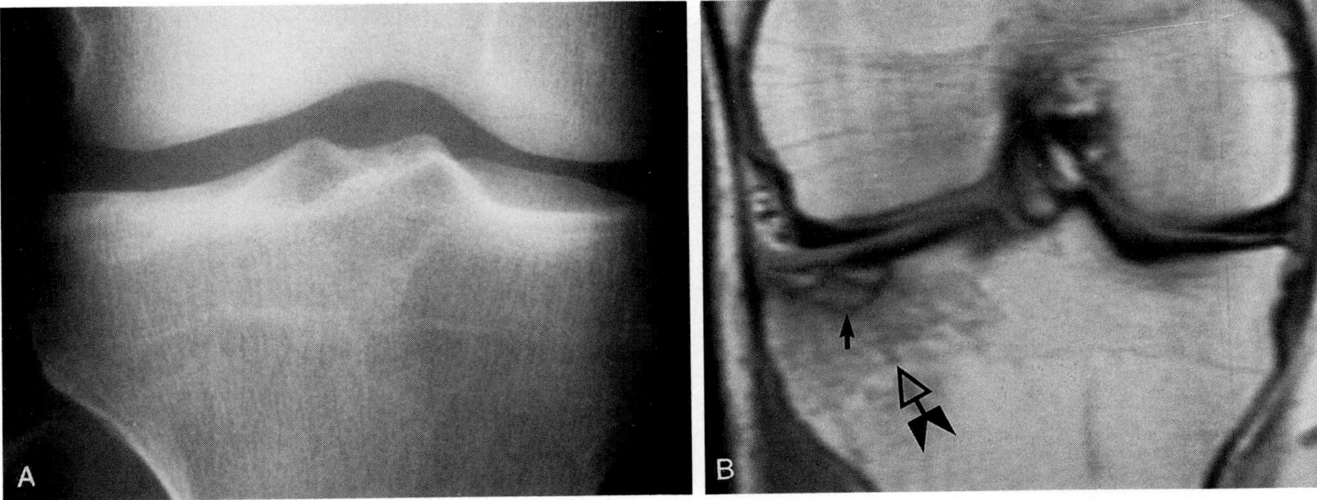

Figure 12–34. Tibial plateau fracture (Hohl's type I) in a 37-year-old man. An occult fracture of the lateral tibial plateau noted on the radiograph *(A)* is well depicted on the coronal T1-weighted image (TR, 1500 ms; TE, 20 ms). *(B)* A wedge-shaped nondisplaced fracture *(arrow)* and an associated poorly defined zone of edema are visualized *(flagged arrow)*. Note that the overlying articular cartilage is intact.

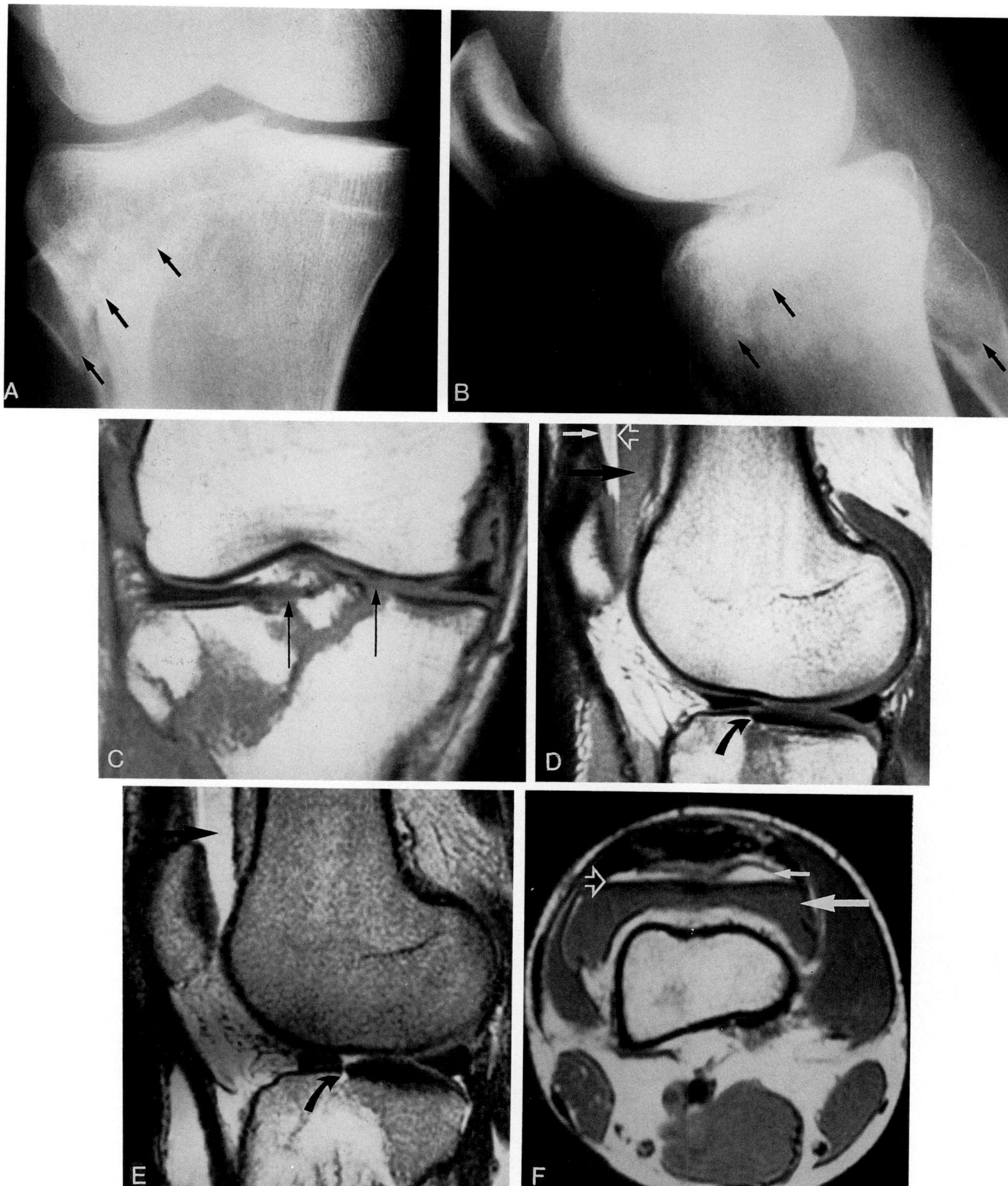

Figure 12–35. An acute tibial plateau fracture (Hohl's type III). *A* and *B*, Initial radiographs show oblique radiolucent fracture lines involving the lateral tibial plateau and proximal fibula *(arrows)*. *C*, Coronal T1-weighted image (TR, 1000; TE, 20 ms) shows a huge wedge fracture area of the lateral tibial plateau, extending to the articular cartilage adjacent to the tibial spine *(long arrows)*. Many displaced loose fragments are noticeable. *D* and *E*, Sagittal MR images show a large zone of presumed medullary edema and hemorrhage, which has a low signal intensity on T1-weighted sequences *(D)* (TR, 1800 ms; TE, 20 ms) and a high signal intensity on T2-weighted images *(E)* (TR, 1800 ms; TE, 80 ms). A step-off subchondral bone is identified, with synovial fluid leaking through the defect *(curved arrow)*. Lipohemarthrosis with fat-serum level (fat, *small arrow*; serum, *thick arrow* in *D*, *E*, and *F*) is identified in the distended suprapatellar bursa. Note the changes on T1- and T2-weighted images in signal intensity due to the different relaxation time and the chemical shift artifact (open arrow in *D*, *E*, and *F*) from fat-serum level. *F*, Axial T1-weighted image.

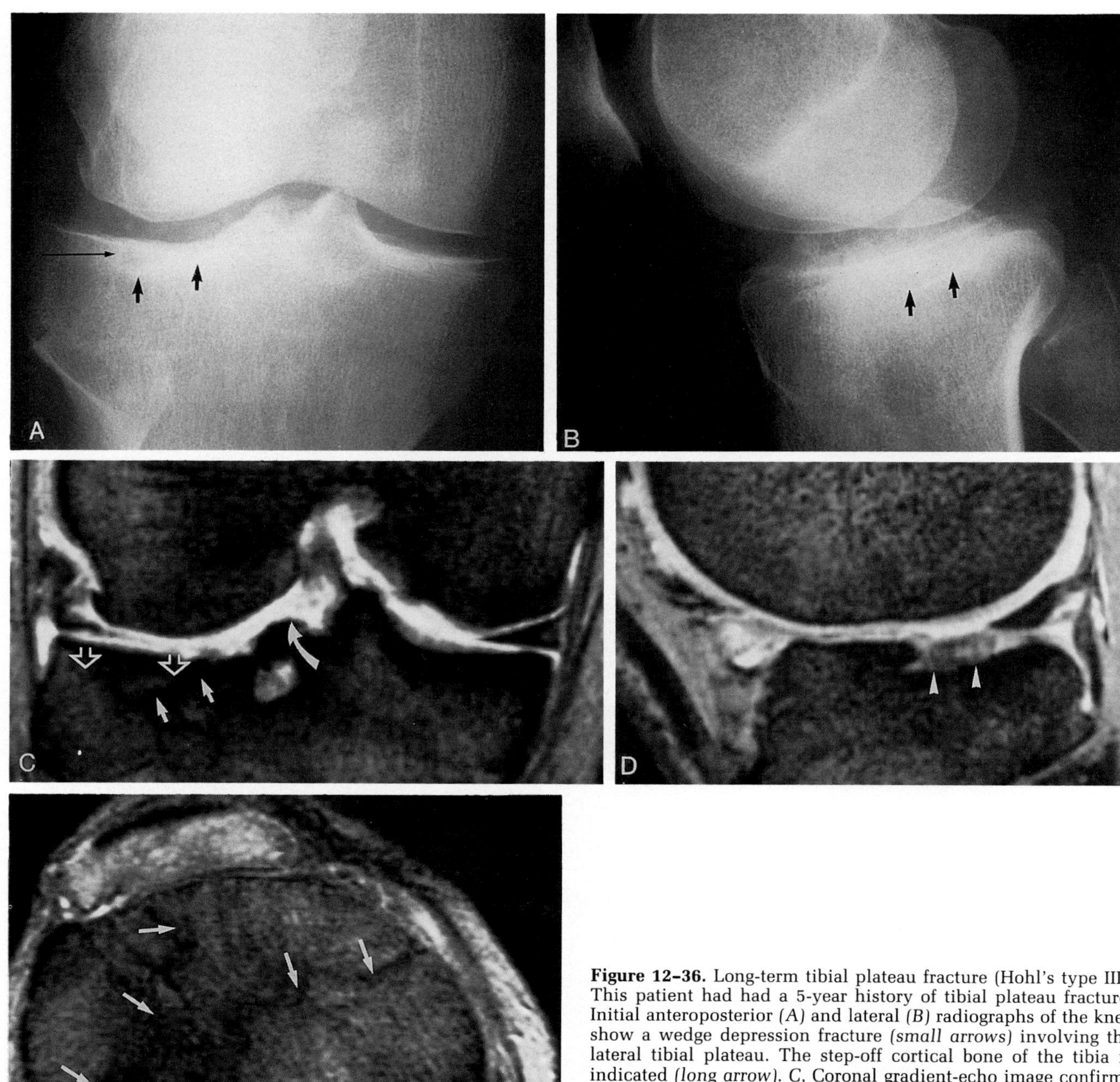

Figure 12–36. Long-term tibial plateau fracture (Hohl's type III). This patient had had a 5-year history of tibial plateau fracture. Initial anteroposterior *(A)* and lateral *(B)* radiographs of the knee show a wedge depression fracture *(small arrows)* involving the lateral tibial plateau. The step-off cortical bone of the tibia is indicated *(long arrow)*. *C,* Coronal gradient-echo image confirms the presence of a wedge depressed fracture of the lateral tibial plateau, involving the tibial spine *(curved arrow)*. A large geode shows high signal intensity on the gradient-echo image. Note that the lateral plateau is depressed down into the metaphysis *(open arrows)*. Comminuted fracture line is indicated *(short arrows)*. *D,* Sagittal gradient-echo image shows denuded articular cartilage being driven into the plateau *(arrowheads)*. *E,* Comminuted fracture lines are better appreciated on the axial image *(arrows)*.

fied as an area of high tracer uptake on bone scan. MR imaging is more specific than scintigraphy in the diagnosis of stress fracture.[51]

The medial and posterior proximal tibia is a common site of stress fracture of the knee.[42] The metaphysis and proximal diaphysis are frequently involved with sparing of the epiphysis.[14] Mink and Deutsch[33] classified stress fractures into two types—linear or amorphous—according to their MR imaging patterns. The linear type appears as a straight or serpiginous fracture line of low signal intensity in the medullary bone perpendicular to the adjacent cortex regardless of pulse sequence (Fig. 12–40). In the acute or subacute phase, the linear component is surrounded by a poorly defined area of low signal intensity on T1-weighted sequences and high signal

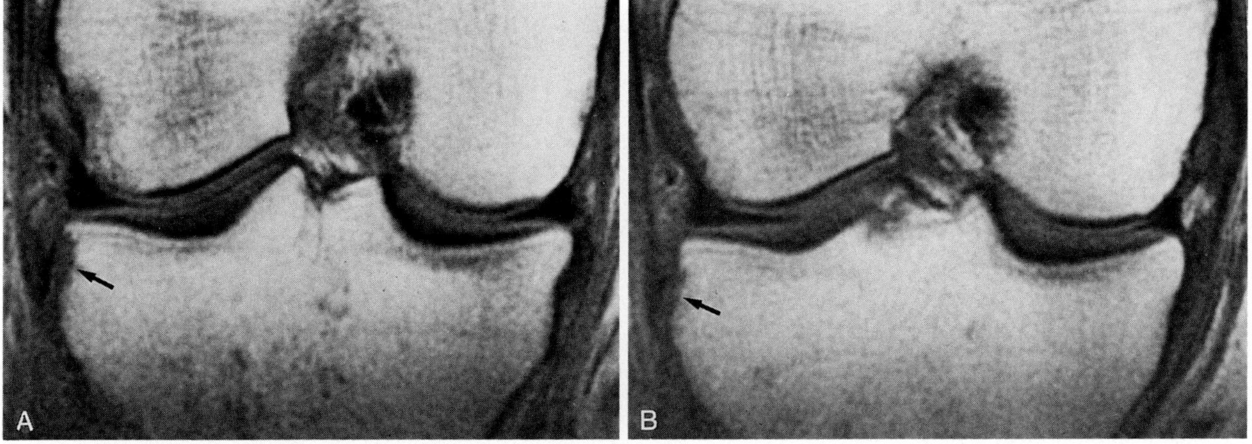

Figure 12–37. Segond fracture. *A* and *B*, Coronal proton density weighted images show focally edematous marrow in the lateral tibial rim, suggesting an injury of the lateral capsular junction *(arrow)*.

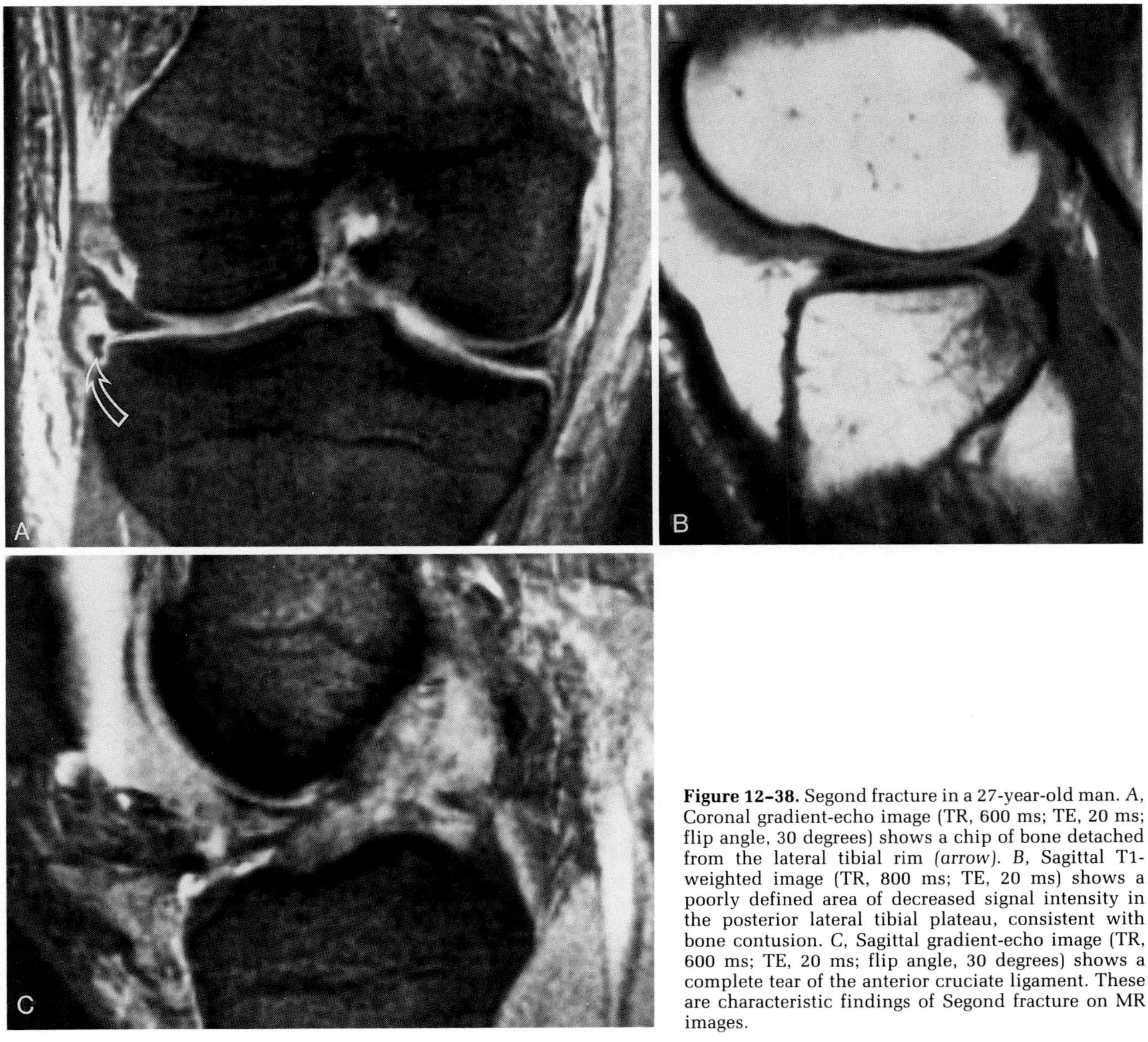

Figure 12–38. Segond fracture in a 27-year-old man. *A*, Coronal gradient-echo image (TR, 600 ms; TE, 20 ms; flip angle, 30 degrees) shows a chip of bone detached from the lateral tibial rim *(arrow)*. *B*, Sagittal T1-weighted image (TR, 800 ms; TE, 20 ms) shows a poorly defined area of decreased signal intensity in the posterior lateral tibial plateau, consistent with bone contusion. *C*, Sagittal gradient-echo image (TR, 600 ms; TE, 20 ms; flip angle, 30 degrees) shows a complete tear of the anterior cruciate ligament. These are characteristic findings of Segond fracture on MR images.

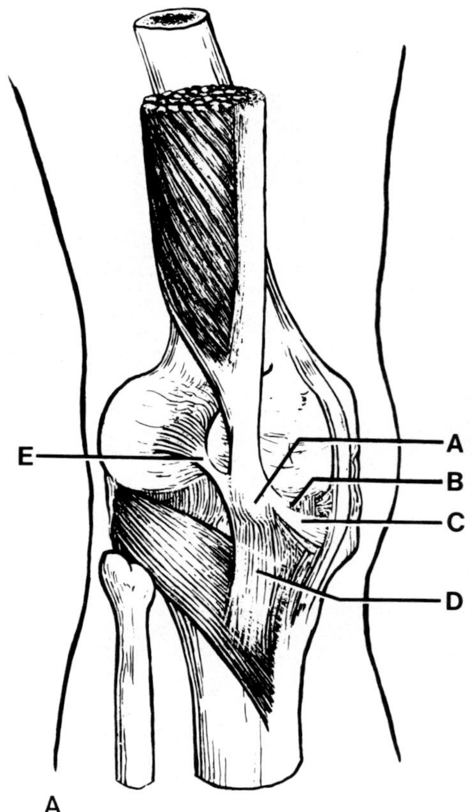

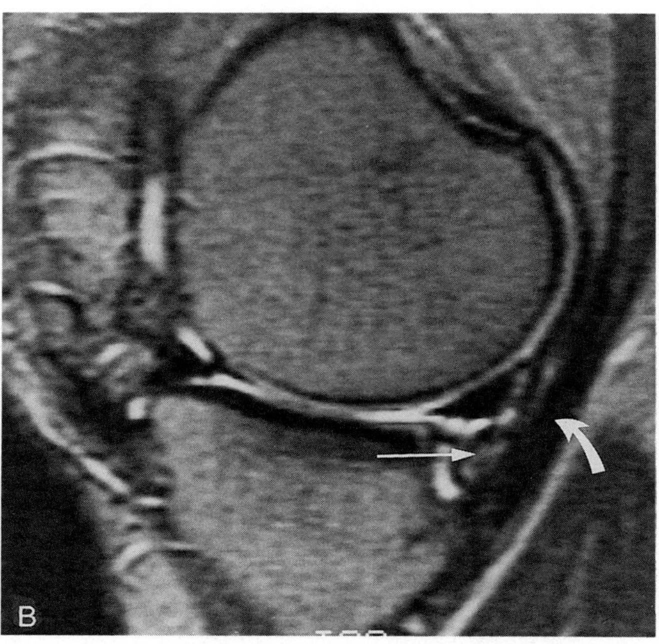

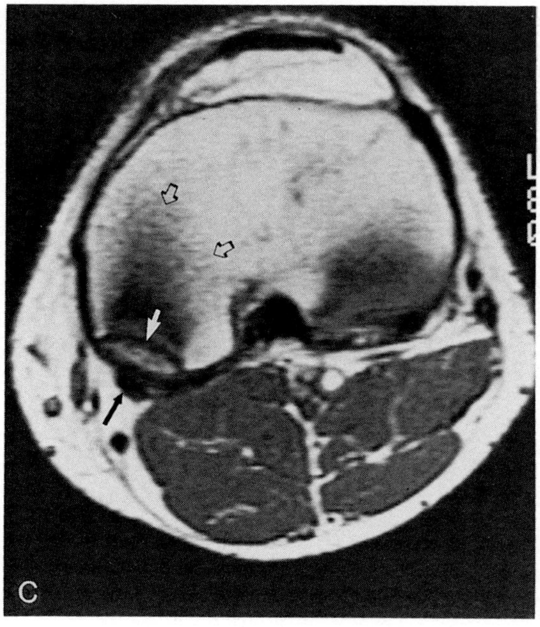

Figure 12–39. *A*, Five distal expansions of the semimembranosus tendon (A, central tendon insertion; B, attachments to the deep capsule and the posterior horn of the medial meniscus [partially obscured]; C, anterior limb inserting deep to the medial collateral ligament; D, distal fibrous expansion over the popliteus muscle; E, lateral contribution to the oblique popliteal ligament). *B* and *C*, Avulsion of the posteromedial tibial plateau by the semimembranosus tendon in a 31-year-old man. This patient injured his left knee by external rotation while playing basketball. *B*, Sagittal T2-weighted image (TR, 2000 ms; TE, 80 ms) through the semimembranosus tendon *(curved arrow)* shows a bony fragment *(long white arrow)* contiguous to the tendinous insertion, surrounded by material of high signal intensity. *C*, T1-weighted axial image (TR, 600 ms; TE, 25 ms) shows a crescentic zone of low signal intensity *(white arrow)* that is consistent with a nondisplaced fracture, contiguous to the central semimembranosus insertion *(black arrow)*. The adjacent areas of low signal intensity are related to partial volume averaging of the menisci and subchondral compact bone *(open arrows)*. (From Yao L, Lee JK: Avulsion of the posteromedial tibial plateau by the semimembranosus tendon: Diagnosis with MR imaging. Radiology 172:513, 1989. © Radiological Society of North America.)

intensity on T2-weighted images, which represents marrow edema.[33,51] Resolution of the lesion on MR imaging occurs as the patient's symptoms improve. The amorphous type appears as a globular region within a geographic area of low signal intensity on T1-weighted sequences with some portions of the lesion increasing in signal on T2-weighted images, corresponding to marrow edema and hyperemia.[33] As mentioned earlier, an amorphous stress fracture is similar to a bone contusion on MR imaging, with lack of a linear component.

Occasionally, in cases of stress fracture, the bone marrow abnormalities seen on MR imaging may extend to the cortex, the periosteum, and the surrounding soft tissues. These findings should not be misinterpreted as infection or malignancy.[51] The

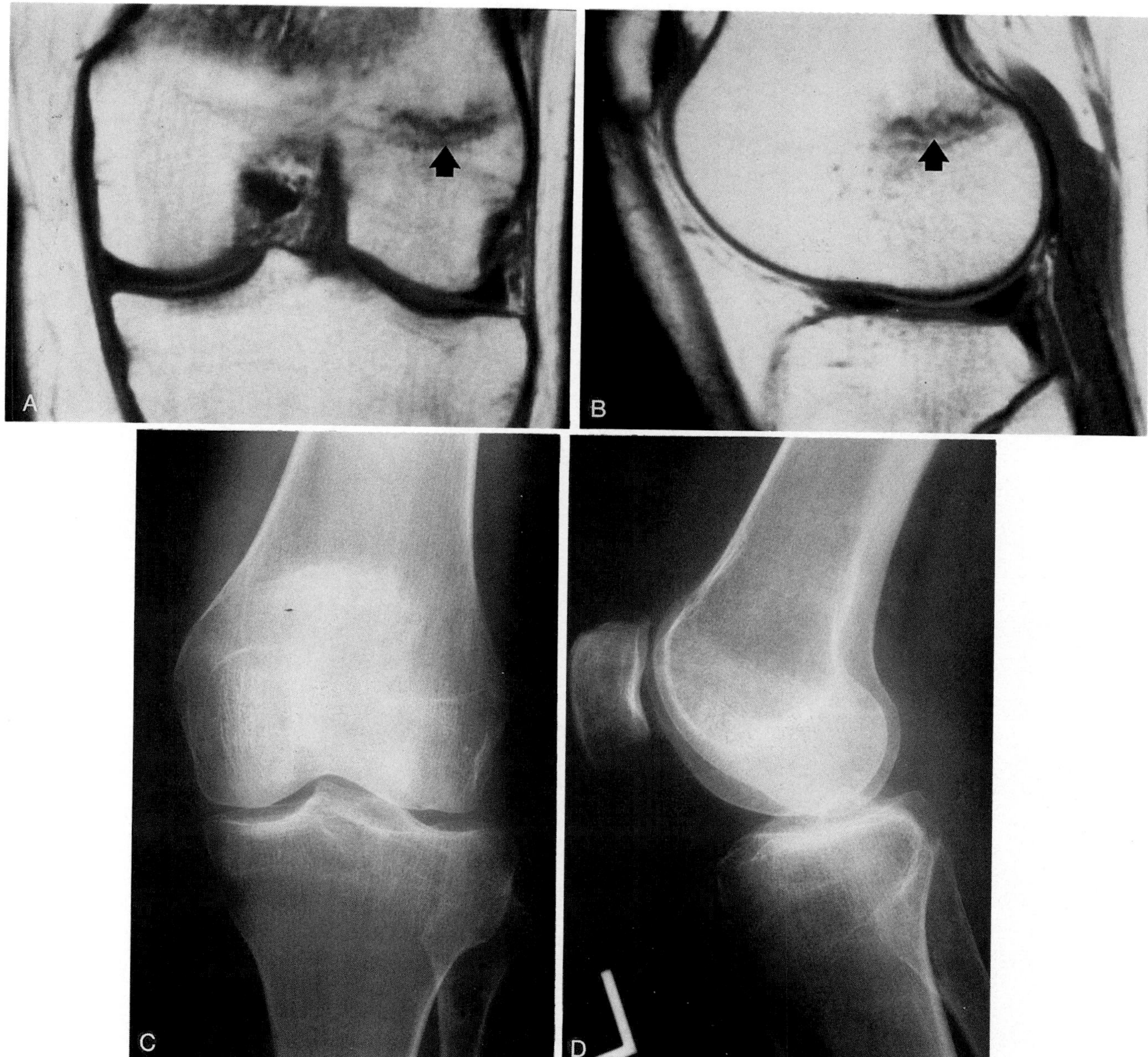

Figure 12–40. Stress fracture in a 36-year-old woman. Coronal *(A)* and sagittal *(B)* T1-weighted images show a linear fracture line of low signal intensity in the metaphysis of the lateral femoral condyle *(arrow)*. No corresponding lesion is seen on anteroposterior *(C)* and lateral *(D)* radiographs of the knee.

lack of a soft tissue mass, the marked marrow edema, and the clinical history suggest the diagnosis of a stress fracture.

OSTEONECROSIS AND RELATED DISORDERS

Medullary Infarcts

Medullary infarcts are seen frequently in patients on steroid therapy. Magnetic resonance imaging allows the depiction of medullary infarcts before they are evident on radiography and bone scans. T1-weighted images show a serpentine border of reactive sclerotic bone of low signal intensity surrounding a core of yellow marrow of high signal intensity (Fig. 12–41).[36] Medullary infarcts may be difficult to differentiate radiographically from an enchondroma. On MR imaging, an enchondroma lacks a serpentine border, and the central core is of decreased signal intensity on T1 weighting with increased signal intensity on T2 weighting.

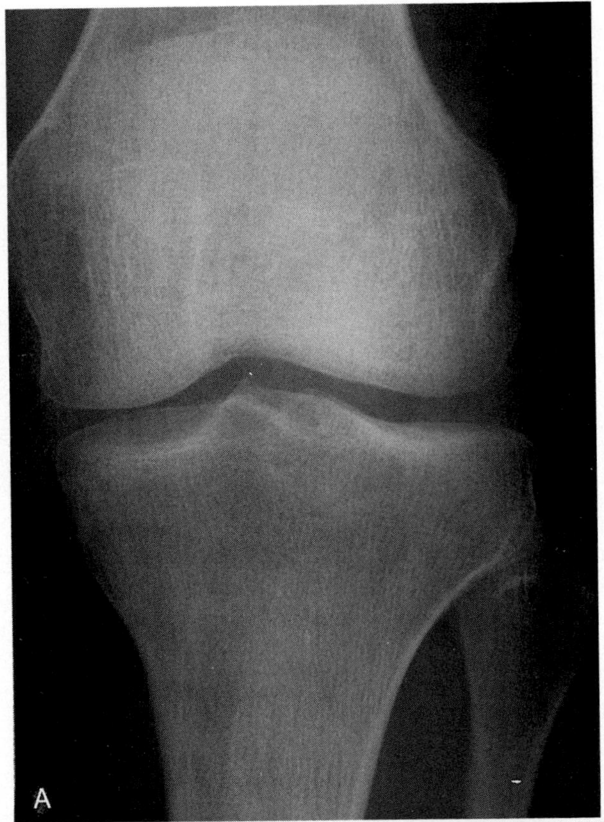

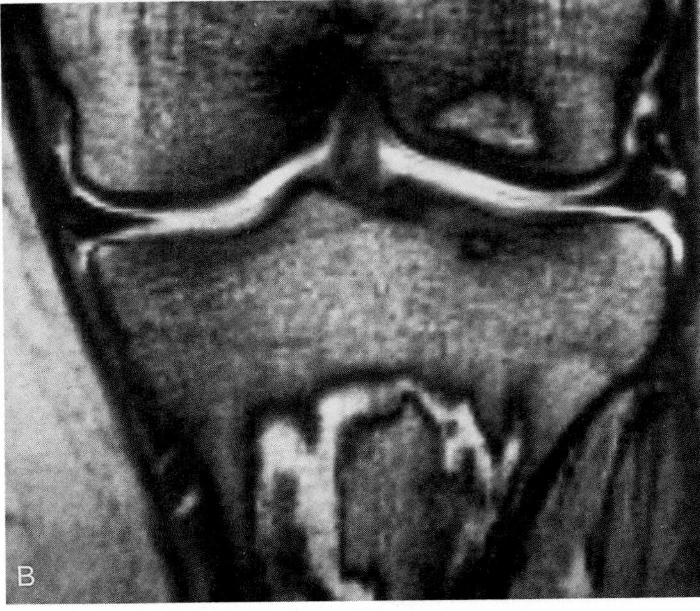

Figure 12–41. Medullary bone infarct in a 19-year-old woman. *A*, Radiograph shows suspicious patchy areas of decreased mineralization in the proximal tibia. *B*, Coronal gradient-echo image shows multifocal, serpentine, inhomogeneous areas of reactive sclerotic bone of low signal intensity surrounding cores of high signal intensity.

Spontaneous Osteonecrosis

Spontaneous osteonecrosis of the knee usually affects the medial femoral condyle. It typically occurs in middle-aged and elderly persons and is characterized by acute localized pain of sudden onset that persists at rest. Spontaneous osteonecrosis may be due to several factors, including trauma with microfractures and subsequent osteonecrosis or primary vascular insufficiency, or it may be a sequela of meniscal tears or severe chondromalacia. Most patients can be treated conservatively with non–weight-bearing therapy. The prognosis varies from complete recovery to progressive joint collapse. The prognosis is relatively poor when the lesion is larger than 5 cm^2 or involves more than 40 per cent of the width of the condyle.[35]

Magnetic resonance imaging allows assessment

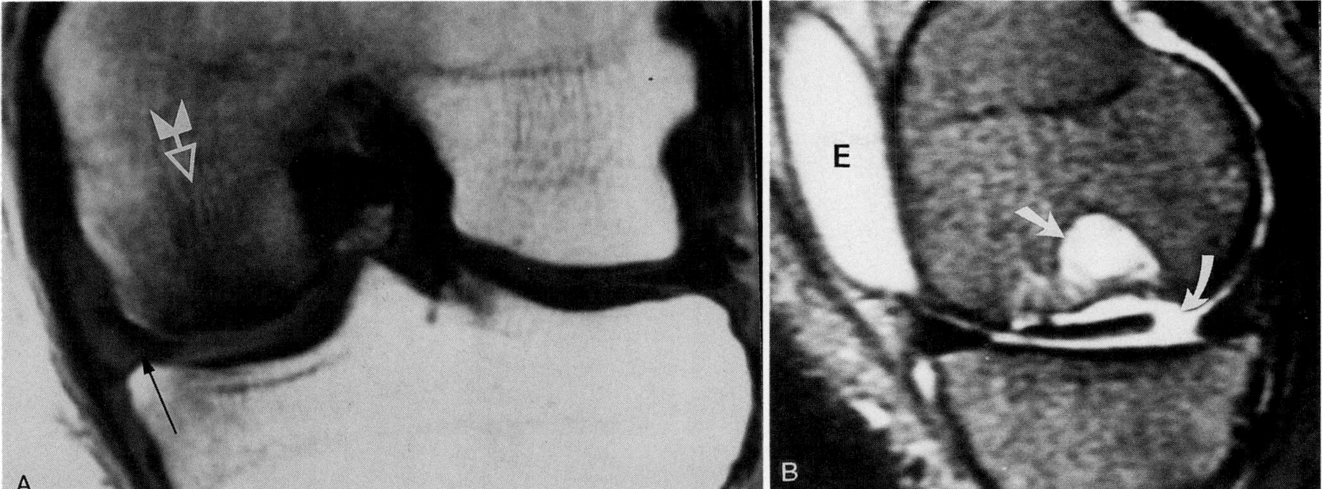

Figure 12–42. Coronal (TR, 600 ms; TE, 20 ms) *(A)* and sagittal (TR, 2200 ms; TE, 80 ms) *(B)* images showing spontaneous osteonecrosis with geode formation of the medial femoral condyle in a 60-year-old man. Extensive subchondral sclerosis of low signal intensity is identified *(flagged arrow)*. A large geode of high signal intensity on the T2-weighted image is seen within the sclerotic area *(straight arrow)*. Synovial fluid is noted communicating with the subchondral bone through a gap in the chondral surface *(curved arrow)*. The torn posterior horn of the medial meniscus is identified, which is a common finding with osteonecrosis *(long arrow)*. E, Associated effusion.

of the extent of involvement. T1-weighted images are best for evaluating the extent of the osteonecrotic lesion, showing discrete, well-marginated, areas of low signal intensity in the subchondral bone (Fig. 12–42).[7] Other disorders simulating osteonecrosis of the knee on MR imaging include bone contusion, osteochondritis dissecans, osteoarthritic changes with subchondral sclerosis, subchondral cyst formation, and transient marrow edema. Its typical location, clinical symptoms, relative lack of loose bodies, and late onset of cartilaginous erosions may be valuable for distinguishing spontaneous osteonecrosis from other disorders.

Osteochondritis Dissecans

Paget in 1870 first described osteochondritis dissecans as "quiet necrosis."[32] These lesions are defects or fragmentation of the subchondral bone and are of uncertain cause. The proposed causes of osteochondritis dissecans include trauma, ischemic necrosis, and ossification anomalies. Osteochondritis dissecans frequently affects the knee during the teenage years. The intercondylar aspect of the medial femoral condyle of the knee is frequently involved. The clinical management of this disease requires understanding of the mechanical stability of the osteochondral fragments. In adults, inappropriate management of the disease may result in early degenerative arthritis. A loose in situ fragment is a ballotable fragment with intact overlying articular cartilage that requires surgical removal or internal fixation.[32] A grossly loose fragment is a displaced fragment with disrupted overlying articular cartilage and is usually resected. Stable osteochondritis dissecans is usually treated conservatively. However, firm or partial attachment of fragments cannot be determined radiographically and clinically. Bone scintigraphy may assist in the distinction between loose and stable lesions. Magnetic resonance imaging permits diagnosis of loosening when both radiography and bone scintigraphy are indeterminate (Fig. 12–43). It also permits the determination of the displacement, size, and location of the fragment; these factors are essential for preoperative planning.

Magnetic resonance imaging is 92 per cent sensitive and 90 per cent specific in distinguishing stable from loose lesions.[32] Both loose and stable lesions are imaged as areas of low to intermediate signal intensity on the T1 weighted images. The presence of interface fluid, which is of intermediate signal intensity on T1 weighting and of high signal intensity on T2 weighting, between the fragment and its parent bone is a reliable sign of loosening (Figs. 12–44, 12–45).[15,32] The signal intensity of the fragment is often inhomogeneous and increases to some extent with T2 weighting, regardless of stability. Occasionally the linear inhomogeneous signal from the fragment may be mistaken for interface fluid. Cystic lesions beneath the fragments are found only in unstable lesions, according to one report.[15] Osteoarthritis can occur as a secondary change to osteochondritis dissecans (Fig. 12–46).

OSTEOMYELITIS

Acute osteomyelitis is commonly caused by hematogenous spread of *Staphylococcus aureus*, and it

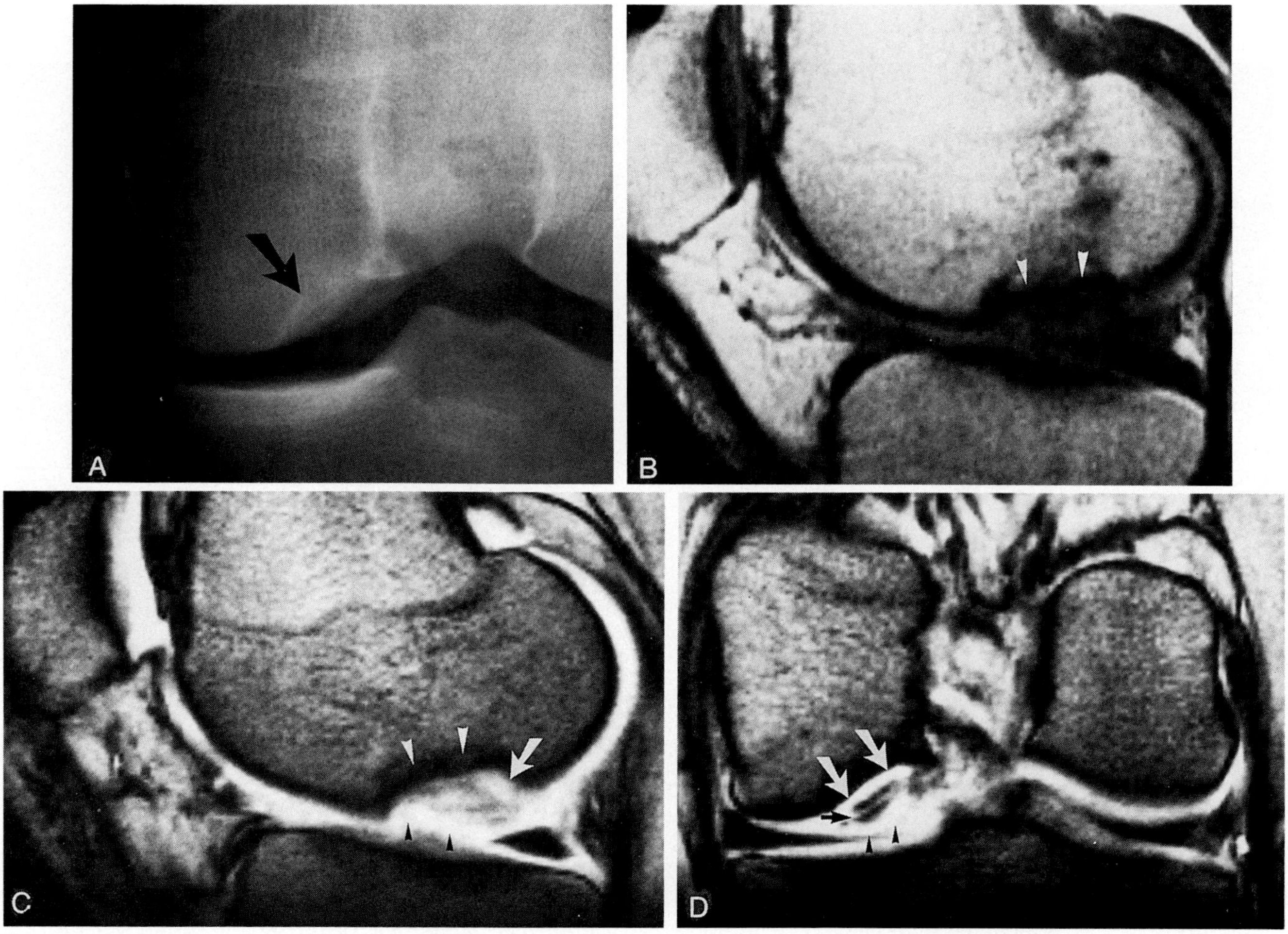

Figure 12–43. Osteochondritis dissecans with partially attached fragment and frayed cartilage in a 36-year-old woman. *A*, Radiograph shows medial condyle defect *(arrow)*. *B*, Sagittal T1-weighted image shows decreased signal intensity of the osteochondral defect *(arrowheads)*, which shows an inhomogeneous increase in signal intensity on gradient-echo images. *C* and *D*, Parent bone-fragment interface shows high signal intensity fluid on gradient-echo images, indicating an unstable fragment *(white arrows)*. A linear region of high signal intensity is found within the fragment, separating the fragment into two pieces *(black arrow* in *D)*. The overlying articular cartilage appears irregular *(black arrowheads)*. The subchondral bone plate appears intact *(white arrowheads* in *B* and *C)*. This patient has had a history of left knee pain for several years. Arthroscopy confirmed the presence of a 2 × 1-cm fragment in the intercondylar aspect of the medial femoral condyle. The fragment had only a small attachment and consisted of two pieces, which were removed subsequently.

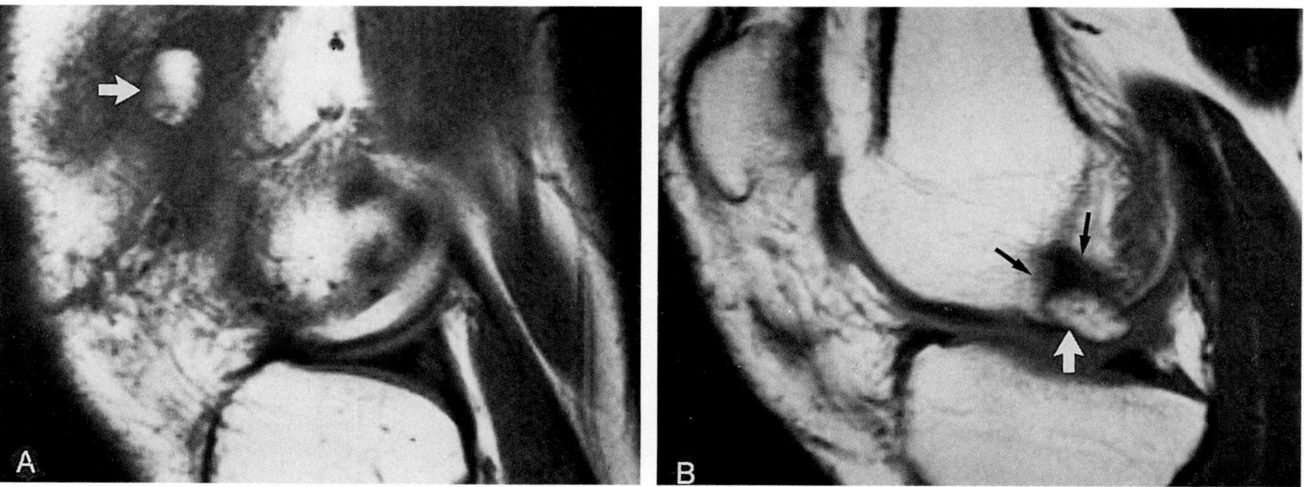

Figure 12–44. Loose body and loose fragment in osteochondritis dissecans in a 32-year-old man. *A*, Sagittal T1-weighted image (TR, 800 ms; TE, 20 ms) shows a loose body in the lateral suprapatellar bursa (arrow). *B*, A 1.5-cm oval fragment is seen in the lateral aspect of the medial femoral condyle *(white arrow)*. The fragment is displaced and appears as a thick area of low signal intensity between the margins of the parent bone, suggesting a loosening of the fragment *(black arrows)*. This patient was treated by arthroscopic removal of the loose body and débridement of the subchondral defect.

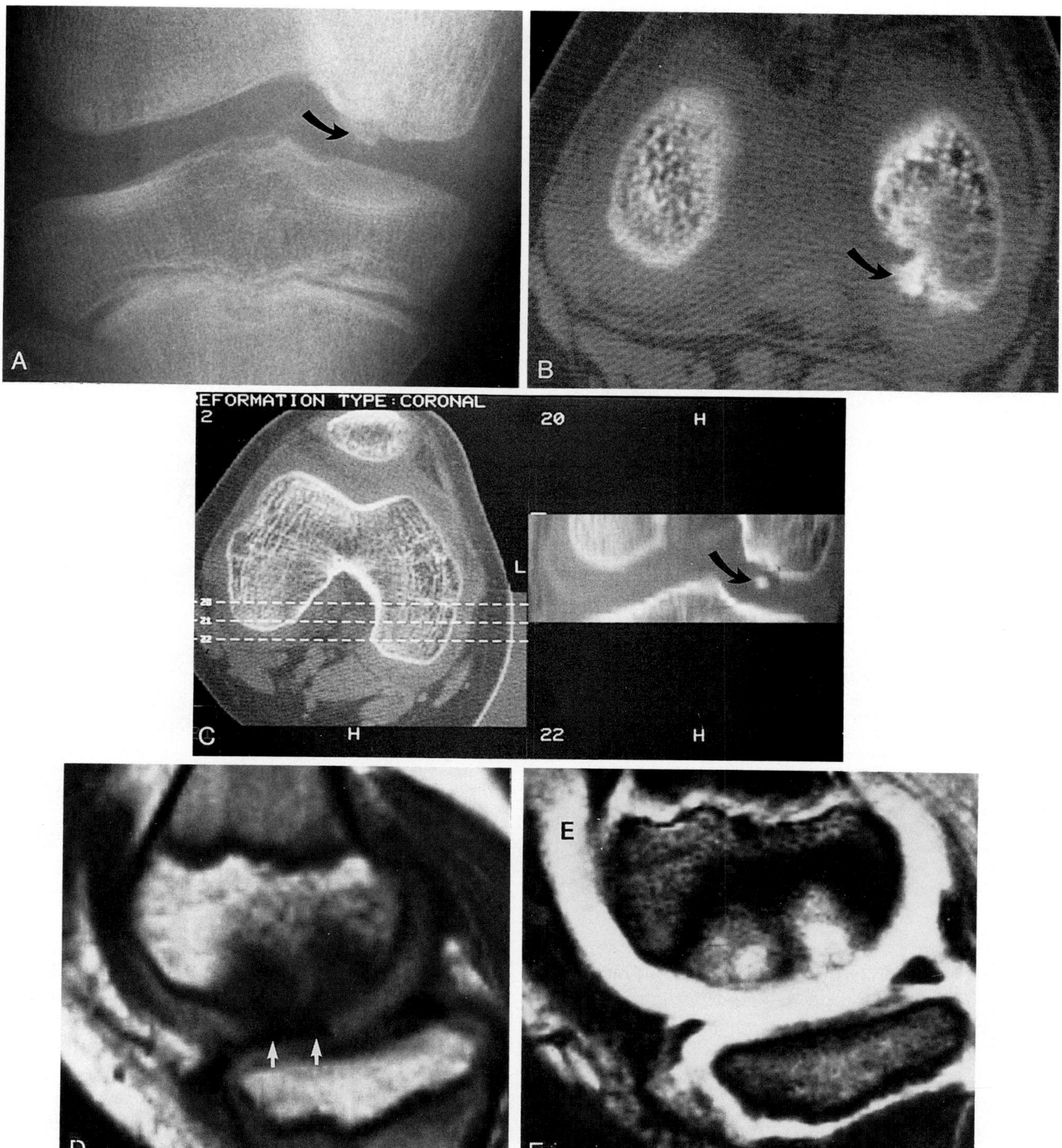

Figure 12–45. Osteochondritis dissecans with displaced fragment. *A,* Initial radiograph shows a bony defect in the medial femoral condyle *(curved arrow).* *B* and *C,* Axial and reformatted coronal computed tomography scans confirm the presence of a dislodged osteochondral body in the posterior medial femoral condyle *(curved arrow).* Sagittal T1-weighted *(D)* and gradient-echo *(E)* images reveal two large cavities filled with synovial fluid, which enters through the disrupted articular cartilage and bone plate *(arrows).* E, Associated effusion.

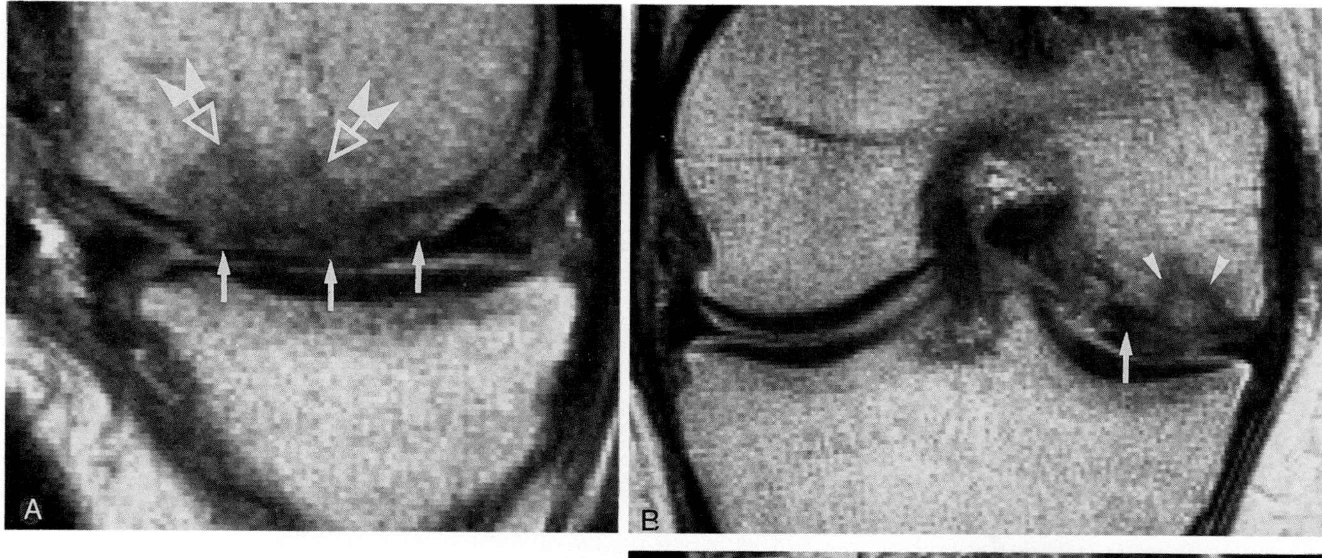

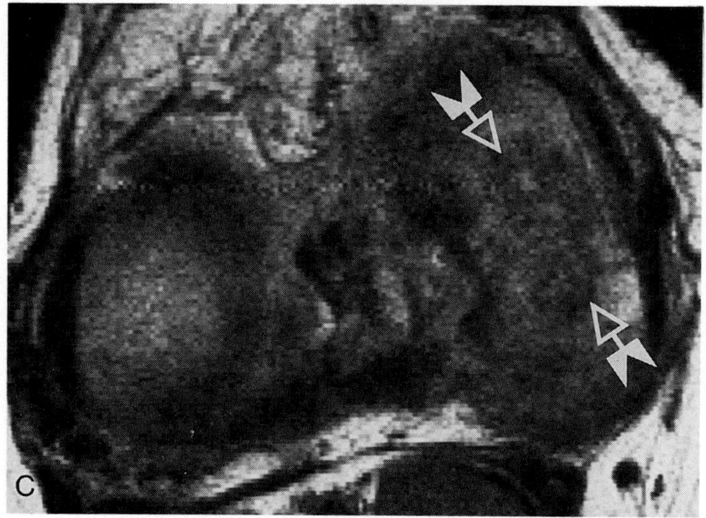

Figure 12–46. Degenerative arthritis secondary to osteochondritis dissecans in a 34-year-old man who was treated for osteochondritis dissecans at age 16 with insertion of pins, which were later removed. He developed recurrent medial knee pain 1 year earlier. Sagittal *(A)*, coronal *(B)*, and axial *(C)* proton density weighted images show diffuse irregularity and ulceration of the articular cartilage of the medial femoral condyle *(arrows)*, corresponding to fibrillation and focal ulceration at arthroscopy. The osteochondritic defect with a well-corticated margin is identified *(arrowheads)*. Note the extensive subchondral sclerosis of low signal intensity *(flagged arrows)*.

frequently affects long bones. Common findings in osteomyelitis, such as periosteal reaction or permeated lucencies, cannot be detected radiographically for at least 7 to 14 days after the onset of infection, or until 35 to 40 per cent of the bone destruction has occurred. Radionuclide scintigraphy is sensitive but nonspecific and is inadequate for differentiating bone from soft tissue infection. Scintigraphy also lacks the spatial resolution for evaluating the extent of infection in preparation for surgery. Computed tomography depicts sequestra, cloacae, sinus tracts, foreign bodies, and intramedullary and soft tissue gas earlier than radiography.[54]

In osteomyelitis, the increase in intramedullary water that results from edema, exudate, hyperemia, and ischemia results in low signal intensity on T1-weighted and high signal intensity on T2-weighted and STIR images (Fig. 12–47).[6,12,18,54] Osteomyelitis is most conspicuous on T1-weighted or STIR sequences owing to the high contrast with normal yellow marrow. The margin of the lesion changes with time, going initially from poorly defined to more distinct, which reflects the resolution of both inflammatory exudate and secondary ischemia. Healing of an adult bone usually results in fibrosis and irregular overgrowth of the bone, which has a low proton density and a short T2 value. In chronic osteomyelitis with active infection, the cortical bone of low signal intensity or the cancellous bone of the sequestrum of high signal intensity is surrounded by exudate of low signal intensity on T1-weighted sequences and high signal intensity on T2-weighted images. A Brodie's abscess exhibits low signal intensity on T1-weighted images and very high signal intensity on T2-weighted images and is outlined by a thick rim of signal void (Fig. 12–48). It should be noted that osteomyelitis with an infectious tract and surrounding purulent fluid can easily be mistaken for Ewing's sarcoma.

Tang and colleagues reported that MR imaging

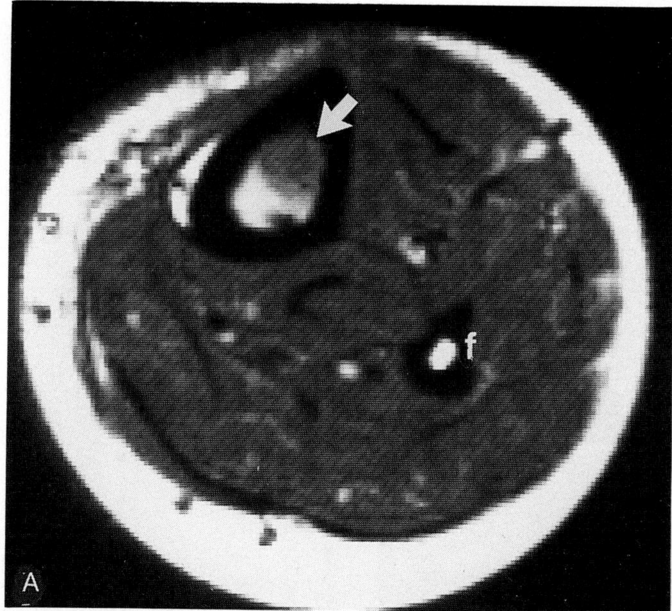

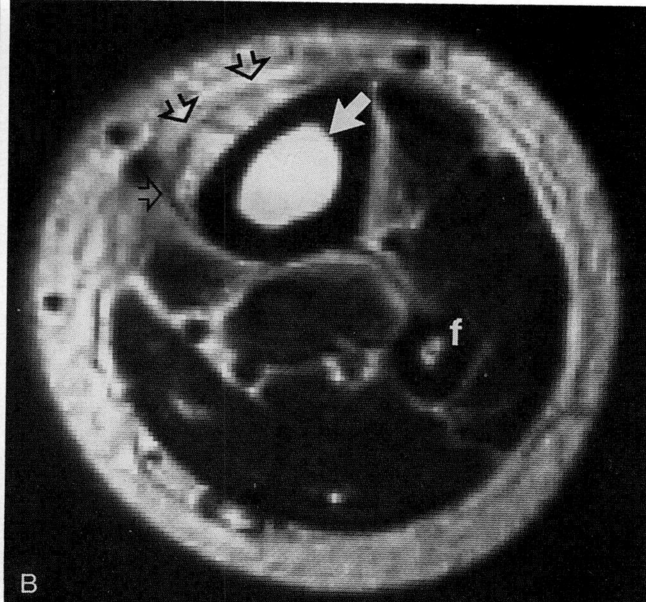

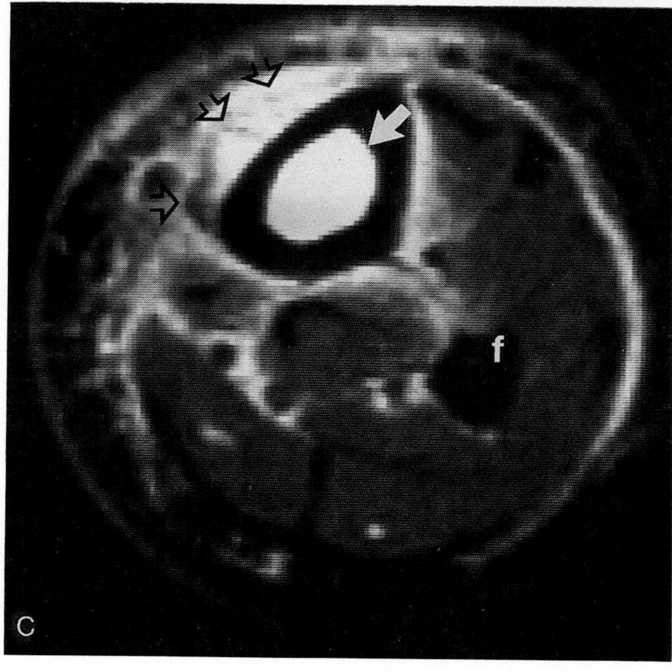

Figure 12–47. Acute hematogenous osteomyelitis of the tibia in a 20-year-old woman. *A*, Axial T1-weighted image (TR, 500 ms; TE, 30 ms) shows low signal intensity of the tibial marrow *(solid arrow)*, which is consistent with infection. The normal fibular marrow (f) is of high signal intensity. *B*, Axial T2-weighted image (TR, 2000 ms; TE, 80 ms) shows the infected tibial marrow *(solid arrow)* with higher signal intensity than the normal fibular marrow (f). The periosteum (dark line, *open arrows)* is elevated by pus (surgically proved). The adjacent cortex of the bone is spared, and no abnormal signal intensity is seen. *C*, Axial short tau inversion-recovery (STIR) image (TR, 1500 ms; TE, 30 ms; inversion time, 100 ms) shows the infected tibial marrow to have high contrast *(solid arrow)* and the inflamed surrounding muscle and subcutaneous tissue to be of high signal intensity. Normal fibular marrow (f) and subcutaneous fat are dark in this sequence. The dark line indicative of an elevated periosteum is noted *(open arrows)*. (From Erdman WA, Tamburro F, Jayson HT, et al: Osteomyelitis: Characteristics and pitfalls of diagnosis with MR imaging. Radiology 180:533, 1991. © Radiological Society of North America.)

is superior to CT because it not only depicts sequestra and sinus tracts but also identifies foci of active infection in areas of chronic osteomyelitis complicated by surgical intervention or fracture.[54] Beltran and co-workers reported that MR imaging is more accurate for the detection of infection involving bone and soft tissue than are plain radiography and CT.[6] In an animal study, Chandnani and associates reported that MR imaging is more sensitive than CT in the detection of osteomyelitis (94 per cent versus 66 per cent), is almost as specific as CT in the exclusion of osteomyelitis (93 per cent versus 97 per cent), and has an overall accuracy greater than CT (93 per cent versus 80 per cent).[12]

JOINT EFFUSIONS

Because the patient is supine with knee extended and slightly rotated externally during routine MR imaging examination of the knee, the distribution of knee joint effusions seen on MR images is different from that seen on radiographs.[34] With the knee in slight flexion, fluid accumulates behind the femoral condyles and the posterior cruciate ligament owing to relaxation of the posterior capsule and the gastrocnemius muscle. With the knee in full extension, the gastrocnemius muscle displaces the fluid into the intercondylar notch and the suprapatellar bursa. Effusions, which appear as areas of intermedi-

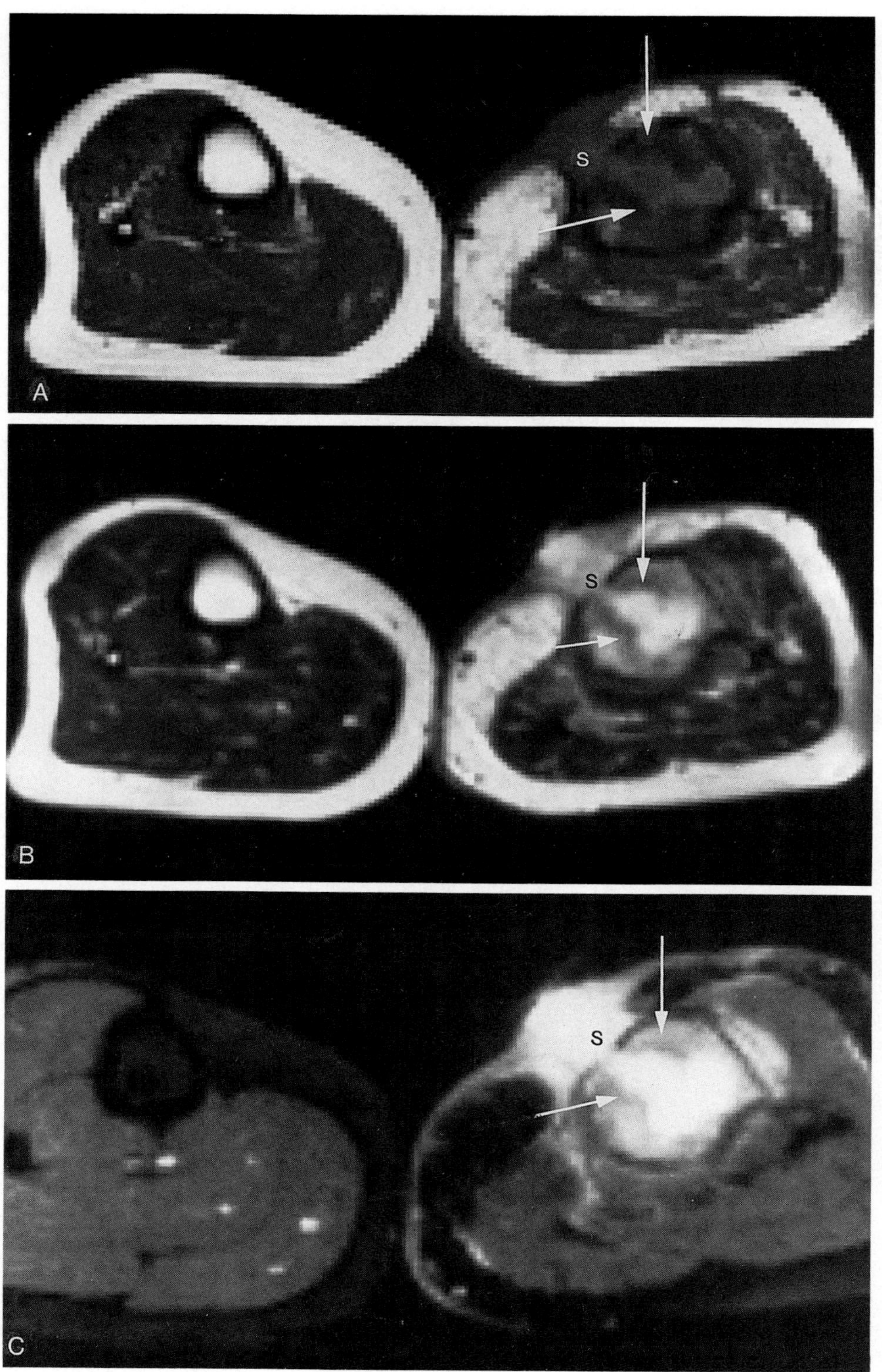

Figure 12–48. Rim sign of chronic osteomyelitis. This 38-year-old woman had had a 28-year history of draining sinus. *A*, Axial T1-weighted (TR, 500 ms; TE, 30 ms) T2-weighted (TR, 2000; TE, 80); and C, short tau inversion-recovery (STIR) (TR, 1500 ms; TE, 30 ms; inversion time, 100 ms) images show a dark rim around the area of histopathologically proved chronic active infection *(arrows)*. The rim is presumed to be fibrous tissue and is dark in all sequences. Disruption of cortical bone at the sinus tract (S), cortical remodeling, and paucity of soft tissue inflammatory change are further characteristics of chronic traumatic osteomyelitis. (From Erdman WA, Tamburro F, Jayson HT, et al: Osteomyelitis: Characteristics and pitfalls of diagnosis with MR imaging. Radiology 180:533, 1991. © Radiological Society of North America.)

ate signal intensity on T1-weighted images and of high signal intensity on T2-weighted and gradient-echo images, are distributed within the suprapatellar bursa, intercondylar notch, and posterior recess. Magnetic resonance imaging is unable to distinguish between inflammatory and noninflammatory effusions.

Any fracture that extends to an intracapsular surface of a joint may allow marrow fat and blood to leak into the joint space and produce a lipohemarthrosis. Radiography taken with a horizontal beam can detect a fat-blood level owing to the differences that are reflected as variations in specific gravity radiographic density. A lipohemarthrosis is a valuable sign of a fracture even when a fracture is not seen radiographically. Magnetic resonance imaging can detect both fat-serum levels and serum-cell levels with the use of both T1- and T2-weighted spin-echo sequences owing to the differences in relaxation time of the different fluid components (Fig. 12–49). Four distinct signal bands may be seen in knees with a lipohemarthrosis (see Fig. 12–35)[24]: (1) a superior band of floating fat appears as a homogeneous region of higher signal intensity on T1 weighting, decreasing in signal intensity on T2 weighting; (2) a thin band of signal void represents chemical shift artifact and occurs at the interface of serum and floating fat when the frequency encoding gradient-echo is in an anteroposterior direction; (3) a central band of serum that assumes a nondependent position and appears as a region of low signal intensity on T1-weighted images and of high signal intensity on T2-weighted images; (4) an inferior band of red blood cells that assumes a dependent position (hematocrit effect) and is of low signal intensity on T1- and T2-weighted images.

Hemarthrosis often occurs after trauma and may indicate the presence of an anterior cruciate ligament tear (72 per cent) or an intracapsular bony fracture. Nontraumatic lesions such as PVNS, hemophilia, and intracapsular tumors also may produce hemarthrosis. Magnetic resonance imaging can detect the serum-cell level of a hemarthrosis and its causal factors.

Pneumarthrosis or an air-fluid level in joints can be found in patients after arthroscopic or arthrographic examinations, open trauma, or any invasive procedures about the joint. Air is seen as a layer of signal void on all pulse sequences and is in a nondependent position (Fig. 12–50).

CYSTIC LESIONS ABOUT THE KNEE

Synovial Cysts

A synovial cyst is an accumulation of fluid within a synovial membrane-lined cavity. Abnormal production of synovial fluid may result in distention of a synovial bursa or herniation of a distended joint capsule secondary to injury or arthritis of the knee. The knee has the largest synovial membrane of any joint in the body and is the most common site for the formation of synovial cysts or other cystic lesions related to the synovium. Synovial cysts can involve any bursa, with the gastrocnemio-semimembranosus bursa involved most commonly (Fig. 12–51).

Infrapatellar, Suprapatellar (Antefemoral), and Prepatellar Cysts. Bursae inferior to, superior to, and in front of the patella are potential sites for cyst formation (Figs. 12–52, 12–53; see also Figs. 12–21, 12–45, 12–49). The suprapatellar bursa is part of the normal knee joint capsule and is a common site for the collection of knee joint effusion; only rarely is it separated into a separate bursa by an intact suprapatellar plica.

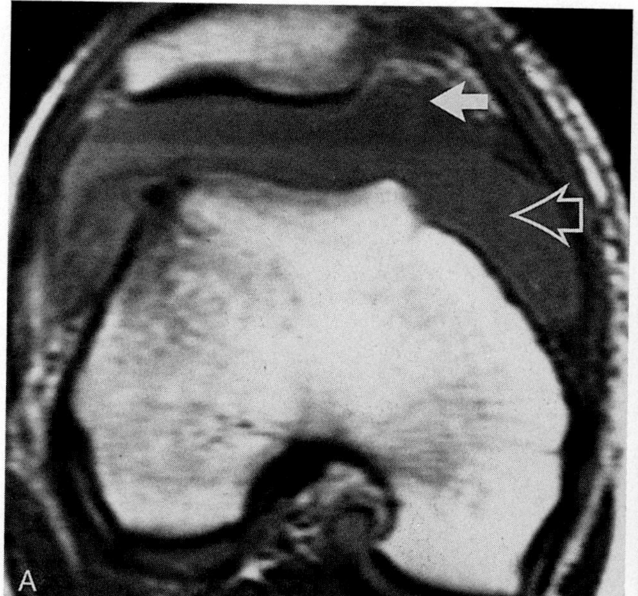

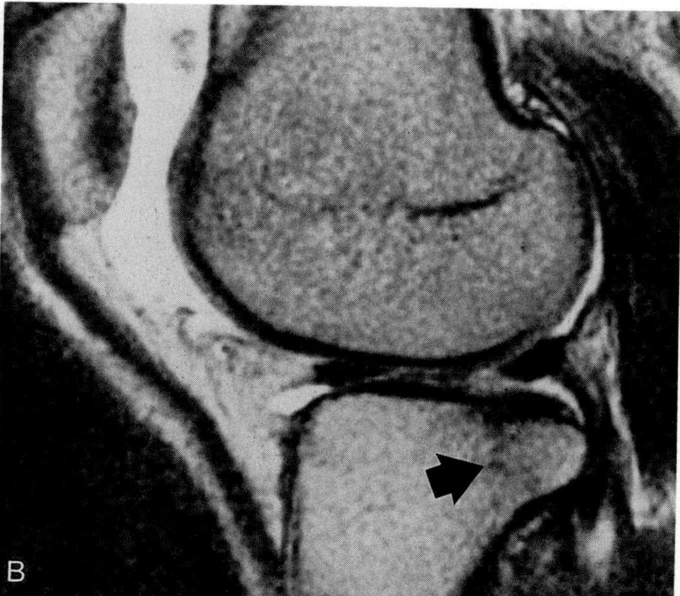

Figure 12–49. Hemarthrosis in a patient after tibial plateau fracture. *A*, Axial T1-weighted image (TR, 800 ms; TE, 20 ms) shows joint effusion with a serum-cell level. The floating layer is serum *(solid arrow)*, whereas the red blood cells are in a dependent position *(open arrow)*. Lateral patellar dislocation also is identified. *B*, Both layers become of high signal intensity on T2-weighted image (TR, 1800 ms; TE, 80 ms). Contusion in the posterior tibial plateau is noted *(arrow)*.

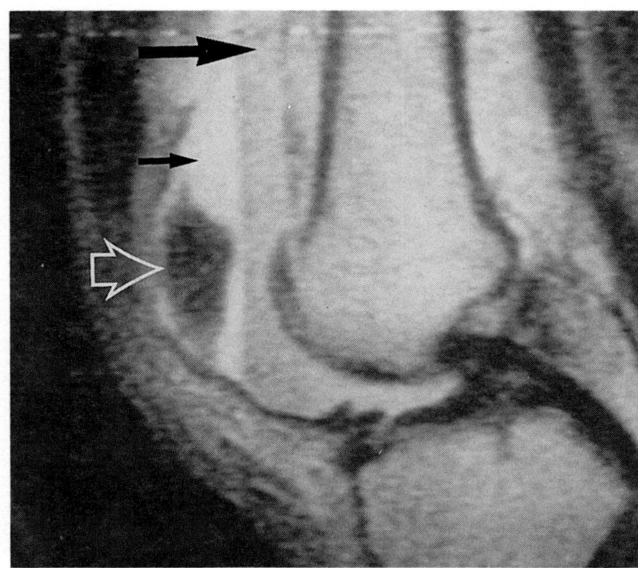

Figure 12–50. Pneumarthrosis and lipohemarthrosis. Sagittal T2-weighted image (TR, 2000 ms; TE, 60 ms) shows signal-void layer of air *(open arrow)* within the fat layer, which is of high signal intensity *(small arrow)*. The dependent position is blood *(large arrow)*.

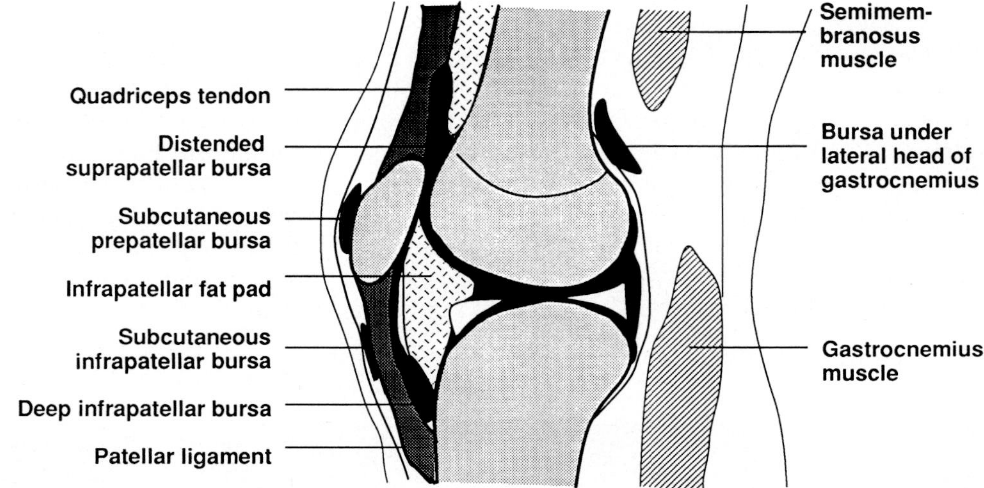

Figure 12–51. Anatomic locations of the bursae of the knee.

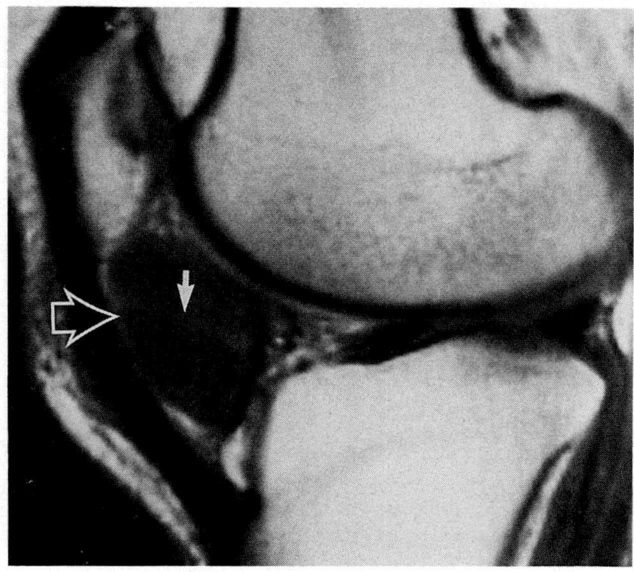

Figure 12–52. Infrapatellar cyst in a 34-year-old woman. Sagittal T1-weighted image (TR, 800 ms; TE, 20 ms) shows a well-defined cyst *(open arrow)* with septum *(small arrow)* in the infrapatellar bursa.

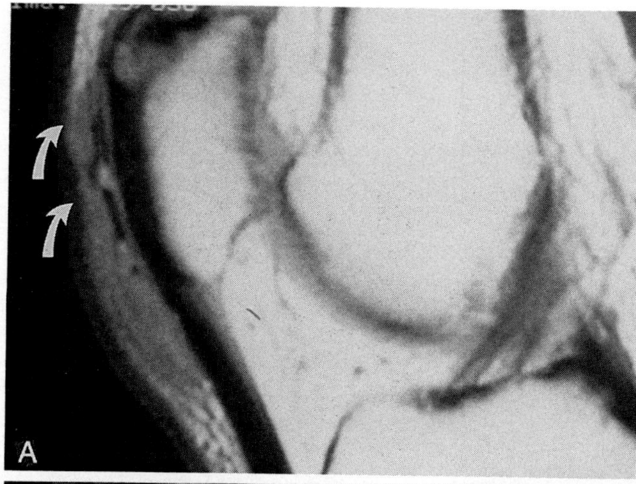

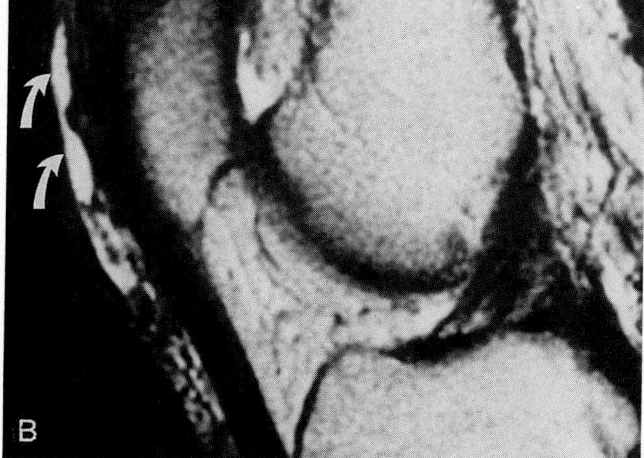

Figure 12–53. Prepatellar bursitis in a 64-year-old man. Sagittal images (*A*, TR, 2000 ms; TE, 20 ms; *B*, TR, 2000 ms; TE, 80 ms) show a lobulated, encapsulated fluid collection lying between the skin and the patella (*arrows*).

Housemaid's knee is a descriptive term for prepatellar bursitis, which is produced by trauma to the soft tissues of the prepatellar region from repeated kneeling. A poorly defined fluid collection is noted on MR images in prepatellar bursitis, which eventually forms a well-defined margin with a thin surrounding capsule (Fig. 12–53).[3]

Popliteal Cysts (Baker's Cysts). Accumulation of fluid within the gastrocnemio-semimembranosus bursa is called a popliteal cyst. This bursa is located in the middle third of the popliteal fossa, in the cleft between the gastrocnemius and semimembranosus muscles. Although popliteal cysts are the most common synovial cyst, only a small proportion of these cysts are diagnosed clinically. They frequently occur in patients with rheumatoid arthritis but also occur with osteoarthritis, juvenile rheumatoid arthritis, psoriatic arthritis, infection, PVNS, trauma, meniscal tears, gout, and Reiter's disease. They can also form without a known cause. On MR imaging, uncomplicated popliteal cysts in a typical location generally are of low signal intensity on T1-weighted images and of high signal intensity on T2-weighted images, whereas a short T1 and long T2 may be seen in dissecting popliteal cysts owing to hemorrhage producing high signal intensity on T1-weighted images (Figs. 12–54, 12–55).[28] Clinically, a ruptured popliteal cyst may mimic thrombophlebitis. Arthrography has traditionally been employed to search for the underlying intra-articular disease and to identify a dissecting or ruptured popliteal cyst; however, it is

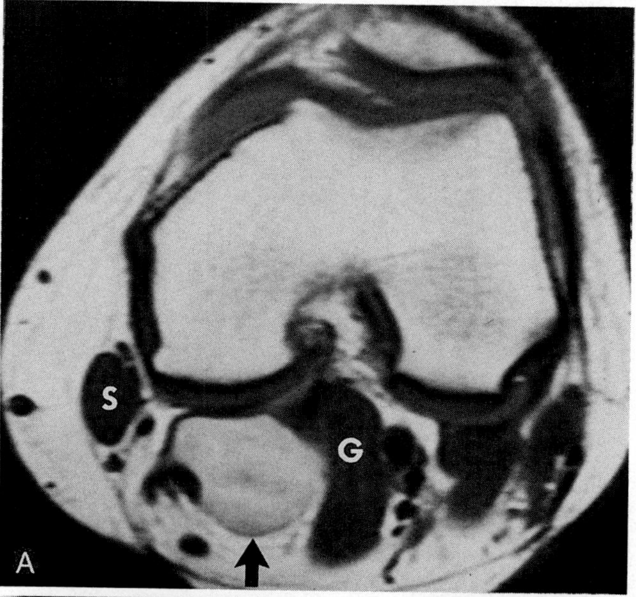

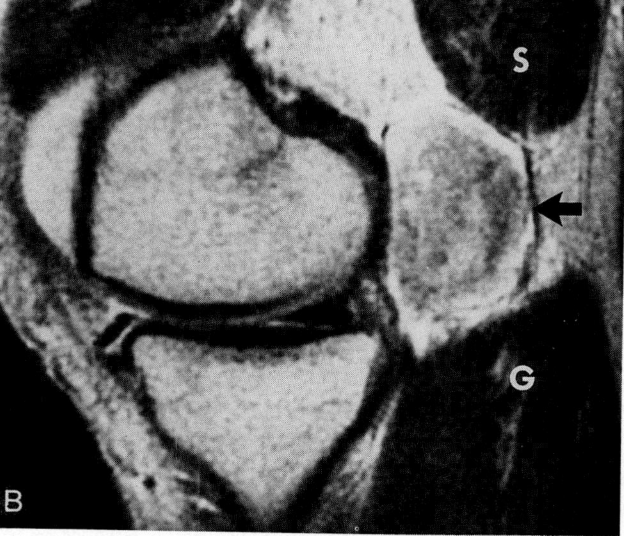

Figure 12–54. Hemorrhagic Baker's cyst in a 20-year-old woman with a tibial plateau fracture. *A*, Axial image (TR, 500 ms; TE, 20 ms) shows an inhomogeneous cyst (*arrow*) lateral to the medial head of the gastrocnemius muscle. *B*, Sagittal image (TR, 1900 ms; TE, 80 ms) shows a hemosiderin deposit of low signal intensity within the cyst (*arrow*). S, Semimembranosus muscle; G, gastrocnemius muscle.

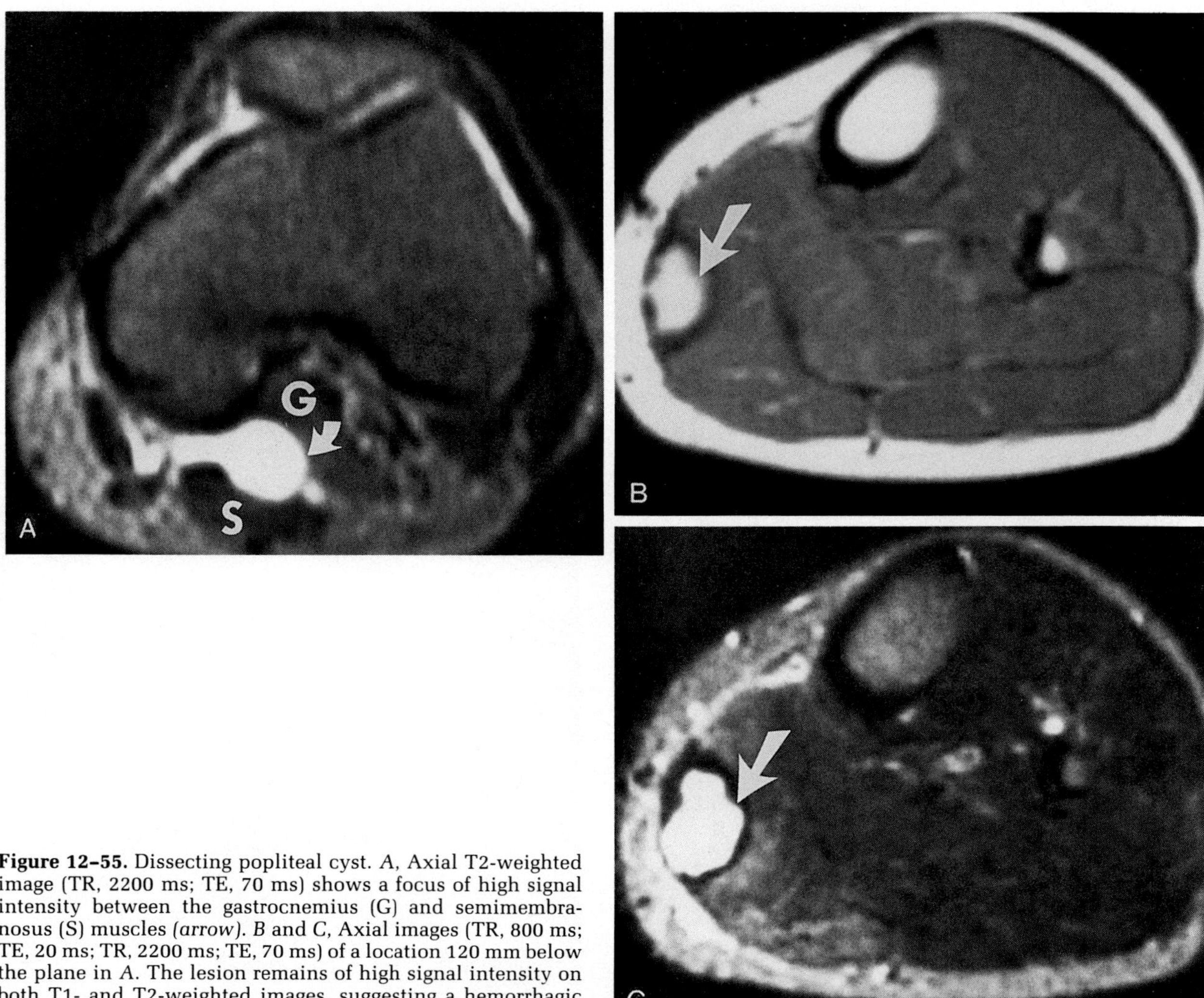

Figure 12–55. Dissecting popliteal cyst. *A*, Axial T2-weighted image (TR, 2200 ms; TE, 70 ms) shows a focus of high signal intensity between the gastrocnemius (G) and semimembranosus (S) muscles *(arrow)*. *B* and *C*, Axial images (TR, 800 ms; TE, 20 ms; TR, 2200 ms; TE, 70 ms) of a location 120 mm below the plane in *A*. The lesion remains of high signal intensity on both T1- and T2-weighted images, suggesting a hemorrhagic dissecting popliteal cyst *(arrow)*.

of limited value when the cyst does not communicate with the knee joint. Magnetic resonance imaging may be useful in this situation. Popliteal cysts may contain loose bodies, which can easily be recognized with MR imaging (Fig. 12–56).

Tibial Collateral Ligament Bursa. This bursa separates the tibial collateral ligament from the deep capsular ligament, and it allows for movement between these structures. Clinically, inflammation of this bursa may cause isolated pain in the medial joint line in the absence of mechanical symptoms. Lee and Yao[27] reported seven cases of tibial collateral ligament bursitis that appeared as vertically elongated, well-defined fluid collections adjacent to the tibial collateral ligament (Figs. 12–57, 12–58). The bursitis may mimic meniscocapsular separation or tears of the medial collateral ligament. It is important to recognize the focal location of fluid at the site of the tibial collateral ligament bursa, in contrast to other abnormalities, such as collateral ligament tears or meniscocapsular separation, which are spread along the course of the ligament superiorly and inferiorly or along the course of the meniscus anteriorly and posteriorly.

Anserine (Anteromedial) Cysts. The axial and most medial sagittal MR images are the best views to trace the pes anserinus tendons (sartorius, gracilis, and semitendinosus) to their tibial insertions where the anserine bursa is located and anteromedial cysts form (Fig. 12–59). In contrast to other cysts, anteromedial cysts occur clinically as firm, tender, cystic masses owing to the tight space beneath the pes anserinus.[37]

Tibiofibular Cysts. Cyst formation in the proximal tibiofibular joint is uncommon. In 10 per cent of adults, this cyst communicates with the joint space.[37] The cyst may cause erosion of bone and may mimic a tumor (Fig. 12–60).

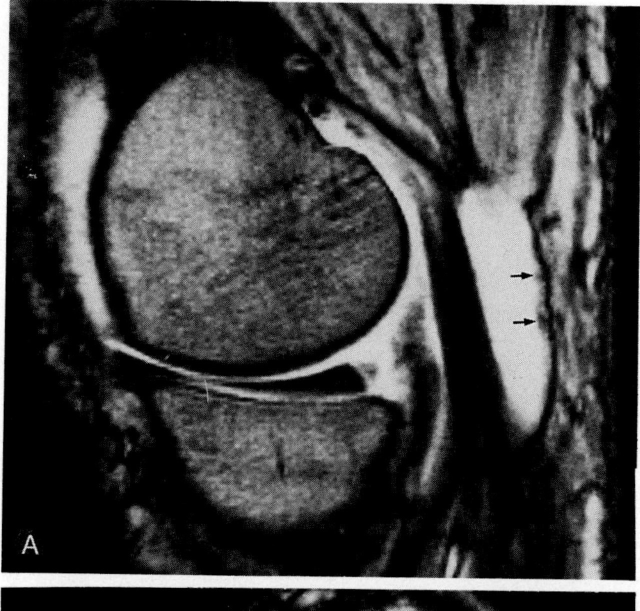

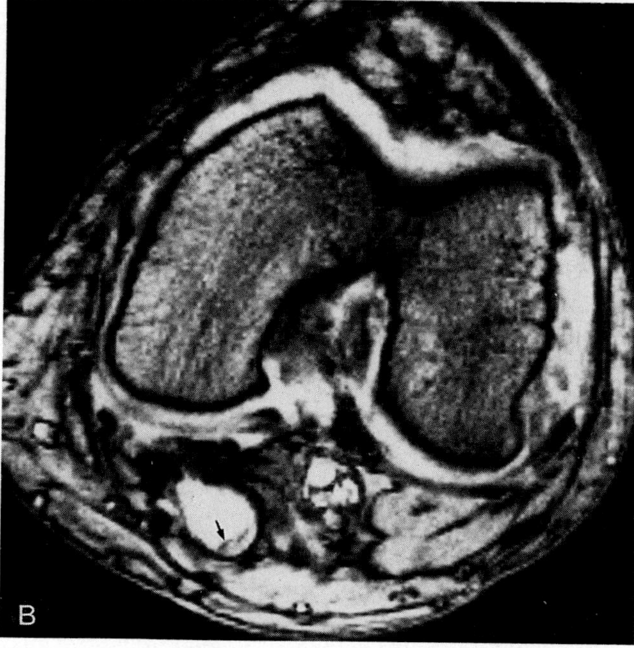

Figure 12-56. Popliteal cyst with loose bodies. Sagittal image (TR, 1000 ms; TE, 30 ms; flip angle, 75 degrees) *(A)* and axial image (TR, 1000 ms; TE, 30 ms; flip angle, 75 degrees) *(B)* show several punctate areas of signal void along the wall of the popliteal cyst, suggesting multiple loose bodies *(arrows)*.

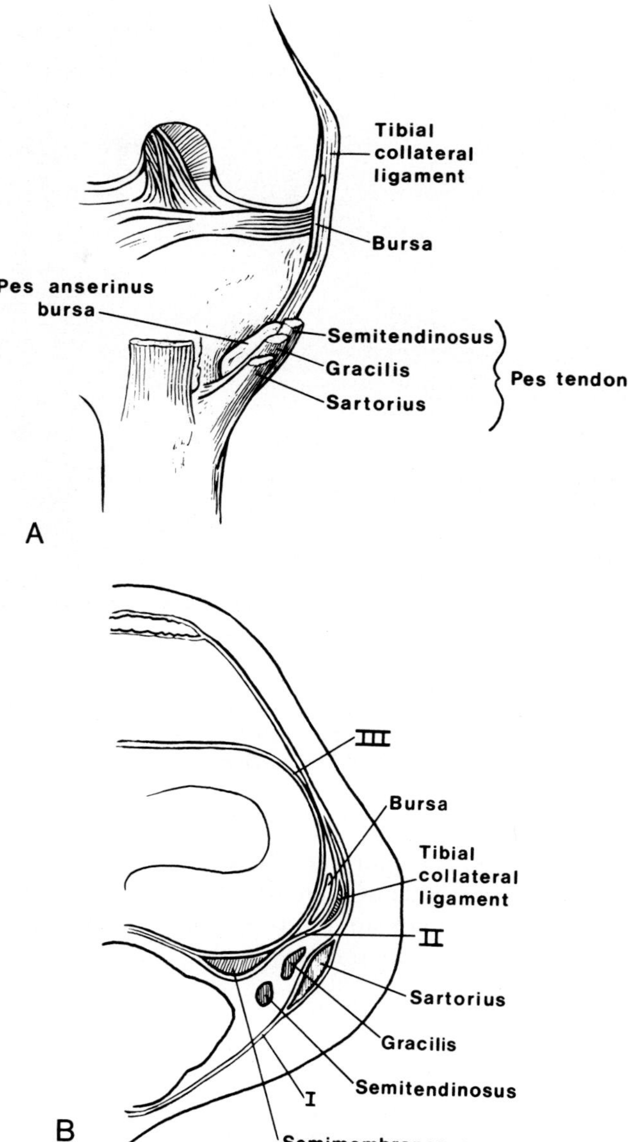

Figure 12-57. *A*, Posterior view of the medial aspect of the knee shows the tibial collateral ligament bursa and pes anserinus bursa. *B*, Axial view of the medial aspect of the knee. Separation of layer III from layer II is exaggerated to illustrate the interposition of the tibial collateral ligament bursa. (From Lee JL, Yao L: Tibial collateral ligament bursa: MR imaging. Radiology 178:855, 1991. © Radiological Society of North America.)

Ganglion Cysts

Ganglions around the knee usually contain jellylike viscous fluid without a synovial lining. Ganglion cysts develop after primary cellular hyperplasia with associated mucin secretion and secondary cystic degeneration of the connective tissue. Clinically, although both meniscal and ganglion cysts frequently occur as palpable masses around the knee, ganglion cysts are not associated with meniscal tears. Moreover, ganglion cysts may or may not communicate with the joint. On MR imaging, ganglion cysts are often found to be multiloculated and are of low signal intensity on T1-weighted images and of high signal intensity on T2-weighted images (Fig. 12-61).[9] Septations, when present, are best seen on T2-weighted images. Meniscal cysts may be recognized by their location within the meniscus or by their extension from a well-defined meniscal tear. The distinction between ganglions and meniscal cysts has important surgical implications. The un-

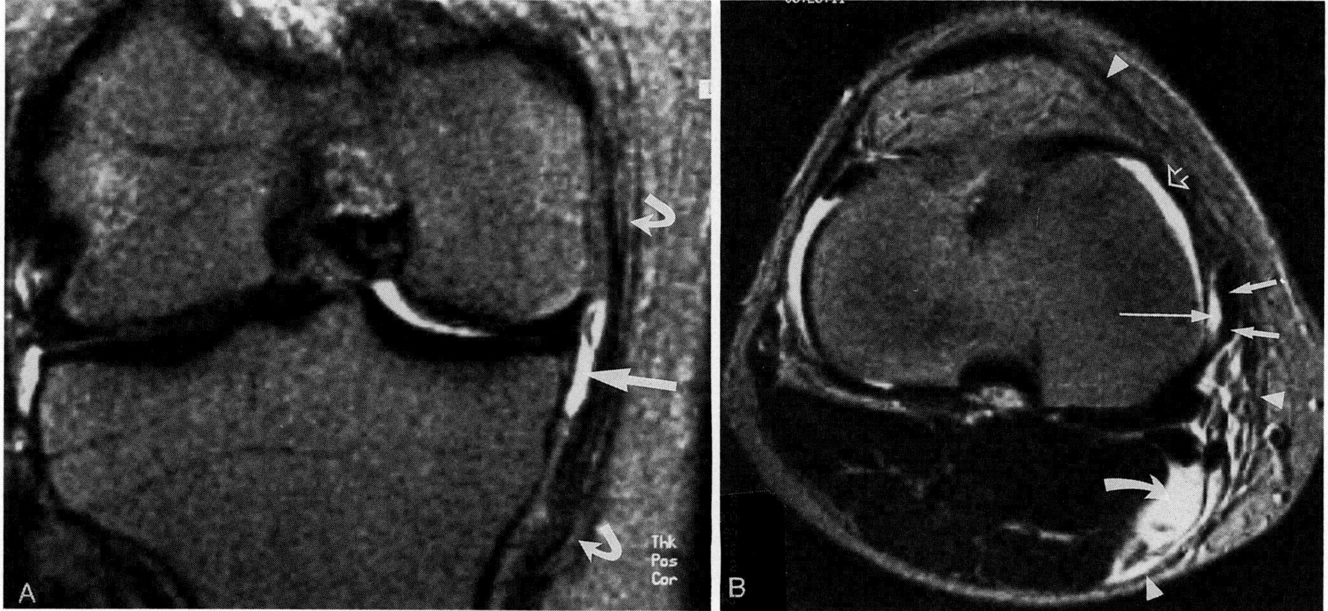

Figure 12–58. Acute tibial collateral ligament strain in a 25-year-old man. *A,* Coronal T2-weighted image (TR, 1800 ms; TE, 80 ms) shows a well-defined elongated fluid collection *(straight arrow)* with one internal separation superiorly. The tibial collateral ligament is somewhat thickened, consistent with the history of acute injury in this case *(curved arrows)*. *B,* Axial T2-weighted image (TR, 1800 ms; TE, 80 ms) shows fluid in the tibial collateral ligament bursa *(long arrow)* adjacent to the tibial collateral ligament *(short arrows)*. Layer I *(arrowheads)* and layer III *(open arrow)* are indicated. A ruptured Baker's cyst is also seen *(curved arrow)*. (From Lee JL, Yao L: Tibial collateral ligament bursa: MR imaging. Radiology 178:885, 1991.© Radiological Society of North America.)

derlying meniscal tear must be repaired or excised to prevent recurrence of a meniscal cyst, whereas ganglion removal does not require an intra-articular approach (also see Chapter 11).

PITFALLS

Pitfalls for menisci and ligaments have been discussed in Chapter 11. This section introduces normal variants and artifacts that may mimic bone pathology.

Normal Depression. A smooth depression on the anterior portions of both femoral condyles adjacent to the anterior horns of both menisci can mimic bone impaction or an osteochondral defect. This pitfall can be recognized confidently by identifying an intact overlying cartilage, by a lack of underlying bone contusion or sclerosis, and by its characteristic location.

Normal tibial depressions can mimic osteochondral defects. These normal depressions, which are related to the attachments for cruciate ligament, are seen in 100 per cent of knees posteromedially and 95 per cent of knees anterolaterally on routine coronal MR images. These depressions are 4 to 10 mm in size at the posterior cruciate ligament insertion and 1 to 6 mm beneath the anterior attachment of the lateral meniscus (Fig. 12–62).

Physis. A normal epiphysis or old physeal scar can mimic a bone fracture (Fig. 12–63). The clinical history and the characteristic appearance of the physis at various stages of maturation are useful clues for their differentiation.

Metallic Artifacts. A postoperative knee may demonstrate multiple artifacts of low signal intensity that can mimic loose bodies. These artifacts are created by the susceptibility effect from microscopic pieces of metallic surgical instruments. These microscopic metallic substances cannot be detected on radiography or CT but are especially pronounced on gradient-echo MR images (Fig. 12–64).

Red Marrow. Residual red marrow is not uncommonly seen in the distal femur or proximal tibia of healthy adults. It may appear to be very extensive and may simulate an infiltrative marrow disease on T1-weighted images. However, normal red marrow should be considered when it spares the epiphysis and patella, and signal intensity does not increase significantly on T2-weighted images (Fig. 12–65). The patient's clinical history and laboratory status are valuable clues in establishing a correct diagnosis.

Bipartite Patella. Bipartite or multipartite patella (Fig. 12–66) is the best example of a developmental variant involving multiple accessory ossification centers in a sesamoid bone. This variant has little or no clinical significance and represents only an incidental finding on MR imaging. However, the

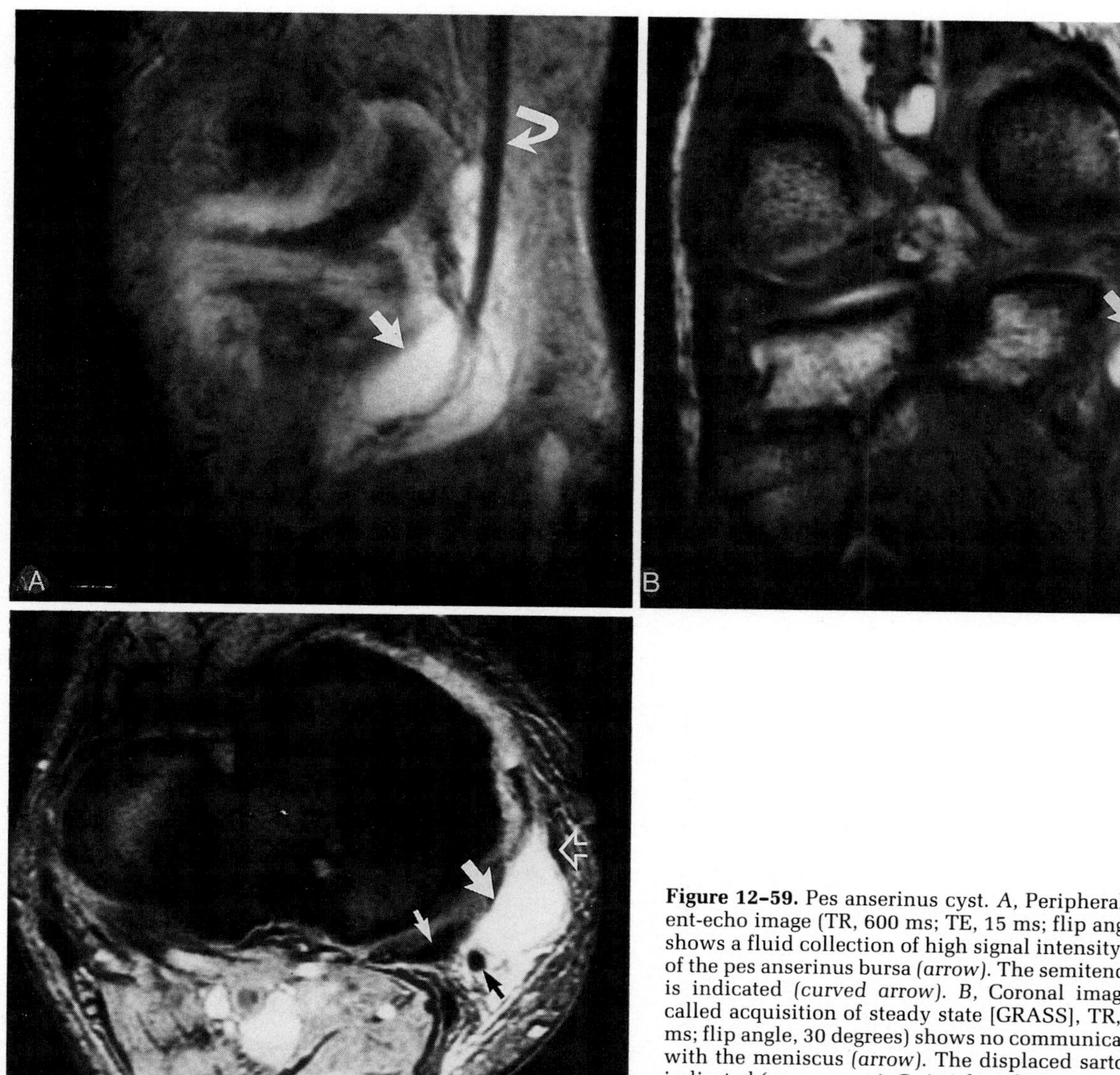

Figure 12–59. Pes anserinus cyst. *A*, Peripheral sagittal gradient-echo image (TR, 600 ms; TE, 15 ms; flip angle, 30 degrees) shows a fluid collection of high signal intensity at the location of the pes anserinus bursa *(arrow)*. The semitendinosus tendon is indicated *(curved arrow)*. *B*, Coronal image (gradient-recalled acquisition of steady state [GRASS], TR, 35 ms; TE, 15 ms; flip angle, 30 degrees) shows no communication of the cyst with the meniscus *(arrow)*. The displaced sartorius tendon is indicated *(open arrow)*. *C*, Axial gradient-echo image (TR, 500 ms; TE, 15 ms; flip angle, 30 degrees) shows the location of the cyst *(heavy arrow)* in relation to the sartorius *(open arrow)*, semimembranosus *(small white arrow)*, and gracilis tendons *(black arrow)*.

Figure 12–60. Popliteal tendon sheath cyst. Posterior coronal image (gradient-recalled acquisition of steady state [GRASS], TR, 35 ms; TE, 15 ms; flip angle, 30 degrees) shows a fluid collection of high signal intensity along the oblique course of the popliteal tendon *(arrow)*,

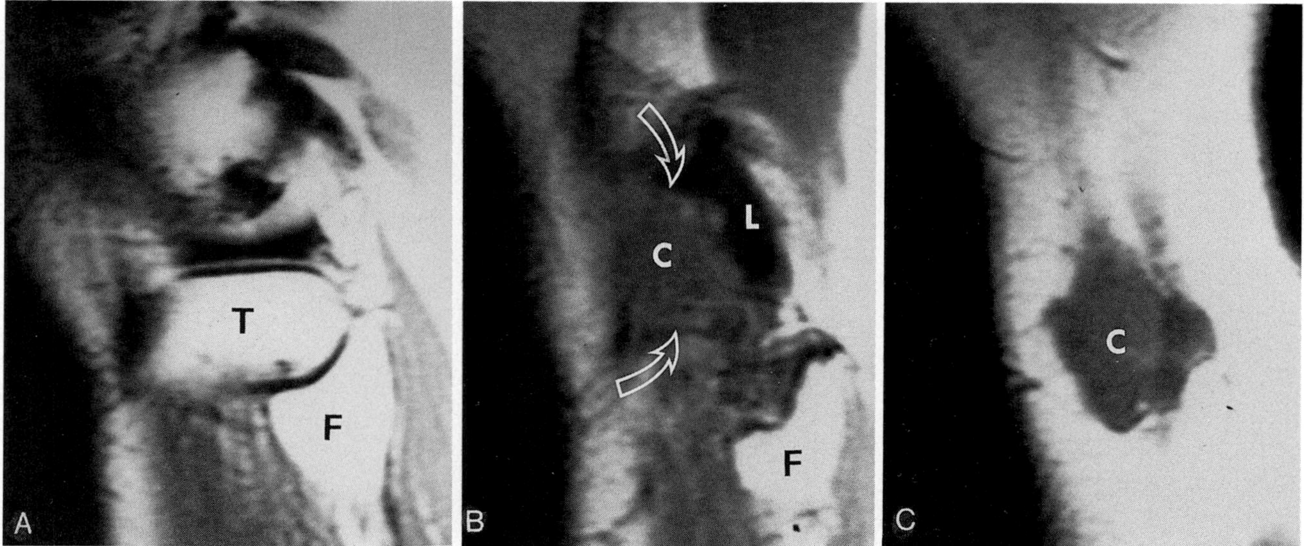

Figure 12–61. Ganglion cyst in a 53-year-old woman. *A, B,* and *C,* Peripheral sagittal consecutive T1-weighted images show a multiloculated cyst *(arrows)* in the lateral aspect of the knee joint, anterior to the fibular collateral ligament. Note that the lateral meniscus is intact. T, Tibia; F, fibula; L, femorofibular ligament; C, ganglion cyst.

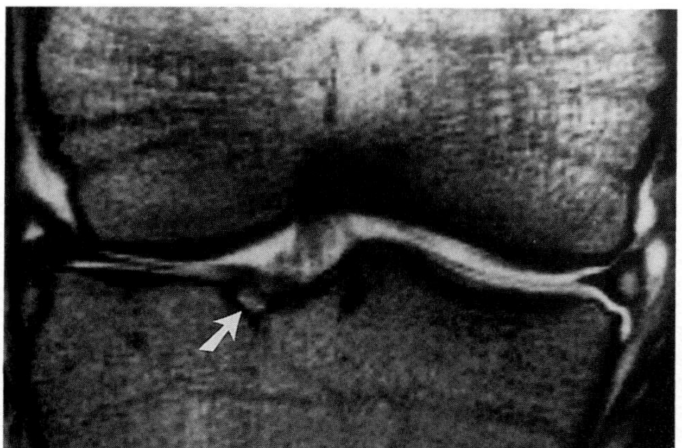

Figure 12–62. Normal tibial depression simulating osteochondral defect. This depression measured 6 mm *(arrow).* Such lesions are consistently located in the anterolateral portion of the tibial plateau beneath the anterior attachment of the lateral meniscus. Sagittal gradient-echo image (TR, 1000 ms; TE, 30 ms; flip angle, 75 degrees).

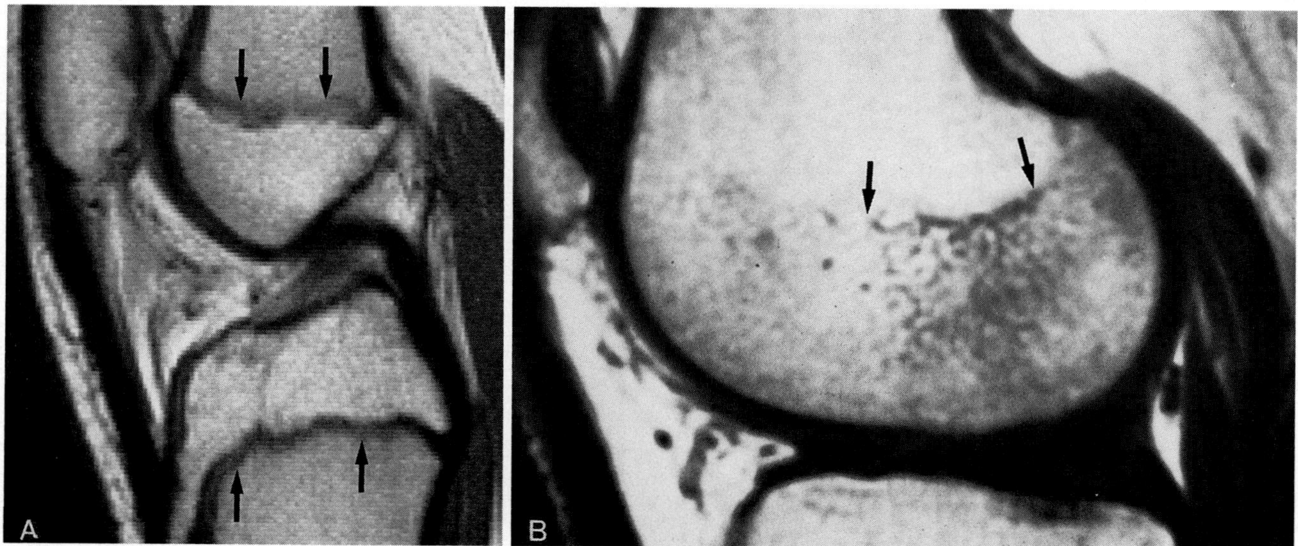

Figure 12–63. Normal physeal line of low signal intensity *(arrows)* in a 10-year-old boy *(A)* and a 72-year-old man *(B).* This physeal line should not be mistaken for a bone fracture.

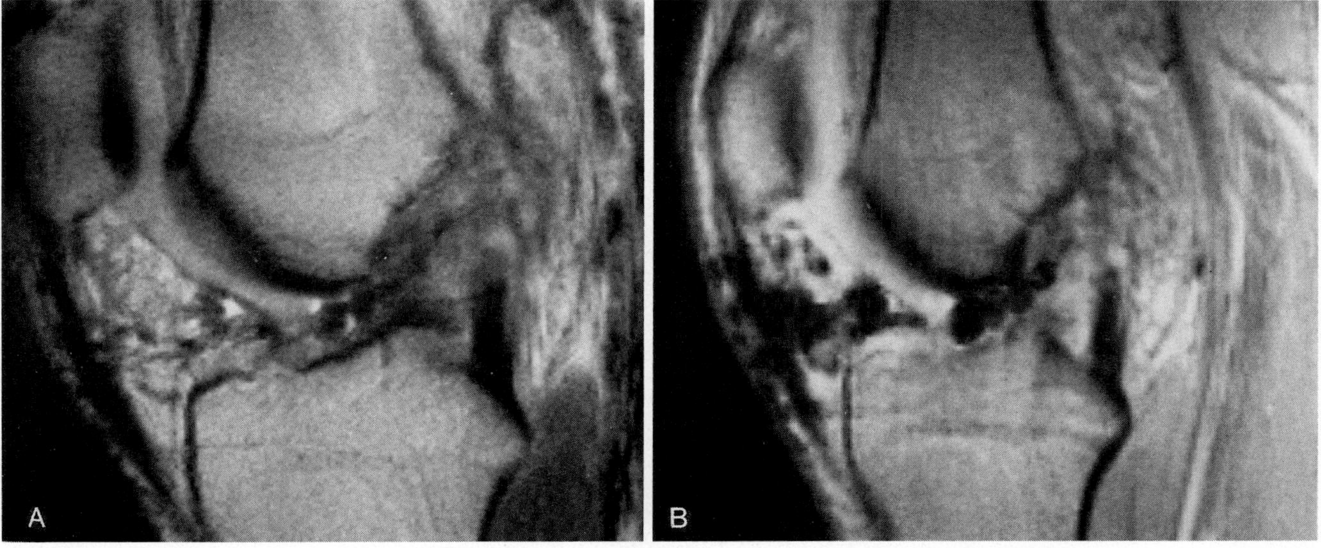

Figure 12–64. Metallic artifacts simulating loose bodies in a postoperative knee. Sagittal T1-weighted image (TR, 800 ms; TE, 20 ms) and *B*, gradient-echo image (TR, 600 ms; TE, 20 ms; flip angle, 30 degrees). Note that the multiple small artifacts of low signal intensity are more pronounced on the gradient-echo image (*B*).

bone fragments of bipartite patella can be mistaken for patellar fracture. Bipartite patella usually involves the superolateral aspect of the patella, in contrast to patellar fracture, which tends to involve the midportion (Fig. 12–67). In bipartite patella, the fragments can form a normal patella if they are replaced.

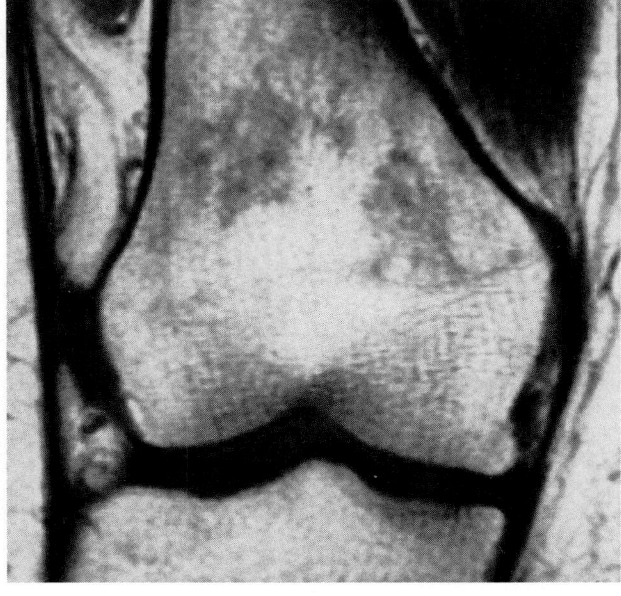

Figure 12–65. Normal bone marrow in a 26-year-old woman. Coronal T1-weighted image (TR, 600 ms; TE, 20 ms) shows residual red marrow of low signal intensity in the femur, in contrast to the marrow fat of high signal intensity. Occasionally, this pattern may mimic a marrow disorder. The epiphysis contains fatty marrow. The areas of low signal intensity in the epiphysis are trabecular bone.

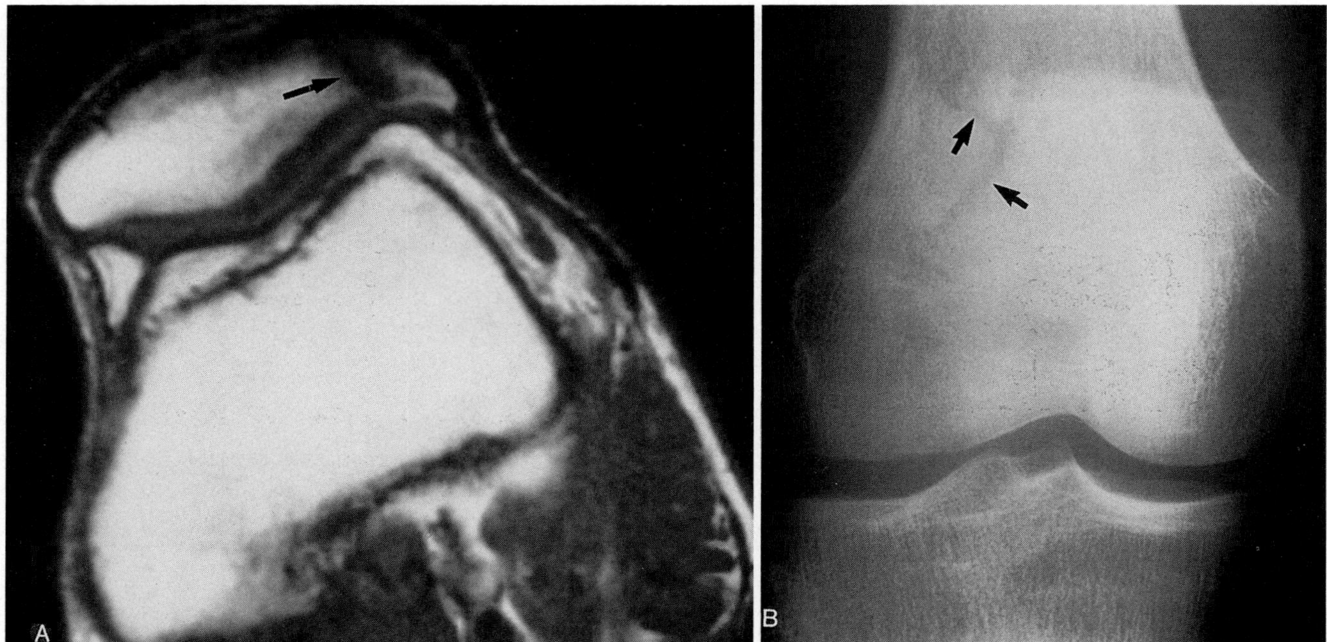

Figure 12–66. Tripartite patella simulating a patellar fracture. *A*, Axial T1-weighted image shows a line in the lateral facet separating the patella into two fragments *(arrow)*. *B*, Radiograph reveals three well-corticated osseous densities at the superolateral margin of the patella, consistent with a tripartite patella *(arrows)*. Note that these fragments do not fit with the parent bone, as they would in a patellar fracture.

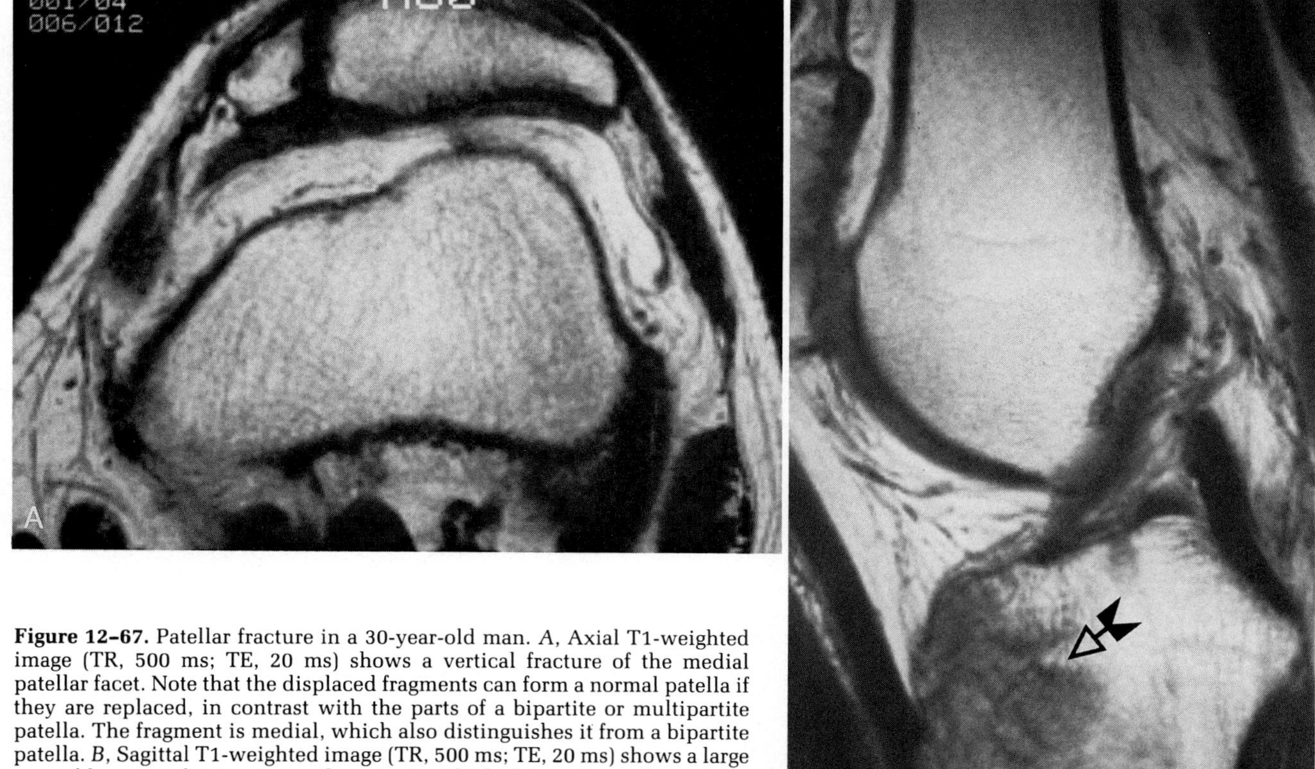

Figure 12–67. Patellar fracture in a 30-year-old man. *A*, Axial T1-weighted image (TR, 500 ms; TE, 20 ms) shows a vertical fracture of the medial patellar facet. Note that the displaced fragments can form a normal patella if they are replaced, in contrast with the parts of a bipartite or multipartite patella. The fragment is medial, which also distinguishes it from a bipartite patella. *B*, Sagittal T1-weighted image (TR, 500 ms; TE, 20 ms) shows a large area of low signal intensity in the anterior tibial plateau *(arrow)*, resulting from a direct blow to the anterior aspect of the knee.

References

1. Ahlbäck S: Osteoarthritis of the knee: A radiographic investigation. Acta Radiol (Stockh) Suppl 277:7, 1968.
2. Arnold WD, Hilgartner M: Hemophilic arthropathy: Current concepts of pathologenesis and management. J Bone Joint Surg [Am] 59:287, 1977.
3. Bellon EM, Sacco DC, Steiger DA, et al: Magnetic resonance imaging in "housemaid's knee" (prepatellar bursitis). Magn Reson Imaging 5:175, 1987.
4. Beltran J: Meniscal tears: MR demonstration of experimentally produced injuries. Radiology 158:691, 1986.
5. Beltran J, Caudill JL, Herman LA, et al: Rheumatoid arthritis: MR imaging manifestations. Radiology 165:153, 1987.
6. Beltran J, Noto AM, McGhee RB, et al: Infection of the musculoskeletal system: High-field-strength MR imaging. Radiology 164:449, 1987.
7. Björkengren AG, Airowaih A, Lindstrand A, et al: Spontaneous osteonecrosis of the knee: Value of MR imaging in determining prognosis. AJR 154:331, 1990.
8. Brook A, Corbett M: Radiological changes in early rheumatoid arthritis. Ann Rheum Dis 36:71, 1977.
9. Burk DL, Dalinka MK, Kanal E , et al: Meniscal and ganglion cysts of the knee: MR evaluation. AJR 150:331, 1988.
10. Chan WP, Lang P, Stevens MP, et al: Osteoarthritis of the knee: Comparison of radiography, CT, and MR imaging to assess extent and severity. AJR 157:799, 1991.
11. Chandnani V, Ho C, Chu P, et al: Knee hyaline cartilage evaluated with MR imaging: A cadaveric study involving multiple imaging sequences and intraarticular injection of gadolinium and saline solution. Radiology 178:557, 1991.
12. Chandnani VP, Beltran J, Morris CS, et al: Acute experimental osteomyelitis and abscesses: Detection with MR imaging versus CT. Radiology 174:233, 1990.
13. Conway WF, Hayes CW, Loughran T, et al: Cross-section imaging of the patellofemoral joint and surrounding structures. Radiographics 11:195, 1991.
14. Daffner RH, Salutano M, Gehweiler JA: Stress fractures in runners. JAMA 247:1039, 1982.
15. DeSmet AA, Fisher DR, Graf BK, et al: Osteochondritis dissecans of the knee: Value of MR imaging in determining lesion stability and the presence of articular cartilage defects. AJR 155:549, 1990.
16. Deutsch A, Resnick D, Dalinka M, et al: Synovial plicae of the knee. Radiology 141:627, 1981.
17. Dorwart RH, Genant HK, Johnston WH, et al: Pigmented villonodular synovitis of synovial joints: Clinical, pathologic, and radiologic features. AJR 143:877, 1984.
18. Erdman WA, Tamburro F, Jayson HT, et al: Characteristics and pitfalls of diagnosis with MR imaging. Radiology 180:533, 1991.
19. Genant HK: Roentgenographic aspects of calcium pyrophospate dihydrate crystal deposition disease (pseudogout). Arthritis Rheum 19:307, 1976.
20. Goodfellow J, Hungerford D: Patellofemoral joint mechanics and pathology. II Chondromalacia patellae. J Bone Joint Surg [Br] 58:291, 1976.
21. Hayes, CW, Sawyer RW, Conway WF: Patellar cartilage lesions: In vivo detection and staging with MR imaging and pathologic correlation. Radiology 176:479, 1990.
22. Hohl M: Tibial condylar fractures. J Bone Joint Surg [Am] 49:1455, 1967.
23. Jelinek JS, Kransdorf MJ, Utz JA, et al: Imaging of pigmented villonodular synovitis with emphasis on MR imaging. AJR 152:337, 1989.
24. Kier R, McCarthy SM: Lipohemarthrosis of the knee: MR imaging. J Comput Assist Tomogr 14:395, 1990.
25. König H, Sieper J, Wolf K-J: Rheumatoid arthritis: Evaluation of hypervascular and fibrous pannus with dynamic MR imaging enhanced with Gd-DTPA. Radiology 176:473, 1990.
26. Kottal RA, Volger JB, Matamoros A, et al: Pigmented villonodular synovitis: A report of MR imaging in two cases. Radiology 163:551, 1987.
27. Lee JL, Yao L: Tibial collateral ligament bursa: MR imaging. Radiology 178:855, 1991.
28. Lieberman JM, Yulish BS, Bryan PJ, et al: Magnetic resonance imaging of ruptured Baker's cyst. J Can Assoc Radiol 39:295, 1988.
29. Lupi L, Bighi S, Cervi PM, et al: Arthrography of the plica syndrome and its significance. Europ J Radiol 11:15, 1990.
30. Lynch TC, Crues J III, Morgan FW, et al: Bone abnormalities of the knee: Prevalence and significance at MR imaging. Radiology 171:761, 1989.
31. McDuffie FC: Morbidity impact of rheumatoid arthritis in society. Am J Med 78:1, 1985.
32. Mesgarzadeh M, Sapega AA, Bonakdarpour A, et al: Osteochondritis dissecans: Analysis of mechanical stability with radiography, scintigraphy, and MR imaging. Radiology 165:775, 1987.
33. Mink JH, Deutsch A: Occult cartilage and bone injuries of the knee: Detection, classification, and assessment with MR imaging. Radiology 170:823, 1989.
34. Mink JH, Deutsch AL: MRI of the Musculoskeletal System: A Teaching File. New York, Raven Press, 1990.
35. Muheim G, Bohne WH: Prognosis in spontaneous osteonecrosis of the knee: Investigation by radionuclide scintigraphy and radiography. J Bone Joint Surg [Br] 52:605, 1970.
36. Munk PL, Helms CA, Holt RG: Immature bone infarcts: Findings on plain radiographs and MR scans. AJR 152:547, 1989.
37. Murayama S, Hines MR, Manzo RP, et al: MR imaging of synovial cysts of the knee. Appl Radiol 3:27, 1991.
38. Myers BW, Masi AT: Pigmented villonodular synovitis and tenosynovitis: A clinical epidemiologic study of 166 cases and literature review. Medicine 59:223, 1980.
39. National Center for Health Statistics: Basic data on arthritis knee, hip and sacroiliac joints in adults 25–74 years, United States 1971–1975. In Vital and Health Statistics. Series 11, no. 123 Washington, DC, DHEW publication no. (PHS) 79-1661.
40. Patel D: Arthroscopy of the plica, synovial fold and their significance. Am J Sports Med 6:217, 1978.
41. Reiser M, Bongartz G, Erlemann R, et al: Magnetic resonance in cartilaginous lesions of the knee joint with three-dimensional gradient-echo imaging. Skeletal Radiol 17:465, 1988.
42. Resnick D, Niwayama G: Diagnosis of Bone and Joint Disorders. 2nd ed. Philadelphia, WB Saunders, 1988.
43. Sabiston CP, Adams ME, Li DKB: Magnetic resonance imaging of osteoarthritis: Correlation of gross pathology using an experimental model. J Orthop Res 5:164, 1987.
44. Sack KE, Genant HK: Radiologist's guide to the laboratory in diagnosing rheumatic diseases. Radiology 139:585, 1981.
45. Segond P: Recherches cliniques et experimentales sur les epanchements sanguins du genou par entorse. Orog Med 7:297, 1879.
46. Senac MO, Deutsch D, Bernstein BH, et al: MR imaging in juvenile rheumatoid arthritis. AJR 150:873, 1988.
47. Shahriaree H: Chondromalacia. Contemp Orthop 11:27, 1985.
48. Shellock FG, Mink JH, Deutsch AL, et al: Patellar tracking abnormalities: Clinical experience with kinematic MR imaging in 130 patients. Radiology 172:799, 1989.
49. Sisk, TD: Knee injuries. *In:* Crenshaw A (ed): Campbell's Operative Orthopaedics. St. Louis, CV Mosby, 1987.
50. Spritzer CE, Dalinka MK, Kressel HY: Magnetic resonance imaging of pigmented villonodular synovitis: A report of two cases. Skeletal Radiol 16:316, 1987.
51. Stafford SA, Rosenthal Dl, Gebhardt MC, et al: MRI in stress fracture. AJR 147:553, 1986.
52. Steinbach LS, Neumann CH, Stoller DW, et al: MRI of the knee in diffuse pigmented villonodular synovitis. Clin Imaging 13:305, 1989.
53. Stoller DW: Magnetic Resonance Imaging in Orthopaedics and Rheumatology. Philadelphia, JB Lippincott, 1989.
54. Tang JS, Gold RH, Bassett LW, et al: Musculoskeletal infection of the extremities: Evaluation with MR imaging. Radiology 166:205, 1988.
55. Vallet AD, Marks PH, Fowler PJ, et al: Occult posttraumatic osteochondral lesions of the knee: Prevalence, classification, and short-term sequelae evaluated with MR imaging. Radiology 178:271, 1991.
56. Weber WN, Neumann CH, Barakos J, et al: Lateral tibial rim (Segond) fractures: MR imaging characteristics. Radiology 180:731, 1991.
57. Weissman BNW, Sledge CB: Orthopaedic Radiology. Philadelphia, WB Saunders, 1986.
58. Yao L, Lee JK: Avulsion of the posteromedial tibial plateau by the semimembranosus tendon: Diagnosis with MR imaging. Radiology 172:513, 1989.
59. Yao L, Lee JK: Occult intraosseous fracture: Detection with MR imaging. Radiology 167:749, 1988.
60. Yulish BS, Lieberman JM, Newman AJ, et al: Juvenile rheumatoid arthritis: Assessment with MR imaging. Radiology 165:149, 1987.
61. Yulish BS, Lieberman JM, Strandjord SE, et al: Hemophilic arthropathy: Assessment with MR imaging. Radiology 164:759, 1987.
62. Yulish BS, Montanez J, Goodfellow DB, et al: Chondromalacia patellae: Assessment with MR imaging. Radiology 164:763, 1987.

13 THE ANKLE AND FOOT

Wing P. Chan, Charles Peterfy, Scott J. Erickson, and Clyde A. Helms

IMAGING TECHNIQUE

Optimal magnetic resonance (MR) images are obtained with the feet or ankles imaged separately using an extremity coil with signal-to-noise characteristics able to support a relatively small (12–16 cm) field of view. The head coil can be used to image both feet simultaneously to allow comparison of one foot with the other but at some cost in terms of spatial resolution. Patients are imaged in the supine position with the feet supported at 90 degrees using foam pads. The feet may be secured with adhesive tape to minimize motion.

At the University of California, San Francisco, routine MR imaging of the ankle is done at a field strength of 1.5 T and includes the following sequences: in the sagittal and coronal planes, (1) T1-weighted spin-echo images with a repetition time (TR) of 600 ms and an echo time (TE) of 20 ms and (2) T2*-weighted multislice gradient-echo images with a TR of 600 ms, a TE of 15 ms, and a flip angle of 30 degrees; in the axial (plantar) plane, (3) proton density–weighted images with a TR of 2000 ms and a TE of 20 ms and (4) T2-weighted spin-echo images with a TR of 2000 ms and a TE of 60 ms (Tables 13–1, 13–2). Other imaging parameters include a 12- to 16-cm field of view and a 256 × 192 acquisition matrix. At such small fields of view, a "no phase wrap" feature must be employed to eliminate aliasing artifact. The slice thickness is 4 mm in the sagittal and coronal planes and 5 mm in the axial planes. We have replaced the conventional T2-weighted axial imaging with a high-resolution, T2-weighted, fast spin-echo sequence (a modification of Rapid Acqui-

Table 13–1. PROTOCOLS OF MAGNETIC RESONANCE IMAGING OF THE ANKLE AT THE UNIVERSITY OF CALIFORNIA, SAN FRANCISCO

Plane	TR (ms)	TE (ms)	FOV (cm)	Slice (mm)	Skip (mm)	NEX	Matrix (pixel)	Flip Angle (degrees)	Time (min)
Sagittal (localizer) (SE)	600	20	20	4	1	1	192		2
Sagittal (GRE)	600	15	20	4	0	2	192	30	4
Axial (SE)	2000	20, 60	20	5	1	2	192		13
Coronal (GRE)	600	20	16	4	1	2	192	30	4
Axial* (FSE)	3500	30, 90	10	4	1	2	256		7
Coronal* (FSE)	3500	30, 90	10	4	1	2	192		6

TR, Repetition time; TE, echo time; FOV, field of view; Skip, interslice gap; NEX, number of excitations; GRE, gradient-recalled echo; SE, spin echo; FSE, fast spin echo; *optional sequence.

Table 13-2. PROTOCOLS OF MAGNETIC RESONANCE IMAGING OF THE FOOT AT THE UNIVERSITY OF CALIFORNIA, SAN FRANCISCO

Plane	TR (ms)	TE (ms)	FOV (cm)	Slice (mm)	Skip (mm)	NEX	Matrix (pixel)	Flip Angle (degrees)	Time (min)
Sagittal (localizer) (SE)	600	20	20	4	1.0	2	192		4
Sagittal (GRE)	600	15	20	4	0	4	192	30	4
Axial (SE)	2000	20, 60	20	5	1.0	2	192		13
Coronal (GRE)	600	20	20	4	0.5	2	192		4
Axial* (FSE)	3500	30, 90	10	4	1.0	2	256		7
Coronal* (FSE)	3500	30, 90	10	4	1.0	2	192		6

TR, Repetition time; TE, echo time; FOV, field of view; Skip, interslice gap; NEX, number of excitations; GRE, gradient-recalled echo; SE, spin echo; FSE, fast spin echo; *optional sequence.

sition Relaxation Enhanced [RARE]) using a TR of 3500 ms, an effective TE of 90 to 100 ms, an echo-train length of 8, and an imaging matrix of 256 × 256 or 192 × 256. We find this sequence provides an optimal visualization of the ankle ligaments with a minimum investment of imaging time. When the fast spin-echo sequence is used, proton density–weighted imaging is replaced by a T1-weighted sequence.

Magnetic resonance imaging of the foot includes the following sequences: in the sagittal and axial planes (relative to the whole body), (1) T1-weighted spin-echo images with a TR of 600 ms and a TE of 20 ms and (2) T2*-weighted spin-echo images with a TR of 600 ms, a TE of 15 ms, and a flip angle of 30 degrees; in the coronal plane (transverse to the foot), (3) proton density–weighted images with a TR of 2000 ms and a TE of 20 ms and (4) T2-weighted spin-echo images with a TR of 2000 ms and a TE of 60 ms. The coronal plane could, alternatively, be imaged using T1-weighted spin-echo and T2-weighted fast spin-echo sequences, as for axial imaging of the ankle.

A short tau inversion-recovery (STIR) sequence is added in cases of trauma, suspected osteomyelitis, or other conditions affecting the marrow. Frequency-selective fat presaturation is added to T1-weighted spin-echo imaging when gadolinium enhancement is used.

THE TENDONS

Tendons are relatively acellular structures composed primarily of densely arranged collagen fibers embedded in an amorphous ground substance. Much of the tensile strength of collagen derives from its highly organized microstructure: microfibrils of tropocollagen are arranged into fibrils, and the fibrils are then organized into fibers. The regularity of this structure also confers an extremely short T2 relaxation time, such that normal tendons are essentially signalless on all pulse sequences. Increased intratendinous signal intensity on T1-weighted images generally indicates hemorrhage or myxoid degeneration. However, under certain conditions the presence of intermediate signal intensity on short TE sequences can be a normal finding (see the discussion of the "magic angle" phenomenon). Increased signal intensity within a tendon on T2-weighted images, however, is always abnormal and indicative of rupture or active inflammation. Morphologically, normal tendons exhibit smooth contours and uniform thickness. Tendon thickening can result from fibrosis due to chronic tendinitis or mass effect from edema and hemorrhage that accompany partial tears and acute inflammation. Partial tears also can result in focal narrowing of a tendon. Complete discontinuity is indicative of a full-thickness tear and usually is accompanied by retraction of the torn ends, with fluid or hemorrhage in the gap.

Achilles Tendon

The Achilles tendon is one of the most common tendons in the body to rupture spontaneously. Clinical evaluation usually is diagnostic; however, in a small proportion of cases, perhaps 25 per cent,[26] physical examination can be misleading. This is particularly true for partial tears or cases in which hematoma between the torn ends of the tendon produces a false impression of continuity on palpation. In these instances, MR imaging can identify the rupture in essentially 100 per cent of cases. Treatment options for rupture of the Achilles tendon include surgical reapposition of the torn ends and casting in plantar flexion. The latter conservative approach is associated with a higher prevalence of recurrent tendon rupture.[23] However, it is likely that these statistics can be improved by the more appropriate patient selection now possible with MR imaging.

The Achilles tendon is formed by a confluence of the tendons of the medial and lateral heads of the gastrocnemius muscle and the tendon of the soleus muscle. The plantaris tendon, when present, runs adjacent to the medial aspect of the deep surface of the Achilles tendon and inserts with it onto the posterior aspect of the calcaneus. On axial MR images, the normal Achilles tendon appears elliptic with a flattened or mildly concave anterior margin, a convex posterior surface, and rounded medial and lateral margins.[26] Mild anterior lobulations of the tendon are seen occasionally; however, this appearance is usually bilaterally symmetric in normal subjects. On midline sagittal images, the Achilles tendon is seen as a signalless cord extending from the gastrocnemius-soleus muscle complex to the posterior calcaneus. Immediately anterior to the Achilles tendon is the large pre-Achilles fat pad. Excessive plantar flexion of the foot during MR imaging studies should be avoided to prevent buckling and pseudothickening of the tendon.

The Achilles tendon assists in plantar flexion of the foot. Injuries of the tendon occur when the foot is suddenly dorsiflexed against the contracting triceps surae group. This may occur at any age during unaccustomed strenuous effort, such as lifting an unexpectedly heavy object, playing a difficult shot at tennis, or suddenly stretching the tendon during a slip on the floor or stairs. Trained athletes are vulnerable to this injury once muscular exhaustion develops. Nonathletic people who rupture the Achilles tendon usually have a predisposing condition, such as gout, rheumatoid arthritis, systemic lupus erythematosus, hyperparathyroidism, chronic renal failure, diabetes mellitus, or long-term use of steroids. Tendon rupture usually occurs at a site of relatively decreased vascularity approximately 2 to 3 cm above the insertion onto the calcaneus.[26] This is also a region where the most superficial fibers of the Achilles tendon run relatively horizontally.

Axial T2-weighted images provide the best views for depicting abnormalities of the Achilles tendon. Chronic tendinitis produces tendinous thickening without discontinuity (Fig. 13–1). Acute tendinitis and partial tears exhibit focal areas of intermediate signal intensity within the tendon on T1-weighted images and of high signal intensity on T2-weighted images owing to hemorrhage and edema. Differentiation of tendinitis from a partial tear on the basis of imaging criteria alone can be extremely difficult; indeed, the two conditions often coexist. Complete ruptures of the Achilles tendon demonstrate tendon discontinuity with intervening fluid or hematoma (Fig. 13–2). The proximal fragment often is retracted, and the distal fragment is lax and buckled. Sagittal images are useful in estimating the longi-

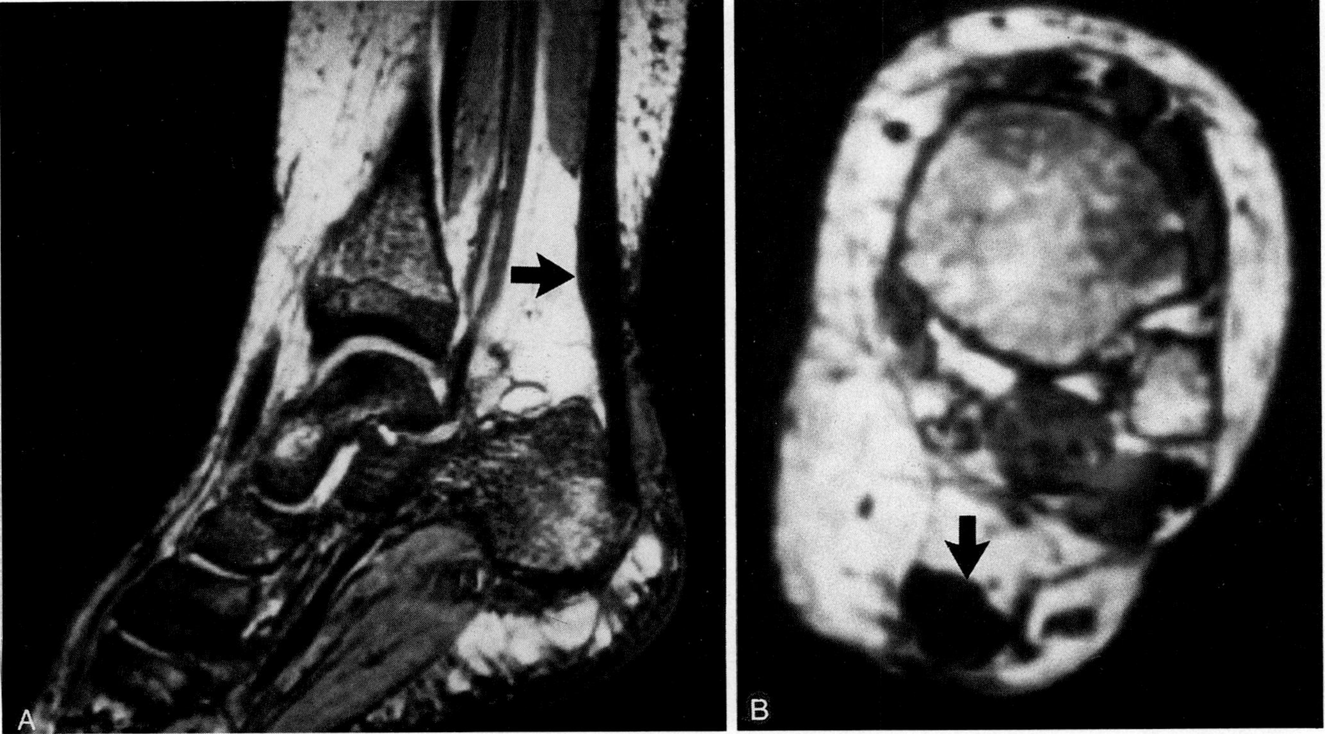

Figure 13–1. Chronic Achilles tendinitis in a 60-year-old woman. Three-dimensional gradient-recalled acquisition of steady state (GRASS) images (*A*: TR, 30 ms; TE, 35 ms; flip angle, 15 degrees; *B*: axial, TR, 90 ms; TE, 33 ms; flip angle, 11 degrees) show fusiform thickening of the Achilles tendon of the left leg (*arrow*). No evidence of increased signal intensity within the tendon is noted.

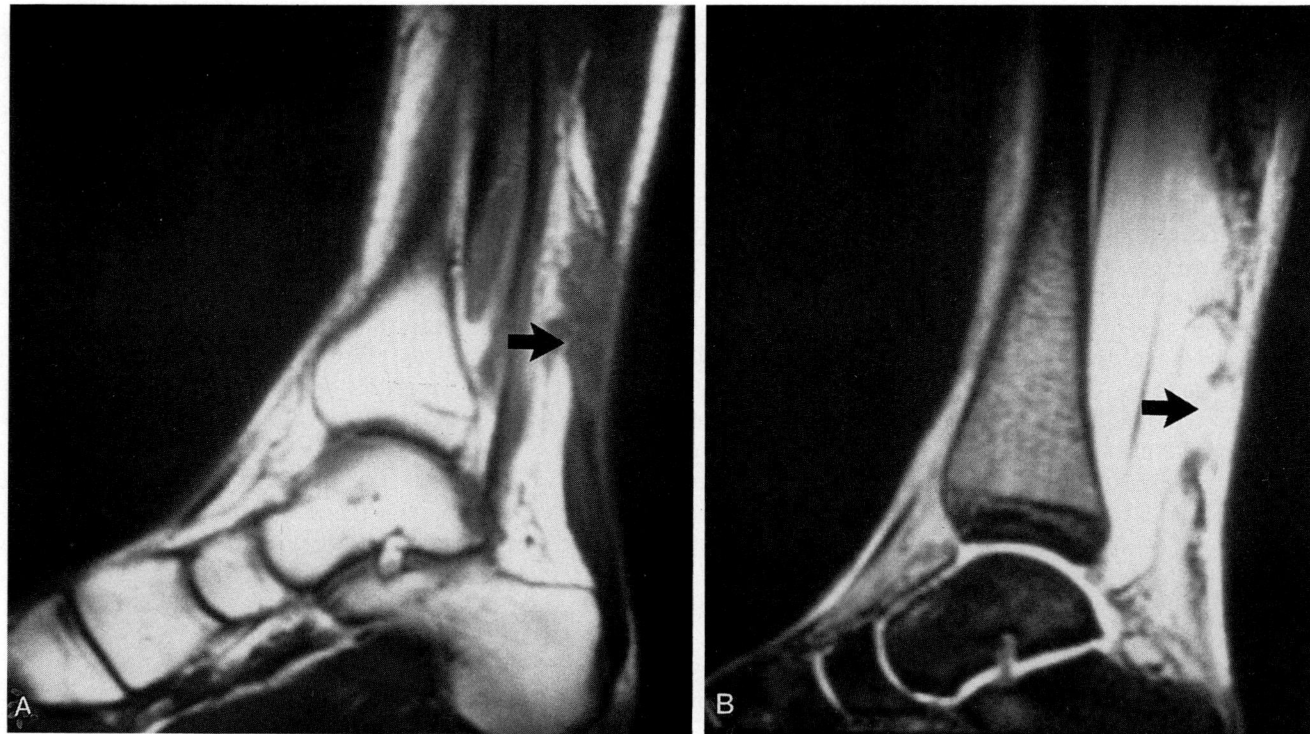

Figure 13–2. Complete rupture of the Achilles tendon in a 23-year-old man. This patient had injured his left leg during a forcible contracture. Sagittal T1-weighted (TR, 500 ms; TE, 20 ms) *(A)* and T2*-weighted *(B)* (TR, 600 ms; TE, 30 ms; flip angle, 30 degrees) images show complete rupture of the Achilles tendon with intervening hematoma *(arrow)*. The site of rupture is 6 cm proximal to the insertion of the Achilles tendon to the calcaneus. There is a 2-cm gap between the proximal and distal torn tendon fibers. This patient subsequently was treated by open Achilles tendon repair.

tudinal distance between the torn ends and for characterizing the edges as smooth or frayed. Because the Achilles tendon has no synovial sheath, fluid collects in a fine, loose fibrous tissue called the *paratenon* just anterior to the tendon (Fig. 13–3). After surgical repair, an increase in signal intensity within or across the tendon indicates rerupture of the tendon (Fig. 13–4). However, a tendon repaired with a polymer of lactic acid implants shows diffuse thickening with streaks of signal within the tendon, which may mimic a tendon rerupture.[20]

Lateral Compartment Tendons

Peroneal Tendons. The lateral compartment of the lower leg contains two muscles: the peroneus longus and the peroneus brevis. Their tendons are contained within a common synovial sheath above the ankle, but then they separate to run deep to the superior and inferior peroneal retinacula and curve posterior to the lateral malleolus (Fig. 13–5). At this level, the tendon of the peroneus longus is posterolateral to that of the peroneus brevis.[18] The two tendons can usually be visualized as separate structures owing to the presence of fat of high signal intensity that surrounds them. The calcaneofibular ligament can be seen deep to the peroneal tendons at the level between the superior and inferior retinacula.[11] The peroneus longus tendon runs inferior to the peroneal tubercle of the calcaneus, present in 45 per cent of normal subjects,[28] and courses along the plantar aspect of the foot to insert on the base of the first metatarsal and medial cuneiform. After curving beneath the lateral malleolus, the peroneus brevis runs a relatively short course to its insertion on the base of the fifth metatarsal.

A groove in the tip of the lateral malleolus accommodates the tendon of the peroneus brevis in 80 per cent of normal persons.[10] Absence or shallowness of this groove predisposes to subluxation or dislocation of the tendon. Vigorous dorsiflexion of the foot with subsequent reflex peroneal contraction also can result in peroneal dislocation by stripping off the overlying superior peroneal retinaculum, allowing the tendons to slide in front of the lateral malleolus. This injury most commonly occurs in skiing, skating, or soccer. Inflammatory arthritides and recurrent mechanical stress are less common causes of dislocation of the peroneal tendons. Clini-

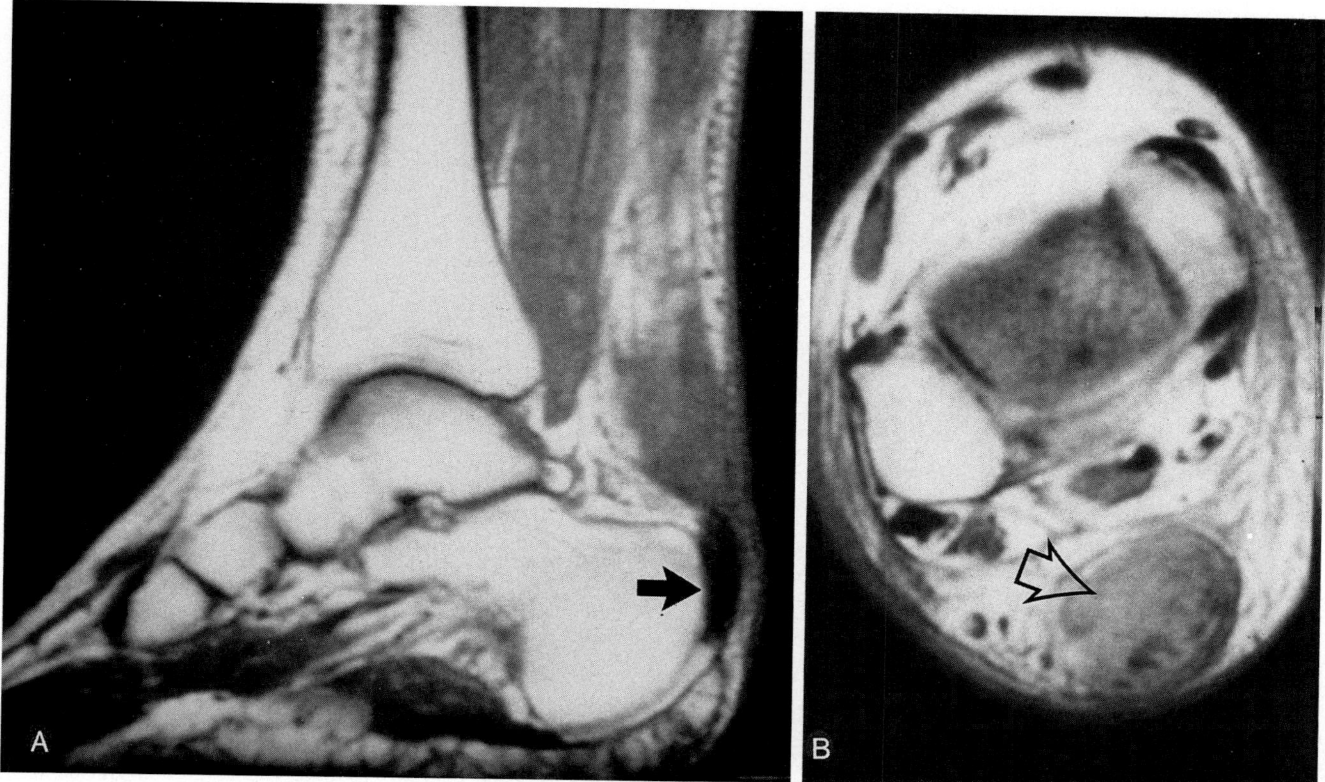

Figure 13–3. Subacute, complete Achilles tendon rupture. *A*, Sagittal T1-weighted image (TR 600 ms; TE, 20 ms) shows a small, residual, torn end of the tendon at the calcaneal insertion *(arrow)*. *B*, Axial T2-weighted image (TR, 2000 ms; TE, 80 ms) shows a hematoma of high signal intensity *(open arrow)* displacing the residual tendon fibers posteriorly.

cally, peroneal dislocation is accompanied by swelling, ecchymosis, and tenderness along the lateral malleolus. Acute peroneal tendon dislocation may require surgical deepening of a shallow or absent fibular groove.

Partial or complete rupture of one or both of the peroneal tendons is rare. Rupture should be considered when lateral ankle pain persists despite an adequate trial of conservative management. Surgical treatment is recommended for chronic lateral ankle instability or tenosynovitis of the peroneal tendon sheath.[15] Complete rupture of peroneal tendons is best seen on T2-weighted axial images, in which the tendon, of low signal intensity, is no longer visualized; in its place, only fluid, of high signal intensity, is seen. The tendon may appear thickened above and below the level of rupture (Fig. 13–6). Partial rupture of the peroneal tendons is similar in appearance to that of other tendons: focal thinning or thickening and increased intratendinous signal locally (Fig. 13–7). Tendinitis tends to produce a more diffusely thickened tendon and is usually associated with fluid within the tendon sheath. In considering tendinitis and rupture of the peroneal tendons, it must be remembered that these tendons under nor-

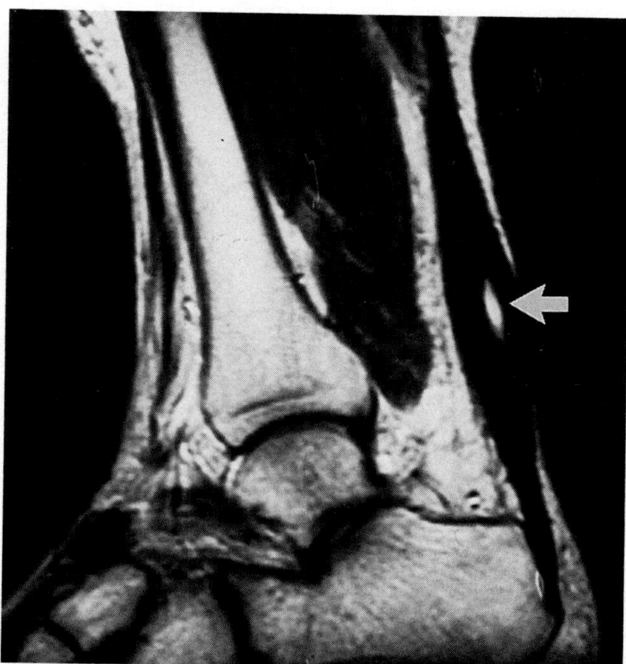

Figure 13–4. Partial rerupture of the Achilles tendon 8 months after surgical repair. Sagittal intermediate T2-weighted images (TR, 1500 ms; TE, 80 ms) show an area of increased signal intensity within the fusiform enlarged Achilles tendon *(arrow)*. Some continuity of fibrous elements is noted.

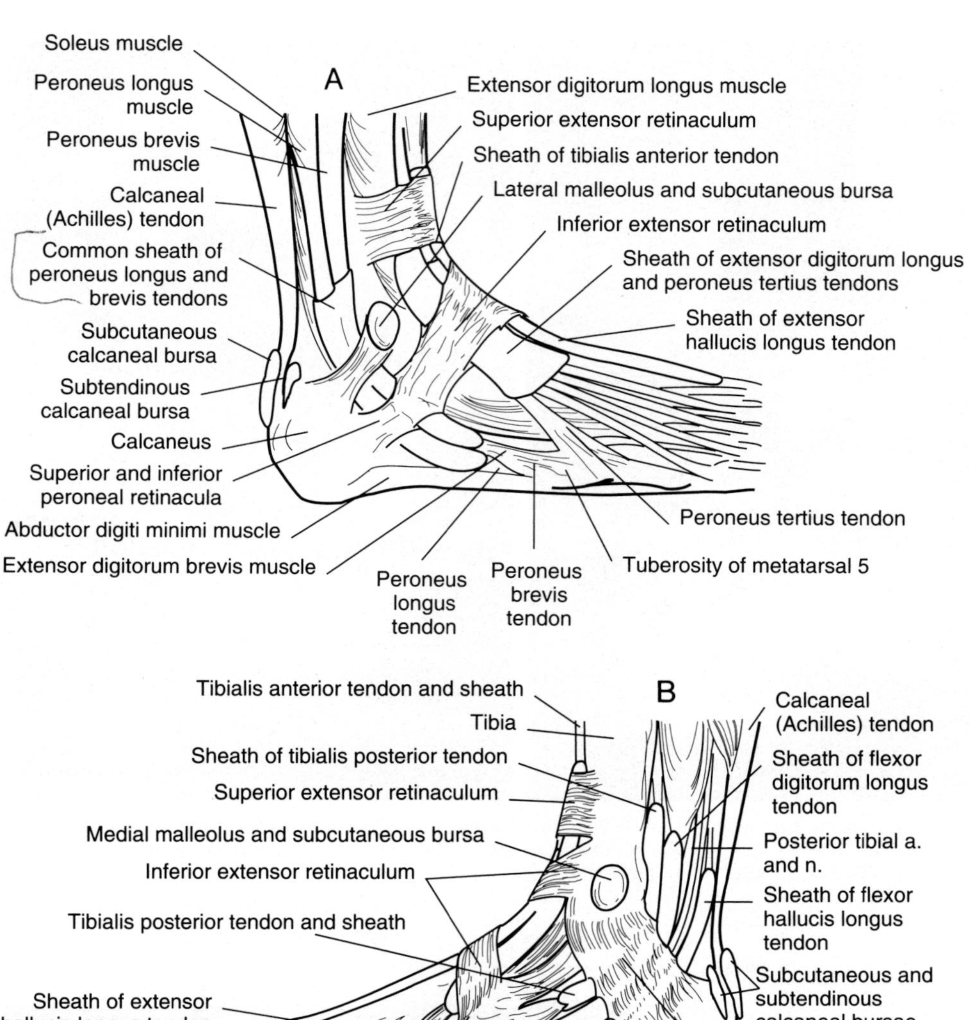

Figure 13–5. Medial *(A)* and lateral *(B)* views of the ankle and foot.

mal circumstances commonly exhibit intermediate intrasubstance signal intensity on short TE sequences (T1-weighted spin-echo, proton density–weighted spin-echo, and short TE gradient-echo images) as they pass through the "magic angle" of 55 degrees (to the static magnetic field) while curving beneath the lateral malleolus[11] (see the discussion of the "magic angle phenomenon").

POSTEROMEDIAL COMPARTMENT TENDONS

In the posteromedial compartment deep to the soleus and gastrocnemius muscles lies a group of three tendons. From medial to lateral, these are the tibialis posterior, the flexor digitorum longus, and the flexor hallucis longus tendons (Fig. 13–8). These can be remembered by the mnemonic "Tom (**t**ibialis posterior), Dick (flexor **d**igitorum longus), and Harry (flexor **h**allucis longus)." Of the three, the tibialis posterior tendon is the most susceptible to injury.

Tibialis Posterior. The tibialis posterior tendon descends posteromedial to the distal third of the tibia, passes just posterior to the medial malleolus, and forms a broad insertion on the plantar surfaces of the navicular; the cuneiforms; and the bases of the second, third, and fourth metatarsals. The muscle-tendon unit serves as a plantar flexor and invertor of the foot, maintains the medial longitudinal arch, and

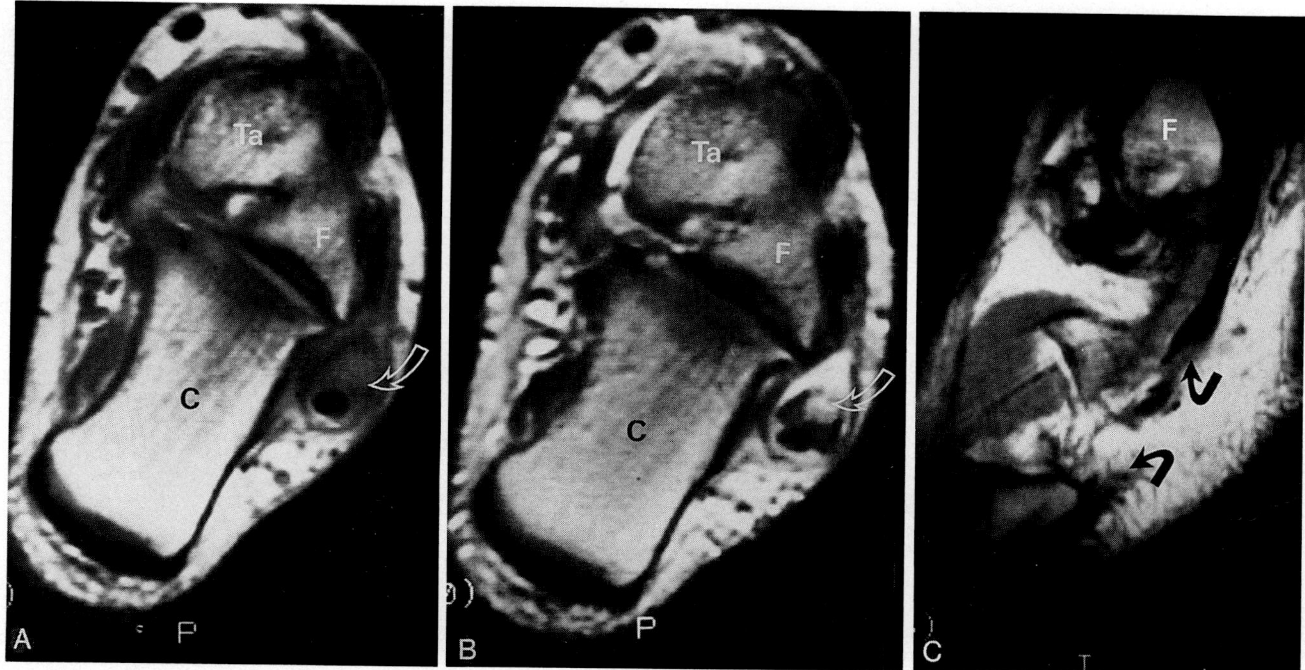

Figure 13-6. Peroneal tendon rupture in a 35-year-old man. Axial T1-weighted (TR, 700 ms; TE, 20 ms) *(A)* and T2-weighted *(B)* (TR, 2000 ms; TE, 80 ms) images show fluid within the peroneal brevis tendon *(open arrow)* and its tendon sheath. Sagittal T1-weighted image (TR, 600 ms; TE, 20 ms) *(C)* shows discontinuity of the peroneal tendon with a large separation between the residual tendon fibers *(solid arrows)*.

is one of the main stabilizers of the hindfoot, preventing valgus (eversion) deformity.

Chronic rupture of the tendon produces painful flatfoot and planovalgus deformities. It most commonly affects women older than 50 years of age.[17] Clinically, the inability to rise up on the toes or to life the heel from the ground when standing only on the affected foot (positive "single-limb heel-rise

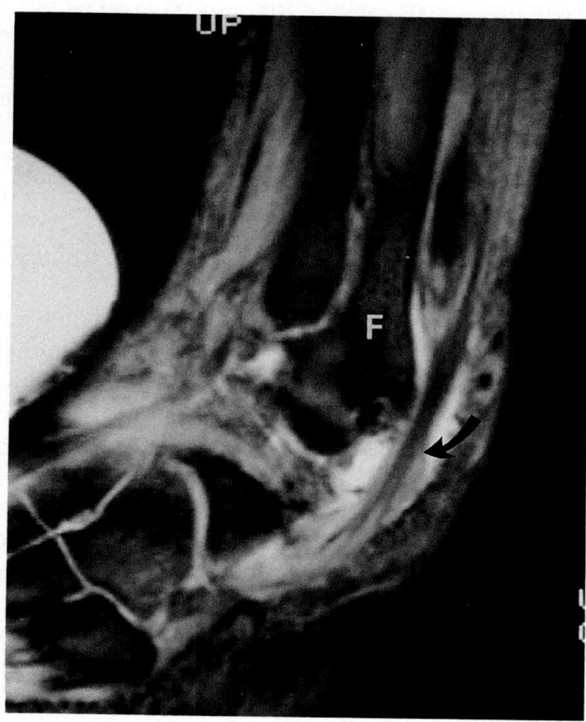

Figure 13-7. Peroneal tendon injury in a 45-year-old woman. Sagittal T2*-weighted image (TR, 520 ms; TE, 15 ms; flip angle, 30 degrees) shows fluid extending along the peroneal tendon. Attenuated tendon fibers are noted *(arrow)*. This may represent a partial rather than a complete rupture of the tendon. F, Fibula.

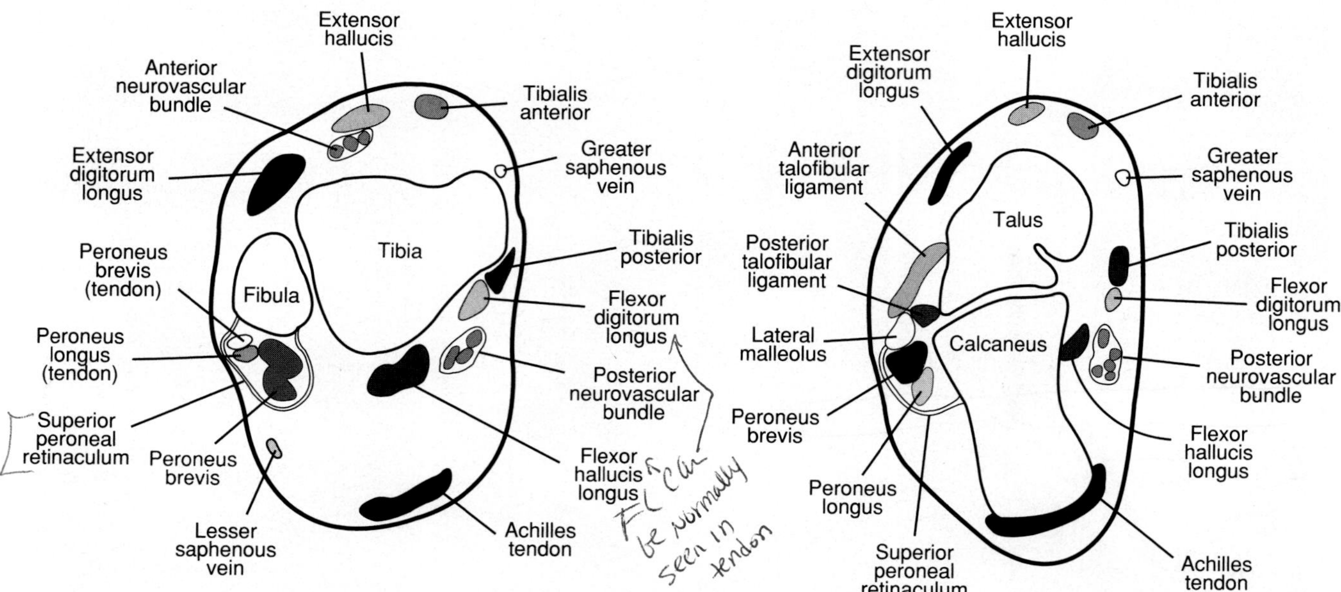

Figure 13–8. Axial planes showing tendons in the ankle.

test") suggests tibialis posterior tendon dysfunction.[15] Tenderness along the course of the tendon supports the diagnosis.

The cause of spontaneous rupture is probably degenerative in nature. Chronic stress on the tendon from supporting the entire weight of the medial longitudinal arch can lead to tenosynovitis and progressive disruption of the tendon fibers. Ruptures tend to occur behind the medial malleolus where the tendon takes a sharp turn and is tightly secured by the flexor retinaculum. This also is a watershed area of decreased vascularity, where the tendon is susceptible to injury and slow to heal. Other factors predisposing to tendon rupture include rheumatoid arthritis, seronegative arthritis, and preexisting flatfoot deformity.

Rupture of the tibialis posterior tendon demands surgery. The surgical approach, however, depends on the site and extent of rupture. Thus, MR imaging can play an important role in guiding therapy.[1,15,27,29]

Except for the intratendinous signal sometimes seen on short TE sequences when the normal tendon curves beneath the medial malleolus[11] (see the discussion of the "magic angle phenomenon"), the normal tibialis posterior tendon is essentially signalless on all pulse sequences. At the level of the medial malleolus, the tendon is approximately two times thicker than the flexor digitorum longus or flexor hallucis longus tendons. However, the thickest segment of the tibialis posterior tendon is found just proximal to its insertion on the navicular bone. Rosenberg and co-workers[27,29] classified rupture of the tibialis posterior tendon into three stages, which are easily differentiated by MR imaging. Type I is partial rupture with tendon hypertrophy and longitudinal splits producing a thickened tendon with foci of high signal intensity on T2-weighted images. Type II tendon rupture also is partial, but with local attenuation rather than enlargement of the tendon. This type of rupture should be suspected when the thickness of the tibialis posterior tendon approaches that of the flexor digitorum longus. Type III rupture is a full-thickness tear, in which frank tendinous discontinuity is seen with fluid and hematoma between the retracted fragments. Fluid is commonly present in the synovial sheath of the injured tendon. However, a small amount of fluid can be normal and should not be mistaken for a pathologic feature.

At least one plane utilizing a T2-weighted sequence is necessary in evaluating tendon pathology, and intratendinous and peritendinous signal intensity changes are best depicted on axial images (Figs. 13–9, 13–10). Sagittal images can show the site of a complete rupture but are most valuable in assessing the degree of retraction of the torn fragments (Fig. 13–11). It should be noted that older tears may not show the characteristic signal intensity changes in the ruptured fibers owing to the formation of granulation tissue within approximately 2 weeks of the initial injury.[6]

Flexor Digitorum Longus Tendon. The flexor digitorum longus tendon lies adjacent to the tibialis posterior tendon, passes medial to the sustentaculum tali in its own canal, and crosses superficially to the flexor hallucis longus tendon beneath the talus to divide into four slips.[15] Each slip passes through a bifurcation of the flexor digitorum brevis

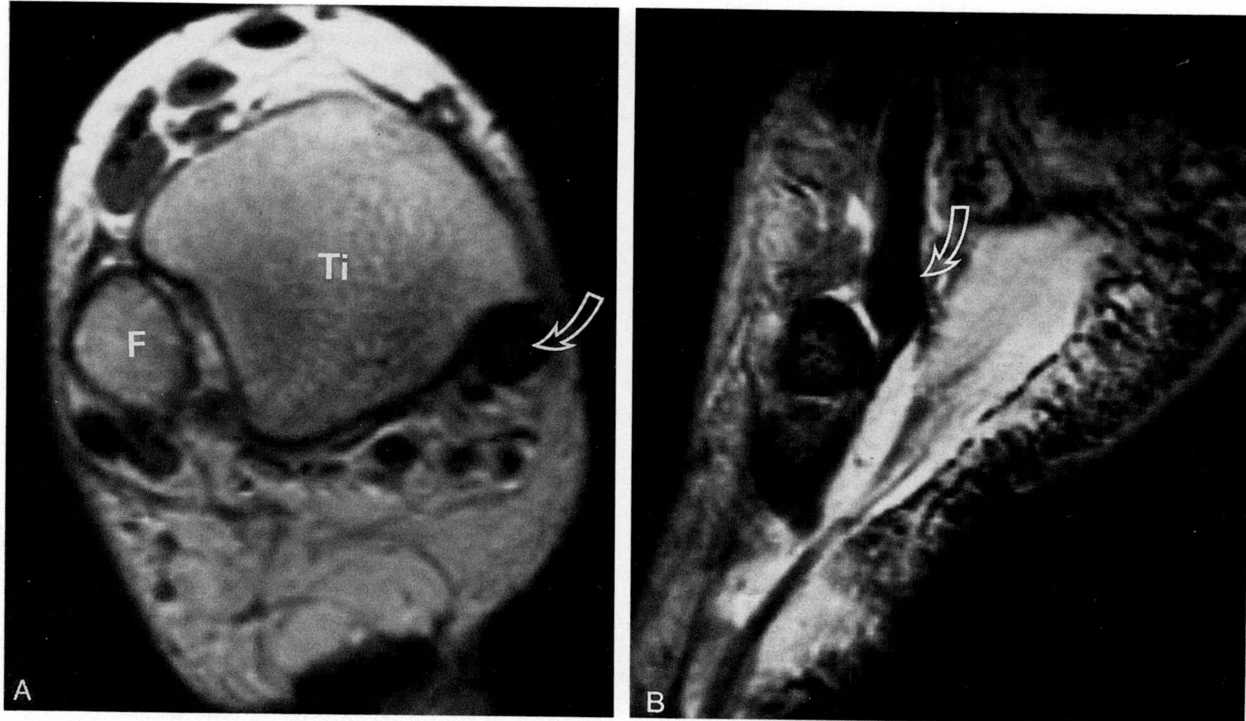

Figure 13–9. Chronic tibialis posterior tendon injury in a 40-year-old woman. This patient had injured her right ankle 18 years earlier. Her medial ankle pain had worsened 2 years ago. MR images (*A*: axial, TR, 2400 ms; TE, 30 ms; *B*: TR, 600 ms; TE, 20 ms; flip angle, 30 degrees) show abnormally enlarged tibialis posterior tendon (*arrow*). Increased signal intensity within the tendon is seen on the proton density–weighted image (*A*). Ti, tibia; F, fibula.

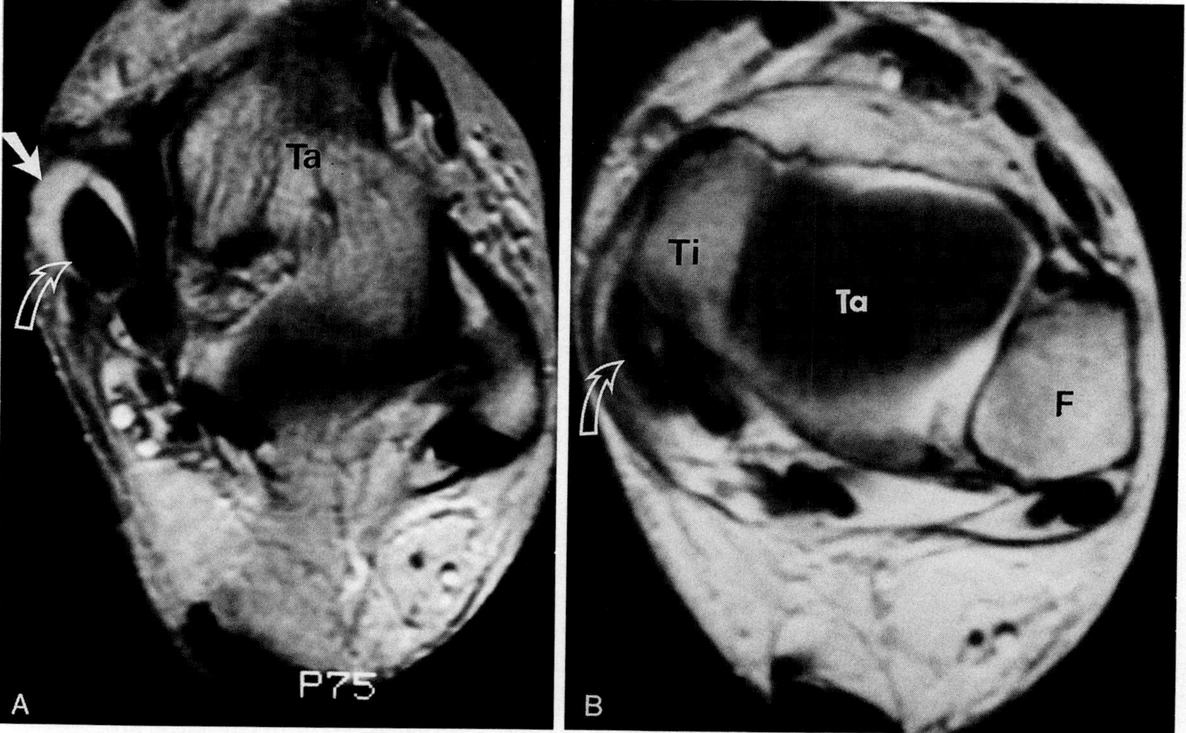

Figure 13–10. Acute tenosynovitis of the tibialis posterior tendon in a 58-year-old woman who suffered from severe ankle pain. *A* and *B*, Axial T2-weighted images (TR, 2000 ms; TE, 80 ms) show an abnormally enlarged posterior tibial tendon (*curved arrow*), which is surrounded by fluid (*straight arrow*). *A*, The tendon is abnormally decreased in size (*curved arrow*), as depicted on the axial image more proximally (*B*). All of these findings suggest severe tendinitis rather than a partial tear of the tendon. Ta, Talus; Ti, tibia; F, fibula.

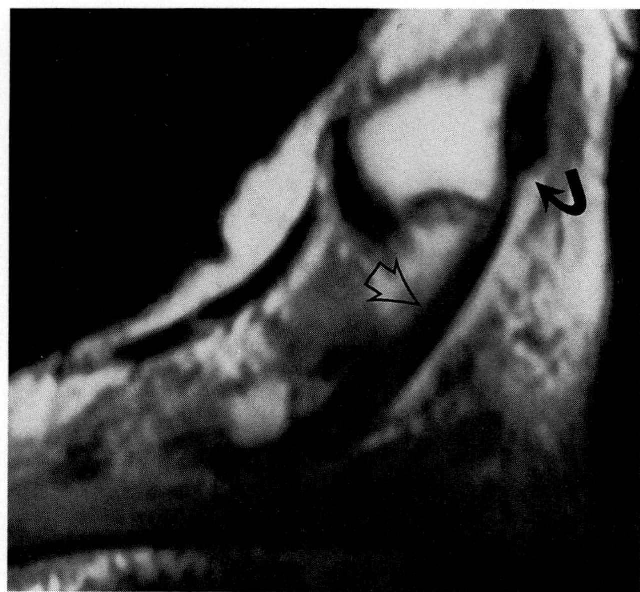

Figure 13–11. Longitudinal rupture of the tibialis posterior tendon in a 50-year-old woman. Sagittal T2-weighted image (TR, 2000 ms; TE, 80 ms) shows proximal retraction of the posterior portion of the posterior tibial tendon *(curved arrow)*. The anterior portion of the tendon tibialis posterior is spared *(arrowhead)*.

tendon at the proximal phalanx of each toe to insert on the plantar aspect of the distal phalanx. The function of this tendon is to flex the ankle and foot in a plantar orientation and actively elevate the arch. Rupture of the tendon results in a characteristic clawfoot deformity. Injuries of this tendon are less common. With MR imaging, axial T2-weighted sequences are best to depict tenosynovitis of this tendon (Fig. 13–12).

Flexor Hallucis Longus Tendon. The flexor hallucis longus tendon lies adjacent to the flexor digitorum longus tendon, passes in its own tunnel beneath the sustentaculum tali, and crosses deep to the flexor digitorum longus tendon to insert on the plantar aspect of the distal phalanx of the great toe. Its functions include flexion of the ankle and great toe, elevation of the arch, and maintenance of foot stability. Rupture of the flexor hallucis longus tendon classically occurs in ballet dancers, and, accordingly, this tendon has been called the Achilles tendon of dance.[15] Injuries to the flexor hallucis longus tendon are rare.

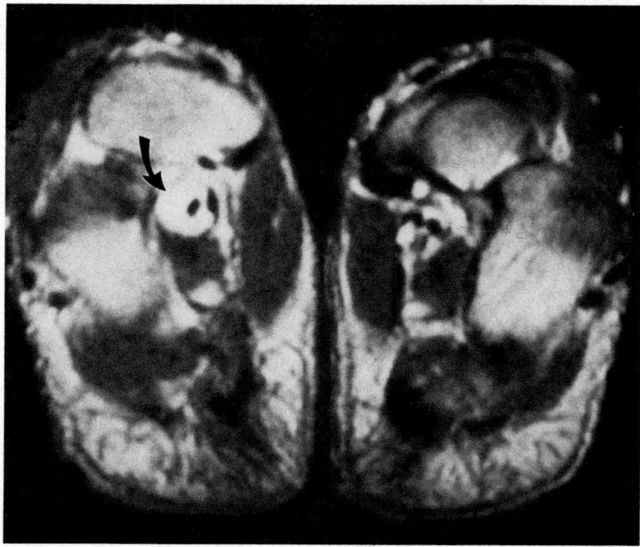

Figure 13–12. Tenosynovitis of the flexor digitorum longus of the right leg. Axial T2-weighted image shows fluid within the tendon sheath *(arrow)*.

Anterior Compartment Tendons

The anterior tendon group consists of the tibialis anterior, extensor hallucis longus, and extensor digitorum longus tendons (from medial to lateral: Tom, Harry, and Dick according to the mnemonic used for the posteromedial group; see Fig. 13–8). Injuries to this group of tendons are uncommon.

Tibialis Anterior Tendon. The tibialis anterior tendon is the most medial of this group and has the largest diameter. It passes deep to the extensor retinaculum along the anteromedial aspect of the foot and inserts on the medial cuneiform and first metatarsal. The muscle-tendon unit serves to invert the foot, supplies two thirds of dorsiflexion power of the foot and ankle, and is an important arch-supporting unit.

Tendinitis of the tibialis anterior tendon occasionally develops when dancers return to vigorous dance after a period of inactivity during which daily warm-up exercises were not performed. As with inflammation of other tendons, tendinitis of the tibialis anterior tendon is generally treated conservatively. Spontaneous rupture of the tendon is most common in the sixth and seventh decades and tends to occur near its insertion on the first tarso-metatarsal joint.[15] Predisposing factors include degenerative arthritis and previous fractures. Rupture may be related to a dorsal exostosis at the first tarso-metatarsal joint or to attrition from rubbing against the edge of the inferior margin of the extensor retinaculum. Rupture is usually complete. The retracted fragments distal to and under the retinaculum are usually easy to palpate. In most patients, the disability is mild with only slight weakness or dorsiflexion and mild footdrop. Nevertheless, surgical repair is occasionally necessary. The MR appearances of rupture and tendinitis of the tibialis anterior tendon are similar to those described for other tendons (Fig. 13–13). Again, T2-weighted axial images are best for visualizing these abnormalities.

Extensor Hallucis Longus Tendon. The extensor hallucis longus tendon runs lateral to the tendon

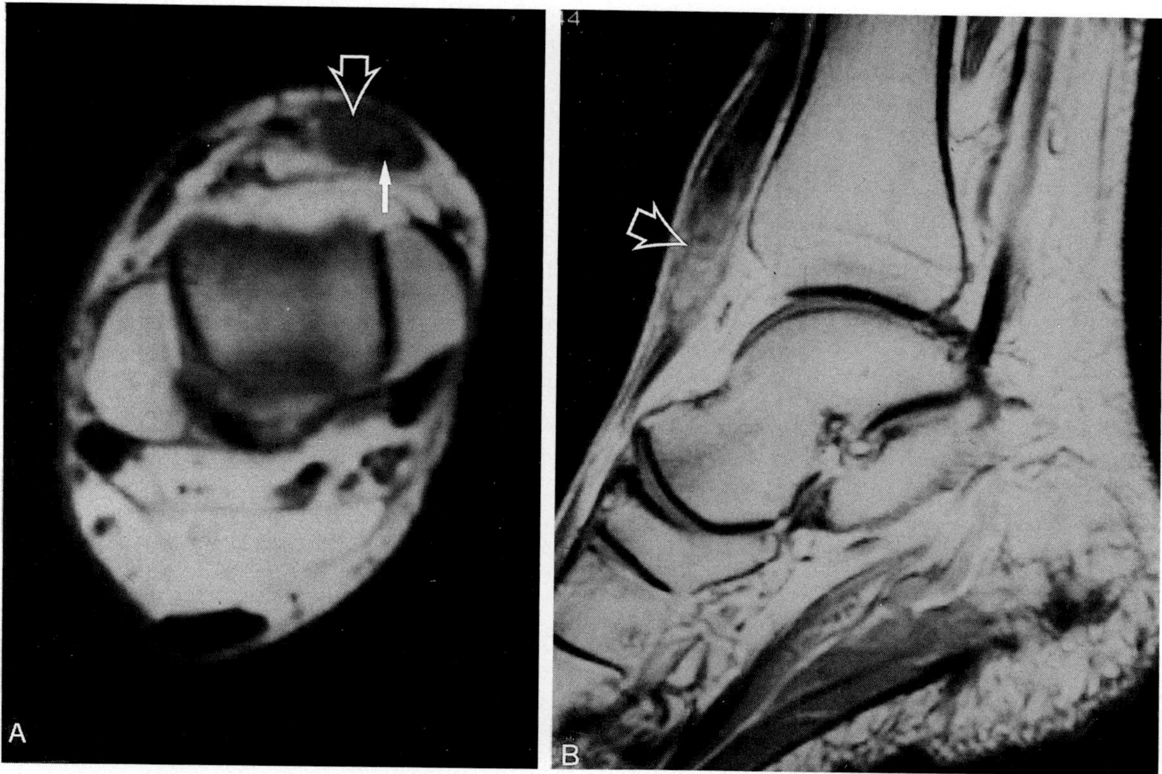

Figure 13–13. Tibialis anterior tendon rupture in a 78-year-old man. MR images (*A*: axial, TR, 800 ms; TE, 20 ms; *B*: sagittal, TR 2200 ms; TE, 80 ms) show a disrupted tibialis anterior tendon with intervening fluid or subacute hemorrhage *(open arrow)* displacing the residual tendon fibers posteriorly *(small arrow)*.

of the tibialis anterior and inserts onto the dorsal aspect of the distal phalanx of the great toe. Its action assists dorsiflexion of the foot. Chronic injury may occur near its insertion or in the forefoot where the tendon is relatively fixed. Acute injury is usually the result of penetrating trauma, such as dropping a sharp knife on the foot while in the kitchen.

Extensor Digitorum Longus and Peroneus Tertius Tendons. The extensor digitorum longus and peroneus tertius tendons run mostly anterolaterally and are difficult to separate. The extensor digitorum longus tendon divides into four slips, each of which inserts onto the dorsal aspect of the middle and distal phalanges of one of the lateral four toes. Fatigue of the extensor digitorum longus muscle with chronic hyperextension of the lateral toes results in a flexion deformity of the distal phalanges due to the unapposed forces of flexor muscles. The peroneus tertius tendon is extremely variable in size and occasionally is absent.[15] When present, it attaches onto the dorsal aspect of the fifth metatarsal.

THE LIGAMENTS

The ankle is stabilized by three sets of ligaments: the complex medial collateral (deltoid) ligament, the distal tibiofibular syndesmosis, and the lateral collateral ligament. Of the three, the lateral collateral ligament is most often injured in ankle sprains. Assessment of the extent of injury has classically relied on clinical evaluation; plain film radiographs (including stress views); and, in some situations, ankle anthrography or peroneal tenography. Normal ligaments exhibit a homogeneous signal void on all MR sequences owing to their abundance of collagen and relatively poor hydration. Discontinuity of a ligament is direct evidence of a complete tear. Associated joint effusion, edema in the adjacent soft tissues, and hemorrhage at the site of ligament interruption are well demonstrated on T2-weighted images. Chronically torn ligaments may appear thickened or attenuated.

The Lateral Collateral Ligament

The lateral collateral ligament of the ankle consists of three components: the anterior talofibular ligament, the calcaneofibular ligament, and the posterior talofibular ligament. Inversion of the supinated foot tightens all three collateral ligaments. Severe inversion stress leads to ligamentous rupture or avulsion of the bony attachments of the lateral collateral ligaments on the fibula. The anterior talofibular is the most common ligament to rupture, followed by the posterior talofibular and calcaneofibular ligaments.[13,15] Although rupture of the anterior talofibular ligament may be an isolated injury (except in the setting of penetrating trauma), calcaneofibular and posterior talofibular ligament ruptures are not found in the presence of an intact anterior talofibular ligament.[13,15] Thus, after inversion injury to the ankle, visualization of an intact anterior talofibular ligament virtually excludes rupture of any of the lateral collateral ligaments.

Anterior Talofibular Ligament. The anterior talofibular ligament extends from the anterior aspect of the lateral malleolus to the lateral neck of the talus (Fig. 13–14). It serves to prevent anterior shift of the talus during dorsiflexion. Rupture of this ligament allows anterior subluxation of the talus and results in lateral rotational instability. The intact anterior talofibular ligament can be seen on axial images virtually 100 per cent of the time, particularly if T2 weighting is used (Fig. 13–15). Accordingly, failure to visualize this ligament is reliable evidence of its rupture (Figs. 13–16, 13–17).

Calcaneofibular Ligament. The calcaneofibular ligament is situated deep to the peroneal tendons and extends obliquely from the anterior tip of the lateral malleolus downward and posteriorly to the lateral surface of the calcaneus (see Fig. 13–14). This ligament is best depicted in the oblique axial plane (Fig. 13–18); however, at least part of this ligament can usually be seen on routine coronal or axial images. The calcaneofibular ligament is extremely resistant to injury (Fig. 13–19), and rupture is indicative of severe trauma to the ankle.

Posterior Talofibular Ligament. The posterior talofibular ligament courses almost horizontally from the posteromedial margin of the lateral malleolus to the lateral aspect of the posterior talar process (see Fig. 13–14). Although its specific function is to prevent posterior shift of the talus during plantar flexion, rupture of the posterior talofibular ligament is invariably accompanied by additional ligamentous ruptures and complete ankle instability or dislocation. The posterior talofibular ligament is best seen in the axial plane at the level at which the relatively round distal fibula becomes concave medially to acquire a comma shape (Fig. 13–20). An intact posterior talofibular ligament can be seen in most cases as a mildly heterogeneous collection of fibers of low signal intensity fanning posteromedially from this concavity in the lateral malleolus. This bony landmark helps distinguish the posterior talofibular ligament from the posterior distal tibiofibular ligament, which is seen higher up, where the fibula is still convex on all borders (Fig. 13–21). The posterior distal tibiofibular ligament, in contrast to the posterior talofibular ligament, is a short, discrete ligament, which in combination with the anterior distal tibiofibular ligament (visible at the same level) and the interosseous membrane between the tibia and the fibula, forms the distal tibiofibular syndesmotic complex (Figs. 13–21, 13–22). Disruption of this syndesmosis (Fig. 13–23) is seen in severe ankle injuries and invariably is associated with collateral ligament rupture or fracture of the malleoli.

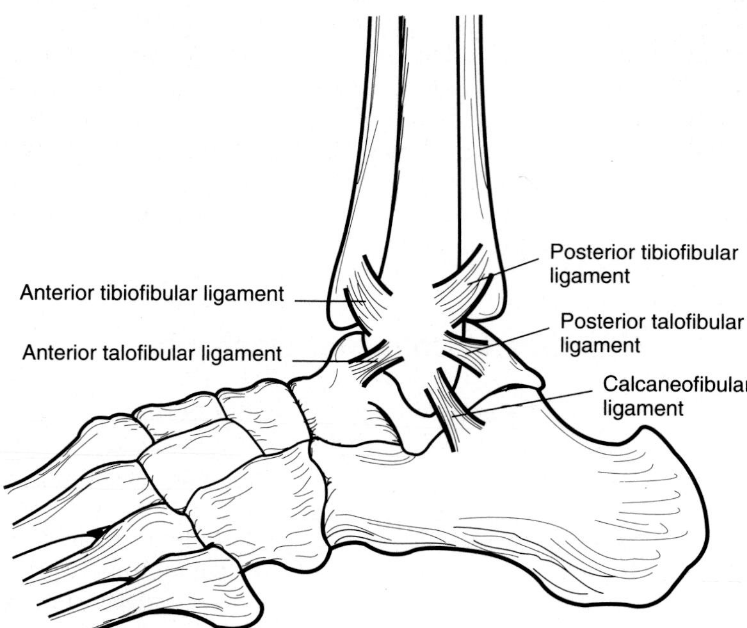

Figure 13–14. Lateral aspect of the ankle showing components of the lateral collateral ligament and the tibiofibular ligament.

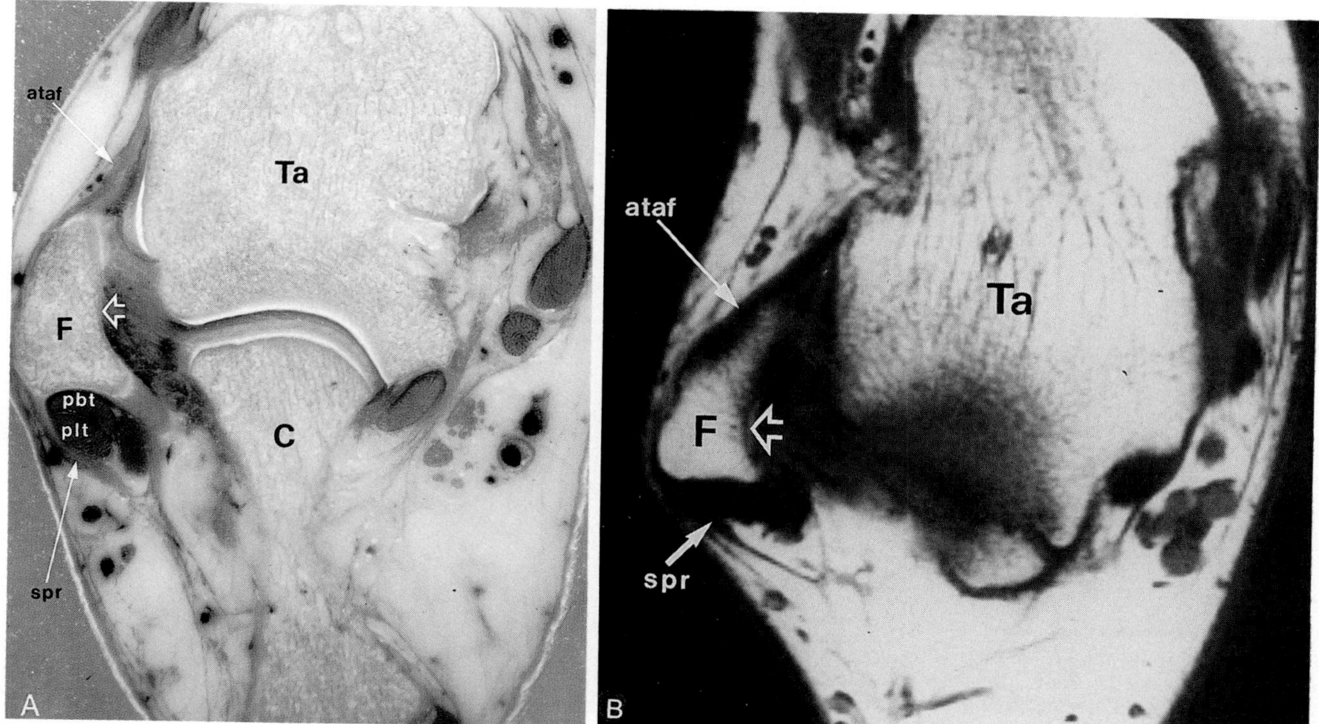

Figure 13–15. A normal anterior talofibular ligament on axial plane. *A*, Cryomicrotome section; *B*, T2-weighted image (TR, 2500 ms; TE, 80 ms). The *open arrowhead* points to the malleolar fossa. Ta, Talus; C, calcaneus; spr, superior peroneal retinaculum; plt, peroneus longus tendon; pbt, peroneus brevis tendon; F, fibula; ataf, anterior talofibular ligament. (From Erickson SJ, Smith JW, Ruiz ME, et al: MR imaging of the lateral collateral ligament of the ankle. AJR 156:131, 1991. © American Roentgen Ray Society.)

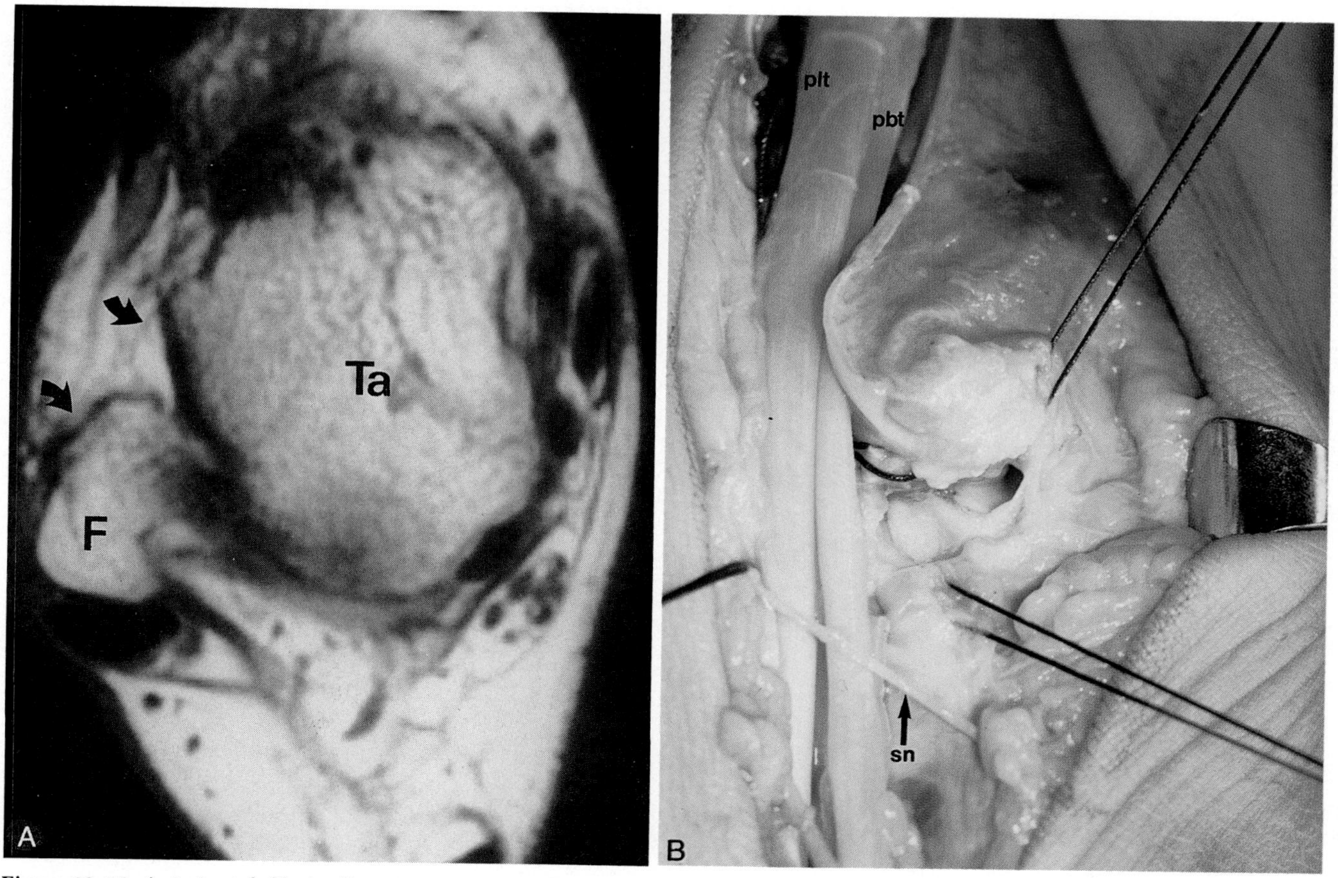

Figure 13–16. Anterior talofibular ligament rupture. *A*, Axial intermediate-weighted image (TR, 2500 ms; TE, 20 ms) shows a large gap between the torn ends of the ligament *(arrows)*. *B*, Intraoperative image shows the torn ends of the ligament (marked by sutures). Ta, Talus; F, fibula; plt, peroneus longus tendon; pbt, peroneus brevis tendon; sn, sural nerve. (From Erickson SJ, Smith JW, Ruiz ME et al: MR imaging of the lateral collateral ligament of the ankle. AJR 156:131, 1991. © American Roentgen Ray Society.)

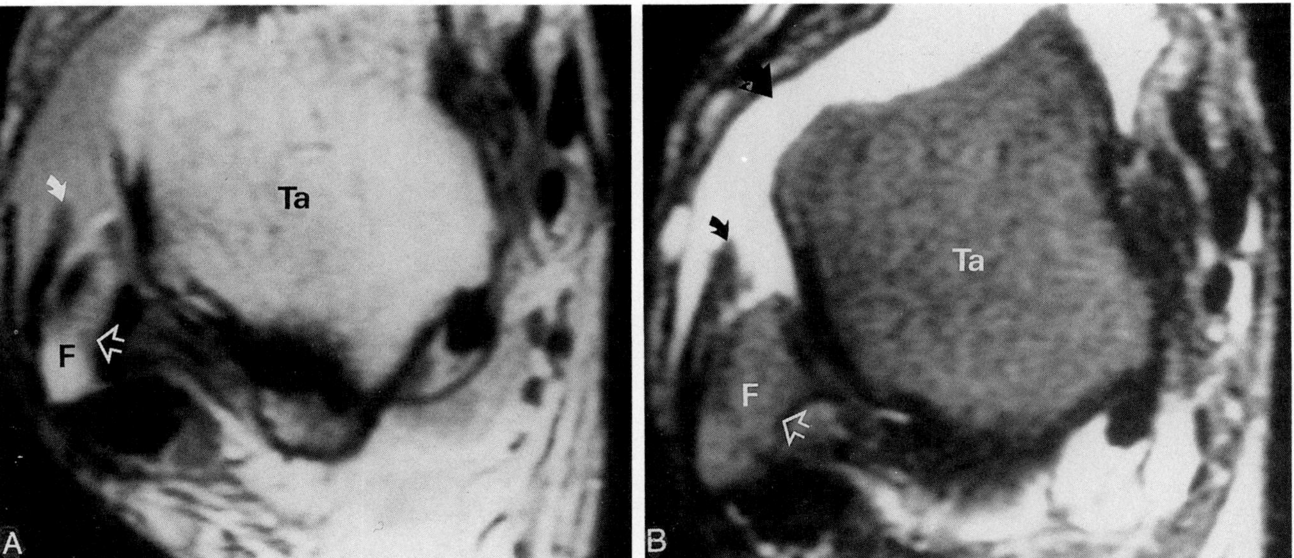

Figure 13–17. Acute anterior talofibular ligament rupture. Axial MR images (*A*: TR, 2500 ms; TE, 20 ms; *B*: TR, 2500 ms; TE, 80 ms) show discontinuity of the ligament *(curved arrow)*. Adjacent large amount of joint fluid is noted *(straight arrow)*. *Open arrowhead* points to the malleolar fossa. TA, Talus; F, fibula. (From Erickson SJ, Smith JW, Ruiz ME, et al: MR imaging of the lateral collateral ligament of the ankle. AJR 156:131, 1991. © American Roentgen Ray Society.)

The Medial Collateral (Deltoid) Ligament

The deltoid ligament is a thick, triangular ligament with many fibers running deep to the flexor retinaculum. The deepest band fans horizontally from the medial malleolus to the talus. Superficial fibers extend anteriorly from the medial malleolus to the navicular bone, medially and inferiorly to the calcaneus, and posteriorly to the sustentaculum tali and dorsal aspect of the talus. The deltoid ligament contributes to ankle stability equally with the anterior talofibular ligament of the lateral collateral ligament group. Both ligaments are brought into tension during plantar flexion, although the strong deltoid ligament is injured far less often than the anterior talofibular ligament.[15] Because of its complexity, the deltoid ligament is difficult to visualize in its entirety in any one plane. Accordingly, its integrity may be difficult to establish by MR imaging alone.

THE BONES

Osteochondritis Dissecans

After the medial femoral condyle, the dome of the talus is the most common site affected by osteochondritis dissecans. Medial talar lesions occur most frequently, and they tend to involve the posterior third of the dome, whereas lateral talar lesions occur anteriorly and usually are more shallow.[8,14] Although the precise cause is debated, some degree of trauma appears to be involved in most cases of osteochondritis dissecans. Patients with medial talar dome lesions usually complain of persistent pain several weeks or months after ankle sprain. Typically, radiographs done at the time of the original injury are negative. In contrast, the majority of the lateral dome lesions are visible on initial radiographs and become manifest sooner. In 1959 Berndt and Harty reproduced these osteochondral fractures in cadaver specimens through two mechanisms: (1) vigorous inversion of the foot with a dorsiflexed ankle resulted in lateral dome fractures; (2) vigorous inversion with a plantar flexed foot and lateral rotation of the tibia resulted in medial dome fractures.[5] Disrupted blood supply to the osteochondral fragment can lead to ischemic necrosis. If the fragment remains attached to the parent bone, revascularization may occur, and the lesion can heal. However, fibrous tissue or cartilage in the undisplaced fracture line can hinder capillary ingrowth and prevent healing. Continued weightbearing may displace the fragment and produce an intra-articular loose body, leading to synovitis and secondary degenerative joint disease.[5]

Pritsch and colleagues categorized osteochondritis dissecans into four arthroscopic grades: grade I, a focal area of cartilage softening, fibrillation, and fissuring without a discernible fracture line; grade II, a breach in the cartilage with a nondisplaced fragment; grade III, a fragment displaced but attached by a flap of articular cartilage; grade IV, a completely

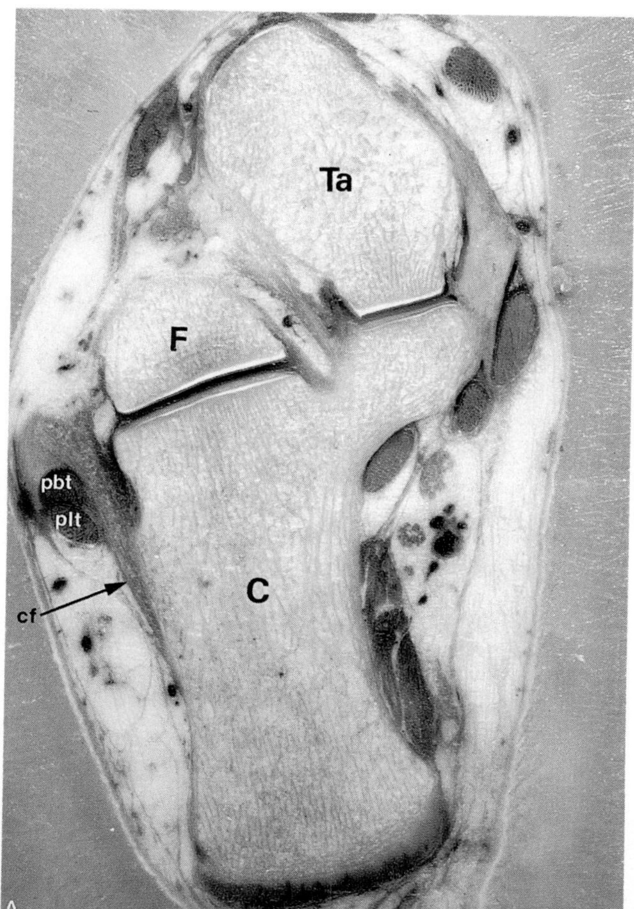

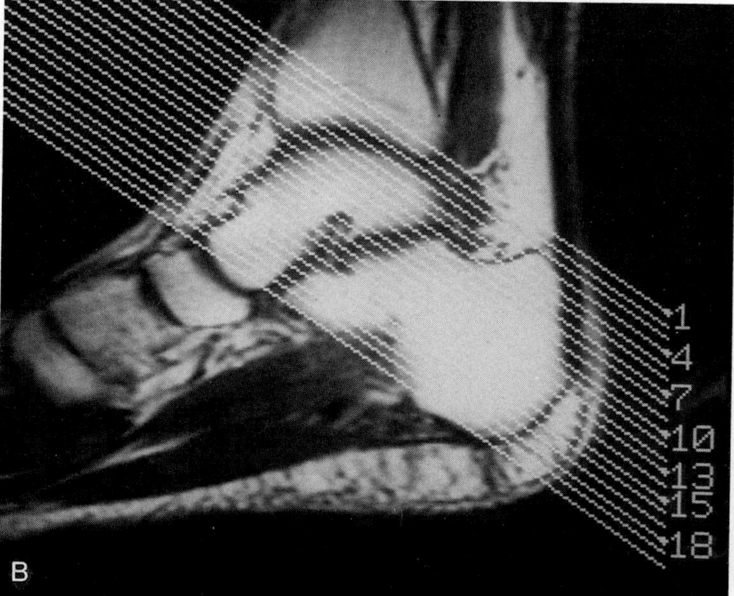

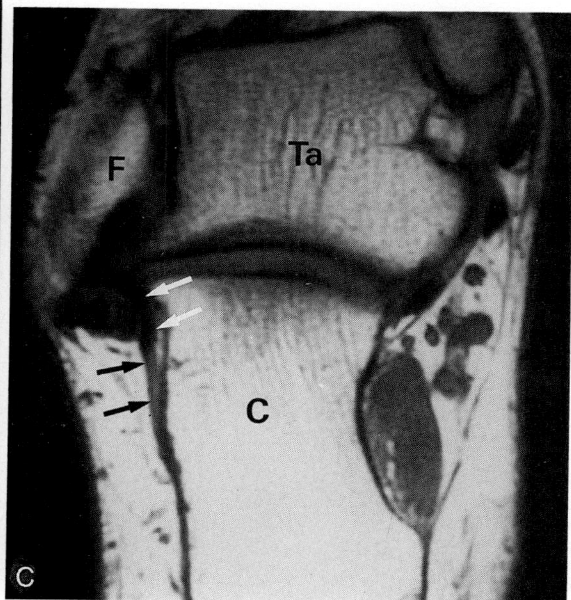

Figure 13–18. Normal calcaneofibular ligament on axial plane. *A*, Cryomicrotome section; *B*, sagittal localizer showing oblique axial plane; *C*, T1-weighted image: TR, 500 ms; TE, 20 ms. *Arrows* point to the course of the calcaneofibular (cf) ligament. Ta, Talus; F, fibula; C, calcaneus; pbt, peroneus brevis tendon; plt, peroneus longus tendon. (From Erickson SJ, Smith JW, Ruiz ME, et al: MR imaging of the lateral collateral ligament of the ankle. AJR 156:131, 1991. © American Roentgen Ray Society.)

detached fragment or loose body within the joint.[25] Detached fragments (grade IV) must be removed surgically and the subchondral bed débrided for optimal results. Partially attached fragments (grades II and III) can often be fixed internally with a screw and allowed to heal with immobilization of the ankle.

Plain radiography, as stated earlier, is relatively insensitive to early osteochondritis dissecans, particularly in cases involving the medial talar dome.

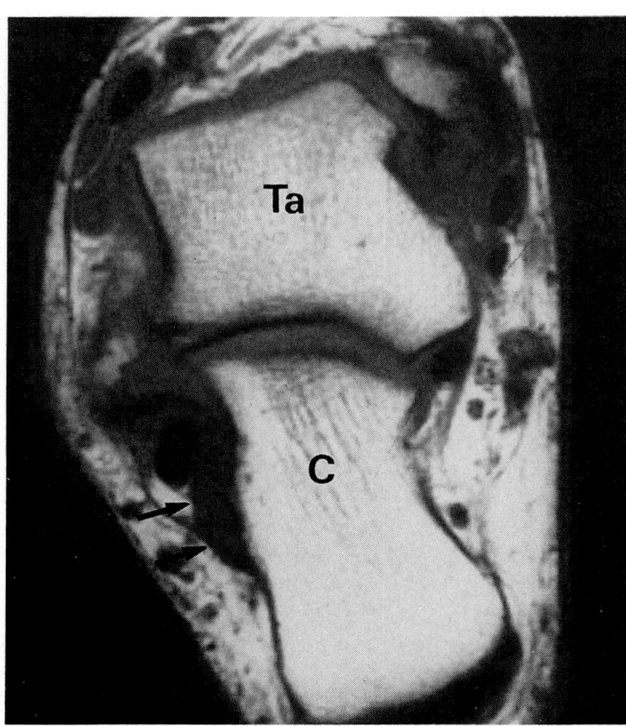

Figure 13–19. Calcaneofibular ligament injury. Oblique axial T1-weighted image (TR, 500 ms; TE, 20 ms) shows abnormal thickening of the ligament *(arrows)*. Ta, Talus; C, calcaneus. (From Erickson SJ, Smith JW, Ruiz ME, et al: MR imaging of the lateral collateral ligament of the ankle. AJR 156:131, 1991. © American Roentgen Ray Society.)

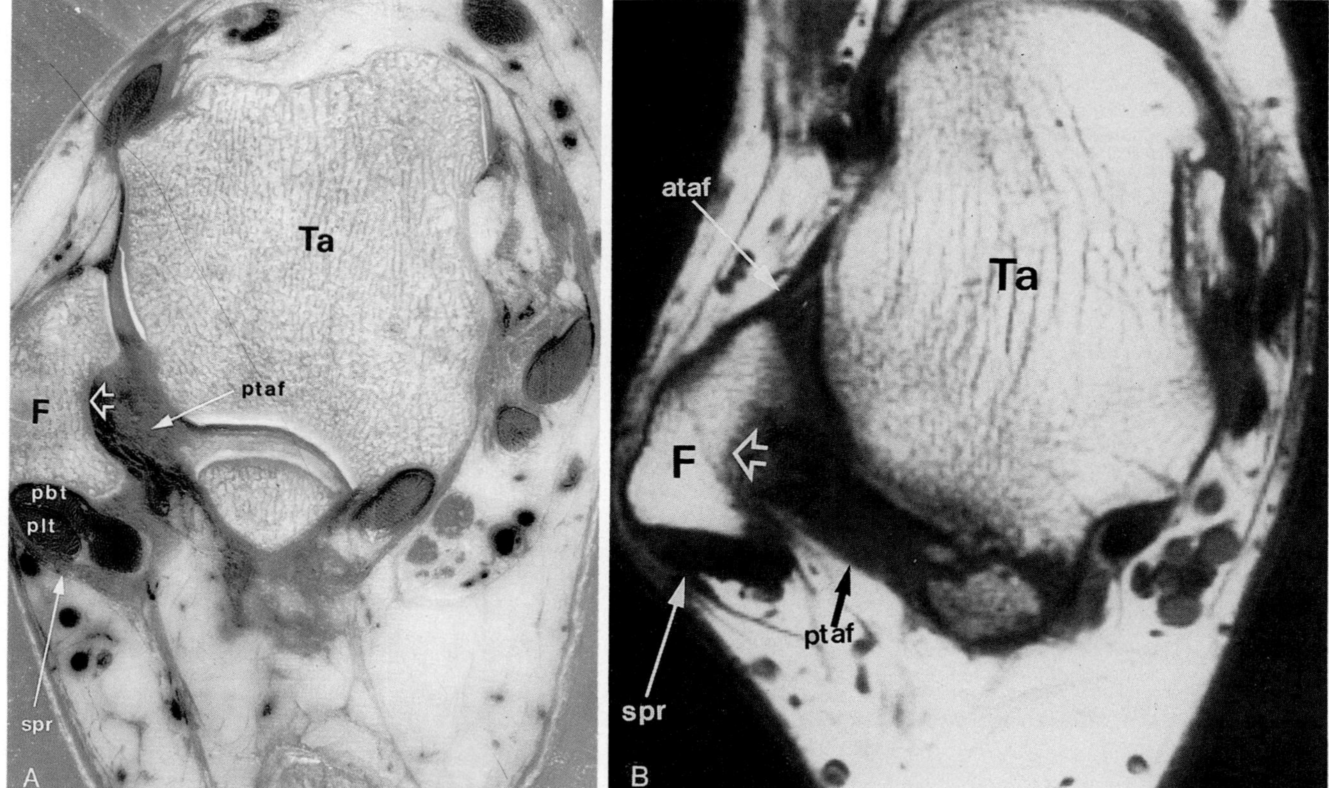

Figure 13–20. Normal posterior talofibular ligament on axial plane. *A*, Cryomicrotome section; *B*, T1-weighted image: TR, 500 ms; TE, 20 ms. *Open arrowhead* points to the malleolar fossa. Ta, Talus; F, fibula; ptaf, posterior talofibular ligament; plt, peroneus longus tendon; pbt, peroneus brevis tendon; spr, superior peroneal retinaculum; ataf, anterior talofibular ligament. (From Erickson SJ, Smith JW, Ruiz ME, et al: MR imaging of the lateral collateral ligament of the ankle. AJR 156:131, 1991. © American Roentgen Ray Society.)

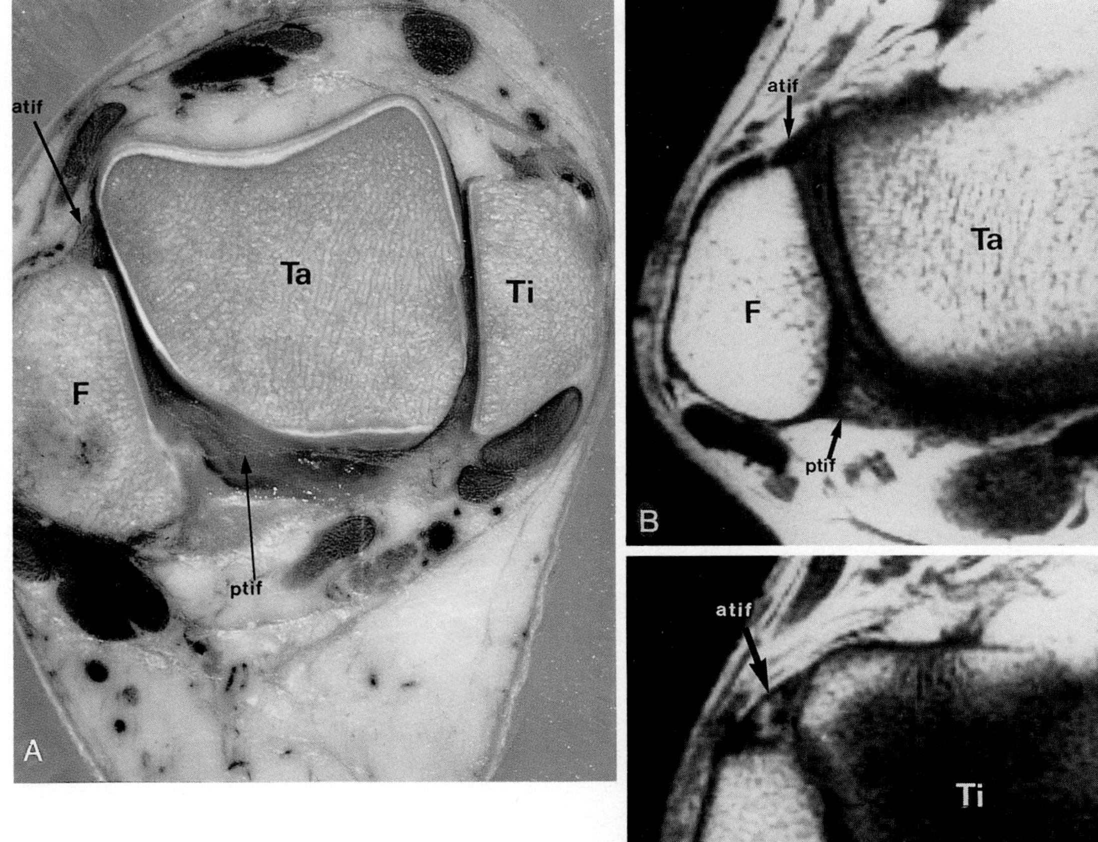

Figure 13-21. Normal tibiofibular ligaments on axial plane. *A*, Cryomicrotome section; *B*, corresponding T1-weighted image (TR, 500 ms; TE, 20 ms); *C*, T1-weighted image (TR, 500 ms; TE, 20 ms) obtained at a level slightly superior to *B*. Ta, Talus, Ti, tibia; F, fibula; atif, anterior talofibular ligament; ptif, posterior talofibular ligament; pbt, peroneus brevis tendon; plt, peroneus longus tendon. (From Erickson SJ, Smith JW, Ruiz ME, et al: MR imaging of the lateral collateral ligament of the ankle. AJR 156:131, 1991. © American Roentgen Ray Society.)

Bone scintigraphy is more sensitive, but it lacks specificity and offers poor spatial resolution. Magnetic resonance imaging is most sensitive in detecting osteochondritis dissecans and can distinguish stable from unstable fragments. Patients who have positive results on bone scintigraphy should be assessed by MR imaging.[2] Accordingly, MR imaging is the most useful noninvasive method for staging osteochondritis dissecans. Nelson and co-workers correlated the MR imaging appearance of osteochondritis dissecans with its arthroscopic grading.[22] In their classification scheme, grade I lesions exhibited uninterrupted overlying articular cartilage on MR imaging (Fig. 13-24).[22] With involvement of the basal layers of the cartilage and vascular subchondral bone (grade II to IV), the defect became filled with blood and fibrin clot. Grade II osteochondral lesions were firmly attached to subchondral bone by a thin band of fibrous tissue of low signal intensity (Fig. 13-25). However, reactive sclerosis can produce a similar dark rim.[22] In grade III lesions, fluid of high signal intensity could be seen in the gap between the fragment and underlying bone (Fig. 13-26). Grade IV lesions showed a complete detachment of the articular cartilage and a fluid-filled defect in the parent bone.

DeSmet and colleagues reported that the most reliable sign of a partially detached fragment is the presence of a thin, irregular line of high signal intensity, similar to fluid, representing loose granulation tissue at the unstable interface on T2-weighted images.[9] Completely detached fragments exhibit a

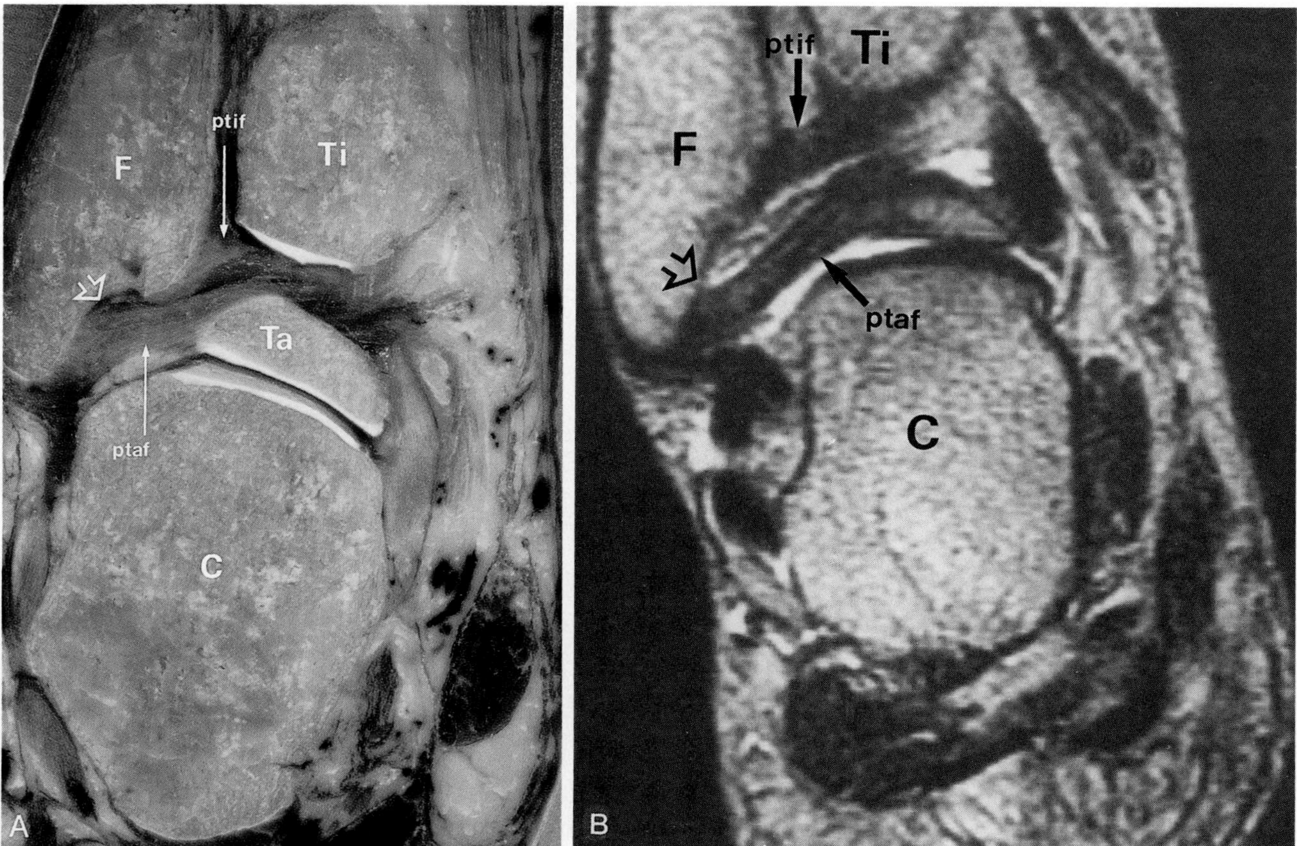

Figure 13-22. Distinction of talofibular from tibiofibular ligaments on coronal plane. *A*, Cryomicrotome section; *B*, T2-weighted image (TR, 2500 ms; TE, 80 ms). F, Fibula; ptif, posterior tibiofibular ligament; ptaf, posterior talofibular ligament; Ta, talus; Ti, tibia; C, calcaneus. (From Erickson SJ, Smith JW, Ruiz ME, et al: MR imaging of the lateral collateral ligament of the ankle. AJR 156:131, 1991. © American Roentgen Ray Society.)

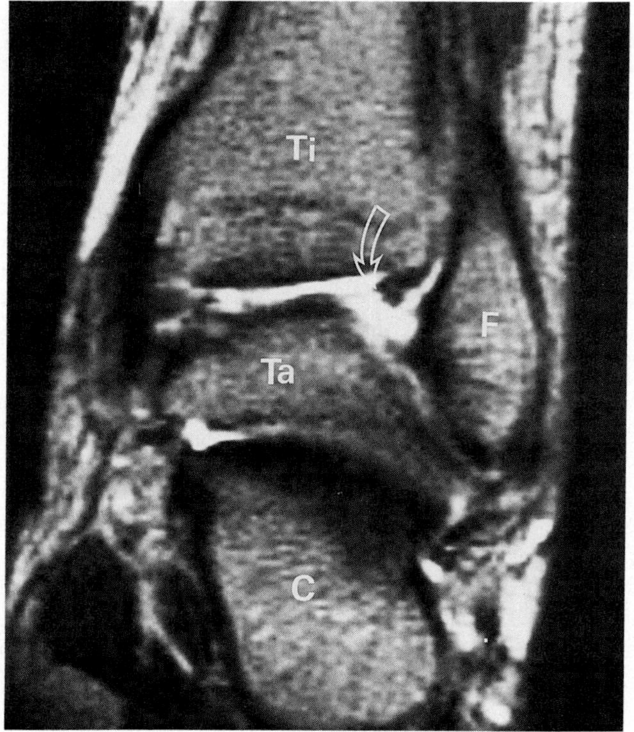

Figure 13-23. Posterior tibiofibular ligament rupture. Coronal T2-weighted image (TR, 2000 ms; TE, 80 ms) shows ligamentous discontinuity *(arrow)*. Ti, Tibia; Ta, talus; F, fibula; C, calcaneus.

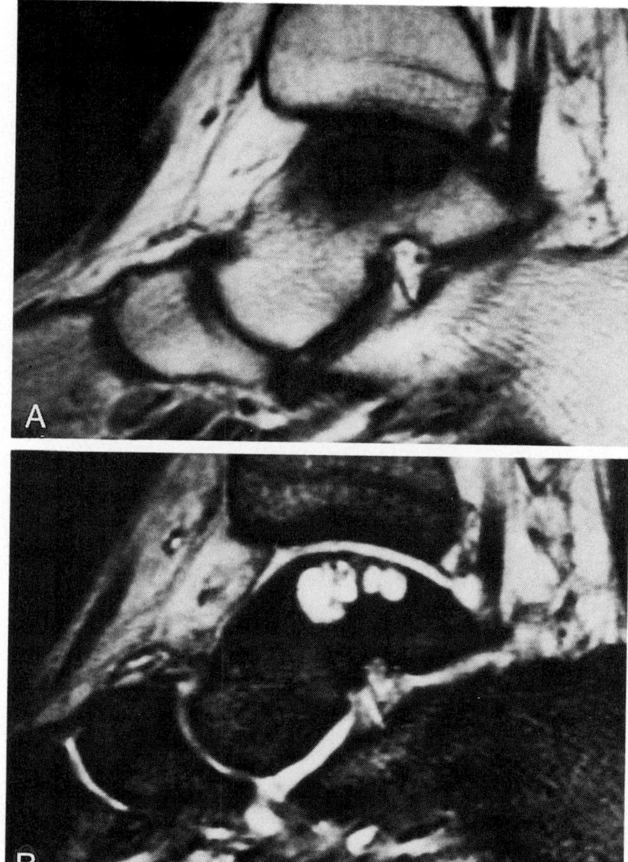

Figure 13-24. Osteochondritis dissecans of the left talus in a 47-year-old man. This patient suffered from chronic ankle pain for 4 years, and conservative therapy had failed. MR images (*A*: sagittal, TR, 600 ms; TE, 20 ms; *B*: sagittal, TR, 600 ms; TE, 15 ms; flip angle, 30 degrees) show a multiloculated fluid collection in the dome of the talus medially. This fluid collection has a well-defined sclerotic border and is consistent with osteochondritis dissecans. No definite adjacent cartilaginous abnormality is noted. This patient was subsequently treated by bone grafting of the lesion to support the subchondral bone and articular cartilage.

smooth fluid line of high signal intensity that encircles the fragment. The signal intensity of the central fragment does not differentiate reliably between loose and healed fragments. A region of low signal intensity centrally can be due to marrow fibrosis or calcification, whereas one of high signal intensity may reflect unabsorbed hemorrhage within the fragment. Displacement of the fragment to form an intra-articular loose body leaves a residual osseous defect in the parent bone. Although a healed osteochondral

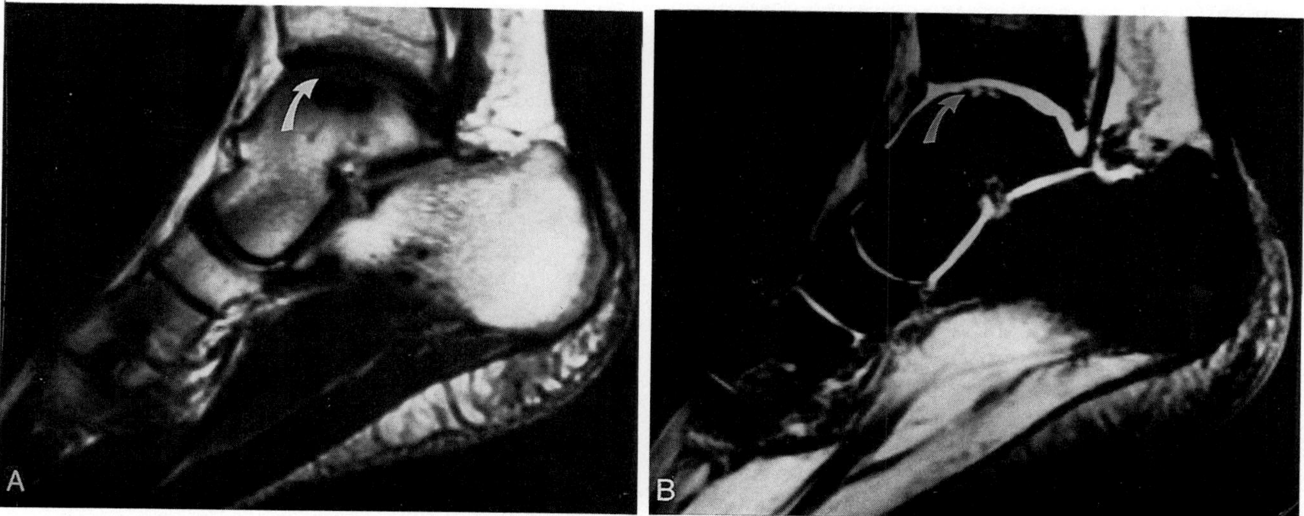

Figure 13-25. Osteochondritis dissecans of the talus. *A*, Sagittal T1-weighted image (TR, 600 ms; TE, 20 ms) shows an osteochondral fragment of low signal intensity from the dome of the talus (*arrow*). *B*, T2*-weighted image (TR, 600 ms; TE, 15 ms; flip angle, 30 degrees) shows the osteochondral fragment and defect with fluid of high signal intensity on the osteochondral cavity.

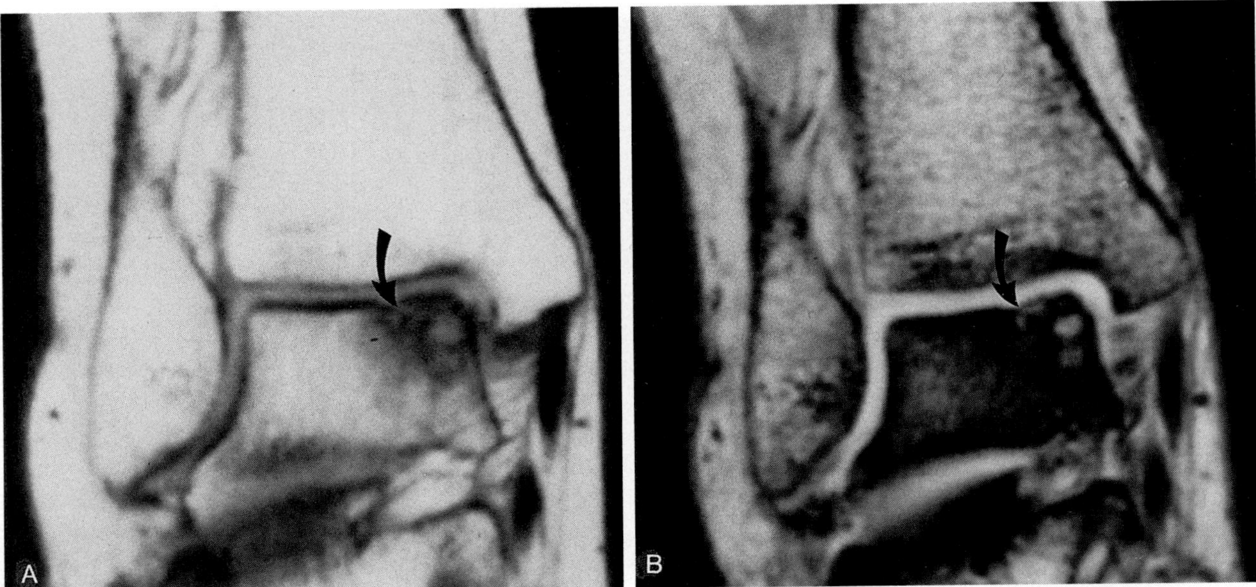

Figure 13-26. Partially separated osteochondral dissecans of the talus. Intermediate-weighted *(A)* and T2-weighted *(B)* coronal images show medial talar subchondral cysts. A cortical defect of the talar dome is noted *(arrow)*. Arthroscopy revealed a partially avulsed flap of the corresponding site.

lesion can have a similar appearance, visualization of such a defect in the talus should raise the possibility of an intra-articular loose body. Thus, in patients with persistent ankle symptoms, MR imaging can identify osteochondral lesions reliably and can evaluate their extent and stability.[2,31]

Stress Fractures

Stress fractures in the foot usually occur in healthy bone after a period of excessive activity. Typically there is no history of any single traumatic incident. One speculation is that bone resorption during the initial phases of remodeling that develops at a site of increased stress leads to focal weakness and increased susceptibility to microfracture in the face of continued exposure to the increased activity. Patients at risk include seasonal athletes and patients who, after prolonged immobilization of the foot in a plaster cast, revert to heavy exercise too rapidly.

Stress fractures can occur at several sites in the appendicular and axial skeleton, but the most common location by far is in the foot, where so-called march fractures are a relatively frequent occurrence. The German military surgeon Breithaupt first described the march fracture in 1855, before the advent of radiography.[15] About 90 per cent of these fractures develop in the second and third metatarsal shafts.[15] The lowest third of the fibula, just at or above the lateral malleolus, is the second most common site for stress fracture. It is particularly common in middle-aged athletes and has been called the "runner's fracture." Calcaneal stress fractures also are common, particularly in diabetics, in whom there often is an accompanying displaced avulsion at the Achilles tendon insertion. Calcaneal stress fractures frequently occur bilaterally.

Diagnosis of stress fractures can be challenging. The fractures are undisplaced, and plain radiographs typically are negative during the initial 10 to 14 days until local callus formation identifies the site of fracture. However, without early protection from weightbearing, stress fractures can go on to overt displacement. Bone scintigraphy is more sensitive than conventional radiography, but it is somewhat nonspecific and lacks spatial resolution. Magnetic resonance imaging is at least as sensitive as scintigraphy, but it offers considerably greater resolution. On MR imaging, stress fractures can be seen as bands of low signal intensity traversing the bone marrow and continuous with the cortex. Hemorrhage or edema of high signal intensity can be seen in the surrounding bone marrow on T2-weighted images (Figs. 13-27, 13-28). Stress fractures must be differentiated from metastases. These different conditions can look identical on bone scintigraphy. However, visualization of the fracture line in the absence of associated mass on MR imaging usually allows a specific diagnosis.

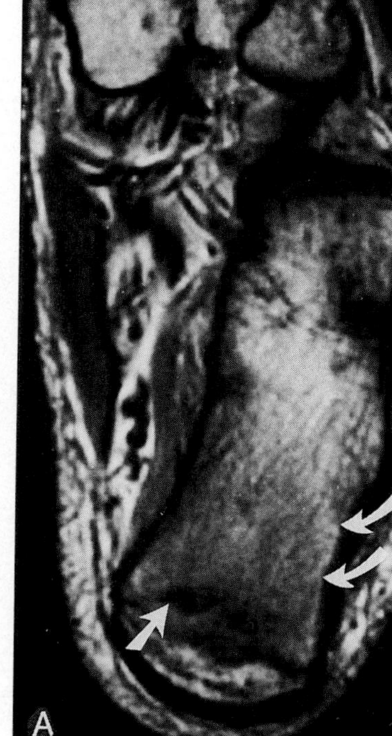

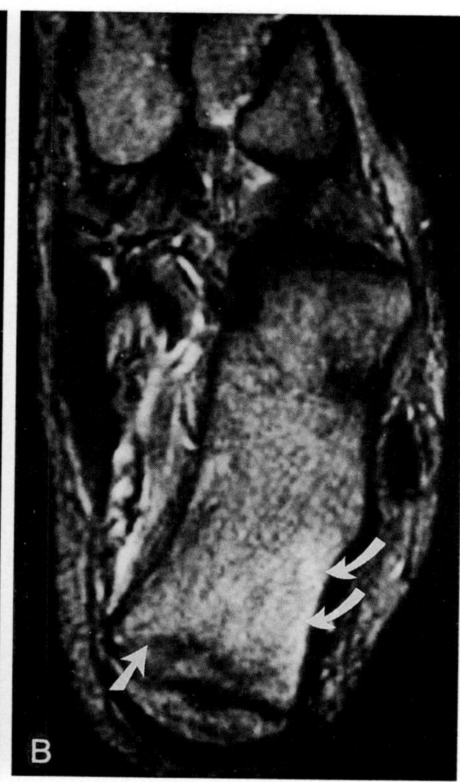

Figure 13–27. Stress fracture of the calcaneus in a 15-year-old boy. This patient, who suffered from heel pain but had no history of acute trauma, had undergone chemotherapy for rhabdomyosarcoma of the infratemporal fossa. Axial MR images (*A*: TR, 1800 ms; TE, 20 ms; *B*: TR, 1800 ms; TE, 80 ms) show a linear area of decreased signal intensity (*arrow*) anterior to the unfused calcaneal apophysis and an adjacent subtle area of marrow edema (*curved arrows*). Edema is of high signal intensity relative to marrow fat on the intermediate image (*A*) and of slightly high signal intensity relative to fat on the T2-weighted image (*B*). Heel pain resolved after rest. (From Kier R, McCarthy S, Dietz MJ, et al: MR appearance of painful conditions of the ankle. Radiographics 11:401, 1991. © Radiological Society of North America.)

MISCELLANEOUS CONDITIONS

Diabetic Foot

Patients with chronic diabetes mellitus invariably develop neuropathic and angiopathic changes in the feet. Often, the first manifestation is infection. The progression of breakdown in the diabetic foot is as follows: Initially, a dermal ulcer develops on the weight-bearing metatarsal heads or over the calcaneal tuberosity; this progressively deepens to form a soft tissue abscess that eventually extends to the bone, producing osteomyelitis; gangrenous auto-amputation can result when there is sufficient vascular compromise.[15] Establishing the site and extent of the infection is important for prognosis and surgical planning. Noninfected neuropathic ulcers and joints (Charcot's joint) can be treated conservatively, but once infection develops, aggressive medical or surgical therapy is required.

Plain radiography is extremely insensitive for detecting osteomyelitis; the films typically remain normal during the first 7 to 15 days of acute infection. Bone scintigraphy with technetium ^{99m}-labeled diphosphonate is a far more sensitive method; however, significant false-positive rates have been reported in diabetic patients.[30]

Magnetic resonance is highly sensitive for the marrow changes in osteomyelitis: on T1-weighted images, fatty marrow of high signal intensity is replaced by exudate of low signal intensity; this exudate becomes bright on T2-weighted sequences (Fig. 13–29). An increase in signal intensity is usually also present in the adjacent soft tissues on T2-weighted images. The most sensitive sequence for detecting marrow edema or infection is STIR, owing to its additive effect on substances exhibiting long T1 and T2 relaxation times (Figs. 13–30, 13–31). In contrast to the findings in osteomyelitis, noninfected neuropathic joints often exhibit low signal intensity on both T1- and T2-weighted images.[3] Although acute neuropathic fractures can mimic osteomyelitis by showing high signal intensity changes on T2-weighted images,[21] the abnormal marrow signal from an ulcer that extends to underlying bone is probably caused by osteomyelitis.[21]

It can be difficult to distinguish an abscess from surrounding edema or cellulitis with MR imaging. Abscesses typically exhibit smooth margins and rim enhancement after the administration of gadopentetate dimeglumine. Cellulitis, in contrast, exhibits diffuse heterogeneous enhancement. Another cause of increased soft tissue signal intensity in patients with diabetes is subaponeurotic edema deep to the

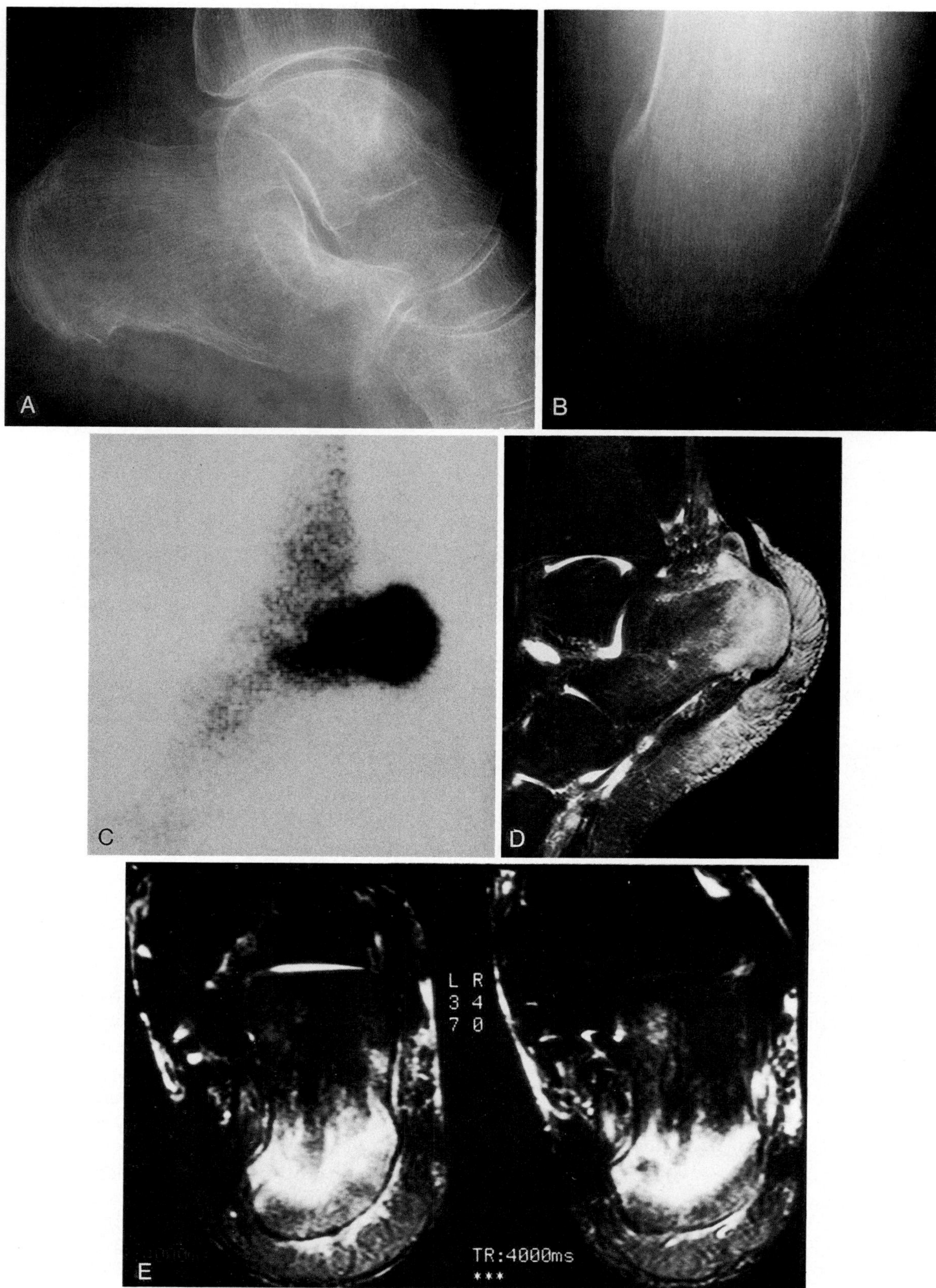

Figure 13–28. Stress fracture in a patient with 1 week of active heel pain. *A* and *B*, Plain radiographs show no evidence of fracture. *C*, Radionuclide bone scan shows hot spot in the calcaneus. *D* and *E*, T2-weighted fast spin-echo images with fat-saturation technique show bone marrow edema of high signal intensity in the calcaneus. Signal intensities of normal bone marrow are suppressed with this technique.

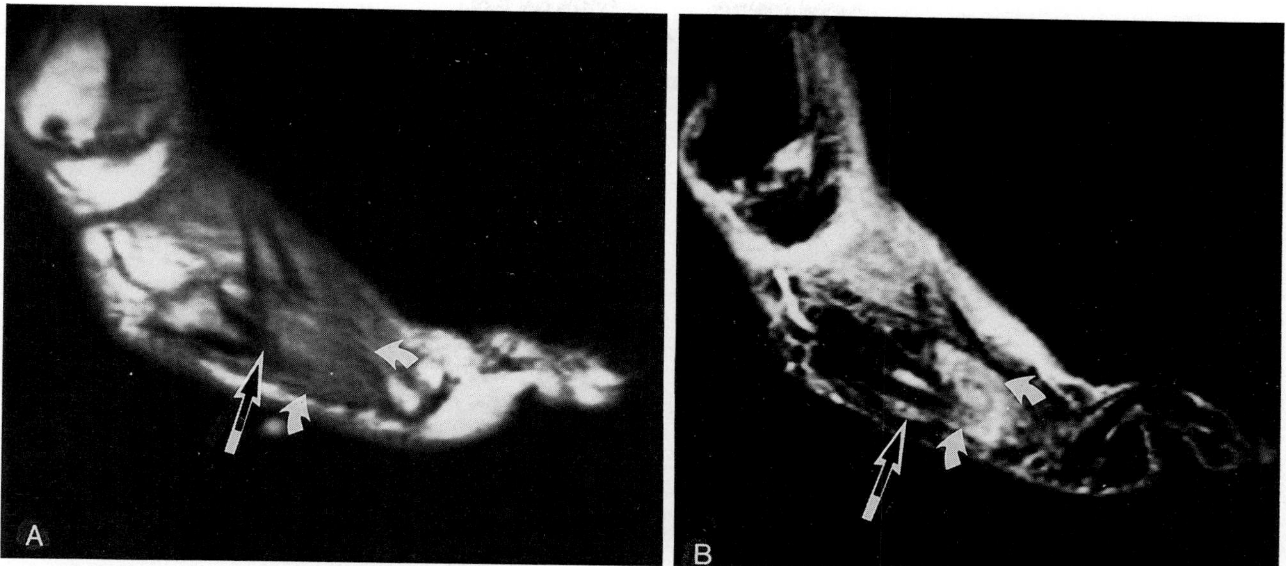

Figure 13–29. Recurrent osteomyelitis and abscesses in a 46-year-old man with diabetic foot. Abnormal bone marrow signal intensity is shown on T1-weighted (*A*: TR, 683 ms; TE, 20 ms) and T2-weighted (*B*: TR, 2016 ms; TE, 100 ms) images whose appearances are consistent with that of osteomyelitis (*straight arrow*). Adjacent soft tissue on the plantar aspect of the foot is of abnormal signal intensity, suggesting abscess (*curved arrows*). (From Yuh WTC, Corson JD, Rezai K, et al: Osteomyelitis of the foot in diabetic patients: Evaluation with plain film, ^{99m}Tc-MDP bone scintigraphy, and MR imaging. AJR 152:795, 1989. © American Roentgen Ray Society.)

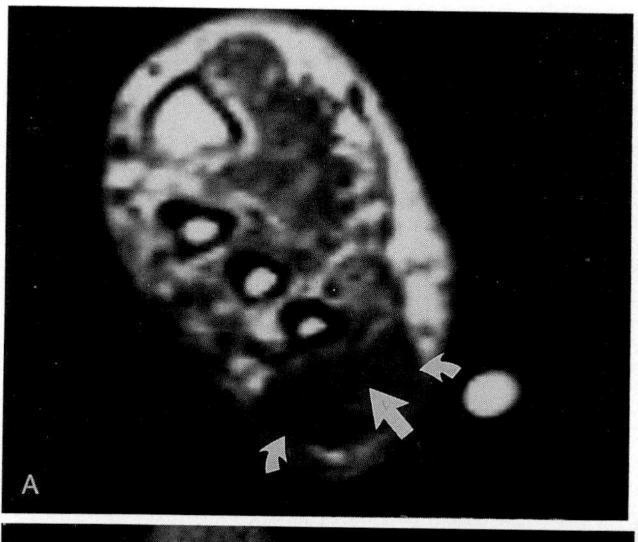

Figure 13–30. Osteomyelitis of fifth metatarsal in a 62-year-old man with diabetic foot. T1-weighted (*A*: axial, TR, 483 ms; TE, 20 ms; *B*: sagittal, TR, 683 ms; TE, 20 ms) and short tau inversion-recovery (STIR) (*C*) images show an abnormal bone marrow signal intensity of the distal metatarsal (*arrow*). Soft tissue abscess (*curved arrows*) adjacent to osteomyelitis also can be seen on T1-weighted (*A*) and STIR (*C*) images, but it is more apparent on the STIR image. Note that the fat signal within normal bone marrow is suppressed on the STIR image (*C*). (From Yuh WTC, Corson JD, Rezai K, et al: Osteomyelitis of the foot in diabetic patients: Evaluation with plain film, ^{99m}Tc-MDP bone scintigraphy, and MR imaging. AJR 152:795, 1989. © American Roentgen Ray Society.)

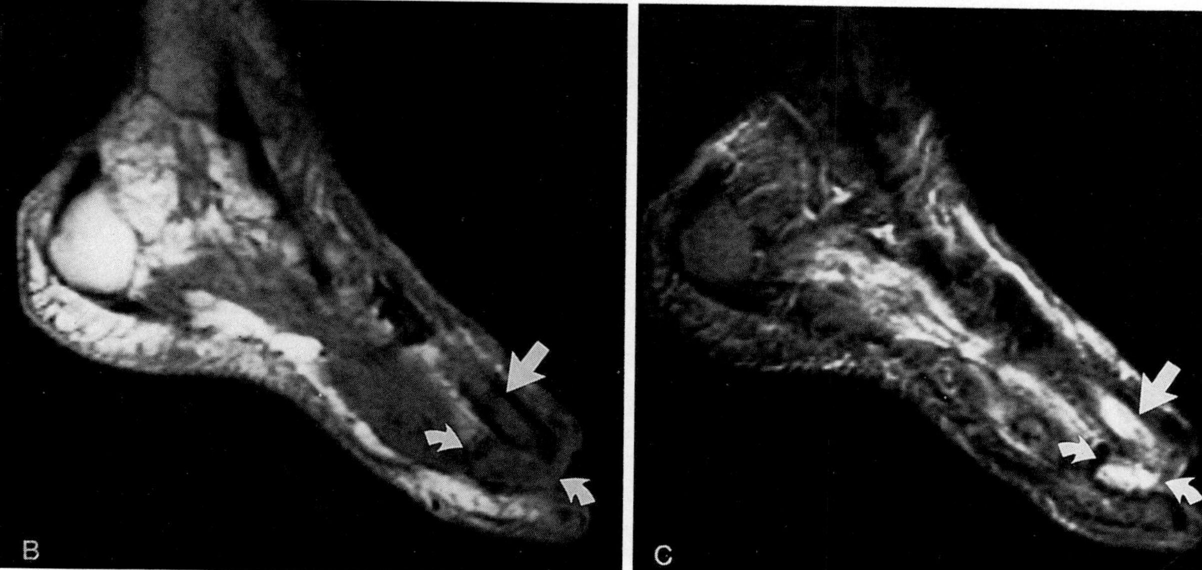

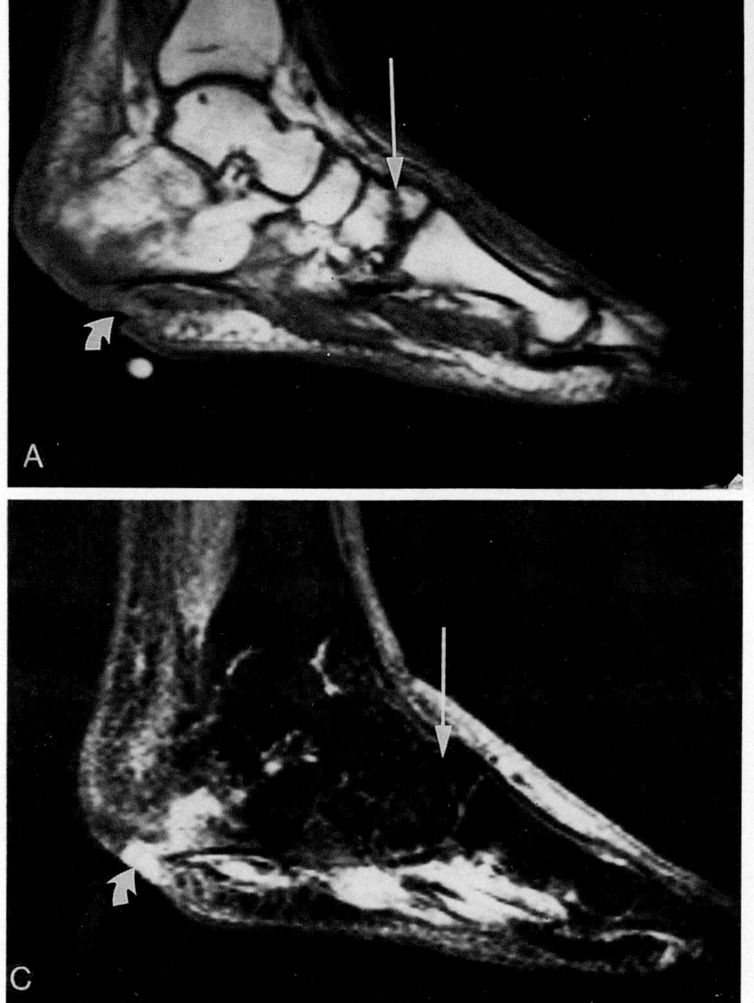

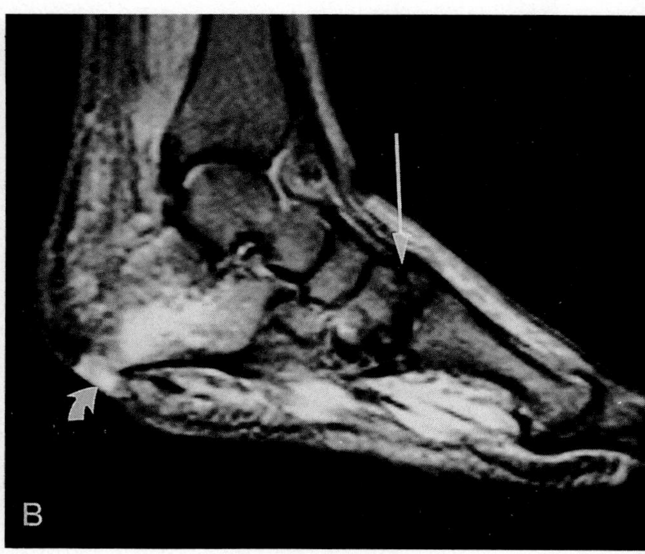

Figure 13–31. Old fracture and osteomyelitis of calcaneus with sinus tract in a 38-year-old man with diabetic foot. MR images (*A*: TR, 583 ms; TE, 20 ms; *B*: TR, 2000 ms; TE, 100 ms; short tau inversion-recovery [STIR]) show abnormal signal intensity of the calcaneus. A sinus tract (*curved arrow*) extends from the open wound into the calcaneus. An old fracture of low signal intensity (*arrow*) is shown on all pulse sequences, presumably due to peripheral neuropathy. (From Yuh, WTC, Corson JD, Rezai K, et al: Osteomyelitis of the foot in diabetic patients: Evaluation with plain film, ^{99m}Tc-MDP bone scintigraphy, and MR imaging. AJR 152:795, 1989. © American Roentgen Ray Society.)

plantar fascia.[21] However, the appearance is nonspecific, and clinical correlation is necessary to eliminate infection. Joint effusions involving the ankle or subtalar joints are a frequent nonspecific finding in diabetic patients. Fluid in the tendon sheaths is indicative of tenosynovitis but not necessarily infection.[3,21] Stress fractures are common in the diabetic foot, and they often are associated with a displaced avulsion of the insertion of the Achilles tendon on the calcaneus.

Tarsal Tunnel Syndrome

The tarsal tunnel is an anatomic space extending from the level of the medial malleolus to the plantar aspect of the navicular bone, containing the medial ankle tendons and posterior tibial neurovascular bundle (Figs. 13–32 to 13–35).[12,16,33] The roof of the tarsal tunnel is formed primarily by the flexor retinaculum, although distally it is bound by the abductor hallucis muscle. The medial surfaces of the talus, sustentaculum tali, and calcaneus form the osseous floor.

Tarsal tunnel syndrome is a compression neuropathy of the posterior tibial nerve and its terminal branches. Causes include conditions that reduce the volume of the tarsal tunnel itself or enlarge the volume of its contents. The most common cause is a ganglion cyst within the tarsal tunnel. Other causes include bony overgrowth associated with tarsal coalition, posttraumatic edema and scarring, and the presence of tumors or aberrant muscles.

Clinical symptoms include burning and paresthesia along the plantar surface of the foot and toes. Reproduction of the pain and paresthesia by percussion along the course of the posterior tibial nerve (Tinel's sign) suggests tarsal tunnel syndrome.[15] Demonstration of decreased nerve conduc-

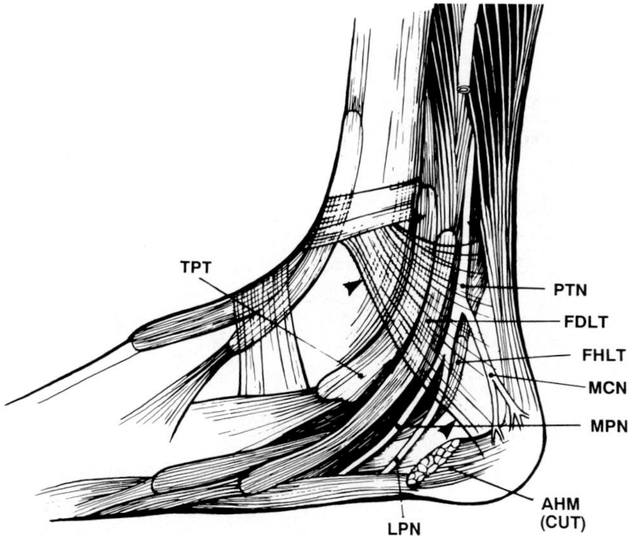

Figure 13–32. Lateral aspect of the ankle showing the contents of the tarsal tunnel. The vascular structures have been excluded. TPT, Posterior tibial tendon; PTN, posterior tibial nerve; FDLT, flexor digitorum longus tendon; FHLT, flexor hallucis longus tendon; MCN, medial calcaneal nerve; MPN, medial plantar nerve; AHM, abductor hallucis muscle; LPN, lateral plantar vein. (From Erickson SJ, Quinn SF, Kneeland JB, et al: MR imaging of the tarsal tunnel and related spaces: Normal and abnormal findings with anatomic correlation. AJR 155:323, 1990. © American Roentgen Ray Society.)

tion confirms the diagnosis; however, normal nerve conduction studies cannot exclude it.

The tarsal tunnel is best evaluated by MR imaging in the axial plane, with the flexor retinaculum seen as a thin band enclosing the tendons of the tibialis posterior, the flexor digitorum longus and flexor hallucis longus muscles, and the posterior tibial nerve and vessels. Thickened flexor retinaculum results in tarsal tunnel syndrome and can easily be detected when comparing it with the asymptomatic foot (Fig. 13–36). A ganglion cyst within the tarsal tunnel is of high signal intensity on T2-weighted images (Fig. 13–37). The medial plantar branch of the posterior tibial nerve can usually be visualized directly adjacent to the flexor hallucis longus tendon.[32]

Unlike carpal tunnel syndrome of the wrist, tarsal tunnel syndrome is often unilateral, with vague and poorly localized symptoms. Magnetic resonance imaging is very helpful in these cases not only in identifying potential causes of the tarsal tunnel syndrome, but also in excluding other reasons for chronic ankle pain, such as tendinitis, osteonecrosis, or ligamentous injury. As with the carpal tunnel syndrome, most cases require surgical decompression by division of the flexor retinaculum.

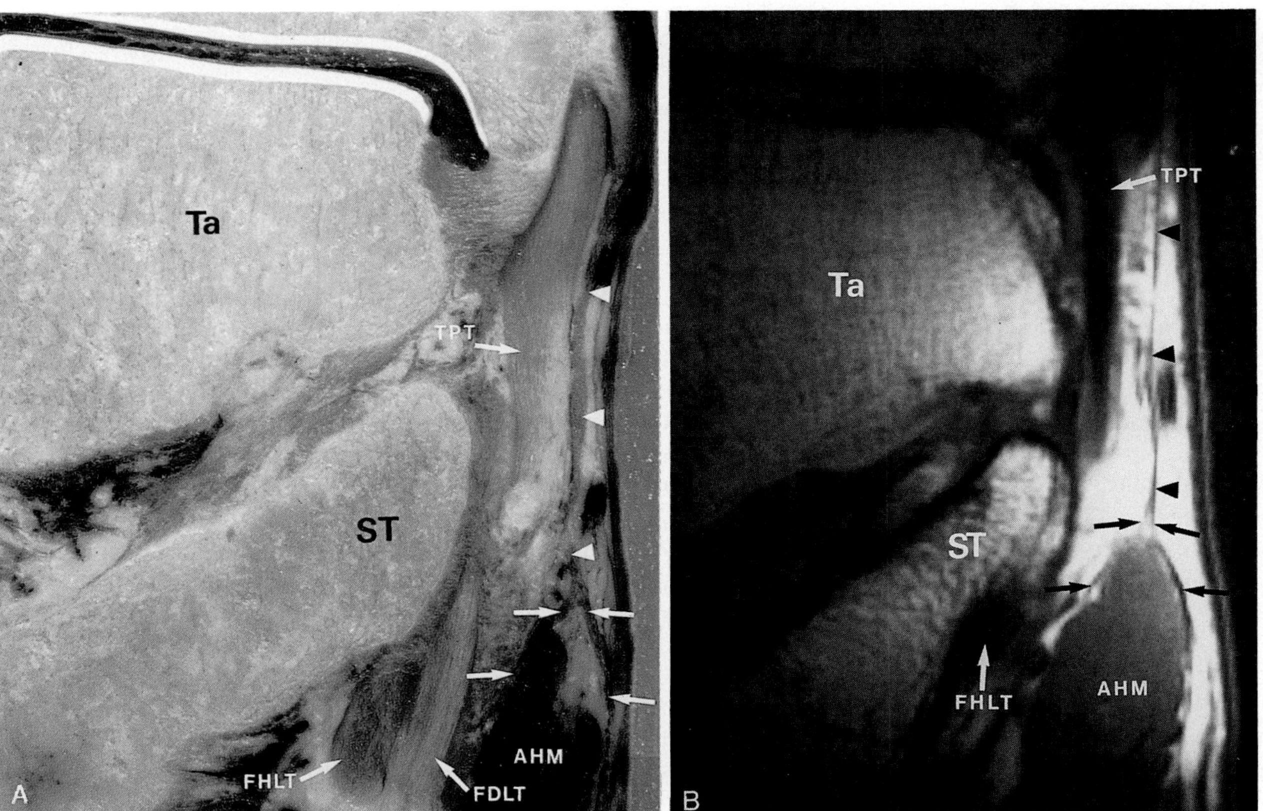

Figure 13–33. Normal tarsal tunnel on coronal plane. *A*, Cryomicrotome section. *B*, T1-weighted image (TR, 500 ms; TE, 20 ms). *Arrows* point to the site of splitting of the flexor retinaculum around the abductor hallucis muscle.

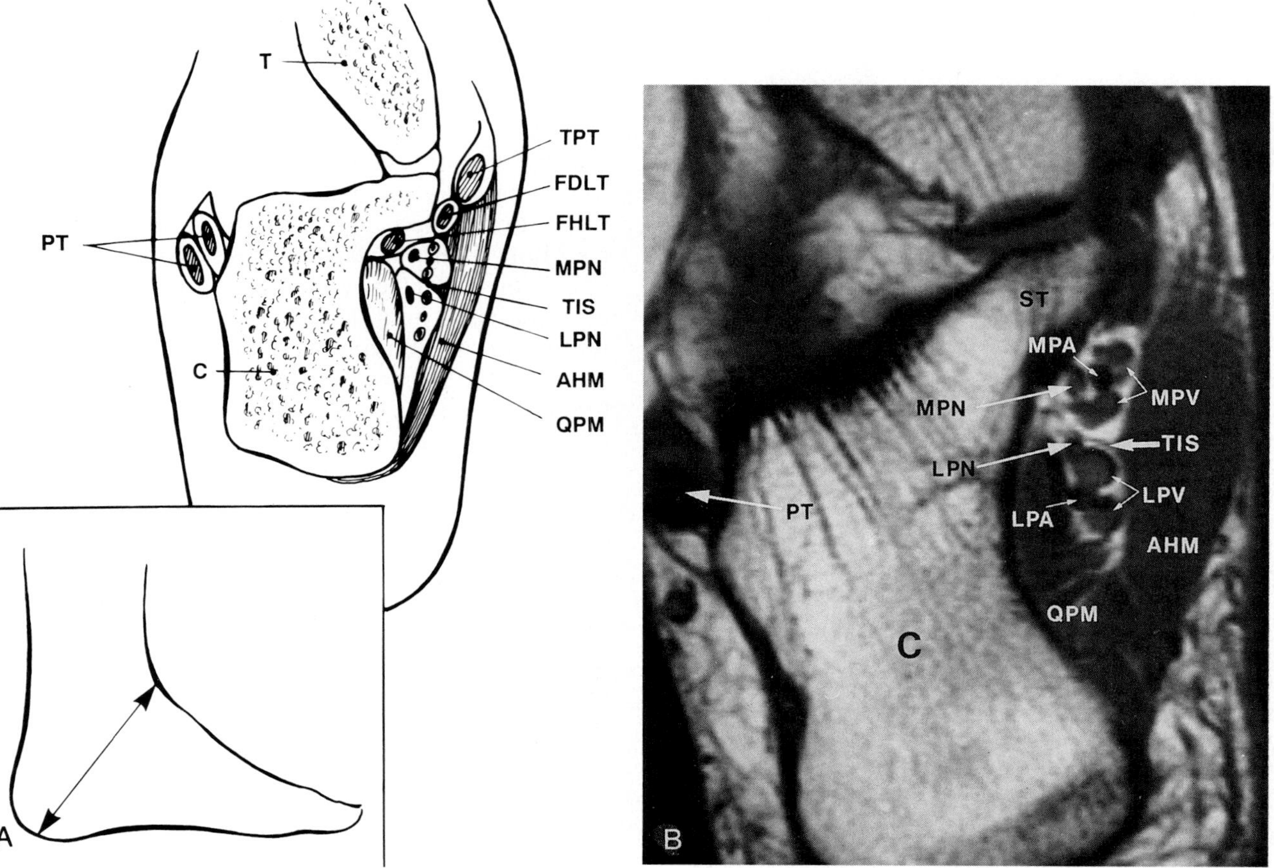

Figure 13-34. Normal tarsal tunnel on oblique axial plane. Drawing (*A*) and T1-weighted image (TR, 500 ms; TE, 20 ms) (*B*) show a transverse interfascicular septum segregating the medial and lateral neurovascular bundles. T, Talus; TPT, posterior tibial tendon; FDLT, flexor digitorum longus tendon; FHLT, flexor hallucis longus tendon; MPN, medial plantar nerve; TIS, transverse interfascicular septum; LPN, lateral plantar nerve, AHM, abductor hallucis muscle; QPM, quadratus plantae muscle; C, calcaneus; PT, peroneal tendons; ST, sustentaculum tali; MPA, medial plantar artery; MPV, medial plantar vein; LPA, lateral plantar artery; LPV, lateral plantar vein. (From Erickson SJ, Quinn SF, Kneeland JB, et al: MR imaging of the tarsal tunnel and related spaces: Normal and abnormal findings with anatomic correlation. AJR 155:323, 1990. © American Roentgen Ray Society.)

Plantar Fasciitis

The plantar aponeurosis consists of three parts: a strong central portion and weaker medial and lateral components. It originates posteriorly from the medial calcaneal tuberosity and divides into five slips as it extends anteriorly to insert into the proximal phalanges. The aponeurosis tightens as the toes are extended. The central portion is important in maintaining the longitudinal arch of the foot.

Plantar fasciitis is a frequent cause of painful heel syndrome. The mechanism has been attributed to repetitive traction on the plantar fascia, resulting in microtears and fascial and perifascial inflammation. Plantar fasciitis is commonly seen in runners and obese patients. In the early stages, patients note heel pain in the morning during the first step onto the floor. In more advanced stages, walking becomes painful, and patients find it uncomfortable even to stand. Treatment of plantar fasciitis is conservative, employing heel pads and oral nonsteroidal anti-inflammatory medication.

Magnetic resonance imaging is useful in demonstrating plantar fasciitis and excluding other causes of heel pain, such as calcaneal stress fracture and medial calcaneal neuritis. The normal plantar fascia is seen as a homogeneous band of low signal intensity with either uniform thickness or minimal tapering along its course in all pulse sequences (Fig. 13-38). Its calcaneal insertion may show a slight flaring of the fascia. Berkowitz and co-workers reported that the mean maximum thickness of the fascia on sagittal and coronal images in asymptomatic patients ranged from 3.0 to 3.5 mm.[4] In patients with plantar fasciitis, this thickness is significantly increased, from 7.4 to 7.6 mm (Fig. 13-39). Frequently there is intrasubstance increased signal intensity of the fascia. Subcutaneous edema is a less common

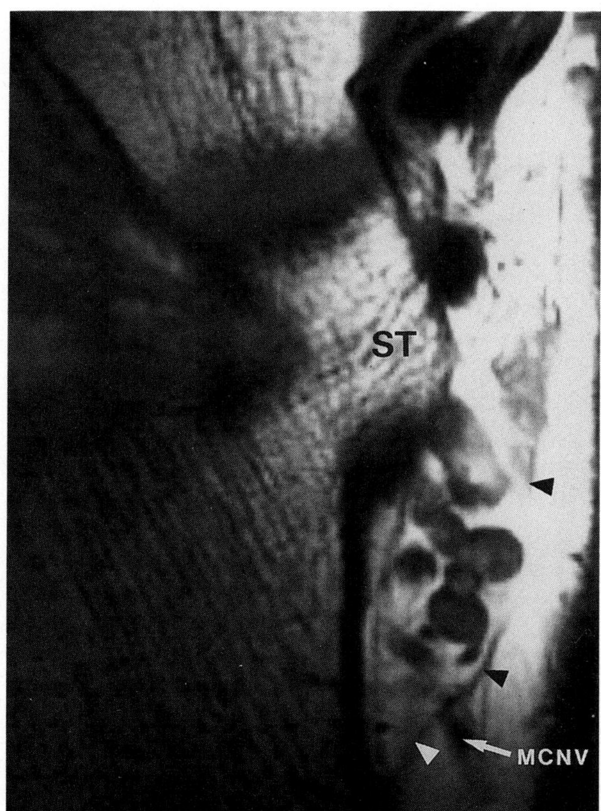

Figure 13–35. Axial MR image (TR, 500 ms; TE, 20 ms) through the inferior aspect of the tarsal tunnel shows the passage of the medial calcaneal neurovascular (MCNV) structures through the flexor retinaculum. ST, Sustentaculum tali. (From Erickson SJ, Quinn SF, Kneeland JB, et al: MR imaging of the tarsal tunnel and related spaces: Normal and abnormal findings with anatomic correlation. AJR 155:323, 1990. © American Roentgen Ray Society.)

finding. Calcaneal bone spurs may develop in symptomatic heels (50 per cent) but can also be found in asymptomatic persons (27 per cent).

Retrocalcaneal Bursitis

Under normal circumstances, body weight is distributed between the midtarsal joints and ligaments. When these mechanics are disturbed, an undue amount of stress is passed to the heel. Normally, the heel pad dissipates this stress. However, if the heel pad mechanism is inadequate, retrocalcaneal bursitis may develop. The retrocalcaneal bursa is a small synovial pouch that separates the distal Achilles tendon from the calcaneus. Patients with retrocalcaneal bursitis typically complain of well-localized pain and tenderness near the insertion of the Achilles tendon on the calcaneus. Clinically, this may be mistaken for Achilles tendinitis. However, MR imaging clearly distinguishes these entities (Fig. 13–40). Use of sports shoes with additional padding usually is adequate treatment for retrocalcaneal bursitis.

Tarsal Coalition

Primary tarsal coalition is a lack of joint formation owing to failure of differentiation and segmentation of primitive mesenchyme. A single gene mutation is responsible, and it is transmitted in an autosomal dominant fashion with variable penetrance. Tarsal coalition is relatively common, with a

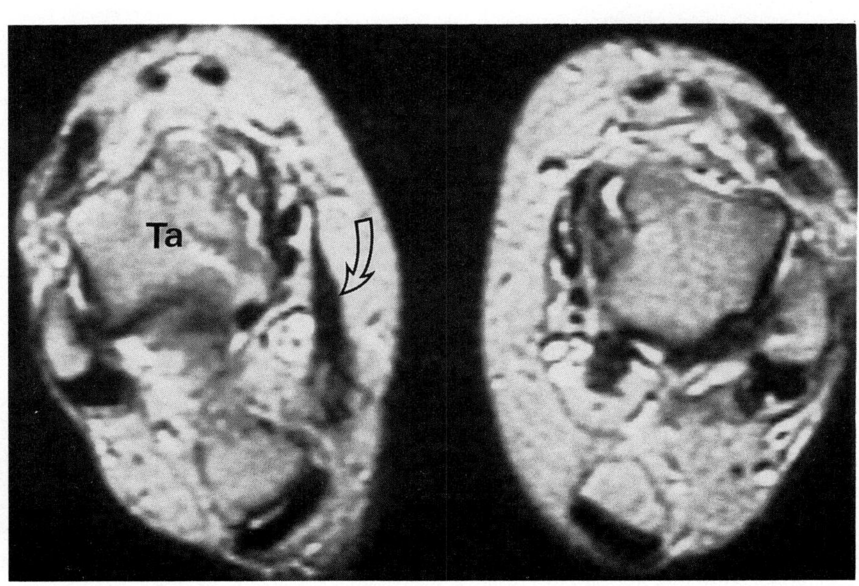

Figure 13–36. Tarsal tunnel syndrome secondary to thickened flexor retinaculum. T2-weighted (TR, 2000 ms; TE, 80 ms) axial images show asymmetric thickening of the flexor retinaculum of the right leg *(arrow)* in comparison to the asymptomatic left side. Ta, Talus.

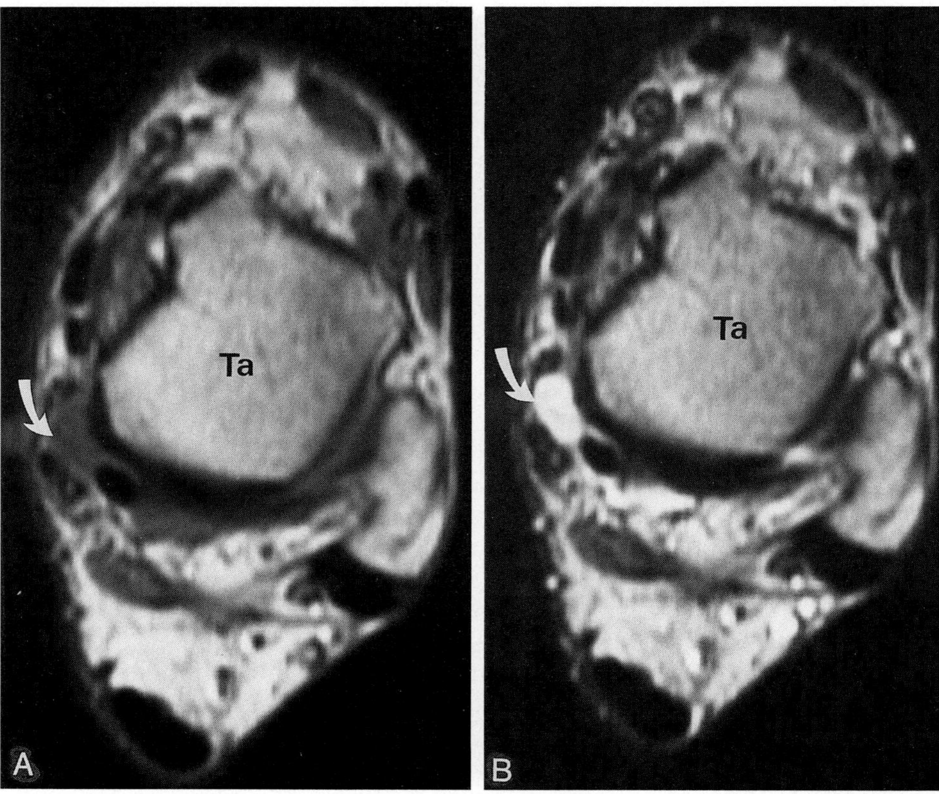

Figure 13–37. Tarsal tunnel syndrome secondary to a ganglion cyst in a 36-year-old man. Axial MR images (*A*: TR, 2000 ms; TE, 20 ms; *B*: TR, 2000 ms; TE, 70 ms) show a soft tissue mass in the tarsal tunnel (*arrow*). Findings at surgery revealed a ganglion cyst compressing the posterior tibial nerve. Ta, Talus.

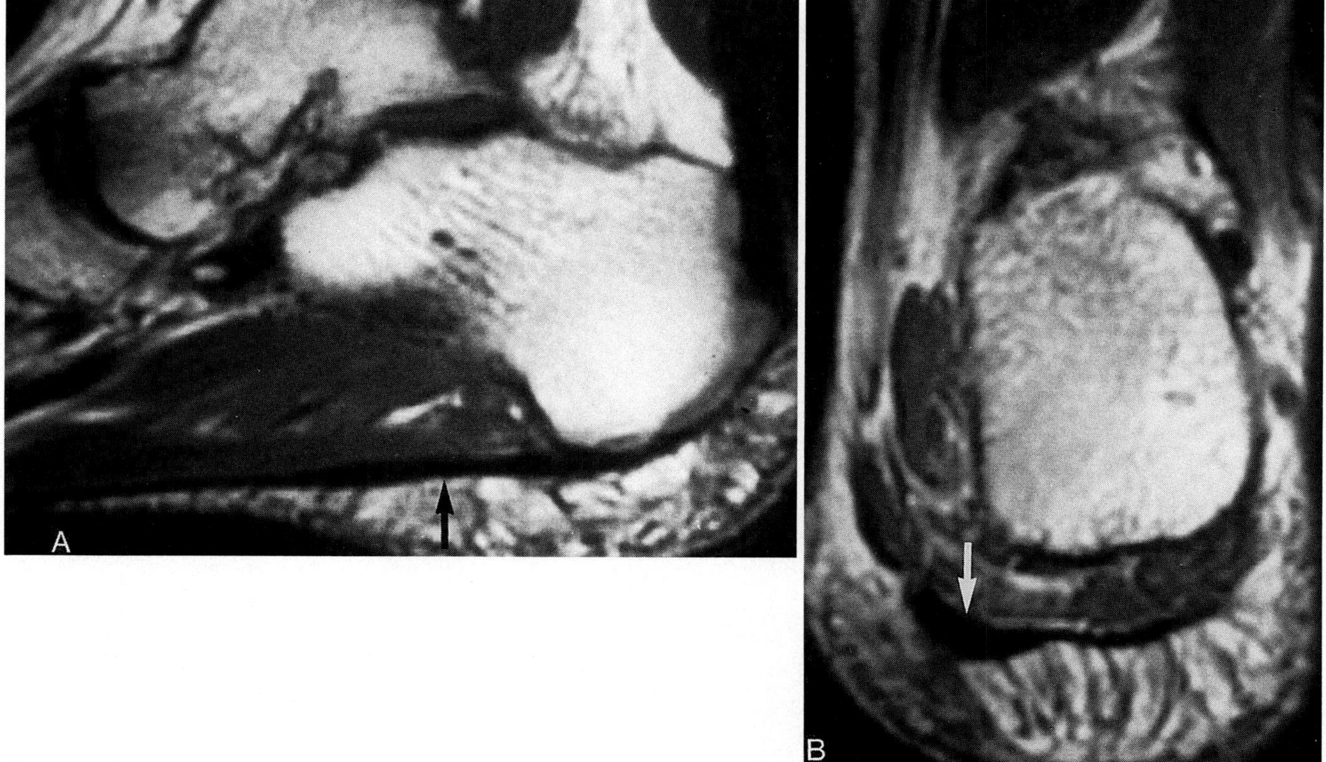

Figure 13–38. Normal plantar fascia in an asymptomatic subject. MR images (*A*: sagittal, TR, 2000 ms; TE, 20 ms; *B*: coronal, TR, 600 ms; TE, 20 ms) show the normal, thin plantar fascia (*straight arrow*). (From Berkowitz JF, Kier R, Rudicel S: Plantar fasciitis: MR imaging. Radiology 179:665, 1991. © Radiological Society of North America.)

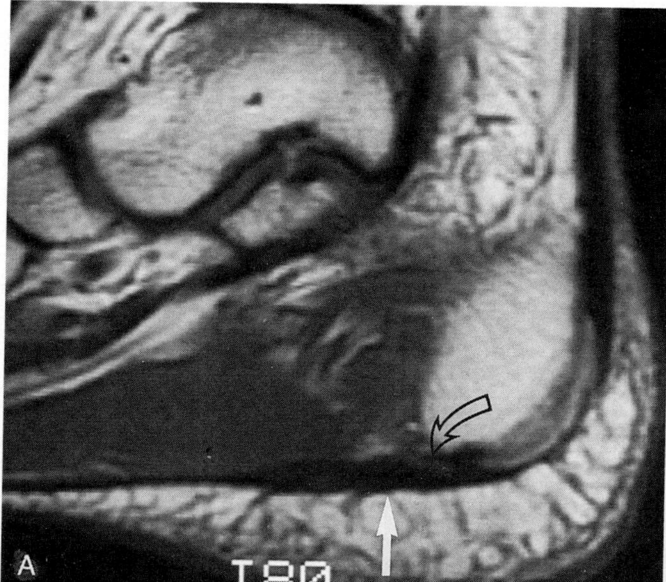

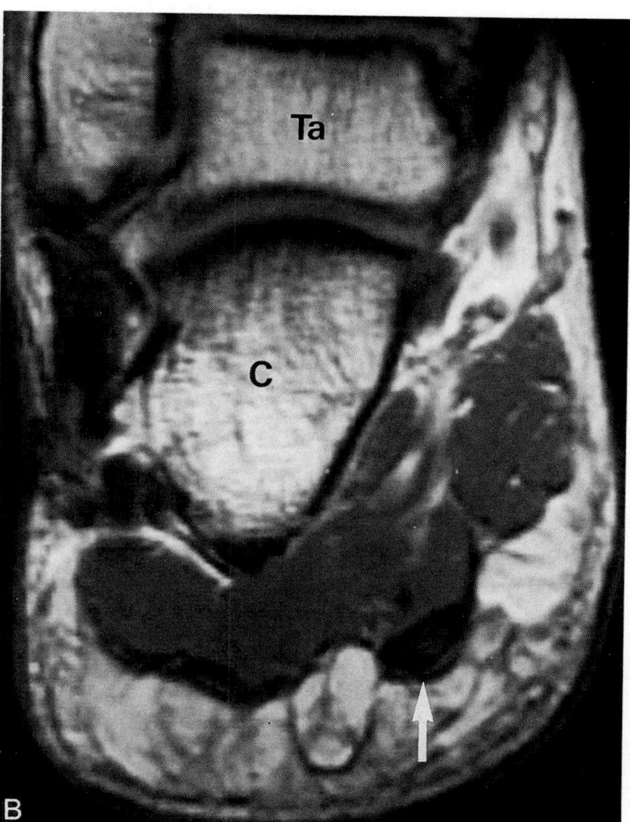

Figure 13–39. Plantar fasciitis in a patient with heel pain. *A*, Sagittal T1-weighted image (TR, 600 ms; TE, 20 ms) shows thickened plantar fascia *(straight arrows)* and calcaneal spur *(curved arrow)*. *B*, The thickened plantar fascia with intrasubstance increased signal intensity is shown on coronal intermediate image (TR, 2200 ms; TE, 20 ms). Ta, Talus; C, calcaneus. (From Berkowitz JF, Kier R, Rudicel S: Plantar fasciitis: MR imaging. Radiology 179:665, 1991. © Radiological Society of North America.)

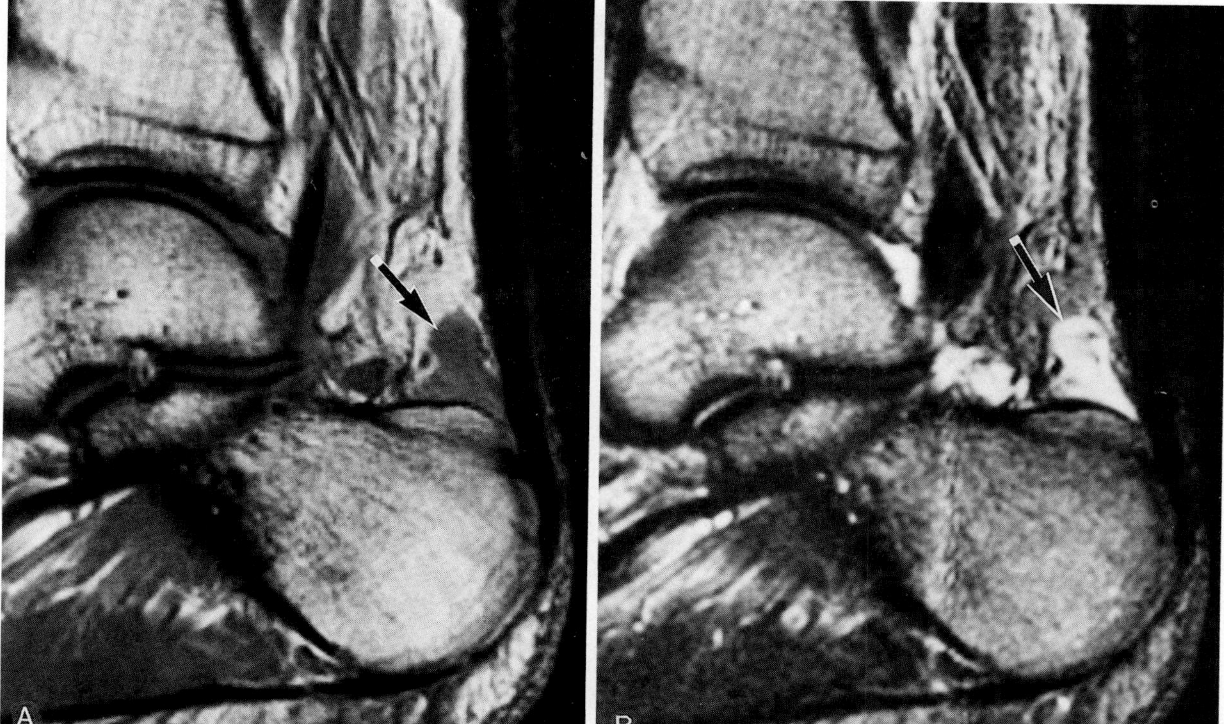

Figure 13–40. Retrocalcaneal bursitis in a 30-year-old man. This patient had been treated conservatively for 6 months for suspected Achilles tendinitis. Sagittal MR images (*A*: TR, 2000 ms; TE, 20 ms; *B*: TR, 2000 ms; TE, 80 ms) show fluid distending the bursa *(arrow)* at the insertion of the normal Achilles tendon into the calcaneus. This patient was subsequently treated by injection of steroids into the bursa, which resulted in an improvement in symptoms. (From Kier R, McCarthy S, Dietz MJ, et al: MR appearance of painful conditions of the ankle. Radiographics 11:401, 1991. © Radiological Society of North America.)

frequency of approximately 1 to 2 per cent.[15] Secondary coalitions resulting from trauma or inflammatory arthritis are less frequent. Talocalcaneal and calcaneonavicular coalitions represent 90 per cent of primary tarsal coalitions, with talonavicular being the third most common.[15] Although these coalitions are isolated findings, coalitions of the more distal forefoot involving the cuboid and cuneiforms are usually syndrome associated, as in otopalatodigital syndrome. Fifty per cent of primary coalitions are bilateral.[15] Presentation is usually between 8 and 16 years of age with subtalar pain and immobility, although the disorder is occasionally also found in adults with tarsal pain. Onset is insidious with ankle sprain, which is a common precipitating factor. Peroneal spastic foot is a classic but uncommon finding. In fact, the clinical picture can be quite misleading, and patients are often mistreated for ankle sprain, tarsal tunnel syndrome, or sinus tarsi syndrome.

Coalitions can be osseous (Fig. 13–41), fibro-osseous (Fig. 13–42), or fibrocartilaginous. Osseous coalitions are actually the least common. Fibro-osseous coalitions exhibit narrowing of the joint space with hypertrophic irregularity of the articular margins but no bony bridging. In fibrocartilaginous coalition, the articulation itself appears open, but a bridging cartilaginous bar is located usually just behind a hypoplastic and mildly irregular sustentaculum tali. Identification of fibrocartilaginous coalitions is important because simple resection of the cartilaginous bar can be curative, whereas osseous and fibro-osseous coalitions require resection and soft tissue interposition to prevent bony regrowth. Early treatment may prevent the development of secondary subtalar arthritis, which once present usually mandates triple arthrodesis.

In light of the prevalence of nonosseous coalitions and the clinical presentation of most cases before skeletal maturation, the poor sensitivity of plain radiography, even with the Harris-Beath view, is not surprising. Computed tomography can demonstrate cortical irregularity of fibro-osseous coalitions and sustentacular deformity in fibrocartilaginous coalition,[19] but the superior soft tissue contrast and multiplanar capability of MR imaging make it an ideal method for evaluating tarsal coalitions. Moreover, MR imaging can be helpful in ruling out other causes of tarsal pain, such as fracture, ligamentous and tendinous injury, and tarsal tunnel and sinus

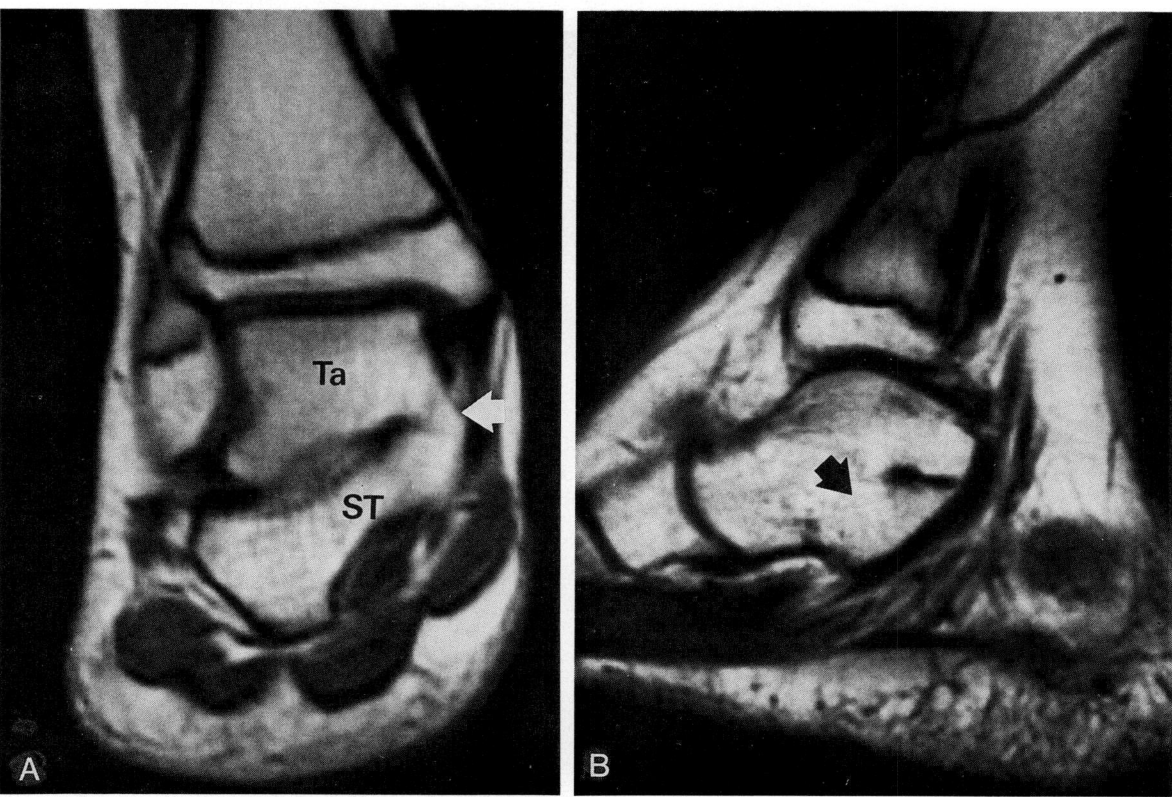

Figure 13–41. Tarsal coalition in a 12-year-old girl who had ankle pain. Coronal *(A)* and sagittal *(B)* T1-weighted (TR, 600 ms; TE, 20 ms) images show fusion *(arrow)* of the talus (Ta) and calcaneus at the level of the sustentaculum tali (ST), which was confirmed at surgery. (From Kier R, McCarthy S, Dietz MJ, et al: MR appearance of painful conditions of the ankle. Radiographics 11:401, 1991. © Radiological Society of North America.)

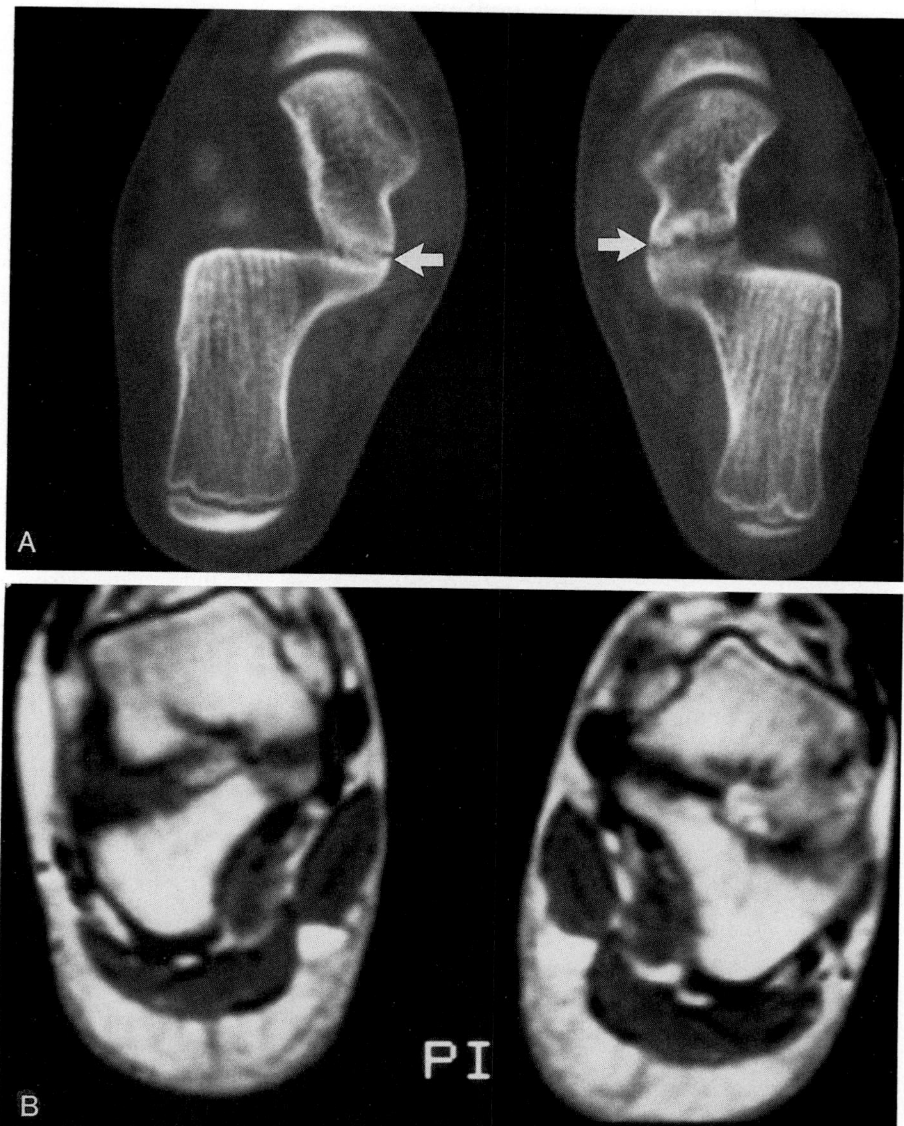

Figure 13–42. Bilateral fibro-osseous coalition of the calcaneo-talar joint. *A*, Axial computed tomography (CT) scan shows irregularity of the articular margins but no bony bridge of the calcaneo-talar joint. *B*, Axial MR T1-weighted images show narrowing of the joint space with fibrotic tissue of low signal intensity of the joint.

tarsi syndromes. The middle (sustentacular) subtalar joint is well visualized in the sagittal and coronal planes. The anterior subtalar joint, which often is continuous with the middle subtalar joint, can be evaluated reliably only in the sagittal plane. The posterior subtalar joint is visible on axial and sagittal images. Calcaneonavicular and talonavicular coalitions also are best demonstrated in the axial and sagittal planes.

PITFALLS

A number of normal variants in MR imaging of the ankle may mimic abnormalities. Awareness of these pitfalls is essential for accurate diagnosis. For example, because of their fanlike configuration, the posterior talofibular and deltoid ligaments have a somewhat frayed and inhomogeneous appearance that can simulate a tear.[24] A small amount of synovial fluid in the shared peroneal tendon sheath is common and should not be mistaken for a longitudinal tear.[24] Fluid in the tendon sheath of the flexor hallucis longus is a normal variant in approximately 20 per cent of the population owing to communication with the ankle joint, and it does not necessarily indicate tenosynovitis. Flow within a normal posterior tibial artery and vein may simulate fluid in the tendon sheath on gradient-echo images (Fig. 13–43). The tibialis posterior tendon normally thickens at its distal insertion; this should not be mistaken for chronic tendinitis.[24] The posterior talofibular and tibiofibular ligaments may mimic loose bodies on sagittal images (Figs. 13–44, 13–45).

Areas of intermediate signal intensity within certain tendons on short TE sequences (T1-weighted

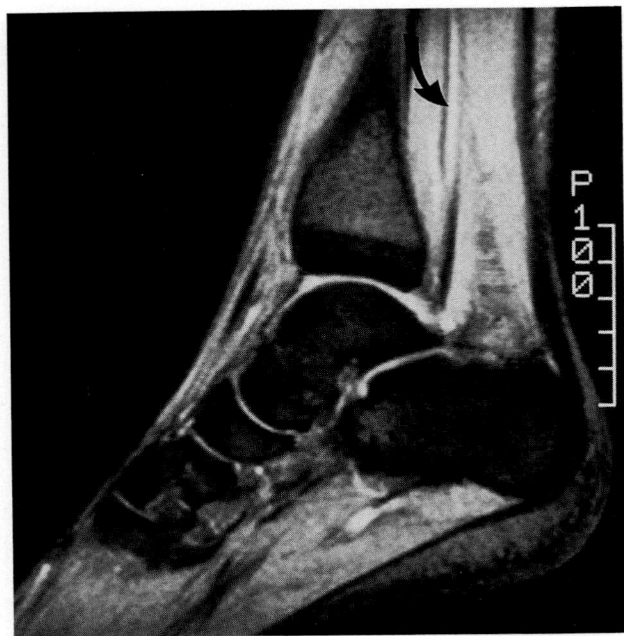

Figure 13–43. Normal posterior tibial artery and vein mimicking fluid in the tendon sheath in a 15-year-old girl. Flow within vessels may appear similar to tenosynovitis on this gradient-echo image (TR, 600 ms; TE, 15 ms; flip angle, 30 degrees).

and proton density–weighted spin-echo sequences and many commonly used gradient-echo sequences) that are due solely to variation of tendinous T2 relaxation with the orientation of the tendon to the static magnetic field (B_0) can mimic tendinitis or rupture (Fig. 13–46). This signal change can be demonstrated in asymptomatic volunteers and has been termed the "magic angle" phenomenon.[11] Because of the structural anisotropy of tendons, their T2 relaxation increases in duration almost one hundredfold (0.25 to 22.0 ms) as their orientation relative to B_0 approaches the magic angle of 55 degrees. Accordingly, under these conditions, signal emerges in otherwise signalless tendons as TEs drop below 30 ms. This phenomenon commonly occurs in the ankle, as several tendons pass through an angle of 55 degrees as they curve into the foot. The tibialis posterior and peroneal tendons are particularly susceptible as they bend around the medial and lateral malleoli, respectively. Disappearance of the tendon signal with longer TE (>60 ms), T2-weighted sequences is extremely helpful in distinguishing magic angle phenomenon from active tendinitis or rupture, both of which exhibit increased signal intensity on T2-eighted images. Reorientation of the tendons by changing foot positioning (plantar flexion) also can reduce this potential artifact but usually is not necessary.

Anomalous leg muscles can be symptomatic, appearing as a soft tissue mass and occasionally being painful. The signal intensity of these muscles is identical to that of the other, normal muscles (Fig. 13–47). Awareness of these entities can yield a definitive diagnosis.[7]

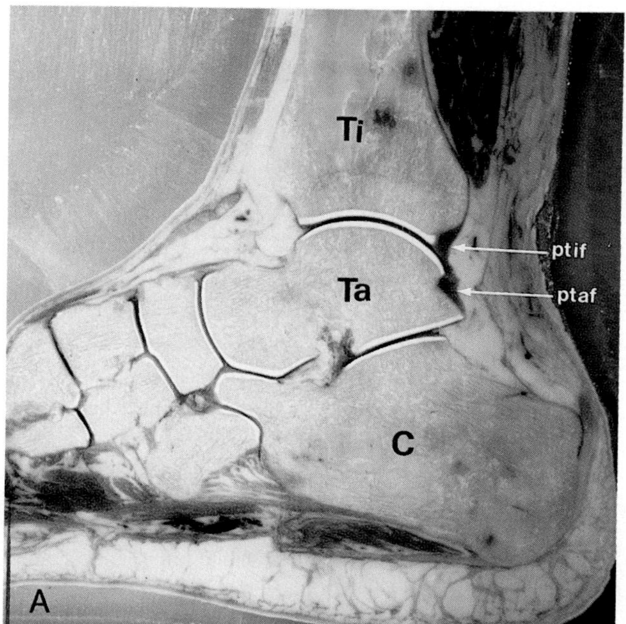

Figure 13–44. Posterior talofibular (ptaf) and tibiofibular (ptif) ligaments mimicking loose bodies. *A*, Cryomicrotome section. *B*, Corresponding intermediate-density image (TR, 2500 ms; TE, 20 ms). *C*, T2-weighted image (TR, 2500 ms; TE, 80 ms) lateral to *B*. Ti, Tibia; Ta, talus; C, calcaneus. (From Erickson SJ, Smith JW, Ruiz ME, et al: MR imaging of the lateral collateral ligament of the ankle. AJR 156:131, 1991. © American Roentgen Ray Society.)

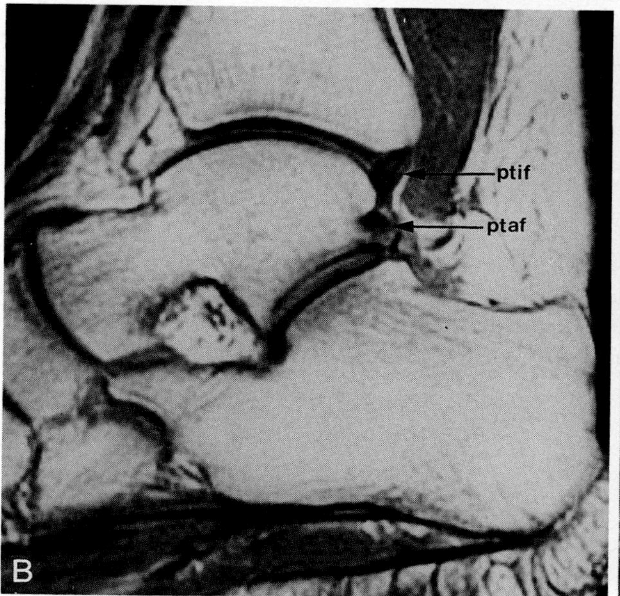

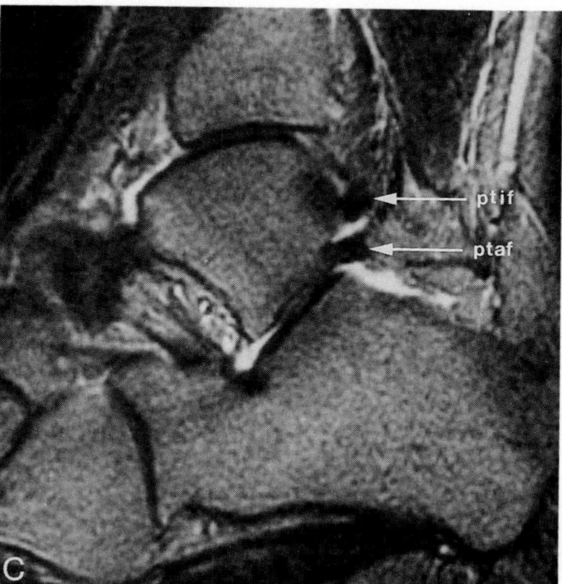

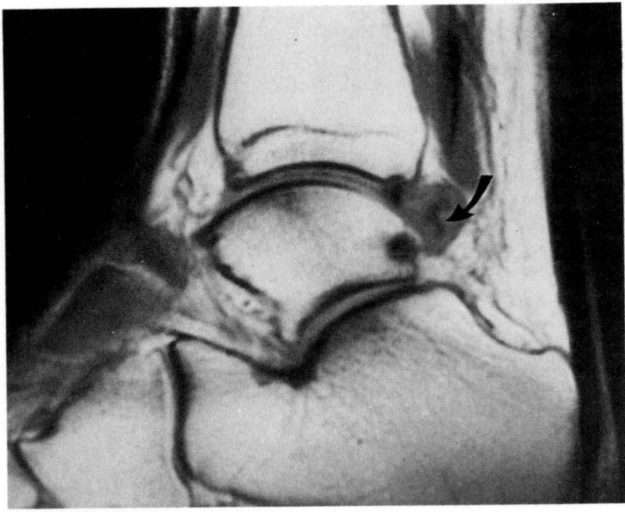

Figure 13–45. Loose body within joint fluid in a patient with detached osteochondritis dissecans *(arrow)*.

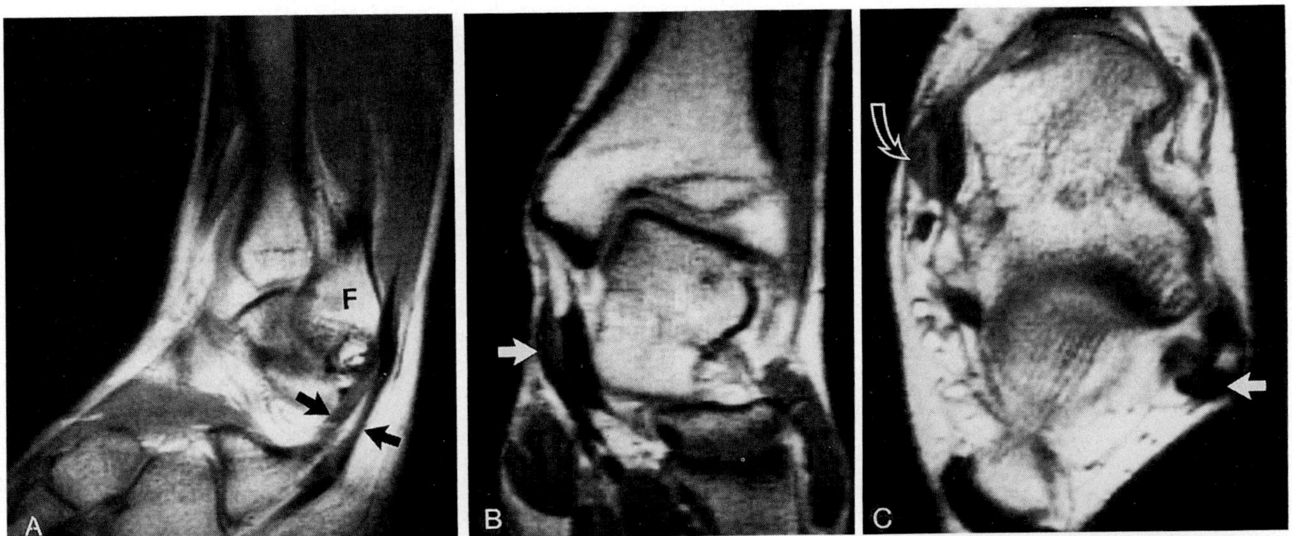

Figure 13–46. Effect of the "magic angle" phenomenon on ankle tendon. T1-weighted images (*A*: sagittal, TR, 753 ms; TE, 30 ms; *B*: coronal, TR, 753 ms; TE, 30 ms; *C* and *D*: axial, TR, 753 ms; TE, 30 ms) show relatively increased signal intensity of the peroneal tendons *(arrow)* and posterior tibial tendon *(curved arrow)*. *E* and *F*, On the T2-weighted image (TR, 1500 ms; TE, 85 ms), all tendon segments are of very low signal intensity, which was independent of orientation. F, Fibula.

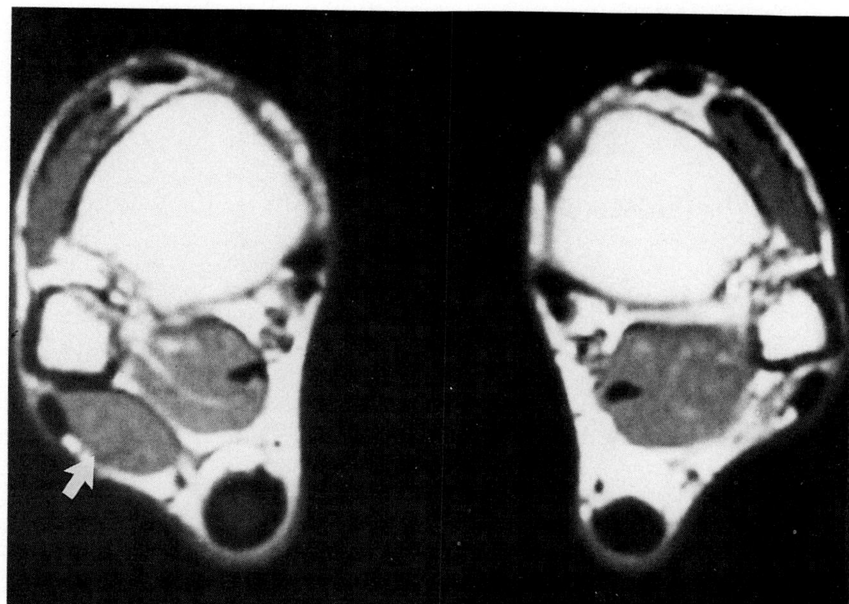

Figure 13–47. Anomalous muscle simulating a soft tissue tumor in a 49-year-old man. T1-weighted axial image (TR, 600 ms; TE, 20 ms) shows a mass posterior to the fibula and adjacent to the peroneal tendon *(arrow)* of the right leg. The mass is of equal signal intensity with muscle. This anomalous muscle should not be mistaken for a pathologic process. Incidentally, abnormal enlargement of the right Achilles tendon with internal signal is noted, suggesting tendinitis there.

References

1. Alexander IJ, Johnson KA, Berquist TH: Magnetic resonance imaging in the diagnosis of the posterior tibial tendon. Foot Ankle 8:144, 1987.
2. Anderson IF, Crichton KJ, Nest C, et al: Osteochondral fractures of the dome of the talus. J Bone Joint Surg [Am] 1989:1143, 1989.
3. Beltran J, Campanini S, Knight C, et al: The diabetic foot: Magnetic resonance imaging evaluation. Skeletal Radiol 19:37, 1990.
4. Berkowitz JF, Kier R, Rudicei S: Plantar fasciitis: MR imaging. Radiology 179:665, 1991.
5. Berndt AL, Harty M: Transchondral fractures (osteochondritis dissecans) of the talus. J Bone Joint Surg [Am] 41:988, 1959.
6. Berquist TH: MRI of the Musculoskeletal System. New York, Raven Press, 1990.
7. Buschmann WR, Cheung Y, Jahss MH: Magnetic resonance imaging of anomalous leg muscles: Accessory soleus, peroneus quartus and the flexor digitorum longus accessorius. Foot Ankle 12:109, 1991.
8. Canale ST, Belding RH: Osteochondral lesions of the talus. J Bone Joint Surg [Am] 62:97, 1980.
9. DeSmet AA, Fisher DR, Burnstein MI, et al: Value of MR imaging in staging osteochondral lesions of the talus (osteochondritis dissecans): Results in 14 patients. AJR 154:555, 1990.
10. Edwards M: The relations of the peroneal tendons to the fibula, calcaneus and cuboideum. Am J Anat 42:213, 1928.
11. Erickson SJ, Cox IH, Hyde JS, et al: Effect of tendon orientation on MR imaging signal intensity: A manifestation of the "magic angle" phenomenon. Radiology 181:389, 1991.
12. Erickson SJ, Quinn SF, Kneeland JB, et al: MR imaging of the tarsal tunnel and related spaces: Normal and abnormal findings with anatomic correlation. AJR 155:323, 1990.
13. Erickson SJ, Smith JW, Ruiz ME, et al: MR imaging of the lateral collateral ligament of the ankle. AJR 156:131, 1991.
14. Flick AB, Gould N: Osteochondritis dissecans of the talus (transchondral fractures of the talus): Review of the literature and new surgical approach for medial dome lesions. Foot Ankle 5:165, 1985.
15. Helal B, Wilson D: The Foot. London, Churchill Livingstone, 1988.
16. Kerr R, Frey C: MR imaging in tarsal tunnel syndrome. J Comput Assist Tomogr 15:280, 1991.
17. Kerr R, Henry D: Posterior tibial tendon rupture. Orthopedics 12:1394, 1989.
18. Kneeland JB, Macrandar S, Middleton WD, et al: MR imaging of the normal ankle: Correlation with anatomic sections. AJR 151:117, 1988.
19. Lee MS, Harcke HT, Kumar SJ, et al: Subtalar joint coalition in children: New observations. Radiology 172:635, 1989.
20. Liem MD, Zegel HG, Balduini FC, et al: Repair of Achilles tendon: Ruptures with a polylactic acid implant: Assessment with MR imaging. AJR 156:769, 1991.
21. Moore TE, Yuh WTC: Abnormalities of the foot in patients with diabetes mellitus: Findings on MR imaging. AJR 157:813, 1991.
22. Nelson DW, DiPaola J, Colville M, et al: Osteochondritis dissecans of the talus and knee: Prospective comparison of MR and arthroscopic classifications. J Comput Assist Tomogr 14:804, 1990.
23. Nistor L: Conservative treatment of fresh subcutaneous rupture of the Achilles tendon. Acta Orthop Scand 47:459, 1976.
24. Noto AM, Cheung Y, Rosenberg ZS, et al: MR imaging of the ankle: Normal variants. Radiology 170:121, 1989.
25. Pritsch M, Horoshovski H, Farine I: Arthroscopic treatment of osteochondral lesions of the talus. J Bone Joint Surg [Am] 68:862, 1986.
26. Quinn SF, Murray WT, Clark RA, et al: Achilles tendon: MR imaging at 1.5 T. Radiology 164:767, 1987.
27. Rosenberg ZS, Cheung Y, Jahss MH, et al: Rupture of posterior tibial tendon: CT and MR imaging with surgical correlation. Radiology 169:229, 1988.
28. Rosenberg ZS, Feldman F, Singson RD: Peroneal tendon injuries: CT analysis. Radiology 161:743, 1986.
29. Rosenberg ZS, Jahss MH, Noto AM, et al: Rupture of the posterior tibial tendon: CT and surgical findings. Radiology 176:489, 1988.
30. Yuh WTC, Corson JD, Baraniewski HM, et al: Osteomyelitis of the foot in diabetic patients: Evaluation with plain film, ^{99m}Tc-MDP bone scintigraphy, and MR imaging. AJR 152:795, 1989.
31. Yulish BS, Mulopulos GP, Goodfellow DB, et al: MR imaging of osteochondral lesions of talus. J Comput Assist Tomogr 11:296, 1987.
32. Zeiss J, Fenton P, Ebraheim N, et al: Normal magnetic resonance anatomy of the tarsal tunnel. Foot Ankle 10:214, 1990.
33. Zeiss J, Saddemi SR, Ebraheim NA: MR imaging of the peroneal tunnel. J Comput Assist Tomogr 13:840, 1989.

14 THE TEMPOROMANDIBULAR JOINT

Wing P. Chan and Clyde A. Helms

Magnetic resonance (MR) imaging of the temporomandibular joint (TMJ) is gaining recognition as a replacement for arthrography and computed tomography (CT) and is becoming the imaging method of choice for suspected internal derangements. It affords an accurate and noninvasive assessment of the disk position and the morphology of the TMJ. This knowledge is of particular value to the surgeon in preoperative staging and in postoperative follow-up studies.

IMAGING PROTOCOLS

At the University of California, San Francisco, the MR imaging protocol for evaluating internal derangement of the TMJ using a field strength of 1.5 T includes T1-weighted spin-echo sequences acquired in the axial (as a localizer) and sagittal planes (Table 14–1). A surface coil of 3 inches in diameter is placed over the diseased joint. This coil improves the signal-to-noise ratio and provides high-resolution images. Alternatively, dual 3-inch coils can image both joints simultaneously in the same time as that required for a unilateral study. This design is useful because up to 80 per cent of patients have bilateral displaced disks.[8,13] However, because usually the most symptomatic joint is treated and the results are satisfactory, it is not necessary to image both joints routinely unless bilateral surgery is being entertained.

A scout axial spin-echo image serves as a localizer for subsequent sagittal images. Axial images are generated with a repetition time (TR) of 600 ms and an echo time (TE) of 20 ms. An axial image obtained at the level of the heads of mandibular condyles is used to achieve optimal angulation for the oblique sagittal or coronal images of the TMJ (Fig. 14–1). Oblique sagittal and coronal images are obtained perpendicular and parallel, respectively, to the condylar head from the scout axial image. Alternatively, straight sagittal images are thought to have an identical resolution if the image thickness is 3 mm or less.[10] Oblique coronal images are used to assess medial or lateral disk displacement.

Sagittal T1-weighted spin-echo images are acquired using a TR of 500 ms and a TE of 20 ms. A closed-mouth position and a partially opened-mouth position are routinely employed at the University of California, San Francisco. Precise localization of the disk position occurs more consistently with a closed-mouth position than with a partially opened-mouth view.[5] Conversely, the partially opened-mouth view allows for better visualization of the disk morphology and signal intensity characteristics.[12,19] This is because the condylar head moves slightly out of the glenoid fossa, allowing the disk, of predominantly low signal intensity, to be separated from the bony cortices of the condylar head and glenoid fossa, also of low signal intensity. Studies are seldom obtained in the full opened-mouth position because the patient cannot maintain a fully opened mouth for a long time, even with a bite block in place, without discomfort and motion. In addition, because disk reduction is a temporary finding and easily diagnosed by clinicians, the full opened-mouth position is not recommended.

Other parameters for sagittal images include a 256 × 192 acquisition matrix, a 12-cm field of view.

Note: All sagittal images in this chapter are oriented with the patient's anterior (face) on the viewer's left, except for Figures 14–2 and 14–13, which are oriented with the patient's anterior on the viewer's right.

Table 14–1. PROTOCOLS OF MAGNETIC RESONANCE IMAGING OF THE TEMPOROMANDIBULAR JOINT AT THE UNIVERSITY OF CALIFORNIA, SAN FRANCISCO

Plane	TR (ms)	TE (ms)	FOV (cm)	Slice (mm)	Skip (mm)	NEX	Matrix (pixel)	Time (min)
Axial (localizer) (SE)	600	20	20	5	1.5	1	192	1~2
Sagittal[1] (SE)	500	20	12	3	0	4	192	4~5
Sagittal[2] (SE)	500	20	12	3	0	4	192	4~5
Sagittal* (FSE)	3500	80	12	3	0.5	4	192	5~6

TR, Repetition time; TE, echo time; FOV, field of view; Skip, interslice gap; NEX, number of excitations; SE, spin echo; FSE, fast spin echo; [1]closed-mouth position; [2]partially opened-mouth position; *optional sequence.

two excitations, and a 3-mm slice thickness. There is no interslice gap. The total imaging time is approximately 6 min. A 12-cm field of view and a 256 × 192 acquisition matrix provide a spatial resolution of 0.63 mm in the phase-encoded direction and of 0.47 mm in the frequency-encoded direction.

In selected cases of joint inflammation, trauma, or suspicion of abnormal joint effusion, multiecho sequences may be employed with a TR of 2000 ms, and TEs of 20 and 80 ms combined with a 256 × 192 acquisition matrix. Conventional T2-weighting is helpful in highlighting joint effusion. However, the scan time is lengthened considerably and the T2-weighting does not add any significant information in demonstrating the disk position and morphology.

Gradient-echo techniques have shown promising results in TMJ imaging when an intermediate flip angle of 20 to 30 degrees is used.[20,23] The major advantage of these techniques is their relative speed of acquisition, which may allow them to replace the time-consuming T2-weighted spin-echo sequences.

Although complementary coronal views have been recommended,[28] we do not routinely employ these views because pure medially and laterally displaced disks are rare.

ANATOMY AND PHYSIOLOGY

The TMJ is a synovial joint formed by the head of the mandibular condyle and bony glenoid fossa and the articular eminence of the temporal bone. The disk is separated from the condyle by a lower joint (or inferior compartment) and from the temporal

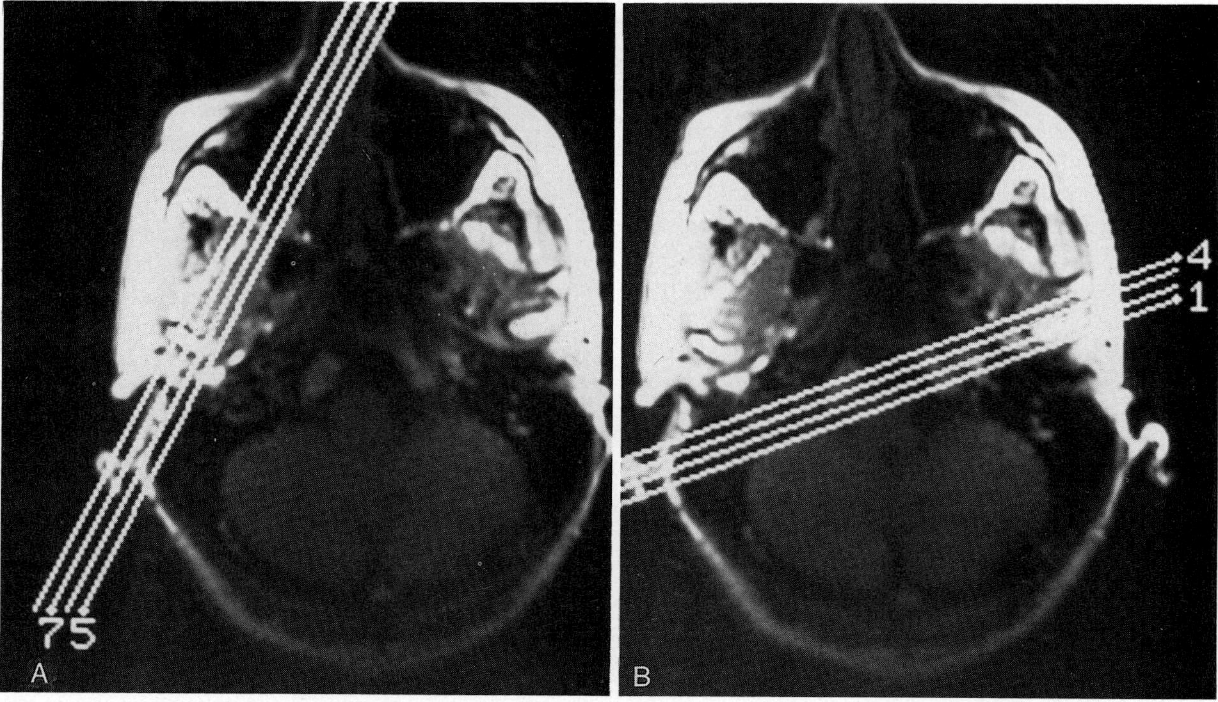

Figure 14–1. Axial scout images (TR, 300 ms; TE, 20 ms) at the level of the external auditory canals. The cursor lines are parallel *(A)* and perpendicular *(B)* to the condylar heads to obtain the oblique sagittal and coronal views, respectively.

bone by an upper joint (or superior compartment). The lower and upper joints are not contiguous except when the disk is perforated. The disk has a bow-tie shape, with a thin intermediate zone connecting a less thick anterior band and a slightly thicker posterior band (Fig. 14–2; see also Fig. 1–33). The intermediate zone is the articular surface of the TMJ disk. The anterior band is attached anteriorly to the superior belly of the lateral pterygoid muscle. The posterior band is attached posteriorly to the bilaminar zone, which consists of loose, vascular connective tissue and nerves that innervate the disk. The disk is firmly attached medially and laterally to the mandibular condyle by collateral ligaments.

Both TMJs function synchronously. In the closed-mouth position, the junction of the posterior band of the disk and the bilaminar zone must be within 10 degrees of the 12 o'clock (or vertical) position relative to the mandibular condyle.[6] When the mouth is opened, the condyle rotates in the articular surface of the disk. The disk moves posteriorly relative to the condyle and then the disk-condyle complex translates forward beneath the eminence of the temporal bone. With closing, these relationships reverse. Another way to judge the position of the disk is by the location of the intermediate zone of the disk, which maintains a constant relationship between the two most closely apposed cortical surfaces of the condyle and the eminence in all degrees of opening (Figs. 14–3, 14–4).[10,19]

A normal disk is flexible and pliable, whereas disks with internal derangements are rigid and inflexible tissue.[14,15,18] The disk in the TMJ is characterized by extensive collagen deposition. The collagen fibrils are densely arranged over the entire disk as well as in the anterior portion of the bilaminar zone, corresponding to the area of low signal intensity seen with MR images in all pulse sequences; an exception is the central portion of the posterior band, where the collagen is less dense and more randomly arranged, consistent with the higher signal intensity seen with T1-weighted images (Fig. 14–5).[6]

DISKS

Internal Derangements

Internal derangement is the most common TMJ disorder and consists of anterior disk displacement. It afflicts 4 to 28 per cent of adults in the United

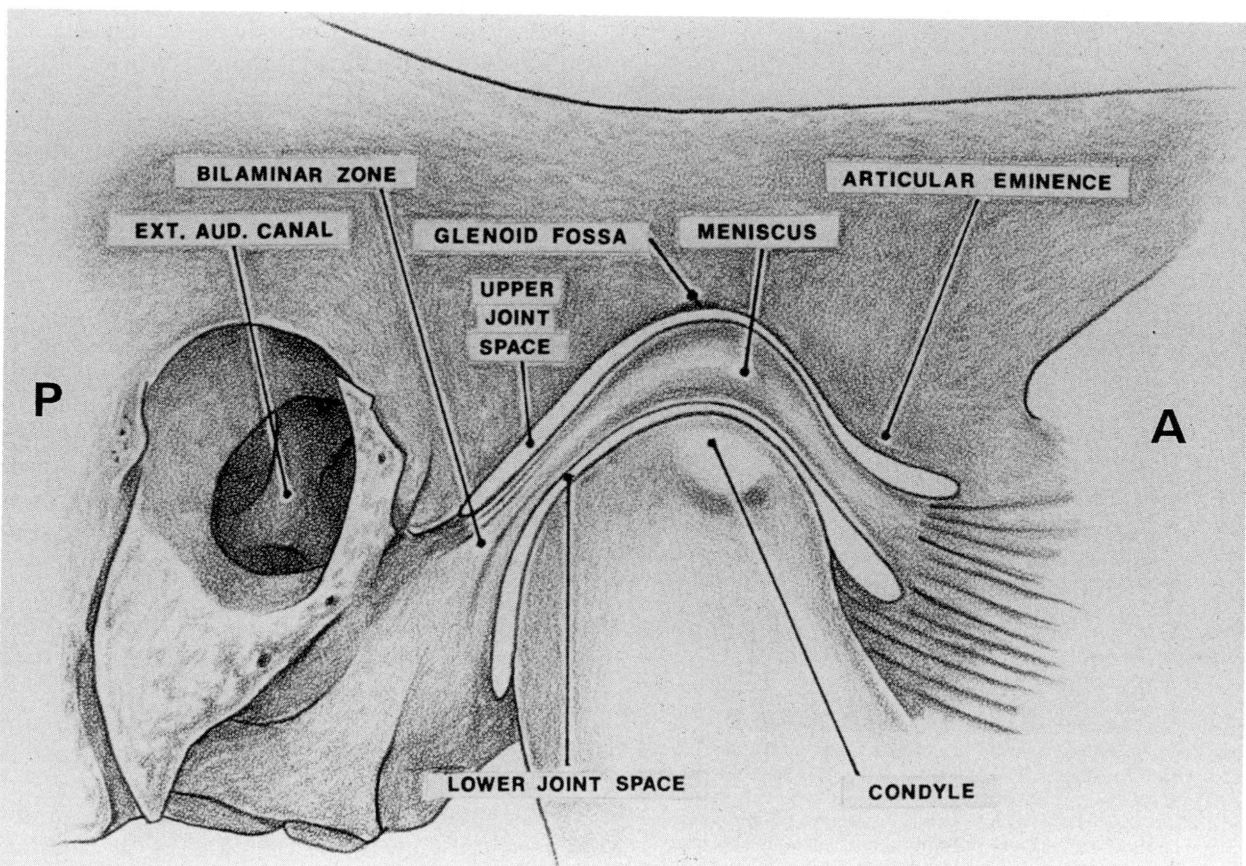

Figure 14–2. Illustration of the normal anatomy of the temporomandibular joint. A, Patient's anterior; P, patient's posterior.

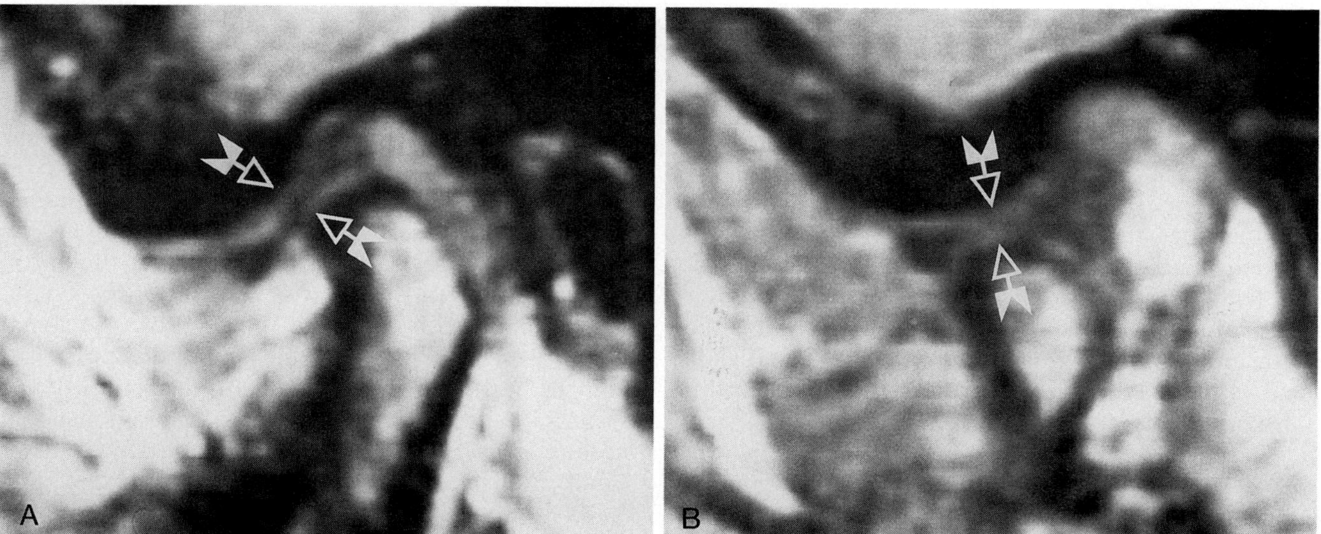

Figure 14–3. Normal disk of the left temporomandibular joint in a 40-year-old woman. T1-weighted (TR, 500 ms; TE, 20 ms) sagittal images (*A*, closed mouth; *B*, partially opened mouth) show the intermediate zone between the two most closely apposed bony surfaces of the condyle and the eminence (*arrows*). The posterior margin of the posterior band is in a vertical position relative to the condyle.

States.[29] Most of these patients are women of child-bearing age. In mild cases, an anteriorly displaced disk can be identified with the closed-mouth position. With mouth opening, the condyle translates under the posterior band of the displaced disk. At the moment when the displaced disk returns to its normal position, a painful and audible "click" may be heard on physical examination. This is clinically termed an anteriorly displaced disk with reduction. At full opening, the condyle and disk are in normal position. With closing, the condyle returns to the fossa, but the disk remains displaced anteriorly.[9] In severe cases, the anteriorly displaced disk can block the anterior translation of the condyle, limiting the mouth opening, and the locked joint is observed with deviation of the jaw (not the disk) to the affected side. The anteriorly displaced disk does not return to its normal position with mouth opening, a condition termed anteriorly displaced disk without reduction.[9] The pain in the joint may be attributed to the neurovascular innervation in the posterior band of the disk and the bilaminar zone.

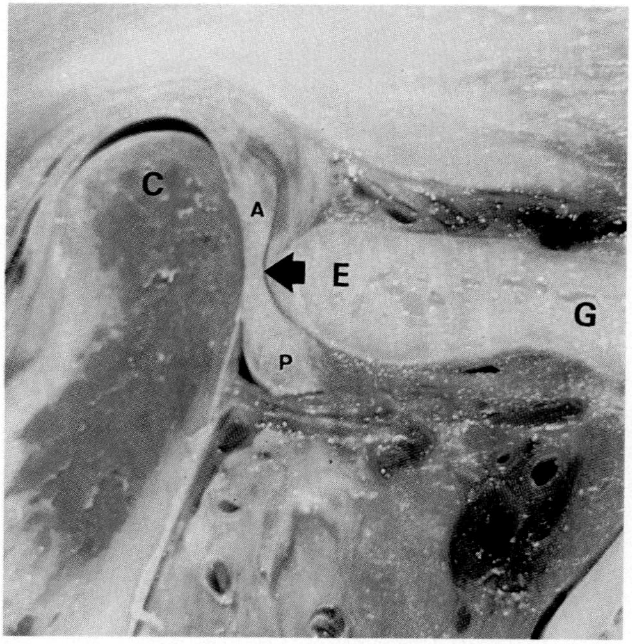

Figure 14–4. Specimens from a cadaver with extreme hyper-opened-mouth position. The condyle (C) is far anterior to the glenoid fossa (G) and the intermediate zone (*arrow*) yet maintains a position between the two most closely apposed bony surfaces. The disk is flexible and pliable and bends around the curved portion of the eminence (E) during mouth opening. A, Anterior band; P, posterior band.

The lateral pterygoid muscle tends to pull the disk anteromedially. This tendency normally is opposed by the elastic force of the bilaminar zone, which maintains the disk in its normal position, as mentioned earlier. A ruptured bilaminar zone results in anteromedial displacement of the disk because the pulling force of the lateral pterygoid muscle is unopposed.[7] This mechanical factor may in part contribute to the anterior (or strictly speaking, anteromedial) disk displacement. Pure medial or lateral displacement of the disk is uncommon. Posterior disk displacement has not been reported. In addition, the shape and position of the disk may be adversely affected by an abnormal bony structure or intra-articular pressure of the joint.

An anteriorly displaced disk is easily identified on a sagittal view of the TMJ (Fig. 14–6). Sagittal T1-weighted spin-echo pulse sequences are essential in

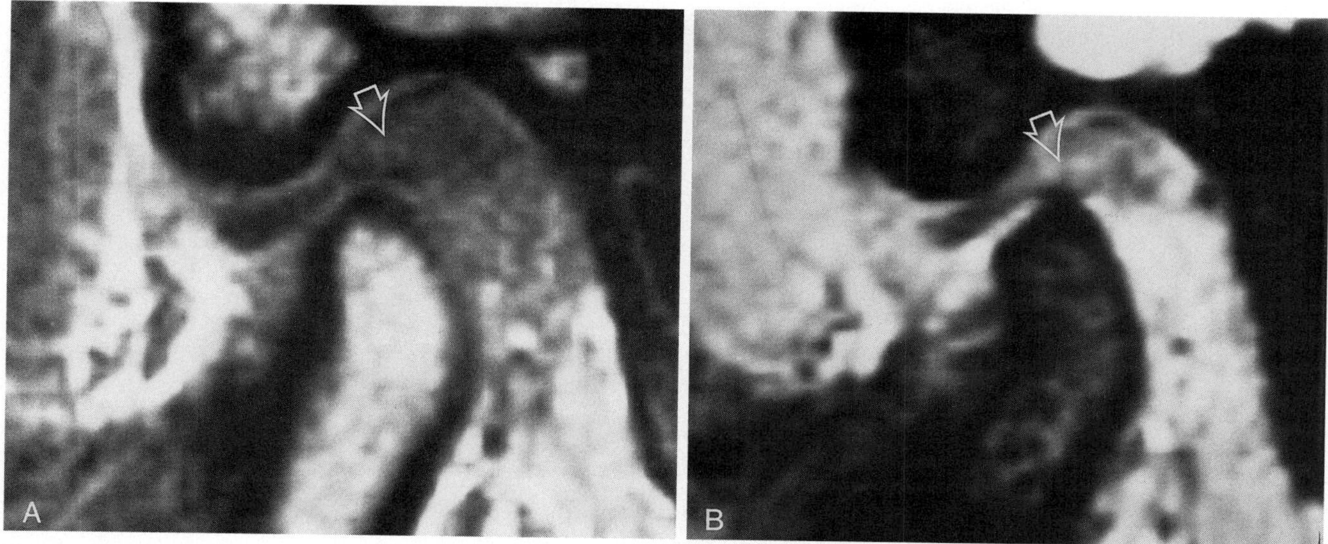

Figure 14–5. Normal disk of the left temporomandibular joint in a 35-year-old woman. Partially opened-mouth sagittal images (*A*: TR, 500 ms; TE, 20 ms; *B*: gradient-recalled acquisition of steady state [GRASS], TR, 33 ms; TE, 15 ms; flip angle, 10 degrees) show the relatively high signal intensity in the center of the posterior band (*arrow*), which is more prominent on the GRASS image. The condyle translates anteriorly beneath the eminence, allowing the disk to be distinguished from the dark cortical surfaces of the condylar head and glenoid fossa. The morphology and signal intensity of the disk are better assessed with this position.

evaluating the morphology and position of the TMJ disk. The anterior band normally is of homogeneous low signal intensity and has a triangular appearance in the sagittal plane with its apex joined to the thinner intermediate zone, which also is of low signal intensity. The central portion of the posterior band normally has a higher signal intensity than that of the anterior band and the intermediate zone (see Fig. 14–5). The inferior and superior margins of the posterior band are uniformly thin and of low signal intensity, whereas the posterior margin is slightly thicker, vertically oriented, linear, and of low signal intensity. Occasionally, this vertically oriented posterior margin is distinct from the central portion of the posterior band, of higher signal intensity, and the bilaminar zone. The bilaminar zone is seen as two dark lines superiorly and inferiorly to a central zone of higher signal intensity. In 95 per cent of asymptomatic persons, the junction of the posterior band of the disk with the bilaminar zone is within 10 degrees from the 12 o'clock position (in relation to the condyle) in the closed-mouth position.[6] This vertical junction of low signal intensity is a diagnostic landmark in assessing anterior disk displacement, which accounts for most of the internal derangements in the TMJ. On T2-weighted spin-echo images, the posterior band is of uniformly low signal intensity without visible contrast between its center and its margins.

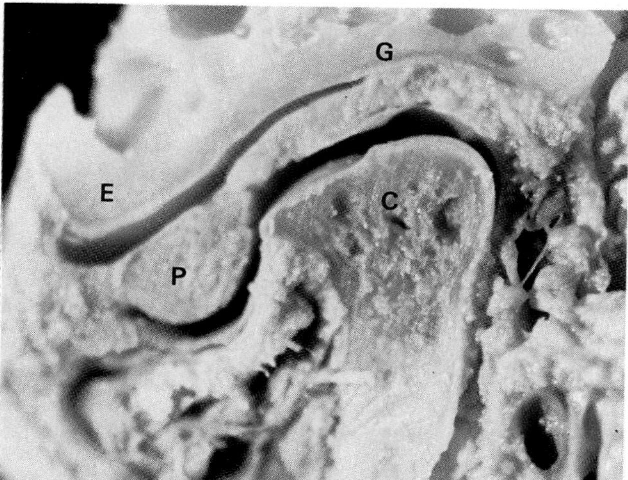

Figure 14–6. Sections from a cadaver with an anteriorly displaced disk. The posterior band (P) is markedly displaced anteriorly in relation to the condyle (C). If the entire disk is located in the anterior recess, the normal translation of the condyle can be blocked. Clinically a locked joint was noted on physical examination. G, Glenoid fossa; E, articular eminence.

In slightly anteriorly displaced disks, the thin intermediate zone is no longer between the closely apposed surfaces of the condyle and the eminence in the closed-mouth position (Fig. 14–7). Using the posterior band for judging the disk position, an anterior disk displacement of 20 degrees can be assessed by inspection. An associated buckling of the disk can be seen when the disk is displaced anteriorly 30 to 40 degrees. In severe cases (80 to 90 degrees of displacement), the disk is usually anterior to the condyle and has a distorted morphology, which commonly does not allow reduction on opening of the mouth.[5]

Helms and co-workers have categorized internal

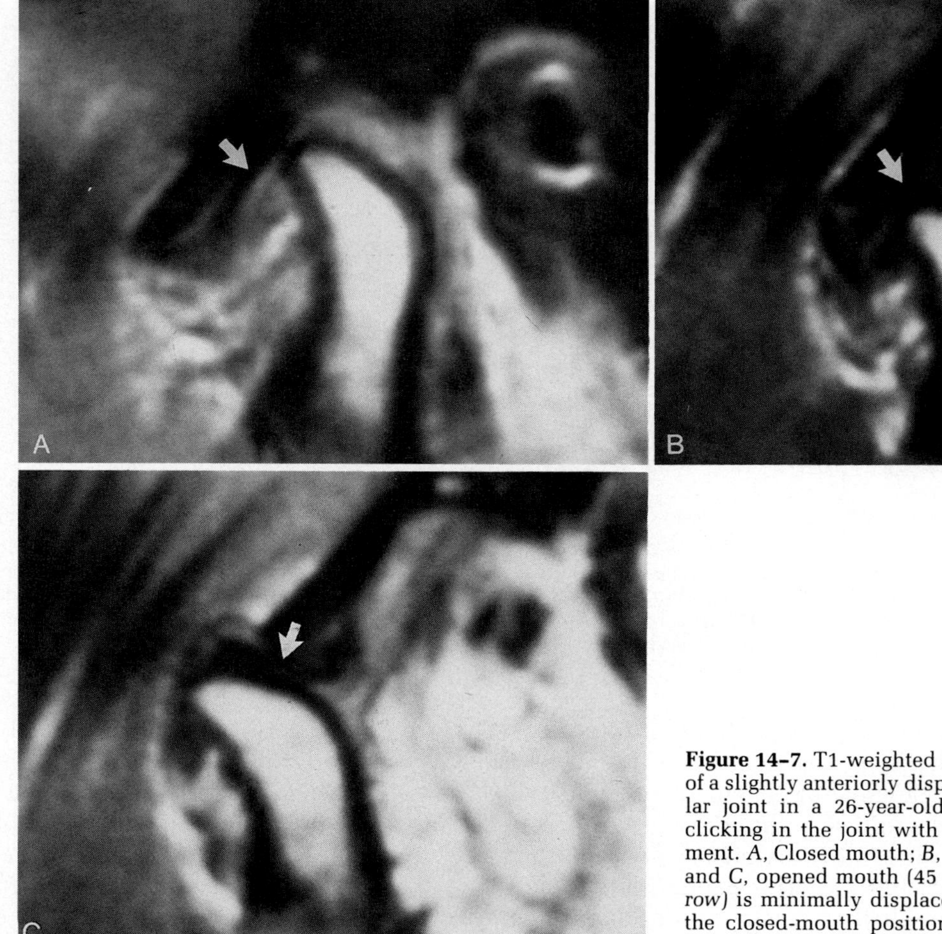

Figure 14–7. T1-weighted sagittal images (TR, 800 ms; TE, 20 ms) of a slightly anteriorly displaced disk of the left temporomandibular joint in a 26-year-old woman. This patient had pain and clicking in the joint with clinically suspected internal derangement. *A*, Closed mouth; *B*, partially opened mouth (10 mm open); and C, opened mouth (45 mm open). The intermediate zone *(arrow)* is minimally displaced anteriorly and is detectable only in the closed-mouth position. If the posterior band were used to judge the proper position, this case would be underestimated.

derangements into two grades of severity according to disk morphology (Fig. 14–8).[11,12] On a normal joint both disk position and morphology are normal. A grade I joint maintains its normal bow-tie configuration, but the disk is displaced anteriorly (Fig. 14–9; see also Fig. 14–18). Disk plication is the preferred treatment for these patients. A grade II joint has a deformed disk that also is displaced anteriorly (Fig. 14–10; see also Fig. 14–17). Disk plication may not relieve patients' symptoms satisfactorily, and more aggressive therapy, such as diskectomy, is required. This grading system correlates well with the severity of the disease process and duration of symptoms. The disk loses its normal hydration as internal derangements progress, resulting in a homogeneous signal of low intensity on T1-weighted images. The higher signal intensity of the central zone of the posterior band is no longer apparent.

Sagittal T2-weighted spin-echo or T2*-weighted gradient-echo images are helpful in detecting effusion (Figs. 14–11, 14–12). Joint effusion is uncommon in asymptomatic persons.[5] Magnetic resonance imaging is limited in detecting disk perforations, although extremely large lesions have been demonstrated in one report.[1] However, this drawback does not bother clinicians because disk perforation is not a surgical indication. In particular cases, arthrography can be performed if disk perforation is really a concern for the clinicians.

The Postoperative Disk

Surgery on the TMJ is indicated when conservative therapy has failed to resolve disk displacements and significant symptoms remain. The surgical procedures that may be performed include disk plication, diskoplasty, diskectomy, diskectomy with placement of a synthetic disk implant, and diskectomy with placement of autogenous tissue (e.g., dermal or cartilage grafts).

Disk plication involves excision of approximately 2 mm of the stretched retrodiskal tissue to allow for repositioning of the displaced disk back on top of the condyle, with the posterior border of the disk then being firmly sutured to the posterior at-

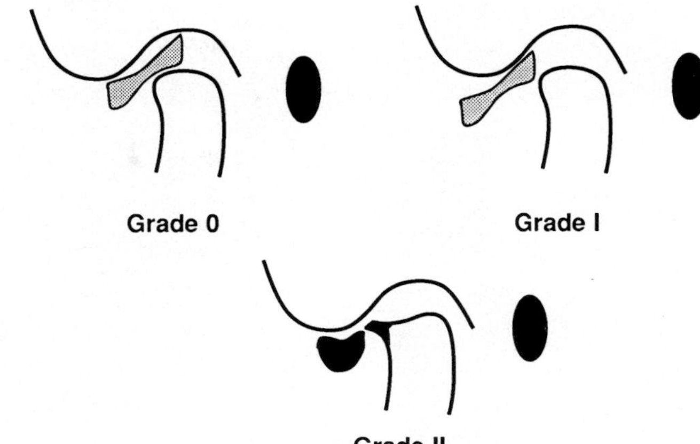

Figure 14–8. Grading system of the severity of the internal derangement. A grade 0 joint has a bow-tie disk in a normal position. A grade I joint has a bow-tie disk that is displaced anteriorly. A grade II joint has a deformed disk that is displaced anteriorly. An anterior osteophyte in the condylar head is frequently noted with grade II joints.

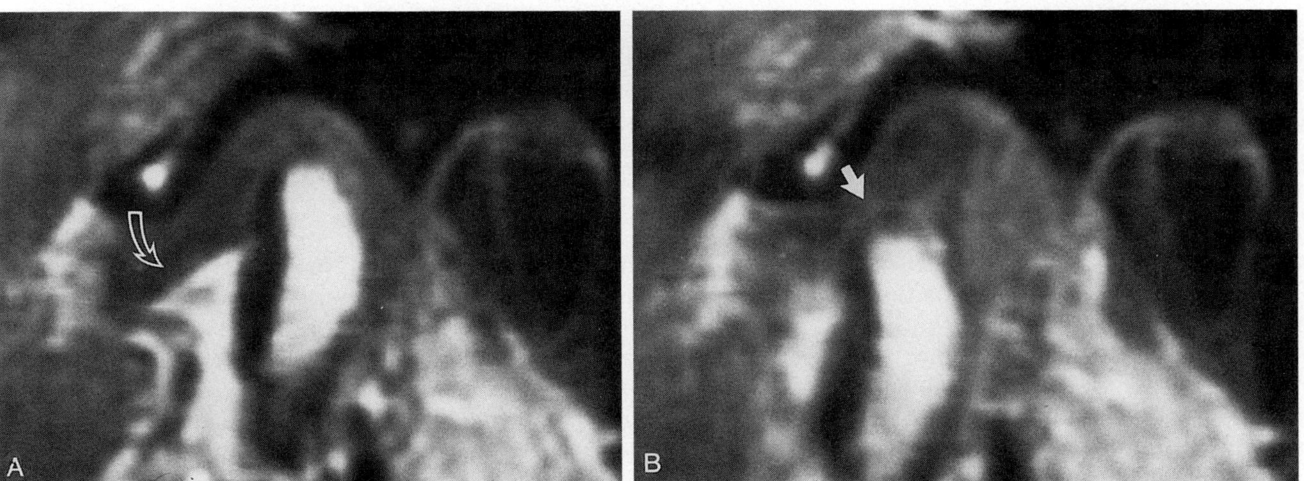

Figure 14–9. Grade I joint of the left temporomandibular joint in a 28-year-old woman. T1-weighted sagittal images (TR, 500 ms; TE, 20 ms) show an anteriorly displaced disk *(curved arrow)* in the closed-mouth position in *A*, which is not detectable in the partially opened-mouth position in *B*. The arrow in *B* points to the intermediate zone, which is in a normal position.

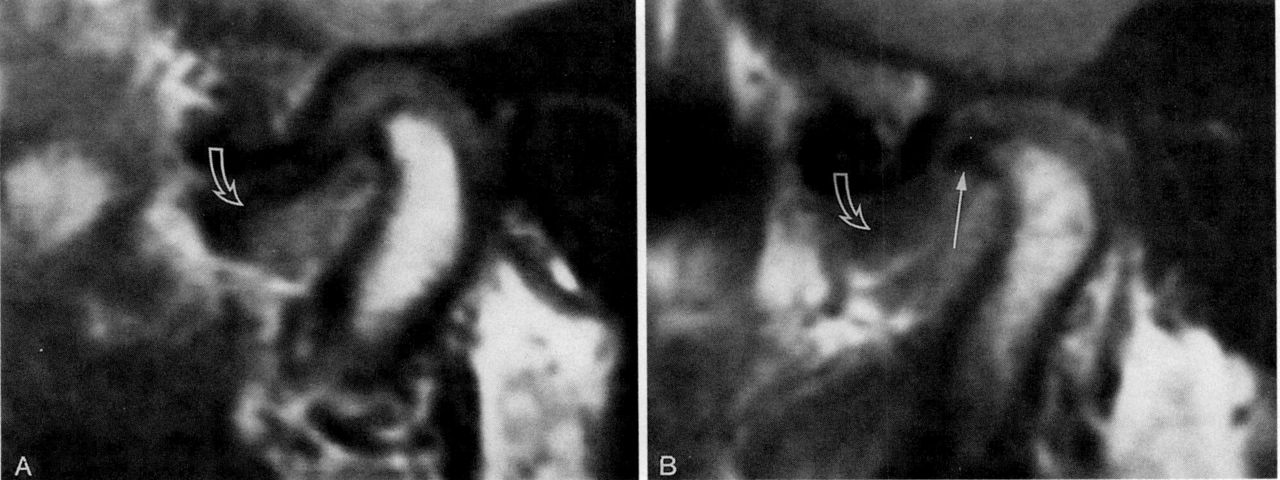

Figure 14–10. Grade I joint (right temporomandibular joint [TMJ]) and grade II joint (left TMJ) in a 42-year-old woman. T1-weighted sagittal images (TR, 600 ms; TE, 20 ms) in the closed-mouth position show bilateral anteriorly displaced disks *(curved arrows)*. The right disk *(A)* is of bow-tie configuration whereas the left disk *(B)* is deformed in shape. A small anterior osteophyte *(straight arrow)* is noted in the left disk.

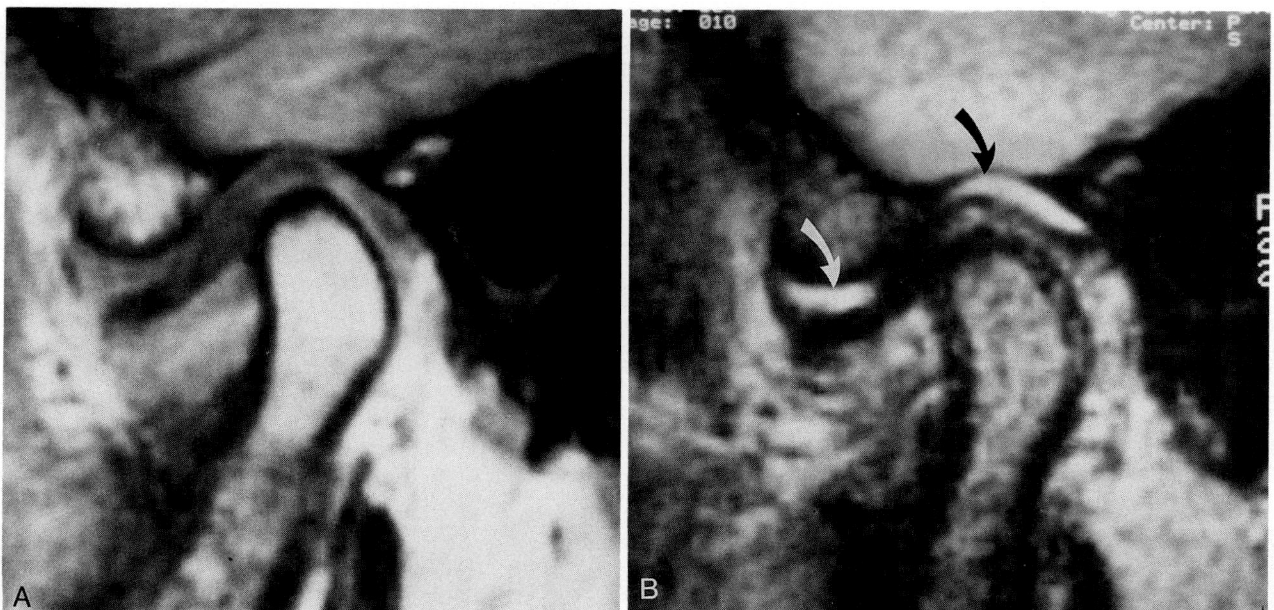

Figure 14–11. Grade I joint with fluid of the left temporomandibular joint. *A*, T1-weighted sagittal image (TR, 600 ms; TE, 20 ms) in the closed-mouth position shows anteriorly displaced disk, which is of normal morphology. *B*, T2-weighted sagittal image (TR, 1500 ms; TE, 60 ms) in a partially opened-mouth position shows upper joint fluid of high signal intensity *(arrows)*.

tachment wall.[2] Magnetic resonance imaging is useful in assessing the disk position after surgical repositioning. Restoration of the normal disk position has a good clinical outcome.

Diskoplasty is a surgical procedure performed to bring the displaced disk back into normal alignment with the mandibular condyle. However, the initial procedure of this technique—loosening the disk from its attachments to the mandibular condyle—results in some immobility of the disk. Magnetic resonance imaging allows the position of the disk to be shown relative to its preoperative position, which is a good means of determinating clinical success after diskoplasty (Figs. 14–13, 14–14).[4] In patients with good clinical outcome, the disks are in a normal or improved position compared with that before surgery. Conversely, in those with poor results, the anteriorly displaced disks showed no improvement in positioning.

Diskectomy can be used in patients with chronic disk displacement with maceration of the disk. Diskectomy with placement of a synthetic disk implant

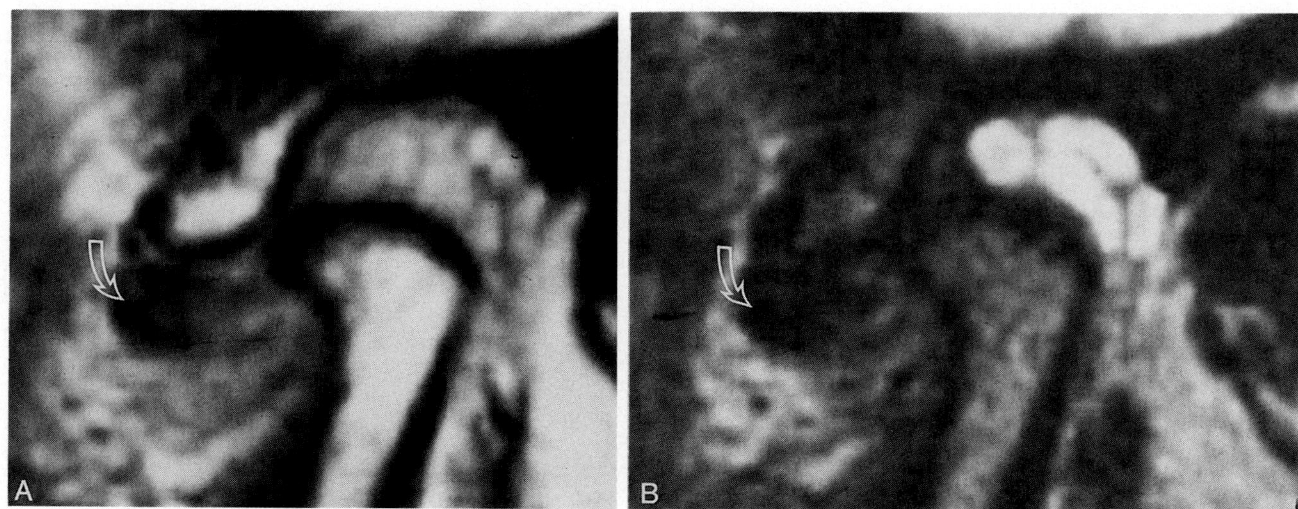

Figure 14–12. Grade II joint with fluid of the left temporomandibular joint in a 40-year-old woman. Sagittal images (*A*: TR, 1000 ms; TE, 20 ms; *B*: TR, 1000 ms; TE, 70 ms) show anteriorly displaced disk in the partially opened-mouth position with deformity *(arrow)*. Mild degenerative joint changes are noted. Lobulated collections of joint fluid involving both compartments and bilaminar zone are observed.

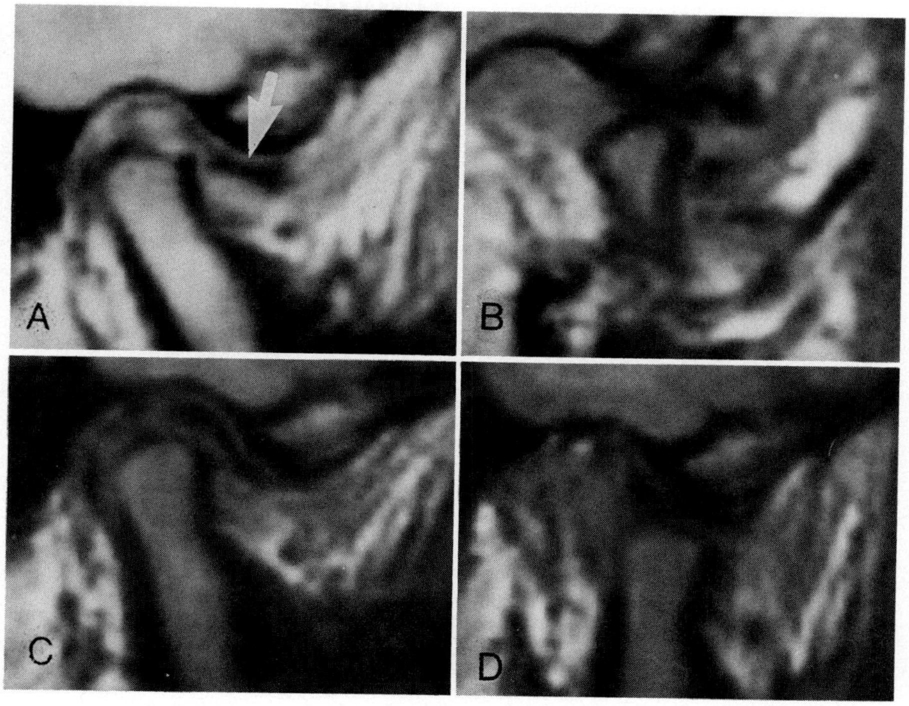

Figure 14–13. Temporomandibular joint after diskoplasty in a patient considered to have a clinically excellent result. T1-weighted sagittal images (TR, 700 ms; TE, 20 ms) show an anteriorly displaced disk *(arrow)* in the closed-mouth position *(A)* and an undisplaced disk in the opened-mouth position *(B)*. In surgery, the disk has been pulled back into its normal position in the closed-mouth view *(C)* and it slides forward minimally in the opened-mouth position *(D)*. Condyle movement is minimally decreased. (From Conway WF, Hayes C, and Campbell R, et al: Temporomandibular joint after meniscoplasty: Appearance at MR imaging. Radiology 180:749, 1991. © Radiological Society of North America.)

(Proplast-Teflon laminate, silicone, Silastic) is designed to function as a disk prosthesis. Complications associated with these implants include a destructive foreign body–type granuloma and avascular necrosis of the mandibular condyle.[27] The destructive granuloma is seen as a soft tissue mass of intermediate signal intensity on both T1- and T2-weighted images (Fig. 14–15). Magnetic resonance imaging is also useful in assessing fracture or displacement of the implant (Fig. 14–16). All of these findings are indications for surgical removal of the prosthesis.

The use of autogenous dermal grafts in TMJ surgery aims to prevent apposition of the bone surfaces

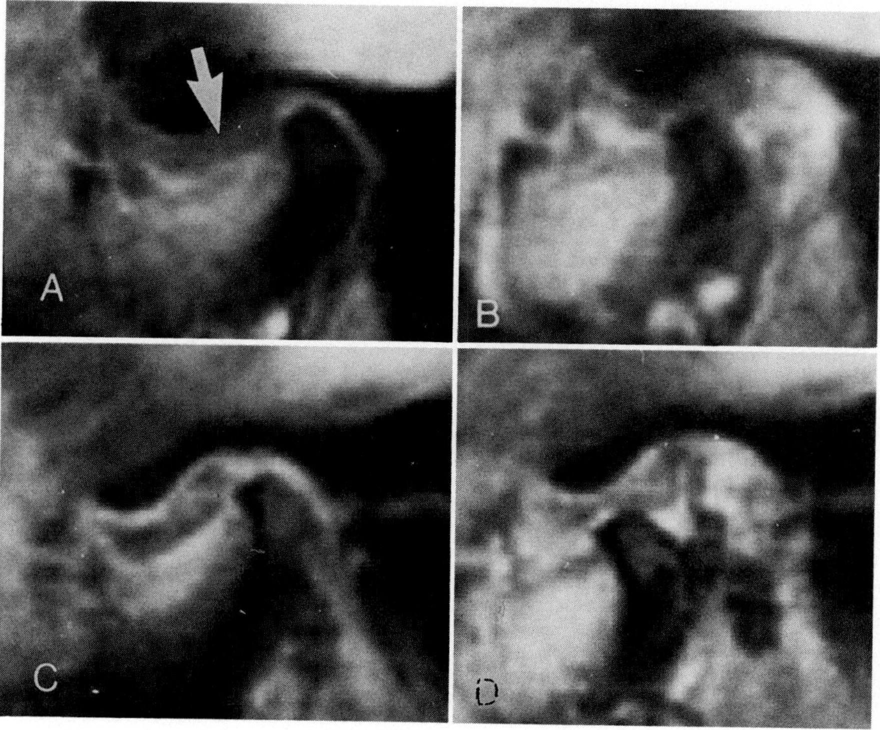

Figure 14–14. The temporomandibular joint after diskoplasty in a patient considered to have a clinically poor result. Fast low-angle shot (FLASH) images (TR, 30 ms; TE, 13 ms; flip angle, 10 degrees) show an anteriorly displaced disk *(arrow)* in the closed-mouth position *(A)* and an undisplaced disk in the opened-mouth position *(B)*. After surgery little change can be seen in the disk position in the closed-mouth view *(C)*, and the disk still becomes reduced in the opened-mouth position *(D)*. The condyle movement is unchanged. (From Conway WF, Hayes C, Campbell R, et al: Temporomandibular joint after meniscoplasty: Appearance at MR imaging. Radiology 180:749, 1991. © Radiological Society of North America.)

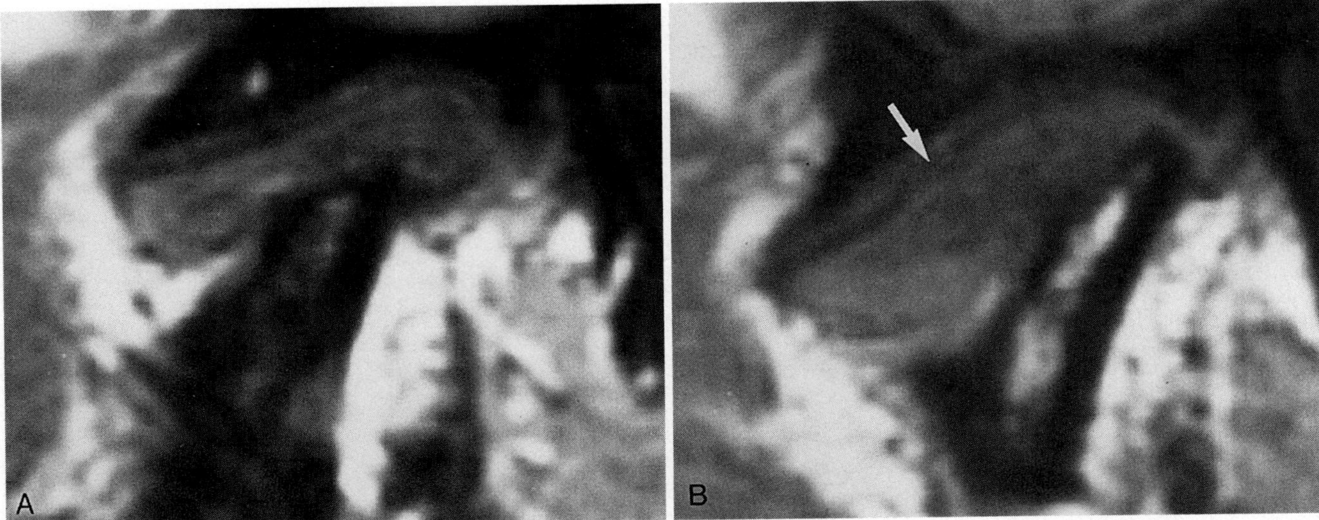

Figure 14–15. Right temporomandibular joint (TMJ) prosthesis and granulation tissue in a 36-year-old woman. This patient had had bilateral TMJ surgery with Proplast-Teflon implants and mandibular osteotomy 4 and 2 years ago, respectively. Bilateral joint pain was noted recently. Consecutive T1-weighted (*A* and *B*, TR, 500 ms; TE, 20 ms) sagittal images in the partially opened-mouth position show a large amount of soft tissue or fluid, or both, within the joint. A linear prosthesis of low signal intensity parallels the glenoid fossa and extends anteriorly to cover the region of the eminence. There is a questionable fracture through the midportion of the prosthesis (*arrow* in *B*). The condylar head shows a pencil-like appearance, which may be secondary to earlier surgery or to erosion. This patient subsequently was treated by arthroplasty with removal of implants and granulation tissue and reconstruction with temporalis fascia flaps. The condylar head revealed severe degenerative changes at surgery. Examination for pathologic conditions confirmed the presence of fragments of prosthetic materials and fibrous connective tissue with foreign body giant cell reaction.

of the mandibular condyle and the temporal bone; apposition of these surfaces produces osteoarthritis. Three dermal graft techniques are used: sandwich, onlay, and patch. The dermal tissue is commonly derived from either the thigh or the buttock. The dermal graft may be placed both above and below the disk, creating a "sandwich," or it can be placed as an "onlay" on the superior surface of the disk. A patch technique aims to repair perforations of the disk. Magnetic resonance imaging does not allow detection of the dermal grafts, but it is useful in detection and evaluation of complications that may follow diskectomy and placement of an allogenic implant. Lieberman and co-workers revealed that little change in disk position was noted when pre- and postoperative MR images were compared.[17] The relationship between the failure to return the disk to the normal position and the clinical outcome is unclear, however.

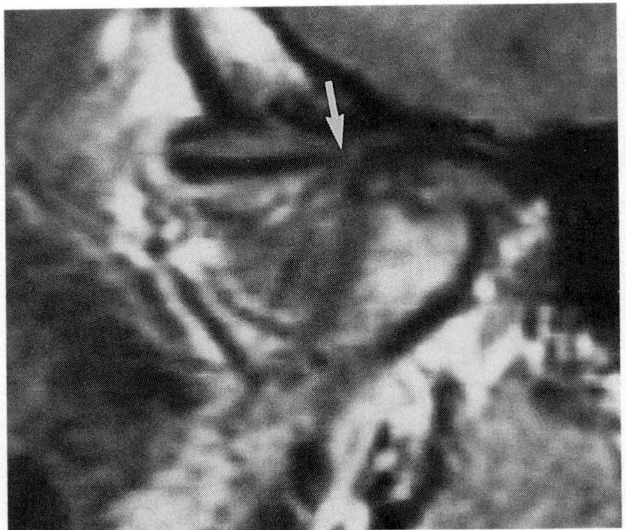

Figure 14–16. Fractured Silastic implant of the right temporomandibular joint (TMJ) in a 31-year-old woman. The patient had TMJ diskectomy and Silastic implant of this joint 3 years earlier. Nonreduction of the disk was suspected recently. T1-weighted (TR, 500 ms; TE, 20 ms) sagittal image in the closed-mouth position shows disruption of the Silastic implant (*arrow*). The irregularly shaped condylar head resulted from a condylar shave at surgery.

BONE

Degenerative Joint Disease

Long-standing disk derangement precedes maceration of the disk and secondary degenerative joint disease.[3,16] Helms and associates, in a study of more than 200 TMJs, reported that 95 per cent of grade II joints (Fig. 14–17), in contrast to 17 per cent of grade I joints (Fig. 14–18), had degenerative joint disease.[12] Degenerative changes in the TMJ initially are manifested as cortical erosion and progress to condylar head flattening with anterior osteophytosis.

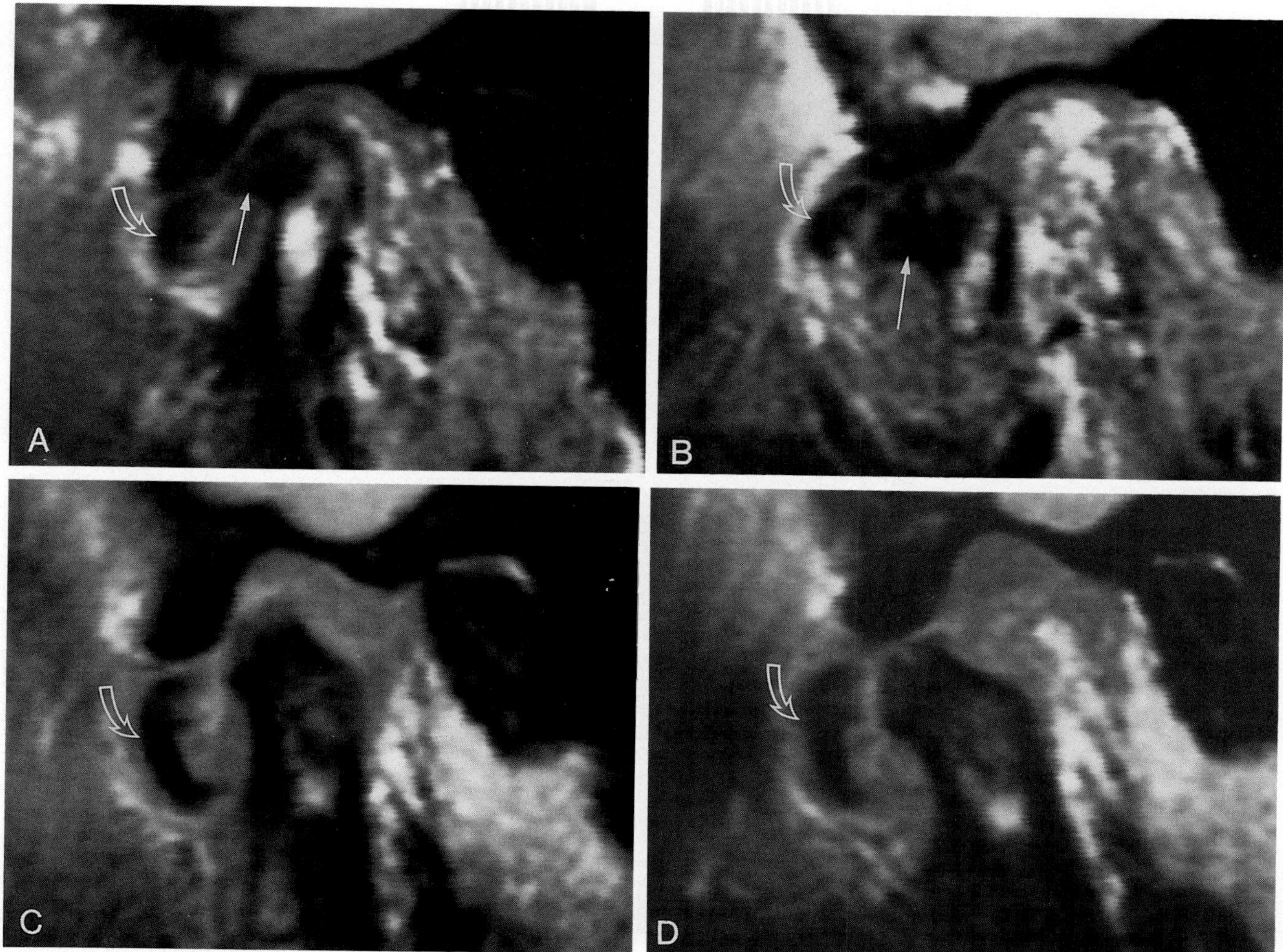

Figure 14–17. Bilateral grade II joint and severe degenerative condyles in a 36-year-old woman. This patient suffered from bilateral temporomandibular joint pain, clicking, and locking. T1-weighted sagittal images (TR, 500 ms; TE, 20 ms) show anterior displacement of the right (*A*, closed mouth; *B*, partially opened mouth) and left (*C*, closed mouth; *D*, partially opened mouth) disks without reduction (*curved arrows*). The disks have abnormally homogeneous low signal intensity, suggesting desiccation. Both disks fail to retain their normal bow-tie configuration. A large anterior osteophyte and an associated area of low signal intensity (sclerotic changes) are present in the right condylar head (*straight arrow* in *A* and *B*). Severe degenerative changes with erosion of the left condylar head are evident.

Avascular Necrosis

Avascular necrosis involving the mandibular condyle has been described in advanced derangement combined with joint inflammation of the TMJ.[22,24–26] Inflammatory arthropathy of the TMJ produces symptoms such as earache, tinnitus, difficult hearing, and fluid in the ear. The inflammatory thickened disk and joint effusion cause increased pressure, compressing the mandibular condyle and resulting in a sensation of joint fullness and limited joint motion. Magnetic resonance imaging features of avascular necrosis include alterations in condylar morphology and marrow signal intensity. Focal, subarticular, or generalized condylar defects are noted. Changes in marrow signal intensity include decreased signal intensity on T1-weighted images and either decreased, variable, or increased signal intensity on T2-weighted images. T2-weighted sequences show subchondral cyst with focal areas of increased subarticular signal intensity, indicating acute avascular necrosis (Fig. 14–19). These pulse sequences also are useful in showing joint effusion, if any, associated with inflammatory arthropathy.

Bone Injury

Bone injury can accompany condylar fractures or dislocation, disk fractures or displacement, and adhesive capsulitis and hematomas. Patients suffer from severe joint pain and limited joint motion. These manifestations may also be accompanied by dysfunction of the contralateral TMJ. Magnetic reso-

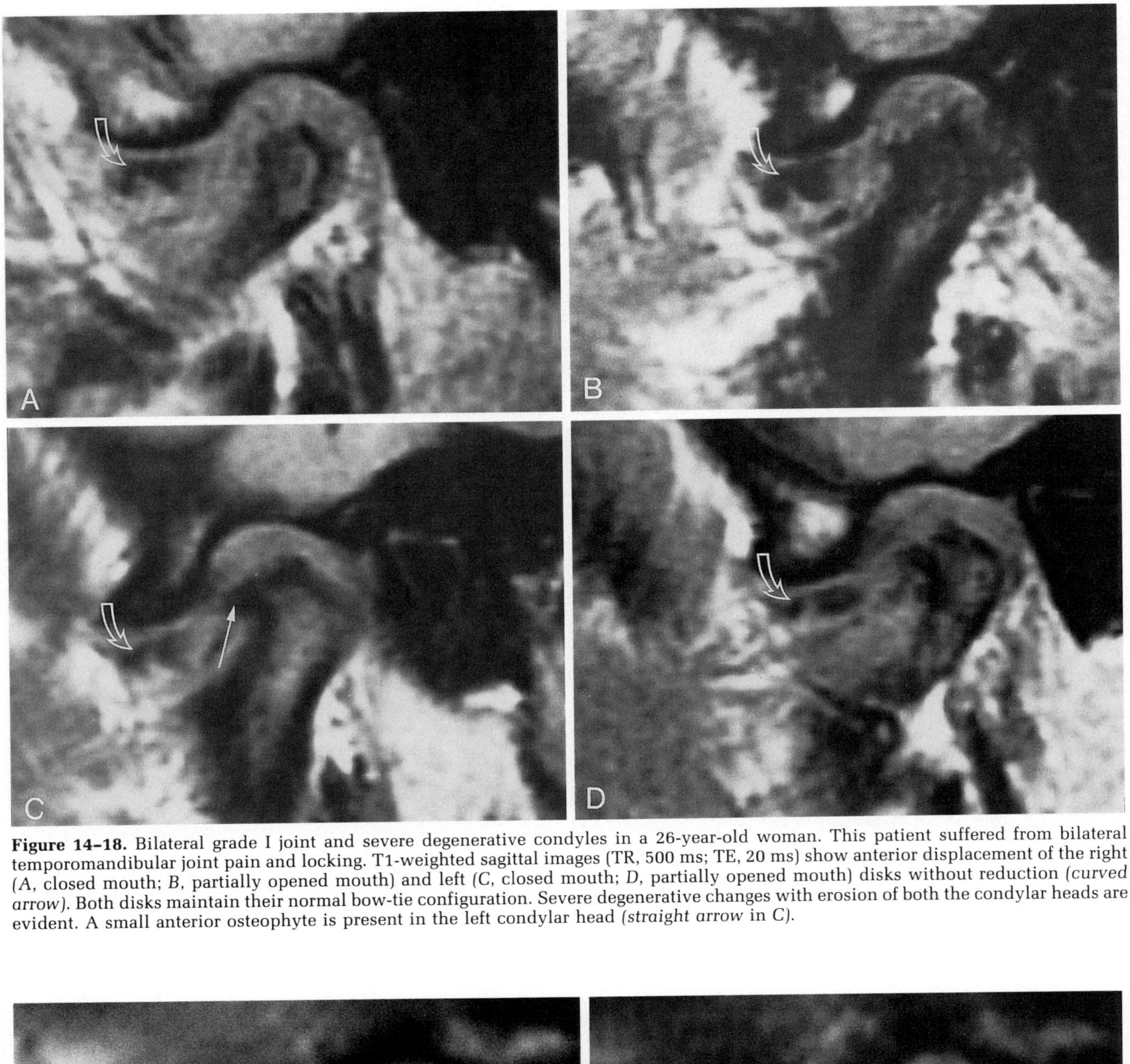

Figure 14–18. Bilateral grade I joint and severe degenerative condyles in a 26-year-old woman. This patient suffered from bilateral temporomandibular joint pain and locking. T1-weighted sagittal images (TR, 500 ms; TE, 20 ms) show anterior displacement of the right (*A*, closed mouth; *B*, partially opened mouth) and left (*C*, closed mouth; *D*, partially opened mouth) disks without reduction (*curved arrow*). Both disks maintain their normal bow-tie configuration. Severe degenerative changes with erosion of both the condylar heads are evident. A small anterior osteophyte is present in the left condylar head (*straight arrow* in *C*).

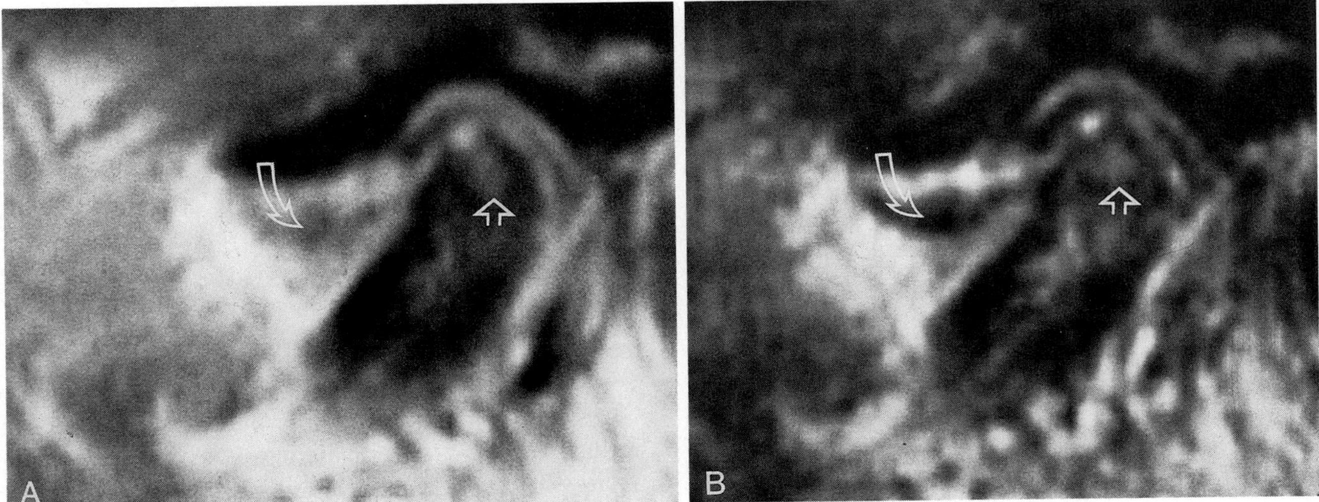

Figure 14–19. Condylar avascular necrosis/osteochondritis dissecans and grade I joint of the left temporomandibular joint in a 54-year-old woman. Sagittal images (*A*: TR 1800 ms; TE, 30 ms; *B*: TR, 1800 ms; TE, 80 ms) in the closed-mouth position show a focal subarticular lesion in the condylar head (*arrowhead*). The lesion is of high signal intensity on T2-weighted images, representing a subarticular edema beneath the eroded articular cartilage. The disk is displaced anteriorly, but it retains a normal configuration (*curved arrow*). (Courtesy Lynne S. Steinbach, M.D., San Francisco, California.)

nance imaging and TMJ radiography and tomography should be performed to assess the type and extent of injury of the craniomandibular apparatus after injury.[21] Complementary sagittal T2-weighted spin-echo or gradient-echo images are recommended in imaging such cases. The radiographs and tomographs may allow identification of fractures that escaped recognition on MR imaging.

References

1. Abbott M, Carvlin MJ, Nowell M, et al: Determination of the sensitivity of MR imaging in the identification of known perforation-like defects of the temporomandibular joint meniscus. Radiology 173(P):100, 1989.
2. Anderson DM, Sinclair PM, McBride KM: A clinical evaluation of temporomandibular joint disk plication surgery. Am J Orthod Dentofacial Orthop 100:156, 1991.
3. Blackwood HJ: Arthritis of the mandibular joint. Br Dent J 115:317, 1963.
4. Conway W, Hayes C, Campbell R, et al: Temporomandibular joint after meniscoplasty: Appearance at MR imaging. Radiology 180:749, 1991.
5. Drace J, Enzmann D: Defining the normal temporomandibular joint: Closed-, partially open, and open-mouth MR imaging of asymptomatic subjects. Radiology 177:67, 1990.
6. Drace J, Young S, Enzmann D: TMJ meniscus and bilaminar zone: MR imaging of the substructure—diagnostic landmarks and pitfalls of interpretation. Radiology 177:73, 1990.
7. Fulmer JM, Harms SE: The temporomandibular joint. Top Magn Reson Imaging 1:75, 1989.
8. Guralnick WC, Kaban LB, Merril RG: Temporomandibular joint afflictions. N Engl J Med 229:123, 1978.
9. Harms SE, Wilk RW. Magnetic resonance of the temporomandibular joint. Radiographics 7:521, 1987.
10. Helms CA: Temporomandibular joint. *In* Berquist, TH (ed): MRI of the Musculoskeletal System. New York, Raven Press, 1990.
11. Helms CA, Doyle GW, Orwig D, et al: Staging of internal derangements of the TMJ with magnetic resonance imaging: Preliminary observations. J Craniomandib Disord 3:93, 1989.
12. Helms CA, Kaban LB, McNeill C, et al: Temporomandibular joint: Morphology and signal intensity characteristics of the disk at MR imaging. Radiology 172:817, 1989.
13. Helms CA, Vogler JB, Morrish RB, et al. Diagnosis of TMJ internal derangements with computed tomography: Review of 200 cases. Radiology 152:459, 1984.
14. Isacsson G, Isberg A, Johansson AS, et al: Internal derangement of the temporomandibular joint: Radiographic and histologic changes associated with severe pain. J Oral Maxillofac Surg 44:771, 1986.
15. Isberg A, Isacsson G: Tissue reactions of the temporomandibular joint following retrusive guidance of the mandible. J Craniomandib Practice 4:143, 1986.
16. Katzberg RW, Keith DA, Guralnick WC, et al: Internal derangements and arthritis of the temporomandibular joint. Radiology 146:107, 1983.
17. Lieberman JM, Bradrick JP, Indresano AT, et al: Dermal grafts of the temporomandibular joint: Postoperative appearance on MR images. Radiology 176:199, 1990.
18. Moffett BC, Johnson LC, McCabe JB, et al: Articular remodeling in adult human temporomandibular joint. Am J Anat 115:119, 1964.
19. Orwig DS, Helms CA, Doyle GW: Optimal mouth position for magnetic resonance imaging of the temporomandibular joint disk. J Craniomandib Disord 3:138, 1989.
20. Rao VM, Vinitski S, Babaria A: Comparison of SE and short TE three-dimensional gradient-echo imaging of the temporomandibular region. Radiology 173:99, 1989.
21. Schellhas KP: Temporomandibular joint injuries. Radiology 173:211, 1989.
22. Schellhas KP: Internal derangement of the temporomandibular joint: Radiologic staging with clinical, surgical, and pathologic correlation. Magn Reson Imaging 7:495, 1989.
23. Schellhas KP, Fritts HM, Heithoff KB, et al: Temporomandibular joint: MR fast scanning. J Craniomandib Pract 6:209, 1988.
24. Schellhas KP, Piper MA, Omile MR: Facial skeleton remodeling due to temporomandibular joint degeneration: An imaging study of 100 patients. AJR 155:373, 1990.
25. Schellhas KP, Wilkes CH, Fritts HM, et al: MR of osteochondritis dissecans and avascular necrosis of the mandibular condyle. AJR 152:551, 1989.
26. Schellhas KP, Wilkes CH: Temporomandibular joint inflammation: Comparison of MR fast scanning with T1- and T2-weighted imaging techniques. AJR 153:93, 1989.
27. Schellhas KP, Wilkes CH, Deeb ME, et al: Permanent Proplast temporomandibular joint implants: MR imaging of destructive complications. AJR 151:731, 1988.
28. Schwaighofer B, Tanaka T, Klein M, et al: MR imaging of the temporomandibular joint: A cadaver study of the value of coronal images. AJR 154:1245, 1990.
29. Solberg WK, Woo MW, Houston JB: Prevalence of mandibular dysfunction in young adults. J Am Dent Assoc 98:25, 1979.

15 MUSCULOSKELETAL NEOPLASM

Philipp Lang, Harry K. Genant, James O. Johnston, and Gordon Honda

Before the advent of cross-sectional imaging techniques, the diagnosis of primary tumors of bone and soft tissue was based on conventional radiography and polytomography. Even today, with the availability of computed tomography (CT) and magnetic resonance (MR) imaging, radiography remains the primary imaging modality for the initial diagnosis of bone and soft tissue neoplasms. Tumors and tumorlike lesions of the long bones frequently can be characterized using conventional radiography. Radiography may, however, be limited in evaluating flat bones, such as the scapula, ribs, vertebrae, and pelvis.[129]

In recent years, MR imaging has evolved as an important diagnostic tool in assessing musculoskeletal neoplasm. It has several features that render it superior to CT in evaluating bone and soft tissue tumors: (1) greater soft-tissue contrast, (2) direct multiplanar imaging capability, (3) lack of beam hardening artifacts from cortical bone, and (4) improved tissue characterization based on signal behavior on different pulse sequences and relaxation parameters.[7] Magnetic resonance imaging is the most accurate imaging method in staging the extent of solitary tumors and tumor-like lesions of bone.[8,45,149] It is more accurate than CT in demonstrating the relationship of tumor to neurovascular bundles.[8,49] Applications for MR imaging in diagnosis and management of musculoskeletal neoplasm are rapidly expanding. In this chapter, we outline current MR imaging techniques for evaluating musculoskeletal neoplasms. The role of MR imaging in lesion detection and staging, lesion characterization, longitudinal assessment of tumors during radiation or chemotherapy, and detection of local recurrence is discussed. The imaging appearance of the most common primary benign and malignant lesions of bone and soft tissue is described. Metastatic disease of bone is discussed in Chapter 16.

IMAGING TECHNIQUES

Patient Selection, Positioning, Coils

Although the presence of vascular clips or cardiac pacemakers precludes MR imaging evaluation of the patient with musculoskeletal neoplasm, most orthopedic implants such as internal fixation devices or prostheses have only minor or no ferromagnetic properties. Torque thus does not present a problem in MR imaging of orthopedic implants. At worst, the implant may cause a localized image artifact resulting from the focal distortion of the magnetic field. Frequently, no artifact or only minor artifacts can be observed (see Chapter 10).

Many patients with musculoskeletal neoplasm fall in the pediatric age group (e.g., patients with osteogenic sarcoma or Ewing's sarcoma). In children, sedation may be indicated to avoid or minimize motion artifacts. Approximately 3 to 4 per cent of all patients studied are claustrophobic. Sedation also is helpful in claustrophobic patients to better tolerate the narrow confines of the MR scanner. Because imaging studies of musculoskeletal tumors frequently are lengthy, it is important to place a patient in a comfortable position in the magnet. Analgesics may be employed when the patient is in pain.

Proper patient positioning and choice of coil are critical for obtaining high-resolution MR imaging studies. In musculoskeletal tumors that may cause skip lesions, such as osteogenic sarcoma, the initial scan should be obtained using the body coil to evaluate the proximal and distal portions of the affected extremity and potentially also the contralateral side. For this purpose, the field of view generally should be large enough to cover a sufficient anatomic area despite the loss in spatial resolution. When the time available for MR examination is limited, this scan may be omitted if the patient has had a bone scan negative for skip lesions before MR imaging. In general, however, it is advantageous to obtain an overview body coil scan even when scintigraphy has been performed, because MR imaging has a higher sensitivity than scintigraphy in detecting tumorlike lesions of bone.[37,69,70,144] Cases have been reported in which transarticular skip metastases were not detected by conventional radiography, CT, and nuclear scintigraphy; MR imaging, however, demonstrated skip lesions.[144,149]

For imaging the immediate area of the tumor, a coil should be chosen that provides optimal spatial resolution, signal-to-noise ratio, and contrast-to-noise ratio for the given anatomic region. Use of the body coil is advantageous in imaging the trunk and the thighs. For the spine, a license-plate–shaped surface coil or a phased-array coil is ideally suited for generating high-resolution images. The shoulder joint is best imaged with a special shoulder coil. Various different designs, such as loop gap resonators or circular surface coils, are available for this purpose. The humerus, elbow, and forearm can be studied with a license plate–shaped surface coil in most instances. The wrist and hand may be imaged with small-diameter surface coils or special wrist coils. The distal femur and the proximal tibia are best imaged with a circumferential knee coil. For evaluating the distal tibia, both the knee coil and the surface coils can be employed. The ankle joint and the foot are generally best imaged with use of a surface coil. A small field of view aids in obtaining an image of good quality with sufficient spatial resolution. However, the field of view must be large enough to include all structures adjacent to the tumor so that soft tissue infiltration, presence or absence of muscle-fat planes, and encroachment of the tumor on neighboring neurovascular bundles can be assessed appropriately.

Pulse Sequences

T1-weighted (short TR, short TE) and T2-weighted (long TR, long TE) spin-echo sequences are most commonly used for evaluating musculoskeletal neoplasm (Table 15–1). On T1-weighted spin-echo

Table 15–1. PULSE SEQUENCES FOR IMAGING OF BONE AND SOFT-TISSUE NEOPLASMS (AT 1.5T).

Pulse Sequence	Acronym	TR (ms)	TE (ms)	Echoes	TI (ms)	Flip Angle (°)	Matrix	NEX	Slice Thickness (mm)	Skip (mm)	Imaging Time (min)[1]
Precontrast											
Spin-echo (T1-weighted)	SE	500–800	20–30	1	—	—	256 × 192 256 × 256	1–2	5	0–5	3–5
Spin-echo (T2-weighted)	SE	2000–2500	80–180	1–4	—	—	256 × 192 256 × 256	1–2	5	0–5	10–20
Gradient-echo ("T1-weighted")	GRE	300–400	15–20	1	—	90	256 × 192 256 × 256	1–2	5	0–5	3–5
Gradient-echo (T2*-weighted)	GRE	500–600	30	2	—	30	256 × 192 256 × 256	1–2	5	0–5	4–7
Short tau inversion recovery	STIR	1500	20–30	1	170–180	—	256 × 192 256 × 256	1–2	5	0–5	5–10
Postcontrast (gadopentetate-dimeglumine)											
Gradient-echo (fast sequential imaging)	GRE	30–60	5–10	1	—	60–90	256 × 128	0.5–1.0	5	—	2–5 sec*
Spin-echo (T1-weighted)	SE	500–800	20–30	1	—	—	256 × 192 256 × 256	1–2	5	0–5	3–5
Gradient-echo ("T1-weighted")	GRE	300–400	15–20	1	—	90	256 × 192 256 × 256	1–2	5	0–5	3–5

TR, Repetition time; TE, echo time; NEX, number of excitations; SE, spin echo; GRE, gradient-recalled echo; STIR, short tau inversion-recovery; [1]unless specified; *per single image; image acquisition without breaks for reconstruction.

Table 15–2. SIGNAL INTENSITIES OF NORMAL AND NEOPLASTIC TISSUES ON DIFFERENT PULSE SEQUENCES

	Spin-Echo T1-Weighted	Spin-Echo T2-Weighted	Gradient-Echo "T1-Weighted"	Gradient-Echo T2*-Weighted	STIR	Spin-Echo T1-Weighted Post-GD-DTPA	Gradient-Echo "T1-Weighted" Post-GD-DTPA
NORMAL TISSUES							
Bone marrow	High	Intermediate to high	High	Low to intermediate*	Low	High	High
Cortical bone	Low	Low	Low	Low	Low	Low	Low
Muscle	Low	Low to Intermediate	Low	Intermediate	Low to intermediate	Low to intermediate	Low to intermediate
Ligaments	Low	Low	Low	Low	Low	Low	Low
Tendons	Low	Low	Low	Low	Low	Low	Low
Vessels	Low	Low	Low, intermediate, high†	Low, intermediate, high†	Low or high	Low	Low
Nerves	Low	Low to intermediate	Low	Intermediate	Low	Low	Low
Subcutaneous fat	High	Intermediate to high	High	Intermediate	Low	High	High
PATHOLOGIC TISSUES							
Intraosseous tumor	Low, intermediate	High, intermediate	Low, intermediate	High‖	High‖	High	High
Extraosseous Tumor	Low, intermediate	High intermediate	Low, intermediate	High‖	High‖	High	High
Fatty tumor portions	High	Intermediate to high	High	Intermediate	Low	High	High
Tumor sclerosis	Low	Low	Low	Low‡	Low	Low	Low
Tumor cysts	Low	High	Low	High	High	Low§	Low§
Hemorrhage—fresh (deoxyhemoglobin)	Low	High	Low	High	High	Low	Low
Hemorrhage—4 weeks (methemoglobin)	High	High	High	High	High	High	High
Hemorrhage—old (hemosiderin)	Low	Low	Low‡	Low‡	Low	Low	Low‡
Peritumorous edema	Low	High	Low	High	High	High	High

*Due to inhomogeneous magnetic susceptibility at trabeculae-marrow interface (see text).
†Depending on flow direction, velocity, phase-encoding direction, distance from entry slice.
‡In many cases with frayed, sometimes cauliflower-like appearance due to inhomogeneous magnetic susceptibility with resultant artifact (in particular with long echo times).
§In selected cases with peripheral enhancement reflecting viable tumor or vessels at cyst margin.
‖Except for tumors that are predominantly fibrous or sclerotic.

images, bone marrow demonstrates high signal intensity (Table 15–2). The majority of tumors are of low signal intensity on this imaging sequence (Table 15–2), thus resulting in good contrast between normal marrow and neoplastic tumor tissue (Fig. 15–1*A*). Because muscle has a signal intensity similar to that of neoplasm on T1-weighted spin-echo images (see Table 15–2), the tumor-muscle interface may not always be well delineated with this sequence (see Fig. 15–1*A*). Fatty tumors (e.g., intraosseous lipoma) or hemorrhagic tumors containing methemoglobin also are of high signal intensity on this pulse sequence and may be difficult to differentiate from normal bone marrow (see Table 15–2).

Gradient-echo sequences with a relatively short TR (200 to 300 ms), a short TE (10 to 15 ms), and a large flip angle (80 to 90 degrees) provide contrast similar to that of conventional T1-weighted spin-echo sequences[33,35,53] (see Tables 15–1, 15–2) and thus can be used to replace T1-weighted spin-echo sequences. These sequences do not, however, significantly reduce imaging time when compared with T1-weighted spin-echo sequences.[35]

On T2-weighted spin-echo images, most neoplasms are of high signal intensity, providing good differentiation from adjacent muscle, which is of intermediate to low signal intensity[56] (see Table 15–2; Fig. 15–2*B*). The signal intensity of most tumors also is higher than that of the adjacent normal marrow. However, some neoplasms may have a signal intensity similar to that of normal marrow on T2-weighting, sometimes rendering distinction difficult (see Table 15–2; Fig. 15–2*B*). This is particularly the case when the images are not heavily T2-weighted. The combination of T1- and T2-weighted spin-echo scans provides good delineation with adequate tissue contrast in most cases of musculoskeletal neoplasm. When more detailed morphologic information is needed (e.g., for planning of biopsy or limb salvage surgery or for monitoring the effects of

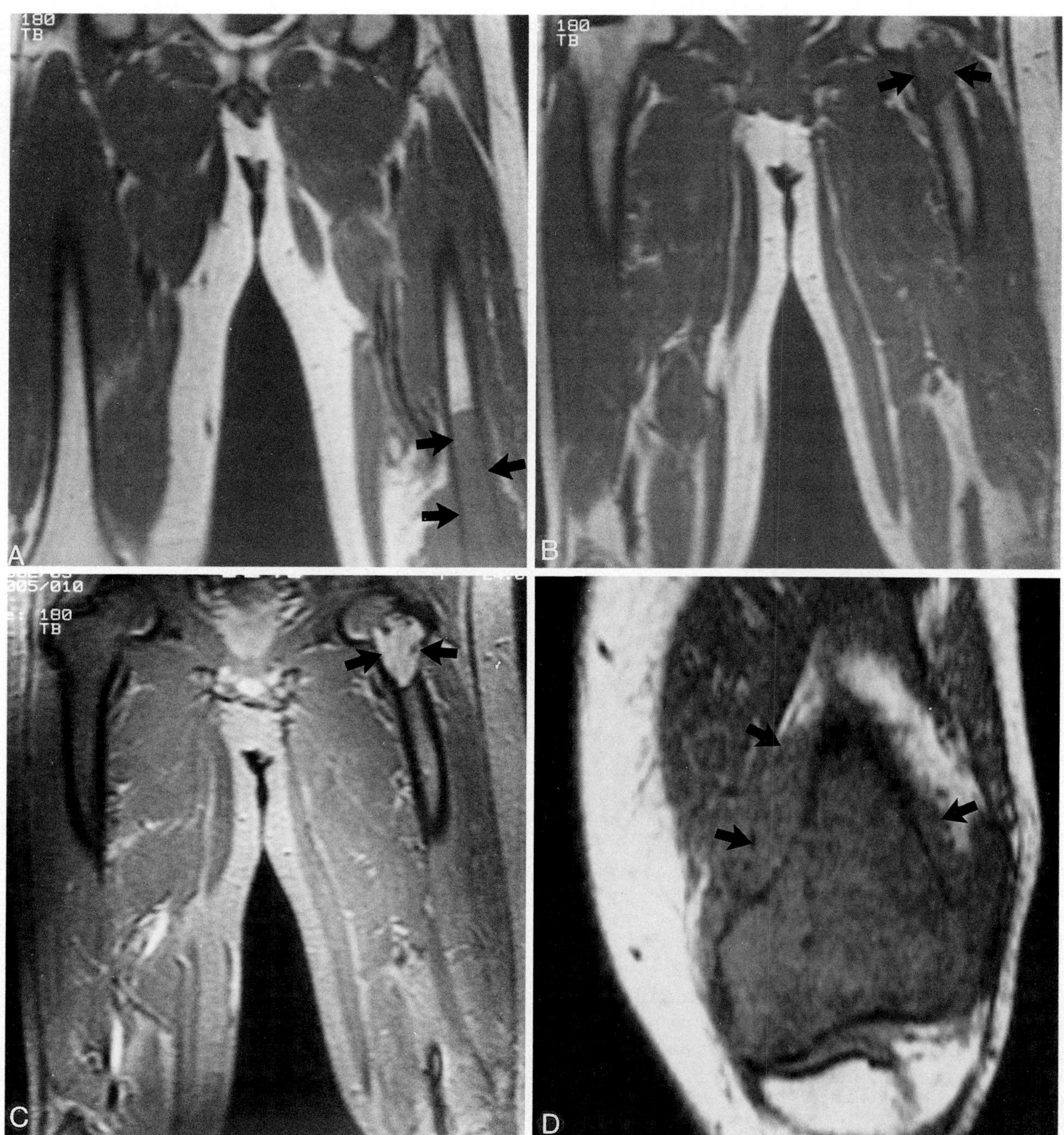

Figure 15–1. Primary lymphoma of bone in a 49-year-old male patient. The symptomatic lesion for which the patient was referred for MR imaging is located in the distal femur on the right side. *A* and *B*, T1-weighted spin-echo scan (TR, 600 ms; TE, 20 ms) obtained with the use of a body coil and a large field of view (48 cm). The lesion, of low signal intensity, is demonstrated in the right distal femur *(arrows)* in *A*. However, a second, asymptomatic lesion of low signal intensity *(arrows)* is shown in the intertrochanteric region of the right femur in *B*. *C*, Short tau inversion-recovery scan (STIR) (TR, 1500 ms; TE, 20 ms; TI, 170 ms). The STIR image confirms the presence of the second lesion *(arrows)* in the intertrochanteric region. The lesion is of high signal intensity on the STIR sequence. *D*, T1-weighted spin-echo scan (TR, 800 ms; TE, 30 ms) obtained with smaller field of view (30 cm) demonstrates the primary lesion of low signal intensity involving the distal femoral metaphysis and epiphysis. The lesion extends into the surrounding soft tissues *(arrows)*.

chemotherapy), these images may be supplemented by other sequences employing different contrast phenomena (see Table 15–1) or by imaging after intravenous administration of gadopentetate dimeglumine (Magnevist).

T2-weighted spin-echo sequences can be replaced by T2*-weighted gradient-echo sequences in evaluating musculoskeletal neoplasms (see Table 15–1). Acquisition of T2*-weighted gradient-echo sequences is more rapid than that of conventional T2-weighted spin-echo sequences. Examination time thus can be shortened when using this

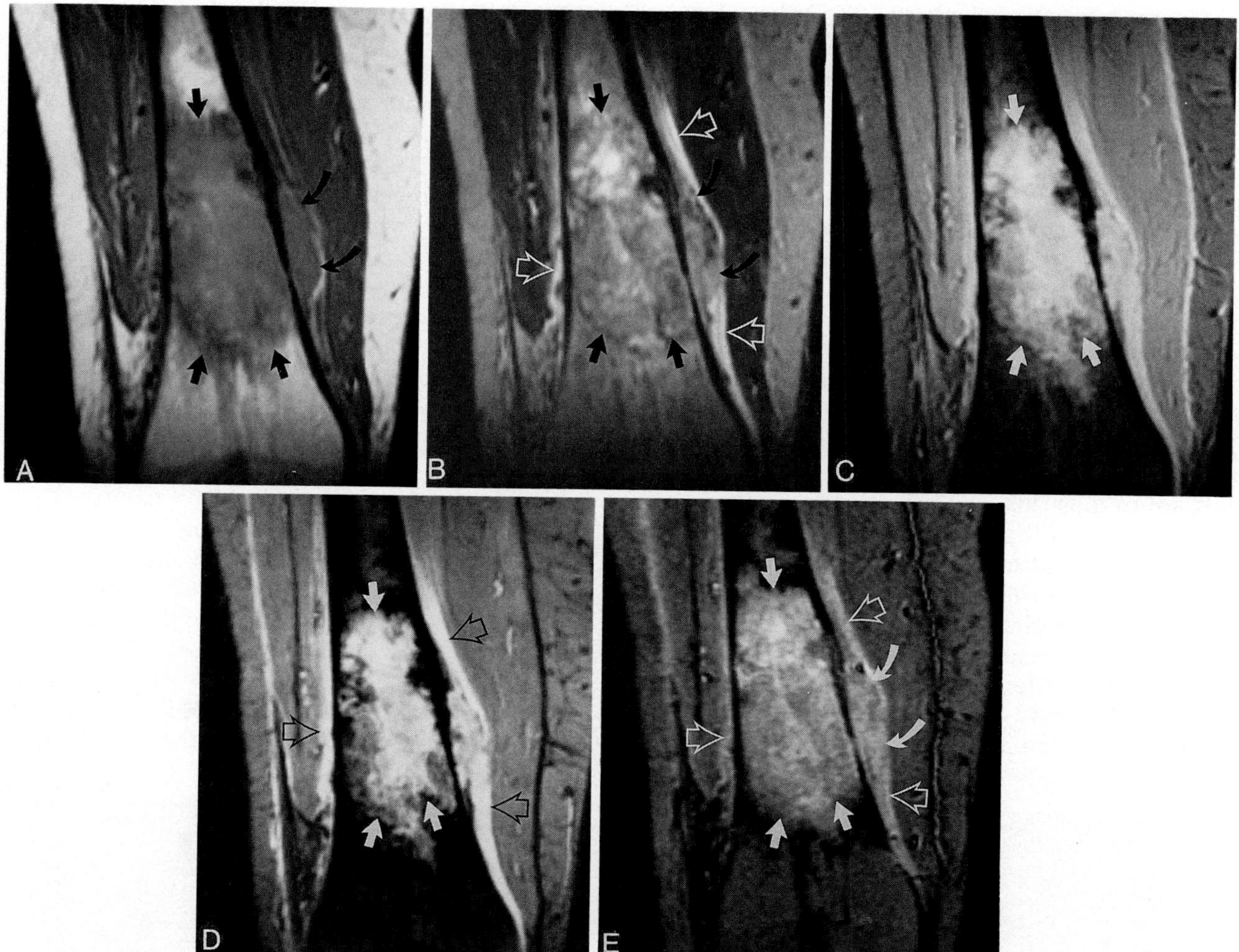

Figure 15–2. Osteogenic sarcoma of the distal femur in a 31-year-old female patient. *A*, T1-weighted spin-echo image (TR, 800 ms; TE, 20 ms). Intraosseous portions of a bone tumor *(straight arrows)* are of low signal intensity, resulting in good contrast to adjacent normal bone marrow, which is of high signal intensity. Extraosseous tumor *(curved arrows)* has a signal intensity similar to that of muscle. The interface between the extraosseous tumor and muscle thus is not as well delineated with this sequence as it is on the T2-weighted spin-echo image in *B*; *B*, T2-weighted spin-echo image (TR, 1600 ms; TE, 80 ms). The tumor increases markedly in signal intensity with this sequence; differentiation between the extraosseous tumor (intermediate to high signal intensity) *(curved solid arrows)* and the adjacent muscle (low signal intensity) is facilitated on T2-weighting. However, the intraosseous tumor *(straight solid arrows)* and the adjacent bone marrow display similar signal intensity, as the sequence is only moderately T2-weighted. Note the perineoplastic edema of high-signal-intensity *(open arrows)* extending along the medial and lateral tumor margin and portions of the normal cortex. *C* and *D*, T2*-weighted gradient-echo image (TR, 600 ms; TE, 14 ms [*C*] and 30 ms [*D*]; flip angle, 30 degrees). The intraosseous tumor *(straight solid arrows)* increases markedly in signal intensity with this imaging sequence, resulting in excellent differentiation from the adjacent normal bone marrow, of low signal intensity. Muscle-tumor contrast on the first echo gradient-echo image *(C)* is not as great as on the T2-weighted spin-echo sequences. The second echo gradient-echo image *(D)* demonstrates perineoplastic edema *(open arrows)* more clearly than the T2-weighted spin-echo image. *E*, Short tau inversion-recovery (STIR) image (TR, 1500 ms; TE, 20 ms; TI, 170 ms). The signal of fatty tissues (i.e., bone marrow and subcutaneous fat) is suppressed with this sequence. The neoplasm *(straight solid arrows)*, however, demonstrates high signal intensity resulting in excellent tumor-marrow contrast. The extraosseous tumor *(curved solid arrows)* also is higher in signal intensity than adjacent muscle, with resultant good tumor-muscle contrast. Perineoplastic edema likewise is of high signal intensity *(open arrows)*. Differentiation between perineoplastic edema and extraosseous tumor is, however, not as good as on the second echo T2*-weighted gradient-echo image.

technique (see Table 15–1, Fig. 15–1*C*). As with spin-echo imaging, T2*-weighted gradient-echo scans can be obtained in dual-echo technique with a relatively short (typically 5 to 15 ms) first echo and a long (typically 25 to 40 ms) second echo, thus providing additional information for tissue characterization (see Fig. 15–1*D*).

Gradient-echo images are, however, more sensitive to artifacts resulting from inhomogeneous magnetic susceptibility. Magnetic susceptibility is defined as the ratio between the induced magnetic field in the tissue and the applied external field; it is a measure of the response of a given tissue to an applied constant field.[46] Magnetic susceptibility effects occur at the interface between different tissues (e.g., dense trabecular bone adjacent to the surrounding bone marrow). Gradient-echo sequences are more sensitive to changes in magnetic susceptibility and resultant T2*-decay than conventional spin-echo sequences, as there is no rephasing of local magnetic field inhomogeneities by a 180 degree-pulse in gradient-echo imaging and T2*-decay is not recovered. Magnetic susceptibility effects and resultant T2*-shortening of bone marrow become greater with an increase in dense trabecular bone, areas of sclerosis, or hemosiderin deposits within the marrow compartment.

Image quality can deteriorate markedly by susceptibility-induced artifacts, in particular when images are generated in the presence of metallic implants. However, artifacts resulting from inhomogeneous magnetic susceptibility in biologic materials are in most instances relatively small and may even aid in identifying certain tissues. A magnetic susceptibility-induced artifact usually leads to a "fraying" of the interface of the object causing the susceptibility variation and the adjacent homogeneous tissue. In hemorrhagic areas with hemosiderin deposits or in osteoblastic regions with significant sclerosis, this may result in a characteristic cauliflowerlike appearance on gradient-echo images (Fig. 15–3). When compared with T2-weighted spin-echo sequences, T2*-weighted gradient-echo sequences provide sufficient contrast for differentiation of tumor from bone marrow or fatty tissue. Distinction of tumor from adjacent muscle may, however, pose a problem on T2*-weighted gradient-echo images, depending on the selected TE, and is generally somewhat inferior to that on T2-weighted spin-echo images[35] (see Fig. 15–2).

Short tau inversion recovery (STIR) sequences suppress the signal of fatty structures such as normal bone marrow (see Table 15–2). Objects with long T1 and T2 relaxation times (e.g., bone tumors), however, are of high signal intensity with this imaging sequence, resulting in excellent tumor-marrow contrast[37,48,49,51,55,66,101,123,138] (see Fig. 15–2*E*). Involvement of the medullary cavity is well demonstrated with STIR imaging owing to the low signal intensity of the marrow fat and the high signal intensity of medullary tumor.[49,66,101,123] Although the findings are still preliminary, this sequence may have greater sensitivity in detecting tumorous lesions of bone and bone marrow than conventional unenhanced spin-echo images.[123] Because the signal of the fat is suppressed with the STIR sequence,

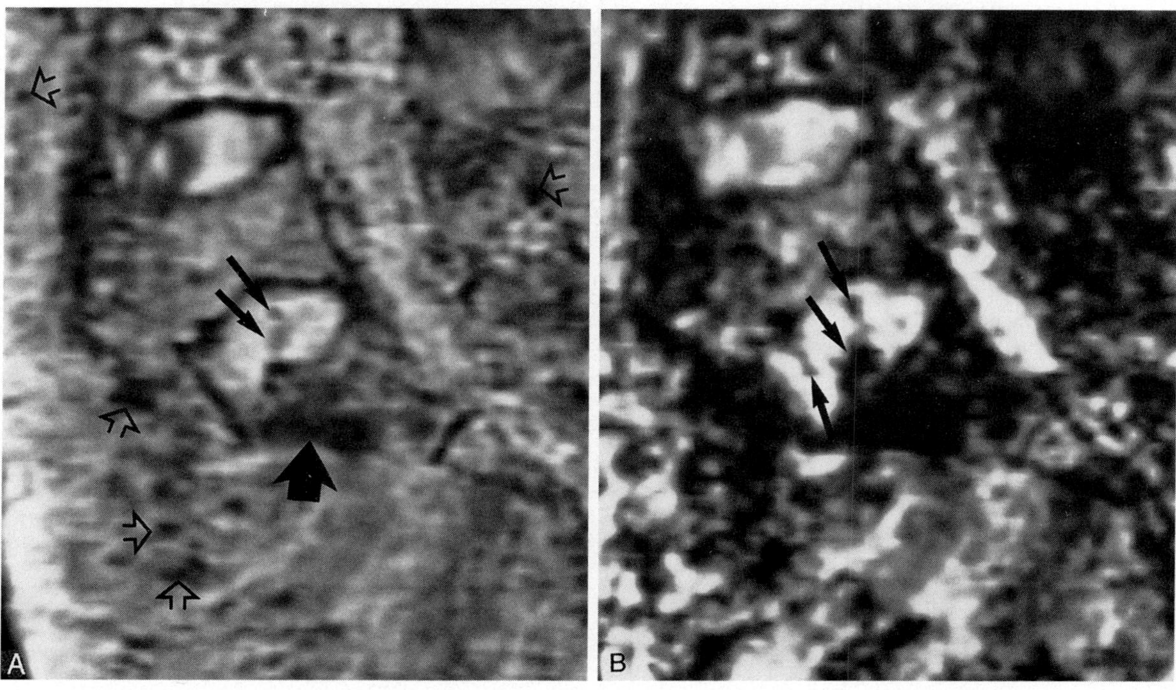

Figure 15–3. Large chondrosarcoma of the pelvis in an 82-year-old female patient (MR images of cadaver specimen). *A*, T1-weighted coronal spin-echo MR image (TR, 800 ms; TE, 30 ms) shows a large heterogeneous tumor mass surrounding and involving the lumbar spine. Note tumor invasion of the sacrum (low signal intensity) (*large solid arrow*). The tumor demonstrates extensive calcified chondroid matrix (low signal intensity) (*open arrows*). Small chondroid matrix calcifications also are seen in the L5–S1 intervertebral disc (*small solid arrows*) (confirmed by histologic examination). *B*, T2*-weighted gradient-echo MR image (TR, 600 ms; TE, 14 ms; flip angle, 30 degrees). The calcified areas are of very low signal intensity on the gradient-echo image and are more apparent than on the spin-echo image. The calcifications within the L5–S1 intervertebral disk appear markedly larger than on the spin-echo image and are more conspicuous (*small solid arrows*). This is attributable to inhomogeneity of magnetic susceptibility at the interface of the calcifications and the adjacent disk matrix, which results in dephasing of spins on gradient-echo sequences. Susceptibility-induced "fraying" of interfaces can be used to identify calcific or sclerotic structures.

muscle frequently is of slightly higher signal intensity than bone marrow on STIR images (see Table 15–2). Nonetheless, normal muscle is markedly lower in intensity than extraosseous tumor allowing good differentiation between normal muscle and extraosseous tumor[49] (see Table 15–2; Fig. 15–2*E*). Peritumorous edema in the muscles and soft tissues adjacent to the lesion exhibits high signal intensity on STIR images[121] (see Table 15–2; Fig. 15–2*E*). These STIR sequences require longer acquisition times than T1-weighted spin-echo images or T2*-weighted gradient-echo sequences and frequently are limited by a relatively low signal-to-noise ratio. Short TR, short TE T1-weighted spin-echo images provide greater spatial resolution than STIR images and therefore depict anatomic details better.[49] Some investigators recommend the combined use of short TR, short TE T1-weighted spin-echo scans and STIR images, thereby capitalizing on the advantages of both (i.e., excellent anatomic detail on T1-weighted spin-echo scans and superior contrast on the STIR sequence).[49]

Fast spin-echo or rapid acquisition relaxation-enhanced (RARE) sequences may also be employed in evaluating musculoskeletal neoplasm.[38,57,58,88] Experience with these sequences in imaging bone tumors is still limited, and tissue contrast remains to be investigated further. However, preliminary experience shows that T2-weighted fast spin-echo techniques provide relatively high signal intensity in fatty marrow and may thereby reduce conspicuity of marrow-based tumors that also are of high signal intensity. This problem can, however, be overcome by the use of a fat-saturated T2-weighted fast spin-echo technique.

Magnetic Resonance Imaging Contrast Media

Magnetic resonance imaging examination of musculoskeletal neoplasms can be supplemented by image acquisition after intravenous administration of MR imaging contrast media.[31-35,56,105,112,118,140] The most frequently employed contrast agent and the only one that is currently approved by the U.S. Food and Drug Administration (FDA) for use in MR imaging of patients is gadopentetate dimeglumine (gadolinium diethylenetriamine pentaaceticacid, Magnevist). This agent is a strongly paramagnetic complex that reduces hydrogen proton relaxation times.[12,13,143] The pharmacokinetic characteristics of gadopentetate dimeglumine resemble those of iodinated contrast agents used in urography and angiography. The standard dose for use in humans is 0.1 mmol per kilogram of body weight.

Other potential MR imaging contrast agents for the enhancement of bone tumors and tumorlike lesions of bone include gadolinium tetraazacyclo-dodecane tetraacetic acid (Gd-DOTA),[140] complex with contrast behavior similar to that of gadopentetate dimeglumine,, and ultrasmall superparamagnetic iron oxide (USPIO) particles.[118]

Sequences that are typically employed after intravenous administration of gadopentetate dimeglumine are T1-weighted spin-echo scans or, alternatively, gradient-echo images with relatively short TRs, short TEs, and large flip angles.[35] On postcontrast scans, most tumors demonstrate marked enhancement, with resultant increase in signal intensity (Fig. 15–4).[3,22,32,33,35,39,40,55,102,105,123,132,138] Only certain benign bone and soft tissue neoplasms, such as lipomas and osteochondromas, may show no increase in signal intensity after intravenous administration of gadopentetate dimeglumine.[33] Fatty tissue and bone marrow may not increase in signal intensity at all or may show only minor degrees of increase on postcontrast images. Muscle generally demonstrates only minor contrast enhancement.[33]

The contrast-to-noise ratio between tumor and bone marrow or fatty tissue on postcontrast T1-weighted images has been reported to be 43 per cent and 37 per cent, respectively, of that in unenhanced T1-weighted spin-echo images.[33] The contrast enhancement of neoplasm has an adverse effect on contrast-to-noise ratio: on precontrast scans, tumor demonstrates low signal intensity and is clearly distinct from bone marrow, which is of high signal intensity. On postcontrast scans, tumor increases in signal intensity and distinction from bone marrow (high signal intensity) is more difficult (see Fig. 15–4*C*). This problem can, however, be overcome with the use of fat saturation sequences. Fat saturation can be achieved with a chemically selective presaturation pulse. After fat saturation, normal bone marrow is of low signal intensity (see Fig. 15–4*D*, *E*). Enhancing tumor, of high signal intensity, is highlighted against the adjacent fatty marrow of low signal intensity (see Fig. 15–4*E*).

The contrast-to-noise ratio between tumor and muscle is 4.5 times greater on postcontrast T1-weighted images than on unenhanced T1-weighted images.[33] However, the contrast-to-noise ratio between tumor and muscle on postcontrast scans is 44 per cent of that found on unenhanced T2-weighted spin-echo images.[33] Despite these limitations, postcontrast scans provide differentiation between viable and necrotic tumor portions, unlike unenhanced imaging studies.[33]

Postcontrast imaging may be supplemented by rapid acquisitions in which images are acquired sequentially after bolus administration of gadopente-

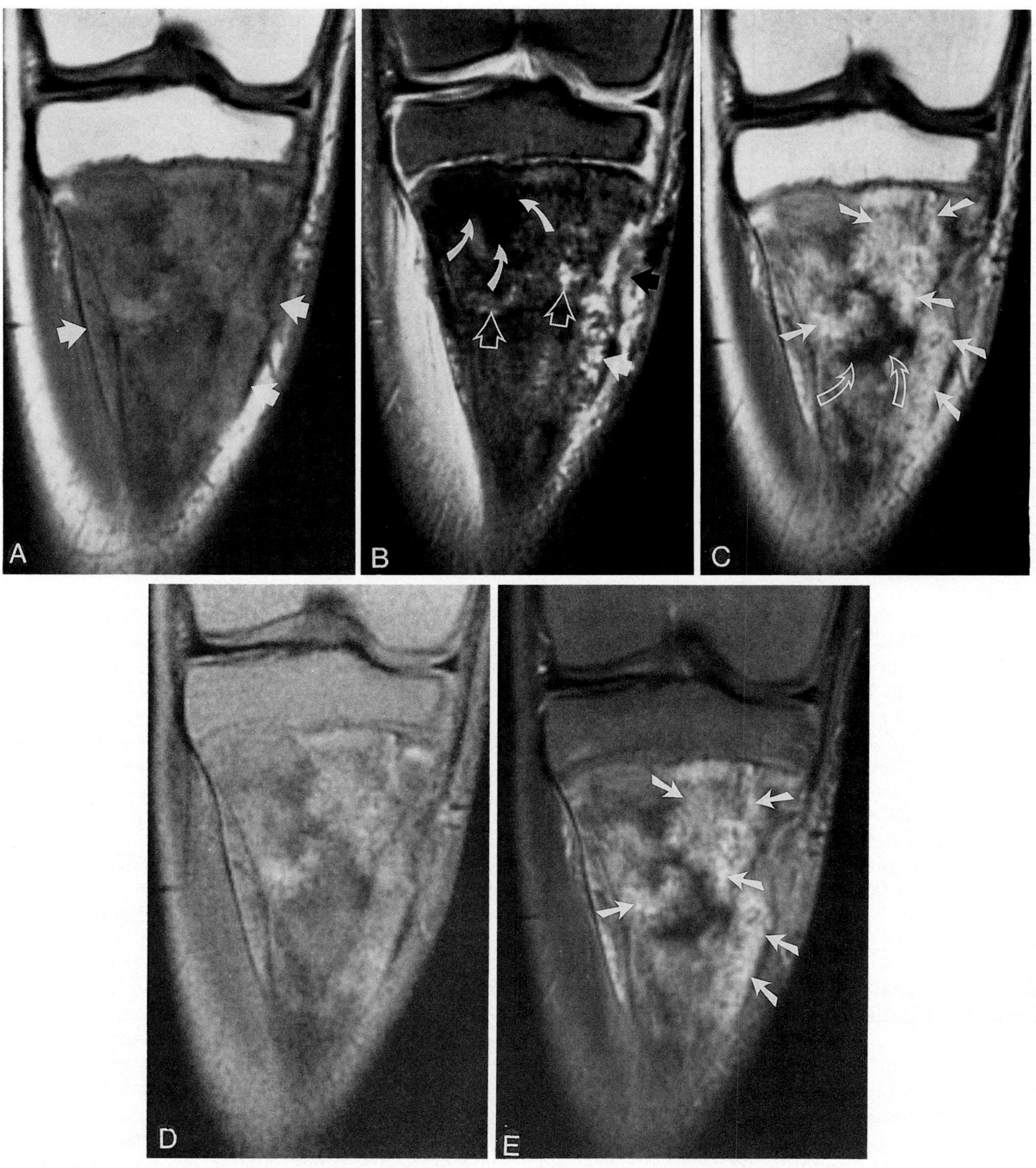

Figure 15–4. Osteogenic sarcoma of the proximal tibia in an 11-year-old male patient. *A*, Precontrast T1-weighted spin-echo MR image (TR, 800 ms; TE, 30 ms) shows a tumor of low signal intensity in the proximal tibia, extending into the soft tissues *(arrows)*. The contrast between metaphyseal tumor (low signal intensity) and fatty marrow (high signal intensity) in the proximal tibial epiphysis is high. *B*, T2*-weighted gradient-echo MR image (TR, 600 ms; TE, 14 ms; flip angle, 30 degrees). The medial portions of the extraosseous tumor *(straight solid arrows)* and portions of the intraosseous tumor increase in signal intensity *(open arrows)*. Note the area of low signal intensity inferior to the physeal plate *(curved solid arrows)*. This probably reflects sclerotic bone. *C*, Postcontrast T1-weighted spin-echo MR image (TR, 800 ms; TE, 30 ms). Marked contrast enhancement is seen in large portions of the tumor *(straight arrows)*. Contrast between enhancing tumor and adjacent fatty marrow (high signal intensity) is lower than on the precontrast image in *A*. Note the area without contrast enhancement in the tibial metaphysis *(curved open arrows)*. This may correspond to a region of nonperfused tumor necrosis, although, in part, it may also reflect an area of sclerosis. *D*, Fat-saturated precontrast T1-weighted spin-echo MR image (TR, 500 ms; TE, 30 ms). Both tumor and normal bone marrow as well as subcutaneous fat demonstrate intermediate to low signal intensity. Tumor-marrow contrast is poor in comparison to the conventional T1-weighted precontrast spin-echo image in *A*. *E*, Fat-saturated postcontrast T1-weighted spin-echo MR image (TR, 500 ms; TE, 30 ms). With this sequence, the enhancing tumor *(arrows)* is highlighted against the normal fatty marrow and subcutaneous fat (low signal intensity). The contrast between marrow and enhancing tumor on the fat-saturated postcontrast T1-weighted MR image is greater than on the conventional postcontrast T1-weighted MR image in *C*.

tate dimeglumine to monitor the dynamics of contrast enhancement.[33,35,76,105,106] Fast gradient-echo sequences are ideally suited for this application. Acquisition time should range between 5 and 10 sec per scan. Acquisition is performed in single slice technique. A single image is generated before administration of the contrast agent. Gadopentetate dimeglumine is then injected in a bolus fashion, and scans are acquired sequentially immediately after the injection[33,35,76,105,106] (Fig. 15–5). Receive and transmit attenuations of pre- and postcontrast images are kept constant to allow quantitative evaluation of signal intensities.

The percentage of increase in signal intensity ($SI^{increase}$) over the baseline intensity is calculated as follows:

$$SI^{increase} = (SI^{max} - SI^{prior}) \times 100 / SI^{prior}$$

where SI^{max} is the maximum signal intensity of the tissue after intravenous application of gadopentetate dimeglumine and SI^{prior} is the signal intensity prior to injection.

The slope of the curve is calculated as the percentage of increase of signal intensity over the baseline value SI^{prior} per minute:

$$\text{Slope} = (SI^{max} - SI^{prior}) \times 100 / (SI^{max} \times T^{max})$$

where T^{max} is the time to reach SI^{max}. Percentage of increase in signal intensity and slope values of the enhancement curves help to differentiate perineoplastic edema from extraosseous tumor.[33] The technique also may improve differentiation between benign and malignant lesions[33] and can be employed to quantify the effects of chemotherapy on neoplastic tissue.[34]

Imaging Planes

Musculoskeletal neoplasms generally should be imaged in at least two planes to delineate the exact dimensions of the lesion. The choice of planes—axial, sagittal, or coronal—depends on the type of tumor, its location, and its extent. The initial localizer scan helps to identify the plane that is best suited to demonstrate both intra- and extraosseous tumor extent. The second plane is designed to further delineate the relationship of the tumor to adjacent structures. In imaging lesions of the distal femur or the proximal tibia, for example, both the coronal and the sagittal planes provide good delineation of the extent of the tumor. The coronal scan is preferred when the neoplasm extends primarily laterally or medially; the sagittal scan is advantageous when the tumor is oriented in an anteroposterior direction. Axial scans are useful to detect neoplastic encroachment or even encasement of the neurovascular bundle (Fig. 15–6). Similarly, sagittal scans also help in this location to demonstrate the spatial relationship between tumor and vascular structures or peripheral nerves.

Additional axial T1- and T2-weighted scans are required when joint involvement is suspected. T1-weighted axial scans help elucidate the infiltration of synovial fat of high signal intensity by tumor of low signal intensity. T2-weighted images are useful in evaluating the integrity of the joint capsule, of low signal intensity, as it is highlighted against neoplasm (high signal intensity) as well as joint fluid or effusion (also high signal intensity).[115]

Oblique scans may be needed when the long axis of the tumor is not parallel to any of the three main planes to avoid partial volume effects.[115] Oblique planes also are useful in upper and lower extremity tumors, when the patient cannot extend the extremity because of pain and the extremity is oriented at an angle (> 5 to 10 degrees) relative to the long axis of the magnet bore.

LOCALIZATION AND STAGING OF LESIONS

Primary bone and soft tissue tumors of a limb girdle or an extremity pose an immediate threat to the affected body part and eventually to life. The choice between local resection, limb salvage surgery, or amputation in primary musculoskeletal tumors depends on the malignant potential of the tumor, its anatomic extent, and its histologic grade.[4,44,87,122] Limb salvage procedures require an accurate preoperative anatomic localization of the lesion to ensure that tumor resection is accomplished with safe margins around the tumor. Similarly, exact delineation of tumor margins is necessary for planning of radiation therapy. Advanced imaging methods are needed for accurate staging of neoplasms and precise localization of the extent of tumors for planning both surgical and radiation therapy.

Conventional radiography remains the initial imaging modality for musculoskeletal neoplasms. Conventional radiography provides valuable information for differential diagnosis of a lesion. It is, however, limited in delineating the intramedullary extent of a musculoskeletal neoplasm and even more so in demonstrating soft tissue involvement[125,126] (Table 15–3).

Nuclear scintigraphy is an ideal screening test for primary neoplasms of bone and soft tissue, for skip lesions, and for more distant metastases involving the axial skeleton, the lungs, or other organ sys-

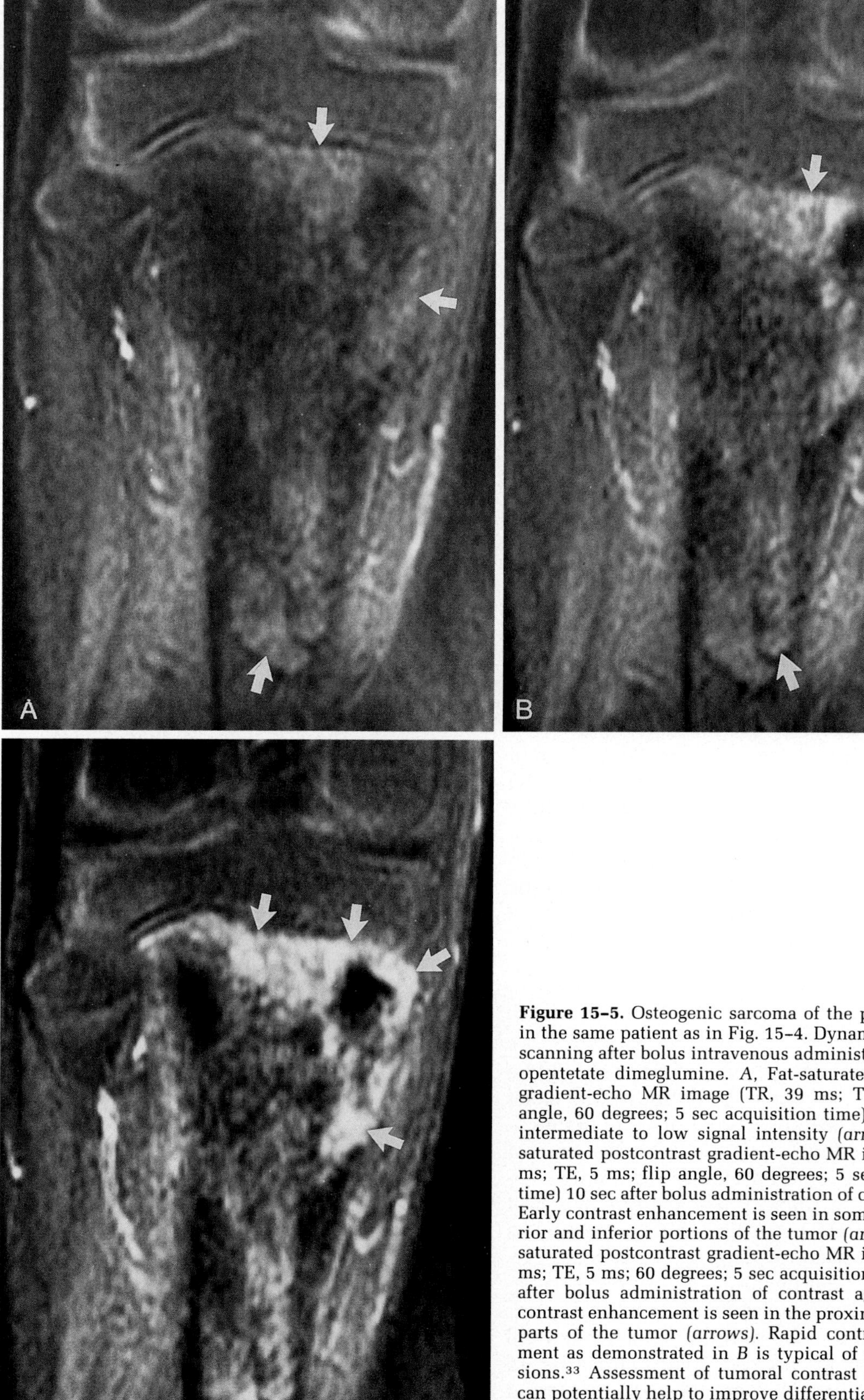

Figure 15–5. Osteogenic sarcoma of the proximal tibia in the same patient as in Fig. 15–4. Dynamic sequential scanning after bolus intravenous administration of gadopentetate dimeglumine. *A*, Fat-saturated precontrast gradient-echo MR image (TR, 39 ms; TE, 5 ms; flip angle, 60 degrees; 5 sec acquisition time). Tumor is of intermediate to low signal intensity *(arrows)*. *B*, Fat-saturated postcontrast gradient-echo MR image (TR, 39 ms; TE, 5 ms; flip angle, 60 degrees; 5 sec acquisition time) 10 sec after bolus administration of contrast agent. Early contrast enhancement is seen in some of the superior and inferior portions of the tumor *(arrows)*. *C*, Fat-saturated postcontrast gradient-echo MR image (TR, 39 ms; TE, 5 ms; 60 degrees; 5 sec acquisition time) 60 sec after bolus administration of contrast agent. Marked contrast enhancement is seen in the proximal and distal parts of the tumor *(arrows)*. Rapid contrast enhancement as demonstrated in *B* is typical of malignant lesions.[33] Assessment of tumoral contrast enhancement can potentially help to improve differentiation between benign and malignant tumors. It may also be used to quantify the effects of chemotherapy in malignant tumors by comparing contrast dynamics before and after chemotherapy.

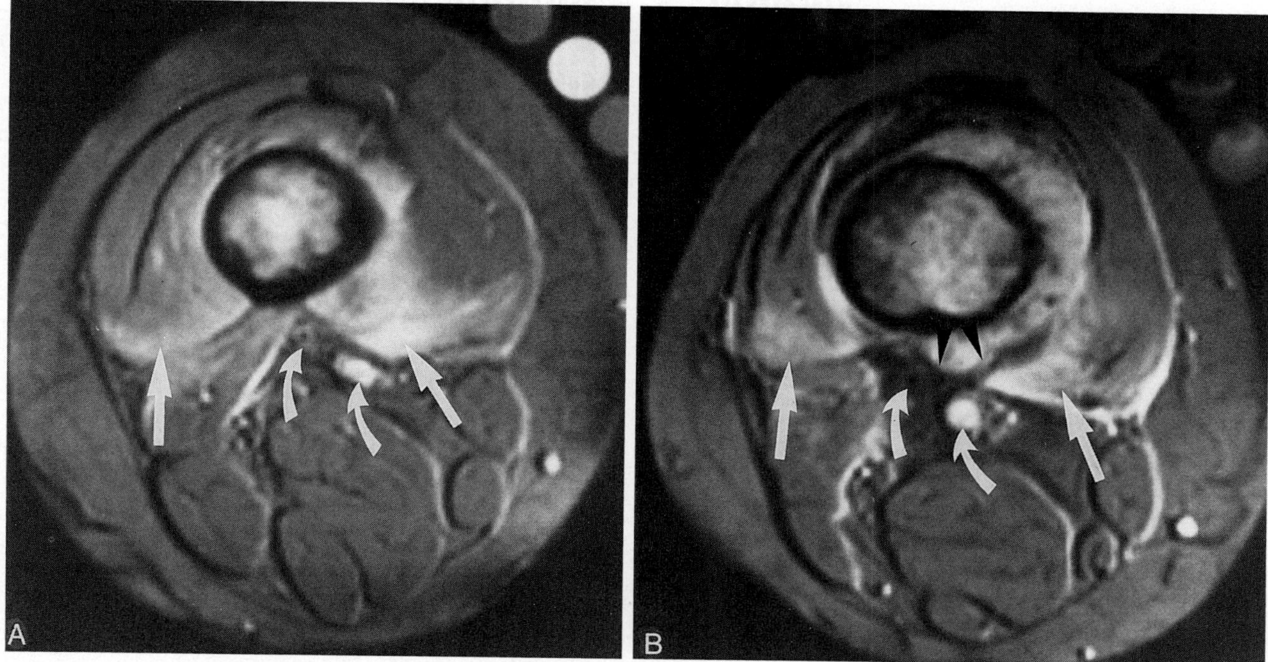

Figure 15-6. Osteogenic sarcoma of the distal femur in the same patient as in Fig. 15-2. *A*, Axial T2*-weighted gradient-echo MR image (TR, 600 ms; TE, 30 ms; flip angle, 30 degrees) demonstrates perineoplastic edema of high signal intensity penetrating the vastus medialis and vastus lateralis muscles *(straight arrows)*. The femoral artery and vein *(curved arrows)* are located immediately adjacent to the perineoplastic edema but are not encased by tumor. *B*, More distal axial T2*-weighted gradient-echo MR image (TR, 600 ms; TE, 30 ms; flip angle, 30 degrees) shows an extraosseous tumor mass protruding posteriorly *(arrowheads)*. Perineoplastic edema also is present at this level *(straight arrows)*. The femoral artery and vein *(curved arrows)* are immediately adjacent to the tumor but not encased by it. The round structures at the upper right corner of the image are fat and water phantoms.

tems (see Table 15-3) as the total body can be imaged in a single session. At present, nuclear scintigraphy of primary bone and soft tissue tumors is performed most frequently using technetium (^{99m}Tc)-diphosphonates such as ^{99m}Tc methylene diphosphonate (MDP).[8,34,100] Other isotopes that have been used in the evaluation of bone and soft tissue tumors include thallium-201 and gallium-67.[103,146] Nuclear scintigraphy is, however, relatively inaccurate in defining the exact extent of the tumor[16,142] (see Table 15-3). Increased bone turnover in the vicinity of the tumor and hyperemia frequently result in a falsely extended uptake pattern at ^{99m}Tc-MDP scintigraphy.[16] Nuclear scintigraphy thus is limited

Table 15-3. RELATIVE ADVANTAGES OF DIFFERENT IMAGING TECHNIQUES IN EVALUATING MUSCULOSKELETAL NEOPLASMS AND NORMAL TISSUES

	Radiography	Nuclear Scintigraphy	Unenhanced CT	Contrast-Enhanced CT	MR Imaging
Cortical bone	+ + +	+	+ + +	+ + +	+ +
Bone marrow	–	+	+	+	+ + +
Soft tissue	+	+	+ +	+ +	+ + +
Intramedullary tumor extent	+	+	+ +	+ +	+ + +
Extramedullary tumor extent	+	+	+ +	+ + +	+ + +
Relationship of tumor to neurovascular bundle	–	–	+	+ +	+ + +
Calcifications	+ +	–	+ + +	+ + +	+ +
Skip lesions	+	+ +	+ +	+ +	+ + +
Differentiation of benign versus malignant	+ + +	+	+	+	+ +
Monitoring of response to chemotherapy or radiation therapy	+	+ +	+ +	+ +	+ + +
Tumor characterization	+ + +	+	+ +	+ +	+ +

–, Not demonstrated; +, moderate; + +, good; + + +, excellent demonstration.

mainly to the detection of metastatic disease arising from a primary neoplasm of bone and soft tissue.[8,23,80,95] Local tumor staging of the primary tumor is performed using cross-sectional imaging techniques such as CT and MR imaging (see Table 15–3).

Intramedullary Tumor Extent

In evaluating the *intramedullary extent* of a primary bone tumor, MR imaging has been reported to be more accurate than CT in multiple studies.[1,8-10,45,56,78,79,95,97,129,133,134,149] Some controversy about these findings existed because many of the CT studies have been performed without using an intravenous contrast agent.[114] In addition, CT scans have been generated with a section thickness greater than that of MR images.[114]

Gillespy and colleagues measured the intramedullary tumor extent on CT and MR images in 17 patients with osteosarcoma of an extremity and correlated it with measurements from macroslides of surgical specimens.[45] Longitudinal intraosseous extension of tumor from the adjacent articular surface was measured on imaging studies and macroslides.[45] The average difference between macroslides and CT measurements amounted to 16.5 mm ± 10.7 mm (15 cases). The average difference between macroslides and MR imaging measurements was 4.9 mm ± 4.3 mm (17 cases). However, the researchers concluded that much of the difference between macroslide and MR imaging values was related to the use of different planes of section for these two methods, because when an identical plane of section was used in a subgroup of five patients, the average difference was 1.8 mm ± 1.6 mm.[45] Similarly, Bloem and colleagues correlated the measurements of intraosseous tumor length from resected tumor specimens with CT and MR measurements.[8] In this study, MR imaging had an almost perfect correlation ($r = 0.99$) with resected tissue, CT had a less substantial correlation ($r = 0.86$), and ^{99m}Tc-MDP scintigraphy had a weak correlation ($r = 0.56$). Magnetic resonance imaging is significantly more accurate than CT scanning in defining the intramedullary extent of neoplasm. However, because most surgeons resect with a safety margin 5 to 6 cm beyond the tumor, both CT and MR imaging are clinically sufficient for evaluating tumor margins.[100,114]

Magnetic resonance imaging is more accurate in demonstrating *epiphyseal involvement* than conventional radiography.[94] Norton and co-workers reported on a series of 15 patients with osteosarcoma; spread to the epiphysis was diagnosed on conventional radiography, on MR imaging, and at autopsy in 9, 12, and 12 patients, respectively, thereby confirming the superiority of MR imaging.[94]

Some controversy exists whether CT or MR imaging is more accurate in delineating *joint involvement*. Bloem and colleagues found that both CT and MR imaging yielded similar results in delineating joint involvement, with sensitivities and specificities in the 90 per cent range.[8] Aisen and co-workers and Zimmer and colleagues,[149] however, both reported MR imaging to be more sensitive in detecting joint involvement in a small number of patients, whereas CT was equal to MR imaging in the majority of cases.

Several studies have reported that conventional radiography and CT are superior to MR imaging in evaluating *cortical destruction*, in depicting *calcification* and *ossification*, and in demonstrating *periosteal or endosteal reaction*.[5,100,125,126,129,149] Dense bone, as found in periosteal reaction or osteoblastic tumor portions, is characterized by low proton density with resultant low signal intensity on all MR imaging sequences. The low amount of signal in these structures is responsible for the reportedly limited delineation on conventional MR imaging sequences. Most of the studies describing the superiority of CT over MR imaging in evaluating cortical bone involvement were, however, performed several years ago at low field strengths and with coil technologies that are outdated at the present time.

Zimmer and co-workers showed that cortical expansion, thinning and destruction, and peri- and endosteal new bone formation all can be identified using MR imaging.[149] More recent studies found that MR imaging is as accurate in the assessment of cortical involvement as CT.[8,63,107] At present it appears, however, that CT is still superior to MR imaging in depicting areas of subtle cortical thickening or thinning. The diagnostic accuracy of MR imaging in detecting subtle areas of cortical thickening or tumor sclerosis may be improved with use of gradient-echo sequences. Focal areas of sclerosis cause local inhomogeneity in magnetic susceptibility, with resultant signal loss on gradient-echo images, because, unlike the case in spin-echo imaging, rephasing is not performed. This signal loss can be used to identify areas of tumor sclerosis (see Fig. 15–3) or periosteal or endosteal reaction.

Involvement of Soft Tissue

Both CT and MR imaging can be employed for evaluating the soft tissue extent of a primary bone tumor or the extent of a primary soft tissue neoplasm. Indirect signs, such as displacement of fat planes, obliteration of fat planes, and general mass effect on neighboring anatomic structures, help in defining the extraosseous tumor extent on CT. However, CT is limited in assessing soft tissue involvement as extraosseous tumor and surrounding soft tissues frequently have similar CT density. The CT study can be amended by the use of intravenous contrast medium, resulting in enhancement of the lesion and improved delineation against the adjacent normal tissue.[81,115,136] However, even with the use of contrast agents, CT is inferior to MR imaging in evaluating the extent of soft tissue lesions in the majority of cases.[97]

Magnetic resonance imaging is significantly superior to unenhanced CT and even CT with contrast enhancement in assessing soft-issue involvement. The main reason is the improved contrast between neoplasm and adjacent normal tissue (e.g., fat and muscle).[1,149] Direct imaging capability in the sagittal and coronal planes also contributes to the superiority of MR imaging over CT for evaluating the extent of soft tissue involvement.[149] Computed tomographic scanning requires mental integration of a large series of transaxial CT scans to estimate the size of the tumor. Multiplanar reconstruction of CT scans is limited by a relatively low in-plane resolution that corresponds to the slice thickness of the original axial CT scan. Direct MR image acquisition in the sagittal, coronal, or oblique planes provides three-dimensional information about the size and shape of tumors that is superior to multiplanar CT reconstructions with respect to spatial and contrast resolution.

Zimmer and co-workers reported MR imaging to be superior to CT in assessing the size of an extraosseous soft-tissue mass in 38 per cent of patients.[149] In detecting soft-tissue calcification, CT is more sensitive than MR imaging[97,136] although MR imaging results may be improved with the use of gradient-echo sequences owing to susceptibility-induced dephasing of spins and local signal loss (see earlier discussion). Zimmer and colleagues demonstrated MR imaging to be better than CT in delineating involvement of fat planes adjacent to the tumor in 7 per cent of patients. In the remaining cases, CT and MR imaging were about equal.[149]

Magnetic resonance imaging also is superior to CT in demonstrating the relationship of the tumor to adjacent *neurovascular bundles*[1,8,97,100,149] (see Fig. 15–6). Vascular structures are seen well on MR images without the need for intravascular contrast agents.[11,86] According to Bloem and colleagues, the sensitivity, specificity, accuracy, positive predictive value, and negative predictive value for CT and MR imaging in assessing involvement of neurovascular bundles are 33 and 100 per cent, 93 and 98 per cent, 82 and 98 per cent, 50 and 91 per cent, and 87 and 100 per cent, respectively.[8]

Differentiation of tumor from normal fat surrounding the neurovascular bundle may be less evident on T1-weighted images after intravenous administration of gadopentetate dimeglumene.[115] On unenhanced T1-weighted MR images, tumor typically demonstrates low signal intensity and thus is clearly distinct from fat, of high signal intensity. On postcontrast images, however, marked contrast enhancement is seen in most tumors resulting in high signal intensity. Contrast between neoplastic tissue and fat (both of high signal intensity) surrounding the neurovascular bundle thus is diminished on MR images after administration of gadopentetate dimeglumine.[33,115]

Magnetic resonance imaging is better than CT in showing the extent of *muscle invasion* and *muscle edema* in 49 per cent of patients[149] (see Fig. 15–6). Bloem and co-workers found a sensitivity, specificity, accuracy, positive predictive value, and negative predictive value for CT and MR imaging in evaluating muscular involvement of 71 and 96 per cent, 93 and 99 per cent, 87 and 98 per cent, 80 and 97 per cent, and 90 and 98 per cent, respectively.[8]

Differentiation between *perineoplastic edema* and *extraosseous tumor* is a well-known problem on CT.[100] On MR imaging, both extraosseous tumor and peritumorous edema that surrounds the tumor in the adjacent reactive zone display similar low signal intensity on T1-weighted images and high signal intensity on T2-weighted images (Figs. 15–7*A*, *B*). Shuman and colleagues compared conventional T2-weighted spin-echo images with STIR sequences in evaluating the extent of soft tissue extremity tumors.[121] They found "peritumoral brightening" in 20 patients; in nine of the 20 patients, this finding was detected only by the STIR sequence and was not evident on T2-weighted spin-echo sequence. However, both peritumorous edema and extraosseous tumor appeared as "peritumoral brightening" on the STIR sequence (see Fig. 15–7*C*). Thus, although the STIR sequence may provide greater lesion conspicuousness than spin-echo images do, it does not afford differentiation between perineoplastic edema and extraosseous tumor.

On conventional post contrast MR images, viable tumor and peritumorous edema also exhibit marked contrast enhancement (Fig. 15–7*E*, *F*). Pettersson and co-workers studied five patients with soft tissue tumors by gadopentetate dimeglumine–enhanced MR imaging and found that both richly vascularized neoplastic tissue and peritumorous

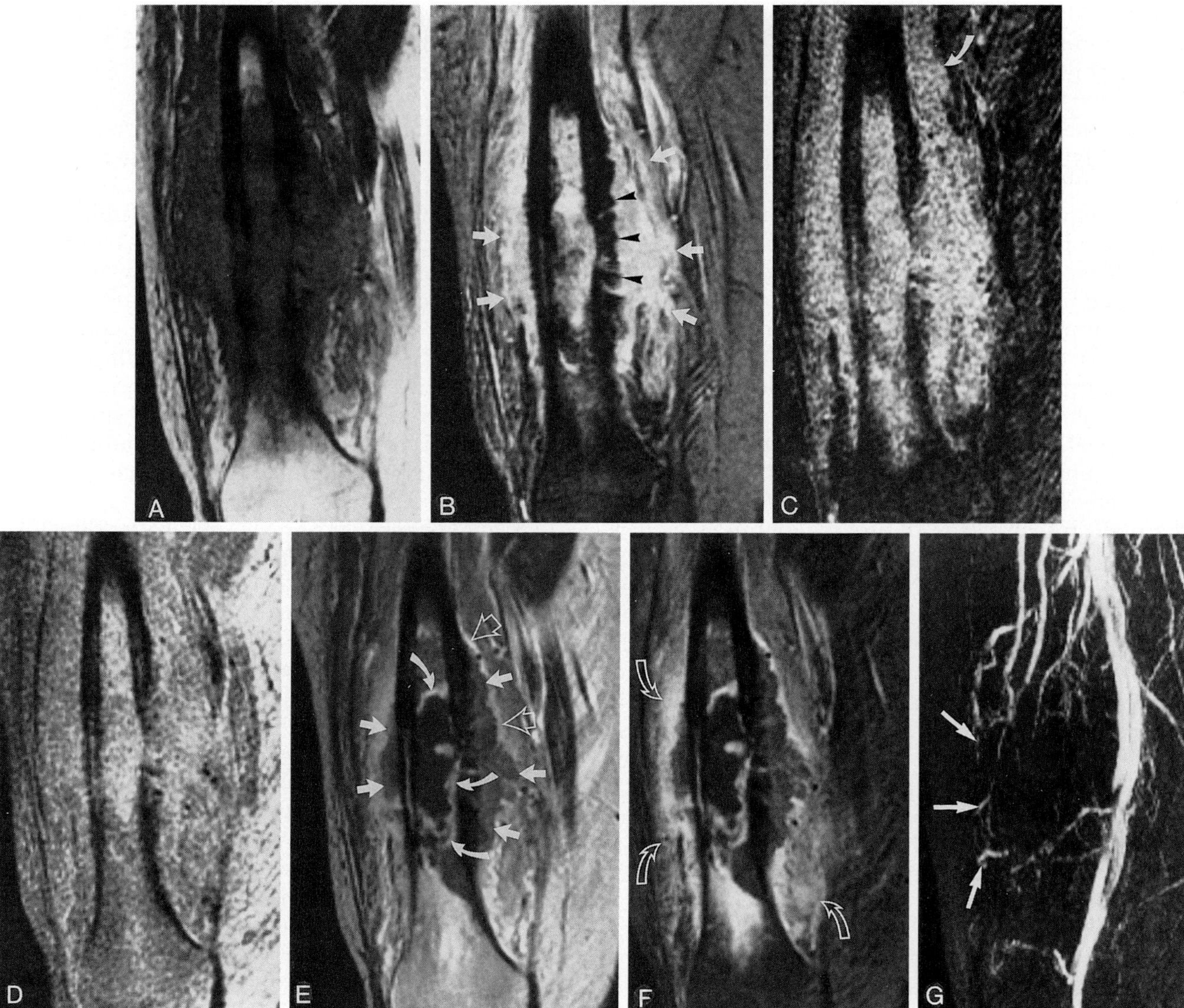

Figure 15–7. Primary lymphoma of the distal femur in a 63-year-old male patient. *A*, Precontrast T1-weighted spin-echo MR image (TR, 800 ms; TE, 30 ms) demonstrates a lesion of low signal intensity in the distal femur. The femoral cortex is disrupted, and the mass is extending into the surrounding soft tissues. *B*, T2*-weighted gradient-echo MR image (TR, 600 ms; TE, 14 ms; flip angles, 30 degrees). The intramedullary portion of the lesion demonstrates high signal intensity. The femoral cortex shows spiculations of low signal intensity *(arrowheads)*. The soft tissue extension also is characterized by high signal intensity *(arrows)*. No clear distinction can be made between the peritumoral edema (high intensity) and the extent of extraosseous tumor (also of high intensity) with this imaging sequence. *C*, Short tau inversion-recovery (STIR) image (TR, 1500 ms; TE, 20 ms; TI, 170 ms). The intra- and extraosseous tumor portions also are of high signal intensity on the STIR sequence. Note the abnormally high signal intensity in the proximal portions of the vastus medialis muscle *(curved arrow)*. By location, this most likely represents perineoplastic edema. STIR similarly does not afford differentiation between perineoplastic edema and extraosseous tumor on the basis of signal intensity. *D*, Precontrast fat-saturated T1-weighted spin-echo MR image (TR, 500 ms; TE, 30 ms) shows the lesion and the adjacent tissues to be of intermediate signal intensity. *E*, Postcontrast T1-weighted spin-echo MR image (TR, 800 ms; TE, 30 ms). The central portions of the intramedullary tumor are of low signal intensity. These corresponded to necrotic tumor tissue at biopsy. The area of low signal intensity is surrounded by an enhancing linear zone of high signal intensity, which represented viable lymphoma at biopsy *(curved solid arrows)*. The soft tissue portion of the tumor *(straight solid arrows)* demonstrates moderate contrast enhancement, resulting in intermediate signal intensity. It is surrounded laterally and superomedially by a markedly enhancing linear zone *(straight open arrows)* which corresponded to perineoplastic edema at biopsy (biopsy specimen taken from superomedial portion). *F*, Postcontrast fat saturated T1-weighted spin-echo MR image (TR, 500 ms; TE, 30 ms). The areas of contrast enhancement are highlighted against fatty structures with this imaging sequence. Marked contrast enhancement of the vastus medialis and vastus lateralis muscles is seen *(curved open arrows)*. These were found to be very edematous but not infiltrated by tumor cells at biopsy. *G*, Two-dimensional time of flight MR angiogram (TR, 50 ms; TE, 8.7 ms; flip angle 45 degrees), coronal maximum intensity projection. Neovascularity is observed surrounding the tumor *(arrows)*. However, almost no vascular branches are seen in the center of the tumor, which is consistent with the nonenhancing area of intramedullary tumor necrosis seen on the postcontrast images.

edema where enhanced.[99] Hanna and colleagues reported that viable tumor, granulation tissue after biopsy, and peritumorous edema were enhanced after the administration of gadopentetate dimeglumine.[55] Similarly, Seeger and associates demonstrated that a streaky enhancement pattern at the periphery of the tumor could represent either perineoplastic edema or infiltration of tumor into adjacent soft tissues.[115] In our own experience, gadopentetate dimeglumine–enhanced MR imaging may, in selected cases be helpful in differentiating between perineoplastic edema and extraosseous tumor (see Fig. 15–7*E, F*). In most patients, however, their distinction is not possible on the basis of MR imaging signal intensity or contrast enhancement alone.

An investigation by Erlemann and co-workers indicates that the two tissues may be distinguished by a fast sequential MR scanning technique after bolus administration of gadopentetate dimeglumine.[33] In their study, viable tumor was found to show a markedly faster contrast enhancement after bolus administration of the contrast agent than did peritumorous edema. Percentage of signal intensity increase over baseline signal intensity and slope values were significantly lower for perineoplastic edema than for viable tumor.[33] The ultimate value of this technique remains to be investigated in a large series of patients with close histologic correlation.

Magnetic resonance imaging after the intravenous administration of gadopentetate dimeglumine is unique in that it provides differentiation between viable and necrotic tumor. Viable tumor is characterized by marked contrast enhancement unless the lesion is hypovascular in nature. Necrotic tumor regions, however, do not demonstrate contrast enhancement and remain low or intermediate in signal intensity on T1-weighted postcontrast scans (see Figs. 15–4, 15–7).[33]

At present, the biopsy site is determined on the basis of clinical findings (e.g., local soft tissue swelling) and conventional radiography, neither of which provides information about local tissue viability. When the biopsy specimen is taken from a necrotic tumor region, frequently the diagnosis is indeterminate. Magnetic resonance imaging after intravenous administration of gadopentetate dimeglumine may thus be of help in identifying viable tumor areas and planning of biopsy.[33]

In imaging spinal soft tissue involvement, CT scans obtained without injection of contrast material may show obliteration of epidural fat, but subtle abnormalities of the spinal canal frequently cannot be demonstrated.[5] Only CT scanning performed after intrathecal injection of ionic contrast material can generate the information that is available with MR imaging.[5] Intrathecal injection of contrast material may, however, induce seizures, headaches, psychic disturbances, and metabolic derangements.[131] Although these side effects have decreased with the introduction of non-ionic contrast media,[71] Nonetheless, CT scanning with intrathecal contrast agent is an invasive procedure that places signifcant stress on the patient. Magnetic resonance imaging demonstrates the relationship among the tumor, the spinal cord, and the subarachnoid space without the need for invasive injections.[5]

CHARACTERIZATION AND SPECIFICITY OF LESIONS

Many criteria employed in conventional radiography and CT for differentiation between benign and malignant lesions are also applicable to MR imaging. Benign lesions typically are sharply demarcated from the adjacent normal marrow and soft tissue.[97,149] Malignant tumors are less well demarcated and have a tendency to infiltrate the adjacent tissues[97] (see Figs. 15–3, 15–6, 15–7). Matrix mineralization is not as well seen on MR images as on CT scans; it appears as a punctate area of low signal intensity.[149] If a sharp, smooth, well-defined border of low signal intensity is seen between the tumor and the adjacent marrow, reflecting tumor capsule, the lesion most likely is benign (Fig. 15–8, see also Fig. 15–11*B*). Such capsules may, however, also be seen in malignant tumors.[149]

Petasnick and co-workers found that most benign soft tissue tumors are characterized by uniform signal intensity on both T1- and T2-weighted images.[97] However, in their series, few benign tumors (e.g., neurofibromas, hemangiomas) had an inhomogeneous signal on T2-weighted images. Most malignant tumors were of uniform intensity on T1-weighted images but were inhomogeneous on T2-weighting. Thus, uniformity of signal on T2-weighting was a reliable indication of benignancy (see Fig. 15–8), but lack of uniformity did not reliably indicate malignancy.[97] In general, morphologic patterns may provide some information about the nature of the lesion that may be used in differentiating between benign and malignant conditions; they are, however, not accurate and not specific.

Zimmer and colleagues measured T1- and T2-relaxation times of normal marrow and of benign and malignant neoplasms.[149] T1-relaxation times for normal marrow, benign tumors, and malignant neoplasms ranged from 195 to 307 ms, 375 to 628 ms, and 404 to 887 ms, respectively.[149] T2-relaxation times for normal marrow, benign tumors, and malignant neoplasms ranged from 80 to 116 ms, 143 to 205 ms, and 126 to 139 ms, respectively.[149] The narrow

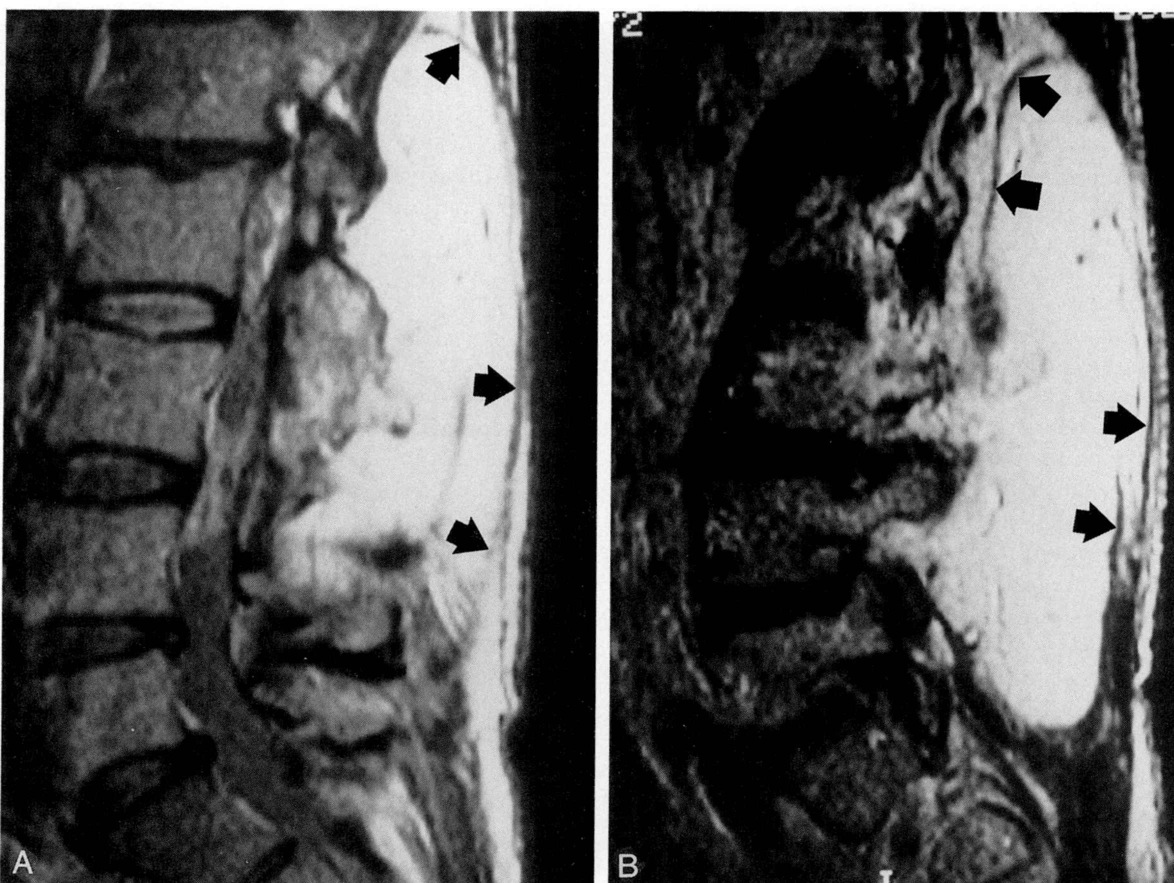

Figure 15-8. Lipoma of the lumbosacral spine (histologically proved). *A*, Sagittal precontrast T1-weighted spin-echo image (TR, 1000 ms; TE, 20 ms) demonstrates a mass of high signal intensity posterior to the lumbar spine. The mass is surrounded by a border of low signal intensity corresponding to the tumor capsule *(arrows)*. *B*, Sagittal precontrast T1-weighted spin-echo image (TR, 1000 ms; TE, 70 ms) at a more lateral site. The mass retains high signal intensity with longer echo time, corresponding to fat within a lipoma. The mass extends into the neural foramina. The tumor capsule is again noted *(arrows)*.

range of T2-relaxation times resulted from the small number of patients studied. T1- and T2-relaxation times demonstrate a significant overlap between benign and malignant lesions and thus are not useful clinically for differentiation between benign and malignant lesions.[97,149]

Similarly, in the same study, no signal intensity differences were observed between benign and malignant lesions.[149] Both exhibited lower signal intensity than normal marrow on T1-weighted spin-echo sequences and inversion-recovery sequences. Both benign and malignant neoplasms demonstrated signal intensity higher than that of normal marrow on T2-weighted spin-echo sequences.[149]

Erlemann and co-workers reported that dynamic MR imaging studies with rapid sequential image acquisition after bolus adminstration of gadopentetate dimeglumine (see Fig. 15-5) can be used to differentiate benign from malignant lesions.[33] In their study, 84.1 per cent of malignant tumors exhibited slope values higher than 30 per cent per minute; 72 per cent of benign tumors showed slopes lower than 30 per cent per minute.[33] However, both benign and malignant tumors demonstrated some overlap using this differential criterion, resulting in an accuracy of approximately 80 per cent with this technique.[33] Ultimately, biopsy is still needed to define the histologic nature of a musculoskeletal neoplasm.

Some of the pathologic tissues that may be found in musculoskeletal neoplasms can be identified on the basis of their signal behavior on MR images. Fat within a tumor produces high signal intensity on T1-weighting and intermediate to high signal intensity on T2-weighted images (see Table 15-2)[100] (see Fig. 15-8). Fibrous tissue demonstrates low signal intensity on all imaging sequences. Similarly, sclerotic bone is characterized by low signal intensity on all sequences (see Table 15-2)[100] (see Fig. 15-3). Cystic, fluid-filled portions of a tumor typically are of low signal intensity on T1-weighted and of high signal intensity on T2-weighted images.[100] The signal intensity of a cyst may, however, be high on T1-weighting when it contains significant amounts of proteinaceous debris.

MONITORING OF THE EFFECTS OF CHEMOTHERAPY AND RADIATION THERAPY

The most frequent skeletal malignancies in children and adolescents are osteosarcoma and Ewing's sarcoma. Before the introduction of chemotherapy, prognosis was poor in these tumors, and more than 70 per cent of the patients died of metastases.[28,68,151] Disease-free survival rate is significantly greater in those patients who respond to chemotherapy than in nonresponders.[8,145] The use of adjuvant chemotherapy has markedly improved 5-year survival rates. The efficacy of chemotherapy usually is assessed by histologic analysis of tumor specimens after resection. Generally, response is considered good when more than 90 per cent of the tumor cells have been devitalized.[65] Preoperative chemotherapy also simplifies the surgical procedure by producing a reduction in tumor size. In tumors that are sensitive to ionizing radiation, radiation therapy has also been shown to help improve survival rates.[141]

It is difficult to assess accurately the response of a tumor to preoperative chemotherapy before surgical resection of the tumor and histologic analysis. However, it is clinically important to know whether a patient responds favorably to a given chemotherapeutic agent or whether the drug should be changed. Clinical signs (e.g., decrease in pain, local heat, and soft tissue mass) often are inadequate in determining the efficacy of preoperative chemotherapy and show poor correlation with the degree of histologic response.[68,145] Similarly, response to radiation therapy may be difficult to evaluate on clinical grounds only. Various imaging modalities, such as conventional radiography, angiography, CT, and scintigraphy, have been employed to assess the efficacy of chemotherapy and radiation therapy. These techniques are only moderately specific in distinguishing responders from nonresponders. Specificity may be as low as 50 per cent because of limitations in delineating the intra- and in particular the extraosseous tumor extent and the inability to demonstrate total tumor destruction.[15,120] Scintigraphy also is limited by a low specificity (i.e., areas of marrow hyperemia, medullary reactive bone, or periosteal new bone may be mistaken for viable tumor).[16]

Magnetic resonance imaging provides both morphologic and quantitative information about response to therapy. Morphologic signs of good response to chemotherapy or radiation therapy are reduction in overall tumor size, improved delineation of muscle and fat planes, and cystic changes[62,96] (Fig. 15–9). Increased calcification at the periphery (see Fig. 15–9 *F–I*) as well as in the central portions also may indicate adequate response to chemotherapy in osteosarcomas.[18,85] Perineoplastic edema decreases with good response to chemotherapy.[62,96] Malignant bone tumors frequently are surrounded by a dark rim consisting of collagenous fibers that are continuous with the periosteum. The dark rim may become thicker and more distinct as the peritumoral edema decreases and as the tumor reponds to chemotherapy or radiation therapy.[96]

Pan and colleagues described four different post-chemotherapy MR imaging patterns. A dark pattern was characterized by predominant areas of low signal intensity on T1- and T2-weighted images.[96] The dark pattern indicated the best response with the least amount of residual viable tumor and the most amount of dense fibrotic scar tissue.[96] A mottled or speckled pattern showed a predominant area of intermediate signal intensity on T1-weighted images and of high signal intensity on T2-weighted images with mottled or speckled regions of low signal intensity. The mottled or speckled pattern also generally was correlated with a low number of viable tumor cells and good histologic response.[96] Mottling corresponded to an uneven distribution of tumor matrix and edematous granulation tissue. Dark speckles reflected scattered islands of hemosiderin.[96] A homogeneous pattern was characterized by a predominant area of intermediate signal intensity on T1-weighted images and of high signal intensity on T2-weighted images with a homogeneous appearance. With this pattern a greater degree of residual viable tumor was found. Increased tumor cellularity with lesser amounts of tumor matrix and granulation tissue accounted for the homogeneous pattern. A cystic pattern was represented by focal areas of low to intermediate signal intensity on T1-weighted images and of high signal intensity on T2-weighted images; this pattern has a distinct multicystic or bubbly appearance. Increased amounts of viable tumor were found with the cystic pattern as the cysts were usually lined by viable tumor cells.[96]

Holscher and colleagues demonstrated a significant correlation between poor histologic response to chemotherapy and increases in signal intensity of the extraosseous tumor component on T2-weighted follow-up MR images compared with pretherapy MR images ($r = 0.57$, $P = 0.02$).[62] Likewise, a decrease in signal intensity of the extraosseous tumor component on T2-weighted follow-up MR images was demonstrated to correlate with good histologic response.[62]

Erlemann and co-workers demonstrated that sequential MR imaging with use of a fast gradient-echo MR imaging sequence after bolus intravenous administration of gadopentetate dimeglumine (see Fig. 15–5) also may be used to assess tumor response to

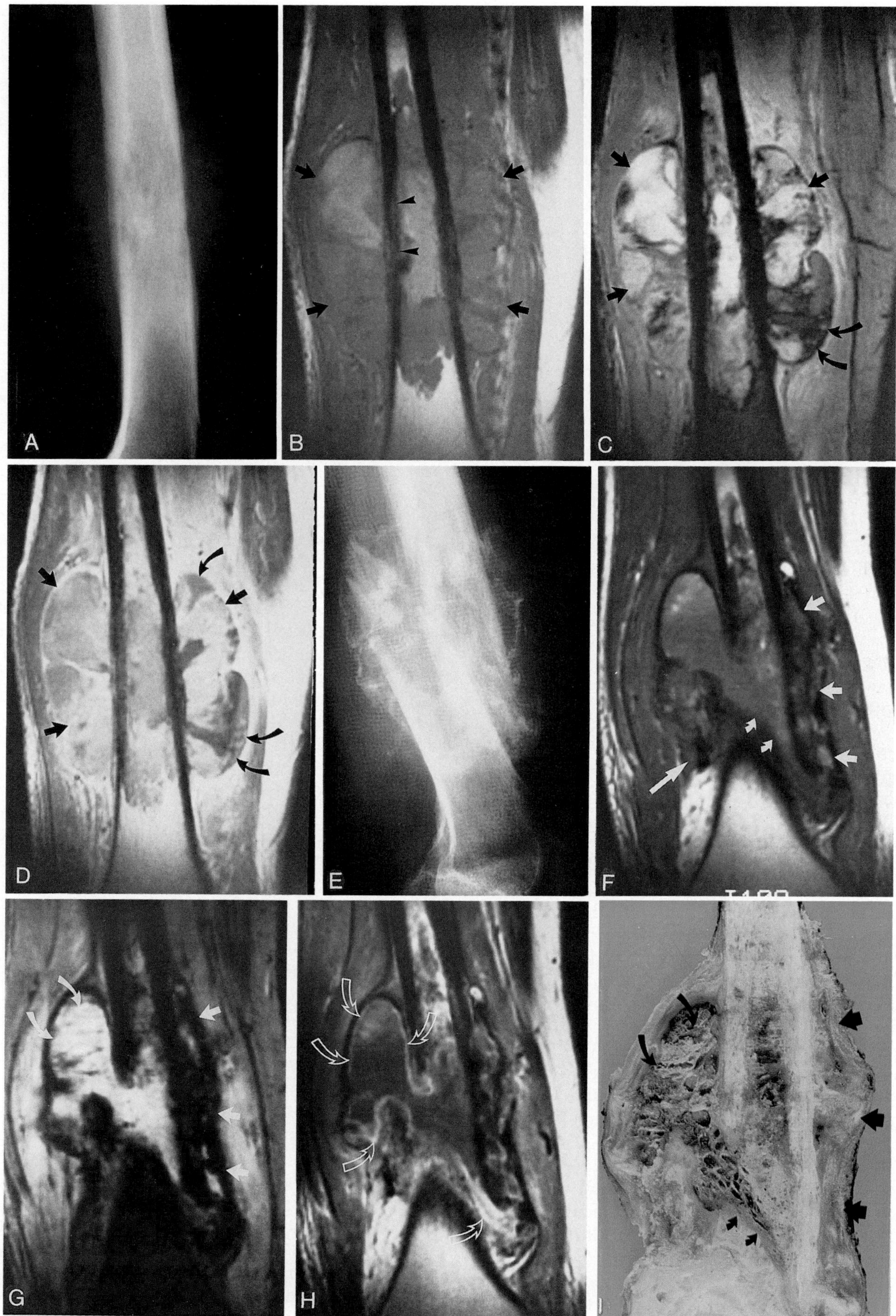

Figure 15–9. See legend on opposite page.

chemotherapy.[34] They found a reduction of the slope of the curve of contrast enhancement over time by at least 60 per cent in responders on follow-up MR imaging after chemotherapy compared with the results in the prechemotherapy study; in non-responders, the reduction was usually less than 60 per cent.[34]

Contradictive results were observed by Erlemann and co-workers[34] and by Holscher and colleagues[62] when changes in tumor volume after chemotherapy were correlated with histologic response. Erlemann and colleagues found that tumor volume did not correlate with histologic response ($P > .05$)[34]; Holscher and co-workers, in contrast, described a statistically significant correlation between reduction in tumor volume after chemotherapy and histologic response ($r = 0.53$, $P \leq 0.03$).[62] This discrepancy may be explained by inadequate techniques applied for tumor volume determination. Both studies employed a relatively crude approach to estimate tumor volume before and after chemotherapy using the formula for elliptical masses: volume $= \pi/6 \times$ height $\times$ width $\times$ depth. The results obtained with this formula represent only an approximation of the true tumor volume. In addition, tumor volume will be artificially elevated using this method in patients in whom the inner portions of a tumor become necrotic after chemotherapy while the outer diameter of the mass remains unchanged.

Lang and co-workers[75] and Heuck and colleagues[60] described a more sophisticated and accurate procedure for quantitative volume determination on the basis of MR images using three-dimensional reconstruction techniques. Conventional two-dimensional MR images are resampled by linear interpolation to generate a uniform resolution in all three dimensions. The resultant binary volume is used as the basis for three-dimensional reconstruction of the structures to be imaged. In bone tumors, the intraosseous tumor, the extraosseous tumor, the adjacent normal bone, and the surrounding soft-tissue can be displayed simultaneously with different colors and transparencies in the final three-dimensional image (Fig. 15–10). The volumes of these tissues can be calculated readily from the binary volume with high diagnostic accuracy.[60,75] To date, this technique has not yet been applied in longitudinal studies to the evaluation of response of primary bone tumors to adjuvant chemotherapy, although it may provide important early, noninvasive information on the efficacy of chemotherapeutic agents.

DISTINGUISHING RECURRENT TUMOR FROM POSTSURGICAL, POSTRADIATION, OR POSTCHEMOTHERAPY CHANGES

Differentiation between recurrent tumor and tissue changes related to surgery, radiation therapy, or chemotherapy poses a difficult diagnostic problem. Recurrence must be suspected when a nodule is observed with low signal intensity on T1-weighted MR images that demonstrates high signal intensity on T2-weighting.[17,47,108,141] Mass effect on surrounding tissues also is suggestive of recurrent tumor. Infiltration of soft tissue and destruction of bone beyond the structural changes that occurred before chemotherapy or radiation therapy are highly indicative of recurrent tumor. In the patient who has

Figure 15–9. Osteogenic sarcoma of the distal femur before *(A–C)* and after *(D–G)* chemotherapy in a 32-year-old patient. *A,* Conventional radiograph demonstrates lytic lesion in the femoral shaft with areas of sclerosis, periosteal reaction, and tumor new bone formation. *B,* Precontrast T1-weighted spin-echo MR image (TR, 800 ms; TE, 30 ms). A lesion of intermediate to low signal intensity is demonstrated in the distal femur. The cortex is disrupted *(arrowheads)*. Marked bilateral soft tissue extension of the tumor *(straight arrows)* is seen. *C,* T2*-weighted gradient-echo MR image (TR, 600 ms; TE, 30 ms; flip angle, 30 degrees). The intra- and extraosseous tumor portions *(straight arrows)* demonstrate intermediate to high signal intensity with this imaging sequence. An extraosseous tumor region of low signal intensity is present inferomedially *(curved arrows)*. *D,* Postcontrast T1-weighted spin-echo MR image (TR, 800 ms; TE, 30 ms). Marked contrast enhancement of the intra- and extraosseous tumor is demonstrated *(straight arrows)*. Only an inferomedial and a small superomedial extraosseous tumor portion do not enhance, which may reflect tumor necrosis *(curved arrows)*. *E,* Conventional radiograph shows pathologic fracture. Marked new bone formation is noted which may be tumor or callus. *F,* Precontrast T1-weighted spin-echo MR image (TR, 800 ms; TE, 30 ms). The patient has suffered a pathologic fracture during chemotherapy *(curved arrows)*. The post-chemotherapy study demonstrates a marked reduction in size of the extraosseous tumor mass medially *(short straight arrows)*. Inferolaterally, the tumor has also significantly decreased in size *(long straight arrow)* when compared with the prechemotherapy MR image in *B*. *G,* T2*-weighted gradient-echo MR image (TR, 600 ms; TE, 30 ms; flip angle, 30 degrees). The medial portions of the extraosseous tumor do not increase in signal intensity with T2*-weighting and are of very low signal intensity, corresponding to dense calcification *(straight arrows)*. The superomedial areas of the extraosseous tumor demonstrate high signal intensity *(curved arrows)* that extends into the medullary cavity, which may reflect tumor necrosis with liquefaction of the necrotic cell mass. *H,* Postcontrast T1-weighted spin-echo MR image (TR, 800 ms; TE, 30 ms). On the postcontrast MR imaging study, only a small linear area surrounding the intra- and extraosseous tumor demonstrates contrast enhancement, reflecting small amounts of residual viable tumor and possibly peritumorous edema *(curved open arrows)*. Most of the tumor does not, however, enhance, indicating extensive tumor necrosis. *I,* Pathologic macrosection confirms extensive tumor necrosis. The superomedial portion of the tumor shows cystic structures, which were filled with liquefied necrotic debris *(large curved arrows)*. Histologically, no viable tumor was found in this region except for a thin peripheral area corresponding to the linear zone of peripheral contrast enhancement. Medially, dense calcification is present *(straight arrows)*. Note the pathologic fracture *(small curved arrows)*.

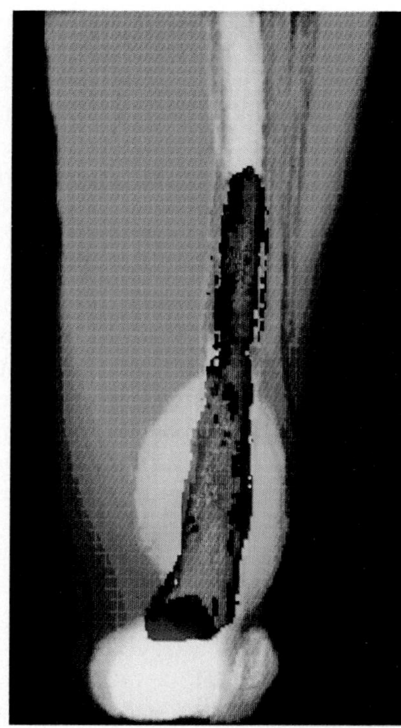

Figure 15–10. Osteogenic sarcoma of the distal femur in an 18-year-old patient. Three-dimensional reconstruction of conventional two-dimensional spin-echo MR images. The normal femoral and patellar bones *(yellow)* are visualized simultaneously with the intraosseous *(blue)* and the extraosseous *(red)* tumor portions. Overlying soft tissue is imaged with green color. The volumes of the intra- and extraosseous tumor can be quantified in cubic centimeters using this technique.

undergone radiation therapy only, however, lesions with high signal intensity on T2-weighting may represent both recurrent tumor and radiation-induced tissue changes.[141]

Areas of low to intermediate signal intensity on T1-weighted sequences without signal intensity increase on T2-weighting and without nodular configuration typically represent postsurgical, postchemotherapy, or postradiation therapy changes. Malignant fibrous histiocytoma may, however, in rare cases exhibit similar signal intensity so that no definite distinction is possible.[108]

Imaging after intravenous administration of gadopentetate dimeglumine may provide some additional information in the patient with suspected recurrent tumor.[73] Tumor and organizing scar or granulation tissue demonstrate marked contrast enhancement. Fibrotic, well-organized scar, however, shows no or only mild contrast enhancement.[73]

Reuther and Mutschler compared CT and MR imaging in detecting recurrent musculoskeletal neoplasm.[108] Sensitivity, specificity, and accuracy for CT and MR imaging were 57.5 per cent and 82.5 per cent, 96.3 per cent and 96.3 per cent, and 85.0 per cent and 92.6 per cent, respectively. Positive and negative predictive values for CT and MR imaging were 52.3 per cent and 75 per cent, and 83.2 per cent and 91.3 per cent.[108] Magnetic resonance imaging thus is significantly more sensitive than CT in detecting recurrent musculoskeletal neoplasm.[108]

Prosthetic devices or other metallic implants are essential for treatment of bone or soft tissue tumors with limb-salvage procedures. In the patient with metallic implants, CT is severely limited in detecting recurrent tumor owing to beam hardening artifacts. Magnetic resonance imaging, however, allows good visualization in the presence of orthopedic implants. When spin-echo sequences are employed, artifact produced by metal is usually limited to the immediate area of the implant. In the patient with a metallic implant, MR imaging, therefore, may allow the diagnosis of recurrent tumor when CT is nondiagnostic.

Choi and associates compared ultrasonography and MR imaging in detecting recurrent tumor in a series of 21 patients.[17] They found a sensitivity and specificity of 100 per cent and 83 per cent, and 79 and 93 per cent for ultrasonography and MR imaging, respectively.[17] Although ultrasonography had a slightly greater sensitivity than MR imaging, this difference was not significant statistically.[17] During the early postoperative period, residual inflammation, edema, blood, and other postoperative changes may obscure recurrent tumor on ultrasonography.[17] In the early postoperative period, MR imaging appears superior to ultrasonography,[17] although hematoma, edema, and other postoperative changes also may represent a diagnostic problem for MR imaging. Ultrasonography is a cost-effective and fast imaging procedure and can even be performed at bedside. In addition, it is not limited by contraindications, such as cardiac pacemakers, that may preclude MR imaging. Both ultrasonography and CT are ideally suited for guiding needle biopsy.[17] Nonferromagnetic needles for MR imaging–guided biopsy have become available, but more development remains to be done in this area.

At present, MR imaging has replaced CT for detecting recurrent musculoskelelal neoplasms. Ultrasonography and MR imaging are characterized by similar sensitivity and specificity. However, ultrasonograms may be more difficult to interpret in the early postoperative period.[17] MR imaging is the imaging modality of choice when ultrasonography is inconclusive.[17] If an initial posttherapy MR imaging study is equivocal, follow-up MR imaging after several weeks or few months—depending on the malignant potential of the tumor—helps to rule out recurrent tumor.

BENIGN BONE NEOPLASMS

Although MR imaging does not provide specific distinction between different tumors, some lesions may demonstrate somewhat singular features. The following pages discuss selected benign and malignant tumors of bone and soft tissue whose MR imaging appearance has been reported in the literature.

Osteochondroma is the most frequently occurring benign bone tumor.[21] Osteochondroma affects typically patients between 10 and 40 years of age. Osteochondroma can occur in any bone that develops by enchondral ossification.[27] The tumor is usually located in the femur, the tibia, the humerus, or the pelvis. Osteochondroma always originates near an epiphyseal line.[27] Its hallmark is the continuity of bone cortex between the normal bone and the cortex of the osteochondroma. The lesion contains a cartilaginous cap. Reports about the MR imaging appearance of osteochondroma have been sparse.[70] On T1-weighted MR images, osteochondroma demonstrates low and intermediate signal intensity; regions of intermediate intensity may reflect cartilaginous tumor tissue. On T2-weighting, areas of mixed high and intermediate signal intensity are observed.[70] Scattered regions of low signal intensity on T1- and T2-weighted images within the lesion correspond to regions of tumor ossification.[70] Osteochondroma usually is well marginated. It may have a mass effect on adjacent soft tissue structures. Soft tissue infiltration, however, is not seen. Magnetic resonance imaging may demonstrate the cartilaginous cap.[70] Although some controversy exists, some researchers believe that a cartilaginous cap measuring more than 3 cm is suggestive of chondrosarcoma.[77]

Aneurysmal bone cyst typically affects older children or young adults, many of whom may have a history of trauma. Histologically, the lesion consists of blood-filled, honeycombed spaces that may be lined with granulation tissue, osteoid, or multinuclear giant cells. Intraosseous lesions distend the cortex as they grow. Extraosseous lesions appear as soft tissue tumor eroding the adjacent bone.[27] They are typically associated with trauma.[27] Aneurysmal bone cysts demonstrate lower signal intensity than that of marrow fat on T1-weighted images.[5,6] Focal areas of high signal intensity on T1-weighted images can be explained by the presence of blood products (i.e., methemoglobin) (Fig. 15–11). The lesion usu-

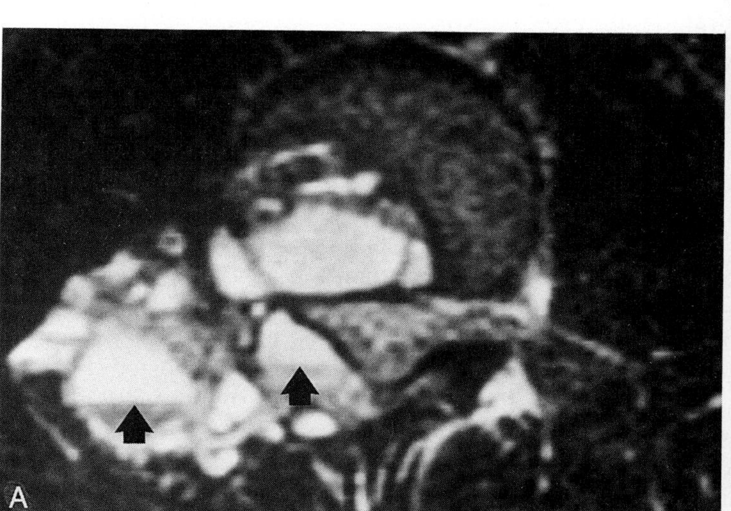

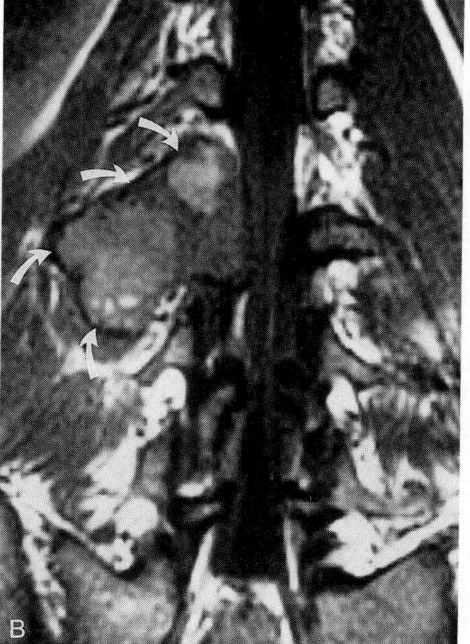

Figure 15–11. Aneurysmal bone cyst of the lumbar spine. *A*, Axial T1-weighted spin-echo MR image (TR, 1000 ms; TE, 20 ms). A large heterogeneous mass is demonstrated extending from the L3 vertebral body laterally into the spinal soft tissues. Multiple fluid levels are visualized *(solid arrows)*. The areas of high signal intensity probably reflect blood products (methemoglobin) within the cystic structures. The lesion is of higher signal intensity than the adjacent muscle, and invades the spinal canal. *B*, Coronal T1-weighted spin-echo MR image (TR, 1000 ms; TE, 20 ms) demonstrates the craniocaudad extent of the mass. The lesion is surrounded by a rim of low signal intensity, which may correspond to an osseous shell *(curved arrows)*.

ally has greater signal intensity than adjacent muscle owing to the presence of blood.[148] On T2-weighting, the tumor is of high signal intensity.[6,148] Aneurysmal bone cysts demonstrate marked contrast enhancement on T1-weighted MR images after intravenous administration of gadopentetate dimeglumine. In addition, aneurysmal bone cysts are characterized by several morphologic features that render the tumor distinct from other bone neoplasms: Aneurysmal bone cysts frequently demonstrate blood-fluid levels (see Fig. 15–11).[6,64] The tumor may have a bubbly appearance with cysts of different signal intensities[6,64] (see Fig. 15–11). Diverticular outpouchings are commonly seen. The tumor may be surrounded by a rim of low signal intensity on all imaging sequences, corresponding to an osseous shell (see Fig. 15–11*B*). The MR imaging appearance of aneurysmal bone cyst is, however, not specific. In particular, fluid-fluid levels also may be observed in chondroblastoma, giant cell tumors, and telangiectatic osteosarcoma.[137]

Hemangioma is a benign vascular tumor that frequently is encountered in the vertebral bodies and in the calvaria. Rarely, hemangioma may occur in tubular and flat bones. Although most vertebral hemangiomas are asymptomatic, they may result in radiculopathy or compressive myelopathy, or both. Histologically, hemangioma consists of multiple vascular channels of various sizes. Most bone hemangiomas are cavernous. Conventional radiography and CT generally permit the diagnosis by showing characteristic images with hypertrophic bone trabeculation in the vertebral body. The MR imaging appearance of vertebral hemangioma is relatively specific. The tumor is of high signal intensity on both T1- and T2-weighted images owing to its fat content and the presence of blood products[25,59,111] (Fig. 15–12). The extraosseous tumor portion does not contain fat and typically is of low signal intensity.[111] Vertebral hemangiomas may occur at multiple levels (see Fig. 15–12). Magnetic resonance imaging provides better definition of subarachnoid space obliteration and spinal cord compression (see Fig. 15–12) than CT in vertebral hemangioma.[59] Hemangioma also can be found in the soft tissues (Fig. 15–13).

Osteoid osteoma accounts for 10 per cent of benign bone tumors.[130] Osteoid osteoma most frequently affects patients between 20 and 30 years of age. The prevalence in men is twice as high as in women. Osteoid osteoma typically causes severe pain. The proximal tibia and femoral neck are frequent sites, but the lesion can affect any bone. The nidus of osteoid osteoma is a circumscribed osteoid focus with or without calcification within a richly vascular osteoblastic connective tissue. Computed

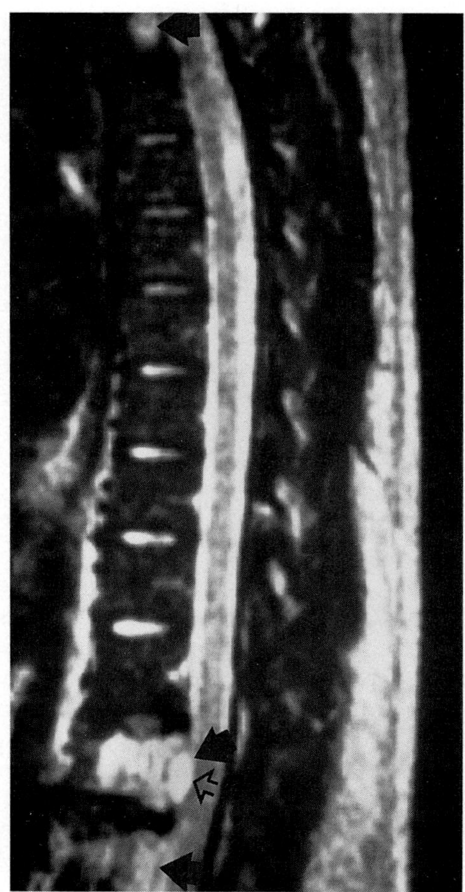

Figure 15–12. Vertebral hemangioma of the thoracic spine. T2-weighted spin-echo image (TR, 2000 ms; TE, 80 ms) demonstrates high signal intensity lesions in the T2, T11, and T12 vertebral bodies (*solid arrows*). The mass at T11 causes spinal cord compression (*open arrow*). The patient was operated on and the hemangioma at T11 and T12 was confirmed histologically.

tomography is useful for localizing the nidus. On CT scans, an osteoid osteoma appears as a lucent focus within surrounding sclerosis.[43] In some cases, a dense calcified center may be seen. On MR imaging, the area of reactive sclerosis is of low signal intensity on both T1- and T2-weighted images.[135] The nidus may be demonstrated as a focal area of high signal intensity on T2-weighted images. However, frequently it may not be depicted by MR imaging, and the MR imaging appearance of osteoid osteoma may be misleading.[135] Correlation of MR imaging findings with radiography and CT is required to avoid erroneous diagnoses.[135]

Giant cell tumors can occur as benign as well as malignant lesions of bone. The tumor contains multinuclear giant cells interspersed in a fibroid stroma. According to Jaffe, giant cell tumor may be subdivided into three grades: Grade 1 consists of normal stromal cells; grade 2 contains moderately atypical stromal cells; and grade 3 is composed of markedly atypical cells that reflect malignancy.[67] Only a small

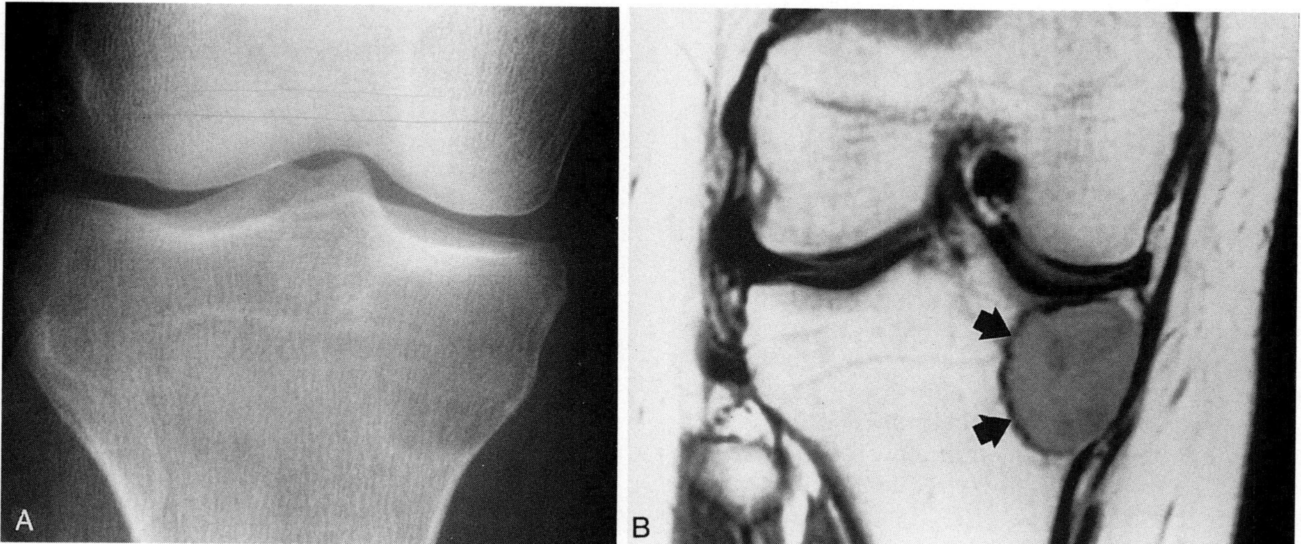

Figure 15–13. Giant cell tumor of the proximal tibia. *A*, Conventional radiograph demonstrates a radiolucent lesion in the metaphyseal end of the proximal tibia, extending to the joint surface. *B*, Coronal T1-weighted spin-echo MR image (TR, 600 ms; TE, 30 ms). The lesion is of intermediate signal intensity with this sequence. It is surrounded by a sharply defined margin of low signal intensity, probably reflecting sclerotic bone *(arrows)*.

percentage of giant cell tumors are grade 3. On radiographs, giant cell tumor is a radiolucent lesion in the metaphyseal end of long bones, extending to the subarticular region. It may also originate in flat bones. On MR imaging, giant cell tumor is well defined with respect to adjacent marrow and cortical bone (Figs. 15–13, 15–14). Areas of homogeneous intermediate or low signal intensity within the tumors are seen on T1-weighted images (see Figs. 15–13, 15–14). T2-weighted images show mixed signal intensity with small inhomogeneous areas of increased signal intensity.[134] However, when hemosiderin is present in a giant cell tumor, a marked decrease in signal intensity of the lesion is observed on both T1- and T2-weighted MR imaging[2] (see Fig. 15–14). This may make evaluation of the integrity of the adjacent cortices difficult.[2] Giant cell tumors also may demonstrate fluid-fluid levels.[137]

Osseous lipoma is a rare benign tumor, intraosseous or parosteal in location. The intraosseous or

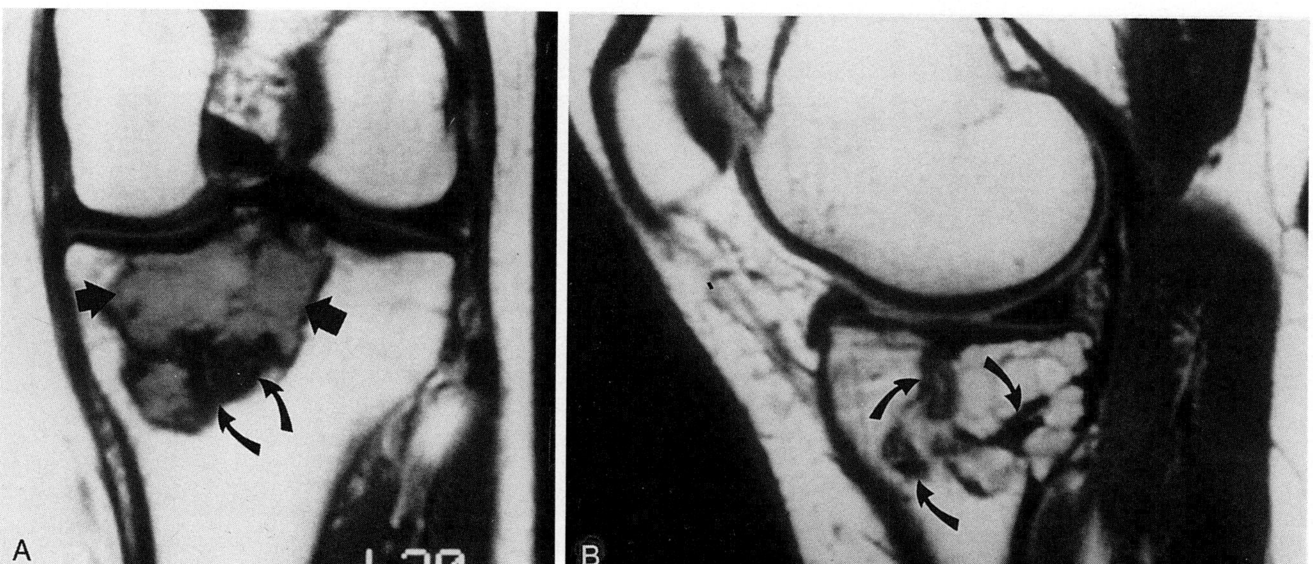

Figure 15–14. Giant cell tumor of the proximal tibia. *A*, Coronal T1-weighted spin-echo MR image (TR, 600 ms; TE, 30 ms). The superior portions of the tumor are of intermediate signal intensity *(straight arrows)*. The inferomedial part of the tumor, however, has low signal intensity *(curved arrows)*. This may correspond to hemosiderin within the tumor. *B*, Sagittal T1-weighted spin-echo MR image (TR, 600 ms; TE, 30 ms) shows multiple foci of very low signal intensity within the tumor mass (intermediate intensity) probably reflecting hemosiderin *(curved arrows)*.

medullary lipoma occurs usually in the metaphysis of a long bone. Frequent sites are the proximal femur and the calcaneus[27] (Fig. 15–15). The MR imaging characteristics are similar to those of soft tissue lipoma. Intraosseous lipoma demonstrates high signal intensity on T1-weighted MR images and intermediate to high signal intensity on T2-weighted images (Fig. 15–15). Cyst formation or signal heterogeneity is possible as a consequence of cell necrosis.[27]

Fibrous defects and *fibrous dysplasia* are frequently found on conventional radiography. Fibrous lesions typically are composed of dense fibrous tissue, resulting in low signal intensity on T1-weighted images. Signal intensity usually is also low on T2-weighting; however, depending on the relative vascularity and other tissue constituents, signal intensity may be intermediate to high on T2-weighted images.[139]

Enchondroma is a benign cartilaginous tumor originating in the medullary cavity. Enchondroma is found most frequently in the fingers and hand. It can, however, also occur in long tubular bones (Figs. 15–16, 15–17). On MR imaging, enchondromas typically are of low signal intensity on T1-weighting and of high signal intensity on T2-weighting. Small, scattered, punctate areas of low signal intensity within the lesion of high signal intensity on T2-weighted images correspond to cartilage calcification and represent a typical finding in enchondroma (see Figs. 15–16, 15–17).

MALIGNANT BONE NEOPLASM

Osteogenic sarcoma is the most frequently occurring skeletal malignancy after multiple myeloma. The prevalence is highest in the first and second decades. A slightly increased rate of occurrence also is found in patients older than 60 years.[27] Osteosarcoma can be subdivided into a number of different types by site and by histologic variation. On radiographs, osteosarcomas produce all varieties of bone

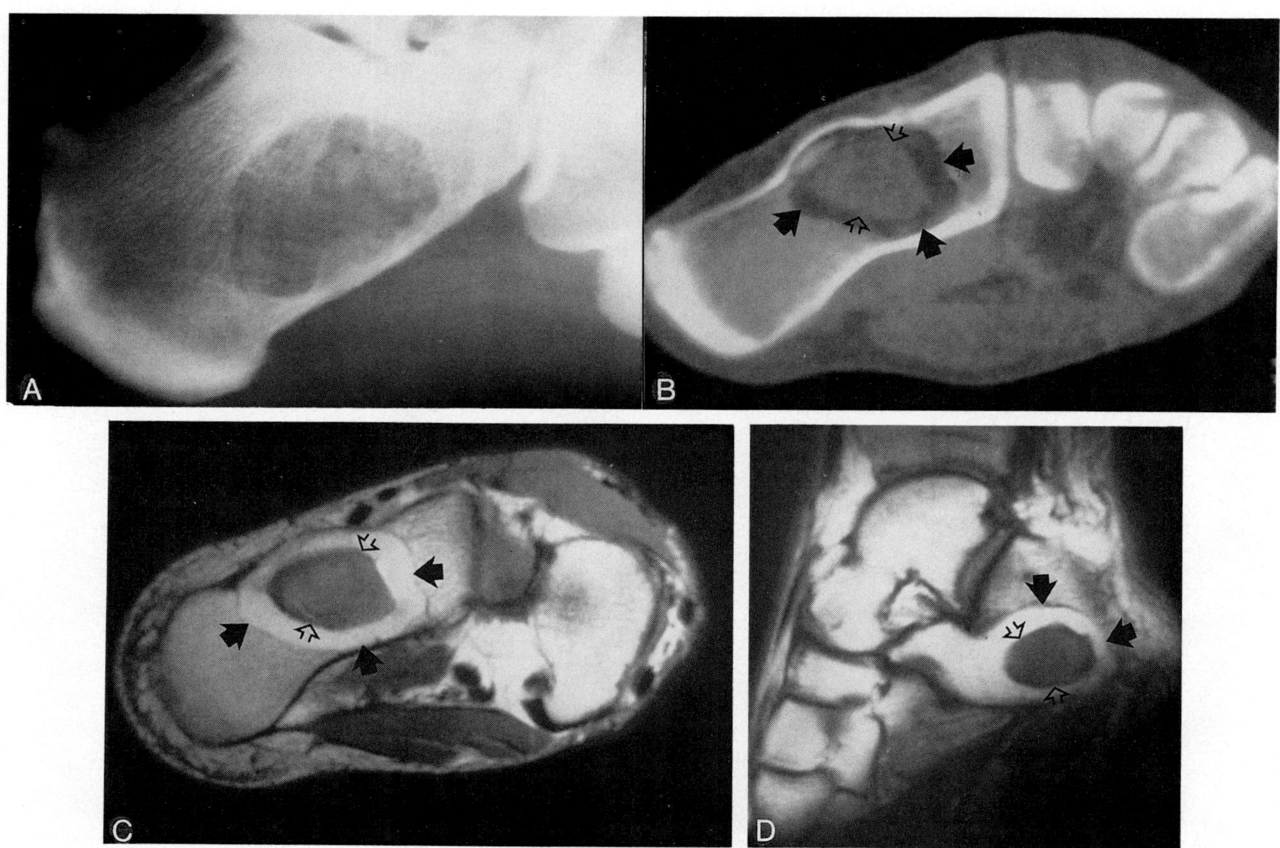

Figure 15–15. Osseous lipoma of the calcaneus. *A*, Conventional radiograph demonstrates a lytic defect with a fine sclerotic margin superiorly. No cortical erosion and no periosteal reaction are seen. *B*, Computed tomographic (CT) scan demonstrates the lytic defect. The defect is of fat density in the periphery *(solid arrows)*. Centrally, however, it is of water density, which may reflect cell necrosis *(open arrows)*. *C* and *D*, Axial *(C)* and coronal *(D)* T1-weighted spin-echo MR images (TR, 800 ms; TE, 30 ms) demonstrate high signal intensity in the periphery of the lipoma *(solid arrows)* that corresponds to fatty tissue and intermediate signal intensity in the center of the lesion *(open arrows)* that corresponds to what was seen on the CT scan in *B*. The central region may reflect an area of cell necrosis and cell debris.

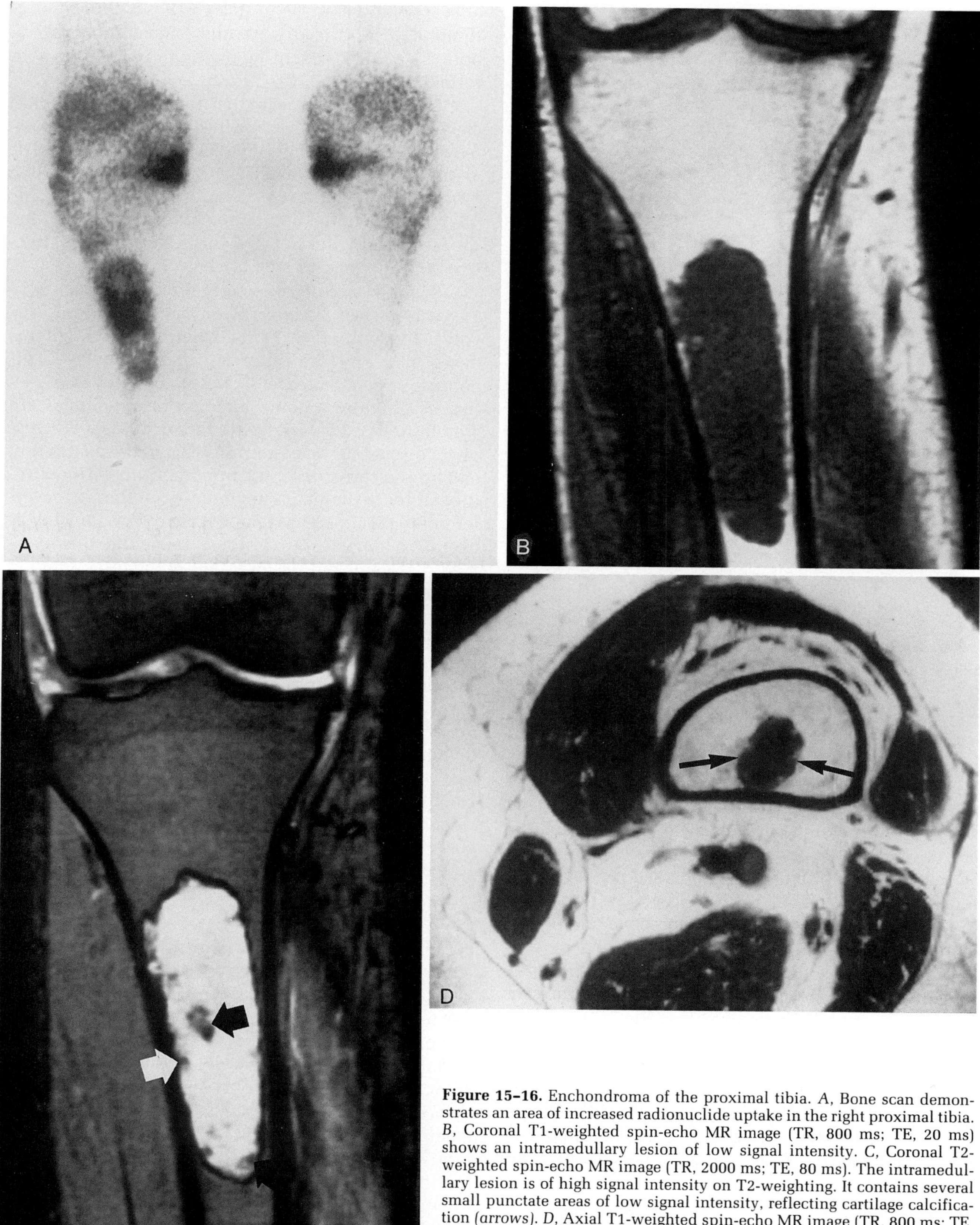

Figure 15–16. Enchondroma of the proximal tibia. *A,* Bone scan demonstrates an area of increased radionuclide uptake in the right proximal tibia. *B,* Coronal T1-weighted spin-echo MR image (TR, 800 ms; TE, 20 ms) shows an intramedullary lesion of low signal intensity. *C,* Coronal T2-weighted spin-echo MR image (TR, 2000 ms; TE, 80 ms). The intramedullary lesion is of high signal intensity on T2-weighting. It contains several small punctate areas of low signal intensity, reflecting cartilage calcification *(arrows)*. *D,* Axial T1-weighted spin-echo MR image (TR, 800 ms; TE, 20 ms) confirms the intramedullary location of the tumor *(arrows)*. Intramedullary location and cartilage calcifications of low signal intensity are findings typical for enchondroma.

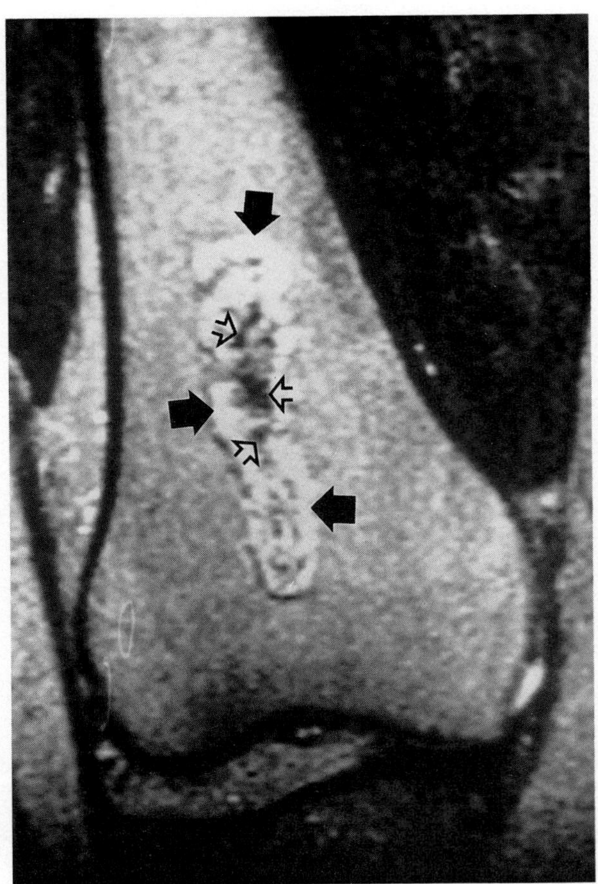

Figure 15-17. Enchondroma of the distal femur. T2-weighted spin-echo MR image (TR, 2000 ms; TE, 120 ms) shows intramedullary lesion with high signal intensity in the distal femur *(solid arrows)*. Scattered punctate areas with low signal intensity correspond to cartilage calcification *(open arrows)*.

change from nearly normal to extremely lytic or dense, to mosaiclike patterns. The MR imaging appearance of osteogenic sarcoma also is characterized by marked variability. The tumor usually is of low signal intensity on T1-weighted images (see Figs. 15-2, 15-4, 15-9). Intra- and extraosseous tumor portions increase in signal intensity on T2-weighting[1,10,133,149] (see Figs. 15-2, 15-4, 15-9). Dense calcified tumor portions are of low signal intensity on both T1- and T2-weighting (see Fig. 15-9*G*; Fig. 15-18). Cartilaginous tumor portions may be of intermediate signal intensity on T1-weighting. Cortical destruction is a frequent finding (see Fig. 15-9*B*). Perineoplastic edema of high signal intensity often is found on T2-weighted, STIR, or postcontrast T1-weighted images[49,121] (see Figs. 15-2, 15-6). The tumor frequently is surrounded by a rim with low signal intensity, reflecting periosteal reaction. Penetration of the thickened periosteum by the extraosseous tumor component may be demonstrated on axial images. Cystic and necrotic changes as well as fluid-fluid levels may be present and are frequent after chemotherapy[62,137] (see Fig. 15-9*E-I*). Intratumoral hemorrhage is of high signal intensity on both T1 and T2 weighting owing to methemoglobin formation (Fig. 15-19).

Chondrosarcoma is the third most common skeletal malignancy after multiple myeloma and osteogenic sarcoma. The median age for chondrosarcoma is 45 years (range, 9 to 70 years). More than half of all patients are older than 40 years[27] (Fig. 15-20). The malignant nature of the lesion is implied on MR imaging by the presence of cortical destruction and tumor necrosis.

Ewing's sarcoma typically affects children 5 to 14 years of age. This sarcoma most commonly involves the femur (Fig. 15-21), then the pelvis, shoulder girdle, and the other leg bones; however, all bones can be affected. The MR imaging characteristics of chondrosarcoma and Ewing's sarcoma are similar to those of osteogenic sarcoma. Magnetic resonance imaging does not allow specific distinction among the three different tumors. Conventional radiography is the imaging modality of choice for differential diagnosis. Magnetic resonance imaging is useful to detect skip lesions, to define the intramedullary and soft tissue extent of the tumor, and to provide posttreatment evaluation.

Fibrosarcoma is a malignant neoplasm of bone that does not form osteoid or chondroid matrix. This type of tumor has an even age distribution from the second to the seventh decades. The tubular bones are involved more frequently in young patients, the flat bones in older patients. On MR images, fibrosarcoma demonstrates low signal intensity on T1 weighing and low or intermediate signal intensity on T2 weighting (Fig. 15-22). Magnetic resonance imaging is particularly useful for evaluating the soft tissue component of the tumor, which is not possible with conventional radiography.

BENIGN SOFT TISSUE NEOPLASMS

Lipomas are common benign neoplasms of soft tissue. They are characterized by high signal intensity on T1-weighted images and intermediate to high signal intensity on T2-weighted images[100] (Figs. 15-8, 15-23). The tumors generally are homogeneous on both sequences. The lesions are well marginated but may contain fibrous septations. Liposarcoma should be suspected when the tumor is inhomogeneous in signal intensity or when the tumor margins are irregular and unsharp (Fig. 15-24).

Intramuscular myxoma is a benign mesenchymal lesion.[29] Intramuscular myxoma occurs mostly in patients 50 to 70 years old. Altered fibroblasts are

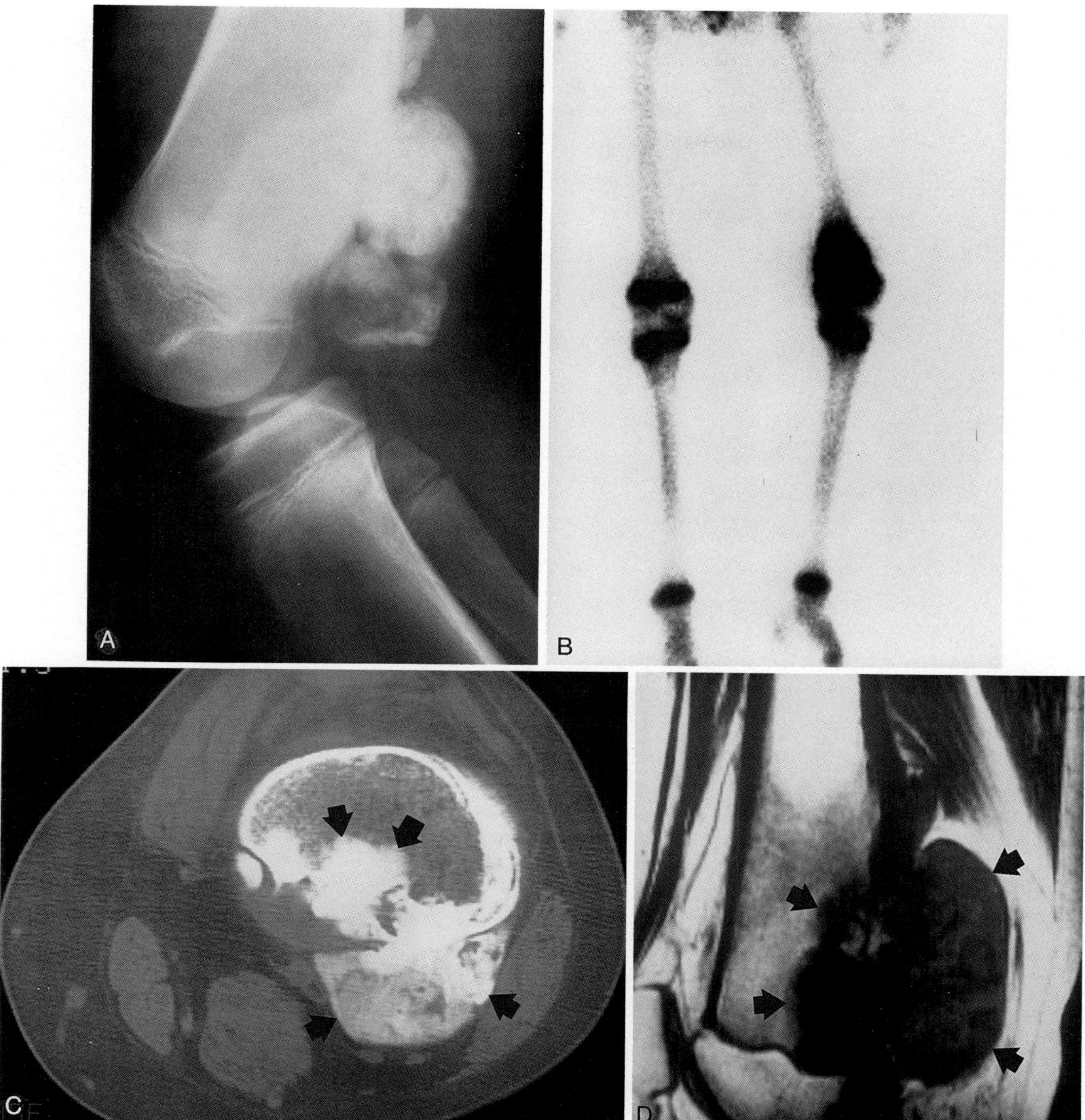

Figure 15–18. Osteogenic sarcoma of the distal femur. *A*, Radiograph shows dense calcification with unsharp margins in the distal femoral metaphysis. A large calcified mass extends posteriorly. *B*, Bone scan demonstrates an area of increased radionuclide uptake in the left distal femur. *C*, Computed tomographic (CT) scan shows densely calcified intra- and extramedullary tumor *(arrows)*. *D*, The mass is of very low signal intensity on a T1-weighted sagittal spin-echo MR image (TR, 600 ms; TE, 30 ms) *(arrows)*. The low signal intensity results in part from the dense calcification.

speculated to produce an excessive amount of mucopolysaccharides and are incapable of forming mature collagen. Most intramuscular myxomas occur in the thigh; the tumor is also found in the shoulder (Fig. 15–25), buttock, arm, and other sites.[29] Some investigators report a high recurrence rate of intramuscular myxoma after surgical resection. Multiple intramuscular myxomas may coexist with fibrous dysplasia of bone.[29] Intramuscular myxoma demonstrates sharp margins on MR imaging[72,98,136] (see Fig. 15–25). Intramuscular myxoma has decreased signal intensity compared with muscle on T1-weighted images (see Fig. 15–25); on T2-weighting the tumor is of marked high signal intensity.[72,98,136]

Text continued on page 433.

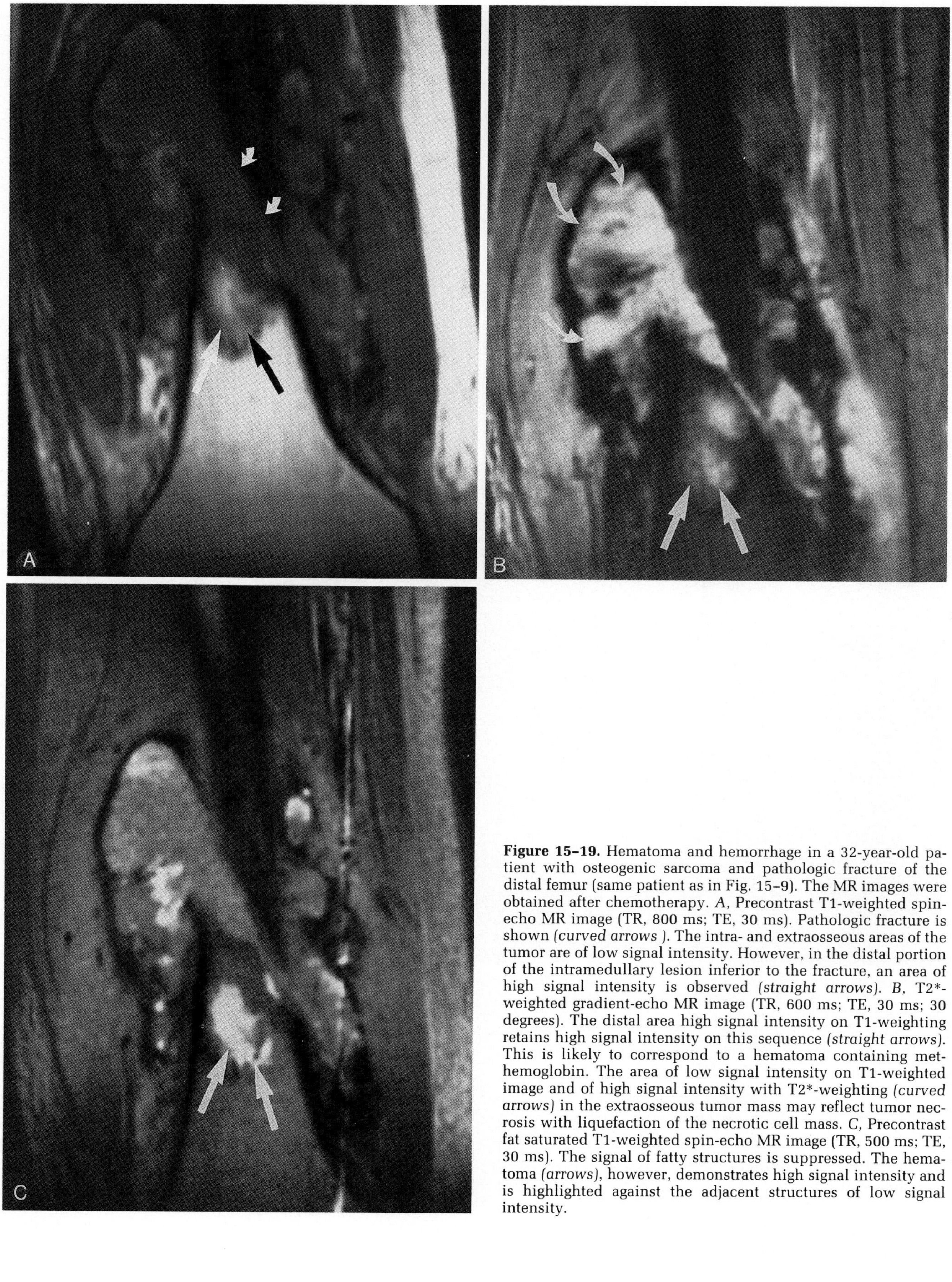

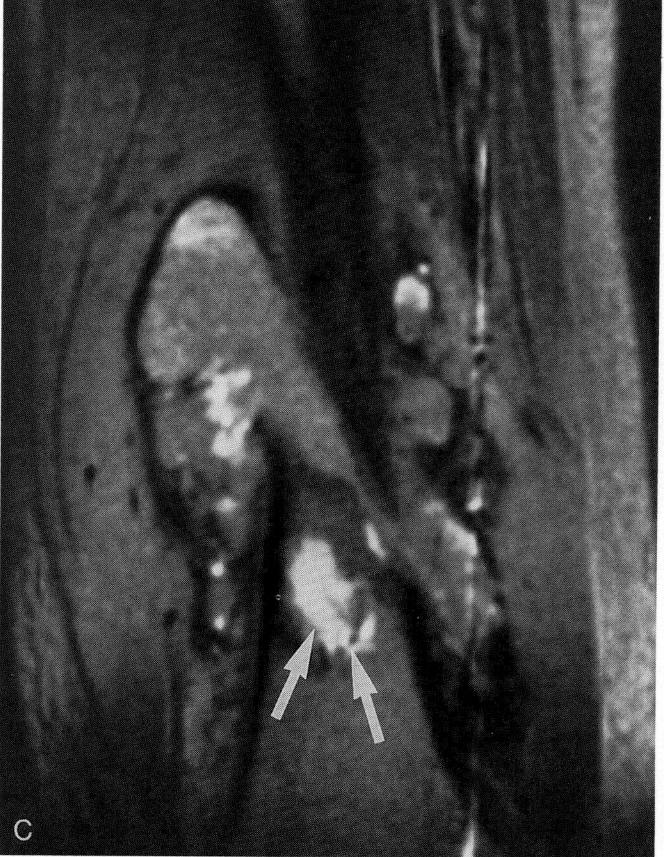

Figure 15–19. Hematoma and hemorrhage in a 32-year-old patient with osteogenic sarcoma and pathologic fracture of the distal femur (same patient as in Fig. 15–9). The MR images were obtained after chemotherapy. *A*, Precontrast T1-weighted spin-echo MR image (TR, 800 ms; TE, 30 ms). Pathologic fracture is shown *(curved arrows)*. The intra- and extraosseous areas of the tumor are of low signal intensity. However, in the distal portion of the intramedullary lesion inferior to the fracture, an area of high signal intensity is observed *(straight arrows)*. *B*, T2*-weighted gradient-echo MR image (TR, 600 ms; TE, 30 ms; 30 degrees). The distal area high signal intensity on T1-weighting retains high signal intensity on this sequence *(straight arrows)*. This is likely to correspond to a hematoma containing methemoglobin. The area of low signal intensity on T1-weighted image and of high signal intensity with T2*-weighting *(curved arrows)* in the extraosseous tumor mass may reflect tumor necrosis with liquefaction of the necrotic cell mass. *C*, Precontrast fat saturated T1-weighted spin-echo MR image (TR, 500 ms; TE, 30 ms). The signal of fatty structures is suppressed. The hematoma *(arrows)*, however, demonstrates high signal intensity and is highlighted against the adjacent structures of low signal intensity.

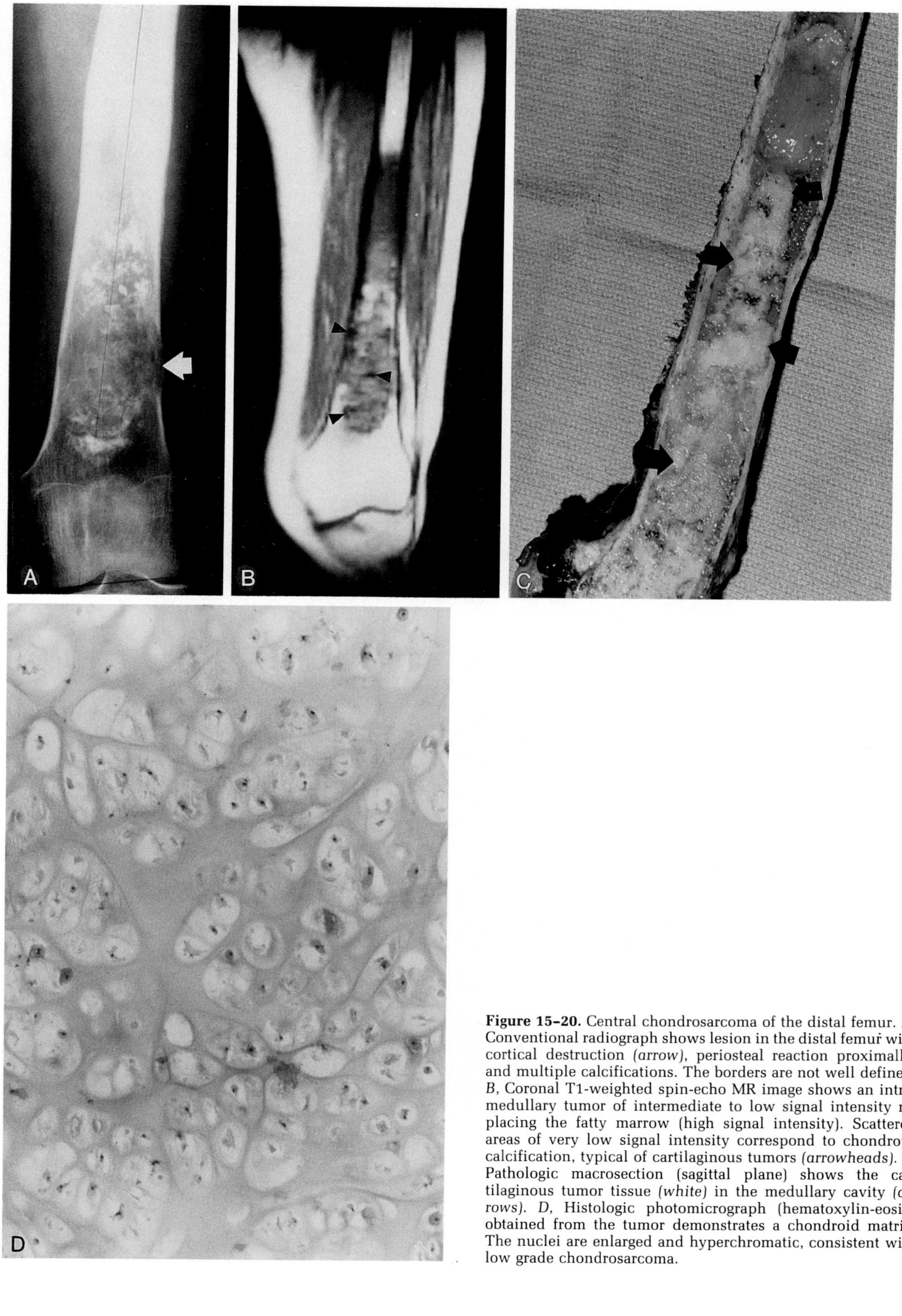

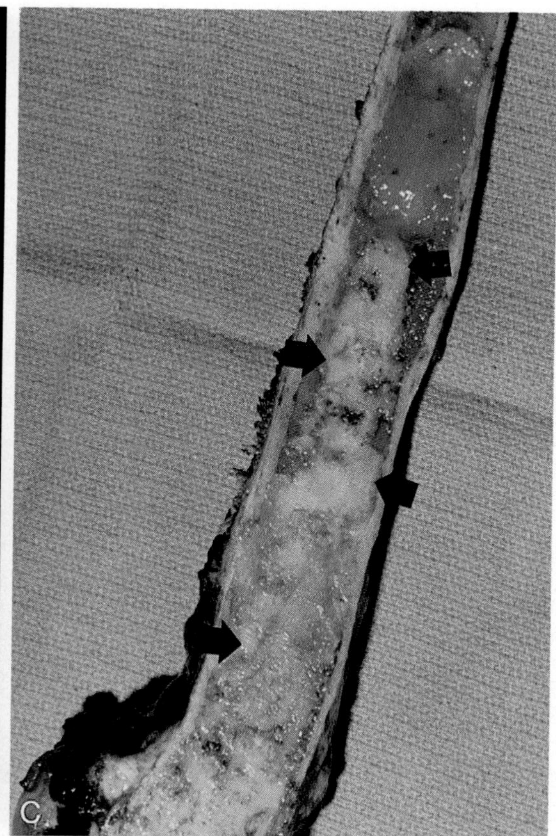

Figure 15–20. Central chondrosarcoma of the distal femur. *A*, Conventional radiograph shows lesion in the distal femur with cortical destruction *(arrow)*, periosteal reaction proximally, and multiple calcifications. The borders are not well defined. *B*, Coronal T1-weighted spin-echo MR image shows an intramedullary tumor of intermediate to low signal intensity replacing the fatty marrow (high signal intensity). Scattered areas of very low signal intensity correspond to chondrous calcification, typical of cartilaginous tumors *(arrowheads)*. *C*, Pathologic macrosection (sagittal plane) shows the cartilaginous tumor tissue *(white)* in the medullary cavity *(arrows)*. *D*, Histologic photomicrograph (hematoxylin-eosin) obtained from the tumor demonstrates a chondroid matrix. The nuclei are enlarged and hyperchromatic, consistent with low grade chondrosarcoma.

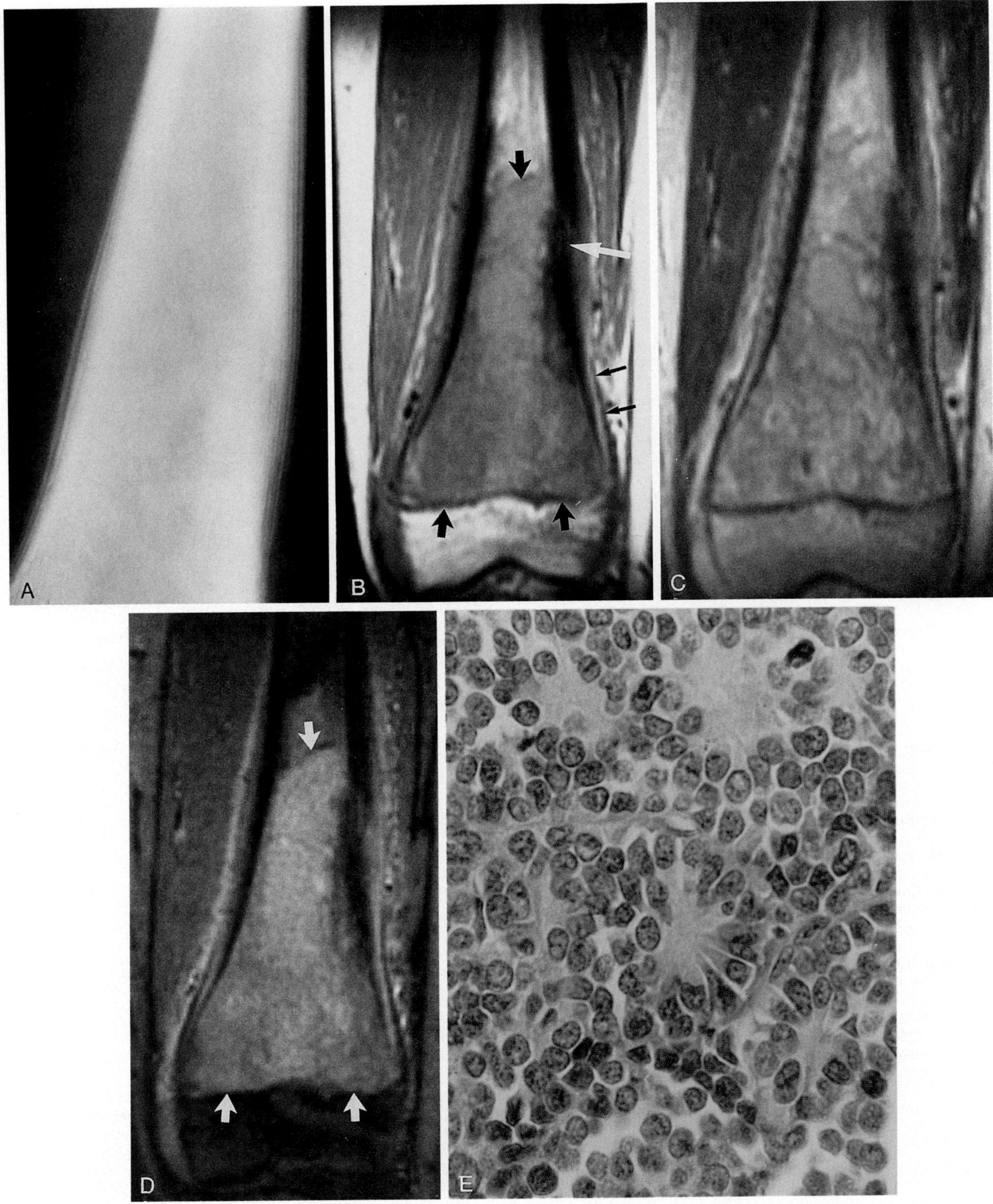

Figure 15–21. Ewing's sarcoma of the distal femur. *A,* Anteroposterior radiograph demonstrates mottled destruction of bone with laminated "onion peel" periosteal reaction. *B,* T1-weighted spin-echo MR image (TR, 800 ms; TE, 20 ms) shows a lesion of low signal intensity *(thick black arrows)* in the distal femoral metaphysis. Cortical destruction is evident medially *(white arrow).* Elevation of the periosteum from the underlying cortex is demonstrated inferiorly *(thin black arrows). C,* T2-weighted spin-echo MR image (TR, 1600 ms; TE, 80 ms). Some portions of the tumor demonstrate an increase in signal intensity with T2-weighting, resulting in a more heterogeneous appearance of the tumor. *D,* Short tau inversion-recovery (STIR) image (TR, 1500 ms; TE, 20 ms; TI, 170 ms). The lesion *(arrows)* is higher in signal intensity than the normal marrow (low signal intensity). *E,* Histologic photomicrograph (hematoxylin-eosin) demonstrates small, round, uniformly sized, and tightly packed tumor cells that form pseudorosettes typical of Ewing's sarcoma.

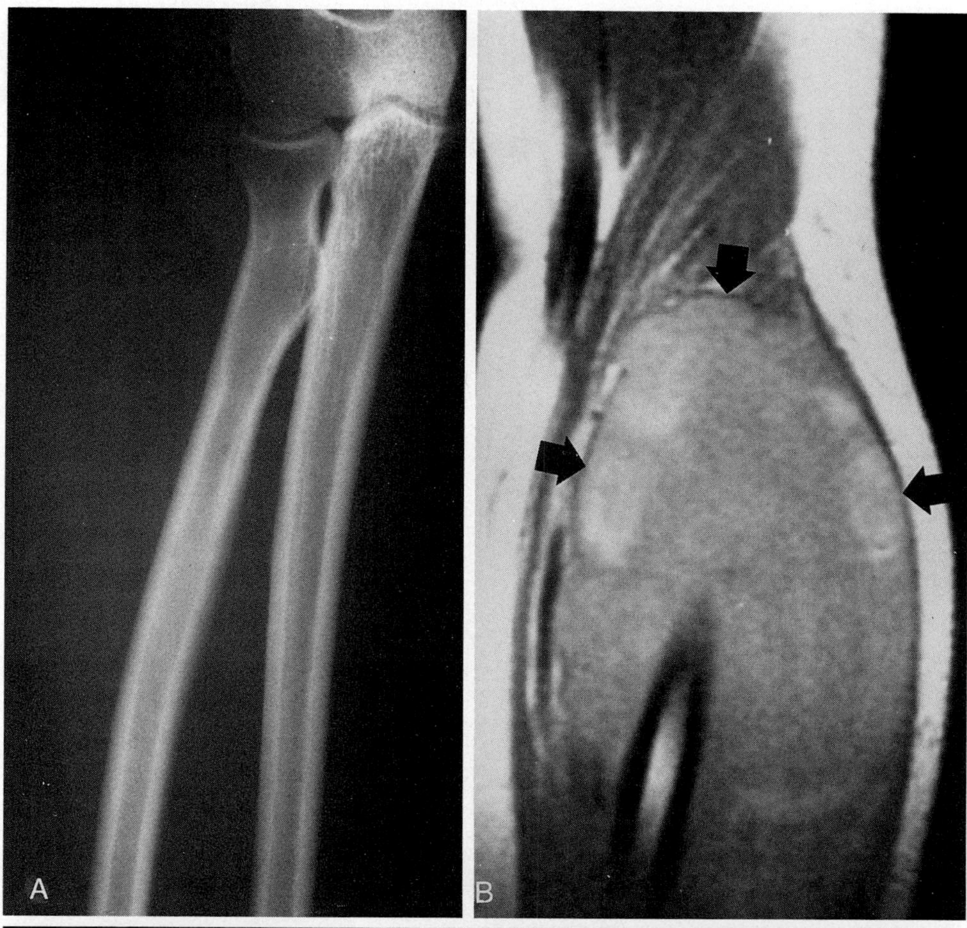

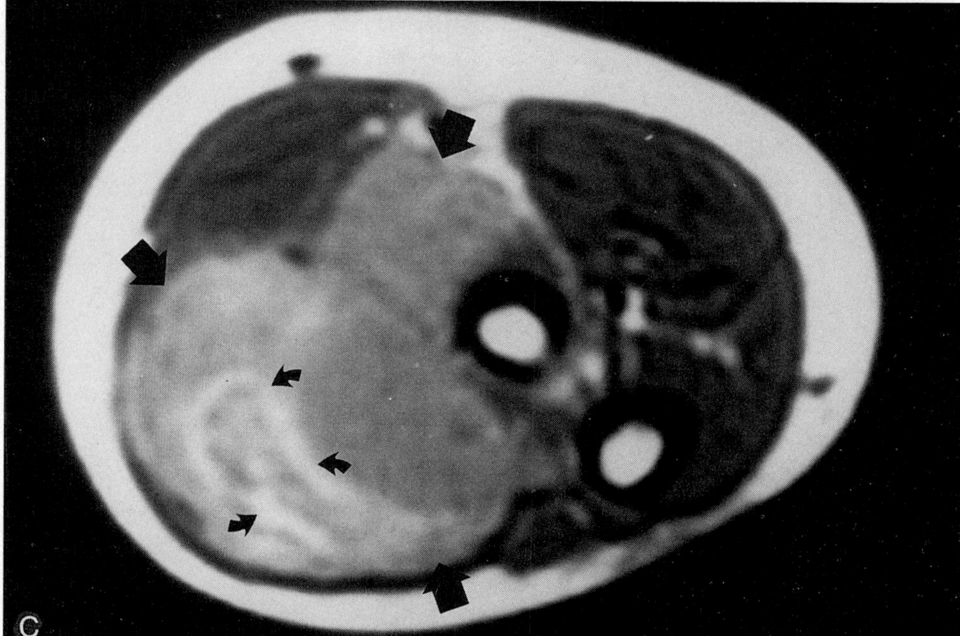

Figure 15-22. Fibrosarcoma of the forearm. *A,* Conventional radiograph demonstrates endosteal resorption and periosteal reaction in the proximal radius. A large soft tissue mass is present. The extent of this mass is difficult to evaluate with radiography. *B,* Coronal T1-weighted spin-echo MR image (TR, 1000 ms; TE, 20 ms) shows an extensive lesion of intermediate to low signal intensity in the soft tissue of the forearm, extending into the cubital fossa *(arrows).* *C,* Axial T1-weighted spin-echo MR image (TR, 1000 ms; TE, 20 ms). The mass *(arrows)* has infiltrated and replaced the brachioradialis, flexor carpi radialis, and extensor carpi ulnaris muscles. The inferior areas of intermediate to high signal intensity may reflect intratumoral bleeding and hemorrhage *(curved arrows).*

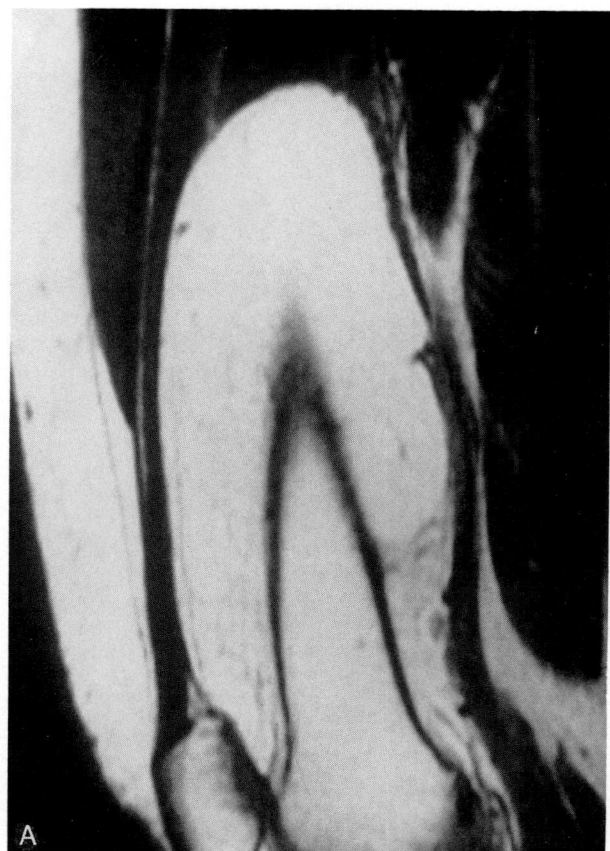

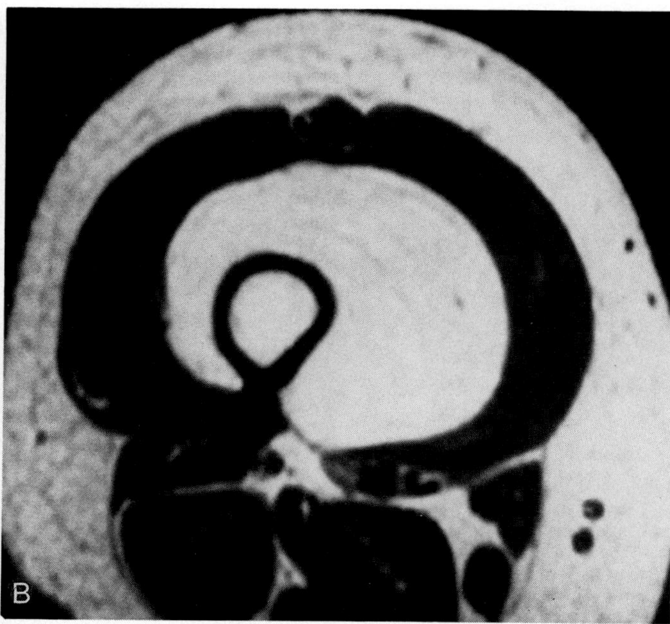

Figure 15–23. Lipoma of the thigh. Sagittal *(A)* and axial *(B)* T1-weighted spin-echo MR images (TR, 800 ms; TE, 20 ms) demonstrate a large soft tissue mass in the thigh. The tumor does not infiltrate the adjacent soft tissue, has high signal intensity similar to that of subcutaneous fat, and is very homogeneous, consistent with a lipoma.

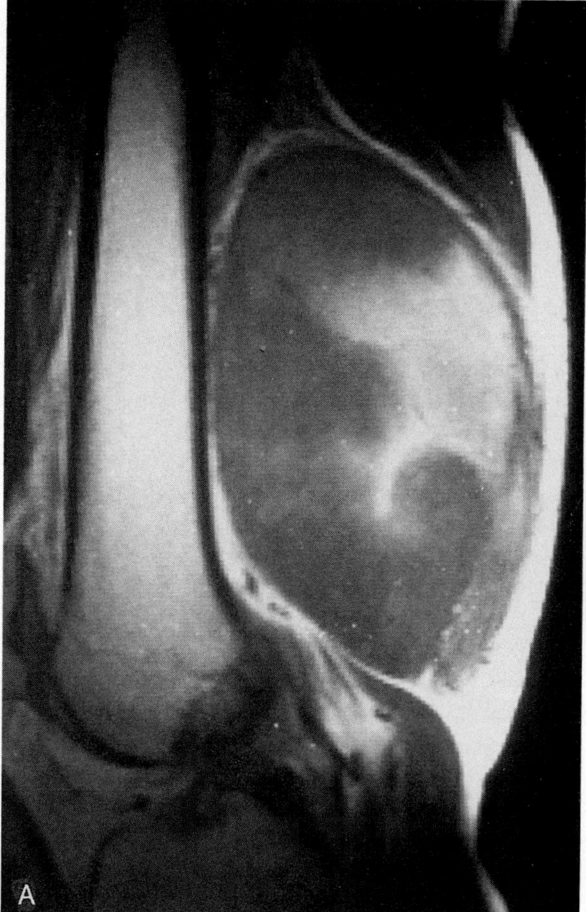

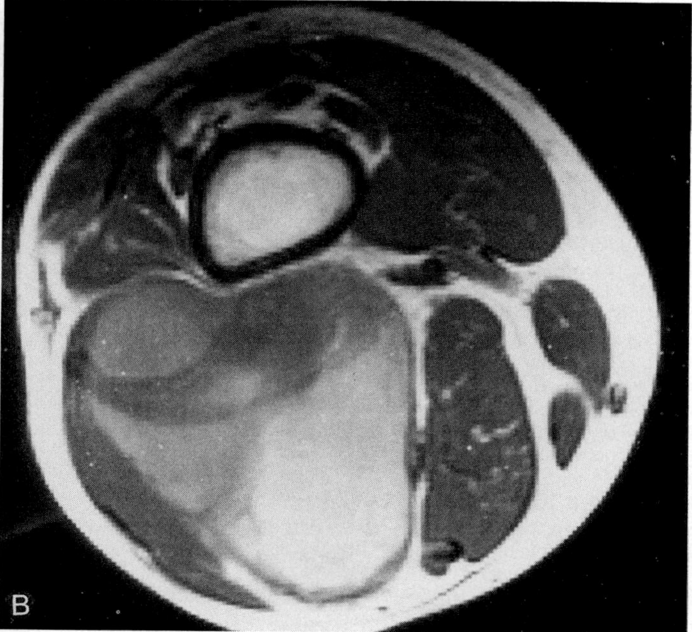

Figure 15–24. Liposarcoma of the thigh. Sagittal *(A)* and axial *(B)* T1-weighted spin-echo MR images (TR, 800 ms; TE, 20 ms) show a large soft tissue mass posterior to the distal femur. In contrast to benign lipoma (Fig. 15–23), the tumor is heterogeneous in signal intensity demonstrating areas of high, intermediate, and low intensity. This should heighten the suspicion for liposarcoma, which was confirmed histologically in this case.

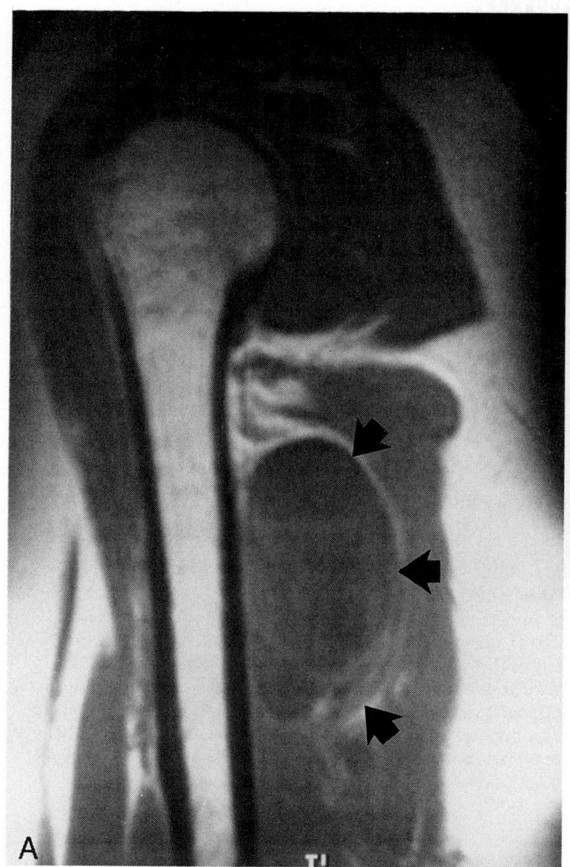

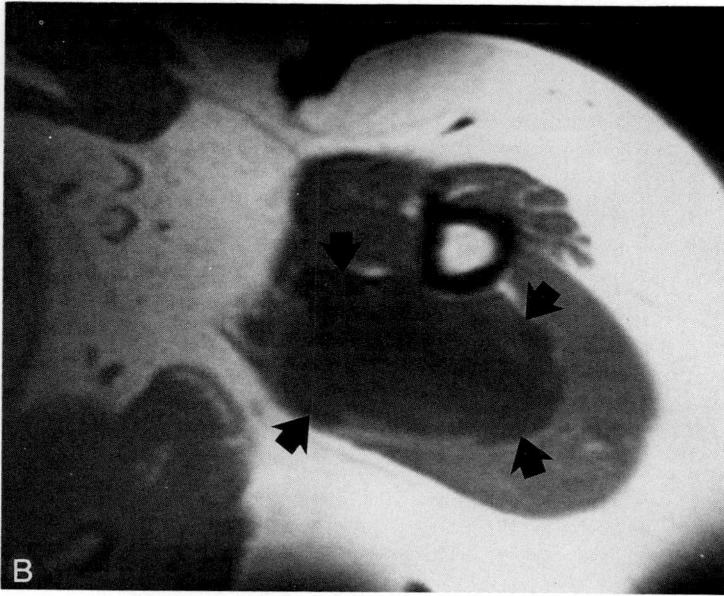

Figure 15–25. Myxoma of the upper arm. Sagittal *(A)* and axial *(B)* T1-weighted spin-echo MR images (TR, 1000 ms; TE, 20 ms) show a tumor of posterior low signal intensity to the humeral shaft *(arrows)*. The tumor, which is of lower signal intensity than the adjacent muscle, has sharp margins and does not infiltrate the adjacent soft tissues.

The imaging pattern on unenhanced MR images is similar to that of cysts. Unlike cysts, however, intramuscular myxomas demonstrate inhomogeneous contrast enhancement in the central portions after intravenous administration of gadopentetate dimeglumine.[98] Occasionally intramuscular myxoma may contain cystic components or calcification.[98]

Extra-abdominal *desmoid tumors* are rare musculoskeletal soft tissue tumors. Patients with desmoid tumors typically range in age from 15 to 40 years.[109] Desmoid tumors are most commonly located in the thighs and buttocks (Fig. 15–26), the shoulder, and the trunk. They demonstrate an aggressive growth pattern with irregular margins (see Fig. 15–26). However, extra-abdominal desmoid tumors do not metastasize. Desmoid tumors are often hypocellular and fibrous.[109,127] More than two thirds of desmoid tumors recur after the initial resection.[109] Because these tumors are composed of fibrous tissue to a significant degree, they usually are of low signal intensity on both T1- and T2-weighted images (see Fig. 15–26). The lesion is inhomogeneous and may demonstrate irregular margins. Bands or streaks of low signal intensity reflect fibrous tissue (see Fig. 15–26). Magnetic resonance imaging is useful for detecting a desmoid tumor, defining the size of the lesion, determining its relationship to adjacent neurovascular bundles, and narrowing the differential diagnostic possibilities before biopsy. Magnetic resonance imaging also is helpful in evaluating the postoperative patient for recurrence of desmoid tumor.

Hemangiomas are a common soft tissue tumor usually found in the skin and subcutaneous tissues.[30] In infants and children, they develop frequently in skeletal muscle.[19,30] Hemangiomas may cause muscular atrophy. Although the tumors are primarily localized lesions, rare cases with involvement of large body portions have been reported (Fig. 15–27). Hemangiomas are subdivided into capillary and cavernous types.[30] On T1-weighted MR images, a hemangioma may be of equal signal intensity to skeletal muscle or may demonstrate slightly greater signal intensity than muscle.[147] Fatty components may cause high signal intensity on T1-weighting. On T2-weighting, the lesion is of marked high signal intensity relative to muscle. In large hemangiomas, dilated vascular structures may be demonstrated on MR images (see Fig. 15–27). Gradient-echo images can be helpful in defining vascular flow more clearly as flowing blood can produce high signal intensity depending on the direction of the phase-encoding

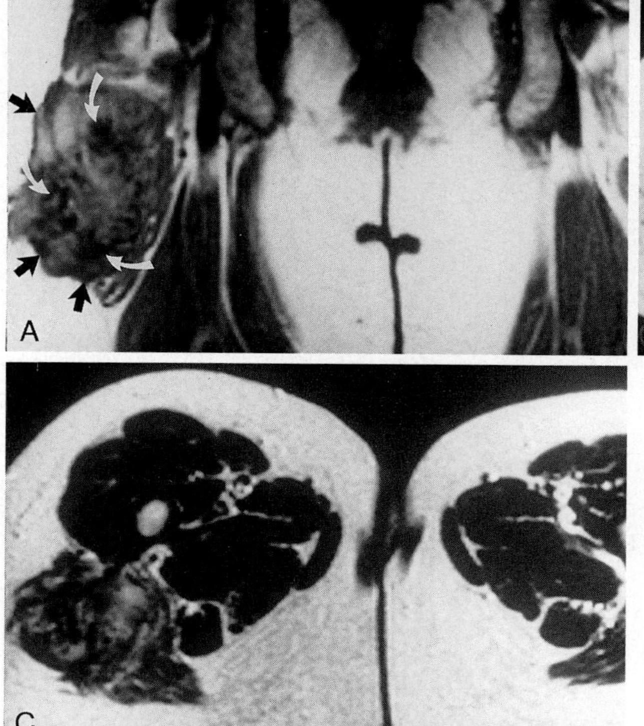

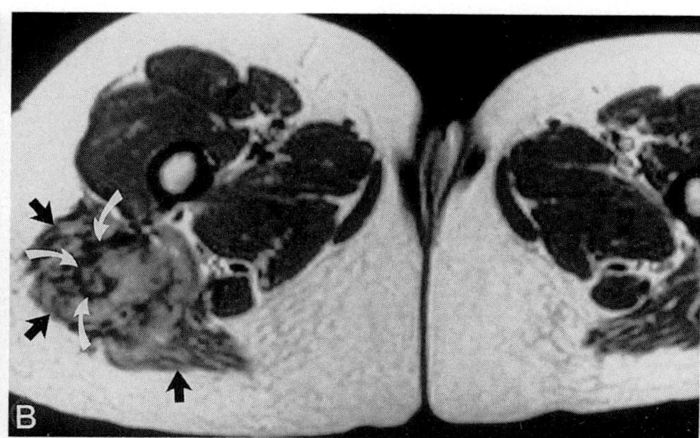

Figure 15–26. Desmoid tumor of the thigh. *A,* Coronal T1-weighted spin-echo MR image (TR, 800 ms; TE, 20 ms) shows a large mass of low signal intensity in the right lateral thigh *(straight arrows).* The mass has very irregular margins. Streaks and bands with very low signal intensity are seen coursing through the lesion *(curved arrows). B,* Axial T1-weighted spin-echo MR image (TR, 800 ms; TE, 20 ms) again shows streaks and bands of very low signal intensity *(curved arrows)* coursing through the lesion *(straight arrows).* The streaks likely reflect fibrous tissue. *C,* Axial T2-weighted spin-echo MR image (TR, 1500 ms; TE, 60 ms). The mass does not increase in signal intensity with longer TR and TE. This suggests that the tumor is composed of fibrous tissue to a significant degree which is typical of desmoid tumors.

gradient.[26] In this setting, MR angiography may be even more helpful in defining the vascular nature of the lesion and in identifying feeding vessels. Fibrous areas with low signal intensity on both T1 and T2 weighting may be present. Muscular atrophy may be depicted on MR images.[14,147]

Angiolipomas and *angiomas* exhibit a heterogeneous pattern of high and low signal intensity on T1- and T2-weighted images[100] (Fig. 15–28). Areas of high signal intensity on both T1- and T2-weighting reflect blood (see Fig. 15–28). As in hemangiomas, multiple serpiginous vessels may be demonstrated on MR imaging (see Fig. 15–28).

Benign nerve sheath tumors include schwannoma (Fig. 15–29) and neurofibroma (Fig. 15–30). Schwannomas typically occur in patients 20 to 50 years old, but have been reported in all age groups.[30] When they arise in the upper extremities, schwannomas usually involve the flexor areas, where the larger nerves are encountered. Neurofibromas are most frequent in young adults 20 to 30 years old.[30] Most peripheral nerve sheath tumors are mildly inhomogeneous on MR scans, being of intermediate or moderately high signal intensity on T1-weighted images (see Fig. 15–29). On T2-weighting, they demonstrate high but heterogeneous signal intensity[124] (see Fig. 15–29). The inhomogeneity is thought to reflect areas of hypo- and hypercellularity[124] or necrosis. The MR imaging appearance of malignant nerve sheath tumors is similar to that of the benign lesions; no specific distinction is possible. Both frequently are sharply demarcated from the surrounding normal tissue.[124] The combined MRI imaging findings of subtle muscle atrophy and a peripheral soft tissue mass in the vicinity of a large nerve trunk suggest a peripheral nerve sheath tumor.

Giant cell tumors of tendon sheath occur most commonly in the tendons of the digits. Histologically, giant cell tumors of the tendon sheath are composed of fibrous tissue, hemosiderin deposits, histiocytes, foam laden macrophages, and giant cells.[30] On T1-weighted images, giant cell tumors are lesions of low signal intensity along the tendon sheath. On T2-weighting, a more inhomogeneous pattern is observed.[119] Regions of low signal intensity correspond to fibrous tissue and hemosiderin deposits.[119] Areas with high signal intensity on T2-weighting likely reflect fluid or inflamed synovium. The lesions usually are well marginated. The signal pattern in conjunction with the location frequently helps to establish the diagnosis.

MALIGNANT SOFT TISSUE NEOPLASMS

Most malignant soft tissue tumors are characterized by indistinct margins.[97] They also frequently demonstrate signal inhomogeneity on T2 weight-

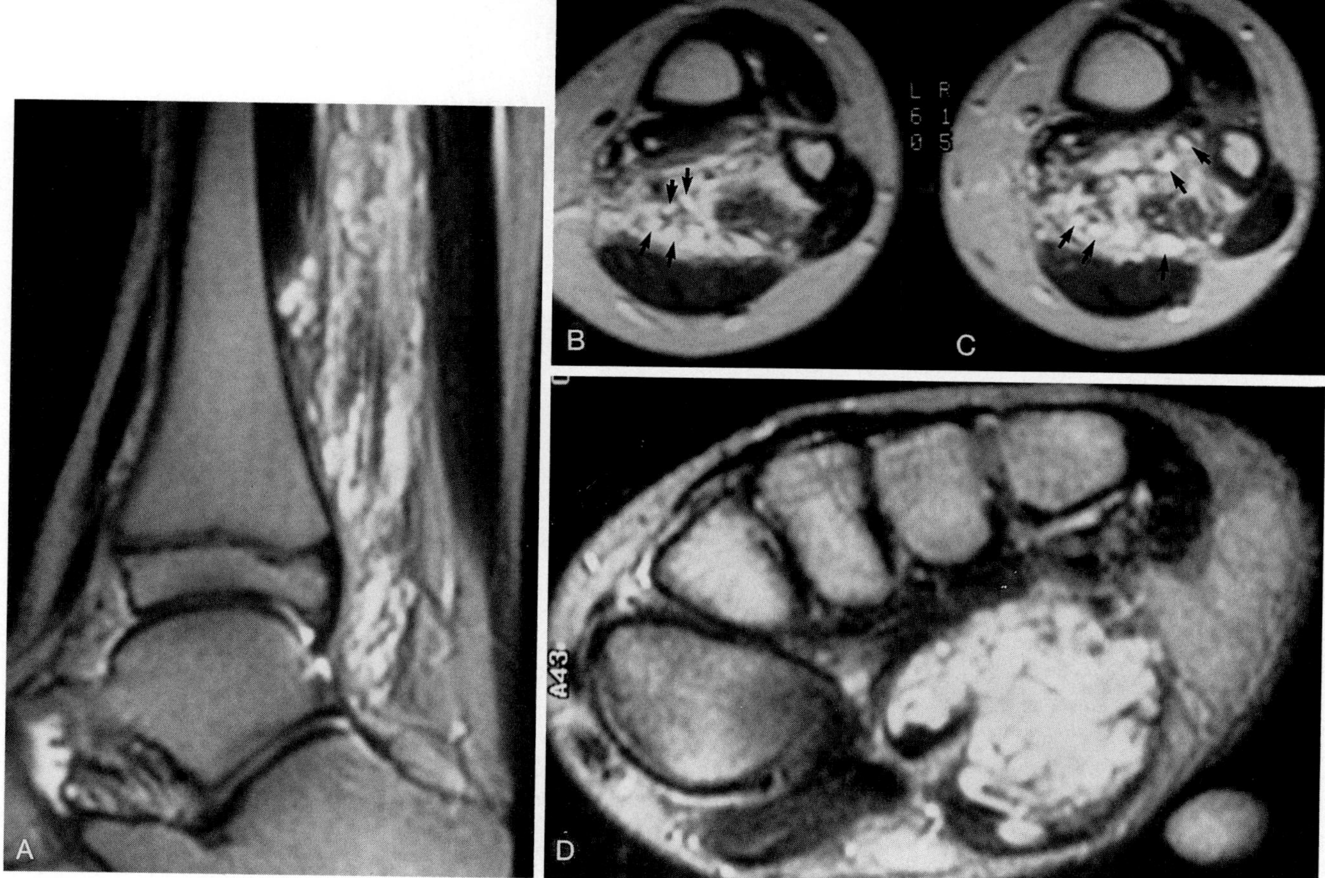

Figure 15-27. Hemangioma in the soft tissues of the calf and foot. *A*, Sagittal T1-weighted spin-echo MR image (TR, 1000 ms; TE, 20 ms) demonstrates a heterogeneous, serpiginous mass of high signal intensity posterior to the distal tibia and anterior to the talus. The high signal intensity is caused by blood products (methemoglobin) in the hemangioma. *B*, and *C*, Axial T1-weighted spin-echo MR images (TR, 1000 ms; TE, 20 ms) at the level of the distal tibia and fibula. The mass has a cavernous or loculated appearance. The round structures of high signal intensity probably correspond to dilated blood-filled vessels *(arrows)*. *C*, Axial T1-weighted spin-echo MR image (TR, 1000 ms; TE, 20 ms) through the foot. The hemangioma extends farther distally along the plantar surface of the foot.

ing.[97] Destruction of adjacent bone and encasement of neurovascular bundles are observed more often with malignant soft tissue tumors than with benign lesions. These features are, however, not specific and can also be observed in benign lesions.[74,83,84,97] Most malignant tumors demonstrate signal intensity higher than that of muscle on T2-weighted images.[97] Some lesions may, however, be of low signal intensity on T2-weighting.

Malignant fibrous histiocytoma is a malignant tumor that can arise from both the bone and the soft tissue. The average age of onset is approximately 50 years.[27] Multiple histologic subtypes are known with predominantly histiocytic, fibromatous, or xanthomatous morphology.[27] Malignant fibrous histocytone typically exhibits intermediate signal intensity on T1-weighted sequences and high signal intensity on T2-weighted images[82,83] (Fig. 15-31). The tumor may, however, also demonstrate low signal intensity on T2-weighted scans when fibromatous tissue predominates histologically.[128,129]

NONNEOPLASTIC LESIONS OF BONE AND SOFT TISSUE

Bone island or *enostosis* is defined as a focus of compact bone within the spongiosa, representing a developmental error in the process of enchondral ossification.[93] It is important to recognize this lesion and not to confuse it with primary or metastatic tumors. Bone islands are diagnosed primarily on the basis of their radiographic appearance.[52] Radiographically, bone islands appear as dense sclerotic lesions with thorny or featherlike radiating spicules. Bone islands are asymptomatic.[52] They may be discovered incidentally on MR images and are characterized by low signal intensity on both T1- and T2-weighted images. The lesion is well defined and located in the marrow cavity. Unlike the case with neoplasms, bone islands do not exhibit cortical destruction, periosteal reaction, soft tissue mass, or peritumoral edema.

In hematoma, the imaging appearance of blood

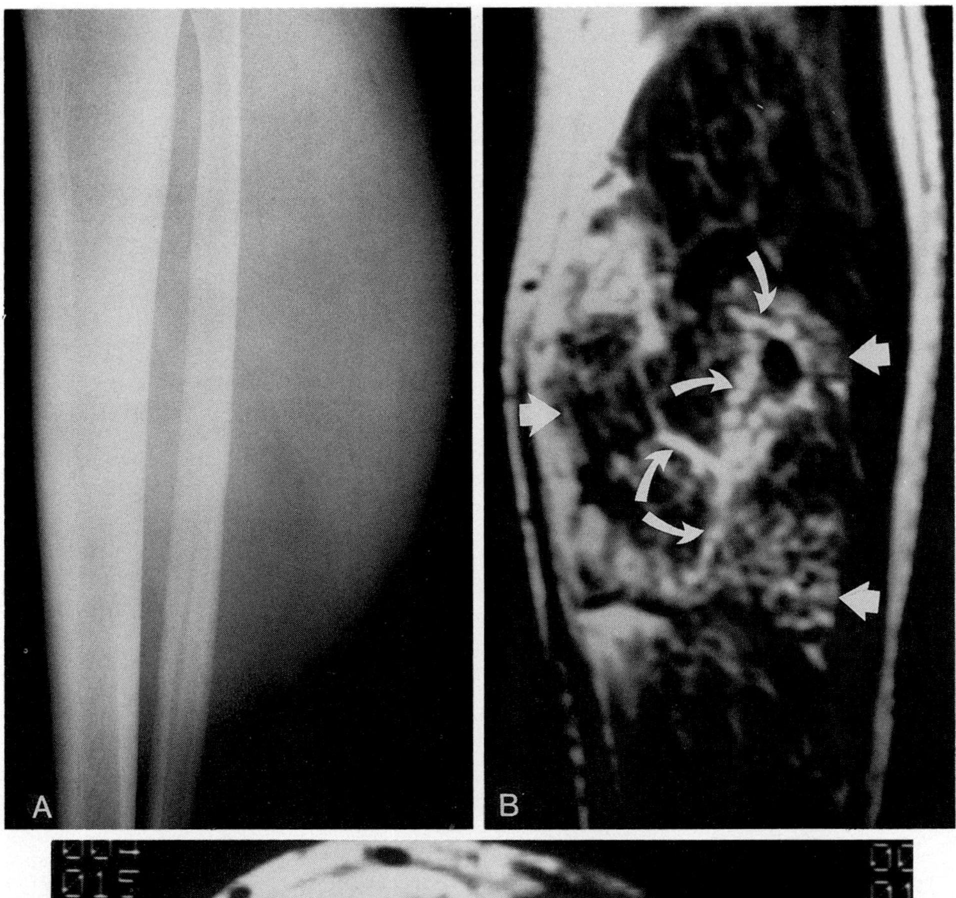

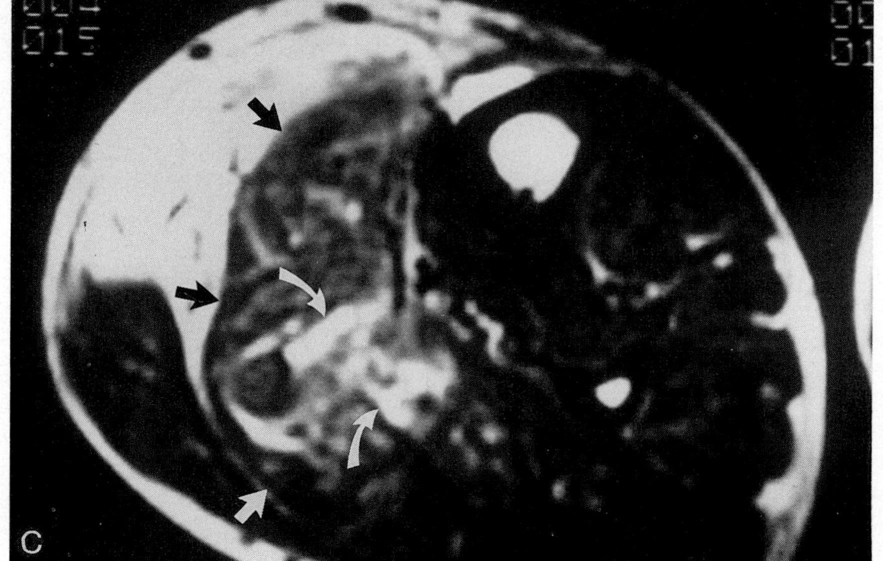

Figure 15–28. Angiolipoma of the forearm. *A*, Conventional radiograph shows a large soft tissue mass. *B*, Coronal T1-weighted spin-echo MR image (TR, 800 ms; TE, 20 ms) demonstrates a large heterogeneous soft tissue tumor with areas of high, intermediate, and low signal intensity *(arrows)*. The areas of high signal intensity have a serpiginous pattern corresponding to blood-filled vascular structures *(curved arrows)*. *C*, Axial T1-weighted spin-echo MR image (TR, 800 ms; TE, 20 ms) shows the heterogeneous mass *(straight arrows)*. Multiple dilated vascular channels of high signal intensity *(curved arrows)* are seen.

depends on the age of the lesion. Fresh hemorrhage, as may be observed after pathologic fracture, is of low signal intensity on T1-weighting and of high signal intensity on T2-weighting in the hyperacute phase. As the amount of intracellular deoxyhemoglobin increases, low signal intensity will be observed on T2-weighted spin-echo images. As deoxyhemoglobin is transformed into methemoglobin, signal of higher intensity is observed in a hematoma on both T1- and T2-weighted images (see Table 15–2)[50,113] (Figs. 15–19, 15–32). When hemosiderin is the predominant blood product in a hematoma, signal intensity is low on both T1 and T2 weighting (see Table 15–2).[50,113] A ring of low signal intensity can be seen surrounding a central area of high signal intensity when hemosiderin has formed in the periphery of a hematoma and methemoglobin is still present in the center of the hematoma. A

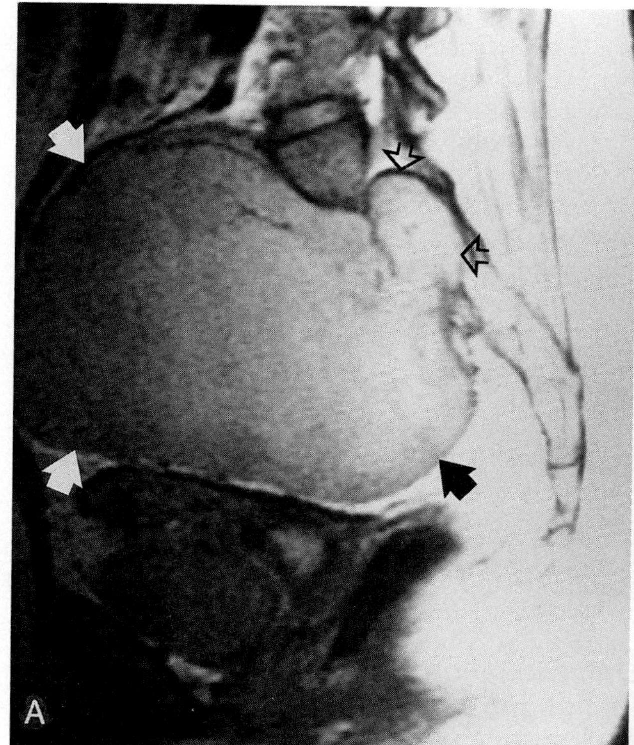

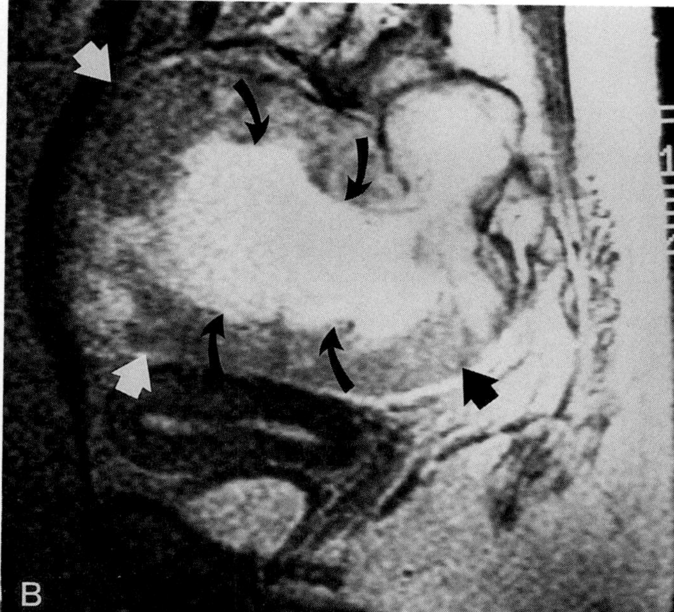

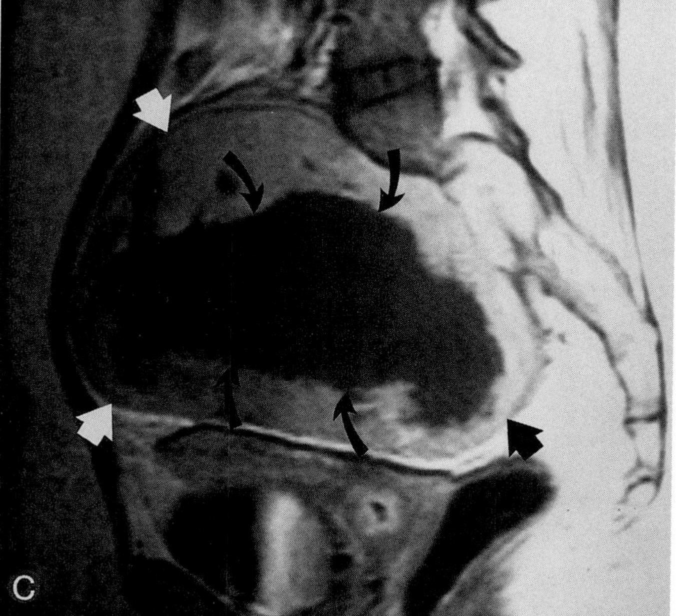

Figure 15–29. Large pelvic schwannoma originating from a lumbar nerve root. *A*, Sagittal precontrast T1-weighted spin-echo MR image (TR, 800 ms; TE, 20 ms) at the level of the lumbosacral spine and pelvis shows a large soft-tissue mass *(solid arrows)*. The mass is of intermediate signal intensity. It originates posteriorly from the L5–S1 neural foramen *(open arrows)*. *B*, Sagittal precontrast T2-weighted spin-echo MR image (TR, 2000 ms; TE, 120 ms). On T2-weighting, the schwannoma demonstrates intermediate signal intensity in the periphery *(straight arrows)* and high signal intensity centrally *(curved arrows)*. It impresses on the uterus. *C*, Sagittal post-contrast T1-weighted spin-echo MR image (TR, 800 ms; TE, 20 ms). The peripheral portions of the schwannoma show contrast enhancement *(straight arrows)*. The central portions do not enhance and remain of low signal intensity, indicating central necrosis *(curved arrows)*.

fraying of the interface between hemosiderin and adjacent normal tissue or even a cauliflowerlike appearance of hemosiderin may be observed on gradient-echo MR images with long echo times (see Table 15–2).

Abscesses are of low signal intensity on T1-weighted images and of high signal intensity on T2-weighted images. When the abscess cavity is filled with necrotic debris, signal inhomogeneity may result on T2-weighted images. An abscess wall of low signal intensity may be observed on precontrast T1-weighted images.[20] The abscess frequently is surrounded by reactive edema of high signal intensity on T2 weighting. After intravenous administration of gadopentetate dimeglumine, marked contrast enhancement is seen in the abscess wall; the inner portions of the abscess do not, however, enhance and remain low in signal intensity on postcontrast T1-weighted MR images.[89]

Cysts are one of the most common causes of soft tissue masses. They can be found in virtually any location in the body, but frequently they occur in close relationship to a joint. Cysts may be caused by distention of the synovial lining of a joint, or they may be distended bursae. They frequently are present adjacent to or communicating with the knee, hip, shoulder, elbow, and wrist joint. Cysts are well marginated, and typically they are of low signal intensity

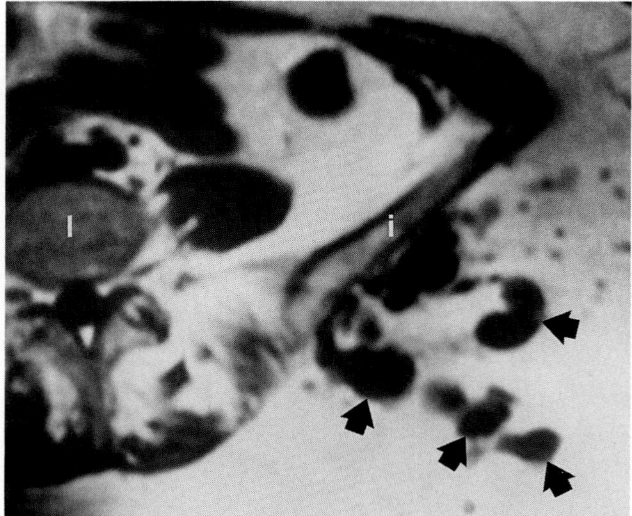

Figure 15–30. Neurofibromatosis. Axial T1-weighted spin-echo MR image demonstrates multiple round neurofibromas in the fat lateral to the iliac wing *(arrows)*. The neurofibromas are of low signal intensity. Note marked atrophy of spinal and pelvic muscles. i, Iliac wing; l, lumbar vertebral body.

on T1-weighted MR images and of high signal intensity on T2-weighted images. Usually the signal intensity on T2-weighted images is homogeneous. However, some signal inhomogeneity may result when cysts are superinfected or when hemorrhage into the cyst occurs.

Ganglion cysts resemble other cysts in their MR imaging appearance. They are found most frequently along tendon sheaths or in juxta-articular locations, but they can also occur intraosseously or in any soft tissue location (Fig. 15–33). Ganglion cysts are composed of myxoid material surrounded by a fibrous capsule. They are of high signal intensity on T2-weighting, unless necrotic material is enclosed in the cysts, which leads to resultant low signal intensity.[36]

MAGNETIC RESONANCE SPECTROSCOPY

A new addition to MR imaging is the use of in vivo MR spectroscopy. In vivo, phosphorus-31 (^{31}P) MR spectroscopy can detect phosphorylated compounds that are present at physiologic concentration, including metabolites such as nucleoside triphosphates (α-, β-, and γ-adenosine triphosphate [ATP] and similar compounds), phosphocreatine (PCr), inorganic phosphate (P^i), and a variety of phosphomonoester and diphosphodiester molecules. Phosphorus-31 MR spectroscopy also is capable of measuring tissue pH in vivo. Although most cells and tissues are able to maintain relatively constant levels of these metabolites through homeostatic processes, tissues such as tumors, which can become hypoxic or ischemic, often show altered levels of certain bioenergetic metabolites and pH.[41,42]

In vivo MR spectroscopy had initially been limited to animal studies using small-bore experimental magnets owing to technical problems inherent in whole-body scanners, such as insufficient field homogeneity. In addition, more sophisticated localization techniques needed to be developed for in vivo application. Technical advances have made possible in vivo acquisition of ^{31}P MR spectra in human bone and soft tissue sarcomas.[24,91,92,104,110,116,117,150] Most studies were in agreement that tumors differed from normal, nonneoplastic tissue by an increase in phosphomonoester and phosphodiester (PDE) levels and in P_i.[24,91,92,104,110,116,117,150]. Tumor metabolism can thus be differentiated from metabolism in normal tissue.

Differentiation between Benign and Malignant Lesions Using Magnetic Resonance Spectroscopy

Negendank and co-workers[91] and Zlatkin and colleagues[150] evaluated to what extent ^{31}P MR spectra may supplement ^{1}H MR imaging in differentiating benign from malignant bone and soft tissue tumors. These investigators found that malignant lesions were distinguished from benign lesions on the basis of significantly higher mean peak ratios of phosphomonoester to β-ATP and of phosphodiester to ATP, a significantly lower mean peak area ratio of PCr to ATP, and a higher mean pH. They concluded that ^{31}P MR spectroscopy may be used to increase diagnostic specificity in bone and soft tissue lesions. However, some controversy exists about these findings: Zlatkin and co-workers reported in the same study that no characteristic spectra were observed for individual tumor types.[150] In both studies, spectra were acquired with the use of surface coil localization techniques, and, consequently, they may have been contaminated by muscle or other soft tissue adjacent to the tumor. In addition, no attempt was made to account for the variation in the sensitivity of the volume, although different coils had been used throughout the patient population.

Ultimately, ^{31}P MR spectroscopy may supplement ^{1}H MR imaging in differentiating between benign and malignant lesions. However, advanced localization techniques such as one-, two-, and three-dimensional chemical shift imaging need to be applied in a larger, well-defined patient population to determine the value of phosphorus spectroscopy in a clinical setting. The three-dimensional chemical

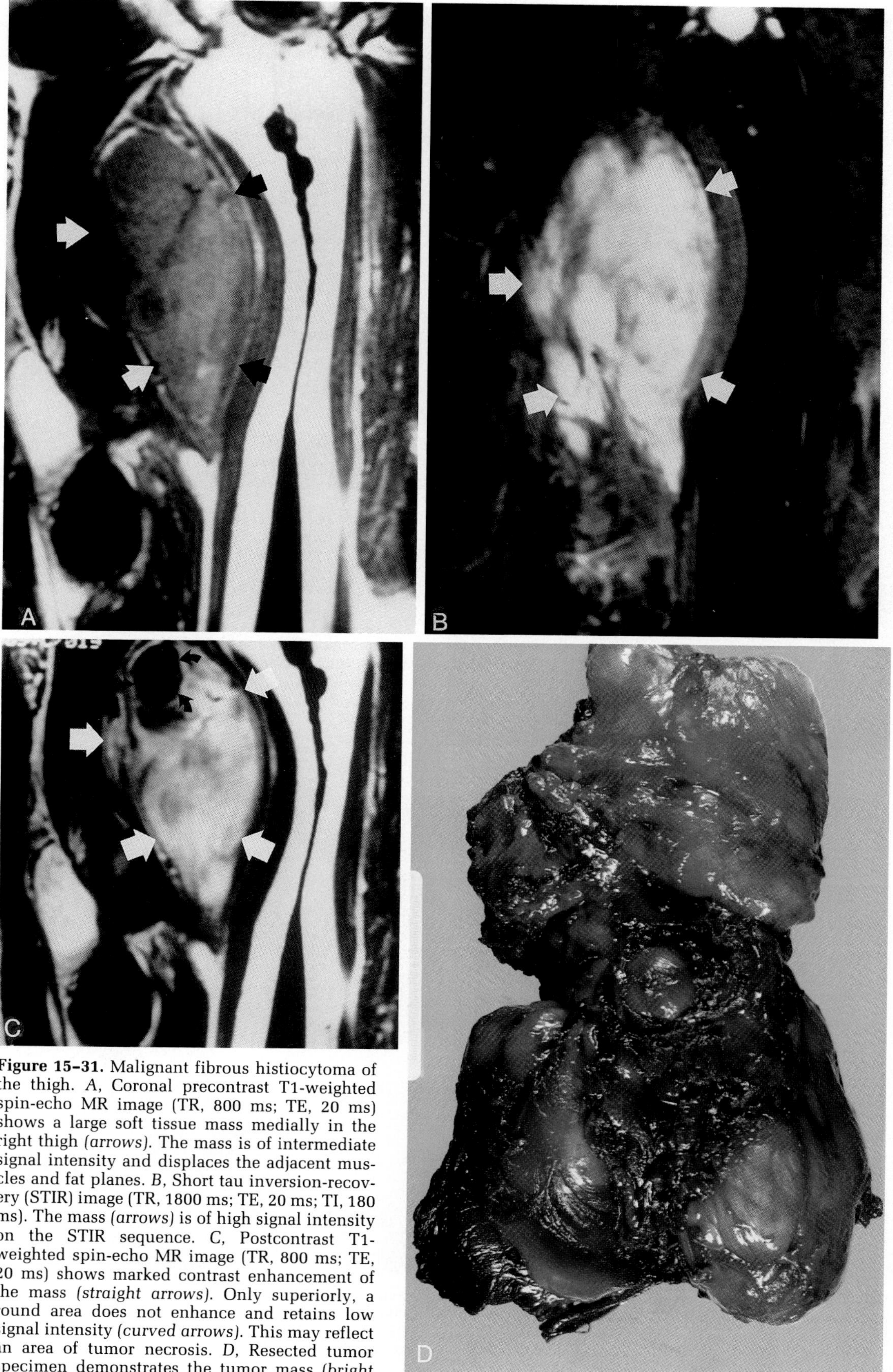

Figure 15–31. Malignant fibrous histiocytoma of the thigh. *A,* Coronal precontrast T1-weighted spin-echo MR image (TR, 800 ms; TE, 20 ms) shows a large soft tissue mass medially in the right thigh *(arrows).* The mass is of intermediate signal intensity and displaces the adjacent muscles and fat planes. *B,* Short tau inversion-recovery (STIR) image (TR, 1800 ms; TE, 20 ms; TI, 180 ms). The mass *(arrows)* is of high signal intensity on the STIR sequence. *C,* Postcontrast T1-weighted spin-echo MR image (TR, 800 ms; TE, 20 ms) shows marked contrast enhancement of the mass *(straight arrows).* Only superiorly, a round area does not enhance and retains low signal intensity *(curved arrows).* This may reflect an area of tumor necrosis. *D,* Resected tumor specimen demonstrates the tumor mass *(bright red)* that is adjacent to normal muscle *(dark red).*

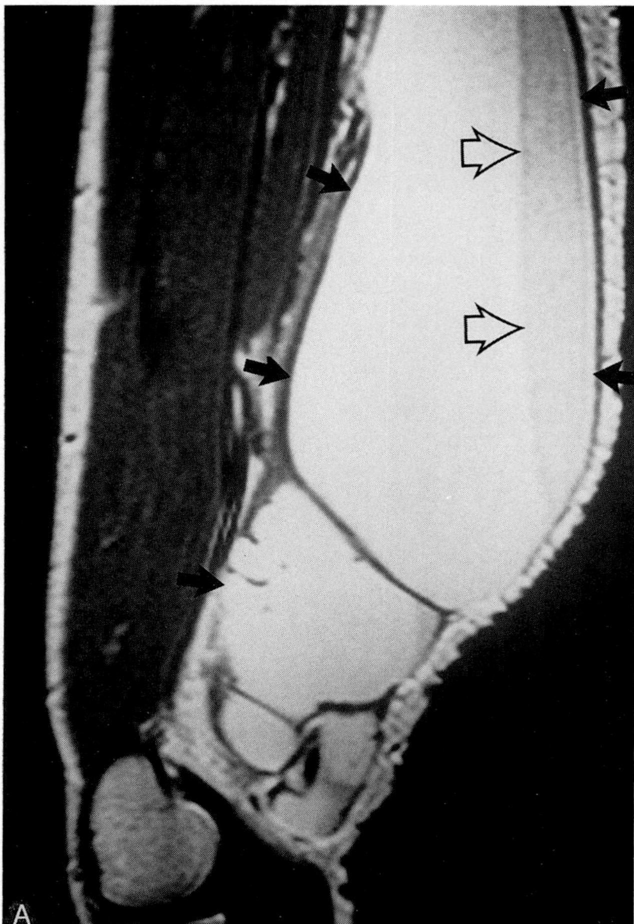

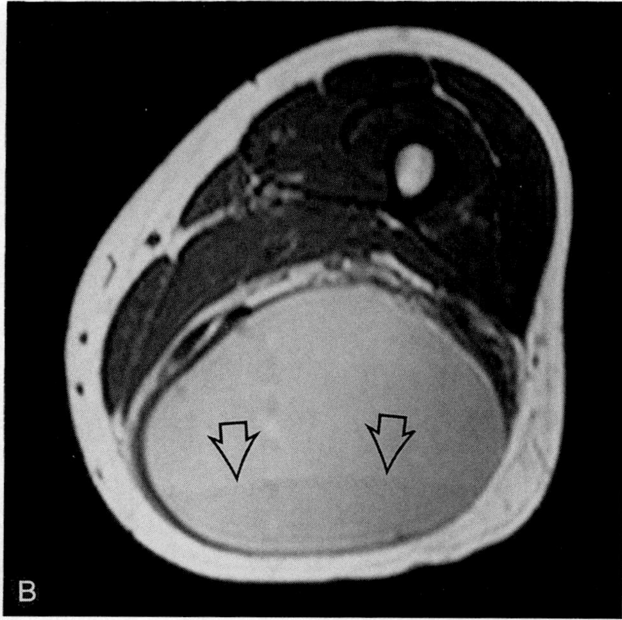

Figure 15–32. Postoperative hematoma of the thigh, 16 days after surgery. Sagittal *(A)* and axial *(B)* T1-weighted spin-echo MR images (TR, 800 ms; TE, 30 ms) show a large postoperative hematoma *(solid arrows)* in the posterior portions of the thigh. The hematoma is of high signal intensity, reflecting the presence of methemoglobin. A fluid-fluid level is seen *(open arrows)*. The more posterior portion *(open arrows)* of the fluid-fluid level is of slightly lower signal intensity and may be explained as cellular debris in the dependent portions of the hematoma or by variations in deoxyhemoglobin, methemoglobin, or hemosiderin content.

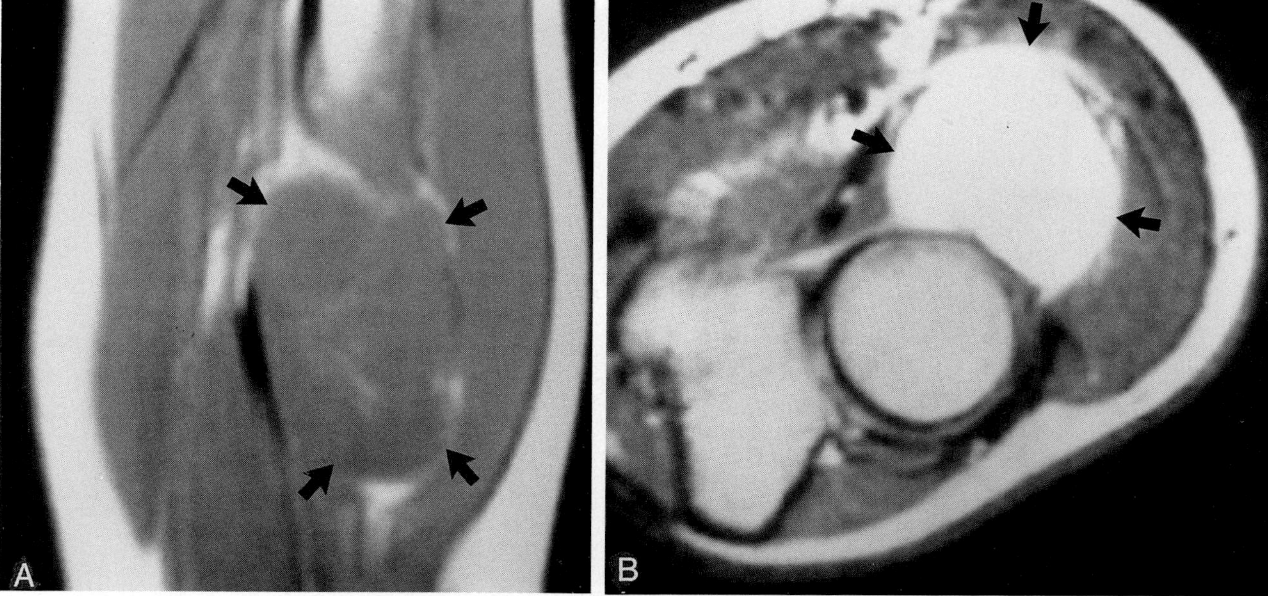

Figure 15–33. Ganglion cyst of the elbow. *A*, Coronal T1-weighted spin-echo MR image (TR, 600 ms; TE, 20 ms) shows a mass *(arrows)* that is of equal intensity to muscle. It displaces but does not infiltrate the adjacent muscle. *B*, Axial T2-weighted spin-echo MR image (TR, 1600 ms; TE, 120 ms). The lesion is of high signal intensity on T2-weighting *(arrows)*. Axial image shows the juxta-articular location of the mass. It is in immediate proximity of the proximal radioulnar articulation, suggesting that it is a ganglion cyst (which was confirmed at surgery).

shift imaging technique can be used to generate ^{31}P metabolite images of a tumor.

Monitoring the Response to Chemotherapy Using Magnetic Resonance Spectroscopy

Preliminary experience also has been reported on the use of ^{31}P MR spectroscopy in assessing the response of bone and soft tissue sarcomas to chemotherapy.[61,92,104,110,116,117] Ross and co-workers observed an increase in P_i, a loss of ATP, and a loss of phosphomonoester in tumor regression.[110] Tumor relapse was accompanied by the reappearance of abnormalities in the spectra. Similar findings had been described by Nidecker and colleagues[92] in a study of one patient after chemotherapy and by Redmond and co-workers.[104] Semmler and associates[116,117] demonstrated changes in the signal intensity of PCr compared with that of P_i after chemotherapy. In a study that involved soft tissue sarcomas exclusively, Dewhirst and colleagues reported that pretherapy tumor pH correlated positively and changes during therapy in pH, ratio of PCr to P_i, ratio of ATP to P_i, and PME signal-to-noise ratio correlated negatively with the percentage of tumor necrosis on the surgical specimen.[24] Hoffer and co-workers found an increase in the ratio of high- to low-energy phosphates—ATP/(P_i + phosphomonoester)—in tumor response in an animal model of reticulum cell sarcoma.[61] However, the opposite pattern of high-energy phosphate decline also can be observed after therapy, often at a faster rate relative to the loss of high-energy phosphates in untreated tumors.[90,110]

The histologic and pathophysiologic bases for increases or decreases in the high- to low-energy phosphate ratios in different tumors remain to be investigated. In the future, accrual of larger clinical data bases in more homogeneous groups of primary bone tumors is necessary to evaluate the clinical use of ^{31}P MR spectroscopy in monitoring response to chemotherapy.

References

1. Aisen AM, Martel W, Braunstein EM, et al: MRI and CT evaluation of primary bone and soft-tissue tumors. AJR 146:749–756, 1986.
2. Aoki J, Moriya K, Yamashita K, et al: Giant cell tumors of bone containing large amounts of hemosiderin: MR-pathologic correlation. J Comput Assist Tomogr 15:1024–1027, 1991.
3. Baierl P, Muhlsteffen A, Haustein J, et al: Comparison of plain and Gd-DTPA-enhanced MR-imaging in children. Pediatr Radiol 20:515–519, 1990.
4. Baker HW: The surgical treatment of cancer. Cancer 43:787–789, 1979.
5. Beltran J, Noto AM, Chakeres DW, Christoforidis AJ: Tumors of the osseous spine: Staging with MR imaging versus CT. Radiology 162:565–569, 1987.
6. Beltran J, Simon DC, Levy M: Aneurysmal bone cysts: MR imaging at 1.5 T. Radiology 158:689–675, 1986.
7. Berquist TH: Magnetic resonance imaging of musculoskeletal neoplasms. Clin Orthop 244:101–118, 1989.
8. Bloem JL, Taminiau AH, Eulderink F, et al: Radiologic staging of primary bone sarcoma: MR imaging, scintigraphy, angiography, and CT correlated with pathologic examination. Radiology 169:805–810, 1988.
9. Bohndorf K, Reiser M, Lochner B, et al: Magnetic resonance imaging of primary tumors and tumor-like lesions of bone. Skel Radiol 15:511–517, 1986.
10. Boyko OB, Cory DA, Cohen MD, et al: MR imaging of osteogenic and Ewing's sarcoma. AJR 148:317–322, 1987.
11. Bradley WG, Waluch V, Lai K, et al: Appearance of rapidly flowing blood on magnetic resonance images. AJR 143:1167–1174, 1984.
12. Brasch RC. Work in progress: Methods for contrast enhancement in NMR imaging and potential applications. Radiology 147:781–788, 1983.
13. Brasch RC, Weinmann HJ, Wesbey GE: Contrast-enhanced NMR imaging: Animal studies using gadolinium-DTPA complex. AJR 142:625–630, 1984.
14. Buetow PC, Kransdorf MJ, Moser RP, et al: Radiologic appearance of intramuscular hemangioma with emphasis on MR imaging. AJR 154:563–567, 1990.
15. Carrasco CH, Charnsangavej C, Raymond K, et al: Osteosarcoma: Angiographic assessment of response to preoperative chemotherapy. Radiology 170:839–842, 1989.
16. Chew FS, Hudson TM: Radionuclide bone scanning of osteosarcoma: Falsely extended uptake patterns. AJR 139:49–54, 1982.
17. Choi H, Varma DGK, Fornage BD, et al: Soft-tissue sarcoma: MR imaging vs sonography for detection of local recurrence after surgery. AJR 157:353–358, 1991.
18. Chuang VP, Benjamin R, Jaffe N, et al: Radiographic and angiographic changes in osteosarcoma after intraarterial chemotherapy. AJR 139:1065–1069, 1982.
19. Cohen EK, Kressel HY, Perosio T, et al: MR imaging of soft-tissue hemangiomas: Correlation with pathologic findings. AJR 15:1079–1081, 1988.
20. Cohen JM, Weinreb JC, Maravilla KR: Fluid collections in the intraperitoneal and extraperitoneal spaces: Comparison of MR and CT. Radiology 155:705–708, 1985.
21. Dahlin DC, Unni KK: Bone Tumors: General Aspects and Data on 8,542 Cases. Springfield, Ill, Charles C Thomas, 1986.
22. Dalinka MK, Zlatkin MB, Chao P, et al: The use of magnetic resonance imaging in the evaluation of bone and soft-tissue tumors. Radiol Clin North Am 28:461–470, 1990.
23. Delbeke D, Powers TA, Sandler MP: Correlative radionuclide and magnetic resonance imaging in evaluation of the spine. Clin Nucl Med 14:742–749, 1989.
24. Dewhirst MW, Sostman HD, Leopold KA, et al: Soft-tissue sarcomas: MR imaging and MR spectroscopy for prognosis and therapy monitoring. Work in progress. Radiology 174:847–853, 1990.
25. Dillon WP, Som PM, Rosenau W: Hemangioma of the nasal vault: MR and CT features. Radiology 180:761–765, 1991.
26. Dumoulin CL: Flow imaging. *In* Budinger TF, Margulis AR, (eds): Medical Magnetic Resonance. A primer-1988. Berkeley, Calif, Society of Magnetic Resonance in Medicine, 1988, pp 85–108.
27. Edeiken J, Dalinka M, Karasick D. Bone tumors and tumor-like conditions. *In* Edeiken J, Dalinka M, Karasick D (eds): Roentgen Diagnosis of Diseases of Bone. Baltimore, Williams & Wilkins, 1990, pp 33–574.
28. Eilber F, Giuliano A, Eckhardt J, et al: Adjuvant chemotherapy for osteosarcoma: A randomized prospective trial. J Clin Oncol 5:21–26, 1987.
29. Enzinger FM: Intramuscular myxoma: A review and follow-up study of 34 cases. Am J Clin Pathol 43:104–110, 1985.
30. Enzinger FM, Weiss SW: Soft tissue tumors. *In* eds. St Louis, CV Mosby, 1988, pp 719–728.
31. Erlemann R, Reiser M, Peters PE, et al: Efficient use of flash sequences in the staging of bone and soft tissue tumors. ROFO 149:178–183, 1988.
32. Erlemann R, Reiser M, Peters PE, et al: Time-dependent changes in signal intensity in neoplastic and inflammatory lesions of the musculoskeletal system following intravenous administration. Radiology 28:269–276, 1988.
33. Erlemann R, Reiser MF, Peters PE, et al: Musculoskeletal neoplasms: Static and dynamic Gd-DTPA-enhanced MR imaging. Radiology 171:767–773, 1989.
34. Erlemann R, Sciuk J, Bosse A, et al: Response of osteosarcoma and Ewing sarcoma to preoperative chemotherapy: Assessment with dynamic and static MR imaging and skeletal scintigraphy. Radiology 175:791–796, 1990.
35. Erlemann R, Vassallo P, Bongartz G, et al: Musculoskeletal neoplasms: Fast low-angle shot MR imaging with and without Gd-DTPA. Radiology 176:489–495, 1990.

36. Feldman F, Singson RD, Staron RB: Magnetic resonance imaging of para-articular and ectopic ganglia. Skel Radiol 18:353–358, 1989.
37. Frank JA, Ling A, Patronas NJ, et al: Detection of malignant bone tumors: MR imaging vs scintigraphy. AJR 155:1043–1048, 1990.
38. Friedburg HG, Wimmer B, Hennig J, et al: Initial clinical experiences with RARE-MR urography. Urology 26:309–316, 1987.
39. Frouge C, Vanel D, Coffre C, et al: The role of magnetic resonance imaging in the evaluation of Ewing sarcoma. A report of 27 cases. Skel Radiol 17:387–392, 1988.
40. Furst G, Pape H, Jurgens T, et al: Use of contrast media in MR tomography of soft tissue sarcomas. Follow-up studies after multimodal therapy. Radiology 29:336–342, 1989.
41. Gadian DG: Nuclear Magnetic Resonance and its Application to Living Systems. New York, Oxford University Press, 1982.
42. Gadian DG: Magnetic resonance spectroscopy as a probe of tumour metabolism. Eur J Cancer 27:526–528, 1991.
43. Gamba JL, Martinez S, Apple J, et al: Computed tomography of axial skeletal osteoid osteomas. AJR 142:769–772, 1984.
44. Gilbert HA, Kagan AR, Winkley J: Management of soft-tissue sarcoma of the extremities. Surg Gynecol Obstet 139:914–918, 1974.
45. Gillespy T 3rd, Manfrini M, Ruggieri P, et al: Staging of intraosseous extent of osteosarcoma: Correlation of preoperative CT and MR imaging with pathologic macroslides. Radiology 167:765–767, 1988.
46. Glazel JA, Lee KH: On the interpretation of water nuclear magnetic resonance relaxation times in heterogeneous systems. J Am Chem Soc 96:970–978, 1974.
47. Glazer HS, Lee JKT, Levitt RG, et al: Radiation fibrosis: Differentiation from recurrent tumor by MR imaging. Radiology 156:721–727, 1985.
48. Golfieri R, Baddeley H, Pringle JS, et al: Primary bone tumors: MR morphologic appearance correlated with pathologic examinations. Acta Radiol 32:290–298, 1991.
49. Golfieri R, Baddeley H, Pringle JS, Souhami R: The role of the STIR sequence in magnetic resonance imaging examination of bone tumours. Br J Radiol 63:251–256, 1990.
50. Gomori JM, Grossman RI, Goldberg HI, et al: Intracranial hematomas: Imaging by high-field MR. Radiology 157:87–93, 1985.
51. Graif M, Pennock JM, Pringle J, et al: Magnetic resonance imaging: Comparison of four pulse sequences in assessing primary bone tumours. Skel Radiol 18:439–444, 1989.
52. Greenspan A, Steiner G, Knutzon R: Bone island (enostosis): Clinical significance and radiologic and pathologic correlations. Skel Radiol 20:85–90, 1991.
53. Haase A: Snapshot FLASH MRI: Applications to T1, T2, and chemical-shift imaging. Magn Reson Med 13:77–89, 1990.
54. Hanna LS, Magill HL, Parham DM, et al: Childhood chondrosarcoma: MR imaging with Gd-DTPA. Magn Reson Imaging 8:669–672, 1990.
55. Hanna SL, Fletcher BD, Fairclough DL, et al: Magnetic resonance imaging of disseminated bone marrow disease in patients treated for malignancy. Skel Radiol 20:79–84, 1991.
56. Harle A, Reiser M, Erlemann R, Wuisman P: The value of nuclear magnetic resonance tomography in staging of bone and soft tissue sarcomas. Orthopäde 18:34, 1989.
57. Hennig J, Friedburg H: Clinical applications and methodological developments of the RARE technique. Magn Reson Imaging 6:391–395, 1988.
58. Hennig J, Friedburg H, Ott D: Fast three-dimensional imaging of cerebrospinal fluid. Magn Reson Med 5:380–383, 1987.
59. Heredia C, Mercader JM, Graus F, et al: Hemangioma of the vertebrae: Contribution of magnetic resonance to its study. Neurologia 4:336–339, 1989.
60. Heuck AF, Steiger P, Stoller DW, et al: Quantification of knee joint fluid volume by MR imaging and CT using three-dimensional data processing. J Comput Assist Tomogr 13:287–293, 1989.
61. Hoffer FA, Taylor GA, Spevak M, et al: Metabolism of tumor regression from angiogenesis inhibition: ^{31}P Magnetic resonance spectroscopy. Magn Reson Med 11:202–208, 1989.
62. Holscher HC, Bloem JL, Nooy MA, et al: The value of MR imaging in monitoring the effect of chemotherapy on bone sarcomas. AJR 154:763–769, 1990.
63. Hudson TM, Hamlin DJ, Enneking WF, Pettersson H: Magnetic resonance imaging of bone and soft-tissue tumors: Early experience in 31 patients compared with computed tomography. Skel Radiol 13:134–146, 1985.
64. Hudson TM, Hanlin DJ, Fitzsimmons JR: Magnetic resonance of fluid levels in aneurysmal bone cyst and in anticoagulated human blood. Skel Radiol 13:267–270, 1985.
65. Huvos AG, Rosen R, Marcove RC: Primary osteogenic sarcoma. Arch Pathol Lab Med 101:14–18, 1977.
66. Ishizaka H, Kurihara M, Heshiki A, et al: MR imaging of the bone marrow using short TI IR. Part 2, Normal and pathological intensity distribution of the bone marrow. Nippon Igaku Hoshasen Gakkai Zasshi 49:134–138, 1989.
67. Jaffe HL: Tumors and Tumorous Conditions of the Bones and Joints. Philadelphia, Lea & Febiger, 1958.
68. Jurgens H, Exner U, Gadner H, et al: Multidisciplinary treatment of Ewing's sarcoma of bone. Cancer 61:23–32, 1988.
69. Kattapuram SV, Khurana JS, Scott JA, el Khoury G: Negative scintigraphy with positive magnetic resonance imaging in bone metastases. Skel Radiol 19:113–116, 1990.
70. Keigley BA, Haggar AM, Gaba A, et al: Primary tumors of the foot: MR imaging. Radiology 171:755–759, 1989.
71. Kiefer SA, Binet EF, David DO, et al: Lumbar myelography with iohexol and metrizamide: A comparative multicenter prospective study. Radiology 151:665–670, 1984.
72. Kilcoyne RF, Richardson ML, Porter BA, et al: Magnetic resonance imaging of soft-tissue masses. Clin Orthop 228:13–22, 1988.
73. Kim EE, Abello R, Holbert JM, et al: Differentiation of Therapeutic Changes from Residual or Recurrent Musculoskeletal Sarcomas Using Gd-DTPA Enhanced MRI. Amsterdam, The Netherlands, Society for Magnetic Resonance in Medicine, 1989.
74. Kransdorf MJ, Jelinek JS, Moser RJ, et al: Soft-tissue masses: Diagnosis using MR imaging. AJR 153:541–547, 1989.
75. Lang P, Genant HK, Steiger P, et al: 3-D reformatting asserts clinical potential in MRI. Diagn Imaging 12:100–105, 1989.
76. Lang P, Stevens M, Vahlensieck M, et al: Rheumatoid arthritis of the hand and wrist: Evaluation of soft-tissue inflammation and quantification of inflammatory activity using unenhanced and dynamic Gd-DTPA enhanced MR imaging. San Francisco, Society of Magnetic Resonance in Medicine, 1991, p 66.
77. Lee JK, Yao L, Wirth CR: MR imaging of solitary osteochondroma: Report of eight cases. AJR 149:557–560, 1987.
78. Lee YY, Van TP: Craniofacial chondrosarcomas: Imaging findings in 15 untreated cases. AJNR 10:165–170, 1989.
79. Lee YY, Van TP, Nauert C, et al: Craniofacial osteosarcomas: Plain film, CT, and MR findings in 46 cases. AJR 150:1397–1402, 1988.
80. Levine E, Lee KR, Neff JR, et al: Comparison of computed tomography and other imaging modalities in the evaluation of musculoskeletal tumors. Radiology 131:431–437, 1979.
81. Lukens JA, McLeod RA, Sim FH: Computed tomographic evaluation of primary osseous malignant neoplasm. AJR 139:45–48, 1982.
82. Mahajan H, Kim EE, Lee YY, Goepfert H: Malignant fibrous histiocytoma of the tongue demonstrated by magnetic resonance imaging. Otolaryngol Head Neck Surg 101:704–706, 1989.
83. Mahajan H, Kim EE, Wallace S, et al: Magnetic resonance imaging of malignant fibrous histiocytoma. Magn Reson Imaging 7:283–288, 1989.
84. Mahajan H, Lorigan JG, Shirkhoda A: Synovial sarcoma: MR imaging. Magn Reson Imaging 7:211–216, 1989.
85. Mail JT, Cohen MD, Mirkin LD, Provisor AJ: Response of osteosarcoma to preoperative intravenous high-dose methotrexate chemotherapy: CT evaluation. AJR 144:89–93, 1985.
86. Mills CM, Brant-Zawadski M, Crooks LE, et al: Nuclear magnetic resonance: Principles of blood flow imaging. AJNR 4:1161–1166, 1983.
87. Morton DL, Eilber FR, Townsend CM, et al: Limb-salvage from multidisciplinary treatment approach for skeletal and soft-tissue sarcoma of the extremity. Ann Surg 184:268–278, 1976.
88. Mulkern RV, Wong ST, Winalski C, Jolesz FA: Contrast manipulation and artifact assessment of 2D and 3D RARE sequences. Magn Reson Imaging 8:557–566, 1990.
89. Muller-Miny H, Reiser M, Erlemann R, Peters PE: The use of gadolinium-DTPA in inflammatory skeletal disease. Radiol Diagn 30:491–495, 1989.
90. Naruse S, Horikawa K, Horikawa Y, et al: Measurements of in vivo ^{31}P magnetic resonance spectra in neuroectodermal tumors for evaluation of the effects of chemotherapy. Cancer Res 45:2429–2433, 1985.
91. Negendank WG, Crowley MG, Ryan JR, et al: Bone and soft-tissue lesions: Diagnosis with combined H-1 and P-31 MR spectroscopy. Radiology 173:181–188, 1989.
92. Nidecker AC, Müler S, Aue P, et al: Extremity bone tumors: Evaluation by ^{31}P MR spectroscopy. Radiology 157:167–174, 1985.
93. Norman A, Greenspan A: Sclerosing dysplasias of bone. *In* Taveras JM, Ferrucci JT (eds): Radiology - Diagnosis, Imaging, Intervention. Philadelphia, J B Lippincott, 1986, pp 1–8.
94. Norton KI, Hermann G, Abdelwahab IF, et al: Epiphyseal involvement in osteosarcoma. Radiology 180:813–816, 1991.
95. O'Flanagan SJ, Stack JP, McGee HM, et al: Imaging of intramedullary tumour spread in osteosarcoma. A comparison of techniques. J Bone Joint Surg 73:998–1001, 1991.
96. Pan G, Raymond AK, Carrasco CH, et al: Osteosarcoma: MR imaging after preoperative chemotherapy. Radiology 174:517–526, 1990.
97. Petasnick JP, Turner DA, Charters JR, et al: Soft-tissue masses of the locomotor system: Comparison of MR imaging with CT. Radiology 160:125–133, 1986.
98. Peterson KK, Renfrew D, Feddersen RM, et al: Magnetic resonance imaging of myxoid containing tumors. Skel Radiol 20:245–250, 1991.
99. Pettersson H, Eliasson J, Egund N, et al: Gadolinium-DTPA enhancement of soft-tissue tumors in magnetic resonance imaging—preliminary clinical experience in 5 patients. Skel Radiol 17:319–323, 1988.

100. Pettersson H, Gillespy T, Hamlin DJ, et al: Primary musculoskeletal tumors: Examination with MR imaging compared to conventional modalities. Radiology 164:237–241, 1987.
101. Porter BA, Olson DO, Stimack GK: STIR Imaging of Marrow Malignancies. New York, Society for Magnetic Resonance in Medicine, 1987, p 146.
102. Prayer L, Imhof H, Stiglbauer R, et al: Gadolinium-DTPA in magnetic resonance tomography. Clinical use—indications. Röntgenblatter 42:493–498, 1989.
103. Ramanna L, Waxman A, Binney G, et al: Thallium-201 scintigraphy in bone sarcoma: Comparison with gallium-67 and technetium-MDP in the evaluation of chemotherapeutic response. J Nucl Med 31:567–572, 1990.
104. Redmond OM, Stack JP, Dervan PA, et al: Osteosarcoma: Use of MR imaging and MR spectroscopy in clinical decision making. Radiology 172:811–815, 1989.
105. Reiser M, Bohndorf K, Niendorf HP, et al: Initial experiences with gadolinium DTPA in magnetic resonance tomography of bone and soft tissue tumors. Radiology 27:467–472, 1987.
106. Reiser M, Bongartz GP, Erlemann R, et al: Gadolinium-DTPA in rheumatoid arthritis and related diseases: First results with dynamic magnetic resonance imaging. Skel Radiol 18:591–597, 1989.
107. Reiser M, Rupp N, Biehl TH, et al: MR in diagnosis of bone tumors. Eur J Radiol 5:1–7, 1985.
108. Reuther G, Mutschler W: Detection of local recurrent disease in musculoskeletal tumors: Magnetic resonance imaging versus computed tomography. Skel Radiol 19:85–90, 1990.
109. Rock MG, Pritchard DJ, Reiman HM, McLeod RA: Extraabdominal demoid tumor. Mayo Clin Tumor Rounds 7:141–147, 1984.
110. Ross B, Helsper JT, Cox IJ, et al: Osteosarcoma and other neoplasms of bone. Magnetic resonance spectroscopy to monitor therapy. Arch Surg 122:1464–1469, 1987.
111. Ross JS, Masaryk TJ, Modic MT, et al: Vertebral hemangiomas: MR imaging. Radiology 165:165–169, 1987.
112. Schorner W, Lang P, Bittner R, Felix et al: Gadolinium-DTPA enhanced MR imaging: Applications in body MRI. Diagn Imaging 4:114–119, 1990.
113. Scott WA, Mark AS, Grossman RI, Gomori JM: Intracranial hemorrhage: Gradient-echo MR imaging at 1.5T. Radiology 168:803–807, 1988.
114. Seeger LL, Eckardt JJ, Bassett LW: Cross-sectional imaging in the evaluation of osteogenic sarcoma: MRI and CT. Semin Roentgenol 24:174–184, 1989.
115. Seeger LL, Widoff BE, Bassett LW, et al: Preoperative evaluation of osteosarcoma: Value of gadopentetate dimeglumine-enhanced MR imaging. AJR 157:347–351, 1991.
116. Semmler W, Gademann G, Bachert BP, et al: In vivo 31-phosphorus spectroscopy of tumors: Pre-, intra- and post-therapy. ROFO 149:369–377, 1988.
117. Semmler W, Gademann G, Bachert BP, et al: Monitoring human tumor response to therapy by means of P-31 MR spectroscopy. Radiology 166:533–539, 1988.
118. Seneterre E, Weissleder R, Jaramillo D, et al: Bone marrow: Ultrasmall superparamagnetic iron oxide for MR imaging. Radiology 179:529–533, 1991.
119. Sherry CS, Harms SE: MR evaluation of giant cell tumors of the tendon sheath. Magn Reson Imaging 7:195–201, 1989.
120. Shirkoda A, Jaffe N, Wallace S, et al: Computed tomography of osteosarcoma after intraarterial chemotherapy. AJR 144:95–99, 1985.
121. Shuman WP, Patten RM, Baron RL, et al: Comparison of STIR and spin-echo MR imaging at 1.5T in 45 suspected extremity tumors: Lesion conspicuity and extent. Radiology 179:247–252, 1991.
122. Simon MA, Enneking WF: The management of soft-tissue sarcomas of the extemities. J Bone Joint Surg [Am] 58:317–327, 1976.
123. Stimac GK, Porter BA, Olson DO, et al: Gadolinium-DTPA-enhanced MR imaging of spinal neoplasms: Preliminary investigation and comparison with unenhanced spin-echo and STIR sequences. AJR 151:1185–1192, 1988.
124. Stull MA, Moser RP, Kransdorf MJ, et al: Magnetic resonance appearance of peripheral nerve sheath tumors. Skel Radiol 20:9–14, 1991.
125. Sundaram M: Radiographic and magnetic resonance imaging of bone and soft-tissue tumors and myeloproliferative disorders. Curr Opin Radiol 3:746–751, 1991.
126. Sundaram M, McDonald DJ: The solitary tumor or tumorlike lesion of bone. Top Magn Reson Imaging 1:17–29, 1989.
127. Sundaram M, McGuire MH, Schajowic ZF: Soft-tissue masses: Histologic bases for decreased signal (short T2) on T2-weighted images. AJR 148:1247–1251, 1987.
128. Sundaram M, McGuire MH, Schajowicz F: Soft-tissue masses: Histologic basis for decreased signal (short T2) on T2-weighted MR images. AJR 148:1247–1250, 1987.
129. Sundaram M, McLeod RA: MR imaging of tumor and tumorlike lesions of bone and soft tissue. AJR 155:817–824, 1990.
130. Swee RG, McLeod RA, Beabout JW: Osteoid osteoma. Radiology 130:117, 1979.
131. Sykes RHD, Wasenaar W, Clark P: Incidence of adverse effects following metrizamide myelography in non-ambulatory and ambulatory patients. Radiology 138:625–627, 1981.
132. Sze G, Krol G, Zimmerman RD, Deck MD: Malignant extradural spinal tumors: MR imaging with Gd-DTPA. Radiology 167:217–223, 1988.
133. Tehranzadeh J, Mnaymneh W, Ghavam C, et al: Comparison of CT and MR imaging in musculoskeletal neoplasms. J Comp Assist Tomogr 13:466–472, 1989.
134. Tehranzadeh J, Murphy BJ, Mnaymneh W: Giant cell tumor of the proximal tibia: MR and CT appearance. J Comput Assist Tomogr 13:282–286, 1989.
135. Thompson GH, Wong KM, Konsens RM, Vibhakar S: Magnetic resonance imaging of an osteoid osteoma of the proximal femur: A potentially confusing appearance. J Pediatr Orthop 10:800–804, 1990.
136. Totty WG, Murphy WA, Lee JKT: Soft-tissue tumors: MR imaging. Radiology 160:135–141, 1986.
137. Tsai JC, Dalinka MK, Fallon MD, et al: Fluid-fluid level: A nonspecific finding in tumors of bone and soft tissue. Radiology 175:779–782, 1990.
138. Uchida M: MR imaging of primary bone and soft tissue tumors. Nippon Igaku Hoshasen Gakkai Zasshi 50:637–648, 1990.
139. Utz JA, Kransdorf MJ, Jelinkek MS, et al: MR appearance of fibrous dysplasia. J Comp Assist Tomogr 13:845–851, 1989.
140. Vanel D, Coffre C, Contesso G, et al: Contribution of gadolinium DOTA in primary tumors of bone and soft tissue. J Radiol 69:735–739, 1988.
141. Vanel D, Lacombe MJ, Couanet D, et al: Musculoskeletal tumors: Follow-up with MR imaging after treatment with surgery and radiation therapy. Radiology 164:243–245, 1987.
142. Watanabe H, Sato T, Hisinuma T, Ogata Y: Comparison of MRI, CT and bone scintigraphy in metastases of experimental neoplasm. Tohoku J Exp Med 163:229–231, 1991.
143. Weinmann HJ, Brasch RC, Press W-R, Wesbey GE: Characteristics of gadolinium-DTPA complex: A potential NMR contrast agent. AJR 142:619–624, 1984.
144. Wetzel LH, Schweiger GD, Levine E: MR imaging of transarticular skip metastases from distal femoral osteosarcoma. J Comput Assist Tomogr 14:315–317, 1990.
145. Winkler K, Beron G, Delling G, et al: Neoadjuvant chemotherapy of osteosarcoma: Result of a randomized cooperative trial (COSS-82) with salvage chemotherapy based on histological tumor response. J Clin Oncol 6:329–337, 1988.
146. Yang CJ, Seabold JE, Gurll NJ: Brown tumor of bone: A potential source of false-positive thallium-201 localization. J Nucl Med 30:1264–1267, 1989.
147. Yuh WTC, Kathol MH, Sein MA, et al: Hemangiomas of skeletal muscle: MR findings in five patients. AJR 149:765–768, 1987.
148. Zimmer WD, Berquist TH, McLeod RA: Magnetic resonance imaging of aneurysmal bone cyst. Mayo Clin Proc 59:633–636, 1984.
149. Zimmer WD, Berquist TH, McLeod RA, et al: Bone tumors: Magnetic resonance imaging versus computed tomography. Radiology 155:709–718, 1985.
150. Zlatkin MB, Lenkinski RE, Shinkwin M, et al: Combined MR imaging and spectroscopy of bone and soft tissue tumors. J Comput Assist Tomogr 14:1–10, 1990.
151. Zucker JM, Henry-Amar M, Sarrazin D, et al: Intensive systemic chemotherapy in localized Ewing's sarcoma in childhood. Cancer 52:1331–1337, 1983.

16 BONE MARROW DISORDERS

Philipp Lang, Harry K. Genant, and Sharmila Majumdar

In evaluating bone marrow, conventional radiography is limited to the assessment of calcific structures associated with bone marrow. Only changes that occur in the cortical bone surrounding the marrow compartment or in the cancellous bone coursing through the marrow can be detected radiographically. Frequently such changes, however, are an indirect reflection of the actual marrow disorder and may be detected only in late stages of the disease, if at all.

Nuclear scintigraphy can provide a functional assessment of bone marrow. Technetium 99m (^{99m}Tc)-methylene diphosphonate scintigraphy serves as a measure of osteoblastic activity and local blood flow[86]; ^{99m}Tc-nanocolloid scintigraphy reflects the activity of the reticuloendothelial system.[19,34,186,228,229] Nuclear scintigraphy using ^{59}Fe can be employed to elucidate iron metabolism in bone marrow.[34,174] However, no matter what isotope is used, nuclear scintigraphy does not provide a detailed morphologic assessment of the marrow compartment owing to its low spatial resolution. In addition, nuclear scintigraphy lacks specificity in evaluating marrow disorders.

Computed tomography (CT) has not proved useful in evaluating marrow disorders because the differences in CT attenuation between normal and pathologic marrow are relatively small. In addition, the dense cortical bone surrounding the marrow compartment frequently causes beam hardening and streak artifacts, which render assessment of the marrow structure difficult.[21,89] Marrow involvement is more likely to be detected by CT scanning when it is asymmetric; however, most myeloproliferative diseases (e.g., lymphoma) are diffuse and symmetric except for rare cases.

Magnetic resonance (MR) imaging represents a unique diagnostic method for noninvasive evaluation of normal bone marrow and marrow disorders. It is characterized by excellent spatial and contrast resolution and affords direct multiplanar image acquisition. This technique has evolved as an important clinical tool for evaluating red and yellow marrow distribution in pediatric and adult patients[33,94,145,151,158,176] and for detecting and staging marrow disorders.[11,22,28,31,54,68-71,76,85,88,105,107,110,111,134,156,159,161,173,188,196,206,211] Magnetic resonance imaging also can be used to monitor the effects of various therapeutic regimens on bone marrow.[8,13,56,66,80,97,101,143,182,198,202-204,212,232] In this chapter, we describe the normal anatomy and physiology of bone marrow. Factors that influence marrow signal intensity and contrast, and MR imaging techniques for assessing bone marrow, also are discussed. An overview of MR imaging of the most important bone marrow disorders is provided. The MR imaging appearance of fractures, osteomyelitis, and bone marrow ischemia is described in Chapters 7, 10, 11, and 12.

FACTORS THAT INFLUENCE SIGNAL INTENSITY AND IMAGE CONTRAST IN THE MAGNETIC RESONANCE IMAGING OF BONE MARROW

Bone marrow in the healthy adult consists of a combination of hematopoietic (red) and fatty (yellow) marrow in conjunction with other cellular components, such as plasma cells, reticuloendothelial cells, osteoblasts, osteoclasts, mast cells, and fibroblasts[47,227] (Fig. 16–1).

The three major tissue constituents of the bone marrow compartment are fat, water, and bone. Fat is characterized by a high proton density, a very short T1 relaxation time, and a moderately short T2 relaxation time.[37,18.] Consequently, fatty tissue demonstrates high signal intensity with T1 weighting and intermediate to high signal intensity with T2 weighting. Water has long T1 and T2 relaxation times resulting in low signal intensity on T1-weighted and high signal intensity on T2-weighted spin-echo images.[84] Cortical and trabecular varieties of bone lack mobile protons and are of low signal intensity on all imaging sequences. Variations in the relative concentration of each component have a direct effect on the imaging appearance of the marrow.

Fatty marrow contains 15 per cent water, 80 per cent fat, and 5 per cent protein.[201,220] Hematopoietic marrow is composed of 40 per cent water, 40 per cent fat, and 20 per cent protein.[201,220] As a result of the greater water content, as well as not yet clearly understood relaxation processes related to the dissimilarity in tissue composition[220] (e.g., protein content and cellularity), red marrow has longer T1 relaxation times than yellow marrow[37,95] and displays lower signal intensity than yellow marrow on T1-weighted spin-echo images (Figs. 16–2, 16–3). On T2 weighting, fatty marrow remains of intermediate to high signal intensity, wherease hematopoietic marrow increases in signal intensity owing to its greater water content and thus longer T2 relaxation time; as a result, it approximates the intensity of fatty marrow[220] (see Fig. 16–2*B*). Differentiation between fatty and hematopoietic marrow is poor on T2-weighted spin-echo MR images in comparison to T1-weighted scans (see Fig. 16–2*B*).

Tissue water exists in different states in bone marrow: bound, structural, and bulk (i.e., free). The three states have different relaxation rates.[7,53,79,91,141] Tissues that have a greater concentration of bound water generally show short T1 and T2 relaxation rates.[144] Bulk or free water has long T1 and T2 relaxation times and causes low signal intensity on T1-weighted and high signal intensity on T2-weighted images. Most pathologic conditions of bone marrow produce an increase in free water with resultant high signal intensity on T2 weighting.[91] In situations in which the free water decreases, signal intensity on T2-weighted images is low.

Inhomogeneity in magnetic susceptibility also may have a profound effect on the MR imaging appearance of bone marrow. Magnetic susceptibility is defined as the ratio between the induced magnetic field in the tissue and the applied external field; it is a measure of the response of a given tissue to an applied constant field. Trabecular bone and the surrounding bone marrow have significantly different magnetic properties. These differences lead to distortions of the magnetic lines of force with resultant local inhomogeneity in the magnetic field. Magnetic susceptibility–generated field inhomogeneities and resultant T2*-shortening become greater with an increase in the concentration of solid inclusions (e.g., trabeculae) in a homogeneous medium (e.g., marrow). Thus, in patients with a dense trabecular network, the trabecular bone induces marked T2*-shortening and signal loss.[193] Gradient-echo se-

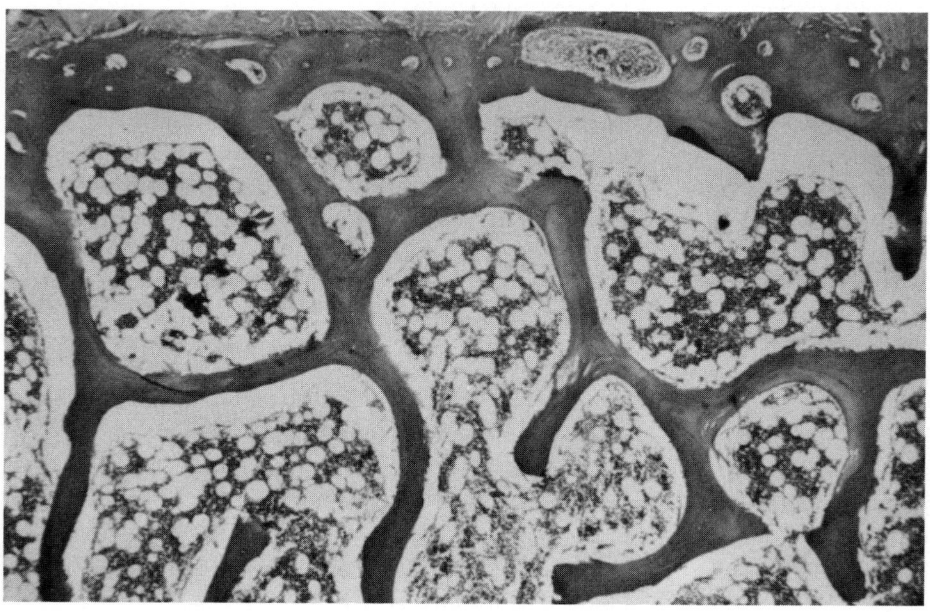

Figure 16–1. Photomicrograph of normal bone marrow (hematoxylin-eosin). Bone marrow is located in the intertrabecular spaces. The multiple round "empty spaces" represent fatty marrow cells (the fat is dissolved during the fixation process). Other cellular components of bone marrow (e.g., hematopoietic, reticuloendothelial, lymphoid, and plasma cells) are interspersed among the fat cells.

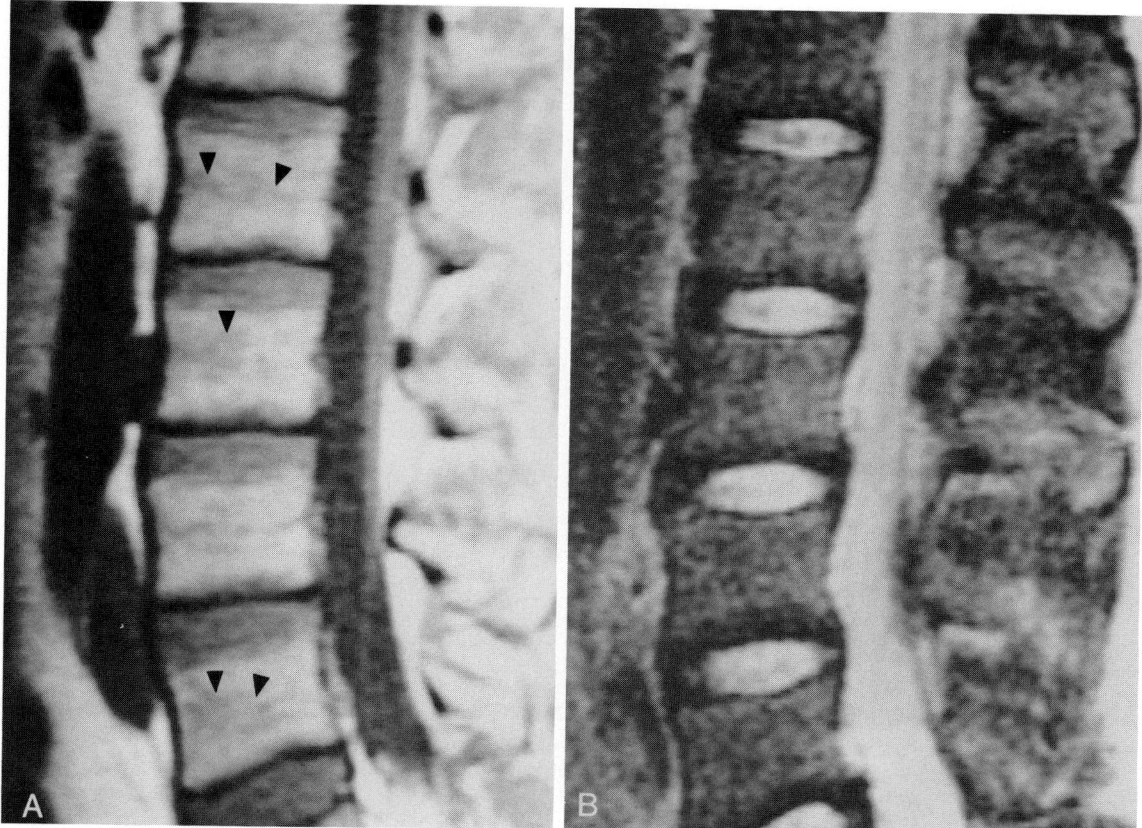

Figure 16–2. Normal lumbar spine in an adult patient. *A*, T1-weighted spin-echo MR image (TR, 800 ms; TE, 20 ms) demonstrates areas of high signal intensity within the vertebral bodies, reflecting fatty bone marrow and areas of intermediate signal intensity corresponding to hematopoietic marrow *(arrowheads)*. Note that hematopoietic marrow is lower in signal intensity than fatty marrow on T1 weighting. *B*, T2-weighted spin-echo MR image (TR, 2000 ms; TE, 80 ms). Fatty marrow and hematopoietic marrow show similar signal intensity, resulting in poor differentiation.

quences are more sensitive to changes in magnetic susceptibility and resultant T2*-decay than conventional spin-echo sequences, as there is no rephasing of local magnetic field inhomogeneities by a 180-degree pulse in gradient-echo imaging and T2*-decay is not recovered (see Fig. 16–3). The T2* effect is increased in areas with more trabecular bone (e.g., epiphysis) over that in areas with less trabecular bone (e.g., diaphysis, metaphysis).[193] A low signal intensity on gradient-echo sequences may thus represent fatty marrow in an area with a high content of trabecular bone and should not be mistaken for hematopoietic marrow[193] (see Fig. 16–3).

Iron is stored in bone marrow in the form of ferritin and hemosiderin. Iron stores exert superparamagnetic effects that also cause inhomogeneity in magnetic susceptibility with resultant dephasing of spins and shortening of T2 and even more so of T2*.[62,63,192] In patients with pathologic conditions that produce increased iron deposition in the marrow, such as hemosiderosis, the bone marrow appears to have low signal intensity on T1-weighted and T2-weighted spin-echo or T2*-weighted gradient-echo sequences (Fig. 16–4).

Fat and water protons have a small difference in resonance frequency (i.e., chemical shift). This resonance difference is approximately 3.5 ppm or 52 to 225 Hz for imaging systems operating in the range of 0.35 T to 1.5 T.[36,231] When the resonance difference occurs in neighboring voxels at tissue interfaces, it can result in a spatial misregistration of resonating spins (i.e., chemical shift artifact). However, intravoxel chemical shift may be used to increase tissue contrast and even to generate fat and water selective images using chemical shift imaging techniques (see following section). In a conventional image, aliphatic and water protons are in phase at the time of the signal readout, and the signal intensity of a given voxel is formed by the sum of signals from aliphatic and water protons in that voxel. When the aliphatic and water protons are in opposed phase (i.e., in an opposed-phase image), however, the net signal intensity of a voxel represents the difference of signals arising from aliphatic and water protons. Images with differing intensity characteristics may be generated depending on the echo times selected, the relative phase of the fat and water signal, and the magnetic field strength.

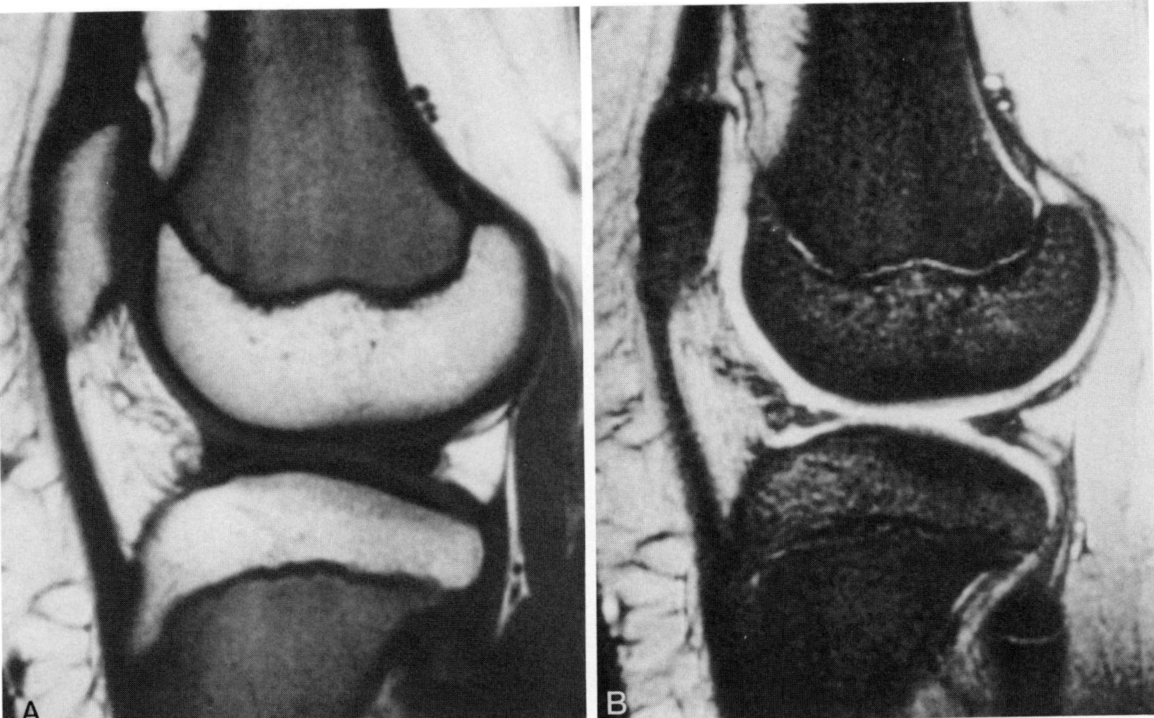

Figure 16–3. Normal hematopoietic marrow in a 9-year-old boy in the distal femoral and the proximal tibial metaphysis. *A*, T1-weighted MR image (TR, 600 ms; TE, 20 ms) shows hematopoietic marrow of intermediate to low signal intensity in the distal femoral and the proximal tibial metaphyses. The distal femoral and the proximal tibial epiphyses show high signal intensity, corresponding to fatty marrow. *B*, T2*-weighted gradient-echo MR image (TR, 500 ms; TE, 30 ms; flip angle, 30 degrees). Both epiphyses and metaphyses show low signal intensity with this imaging sequence. Inhomogeneous magnetic susceptibility at the interface between marrow and trabecular bone causes dephasing of spins with T2*-shortening and resultant low signal intensity in the epiphysis and metaphysis. Hematopoietic and fatty marrow cannot be differentiated on the gradient-echo sequence. Low signal intensity of the epiphyseal fatty marrow on gradient-echo sequences due to inhomogeneous magnetic susceptibility should not be mistaken for hematopoietic marrow.

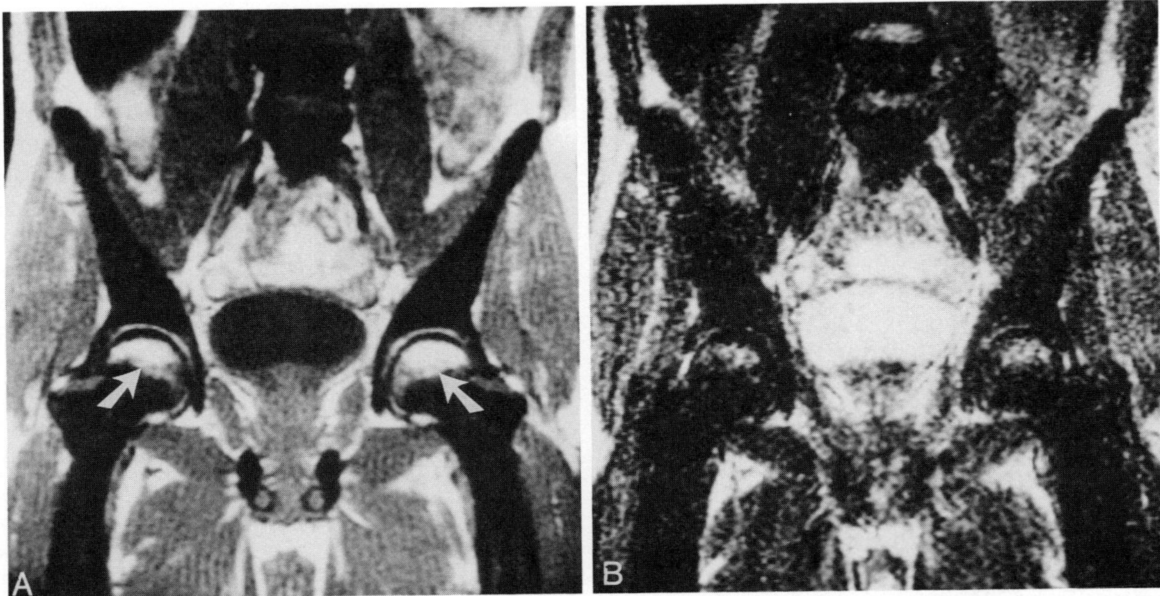

Figure 16–4. Hemosiderosis in a patient with sickle cell anemia. *A*, T1-weighted spin-echo MR image (TR, 800 ms; TE, 20 ms) of the pelvis demonstrates very low signal intensity of the bone marrow in the femurs, the pelvis, and the lumbar spine. Femoral epiphyses are spared *(arrows)*. *B*, T2-weighted spin-echo MR image (TR, 2000 ms; TE, 120 ms). The bone marrow also has very low signal intensity with this imaging sequence. The low signal intensity on both T1- and T2-weighted sequences is caused by intramedullary iron deposits. (Courtesy of Phoebe Kaplan, M.D.)

IMAGING TECHNIQUES

Pulse Sequences

Spin-echo scans with short repetition time (TR) (400 to 800 ms) and short echo time (TE) (15 to 40 ms)—that is, T1-weighted spin-echo images—are employed most frequently for evaluating bone marrow disease. T1-weighted spin-echo sequences have a reasonably short acquisition time (Table 16–1) and provide good image quality, which makes them ideally suited as a screening test for bone marrow disease. Fatty marrow appears bright on T1-weighted spin-echo scans. Neoplastic processes and most other lesions are of low signal intensity on T1-weighted images, resulting in good image contrast. Because the T1 relaxation time of fatty marrow increases with higher field strength to a lesser extent than does water-containing tissue,[102] greater contrast between marrow types may be noted with high field strength than with low field strength.[145] However, because T1 increases overall at higher field strength, a longer TR is needed on a high field magnet.

Proton density images (i.e., images with long TR [2000 to 2500 ms] and short TE [15 to 25 ms] have high signal-to-noise ratios and good image quality. However, proton density images are primarily sensitive to changes in spin density and are relatively insensitive to changes in T1 and T2 relaxation times. Because changes in spin density are relatively small in comparison to changes in T1 and T2 relaxation times in imaging bone marrow, proton density images generally have lower contrast than T1- and T2-weighted spin-echo sequences in imaging pathologic marrow.

T2-weighted spin-echo sequences—i.e., images obtained with long TR (2000 to 2500 ms) and long TE (80 to 180 ms)—have lower red-yellow marrow contrast than T1-weighted spin-echo images (see Fig. 16–2, Table 16–1). Fatty marrow is of intermediate signal intensity on T2-weighted spin-echo images. Hematopoietic marrow increases in signal intensity on T2 weighting over that seen on T1-weighted images and thus approximates the signal intensity of fatty marrow (see Fig. 16–2). However, T2-weighted spin-echo images provide improved specificity in imaging pathologic conditions of bone marrow: An area that is of low signal intensity on T1-weighted images located within fatty marrow (high signal intensity) may represent hematopoietic marrow as well as a pathologic (e.g., neoplastic) lesion. Hematopoietic marrow has a signal intensity similar to that of the surrounding fatty marrow on T2 weighting owing to similar T2 relaxation times (see Fig. 16–2). A neoplastic process, however, in most cases demonstrates signal intensity greater than that of fatty marrow on T2 weighting. However, if a lesion has a T2 relaxation time similar to that of fat or if a sequence with a more mixed weighting is used, differentiation of some pathologic processes from fat may be difficult. T2-weighted spin-echo images are limited by rather long acquisition times (see Table 16–1) and relatively poor signal-to-noise ratios.

Table 16–1. PULSE SEQUENCES FOR IMAGING OF BONE MARROW (AT 1.5 T).

Pulse Sequence	Acronym	TR (ms)	TE (ms)	Echoes	TI (ms)	Flip Angle (°)	Matrix	NEX	Slice Thickness (mm)	Skip (mm)	Imaging Time (min)
Precontrast											
Spin-echo (T1-weighted)	SE	400–800	15–40	1	—	—	256 × 192 256 × 256	1–2	5	0–5	3–5
Spin-echo (T2-weighted)	SE	2000–2500	80–180	1–4	—	—	256 × 192 256 × 256	1–2	5	0–5	10–20
Gradient-echo ("T1-weighted")	GRE	300–400	15–20	1	—	90	256 × 192 256 × 256	1–2	5	0–5	3–5
Gradient-echo (T2*-weighted)	GRE	500–600	30	2	—	30	256 × 192 256 × 256	1–2	5	0–5	4–7
Short tau inversion-recovery	STIR	1500	20–30	1	170–180	—	256 × 192 256 × 256	1–2	5	0–5	5–10
Postcontrast (gadopentetate-dimeglumine)											
Spin-echo (T1-weighted)	SE	400–800	15–40	1	—	—	256 × 192 256 × 256	1–2	5	0–5	3–5
Gradient-echo ("T1-weighted")	GRE	300–400	15–20	1	—	90	256 × 192 256 × 256	1–2	5	0–5	3–5

TR, Repetition time; TE, echo time; NEX, number of excitations; SE, spin echo; GRE, gradient-recalled echo; STIR, short tau inversion-recovery.

Several fat and water selective imaging techniques have been proposed to date.[36,59,104,135,181,210] Most of these techniques are based on the difference in resonance frequency between aliphatic and water protons (i.e., chemical shift). If the difference in resonance frequency between the aliphatic and water protons in tissue is represented by Δf, the signal intensity shows a cyclic oscillation (Fig. 16–5), going through maximal and minimal values of signal intensity at a period given by the following equation:

$$T = 1000/\Delta f \text{ ms}$$

This implies that the time interval between two maximal points in the signal intensity value is T, and the maxima and minima occur at intervals of T/2. The signal is maximal when aliphatic and water protons are in phase; the signal is minimal when they are in opposed phase. For a separation of 3.5 ppm in resonance frequencies at 0.35 T, the time period of such oscillations is 19.2 ms. At 1.5T, the time period of such oscillations is 4.5 ms.

In the Dixon technique,[36,131,231] images of fat and water may be generated by postprocessing two acquired images in which the fat and water protons are in phase and 180 degrees out of phase. These images are produced by simple modifications of the pulse separations in spin-echo sequences. The out-of-phase image is obtained by applying the 180-degree refocus pulse at a time $TE/2 + \Delta t$, where Δt in seconds is given by $1/4\Delta f$, with Δf being the separation between the fat and water peaks in hertz.

The hybrid sequence combines the Dixon technique of chemical shift imaging with a presaturating radiofrequency pulse applied before the excitation pulse.[210]

One description of a three-point Dixon technique[59] involves producing three images, one in which the fat and water are in phase, one in which the fat and water components are 180 degrees out of phase, and one in which the fat and water components are −180 degrees out of phase. The images in which fat and water are 180 degrees out of phase are obtained as before by the application of the 180-degree refocusing pulse at a time $TE/2 + \Delta t$, whereas the image in which fat and water are −180 degrees out of phase is obtained by applying the 180-degree radiofrequency pulse at a time $TE/2 - \Delta t$. In this imaging sequence, it is necessary to calculate, in addition to the magnitude, in- and out-of-phase images, phase-sensitive in-phase images, and 180-degree and −180-degree out-of-phase images. On the basis of the magnitude and phase reconstructions of these three sets of images, fat and water images may be generated.[59,135]

Numerous other techniques exist for generating "fat only" and "water only" images. These include saturation of fat or water by chemically selective radiofrequency pulses followed by "spoiler" gradients,[103,108] selected refocusing of transverse magnetization,[149] and alteration of the polarity of the section select gradient.[64] Fat saturation techniques have been widely applied to the study of bone marrow,[66,67,70,180,181,231] and some investigators believe that they provide improved discrimination between healthy and pathologic bone marrow.[180]

Short tau inversion-recovery sequences (STIR) represent an additional technique to suppress the signal of fatty tissues. Fatty tissue (e.g., bone marrow) is of low signal intensity on a STIR image. However, objects with long T1 and T2 relaxation times (e.g., neoplasm, bone infarcts, infiltrative disorders) produce high signal intensity with this imaging sequence[76,93,165] (Fig. 16–6). Involvement of the medullary cavity is well demonstrated owing to the low signal intensity of the normal marrow and the high signal intensity of the pathologic process. In using STIR sequences, the inversion time is chosen so that it is identical to the cross-over time of fat. After the 180-degree inversion pulse, fat crosses over the zero point before water. If a 90-degree pulse is applied at the time when fat crosses over the zero point, net magnetization of water is tilted to the transverse plane. A subsequent 180-degree pulse refocuses signal from water protons only resulting in effective fat suppression. Because T1 increases with magnetic field strength,[102] longer TR and longer inversion times are needed with high field systems than with low or midfield magnets.

Acquisition times of STIR sequences are longer than those of T1-weighted spin-echo images or T2*-weighted gradient-echo sequences. Also STIR images are frequently limited by a relatively low signal-to-noise ratio. Short TR, short TE T1-weighted spin-

Figure 16–5. Reconverted hematopoietic marrow in the distal femur: echo time–dependent oscillation of yellow and red marrow signal. *A*, Gradient-echo sequence (TR, 250 ms; flip angle, 90 degrees) with TE varying from 10 to 21 ms at 1 ms increments. Plot of the red and fatty marrow signal intensities (in arbitrary units) against the TE. As illustrated in the MR images (*B–E*), the signal intensity of fatty and red marrow oscillates periodically depending on the selected echo time. This oscillation results from the difference in resonance frequency of aliphatic and water protons. *B–E*, Gradient-echo sequence (TR, 250 ms; flip angle, 90 degrees) with TE varying from 10 to 21 ms at 1 ms increments (*B*, 11 ms; *C*, 14 ms; *D*, 16 ms; *E*, 18 ms). The contrast between red and fatty marrow depends heavily on TE. Maximal contrast is obtained when aliphatic and water protons are in opposed-phase at TE 11 ms (*B*) and 16 ms (*D*), thereby canceling their signal. Note the test tubes in *B* containing fat (left, 1) and water (right, 2) solutions. (From Lang P, Fritz R, Majumdar S, et al: Hematopoietic bone marrow in the adult knee: Spin-echo and opposed-phase gradient-echo MR imaging. Skel Radiol 22:95–103, 1993.)

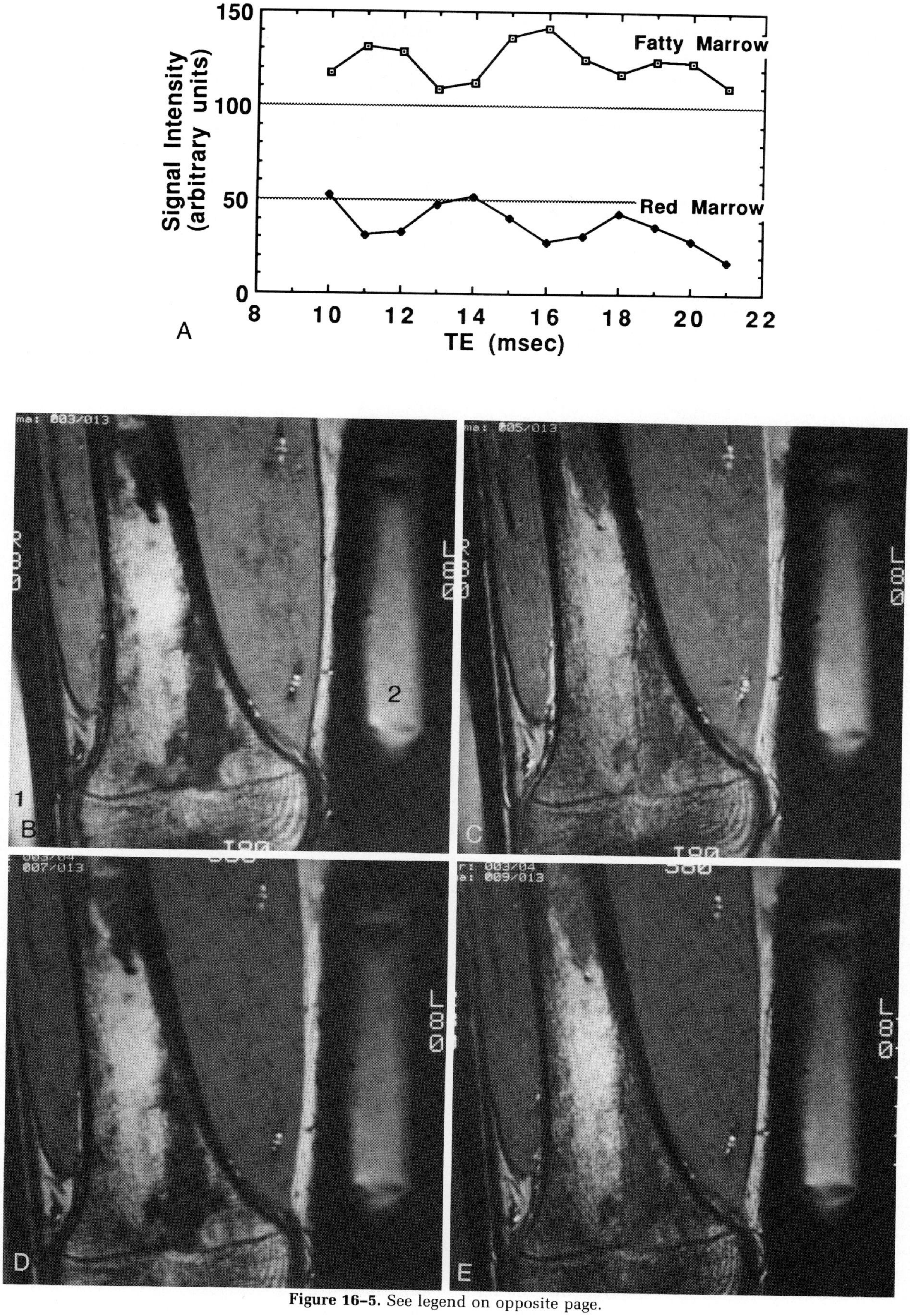

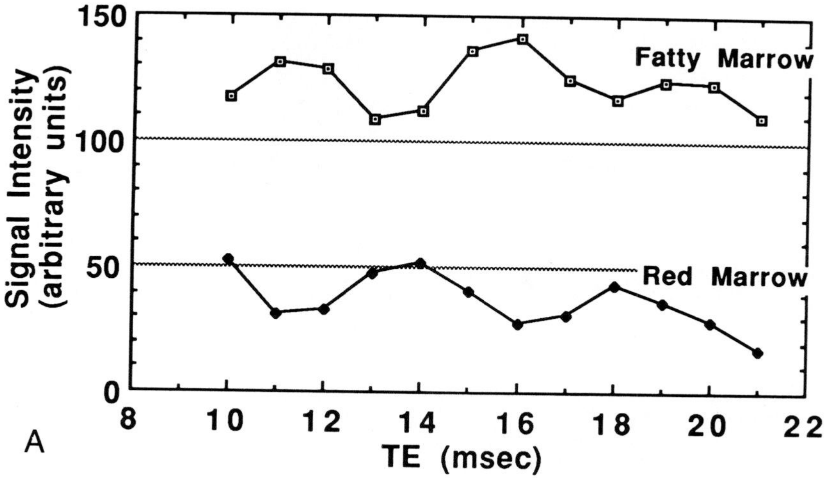

B C D E

Figure 16–5. See legend on opposite page.

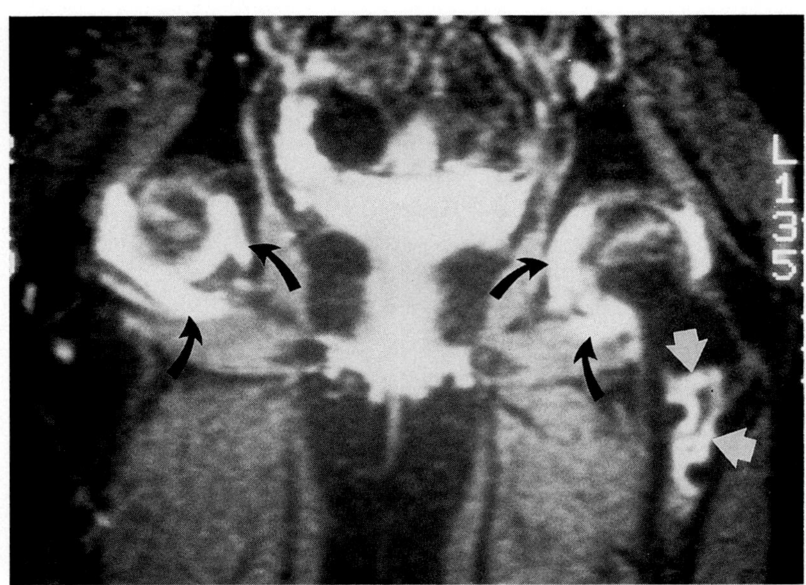

Figure 16–6. Image of sickle cell anemia with bone infarcts and joint effusions. Short tau inversion-recovery (STIR) image of normal bone marrow shows low signal intensity. Bone infarcts *(straight arrows)* and joint effusions *(curved arrows)* demonstrate high signal intensity, resulting in good image contrast.

echo images provide greater spatial resolution than STIR images and therefore a better depiction of anatomic details.[61] Some researchers recommend the combined use of short TR, short TE T1-weighted spin-echo scans and STIR images, thereby capitalizing on the advantages of both (i.e., excellent anatomic detail on T1-weighted spin-echo scans and superior contrast on the STIR sequence).[61]

Gradient-echo sequences with a relatively short TR (200 to 300 ms), a short TE (10 to 15 ms), and a large flip angle (80 to 90 degrees) (T1-weighted) provide contrast similar to that of conventional T1-weighted spin-echo sequences[46,73] (see Table 16–1). They may be used to replace T1-weighted spin-echo sequences. These sequences, however do not significantly reduce imaging time in comparison to T1-weighted spin-echo sequences[46] (see Table 16–1).

Intravoxel chemical shift effects may have a significant impact on the signal intensity and contrast of red and yellow marrow on gradient-echo images. By selecting an appropriate TE, signals can be recorded at a time when water and fat protons are oriented in opposite directions (see earlier discussion), resulting in an opposed-phase gradient-echo image (see Fig. 16–5). In an opposed phase gradient-echo image, voxel brightness represents the net difference between water and fat magnetization. Signal of fatty marrow is only slightly reduced in the opposed-phase gradient-echo image, because fatty marrow contains only a little water.[201,220] Hematopoietic marrow, however, contains approximately equal amounts of water and fat[201,220]; the net difference of water and fat magnetization is therefore markedly lower in hematopoietic marrow than in fatty marrow, resulting in reduced signal intensity on opposed-phase gradient-echo images. Contrast between hematopoietic and fatty marrow is thereby significantly more pronounced on opposed-phase gradient-echo sequences with short TR, short TE, and large flip angles than on in-phase gradient-echo images or T1-weighted spin-echo images[118-120,125] (see Fig. 16–5*B*, *D*). Opposed-phase gradient-echo sequences can easily be implemented in existing scanner protocols.[118-120,125]

T2*-weighted gradient-echo sequences may be used instead of T2-weighted spin-echo sequences in imaging bone marrow (see Table 16–1). T2*-weighted gradient-echo sequences have shorter acquisition times than conventional T2-weighted spin-echo sequences. However, gradient-echo sequences are more subject to artifacts resulting from inhomogeneous magnetic susceptibility than are spin-echo images because no rephasing occurs in gradient-echo imaging by a 180-degree pulse and T2*-decay is not recovered. Because T2*-weighted gradient-echo sequences have longer echo times than gradient-echo sequences with predominantly T1 contrast (T1-weighted) (see Table 16–1), they are more sensititve to inhomogeneity in magnetic susceptibility. Differentiation between hematopoietic and fatty marrow thus may not always be possible owing to spin dephasing and T2* shortening at the interface between trabecular bone and marrow (i.e., the presence of trabecular bone suppresses signal). For the same reason, pathologic conditions (e.g., hemosiderosis) may be obscured on T2*-weighted gradient-echo images (Fig. 16–7).

Fast spin-echo or rapid acquisition relaxation enhanced (RARE) sequences[49,81,82,154] are characterized by very fast image acquisition times. They

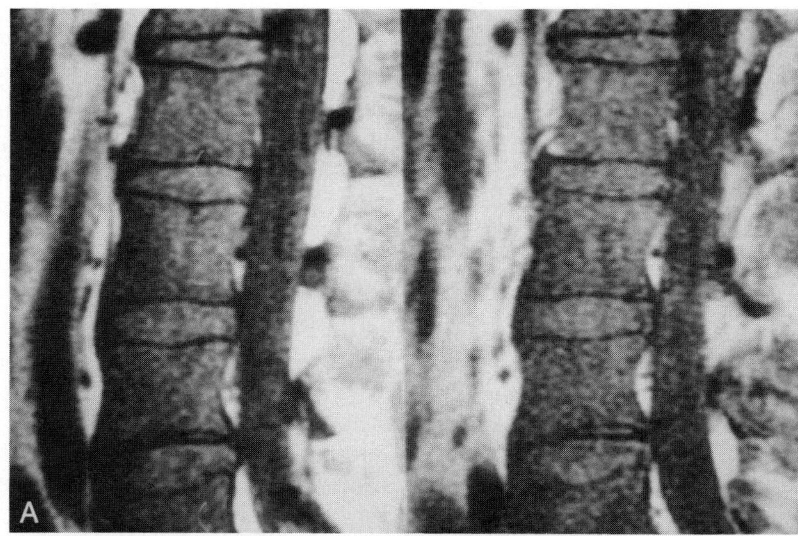

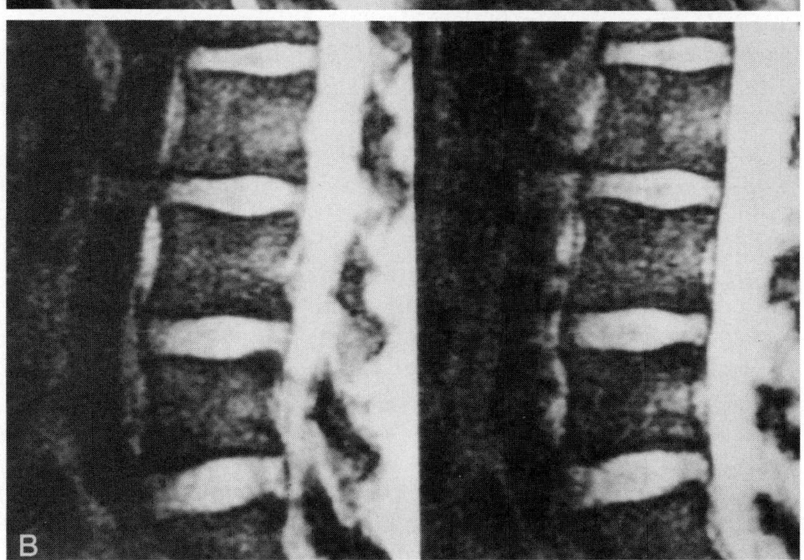

Figure 16–7. Magnetic resonance image of the spine in an adult patient with acquired immunodeficiency syndrome (AIDS). *A*, T1-weighted spin-echo MR images (TR, 1000 ms; TE, 20 ms). The vertebral bodies show mottled low signal intensity. The signal intensity of the vertebral bodies is markedly reduced in comparison to that of a normal spine (see Fig. 16–2). The low signal is thought to be related to increased amounts of storage iron within the bone marrow, probably secondary to the anemia of chronic disease.[57] *B*, T2*-weighted gradient-echo MR images (TR, 600 ms; TE, 35 ms; flip angle, 30 degrees). The vertebral marrow is of uniformly low signal intensity. However, the pathologic low signal intensity identified on the T1-weighted spin-echo image is not readily depicted on the T2*-weighted gradient-echo sequence, as the low signal intensity also may be the result of inhomogeneous magnetic susceptibility at the interface between trabecular bone and marrow, with resultant spin dephasing, T2*-shortening, and signal loss.

may also be used to evaluate pathologic conditions of bone marrow. Preliminary experience indicates that T2-weighted fast spin-echo techniques provide relatively high fatty marrow signal. The conspicuousness of marrow-based lesions that also are of high signal on T2 weighting thus may be decreased. However, this problem can be solved with use of a fat-saturated T2-weighted fast spin-echo technique.

Imaging Coils

In imaging patients with disorders or lesions of bone marrow, a coil should be chosen that affords maximal signal-to-noise ratio for the anatomic region in question. The use of small fields of view are recommended to obtain high spatial resolution. However, the field of view must be large enough to cover the margins of the abnormality completely. In imaging pathologic changes of the bone marrow of the lumbar or thoracic spine, a license plate–shaped surface coil provides both high spatial resolution and good signal-to-noise ratio. However, the fields of view available with the license plate–shaped surface coil are not large enough to allow imaging of both the thoracic and lumbar or the cervical and thoracic spine simultaneously. Thus, when multilevel disease is suspected, imaging should be performed with a phased-array spine coil or, when this is not available, the body coil. The body coil also is useful for imaging the trunk and the thighs. The distal femur and the proximal tibia are best imaged with a circumferential knee coil. For evaluating the distal tibia, both the knee coil and surface coils can be employed. The ankle joint and the foot generally are best imaged using a surface coil. For imaging the shoulder joint, a designated shoulder coil is used.

Various designs, such as loop gap resonators or circular surface coils, are available for this purpose.

Magnetic Resonance Imaging Contrast Media

In selected cases, additional information about pathologic marrow lesions may be obtained by generating MR images after administration of a contrast medium.[24,76,126,191,194,205,208,215,219] The most commonly employed MR imaging contrast agent and the only one that is currently approved by the U.S. Food and Drug Administration is gadopentetate dimeglumine (gadolinium-diethylenetriamine penta-acetic acid). Gadopentetate dimeglumine is a strongly paramagnetic complex that reduces hydrogen proton relaxation times.[15,16,226] The standard dose for use of gadopentetate dimeglumine in humans is 0.1mmol/kilogram of body weight.

Other potential MR imaging contrast agents for imaging bone marrow and bone marrow lesions are gadolinium tetra-azacyclododecane tetra-acitic acid (DOTA),[219] a gadolinium complex with contrast behavior similar to that of gadopentetate dimeglumine; nonionic gadolinium complexes; and ultra-small superparamagnetic iron oxide particles.[194]

Following administration of gadopentetate dimeglumine, T1-weighted spin-echo sequences or gradient-echo images with a short TR, short TE, and large flip angle (T1-weighted) are acquired. These sequences may be supplemented with T1-weighted fat saturation techniques. In the adult, normal fatty marrow and hematopoietic marrow demonstrate typically only subtle contrast enhancement. The degree of enhancement depends on the amount of marrow vascularity. Benign or malignant lesions of the bone marrow will demonstrate marked contrast enhancement in the majority of cases.[5,76,109,128,205,218] Sometimes the areas of contrast enhancement are difficult to differentiate from the high signal intensity of the adjacent marrow. In this situation, marrow signal can be suppressed using a fat saturation technique, resulting in greater contrast between the enhancing lesion and the suppressed marrow, more of low signal intensity.[124,126]

In children, particularly those who are less than 2 years old, marked enhancement can be observed in the bone marrow of the spinal column after intravenous administration of gadopentetate dimeglumine.[208] Enhancement of the bone marrow in children is thought to be due to the unusual prominence of the vasculature, in association with the permeability of the capillary endothelium and the plentiful extravascular space.[208] Although marked and diffuse enhancement of vertebral bodies in adults often is thought to indicate a pathologic marrow state, caution must be used before the same criteria are applied to children.

Magnetic resonance imaging after administration of gadopentetate dimeglumine also may help to differentiate between benign and malignant spinal fracture[26] (see the section on Metastatic Disease). An additional application for MR imaging after gadopentetate dimeglumine administration may be the detection of early marrow ischemia. Cova and associates and Tsukamoto and colleagues performed animal experiments in which they occluded the arterial[24] or the venous[215] supply to the proximal femur. After arterial and venous ligation, they observed a decreased degree of contrast enhancement of the bone marrow compared with the preligation MR images.[24,215] Lang and colleagues performed a study in human subjects in which they obtained pre- and postcontrast MR images in patients who had sustained a fracture of the femoral neck.[126,127] Findings from MR imaging were correlated to superselective digital subtraction angiograms and to clinical-radiographic follow-up data. In patients with impaired bone marrow perfusion, contrast enhancement was not observed in the femoral head on the fractured side; the femoral head on the unaffected side, however, demonstrated normal contrast enhancement.[128,129] In patients with persistent perfusion, uniform contrast enhancement was observed in the femoral head, neck, and shaft on both the fractured and the healthy side.[126,127] Although larger studies are needed to validate the animal experiments and the preliminary studies in human subjects, it appears that MR imaging after intravenous administration of gadopentetate dimeglumine may be employed to detect early bone marrow ischemia.

In addition to static imaging, rapid scans may be obtained in which images are acquired sequentially after bolus administration of the contrast agent. This technique is designed to assess the dynamics of contrast enhancement quantitatively.[44,46,129,171,172] Most commonly, fast gradient-echo sequences are employed for this application. Acquisition time should range between 5 and 10 sec per scan. A single-slice technique is used. A single image is generated before administration of the contrast medium. Gadopentetate dimeglumine is injected in a bolus fashion and sequential scan acquisition is started immediately after the injection.[44,46,129,171,172] Receiver and transmitter attenuations of pre- and postcontrast images are kept constant, thereby permitting quantitative assessment of signal intensities.

The percentage of increase in signal intensity ($SI_{increase}$) over the baseline intensity is calculated as follows:

$$SI_{increase} = (SI_{max} - SI_{prior}) \times 100/SI_{prior}$$

in which SI_{max} is the maximal signal intensity of the tissue after intravenous application of gadopentetate dimeglumine and SI_{prior} is the signal intensity before injection.

The slope of the curve is calculated as the percentage of increase of signal intensity over the baseline value of SI_{prior} per minute:

$$Slope = (SI_{max} - SI_{prior}) - 100/(SI_{max} \times T_{max})$$

in which T_{max} is the time to reach SI_{max}.

At the present time, this technique has not yet been applied to the evaluation of diffuse bone marrow lesions. However, studies on musculoskeletal tumors[44,45] have shown that the percentage of increase in signal intensity and the slope values of the enhancement curves may help to better characterize lesions, to differentiate between benign and malignant lesions, and to monitor response to chemotherapy or radiation therapy. The same may hold true for imaging other pathologic conditions of bone marrow.

AGE-DEPENDENT CHANGES IN THE DISTRIBUTION OF FATTY AND HEMATOPOIETIC MARROW

The distribution of hematopoietic and fatty marrow is a dynamic process that is influenced by biologic age, sex, erythrocyte demand, pathologic marrow conditions, and other factors.[113,220] Knowledge of the usual age distribution of red and yellow marrow is of clinical importance, as hematopoietic marrow is the site of origin of processes that are related to red marrow vascularity[113] (e.g., marrow infarction, osteomyelitis, and metastases) and of lesions that arise from the reticulum cell or its derivatives[113] (e.g., reticulum cell sarcoma, multiple myeloma, Ewing's sarcoma, and other lesions).

Myeloid hematopoiesis commences in the fourth intrauterine month and supplements and eventually replaces blood cell formation by the liver. At the time of birth, bone marrow is fully responsible for red blood cell production, and all bone marrow cavities are actively engaged in hematopoiesis[47,113] Conversion from red to yellow marrow begins in the terminal phalanges of the feet and probably the hands shortly before birth.[42] The conversion from hematopoietic to fatty marrow is a steady, gradual process that progresses from distal to proximal. The replacement occurs first in the diaphysis of the peripheral long bones, then slowly expands centripetally. Marrow conversion is more rapid in the distal long bones than in the proximal long bones, although the overall pattern of conversion starting in the diaphysis is similar.[113] By the age of 12 to 14 years, fat will be present macroscopically in the midshaft of all long bones.

At the time of birth, the cartilaginous epiphyses do not contain hematopoietic marrow. However, hematopoietic marrow is found in varying concentration in the epiphyses once they ossify.[113] By the age of 7 years, fat is observed macroscopically in the distal epiphyses of the distal long bones.[164] With puberty, the epiphyses of the distal long bones are composed of significantly more fat than the femoral and humeral epiphyses, although all of them are predominantly composed of fatty marrow (Fig. 16–8).

An adult pattern of marrow distribution is reached by the age of 25 years. In this pattern, red marrow is found macroscopically in the vertebrae, sternum, ribs, pelvis, skull, and proximal shafts of the femora and humeri.[77,113] The epi-, meta-, and diaphyseal marrow spaces of the more peripheral bones are of uniform high signal intensity on T1 weighting, corresponding to homogeneous fatty marrow. However, variations from this adult pattern can be observed in clinically and hematologically normal patients (Fig. 16–9).

Knowledge of the normal pattern of red and yellow marrow on MR imaging and its age variations in the different parts of the body is essential before pathologic processes in the marrow can be identified.

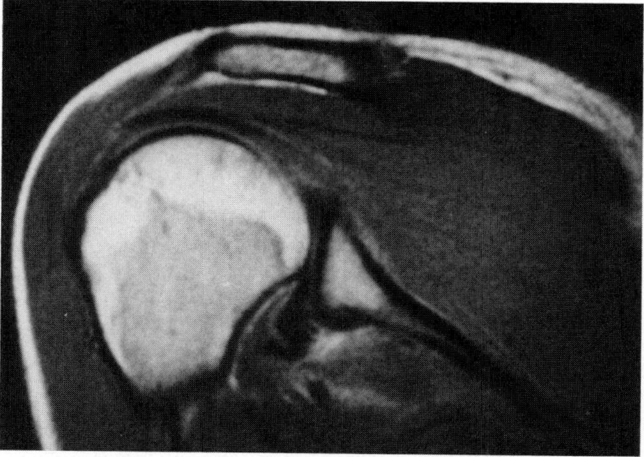

Figure 16–8. Distribution of fatty and hematopoietic marrow in the proximal humeral epiphysis and metaphysis during adolescence (17-year-old patient). An oblique coronal T1-weighted spin-echo MR image (TR, 600 ms; TE, 20 ms) shows intermediate signal intensity in the proximal humeral metaphysis, reflecting hematopoietic marrow. The epiphysis shows high signal intensity, corresponding to fatty marrow.

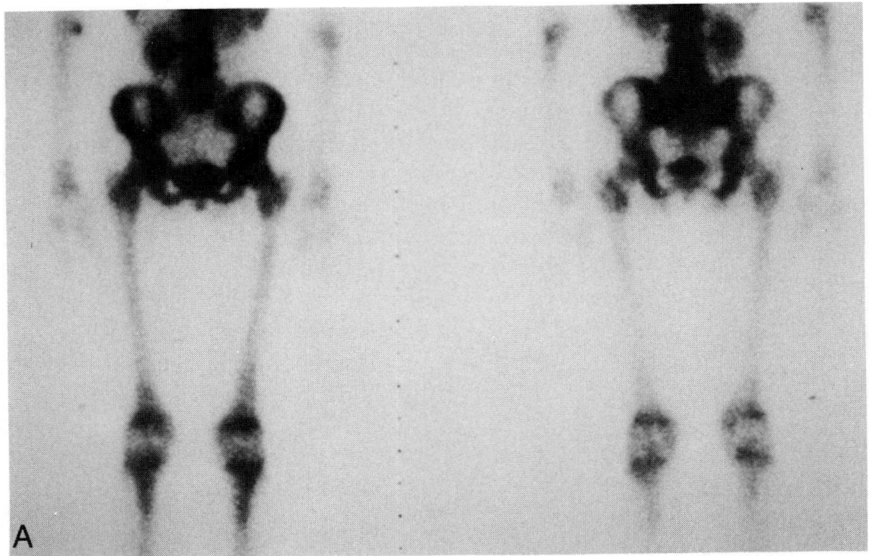

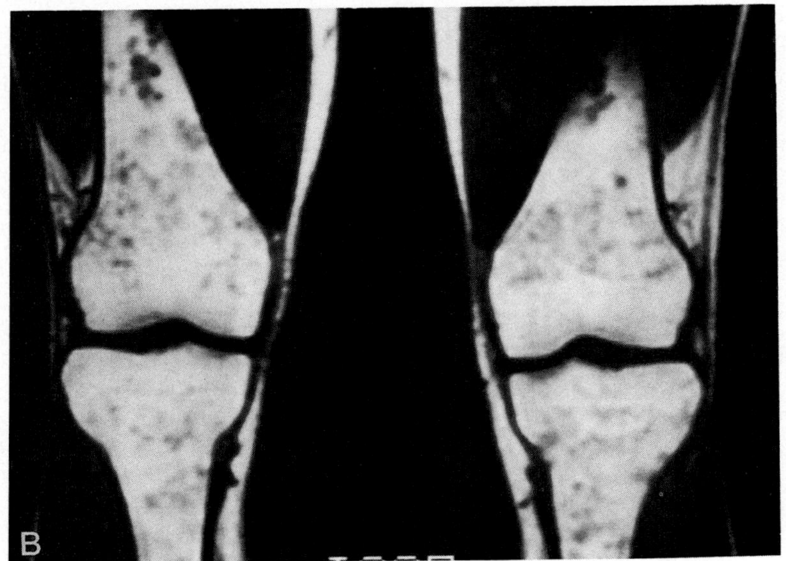

Figure 16–9. Variation from the adult pattern of red and yellow marrow distribution in a clinically and hematologically asymptomatic 32-year-old patient (patient had normal results of red and white cell counts, differential count, and iron studies). *A*, Bone scan is normal. No focal areas of increased radionuclide uptake are seen. *B*, Coronal T1-weighted spin-echo MR image (TR, 800 ms; TE, 20 ms) shows multiple scattered areas of low signal intensity in the left and right femur and tibia. Diaphyses, metaphyses, and epiphyses are involved. Although this patient was clinically and hematologically healthy and remained so on follow-up, such variation from the normal adult pattern of red and yellow marrow distribution should prompt careful clinical and laboratory assessment of the patient to rule out a clinically asymptomatic hematologic disorder or malignancy.

Skull

Okada and colleagues found three different patterns of bone marrow in the skull. In pattern 1, bone marrow in the clivus and calvaria is of uniformly low signal intensity on T1-weighted images. In pattern 2, regions of low and high signal intensity are seen in the marrow spaces of the skull. In pattern 3, uniformly high signal intensity is demonstrated on T1 weighting in the clivus and the calvaria.[158] Pattern 1 marrow is seen in most infants less than 1 year old. The number of patients with pattern 1 marrow decreases rapidly in early childhood, whereas the number of patients with low and high signal intensity (pattern 2) and uniformly high signal intensity (pattern 3) gradually increases with age. Okada and associates did not observe marrow of pattern 1 in either the clivus or the calvaria after the age of 7 years.[158] Most patients had pattern 3 marrow by the age of 15 years.[158] These findings are similar to what has been described by Ricci and co-workers.[176]

Spine

Ricci and colleagues identified four main patterns of marrow distribution in the spine. In pattern 1, the vertebral body is of uniformly low signal intensity on T1-weighted images except for linear areas of high signal intensity that are superior and inferior to the basivertebral vein. In pattern 2, band-like and triangular areas of high signal intensity are found near the end plates and anteriorly and posteriorly at the corners of the vertebral body (see Fig.

16–2). In pattern 3, diffusely distributed areas of high signal intensity are seen on T1 weighting, consisting of either numerous indistinct dots measuring a few millimeters or less (pattern 3a) or relatively well-marginated areas ranging in size from 0.5 to 1.5 cm (pattern 3b).[179]

In the cervical spine, pattern 1 is found predominantly in patients younger than 40 years of age. The majority of patients older than 40 years have a pattern 2 or 3. Most patients with a pattern 3 are older than 50 years.[176]

In the thoracic spine, three fourths of the patients with pattern 1 are younger than 30 years, and approximately 90 per cent of those with pattern 2 are older than 50 years. The age distribution of pattern 3 is not consistent in the thoracic spine.[176]

In the lumbar spine, pattern 1 was again seen in young patients (20 to 30 years of age). Patterns 2 and 3 show an increase in numbers with age and account for more than three fourths of all cases in patients older than 40 years.[176]

Pattern 2, with conversion of hematopoietic to fatty marrow near end plates, may, in part, relate to mechanical stresses and degenerative disk disease.[117,149,150,176]

Overall, there is continued gradual replacement of hematopoietic with fatty marrow that continues until the time of death. In normal elderly patients, it is not uncommon to find that the spine as well as the pelvis appears with marked high signal intensity on T1-weighted spin-echo images, reflecting the predominance of fatty marrow. Large variations exist, however, owing to differences among individuals and responses to stress.

Pelvis

In the pelvis, two major patterns of red and yellow marrow distribution have been described by Ricci and associates.[176] In pattern 1, a small area of high signal intensity is seen in the acetabulum that is superior and medial to the hip joint (Fig. 16–10). In pattern 2, additional regions of high signal intensity are found in the ilium and adjacent to the sacroiliac joint.[176] Pattern 1 is seen predominantly in patients younger than 40 years and pattern 2 occurs mostly in patients older than 40 years.[176]

Dawson and colleagues identified different patterns of bone marrow signal intensity and heterogeneity for four age groups: infant, child, adolescent, and young adult.[30] In the first year of life, red marrow with relatively homogeneous low signal intensity on T1-weighted MR images is seen in the posterior ilium, the anterior ilium, the ischium, the pubis, and the sacrum. At ages 1 to 10 years, marrow with intermediate or slightly increased signal intensity is seen in the acetabulum and the anterior ilium. Bone marrow in the posterior ilium, the ischium, the pubis, and the sacrum has relatively homogeneous intermediate signal intensity. At ages 11 to 20 years, marrow with intermediate signal intensity is observed throughout the pelvis, although marrow with slightly increased signal intensity may again be seen in the anterior ilium or the acetabulum. Marked marrow heterogeneity is not uncommon at this age. This constellation constitutes the "adolescent pattern." At ages 21 to 24 years, marrow with intermediate signal intensity is again seen in the posterior ilium, the ischium, the pubis, and the sacrum. Marrow with

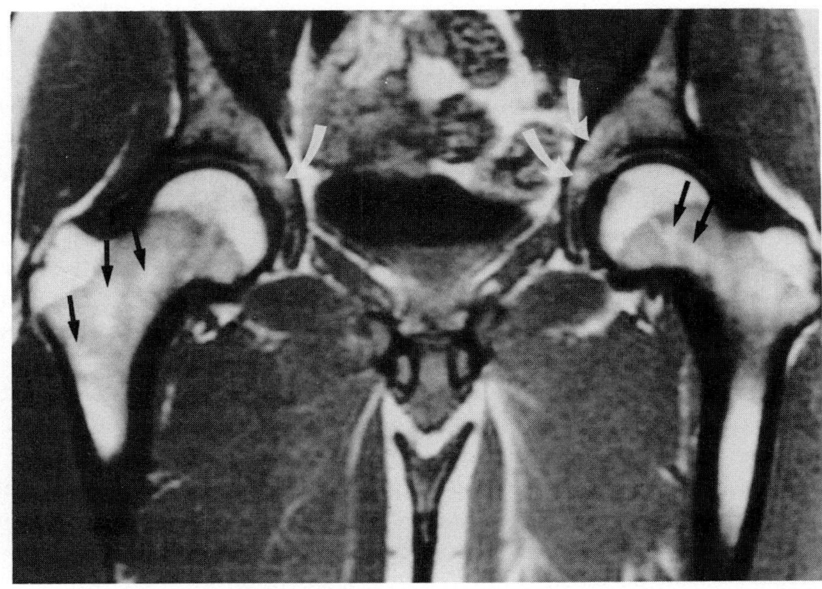

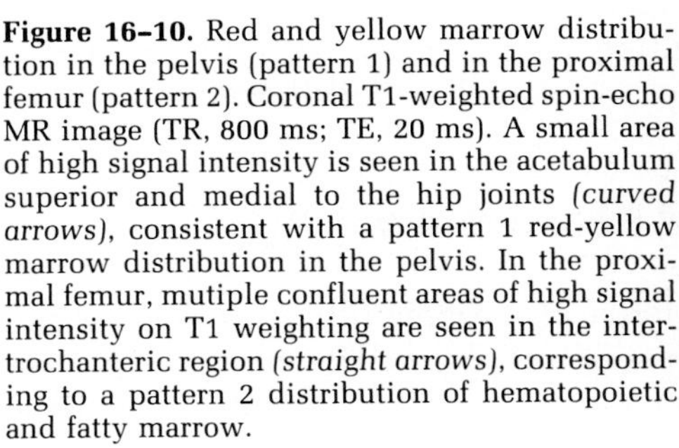

Figure 16–10. Red and yellow marrow distribution in the pelvis (pattern 1) and in the proximal femur (pattern 2). Coronal T1-weighted spin-echo MR image (TR, 800 ms; TE, 20 ms). A small area of high signal intensity is seen in the acetabulum superior and medial to the hip joints *(curved arrows)*, consistent with a pattern 1 red-yellow marrow distribution in the pelvis. In the proximal femur, mutiple confluent areas of high signal intensity on T1 weighting are seen in the intertrochanteric region *(straight arrows)*, corresponding to a pattern 2 distribution of hematopoietic and fatty marrow.

increased signal intensity is seen in the acetabulum and the anterior ilium. Marrow heterogeneity, if present, is most marked in the acetabulum.

Proximal Femur

Four different MR imaging patterns of hematopoietic and fatty marrow have been described in the proximal femur.[176] Pattern 1a demonstrates uniformly high signal intensity on T1-weighted images only in the proximal femoral epiphysis and in the greater and lesser trochanters. In pattern 1b, high signal intensity is observed in a triangular area inferior to the femoral head medially, and fatty marrow extends into a portion of the intertrochanteric region medial to the greater trochanter. In pattern 2, areas of high signal intensity on T1 weighting are seen in the intertrochanteric region, consisting of small foci of high signal intensity that may be partially confluent (see Fig. 16–10). In pattern 3, signal intensity is uniformly high throughout the proximal femur.[176] Approximately 80 per cent of patients with pattern 1 are younger than 50 years. Pattern 2 has a maximal frequency in the middle age group. Approximately 90 per cent of patients with pattern 3 are older than 50 years.[176]

Early conversion of hematopoietic to fatty marrow in the intertrochanteric region may be observed in patients who subsequently develop osteonecrosis of the femoral head.[145] The early conversion is thought to be an effect of decreased vascularity, which may allow the physician to identify patients who are at increased risk for osteonecrosis.[145]

Distal Femur

In children younger than 1 year old, uniformly low signal intensity is observed in the femoral diaphysis and the distal metaphysis.[151] Between the ages of 1 to 5 years, the diaphyseal marrow begins to convert to fatty marrow. The distal metaphyses retain intermediate signal intensity on T1-weighted images, reflecting the presence of hematopoietic marrow. Between the ages of 6 to 10 years, the conversion of red to yellow marrow is completed in the femoral diaphysis. However, the distal metaphysis remains of relatively homogeneous intermediate signal intensity on T1 weighting, reflecting the presence of hematopoietic marrow (see Fig. 16–3). Between the ages of 11 and 15 years, conversion of hematopoietic to fatty marrow begins in the distal femoral metaphysis, seen as areas of signal inhomogeneity and signal intensity slightly greater than that observed between 6 and 10 years. Between 16 and 20 years, the heterogeneity of metaphyseal marrow increases, reflecting continuing conversion of hematopoietic to fatty marrow so that relatively large, well-circumscribed areas of yellow marrow often are seen. Between the ages of 21 and 24, conversion to the adult pattern is observed. The diaphysis and distal metaphysis show increased signal intensity and are more and more uniform, corresponding to homogeneous areas of fatty marrow.[151]

Epiphyses

Jaramillo and co-workers evaluated conversion from hematopoietic to fatty marrow in the epiphyses of patients from birth to 16 months.[94] In their study, patients with epiphyseal marrow of low signal intensity on T1-weighted images, reflecting the presence of hematopoietic marrow, had a mean age of 3.9 months (standard deviation, 3.2 months). Patients with epiphyseal marrow that was of high signal intensity on T1 weighting, corresponding to fatty marrow, had a mean age of 9.6 months (standard deviation, 3.9 months). These researchers concluded that marrow of low signal intensity on T1-weighted images represents normal red marrow in a recently formed ossification center. Epiphyseal marrow shows high signal intensity within a few months of development of the secondary center of ossification.[94] These findings are consistent with what has been reported by Moore and Dawson.[151]

RECONVERSION

Fatty marrow is in a labile state and is capable of relatively fast reconversion to hematopoietic marrow when the demand for red blood cells is increased or the marrow system is replaced.[27] Conditions in which reconversion can be seen include anemia, metastatic disease, multiple myeloma, acquired heart disease with chronic heart failure, lymphoma, and other myeloproliferative disease.[113] Reconversion follows the reverse pattern of physiologic red and yellow marrow conversion and begins proximally, extending more and more distally depending on the degree of erythrocyte demand.[27,113]

Patterns of Reconversion

Five patterns of intramedullary loss of signal intensity reflecting reconverted hematopoietic marrow have been reported in adult patients in the distal femur[120,121] (Fig. 16–11).

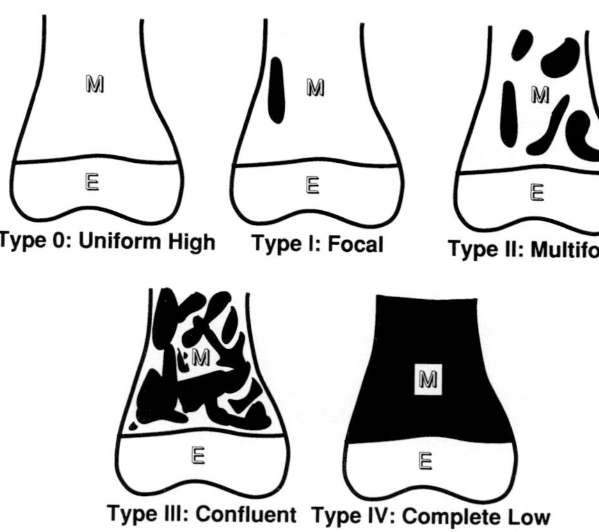

Figure 16–11. Patterns of residual and reconverted hematopoietic bone marrow in the distal femur. M, Metaphysis; E, epiphysis. (From Lang P, Fritz R, Majumdar S, et al: Hematopoietic bone marrow in the adult knee: Spin-echo and opposed-phase gradient-echo MR imaging. Skel Radiol 22:95–103, 1993.)

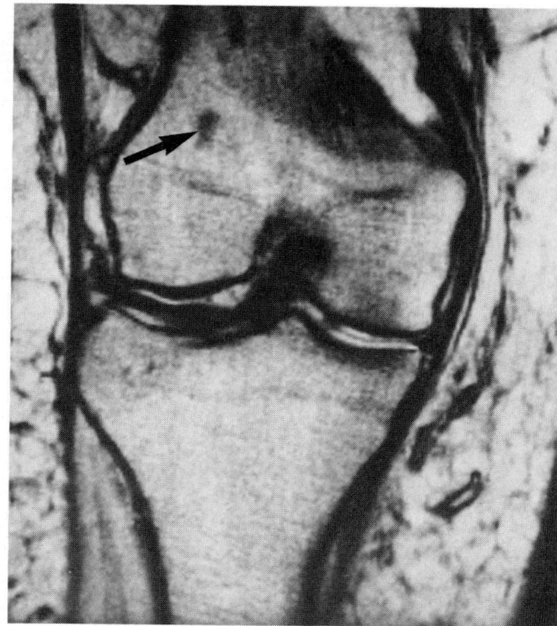

Figure 16–12. Type I pattern of reconverted hematopoietic bone marrow in the distal femur. (From Lang P, Fritz R, Majumdar S, et al: Hematopoietic bone marrow in the adult knee: Spin-echo and opposed-phase gradient-echo MR imaging. Skel Radiol 22:95–103, 1993.) Coronal opposed-phase gradient-echo image (TR, 1000 ms; TE, 30 ms; flip angle, 75 degrees) demonstrates a type I pattern (i.e., focal, punctuate, marrow signal reduction *[arrow]*) corresponding to a small island of hematopoietic bone marrow.

1. Type 0: uniform high signal (i.e., no signal change)
2. Type I: focal, punctate
3. Type II: multifocal without confluence
4. Type III: confluent, patchy
5. Type IV: complete, homogeneous

In type 0 (see Fig. 16–11), the marrow cavity of the distal femur is of uniform, homogeneous high signal intensity on T1 weighting corresponding to fatty marrow without any interspersed red marrow. In type I (Figs. 16–11, 16–12), a single, focal, punctate area of intermediate to low signal intensity is seen on T1-weighted spin-echo images. In type II, several focal areas of signal reduction are found; these areas do not communicate with each other (Figs. 16–11, 16–13). The areas of signal reduction observed in types I and II are usually round or oval (see Figs. 16–12, 16–13). A type III pattern is represented by multiple, confluent areas of low signal intensity that involve a large portion of the marrow space (Figs. 16–11, 16–14). A type IV pattern is characterized by a homogeneous area of intermediate to low signal intensity on T1-weighted spin-echo images that occupies the entire marrow space (see Figs. 16–1, 16–15). Areas of reconverted hematopoietic marrow are of markedly lower signal intensity on opposed-phase gradient-echo sequences than on conventional T1-weighted spin-echo images, resulting in significantly greater contrast between yellow and red marrow than on conventional T1-weighted spin-echo images[120,121] (see Fig. 16–14).

Hashimoto described an anatomy-based classification system for the distribution of hematopoietic

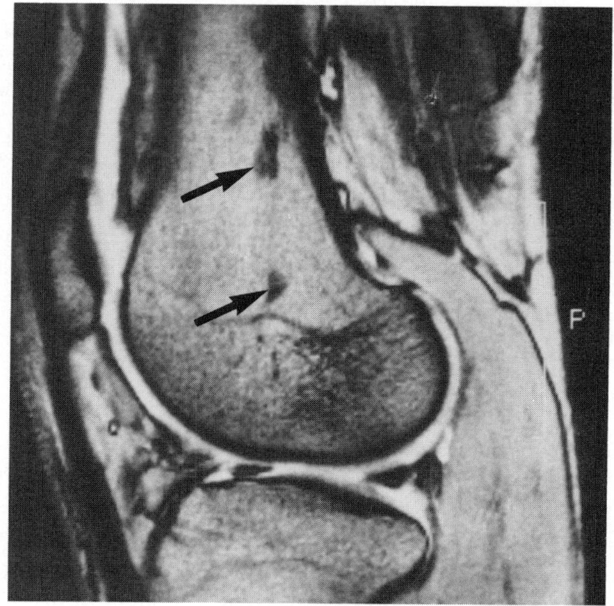

Figure 16–13. Type II pattern of reconverted hematopoietic bone marrow in the distal femur. Sagittal opposed-phase gradient-echo (TR, 1000 ms; TE, 30 ms; flip angle, 75 degrees) image demonstrates type II multifocal hematopoietic marrow of low signal intensity without confluence *(arrows)* located within the fatty marrow of high signal intensity. Patient's clinical and laboratory data were normal. (From Lang P, Fritz R, Majumdar S, et al: Hematopoietic bone marrow in the adult knee: Spin-echo and opposed-phase gradient-echo MR imaging. Skel Radiol 22:95–103, 1993.)

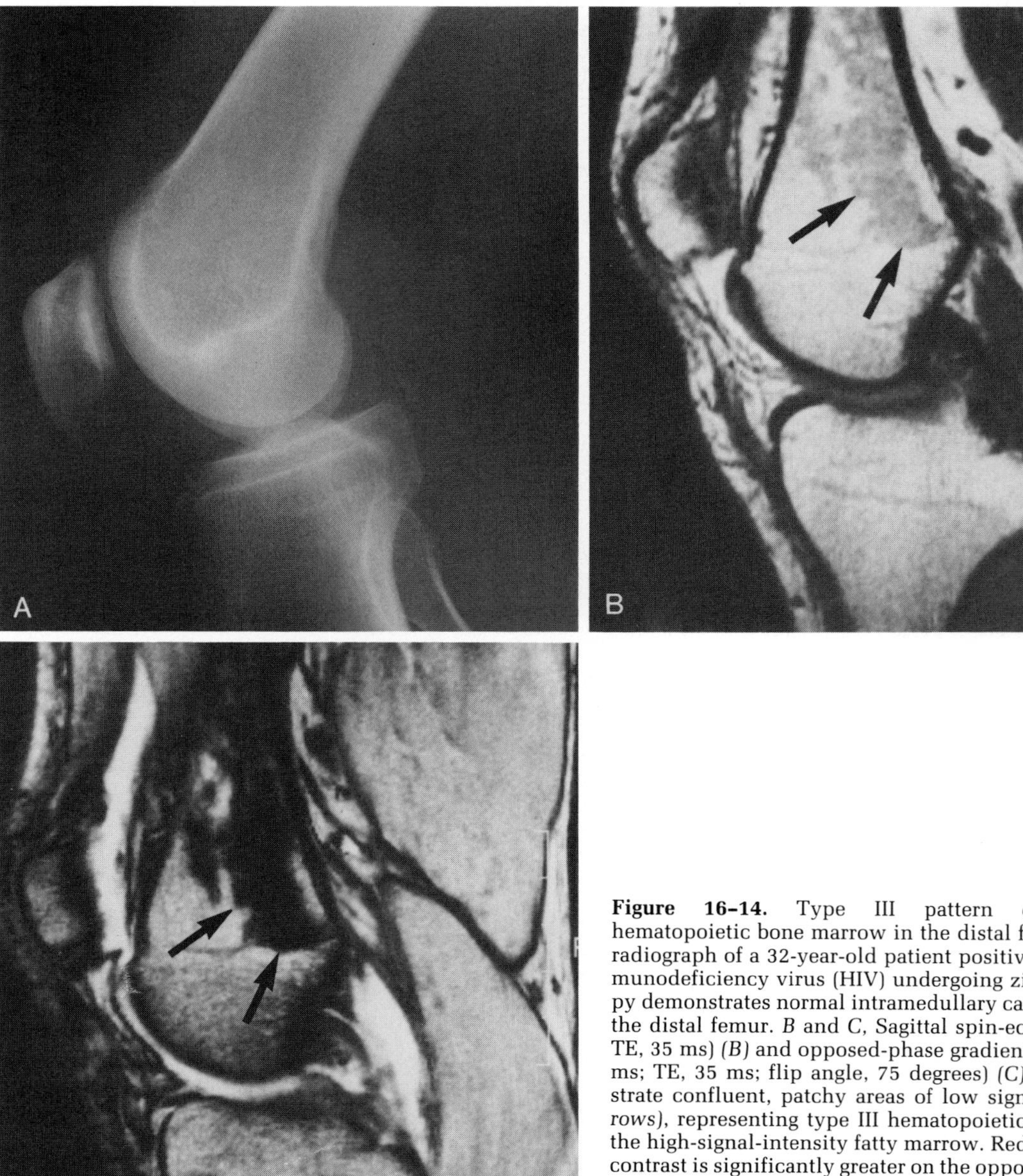

Figure 16–14. Type III pattern of reconverted hematopoietic bone marrow in the distal femur. *A*, Lateral radiograph of a 32-year-old patient positive for human immunodeficiency virus (HIV) undergoing zidovudine therapy demonstrates normal intramedullary cancellous bone in the distal femur. *B* and *C*, Sagittal spin-echo (TR, 500 ms; TE, 35 ms) *(B)* and opposed-phase gradient-echo (TR, 1000 ms; TE, 35 ms; flip angle, 75 degrees) *(C)* images demonstrate confluent, patchy areas of low signal intensity *(arrows)*, representing type III hematopoietic marrow within the high-signal-intensity fatty marrow. Red-yellow marrow contrast is significantly greater on the opposed-phase gradient-echo *(C)* than on the spin-echo image *(B)*. (From Lang P, Fritz R, Majumdar S, et al: Hematopoietic bone marrow in the adult knee: Spin-echo and opposed-phase gradient-echo MR imaging. Skel Radiol 22:95–103, 1993.)

marrow in the femur that differentiates three different types.[77] In class 1, the entire proximal and distal femoral marrow space is occupied by red marrow, corresponding to a type IV signal pattern on MR imaging. In class 2, both red and yellow marrow are found in varying distributions along the femur. In class 3, red marrow is absent in the femoral marrow cavity except in the intertrochanteric region, corresponding to a type 0 MR imaging marrow signal. Anatomic class 1 red marrow distribution was suspected of being pathologic.[77] On the basis of recent MR imaging studies,[33,120,121,195] it may, however, only reflect a physiologic response to increased erythrocyte demand.

Prevalence

Data on the prevalence of areas of hematopoietic marrow in the distal femur in adult patients vary markedly. Deutsch and co-workers found only 10 cases of red marrow in the distal femur in a total of 1400 MR imaging examinations of adult patients (prevalence, 0.7 per cent).[33] However, they only considered patients with types III and IV MR imaging signal changes, although types I and II patterns also must be considered a variation from the reported anatomic adult red-yellow marrow distribution.[77,113] Shellock and colleagues described a prevalence of 15 per cent in patients with symptomatic

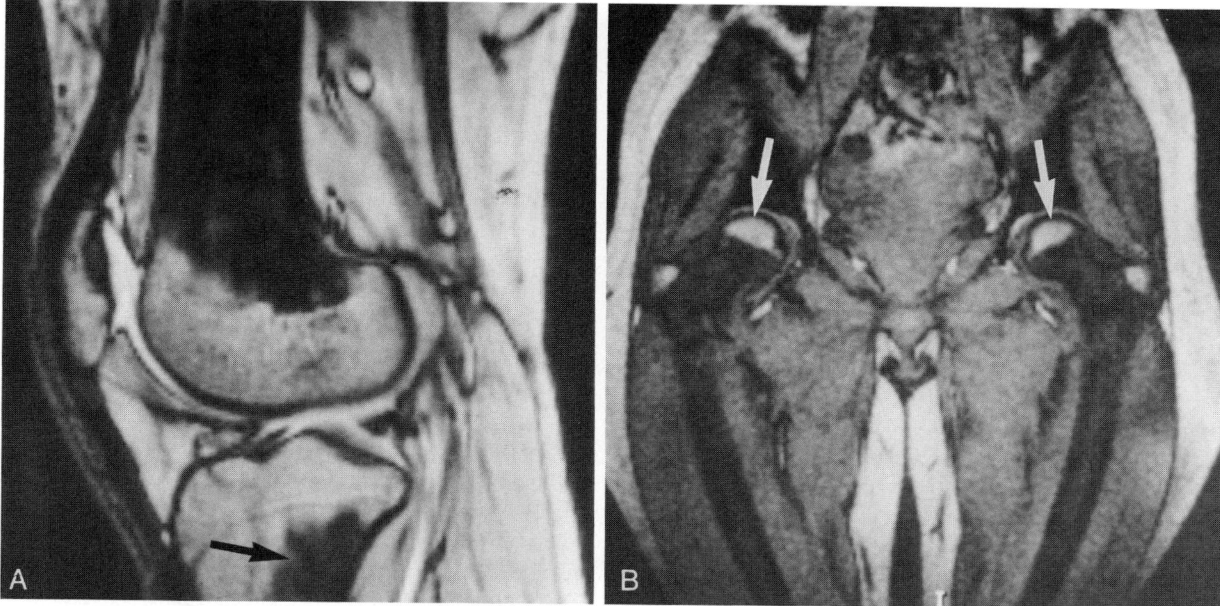

Figure 16–15. Type IV pattern of reconverted hematopoietic bone marrow in the distal femur. *A*, Sagittal opposed-phase gradient-echo image of the knee (TR, 1000 ms; TE, 30 ms; flip angle, 75 degrees) in a 28-year-old female patient with anemia demonstrates a type IV marrow pattern with complete, homogeneous reduction of intramedullary signal of the distal femur. The area of signal intensity change do not extend into the distal epiphysis. The epiphysis shows normal high signal intensity, corresponding to fatty marrow. The patella also is spared. Note involvement of the tibial metaphysis *(arrow)*. *B*, Coronal opposed-phase gradient-echo image of the hip (TR, 1000 ms; TE, 30 ms; flip angle, 75 degrees) in the same patient demonstrates type IV marrow pattern with homogeneous signal reduction, reflecting hematopoietic marrow that extends from the pelvis to the femoral metaphysis and diaphysis. The proximal femoral epiphyses also are spared *(arrows)*. (From Lang P, Fritz R, Majumdar S, et al: Hematopoietic bone marrow in the adult knee: Spin-echo and opposed-phase gradient-echo MR imaging. Skel Radiol. 22:95–103, 1993.)

knee disorders.[195] Lang and associates,[120,121] however, reported a prevalence of areas of hematopoietic marrow in the distal femur of approximately 35 per cent in an adult patient population similar to the one described by Shellock and co-workers.[195] The higher prevalence in the studies by Lang and colleagues[120,121] may be explained by the use of an opposed-phase gradient-echo sequence, which renders red marrow markedly more conspicuous relative to the spin-echo sequences employed by Deutsch and co-workers[33] and Shellock and associates.[105] Small islands of hematopoietic marrow within the fatty marrow may be missed when window and level are not carefully adjusted on conventional spin-echo images or when they are adjusted primarily for visualization of the menisci.

Differential Diagnosis

Because marrow reconversion is encountered relatively frequently[120,121,195] it is important not to confuse it with an infiltrative or myeloproliferative process. The differential diagnosis of residual or reconverted hematopoietic marrow includes primary and metastatic bone tumors, nonneoplastic infiltrative disorders, and lymphoma. Residual or reconverted hematopoietic marrow can be differentiated from most tumorous lesions on the basis of MR imaging signal intensity. Hematopoietic marrow and most tumors demonstrate low signal intensity on T1-weighted images.[220] On T2-weighting, hematopoietic marrow increases in intensity and demonstrates signal intensity similar to (although, in most instances, slightly lower than) that of fatty marrow.[220] Tumors, however, most often show signal intensity greater than that of fatty marrow on T2 weighting.[233]

Tumors or malignant marrow disorders such as lymphomas frequently involve the epiphysis. In hematopoietic reconversion of fatty marrow, however, sparing of the epiphysis typically is seen.[32,120,121,195] In all previous reports on hematopoietic reconversion in the distal femur, femoral and tibial epiphyses and the patella were not involved in any of the patients.[32,120,121,195] The presence or absence of epiphyseal involvement may thus represent an additional useful diagnostic parameter to distinguish reconverted hematopoietic marrow from benign or malignant marrow disorders. Absence of cortical destruction and soft tissue mass further aids in differentiating reconverted red marrow from tumorous lesions.

Residual Red Marrow

Areas of hematopoietic marrow in the distal femur also may be demonstrated on MR imaging in adult patients with normal red and white blood counts, differential counts, reticulocyte counts, serum iron, serum ferritin, and total iron binding capacity.[120,121] Interestingly, they are present primarily in female patients.[120,121] The presence of hematopoietic marrow in the distal femur in healthy women may be related to a substantially greater erythrocyte demand than in men, resulting from menstrual bleeding. Thus, it appears possible that localized areas of hematopoietic marrow in the distal femur may reflect a physiologic variation in red-yellow marrow distribution in women and not an active reconversion. In addition, MR imaging may afford better (i.e., more sensitive) detection of areas of red marrow within the surrounding fatty marrow than anatomic macrosections, which would explain deviations from the reported macroanatomic[77,113] adult red-yellow marrow distribution. Microscopic red marrow has been found in the distal third of the femur and humerus in the adult,[164] although these studies did not quantify the amount of red marrow. Superiority of MR imaging over anatomic macrosectioning in delineating marrow compositional changes has been documented in previous studies.[151]

MYELOID DEPLETION

Myeloid depletion is seen in patients who have had previous radiation therapy or chemotherapy and in patients who have aplastic anemia. The cellularity of the hematopoietic marrow decreases with myeloid depletion, and the fatty marrow expands. Aplastic anemia is a relatively rare disorder in which hypo- or acellularity of the bone marrow results in pancytopenia (i.e., anemia, neutropenia and thrombocytopenia). Several agents or disorders, such as drugs, toxins, hepatitis, and viral infections, that cause aplastic anemia have been identified.[170] However, in approximately half of the cases, aplastic anemia is idiopathic. Pancytopenia is evident in the peripheral blood. Bone marrow biopsy specimens show hypocellular or aplastic marrow with a variable degree of expanding fatty marrow. Small areas of normal marrow may be seen within the aplastic marrow.[170] Because small areas with normal hematopoiesis may coexist with the overall aplastic marrow, bone marrow biopsy results may sometimes be subject to sampling errors.

On MR imaging, high signal intensity of the marrow on T1 weighting is present in the affected areas in patients who had radiation therapy (see also the section on Malignant Marrow Infiltration or Replacement). This is particularly dramatic with sharply defined radiation ports affecting the spine. The radiated levels appear very bright on T1-weighted images, but the adjacent nonirradiated vertebrae demonstrate normal hematopoietic marrow with intermediate signal intensity.

In aplastic anemia, high signal intensity is found on T1-weighted MR images throughout the entire skeleton, including the spine and the pelvis (Fig. 16–16). The changes are most apparent in areas that normally contain hematopoietic marrow. Few scattered, small areas of intermediate signal intensity typically represent nests of residual hematopoietic marrow.[22,105,107,159] On STIR sequences, the vertebral bodies of patients with aplastic anemia are of uniformly low signal intensity, reflecting the presence of fatty marrow.[115]

Hypocellularity and acellularity of the marrow and expansion of fatty marrow in aplastic anemia can also be assessed using relaxation time measurements. In aplastic anemia, T1-relaxation times of the marrow typically are shortened.[179,180,197] Aplastic anemia can be distinguished from normal marrow because of significantly shortened T1 relaxation times ($P < 0.001$).[197] A significant correlation has been described between T1 relaxation and bone marrow cellularity ($r = 0.74$, $P < 0.001$).[197] T2 relaxation times are of no value in the characterization of bone marrow disorders.[197]

In patients treated for and recovering from aplastic anemia, T1-weighted MR images will demonstrate multiple focal areas of low signal intensity, reflecting repopulated hematopoietic marrow within the fatty marrow of high signal intensity (Fig. 16–17).[107,143,159] Histologically, the focal areas of low signal intensity on T1 weighting correspond to hypercellular hematopoietic tissue interspersed with patches of hypocellularity and fatty marrow.[107] Kaplan and associates reported that repopulation begins in the spine (see Fig. 16–17), while the pelvis and proximal femurs retain high signal intensity, corresponding to the presence of fatty marrow.[107] Although no studies have been performed yet in which patients were followed longitudinally during various stages of therapy of aplastic anemia, it appears possible that repopulation of hematopoietic marrow follows a pattern similar to that in marrow reconversion, originating centrally and extending more and more peripherally until a normal pattern of red-yellow marrow distribution is reached.

Several studies have suggested clonal evolution of aplastic anemia to myelodysplastic syndromes and leukemia.[153] Negendank and co-workers observed an unexpected inhomogeneous or diffuse pat-

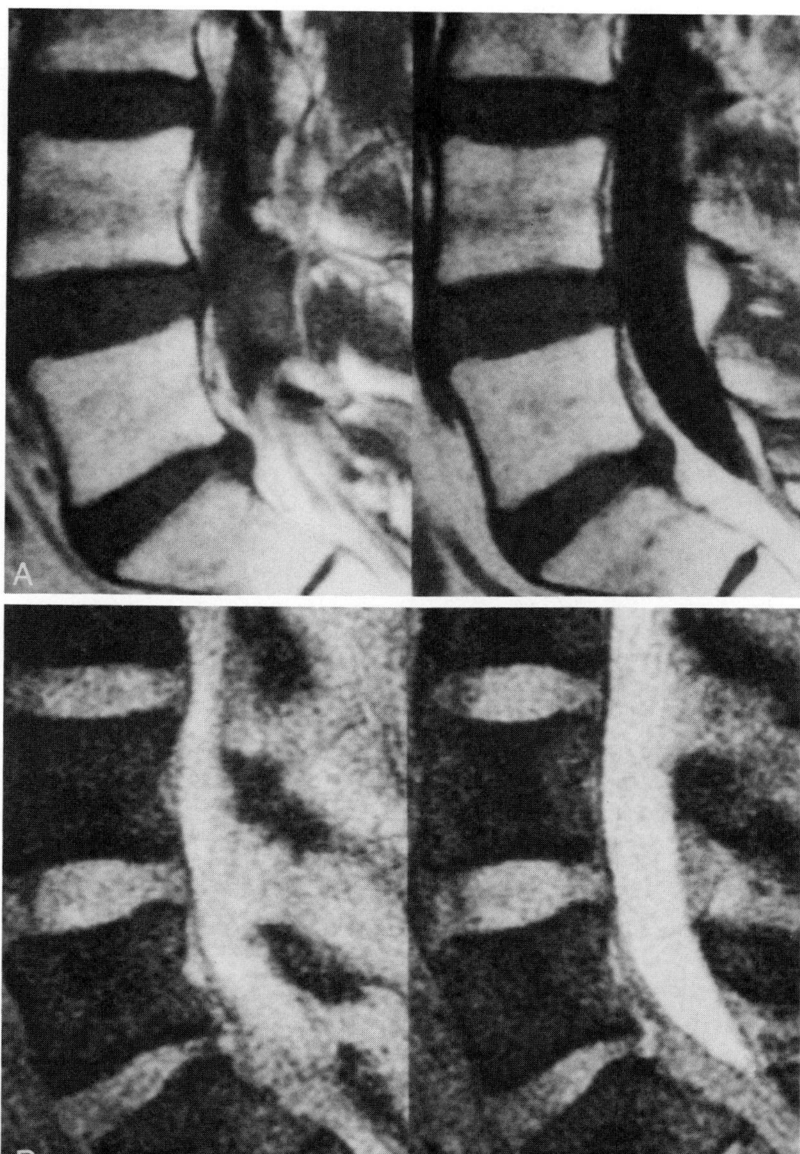

Figure 16–16. Untreated aplastic anemia. *A*, T1-weighted spin-echo MR images (TR, 800 ms; TE, 30 ms) show an abnormally bright signal of the vertebral marrow, indicating hypocellularity of hematopoietic marrow and expansion of fatty marrow. *B*, T2*-weighted gradient-echo MR images (TR, 500 ms; TE, 25 ms; flip angle, 30 degrees). On the gradient-echo image, the vertebral marrow is low in signal owing to inhomogeneity in magnetic susceptibility at the interface between trabecular bone and marrow, with resultant spin dephasing and signal loss. Hypocellularity of hematopoietic marrow and expansion of fatty marrow cannot be diagnosed.

tern of low signal intensity on T1-weighted MR images in several patients with hypocellular marrow biopsies and a clinical diagnosis of aplastic anemia.[156] Subsequent marrow or cytogenetic studies led to diagnoses of hypoplastic myelodysplastic syndromes. Negendank and colleagues concluded that MR imaging is able to detect early clonal disease in patients with aplastic anemia and can distinguish aplastic anemia from hypoplastic myelodysplastic syndromes.[156]

ANEMIAS

Anemias that are not caused by stem cell failure (e.g., hemolytic anemias) typically result in reconversion of fatty to hematopoietic marrow (see the section on Reconversion). Sickle cell anemia affects approximately 0.15 per cent of the black population in the United States.[221] The disease is severe, resulting in significant morbidity and shortened life expectancy. Patients with sickle cell anemia frequently suffer from painful sickle cell crises involving the limbs and abdomen as a consequence of sickling-induced ischemia. The primary mechanism of the anemia is vascular hemolysis.[221] The peripheral blood smear shows markedly distorted red cells, including the characteristic sickle cells.

The multiple repetitive episodes of organ ischemia and in particular bone marrow ischemia result in a host of abnormalities. The medullary spaces usually are widened owing to the marked compensa-

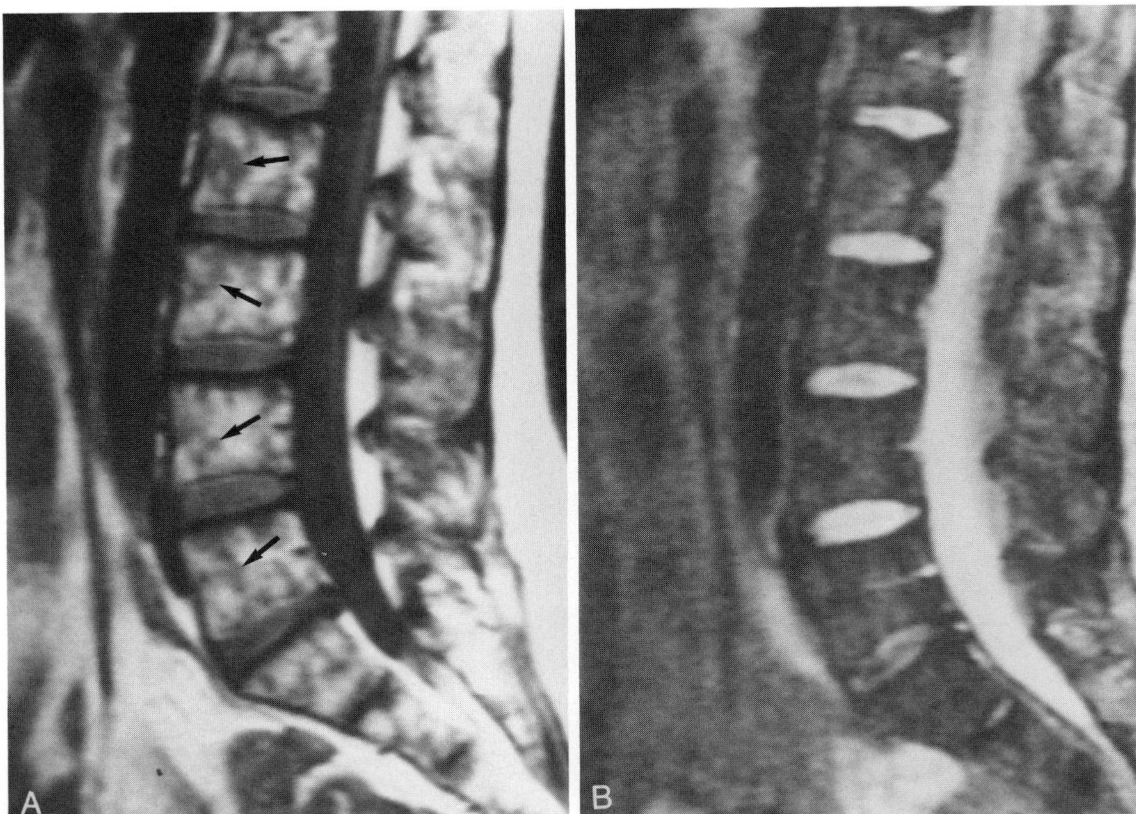

Figure 16–17. Treated aplastic anemia with recovery. *A*, Sagittal T1-weighted spin-echo MR image (TR, 600 ms; TE, 25 ms) of the lumbar spine shows multiple focal areas of low signal intensity *(arrows)*. These areas correspond to repopulated hematopoietic marrow. *B*, T2-weighted spin-echo MR image (TR, 2000 ms; TE, 80 ms). The focal areas of hematopoietic marrow are of approximately equal intensity to that of fatty marrow. (Courtesy of Phoebe Kaplan, M.D.)

tory expansion of hematopoietic bone marrow. The spine frequently takes on a distorted appearance, and the femoral heads commonly show osteonecrosis.

Technetium 99m sulfur colloid bone marrow scans demonstrate increased uptake, consistent with red marrow hyperplasia.[167] On T1-weighted MR images, the signal intensity of the bone marrow is diffusely decreased, reflecting hyperplastic hematopoietic marrow[140,167] (Figs. 16–18, 16–19). Hematopoietic marrow has been found on T1-weighted MR images in the thoracolumbar spine, the proximal femur, and the knee joint of adult patients with sickle cell disease.[140,167] Hematopoiesis may even occur in the epiphyses once an ossification center has formed (Fig. 16–20). On T2 weighting, hematopoietic and fatty marrow are approximately of equal signal intensity, so that the expansion of the red marrow component in sickle cell anemia is more difficult to appreciate.[140,167] When the patients develop secondary hemosiderosis, "black marrow"[167] with markedly decreased signal intensity on T1-weighted MR images can be observed (see Fig. 16–4).

During painful sickle cell crises, different MR imaging patterns can be observed.[142,170] Although conventional radiographs are unchanged, T1-weighted MR images frequently demonstrate one or more focal areas of low signal intensity in the affected joint. On T2 weighting, these areas increase markedly in signal intensity and are of high signal intensity compared with the adjacent fatty marrow, reflecting bone marrow edema or hyperemia and possibly early bone marrow infarction[140,167] (Figs. 16–20, 16–21).

In late stages of sickle cell disease, manifest bone infarcts can be seen in both areas of fatty marrow (e.g., the proximal femoral epiphyses) and areas of hematopoietic marrow (e.g., the proximal femoral metaphysis).[140,167-169] Bone marrow infarcts show inhomogeneous low signal intensity on T1-weighted images. With T2 weighting, bone infarcts may be of high signal intensity,[140,168,169] which may be caused by reactive granulation tissue that is permeating the infarct. Bone infarcts may, however, also retain low signal intensity on T2-weighted images when they are old and fibrous and sclerotic tissue predominates.[167] Magnetic resonance imaging is unique in that it allows differentiation between acute and old bone marrow infarction.[167]

In evaluating patients with painful sickle cell

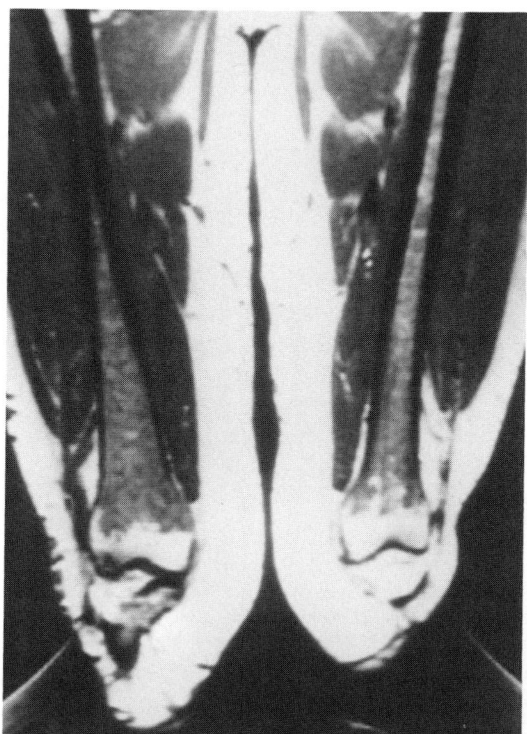

Figure 16–18. Coronal T1-weighted spin-echo MR image (TR, 1000 ms; TE, 20 ms) of a 32-year-old patient with sickle cell anemia shows compensatory hematopoietic marrow with intermediate signal intensity in the femoral diaphysis and the distal femoral metaphysis. Only the distal femoral epiphysis contains fatty marrow of high signal intensity.

crisis, the symptomatic joints should be imaged with T1- and T2-weighted scans. Typically, the pelvis and hip joints, the distal femur and knee joint, and the tibia and the ankle are evaluated for patterns of acute or chronic infarction.

In other forms of anemia (e.g., anemia associated with chronic renal disease), MR imaging can be used to evaluate response to treatment with erythropoietin.[100] Changes in image contrast and T1 relaxation times can be observed before any response is seen in the hemoglobin concentration in the peripheral blood.[100] The signal intensity of the bone marrow decreases and the T1 relaxation time increases with erythropoietin treatment, which reflects the expansion of hematopoietic marrow.[100]

BENIGN MYELOPROLIFERATIVE DISORDERS

Benign myeloproliferative diseases include polycythemia vera and myelofibrosis. Polycythemia vera is a rare benign monoclonal myeloproliferative disorder of the pluripotent stem cell that is marked by increased autonomous production of predominantly erythrocytes but also thrombocytes and granulocytes.[155] Red blood cell counts and hematocrit readings are increased; the morphology of the peripheral blood cells is, however, normal. The increase in red blood cell mass may slow local blood flow, resulting in ischemia and hemorrhage. Thromboembolic events are a common complication in polycythemia vera. Bone marrow biopsy demonstrates hypercellular marrow. However, marrow cellularity also may be normal, or marrow fibrosis may be present in selected biopsy samples.[115]

As the disease progresses, the patients develop splenomegaly. The hematopoietic marrow extends into the humerus and femur. Extramedullary hematopoiesis may commence in the liver and in the spleen. In late stages, polycythemia vera transforms into myelofibrosis.[155] The bone marrow becomes filled with fibrotic tissue, and extramedullary hematopoiesis increases further. Occasionally, polycythemia vera may transform into an acute or chronic myelocytic leukemia.[115]

Bone marrow failure and pancytopenia develop with increasing myelofibrosis, because extramedullary hematopoiesis cannot compensate completely for the loss of medullary function.[155] The patient

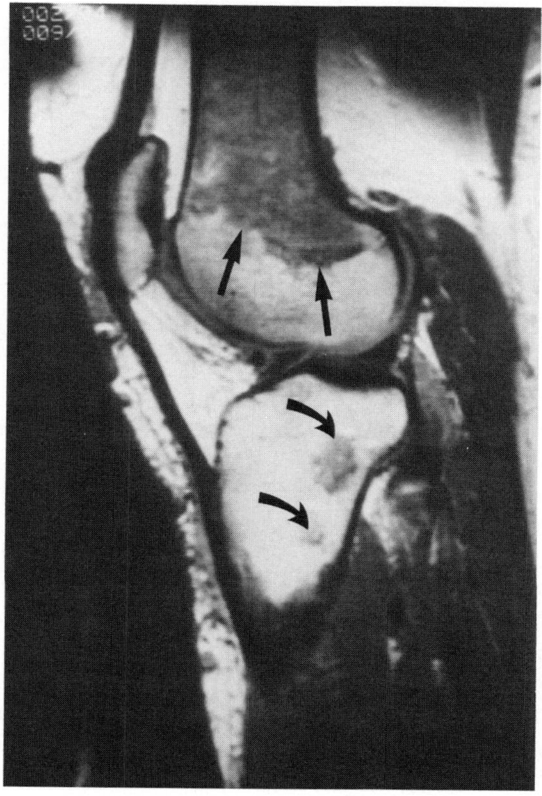

Figure 16–19. Sagittal T1-weighted MR image (TR, 800 ms; TE, 30 ms) of a 38-year-old patient with sickle cell anemia shows hematopoietic marrow extending into the distal femoral metaphysis *(straight arrows)*. Scattered areas with intermediate signal intensity, consistent with hematopoietic marrow, also are seen in the proximal tibial metaphysis *(curved arrows)*.

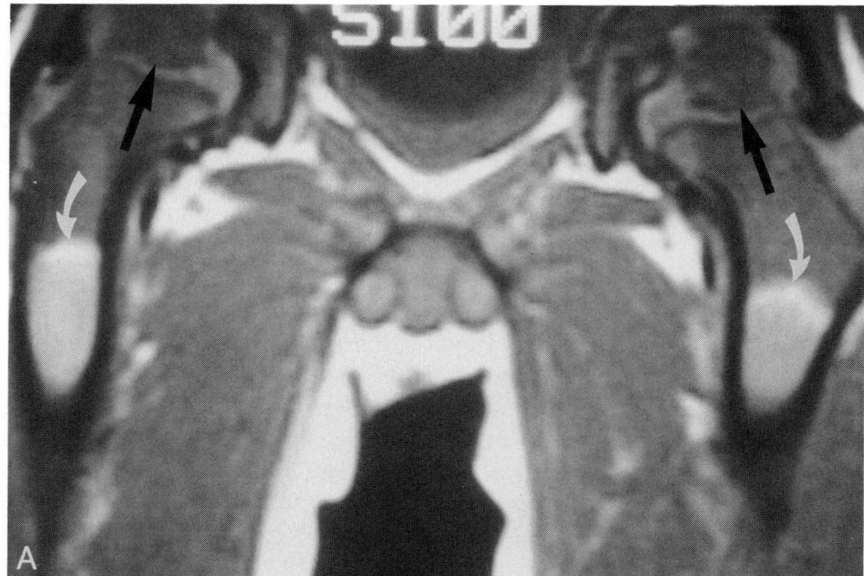

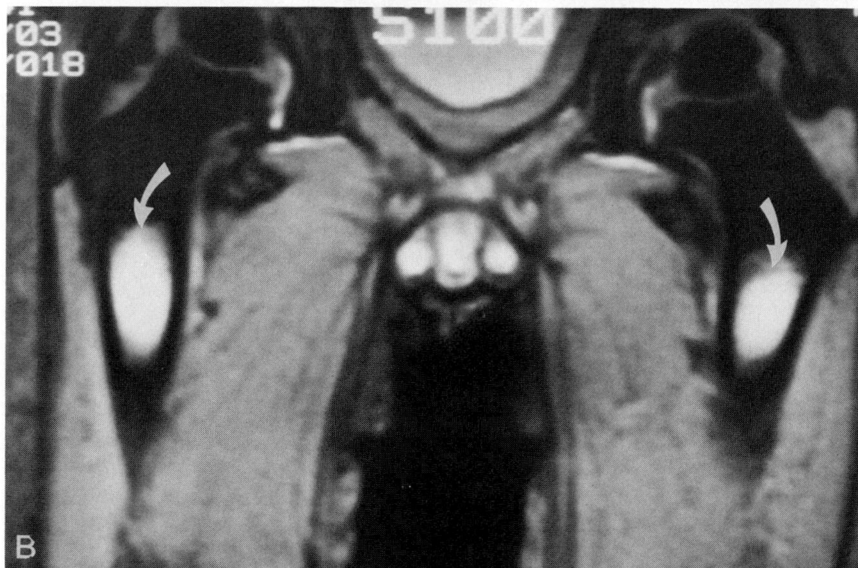

Figure 16–20. Images of sickle cell anemia—epiphyseal hematopoiesis in a 10-year-old patient. *A*, Coronal T1-weighted spin-echo MR image. Hematopoietic marrow with intermediate signal intensity is shown in the proximal femoral metaphyses as well as in the ossified portions of the epiphyses *(straight arrows)*. Under normal conditions, the ossified portions of the epiphyses contain fatty marrow. The presence of epiphyseal hematopoietic marrow results from increased red blood cell demand in sickle cell anemia. Additional distal areas with abnormally high signal intensity reflect hemorrhagic bone infarcts in the femoral shafts *(curved arrows)*. *B*, Coronal T2*-weighted gradient-echo MR image (TR, 600 ms; TE, 14 ms; flip angle, 30 degrees). The proximal femoral epiphysis and the metaphysis retain low signal intensity with this imaging sequence. Again noted are areas of hemorrhagic bone infarcts in the femoral shafts with high signal intensity *(curved arrows)*.

ultimately succumbs to hemorrhage, infection, or liver failure.[155]

Polycythemia vera has to be differentiated from other forms of polycythemia (i.e., increase in red cell mass) resulting from an "external" stimulus (e.g., hypoxia, neoplasm, or paraneoplastic syndromes). Myelofibrosis can occur independently of polycythemia vera. It can be caused by a neoplasm as well as an infection.

The most striking finding on MR imaging in polycythemia vera is partial or even complete reconversion of fatty to hematopoietic marrow. The red-yellow marrow distribution differs from the normal pattern. The epiphyses also may show hematopoiesis. The reconversion is documented on T1-weighted MR images by the presence of zones of intermediate to low signal intensity consistent with the presence of hematopoietic marrow in areas in which fatty marrow is expected. On T2 weighting, the signal intensity of the marrow varies depending on the amount of cellular tissue and tissue water as well as on the degree of myelofibrosis. As long as hypercellularity of hematopoietic marrow predominates, the areas of hematopoietic marrow are approximately of equal signal intensity to fatty marrow on T2 weighting. Once myelofibrosis develops, patchy small areas with low signal intensity, reflecting focal tissue fibrosis, or confluent homogeneous areas with low signal intensity are seen on T1-weighted MR images (Fig. 16–22).[106] On T2 weighting, these areas do not increase in signal intensity.

Similar to the results in leukemia, marrow in polycythemia vera demonstrates prolonged T1 relaxation times.[96,197] T2-relaxation times are nor-

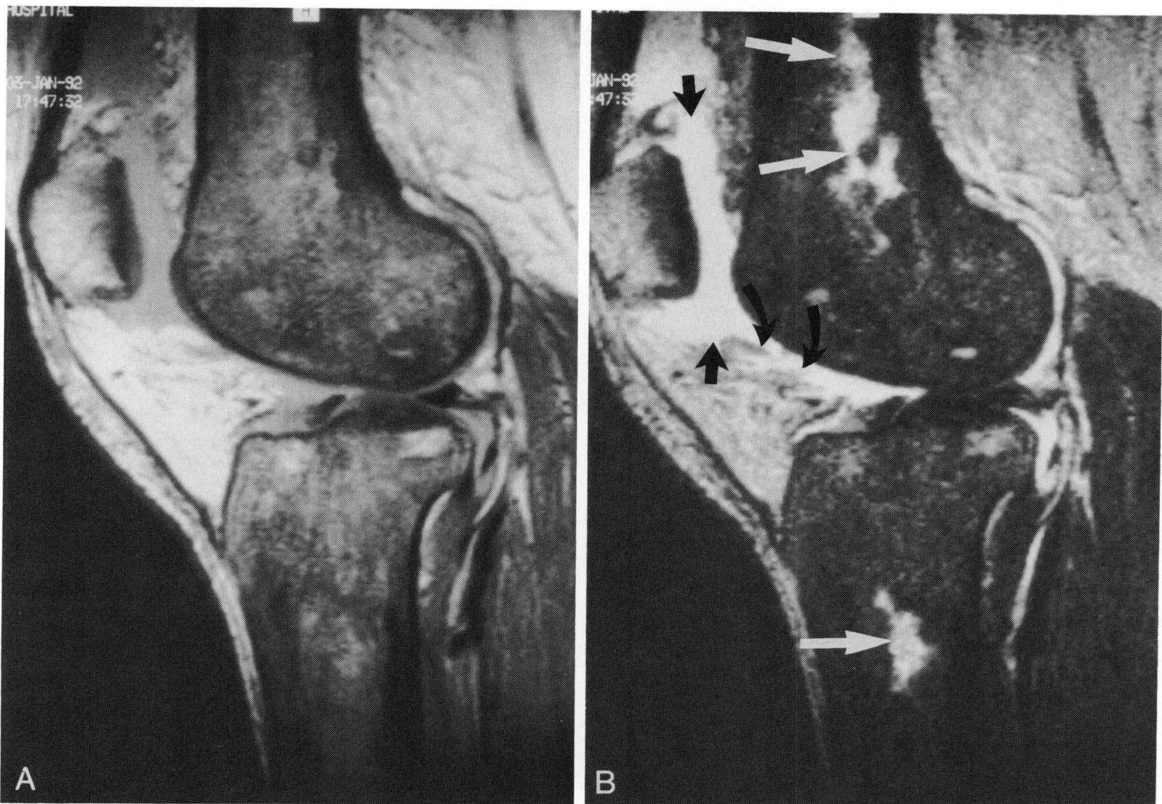

Figure 16–21. Sickle cell crisis with medullary bone infarcts. *A*, Sagittal T1-weighted spin-echo image (TR, 1000 ms; TE, 20 ms) shows abnormal heterogeneous low signal intensity of femoral and tibial marrow. The low signal intensity is caused by hematopoietic marrow. The signal heterogeneity suggests the presence of bone infarcts. *B*, Sagittal T2*-weighted gradient-echo MR image (TR, 500 ms; TE, 20 ms; flip angle, 30 degrees). Areas with high signal intensity that reflect bone marrow edema and infarcts are shown in the femur and in the tibia *(white straight arrows)*. Hoffa's fat pad is irregularly shaped *(curved arrows)*, consistent with synovitis. A joint effusion is present *(black straight arrows)*.

mal.[96,203] Kaplan and associates correlated MR imaging findings and clinical-laboratory parameters in patients with polycythemia vera and patients with myelofibrosis.[106] They found that the presence of hematopoietic marrow of abnormally low-signal-intensity in the femoral capital epiphysis and the greater trochanter correlated well with clinical-laboratory disease parameters of polycythemia vera and myelofibrosis. Splenic volume, as measured from MR images, was significantly greater in the myelofibrosis group than in the polycythemia vera group ($P < 0.001$).[106]

MALIGNANT MARROW INFILTRATION OR REPLACEMENT

Marrow infiltration or marrow replacement is seen with infiltrative neoplastic disorders such as leukemia, lymphoma, multiple myeloma, and metastatic disease. Metastatic disease is discussed in detail in the following section. Some of these lesions (e.g., leukemia and multiple myeloma) are thought to originate in the hematopoietic marrow. The location of other lesions is largely dependent on local vascular anatomy. Lesions that spread in a hematogenous pattern favor the richly vascularized hematopoietic marrow over the relatively sparsely vascularized fatty marrow.

Different MR imaging patterns of bone marrow infiltration have been described.[11] In general, the lesions are of low signal intensity relative to adjacent fatty marrow on T1-weighted MR images. They may be diffuse and uniform (Fig. 16–23), diffuse and nonuniform (Fig. 16–24), or patchy (Fig. 16–25) in appearance.[11] The signal intensity of infiltrative marrow lesions on T2-weighted MR images shows marked variation depending on the cellularity; the degree of vascularization; the water content; and the presence of granulation tissue, fibrosis, or hematoma.[233] Leukemia and lymphoma as well as other infiltrative lesions of the bone marrow frequently demonstrate high signal intensity on T2 weighting[11,85,105] (see Fig. 16–23). However, despite the variability in morphologic appearance and signal intensity, MR imaging does not provide specific

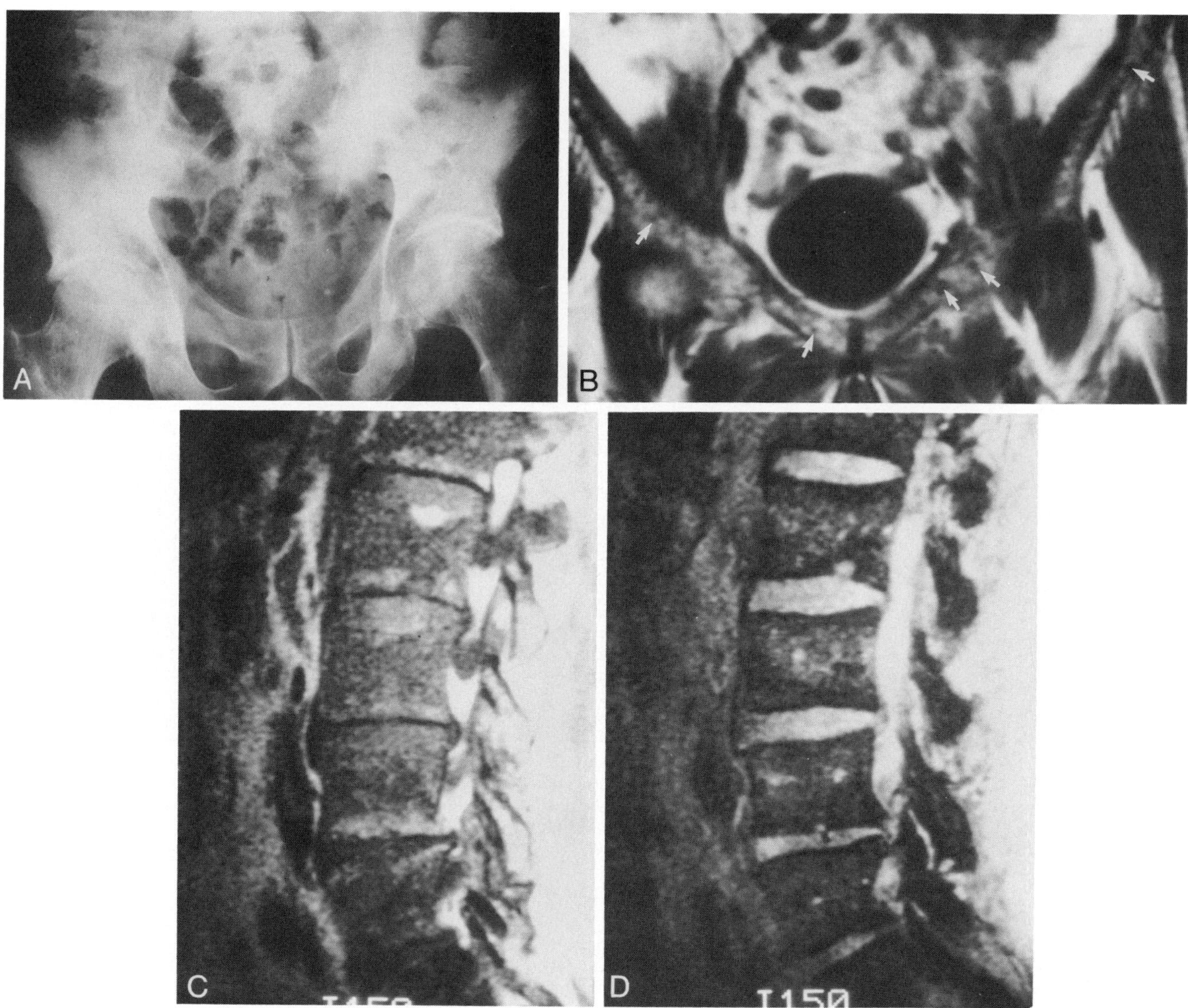

Figure 16-22. Myelofibrosis in the pelvis and lumbar spine. *A*, Conventional radiograph shows homogeneous increased density of the iliac bones, femoral heads, and lower lumbar vertebrae. Mottled increased density is seen in the superior and inferior rami of the os pubis and the ramus of the os ischium. *B*, Coronal T1-weighted spin-echo MR image (TR, 800 ms; TE, 30 ms) of the pelvis shows multiple patchy foci with low signal intensity *(arrows)* in the left and right iliac bones *(arrows)*. *C*, Sagittal proton density–weighted spin-echo MR image (TR, 1000 ms; TE, 40 ms) shows confluent homogeneous reduction in marrow signal intensity throughout the lumbar spine. *D*, Sagittal T2*-weighted gradient-echo MR image (TR, 600 ms; TE, 30 ms; flip angle, 30 degrees). The marrow retains low signal intensity on T2*-weighting.

distinction among different infiltrative marrow disorders (e.g., leukemia, lymphoma, multiple myeloma). Magnetic resonance imaging shows substantial overlap not only among different tumor types but also between benign and malignant lesions.[76,157,159,163,196,233] Similarly, differentiation between malignant infiltration of the marrow and hematopoietic marrow in marrow reconversion may be difficult, in particular when the abnormality is diffuse and uniform. Only occasionally may the MR imaging appearance correspond to the histologic composition of the lesion. Foci of decreased signal intensity within diffuse abnormal marrow of low signal intensity may reflect the nodular histologic pattern of Hodgkin's disease.[159]

Multiple myeloma is difficult to diagnose using conventional radiography and nuclear scintigraphy. On conventional radiography, multiple myeloma may cause osteopenia, which is, however, nonspecific. Because multiple myeloma causes marrow replacement rather than increased bone turnover, nuclear scintigraphy findings often are false negative.[28,48,214] Magnetic resonance imaging frequently detects multiple myeloma in the bone marrow when radiographic and nuclear scintigraphic findings are negative.[28,51]

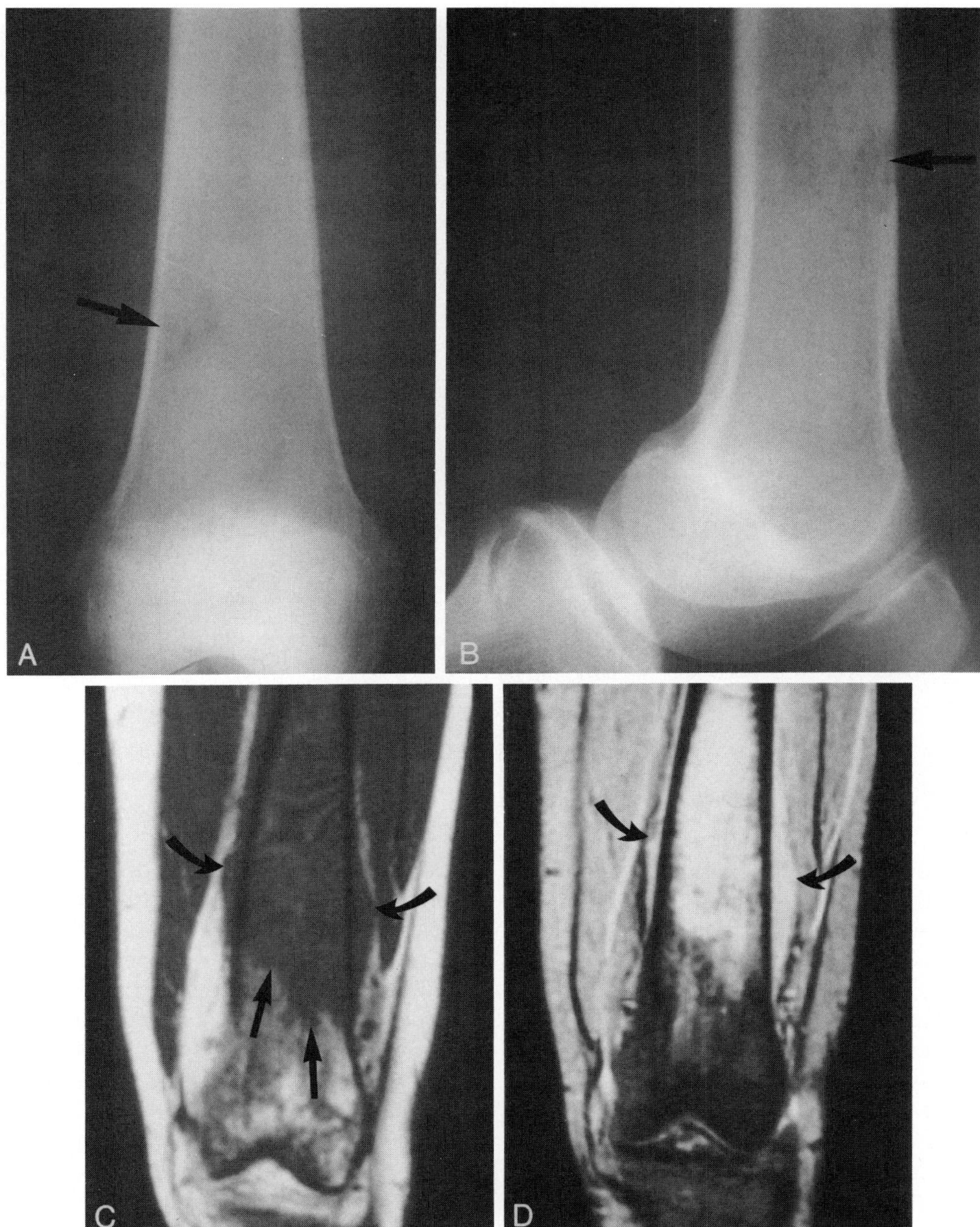

Figure 16–23. Chronic myeloid leukemia. *A* and *B*, Conventional radiographs in anteroposterior (*A*) and lateral (*B*) projections show a lytic defect in the distal femur (*arrow*). *C*, T1-weighted spin-echo MR image (TR, 600 ms; TE, 30 ms) shows a diffuse uniform lesion of low signal intensity in the femoral metaphysis, extending into the diaphysis (*straight arrow*). Note the extraosseous extension (*curved arrows*). *D*, T2*-weighted gradient-echo MR image (TR, 500 ms; TE, 30 ms; flip angle, 30 degrees). The lesion shows high signal intensity with T2*-weighting and appears very homogeneous. The extraosseous extension (*curved arrows*) also shows high signal intensity.

T1-weighted MR images demonstrate replacement of fatty marrow with tissue of intermediate to low signal intensity. On T2 weighting, multiple myeloma shows intermediate to high signal intensity. On STIR images, multiple myeloma is of high signal intensity. Occasionally, differentiation between multiple myeloma and normal hematopoietic marrow may be difficult on T1-weighted MR images. However, knowledge of the age-related distribution of red and yellow marrow helps in this differentiation. In addition, signal intensity higher than that of fatty marrow on T2-weighted images suggests an infiltrative marrow process.[28,51] Nonetheless, Libshitz and associates indicate that MR imaging may be characterized by a high rate of false-negative results of up to 50 per cent in imaging multiple myeloma.[133]

Despite its low specificity in differentiating infiltrative marrow disorders, MR imaging is

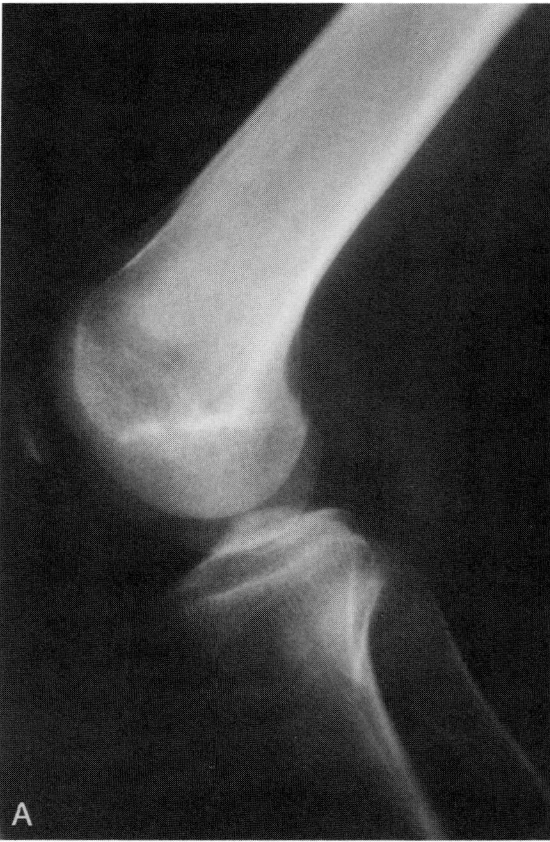

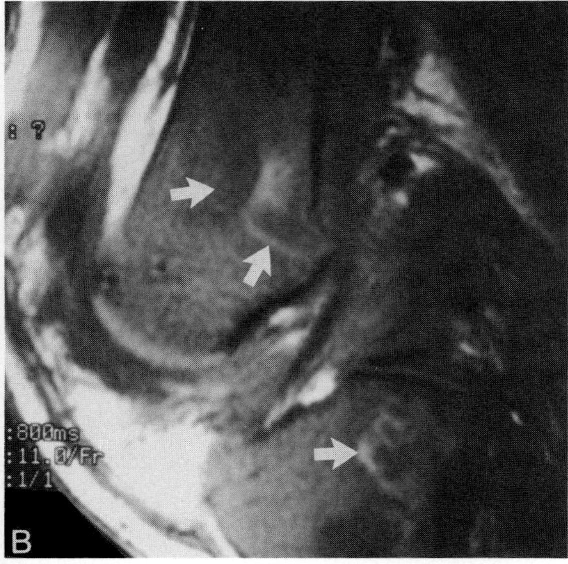

Figure 16–24. Chronic myeloid leukemia with bone infarcts. *A*, Conventional radiograph appears normal except for a subtle area of sclerosis in the distal femur. *B*, T1-weighted spin-echo MR image (TR, 600 ms; TE, 30 ms) shows complete replacement of fatty marrow by leukemic marrow with low signal intensity in the metaphyses and epiphyses. The leukemic marrow is, however, not uniform. Irregular areas of even more decreased as well as increased signal intensity are seen in the femur and in the tibia, corresponding to bone infarcts *(arrows)*.

ideally suited to define the extent of such lesions.[1,12,58,163,231] Magnetic resonance imaging is significantly more accurate than any other imaging modality in defining the exact dimensions of an infiltrative marrow process[1,12,58,163,231] and thereby lends itself to planning of biopsy.[28,111,196]

Hoane and colleagues evaluated 109 patients with lymphoma by MR imaging and blind iliac crest biopsy.[85] Ten patients had positive results on biopsies and negative MR image examinations; eight of these had microscopic infiltration (less than or equal to 5 per cent) with tumor. Biopsy thus better detected low-grade microscopic involvement. However, MR imaging detected marrow tumor either in the crests or elsewhere in 33 per cent of patients with negative biopsies. Marrow involvement was confirmed in 60 per cent of these patients by clinical methods.[85] Therefore, up to one third of the patients evaluated with routine biopsies may have occult marrow tumor detectable by MR imaging.[85] In patients with negative marrow biopsies, especially those with Hodgkin's disease or intermediate to high-grade non-Hodgkin's lymphomas, MR imaging scans found focal lesions distant from the crests.[85] Optimal staging of infiltrative marrow disorders incorporates both biopsy and MR imaging.

Several investigators have measured T1 and T2 relaxation times in different infiltrative marrow disorders.[98,99,177,179,198,213] Most studies agreed that there are no significant differences in T1 and T2 relaxation times between benign and malignant bone marrow lesions.[14,20,231] In comparison to normal tissue, both benign and malignant tumor cells have prolonged T1 relaxation times and variable T2 relaxation times.[14,20,231] Only Sugimura and colleagues observed significant differences between benign and malignant lesions in the ratios of T1 to T2 and in the T1 ratios and T2 ratios, which they defined as T1 ratio = (T1 for affected vertebrae / T1 for normal vertebrae) and T2 ratio = (T2 for affected vertebrae / T2 for normal vertebrae).[206] These findings have not, however, been confirmed in follow-up studies as yet.

Leukemic marrow,[98,99,179,213] similar to that in lymphoma[177,198] and myeloma, is characterized by a prolongation of T1 relaxation times in comparison to those for normal marrow. The prolonged T1 relaxation is the cause for the low signal intensity of leukemic marrow on T1-weighted MR images. The cause of the T1 prolongation and the resultant low signal intensity is not exactly known, although an increase in cellularity[105,157,159] due to tumor invasion of marrow and an increase in total water content have been discussed.[92,98,213] T2 relaxation changes are less specific and do not reach statistical significance when compared with those from age-matched control subjects.[152]

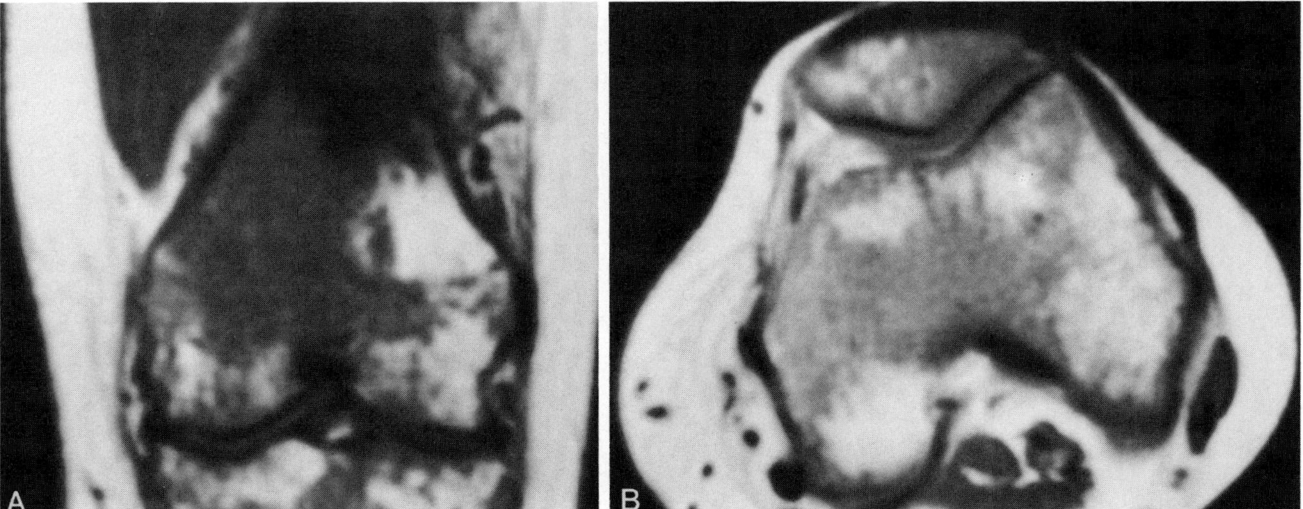

Figure 16–25. Chronic myeloid leukemia. Coronal *(A)* and axial *(B)* T1-weighted spin-echo MR images (TR, 600 ms; TE, 20 ms) show patchy leukemic infiltration with low signal intensity in the fatty bone marrow.

Jensen and co-workers used volume-selective MR proton spectroscopy to investigate the hematopoietic (iliac bone) and fatty bone marrow (tibia) in patients with leukemia and polycythemia vera.[98] Selective measurements of the T1 and T2 relaxation times for the water and fat resonances in the bone marrow spectra were obtained. Significant differences could be detected in the spectral patterns from iliac bone marrow in patients with leukemia and in healthy normal controls. The water content was higher in the iliac bone marrow spectra of the leukemic patients than in the normal subjects, which was thought to indicate an increase in the amount of hematopoietic tissue and a corresponding decrease in the marrow fat content. The T1 relaxation times of the water resonance in the spectra from the iliac bone marrow of the leukemic patients were significantly prolonged at diagnosis compared with those of the normal control subjects and of the patients with polycythemia vera. After chemotherapeutic induction of remission, the spectra from the iliac bone marrow in the patients with leukemia resembled normal spectra.[98] Similar findings were reported by Irving and co-workers, who found an increase in water in the marrow of patients with leukemia over that in patients who have normal marrow.[92] In the future, MR spectroscopy may thus be used to diagnose involvement of marrow by leukemia and potentially to monitor response to therapy.

Marrow Changes after Chemotherapy or Radiation Therapy

Before chemotherapy, leukemic or lymphomatous marrow is of intermediate to low signal intensity on T1-weighted MR images and of variable, often elevated, signal intensity on T2 weighting.[11,85,105] With the induction of chemotherapy, the signal intensity of the involved marrow increases on T1-weighted MR images and then returns to normal on remission.[8,72,152,212] Similarly, findings on T2-weighted images normalize with response to chemotherapy. The high signal intensity on T1-weighted MR images in patients who are in remission results from a decrease in tumor cells infiltrating the marrow and from an increase in the fat fraction of the marrow, reflecting a return of fatty marrow.[56,66,143,182] The increase in fat fraction in remission has been documented with the use of chemical shift imaging techniques[56,66,143] as well as with the use of quantitative dual-energy body CT correlated with MR imaging.[182] In cases of relapse, sharp decreases in marrow fat fraction are seen with quantitative chemical shift imaging techniques.[56] T1-weighted MR images in relapse show patchy or diffuse areas of decreased signal intensity in marrow that had had normal high signal intensity during remission.

Relaxometry studies also may be employed to evaluate patients for remission and to detect relapse. Before the induction of chemotherapy, leukemic and lymphomatous bone marrow will show markedly prolonged T1 relaxation times[98,99,177,179,198,213] With remission, T1 relaxation times of the marrow decrease significantly toward or into the normal range.[80,97,101,152] With relapse, T1 relaxation times will show again marked prolongation in comparison to normal marrow or to the T1 relaxation times during remission.[80,97,101]

Radiation-induced changes in the bone marrow depend on the radiation dose and the time course of radiation treatment. Hematopoietic marrow decreases and fatty marrow increases after radiation.[207]

Knowledge of the chronologic evolution of bone marrow changes during and after radiation therapy is essential in differentiating normal postradiation changes from other marrow abnormalities. During the first 2 weeks of radiation therapy, no definite change in the appearance of the marrow is observed on spin-echo images.[204] Stevens and colleagues described, however, an increase in signal intensity on STIR images, apparently reflecting early marrow edema and necrosis.[204] Between weeks 3 and 6, the marrow shows an increasingly heterogeneous signal intensity and prominence of the signal from central marrow fat, shown best on T1-weighted images.[204] Different MR imaging patterns have been described for the late phase after radiation therapy. Yankelvitz and colleagues[232] and Starz and associates[202] found homogeneous fatty replacement with high signal intensity in the vertebral bodies on T1-weighted MR images. Stevens and researchers described two different marrow signal patterns 6 weeks to 14 months after therapy.[204] These consist of either homogeneous fatty replacement or a band pattern of peripheral intermediate signal intensity, possibly representing hematopoietic marrow, surrounding the central marrow fat.[204] The presence of focal marrow lesions or soft tissue edema during the course of radiation therapy should raise the possibility of the presence of a pathologic process other than radiation change.[204] Magnetic resonance imaging can detect radiation-induced marrow changes as early as 2 weeks after starting therapy and can be used to assess radiation-induced changes of the bone marrow.

Osteonecrosis is a potential complication of radiation therapy. Magnetic resonance imaging is ideally suited to detect radiation-induced osteonecrosis in early stages[142,146] and to define the extent of the areas of necrosis.[116,123]

Bone Marrow Transplantation

Bone marrow transplantation has become a treatment option for several different marrow disorders. After transplantation, T1-weighted MR images of vertebral marrow show a characteristic band pattern consisting of a peripheral zone of intermediate signal intensity and a central zone of bright signal intensity.[203] Reciprocal changes are seen on STIR sequences (i.e., the peripheral zone is of high signal intensity and the central zone is of low signal intensity).[203] On histologic examination, the central zone corresponds to fatty marrow; the peripheral zone corresponds to a zone of regenerating hematopoietic cells.[203] Two cases of bone marrow transplantations have been described that demonstrated uniformly low signal intensity on both T1- and T2-weighted MR images.[13] The cause of this signal pattern could not be determined. Marrow hypercellularity, sclerosis, and fibrosis were discussed as possible explanations.[13] Although some variability in the MR imaging appearance of bone marrow seems to occur after transplantation, which requires further study, MR imaging may be useful in screening for residual marrow disease and determining marrow engraftment. Future investigations should determine whether MR imaging can be used to differentiate repopulation with normal cells from repopulation with malignant cells in cases of recurrence.

Stevens and co-workers also determined posttransplantation T1 and T2 relaxation times of the vertebral marrow.[203] They observed no statistically significant trends in relaxation times.[203] Smith and colleagues studied patients with Hodgkin's disease.[198,199] The pretransplantation T1 relaxation times of the vertebral marrow were prolonged. After bone marrow transplantation, normalization of the T1 relaxation times was seen, consistent with good response to treatment.[198,199] Interestingly, a significant correlation also was shown between the T2 relaxation rate and the numbers of colony-forming units of the granulocyte and macrophage line in marrow cultured after transplantation.[198] Although further work is still needed, MR relaxometry may become an additional useful tool to evaluate response to bone marrow transplantation.

METASTATIC DISEASE

Conventional radiography is the initial imaging modality for detecting, staging, and follow-up of tumor in patients with metastatic bone disease. Conventional radiography is not, however, suited to provide a survey of the entire skeleton, which is needed for accurate staging of the extent of metastatic disease. Nuclear scintigraphy has a higher sensitivity than conventional radiography in detecting skeletal metastases. Nuclear scintigraphy remains, at present, the imaging and screening modality of choice for skeletal metastases, as it provides total body images (Table 16–2).

Magnetic resonance imaging also is limited in screening for metastases because it likewise cannot yield images of the total body. However, using a phased-array spine coil or a body coil, or both, and a large field of view, MR imaging can be employed to evaluate the spine (Fig. 16–26), the pelvis, and the upper and lower extremities for metastatic disease[3,4,9,23,28,48,60,200] (see Table 16–2). Advantages of MR imaging over nuclear scintigraphy in imaging metastatic disease include greater sensitivity in detecting lesions[3,4,9,48] and improved specificity in

Table 16–2. COMPARISON OF NUCLEAR SCINTIGRAPHY AND MR IMAGING IN THE EVALUATION OF METASTATIC DISEASE OF BONE

	Nuclear Scintigraphy	MR Imaging
Total body screening	Yes	No (only spine, extremities)
Sensitivity	+	+ +
Specificity	–	+
Imaging complex bony anatomy	–	+ +
Evaluation of extent of lesion	–	+ +
Imaging multiple myeloma	–	+

–, Insufficient; +, good; + +, excellent.

differentiating metastases from other pathologic conditions such as osteoarthritis or spinal osteochondrosis (see Table 16–2). Magnetic resonance imaging also may be superior to nuclear scintigraphy when metastatic disease is suspected in areas with complex bony anatomy, such as the craniovertebral junction, the sternum, and the sacrum, which are difficult to image with non–cross-sectional techniques.[144] Additionally, MR imaging may be helpful in demonstrating the extent of a metastatic lesion, such as in the femur, where prophylactic pinning may be justified.

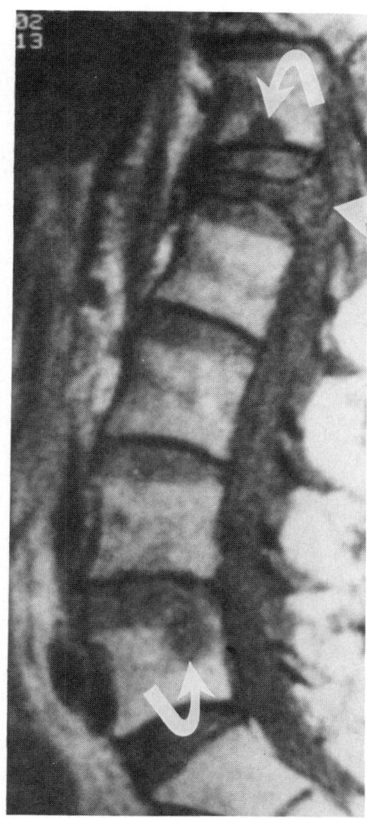

Figure 16–26. Metastatic disease of the spine (colon carcinoma). Sagittal T1-weighted spin-echo MR image (TR, 800 ms; TE, 30 ms) shows collapse of the L1 vertebral body. The L1 vertebral body is of low signal intensity and bulges posteriorly in a convex fashion (*straight arrow*). Lesions of low signal intensity corresponding to other metastases are seen in the T12 and the L5 vertebral bodies (*curved arrows*). The signal pattern of the vertebral bodies L2 to L4 also is inhomogeneous, suggesting possible metastatic involvement.

Magnetic Resonance Imaging Protocols for Metastatic Disease

T1-weighted spin-echo sequences with a short TR and TE or alternatively gradient-echo sequences with a short TR and TE and a large flip angle (80 to 90 degrees) are ideally suited for screening purposes because they allow short acquisition times. When metastases have been identified, T2-weighted spin-echo sequences or T2*-weighted gradient-echo sequences can be obtained for more detailed assessment of the metastatic lesion (Figs. 16–27, 16–28). T2- or T2*-weighted images are particularly useful for evaluating spinal cord involvement. STIR sequences provide higher contrast between metastases (high signal intensity) and normal marrow (low signal intensity) than T2-weighted spin-echo images.[205] In addition, STIR sequences may be more sensitive in detecting metastatic disease than conventional spin-echo images.[205] Some investigators recommend the combined use of T1-weighted spin-echo images and STIR images rather than T1- and T2-weighted spin-echo sequences.[61,205]

Intradural metastases frequently are not demonstrated on unenhanced MR images unless they cause a significant mass effect.[114] On T1-weighted images after intravenous administration of gadopentetate dimeglumine, intradural metastases demonstrate significant contrast enhancement.[205] Contrast enhanced T1-weighted spin-echo scans thus are required when intradural spread is suspected clinically.[205] However, lumbar puncture and cytologic examination of cerebrospinal fluid constitute the primary diagnostic test for intradural metastasis.

Magnetic Resonance Imaging Patterns of Metastatic Disease

Skeletal metastases typically are of low signal intensity on T1-weighted images and of high signal intensity on T2-weighted images (see Fig. 16–27). Four different patterns of metastatic marrow involvement have been described.[2] In type I, metastases are focal and are of low signal intensity (less

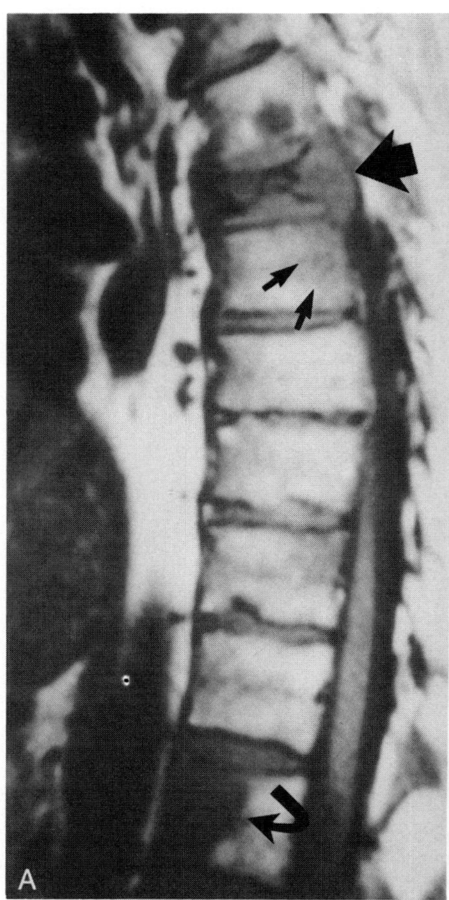

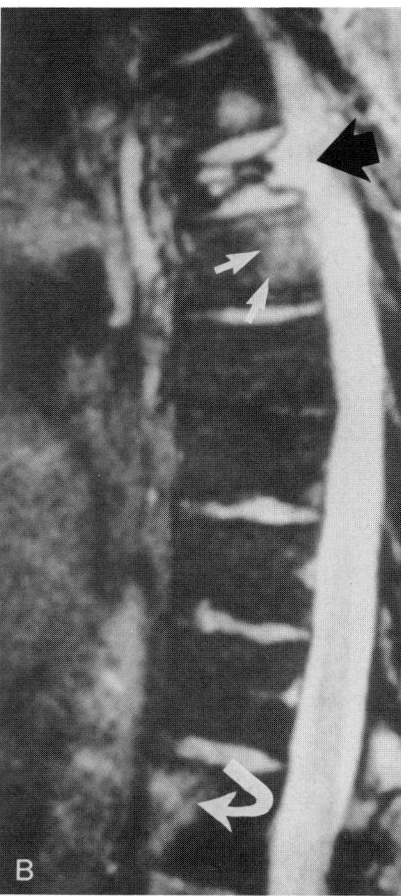

Figure 16–27. Metastatic disease of the spine (breast carcinoma). Type I signal pattern (i.e., focal, low signal intensity on T1 weighting and high signal intensity on T2 weighting). *A*, Sagittal T1-weighted spin-echo MR image (TR, 800 ms; TE, 30 ms) demonstrates the pathologic fracture and collapse of the T6 vertebral body, which is of pathologic low signal intensity and extends posteriorly into the spinal canal *(large straight arrow)*. A sharply marginated lesion of low signal intensity also is shown anteriorly in the T12 vertebral body *(curved arrow)*. Note the subtle decrease in signal intensity in the posterior portion of the T7 vertebral body *(small straight arrows)*. *B*, Sagittal T2*-weighted gradient-echo MR image (TR, 500 ms; TE, 30 ms; flip angle, 30 degrees). The T6 vertebral body shows high signal intensity with T2*-weighting *(large straight arrow)*. The posterior portions of the T7 vertebral body increase with this sequence, suggesting metastatic involvement *(small straight arrows)*. The lesion in the T12 vertebral body also is of high signal intensity *(curved arrow)*. Multilevel involvement is typical of metastatic disease.

than that of normal marrow) on T1 weighting and of high signal intensity (greater than that of normal marrow) on T2 weighting (see Fig. 16–27). Type I lesions frequently represent lytic metastases or multiple myeloma. Type II lesions also are focal and of low signal intensity on both T1 and T2 weighting. This MR imaging pattern reflects blastic or sclerotic metastases. Bone islands and benign vertebral sclerosis secondary to osteochondrosis can cause a similar imaging appearance. Conventional radiography is helpful to avoid confusion of metastases with benign changes. In type III, diffuse inhomogeneous signal is seen in the vertebral body (see Fig. 16–28). In type IV, diffuse homogeneous areas of low signal intensity on T1 weighting and of high signal intensity on T2 weighting are demonstrated (see Fig. 16–23). This pattern can be observed in lymphoma or other infiltrative marrow disorders.[2]

Sensitivity

Magnetic resonance imaging is significantly more sensitive than nuclear scintigraphy in detecting metastatic disease of bone.[3,4,9,48] Algra and colleagues analyzed bone scintigrams and MR images of the spine in 71 patients with histologically proved skeletal metastases in a double-blind, prospective study.[3] Nuclear scintigraphy permitted identification of 499 abnormal vertebrae, and MR imaging identified 818 abnormal vertebrae.

Avarami and associates studied 40 patients with known primary tumor suspected of having spinal metastatic disease with MR imaging of the thoracic and lumbosacral spine.[4] Conventional radiographs and CT scans of the spine were normal in all patients. Radionuclide bone scans were equivocal. In 21 patients, focal or diffuse vertebral MR imaging abnormalities were detected. Malignancy was confirmed, using needle biopsy, in the 21 patients with MR imaging abnormality.

At present, nuclear scintigraphy remains the primary screening modality for metastatic disease of bone because it is cost-effective and provides an overview of the entire skeleton. However, when radionuclide bone scans are equivocal in the patient with known primary tumor and suspected metastatic disease of bone, MR imaging is the modality of choice owing to its excellent sensitivity in detecting skeletal metastases.

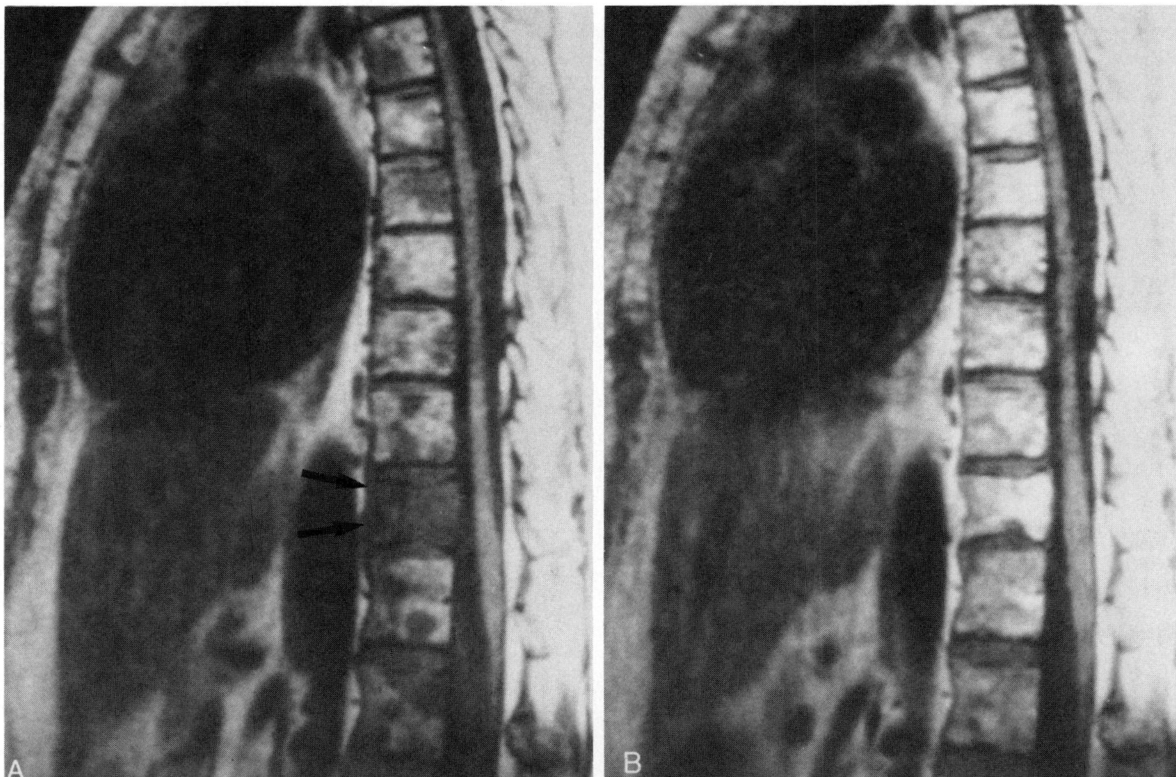

Figure 16–28. Metastatic disease of the spine (breast carcinoma). Type III signal pattern (i.e., inhomogeneous areas). *A*, Sagittal T1-weighted spin-echo MR image (TR, 1000 ms; TE, 20 ms) shows multiple foci of low signal intensity in the vertebral bodies of the thoracic spine. Overall, the areas are inhomogeneous in appearance, although some of them are more focal. A pathologic fracture of the T10 vertebral body *(arrows)* is seen. *B*, Sagittal intermediate-weighted spin-echo MR image (TR, 1000 ms; TE, 80 ms). The lesions increase in signal intensity with longer echo time.

Differential Diagnosis of Metastatic Disease

Differentiation between *vertebral osteochondrosis* and metastatic disease is not possible with radionuclide bone scans in the majority of patients.[214] Magnetic resonance imaging is superior to nuclear scintigraphy in this setting owing to its superior spatial and contrast resolution. Degenerative changes typically involve the facet joints and vertebral end plates with hypertrophic spurring. Metastatic disease usually affects the vertebral body or the pedicle. Cortical destruction is a hallmark of metastatic disease on MR images. Evidence of soft tissue mass also is highly suggestive of metastatic disease, although hematoma and edema in benign traumatic compression fracture can occasionally mimic a malignant soft tissue mass. Irregularities of the vertebral end plates and multilevel disease are more frequently found in metastatic disease, although they also can be observed in spinal osteochondrosis.

Other conditions that may be confused with vertebral metastases on MR imaging include vertebral osteomyelitis, focal fatty deposits, hemangioma, and benign compression fracture. In *vertebral osteomyelitis*, T1-weighted MR images demonstrate decreased signal intensity of the involved intervertebral disk, loss of disk height, and decreased signal intensity in the marrow spaces of the involved vertebral bodies.[148] On T2 weighting, a marked increase in signal intensity is observed in the intervertebral disk and in the involved vertebral marrow.[148] Unlike in metastatic disease, involvement of the intervertebral disk is a persistent finding in vertebral osteomyelitis and represents an important differential criterion.[148] Clinical and laboratory findings further aid in differentiating vertebral osteomyelitis from metastatic disease.

Focal fatty deposits in the vertebral bodies are characterized by high signal intensity on T1 weighting and intermediate to high signal intensity on T2 weighting.[75] Typically they are round or oval and homogeneous, with a relatively sharp interface to the adjacent marrow[75] (see the section on Other Marrow Diseases or Disorders That Affect the Marrow).

Vertebral hemangioma is characterized by high signal intensity on both T1- and T2-weighted MR images, reflecting the presence of blood products and fat[83,185] (see also Chapter 15).

Differentiation between a *benign* and a *patho-*

logic vertebral compression fracture represents another diagnostic challenge. Convex anterior and posterior contours of the compressed vertebrae are features that favor pathologic fracture[6] (see Fig. 16–26). Baker and colleagues studied patients with benign and pathologic compression fractures with T1- and T2-weighted spin-echo as well as STIR sequences.[6] Chronic benign compression fractures were characterized by homogeneous, isointense signal intensity compared with that of normal vertebral bodies on T1-weighted, T2-weighted, and STIR images except for small areas of sclerosis[6] (Fig. 16–29). Acute benign fractures demonstrated inhomogeneous low signal intensity on T1-weighted spin-echo images and inhomogeneous high signal intensity on T2-weighted spin-echo and STIR scans.[6] Pathologic fractures, by contrast, showed homogeneous replacement of vertebral marrow with low signal intensity on T1-weighted images and high signal intensity on T2-weighted and STIR sequences[6] (see Figs. 16–26 to 16–28). Generally, in simple or osteoporotic fractures, altered MR imaging signal follows closely the distribution of the fractures, but in metastatic disease it does not. On the basis of these data, MR imaging may provide differentiation between benign and malignant vertebral compression fracture.

Preliminary studies also indicate that MR imaging after intravenous administration of gadopentetate dimeglumine may provide additional information to differentiate between benign and pathologic vertebral fracture.[26] In both benign and malignant vertebral collapse, distinct contrast enhancement is observed. However, linear enhancement parallel to the fracture line is more frequent in benign, osteoporotic fracture (Fig. 16–30), whereas a patchy enhancement pattern is more typical of malignant vertebral collapse.[26]

T1 and T2 relaxation times also may be used to distinguish benign from malignant vertebral lesions.[206] The clinical utility of relaxation time measurements awaits to be investigated in larger studies.

BONE MARROW EDEMA

Bone marrow edema is a frequent finding in MR imaging of the skeleton and can be seen in association with a wide variety of disorders, such as transient osteoporosis,[10,162,230] osteonecrosis,[216] reflex sympathetic dystrophy,[187,190] bone contusion or fractures,[32,50,126,127,222] osteomyelitis and infection,[43,112,148] and neoplasm[189] (see also Chapters 10–12). The pathophysiologic principles underlying bone marrow edema are not yet completely under-

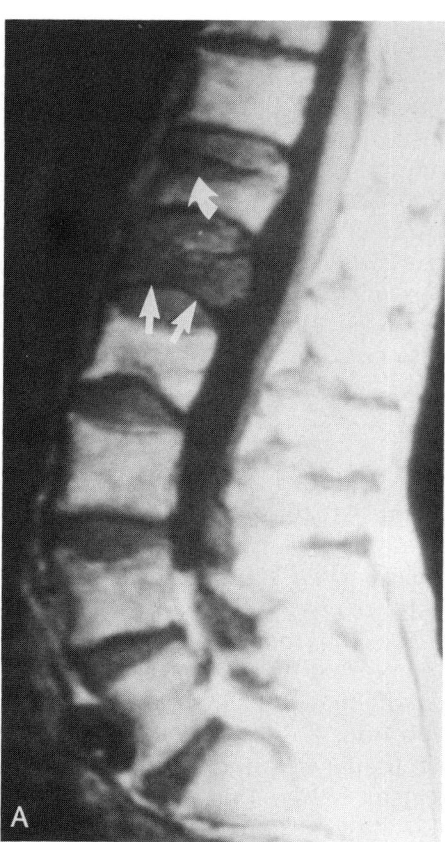

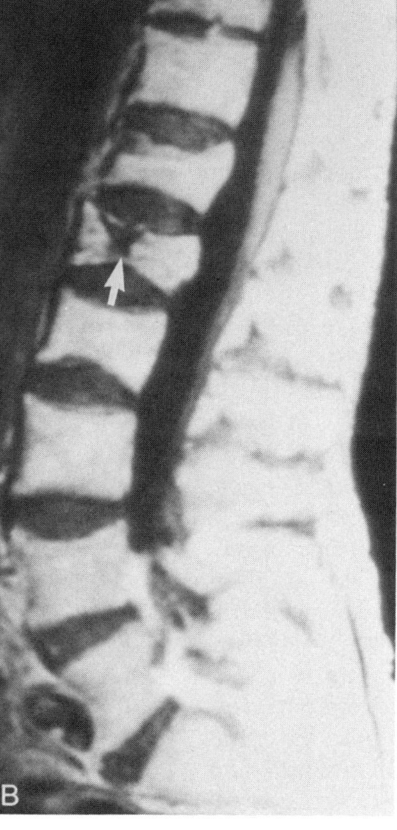

Figure 16–29. Benign osteoporotic compression fracture. *A*, Sagittal T1-weighted MR spin-echo image (TR, 800 ms; TE, 20 ms) obtained 3 days after the injury. The L1 vertebral body demonstrates low signal intensity and wedge deformity *(straight arrows)*. The superior endplate of the T12 vertebral body also is deformed and shows low signal intensity *(curved arrow)*. *B*, Sagittal T1-weighted spin-echo MR image (TR, 800 ms; TE, 20 ms) obtained 6 months after the injury. Chronic benign compression fracture of the T12 vertebral body shows homogeneous signal intensity of equal intensity to that of normal vertebral bodies. Small area with low signal intensity *(straight arrow)* reflects residual sclerosis following fracture. (Courtesy of William Dillon, M.D.)

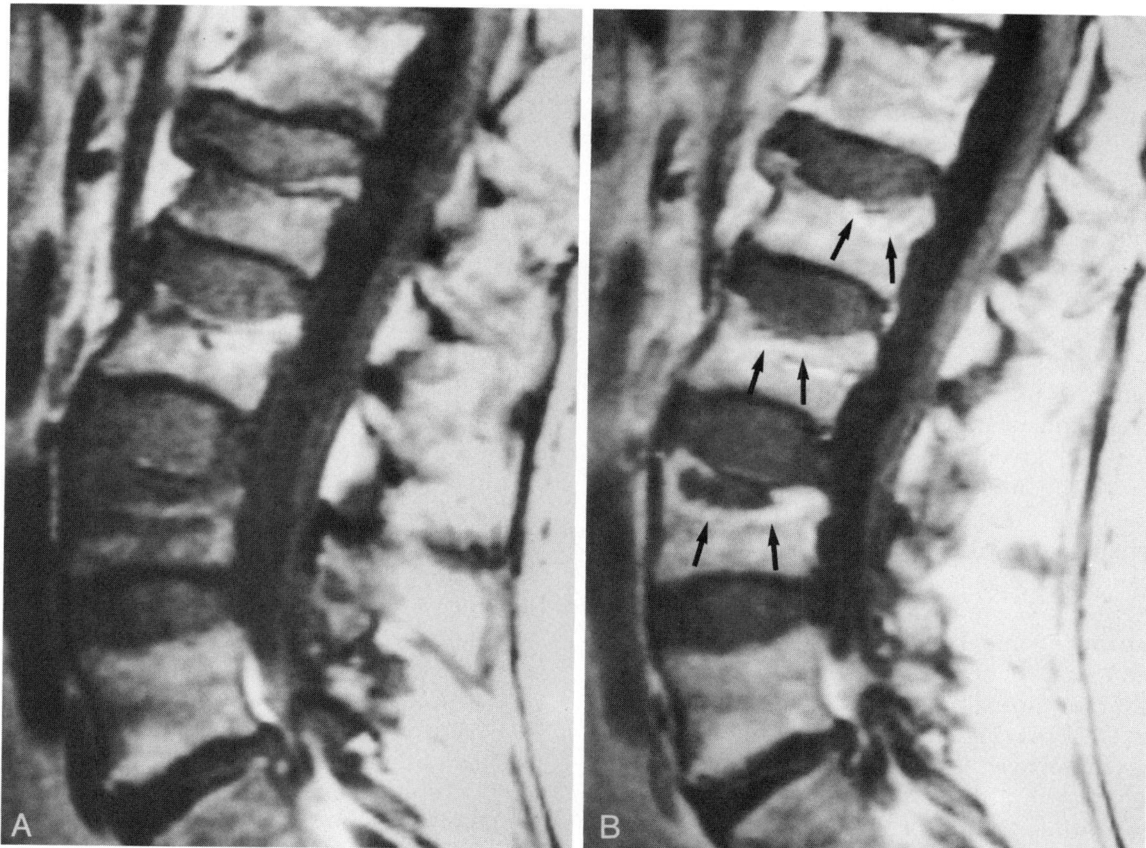

Figure 16–30. Benign osteoporotic compression fractures—gadolinium enhancement pattern. *A*, Precontrast sagittal T1-weighted spin-echo MR image (TR, 800 ms; TE, 20 ms) shows multiple compression fractures involving the L4, L3, and L2 vertebral bodies. *B*, Postcontrast sagittal T1-weighted spin-echo MR image (TR, 800 ms; TE, 20 ms) shows linear enhancement *(arrows)* paralleling the fracture lines in the fractured vertebral bodies. This is typical of a benign, osteoporotic fracture.

stood. Some researchers speculate that extracellular water increases in the interstitium and that hypervascularity and hyperperfusion contribute to this phenomenon.[220] Increased capillary permeability also may play a role in the pathophysiology of bone marrow edema.

In bone marrow edema, areas of low signal intensity are demonstrated in the marrow cavity of high signal intensity on T1-weighted images. These areas are of high signal intensity on T2 weighting. The signal abnormality usually is relatively focal and frequently is not sharply marginated. The alteration often is homogeneous in appearance. The signal alterations may be somewhat striking. They can, however, also be subtle. Chemical shift imaging techniques producing water-selective images help to highlight areas of bone marrow edema.[189]

Transient or migratory osteoporosis is an uncommon disorder of uncertain causation.[175] A common site for transient osteoporosis is the hip. Conventional radiographs usually demonstrate a focal area of osteoporosis (e.g., in the femoral head and neck).[90] Radionuclide uptake is increased[55] (see Fig. 10–13). Clinical and radiographic findings regress spontaneously within 6 to 12 months without any obvious morphologic defects.[175]

On MR imaging, diffuse signal abnormalities are observed that are typical for bone marrow edema (see Figs. 10–13, 10–14).[10,65,78,217,230] The bone marrow in the abnormal areas demonstrates low signal intensity on T1-weighted images and high signal intensity on T2-weighted images (see Figs. 10–13, 10–14). The MR imaging findings are thought to reflect transient bone marrow edema associated with transient osteoporosis[230] (see Figs. 10–13, 10–14). Interestingly, the MR imaging changes regress completely after 6 to 10 months,[10] which is important for differentiating transient osteoporosis from osteonecrosis.

Turner and associates described an imaging pattern consistent with bone marrow edema in the femoral head and neck of patients who later on developed osteonecrosis of the femoral head[216] (see Fig. 10–4). Focal signal abnormalities in the femoral epiphysis typical for osteonecrosis were not seen at the time of the MR imaging studies. Nonetheless, the patients had clinically and radiographically manifest osteonecrosis within a few months after the MR imaging study that showed bone marrow edema.[216] The pathophysiologic mechanism for bone marrow edema in patients with early stages of osteonecrosis is not known. However, venous congestion has been speculated to play a role in the pathophysiology of osteonecrosis[215] and may well be responsible for the presence of bone marrow edema.

Reflex sympathetic dystrophy and transient osteoporosis may be related disorders.[175] The pathophysiologic basis for the presence of bone marrow edema in reflex sympathetic dystrophy is not known and requires further study.

Traumatic and stress fractures both produce signal patterns consistent with bone marrow edema. On T1 weighting, areas with a signal intensity lower than that of the adjacent marrow are demonstrated. On T2-weighted images, these areas increase in signal intensity and are of high signal intensity when compared with normal marrow[32,50,126,127,222] (see Fig. 10–23). These areas are thought to correspond to bone marrow edema. However, quite possibly, acute intramedullary hemorrhage also may contribute to this signal pattern. The same signal pattern can be seen on MR imaging when conventional radiography, polytomography, and CT show no fracture and the trabecular architecture appears intact on these imaging studies.[32] In this instance, bone marrow edema demonstrated by MR imaging may indicate trabecular microfractures that are too small to be detected by radiography and CT.

METABOLIC DISORDERS

Gaucher's Disease

Gaucher's disease is a metabolic disorder characterized by increased abnormal storage of complex lipids in the reticuloendothelial system. Biochemically, there is decreased activity of the lysosomal hydrolase acid beta-glucosidase.[18] The cells typical for Gaucher's disease show marked accumulation of galactoside cerebrosides, glucoside cerebrosides, and polycerebrosides within the cytoplasm of the reticulum cells.[17] Three genetic variants of Gaucher's disease are known, corresponding to the clinical types of the disease.[17] Type I is the so-called chronic nonneuropathic or adult form; it is the most common form of Gaucher's disease. Type II is the infantile form; it is rare. Type III develops during childhood and also is named the juvenile or subacute form. Gaucher's disease affects liver, spleen, skin, bone and bone marrow, and potentially other organ systems. Gaucher's cells packing the bone marrow can cause dull bone pain. Acute bone infarction can occur and will cause crises of acute bone pain, tenderness, and temperature elevation.[132] These crises are most common in the femur. The "Gaucher crises" may simulate osteomyelitis. On radiographs, patients with Gaucher's disease have osteopenia, osteolytic lesions, pathologic fractures, and bone infarcts.

On MR imaging, homogeneous and inhomogeneous reduction in marrow signal intensity on both T1- and T2-weighted images is seen[25,74,130,183] (Fig. 16–31). The extent of marrow packing can be evaluated with MR imaging. Marrow involvement typically follows the distribution of hematopoietic marrow and progresses from proximal to distal in the appendicular skeleton, with a tendency to spare the epiphyses unless bone involvement is severe. Horev and associates reported areas of high signal intensity on both T1- and T2-weighted images in patients with a Gaucher's disease crisis.[87] The signal pattern was thought to reflect intramedullary hemorrhage.[87]

Magnetic resonance imaging is particularly useful for assessing avascular complications associated with Gaucher's disease. Bone infarcts can be identified readily. Acute and subacute bone infarcts appear as circumscribed areas of low signal intensity on T1-weighted images and of high signal intensity on T2-weighted images[25] (see Fig. 16–31). The high signal intensity may reflect edema, hyperemia, or granulation tissue. Chronic infarcts demonstrate low signal intensity on both T1- and T2-weighted images.[130,183]

The diagnosis of osteomyelitis in the presence of Gaucher's disease is difficult if not impossible with conventional radiography and nuclear scintigraphy, including radiogallium scans.[209] Magnetic resonance imaging does not provide clear differentiation between painful Gaucher's crisis and pyogenic osteomyelitis either.[25]

Amyloidosis

Amyloidosis is the accumulation and infiltration of body tissues by a protein polysaccharide complex. Two forms of amyloidosis, primary and secondary, need to be differentiated. Primary amyloidosis is idiopathic. Secondary amyloidosis occurs in chronic disease. Amyloid deposits can be seen in and around the joints or within the marrow spaces. Osteolytic lesions frequently are present in the proximal humerus and the proximal femur (see Fig. 10–29). The osteolytic lesions may have a sclerotic border. On MR imaging, amyloid typically is of low signal intensity on T1-weighted MR images (see Fig. 10–29) and of low to intermediate signal intensity on

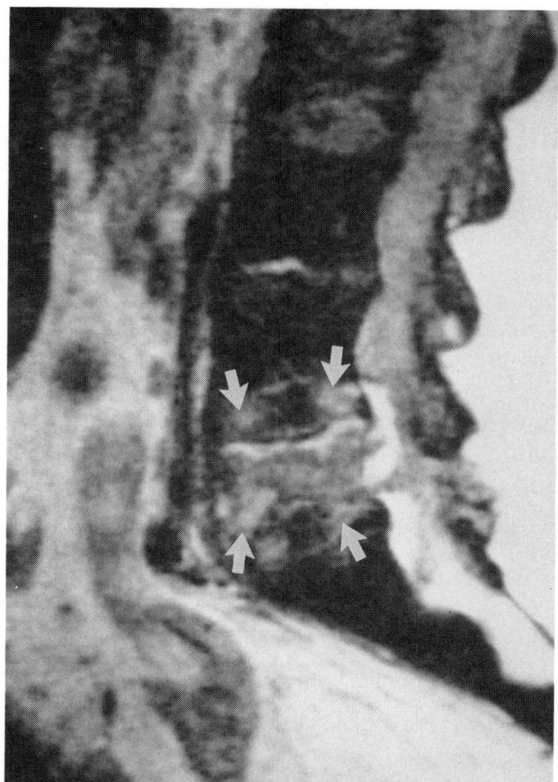

Figure 16–31. Gaucher's disease. Sagittal T2-weighted spin-echo MR image (TR, 1800 ms; TE, 120 ms) shows homogeneous low signal intensity in the marrow of the sacrum, of the L3 vertebral body, and partially of the L4 vertebral body, corresponding to marrow packing with Gaucher cells. Note areas of high signal intensity *(arrows)* in the L5 vertebral body and the inferior portions of the L4 vertebral body. These areas may reflect acute bone infarction with edema, hyperemia, or granulation tissue.

T2-weighted images. Lysis of cortical bone, progressive joint destruction, and pathologic fracture may occur as the intramedullary amyloid deposits expand.

OTHER MARROW DISEASES OR DISORDERS THAT AFFECT THE MARROW

Focal Fatty Deposition

Focal fatty deposits in areas of hematopoietic marrow are a frequent finding on MR images.[52,75] Hajek and co-workers found areas of focal fatty deposition in the lumbar spine of 60 per cent of the patients enrolled in their study.[75] On MR imaging, focal fatty deposits demonstrate high signal intensity on T1-weighted images. On T2 weighting, the signal intensity of focal fatty deposits is intermediate to high, similar to that of fatty marrow.[52,75] The prevalence of focal fatty deposits increases with age; no gender differences have been observed.[52] The prevalence of focal fatty deposits is increased in patients with scoliosis and osteoporosis, but also after cytostatic therapy of different malignomas.[52] Focal fatty deposits represent a normal phenomenon and should not be misinterpreted as a pathologic condition.[52,75] These deposits can be differentiated from metastases, because most metastases are of low signal intensity on T1-weighted images. Absence of other findings, such as vertebral destruction and collapse or mass effect, aids in the differential diagnosis.

Hemangioma

Hemangioma is a benign vascular tumor frequently seen in the vertebral bodies and in the skull. Most vertebral hemangiomas are asymptomatic; however, they may result in radiculopathy or compressive myelopathy, or both. Hemangioma consists histologically of multiple vascular channels. On conventional radiography and CT, hypertrophic bone trabeculation is demonstrated in the vertebral body. On MR imaging, hemangioma is of high signal intensity on both T1- and T2-weighted images owing to the fat content and the presence of blood products[35,83,185] (see Fig. 15–12). The extraosseous tumor portion does not contain fat and typically is of low signal intensity.[185] Vertebral hemangiomas may occur at multiple levels (see Fig. 15–12).

Vertebral End Plate Changes

Degenerative disk disease frequently is associated with signal changes in the vertebral end plates.[149,150] Modic and colleagues have identified different patterns of signal alterations in the vertebral end plates.[150] In type 1, decreased signal intensity is seen in the end plates adjacent to the degenerated intervertebral disk on T1-weighted spin-echo images, and increased signal intensity is present on T2-weighted images, corresponding to fissuring of the end plates and vascularized fibrous tissue. In type 2, the signal intensity of the end plate is increased on T1-weighted images and is isointense or slightly increased on T2-weighted images. Type 2 corresponds histologically to yellow marrow replacement.[150] The signal changes of the end plates can be distinguished easily from benign or malignant marrow disorders, because they are always associated with intervertebral disk disease and degeneration and because they are oriented parallel to the end plates. They are likely to be caused by changes in biomechanical forces in the degenerated motion segment.[117,122]

Osteopetrosis

Several different forms of osteopetrosis have been described. The benign or autosomal dominant form is associated with normal life expectancy.[38] The autosomal recessive malignant infantile form is characterized by multiple systemic manifestations, such as anemia, hepatosplenomegaly, and cranial nerve palsies.[38] An intermediate form with recessive inheritance is known.

In the benign form of osteopetrosis, the changes may be limited to osteosclerosis, and the bone marrow may have normal appearance on MR imaging.[166] In the malignant autosomal recessive form, the osteopetrotic bone marrow is of low signal intensity on both T1- and T2-weighted images.[39-41,160,166] Patterns of abnormal distribution of bone marrow appear to be age dependent.[41] In patients younger than 1 year, marrow stores are primarily in the skull base and at the ends of the long bones. In patients 3 to 5 years old, marrow stores shift to the diaphyseal regions of long bones. These marrow spaces usually are less opaque than the surrounding sclerotic bone on conventional radiographs and may be of higher signal intensity than sclerotic bone on T2-weighted MR images.[41]

Acquired Immunodeficiency Syndrome

Some studies have reported abnormalities in the signal intensity of the vertebral bone marrow in patients with acquired immunodeficiency syndrome (AIDS).[57] The vertebral bodies may be of abnormal homogeneous or inhomogeneous low signal intensity on T1-weighted spin-echo images[57] (see Fig. 16–7*A*). On T2*-weighted gradient-echo images, the pathologic low signal intensity identified on T1-weighted spin-echo images may not be depicted readily, because low vertebral signal intensity also may be caused by inhomogeneous magnetic susceptibility at the interface between trabecular bone and marrow, with resultant spin dephasing, T2*-shortening, and signal loss (see Fig. 16–7*B*).

On histologic examination of bone marrow biopsies, increased amounts of stainable iron were found in AIDS patients with decreased signal intensity on T1-weighted spin-echo images.[57] The cause of the increased iron is not completely understood. Possible explanations include increased retention of iron within the bone marrow due to anemia of chronic disease or iron overload after multiple blood transfusions.[57]

Hemochromatosis

Hemochromatosis is an inherited disorder of iron metabolism resulting in excessive body iron. Clinical manifestations of hemochromatosis are cirrhosis, grayish pigmentation, and diabetes mellitus. The mode of inheritance is autosomal recessive. The iron overload appears to be due to an increased absorption of dietary iron. Secondary forms of hemochromatosis or so-called hemosiderosis can be seen in patients who require long-term transfusion therapy and in patients who have chronic hemolytic anemias. On MR images, the iron overload of the bone marrow results in low marrow signal on both T1- and T2-weighted images (Fig. 16–32). Concurrent low signal intensity on both T1 and T2 weighting can be observed in the liver and spleen (Fig. 16–32).

OSTEOPOROSIS AND IMAGING OF TRABECULAR STRUCTURE

Several reports indicate the potential use of MR imaging as a means of assessing bone mineral density and perhaps even bone structure without ionizing radiation.[29,37,136,137,139,193]

With increasing age, T1 and T2 relaxation times of the vertebral marrow have been demonstrated to decrease progressively[37,178]; histologically, loss of vertebral bone mineral with concomitant decrease in hematopoietic marrow and increase in fatty marrow occurs. Fatty marrow expands into the widened marrow spaces as trabecular rarefaction progresses. Consequently, the T1-relaxation times of vertebral marrow become similar to those of fat, which has relatively short T1 and T2 relaxation times. Because bone mineral density and amount of intertrabecular fatty marrow are inversely related, it has been speculated that MR relaxation parameters may in the future potentially be used to assess bone mineral density.[37] However, the amount of fatty marrow also depends on a large number of other factors, such as red cell demand, so that this approach does not appear promising at the present time.

Other recent studies have focused on the influence of trabecular bone on marrow signal intensity. Davis and colleagues performed experiments in which they immersed bone powder in water and cottonseed oil; as the bone surface-to-volume ratio increased, they observed a significant decrease in T2*-relaxation for both water and oil.[29] Similarly, Rosenthal and associates have measured a reduction

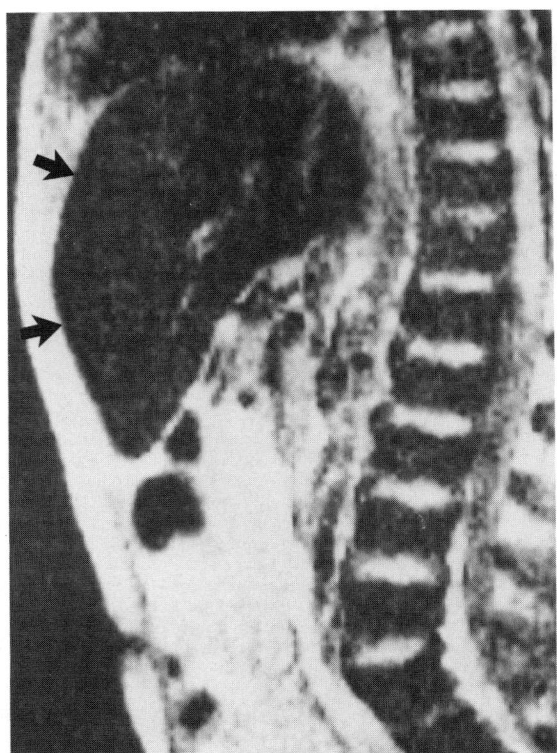

Figure 16–32. Hemochromatosis. Sagittal T2-weighted MR image (TR, 1600 ms; TE, 120 ms) of the lumbar spine shows generalized low signal intensity of the vertebral marrow due to iron overload. Note low signal intensity of the liver *(arrows)*, which also is caused by excessive iron storage.

in the T2* of water present in the trabecular spaces compared with that of extratrabecular water using specimens of excised human vertebrae.[184] Their results suggest that the presence of trabecular bone in the marrow space causes T2*-shortening and thus signal loss.

The influence of trabecular density on MR imaging signal intensity in gradient-echo images can be reproduced in vivo in imaging the knee joint.[193] Fatty marrow demonstrates high signal intensity on gradient-echo MR images in the diaphysis with its relatively low trabecular density. In areas of high trabecular density, such as the epiphysis, however, low signal intensity is observed on gradient-echo MR images, resulting from more pronounced T2*-shortening at the bone–bone marrow interface.

Bone density and T2* relaxation have been correlated directly in vitro and in vivo. Majumdar and colleagues found a strong correlation between the vertebral bone density determined by quantitative CT and the inverse of T2* relaxation (1/T2*) ($r = 0.92$; $P \leq 0.0001$) in an in vitro study[137,139] (Fig. 16–33). In a second experiment, Majumdar and colleagues also correlated elastic modulus of trabecular bone specimen with 1/T2*.[138] In these preliminary

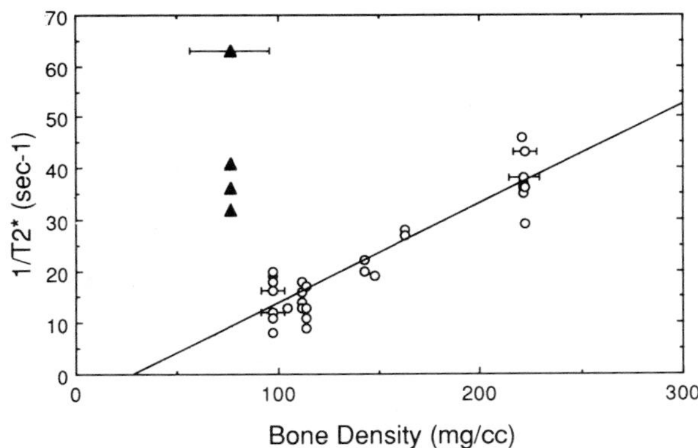

Figure 16–33. Variation of 1/T2* as a function of trabecular density. Bone mineral density was determined with quantitative computed tomography. 1/T2* was measured using a gradient-echo sequence with echo times varying from 10 to 50 ms. The correlation between 1/T2* and bone mineral density is high ($r = 0.92$, $P < 0.0001$). (Triangle: specimen that disintegrated during evacuation with a vacuum pump—excluded from statistical analysis.) (From Majumdar S, Thomasson D, Shimakawa A, et al: Quantitation of the susceptibility difference between trabecular bone and bone marrow: Experimental studies. Magn Reson Med 22:111, 1991.)

studies, the correlation of 1/T2* with elastic modulus ($r = 0.98$) was greater than that between 1/T2* and bone density ($r = 0.82$) determined by quantitative CT.[138]

Wehrli and co-workers obtained similar results using MR interferometry.[224] They found that T2* in healthy persons increases slightly with age.[224] In patients with osteoporosis, however, T2* values are significantly prolonged, which is likely to be caused by an enlargement of the intertrabecular space.[224]

An additional new MR-based modality for evaluating trabecular microarchitecture is MR microscopy. Wehrli and colleagues were able to generate microscopic images with sufficient spatial resolution to demonstrate individual trabeculae.[223,225] The in-plane resolution of these images ranged from 33 × 33 μm to 66 × 66 μm. The study had, however, been performed at a field strength of 9.4T—significantly higher than what is currently available for in vivo studies in humans. In addition, the marrow had been removed from the specimen to avoid chemical shift artifact.

T2* decay may be influenced not only by trabecular density but also by trabecular geometry, as is suggested by the greater correlation of 1/T2* with elastic modulus than in bone density studies.[138] Magnetic resonance imaging may thereby provide unique information not only on trabecular density but also on trabecular structure and architecture. The information provided by quantitative MR imaging may be useful in the future in assessing bone strength and predicting fracture risk. However, both techniques require significant further development before they can be used clinically.

References

1. Aisen AM, Martel W, Braunstein EM, et al: MRI and CT evaluation of primary bone and soft-tissue tumors. AJR 146:749–756, 1976.
2. Algra P, Bloem JL, Arndt JW, et al: Sensitivity for MRI of Vertebral Metastasis: A Comparison of MRI and Bone Scintigraphy. Amsterdam,, Society for Magnetic Resonance in Medicine, 1989, p 122.
3. Algra PR, Bloem JL, Tissing H, et al: Detection of vertebral metastases: Comparison between MR imaging and bone scintigraphy. Radiographics 11:219–232, 1991.
4. Avrahami E, Tadmor R, Dally O, Hadar H: Early MR demonstration of spinal metastases in patients with normal radiographs and CT and radionuclide bone scans. J Comput Assist Tomogr 13:598–602, 1989.
5. Baierl P, Muhlsteffen A, Haustein J, et al: Comparison of plain and Gd-DTPA-enhanced MR-imaging in children. Pediatr Radiol: 20:515–519, 1990.
6. Baker LL, Goodman SB, Perkash I, et al: Benign versus pathologic compression fractures of vertebral bodies: Assessment with conventional spin-echo, chemical-shift, and STIR MR imaging. Radiology 174:495–502, 1990.
7. Bakker CJG, Viend J: Multi-exponential water proton spin-lattice relaxation in biological tissues and its implication for quantitative NMR imaging. Phys Med Biol 29:509–518, 1984.
8. Bentz M, Dohner H, Guckel F, et al: Assessment of bone marrow infiltration by magnetic resonance imaging in patients with hairy cell leukemia treated with pentostatin or alpha-interferon. Leukemia 5:905-907, 1991.
9. Benz BG, Gross FW, Widemann B, Linden A: Bone marrow metastases of neuroblastoma: MRT in comparison with bone marrow cytology and mIBG scintigraphy. ROFO 152:523–527, 1990.
10. Bloem JL: Transient osteoporosis of the hip: MR imaging. Radiology 167:753–755, 1988.
11. Bohndorf K, Benz BG, Gross FW, Berthold F: MRI of the knee region in leukemic children. Part I. Initial pattern in patients with untreated disease. Pediatr Radiol 20:179–183, 1990.
12. Bohndorf K, Reiser M, Lochner B, et al: Magnetic resonance imaging of primary tumors and tumor-like lesions of bone. Skel Radiol 15:511–517, 1986.
13. Boothroyd AE, Sebag G, Brunelle F: MR appearances of bone marrow in children following bone marrow transplantation. Pediatr Radiol 21:291–292, 1991.
14. Brady TJ, Gebhardt MC, Pykett IL, et al: NMR imaging of forearms in healthy volunteers and patients with giant cell tumor of bone. Radiology 144:549–552, 1982.
15. Brasch RC: Work in progress: Methods for contrast enhancement in NMR imaging and potential applications. Radiology 147:781–788, 1983.
16. Brasch RC, Weinmann HJ, Wesbey GE: Contrast-enhanced NMR imaging: Animal studies using gadolinium-DTPA complex. AJR 142:625–630, 1984.
17. Carter HE: Sphingolipids. *In* Page IH (ed): Chemistry of Lipids as Related to Atherosclerosis. Springfield, Ill, Charles C Thomas, 1958, pp 82–95.
18. Carter HE, Glick FJ, Norris WP, Phillips GE: Biochemistry of the sphingolipids. III. Structure of sphingosin. J Biol Chem 170:285–289, 1947.
19. Cartia GL, Ciambellotti E, Coda C: Bone marrow scintigraphy with ^{99m}Tc-nanocolloid. A complement to bone scintigraphy with ^{99m}Tc-MDP in oncologic diagnosis. Minerva Med 82:715–721, 1991.
20. Cherryman GR, Smith FW: NMR scanning for skeletal tumours. Lancet 1:1403–1404, 1984.
21. Coffre C, Vanel D, Contesso G, et al: Problems and pitfalls in the use of computed tomography for local evaluation of long bone osteosarcoma. Skel Radiol 13:147–153, 1985.
22. Cohen MD, Klatte EC, Baehner RA, et al: Magnetic resonance imaging of bone marrow disease in children. Radiology 151:715–718, 1984.
23. Colman LK, Porter BA, Redmond JIII, et al: Early diagnosis of spinal metastases by CT and MR studies. J Comput Assist Tomogr 12:423–426, 1988.
24. Cova M, Kang YS, Tsukamoto H, et al: Bone marrow perfusion evaluated with gadolinium-enhanced dynamic fast MR imaging in a dog model. Radiology 179:535–539, 1991.
25. Cremin BJ, Davey H, Goldblatt J: Skeletal complications of type I Gaucher disease: The magnetic resonance features. Clin Radiol 41:244–247, 1990.
26. Cuenold CA, Laredo JD, Chicheportiche V, et al: Vertebral collapses: Distinction between porotic and malignant causes on MR images before and after Gd-DTPA enhancement. Radiology 177 (P): 240, 1990.
27. Custer RP: Studies on the structure and function of bone marrow. I. Variability of the hematopoietic pattern and consideration of method for examination. J Lab Clin Med 17:951–959, 1932.
28. Daffner RH, Lupetin AR, Dash N, et al: MRI in the detection of malignant infiltration of bone marrow. AJR 146:353–358, 1986.
29. Davis CA, Genant HK, Dunham JS: The effects of bone on proton NMR relaxation times of surrounding liquids. Invest Radiol 21:472–477, 1986.
30. Dawson KL, Moore SG, Rowland JM: Age-related marrow changes in the pelvis: MR and anatomic findings. Radiology 183:47–51, 1992.
31. Decho T, Horstmann G, Randzio G: A comparative study between bone marrow scintigraphy and magnetic resonance tomography in oncologic patients. ROFO 154:300–305, 1991.
32. Deutsch AL, Mink JH: Occult fractures of the proximal femur: MR imaging. Radiology 170:113–116, 1989.
33. Deutsch AL, Mink JH, Rosenfelt FP, Waxman AD: Incidental detection of hematopoietic hyperplasia on routine knee MR imaging. AJR 152:333–336, 1989.
34. Dewes W, Ruhlmann J, Loos U, et al: MR tomography in lymphatic and leukemic bone marrow infiltrations. A comparison with scintigraphy and conventional x-ray diagnosis. ROFO 147:654–661, 1987.
35. Dillon WP, Som PM, Rosenau W: Hemangioma of the nasal vault: MR and CT features. Radiology 180:761–765, 1991.
36. Dixon WT: Simple proton spectroscopic imaging. Radiology 153:189–194, 1984.
37. Dooms GC, Fisher MR, Hricak H, et al: Bone marrow imaging: Magnetic resonance studies related to age and sex. Radiology 155:429–432, 1985.
38. Edeiken J, Dalinka M, Karasick D: Dysplasias. *In* Edeiken J, Dalinka M, Karasick D (eds): Edeiken's Roentgen Diagnosis of Diseases of Bone. Baltimore, Williams & Wilkins, 1981, pp 1461–1744.
39. Elster AD, Theros EG, Key LL, Chen MY: Cranial imaging in autosomal recessive osteopetrosis. Part I. Facial bones and calvarium. Radiology 183:129–135, 1992.
40. Elster AD, Theros EG, Key LL, Chen MY: Cranial imaging in autosomal recessive osteopetrosis. Part II. Skull base and brain. Radiology 183:137–144, 1992.
41. Elster AD, Theros EG, Key LL, Stanton C: Autosomal recessive osteopetrosis: Bone marrow imaging. Radiology 182:507–514, 1992.
42. Emery JL, Follett GF: Regression of bone marrow haematopoiesis from the terminal digits in the foetus and infant. Br J Haematol 10:485–488, 1964.
43. Erdman WA, Tamburro F, Jayson HT, et al: Osteomyelitis: Characteristics and pitfalls of diagnosis with MR imaging. Radiology 180:533–539, 1991.
44. Erlemann R, Reiser MF, Peters PE, et al: Musculoskeletal neoplasms: Static and dynamic Gd-DTPA-enhanced MR imaging. Radiology 171:767–773, 1989.
45. Erlemann R, Sciuk J, Bosse A, et al: Response of osteosarcoma and Ewing sarcoma to preoperative chemotherapy: Assessment with dynamic and static MR imaging and skeletal scintigraphy. Radiology 175:791–796, 1990.
46. Erlemann R, Vassallo P, Bongartz G, et al: Musculoskeletal neoplasms: Fast low-angle shot MR imaging with and without Gd-DTPA. Radiology 176:489–495, 1990.
47. Erslev AJ: Medullary and extramedullary blood formation. Clin Orthop 52:25–36, 1967.
48. Federico M, Magin RL, Swartz HM, et al: Detection of bone marrow involvement in patients with cancer. Tumori 75:90–96, 1989.
49. Friedburg HG, Wimmer B, Hennig J, et al: Initial clinical experiences with RARE-MR urography. Urologe 26:309–316, 1987.
50. Froelich JW: Imaging of fractures: Stress and occult (editorial). J Rheumatol 18:4–6, 1991.
51. Fruehwald FXJ, Tscholakoff D, Schwaighofer B, et al: Magnetic resonance imaging of the lower vertebral column in patients with multiple myeloma. Invest Radiol 23:193–199, 1988.
52. Fruhwald F, Fruhwald S, Hajek PC, et al: Focal fatty deposits in spinal bone marrow—MR findings. MRI of focal fatty deposits. ROFO 148:75–78, 1988.
53. Fullerton GD, Potter JL, Dornbluth NC: NMR relaxation of protons in tissues and other macromolecular water solutions. Magn Reson Imaging 1:209–228, 1982.
54. Gamillscheg A, Urban C, Ranner G, et al: Malignant bone marrow infiltration in nuclear magnetic resonance tomography. Pediatr Radiol 24:103–111, 1989.
55. Gaucher A, Colomb JN, Naoun AR, et al: The diagnostic value of Tc-99m diphosphonate bone imaging in transient osteoporosis of the hip. J Rheumatol 6:574–583, 1979.
56. Gerard EL, Ferry JA, Amrein PC, et al: Compositional changes in vertebral bone marrow during treatment for acute leukemia: Assessment with quantitative chemical shift imaging. Radiology 183:39–46, 1992.
57. Geremia GK, McCluney KW, Adler SS, et al: The magnetic resonance hypointense spine of AIDS. J Comput Assist Tomogr 14:785–789, 1990.
58. Gillespy TIII, Manfrini M, Ruggieri P, et al: Staging of intraosseous extent of osteosarcoma: Correlation of preoperative CT and MR imaging with pathologic macroslides. Radiology 167:765–767, 1988.

59. Glover GH: Multipoint Dixon technique for water and fat proton and susceptibility imaging. J Magn Reson Imaging 1:521–530, 1991.
60. Godersky JC, Smoker WR, Knutzon R: Use of magnetic resonance imaging in the evaluation of metastatic spinal disease. Neurosurgery 21:676–680, 1987.
61. Golfieri R, Baddeley H, Pringle JS, Souhami R: The role of the STIR sequence in magnetic resonance imaging examination of bone tumours. Br J Radiol 63:251–256, 1990.
62. Gomori JM, Grossman RI: Mechanisms responsible for the MR appearance and evolution of intracranial hemorrhage. Radiographics 8:427–440, 1988.
63. Gomori JM, Grossman RI, Goldberg HI, et al: Intracranial hematomas: Imaging by high-field MR. Radiology 157:87–93, 1985.
64. Gomori JM, Holland GA, Grossman RI, et al: Fat suppression by section select gradient reversal on spin-echo imaging. Radiology 168:493–495, 1988.
65. Grimm J, Higer HP, Benning R, Meairs S: MRI of transient osteoporosis of the hip. Arch Orthop Trauma Surg 110:98–102, 1991.
66. Guckel F, Brix G, Semmler W, et al: Proton chemical shift imaging of bone marrow for monitoring therapy in leukemia. J Comput Assist Tomogr 14:954–959, 1990.
67. Guckel F, Brix G, Semmler W, et al: Systemic bone marrow disorders: Characterization with proton chemical shift imaging. J Comput Assist Tomogr 14:633–642, 1990.
68. Guckel F, Dohner H, Knauf W, et al: MR tomography detection of bone marrow infiltration by malignant lymphomas. Onkologie 1:34–37, 1989.
69. Guckel F, Heilig B, Semmler W, et al: Magnetic resonance tomographic imaging of bone marrow involvement in angioimmunoblastic lymphadenopathy. ROFO 153:345–347, 1990.
70. Guckel F, Semmler W, Brix G, Bachert BP, et al: Bone marrow changes in Hodgkin's disease: MR tomography and chemical shift imaging. ROFO 150:670–673, 1989.
71. Guckel F, Semmler W, Dohner H, et al: NMR tomographic imaging of bone marrow infiltrates in malignant lymphoma. ROFO 150:26–31, 1989.
72. Guckel F, Semmler W, Knauf W, et al: NMR-tomographic quantification of infiltrative bone marrow changes and therapeutic follow-up of patients with hairy cell leukemia. ROFO 152:595–600, 1990.
73. Haase A: Snapshot FLASH MRI: Applications to T1, T2, and chemical-shift imaging. Magn Reson Med 13:77–89, 1990.
74. Hainaux B, Christophe C, Hanquinet S, et al: Gaucher's disease. Plain radiography, US, CT and MR diagnosis of lungs, bone and liver lesions. Pediatr Radiol 22:78–79, 1992.
75. Hajek PC, Baker LL, Goobar JE, et al: Focal fat deposition in axial bone marrow: MR characteristics. Radiology 162:245, 1987.
76. Hanna SL, Fletcher BD, Fairclough DL, et al: Magnetic resonance imaging of disseminated bone marrow disease in patients treated for malignancy. Skel Radiol 20:79–84, 1991.
77. Hashimoto M: Pathology of bone marrow. Acta Haematol 27:193–216, 1962.
78. Hauzeur JP, Hanquinet S, Gevenois PA, et al: Study of magnetic resonance imaging in transient osteoporosis of the hip. J Rheumatol 18:1211–1217, 1991.
79. Hazelwood CF, Nichols BL, Chamberlain NF: Evidence for existence of a minimum of two phases of ordered water in skeletal muscle. Nature 222:747–750, 1969.
80. Henkelman RM, Messner H, Poon PY, et al: Magnetic resonance imaging for monitoring relapse of acute myeloid leukemia. Leuk Res 12:811–816, 1988.
81. Hennig J, Friedburg H: Clinical applications and methodological developments of the RARE technique. Magn Reson Imaging 6:391–395, 1988.
82. Hennig J, Friedburg H, Ott D: Fast three-dimensional imaging of cerebrospinal fluid. Magn Reson Med 5:380–383, 1987.
83. Heredia C, Mercader JM, Graus F, et al: Hemangioma of the vertebrae: Contribution of magnetic resonance to its study. Neurologia 4:336–339, 1989.
84. Herfkens R, Davis P, Crooks L, et al: Nuclear magnetic resonance imaging of the abnormal live rat and correlation with tissue characteristics. Radiology 141:211–218, 1981.
85. Hoane BR, Shields AF, Porter BA, Shulman HM: Detection of lymphomatous bone marrow involvement with magnetic resonance imaging. Blood 78:728–738, 1991.
86. Hoffer PB, Genant HK: Use of bone scanning agents in the evaluation of arthritis. Semin Nucl Med 6:121–137, 1976.
87. Horev G, Kornreich L, Hadar H, Katz K: Hemorrhage associated with "bone crisis" in Gaucher's disease identified by magnetic resonance imaging. Skel Radiol 20:479–482, 1991.
88. Hosten N, Sander B, Schorner W, et al: Magnetic resonance tomographic screening studies of the bone marrow with gradient echo sequences: I. The contrast relations of phase-identical and phase-shifted gradient echo sequences. Studies on probands and pathological-anatomical preparations. ROFO 154:614–620, 1991.
89. Hudson TM, Hamlin DJ, Enneking WF, et al: Detection of bone marrow metastases using quantitative computed tomography. Skel Radiol 13:134–146, 1985.
90. Hunder GG, Kelly PJ: Roentgenologic transient osteoporosis of the hip: A clinical syndrome? Ann Intern Med 68:539–552, 1968.
91. Inch WR, McCredic JA, Geiger C, et al: Spin-lattice relaxation times for mixtures of water and gelatin or cotton, compared with normal and malignant tissue. JCNI 53:689–690, 1974.
92. Irving MG, Brooks WM, Brereton IM, et al: Use of high resolution in vivo volume selected ^{1}H-magnetic resonance spectroscopy to investigate leukemia in humans. Cancer Res 47:3901–3906, 1987.
93. Ishizaka H, Kurihara M, Heshiki A, et al: MR imaging of the bone marrow using short TI IR. I. Normal and pathological intensity distribution of the bone marrow. Nippon Igaku Hoshasen Gakkai Zasshi 49:128–133, 1989.
94. Jaramillo D, Laor T, Hoffer FA, et al: Epiphyseal marrow in infancy: MR imaging. Radiology 180:809–812, 1991.
95. Jenkins JPR, Stehling M, Sivewright G, et al: Quantitative magnetic resonance imaging of vertebral bodies: A T1 and T2 study. Magn Reson Imag 7:17–23, 1989.
96. Jensen KE, Grube T, Thomsen C, et al: Prolonged bone marrow T1-relaxation in patients with polycythemia vera. Magn Reson Imaging 6:291–292, 1988.
97. Jensen KE, Grundtvig SP, Thomsen C, et al: Magnetic resonance imaging of the bone marrow in patients with acute leukemia during and after chemotherapy. Changes in T1 relaxation. Acta Radiol 31:361–369, 1990.
98. Jensen KE, Jensen M, Grundtvig P, et al: Localized in vivo proton spectroscopy of the bone marrow in patients with leukemia. Magn Reson Imaging 8:779–789, 1990.
99. Jensen KE, Sorensen PG, Thomsen C, et al: Prolonged T1 relaxation of the hematopoietic bone marrow in patients with chronic leukemia. Acta Radiol 31:445–448, 1990.
100. Jensen KE, Stenver D, Jensen M, et al: Magnetic resonance imaging of the bone marrow following treatment with recombinant human erythropoietin in patients with end-stage renal disease. Int J Artif Organs 13:477–481, 1990.
101. Jensen KE, Thomsen C, Henriksen O, et al: Changes in T1 relaxation processes in the bone marrow following treatment in children with acute lymphoblastic leukemia. A magnetic resonance imaging study. Pediatr Radiol 20:464–468, 1990.
102. Johnson JA, Herfkens RJ, Brown MA: Tissue relaxation time: In vivo field dependence. Radiology 156:805–810, 1985.
103. Joseph PM: A spin echo chemical shift MR imaging technique. J Comput Assist Tomogr 9:651–658, 1985.
104. Joseph PM, Shetty A: A comparison of selective saturation and selective echo chemical shift imaging techniques. Magn Reson Imaging 6:421–430, 1988.
105. Kangarloo H, Dietrich RB, Taira R, et al: MR imaging of bone marrow in children. J Comput Assist Tomogr 10:205–209, 1986.
106. Kaplan KR, Mitchell DG, Steiner RM, et al: Polycythemia vera and myelofibrosis: Correlation of MR imaging, clinical, and laboratory findings. Radiology 183:329–334, 1992.
107. Kaplan PA, Asleson RJ, Klassen LW, Duggan MJ: Bone marrow patterns in aplastic anemia: Observations with 1.5 T MR imaging. Radiology 164:441–444, 1987.
108. Keller PJ, Hunter WW, Schmalbrock P: Multisection fat-water imaging with selective presaturation. Radiology 164:539–541, 1987.
109. Kimura F, Kim KS, Friedman H, et al: MR imaging of the normal and abnormal clivus. AJR 155:1285–1291, 1990.
110. Kirsch DL, Colletti PM, Zee C-S, et al: MRI of the skull. Magn Reson Imaging 8:217–222, 1990.
111. Knauf WU, Guckel F, Dohner H, et al: Detection of bone marrow infiltration by non-Hodgkin's lymphoma—comparison of histological findings, analysis of gene rearrangements, and examination by magnetic resonance imaging. Klin Wochenschr 69:345–350, 1991.
112. Koenker RM, DeLuca SA: Vertebral osteomyelitis. Am Fam Physician 39:169–174, 1989.
113. Kricun ME. Red-yellow marrow conversion: Its effect on the location of some solitary bone lesions. Skel Radiol 14:10–19, 1985.
114. Krol G, Sze G, Malkin M, Walker R: MR of cranial and spinal meningeal carcinomatosis: Comparison with CT and myelography. AJNR 9:709–714, 1988.
115. Kusumoto S: Magnetic resonance imaging of the bone marrow in patients with aplastic anemia and myelodysplastic syndrome. Rinsho Ketsueki 33:423–429, 1992.
116. Lang P: Magnetic Resonance Imaging in Avascular Necrosis of the Femoral Head. Stuttgart, Enke, 1990.
117. Lang P, Chafetz N, Genant HK, Morris JM: Lumbar spinal fusion. Assessment of functional stability with magnetic resonance imaging. Spine 15:581–588, 1990.
118. Lang P, Fritz R, Majumdar S, Genant HK: Hämatopoetische Knochenmarkshyperplasie: mögliche Ursache von Fehlinterpretationen in der MR-Tomographie. Wiesbaden, Springer-Verlag, 1991, p 240.
119. Lang P, Fritz R, Majumdar S, Genant HK: Hematopoietic bone marrow

hyperplasia of the knee: Use of out-of-phase gradient-echo MR imaging. *In* Lemke HU, Rhodes ML, Jaffe CC, et al. (eds): Computer Assisted Radiology [Computergestützte Radiologie]. Berlin, Springer-Verlag, 1991, pp 36–39.
120. Lang P, Fritz R, Majumdar S, et al: Hematopoietic bone marrow in the adult knee: Spin-echo and opposed-phase gradient-echo MR imaging. Skel Radiol 22:95–103, 1993.
121. Lang P, Fritz R, Vahlensieck M, et al: Residual and reconverted hematopoietic bone marrow in the distal femur. Spin-echo and opposed-phase gradient-echo MRT. ROFO 156:89–95, 1992.
122. Lang P, Genant HK, Chafetz N, et al: Magnetic resonance tomography in the determination of the functional stability of posterolateral lumbar spondylodeses. ROFO 147:420, 426, 1987.
123. Lang P, Jergesen HE, Moseley ME, et al: Avascular necrosis of the femoral head: High-field-strength MR imaging with histologic correlation. Radiology 169:517–524, 1988.
124. Lang P, Li J, Stevens M, et al: Rheumatoid Arthritis of the Wrist: Comparison of Unenhanced and Contrast-enhanced Spin-echo, Gradient-echo, and Fat Saturation Sequences. Berlin, Society for Magnetic Resonance in Medicine, 1992.
125. Lang P, Majumdar S, Fritz R, et al: Bone Marrow MR Imaging: Assessment of Contrast Phenomena in Hematopoietic Hyperplasia. Vienna, European Congress of Radiology, 1991, p 1768.
126. Lang P, Mauz M. Schörner W, et al: Acute femoral neck fracture: Assessment of femoral head perfusion using gadopentetate dimeglumine enhanced MR imaging. AJR (in press).
127. Lang P, Schwetlick G, Langer M, et al: Acute femoral neck fracture: Unenhanced and Gd-DTPA enhanced MR imaging. Magn Reson Imaging 8:12, 1990.
128. Lang P, Schwetlick G, Mauz M, et al: Avascular necrosis of the hip: Unenhanced and Gd-DTPA enhanced MR imaging. Washington, DC, 90th Annual Meeting of the American Roentgen Ray Society, 1990.
129. Lang P, Stevens M, Vahlensieck M, et al: Rheumatoid arthritis of the hand and wrist: Evaluation of soft-tissue inflammation and quantification of inflammatory activity using unenhanced and dynamic Gd-DTPA enhanced MR imaging. Society of Magnetic Resonance in Medicine, San Francisco, 1991, p 66.
130. Lanir A, Hadar H, Cohen I, et al: Gaucher's disease: Assessment with MR imaging. Radiology 161:239–244, 1986.
131. Lee JKT, Dixon WT, Ling D, et al: Fatty infiltration of the liver: Demonstration by proton spectroscopic imaging. Radiology 153:195–201, 1984.
132. LeRoy JG: The oligosaccharidoses (formerly mucolipidoses). *In* Emery AEH, Rimoin DL (eds): Principles and Practice of Medical Genetics Edinburgh, Churchill Livingstone, 1983, pp 1348–1359.
133. Libshitz HI, Malthouse SR, Cunningham D, et al: Multiple myeloma: Appearance at MR imaging. Radiology 182:833–837, 1992.
134. Linden A, Zankovich R, Theissen P, et al: Bone marrow scintigraphy and magnetic resonance tomography in malignant lymphomas: comparison with histologic results. Nuklearmedizin 28:166–171, 1989.
135. Lodes CC, Felmlee JP, Ehman RL, et al: Proton MR chemical shift imaging using double and triple phase contrast acquisition methods. J Comput Assist Tomogr 13:855–861, 1989.
136. Majumdar S: Quantitative study of the susceptibility difference between trabecular bone and bone marrow: Computer simulations. Magn Reson Med 22:101–110, 1991.
137. Majumdar S, Genant HK: Quantitation of susceptibility effects in trabecular bone and their correlation with bone density. Society for Magnetic Resonance Imaging. 8th Annual Meeting, Printed Program Supplement 1990, p 22.
138. Majumdar S, Keyak J, Lee I, et al: Relationship between the mechanical properties of trabecular bone and intratrabecular marrow relaxation time T2*. Book of Abstracts, Berlin, Society for Magnetic Resonance in Medicine, 1992, p 1301.
139. Majumdar S, Thomasson D, Shimakawa A, Genant HK: Quantitation of the susceptibility difference between trabecular bone and bone marrow: Experimental studies. Magn Reson Med 22:111–127, 1991.
140. Mankad VN, Yang YM, Williams JP, Brogdon BG: Magnetic resonance imaging of bone marrow in sickle cell patients. Am J Pediatr Hematol Oncol 10:344–347, 1988.
141. Mansfield P, Morris PG: Water in biological systems. *In* NMR Imaging in Biomedicine. New York, Academic Press, 1982, pp 10–31.
142. Markisz JA, Knowles JR, Altchek DW, et al: Segmental patterns of avascular necrosis of the femoral heads: Early detection with MR imaging. Radiology 162:717–720, 1987.
143. McKinstry CS, Steiner RE, Young AT, et al: Bone marrow in leukemia and aplastic anemia: MR imaging before, during, and after treatment. Radiology 162:701–707, 1987.
144. Mitchell DG, Burk DL, Vinitski S, Rifkin M: The biophysical basis of tissue contrast in extracranial MR imaging. AJR 149:831–837, 1987.
145. Mitchell DG, Rao VM, Dalinka M, et al: Hematopoietic and fatty marrow distribution in the normal and ischemic hip: New observations with 1.5T MR imaging. Radiology 161:199–202, 1986.
146. Mitchell DG, Rao VM, Dalinka MK, et al: Femoral head avascular necrosis: Correlation with MR imaging, radiographic staging, radionuclide imaging, and clinical findings. Radiology 162:709–715, 1987.
147. Mitchell DG, Vinitski S, Rifkin MD, Burk DL: Sampling brand width and fat suppression: Effects on long TR/TE MR imaging of the abdomen and pelvis at 1.5T. AJR 153:419–425, 1989.
148. Modic MT, Feiglin DH, Piraino DW, et al: Vertebral osteomyelitis: Assessment using MR. Radiology 157:157–166, 1985.
149. Modic MT, Ross JS: Magnetic resonance imaging in the evaluation of low back pain. Orthop Clin North Am 22:283–301, 1991.
150. Modic MT, Steinberg PM, Ross JS, et al: Degenerative disk disease: Assessment of changes in vertebral body marrow with MR imaging. Radiology 166:193–199, 1988.
151. Moore SG, Dawson KL: Red and yellow marrow in the femur: Age-related changes in appearance at MR imaging. Radiology 175:219–223, 1990.
152. Moore SG, Gooding CA, Brasch RC, et al: Bone marrow in children with acute lymphocytic leukemia: MR relaxation times. Radiology 160:237–240, 1986.
153. Mufti GJ, Hamblin TJ, Lee-Potter JP: The aplasia-leukemia syndrome. Aplastic anemia followed by dyserythropoiesis, myeloproliferative syndrome, and acute myeloid leukemia. Acta Haematol 69:349–356, 1983.
154. Mulkern RV, Wong ST, Winalski C, Jolesz FA: Contrast manipulation and artifact assessment of 2D and 3D RARE sequences. Magn Reson Imaging 8:557–566, 1990.
155. Murphy S: Polycythemia vera. *In* Williams W, Beutler E, Erslev A, et al (eds): Hematology. New York, McGraw-Hill, 1989, pp 185–196.
156. Negendank W, Weissman D, Bey TM, et al: Evidence for clonal disease by magnetic resonance imaging in patients with hypoplastic marrow disorders. Blood 78:2872–2879, 1991.
157. Nyman R, Rehn S, Glimelius B, et al: Magnetic resonance imaging in diffuse malignant bone marrow diseases. Acta Radiol 28:199–205, 1987.
158. Okada Y, Aoki S, Barkovich AJ, et al: Cranial bone marrow in children: Assessment of normal development with MR imaging. Radiology 171:161–164, 1989.
159. Olson DO, Shields A, Scheurich CJ, et al: Magnetic resonance imaging of the bone marrow in patients with leukemia, aplastic anemia, and lymphoma. Invest Radiol 21:540–546, 1986.
160. Otsuka N, Fukunaga M, Ono S, et al: Bone marrow scintigraphy and MRI in a patient with osteopetrosis. Clin Nucl Med 16:443–445, 1991.
161. Papavasiliou C, Trakadas S, Gouliamos A, et al: Magnetic resonance imaging of marrow heterotopia in haemoglobinopathy. Eur J Radiol 8:50–53, 1988.
162. Pay NT, Singer WS, Bartal E: Hip pain in three children accompanied by transient abnormal findings on MR images. Radiology 171:147–149, 1989.
163. Pettersson H, Gillespy T, Hamlin DJ, et al: Primary musculoskeletal tumors: Examination with MR imaging compared to conventional modalities. Radiology 164:237–241, 1987.
164. Piney A: The anatomy of the bone marrow. With special reference to the distribution of the red marrow. Br J Med 2:792–795, 1922.
165. Porter BA, Olson DO, Stimack GK: STIR Imaging of Marrow Malignancies. New York, Society for Magnetic Resonance in Medicine, 1987, p 146.
166. Rao VM, Dalinka MK, Mitchell DG, et al: Osteopetrosis: MR characteristics at 1.5T. Radiology 161:127–130, 1986.
167. Rao VM, Fishman M, Mitchel DG, et al: Painful sickle cell crisis: Bone marrow patterns observed with MR imaging. Radiology 161:211–215, 1986.
168. Rao VM, Mitchell DG, Rifkin MD, et al: Marrow infarction in sickle cell anemia: Correlation with marrow type and distribution by MRI. Magn Reson Imaging 7:39–44, 1989.
169. Rao VM, Mitchell DG, Steiner RM, et al: Femoral head avascular necrosis in sickle cell anemia: MR characteristics. Magn Reson Imaging 6:661–667, 1988.
170. Rappaport J, Bunn H: Bone marrow failure, aplastic anemia, and other primary bone marrow disorders. *In* Braunwald E, Isselbacher K, Petersdorf R, et al, (eds): Harrison's Principles of Internal Medicine. New York, McGraw-Hill, 1991, pp 1567–1571.
171. Reiser M, Bohndorf K, Niendorf HP, et al: Initial experiences with gadolinium DTPA in magnetic resonance tomography of bone and soft tissue tumors. Radiologe 27:467–472, 1987.
172. Reiser M, Bongartz GP, Erlemann R, et al: Gadolinium-DTPA in rheumatoid arthritis and related diseases: First results with dynamic magnetic resonance imaging. Skel Radiol 18:591–597, 1989.
173. Reiser M, Kaiser W, Zeitler E: Bone marrow diseases in children. Significance of MRI in comparison to other imaging modalities. Ann Radiol 30:511, 1987.
174. Reske SN: Recent advances in bone marrow scanning. Eur J Nucl Med 18:203–221, 1991.
175. Resnick D, Niwayama G: Transient osteoporosis of the hip. *In* Resnick D, Niwayama G (eds): Diagnosis of Bone and Joint Disorders, Philadelphia, W. B. Saunders, 1981, p 1653.

176. Ricci C, Cova M, Kang YS, et al: Normal age-related patterns of cellular and fatty bone marrow distribution in the axial skeleton: MR imaging study. Radiology 177:83–88, 1990.
177. Richards MA, Webb JA, Jewell SE, et al: Low field strength magnetic resonance imaging of bone marrow in patients with malignant lymphoma. Br J Cancer 57:412–415, 1988.
178. Richards MA, Webb JA, Jewell SE, et al: In-vivo measurement of spin lattice relaxation time (T1) of bone marrow in healthy volunteers: The effects of age and sex. Br J Radiol 61:30–33, 1988.
179. Roberts N, Smith SR, Percy DF, et al: The quantitative study of lumbar vertebral bone marrow using T1 mapping and image analysis techniques: Methodology and preliminary results. Br J Radiol 64:673–678, 1991.
180. Rosen BR, Fleming DM, Kushner DC, et al: Hematologic bone marrow disorders: Quantitative chemical shift MR imaging. Radiology 169:799–804, 1988.
181. Rosen BR, Wedeen VJ, Brady TJ: Selective saturation NMR imaging. J Comput Assist Tomogr 8:813–818, 1984.
182. Rosenthal DI, Hayes CW, Rosen B et al: Fatty replacement of spinal bone marrow due to radiation: Demonstration by dual energy quantitative CT and MR imaging. J Comput Assist Tomogr 13:463–465, 1989.
183. Rosenthal DI, Scott JA, Barranger J, et al: Evaluation of Gaucher's disease using magnetic resonance imaging. J Bone Joint Surg 68A: 802–808, 1986.
184. Rosenthal H, Thulborn KR, Rosenthal DI, et al: Magnetic susceptibility effects of trabecular bone on magnetic resonance imaging of bone marrow. Invest Radiol 25:173–178, 1990.
185. Ross JS, Masaryk TJ, Modic MT, et al: Vertebral hemangiomas: MR imaging. Radiology 165:165–169, 1987.
186. Ruther W, Hotze A, Moller F, et al: Diagnosis of bone and joint infection by leucocyte scintigraphy. A comparative study with ^{99m}Tc-HMPAO-labelled leucocytes, ^{99m}Tc-labelled antigranulocyte antibodies and ^{99m}Tc-labelled nanocolloid. Arch Orthop Trauma Surg 110:26–32, 1990.
187. Sarrat P, Acquaviva PC, Lafforgue P, et al: Algodystrophy of the femoral head. Contribution of new imaging methods. J Radiol 69:495–500, 1988.
188. Schicha H, Franke M, Smolorz J, et al: Diagnostic strategies and staging procedures for Hodgkin's lymphoma: Bone marrow scintigraphy and magnetic resonance imaging. Recent Results Cancer Res 117:112–119, 1989.
189. Schick F, Bongers H, Aicher K, et al: Subtle bone marrow edema assessed by frequency selective chemical shift MRI. J Comput Assist Tomogr 16:454–460, 1992.
190. Schimmerl S, Schurawitzki H, Imhof H, et al: Sudeck's disease—MRT as a new diagnostic procedure. ROFO 154:601–604, 1991.
191. Schorner W, Lang P, Bittner R, Felix R: Gadolinium-DTPA enhanced MR imaging: Applications in body MRI. Diagn Imaging 4:114–119, 1990.
192. Scott WA, Mark AS, Grossman RI, Gomori JM: Intracranial hemorrhage: Gradient-echo MR imaging at 1.5T Radiology 168:803–807, 1988.
193. Sebag GH, Moore SG: Effect of trabecular bone on the appearance in marrow in gradient echo imaging of the appendicular skeleton. Radiology 174:855–859, 1990.
194. Seneterre E, Weissleder R, Jaramillo D, et al: Bone marrow: Ultrasmall superparamagnetic iron oxide for MR imaging. Radiology 179:529–533, 1991.
195. Shellock FG, Morris E, Deutsch AL, et al: Hematopoietic bone marrow hyperplasia: High prevalence on MR images of the knee in asymptomatic marathon runners. AJR 158:335–338, 1992.
196. Shields AF, Porter BA, Churchley S, et al: The detection of bone marrow involvement by lymphoma using magnetic resonance imaging. J Clin Oncol 5:225–230, 1987.
197. Smith SR, Williams CE, Davies JM, Edwards RH: Bone marrow disorders: Characterization with quantitative MR imaging. Radiology 172: 805–810, 1989.
198. Smith SR, Williams CE, Edwards RH, Davies JM: Quantitative magnetic resonance imaging in autologous bone marrow transplantation for Hodgkin's disease. Br J Cancer 60:961–965, 1989.
199. Smith SR, Williams CE, Edwards RH, Davies JM: Quantitative magnetic resonance studies of lumbar vertebral marrow in patients with refractory or relapsed Hodgkin's disease. Ann Oncol 2:39–42, 1991.
200. Smoker WRK, Godersky JC, Knutzon RK, et al: The role of MR imaging in evaluating metastatic spinal disease. AJR 149:1241–1248, 1987.
201. Snyder WS, Cook MJ, Nasset ES, et al: Report of the Task Group on Reference Man. New York, Pergamon Press, 1975.
202. Starz I, Einspieler R, Poschauko H, et al: MR study of bone marrow changes of the spine following radiotherapy. Röntgenblatter 43:355–358, 1990.
203. Stevens SK, Moore SG, Amylon MD: Repopulation of marrow after transplantation: MR imaging with pathologic correlation. Radiology 175:213–218, 1990.
204. Stevens SK, Moore SG, Kaplan ID: Early and late bone marrow changes after irradiation: MR evaluation. AJR 154:745–750, 1990.
205. Stimac GK, Porter BA, Olson DO, et al: Gadolinium-DTPA-enhanced MR imaging of spinal neoplasms: Preliminary investigation and comparison with unenhanced spin-echo and STIR sequences. AJR 151:1185–1192, 1988.
206. Sugimura K, Yamasaki K, Kitagaki H, et al: Bone marrow diseases of the spine: differentiation with T1 and T2 relaxation times in MR imaging. Radiology 165:541–544, 1987.
207. Sykes MP, Chu FCH, Wilkerson WG: Local bone marrow changes secondary to therapeutic irradiation. Radiology 75:919–924, 1960.
208. Sze G, Bravo S, Baierl P, Shimkin PM: Developing spinal column: Gadolinium-enhanced MR imaging. Radiology 180:497–502, 1991.
209. Sziklas JJ, Negrin JA, Rosshirt W, et al: Diagnosing osteomyleitis in Gaucher's disease. Observations on two cases. Clin Nucl Med 16:487–489, 1991.
210. Szumowski J, Eisen JK, Vinitski S, et al: Hybrid methods for chemical shift imaging. Magn Reson Med 9:379–388, 1989.
211. Tesoro TJ, Balzarini L, Ceglia E, et al: Magnetic resonance imaging in the initial staging of Hodgkin's disease and non-Hodgkin lymphoma. Eur J Radiol 12:81–90, 1991.
212. Thompson JA, Shields AF, Porter BA, et al: Magnetic resonance imaging of bone marrow in hairy cell leukemia: Correlation with clinical response to alpha-interferon. Leukemia 1:315–316, 1987.
213. Thomsen C, Sorensen PG, Karle H, et al: Prolonged bone marrow T1-relaxation in acute leukaemia. In vivo tissue characterization by magnetic resonance imaging. Magn Reson Imaging 5:251–257, 1987.
214. Thrall JH, Ellis BI: Skeletal metastases. Radiol Clin North Am 25: 1155–1170, 1987.
215. Tsukamoto H, Kang YS, Jones LC, et al: Evaluation of marrow perfusion in the femoral head by dynamic magnetic resonance imaging. Effect of venous occlusion in a dog model. Invest Radiol 27:275–281, 1992.
216. Turner DA, Templeton AC, Seizer P, et al: Femoral capital osteonecrosis: MR findings of diffuse marrow abnormalities without focal lesions. Radiology 171:135–140, 1989.
217. Urbanski SR, de Lange EE, Eschenroeder HJ: Magnetic resonance imaging of transient osteoporosis of the hip. A case report. J Bone Joint Surg [Am] 73:451–455, 1991.
218. Vande Berg B, Malghem J, Labaisse MA, et al: Avascular necrosis of the hip: Comparison of contrast-enhanced and non-enhanced MR imaging with histologic correlation. Radiology 182:445–450, 1992.
219. Vanel D. Coffre C, Contesso G, et al: Contribution of gadolinium DOTA in primary tumors of bone and soft tissue. J Radiol 69:735–739, 1988.
220. Vogler JB, Murphy WA: Bone marrow imaging. Radiology 168:679–693, 1988.
221. Waterbury L: Anemia. *In* Barker LR, Burton JR, Zieve PD (eds): Ambulatory Medicine. Baltimore, Williams & Wilkins, 1991, pp 549–563.
222. Weber WN, Neumann CH, Barakos JA, et al: Lateral tibial rim (Segond) fractures: MR imaging characteristics. Radiology 180:731–734, 1991.
223. Wehrli FW, Chung J, Kugelmass SD, et al: Relationship between Young's modulus and morphometric parameters in human trabecular bone studied by NMR microscopy. Book of Abstracts. Berlin, Society for Magnetic Resonance in Medicine, 1992, p 973.
224. Wehrli FW, Ford JC, Attie M, et al: Trabecular structure: Preliminary application of MR interferometry. Radiology 179:615–621, 1991.
225. Wehrli FW, Wehrli SL, Williams J et al: Anisotropy of trabecular microstructure studied by high-field NMR microscopy and linewidth measurements. Book of Abstracts. New York, Society of Magnetic Resonance in Medicine, 1990, p 127.
226. Weinmann HJ, Brasch RC, Press W-R, et al: Characteristics of Gadolinium-DTPA complex: A potential NMR contrast agent. AJR 142: 619–624, 1984.
227. Weiss L: The histophysiology of bone marrow. Clin Orthop 52:13–23, 1967.
228. Widding A, Smolorz J, Franke M, et al: Bone marrow investigation with technetium-99m microcolloid and magnetic resonance imaging in patients with malignant myelolympho-proliferative diseases. Eur J Nucl Med 15:230–238, 1989.
229. Widding A, Stilbo I, Hansen SW, et al: Scintigraphy with nanocolloid Tc-99m in patients with small cell lung cancer, with special reference to bone marrow and hepatic metastasis. Eur J Nucl Med 16:717–719, 1990.
230. Wilson AJ, Murphy WA, Hardy DC, et al: Transient osteoporosis: Transient bone marrow edema? Radiology 167:757–760, 1988.
231. Wismer GL, Rosen BR, Buxton R, et al: Chemical shift imaging of bone marrow: Preliminary experience. AJR 145:1031–1037, 1988.
232. Yankelevitz DF, Henschke CI, Knapp PH, et al: Effect of radiation therapy on thoracic and lumbar bone marrow: Evaluation with MR imaging. AJR 157:87–92, 1991.
233. Zimmer WD, Berquist TH, McLeod RA, et al. Bone tumors: Magnetic resonance imaging versus computed tomography. Radiology 155:709–718, 1985.

17 DISORDERS OF SKELETAL MUSCLE

Wing P. Chan, James L. Fleckenstein, Gin-Chung Liu, and Harry K. Genant

Magnetic resonance (MR) imaging is well known for its ability to delineate soft tissue structures. It allows not only the localization of soft tissue abnormalities but also the assessment of changes in tissue composition and of biochemical defects. Magnetic resonance imaging of muscle usually is performed to determine the extent of a focal disease process when selective testing or treatment is being considered. It can minimize the problems of sampling errors in procedures traditionally performed blindly, such as biopsy, electromyography, and MR spectroscopy.

This chapter discusses both clinical and research aspects in MR imaging of pathologic conditions of skeletal muscle.

IMAGING TECHNIQUES

The main goal in imaging abnormal skeletal muscle is the delineation of changes in the fat-water composition. Magnetic resonance imaging allows the delineation of changes with optimized protocols. Skeletal muscle is of low signal intensity on T2-weighted spin-echo sequences. On T1-weighted images, fat is of high signal intensity because of its short T1 relaxation time. On T2-weighted images, fat shows relatively high to intermediate signal intensity because it has a long T2 relaxation time. Increased tissue water results in increased spin density and T1 and T2 relaxation times, which are manifested by decreased and increased signal intensity, respectively, on heavily T1- and T2-weighted spin-echo images. In general, edema fluid becomes more intense than fat on long TR, long TE spin-echo sequences. However, it is important to note that mild edematous changes may be indistinguishable from mild fatty changes at 0.35 T if suboptimal imaging pulse sequences are employed.[22] The T2 relaxation time of fat is about 60 ms, which is twice that of normal muscle (about 30 ms). When edema is extensive enough to cause the muscle T2 time to exceed 60 ms, edematous muscle can be distinguished from fat.[22] If edema is mild and the T2 time is only slightly increased (between 30 and 60 ms), the edematous change may be indistinguishable from mild fatty changes. Hence, fat suppression sequences have been developed.

The detection of muscle edema is enhanced by using fat suppression techniques, such as a T2-weighted chemical shift sequence or short tau inversion-recovery (STIR) sequence. STIR imaging not only is highly sensitive to edematous muscle but also provides some specificity by suppressing the signal from fat. However, the drawbacks of STIR imaging include prominent motion artifacts (especially at high field strength), relatively low signal-to-noise ratios, and relatively long scan times. To obtain images sensitive to edema and short scan times, various parameters can be altered.[14]

Sequences using a short repetition time (TR) and a short echo time (TE) optimize the contrast between muscle and fat. If imaging of edemalike processes (e.g., early denervated, exercised, or necrotic muscles) is the goal, T2-weighted or STIR sequences are necessary to detect associated edematous changes. In lesions such as polymyositis, both T1- and T2-weighted (or STIR) sequences are required to sensitively detect edemalike processes. The sequelae of the primary disease process, edema, and fatty changes must be estimated. In fact, if STIR and T1-weighted spin-echo images are obtained, the time-consuming T2-weighted sequence frequently is unnecessary.

A locator sequence in a longitudinal plane, typically the coronal plane, is helpful for orientation of the extent of disease and for identification of palpable bone landmarks, such as the large joints. The axial plane provides sufficient information on the size of muscle groups and tissue characterization and has the advantage of showing all the muscles, especially in both limbs. T1-weighted images are generated with a TR of 500 to 800 ms and a TE of 20 to 30 ms. T2-weighted images are generated with a TR of 2000 to 3000 ms and a TE of 60 to 80 ms. STIR images are acquired by using a TR of 1500 ms, a TE of 30 ms, and an inversion time (TI) of 100 to 160 ms. When fine anatomic detail is needed, the smallest appropriate coil and pixel size should be used to maximize imaging efficiency. A smaller coil improves signal-to-noise ratios and, therefore, gives images with high spatial resolution that do not require additional imaging time. Conversely, when tissue characterization is more important than fine anatomic detail, large voxels can be used to maximize the signal-to-noise ratio. The combination of small coils and large voxels reduces the number of excitations and phase-encoding steps.[22] The end result is the potential for relatively fast, economical imaging for tissue characterization.

EXERCISE, IMMOBILIZATION, AND AGING

Hypertrophy is a common response in muscles subject to long-term physical training. Slow-twitch type I fibers respond to low-intensity dynamic exercise, such as marathon running or swimming, which rely upon oxidative metabolism.[2] Fast-twitch type II fibers respond to short bursts of high-intensity activity, such as weight lifting, which relies on anaerobic glycolysis.[2] Studies using animal models[6] suggest that MR imaging can discriminate between muscles having different proportions of fiber types. The rat soleus muscle, which is composed almost exclusively of type I fibers, has longer T1 and T2 relaxation times than the gastrocnemius muscle, which has a greater proportion of type II fibers.

The transient reaction of skeletal muscle to exercise includes consumption of phosphocreatine, a rise in the level of inorganic phosphate, and a decrease in pH.[11,41,47,48] These physiologic changes can be reversed within a few minutes. An increase in water content accompanies these processes. Extracellular water is increased at low-intensity exercise, and intracellular water is increased at maximal-intensity exercise.[22] Fleckenstein and colleagues reported an increase in T1 and T2 relaxation times after acute exercise[17] (Fig. 17–1). Sjogaard and Saltin[54] found that the increase in water content is mostly extracellular in nature at low signal intensity, which probably contributes to the increase in signal intensity of the exercised muscles as intracellular water in muscle has a short T2 and a long T1.[3]

Exercise-enhanced MR studies are used to probe dynamic muscle physiology and pathophysiology. Included applications of exercise enhancement are a verification of muscle boundaries and assessment of intersubject variations in normal anatomy[15,31] and pathophysiology.[19]

Disuse atrophy involving predominantly type II fibers is a common response to any prolonged period of immobilization (Fig. 17–2).[2] Atrophy of type II fibers also is a common response to various degenerative and denervative processes. Acquired contractures frequently result from a combination of immobility, severe muscle atrophy, and replacement fibrosis.

GENERAL MUSCLE REACTIONS TO DISEASES

Despite the wide variety of diseases that may involve skeletal muscle, the response of muscle to disease is relatively restricted. Denervation, myonecrosis, fiber size alteration (i.e., atrophy and hypertrophy), and connective tissue changes (i.e., fatty replacement and fibrosis) are the dominant histopathologic findings.[32]

Denervation

Denervation is the major histopathologic finding in neurogenic muscle disease. Denervated muscle is characterized by angular, atrophic fibers, and groups of atrophic fibers are pathognomonic.[2] The residual innervated fibers may show compensatory hypertrophy. Type II fibers appear to atrophy at a faster rate than type I fibers. The longer the denervation lasts, the more atrophic a fiber becomes. All fiber types atrophy most rapidly within the first 2 to 3 months after denervation and degenerate quite slowly thereafter until a stationary state is reached.[2] Chronically denervated fibers may be reduced in caliber up to 10 to 20 per cent of the original size, and the muscle may show a corresponding increase in interstitial connective tissue.[2] In severely atrophied muscle,

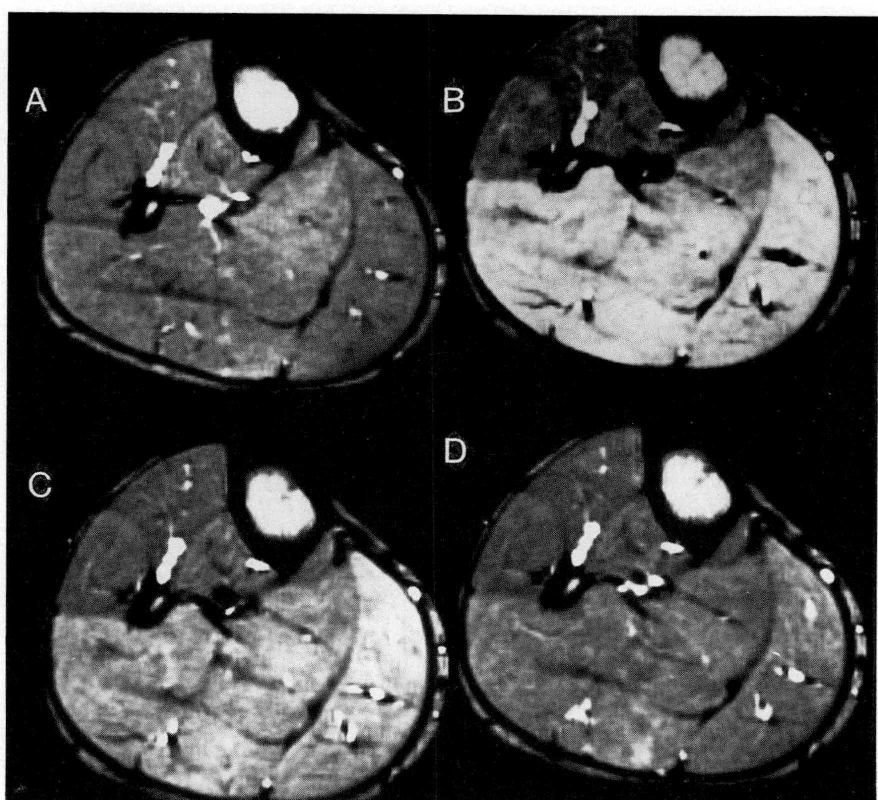

Figure 17–1. Acute effects of exercise on the MR imaging appearance of leg muscle. T2*-weighted gradient-echo images (TR, 500 ms; TE, 30 ms; flip angle, 30 degrees) before *(A)* and for the first *(B)*, second *(C)*, and third *(D)* 4-minute periods after plantar flexion. The increase of signal intensity in stressed calf muscles is normal and transient. (From Fleckenstein JL, Weatherall PT, Parkey RW, et al: Sports-related muscle injuries: Evaluation with MR imaging. Radiology 172:793, 1989.)

both adipose and fibrous replacement occur. Denervation may be reversible for several months and possibly up to a year after onset.

Magnetic resonance imaging allows the identification of the subacute phase of denervated muscles before fatty deposition.[50] In an animal model,[46] prolongation of muscle relaxation times is evident in the first few weeks after denervation and is proportional to a diminution in fiber size and an attendant increase in the extracellular water content. Prolonged proton T1 and T2 relaxation times during this early stage are reflected on STIR images by high signal intensity. The end-stage appearance of denervation is fatty change, as characterized by T1 shortening (Figs. 17–3, 17–4). On T2-weighted images, fatty changes and atrophic muscles also display high signal intensity but are of lesser intensity than edematous muscle. It is important to note that atrophic muscle with increased signal intensity on any sequence should direct attention to the nearby nerves (see Figs. 17–3, 17–4).

Myonecrosis

The most severe consequence of skeletal muscle injury is fiber necrosis. Because injury is usually focal, segmental necrosis is a common result. Segmental necrosis is followed by regeneration, but the full expression of mature fiber type is not achieved until innervation is reestablished.[2] The skeletal muscle cell does have considerable regenerative capacity, provided that viable satellite cells persist, but with sustained or repetitive injury, fibroblastic proliferation may exceed regeneration, and interstitial fibrosis develops.[2] For example, infarcted muscle resulting from ischemia undergoes all the changes of necrosis and regeneration. If the blood supply is not restored fairly rapidly, the repair will be by fibrosis and not by regeneration.

Clinically, elevation of the serum creatine kinase concentration indicates myonecrosis. However, laboratory testing cannot distinguish between a small amount of enzyme release from a large volume of muscle and a large amount of enzyme release from a small volume of tissue. In addition, the enzyme elevation is highly dependent on the timing of specimen collection.

Myonecrosis appears as areas of high signal intensity on proton-weighted, T2-weighted, and STIR sequences but usually is not apparent on T1-weighted images unless it is very severe.[23,59] Necrotic muscle, as seen on MR images, may persist for far longer than is indicated by serum creatine kinase levels.[17] Drug overdose and crush injury are common conditions for which imaging necrotic tissue is

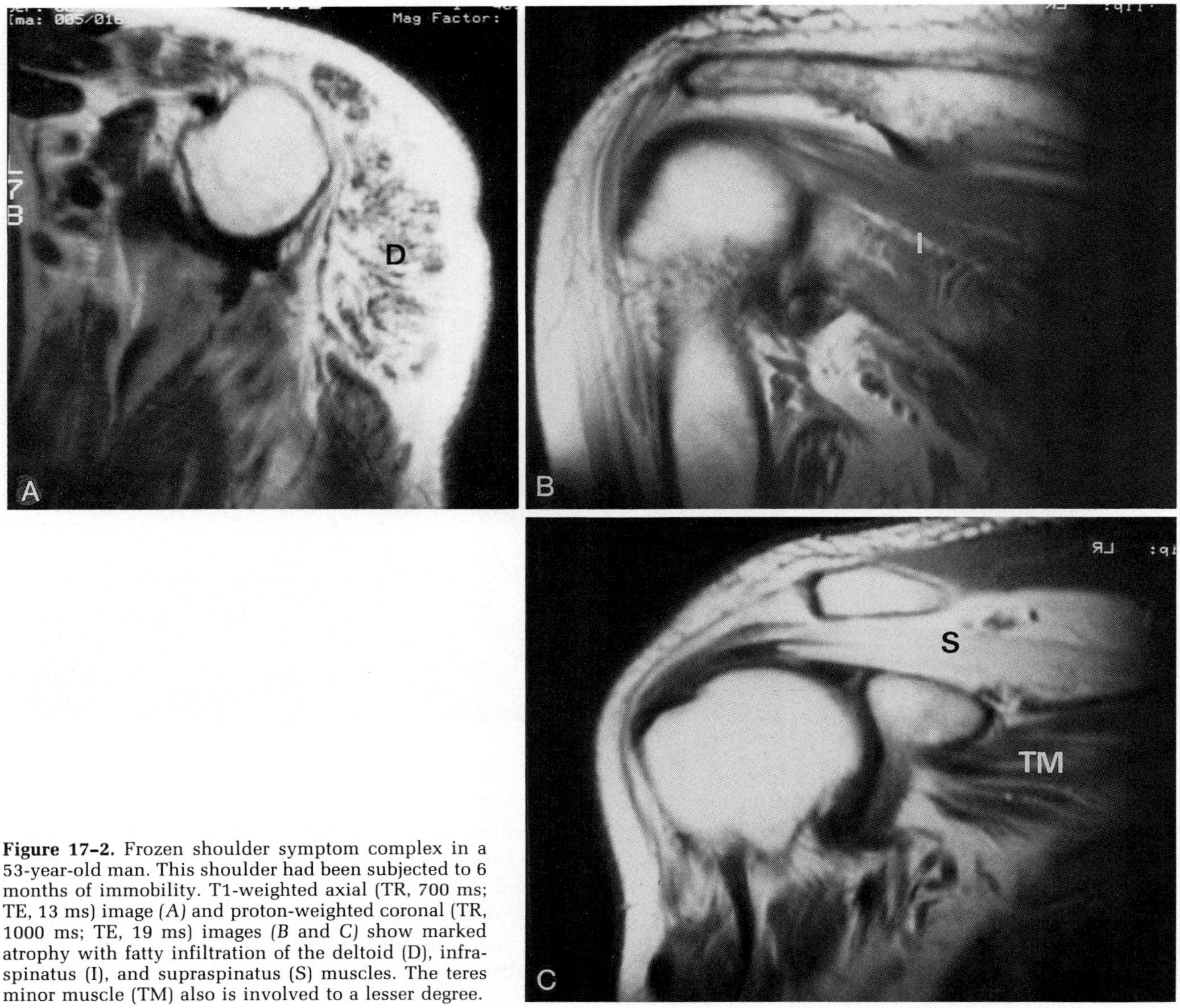

Figure 17–2. Frozen shoulder symptom complex in a 53-year-old man. This shoulder had been subjected to 6 months of immobility. T1-weighted axial (TR, 700 ms; TE, 13 ms) image *(A)* and proton-weighted coronal (TR, 1000 ms; TE, 19 ms) images *(B* and *C)* show marked atrophy with fatty infiltration of the deltoid (D), infraspinatus (I), and supraspinatus (S) muscles. The teres minor muscle (TM) also is involved to a lesser degree.

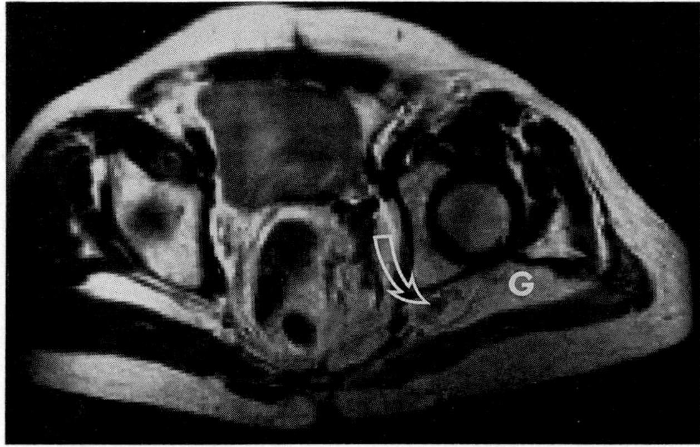

Figure 17–3. Denervation with associated muscle atrophy. Axial T2-weighted (TR, 2000 ms; TE, 60 ms) image of the pelvis shows an atrophic gluteus maximus (G). Note the nodular and enlarged sciatic nerve *(arrow)*, which has been infiltrated by prostate carcinoma. (From Fleckenstein JL, Weatherall PT, Bertocci LA, et al: Locomotor system assessment by muscle magnetic resonance imaging. Magn Reson Q 7:79, 1991.)

requested (Fig. 17–5). An interesting feature of such conditions is that some muscles are severely affected, whereas nearby muscles are spared. Similarly, focal patterns also are the rule in postexertional muscle necrosis, which occurs frequently in the weekend athlete (i.e., one who is unaccustomed to intense physical activity). Muscle strains are another form of exertion-related injury that have MR imaging features indistinguishable from necrosis, but muscle strains are not usually considered a form of necrosis. Inflammatory conditions of the muscles similarly show necrosis histopathologically. In these conditions, focal involvement is the rule, and tissue alterations occur with striking symmetry. Magnetic resonance imaging also allows the detection of acute necrosis before development of atrophy in patients with phosphofructokinase and myophosphorylase deficiency.[20]

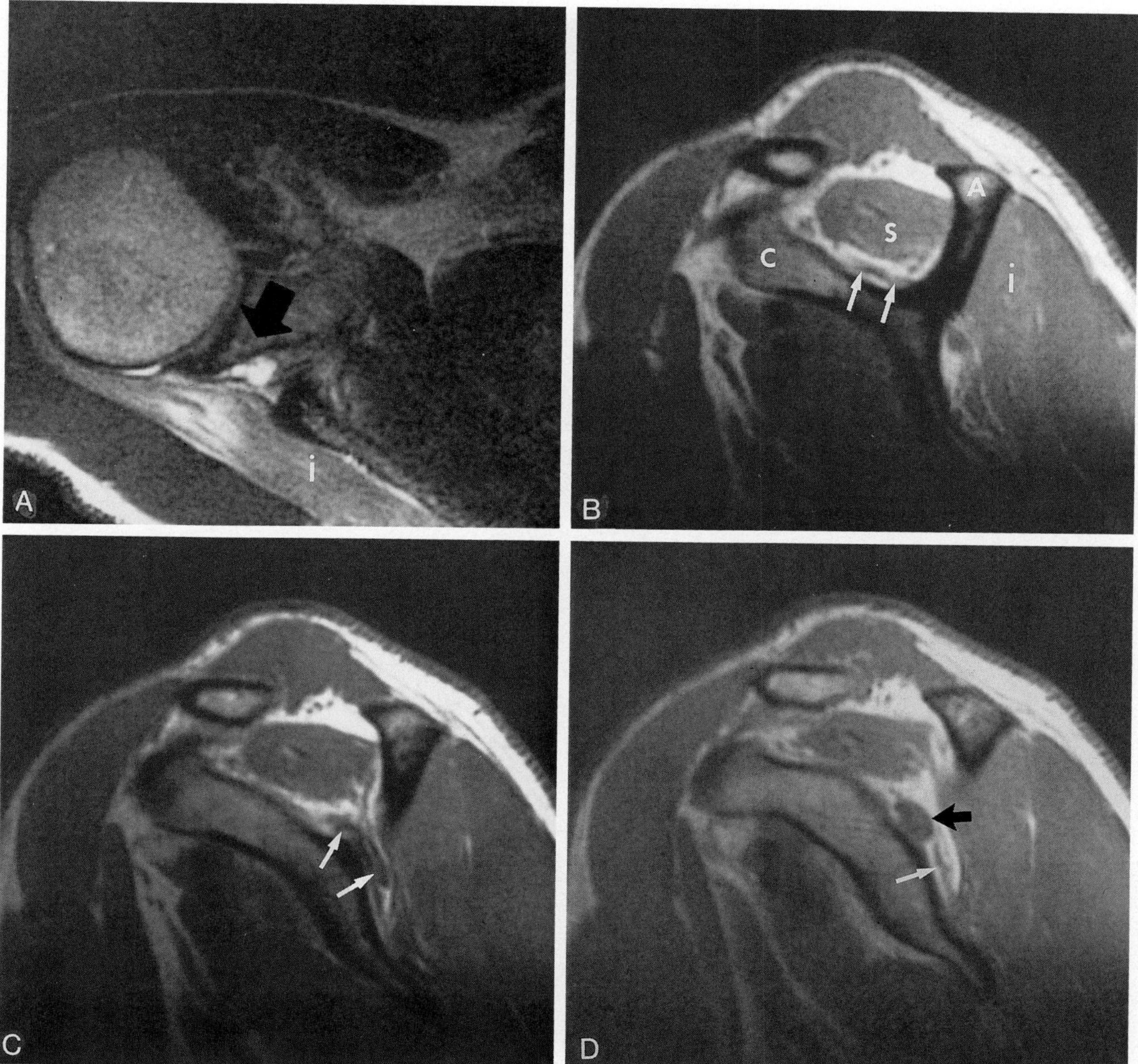

Figure 17–4. Denervation of the infraspinatus muscle due to a ganglion cyst in the spinoglenoid notch that has entrapped the suprascapular nerve. *A*, T2-weighted axial (TR, 2000 ms; TE, 60 ms) image shows a 1-cm mass of high signal intensity *(arrow)* adjacent to the distal suprascapular nerve. Atrophy and increased signal intensity of the infraspinatus muscle (i) is noted. *B*, *C*, and *D*, Sequential medial to lateral T1-weighted oblique sagittal (TR, 800 ms; TE, 20 ms) images show the course of the suprascapular nerve *(white arrows)* beneath the supraspinatus muscle in the supraspinatus fossa in *B*, through the spinoglenoid notch in *C*, and beneath the infraspinatus fossa in *D*. The ganglion cyst is seen laterally in the spinoglenoid notch *(black arrow* in *D)*. A, acromion; C, coracoid process; S, supraspinatus muscle; i, infraspinatus muscle. (From Fritz RC, Helms CA, Steinbach LS, et al: Suprascapular nerve entrapment: Evaluation with MR imaging. Radiology 182:437, 1992.)

Atrophy, Hypertrophy, and Pseudohypertrophy

Atrophy (shrinkage) implies diminution in the muscle fiber size and is the hallmark of end-stage muscle disease. The most common causes of atrophic muscle fibers are denervation and disuse. Metabolic disturbances of the muscle cells (e.g., malnutrition, hyperthyroidism, Cushing's syndrome) that cause a decrease of the anabolic-catabolic ratio also can lead to shrinkage.[7] In Duchenne's, Becker's, and limb girdle muscular dystrophies, a spectrum of fiber sizes is seen, with atrophic, hypertrophic, and normal-sized fibers scattered in random relationship. Spinal muscular atrophy tends to be manifested by diffuse atrophy with increased subcutaneous fat.

Hypertrophy is defined as the increased girth of

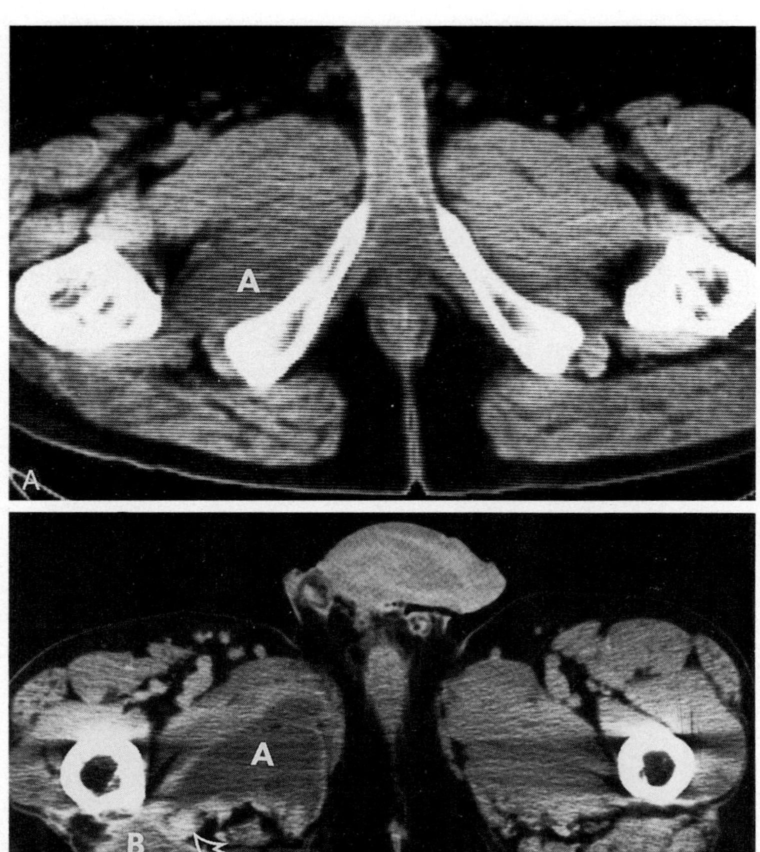

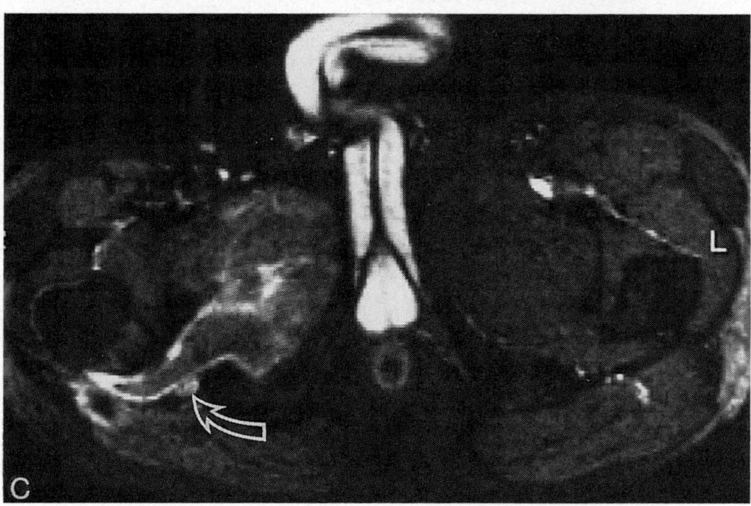

Figure 17–5. Myonecrosis caused by a drug overdose in a 34-year-old man 1 week after he attempted suicide by diphenhydramine overdose. Severe rhabdomyolysis and sciatic nerve palsy were present clinically. *A*, Precontrast computed tomography (CT) image shows faintly decreased density in the right adductor magnus (A). *B*, A postcontrast CT image obtained slightly inferiorly shows peripheral enhancement about the low-density adductor magnus (A) and biceps femoris (B) muscles. Note the enhanced and enlarged sciatic nerve *(arrow)*. *C*, Short tau inversion-recovery (STIR) MR image shows all of these findings. (From Fleckenstein JL: Magnetic resonance imaging and computed tomography of skeletal muscle pathology. *In* Bloem JL, Sartoris DJ [eds]: MRI and CT of the Musculoskeletal System. ©1992, the Williams & Wilkins Co., Baltimore.)

muscle fibers, which is mainly due to an absolute increase of the number or the size of myofibrils, or both. True hypertrophy may result from heavy or excessive use of the muscle or muscle group (i.e., physiologic hypertrophy) or may be an expression of a disease of the muscle fibers, such as hypothyroid myopathy (i.e., pathologic hypertrophy). Physiologic hypertrophy in physical training has been described earlier. Pathologically hypertrophic fibers are especially common in Duchenne's dystrophy, limb girdle dystrophy, and myotonic dystrophy; less common in partial denervation (i.e., chronic neuropathies); and rare in inflammatory myopathies. In progressive muscular dystrophy, a number of more or less hypertrophic fibers constantly exist in unaffected muscles as a compensatory response to the weakness in atrophic diseases.[7]

The distinction between pseudohypertrophy and true hypertrophy cannot be made absolutely because both conditions commonly coexist (e.g., progressive muscular dystrophy). Pseudohypertrophy may be defined as an increase in muscle volume that may not be related to the increase in number or size of myofibrils. Magnetic resonance imaging can detect pseudohypertrophy due to fatty replacement. As a rule, microscopic examination alone permits discrimination between hyperplasia and hypertrophy of muscle fibers.[32]

Fatty Infiltration and Fibrosis

Mesenchymal abnormalities common in muscle lesions include fat deposition and fibrosis. Microscopically, fatty deposits occur early in many pathologic conditions but are readily apparent on imaging studies only in later stages. On MR images, subacute hematoma and fat may have a similar signal intensity on spin-echo images, but chemical shift techniques for fat suppression can distinguish between blood and fat.

Recurrent or severe injury, regardless of cause, may result in fibroblast proliferation and subsequent fibrosis. However, only a few cases of fibrosis of muscles have been diagnosed with the use of MR imaging.[55]

DISORDERS OF THE NEUROMUSCULAR UNIT

Primary neurogenic disorders of locomotor dysfunction can be categorized on the basis of whether the lesion lies within anterior or lateral horn cells, axons, or in the neuromuscular junction. Early histopathologic findings are characterized by denervation atrophy and, when reinnervation occurs, by fiber angulation and fiber-type grouping.[32] Magnetic resonance imaging features of denervated muscles have been described earlier. In the early stage, fiber atrophy with resultant increase in extracellular water results in prolongation of T1 and T2 relaxation times. These characteristics are in contrast to the end-stage appearance of denervation, in which T1 shortening from fat deposition is the dominant MR imaging feature.

Diseases of Motor Neurons

Motor neuron disease is a family of disorders, characterized by progressive, widespread degeneration of motor nerve cells. Onset is variable, usually in middle to old age, with a male predominance. The disease affects the anterior horn cells, brainstem nuclei, and Betz's cells in the cerebral cortex. The cause of motor neuron disease remains unknown.

An example of the combination of upper and lower motor neuron disease is amyotrophic lateral sclerosis, which describes a form in which corticospinal tract signs predominate. Fatty infiltration of the tongue may be a useful clue to the diagnosis.[8] Fatty infiltration of the lower limb is mild, highly selective, and symmetric in amyotrophic lateral sclerosis.

An example of lower motor neuron disease is spinal muscular atrophy, which implies a group of inherited disorders characterized by degeneration of anterior horn cells and progressive muscle weakness. It can be categorized according to the age of onset and reserved motor function.[2] In the severe type, the onset of muscle weakness appears before 6 months. The disease is fatal by 3 years. In the intermediate type, the onset is usually before 3 years and the average age of death exceeds 10 years. The mild type is a relatively benign condition, and the onset is insidious, between 2 and 50 years of age. Muscular atrophy and fatty infiltration of the thigh are symmetric (Fig. 17–6).[42]

Diseases of the Neuromuscular Junction

Common disorders of neuromuscular transmission include the Eaton-Lambert syndrome, botulism, tick paralysis, and myasthenia gravis. Biopsy in such patients may reveal changes compatible with structurally denervated muscle, which are presumably the result of physiologic denervation of the fibers secondary to disrupted neuromuscular transmission.[27]

MUSCULAR DYSTROPHIES

The muscular dystrophies are a group of primary myopathies that commonly have a hereditary cause and are characterized by a progressive noninflammatory degeneration of skeletal muscles. Common manifestations are muscle weakness and wasting, elevated serum enzyme levels, and myopathic patterns on electromyography.

Duchenne's Dystrophy

Duchenne's dystrophy is the most common and the most severe form of the muscular dystrophies in children. The disease is inherited as an X-linked

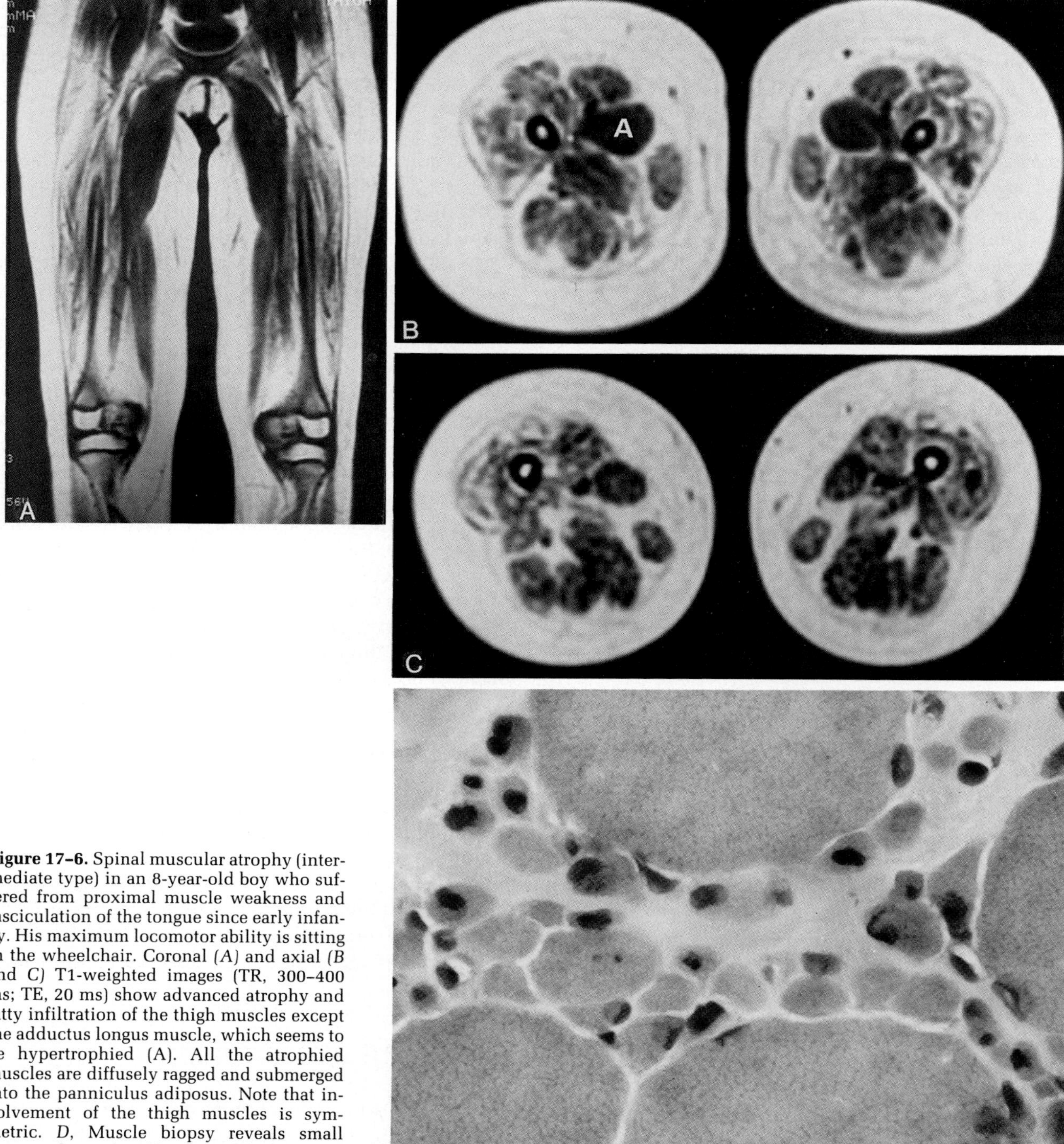

Figure 17–6. Spinal muscular atrophy (intermediate type) in an 8-year-old boy who suffered from proximal muscle weakness and fasciculation of the tongue since early infancy. His maximum locomotor ability is sitting in the wheelchair. Coronal *(A)* and axial *(B* and *C)* T1-weighted images (TR, 300–400 ms; TE, 20 ms) show advanced atrophy and fatty infiltration of the thigh muscles except the adductus longus muscle, which seems to be hypertrophied (A). All the atrophied muscles are diffusely ragged and submerged into the panniculus adiposus. Note that involvement of the thigh muscles is symmetric. *D,* Muscle biopsy reveals small groups of atrophic fibers (modified Gomori trichrome stain, ×200).

recessive trait, predominantly in males, with an onset of muscular weakness at age 4 to 6 years; 95 per cent of patients are confined to a wheelchair by the age of 12 years; and death eventually occurs from respiratory or cardiac failure between the ages of 10 and 20 years.[13] Duchenne's dystrophy is characterized by an initial symmetric and selective involvement of the proximal pelvic girdle muscles, followed by the calf muscles and the proximal shoulder girdle muscles. Pseudohypertrophy of the calves accounts for 80 per cent of cases. Pseudohypertrophy is attributable to an excess of adipose and connective tissue interspersed between the degenerated muscle bundles.

Duchenne's dystrophy has been studied qualitatively and quantitatively by MR imaging.[9,36–40,49] The delineation of the fat-muscle interface of the affected muscle is more apparent on short TR and short TE pulse sequences. The pattern of fatty infiltration appears as featherlike fronds interspersed between the diseased muscles on these sequences. However, the outlines of the individual muscles are not seen distinctly in advanced fatty infiltration. Longer TR and TE sequences do not offer any advantage in delineating fatty infiltration, and in addition they prolong scanning time. The intermuscular septa are outlined as low-signal curvilinear lines in a fat saturation imaging technique.[9] The sartorius and gracilis muscles are relatively spared (Fig. 17–7).[42] The posterior tibialis muscle also is less involved (Fig. 17–8).[9] The degree of increased signal intensity of the affected muscle groups parallels the severity of the fatty infiltration. The degree of fatty infiltration of the diseased muscles is an index of the decline in muscle strength.[9] Hence, MR imaging is valuable in identifying the extent and severity of this condition, thereby providing a powerful clinical investigational procedure for therapeutic interventions and monitoring.

Becker's Dystrophy

Becker's dystrophy is a comparatively benign variant of a sex-linked muscular dystrophy. The clinical course represents the most dependable means of distinguishing between Duchenne's and Becker's dystrophy. The Becker variant is rare, and cardiac involvement is much less frequent. The disease has a later onset, with less severe contractures and a slower course. Similarly, atrophied muscles with fatty infiltration are the predominant MR imaging features (Fig. 17–9).

Myotonic Dystrophy

Myotonia is a clinical phenomenon characterized by a failure of relaxation after a strong contracture. Myotonic dystrophy (Steinert's disease) is an autosomal dominant disorder causing diffuse weakness, which is most prominent in the distal limb muscles, the neck, and the face. The initial symptoms usually appear between the ages of 15 and 40 years. It is one of the more common myopathies and the most frequent of the myotonias. The disease not only is limited to skeletal muscle but also affects myocardial and visceral smooth muscle. Magnetic resonance imaging shows selective muscle atrophy with fatty infiltration (Fig. 17–10) as well as occasional edema like changes.

Facioscapulohumeral Dystrophy

Facioscapulohumeral dystrophy is the mildest form of muscular dystrophy. It is characterized by an onset of any age, expression in either sex, and initial involvement of the facial and shoulder girdle muscles, followed by spread to the pelvic girdle muscles. Muscle hypertrophy and joint contractures are rare. Magnetic resonance images show fatty infiltration of muscle with a severity less than that in the Duchenne's and limb girdle dystrophy groups.

Limb Girdle Dystrophy

Limb girdle dystrophy is a heterogeneous group of diseases with a distribution of muscular involvement essentially similar to that in facioscapulohumeral dystrophy but without facial involvement. It is characterized by an onset usually in the second to third decade; expression in either sex; and involvement of the shoulder and the pelvic girdle muscles, with pseudohypertrophy of the calves and vastus lateralis in one third of patients. The disease ranges in extent from severe and symmetric to moderate and asymmetric, as displayed on MR images.[42]

METABOLIC MYOPATHIES

The local muscle glycogen is the main energy source for skeletal muscle contraction. Glycogenoses or glycogen storage diseases are a group of disorders characterized by deranged metabolism of glycogen, glucose, or both. Muscle symptoms are predominantly manifested in types II (Pompe's disease) (Fig. 17–11), V (McArdle's disease), VII (Tarui's disease) (Fig. 17–12), and X (diSant' Agnese's disease), whereas in other types, symptoms of liver dysfunction and hypoglycemia dominate the clinical picture.

There is no decrease in muscle pH in patients with muscular glycogenosis because of the failure of breakdown of glycogen, which results in an inability to produce lactate during exercise.[11,48] Lactate is a major osmotically active product of glycogenolysis, and its absence in patients with glycogenosis may be associated with impaired water translocation from the vascular to the extravascular space.[19] Therefore, negligible variation in exercise-induced muscular enhancement on T2-weighted images may reflect

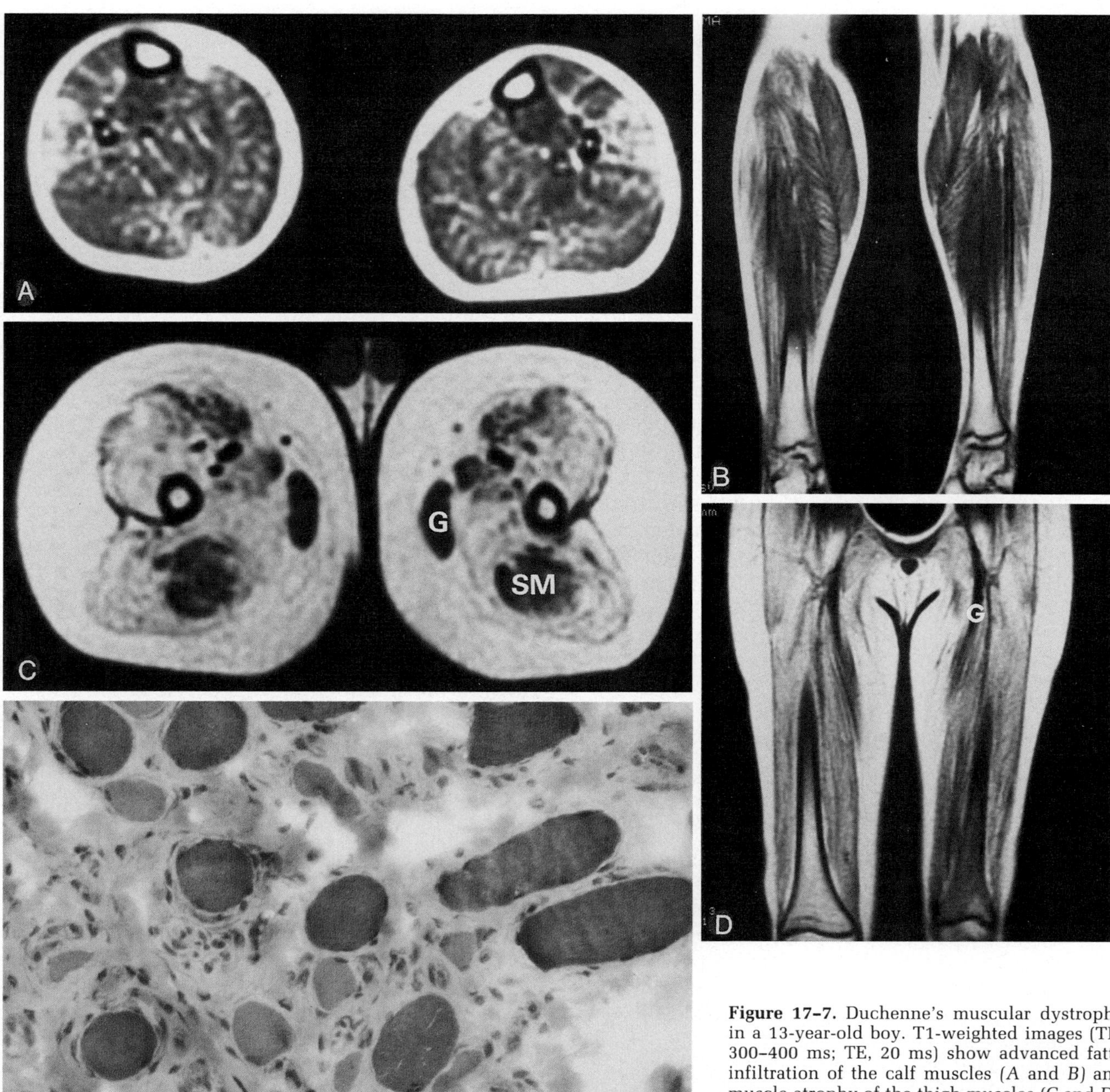

Figure 17-7. Duchenne's muscular dystrophy in a 13-year-old boy. T1-weighted images (TR, 300–400 ms; TE, 20 ms) show advanced fatty infiltration of the calf muscles (*A* and *B*) and muscle atrophy of the thigh muscles (*C* and *D*), with selective bilateral preservation of the gracilis (G) and semimembranosus muscles (SM). Note that involvement of the thigh and calf muscles is symmetric. *E*, Muscle biopsy reveals variation in fiber size and focal muscle fiber necrosis and phagocytosis (hematoxylin and eosin stain, ×50).

the impairment of muscle glycogenolysis.[20] Fleckenstein and colleagues[18] proposed that a glycogenolysis-mediated effect (i.e., a lactate-mediated increase in osmotic strength establishing a gradient for water translocation) is the critical mediator of exercise-induced changes in signal intensity; de Kerviler and co-workers[34] hypothesized that muscle perfusion mediates these changes.

MYOSITIS

The term *myositis* implies an inflammatory reaction to muscle of a type usually related to the presence of infection. It may be acute, subacute, or chronic, and it may involve one or many muscles. A causative agent (including pyogenic bacterium, virus, parasite, or spirochete) may or may not be iden-

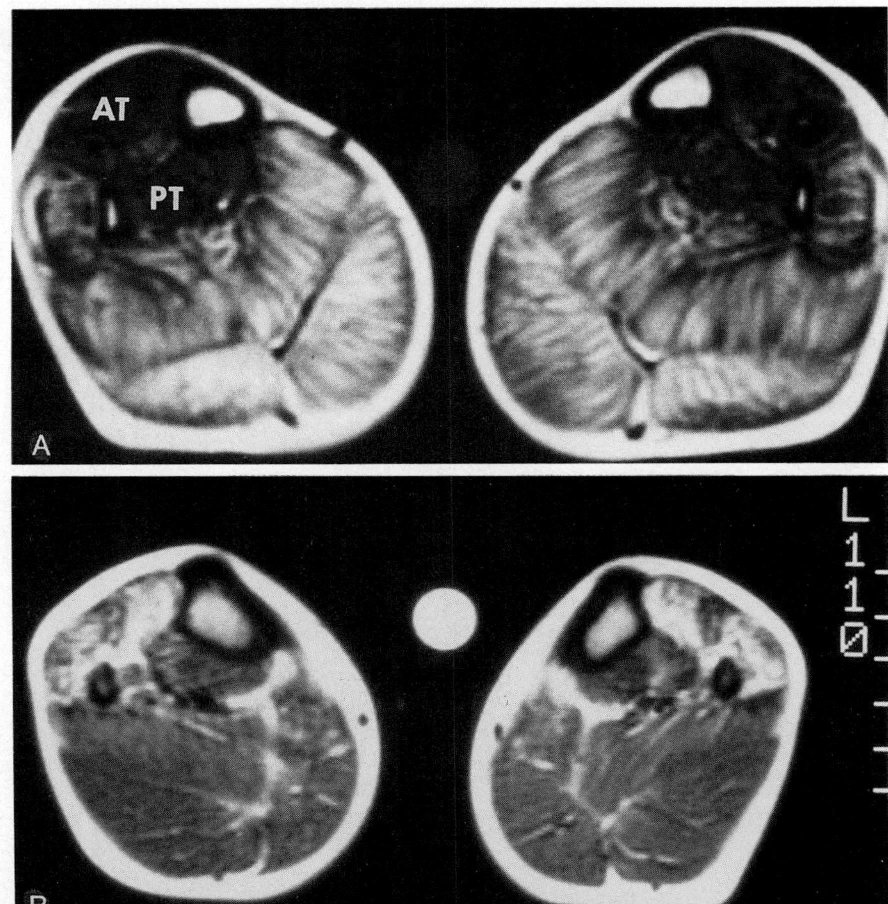

Figure 17-8. Duchenne's muscular dystrophy in two patients. *A* (patient 1), T1-weighted axial image (TR, 500 ms; TE, 20 ms) shows advanced fatty infiltration of the posterior and lateral calf muscles. The anterior (AT) and posterior tibialis (PT) muscle is relatively spared. *B* (patient 2), T1-weighted axial image (TR, 500 ms; TE, 20 ms) shows fatty infiltration predominates in the anterior tibialis muscle, whereas other calf muscles are relatively spared.

tified in the inflammatory lesion. If no cause is identified, the inflammatory reaction is categorized as idiopathic. The inflammatory lesion in idiopathic myositis is believed to be due, at least in part, to an altered immunologic reaction.[32]

Pyogenic Bacterial Myositis (Pyomyositis)

Pyogenic bacterial myositis (pyomyositis) denotes a bacterial infectious disease of the skeletal muscle; the thighs, calves, and buttocks are involved predominantly. It occurs more frequently in males than in females and is common in tropical countries. Skeletal muscle can be infected via three routes: by blood stream infections (e.g., septicemia), by direct extension (e.g., infectious arthritis), and by open traumatic wounds (e.g., injection). The most commonly cultured pathogen is *Staphylococcus aureus;* streptococcal species are less common.[32] Even if no infective agent can be identified, infection cannot be excluded.

Onset is often insidious, with fever, chills, and sweating followed by ill-defined local pain and swelling of the affected muscle or groups of muscles. Leukocytosis and elevated erythrocyte sedimentation rate are frequent but inconstant laboratory findings, whereas concentrations of creatine kinase are normal and lactate dehydrogenase levels are high secondary to anaerobic metabolism. In the early stage, the lesion histologically reveals edematous and cellular inflammatory changes, but muscle fibers themselves are intact. An abscess forms as the process continues. Muscle architecture is obliterated and replaced by purulent exudate. A great increase in cellularity occurs within the few days of infection. Muscle abscesses become encapsulated by fibroblastic connective tissue. Extensive fibrosis results in contracture, which is noted in the advanced stage.[32] Clinically, fluctuation of the infected tissue, which occurs within 5 to 10 days, indicates the formation of an abscess.[32] In addition to antibiotic therapy, evacuation by surgical drainage is indicated in the treatment of an abscess. Delayed treatment may cause death.

Magnetic resonance imaging is helpful in differentiating other pathologic processes from pyo-

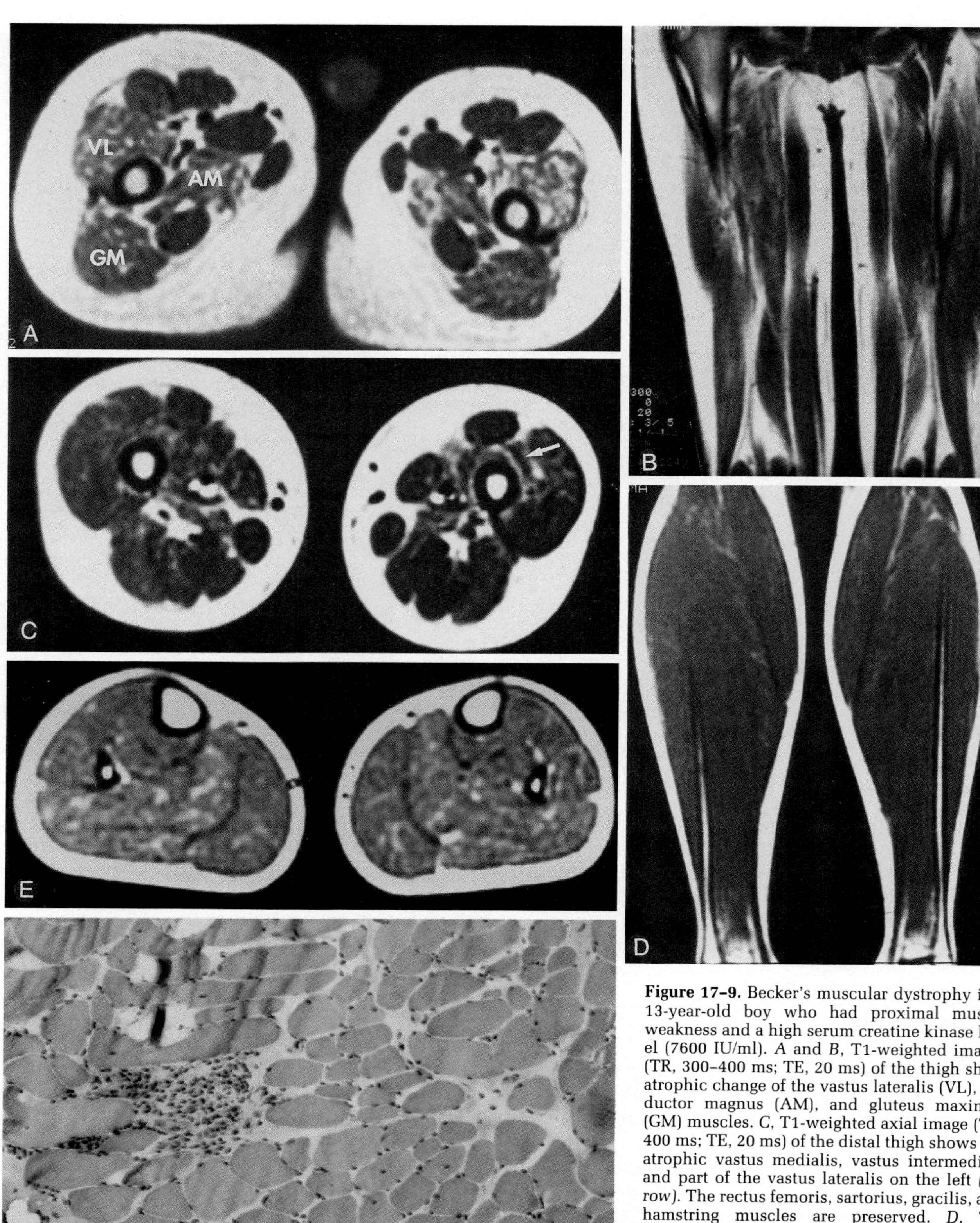

Figure 17–9. Becker's muscular dystrophy in a 13-year-old boy who had proximal muscle weakness and a high serum creatine kinase level (7600 IU/ml). *A* and *B*, T1-weighted images (TR, 300–400 ms; TE, 20 ms) of the thigh show atrophic change of the vastus lateralis (VL), adductor magnus (AM), and gluteus maximus (GM) muscles. *C*, T1-weighted axial image (TR, 400 ms; TE, 20 ms) of the distal thigh shows the atrophic vastus medialis, vastus intermedius, and part of the vastus lateralis on the left (*arrow*). The rectus femoris, sartorius, gracilis, and hamstring muscles are preserved. *D*, T1-weighted coronal image (TR, 300 ms; TE, 20 ms) and axial image (*E*) (TR, 400 ms; TE, 20 ms) of the calf show no evidence of atrophy or fatty infiltration of the calf muscles. The severity of the disease is milder than that in Duchenne's muscular dystrophy. *F*, Muscle biopsy of the biceps reveals variation in fiber size and focal muscle fiber necrosis and phagocytosis (hematoxylin and eosin stain, ×50).

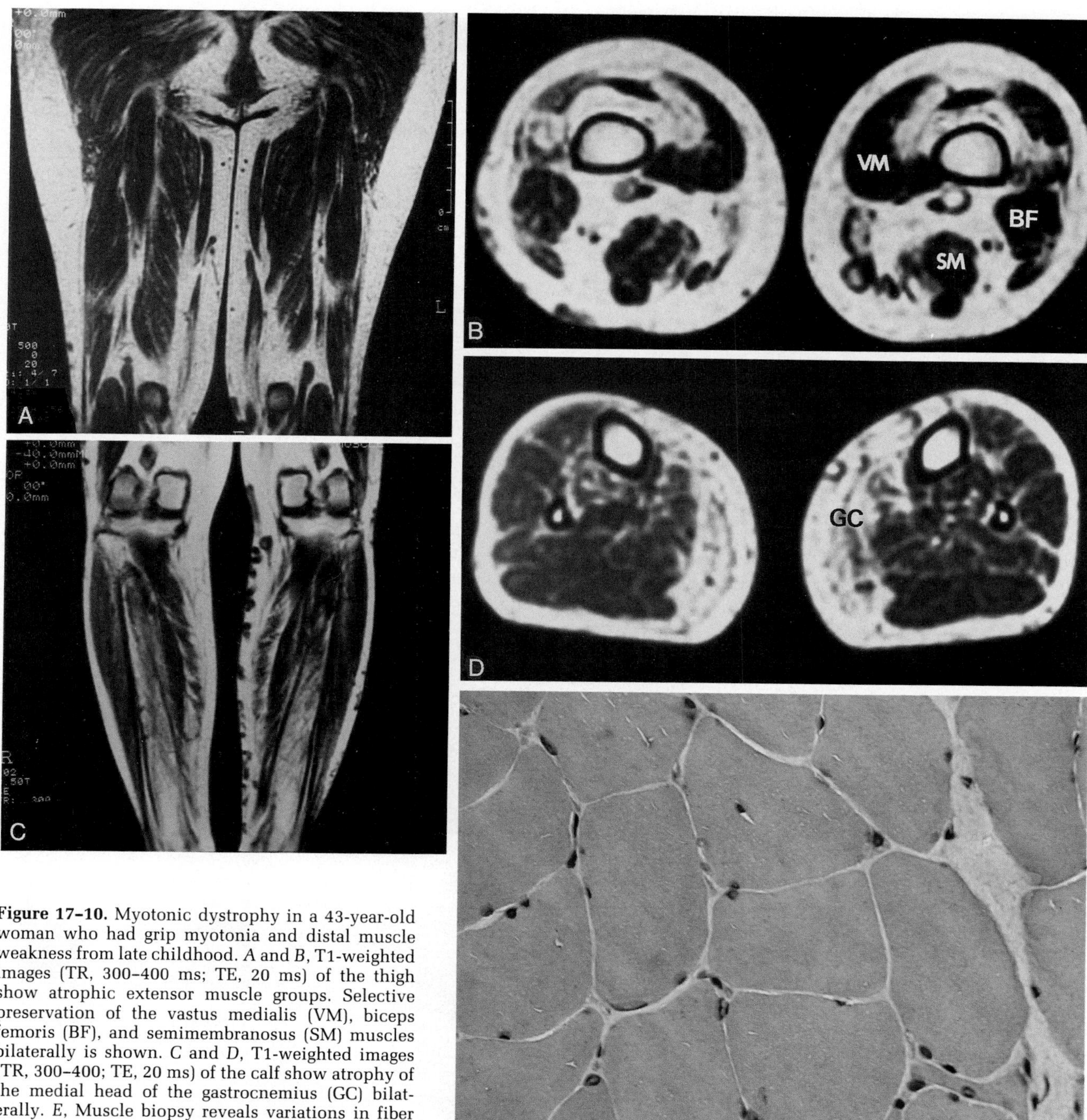

Figure 17–10. Myotonic dystrophy in a 43-year-old woman who had grip myotonia and distal muscle weakness from late childhood. *A* and *B*, T1-weighted images (TR, 300–400 ms; TE, 20 ms) of the thigh show atrophic extensor muscle groups. Selective preservation of the vastus medialis (VM), biceps femoris (BF), and semimembranosus (SM) muscles bilaterally is shown. *C* and *D*, T1-weighted images (TR, 300–400; TE, 20 ms) of the calf show atrophy of the medial head of the gastrocnemius (GC) bilaterally. *E*, Muscle biopsy reveals variations in fiber size and internal nuclei and a small angular fiber (hematoxylin and eosin stain, ×100).

myositis, in outlining the extent of involvement, and in localizing drainable fluid collections.[58] Other conditions that may produce symptoms similar to those of pyomyositis include septic arthritis, osteomyelitis, thrombophlebitis, and polymyositis. Septic arthritis typically is accompanied by joint effusion, which is easily demonstrated on MR images. However, joint effusion is not a feature of pyomyositis. Conventional radiographs help to diagnose osteomyelitis. In thrombophlebitis, markedly increased signal intensity in the intermuscular fascia is seen, with relative sparing of the muscles, owing to the obstructive edema.[58] Involved vessels also may show increased signal intensity owing to their slower flow or thrombosis. Bilateral symmetric involvement is uncommon in pyomyositis but more common in polymyositis. In addition, the absence of elevated levels of serum muscle enzymes favors pyomyositis.

On T1-weighted images, infected muscle has a

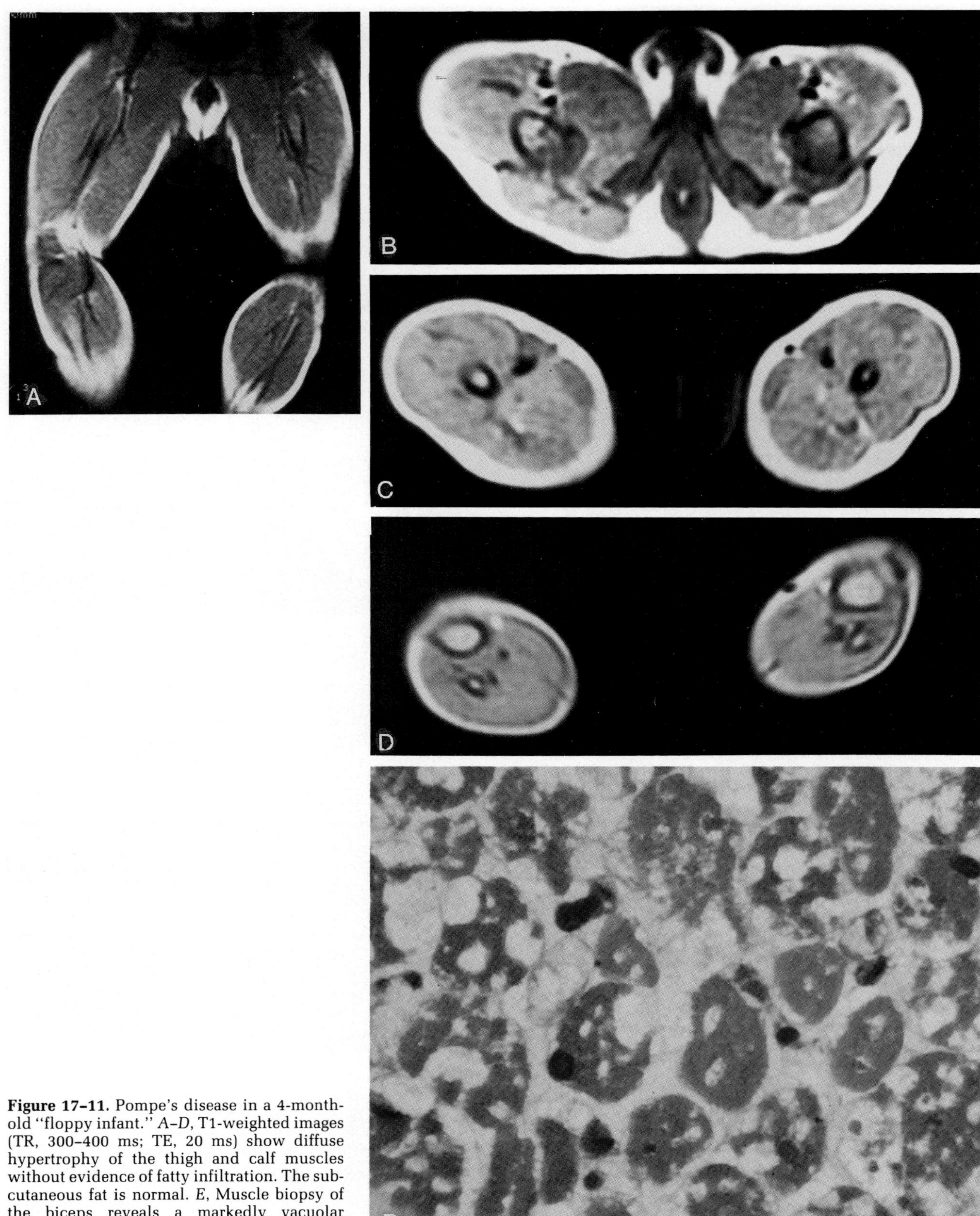

Figure 17–11. Pompe's disease in a 4-month-old "floppy infant." *A–D*, T1-weighted images (TR, 300–400 ms; TE, 20 ms) show diffuse hypertrophy of the thigh and calf muscles without evidence of fatty infiltration. The subcutaneous fat is normal. *E*, Muscle biopsy of the biceps reveals a markedly vacuolar myopathy.

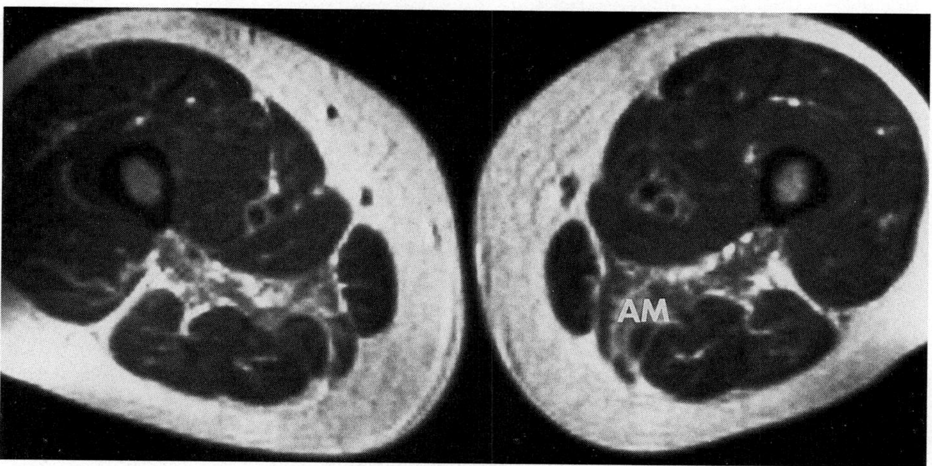

Figure 17–12. Tarui's disease (phosphofructokinase deficiency) in a 48-year-old man who had recurrent subclinical rhadomyolysis. MR imaging of the thighs was performed as a screening test for muscle abnormalities. Proton-weighted axial image (TR, 2000 ms; TE, 30 ms) shows fatty deposition and focal diminution of the adductor magnus (AM) bilaterally. This case suggests that severe muscle necrosis may lead to atrophy and fatty replacement. (From Fleckenstein JL, Peshock RM, Lewis SF, et al: Magnetic resonance imaging of muscle injury and atrophy in glycolytic myopathies. Muscle Nerve 12:849, 1989. © John Wiley & Sons, Inc., a division of John Wiley & Sons, Inc.)

signal intensity similar to that of normal tissue with or without a higher signal intensity rim around a central focus (Figs. 17–13, 17–14). The presence of this rim may be peculiar to muscle abscesses.[16] On T2-weighted or STIR images, the corresponding rim is of low signal intensity and surrounds the central focus, which is of very high signal intensity. The physical basis of the rim is unknown. In contrast, a study of intramuscular abscess in laboratory animals[45] found that the capsular rim is inapparent on T1-weighted images, but improved detection is found after the infusion of gadolinium, which significantly enhances this vascularized rim in the periphery of muscle abscess. The circular enhancement of the abscess rim corresponds to the inflammatory zone with abundant white blood cells, whereas the necrotic center of abscess is weakly enhanced. The necrotic portion can be filled partially with contrast material over time and becomes more homogeneous between 25 to 50 minutes after administration.[45]

Multiple lesions are a frequent feature in pyomyositis (Fig. 17–15) and were encountered in 43 per cent of patients in one series.[10] However, multiple sites of involvement are not a specific characteristic of pyomyositis because polymyositis, Kaposi's sarcoma, and lymphoma also have this feature.

In pyomyositis, abnormality is typically restricted to the muscle and less frequently involves the subcutaneous tissue. In patients with acquired

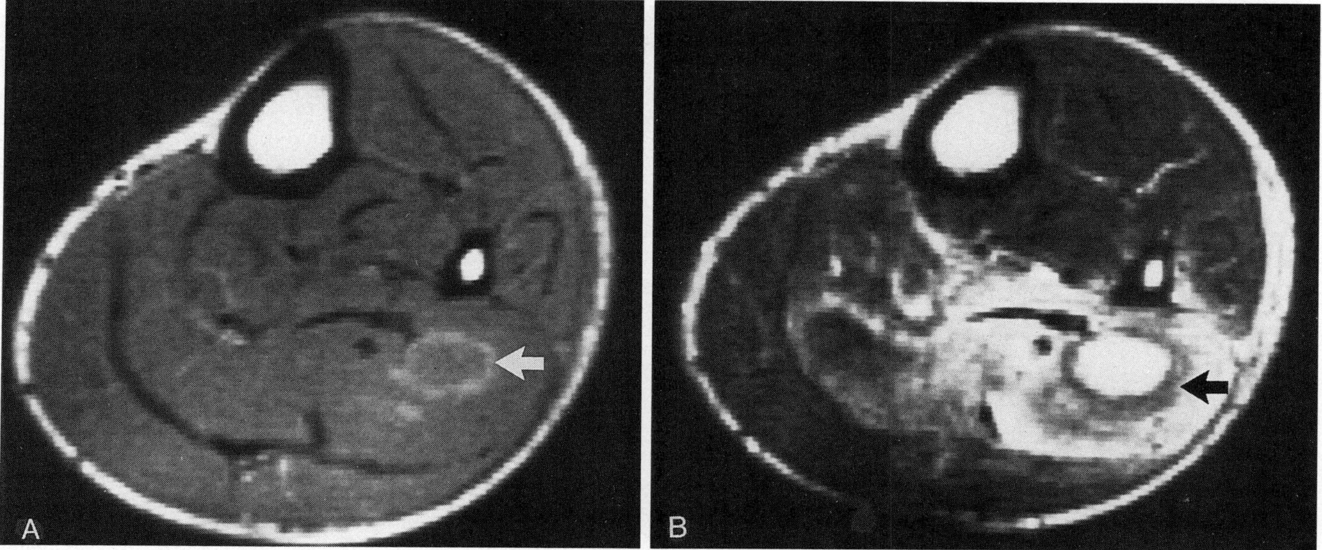

Figure 17–13. Pyomyositis in patients with acquired immunodeficiency syndrome (AIDS). *A*, T1-weighted image (TR, 500 ms; TE, 30 ms) shows a rim *(arrow)* of increased signal intensity in the left soleus muscle. *B*, This rim is closely associated with a band of decreased signal intensity *(arrow)* on the T2-weighted image (TR, 2000 ms; TE, 90 ms). At aspiration, the central area of high signal intensity was found to consist of pus. (From Fleckenstein JL, Burns DK, Murphy FK, et al: Differential diagnosis of bacterial myositis in AIDS: Evaluation with MR imaging. Radiology 179:653, 1991.)

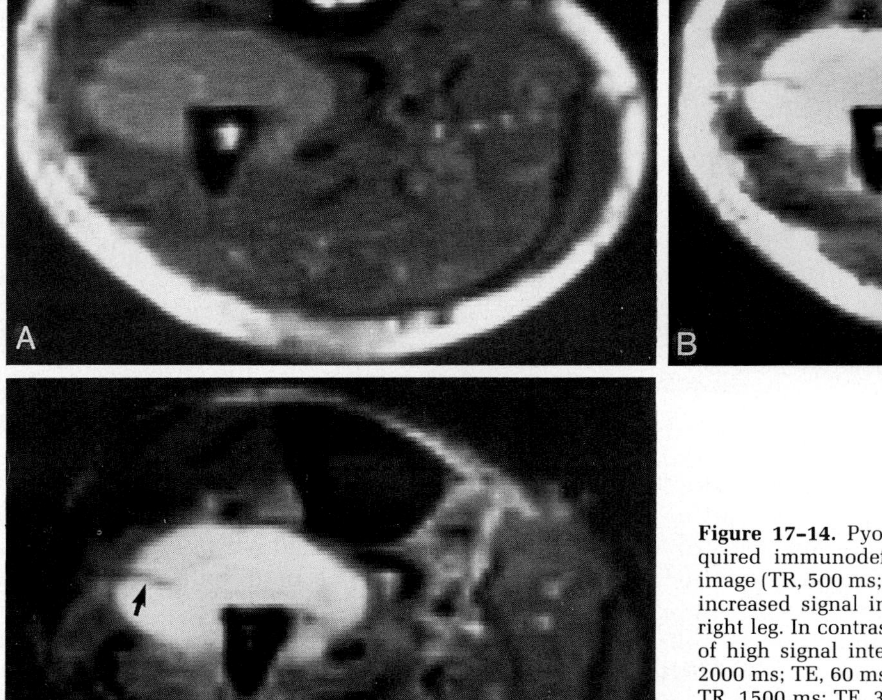

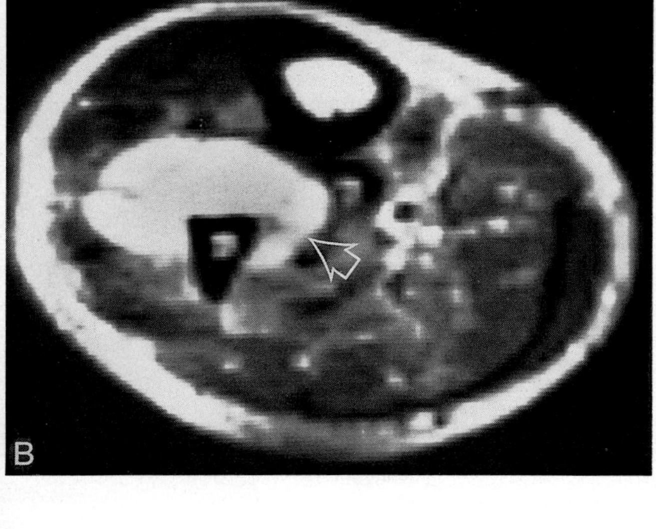

Figure 17–14. Pyomyositis and abscess in a patient with acquired immunodeficiency syndrome (AIDS). *A*, T1-weighted image (TR, 500 ms; TE, 30 ms) shows a region of homogeneously increased signal intensity in the anterior compartment of the right leg. In contrast to the case in Figure 17–13, there is no rim of high signal intensity in this case. On T2-weighted (*B*: TR, 2000 ms; TE, 60 ms) and short tau inversion-recovery (STIR) (*C*: TR, 1500 ms; TE, 30 ms; TI, 100 ms) images, the central area of the muscle abscess shows very high signal intensity consistent with frank pus, which was found at aspiration. The process readily crosses the interosseous membrane, involving the posterior compartment (*arrow* in *B*) and intracompartmental fascial boundaries (*arrow* in *C*). (From Fleckenstein JL, Burns DK, Murphy FK, et al: Differential diagnosis of bacterial myositis in AIDS: Evaluation with MR imaging. Radiology 179:653, 1991.)

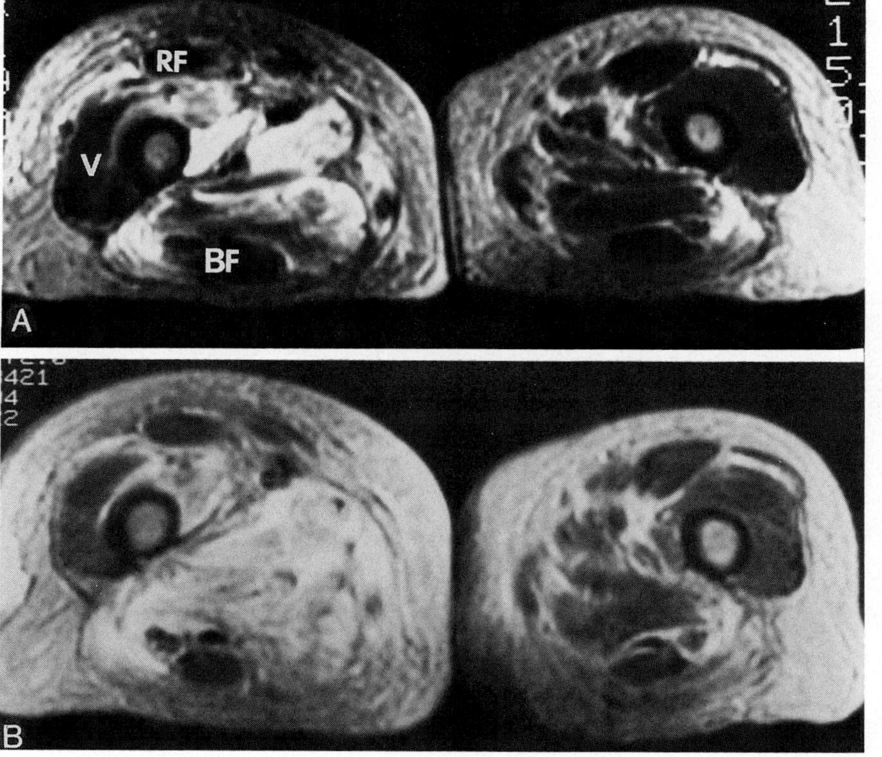

Figure 17–15. Recurrent pyomyositis in a 44-year-old woman who suffered from swelling and pain of the right leg. This patient had renal failure secondary to diabetes mellitus and was on hemodialysis. *A* (initial study), Axial T2-weighted image (TR, 2500 ms; TE, 80 ms) shows markedly increased signal intensity within muscles and intramuscular septa of the right leg. Only parts of the rectus femoris (RF), vastus lateralis and intermedius (V), and biceps femoris (BF) muscles are spared. The left leg also is involved to a lesser degree. No focal fluid collection to suggest a focal abscess is identified. No evidence of osteomyelitis is noted. *B* (10 days after the initial examination but before antiobiotic treatment), Axial T2-weighted image (TR, 2500 ms; TE, 80 ms) shows progressive swelling of the muscles of the right leg. Interval development of more pronounced involvement of the right vastus lateralis muscle. Although no organisms were documented on a culture of the biopsy material, this patient was treated with two different courses of antibiotic therapy with apparent resolution after this MR study.

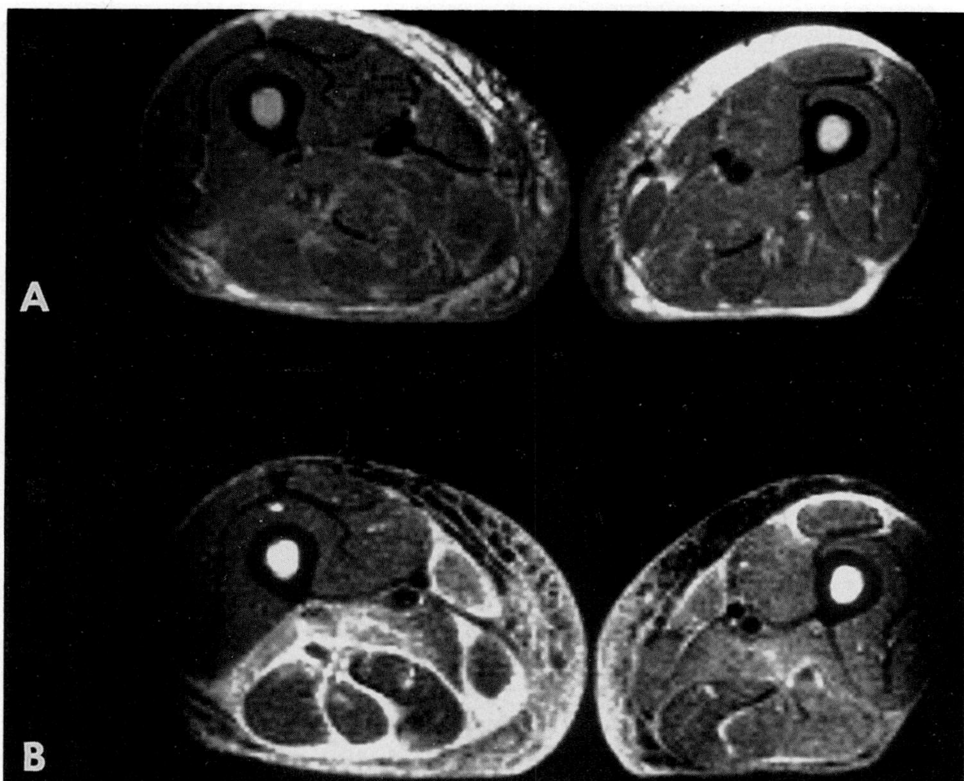

Figure 17–16. Lymphoma-associated lymphedema in a patient with acquired immunodeficiency syndrome (AIDS). This patient had unilateral leg swelling associated with distal cellulitis. Cellulitis improved in response to intravenously administered vancomycin hydrochloride, but swelling persisted. MR imaging was performed at that time. *A*, T1-weighted image (TR, 500 ms; TE, 30 ms) shows skin thickening, subcutaneous tissue expansion, and a streaky, trabecular pattern of low signal intensity that permeates the subcutaneous fat. *B*, Short tau inversion-recovery (STIR) image (TR, 1500 ms; TE, 30 ms; TI, 100 ms) obtained at the same level shows an abnormal increase in signal intensity in the subcutaneous fat and a variable increase in signal intensity in the muscles. (From Fleckenstein JL, Burns DK, Murphy FK, et al: Differential diagnosis of bacterial myositis in AIDS: Evaluation with MR imaging. Radiology 179:653, 1991.)

immunodeficiency syndrome (AIDS) or with positive results for human immunodeficiency virus (HIV), subcutaneous tissue changes on MR images can be present in dermal infections (e.g., carbunculosis), Kaposi's sarcoma, and lymphedema (Fig. 17–16).[16] Lymphatic obstruction is common in Kaposi's sarcoma and may occur in the presence of minimal skin lesions.[57] Subcutaneous tissue changes on MR images include skin thickening, subcutaneous tissue expansion, and a streaky trabecular pattern of signal alterations that permeates the subcutaneous tissue.[16] These streaks are of low signal intensity on T1-weighted images and of high signal intensity on T2-weighted or STIR images. Associated fluid collections are noted as well.

In differentiating pyomyositis from cellulitis, Yuh and colleagues suggested that the disproportionate involvement of muscle compared with subcutaneous tissue favors the diagnosis of pyomyositis.[58] In cellulitis, signal intensity changes in muscle are relatively slight because there is limited lymphatic drainage in muscle compared with that in subcutaneous fat. Occasionally, MR imaging may not reliably differentiate more advanced stages of cellulitis from pyomyositis, particularly when involvement is extensive (Fig. 17–17). Magnetic resonance imaging patterns may be more confusing when the two entities coexist. In addition, differentiation between abscess and tumor is not easily made on MR images.

Other Types of Myositis

Viral infection of muscle is extremely rare. In one example of MR findings,[29] differentiation of affected and less affected muscle was seen earlier on the T2-weighted fat-suppressed sequence. Skeletal muscle affected by primary tuberculous infection is extremely rare. Fungus infection of muscles can be caused by direct extension from a neighboring focus, such as from actinomycosis of the skin. The formation of abscesses and fistulas is the rule. Parasitic myositis is rare and is seldom fatal; trichinosis is the best known example.

Polymyositis and Dermatomyositis

Polymyositis is an inflammatory myopathy of unknown cause. When a characteristic skin rash accompanies the polymyositis, the term *dermatomyositis* is applied. The frequency of polymyositis in the inflammatory myopathies is about 35 to 60 per cent, and that of dermatomyositis is about 15 to 35 per cent.[6] The disease can occur in any age,

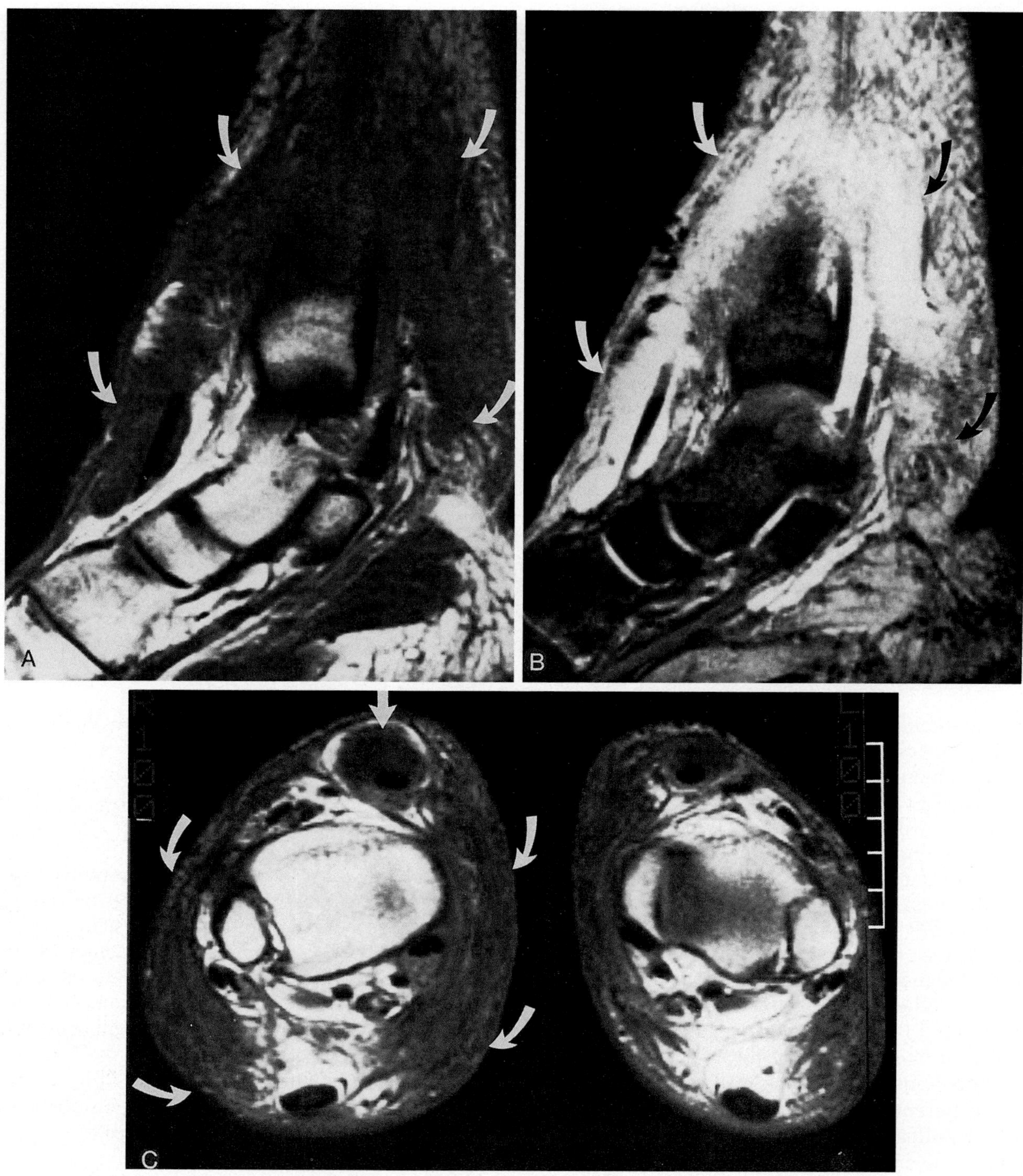

Figure 17–17. Extensive granulomatous inflammation of both legs in a 39-year-old man. T1-weighted sagittal image *(A)* shows large areas of low signal intensity *(curved arrows)*, which become very bright on the T2*-weighted image *(B)* of the low leg. Axial T1-weighted image *(C)* shows extensive involvement of the subcutaneous fat bilaterally of both legs. Note no evidence of osteomyelitis. Fluid collection within the anterior tibial tendon sheath *(straight arrow* in *C)* also is shown bilaterally. An atypical organism, *Mycobacterium smegmatis*, was obtained for culture subsequently.

but its frequency is greater after the age of 40 years. A variety of collagen vascular diseases including rheumatoid arthritis, systemic lupus erythematosus, progressive systemic sclerosis, polyarteritis nodosa, and Sjögren's syndrome may be complicated by the development of polymyositis. There is an increased association of polymyositis-dermatomyositis with neoplasia, with tumors of the ovary and stomach found to be more common. The reported frequency ranges from 7 to 34 per cent[5,56] and even higher

(71 per cent) in men over the age of 50 years.[53] Nearly 70 per cent of patients improve after treatment. The overall mortality rate is between 15 and 25 per cent.

Five major clinical criteria may be used to define polymyositis-dermatomyositis.[6] They are briefly described as follows: (1) symmetric weakness of the limb-girdle muscles and anterior neck flexors, progressing over weeks to months; (2) muscle biopsy evidence of necrosis of types I and II fibers, regeneration with basophilia, perivascular and interstitial inflammation (predominantly lymphocytic), and fiber atrophy in a perifascicular distribution; (3) elevation of skeletal muscle enzymes in the serum, particularly creatine kinase; (4) positive electromyographic findings; and (5) positive dermatologic features. Definitive diagnosis is made for dermatomyositis if three or four criteria (plus the rash) are present and for polymyositis if four criteria (without the rash) are present. The classic rash is a lilac discoloration of the eyelids with periorbital edema, for example.

Magnetic resonance imaging is useful in determining the distribution and extent of the involved muscles. Polymyositis-dermatomyositis is characterized by proximal and symmetric atrophy and high degree of selective involvement, usually with sparing of the gracilis and sartorius muscles for unknown reasons (Figs. 17–18, 17–19). Compensatory hypertrophy often does not occur in polymyositis-dermatomyositis, and this feature may prove to be a means of distinguishing these diseases from other diseases or dystrophies. The affected muscle is of high signal intensity on T1-weighted images, suggesting fatty change from degenerated fibers. The fat replacement and atrophy are thought to correlate significantly with the clinical activity of disease.[33] STIR images are helpful in demonstrating the edema of high signal intensity, which suggests the presence of foci of relatively active disease (Fig. 17–20; see also Fig. 17–19).[22]

Childhood Dermatomyositis

Childhood dermatomyositis differs from the adult variety in that the primary target of the disease process is blood vessels, and therefore the disorder is actually a systemic vasculopathy. The earliest manifestation is of proximal lower extremity weakness, followed by proximal upper limb weakness. Contractures frequently occur. Typically, the skin lesion appears as discoloration of the upper lids and malar aspect of the face. The course of the disease varies,

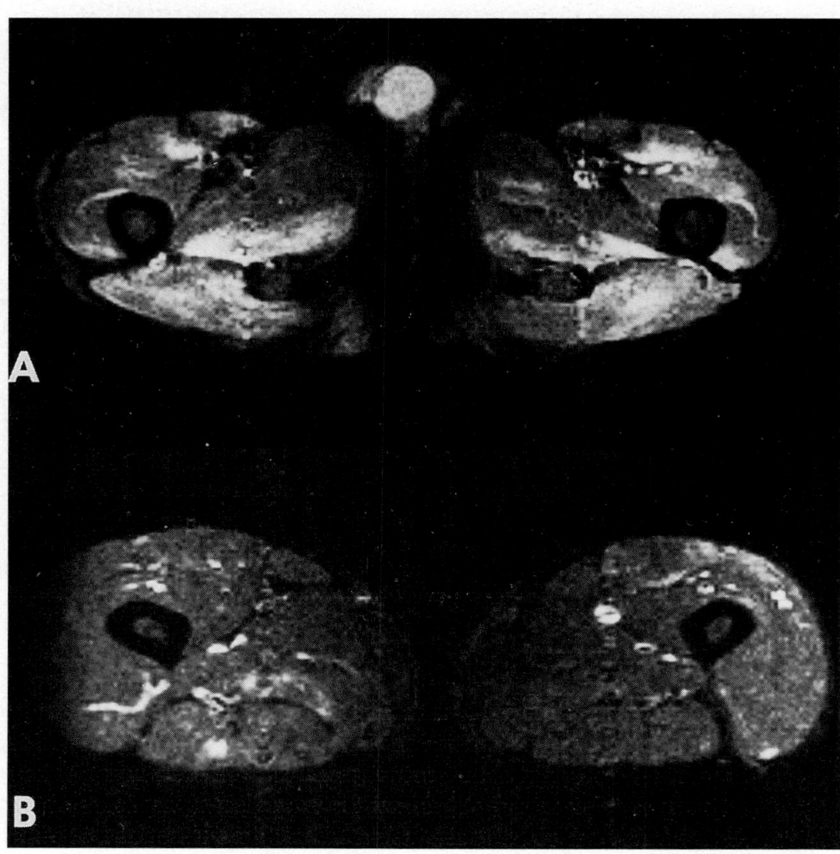

Figure 17–18. Acute idiopathic polymyositis in a 28-year-old man. This previously healthy adult developed rapidly progressive proximal weakness and rhabdomyolysis. Short tau inversion-recovery (STIR) images show strikingly symmetric involvement of the proximal thigh muscles *(A)*, but less severe abnormality is seen 10 cm more distally *(B)*. (From Fleckenstein JL: Magnetic resonance imaging and computed tomography of skeletal muscle pathology. *In* Bloem JL, Sartoris DJ [eds]: MRI and CT of the Musculoskeletal System. © 1992, the Williams & Wilkins Co., Baltimore.)

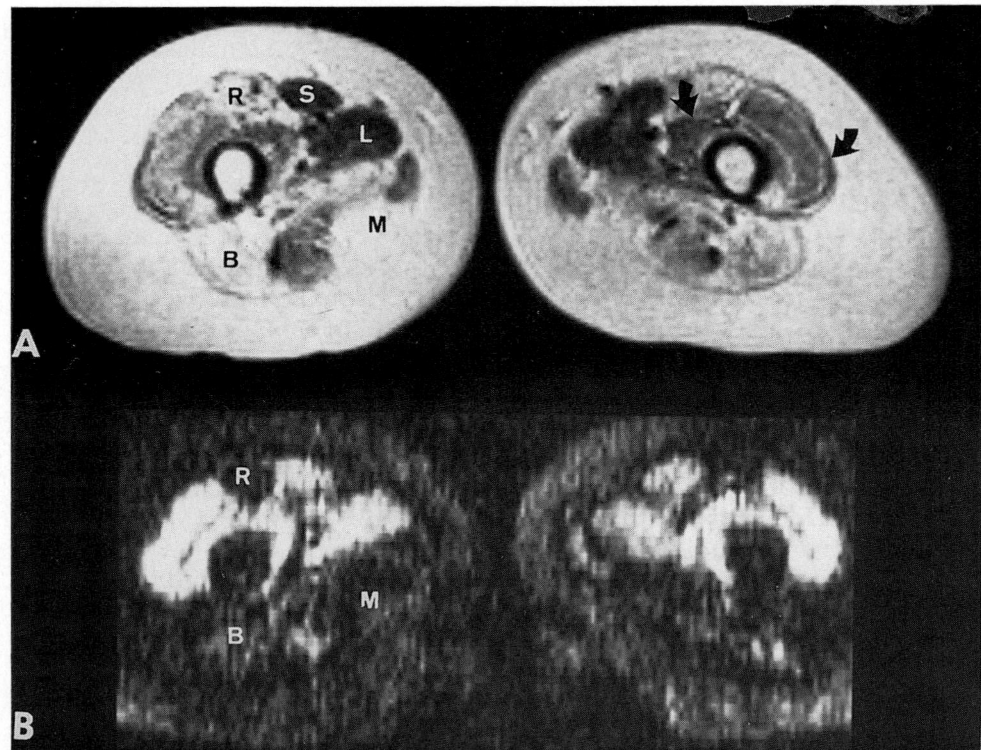

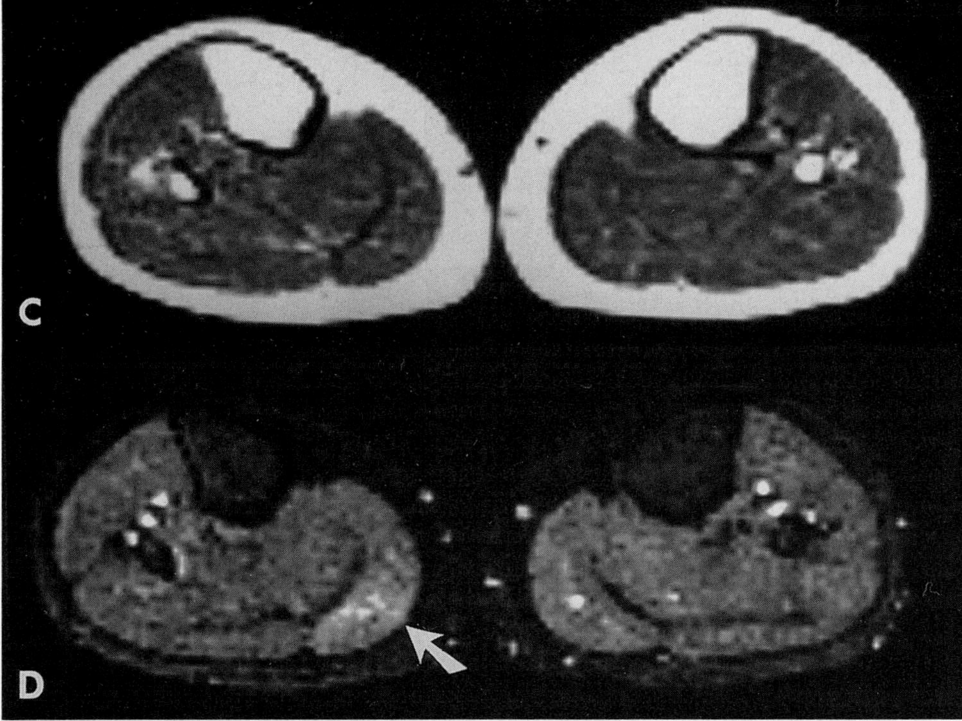

Figure 17–19. Chronic idiopathic polymyositis in a 37-year-old woman. This patient had a history of polymyositis for 17 years before this MR imaging study. *A*, A spin-echo image (TR, 2000 ms; TE, 30 ms) of the thigh shows symmetric fatty infiltration of the rectus femoris (R), adductor magnus (M), and biceps femoris (B) muscles. The sartorius (S), adductor longus (L), and vastus muscles are spared (*arrow*). *B*, A 1-min fast short tau inversion-recovery (STIR) image at the same level shows suppressed signal intensity where fat predominates. The muscle with edema and inflammation is seen to have very high signal intensity. *C*, An image (TR, 2000 ms; TE, 30 ms) of the same patient's calves shows no abnormality. *D*, A STIR image shows a single focus of increased signal intensity in the medial head of the right gastrocnemius muscle (*arrow*), suggesting foci of relatively active disease process there. (From Fleckenstein JL: Magnetic resonance imaging and computed tomography of skeletal pathology. *In* Bloem JL, Sartoris DJ [eds]: MRI and CT of the Musculoskeletal System. © 1992, the Williams & Wilkins Co., Baltimore.)

with spontaneous remission occurring in some cases. In addition, there is no increased incidence of neoplasia with childhood dermatomyositis as there is with the adult variety.

Hernandez and colleagues have reported that the involved muscles in childhood dermatomyositis have a signal intensity that is indistinguishable from that of uninvolved muscles on T1-weighted images.[30] On T2-weighted images, the involved muscles are of high signal intensity, presumably because of the accumulation of extracellular water (Fig. 17–21). Because vasculitis resulting in microinfarc-

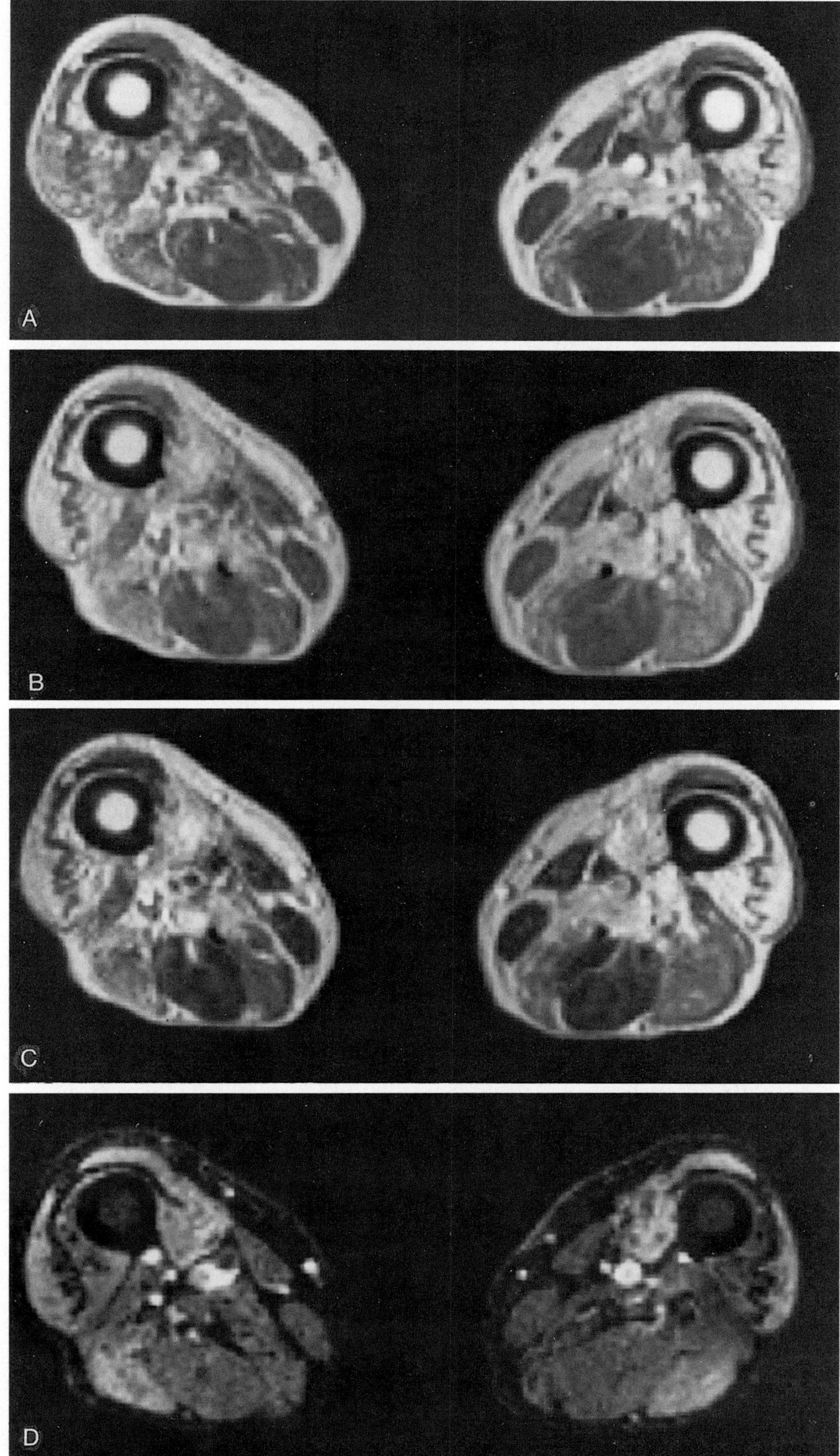

Figure 17–20. Chronic polymyositis. Spin-echo images (*A*: TR, 500 ms; TE, 30 ms; *B*: TR, 2000 ms; TE, 30 ms; *C*: TR, 2000 ms; TE, 60 ms) show the extensive areas of high signal intensity throughout the thigh muscles, indicating fatty infiltration. *D*, In the short tau inversion-recovery (STIR) image, the signal from fat is suppressed, while regions of edema are of high signal intensity, suggesting foci of relatively active disease process there.

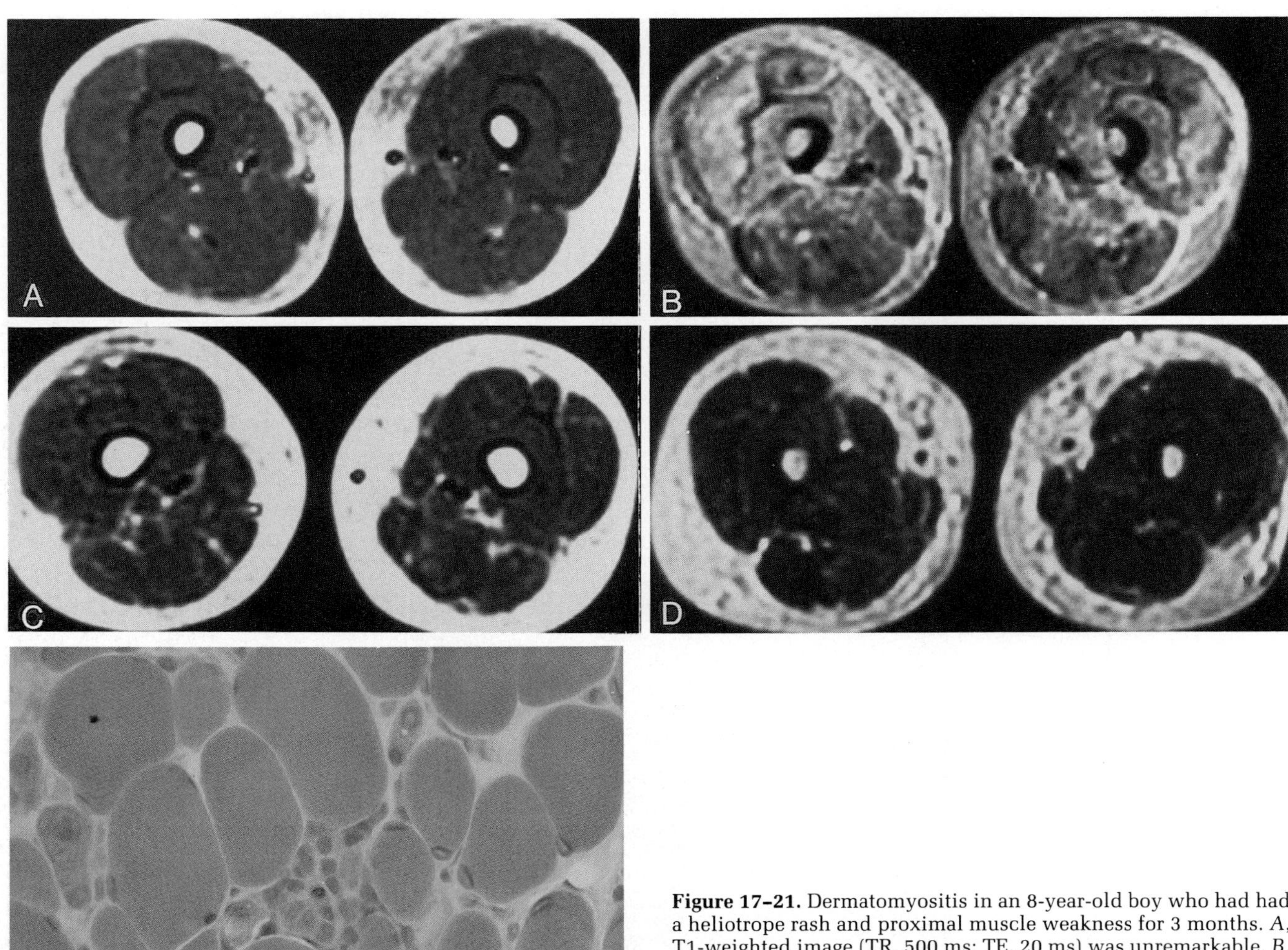

Figure 17–21. Dermatomyositis in an 8-year-old boy who had had a heliotrope rash and proximal muscle weakness for 3 months. *A*, T1-weighted image (TR, 500 ms; TE, 20 ms) was unremarkable. *B*, T2-weighted image (TR, 1500 ms; TE, 90 ms) shows high signal intensity throughout the muscles of the thigh, indicating active edema and inflammation. Follow-up MR images (*C*: TR, 500 ms; TE, 20 ms; *D*: TR, 1500 ms; TE, 90 ms) 10 months later in a state of remission show apparent normal signal intensities on both sequences. *E*, Muscle biopsy reveals variations in fiber size and focal fiber necrosis with phagocytosis (hematoxylin and eosin stain, ×100).

tion is a dominant pathologic feature of childhood dermatomyositis, an increased water content of infarcted muscle may be responsible for the MR imaging signal intensity changes. However, the diseased muscles in adult-type dermatomyositis are of increased signal intensity on T1-weighted images.[33] Hernandez and colleagues proposed that their discrepancy could be related to the chronicity of the disease.[30]

EXERTIONAL MUSCLE INJURIES

Strains and Ruptures

Muscle pain that begins during exercise is referred to as muscle strain. The majority of muscle strain injuries are thought to be the result of eccentric actions.[51,52] Muscles that cross two joints and those that have a high proportion of fast-twitch, type II muscle fibers (e.g., the biceps, rectus femoris, medial gastrocnemius) most often are involved in this type of injury.[28] Involvement of the rectus femoris is most common and accounted for 38 per cent of muscle injuries in a sonographic study.[24] Clinically, a strain is considered mild (first-degree) if weakness is not present, indicating no myofascial disruption. In a moderate (or second-degree) strain, weakness is associated with a variable degree of separation of muscle from tendon or fascia. Myofascial separation is complete in severe (or third-degree) strain, resulting in a lack of muscle function.[21]

In healthy subjects, acutely strained muscles often show focal abnormalities in signal intensity on

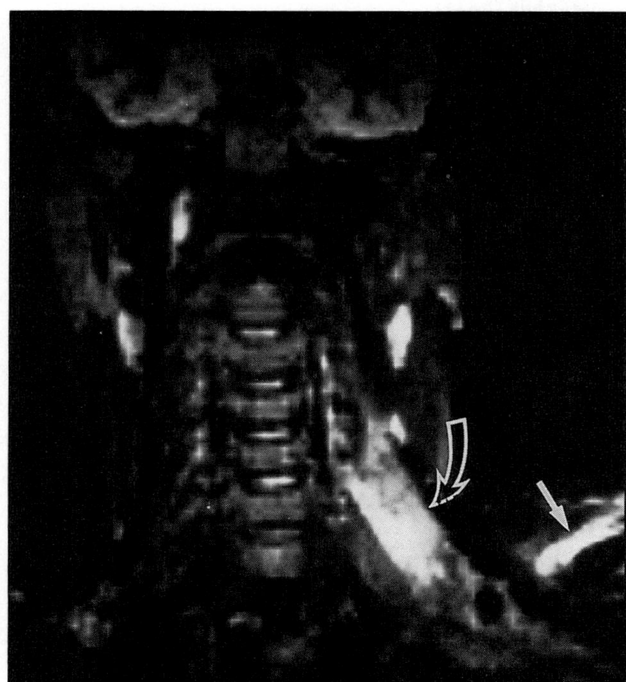

Figure 17–22. Cervical strain in a woman after a motorcycle accident. Coronal short tau inversion-recovery (STIR) image shows marked and diffuse high signal intensity in the left scalenus anterior muscle *(curved arrow)* as well as the left supraspinatus muscle *(straight arrow)*. The other areas of high signal intensity in the neck represent flowing venous blood. (From Fleckenstein JL, Weatherall PT, Bertocci LA, et al: Locomotor system assessment by muscle magnetic resonance imaging. Magn Reson Q 7:79, 1991.)

MR images. T2-weighted spin-echo, T2*-weighted gradient-echo, and STIR images (Fig. 17–22) show distinct high signal intensity in an acutely strained muscle compared with normal muscle, whereas there are no obvious signal intensity differences on T1-weighted spin-echo images. From the second day after injury, a rim of high signal intensity may be found surrounding the strained muscle (Fig. 17–23). Later on, abnormalities of signal intensity may become more diffuse and homogeneous before returning to normal.[23] However, muscle strains have MR imaging features that are indistinguishable from those of necrosis unless avulsed fibers are visible. Differentiation of muscle strains from hematoma is important because muscle strain is better treated by massage, whereas hematoma is better treated by drainage. Hematoma may have a variable MR appearance depending on the age and volume of the blood.

Paramedial spinal muscle strains have MR imaging features similar to those of other exertional strain (see Fig. 17–22). Magnetic resonance imaging helps to document the presence, location, and severity of the strained muscle. This may help to identify muscle injury as a cause of symptoms in patients with whiplash and on-the-job injuries.

Some research efforts have focused on the myotendinous junction as the weak point in the myotendinous unit during passive stretch (Fig. 17–24).[43,44]

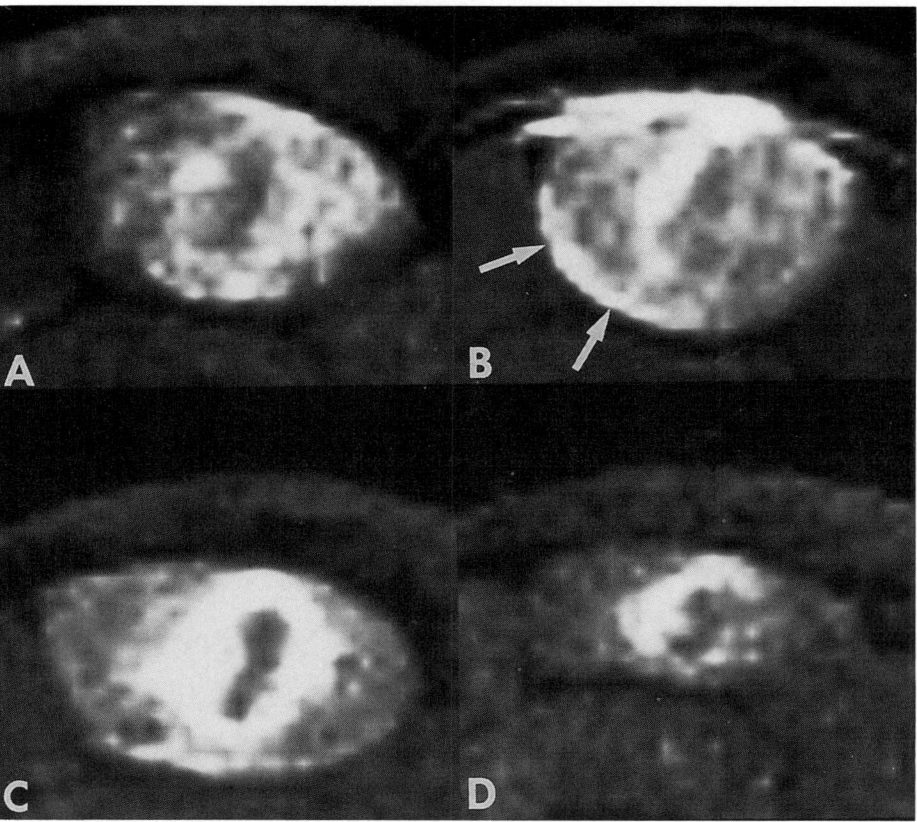

Figure 17–23. Healing of a muscle strain in a patient (a weekend athlete) who had had acute thigh pain while running for a ball. Serial axial short tau inversion-recovery (STIR) images show the course of the rectus femoris strain at 18 hours *(A)*, 36 hours *(B)*, 72 hours *(C)*, and 12 days *(D)* after the injury, when he was asymptomatic. Note the evolution of concentric heterogeneity over time, with a thin rim of very high signal intensity adjacent to the intermuscular fascia at 36 hours *(arrows)*. (From Fleckenstein JL, Weatherall PT, Parkey RW, et al: Sports-related muscle injuries: Evaluation with MR imaging. Radiology 172:793, 1988.)

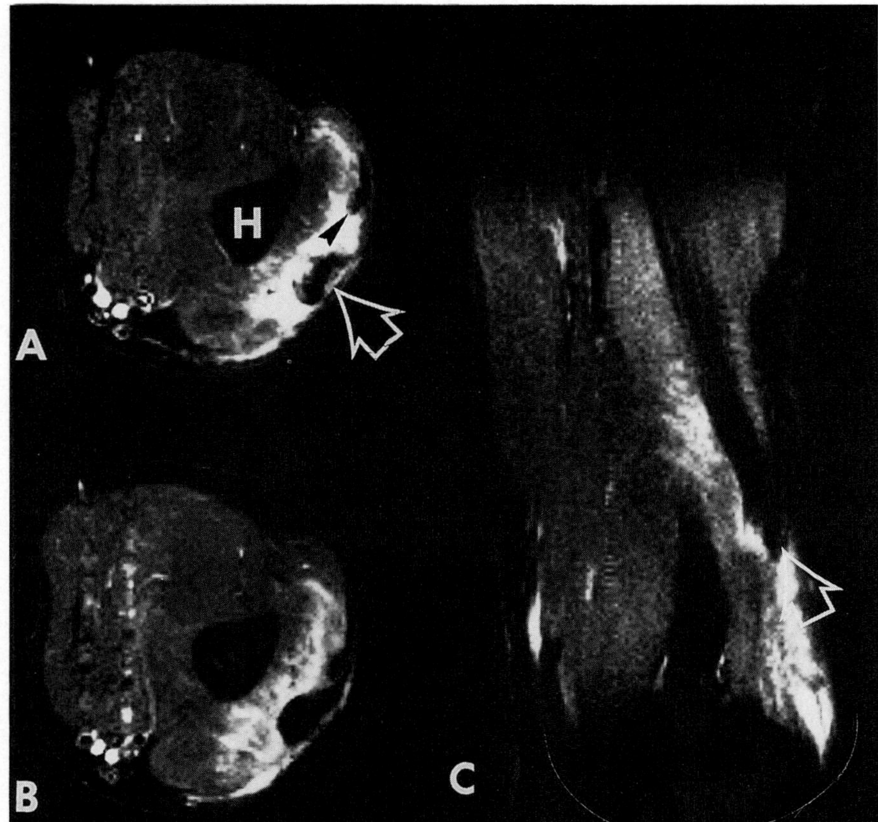

Figure 17–24. Triceps rupture. Short tau inversion-recovery (STIR) images (*A* and *B*, contiguous axial; *C*, coronal; TR, 1500 ms; TE, 30 ms; TI, 100 ms) obtained at the level of the distal triceps muscles show extensive muscular edema and fluid intervening between the retracted tendon *(open arrow)* and the humerus (H). Discontinuity is evident between the remaining intact tendon *(solid arrowhead)* and the retracted portion. Retraction of the myotendinous portion superiorly *(open arrow)* is well shown on the coronal image *(C)*. (Reprinted from Fleckenstein JL, Shellock FG: Exertional muscle injuries: Magnetic resonance imaging evaluation. Top Magn Reson Imaging 3:50, with permission of Aspen Publishers, Inc., © 1991.)

This junction is thought to be a frequent point of rupture. This is of clinical significance because complete rupture is an indication for surgical tendinous repair, whereas conservative treatment is usually sufficient for small fascial injury.

Delayed Onset Muscle Soreness

Muscle pain that occurs hours to days after exercise is referred to as delayed onset muscle soreness. Other symptoms may include joint stiffness and swelling. These symptoms are usually localized to the myotendinous junction, increase in intensity within the first 24 hours after exercise, peak in 2 to 5 days, and disappear 7 to 10 days after exercise.[51,52] Several investigators have postulated that increased intramuscular fluid pressure, inflammation, or damage to the corresponding connective tissue is responsible for symptoms.[4,25,26]

Damaged muscles are depicted on T2-weighted spin-echo, T2*-weighted gradient-echo, or STIR MR images as areas of increased signal intensity, corresponding to increased T2 relaxation times, that may persist up to 4 weeks[23] or more[51,52] after the disappearance of symptoms (Fig. 17–25). Changes in relaxation times also are noted before the onset of pain and elevation of the serum creatine kinase levels.[23] In severe cases, rim patterns similar to those seen in strains may be noted in the periphery of muscle or in the perifascial and intermuscular spaces, closely apposed to the injured muscle.[23]

Chronic Muscle Overuse Syndromes

Muscle pain that occurs in patients with recreation- or occupation-related muscle activity is referred to as chronic muscle overuse syndrome. Joint stiffness is a frequent symptom as well. Examples of recreational activities are those performed by runners, tennis players (Fig. 17–26), and baseball pitchers. Examples of occupation-related repetitive muscular activities are those carried out by typists, waitresses (Fig. 17–27), and musicians. However, chronic muscle overuse syndromes are poorly understood, and they occasionally are confused with delayed onset muscle soreness.

Sequelae of Muscle Injuries

Whether or not intense exercise is resumed, a previously strained or partially torn muscle may predispose to a completely torn muscle. Fatty replacement of traumatized muscle is rare, but it occurs

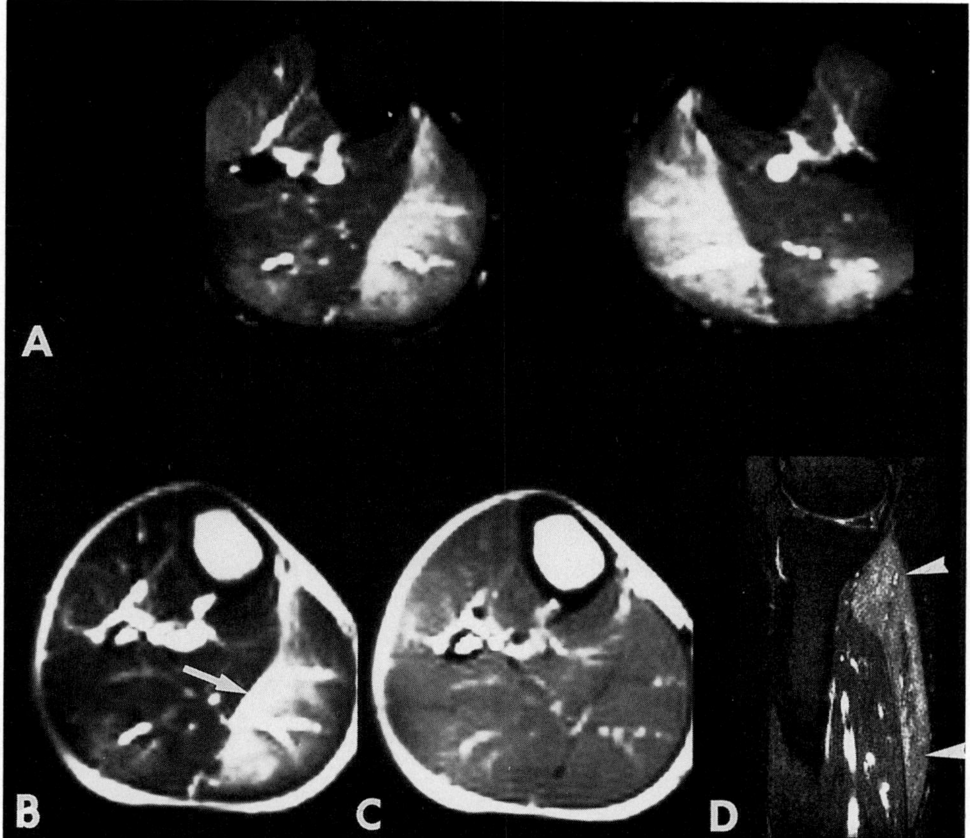

Figure 17–25. Delayed onset muscle soreness. MR imaging obtained in a volunteer 48 hours after exercise of both legs. Short tau inversion-recovery (STIR) *(A)* and T2-weighted *(B*: TR, 2000 ms; TE, 60 ms) axial images show markedly increased signal intensity of the medial head of the gastrocnemius, especially adjacent to the deep fascial boundary *(arrow)*. *C*, T1-weighted axial image (TR, 500 ms; TE, 30 ms) shows no evidence of signal alteration. *D*, Sagittal STIR image shows the longitudinal extent of the abnormality *(arrowheads)*. (From Fleckenstein JL, Weatherall PT, Parkey RW, et al: Sports-related muscle injuries: Evaluation with MR imaging. Radiology 172:793, 1988.)

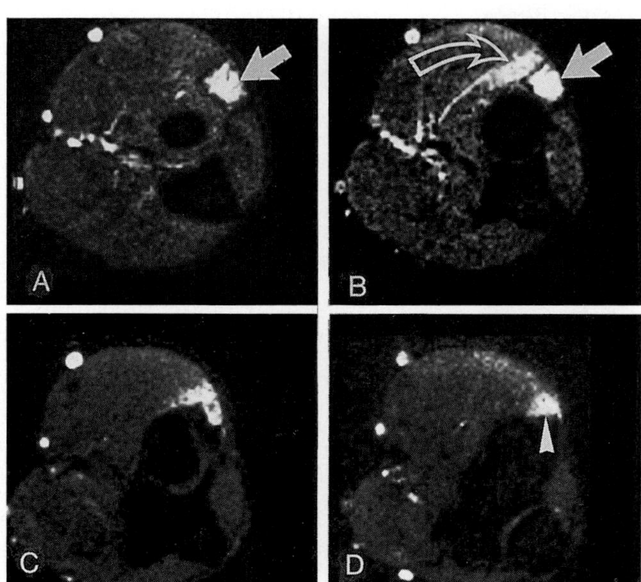

Figure 17–26. Chronic muscle overuse syndrome in a patient with "tennis elbow." The patient stopped playing tennis owing to increased exertional pain over the lateral humeral epicondyle. Short tau inversion-recovery (STIR) axial images proceeding from distal *(A)* to proximal *(D)* show focal increased signal intensity in the extensor digitorum communis *(straight arrow)* and extensor carpi radialis brevis *(curved arrow)*, where they originate from the lateral epicondyle *(arrowhead)*. (From Fleckenstein JL, Weatherall PT, Bertocci LA, et al: Locomotor system assessment by muscle magnetic resonance imaging. Magn Reson Q 7:79, 1991.)

with regularity in patients with myopathies. Fibrosis and muscle ossification are not uncommon complications in muscle tears and contusion.

In the compartment syndrome, edema or hemorrhage occurs within intact fascial boundaries, and increases in muscle pressure impair blood and oxygen delivery.[21,22] This may be an indication for immediate surgical decompression. Patients who have deep venous thrombosis of the lower extremities or compartment syndrome may have painful swollen legs as the presenting feature, and a correct clinical diagnosis is occasionally hard to make. Magnetic resonance imaging allows evaluation of the status of the vessels, subcutaneous tissues, and bone marrow in addition to the muscles. Muscle perfusion in injured muscles in a compartment syndrome can be assessed by using gadopentetate dimeglumine (Fig. 17–28). Exercise-enhanced MR imaging improves assessment of chronic compartment syndromes, the diagnostic criterion for which is progressive elevation of the muscle proton T1 relaxation time during recovery after exercise.[1]

Another complication of muscle injuries is myositis ossificans. This is a nonneoplastic heterotopic proliferation of bone and cartilage in the skeletal muscle. It can be categorized into a localized form

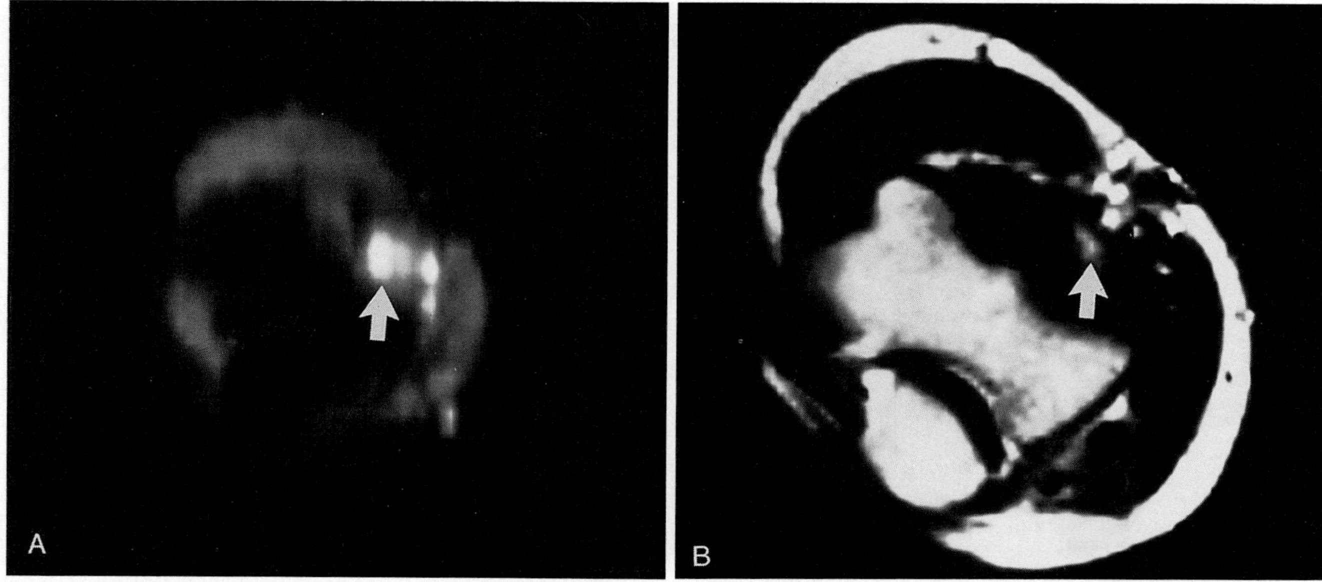

Figure 17–27. Chronic muscle overuse syndrome in a waitress with "waitress elbow." This patient complained of progressive pain near her elbow for several days that was exacerbated by her holding a cocktail tray. Ultimately she was forced to stop waitressing. *A*, A 1-minute, low-resolution fast short tau inversion-recovery (STIR) (TR, 1500 ms; TE, 30 ms; TI, 100 ms) scout sequence provided 16 1-cm slices, one of which showed a small focus of increased signal intensity in the distal brachialis muscle *(arrow)*. *B*, A spin-echo (TR, 1000 ms; TE, 80 ms) image confirms the focus of edema *(arrow)*. Both the MR imaging abnormality and the symptoms resolved during 1 week off work. (Reprinted from Fleckenstein JL, Shellock FG: Exertional muscle injuries: Magnetic resonance imaging evaluation. Top Magn Reson Imaging 3:50, with permission from Aspen Publishers, Inc., © 1991.)

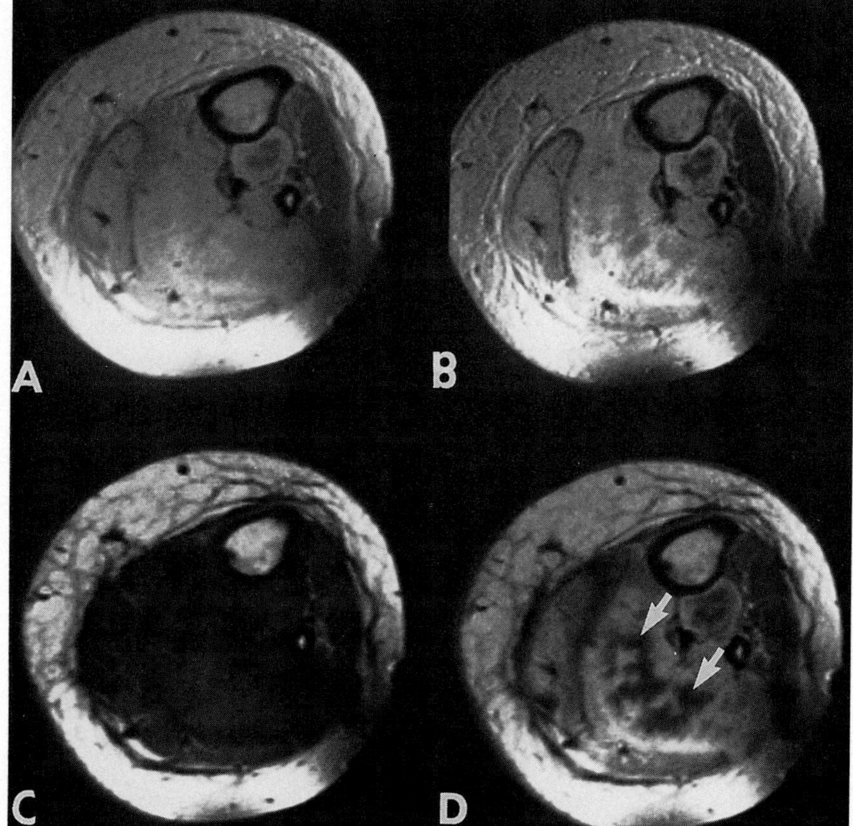

Figure 17–28. Sequelae of exertional muscle injury: compartment syndrome in a 26-year-old psychotic patient. This patient complained of swelling of the thigh 3 days after being discharged from the psychiatric emergency department. She was treated for deep venous thrombosis on the basis of an equivocal nuclear scintigraphic examination. It was later found that she had been attacked by several men and that her legs had been violently compressed during the struggle. Clinical evidence of compartment syndrome led to fasciotomy, after which this imaging study was performed. Edematous proximal leg muscles are evident by an increased signal intensity on *A* (TR, 2000 ms; TE, 30 ms) and *B* (TR, 2000 ms; TE, 60 ms) sequences and an absence of marked abnormality on the *C* (TR, 500 ms; TE, 30 ms) sequence. The TR, 500 ms; TE, 30 ms sequence was repeated after an infusion of gadolinium *(D)*. Failure of edematous-appearing muscle to become enhanced *(arrows)* suggests a focal perfusion. After 5 days, a subsequent MR imaging showed defective focus of very high signal intensity in the same region on a T1-weighted sequence, indicating hemorrhage (not shown). Whether this focus was originally due to blood or was nonviable muscle, which subsequently bled, is unknown. (From Fleckenstein JL, Weatherall PT, Bertocci LA, et al: Locomotor system assessment by muscle magnetic resonance imaging. Magn Reson Q 7:79, 1991.)

that involves a single muscle or group of muscles after trauma and a generalized form that is unrelated to trauma and consists of a progressive and widespread ossifying process in many skeletal muscles. Traumatic myositis ossificans also has been termed traumatic ossifying myositis, ossifying hematoma, calcified hematoma, and traumatic periosteal bone formation.[32] The ossifications that appear in the muscle may develop after a direct severe blow to a muscle (myositis ossificans traumatica) or may arise after repeated minor trauma (myositis ossificans circumscripta). Myositis ossificans traumatica is commonly seen as a painful mass in young patients within a period of 1 to 4 weeks after trauma. The anterior thigh muscles and brachialis of the upper arm are most frequently involved. Myositis ossificans circumscripta may be entirely asymptomatic.

The initial injury is associated with hemorrhage of variable degree; inflammatory changes and extensive proliferation of the intramuscular connective tissue then follow; and soon, islands of bone and cartilage arise in these thickened connective tissue septa. The muscle fibers are not primarily involved in the process of ossification but are compressed in the fibrosing and calcifying tissue.[32] In advanced ossified lesions, the bone may be formed directly from elements of the fibrous tissue or displaced osteoblasts, or it may arise in islands of cartilage. The newly formed bone is of the cancellous type and resembles the bone of callus formation in healing fractures. In addition, calcification of tendons and myotendinous junctions is a common posttraumatic complication.

The MR imaging appearance of myositis ossificans varies with disease progression and relates to histologic changes.[35] Magnetic resonance imaging is useful for defining the extent of soft tissue abnormalities and for excluding the possibility of abnormalities of the adjacent cortical bone and bone marrow.[12] The soft tissue shows diffuse high signal intensity on T2-weighted images except for the ossified portion, which is of low signal intensity (Fig. 17-29). The absence of abnormality in an adjacent cortex or bone marrow is the cue to exclude neoplasms.

OTHER MUSCLE TRAUMA

Crush Syndrome

Massive injuries to muscle usually accompany severe bodily trauma. Crush syndrome, unlike exertional muscle injuries, is a life-threatening condition. Large amounts of myoglobin spill into the serum and cause renal failure. This condition is worsened by the circulatory collapse (shock) that may be attendant in crush syndrome. Cardiac complications, particularly conduction disorders, may result from increased entrance of potassium from injured muscle into the blood stream.[32]

Burns, Including Electrical Injuries

Regional myonecrosis occurs at the site of burns. Liberation of myoglobin, as in the crush syndrome, may lead to fatal myoglobinuria and lower nephron nephrosis. High-voltage current, including lightning, may cause severe muscle injuries (Fig. 17-30). The damage often is seen at the site of and along the pathway of the current. Because the path is frequently unpredictable, it may be very important to identify necrotic muscle preoperatively so that infection and sepsis can be avoided. For example, weakness and wasting around the shoulder and upper arm may develop after electrical burns to the thumb and index fingers. The weakness may be immediate or delayed for a few months, and atrophy can progress for a few months before regression.[32]

Iatrogenic Trauma

The site of intramuscular injections is of high signal intensity on T2-weighted or STIR images. In infants, hematoma in the anterior thigh muscle may occur in muscle trauma from intramuscular injections (Fig. 17-31). In addition, muscle biopsy may leave a mark for a certain period of time. Magnetic resonance imaging can document the site of biopsy, which should not be mistaken for a disease process.

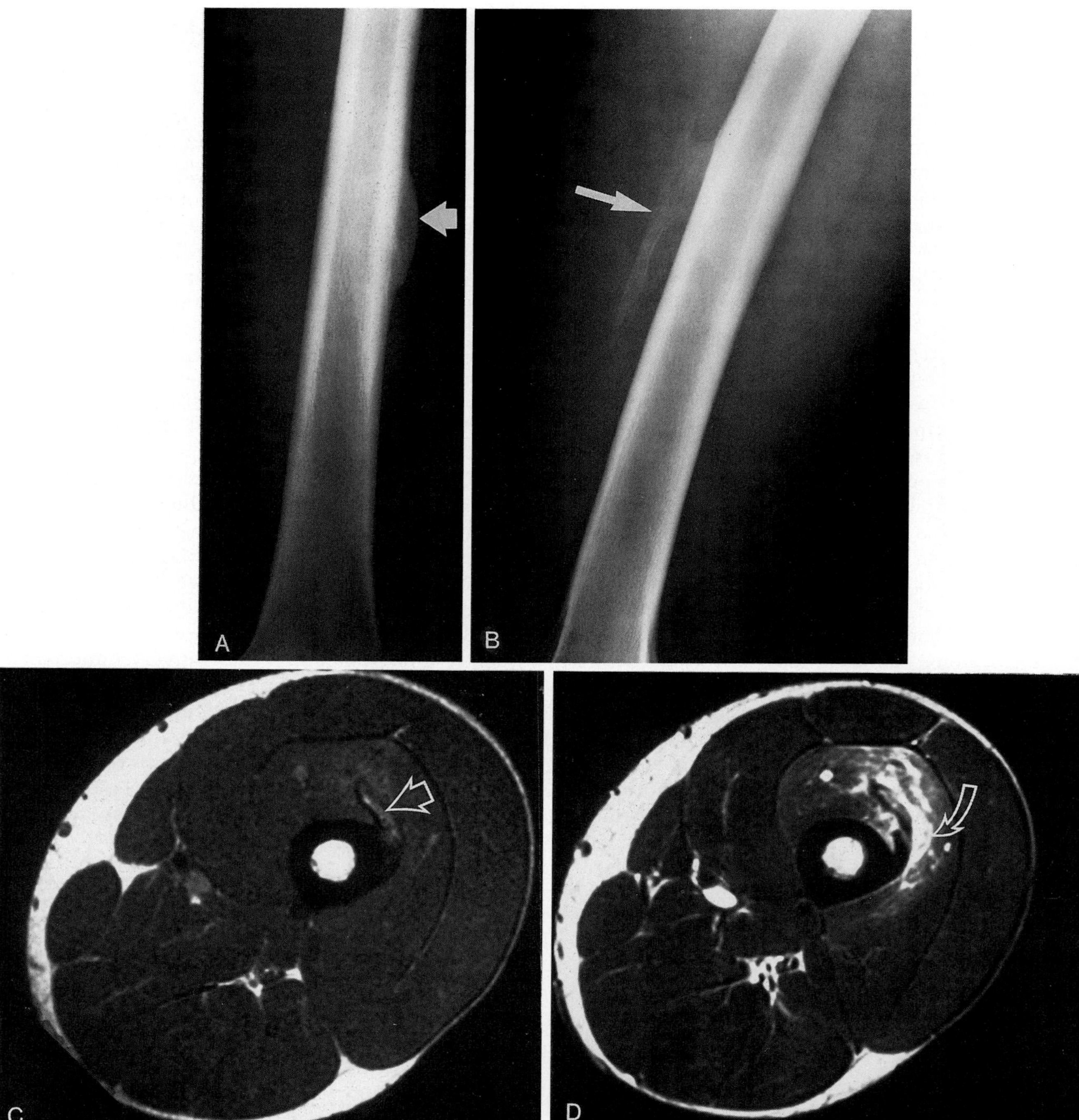

Figure 17–29. Sequelae of muscle contusion: myositis ossificans. This patient complained of persistent pain after a blunt blow to the thigh 4 months previously. *A* and *B*, Radiographs confirm the presence of an ossified subperiosteal hematoma (*arrow* in *A*) and associated myositis ossificans (*arrow* in *B*). *C*, T1-weighted axial image (TR, 600 ms; TE, 20 ms) shows the contiguity of both lesions (*open arrow*). *D*, T2-weighted axial image (TR, 2000 ms; TE, 80 ms) shows related extensive edema in the vastus intermedius muscle (*curved arrow*). The causative relationship between the muscle edema and the spicule of bone cannot be determined, but rehabilitation efforts were directed to the appropriate muscle. (From Fleckenstein JL, Weatherall PT, Bertocci LA, et al: Locomotor system assessment by muscle magnetic resonance imaging. Magn Reson Q 7:79, 1991.)

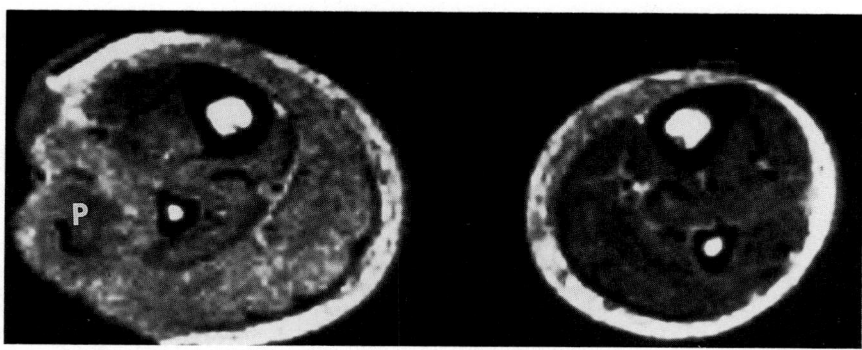

Figure 17-30. Thermal myonecrosis in a patient who had grounded high-voltage wires. Moderately T2-weighted axial image (TR, 2000 ms; TE, 60 ms) shows marked swelling of the peroneus longus muscle (P). The adjacent subcutaneous fat shows the extensive loss of architecture that resulted from fasciotomy. This patient underwent amputation subsequently. (From Fleckenstein JL: Magnetic resonance imaging and computed tomography of skeletal muscle pathology. *In* Bloem JL, Sartoris DJ [eds]: MRI and CT of the Musculoskeletal System. © 1992, the Williams & Wilkins Co., Baltimore.)

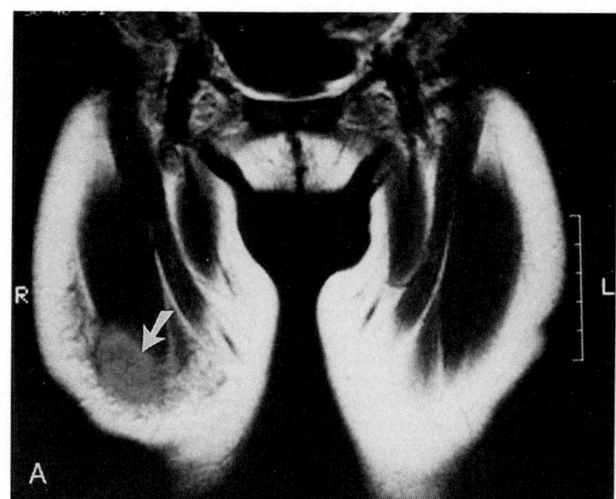

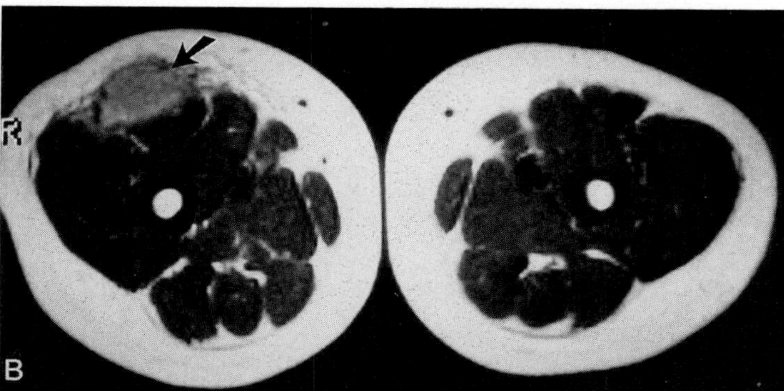

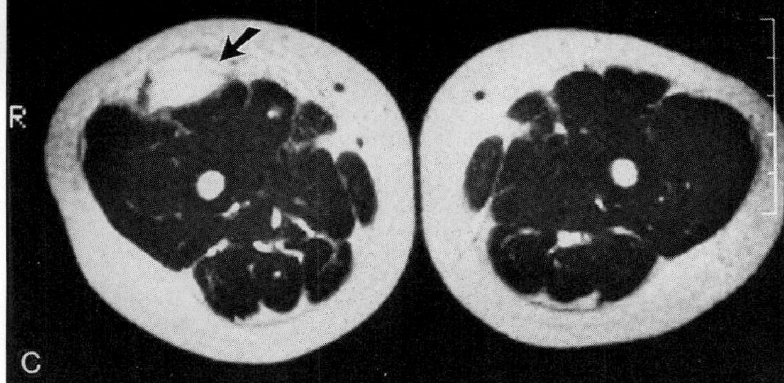

Figure 17-31. Intramuscular injection in an 18-month-old infant. *A–C,* T1-weighted coronal *(A)* and axial *(B)* images (TR, 750 ms; TE, 40 ms) show increased signal intensity (relative to other muscles) in the rectus femoris muscle *(arrow),* which becomes very bright on the T2-weighted image *(C).* This infant had a history of recent intramuscular injection. The abnormal signal intensity may be due to the drug itself or to the hematoma.

References

1. Amendola A, Rorabeck CH, Vellett D, et al: The use of magnetic resonance imaging in exertional compartment syndromes. Am J Sports Med 18:29, 1990.
2. Anderson JR: Atlas of Skeletal Muscle Pathology. Lancaster, MTP Press, 1985.
3. Archer BT, Fleckenstein JL, Bertocci LA, et al: Effect of perfusion on exercised muscle: MRI evaluation. JMRI 2:407, 1992.
4. Armstrong RB: Mechanisms of exercise-induced delayed onset muscular soreness: A brief review. Med Sci Sports Exerc 16:529, 1984.
5. Arundell FD, Wilkinson RD, Haserick JR: Dermatomyositis and malignant neoplasma in adults. Arch Dermatol 82:772, 1960.
6. Bohan A, Peter JB: Polymyositis and dermatomyositis. N Engl J Med 292:344, 1975.
7. Carpenter S, Karpati G: Pathology of Skeletal Muscle. New York, Churchill Livingstone, 1984.
8. Cha CH, Patten BM: Amyotrophic lateral sclerosis: Abnormalities of the tongue on magnetic resonance imaging. Ann Neurol 25:468, 1989.
9. Chan WP, Miller RG, Lang P, et al: MR imaging of Duchenne muscular dystrophy. Magn Reson Imaging 1:233, 1991.
10. Chiedozi LC: Pyomyositis: A review of 205 cases in 112 patients. Am J Surg 137:255, 1979.
11. Duboc D, Jehenson P, Tran-Dinh S, et al: Phosphorus NMR spectroscopy study of muscular enzyme deficiencies involving glycogenosis and glycolysis. Neurology 37:663, 1987.
12. Ehara S, Nakasato T, Tamakawa Y, et al: MRI of myositis ossificans circumscripta. Clin Imaging 15:130, 1991.
13. Emery AEH: Duchenne Muscular Dystrophy. Oxford, Oxford University Press, 1987.
14. Fleckenstein JL, Archer BT, Barker BA, et al: Fast short-tau inversion-recovery MR imaging. Radiology 179:499, 1991.
15. Fleckenstein JL, Bertocci LA, Nunnally RL, et al: Exercise-enhanced MR imaging of variations in forearm muscle anatomy and use: Importance in MR spectroscopy. AJR 153:693, 1989.
16. Fleckenstein JL, Burns DK, Murphy FK, et al: Differential diagnosis of bacterial myositis in AIDS: Evaluation with MR imaging. Radiology 179:653, 1991.
17. Fleckenstein JL, Canby RC, Parkey RW, et al: Acute effects of exercise on MR imaging of skeletal muscle in normal volunteers. AJR 151:231, 1988.
18. Fleckenstein JL, Haller RG, Bertocci LA, et al: Glycogenolysis, not perfusion, is the critical mediator of exercise-induced muscle modifications on MR images. Radiology 183:25, 1992.
19. Fleckenstein JL, Haller RG, Lewis SF, et al: Absence of exercise-induced MRI enhancement of skeletal muscle in McArdle's disease. J Appl Physiol 71:961, 1991.

20. Fleckenstein JL, Peshock RM, Lewis SF, et al: Magnetic resonance imaging of muscle injury and atrophy in glycolytic myopathies. Muscle Nerve 12:849, 1989.
21. Fleckenstein JL, Shellock FG: Exertional muscle injuries: Magnetic resonance imaging evaluation. Top Magn Reson Imaging 3:50, 1991.
22. Fleckenstein JL, Weatherall PT, Bertocci LA, et al: Locomotor system assessment by muscle magnetic resonance imaging. Magn Reson Q 7:79, 1991.
23. Fleckenstein JL, Weatherall PT, Parkey RW, et al: Sports-related muscle injuries: Evaluation with MR imaging. Radiology 172:793, 1989.
24. Fornage BD, Touche DH, Segal P, et al: Ultrasonography in the evaluation of the muscular trauma. J Ultrasound Med 2:549, 1983.
25. Friden J, Sfakianos PN, Hargens AR: Muscle soreness and intramuscular fluid pressure: Comparison between eccentric and concentric load. J Appl Physiol 61:2175, 1986.
26. Fritz VK, Stauber WT: Characterization of muscles injuries by forced lengthening. II. Proteoglycans. Med Sci Sports Exerc 20:354, 1988.
27. Galdi AP: Diagnosis and management of muscle disease. New York, Spectrum Publications, 1984.
28. Garrett WE: Muscle strain injuries: Clinical and basic aspects. Med Sci Sports Exerc 22:436, 1990.
29. Hernandez RJ, Keim DR, Chenevert TL, et al: Fat-suppressed MR imaging of myositis. Radiology 182:217, 1992.
30. Hernandez RJ, Keim DR, Sullivan DB, et al: Magnetic resonance imaging appearance of the muscles in childhood dermatomyositis. J Pediatr 117:546, 1990.
31. Jeneson JAL, Taylor JS, Vigneron DB, et al: 1H MR imaging of anatomical compartments within the finger flexor muscles of the human forearm. Magn Reson Med 15:491, 1990.
32. Kakulas BA, Adams RD: Diseases of Muscle. Philadelphia, Harper & Row Publishers, 1985.
33. Kaufman LD, Gruber BL, Gerstman DP, et al: Preliminary observations on the role of magnetic resonance imaging for polymyositis and dermatomyositis. Ann Rheum Dis 46:569, 1987.
34. de Kerviler E, Leroy-Willig A, Jehenson P, et al: Exercise-induced muscle modifications: Study of healthy subjects and patients with metabolic myopathies with MR imaging and P-31 spectroscopy. Radiology 181:259, 1991.
35. Kransdorf MJ, Meis JM, Jelinek JS: Myositis ossificans: MR appearance with radiologic-pathologic correlation. AJR 157:1243, 1991.
36. Kuriyama MK, Hayakawa K, Konishi Y, et al: MR imaging of myopathy. Comput Med Imaging Graphics 13:329, 1989.
37. Lamminen AE, Tanttu JI, Sepponen RE, et al: Magnetic resonance of diseased skeletal muscle: Combined T1 measurement and chemical shift imaging. Br J Radiol 63:591, 1990.
38. Liu G-C, Jaw T-S, Jong T-J, et al: MR grading system of Duchenne muscular dystrophy: Correlation with clinical staging. Radiology 181(P):200, 1991.
39. Matsumara K, Nakano I, Fukuda N, et al: Proton spin-lattice relaxation time of Duchenne dystrophy skeletal muscle by magnetic resonance imaging. Muscle Nerve 11:97, 1988.
40. Matsumura K, Nakano I, Fukuda N, et al: Duchenne muscular dystrophy carriers. Neuroradiology 31:373, 1989.
41. Mole PA, Coulson RL, Caton JR, et al: *In vivo* ^{31}P NMR in human muscle: Transient patterns with exercise. J Appl Physiol 59:101, 1985.
42. Murphy WA, Totty WG, Carroll JE: MRI of normal and pathologic skeletal muscle. AJR 146:565, 1986.
43. Nikolaou PK, Macdonald BL, Glisson RR, et al: Biomechanical and histochemical evaluation of muscle after controlled strain injury. Am J Sports Med 15:9, 1987.
44. Nikolaou PK, Macdonald BL, Glisson RR, et al: The effect of architecture on the anatomical failure site of skeletal muscle (abstract). Trans Orthop Res Soc 11:228, 1986.
45. Paayanen H, Grodd W, Revel D, et al: Gadolinium-DTPA enhanced MR imaging of intramuscular abscesses. Magn Reson Imaging 5:109, 1987.
46. Polak JF, Jolesz FA, Adams DF: NMR of skeletal muscle differences in relaxation parameters related to extracellular/intracellular fluid spaces. Invest Radiol 23:107, 1988.
47. Radda GK, Rajagopalan B, Taylor DJ: Biochemistry in vivo: An appraisal of clinical magnetic resonance spectroscopy. Magn Reson Q 5:122, 1989.
48. Ross BD, Radda GK, Gadian DG, et al: Examination of a case of suspected McArdle's syndrome by ^{31}P nuclear magnetic resonance. N Engl J Med 304:1338, 1981.
49. Schreiber A, Smith WL, Ionasescu V, et al: Magnetic resonance imaging of children with Duchenne muscular dystrophy. Pediatr Radiol 17:495, 1987.
50. Shabas D, Gerard G, Rossi D: Magnetic resonance imaging examination of denervated muscle. Comput Radiol 11:9, 1987.
51. Shellock FG, Fukunaga T, Mink JH, et al: Acute effects of exercise on MR imaging of skeletal muscle: Concentric versus eccentric actions. AJR 156:765, 1991.
52. Shellock FG, Tetsuo F, Mink JH, et al: Exertional muscle injury: Evaluation of concentric versus eccentric actions with serial MR imaging. Radiology 179:659, 1991.
53. Shy GM: The late onset myopathy: A clinicopathologic study of 131 patients. World Neurol 3:146, 1962.
54. Sjogaard G, Saltin B: Extra- and intracellular water spaces in muscle of man at rest and with dynamic exercise. Am J Physiol 243:R271, 1982.
55. Whyte AM, Lufkin RB, Bredenkamp J, et al: Sternocleidomastoid fibrosis in congenital muscular torticollis: MR appearance. J Comput Assist Tomogr 13:163, 1989.
56. Williams RC: Dermatomyositis and malignancy: A review of the literature. Ann Intern Med 50:1174, 1959.
57. Witte MH, Fiala M, McNeill GC, et al: Lymphangioscintigraphy in AIDS-associated Kaposi's sarcoma. AJR 155:311, 1990.
58. Yuh WTC, Schreiber AE, Montgomery WJ, et al: Magnetic resonance imaging of pyomyositis. Skeletal Radiol 17:190, 1988.
59. Zagoris RJ, Karstaedt N, Koubek TD: MR imaging of rhabdomyolysis. J Comput Assist Tomogr 10:268, 1986.

INDEX

Note: Page numbers in *italics* refer to illustrations; page numbers followed by t refer to tables.

B

C

D

G

J

K

L

M

N

O

T

U

V

W

Y

Z